A BIRD ATLAS OF KENYA

A Bird Atlas of Kenya

By
ADRIAN LEWIS
African Birding Ltd, Nairobi, Kenya
DEREK POMEROY
Department of Zoology, Makerere University, Kampala, Uganda

A.A.BALKEMA / ROTTERDAM / BROOKFIELD / 1989

Photograph on the cover: Crowned Crane *Balearica pavonina* by John Karmali F.R.P.S.

CIP-DATA KONINKLIJKE BIBLIOTHEEK, DEN HAAG

Lewis, Adrian

A bird atlas of Kenya / by Adrian Lewis, Derek Pomeroy. – Rotterdam (etc.): Balkema. – Ill.
With index, ref.
ISBN 90 6191 716 6 bound
SISO af.m-keni 598.33 UDC 598.2(676.2)
Subject heading: birds; Kenya.

Published by

A.A.Balkema, P.O.Box 1675, 3000 BR Rotterdam, Netherlands

A.A.Balkema Publishers, Old Post Road, Brookfield, VT 05036, USA

ISBN 90 6191 716 6

Printed in the Netherlands

Contents

List of plates

Preface

The publication of this atlas is the culmination of a major cooperative effort that lasted over six years and involved over two hundred contributors, the majority of whom are residents of Kenya. Its coverage is certainly not complete, but we believe nevertheless that it gives a fair indication of Kenya's avifauna, especially during the main atlas period of 1970-1984.

Birdwatchers come in all shapes and sizes and their degree of scientific interest and knowledge varies equally. Consequently we have tried to make the atlas comprehensible to all whilst aiming at scientific accuracy. Words which may be less familiar to some are included in a glossary.

There is a great temptation when obtaining a work of this type to plunge directly into the maps and the species accounts. But we strongly urge readers also to take note of both the main introduction, and the introductions to the individual families of birds, since these will certainly provide a greater understanding of our subject.

Although work on this atlas is complete, it is only a start. There is still plenty to discover, there are still many areas that are poorly known ornithologically, and the avifauna of Kenya is changing.

Adrian Lewis
Derek Pomeroy

Acknowledgements

This atlas could not exist without the efforts and ready cooperation of the many people mentioned here. We are deeply grateful and indebted to them all.

The project's staunchest supporter has without a doubt been Mrs Mary Sinclair, who began by contributing records, and then went on to assist A.D.L in extracting data from the whole of the specimen collection at the National Museum, Nairobi, in transferring over 20 000 records from the atlas' checklists to its maps, and in transferring data from the East African Natural History Society's Nest Record Scheme. She also typed sections of early drafts and the first draft of the bibliography, as well as compiling and typing the first draft of the index. Finally, she worked with A.D.L. on the proof-reading of the entire typescript, prior to its submission to our publisher. Mary has been a source of cheerful encouragement and positive thinking throughout, always ready to take on more work. This book owes its existence to her, more than to any other single contributor: we cannot thank her enough.

Dr David Pearson first drew A.D.L.'s interest to the atlas project in 1981, and has been a source of much help and discussion throughout. In particular, he produced a summary of the Afrotropical ranges of all Kenya's Palearctic migrants specifically for this study, and this forms the basis of the Afrotropical sections in these species' texts. This summary also contains much information on these birds' distributions and seasonalities in Kenya. It is the source of all unreferenced data on these topics in their accounts. David also produced very useful comments on the penultimate draft of the species accounts, and contributed many records from obscure and poorly known areas of Kenya.

The project's computing, including the writing of a suitable computer program (see Appendix), the input of the data and the printing of the maps ready for the publisher, has been handled entirely by Dr & Mrs A.R.D. Taylor, assisted by Dr D. Widdowson. This has taken a tremendous work-load from our shoulders, for which we are extremely grateful. A.M. Low, L.M. Widdowson and P.A. Wootton very kindly undertook the proof-reading of the data on the computer against those on our base maps.

The photographs that illustrate this book have been generously donated by J.S. Karmali FRPS, D.K. Richards FRPS and R.S. Daniell ARPS.

The birdwatchers whose observations have provided the bulk of the data shown on the maps have been superbly cooperative throughout. We are especially grateful to them for their patience and forbearance in the face of A.D.L.'s detailed and often sceptical cross-examination of their records. They are: F. Alexander; A.M. Alio; R.G. Allan; the late Miss P.M. Allen; E. Allsop; D. Angwin; E.B. Angwin; E.H. Arensen; J.S. Ash; I. Ayres; D.E.G. Backhurst; G.C. Backhurst; A. Barcas; D.S. Barnett; the late T.J. Barnley and Mrs J.G. Barnley; J. Bartram; J.S.S. Beesley; L. Bennun; Mr & Mrs D.F.G. Blaxall; D.J. & A.S. Brass; H.M.H. Braun; C. Briffett; N. Brightman; H.A. & P.L. Britton; C.H. Brown; F.N. Bruce-Miller; Mr & Mrs P. Bruce-Miller; K. Burton; R. & J.M.S. Caldwell; A.C. Cameron; A.L. & T. Campbell; D. Campbell; K.L. Campbell; M.S. & D. Candy; L. Carnegie; N. Carr-Hartley; D.K. Chepus; J. Chism; J. Chittick; J.G. Collins; M.A.C. Coverdale; D. Craig; C. & H.J.S. Crees; E. Culwick; D. Curtis; J. Darlington; F.A.D. Davey; G.M. & P.R.A. Davey; R.A. Davey; J. Dawson; S. Deverell; C.F. Dewhurst; A.W. Diamond; J.P. Dittami; D. Duncan; J. & H.F. Eggers; C.C.H. Elliott; R. Elliott; J.H. Fanshawe; F. & J. Farnsworth; P.C. Faull; C.C. & V.C. Fayad; I.J. Ferguson-Lees; B. Fletcher; H.D. & P.C. Fletcher; J.R. Forbes; C. Foxall; P.M. & I.W. Francombe; J.D. Gerhart; C.M. & N. Gichuki; H. Gomez de Silva; J. & W.M. Good; M.E.J. Gore; L.V.J. Grant; L.M. & R. Greenham; A. & R.M. Greenshields; J. & T. Grumbley; G.G. Gynn; V. Haas; D.P.

Harding; I.W. Hardy; J. Harjula; J.B. & L. Hartley; B.J. Hayes; J. Hayes; J. Heap; S. & B.R. Heath; M.S. & P. Hemphill; B. & A.J. Holcombe; V. Holmgren; A. & J.D. Homfray; J.F.M. Horne; D. & G.C. Irvine; H.A. Isack; A.G. Johnston; P.J. Johnston; L. Jorgensen; K. Kachalon; D.F. Kent; The Kenya Exploration Society; P.C. Lack; H.F. Lamprey; M. Le Marchant; W. Lemke; P. Le Pelley; L. Lind; A.-M. Lohding; E. Lucas; M. Maas; Y. Malcolm-Coe; I.M. Marshall; F.J. McCartney; R. McCartney; R.A.M. McVicker; B.S. Meadows; H. Merriman; M.H. Miller; A. & A.C. Mills; J.E. Miskell; A. Moody; C. Mortensen; D. Murdoch; C. Naesheim; Nairobi School Junior Wildlife Club; J.R. Ness; F.A. Ng'weno; P. Nickalls; G. Nikolaus; T.C. Noad; E. O'Loan; D.K. Olson; R.M. Osborn; M. Palmer; B.T. Parsons; G.N. Paterson; D.J. Pearson; P. Peltonen; D. Percival; M. Pickford; W.J. Plumb; B. Prentice; J. Prettyjohn; O.A.E. Rasa; C. Reid; F.S. & G.D.F. Reid; H.-U. Reyer; D.K. Richards; C.A. & A.A. Riley; E.J. Roberts; T. Romig; R. Rose; C. Ryall; S. Sassoon; D. Schmidl; T. & R.E.M. Scholtens; P.A. Scott; D. Sheppard; L.L. Short; K.-H. Siebenrock; L.A. Siemens; S. Simonson; S.P. Simpkin; M. & K.L.G. Sinclair; M.E. Smalley; P.S. Soorae; P.J. Squelch; T. Stevenson; F. & C.G. Taylor; B. Tengecho; N.R. Thomas; S.L. & M.C. Tremlett; S. Trevor; D.P.C. Trump; E.H. Tucker; G. Tudor; D.A. Turner; G. Vanderstichelen; R.W.M. Watson; M.L. Watts; J. Wood; P.A. Wootton; R. Wouters; Mr & Mrs C. Ziegler.

The following organisations and individuals have generously donated funds to subsidise the project's expenses during the six years of its existence: The WWF Regional Office for East and Central Africa; The East African Natural History Society; Mrs F. Alexander; Mr A.L. Archer; the late T.J. Barnley and Mrs J.G. Barnley; Mr & Mrs A.S. Brass; Mr & Mrs D. Candy; East African Ornithological Safaris Ltd.; Flamingo Tours Ltd.; Mr & Mrs P.C. Fletcher; Mrs J.B. Hartley; Hogg Robinson Kenya Ltd.; Mr J.S. Lee; Phoenix of E.A. Assurance Co. Ltd.; Mr J.D.M. Silvester; Mr P.J. Squelch; The Old Mutual; Tonbridge School Natural History Society c/o Mr & Mrs A.J. Holcombe; Mr & Mrs R.C. Wootton.

As well as the provision of bird records, many people have helped us in other ways, e.g. in the location of obscure place names, in handling the mountain of paperwork that this project has generated and in the storing of vital papers, in the manipulation of data for use in the environmental overlays and in atlas-derived publications, and with transport and warm hospitality during many data-gathering expeditions. Of particular note here are I.J.P.Loefler and T.S. Sembhi, who made their aircraft and fuel available for sorties to remote areas: this resulted in several notable trips to poorly known areas along the borders with Sudan and Ethiopia.

Others who provided various aid are: E. Abe; D.E.G. Backhurst; the late T.J. Barnley and Mrs J.G. Barnley; S. Bleazard; Sir M. Blundell; K.R. Bock; D.J. & A.S. Brass; H.M.H. Braun; K. Burton; A.B. Carles; D.K. Chepus; M.P. Clifton; W. Colley; J. & H.F. Eggers; P.S. Frere; S. & B.R. Heath; P.J. Johnston; J. Kisipolo; N.A. Larkman; V.K. Loefler; A. Mutere; C.O. Oswala; the late Mr N. Palmer; F.S. & G.D.F. Reid; V. Richards; E.J. & N. Roberts; P. Robertshaw; K.L.G. Sinclair; K. Smith; W.J.A. Stanbury; T. Stevenson; B. Tengecho.

The East African Natural History Society allowed us access to the data of its Nest Record Scheme, and to numerous unpublished bird checklists in its possession. Its Ornithological Sub-Committee (OS-C) donated 66 copies of their *Check-list of the Birds of Kenya* for use as the atlas' basic data files.

We are very grateful to Mrs C.M. Gichuki for her wholehearted cooperation while we worked in the Department of Ornithology at the National Museums of Kenya, Nairobi. D.A. Turner read through our text, and provided telex facilities via East African Ornithological Safaris Ltd. R.G. Allan provided detailed comments on our accounts of the three species of *Quelea*. B.S. Meadows kindly provided details of surface water types and their distribution in Kenya.

J.S. Ash, a meticulous correspondent, sent numerous and very valuable data from the poorly known squares on the borders with Uganda, Ethiopia and Somalia, as did J.S.S. Beesley for

Tanzania, and G. Nikolaus for Sudan. J.D. Gerhart, G.J. Morel, N.J. Skinner and J.-P. van de Weghe provided various other extralimital ornithological information. P.C. Lack was the source of much essential discussion, and the late Mr T.J. Barnley took great trouble to send much original information from the Kitale area.

D.J. Brass generously offered her typing services when the manuscript built up to an alarming degree in the early stages of the writing of the text, and C.L. Brown, S. Cameron, B.A. Partridge, M. Price and J. Rollins followed suit, handling large amounts of material later on in the scheme. J.B. Hartley typed the penultimate draft, and took on the formidable task of typing the final draft camera-ready; she also administered the project's finances in later years.

Mr A.T. Balkema, of A.A. Balkema Publishers, has been most enthusiastic and encouraging throughout, and has provided invaluable expertise in the fields of layout and production.

Finally, A.D.L. would like to thank his wife Juliet for her forbearance, encouragement and help during six years of continuous and often frantic work.

Introduction

THIS ATLAS

The idea of a bird atlas is to show distributions in some detail. The first country in tropical Africa to have such an atlas was the Gambia (Jensen & Kirkeby 1980), and the second was the Sudan (Nikolaus 1987). Kenya is the third. There are also atlases for Tunisia (Thomsen & Jacobsen 1979) and Natal (Cyrus & Robson 1980). Soon there will be more (Fig. 1); within the next few years the whole of southern Africa will have been covered (Hockey & Ferrar 1985). Eventually, we hope, there will be an atlas of African birds, as one is planned for Europe.

This atlas originated with a proposal of the Ornithological Sub-Committee (OS-C) of the East African Natural History Society (EANHS) (OS-C 1979; Pearson 1979b). Other schemes for eastern Africa had already been initiated in Ethiopia, Somalia, Sudan and Zambia. In 1980, at the Fifth Pan-African Ornithological Congress in Malawi (PAOC V), various suggestions were made as to how African atlases might be standardised (Ash & Pomeroy 1981). One proposal was that the basic unit should be squares of half by half a degree, i.e. quarter square degrees (QSDs). Data for East Africa were requested on this basis, indicating months of records where possible, together with indications of breeding. Turner (1981) reported on the early progress of this scheme, and published a map of coverage for Kenya and Tanzania.

By late 1981, however, it was apparent that the relatively few observers in Tanzania and Uganda would make it difficult for an East African atlas to be accomplished with any rapidity. Meanwhile, the predictably faster accumulation of data for Kenya had already reached an advanced stage. Hence it was agreed to institute separate atlas schemes for each of the East African countries, to be co-ordinated by individuals rather than by the OS-C as a whole. This book is the product of the Kenyan scheme. A similar atlas is planned for Uganda (Carswell & Pomeroy 1984), whilst N.E. Baker is compiling data for Tanzania.

The primary purpose of the *Bird Atlas of Kenya* is to document the distributions of birds in Kenya as thoroughly as practicable, particularly for the period 1970-1984. For 871 of the species, this information is presented as maps which in most cases considerably amplify the accounts given by Britton (1980a). Further, they greatly facilitate analyses of distributions (see p. 13). The unmapped species are those with few records and/or squares (generally five or less); the squares where these species have occurred are listed.

Once the basis for a Kenyan atlas had been agreed, it was decided that the mapping should continue to be based on the QSD, rather than some multiple of a kilometre grid, which is almost universal in Europe (see for example Udvardy 1981). On the QSD system, Kenya is covered by 228 squares and part-squares, but had squares of 10 km been used, there would have been nearly 30 times as many. Clearly we could not have hoped to achieve more than a very scanty coverage on that scale.

The use of a coarser grid has an unexpected advantage. Many older records have only vague localities, such as 'Machakos' or 'Chalbi Desert'. We can usually place these quite adequately in the appropriate QSD but could only rarely have done so with 10 km squares, and such records would thus have been lost to the atlas. On the other hand, the QSD is really too large a unit for species with very specialised requirements.

At the inception of the atlas for Kenya, we decided to discard the use of monthly data, used in some schemes in southern Africa, but classed as optional at PAOC V (Ash & Pomeroy 1981).

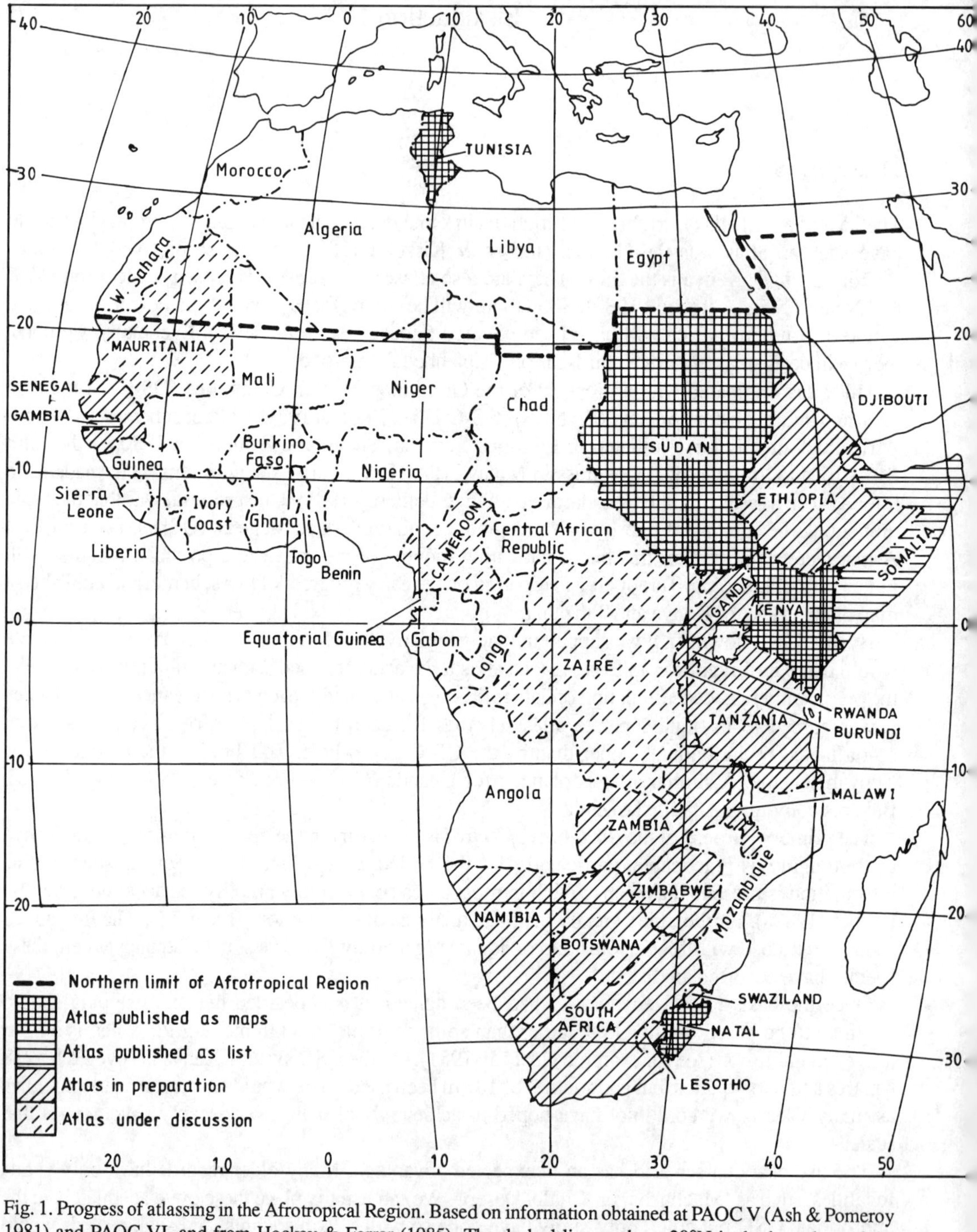

Fig. 1. Progress of atlassing in the Afrotropical Region. Based on information obtained at PAOC V (Ash & Pomeroy 1981) and PAOC VI, and from Hockey & Ferrar (1985). The dashed line at about 20°N is the boundary of the Afrotropical Region (Cramp & Simmons 1977). Note that Senegal and Gambia are shown separately here, whilst in the text they are referred to as Senegambia.

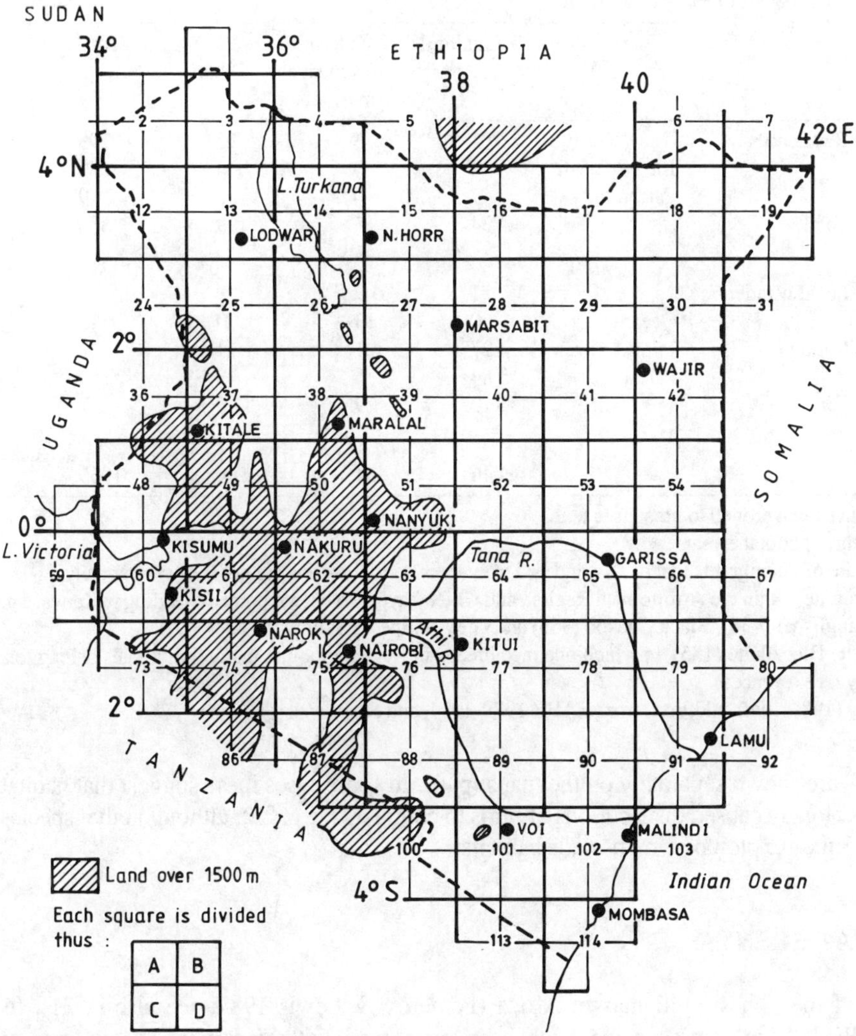

Fig. 2. Kenya base map, showing Quarter-square-degree (QSD) numbers, the main towns and rivers, and land over 1500 m in altitude.

Instead, we concentrated on presence and on breeding records, both for recent records and those prior to 1970.

The human population of Kenya is increasing by about 4% annually, and this, together with the adoption of new land-use practices and people's rising expectations, is leading directly and indirectly to many habitat modifications, of which the most striking are deforestation and desertification. Conversion of native forest to plantation, or forest and woodland to arable land, and improved agricultural techniques, naturally result in extensive changes in bird populations (Carlson 1986a; Pomeroy & Muringo 1980).

Not all changes are of recent occurrence, however. The settlement of the highlands by Europeans in the early decades of the present century was accompanied by widespread bush clearance and deforestation (M. Blundell *in litt.*). Where they differ from records for the period 1970-1984,

Table 1. The major components of Kenya's avifauna

			Landbirds	Waterbirds Exclusively marine	Waterbirds Others	Totals
Breeding species[a]	Residents[b]		701	5	93	799
	Migrants[c]	Intra-African[d]	26	0	4	30
		Palearctic[e]	7	0	2	9
	Total		734	5	99	838
Non-breeding species	Local wanderers		1	9	9	19
	Migrants	Oceanic	0	13	0	13
		Intra-African	27	1	4	32
		Palearctic	84	6	73	163
	Total		112	29	86	227
Overall totals			846	34	185	1065

a: not all of these have been proved to breed; see text.
b: found in Kenya throughout the year.
c: vagrants, which are often migrants that have wandered from their normal routes, are not shown separately.
d: species which migrate within the Afrotropical Region, and which have also been recorded breeding in Kenya, e.g. Chestnut-bellied Kingfisher (436), Black Cuckoo (366) and Grey-rumped Swallow (561).
e: species such as the Black Kite (135), in which one race breeds in Kenya although many individuals belong to a different, migratory race (or races).
(Based on Pomeroy (1981), with additions from EABR 1979-1984, and *Scopus* 9).

pre-1970 records are shown separately on the maps (p. 6). In some cases these suggest that habitat modification has already caused distributions of birds to contract (see p. 22); although other species have expanded as they exploit new man-made habitats.

THE AVIFAUNA OF KENYA

Kenya has one of the richest avifaunas in Africa (Pomeroy & Lewis 1987; see also p. 31). In compiling the *Bird Atlas of Kenya* we have been concerned with documenting our present knowledge of this avifauna and in showing how that information can be interpreted and used, for example in promoting conservation of endangered species and their habitats, and in ecological studies.

The area of Kenya is some 582 600 km², which is close to the average size of African countries. Within this area are found an unusual diversity of habitats, which are well described and illustrated by Britton (1980a). Lind & Morrison (1974) give more details of the different vegetation types, whilst the rangelands, which make up so much of the country, are fully described by Pratt & Gwynne (1977).

Kenya is predominantly a dry country, only about a quarter of the land being capable of supporting rain-fed agriculture; this is mostly at the coast, in the highlands of the south and west, and near Lake Victoria (Fig. 2). The same areas naturally support most of the human population, estimated in 1986 as about 20 millions. Away from the highlands, there are only two permanent rivers of any size, the Athi-Galana-Sabaki system and the Tana, and both of these arise in the highlands. Increasingly the flow in these rivers is being affected by hydroelectric dams and irrigation schemes.

The wide range of habitats in Kenya is a reflection of the great altitudinal range and distinct regional patterns of rainfall. Consequently most species have well-defined distributions. The very few that occur throughout the country include the Laughing Dove (333)*, Drongo (566) and Grey-headed Sparrow (990).

Collection of records for this atlas mainly ceased at the end of 1984, although data from some poorly known areas (mainly in the far north) continued to be accepted until late 1985. By this latter time, 1065 species were included on the Kenyan list. These comprise the 1053 mentioned by Britton (1980a), 13 added since then (EABR; *Scopus* 9:53-54, 110-111) and one deleted (Parker 1984). Of this total of 1065 some 838 (79%) are believed to breed in Kenya (Table 1), but as yet not all have been proved to do so. For example, there are no Kenyan breeding records for several species of cisticola (*e.g.* Boran (728) and Foxy (731)), or for various forest birds such as the Cameroon Sombre Greenbul (614) or Shrike Flycatcher (795). Since it is unlikely that any of these species is migratory, it must be assumed that they breed locally. The north and east of the country,including the coast, have relatively few breeding records, largely because of lack of observation (see p. 10, and as examples, Crested Lark (543) and Superb Starling (890)).

Many of the non-breeding species are Palearctic migrants (Table 1; p. 28), although others are birds which breed elsewhere in Africa. The latter are usually referred to as intra-African migrants.

COLLECTION AND COMPILATION OF DATA

Once the Kenya Atlas scheme had been launched, data accumulated rapidly. Over 8000 records came in 1981 (the first year), 22 000 more in 1982, a further 5000 in 1983 and 3500 in 1984. Most of our methods have already been reported (Lewis & Pomeroy 1982ab, 1983ab, 1984). Below we summarise the most important aspects.

Sources of data

(a) Published records. The literature on the birds of Kenya is extensive, so the task of abstracting records was considerable. The primary sources of reference were Jackson (1938), Mackworth-Praed & Grant (1957, 1960), Hall & Moreau (1970), Snow (1978), Britton (1980a) and Brown *et al.* (1982). Jackson himself was extremely thorough and it was not generally considered necessary to re-check references which he quotes. Numerous other sources were checked however; they all feature in the bibliography.

In the periodical literature, many important papers, especially for earlier years, have appeared in *The Ibis* (abstracted back to 1894), but large amounts of material were also found in the *Bulletin of the British Ornithologists' Club* (abstracted back to 1930), the *EANHS Bulletin* (abstracted in total), the *Journal of the East African Natural History Society and National Museum* (abstracted in total) and *Scopus* (abstracted in total), with lesser amounts from *Alauda* (abstracted back to 1960), *Ardea* (back to 1960), *The Auk* (back to 1960), the *East African Wildlife Journal* (later the *African Journal of Ecology)* (abstracted in total), *Le Gerfaut* (back to 1960), *Journal für Ornithologie* (back to 1970), *The Ostrich* (back to 1950) and *Die Vogelwarte* (back to 1969).

As with other sources of information, exact localities were not always easy to determine, despite reference to the comprehensive gazetteers of the United States Board on Geographic Names (1964),

* The numbers following species names in this book, for example Common Bulbul (609), refer to their serial numbers in the main part of the atlas, where scientific names are given.

Britton (1980a) and Jackson (1938). Furthermore, reference is commonly made to areas greater in size than a QSD, thus: 'common in Machakos District'. In such instances, the species was only entered in the QSD actually containing the settlement named, while records like 'common in Tsavo West' were only entered in the QSD containing most of that park, *i.e.* QSD 89C. It should be stressed however that we followed a policy of 'if in doubt, leave it out', and no unusual records (those of uncommon species or birds apparently out of range) were mapped in this way.

(b) Museum skins. The whole collection of bird skins in the National Museums of Kenya at Nairobi was examined; this entailed checking more than 22 000 specimens. Here again, some localities could not be determined with certainty and those records were omitted.

(c) Contributions received direct from observers. This was by far the greatest source of records, yielding in excess of 30 000 by Dec. 1984. More than 200 people contributed (they are acknowledged on p. XI), several providing more than 5000 records, although naturally there were many duplicates. In contrast to the first two categories, most of these records refer to the recent period, 1970-1984. The publication in 1981 of the *Check-list of the Birds of Kenya* (by the Ornithological Sub-Committee of the EANHS) greatly simplified the process, observers being supplied with copies into which they entered their records. The procedures for vetting records are given in the next section.

(d) Planned safaris. The authors and other observers made a number of visits to poorly covered areas specifically for the purpose of obtaining atlas records. More commonly, contributors modified their routes whilst travelling through Kenya, making additional stops purposely to seek records in various places. In addition, David Pearson has covered much of Kenya whilst recording the seasonal distributions of Palearctic migrants (Pearson 1978, 1982, 1984b). Despite these efforts, substantial parts of eastern Kenya, and to a lesser extent the north-west, were scantily covered (see p. 9).

(e) Nest record cards were a major source of information on breeding. The many thousands of cards for Kenya were checked up to the end of 1983, and appropriate information was abstracted. Queries were resolved by corresponding with the observers.

(f) The East African Ringing Scheme. Ringing in Kenya has been concentrated at several places, including Nairobi 75B, Magadi 75C, Nyanza and the coast. Of particular importance has been the ringing station at Ngulia Lodge in Tsavo West National Park 101A (Pearson & Backhurst 1976b; Backhurst & Pearson 1977ab-1984; Backhurst *et al.* 1986). Several Palearctic migrants have been recorded mainly or even exclusively from this site, for example River Warbler (688) and Savi's Warbler (689). In addition, the East African Bird Ringing Reports were abstracted (Blencowe 1960b, 1962d; Smart 1967; Backhurst 1968 annually to 1974, 1977, 1981b).

(g) Comparatively little sea-watching has been done off Kenya's coast, so that pelagic species are the least known sector of the avifauna. But observations such as those of Britton (1981), Taylor (1982a),Pearson & Coverdale (1984), and the Hemphills' (EABR 1982-1985), show that increased efforts would be rewarded.

Categories of data

The atlas maps use symbols to indicate the nine categories of records. These are set out in Table 2. A record in any recent category (on or since 01 Jan. 1970) was considered to supersede an earlier

Table 2. Mapping symbols

Status*	Earlier (or pre) records (up to & including 31 Dec 1969)	Recent (or post) records (on or since 01 Jan 1970)
Presence		
Probable breeding		
Confirmed breeding		

Combined records, occurring only where an Earlier record has a higher status than a Recent one (see text) are:

Recent presence and earlier probable breeding	
Recent presence and earlier confirmed breeding	
Recent probable breeding and earlier confirmed breeding	

* See Table 3 for definitions of breeding status.

Table 3. Categories of observations considered as indicating *probable breeding* or *confirmed breeding*, based upon the European grades as set out in Sharrock (1976)

Probable breeding	Confirmed breeding
Pair observed in suitable habitat in breeding season	Distraction display or injury feigning
Evidence of a permanent territory (e.g. song) on at least two occasions a week or more apart, at the same place	Used nest or egg shells found
	Nest with eggs
	Adult incubating
Egg in oviduct (museum specimen)	Adult entering or leaving nest hole (or similar site) in circumstances indicative of an occupied nest
Courtship and display by non-migratory species	Adult carrying faecal sac
Visiting probable nest site	Adult carrying food for young
Nest building activities, including hole excavation	Recently fledged young of nidicolous species, or downy young of nidifugous species
Agitated reaction of adult to observer	Adult feeding young (only for non-migratory species)
Brood patch (on bird in hand)	

Note that attempts at breeding which are known to have failed prior to egg-laying are not included in either category.

record of equal or lower status. Thus, if a bird was present in a particular QSD in both 1965 and 1984, it is shown only as a recent record. But had the 1965 record been one of probable breeding whilst that of 1984 was only a 'presence' record, then the combined symbol would be 'recent presence and earlier probable breeding' (see Table 2). For species whose squares are few enough simply to be listed, rather than illustrated by a map, earlier records are termed 'pre', whilst recent ones are termed 'post'. Pre-1970 (*i.e.* earlier) records that were superseded by recent ones were not usually retained in our files, unless the species concerned is rare in Kenya.

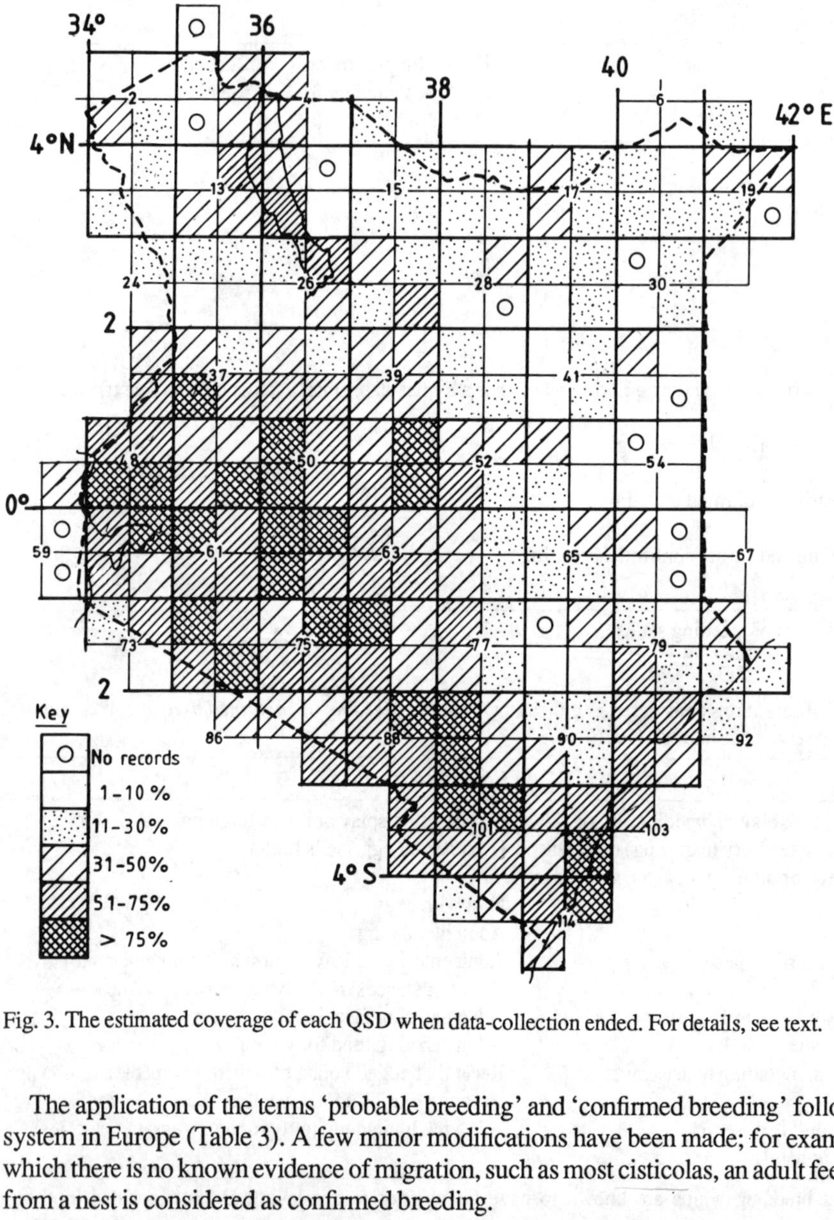

Fig. 3. The estimated coverage of each QSD when data-collection ended. For details, see text.

The application of the terms 'probable breeding' and 'confirmed breeding' follows a widely used system in Europe (Table 3). A few minor modifications have been made; for example in species for which there is no known evidence of migration, such as most cisticolas, an adult feeding young away from a nest is considered as confirmed breeding.

Where necessary, we corresponded with observers to clarify the categories of their records, or concerning any unusual records or those with insufficient detail. Records of rarities or of species seen in unusual localities (category 'S' of the EABR (see for example EABR 1983, p. 107)) are included only if they have been accepted by the Ornithological Sub-Committee of the EANHS. Other records of less common species were adjudicated by ourselves, basing our judgement on (a) ranges accepted by Britton (1980a), (b) the premise that distributions invariably follow some pattern, related to the species' ecology, and (c) our personal experience.

The atlas shows the broad patterns of distributions of birds on a national scale, rather than specific localities. In a small proportion of cases, records which could not be located precisely were entered in the square where we thought it most likely that they had been made. This happened more frequently in the drier parts of Kenya, where topographical maps show so little detail that the observer himself often experiences difficulty in knowing his precise location.

Although the title of this book is *A Bird Atlas of Kenya*, a few of the observations included on the maps were made just outside the country, although within a square that is also partly Kenya (examples of such QSDs are 2C, 31A, 48A and 86B). This is because many of the areas along Kenya's borders are poorly covered by birdwatchers and by including records from border squares, regardless of which side of the border they come from, the patterns of distribution become clearer. However, this procedure was not followed when (a) the species concerned was rare in Kenya or new to the Kenyan avifauna, or (b) the record came from a habitat not found in the Kenyan part of the square.

Full information has been kept on all unusual records, so that if necessary re-assessment will be possible in the future. All records relating to the Atlas, except those derived from published sources, are deposited in the National Museum, Nairobi, and the Library of the British Museum (Natural History) at Tring. Copies of the computer diskettes used to produce the maps are also held by both of these institutions.

Methods of compilation

Data from all sources, after checking where necessary, were entered into copies of the *Check-list of the Birds of Kenya* (OS-C 1981), using one column for each QSD. For the less usual records, the entries were initialled for easy reference to their sources. In turn, the data from the check-lists were used to prepare draft maps for each species: there were thus two complete sets of data in addition to the original records from which they were compiled.

The maps in this atlas are reproduced from originals made on a BBC Microcomputer Model B, using a programme by A.R.D. Taylor (see Appendix). To ensure accuracy, the computer maps have been carefully checked against the hand-entered originals. There were two major advantages in entering the data into the computer: firstly in producing the maps, and secondly in analysing the distributions (see p. 13).

Coverage

Figure 3 shows the estimated coverage for each QSD, based on the method of Pomeroy & Lewis (1983). Briefly this was as follows. We assumed that the number of species recorded from a square reflected both habitat diversity and observer effort. Two measures of habitat diversity were used: the altitudinal range within the square, r, and the number w of types of water body it contained. Observer effort, e, was scored on a 10-point scale, with a maximum of 5 for effort by observers resident in the square and a further 5 for visiting birdwatchers. Using updated data for 33 comparatively well-known QSDs, we derived a multiple linear regression to predict the number of species, S:

$$S = 255.5 + 0.033r + 24.6w + 19.2e$$

The equation was highly significant statistically ($P \ll 0.001$).

We then proceeded to estimate the maximum number of species to be expected from each QSD on the assumption of maximum observer effort (i.e., e was set to 10). Then the observed numbers of species for each QSD were expressed as percentages of the estimated maxima, and the results plotted to give Figure 3. No allowance was made for part squares having a smaller area.

The highest estimated total of possible species for any one QSD is 643, shared by 36D and 63A; they include Mts Elgon and Kenya respectively. The highest actually achieved were 618 in 75B (Nairobi: 100% of estimate) and 552 in 62A (Nakuru: 90%).

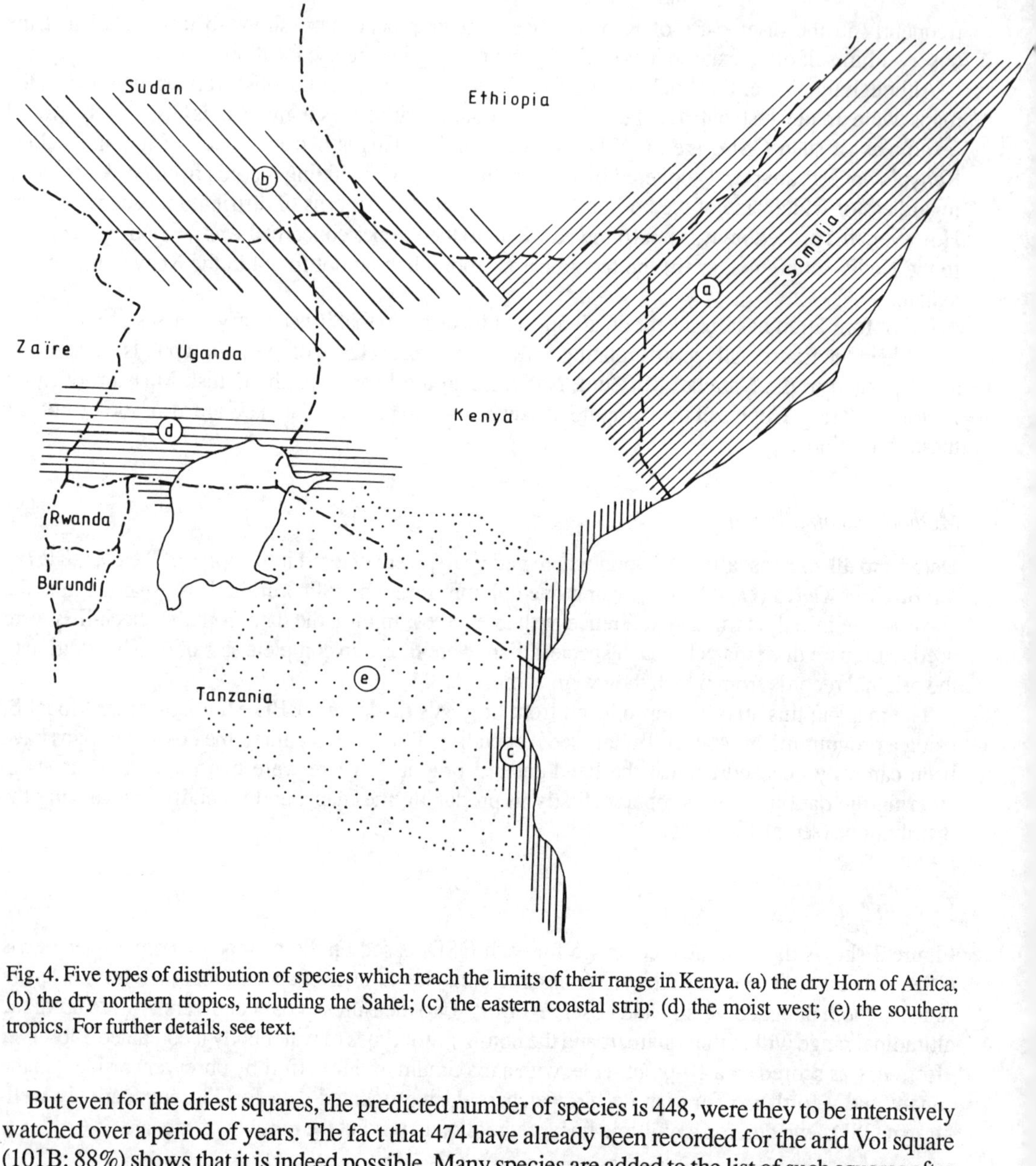

Fig. 4. Five types of distribution of species which reach the limits of their range in Kenya. (a) the dry Horn of Africa; (b) the dry northern tropics, including the Sahel; (c) the eastern coastal strip; (d) the moist west; (e) the southern tropics. For further details, see text.

But even for the driest squares, the predicted number of species is 448, were they to be intensively watched over a period of years. The fact that 474 have already been recorded for the arid Voi square (101B: 88%) shows that it is indeed possible. Many species are added to the list of such squares after rare but spectacular spells of prolonged rain, such as much of Kenya experienced in 1977.

Only 13 QSDs completely evaded observation. Luckily, they are widely scattered, principally in the east where the topography and vegetation are comparatively uniform. Two others (59BD) are in Lake Victoria. We believe that despite these gaps the *patterns* of distribution for all but a few species are clear from the maps. Nevertheless, it is also true that from our estimates, a coverage of more than 50% was only achieved in 72 squares, almost all of them in the highlands and coastal regions. Thus, Kenya's highland avifauna is in general well covered, whilst the dryland squares, almost equal in richness and diversity, are not. Compare, for example, the distribution of the Tawny-flanked Prinia

(743), for which coverage is probably complete or nearly so, with its equally conspicuous dry country relative, the Pale Prinia (744), where large gaps are apparent. For less conspicuous species, the dry country coverage is apt to be poor: see for instance Red-naped Bush Shrike (848) and Grosbeak Canary (1060).

Breeding data for dry country species are particularly sparse. It is obvious for example that Williams' Bush Lark (535) must breed in Kenya, since it is endemic, yet we do not have a single breeding record. It is in fact difficult to obtain breeding records in arid areas, because a majority of the species there only breed after rains, which are sporadic, and which also make many such areas impassable by road. Such coverage as there is for dry country species' breeding is largely confined to the more accessible places like the Tsavo 89C 101AB and Amboseli 88C National Parks, or the few localities where birdwatchers have been resident, such as Baringo 50A and Wajir 42A. Thus for species with a wide distribution, such as the Little Bee-eater (442), the apparent lack of breeding in the north could be real, but might equally well reflect the lack of observers. This source of bias must be borne in mind when interpreting maps of dry country species.

INTERPRETATION OF THE MAPS

Our present knowledge of the requirements of most bird species is very limited, yet such information can be of vital importance in understanding their distributions and conservation needs. Conversely, by attempting to analyse distributions we may gain insights into the species' requirements.

Distributions can be considered at two levels. Firstly there is the species' *range*, which is the overall geographical area within which it occurs. The maps in Brown *et al.* (1982) and Mackworth-Praed & Grant (1957, 1960) are of this type. Secondly, there are the particular localities within that range where the species is found: these are largely a reflection of its habitat requirements. The maps in this atlas, whilst showing more detail than the overall range, are not of large enough scale to show individual localities. As an example, the Common Bulbul (609) has a range which includes almost the whole of the Afrotropical Region and extends into the Palearctic. But even within Kenya it is absent from some parts, notably the montane zones and open grasslands, whilst in the drylands it is largely confined to wooded habitats along watercourses.

Types of distribution

Several attempts have been made to analyse the distributions of birds in the Afrotropical Region as a whole (notably Moreau 1966, Crowe & Crowe 1982), or substantial parts of it (*e.g.* Diamond & Hamilton 1980).

Many types of distribution are represented on the maps in this atlas. Firstly, on a global scale, Kenya represents only a very small part of the world range of such species as the Osprey (141) and the Barn Owl (381). A larger number of species occur throughout most of the Afrotropical Region, *e.g.* the Hamerkop (44), the Didric Cuckoo (374) and the Drongo (566). Some Palearctic species also occur throughout the Afrotropical Region during the northern winter: the Wood Sandpiper (263) and Eurasian Swallow (553) are in this category. At the other extreme are Kenya's most localised endemic species, *e.g.* the Sokoke Scops Owl (383) and Clarke's Weaver (959), both restricted to the Sokoke-Arabuko Forest 102B (see p. 28).

Species at the edge of their range

There are a large number of species that occupy discrete areas of the Afrotropical Region, and many of these reach the edge of their range in Kenya (see p. 31). Five categories can be recognised (Fig. 4).

a) Birds originating in the Horn of Africa, *i.e.* Ethiopia, Somalia and perhaps north-eastern Kenya, and having their western or south-western limits in Kenya or north-eastern Tanzania. Examples include Somali Bee-eater (451), Pale Prinia (744), Golden Pipit (824), Shelley's Starling (888) and Golden-breasted Starling (893).

b) Species that occur from the west of Africa across the northern tropics, in some cases along the arid southern edge of the Sahara (the Sahel), to reach the south-eastern extremity of their range in north-west Kenya. Examples include Fox Kestrel(157), Abyssinian Roller (453), Red-breasted Wheatear (637) and Yellow-billed Shrike (868).

c) Species confined to the east coast of Africa, extending only limited distances inland, *e.g.* Mangrove Kingfisher (440), East Coast Batis (797), Black-breasted Glossy Starling (877), Mouse-coloured Sunbird (936) and Zanzibar Red Bishop (973).

d) Species from the west African equatorial forests or adjacent lowlands which extend eastwards to Kakamega and other western forests. Examples are Grey Parrot (346), Grey-throated Barbet (489), Splendid Glossy Starling (882), Green-throated Sunbird (933) and Red-headed Malimbe (968).

e) Species which, widespread in southern Africa, have their northern limits in southern Kenya. Examples are Rufous-bellied Heron (33), Dickinson's Kestrel (154), Brown-hooded Kingfisher (437), Green-capped Eremomela (766) and Southern Red Bishop (971).

Ecological factors affecting distributions

Within its overall range, the distribution of a species reflects its ecological requirements. In some cases these are obvious – Pel's Fishing Owl (388) is virtually restricted to large rivers, the Scarlet-tufted Malachite Sunbird (928) to the montane zones and the Mountain Greenbul (615) to highland forests.

However, amongst the more widespread species, factors that limit their distributions are often less obvious. Indeed, the requirements of the majority are simply not known, for even though we can say that the Buff-bellied Warbler (749) is 'characteristic of acacia canopies in bushed, wooded and forest edge environments', it is not the trees themselves that it needs (except for nest sites), but certain of the insects that live on them. There are likely to be several hundred species of insects to be found on a tree such as *Acacia xanthophloea*, only comparatively few of which will be taken by the warbler. The almost total lack of knowledge at this level prevents us from achieving a better understanding of birds' distributions.

One of the more puzzling patterns of bird distribution in Kenya, seen in a number of species, consists of two more or less separate populations, one at the coast and the other in the highlands. It is true that these are the wettest areas of Kenya (see Fig. 6) but in fact rainfall at the coast is substantially lower than in the highlands, and, because it is hotter and has a fairly severe dry season, the coast is much less humid (Fig. 7). Furthermore, the structure and composition of coastal forests is very different from those in the highlands. In many species with both coastal and highland populations, two races are involved, suggesting that they have begun to evolve differences in the two areas in response to the different environments. Examples are the Speckled Mousebird (425), Black-headed Batis (798) and Black-winged Red Bishop (974).

Nevertheless, there are several environmental variables which are well known to influence distribution, albeit indirectly in most cases. For aquatic species, the extent and type of water bodies are obviously important. Similarly, many bird species, both aquatic and terrestrial, are restricted to a certain range of altitudes. It is extremely unlikely that they are actually responding to altitude itself, but to a large extent altitude determines temperature. In turn, temperature controls many aspects of plant growth, and all birds depend, directly or indirectly, upon plants. So by relating bird distributions to altitude, we are taking a short cut: and altitude is familiar and easy to determine.

Another major influence, particularly on terrestrial species, is that of rainfall. Some species, such as sandgrouse and doves, need water to drink, but the main effects of rainfall are again indirect, especially through plants. Three aspects of rainfall are important: its amount, its seasonality and its reliability. The last of these is quite closely correlated with the first: the lower the mean rainfall, the less reliable it is.

Taken together, temperature and rainfall are important in determining the type of natural vegetation to be found in an area; a third factor is soil type, which can modify the vegetation considerably. Increasingly too, man himself is deciding upon the type of vegetation he wants on his farm or in his garden or pond, although he too is constrained to some extent by temperature, rainfall and soil.

We have already mentioned man's impact upon the distributions of birds through such activities as deforestation (see p. 3). But of course some birds have spread as a result of man's activities. A good example is the Little Swift (424) which, formerly confined to the vicinity of cliffs for breeding, has increased both its range and numbers by nesting on permanent buildings. Also spreading is the House Sparrow (992), which was introduced to the coast by man early in this century. Its numbers remained low for many years but in the early 1980s it began spreading quite rapidly inland, apparently following the road from Mombasa 114B towards Nairobi 75B (Lewis 1983c).

Relating distributions to climate and vegetation

In preparing the text for each species, we used a series of transparent overlays, which are reproduced as Figures 5-11. Three of them, for altitude, rainfall and wooded habitats, also appear on the overlay accompanying this book. They are designed to fit over the atlas maps of individual species, to enable the reader to examine correlations of environment and distribution at first hand.

Before proceeding, we would add a word of caution. The existence of a correlation between, for example the distribution of the Dusky Turtle Dove (332) and areas of high altitude, does not *prove* that the dove prefers high altitudes. In fact, as we suggested earlier, it is the plants which provide the dove's food that are likely to matter most to it, and the distributions of the plants may quite possibly be determined by temperature: many highland species are unable to thrive at the higher temperatures of lowland areas.

We considered five environmental factors in relation to the distributions of birds in Kenya: altitude, rainfall, moisture, forests and woodlands, and, for aquatic species, types of water bodies.

Altitude

The altitude overlay (reproduced also as Fig. 5) shows the approximate *average* altitude of each QSD. The information was taken from the 1:250 000 topographical maps of the Survey of Kenya. Clearly, average values have more meaning for parts of the country where the altitudinal range within a QSD is relatively small, as in the east and much of the north, than for mountainous regions. The most extreme case is QSD 63A, which contains Mt. Kenya and has a range exceeding 3500 m. But despite being a rather coarse measure, average altitudes vary sufficiently for the general pattern of the highlands, the eastern plateau and the coastal plains to show clearly.

A good example of a species whose distribution correlates clearly with altitude is the Singing Bush Lark (527), which is mainly found above 400 m but below 1800 m. In contrast, the Robin Chat (670) is a typical highland species, rarely seen below 1500 m.

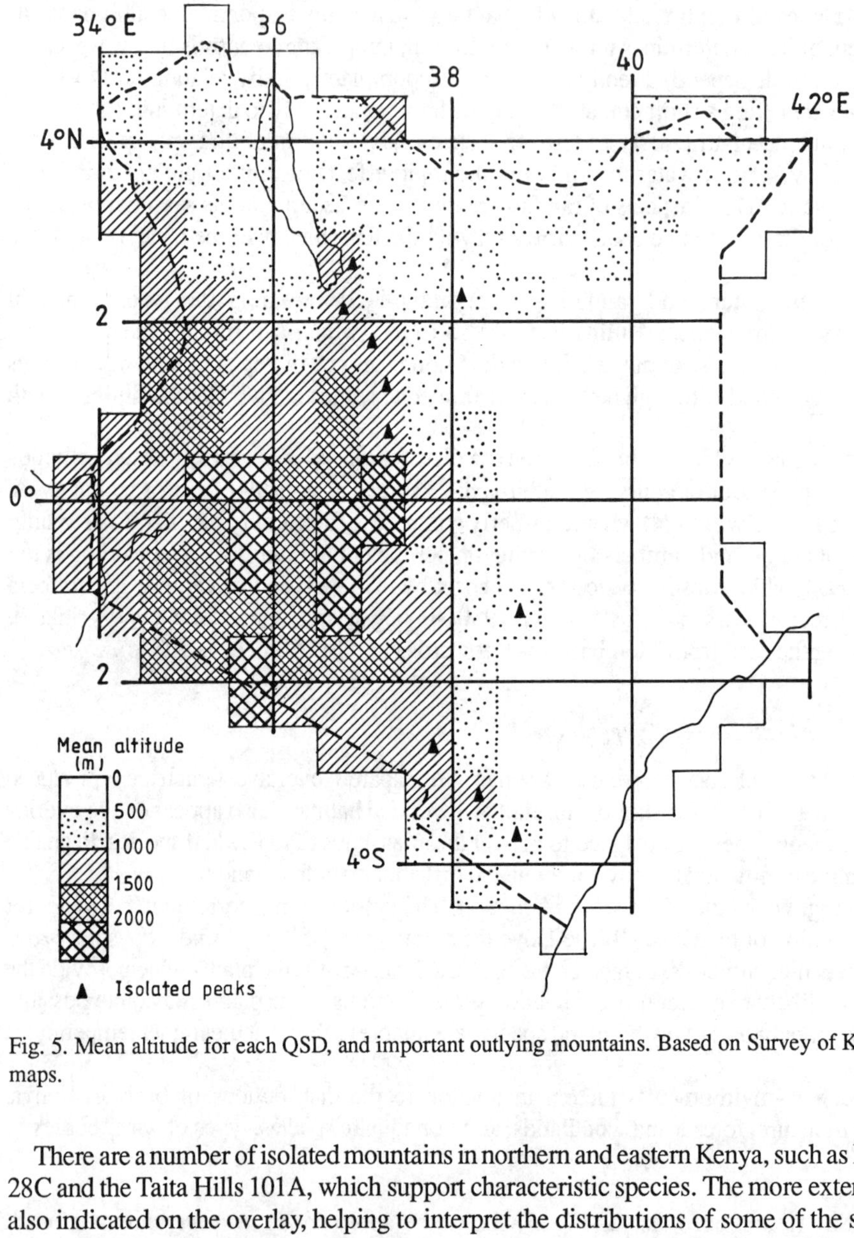

Fig. 5. Mean altitude for each QSD, and important outlying mountains. Based on Survey of Kenya topographical maps.

There are a number of isolated mountains in northern and eastern Kenya, such as Mt Marsabit 27D 28C and the Taita Hills 101A, which support characteristic species. The more extensive of these are also indicated on the overlay, helping to interpret the distributions of some of the species occurring near their summits. Examples are the Eastern Double-collared Sunbird (922) and the Montane White-eye (938).

Rainfall

The *National Atlas of Kenya* (Survey of Kenya 1970) provides a map showing mean annual rainfall for the whole country, and this was used as a basis for Figure 6 and the corresponding overlay. There is a distinct similarity between the patterns of Figures 5 and 6: but there are important differences too. Most notable is the zone of higher rainfall along the coast.

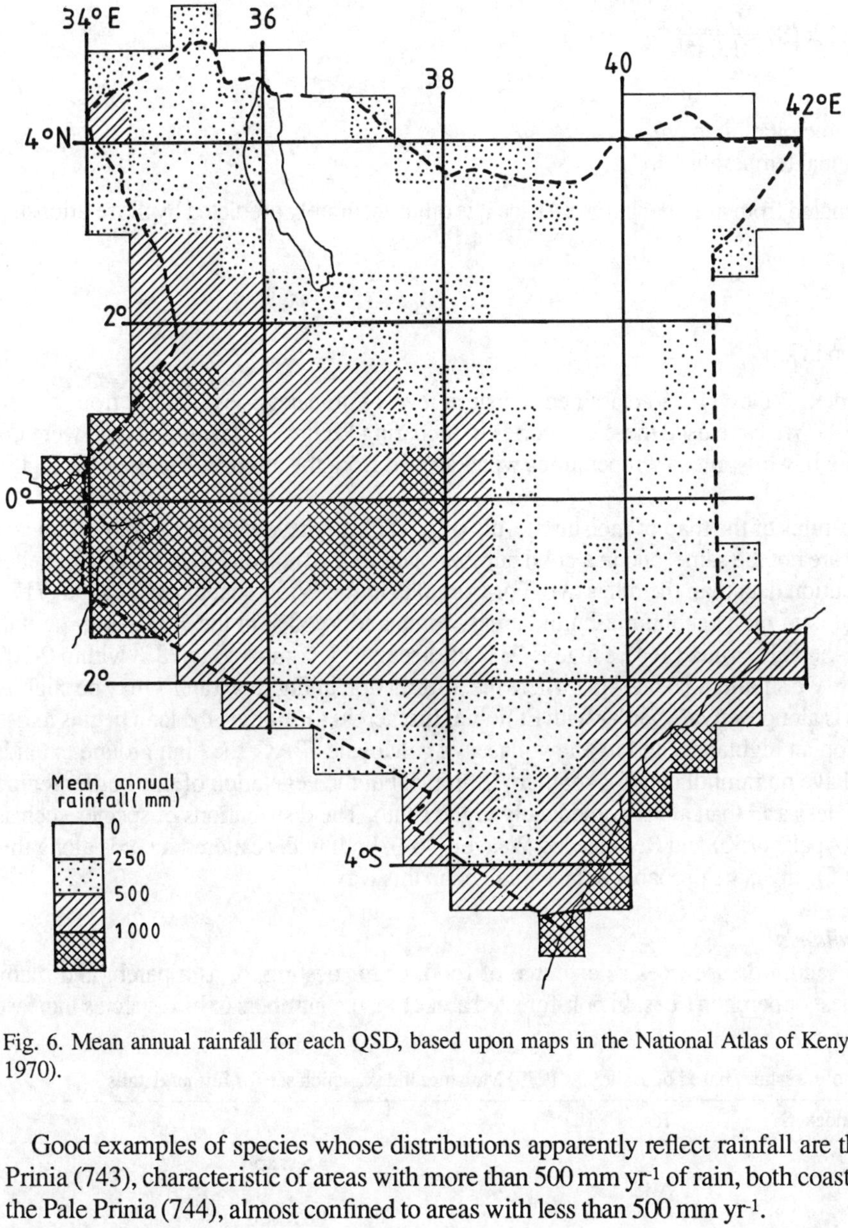

Fig. 6. Mean annual rainfall for each QSD, based upon maps in the National Atlas of Kenya (Survey of Kenya 1970).

Good examples of species whose distributions apparently reflect rainfall are the Tawny-flanked Prinia (743), characteristic of areas with more than 500 mm yr^{-1} of rain, both coastal and inland; and the Pale Prinia (744), almost confined to areas with less than 500 mm yr^{-1}.

Moisture index

The indirectness of the effects of both altitude and rainfall on bird distributions has led us to consider another measure, namely a moisture index. Many studies have shown that plant growth and production are closely correlated to both temperature and rainfall (see for example Leith 1972). These two parameters contribute to estimates of evapo-transpiration, together with estimates of radiation and wind, but the latter are rarely available. We therefore used the simpler moisture index. One of the more generally accepted versions is Bailey's Moisture Index (Bailey 1979), which takes the form:

$$\text{Moisture Index (S)} = \frac{0.18P}{1.045^t}$$

where

P = annual precipitation in cm

t = mean annual temperature in °C,

t can also be estimated from altitude; in East Africa it is quite accurately predicted by the relationship (Kenworthy 1966):

$$t = 30.2 - 6.5E$$

where

E = elevation in km.

Based on this index, Bailey (1979) classified moisture levels into categories ranging from arid to perhumid (Table 4). We have used these categories in preparing Figure 7. The rainfall data were the same as for Figure 6, whilst mean temperatures were derived from the altitude data used for Figure 5.

The general features of the map of moisture levels (Fig. 7) clearly resemble those for rainfall and altitude, but they are not the same, and in a good number of cases the moisture map provides a better 'fit' to the distribution data than the other two. For example, in the Wing-snapping Cisticola (717), 90% of the range is in moist subhumid-humid areas, compared to 63% in the 1000+ mm rainfall category. And in the Pale Prinia (744), range is 93% arid-semiarid, compared to 78% within 0-500 mm rainfall areas. We should note one area where the information in Figures 6 and 7 may be slightly misleading. This is along Kenya's border with Ethiopia. Just across the border, the land begins to rise towards the Ethiopian highlands, and in a few places, *e.g.* Moyale 17AC, the high ground extends into Kenya. We have no rainfall data for southern Ethiopia, but the vegetation of this border region suggests that it is less arid than areas immediately to the south. The distributions of species such as Yellow-breasted Apalis (750) and Red-headed Weaver (969), both with scattered records along this northern border of Kenya, are probably to be explained in this way.

Forests and woodlands

Many species of landbirds use trees as a source of food, or for nesting, or as a perch: and many waterbirds also nest or perch in trees. In non-forested areas too, the numbers of bird species increase

Table 4. Categories of moisture, based on Bailey's (1979) Moisture Index, which see for further details

Value of Moisture Index, S	Term	
		Most arid
1.2	Arid	. . . .
		Less arid
2.5	. . . .	
	Semiarid	
4.7	. . . .	
	Dry subhumid	
6.4	. . . .	
	Moist subhumid	
8.7	. . . .	
	Humid	
16.2	. . . .	
	Perhumid	

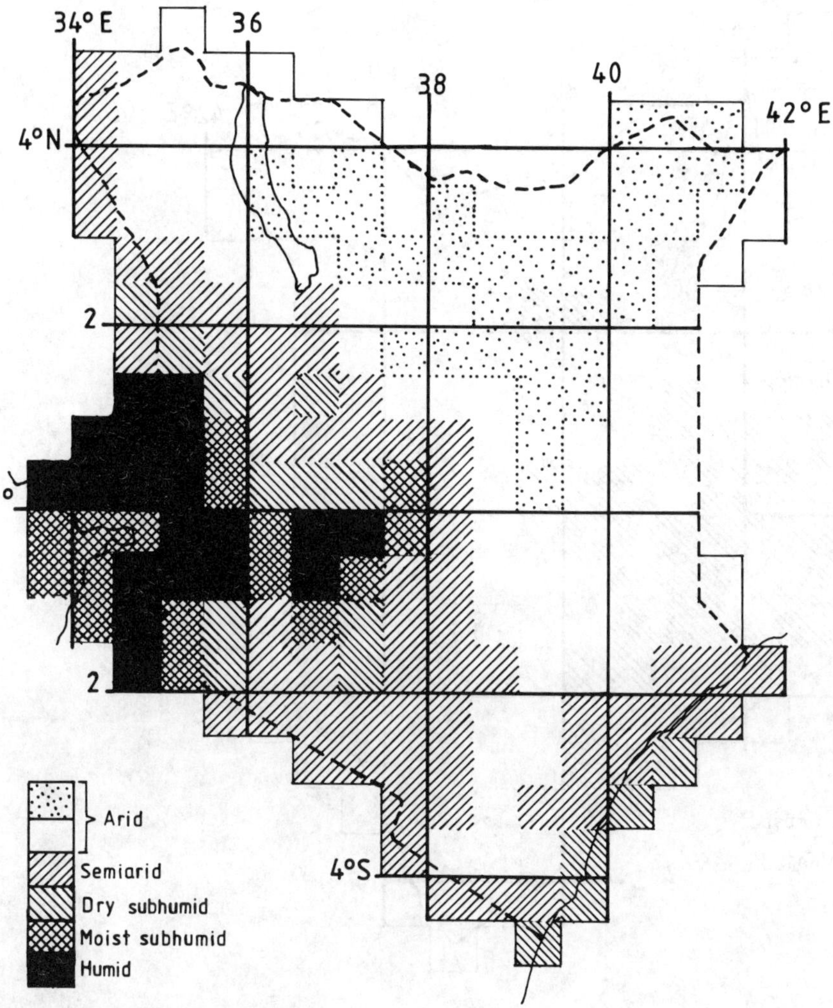

Fig. 7. Values of Bailey's Moisture Index for each QSD, determined as described in the text. See also Table 4.

in proportion to the amounts of woody vegetation – that is, trees and shrubs (Pomeroy & Tengecho 1986). However, by far the greatest bulk of woody vegetation is to be found in forests, which in the tropics are extraordinarily rich in insect species. These in turn support a rich diversity of birds, almost all of them breeding in the forest. (In contrast, whilst semiarid woodlands may support as many species of birds as a forest, far fewer of them breed in any particular woodland (Pomeroy & Lewis 1987).)

In Figures 8 and 9, forests and woodlands are shown separately, but are combined on the overlay. Ecologists generally define forests as places where the canopies of the trees form a continuous layer, and indeed canopies of individual trees commonly overlap each other. We took the distribution of forests mainly from the 1:250 000 topographical maps of the Survey of Kenya, supplemented by our own field knowledge and that of several colleagues. The minimum extent of forests considered for Figure 9 and the overlay was not defined precisely but was never less than 2 km^2 and usually more than 10 km^2. Several types of forest were included.

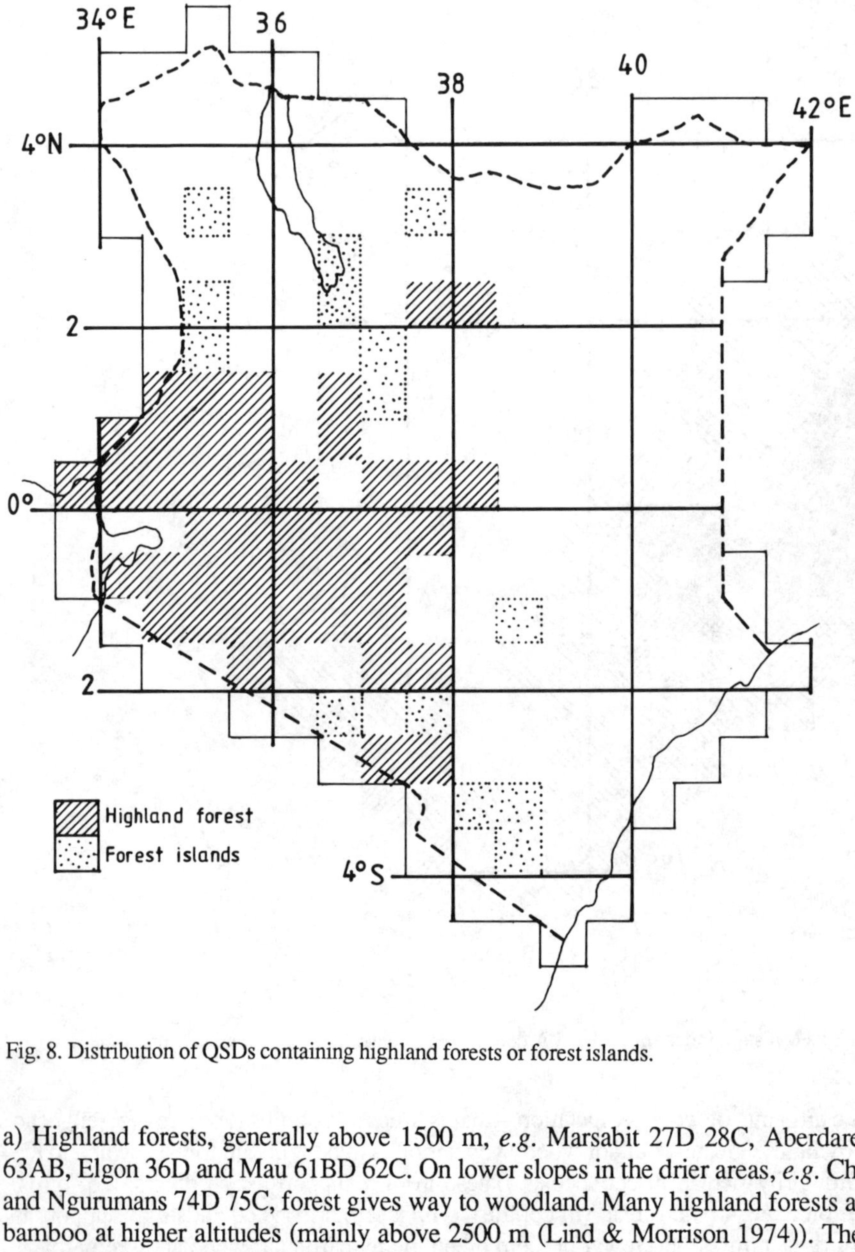

Fig. 8. Distribution of QSDs containing highland forests or forest islands.

a) Highland forests, generally above 1500 m, *e.g.* Marsabit 27D 28C, Aberdares 62BD, Kenya 63AB, Elgon 36D and Mau 61BD 62C. On lower slopes in the drier areas, *e.g.* Cheranganis 37CD and Ngurumans 74D 75C, forest gives way to woodland. Many highland forests are dominated by bamboo at higher altitudes (mainly above 2500 m (Lind & Morrison 1974)). The only published studies of highland forest birds in Kenya are those of Zimmerman (1972) and Carlson (1986a).

b) Island forests are of great interest and importance (see for example Diamond & Keith 1980). Higher altitude forests are found on most of the isolated peaks shown on Figure 5; medium altitude forest islands include the Chyulu Hills 88D and Mt Endau 77B. Groundwater from the Chyulu Hills supports an important forest at nearby Kibwezi 88B. Each of the forests in this category is separated by a considerable distance from the others and from the main highland forest blocks. The significance of this in terms of species distribution is discussed on page 14.

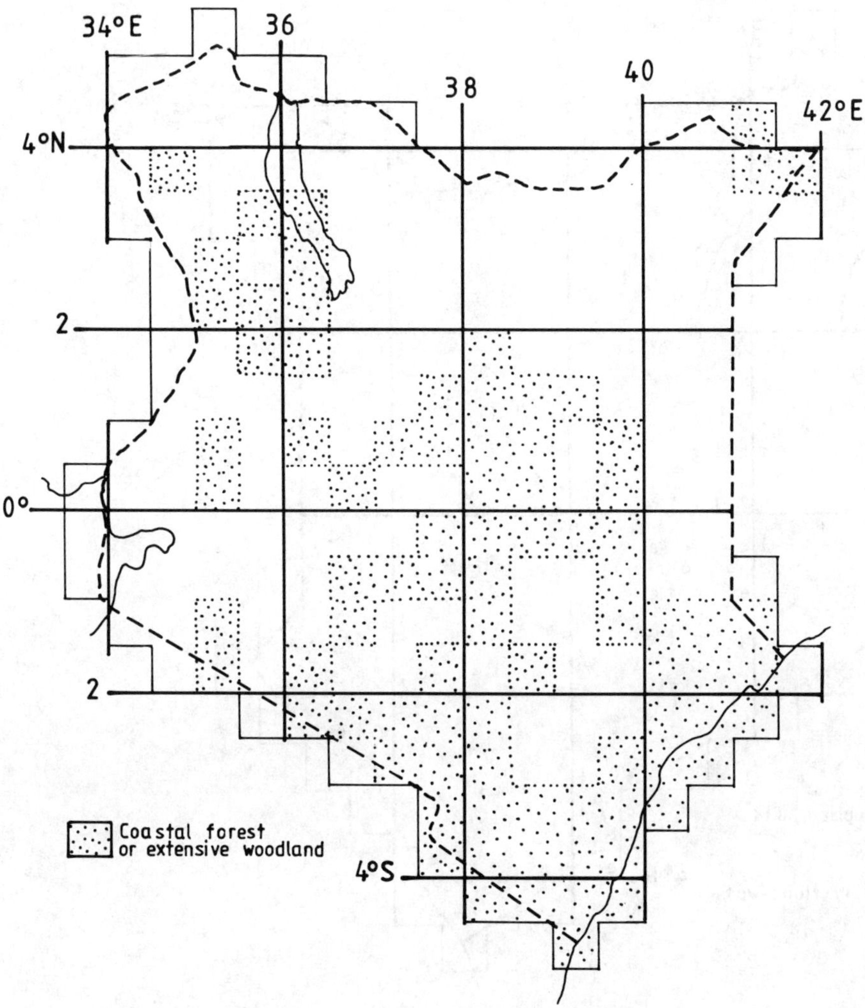

Fig. 9. Distribution of QSDs containing lowland (coastal) forest or extensive woodland.

c) Lowland forests in Kenya are restricted to the coastal strip, and the lower reaches of the Athi and especially Tana rivers. They give way to woodlands on their inland, drier sides. The Sokoke 102B and neighbouring coastal forests were important refuges during the dry glacial periods, and support two endemic species (Britton & Zimmerman 1979).

d) Woodlands. We follow Pratt & Gwynne (1977) in defining a woodland as a place dominated by trees whose canopies cover more than 20% of the land area. The trees are not necessarily tall – some species of *Commiphora* and *Acacia* are only 4-5 m high. Unlike forests, the ground layer of woodlands is typically dominated by grasses. All non-forested squares contain some woodland, particularly along watercourses, but in Figure 8, and the corresponding overlay, we have shown only those QSDs where there is a substantial area. Good examples of extensive woodlands are those along the Turkwell river in northern Kenya 37A 14C, and south of the Athi river in the Kiboko-Voi 88B 101B region. Studies of woodland birds include Lack (1985) and Pomeroy & Muringo (1984).

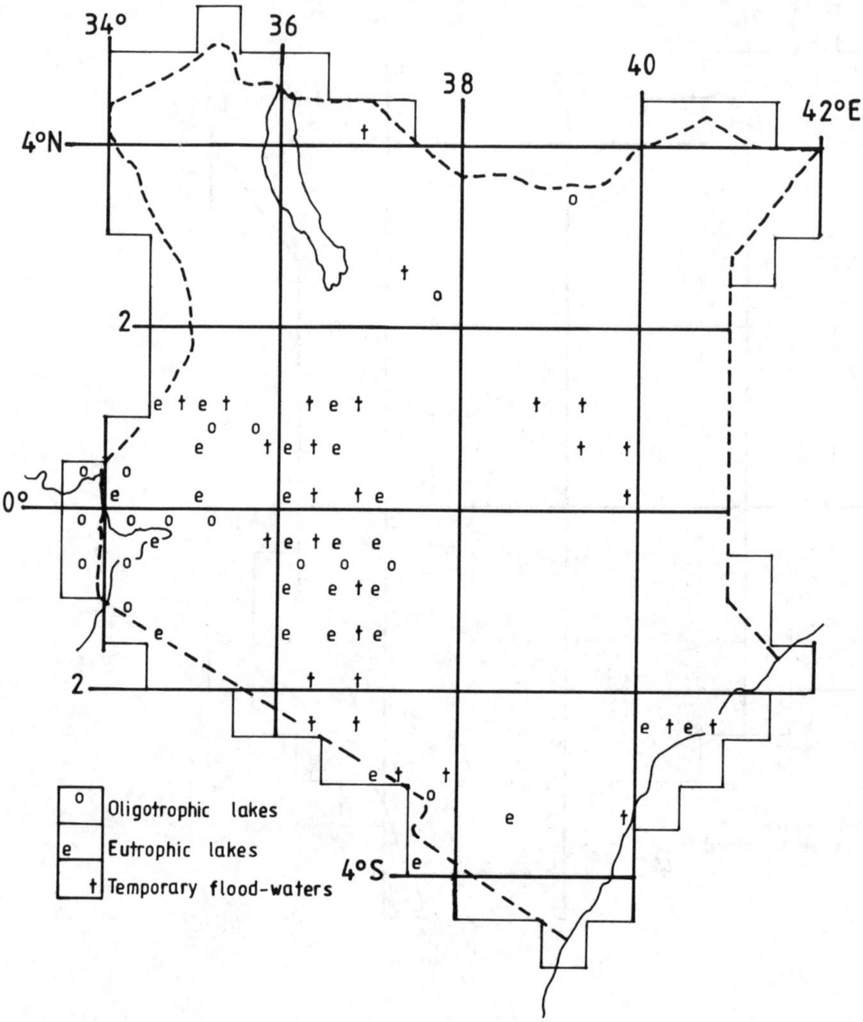

Fig. 10. Distribution of QSDs containing freshwater lakes and temporary flood-waters.

e) Exotic plantations are not distinguished in Figure 8 because all of the larger ones occur in areas where some natural forest remains. So far, only one brief study of these potentially important forests has been made (Carlson 1986a).

Aquatic habitats

Most if not all of the major types of aquatic habitats of tropical Africa are to be found in Kenya: certainly there is a rich diversity, and more than 200 waterbird species have been recorded (Table 1).

a) Marine. The Kenya coast is dominated by two major habitats: sandy beaches and mangrove forests. There is a fringing coral reef along much of the coast, and some offshore rocky islands. Each of these has its characteristic fauna. There are however no breeding sites of pelagic species such as boobies or frigatebirds.

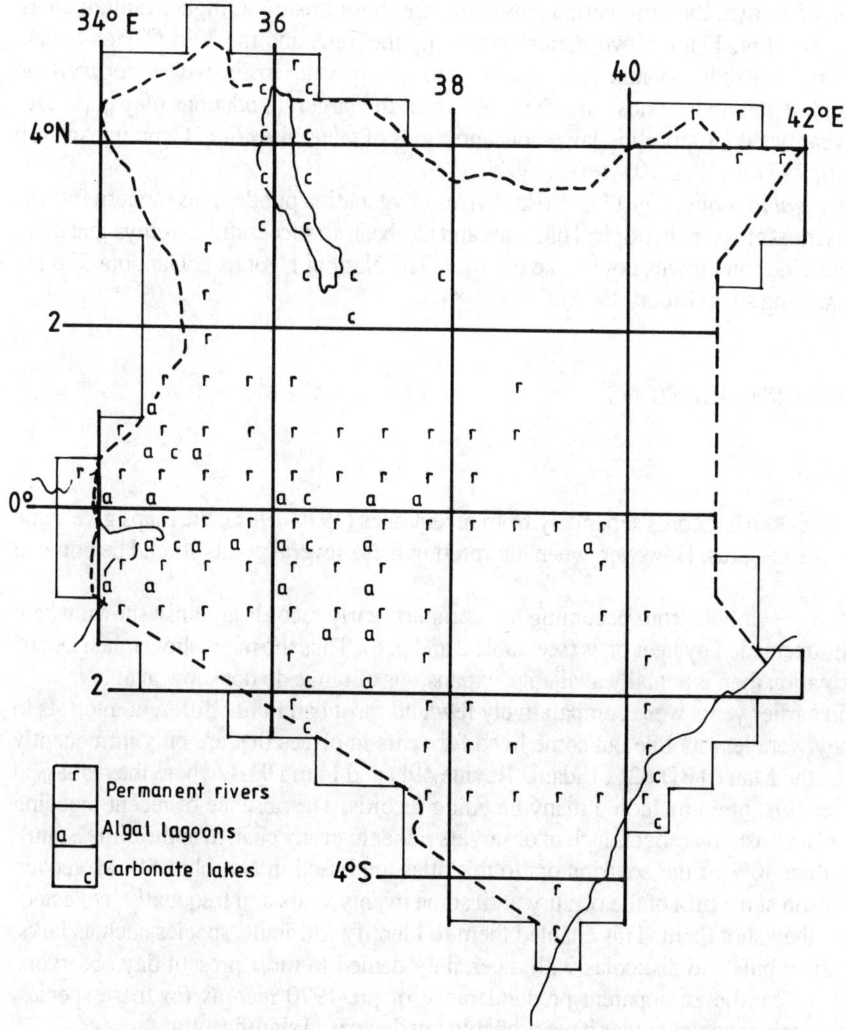

Fig. 11. Distribution of QSDs containing other important water bodies.

The largest concentrations of shorebirds are associated with tidal creeks and estuaries, notably Mida Creek 102B, the Sabaki estuary 103A and the Lamu archipelago 91B 92A, although much of the latter is fringed with mangrove.

b) Inland waters. The numerous *lakes* in Kenya vary greatly in size and composition. Lake Victoria consists largely of water from rain falling directly on to its surface, and consequently it has a low concentration of nutrients. The flora and fauna are correspondingly sparse: it is said to be *oligotrophic* and is one of several such lakes in Kenya (Fig. 10).

Many of the smaller lakes have a higher nutrient status: these *eutrophic* lakes such as Naivasha 62C and Baringo 50A are more productive. A few have very high concentrations indeed of nutrients, especially carbonates, and are generally referred to as 'soda lakes', although the term *carbonate* lake is more appropriate. Well known examples are Turkana, Bogoria 50C, Nakuru 62A and Magadi 75C (Fig. 11). There is an extensive literature on some of these (*e.g.* Vareschi & Jacobs 1985).

In the drier parts of Kenya, the only permanent rivers are those arising in higher rainfall areas. There are few of these (Fig. 11), the two major ones being the Tana and the Athi-Galana-Sabaki system. Occasional periods of heavy rain in the drier regions can produce extensive *temporary flood water* in hollows and depressions almost anywhere. A few of the larger inundations may persist for months or even a year or two, supporting large concentrations of waterbirds (*e.g.* Cunningham-van Someren & Richards 1977ab) (Fig. 10).

Finally we show *algal lagoons* (Fig. 11), also known as oxygenation ponds. This modern form of sewage treatment is most efficient in tropical climates and has been so successful in Kenya that more than 30 of the larger cities and towns now make use of it. The Nairobi lagoons at Dandora 75B are well-known birdwatching sites (Moore 1982).

ANALYSING BIRD DISTRIBUTIONS

Changes of distribution

By showing early (pre-1970) records separately from recent ones (1970-1985), the maps give some indications of changes in status. However, when interpreting these several points should be borne in mind.

i) To prevent map symbols from becoming too complex, early records are only shown where they have not been superseded by later ones (see Table 2 and p. 6). Thus the maps show much less of the pre-1970 information than is actually available; expansions of range do not show at all.

ii) Observers in earlier years were comparatively few and most had rather different interests to those of today. They were less mobile but some lived for years in places that are only infrequently visited now, such as the Mau 61BD 62C, Eldama Ravine 49D and Lamu 91B, where they amassed large numbers of records, often including many breeding records. The absence of recent breeding records from such places usually reflects lack of observers rather than any change in the birds' status. By contrast, more than 40% of the contributors to this atlas are based in Nairobi 75B. Moreover, observers from Jackson at the turn of the century until some twenty years ago frequently 'collected' their birds – that is, they shot them. This enabled them to identify 'difficult' species such as larks, alethes, nightjars, greenbuls and cisticolas with a certainty denied to most present-day observers. Here again, the result can be an apparent preponderance of pre-1970 records for these species, although in most cases their actual ranges have probably not decreased significantly.

iii) Many forest species have obviously declined with the deforestation of large areas of Kenya, most especially the highlands, yet the maps give little indication of this because the QSD is a large unit. Most QSDs in the highlands have much less forest now than a hundred years ago, although they still retain some fragments.

Whilst bearing these facts in mind, we can still discern some probable changes. Despite the large size of a QSD, the maps do suggest losses resulting from agricultural practices.

More detailed studies provide evidence for changes in distribution which, whilst not yet apparent on atlas maps, are likely to be seen in future. Carlson (1986a) found that of 18 species occurring at densities of 1.0 birds ha^{-1} or more in natural forest, only six were found at all in adjacent coniferous plantations, and all at much lower densities. Overall, the natural forest supported about 44 birds ha^{-1}, whilst there were fewer than 11 birds ha^{-1} in the plantations. Interestingly, whilst Palearctic species made up a quarter of the plantation birds, none was found in the natural forest (see p. 28).

Changes taking place in dry woodlands cleared for cultivation were also pronounced, though less drastic than in forests (Pomeroy & Muringo 1984). Of the 15 commonest species in a natural woodland, 14 also occurred in adjacent cultivation, but every one was less common. However, the total densities of birds in the two habitats were similar because some species, notably doves and weavers, were abundant in the cultivation. Lack (1980) described the changes in bird populations in a part of Tsavo East National Park 101B where most of the woody vegetation had been destroyed by elephants *Loxodonta africana*.

But whilst many of the richer terrestrial habitats are declining, some interesting new aquatic habitats are being created, supporting a remarkable diversity of birds. Pomeroy & Tengecho (1982a) studied three of the many small dams east of Nairobi 75B, and found 59 species of waterbirds frequenting them. Oxygenation ponds, such as Dandora near Nairobi, are richer still.

Some well documented changes in bird distributions are hard to explain. No less than 52 of the 618 species recorded in the Nairobi QSD (75B) have not been reported since 1969. It is possible that some of these apparent losses are attributable to habitat change, such as Kenya Crested Guineafowl (178) and Retz's Helmet Shrike (872). But why should the Ashy Flycatcher (785) or Doherty's Bush Shrike (856) no longer be seen in the area?

Habitat changes will continue to occur, and an important function of this atlas is to record bird distributions in a particular period so that future changes can be assessed.

Degree of mobility

The extent to which birds wander outside the breeding season varies considerably. Good examples are to be found amongst the bee-eaters. The White-fronted (441), a highland species, has only once been recorded over 200 km from its breeding range, whereas its congener the White-throated (450), a bird of rather lower and drier areas, is an intra-African migrant.

Evidence for non-dispersion can often be discerned from the maps, even when breeding records are few. Thus the seven breeding squares of the Mountain Greenbul (615) span its total Kenyan range quite well.

Extensive movements are apparent in many aquatic species. The Great White Egret (39) has only bred at five sites in Kenya in recent years, and the Woolly-necked Stork (48) is not known to breed in Kenya at all, yet both species have been recorded over a wide area. Scavengers also travel far: thus it is probable that all breeding colonies of the Marabou (50) in Kenya are known, yet the map shows that the species has been recorded in almost every part of the country.

Species abundance

A few European atlas projects have combined records of distribution with estimates of abundance (Udvardy 1981). The PAOC V meeting recommended that species abundance should be recorded as supplementary data where possible (Ash & Pomeroy 1981). No attempt was made to do this in the case of the Kenya atlas however. There are relatively few data on bird populations in Kenya, although attempts at estimating numbers have been made for a few species of particular interest: thus the Sokoke Scops Owl (383) is believed to number from 1300-1500 pairs (Britton & Zimmerman 1979). Brown (1973a) has given estimates for the population of the Lesser Flamingo (60) throughout Africa, and when the Great White Pelican (16) bred at Lake Elmenteita 62A in 1968, he counted about 1150 pairs, which was the total breeding population in Kenya. From 1979 to 1983, the average breeding population of Marabou Storks (50) in Kenya was estimated at about 265 pairs (Pomeroy 1986).

Regular counts of some waterfowl species have been made for a number of years (Meadows 1979, 1980). Apart from various estimates of flamingo populations between 1954 and 1978 (Vareschi 1978 and Vareschi & Jacobs 1985), these are the only counts in Kenya over a sufficient period for trends to be detected.

Symbol	Method	References
○	Mist-netting	Britton (1978), Britton & Zimmerman (1979); Zimmerman (1972)
⊙	Line transect	Carlson (1986a)
⊗	Plots, transects	Lack (1980: Table 6.5)
●	Line transects	Pomeroy & Muringo (1984); Pomeroy (unpublished)

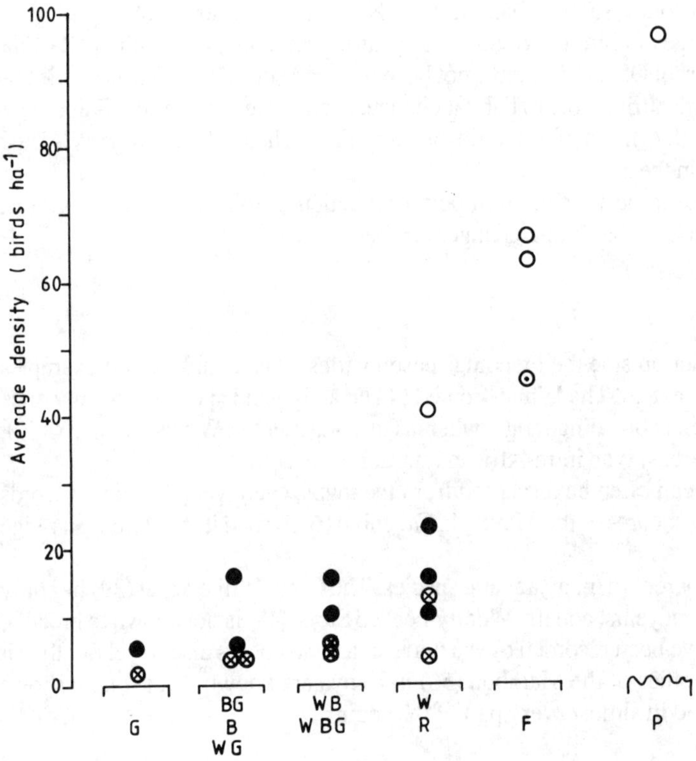

Fig. 12. Estimated densities of bird populations (all species combined) in various natural habitats in Kenya. All are average values, based on series of counts. The habitats are categorised in increasing order of vegetation amount, thus: G (grassland), BG (bushed grassland), B (bushland), WG (wooded grassland), WB (wooded bushland), WBG (wooded bushed grassland), W (woodland), R (riverine forest), F (forest). Exact definitions of these categories are given by Pratt & Gwynne (1977). At the right, P = Papyrus Swamp. As explained in the text, the various data are only rather generally comparable, because of the variety of techniques employed.

Estimates of population density (birds per unit area) have been made for a variety of terrestrial habitats (Fig. 12). The wide range of values probably reflects true differences, despite the fact that several methods were used in deriving them. It is evident that bird populations increase with the amount of woody vegetation (see also Pomeroy & Tengecho 1986). The high figure for papyrus may well be attributed to the high productivity for which this plant is renowned (Jones 1983).

We can safely assume from Figure 12 that, for Kenya as a whole, there must be at least 10 birds ha^{-1}, which would give a population for the whole country of rather more than five hundred million birds.

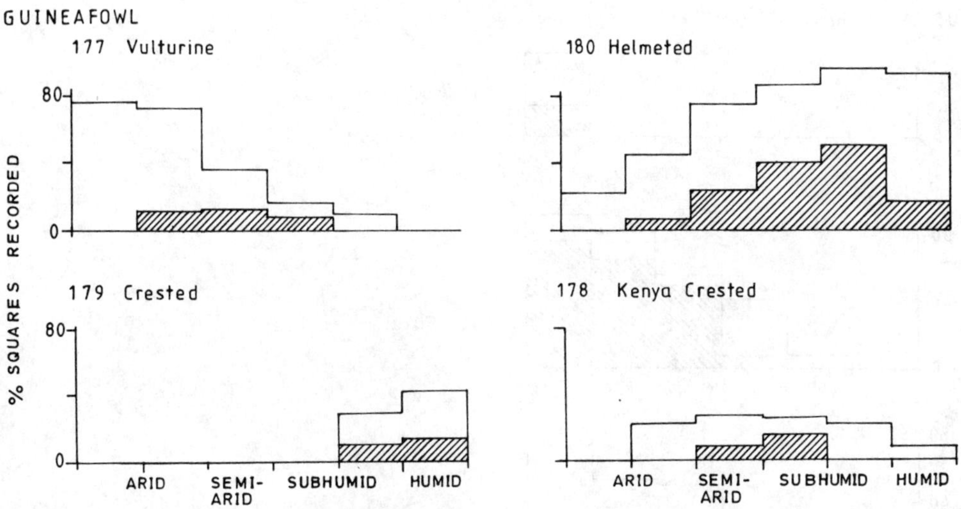

Fig. 13. The proportion of QSDs occupied by all four species of guineafowl in each of the six moisture categories. The shaded areas represent breeding records (including probable breeding). See text for further details.

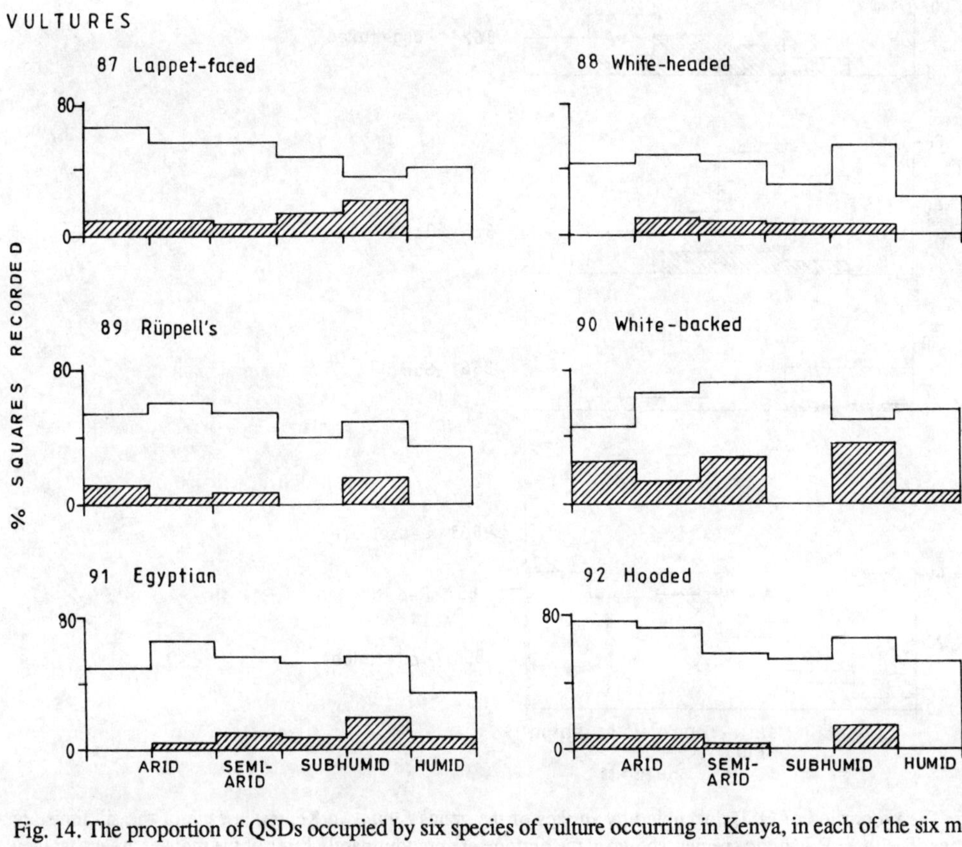

Fig. 14. The proportion of QSDs occupied by six species of vulture occurring in Kenya, in each of the six moisture categories.

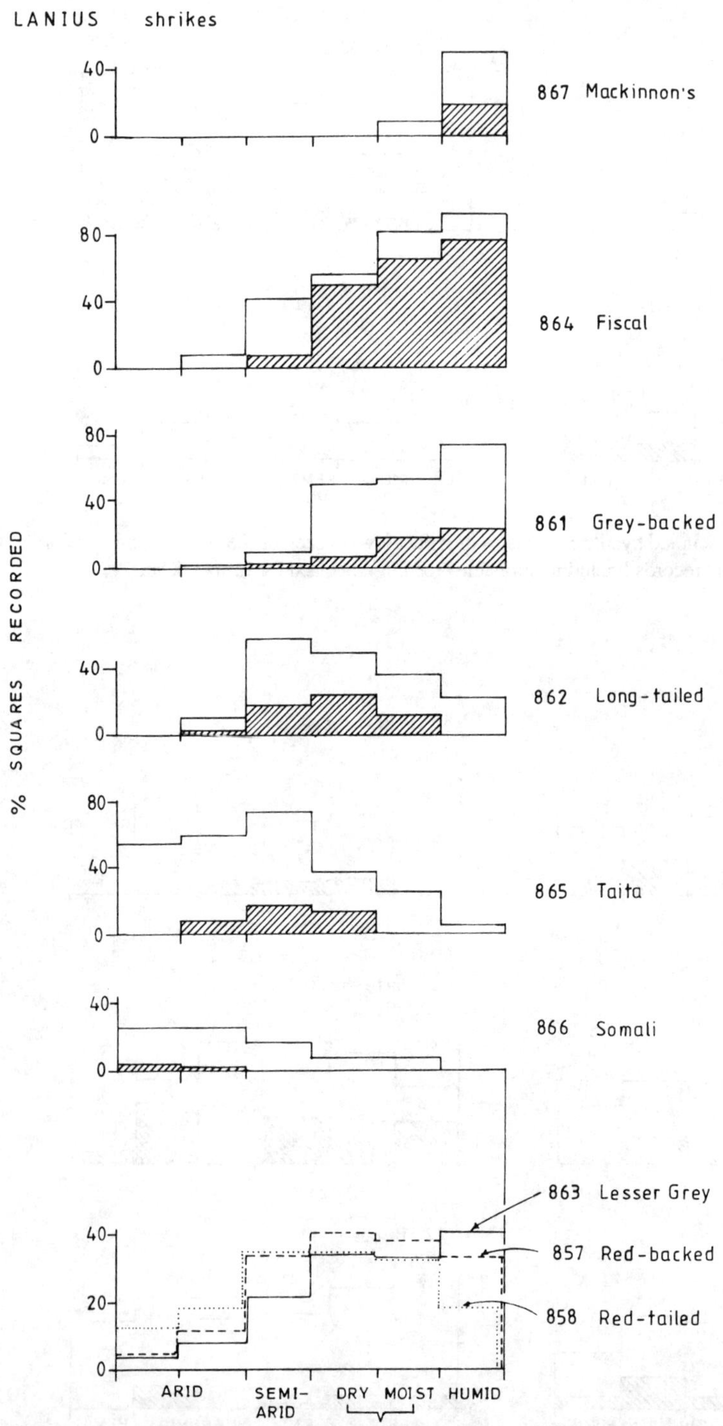

Fig. 15. The proportion of QSDs occupied by shrikes of the genus *Lanius* in Kenya, in each of the six moisture categories. The three Palearctic species, shown at the bottom, are predominantly birds of the moister areas, but they differ ecologically from the resident species.

For closely related species, especially those with similar values for detectability and observer coverage (p. 34), the numbers of QSDs give a rough guide to relative abundance. Thus the Wood Sandpiper (263), recorded from 103 QSDs, is certainly much more numerous than the Terek Sandpiper (264), observed in only 25. Similarly, the Nubian Woodpecker (512: 144 QSDs) can safely be assumed to be much more numerous than the Brown-backed (523: 17 QSDs).

Ecological differences between species

The compilation of all records on computer diskettes greatly simplified analyses of distributions in relation to the variables mapped in Figures 5-11. To illustrate this, we have selected just one variable – moisture – to show how closely related species may be compared.

Figure 13 represents the percentage occurrence of the four Kenyan guineafowls in each of the six moisture categories (Fig. 7). Thus, the Vulturine Guineafowl (177) has been recorded in 27 of the 34 most arid QSDs, shown in the figure at 78%. In these examples, all squares have been included in the analysis, even those with no records. The omission of empty squares, and those with very few records, sometimes helps to elucidate distribution patterns (see Pomeroy, Lewis & Mutere 1982). The figure also shows the percentages of squares with breeding records. Pre- and post-1970 records have been combined.

The four species of guineafowl clearly differ quite markedly in their preferences for moist or arid environments. The contrast between the Vulturine and Helmeted (180) is conspicuous, especially when allowance is made for the scarcity of observers in arid areas. The Kenya Crested (178) and Crested (179) are allopatric, and differ from the other two species in frequenting denser thickets, and forests. Hence although the four species have essentially similar diets, it is unusual to find more than one in any particular place. In this way we assume they avoid competition.

In marked contrast are six vulture species (Fig. 14). Vultures travel great distances (Pennycuick 1972) and, unlike the largely sedentary guineafowl, all species can occur together. But they differ markedly in their feeding behaviour (Kruuk 1967; Houston 1975). For instance, the Lappet-faced (87) is a carcass-opener, the African White-backed (90) feeds manly on viscera whilst the Hooded (92) picks the bones.

Other cases are less simple. Eleven species of *Lanius* (shrikes) have been recorded in Kenya, five being Palearctic migrants, two of which are rare. The nine commoner species are represented in Figure 15, with the Afrotropical species arranged according to their frequency along the moist-arid gradient. The differences between the extremes – Mackinnon's (867) and Somali (866) Fiscals – are very clear. But in between there is considerable overlap, particularly with the Fiscal (864), Grey-backed Fiscal (861) and Long-tailed Fiscal (862). The last two of these are allopatric, but the Fiscal and the Grey-backed occur together in many parts of eastern Africa, including much of western Kenya.

At present, we understand rather few of the reasons why particular species occur where they do, but analyses such as those we have described can help. We expect closely related species to have similar requirements, which might lead to competition between them if they occurred together. Their similarities may be the reason for many members of superspecies being allopatric, as in the case of the Crested and Kenya Crested Guineafowls, and the Grey-backed and Long-tailed Fiscals. But it may also happen that two species with very similar requirements can occur together so long as their numbers are low, allowing enough resources for both. As yet we simply do not know in most cases. But anyone who has watched a vulture party at the height of its frenzied activity can see that there are certainly some instances where species do compete – vigorously.

Further examples are given in Pomeroy, Lewis & Mutere (1982) and Pomeroy (in press, c).

Restricted distributions: endemic species

Whilst most of Kenya's species occur in other countries too, a few are confined to Kenya, and are known as Kenya *endemics*. In addition, a good number of species have races restricted to some part of Kenya. Clearly endemic species and races have limited distributions, a situation which can arise in one of two ways. They may be the surviving relicts of a formerly widespread species, or they may have evolved where we find them today and never spread far from their origin.

There are several reasons why the range of a species may expand or contract – but the biggest cause today is man. Until the recent past, however, changing climate was probably the primary cause. During the past million years or so, the global climate has undergone a series of cold 'glacial' periods, which were comparatively dry, and alternated with milder, wetter periods – the interglacials. Many authorities believe that we are at present living in an interglacial period, rather than that the 'ice ages' are over.

During the past few years, evidence has been accumulating to show that tropical Africa was profoundly affected by these climatic changes. The greatest impact was probably on forests, which shrank in the glacial periods, and expanded in the interglacials. Until recently, when man began the process of deforestation, Africa's forests were probably as extensive as they have been at any time in the past million years, and were certainly much more widespread than 10 000 years ago (Diamond & Hamilton 1980). During the last glacial, there was almost no forest in most of the areas of Kenya where, in 1850, forests were extensive: the highlands, Aberdares 62BD, Mts Elgon 36D and Kenya 63A, and the west, including Kakamega 48D, Nandi 49C and much of the Lake Victoria basin. The only significant area of forest in Kenya that did survive this cold, dry period was at the coast; other forest islands survived in western Uganda, and on several of the lower mountains in Tanzania, most notably the Usambaras.

Those isolated forest refuges which survived the glaciations all contain endemic species – ones that have not subsequently spread as the forests have spread in the past 10 000 years. (See Stuart (1981) for a detailed review of this subject.) There are six endemic species of birds in the Usambaras, and four in the coastal forests of East Africa (two of them endemic to Kenya: the Sokoke Scops Owl (383) and Clarke's Weaver (959)). But as the climate warmed, the refuges acted as centres of recolonisation. The forests of western Kenya were largely recolonised from Uganda and Zaire. About 35 species reach their eastern limit in the Kakamega-Nandi 48D 49C Forests.

There are five other endemic species in Kenya, namely the Violet Wood Hoopoe (460), Williams' Bush Lark (535), Hinde's Pied Babbler (600), the Aberdare Cisticola (723) and the Tana River Cisticola (726). Whether these arose in the interior and failed to spread, or are the last survivors of once widespread species, is a matter for speculation.

Migrants

Nearly 80% of Kenya's bird species breed in Kenya, or are thought to do so (Table 1). Most of the remainder are species that breed in the Palearctic Region during the northern summers but spend their winters in the Afrotropical. To do so, they either cross or skirt the Sahara and other deserts on their annual migrations. Others, the intra-African migrants, come to Kenya as non-breeding visitors from within the Afrotropical Region. Amongst the species that breed in Kenya, many, perhaps half, make regular movements away from their breeding sites for part of the year.

Palearctic migrants from western Europe mostly fly south or south-west to West Africa; few cross the equatorial forests. In contrast, most Palearctic migrants in eastern Africa have come from eastern Europe or from western, central or northern Asia, many from as far east as Siberia. A few originate in the Indian sub-continent, part of the Oriental Region. Many cross the equator and some continue

Table 5A. Percentages of Palearctic migrants in various habitats. Britton (1947f) considered that in SW Kenya, southern passage was mainly in Oct and Nov; birds present in Dec-Jan were presumed to be overwintering. Data were collected in various ways and many sample sites are atypical or unrepresentative. Hence only rather general conclusions can be drawn.

	Locality	QSD	Oct-Nov	Dec-Jan	Feb-Apr	Reference
Areas not extensively affected by man						
Evergreen forest	Kinale	62D	-	-	0	Carlson 1986a
Evergreen forest	Kakamega	48D	-	1	-	Britton 1974f
Evergreen forest edge	Mt Elgon	36D	-	9	-	Britton 1974f
Riverine woodland	Tsavo E NP	101B	-	10	-	Ulfstrand 1973
Moist acacia woodland	Naivasha	62C	-	6	-	Ulfstrand 1973
Dry mixed woodland	Masalani	89A	(0)	12	5	Pomeroy (unpubl.)
Dry acacia woodland	Olturot	27A	5	4	3	Pomeroy (in press, b)
Moist	Kisumu	60B	(22)	(13)	10	Britton 1974f*
Dry	Nairobi NP	75B	-	5	-	Ulfstrand 1973
Dry	Amboseli NP	88C	-	4	-	Ulfstrand 1973
Dry Various wooded	Tsavo E NP	101B	10	4	15	Lack 1980, 1983
Dry and bushed areas	Tsavo E NP	101B	-	12	-	Ulfstrand 1973
Dry	Magadi	74D	-	12	-	Ulfstrand 1973
Dry	Olorgesaille	75C	-	14	-	Ulfstrand 1973
Arid	Olturot	27A	5	4	3	Pomeroy (in press, b)
Lakeshore	Nakuru NP	62A	-	60	-	Britton 1974f
Papyrus swamp edge	Kisumu	60B	2	5	1	Britton 1974f*
Areas extensively affected by man						
Coniferous forest plantation	Kinale	62D	-	-	(23)	Carlson 1986a
Lantana-dominated thicket	Kisumu	60B	(53)	(24)	(16)	Britton 1974f*
Arable farm	Naivasha	62C	-	10	-	Ulfstrand 1973
Moist	Nairobi	75B	-	9	-	Ulfstrand 1973
Moist	Kisumu	60B	6	7	4	Britton 1974f*
Dry Cultivations	Masalani	89A	(0)	10	2	Pomeroy (unpubl.)
Dry	Voi	101B	-	4	-	Ulfstrand 1973

Figures in parentheses are based upon counts of less than 50 birds.
*based upon netted species (his Table 1); other references to this paper are from his Tables 4 and 5.

Table 5B. Summary of data in Table 5A. Data on forests and aquatic habitats (papyrus, lakeshore) have been omitted, as have those based upon small samples. These figures can be criticised on various grounds but they give a first approximation: all are percentages.

	Oct-Nov	Dec-Jan	Feb-Apr
Areas not extensively affected by man	7*	10	7
Areas extensively affected by man	6*	8	3*

* based upon rather few data (*cf.* Table 5A).

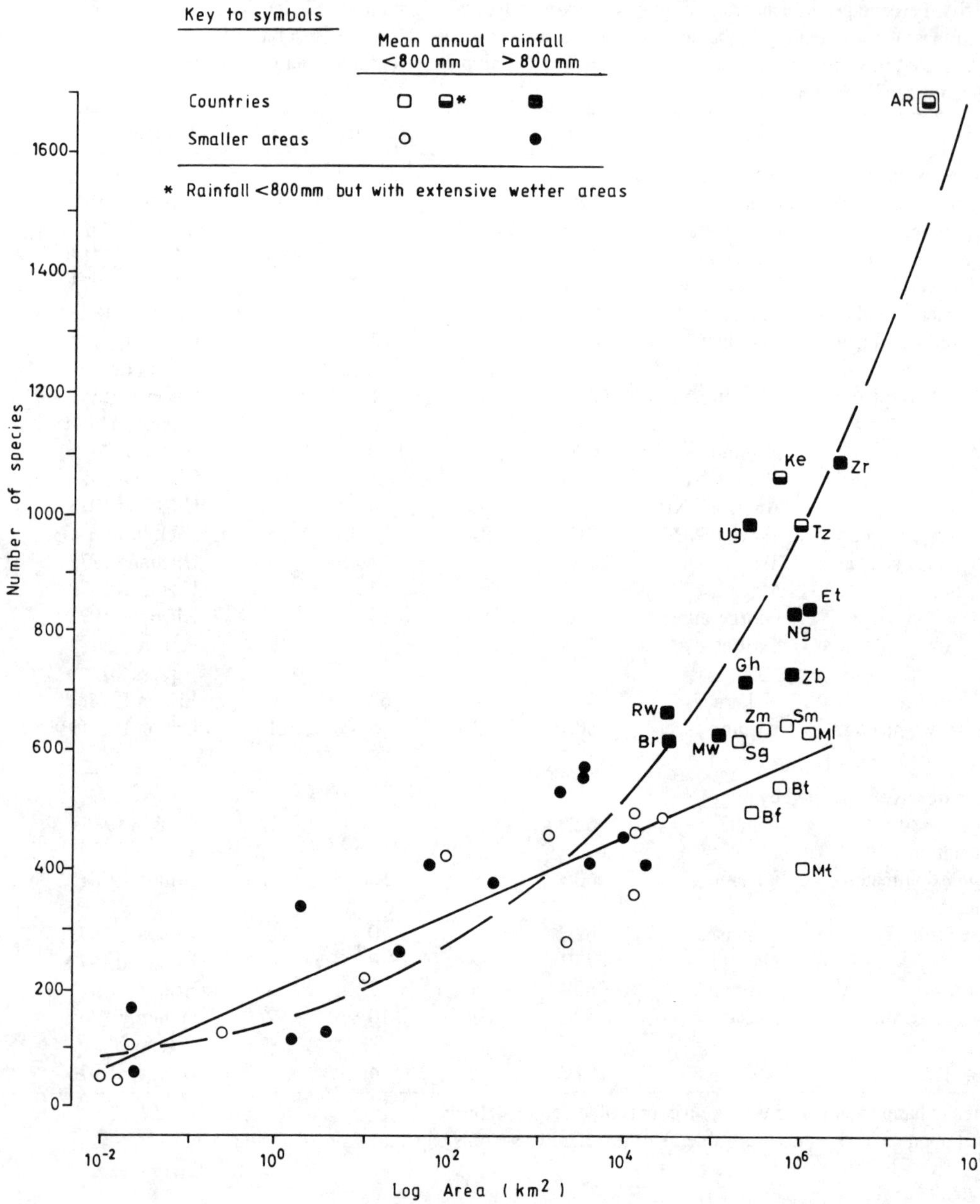

Fig. 16. The relationship between species number and area in the Afrotropical Region. The number of species increases with area: the curve is for moist areas and the straight line for dry ones. All data are for tropical Africa, except 'AR' (see below). The equations of the regression lines, which can be used to predict values expected for any area, are as follows (A = area, S = numbers of species):
moist areas : $\log_{10}S = 0.125 \log_{10}A + 2.231$
dry areas : $S = 65.4 \log_{10}A + 187.1$
For further details, see Pomeroy & Lewis (1987). The countries are indicated by their initials, thus Botswana (Bt), Burkino Faso (BF), Burundi (Br), Ethiopia (Et), Ghana (Gh), Kenya (Ke), Malawi (Mw), Mali (Ml), Mauritania (Mt), Nigeria (Ng), Rwanda (Rw), Senegambia (Sg), Somalia (Sm), Tanzania (Tz), Uganda (Ug), Zaire (Zr), Zambia (Zb) and Zimbabwe (Zm). AR indicates the total Afrotropical Region. (Redrawn, with permission, from Pomeroy & Lewis (1987)).

southwards as far as Mozambique, Zimbabwe and South Africa. Taxonomically, the Palearctic migrants are very diverse. Amongst the longest distance migrants are some of the swallows, terns and waders. The swallows and warblers probably make up the greatest numbers of individual birds. In some habitats, these migrants outnumber their Afrotropical relatives. This is strikingly so in the case of waders and swallows, and also of falcons, where most of the resident species are quite rare; some Palearctic falcons, such as the Hobby (152) and Lesser Kestrel (160), occur in flocks of hundreds or even thousands (see, for example, Smalley (1983)).

In terms of habitats, the Palearctic species frequently dominate shorelines, both marine and inland, and they can at times make up a large proportion of the birds in bushland habitats, especially in eastern Kenya, including Tsavo 89C 101 (Lack 1983). Elsewhere they are usually less significant and they scarcely penetrate forests at all (see Table 5A, with a summary in Table 5B). 'Oversummering' by Palearctic birds – that is, staying in Africa during the northern summer, rather than migrating north to breed – is common in immatures of most species of terns and waders (*e.g.* Pearson & Britton 1980), but oversummering by passerines such as the Northern Wheatear (639) in Tsavo is exceptional.

Whilst the movements of Palearctic species in Africa have been extensively studied and reviewed by Moreau (1972) and Curry-Lindahl (1981), far less is known about intra-African migrants. To take just two examples, the Spotted Ground Thrush (681) is apparently a non-breeding migrant to Kenya's coastal lowlands but the location of its breeding range is unknown; and the Grey-rumped Swallow (561) disappears completely from some areas for several months at a time and is believed to migrate extensively: but nobody knows where it goes.

There is almost no knowledge of the movements of the many dry country species which leave their breeding areas after the young are fledged; they are often referred to as 'wanderers'. The only significant exception is the Red-billed Quelea (986) (Ward 1971; Allan 1983), which has been shown to make predominantly north-south movements from southern Somalia, through eastern Kenya (where it often breeds) and into eastern Tanzania.

Peter Lack (pers. comm.) points out that Palearctic species can be considered as Afrotropical species which move out to breed. This is almost certainly how they came to colonise the northern Palearctic regions after the ice ages. There are also a number of Kenyan species which move out to breed but stay within the Afrotropical Region. Examples include several bee-eaters and nightjars, and a number of aquatic birds like the Great White Pelican (16). Whilst some of these species have been recorded breeding in Kenya, most of the population seen here are migrants.

KENYA IN AN AFRICAN CONTEXT

Comparison of avifauna

Kenya is well known for the numbers of bird species that have been recorded here. Only one African country, Zaire, has more species recorded, but clearly this is partly because Zaire is much larger – some four times – than Kenya. Similarly, one would not expect so many species in a small country, such as Burundi, about one twentieth the size of Kenya.

Figure 16 shows the numbers of species in various countries of tropical Africa, related to their sizes. In order to present a clearer perspective, some smaller areas such as National Parks are included too. The countries as well as the smaller areas are divided into two categories, moist and dry, on the basis of their annual rainfall, those with a modal rainfall averaging more than 800 mm yr^{-1} being designated moist. Two generally 'dry' countries, Kenya and Tanzania, are nevertheless included with the moist category because they have extensive moist areas, unlike the other dry countries.

For dry areas, species numbers rise progressively with size. The same is true for moist areas except that for the largest areas – whole countries – the slope is much steeper. These relationships can be described mathematically, as shown by the two lines in Figure 16. The upper, curved line which is for moist areas, has two countries well above it – Kenya and Uganda. On this basis we can say that Kenya and Uganda have the richest avifaunas in tropical Africa, relative to their size.

Species richness is strongly related to latitude (Pomeroy & Lewis 1987) and this is one reason for Kenya's having so many species. Another reason is that Kenya lies at the conjunction of several sub-regions which have been important centres of speciation within the Afrotropical Region (Crowe & Crowe 1982; Diamond & Hamilton 1980). One distinct avifauna originated in the arid area of Somalia, southern Ethiopia and northern Kenya. A second sub-region is centred on the equatorial forests of the Congo basin, and a third stretches across the Sahel, reaching north-western Kenya (*cf.* Fig. 4).

Other African atlases

We return finally to the opening paragraph of this introduction, where we mentioned other atlases and, in Figure 1, showed the position for the whole continent in 1987. Planned or existing atlases by then covered just over half of the area of Africa. Ultimately, perhaps, there will be an atlas for the whole of Africa. For such an idea to become a reality, there is clearly a need for some degree of uniformity. This is why PAOC V suggested that the unit of mapping should be a square of half-degree side, the quarter square degree (QSD). Some atlases, including Natal (Cyrus & Robson 1980), use squares of quarter-degree side (16 to a degree square), but this of course is compatible with the QSD.

Other recommendations of the PAOC V meeting were that breeding records should be noted, and that records prior to 01 January 1970 should be kept separately, as in this atlas. Optimal additional features might be data on seasonality (some atlases show or are planned to show monthly occurrence), and on abundance.

Atlassing in southern Africa is well advanced and there is a high degree of coordination between the 12 schemes (Hockey & Ferrar 1985). Now, with the publication of this atlas, and others planned, eastern Africa will be the next part of the continent to be covered. We hope this will stimulate those in the west and north to initiate schemes in these important areas, which are largely blank on Figure 1.

A GUIDE TO THE SPECIES ACCOUNTS IN THIS BOOK

1. *Bird names and sequence*

Scientific and English nomenclature in this atlas, and the sequence in which the families are presented, are exactly as in Britton (1980a). We gave considerable thought to the question of the systematic order within families, particularly in view of the thorough research which went into Britton (1980a). The overriding consideration in our choice of an order has been the juxtaposition of closely related species, to facilitate comparisons.

Britton's (1980a) adoption of the alphabetical arrangement of genera within families, and species within genera – following Lack (1968) – leads to an unsatisfactory sequence. In the larger genera, such as *Ploceus* and *Cisticola*, species which are clearly close relatives become widely separated, whilst more distantly related species come together. This is particularly unsatisfactory for an atlas, where one wants to compare the distributions of members of superspecies, *e.g.* Scarlet-chested and Hunter's Sunbirds (930, 931), or the Wailing, Ashy and Tana River Cisticolas (724-726).

This comparison of closely related forms was a prime objective of Hall & Moreau (1970) and Snow (1978) in their Afrotropical atlases, and we use their sequences here. Sequences for Palearctic species, which were not covered by these works, have been taken from Voous (1977) and, for the few marine species which he did not cover, from Harrison (1983).

Thus we have introduced our own series of species numbers, but the Britton numbers and Mackworth-Praed & Grant numbers for each species are given too (see p. 34), to facilitate cross-reference.

This atlas is primarily concerned with distributions at species level, but subspecies are sometimes mentioned (see p. 36).

2. *Geographical abbreviations*

Geographical terms are abbreviated thus:

central	C
east (-ern)	E
north (-ern)	N
south (-ern)	S
west (-ern)	W
Lake	L
River(s)	R
Mount(s)	Mt(s)
National Park	NP
Game Reserve	GR
Quarter Square Degree	QSD

3. *Abbreviation of standard references*

Frequently cited references are abbreviated as follows:

AGH	Andrews, Groves & Horne 1975
AM	Ash & Miskell 1983
Ba	Backhurst, Britton & Mann 1973
Be	Betts 1966
Br	Britton 1980a
BUN	Brown, Urban & Newman 1982
EABR	East African Bird Report, *Scopus*. Years are given if three or less issues are quoted.
HM	Hall & Moreau 1970
HH	Hopson & Hopson 1975
Ja	Jackson 1938
La	Lack, Leuthold & Smeenk 1980
MPG	Mackworth-Praed & Grant 1957, 1960
Se	Sessions 1966
Sn	Snow 1978
St	Stevenson 1980

4. *Localities*

Localities are followed by their QSD number (see Fig. 2 and the overlay), *e.g.* Nairobi 75B, Kakamega Forest 48D. Precise localities are given in the gazetteer to Britton (1980a), and in the United States Board on Geographic Names (1964).

5. *Family accounts*

Each family is introduced by summarising its representation in Kenya, related to the Afrotropical Region and the world. The numbers of species quoted for each family are approximate, since authorities sometimes disagree. Afrotropical numbers exclude Madagascar and oceanic islands such as the Seychelles.

Certain general characteristics of each family may be noted, particularly where these affect the interpretation of the maps *e.g.* whether the species are easily observed or skulking, and whether they are easily identified (see *detectability* and *coverage*, below). These factors are in turn used to discuss the significance of features on the maps such as (a) clusters of pre-1970 squares, which may signify a contraction of range, and (b) absence or paucity of breeding squares, which may indicate migratory behaviour.

6. *The titles of the species accounts*

The following example can be used:

609 COMMON BULBUL *Pycnonotus barbatus*
B732 / MPG741-744 / D3C3

The top line consists of the sequence number for this atlas (609), and the English (capitals) and scientific (italics) names: full details of taxonomic status, including authority and races, are given by Britton (1980a).

The second line consists of the sequence numbers from Britton 1980a (given as B) and Mackworth-Praed & Grant 1957, 1960 (given as MPG), and the Detectability (D) and Coverage (C) numbers for the species.

These DC numbers require some elucidation here. Each is accorded a rating on a four point scale (Table 6). These ratings are intended to aid interpretation of the maps and should also be considered when comparing species. Ease of detection clearly influences the completeness of a species' map: nocturnal species, notably owls and nightjars, are harder to detect than birds like plovers and woodpeckers, although the nocturnal species might be just as numerous. By detectability we also imply that one not only has to see the species but also to identify it, so that pygmy crakes (185-189) and the smaller honeyguides (504-506) generally have low detectability – and nightjars (except such as the Pennant-winged (410)) are given low ratings.

Coverage refers to observers: obviously even a rare bird has a good chance of being spotted in Nairobi National Park 75B, but it is virtually certain to be missed if it lives at Fudua 90B, which is north-west of Garsen in a very poorly known area. In general, the remarks made earlier (p. 10) on dry country species apply here – mostly they live in poorly covered areas.

The difference in meaning of detectability and coverage should become clear from the following examples. The Nubian Shrike (859) is a distinctive species (D3), but it is a migrant which only rarely reaches Kenya: there are reports of only four birds so far, and it is quite possible that others have gone undetected in Kenya's poorly known northern areas *i.e.* our map's coverage of the bird's occurrence may be poor (C0). The same applies to the Swallow-tailed Bee-eater (445), for which there is only a single, old record. On the other hand, the Little Grey Greenbul (611), whilst obscure and difficult to identify (D1), is apparently restricted to the Kakamega and Malaba Forests of 48D, so that its range is known very well (C3).

Maps are not given for species with very restricted distributions in Kenya. Instead, their square numbers are listed, together with the classes of record (see p. 6) which occur in them *e.g.*:

Table 6. Categories of detectability and coverage

	Detectability		Coverage
D3	Very well recorded: conspicuous and easily identified, active and/or vocal throughout day e.g. Black Kite (135), Hoopoe (458), Common Bulbul (609)	C3	Good: very well known and at least fairly common species e.g. Marabou Stork (50), Fish Eagle (134), Eurasian Swallow (553)
D2	Well recorded: species which require reasonably good views for identification or are often identified by their calls; fairly easily seen e.g. larger falcons, most waders and swifts, the Red-chested Cuckoo (365), tchagras	C2	Fair: fairly common but frequently overlooked, or uncommon but fairly distinctive e.g.Grasshopper Buzzard (120), Spotted Thicknee (275), Tree Pipit (817), Retz's Helmet Shrike (872); or fairly common but in rather restricted habitats such as dams, ponds, e.g. Jacana (211), Green Pigeon (340)
D1	Poorly recorded: hard to identify and/or skulking and/or nocturnal e.g. many nightjars, crakes, cisticolas, the African Finfoot (202)	C1	Inadequate: many of these species are relatively uncommon, or the less conspicuous dry country birds; may also be difficult to identify; e.g. Blue Quail (175), Barred Owlet (390), Gambaga Flycatcher (781)
DO	Seriously under-recorded: secretive and often very difficult to identify, including some crepuscular/nocturnal species, e.g. Pintail Snipe (252), small members of *Indicator*, Friedmann's Bush Lark (536), Rock-loving Cisticola (733)	CO	Poor: either low detectability and/or in remote areas (see coverage map) or very rare e.g. *Sarothrura* crakes, small *Indicator* species, Friedmann's Bush Lark (536), Collared Lark (530), Pygmy Sunbird (903)

491 GREEN TINKERBIRD *Pogoniulus simplex*
B554 / MPG591 / D2C3
NO MAP: post presence 102BD 114ABC;
post confirmed 103A

Species whose presence in Kenya has been mentioned in reputable publications, but later doubted or withdrawn, do not have a sequence number on this atlas. They are clearly labelled *e.g.:*

– DWARF HONEYGUIDE *Indicator pumilio*
B568 / MPG- / DOCO
RECORDS DELETED

7. *The maps*

The maps are the central feature of the atlas: and are provided for 871 of the 1065 species that constitute the Kenyan avifauna. The mapping symbols are shown in Table 2 (see also p. 6).

8. *The Afrotropical section of each species' text*

This paragraph briefly describes the species' distribution within the Afrotropical Region. Many of Kenya's birds are at or near the limits of their range (see p. 11), so that their distribution within sub-Saharan Africa aids in the interpretation of their distribution in Kenya. A final sentence in this section stresses if Kenya lies at the limit of the range, *e.g.* Honey Buzzard (137).

We have taken the Afrotropical Region to be the area south of the Palearctic as defined by Cramp & Simmons (1977): this is shown in Figure 1. In these range descriptions, E Africa refers to the eastern side of the continent, while East Africa means Kenya, Uganda and Tanzania.

In this section, the data for Afrotropical species have been extracted from Hall & Moreau (1970) and Snow (1978), with additional data for Somalia from Ash & Miskell (1983), while material on marine and other aquatic species has been taken from Harrison (1983). There is, however, no text of comparable accuracy on the distribution of Palearctic migrants within the Afrotropical Region during the northern winter. David Pearson (pers. comm.) provided these data, together with notes on each Palearctic species' overall distribution and seasonality in Kenya.

9. *The Kenyan section of each species' text*

Here we provide a summary of the species' occurrence in Kenya, including comments upon its map. The summary consists of data derived from the map, the literature, and personal experience. Map interpretation included extensive use of the various overlays, and environmental data from this source are often quoted as percentages *e.g.* 'Range is 94% within 500+ mm rainfall areas', which means that 94% of the squares in which the species is recorded have a modal rainfall of at least 500 mm per year. In the case of very localised species, these environmental data are presented in terms of actual numbers of squares *e.g.* 'four of five squares are in 500+ mm rainfall areas'. Data from the altitude overlay are quoted only if they amplify or contradict altitudinal ranges given in the literature (*e.g.* Red-winged Starling (885)).

Important records for which the exact QSD is uncertain (*e.g.* Palm Nut Vulture (86) in 37C) are mentioned in the text with an indication about plotting. Records that are too imprecisely located to plot at all are also mentioned (*e.g.* Yellow-billed Shrike (868) in 25-13D), as are potentially important but unconfirmed reports (*e.g.* Broad-billed Prion (7) off Shimoni 114C).

Apparent expansions (*e.g.* Somali Bee-eater (451)) or contractions (*e.g.* Marsh Tchagra (843)) of range are discussed, often with reference to factors such as the possible effects of observer frequency in remote areas or the relative absence of specimen collecting in recent years (*e.g.* Bristle-crowned Starling (884)).

Closely related species are discussed in some detail, since their inter-relationships often amplify the distributions seen on the maps, for example in the case of allopatric members of superspecies, where various birds occupy mutually exclusive ranges (*e.g.* the East Coast and Chin-spot Batis (797, 799), and the Black-backed, Northern and Pringle's Puffbacks (836-838)). Races are mentioned if they amplify the map's features (*e.g.* Speckled Mousebird (425) and Black-headed Batis (798)), or if they are readily identifiable in the field (*e.g.* Lilac-breasted Roller (455) and Black and White Mannikin (1042)). Easily identifiable morphs are also referred to (*e.g.* Little Egret (35) and Augur Buzzard (119)).

We make frequent references to Ash & Miskell's (1983) *Birds of Somalia: their habitat, status and distribution*, because areas to the east of 40°E are amongst the most poorly known ornithologically in Kenya. Hence, despite low observer frequencies in these areas of both countries, any analogy that can be drawn between a bird's range in the far east of Kenya and adjacent areas of south Somalia helps to demonstrate the validity of the distributions shown on our maps (*e.g.* Red-chested Cuckoo (365) and Black-breasted Glossy Starling (877)).

Although we did not collect seasonal data for this atlas (see p. 1), we have tried to summarise those existing in the literature, and in the unpublished notes of ourselves and others. Seasonality of most Palearctic species is reasonably well known, whereas that of most Afrotropical birds is not.

We have mentioned even the slightest evidence of movements or wandering in Afrotropical species, sometimes including examples from outside Kenya, as this topic is so little known here and, indeed, in Africa in general. Any mentions of residential status are included, unless the bird's family is known to be essentially sedentary, *e.g.* the Turacos Musophagidae, Babblers Timaliidae and Bush Shrikes Malaconotidae.

If not in chronological order, references are listed in their order of importance to the point being made. In most cases, the criterion for the use of a reference from the literature has been its mention of aspects of a species in Kenya. In addition to the sources quoted in the text, the bibliography also contains references to writings on the birds of Kenya that are not sufficiently specific to be included in any species or family account.

10. *Opportunities for further observations on birds in Kenya*

This atlas attempts to provide an explanatory overview of the birds of Kenya. We hope that it will enable interested observers to derive greater satisfaction from their birdwatching by seeing the significance of their observations in a broader context.

But we have only just begun the fascinating study that is ornithology in Kenya. There are a multitude of ways in which relatively simple observations can add substantially to what is known.

Here are some suggestions:

a) Studies of single species. This atlas provides a broad summary of what is known about each species in Kenya, directly via the text and indirectly via the references cited. How little that is known about many birds, especially common ones, may come as a surprise: J.P.Dittami's (1983) studies of the Blue-eared Glossy Starling (879) at Nakuru 62A are a classic case. Such topics provide almost unlimited opportunity for constructive observations, ranging from the activities of the Common Bulbuls (609) in your garden, to those of the Shovelers (78) wintering on Lake Turkana. Published examples of this type of study include: Bennun 1985, Brown 1972a, Candy 1984, Lewis 1982d, Meadows 1977a, Pearson 1981a and Sessions 1972.

b) Studies of single areas. Rather than concentrating on one species, you may prefer to look at some or all of the birds in a defined area, *e.g.* your garden, a stretch of lakeshore, a section of forest edge *etc*. Bearing in mind the very little that is known about the movements of birds, particularly African species, within Kenya, simple periodic counts of individuals can easily produce valuable data *e.g.* J.P.Dittami's study already referred to, Stevenson's (1980) at Lake Baringo 50A, and D.E.P.'s data (presented in this atlas) from the Ruaraka-Kahawa-Ruiru area of 75B. Records of rainfall and the state of the vegetation (*i.e.* lush, in flower, dessicated etc.) are useful here, and the study site should be visited regularly. Other examples of this type of study are: Britton 1978, Britton & Zimmerman 1979, Carlson 1986a, Dittami 1981, Lack 1985, Pearson 1984d and Pomeroy 1982.

c) Testing specific statements in this atlas. Many points in this atlas can be tested by simple observations. Is the apparent paucity of breeding of the Hamerkop (44) in the coastal lowlands real? And what of that of the Silvery-cheeked Hornbill (465) in Kenya as a whole? Do both the Yellow and Montane White-eyes (937, 938) occur on Mt Nyiru 26D and , if they do, are they alongside each other or in some way segregated? Can anyone find a Shoebill (-) in the Winam Gulf 60AB, or a truly wild Mallard (-) anywhere? Does the White-winged Apalis (752) still exist along the lower Tana River?

d) Adding to this atlas. Whereas the total number of records needed to complete this atlas may be as high as 135 000 (see p. 9), we have amassed less than 43 000 : there is plenty of opportunity for further atlassing work, whether as expeditions to poorly covered, remote areas, or as general birdwatching on better known ground. When recording a 'new' bird for an atlas square, the following questions should be borne in mind. Is the species in an area where it is to be expected, both in terms of its habitat (and seasonality, if known), and of nearby QSDs with other records? If not, then see the points made in the next section.

e) Recording of rarities and vagrants. This atlas will tell you whether any bird that you see is rare in a local or national context. When faced with an apparent rarity, remember that *details* are essential. Firstly, accuracy of identification: what features distinguished your bird with certainty from other, similar species? Photographs are invaluable, as is the corroboration of other observers. Where did you see the bird, and what was the date? How good were the lighting conditions, for how long was the bird in view, and approximately how far were you from it? Was the area uncharacteristically lush or arid, and were there harsh weather conditions elsewhere that might have caused the bird to wander? Details of rare birds should be sent to the East African Bird Report, an annual publication of the journal *Scopus*, c/o D.A. Turner, P.O.Box 48019, Nairobi, Kenya.

Species accounts

Family Struthionidae: Ostrich

A monotypic family, relatively recently extinct in Arabia, and now surviving only in Africa, mainly S of the Sahara (Sn; BUN). The Ostrich is Kenya's only flightless bird. It avoids large mammalian predators by means of its speed and keen eyesight.

1 OSTRICH *Struthio camelus*
B1 / MPG1 / D3C3

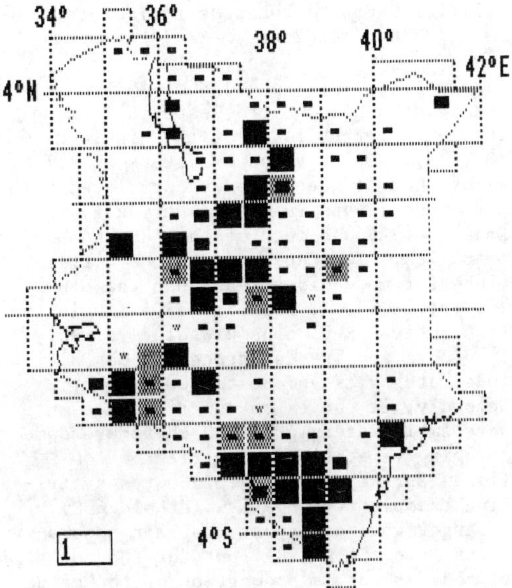

Afrotropical: in two distinct areas. In the N, across the N tropics between 10° and 20°N, and S from Sudan, Ethiopia and Somalia, through Kenya and NE Uganda, to E and S Tanzania. There is an isolated population S of 15°S.

Kenyan: locally common in dry grasslands, including those with bush and trees, mainly below 2000 m (Br; Ja; MPG; Be; Se; La). Range is throughout our rainfall categories, but in only 22% of the 1000+ mm squares. Paucity of records in 63-65 and 76-78 resembles the distribution of Secretary Bird (85), Kori Bustard (209), Crowned Plover (220), the vultures, and Chestnut-bellied Sandgrouse (318). These areas are partly cultivated in the W, and covered in dense bush further E. The former tends to exclude terrestrial species, particularly large ones, while the latter reduces their range of vision. Apparent absence from the poorly known E borders compares with status in adjacent S Somalia (AM). But virtual absence from even semi-arid areas of Kenya's coastal strip, presumably due to dense habitats, is contrary to the Somali range N of the equator (AM). Br considers this species resident, and reports of this status come from 74 (Be) and 74B 75 88AD 89C 100B 101A (DEP pers. obs.).

The Ostrich is not endangered in Kenya. The use of its eggs for food or tourist souvenirs is illegal, and of negligible extent. It is obviously expelled by widespread cultivation (particularly smallholders) and settlement (the probable cause of the lack of post-1969 records from 61CD: see Se), but this is not a factor over most of its range. The highest density known (0.6-0.8 birds per km^2) is attained in the protection of Nairobi NP 75B (BUN). Attempts at farming early in this century did not persist (see Ja's (p.1300) mention of paddocks at Theki 76C).

The Somali Ostrich *S.c. molybdophanes* of N and E Kenya and the Maasai Ostrich *S.c. massaicus* of more SW areas may be approaching or at species status in view of (a) interbreeding difficulties when *molybdophanes* was introduced into the range of *massaicus* in Nairobi NP 75B (A D Forbes-Watson pers. comm. to T Stevenson; Reynolds 1972a; Ng'weno 1972), although fertile hybrids were produced (BUN), and (b) their apparently differing habitat preferences (*molybdophanes* entering more densely bushed areas: Ja; MPG; Sn; BUN; Lack 1985). Other references: Leuthold 1970b, 1977; Hurxthal 1979.

Family Podicipedidae: Grebes

20 species of waterbird, with representatives world-wide apart from Antarctica. The three on the Afrotropical mainland all belong to species that also occur in the Palearctic. All three breed in Kenya.

2 LITTLE GREBE *Tachybaptus ruficollis*
B4 / MPG4 / D2C2

Afrotropical: widespread S of 15°N.
Kenyan: a common bird, especially numerous on the rift lakes of the highlands (Br; Ja). It is found on waters of all kinds, but is not markedly correlated with permanent rivers in dry country e.g. the Daua 7C 19AB and much of the Tana 64AB 65ABD 78B. Two factors make the Little Grebe far more widespread than the other two grebes (3,4), both in Kenya and in the Afrotropical Region as a whole. Firstly, it occurs and breeds throughout our altitude categories and up to 3000 m (Br; Se). Secondly, it can exist and reproduce on very small waters provided they have some fringing vegetation and a rich bottom flora (Sn; Reynolds (1975a) notes a striking example at Bangali waterhole 65C).

Larger waters support its main populations (*e.g.* EABR 1981) but, at the onset of heavy rains, a proportion of the birds on these rapidly spreads to floods and other seasonal waters (*e.g.* EABR 1983), where breeding occurs if these temporary waters are sufficiently long-lived (Brown & Britton 1980; BUN). This regular movement clearly contrasts with the limited wandering of the other two grebes.

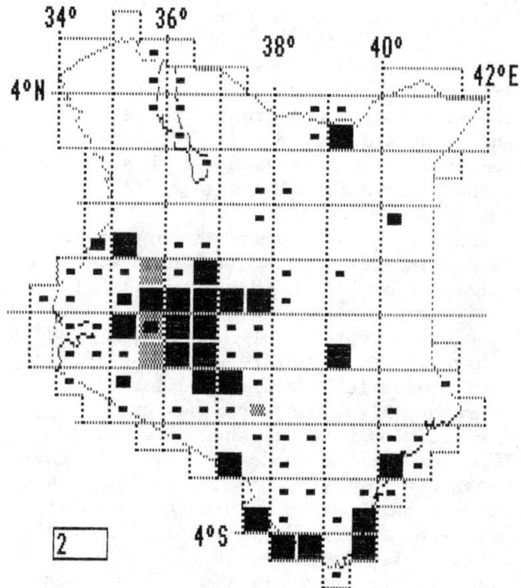

3 BLACK-NECKED GREBE *Podiceps nigricollis*
B3 / MPG3 / D2C2

Afrotropical: a highland bird of Ethiopia, Kenya and N Tanzania, also widespread S of 20°S.
Kenyan: breeds on lakes in the rift valley highlands (all three breeding squares above 1500 m), but can move to dams, sewage ponds and waterholes at other seasons. It is generally in small numbers (Br), but large concentrations are known from moderate to high altitude soda lakes in the rift *i.e.* L Bogoria 50C (particularly), L Nakuru and L Elmenteita 62A (EABR). Wandering to lower altitudes 13B 27D 88B 89C 100D 101B is probably in response to periods of high rainfall (Br; HH; La; EABR). This wandering is more extensive than that

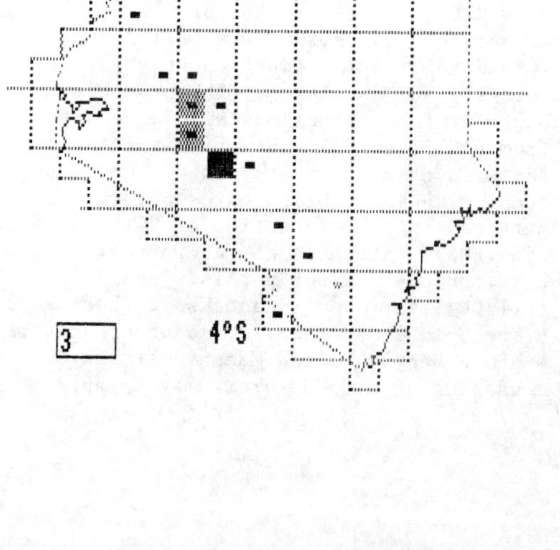

of the Great Crested Grebe (4), but the Black-necked is similarly absent from L Victoria.

4 GREAT CRESTED GREBE *Podiceps cristatus*
B2 / MPG2 / D3C2

Afrotropical: while there are scattered records from many areas, particularly in the S tropics, this grebe is mainly concentrated on highland lakes of Ethiopia and East Africa, and S of 25°S. Distribution in Kenya represents the E limit of the range.

Kenyan: locally common on larger water bodies at 1500-3000 m (Br; Se). The breeding population is centred on L Nakuru 62A (since the introduction of fish there in 1960: BUN) and L Naivasha 62C, although a marked population decline has occurred at the latter due to the introduction of gill-nets for fishing (Brown & Britton 1980; J Hayes *in litt.*; *cf.* White-backed Duck (84)). Small numbers in other areas (*e.g.* Se; St; EABR) include birds dispersing at the onset of heavy rains, but this dispersal is more limited than that of the other two grebes (2,3). Ja notes occurrence on L Turkana (400 m) but omits localities. The record in 4A (J S Ash *in litt.*) presumably represents the S limit of birds from the Ethiopian highlands. This grebe is notably absent from L Victoria, despite apparently suitable habitat (Ja; BUN; *cf.* Black-necked Grebe (3)).

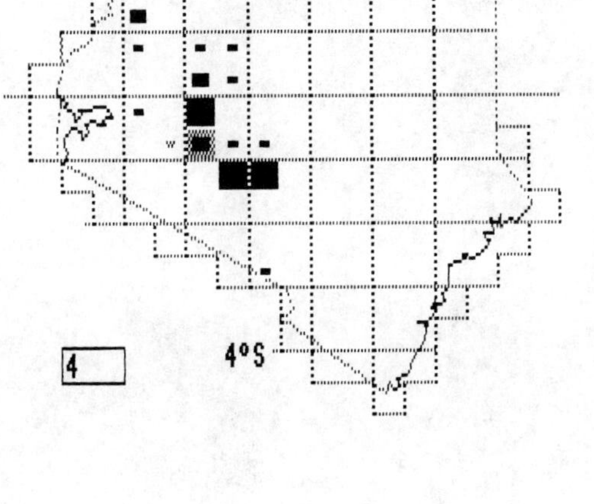

Family Diomedeidae: Albatrosses

The 13 species of albatross are pelagic birds of the S oceans and N Pacific. Seven are known from Afrotropical waters. One is recorded off the Kenya coast as a vagrant but, in view of our sparse knowledge of the pelagic avifauna, it may well be commoner than this and other species can be expected. Due to the cost of boat hire, pelagic birdwatching off Kenya is an expensive exercise and, in addition to this, it is not uncommon to find little of interest in the course of a day's search. Observers frequently offshore, *i.e.* professional game fishers, undoubtedly see notable species but these records are usually not sufficiently substantiated.

5 BLACK-BROWED ALBATROSS *Diomedea melanophris*
B5 / MPG- /D1CO
NO MAP: post presence 114C; pre presence 114B

Afrotropical: breeds in the far S oceans and normally extends N only to the S tropic, but wanderers are known to the N Atlantic and N along the E African coast to Kenya.

Kenyan: the two records are a bird at Mombasa 114B on 30 June 1955 (Bednall 1956) and an adult off Shimoni 114C in Oct 1978 (EABR 1982). Immatures off Shimoni 114C on 12 Sept and 10 Oct were either this species or the Grey-headed Albatross *D. chrysostoma* which is, however, rare N of 35°S (EABR 1983). Other records of small albatrosses may refer to this species or possibly to the Yellow-nosed Albatross *D. chlororhynchos*, which also occurs in the S Indian Ocean and which is the commonest albatross off SE Africa from May to Sept (BUN; Harrison 1983).

Family Procellariidae: Prions, Petrels, Shearwaters

55 species, with representatives throughout the world's oceans. 28 (51%) are known from Afrotropical waters. Five have occurred along Kenya's coasts as vagrants or non-breeding visitors, and there is one striking inland record. Other reference: Campbell 1971b.

- GIANT PETREL *Macronectes giganteus*
B- / MPG15 / D1CO
RECORD WITHDRAWN

Afrotropical: this species (now subdivided: see for *e.g.* Harrison 1983) occurs throughout the oceans S of the tropics, N to Angola and Mozambique in S Africa.

Kenyan: a record from Shimoni 114C (Mann 1976) lacked substantiation and so Br excluded this petrel from the East African avifauna. This record was subsequently withdrawn by the observer, so that the only certain occurrence for the tropical Indian Ocean is one collected at 7°57'S off Tanzania (Britton 1982a).

6 PINTADO PETREL *Daption capense*
B- /MPG- /D2CO
NO MAP: post presence 114B

Afrotropical: throughout the world's oceans S of the tropics, N to Namibia and Mozambique in S Africa. Wanderers in the Indian Ocean have reached N to Kenya, the Arabian Sea and India (Sinclair 1981; Harrison 1983).

Kenyan: the single record is of a bird 10 km off Mombasa 114B on 26 Sept 1974 (Sinclair 1981). Sept is the peak month for occurrences off S Africa (BUN).

7 BROAD-BILLED PRION *Pachyptila vittata*
B- / MPG14 / DOCO
NO MAP: post presence 91C 103A

Afrotropical: breeds on S hemisphere islands and is found throughout the world's oceans S of the tropics, with dispersal N to Angola, the E seaboard of Africa and the Indian Ocean.

Kenyan: the two records are of a bird off Ras Ngomeni 91C on 10 Sept 1983 (Coverdale & Pearson 1984), and an adult male found dead at Watamu 103A on 16 Aug 1984 (EABR 1984; National Museum Collection, Nairobi). These compare with the Jul-Aug seasonality of records from S Africa (BUN), SE Tanzania (a single possible sighting in Jul: Br), Kenya (a probable *Pachyptila* sp. off Shimoni 114C on 25 Aug 1981: EABR 1981) and Somalia (a wreck along the S coast during Aug 1979: Ash 1983).

8 JOUANIN'S PETREL *Bulweria fallax*
B6 / MPG13A / DOCO
NO MAP: post presence 103A

Afrotropical: although its breeding grounds are unknown, this petrel is endemic to the NW Indian Ocean, SW to Somalia (AM) and Kenya.

Kenyan: the two records, on strikingly coincident dates, are both from Malindi 103A: a female captured offshore in floating seaweed on 13 Dec 1953 (Jouanin 1957), and an exhausted bird caught on 10 Dec 1985 (Boothroyd 1986; National Museum Collection, Nairobi).

9 WEDGE-TAILED SHEARWATER *Puffinus pacificus*
B- / MPG10 / DOCO
NO MAP: post presence 91C 103A

Afrotropical: breeds on islands in the Indian Ocean and is known as a non-breeding visitor along the whole E seaboard of Africa. It is also widespread in the Pacific.

Kenyan: an uncommon species that probably occurs regularly. Subsequent to sightings that have remained unconfirmed (Brown 1974a; EABR; Br), the only definite records have been of one or possibly more off Ras Ngomeni 91C on 26 Apr 1983 (Pearson & Coverdale 1984), and one off Watamu 103A on 30 Dec 1983 (EABR 1983). All unconfirmed records are Aug-Feb, which compares with Sept-Mar seasonality for the Indian Ocean seaboard of Africa noted by BUN. The confirmed late Apr record seems more similar to recent Apr-Jul sightings off Somalia (Ash 1983).

10 AUDUBON'S SHEARWATER *Puffinus lherminieri*
B7 / MPG9 / D1CO

Afrotropical: breeds on islands in the Indian Ocean and on unknown sites off Arabia, and is known as a non-breeding visitor along the E seaboard of Africa S to S Africa. Other populations exist in the Pacific and off Central America.

Kenyan: well known to offshore fishermen, this shearwater almost certainly occurs quite frequently and has been seen in large numbers (Brown 1973b; Cunningham-van Someren 1974; EABR 1977, 1983), but substantiated records are few. Most records are unassigned racially, but two examples of *P.l. bailloni* (the Mascarene islands race) have occurred. One was a remarkable vagrancy, found 550 km inland at the foot of Limuru TV mast 75B on 18 Oct 1963, during a period of fine weather and only weak coastal breezes (Spratt 1963; Br; N Palmer *in litt.*). The other was found moribund on the reef at Malindi 103A on 13 Jan 1981 (EABR 1981). All records are Oct-Mar, which agrees with Nov sightings, thought to be *bailloni*, off S Somalia (AM).

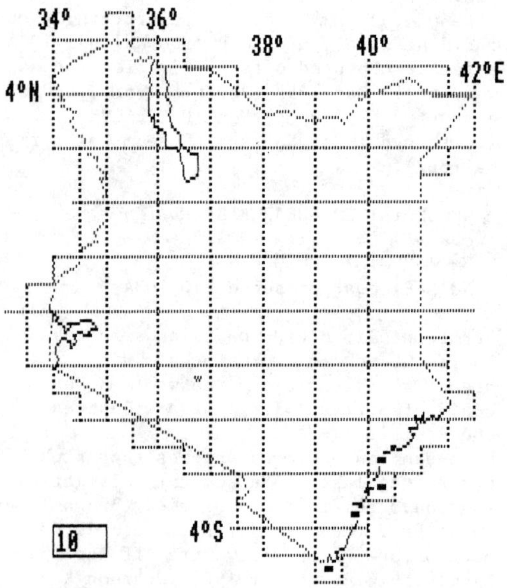

Family Hydrobatidae: Storm Petrels

About 20 species, with representatives throughout the sub-Arctic oceans. Nine are known from Afrotropical waters. Three have occurred off Kenya as non-breeding visitors or vagrants.

11 WILSON'S STORM PETREL *Oceanites oceanicus*
B- / MPG7 / D1C2

Afrotropical: breeds in the Antarctic and nearby latitudes, and then migrates into the N hemisphere. It occurs along both seaboards of Africa, where it is the commonest storm petrel.

Kenyan: as with the Wedge-tailed Shearwater (9), a series of unconfirmed sightings, often by offshore fishermen, presaged the first certain record in Apr 1980 (Britton 1981). There have been several subsequent records (Taylor 1982a; EABR 1982, 1983), including daily onconfirmed sightings by fishermen in Dec. It may well be a regular visitor during post-breeding migration Apr-Sept, while records in other months could reflect non-breeders temporarily resident in the tropics (Taylor 1982a; Harrison 1983).

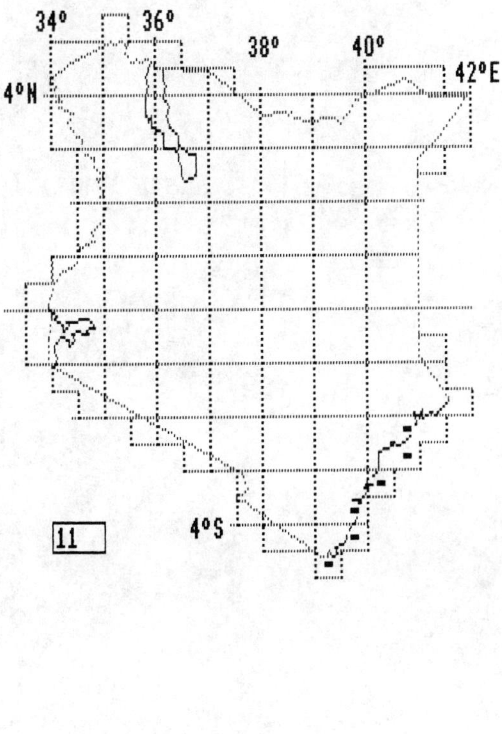

12 LEACH'S STORM PETREL *Oceanodroma leucorhoa*
B8 / MPG6 / D1C0
NO MAP: post presence 102D; pre presence 114B

Afrotropical: breeds in the N Atlantic and Pacific Oceans and then migrates S to the equator and S tropics for the N winter. Thus it is regular on the W African seaboard but very rare in the Indian Ocean, where Harrison (1983) quotes occurrences only from Kenya and the Persian Gulf.

Kenyan: the only certain records are of a tideline corpse near the mouth of the Tiwi R 114B on 8 Feb 1967 (Parsons 1969), and a single bird seen off Mtwapa Creek 102D on 4 Oct 1981 (Taylor 1982a). Earlier unconfirmed records (Cunningham-van Someren 1974; EABR 1978) were considered probable by Br but largely discounted by Taylor (1982a). Harrison (1983) suggests Indian Ocean occurrences to represent migration N into the "wrong" ocean *i.e.* the Indian Ocean instead of the Atlantic, but Cramp & Simmons (1977) suggest that Indian Ocean birds originate in the Pacific.

13 MATSUDAIRA'S STORM PETREL *Oceanodroma matsudairae*
B- / MPG- /D0C0
NO MAP: post presence 114B

Afrotropical: breeds only on the Volcano Isles, S of Japan, and then migrates W into the equatorial Indian Ocean, where it is most common in the W between June and Sept (Bailey *et al*. 1968).

Kenyan: the two records are of single birds off Mombasa 114B on 5 Jul and 22 Aug 1981 (Taylor 1982a). A sight record E of 92A on 16 Sept 1963 is unconfirmed (Bailey *et al*. 1968). The regular presence of this species in the equatorial W Indian Ocean must make occurrences off Kenya likely, so that the few records probably reflect identification difficulties and the generally scanty observer coverage of pelagic species.

Family Phaethontidae: Tropicbirds

Three species which inhabit the tropical oceans, extending significantly beyond the tropics in the Pacific. All three are known in Afrotropical waters. Two have occurred off Kenya, one as a reasonably regular non-breeding visitor and the other apparently as a vagrant.

14 RED-TAILED TROPICBIRD *Phaethon rubricauda*
B10 / MPG17 / D2CO
NO MAP: pre presence 80C

Afrotropical: breeds on islands in the Pacific and the S tropics of the Indian Ocean. Hence it is to be expected off the S tropical E African seaboard, but it is a far more strictly oceanic species than the White-tailed Tropicbird (15) and is very rarely encountered inshore (BUN; Harrison 1983).

Kenyan: the sole record is of a bird seen near Kiunga 80C, at some time between mid-Jul and mid-Sept 1961 (Fogden 1964). There are no records from either Tanzania (Br) or Somalia (AM).

15 WHITE-TAILED TROPICBIRD *Phaethon lepturus*
B9 / MPG18 / D2CO
NO MAP: post presence 102D 103A 114C

Afrotropical: breeds in most of the world's tropical oceans, including the Indian Ocean islands, and occurs in the S tropics along both seaboards of Africa.

Kenyan: a non-breeding visitor now seen fairly often off Kenya's S coast Aug-Mar, particularly in the Pemba Channel off Shimoni 114C, where it is regular in small numbers (Br; EABR). The northernmost records are singles inshore at Mambrui 103A on 2 Feb 1927 (Ja) and Malindi 103A on 11 Nov (EABR 1983), and there are no records from Somalia (AM).

Family Pelecanidae: Pelicans

Seven or eight species, with members in all the tropical and temperate zones of the world's zoogeographical regions. Two very different species occur in the Afrotropical Region. Both breed in Kenya.

16 GREAT WHITE PELICAN *Pelecanus onocrotalus*
B11 / MPG31 / D3C3

Afrotropical: widespread S of 15°N, apart from the equatorial lowland forest areas of the W, and arid E Somalia. Breeding colonies are scattered throughout but the largest, which house the greater part of the African population, are on the E African rift lakes. Large scale movements are known, but there is no evidence for birds of Palearctic origin in E Africa (Br).

Kenyan: a highly gregarious pelican of large water bodies that is most numerous on lakes, particularly those in the rift valley *e.g.* 5000 normally at L Nakuru 62A and a peak count of 35 000 there (Br; Bartholomew & Pennycuick 1973). Smaller numbers are found on lakes elsewhere, and on extensive floodwaters and large, sluggish rivers. This bird is scarce at the coast, and of only transient occurrence on waters lacking adequate fish. The single Kenyan breeding colony, which has held up to 8000 pairs, started in 1968 on small rocky islands in L Elmenteita 62A, after the introduction of fish into L Nakuru 62A had created a food source. This colony is on an accessible and insecure site (Brown *et al*. 1973; Brown 1973a; Brown & Britton 1980; EABR 1982).

White Pelicans frequently make long diurnal movements in large flocks between the various rift lakes, with single flights of over 100 km known in Kenya (Pennycuick 1972; Lewis 1982a). BUN give a potential for 400-500 km travel per day. Juveniles ringed at L Shala in S Ethiopia made the 915 km journey to the L Nakuru area in as little as 23 days (Backhurst 1977). Other reference: Guillet & Furness 1985.

17 PINK-BACKED PELICAN *Pelecanus rufescens*
B12 / MPG32 / D2C2

Afrotropical: widespread S of 15°N, apart from much of the SW.

Kenyan: this species exhibits marked contrasts with the White Pelican (16). Firstly, the Pink-backed often utilises much smaller water bodies, and can be found on any aquatic habitat that holds adequate fish except the open sea (Br; BUN). It is thus

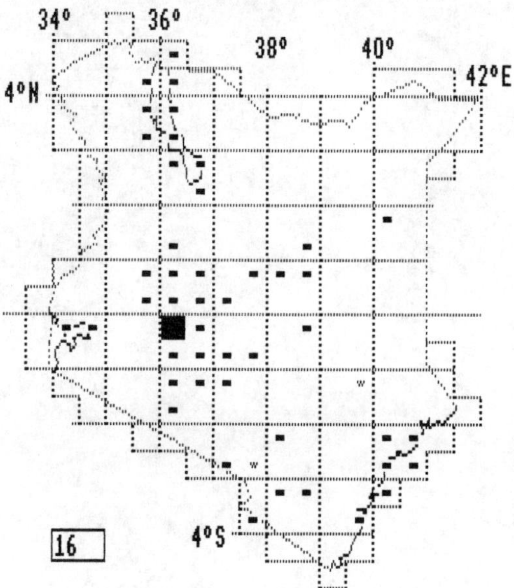

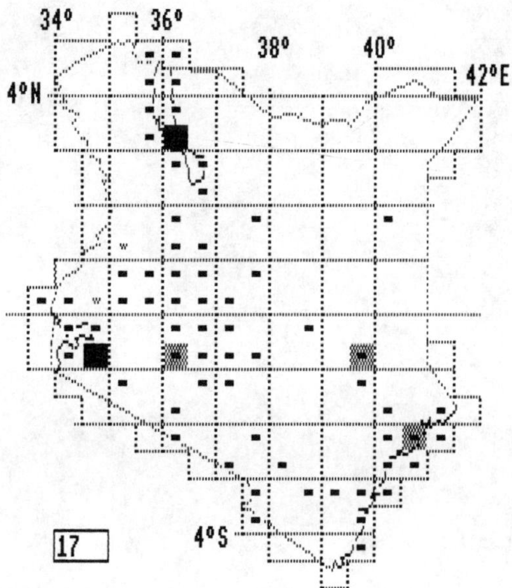

more widespread than the White Pelican and, at QSD scale, this distinction can best be seen on the maps in the W 48 49 60, and also in the coastal strip and along the lower Tana R 65D 78B 79C. It is also far less gregarious than the White Pelican, usually being encountered in groups of 30 or less, which disperse widely from their few breeding sites (breeding sites = 7% of the species' total squares) and perform considerable local movements (Sn; Br). Finally, the two birds differ significantly in their breeding behaviour. The Pink-backed nests almost invariably in trees, and its smaller, more numerous colonies are less prone to disturbance, even to the extent of being sited in the vicinity of human activity. The best known in Kenya is at Rakewa 60D, which has been used by up to 250 pairs for over 200 years, and which is protected by the area's Luo people (Burke & Brown 1970; Brown & Britton 1980). See Greater Cormorant (21).

Family Sulidae: Boobies, Gannets

Depending on authority, this family contains either six or nine species. Representatives are found throughout tropical and sub-tropical oceans, and also in the N Atlantic. Five are known in Afrotropical waters (BUN). Two occur along the Kenya coast as apparently regular, non-breeding visitors, while a third is more strictly oceanic and only a very rare straggler inshore. Records of a fourth species are unsatisfactory.

18 RED-FOOTED BOOBY *Sula sula*
B16 / MPG20 /D1C0
NO MAP: post presence 102D 103A 114BC

Afrotropical: although this booby is found in most areas of the world's tropical oceans and breeds on islands in the Indian Ocean, it is very rare in African coastal waters, with records for the E seaboard of only stragglers to S Somalia (Ash 1983) and Kenya.

Kenyan: the first African record was an immature captured 8 km off Kilifi 102D on 24 Oct 1976 (Britton 1977d), while the second was a corpse at Shanzu 114B on 20 Mar 1978 (EABR 1978). These were followed by an immature off Malindi 103A on 16 Nov 1980 (EABR 1980), and an individual off Shimoni 114C on 24 Jan 1983 (EABR 1984).

19 MASKED BOOBY *Sula dactylatra*
B14 / MPG22 / D2C1

Afrotropical: found over most of the tropical oceans. It breeds on Indian Ocean islands and regularly occurs along Africa's E seaboard from Somalia to N Mozambique, but is absent from most of the SE Atlantic and thus a rare vagrant to Africa's W coast.

Kenyan: regular small numbers off the Kenyan and Tanzanian coasts at all seasons are thought to originate from the breeding colony on Latham Island, SE of Zanzibar (Br). However, since 1976, all Kenyan records have been in the period Aug-Jan (EABR).

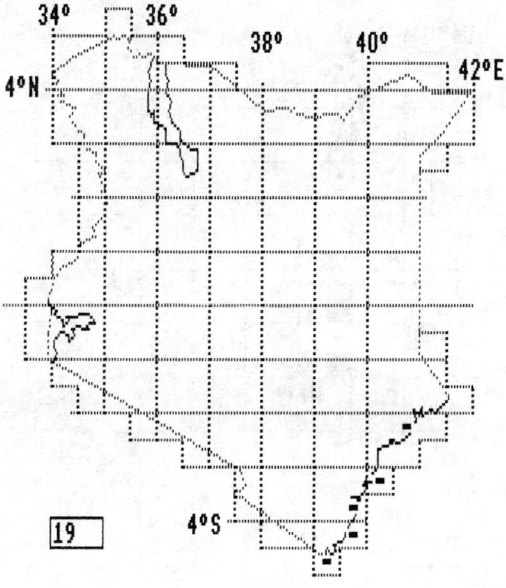

20 BROWN BOOBY *Sula leucogaster*
B15 / MPG19 / D1C1
NO MAP: post presence 102D 103A 114C

Afrotropical: occurs in most of the tropical oceans. It breeds on islands in the N and NE Indian Ocean and is known along Africa's E seaboard S to Kenya and probably (although there is no certain record) Tanzania. Off W coasts, it extends from the Gulf of Guinea to the N tropic.

Kenyan: recent records suggest this booby to be a regular non-breeding visitor in very small numbers off the coast during Dec-Feb and Aug-Sept (Br; Brown 1973b; EABR).

- GANNET *Sula bassana*
B13 / MPG24 / D2C0
RECORDS UNSATISFACTORY

Afrotropical: the Northern Gannet is found in the N Atlantic S to Senegambia, while the Cape Gannet (which is sometimes treated as a separate species) breeds in S Africa and extends N to the Gulf of Guinea and, on the E seaboard, to Tanzania.

Kenyan: Ja reports sight records from Mombasa 114B and Vanga 114C, but these are not accepted because of the possibility of confusion with the more common Masked Booby (19) and the lack of recent sightings. It may be a regular visitor Jul-Sept to Tanzanian waters (Br).

Family Phalacrocoracidae: Cormorants

About 28 species, with members in all of the world's zoogeographical regions. Six (21%) are found in the Afrotropical Region, and two are resident breeding birds in Kenya.

21 GREATER CORMORANT *Phalacrocorax carbo*
B18 / MPG25 /D3C3

Afrotropical: predominantly E and S Africa from Sudan, Ethiopia and Somalia (see Ash 1983) S through E Africa to Zaire, Malawi, Mozambique and most of the continent S of 15°S. There are also populations in the L Chad basin and along the W coast N of 10°N.

Kenyan: essentially a bird of large inland waters, especially lakes, with non-vegetated or rocky shores (Br; *cf.* Long-tailed Cormorant (22) and Darter (23)). Range is in 71% of the squares with larger lakes, the most notable exception being L Magadi 75C 87A, which lacks suitable fish. All post-1969 breeding records are from large lakes in the rift valley. It can wander to smaller waters (*e.g.* Se; La; EABR 1981), but this tendency is far less developed than in the Long-tailed Cormorant and the Greater is much less common at the coast (*cf.* MPG). As noted by Sn, the relationship of this species to the Long-tailed Cormorant is similar to that of the White Pelican (16) to the Pink-backed Pelican (17).

22 LONG-TAILED CORMORANT *Phalacrocorax africanus*
B17 / MPG27 / D3C3

Afrotropical: widespread over most of Africa S of 17°N.

Kenyan: widespread and locally common on a great variety of waters that usually have well developed fringing vegetation and/or partially inundated trees (Br; Ja). Occurrence at the coast is on the sheltered waters of creeks, mangrove swamps etc., rather than the open sea. It differs from the Greater Cormorant (21) in being more associated with permanent rivers (present in 55% of the squares that contain them, against the Greater's 34%), and by occurring on many smaller waters away from lakes (26 non-lake squares, as against the Greater's 13). The Long-tailed Cormorant breeds alongside herons, storks, ibises and darters in established heronries, as at L Jipe 100D (actual colony just inside Tanzania)(Lack 1977b; Turner 1978) and Garsen 91A (Coverdale *et al.* 1983). It also undertakes poorly known local movements (*e.g.*Se) and, again in contrast to its congener, can rapidly establish breeding colonies on temporary inundations (Sn; Br).

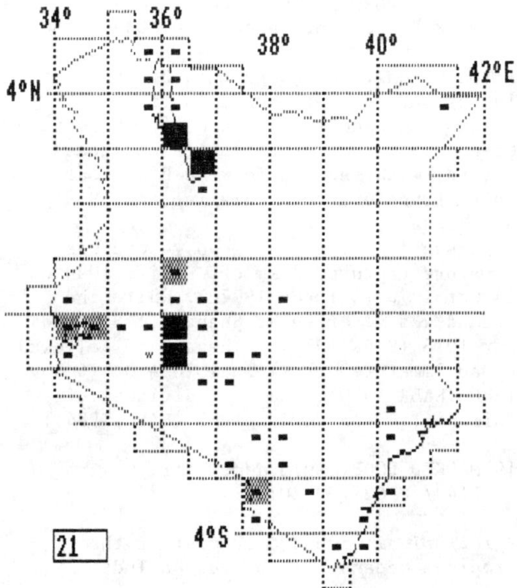

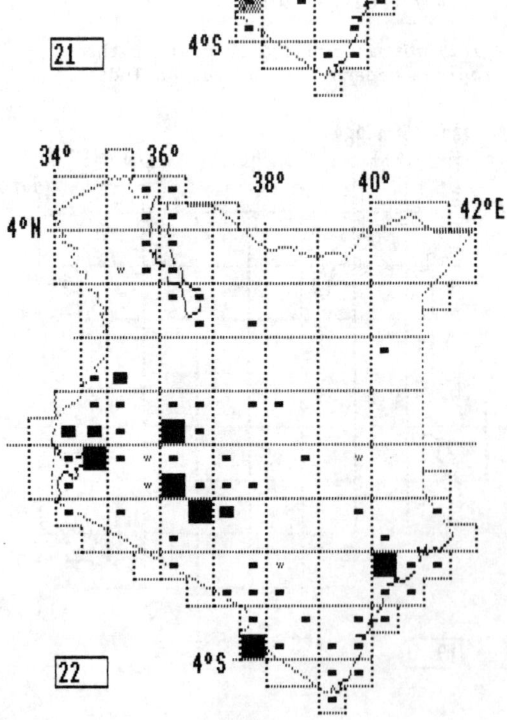

Family Anhingidae: Darters

Four species, distributed throughout the tropical and warm temperate regions of the continents, including the extreme S Palearctic. One occurs in the Afrotropical Region, and is resident in Kenya.

23 DARTER *Anhinga rufa*
B19 / MPG28 / D2C2

Afrotropical: widespread over most areas S of 15°N.

Kenyan: a bird of still or sluggishly moving waters with fringing vegetation, including coastal inlets (Br). It performs considerable but poorly known local movements in response to rainfall (MPG; Br; *e.g.* St), and usually breeds in colonies of herons, storks and ibises (Brown & Britton 1980; *e.g.* North 1959b, Coverdale *et al.* 1983).

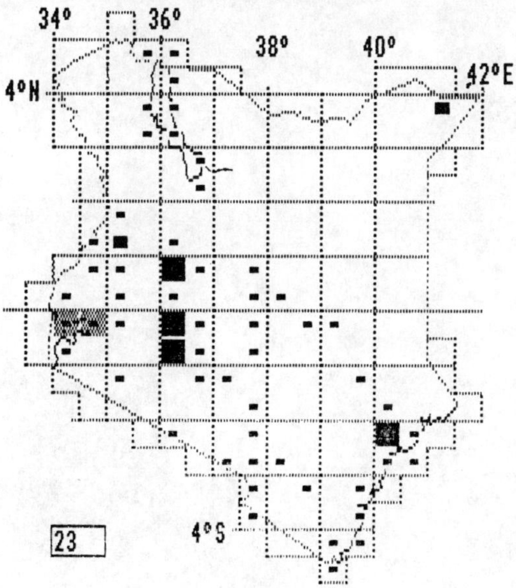

Family Fregatidae: Frigatebirds

Five species, with representatives throughout the tropical and subtropical oceans. Four are known in Afrotropical waters, and two occur as non-breeding visitors along Kenya's coast. While members of this family are amongst the most distinctive of seabirds, their identification to species level is notoriously difficult. New information published by Harrison (1983) may alleviate this problem (see Ash 1985b).

24 GREATER FRIGATEBIRD *Fregata minor*
B21 / MPG29 / DOCO
NO MAP: post presence 80C 103A; pre presence 91D

Afrotropical: breeds on Indian Ocean islands and occurs off the E seaboard of Africa from Somalia S to the Cape, though it is rare inshore. It is also found in the Pacific and the SW Atlantic.

Kenyan: sight records of frigatebirds have fallen into three categories. Close views of individuals that are entirely dark indicate adults of either this species, the Magnificent Frigatebird *F. magnificens* or the Ascension Frigatebird *F. aquila* but, since the first two are unknown in the Indian Ocean, the Greater is by far the most probable. A record has been accepted on this basis (EABR 1981). There has also been a recent sight record of a Lesser Frigatebird (25). Most reports, however, have been of birds with varying amounts of white below that have not been satisfactorily identified to species level: these are thus *Fregata* spp.

The confirmed records are one collected at Kipini 91D on 31 Aug 1951 (Forbes-Watson 1972a), a bleached skull found near Kiunga 80C on 29 Aug 1971 (Britton 1974c), and an all dark bird (referred to above) at Watamu 103A on 13 Aug 1981 (EABR 1981). All dark birds off Shimoni 114C on 8 Sept and 9 Oct 1981 were probably this species (EABR 1981). This Aug-Oct seasonality compares with Aug-Nov birds off SE Somalia (Ash 1985b).

Fregata spp. have occurred in 91BC 102B 103A 114BC, in all months except May and Jul (Ja; Stoneham 1944a; Ng'weno 1971; Mann & Britton 1971ab; Cunningham-van Someren 1972ab; Forbes-Watson 1972a; Britton 1974c; Brown 1975a; Br; EABR). Seven or eight 60 km off Shimoni 114C on 24-25 Mar 1974 suggests that they may be more common at sea than is usually thought (Britton 1974c).

25 LESSER FRIGATEBIRD *Fregata ariel*
B20 / MPG30 / DOCO
NO MAP: post presence 103A

Afrotropical: breeds on Indian Ocean islands, and disperses N to the coasts of India and Somalia (Harrison 1983; AM : but see Ash 1985b). There is a single record from Kenya, and two from Tanzania (Br; EABR 1983). It also occurs in the Pacific and SW Atlantic.

Kenyan: the single record concerned an adult male off Watamu 103A in Jan 1980 (Grant 1985; EABR 1983).

Family Ardeidae: Bitterns, Herons, Egrets

62 species, with representatives throughout the tropical and temperate regions of the world. 21 (34%) occur in the Afrotropical Region and, of these, 18 (86%) are found in Kenya. Most species breed in Kenya, but populations of some are augmented by migrants from the Palearctic. The only totally non-breeding, regular visitor is the Madagascar Squacco Heron (32). Many of these species breed colonially, often alongside birds of other families, in long-established heronries: see Kisumu 60B (Ja; Campbell 1972b; Harper 1974; Parsons 1976, 1977; Swegen 1984); Garsen 91A (North 1959b; Britton 1974e; Coverdale *et al*. 1983); L Jipe 100D (actual heronry just inside Tanzania)(Lack 1977b; Turner 1978). Coverage of breeding sites for the larger and less skulking species is probably reasonably complete, so that breeding:presence square ratios are meaningful.

\- BITTERN *Botaurus stellaris*
B22 / MPG52 / D2C0
RECORDS UNSATISFACTORY

Afrotropical: Palearctic migrants have been recorded S to Nigeria, N Zaire, Sudan and Ethiopia, while a resident race occurs S of about 8°S.

Kenyan: records from Amboseli 88C (Lowis 1974), Naivasha 62C (Wood 1974) and Nairobi 75B (Norris 1974) could conceivably have been Palearctic birds S of their normal wintering range, but these sightings have not been fully substantiated (Br). The range for Kenya shown by BUN is erroneous.

26 LITTLE BITTERN *Ixobrychus minutus*
B23 / MPG50 / D1C2

Afrotropical: both the Palearctic nominate race and the resident *I.m. payesii* are locally distributed over most areas S of 15°N.

Kenyan: a skulking and secretive but not uncommon bird of dense aquatic vegetation (Br; Ja; MPG; St). Most sightings are reported (hence C2) because it is infrequently seen but distinctive. All four breeding squares are above 1000 m. The few records of the Palearctic race range from Nyanza 60 to the coast and N to Turkana, Nov to early May but mostly at times of passage (Ba; EABR). Birds undergoing nocturnal movements during the autumn at Ngulia Lodge 101A have included singles of *payesii* and numbers of the Palearctic race, as well as flocks of up to 14 of unknown race (EABR).

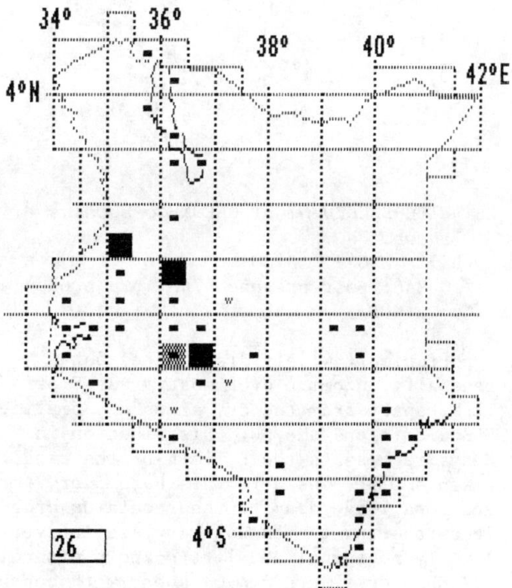

27 DWARF BITTERN *Ixobrychus sturmii*
B24 / MPG51 / D1C2

Afrotropical: widespread S of 15°N, apart from the arid areas of N Somalia and the SW.

Kenyan: an uncommon species that ranges to even the smallest water bodies, usually skulking in dense vegetation (Br; Ja; St). See Little Bittern (26) for the logic behind C2. Birds breeding in Kenya may be resident, but this species is a known intra-African migrant and equatorial populations are most probably augmented by birds from the S (and possibly N) tropics during their non-breeding seasons *e.g.* individuals on floodwaters during May-Jul (Britton & Britton 1976b; Br; BUN). Presumed migrants have occurred at Ngulia Lodge 101A in Nov-Jan and Apr (EABR).

(map overleaf)

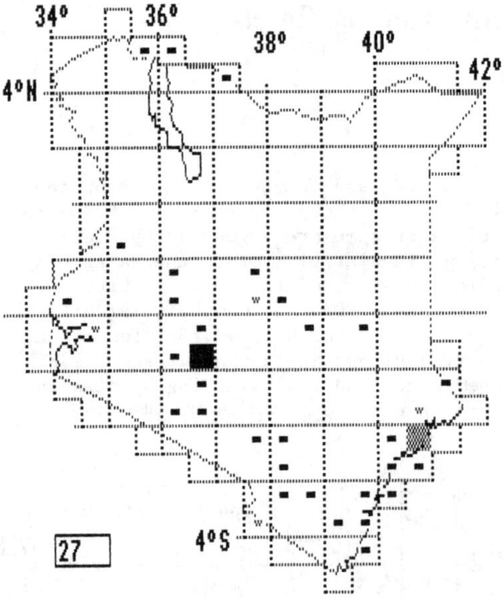

28 WHITE-BACKED NIGHT HERON *Gorsachius leuconotus*
B39 / MPG49 / D1C0
NO MAP: post presence 74A; pre presence 63C 100B

Afrotropical: widely distributed but generally uncommon over most areas S of 12°N, apart from the dry areas of Somalia, E Ethiopia and the SW. Distribution in Kenya represents the E limit of the range.

Kenyan: Br considers this heron very rare and localised, despite the recording problems inherent in its solitary, secretive, largely nocturnal and little-known nature. It inhabits rivers and marshes, with dense fringing vegetation in which it spends the day (MPG; Br). The three precisely located records are from Taveta 100B (Ja), Murang'a 63C (Br) and near Governor's Camp, Maasai Mara 74A (three individuals on 11-12 Jul 1978: EABR 1978). There is also a specimen in the National Museum, Nairobi, collected on 26 Aug 1960 on a river, possibly the Mara, near Narok 74B. Williams (1968) mentions the Mara R in Maasai Mara 74AC. BUN quote May-June egg-laying dates for Kenya, but omit actual localities: we cannot trace the source of this record.

29 NIGHT HERON *Nycticorax nycticorax*
B40 / MPG48 / D2C2

Afrotropical: widespread S of 15°N, with the exception of much of the equatorial lowland forest of the W, and of the dry areas of E Somalia and the SW. Palearctic migrants penetrate S at least to Cameroon (BUN) and Mozambique (Br).

Kenyan: a gregarious bird found in dense fringing vegetation around water bodies of all types (Br; Ja; MPG). It is largely nocturnal and somewhat local, but far commoner and more easily detected than the White-backed Night Heron (28). Breeding is colonial, usually alongside other species and often in established heronries. At least local movements to breeding sites must occur (breeding squares = 17% of its total), but details are lacking. The regular occurrence of Palearctic migrants is quite possible, but indications are so far restricted to birds of uncertain origin attracted to the Ngulia Lodge lights 101A on the night of 26 Dec 1973 (Backhurst & Pearson 1977b), and in unsuitable habitats at times of passage *e.g.* on Central Island, L Turkana 14C during Apr (Br).

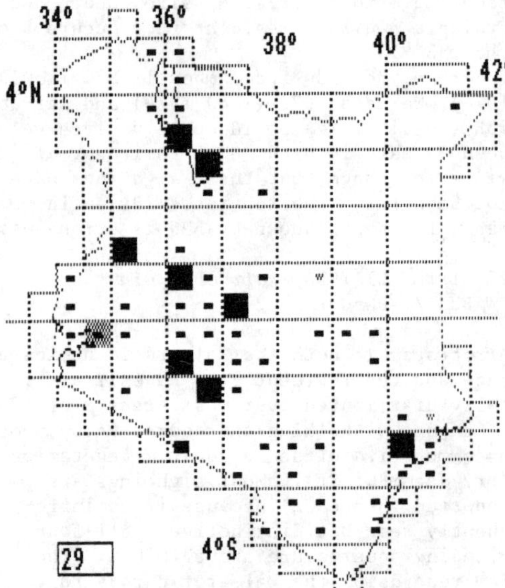

30 CATTLE EGRET *Bubulcus ibis*
B32 / MPG42 / D3C3

Afrotropical: widespread S of 15°N, apart from the interior of the arid SW. This is a pronounced intra-African migrant, apparently breeding in only relatively restricted areas of the continent (Sn). While certainly an abundant non-breeding migrant to S Somalia (AM), the relatively few breeding birds in E Africa also suggest that most of the large numbers here breed elsewhere, and there are ringing recoveries from S Africa (Backhurst 1969b).

Kenyan: a common, approachable, gregarious egret. It is a habitual commensal of large mammals and, like the Black-headed Heron (43), can often be found far from water on dry grasslands. This egret also feeds at active cultivation (MPG). Breeding is colonial, in established heronries, alongside the Long-tailed Cormorant (22) (Brown & Britton 1980), or by itself. A recently discovered colony at N Kinangop 62D (Siemens 1984) is unique in Kenya by virtue of its position in a village, but this type of situation is common in the N tropics (Sn). While most Kenyan birds may be non-breeding visitors, the only indications of movements are a regular Mar-Apr N passage at Mau Narok 61D (Se), and an Oct-May peak in Tsavo E 89D 101B (La).

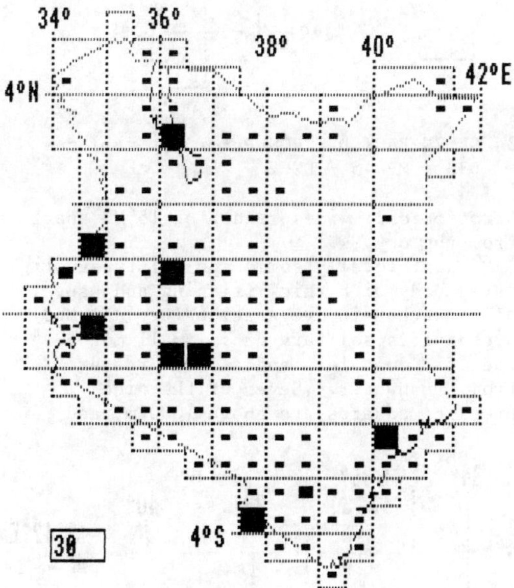

31 SQUACCO HERON *Ardeola ralloides*
B30 / MPG43 / D2C2

Afrotropical: widespread S of 15°N, avoiding only the interior of the arid SW. As in the Grey (40) and Purple (41) Herons, Palearctic migrants probably greatly outnumber resident populations in the N tropics and *vice versa* further S (Sn; BUN).

Kenyan: locally common on lakes and sluggish water bodies with fringing or floating vegetation: coastal records do not refer to the open shore (Br; MPG; St). Breeding can take place solitarily, in small colonies or alongside other species in established heronries. Local movements (Sn) doubtless occur, but no details are known. Palearctic migrants penetrate Africa at least S to Zaire, but there is no direct evidence for them in Kenya (Br). There are occasional nocturnal arrivals at Ngulia Lodge 101A during the autumn (Backhurst & Pearson 1977b). See Madagascar Squacco Heron (32).

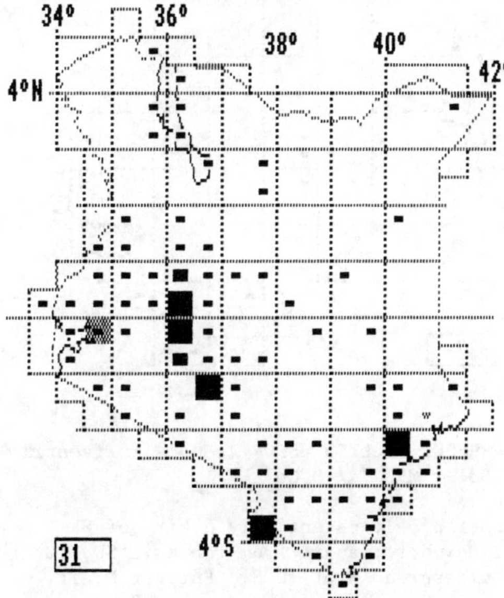

32 MADAGASCAR SQUACCO HERON *Ardeola idae*
B29 / MPG44 / D1C1

Afrotropical: breeds in Madagascar Oct-Mar, and is a regular non-breeding migrant during other months to E Africa, from Mozambique and Zimbabwe N to E Zaire and S areas of Uganda, Kenya and Somalia (Turner 1980; Ash 1983).

Kenyan: this species forms part of a species-group with the Squacco Heron (31), and the two frequent broadly similar habitats except for the Madagascar's apparent preference for more open situations in its non-breeding range (Sn; Turner 1980). Their maps, however, are distinctly different, with the Squacco more widespread in the W and N, while the distribution of the present species reflects its penetration into Kenya as a migrant from the S. Most records from the interior are in or E of the rift, N to the N and NE peripheries of the C highlands 50A 51B 52C, while the trend along the coastal strip culminates at 2-3°N on the Somali coast (Ash 1983). Dates are mostly May-Oct (extremes Apr-Dec) with up to 25 birds in one place, and there is a regular S passage along the coast Aug-Oct (Turner 1980; EABR).

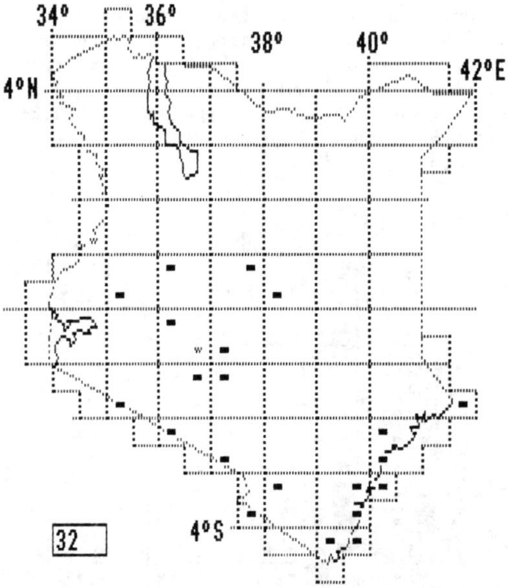

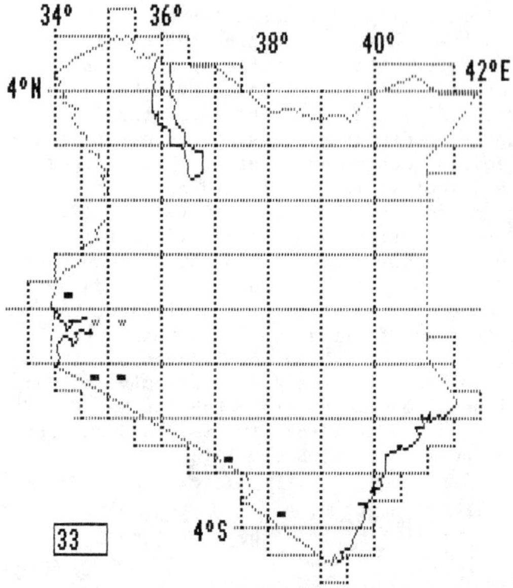

33 RUFOUS-BELLIED HERON *Ardeola rufiventris*
B31 / MPG47 / D2CO

Afrotropical: essentially a bird of S Africa where, apart from the arid SW, it is widespread S of 10°S. Further N, it extends E of the Congo basin to SE Zaire, the W rift, Tanzania, SW Uganda and, on the NE limit of the range, the SW borders of Kenya.

Kenyan: the few records of this marsh and inundation species are: (a) one near Songhor 61A June 1949 (D K Bednall *in litt.*), (b) one in papyrus on the S shores of the Winam Gulf 60B Aug or Sept 1949 (D K Bednall *in litt.*), (c) two at Amboseli 88C 15 Oct 1973 (Binford & Zimmerman 1974), (d) one at Salt Lick Lodge 101C 9 Nov 1978 (EABR 1980), (e) one in Mara GR 74A 24 Jul 1980 (EABR 1980), (f) one near Thika 76A 3 Jul 1983 (EABR 1983), (g) 3-6 near Sio Port 48C 25 May and one there 21 Sept 1983 (EABR 1983) and (h) one in NW Mara GR 73B 19 Aug 1983 (EABR 1983). In view of the single Apr-May breeding record for E Africa (Uganda)(Brown & Britton 1980), this restriction of records to W and SW Kenya may indicate post-breeding dispersal of wanderers E and NE of their normal range. Virtual absence from the apparently suitable papyrus swamps of W Kenya remains unexplained (Br).

34 GREEN-BACKED HERON *Butorides striatus*
B33 / MPG45 / D2C2

Afrotropical: widespread S of 15°N, apart from the dry SW.

Kenyan: locally common on all types of water body with thick fringing and especially overhanging vegetation (Br; Ja; MPG). Breeding is solitary or in small monospecific colonies, and all records are from freshwater. Seven of the nine breeding squares are above 1000 m, and

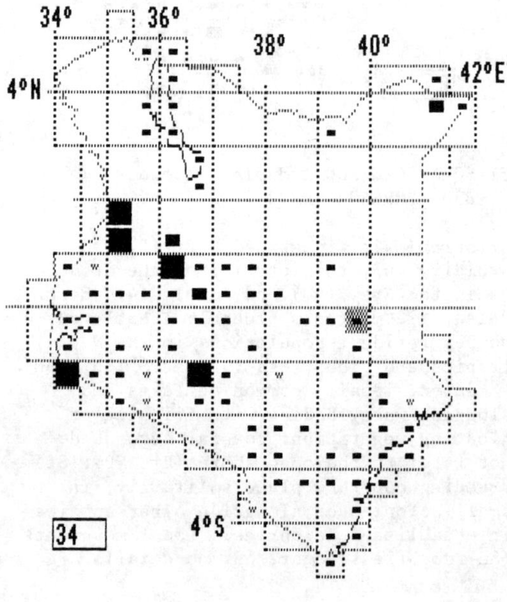

there are none in the coastal lowlands: these patterns are likely to reflect observer frequencies and the skulking nature of this species. Birds around Ngulia Lodge 101A at dawn in Dec may have been nocturnal migrants (Backhurst & Pearson 1977b), but this heron is otherwise considered resident (Br; Be).

35 LITTLE EGRET *Egretta garzetta*
B36 / MPG40 / D3C3

Afrotropical: widespread S of 15°N, apart from the dry SW, the Congo basin and some areas to the N of the Gulf of Guinea. Palearctic migrants penetrate S to the equator, possibly mostly in the W (BUN), and there is no evidence for them in E Africa (Br).

Kenyan: locally common on all types of waters, breeding on rocky islands (L Turkana 14C 26B) and in established heronries. As with the Great White Egret (39), the small number of breeding sites (breeding squares = 8% of its total) means that extensive movements must occur, but no direct evidence for these or for the occurrence of Palearctic migrants exists (Br).

There is pronounced colour dimorphism, with white birds greatly predominating everywhere (*e.g.* St) except along the coast S of Mombasa 114B (Br). Dark birds are confined to the coast (mainly S of Mombasa) and the rift valley (especially at L Turkana)(D J Pearson *in litt.*).

The Little Egret forms part of a superspecies with the African Reef Heron (36) (Sn).

36 AFRICAN REEF HERON *Egretta gularis*
B37 / MPG41 / D2C1

Afrotropical: predominantly marine, along the W coast between W Sahara and Gabon, and along the E coast from the Red Sea S to NE Tanzania. Wanderers inland have occurred in Sudan, Ethiopia, Uganda and Kenya.

Kenyan: the concentration of records along the coast and the rift valley lakes reflects birds moving S along these pathways from their breeding areas in coastal Ethiopia and N Somalia. Rather than a regular post-breeding dispersal, however, the rather erratic occurrences in Kenya suggest wanderers slightly S of their normal non-breeding range (Br), which may extend as close to Kenya as 1°30'N on the Somali coast (AM).

Britton (1975) knew of only four previous coastal records, perhaps due to problems of identification (see *e.g.* North 1966; Forbes-Watson 1966), but the species is now well known there, particularly in Mida Creek 102B, occurring erratically in most months of the year (EABR). Inland, in addition to brief occurrences, single birds have been resident at L Baringo 50A for 16 months (St) and at L Nakuru 62A for five months (EABR 1982). Small numbers have been reported at L Turkana during much or all of the year (Forbes-Watson 1966; Owre 1967; Br), though their possible breeding there remains controversial. See Little Egret (35).

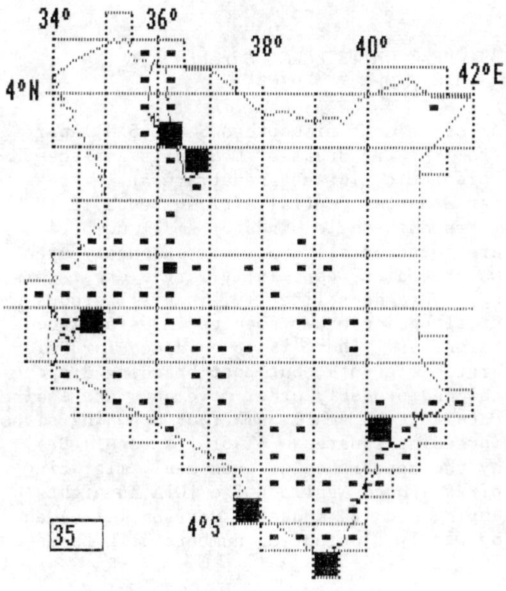

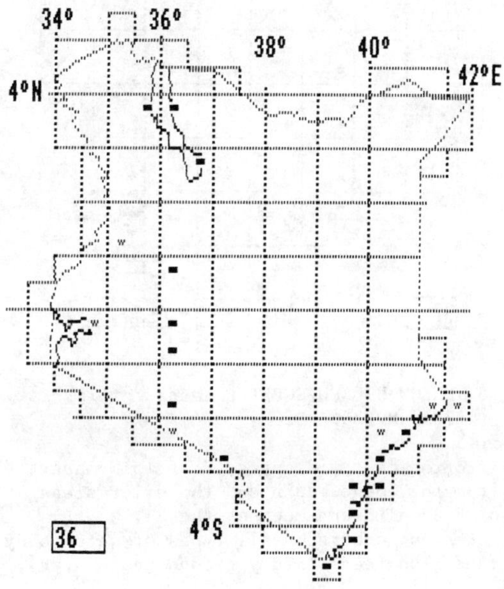

37 BLACK HERON *Egretta ardesiaca*
B35 / MPG39 / D2C2

Afrotropical: widespread but local over most of Africa S of 15°N, apart from much of the Congo basin and the dry areas of E Somalia and the SW.

Kenyan: a local bird of reasonably open lake and marsh shores and coastal creeks, usually singly or in small flocks (Br; Ja). Although formerly also at Kisumu 60B (Ja), breeding is now apparently restricted to the Garsen heronry 91A. Extensive post-breeding dispersal and/or wandering must occur, as is shown by seasonal concentrations of hundreds at L Jipe 100D (Forbes-Watson & Turner 1973) and L Bilisa 91A (EABR 1977).

The map shows two broad areas of occurrence that appear to be linked to seasonality. Birds in the coastal lowlands and at L Jipe are present at least locally during all months (EABR). N of Amboseli 88C and Shombole swamp 87A, however, there are no recent breeding records and dates mostly agree with the Oct-May seasonality recorded in Somalia (AM) and E Uganda (EABR), where the bird is also a non-breeding visitor.

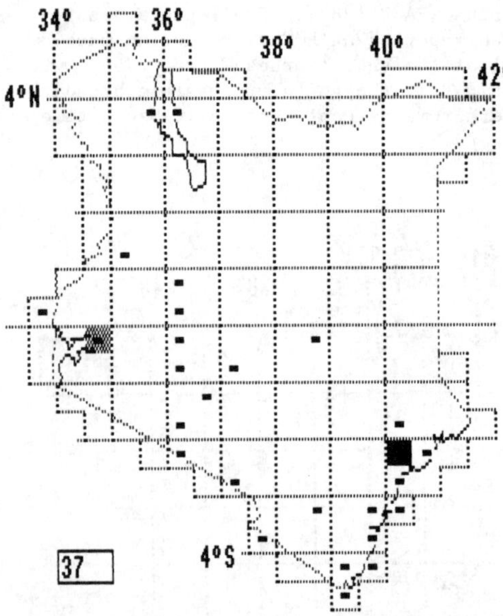

38 YELLOW-BILLED EGRET *Egretta intermedia*
B38 / MPG38 / D2C2

Afrotropical: widespread S of 15°N, apart from the Congo basin and the drier areas of E Somalia and much of the SW.

Kenyan: occurs on all types of water body (and also feeds on dry ground near water), but generally in less open situations than the other large egrets (Br; BUN). It is usually solitary, or in small numbers. Breeding is colonial, alongside other species, and often in established heronries. The low number of breeding sites (breeding squares = 10% of its total) indicate at least local movements (*cf.* Little (35) and Great White (39) Egrets). All records over a five-year period at Mau Narok 61D were of passage migrants during May (Se), and irregular absences of several months duration occur at L Baringo 50A (St). It becomes slightly more common in Tsavo E 101B during the rains (La).

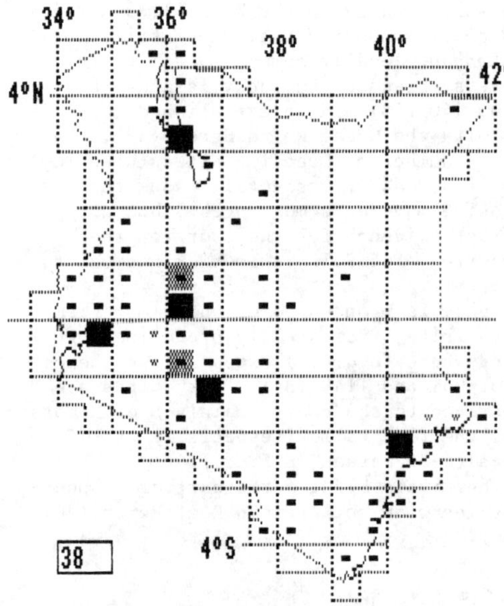

39 GREAT WHITE EGRET *Egretta alba*
B34 / MPG37 / D3C3

Afrotropical: widespread S of 15°N, apart from the dry areas of E Somalia and the SW. Palearctic migrants penetrate at least as far S as the Central African Republic (BUN).

Kenyan: single birds or small numbers are found on all types of reasonably open water bodies, especially larger ones. This egret breeds solitarily, in small groups or alongside other species in established heronries. There is no evidence for Palearctic migrants, but post-breeding dispersal and probably other movements are indicated by the small number of breeding sites (breeding squares = 6% of its total)(Br), by the appearance of presumably migrating birds around Ngulia Lodge 101A on nights during Dec (Backhurst & Pearson 1977b), and by widely fluctuating numbers at L Baringo 50A (St).

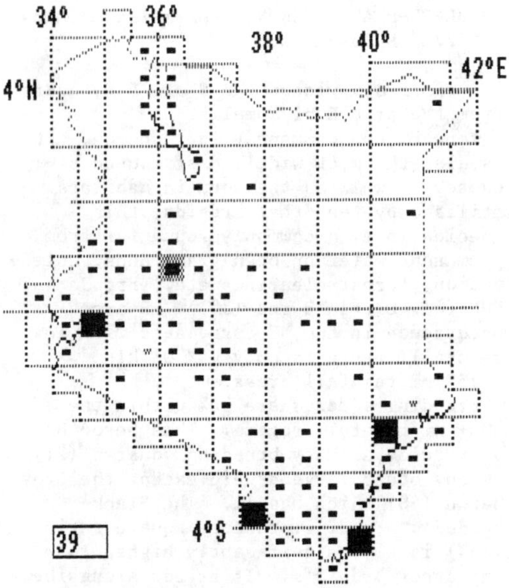

40 GREY HERON *Ardea cinerea*
B25 / MPG33 / D3C3

Afrotropical: widespread S of 15°N. Birds in the N tropics are mainly Palearctic migrants with minor numbers of Afrotropical residents, while those further S are primarily Afrotropical (*cf.* Squacco (31) and Purple (41) Herons)(Sn; BUN).

Kenyan: a usually solitary species, mainly associated with aquatic habitats including the open seashore, but also found on dry grassland (Br; Ja). Kenya is situated at latitudes where both Palearctic and resident birds could occur (see above). There is no evidence for the former (Br), but the following points demonstrate movements of some type. The few pairs at each breeding colony appear disproportionately small in relation to the overall population, suggesting that many birds breed elsewhere (Br). It is an occasional visitor to Mau Narok 61D only during the N winter (Se). Occurrence at L Baringo 50A may be in all months, but it is sometimes absent June-Aug: increases in numbers during the N winter are followed by sudden decreases in Apr and May (St), though whether this reflects rains-induced dispersal or Palearctic migration is uncertain. Other references: Skinner 1972; Irvine & Irvine 1978.

41 PURPLE HERON *Ardea purpurea*
B28 / MPG36 / D2C2

Afrotropical: widespread S of 15°N, apart from the arid areas of E Somalia and the SW. There is a possible division of populations similar to that of the Squacco (31) and Grey (40) Herons. Thus the great majority of breeding sites are in the S tropics of E and SE Africa, so that the numerous non-breeding birds seen in other areas may well be migrants from the Palearctic (Sn).

Kenyan: a skulking and usually solitary heron that frequents waters of any type, including seasonal floods, provided that

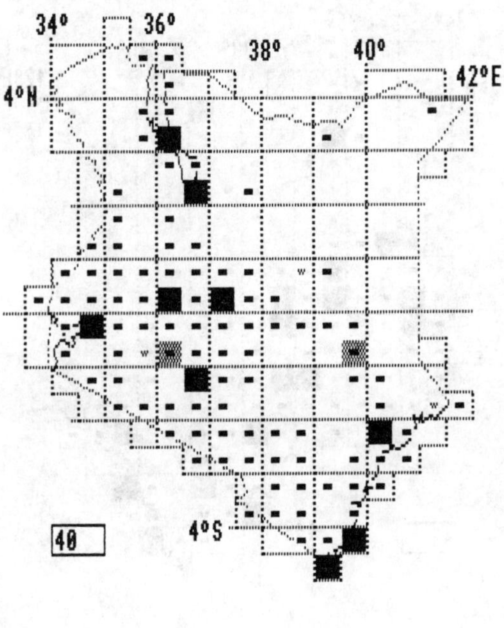

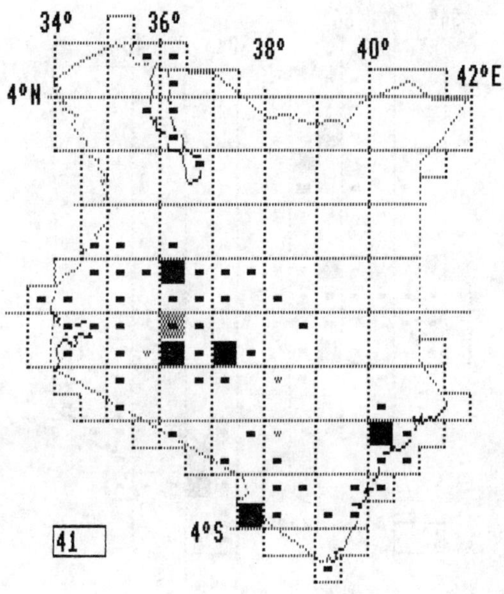

they have reedbeds or other dense fringing vegetation (Br; Ja; MPG; HH; St). It breeds both in established heronries and at other sites. As with the Grey Heron, Kenya's geographical position makes the presence of Palearctic birds likely, but there is as yet no definite evidence (Br) and these migrants may mostly winter in W Africa (BUN) or to the N of Kenya in the E. Br, Se and La mention evidence of at least local movement, and this is also indicated by the restriction of breeding to 10% of the species' squares.

42 GOLIATH HERON *Ardea goliath*
B26 / MPG35 / D3C3

Afrotropical: widespread S of 15°N, apart from the E Somalia and SW arid areas.

Kenyan: usually solitary or in pairs, though flocks are known *e.g.* EABR 1981 (Br; Ja; MPG). It is almost always localised in the vicinity of permanent water, whether fresh, salt or carbonate, even fishing in the running tide (BUN). All breeding records are from rift valley lakes and Tsavo E 89D 101B. It does not use the established heronries mentioned in this family's introduction, and the colonies of approximately 20 pairs each at L Baringo 50A are an exception to its usually solitary breeding (Brown & Britton 1980; St). Breeding squares are only 10% of its total and movements obviously occur, but no details are known. Other reference: Skinner 1972.

43 BLACK-HEADED HERON *Ardea melanocephala*
B27 / MPG34 / D3C3

Afrotropical: widespread S of 15°N, apart from the arid E of Somalia.

Kenyan: the commonest large heron. It is also the most widely distributed because, as well as the aquatic habitats utilised by the other ardeids, this species is also commonly found far from permanent water on open grasslands, cultivation, forest clearings etc. (Br; Ja; BUN)(*cf.* Cattle Egret (30)). Sparse occurrence in the NE correlates with low rainfall: range is only 15% within 0-250 mm rainfall areas.

Breeding squares are 96% within the 500+ mm rainfall regime. This heron has over twice as many breeding squares (21) as any other large ardeid except the Grey Heron (40), which has 11. The Black-headed's breeding:presence square ratio (19%) is also significantly higher than any other ardeid's. It breeds alongside other species in established heronries, *e.g.* at Kisumu 60B (see Parsons 1977) and L Jipe 100D but, again like the Cattle Egret, it also commonly has monospecific colonies in or near human centres *e.g.* Nairobi railway yards 75B (North 1963), in Njoro High Street 61B (Standish King 1972b; Cunningham-van Someren 1982), and at Malindi 103A (Brown & Britton 1980).

Br notes that movements to nest sites must occur, but that this bird may be mainly resident in many areas. DEP (pers. obs.) found it resident in 75, but also noted regular Nov-Feb influxes to the

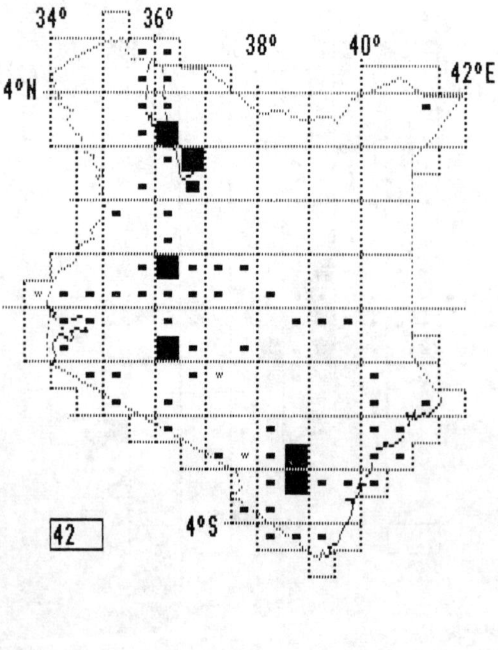

42

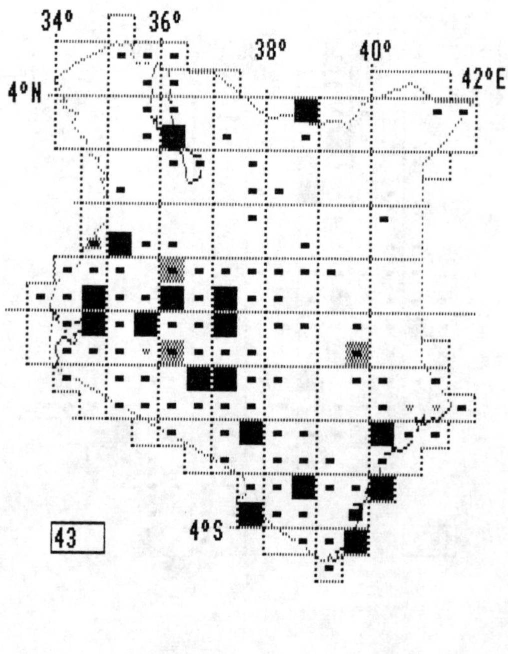

43

Ruaraka-Kahawa-Ruiru area of the square. This seasonality coincides with Nov-Mar influxes to the Aruba area of 101B (La). Some movement into dry areas after the rains has been reported from semiarid Baringo 50A (St) and arid Tsavo E 101B (La).

Family Balaenicipitidae: Shoebill

A monotypic family, endemic to papyrus and *Miscanthidium* swamps in E tropical Africa, where the total population may be as low as 1500 individuals (BUN).

- SHOEBILL *Balaeniceps rex*
B41 / MPG54 / DICO
RECORDS UNSATISFACTORY

Afrotropical: endemic to papyrus swamps from the Central African Republic, Sudan and W Ethiopia S through Uganda, W Tanzania and Rwanda to SE Zaire, Zambia and Botswana. Presence in W Kenya requires confirmation.

Kenyan: a photograph, supposedly taken in the Yala Swamp 48C, the most suitable habitat for this species in Kenya, constituted the sole confirmed record (Mann 1976; Britton 1978; Br), but this has now been admitted as a hoax (Parker 1984).

The remaining reports are those of Johnston (1902), Preston (1976) and Britton & Britton (1985). The first involves only hearsay evidence from colonial settlers in 1899 of occurrence along the shores of NE L Victoria S to the Winam Gulf 60AB. Despite searching, Johnston himself never found it. Preston's records are of one at Fourteen Falls 76A in June 1971 and two at Shombole Swamp 87A in Apr 1976. Despite the bird's singular appearance, these are not convincing due to lack of details, the extreme unlikelihood of this normally sedentary species so far E of its usual range, and the absence of corroboration from observers already familiar with the bird. Finally, Britton & Britton (1985) did not find this bird during a day inside the then undisturbed Yala Swamp 48C in 1969, although the local Luo people said that occasional (as opposed to territorial or sedentary) individuals occurred there. From reports of extensive swamp drainage at Yala, Britton & Britton (1985) consider the sole suitably extensive habitat for this species in Kenya to have been destroyed.

Plate 1. The Ostrich, endemic to Africa and Kenya's only flightless bird, in typically open habitat (J.S. Karmali F.R.P.S.). *See page 41*.

Plate 2. The Cattle Egret's characteristic affinity for large mammals (J.S. Karmali F.R.P.S.). *See page 56.*

Plate 3. Black Heron 'umbrella-ing': this species habitually mantles its head with its wings while fishing, probably to cut down surface reflection from the water in order to see its prey more easily (J.S. Karmali F.R.P.S.). *See page 60.*

Plate 4. The Hamerkop, sole member of an Afrotropical endemic family (J.S. Karmali F.R.P.S.). *See page 65*.

Family Scopidae: Hamerkop

A monotypic family, endemic to the Afrotropical Region and widespread in Kenya.

44 HAMERKOP *Scopus umbretta*
B42 / MPG53 / D3C3

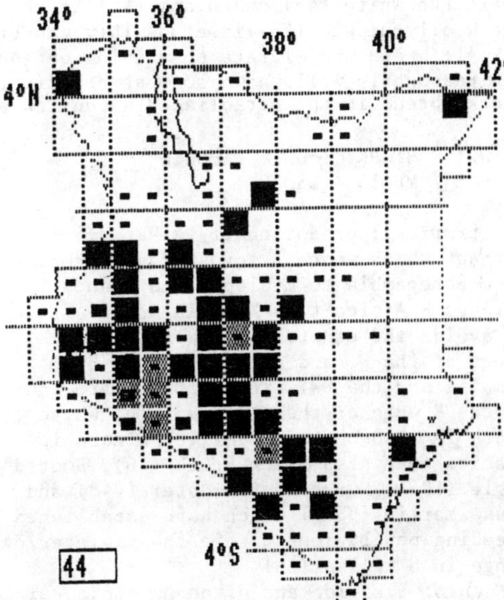

Afrotropical: widespread S of 15°N.

Kenyan: provided there are shallows and adequate prey, this reasonably common species occurs singly or in pairs on all types of still or slow-moving waters, in environments ranging from forest to semi-desert (Br; Ja; MPG). It is predictably more frequent in wetter country, however: range is 64% within 500+ mm rainfall areas, and the bird is scarce at L Turkana (HH).

Breeding squares are 77% above 1000 m and virtually absent from the apparently suitable coastal high rainfall area and the Tana R. Absence from the latter may reflect infrequency of observers in some areas, but the Hamerkop's large and conspicuous nest ought to ensure good coverage of breeding in the better known coastal lowlands, so that this apparent absence of nesting may be real. If this is true, then most birds in the coastal lowlands are non-breeding visitors and the species is more mobile than is commonly realised. Breeding is known from coastal squares in S Somalia (AM).

Br thinks large scale movements unlikely, but notes wandering to concentrations of food in, amongst other types, seasonal waters. BUN also reports dispersal to small waters during the rains. An apparent Oct-Nov seasonality occurs at L Turkana (HH). Be found this bird resident in 74, and DEP (pers. obs.) notes similar status in 75BD 89A.

Family Ciconiidae: Storks

The 17 storks have representatives throughout the world's zoogeographical regions, including eight species (47%) in the Afrotropical. All of these eight also occur in Kenya, where the White (45) and Black (46) Storks are non-breeding visitors from the Palearctic, the Woolly-necked (48) is a possibly non-breeding, intra-African migrant, and the remainder all breed and migrate to varying extents. As with some of the herons, breeding sites are reasonably well known so that breeding:presence square ratios are meaningful. Some storks breed in the established heronries already mentioned.

45 WHITE STORK *Ciconia ciconia*
B45 / MPG55 / D3C3

Afrotropical: predominantly a Palearctic migrant which winters across the N tropics from Senegambia to Ethiopia and thence S through E Africa to N Namibia and S Africa. It avoids the equatorial lowland forest areas of the W, and the arid areas of Somalia and the SW. Kenyan birds are thus on the E edge of their range. The White Stork is also one of the five Palearctic species (see also Black Stork (46), Booted Eagle (124), Eurasian Bee-eater (448) and House Martin (563)) which have established breeding populations within their wintering range in S Africa (Sn).

Kenyan: a common and often numerous Palearctic migrant and winter visitor inland, mainly late Oct-early May (Br; Ja). Although flocks of 1000+ can occur in Tsavo E 101B (EABR 1981), it is still unrecorded in all coastal squares (Br).

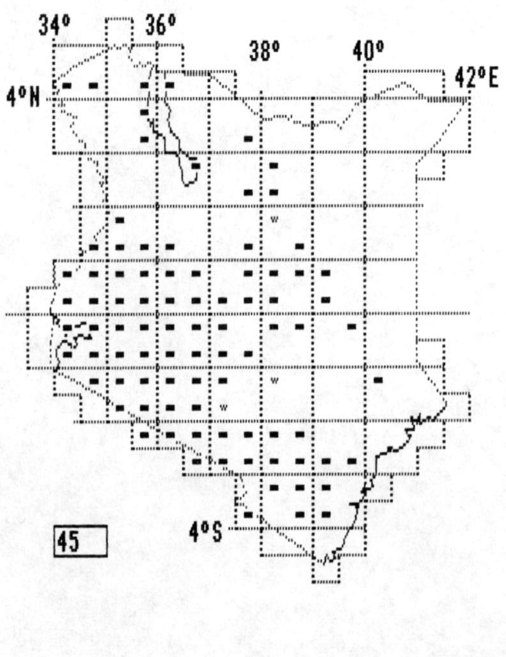

Absence from E and NE Kenya is apparently real, since it has not been recorded by observers resident at Rhamu 19A or Wajir 42A, and there is only a single record for S Somalia (AM).

Wintering is mainly above 1600 m on grasslands and cultivation of S and C Kenya (Br). As with Abdim's Stork (47), migration is principally W of the rift valley, apparently mainly through the L Victoria basin: Nov-Dec passage is heavier than that in Mar-Apr (Britton & Britton 1976b; Br; HH; Lewis & Pearson 1981a; EABR 1983).

Oversummering in small numbers is common (Stoneham 1944b; Se; St; EABR). Oversummering range is mainly W of or in the rift valley, 69% above 1500 m, 85% moist subhumid-humid, and 92% within 500+ mm rainfall areas. Newly fledged young with two adults at L Nakuru 62A in Jul 1982 were probably only about two months old and possibly bred in Kenya (Dittami & Haas 1982), though the day lengths and ambient temperatures of this equatorial regime would differ substantially from those of their S African and Palearctic breeding areas (Sn; *cf.* equatorial breeding of Common Sandpiper (265)).

Numerous ringing recoveries show movements between Kenya and South Africa, Algeria, Greece, East and West Germany, Poland and the USSR (Backhurst: E A Bird Ringing Reports).

46 BLACK STORK *Ciconia nigra*
B47 / MPG56 / D2C2

Afrotropical: occupies two distinct areas. Migrants from the Palearctic winter across the N tropics from Senegambia to Ethiopia, S to Uganda, Kenya and N Tanzania. There is a stable breeding population in the S, from S Africa N to Zambia. Distribution in Kenya represents the E limit of the range.

Kenyan: this species contrasts with the White Stork (45) by occurring singly or in pairs and overall in very small numbers, by predominantly feeding near small waters and along water courses, and by showing no tendency for oversummering (Br; Se; HH; La). Range is 72% within 500+ mm rainfall areas. The concentrations of squares in C Kenya and the SE 89 101AB (La) show its principal wintering areas Nov-Apr (Br; EABR). It may have become generally more common during this century (Ba), but it remains scarce in W Kenya.

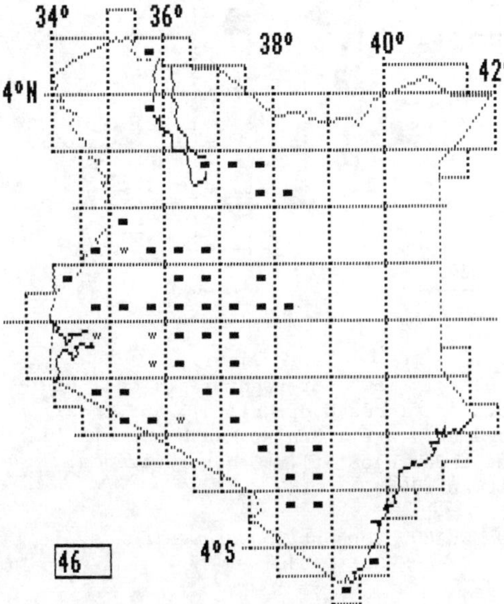

46

47 ABDIM'S STORK *Ciconia abdimii*
B44 / MPG58 / D2C2

Afrotropical: a trans-equatorial intra-African migrant which breeds during the rains in the N tropics, in a zone extending from Senegambia and Ivory Coast E to Sudan and Ethiopia, with a small S outlier in the N of the L Victoria basin. After breeding, it migrates S across the equator to S Kenya, SW to Angola and S to S Africa, where it is present during the S rainy seasons. Its crossings of the equator also coincide with wet seasons.

Kenyan: a highly migratory stork that feeds in open, dry country and can flock in thousands (Br). The map illustrates three main phenomena. The breeding on cliff sites in the Lokitaung 3D area (Blencowe 1962a) represents the S edge of the main N tropics breeding range (Kahl 1968). The breeding in W Kenya 48CD 60BD is the E edge of the equatorial L Victoria basin breeding population, which nests in trees, singly or in small colonies, in highly populated areas that include town centres *e.g.* Kakamega 48D and Kisumu 60B

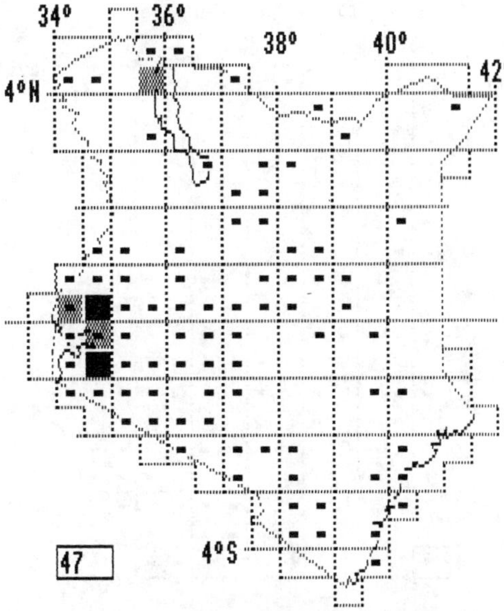

47

(North 1940; Kahl 1968; Britton & Britton 1976b; Maas 1979).

The rest of the map points refer to birds that are predominantly migratory, with some non-breeding *i.e.* 'wintering' visitors Oct-May. It is possible that the bulk of migration between the N and S tropics takes place through the L Victoria basin, *i.e.* avoiding the lowland forests of the Congo basin, and thus migrants are far more abundant in W Kenya than E of the rift valley and at the coast (Br; BUN). Heavy S passage with flocks of up to 500 birds is recorded Nov-Dec from Nyanza 48 60 (Britton & Britton 1976b), L Turkana (HH), and Mau Narok 61D (Se). N passage in W Kenya Mar-May is less intense, with flocks usually below 50 but exceptionally up to 900 (Britton & Britton 1976b; Lewis & Pearson 1981a; EABR 1983). Kenya is at the N extremity of the S tropics non-breeding range and most individuals move further S, but 'wintering' birds still occur in hundreds in S Kenya (*e.g.* in Tsavo E 101B: La) and in smaller numbers N of the equator, Dec-Feb (Br).

48 WOOLLY-NECKED STORK *Ciconia episcopus*
B46 / MPG57 / D3C2

Afrotropical: widespread S of 15°N, apart from the E Somalia and SW arid areas. Few breeding localities are known, but it is an intra-African migrant although its movements are in general poorly understood.

Kenyan: at best locally common, typically near water and in pairs or small groups, up

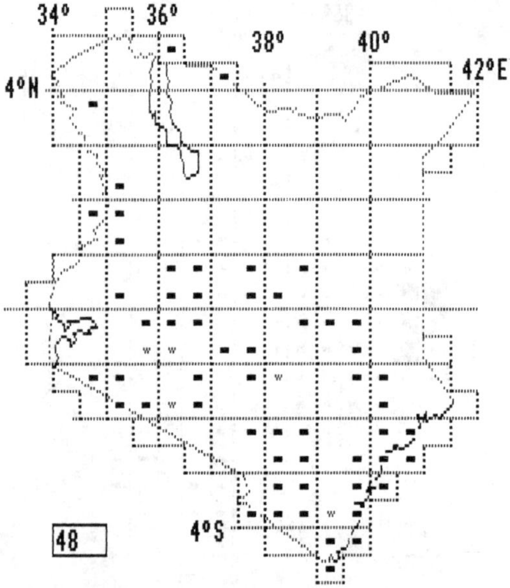

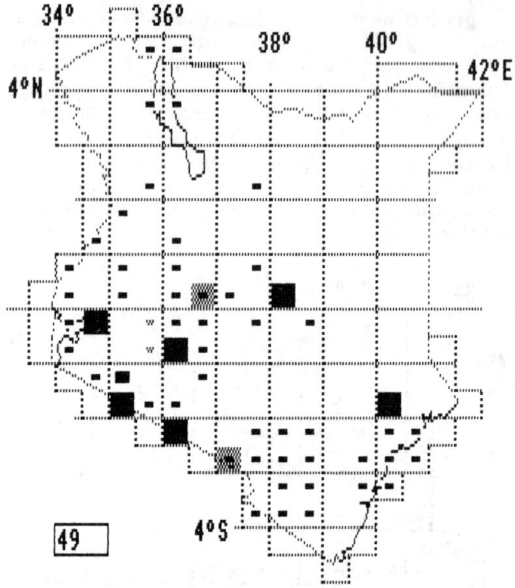

to 3000 m (MPG; Br; Ja; Se). Occurrence is in two distinct regions. It is reasonably common and present in all months along the coast, and there has been the suggestion that it breeds there, undiscovered, in mangroves (Br; Kahl 1968). Small numbers are also found throughout the year in Tsavo E 101B (La). Elsewhere, this stork is an erratic wanderer or intra-African migrant (Br; Se; St; EABR).

Range is only 4% within 0-250 mm rainfall areas and thus absent from the driest areas of the NE, including L Turkana (a single QSD on the Ethiopian border). It is uncommon in the 1000+ mm rainfall country of W Kenya, and there is no record from oligotrophic L Victoria.

49 SADDLE-BILLED STORK *Ephippiorhynchus senegalensis*
B48 / MPG60 / D3C2

Afrotropical: widespread S of 15°N, excepting the drier areas of E Somalia and the SW.

Kenyan: the Saddle-billed Stork is found singly or in pairs, always thinly distributed (Br; Ja; MPG). It frequents rivers, lakes and swamps with open water, but there are few records from L Turkana (HH) and none from the coast S of Malindi 103A. As stated by Br, it can occur up to 3000 m (*e.g.* Se) and breed up to 1900 m (*e.g.* at L Naivasha 62C), but range is mostly lower than this, with 65% of the squares below 1500 m. This species nests solitarily, and our coverage of breeding areas is undoubtedly low. Br reports seasonal visitors to floods N of Malindi, and local movements are known at Mau Narok 61D (Se) and L Baringo (St). Other reference: Pitman 1965b.

50 MARABOU *Leptoptilos crumeniferus*
B49 / MPG61 / D3C3

Afrotropical: widespread S of 15°N, although rare in the arid areas of E Somalia and much of the SW. Intra-African movements occur.

Kenyan: the widespread distribution of this sometimes locally numerous species reflects its utilisation of both aquatic and dry habitats, its propensity to wander widely like a vulture in search of food (Pomeroy 1978b; *e.g.* Se), and its apparently often substantial movements to and from its few breeding sites. Thus it appears that Marabous could occur in any of our squares, but there are unexplained absences *e.g.* along the coastal strip S of Malindi 103A where we have found only a single record, from Mombasa 114B. The species is less common in areas with below 250 mm rainfall (3% of its squares), though this may be partly due to lack of observers.

The Marabou was rare in most of E Africa early this century (see Ja), with only small numbers at well scattered localities, but it is more numerous now due to the great increase of concentrated human wastes at sites like fish-processing factories (*e.g.* at Baringo 50A (St) and formerly at Naivasha 62C (Brown 1973a), abattoirs and

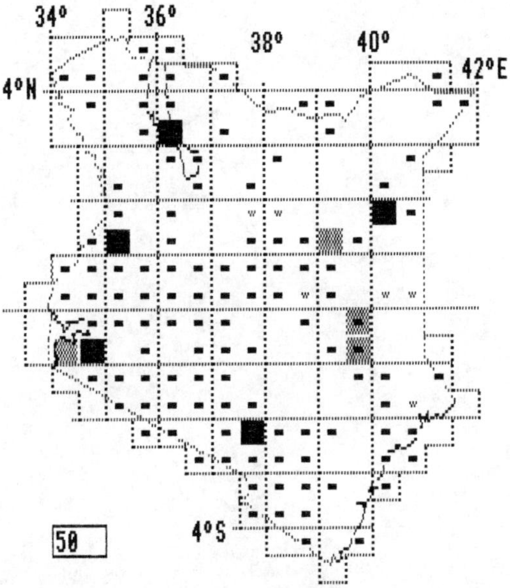

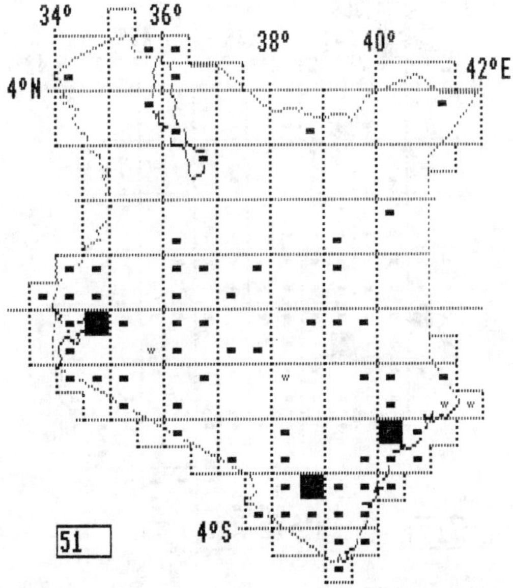

refuse tips (Pomeroy 1978b; Brown 1973a). Absence from the rubbish dumps at Mombasa 114B and Kisumu 60B remains unexplained (Br). In addition to refuse, Marabous visit carrion with vultures (Pomeroy 1977), and opportunistically take live vertebrate (Brown 1963b; Brown 1973a) and invertebrate prey.

This stork breeds colonially, often in or near human settlements *e.g.* at Kitale 37C, Wajir 42A and Kiboko 88B, and in Kenya its colonies are monospecific. These few colonies are quite well known (North 1939, 1943; Pomeroy 1975, 1979; Archer 1978), and our coverage of breeding may well be complete. The scarcity of these sites (7% of the species' squares) may partly reflect the need for a reliable dual source of food *i.e.* carrion/refuse for the bulk of the diet, and small vertebrates *e.g.* fish, frogs to provide calcium for the nestlings (Kahl 1968): refuse tips near permanent swamps would seem ideal.

Seasonal movements certainly occur, *e.g.* a marked increase at L Baringo 50A Dec-May (St), and are consistent with movements of individual birds N and S to or across the equator to avoid dry weather (Pomeroy 1978ab). Other references: Merz 1980; Pomeroy 1986.

51 OPEN-BILLED STORK *Anastomus lamelligerus*
B43 / MPG59 / D3C2

Afrotropical: widespread S of 15°N, avoiding only the E Somalia and SW arid areas. Most breeding colonies appear to be in E-C Africa, approximately between Namibia and Uganda-Kenya, so that much intra-African movement presumably occurs (Sn).

Kenyan: a gregarious, locally abundant and usually aquatic species that is most frequent in the L Victoria basin and in the coastal lowlands (Br; Ja), where the map shows full coverage. Breeding is in monospecific colonies or in established heronries, and the three sites are also in these two regions. Widespread distribution over the rest of the country indicates this bird to be a marked wanderer and intra-African migrant which, during its movements, may temporarily appear anywhere, even far from water. It is, however, uncommon in the highlands, with range 77% below 1500 m (Br; Ja).

Three notes of movement are in the Oct-Dec period: most records at L Baringo 50A have been of Oct-Dec passage birds (St), regular S movements in Nov-Dec are known at L Turkana (HH), and E movement occurs over Mau Narok 61D (at 3000 m) in Nov (Se).

52 YELLOW-BILLED STORK *Mycteria ibis*
B50 / MPG62 / D3C3

Afrotropical: widespread S of 15°N, except for the arid E of Somalia.

Kenyan: this common stork frequents all types of aquatic habitat, though generally preferring larger waters, and can feed even in turbid conditions (Br; BUN). It breeds colonially, usually alongside other species (*e.g.* North 1939; Parsons 1976; Swegen 1984).

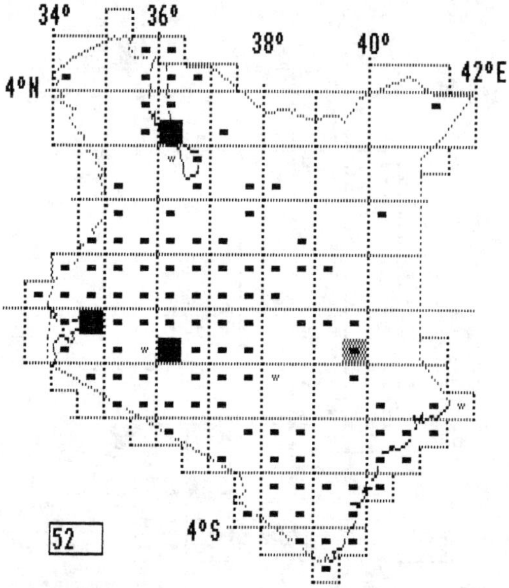

Kahl (1968) knew of only one Kenyan nesting site and presumed that many more exist, but the Yellow-billed Stork is well known and our coverage of the breeding of this conspicuous species is probably good. Breeding is in only 4% of its squares, and there are no breeding records from either Uganda (Br) or Somalia (AM). Hence numerous at least local movements must occur, and flocks of hundreds accumulate irregularly and reside for a few months at L Baringo 50A (St). Br considers few individuals wholly resident.

Family Threskiornithidae: Ibises, Spoonbills

28 species, with representatives in the warmer areas of all the world's zoogeographical regions. Ten (36%) occur in the Afrotropical Region. Of the six found in Kenya, the Eurasian Spoonbill (58) is a rare vagrant from the Palearctic, the Glossy Ibis (56) breeds locally but is possibly augmented by Palearctic migrants, and the remainder are all locally breeding Afrotropical residents. Breeding:presence square ratios are meaningful for the non-forest or woodland dwelling members of this family, all of which utilise established heronries to some extent.

53 SACRED IBIS *Threskiornis aethiopica*
B54 / MPG63 / D3C3

Afrotropical: widespread S of 17°N, apart from the arid SW.

Kenyan: the wide distribution of this common ibis reflects wandering to feed on moist ground near or far from all types of water, including small temporary pools (Br; Ja; BUN). It is also a scavenger around human settlements. Breeding is colonial, often alongside other species and in established heronries. Hundreds of pairs have used the Kolal ibisery 60B (Parsons 1976, 1977). As with many of the herons and storks, at least local movements must occur to its small number of breeding sites (12% of its total squares). St reports fluctuations in abundance at L Baringo 50A, and Se notes visitors around 3000 m at Mau Narok 61D. This ibis is apparently a non-breeding visitor to Somalia (AM).

54 HADADA *Bostrychia hagedash*
B51 / MPG65 / D3C2

Afrotropical: widespread S of 15°N, with the exceptions of the drier areas of E Somalia and the SW of the continent N to N Angola.

Kenyan: locally common where mainly permanent waters are near open forest or woodland, including coastal shores and mangroves (Br; Ja; MPG). Range is 77% within 500+ mm rainfall areas and only 16% arid. It is noticeably absent from apparently suitable wooded habitats around L Turkana and on Mt Marsabit 27D 28C, and occurrence in the arid NE and E is confined to locally moist situations *e.g.* waterholes at Wajir 42A, and riverine vegetation around the Daua 19A, Ewaso Ng'iro 40D 52B and Tana 64B 65ABD 78B 79C. It is uncommon around rivers in arid Tsavo E 101B (La).

Breeding is solitary. The Hadada's

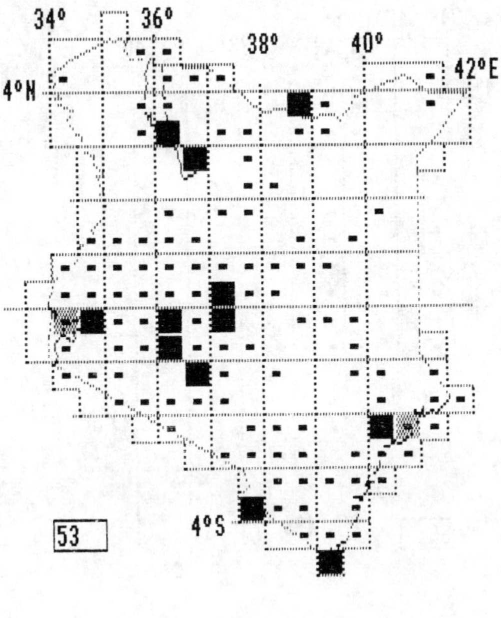

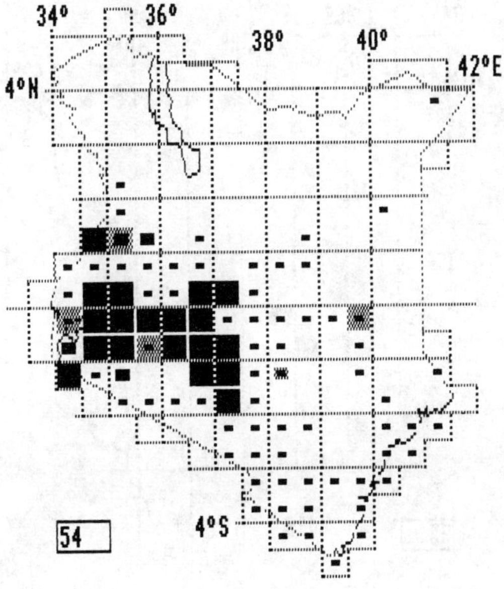

breeding squares are 90% above 1000 m, and completely absent from the seemingly suitable coastal lowlands. There are two possible explanations for this. Coastal breeding may have escaped detection, but this appears unlikely in view of this bird's vocal and conspicuous character, and the reasonably high frequency of observers (including residents) in the coastal strip. Alternatively, and contrary to Br's view of this as a mostly resident species, the coastal lowlands may be an off-season area into which birds move from relatively distant breeding grounds. This view parallels this ibis' apparent status as a non-breeding visitor to inland and coastal S Somalia (AM). Reports of resident status in the highlands come from Narok 74 (Be), Mau Narok 61D (Se) and 75B (DEP pers. obs.). Other reference: Gough 1952.

55 GREEN IBIS *Bostrychia olivacea*
B52 / MPG66 / D1C2

Afrotropical: a forest species with a remarkable fragmented distribution. In the W it is found in the lowlands around the Gulf of Guinea, and on islands in the Gulf. Further E, it is known from E Zaire, NE Tanzania and the highlands of Kenya.

Kenyan: a poorly known and local inhabitant of dense montane forests at 2000-3700 m on the Aberdare Mts 62BD, Mt Kenya 51C 63AB and the Nyambeni Hills 51D (Br; Ja; MPG). It replaces the similar Hadada (54) at altitudes above the latter's range, and is seldom identified because it is resident in dense habitats, utters a Hadada-like (though in fact distinctive) call, and makes crepuscular flights between roosts and feeding grounds.

Most records are from Mt Kenya 63A (EABR) but this reflects observer frequency and it is probably as common on the more poorly covered Aberdare Mts 62BD (*e.g.* Parker 1982). Status W of the rift valley is uncertain. Although Zimmerman (1972) thought early records from Nandi 49C (Meinertzhagen 1937) and Mt Elgon 36D (Ja) feasible, these have never been confirmed despite extensive later coverage of the areas (and of the Kakamega Forest 48D) by numerous experienced observers. Be described it as resident around 2500 m in the Mau forests 61BD 62C, but Se could find only the Hadada in these areas. The Green Ibis breeds solitarily near streams, and the only confirmed breeding records were in 1910 (Ja), 1959 and 1981 (Parker 1982).

56 GLOSSY IBIS *Plegadis falcinellus*
B53 / MPG68 / D3C2

Afrotropical: throughout S of 15°N, apart from the arid SW. Breeding of the resident Afrotropical population appears to be mainly in the S and E, while Palearctic migrants penetrate S to Niger, Chad, Sudan and probably the equator (Sn; BUN).

Kenyan: a rather local bird of swampy habitats inland, and brackish coastal lagoons (Br; MPG). It breeds colonially in the Garsen heronry 91A, and formerly at

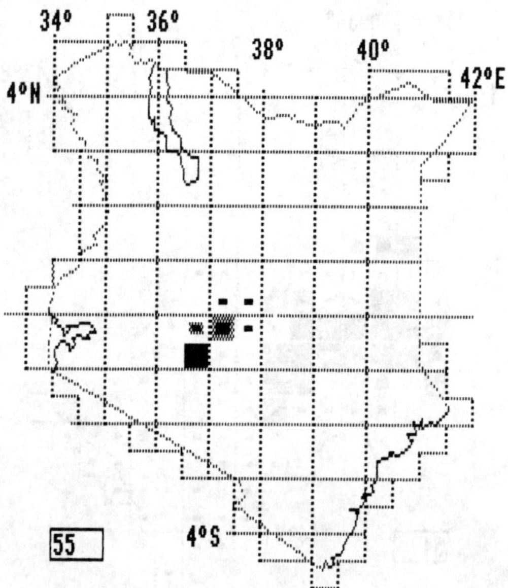

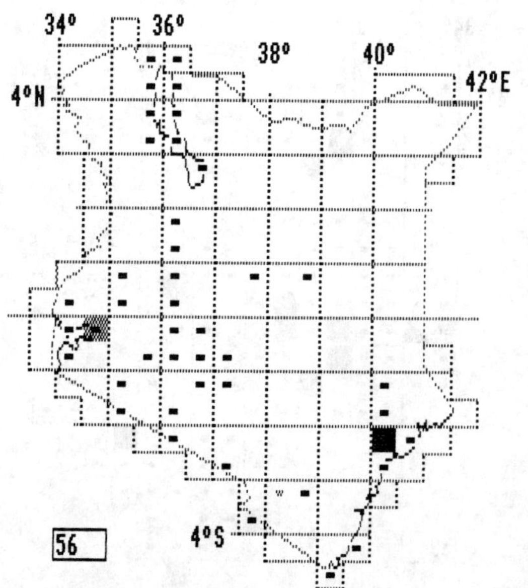

the old Kisumu heronry 60B (Ja). While present throughout the year on L Nakuru 62A and L Naivasha 62C, numbers at many other localities *e.g.* L Turkana (MPG), L Baringo 50A (St) and W Kenya (Br) increase during the N winter, suggesting addition of Palearctic migrants (Br). This may be the case, particularly for birds at L Turkana, but small influxes in Tsavo E 101B are June-Aug (La), and most Kenyan birds are more likely to be intra-African migrants from either the N or S tropics (Br).

57 AFRICAN SPOONBILL *Platalea alba*
B55 / MPG70 / D3C3

Afrotropical: widespread S of 17°N, with the exceptions of most of the SW of the continent from the Congo basin S to S Africa, and arid E Somalia.

Kenyan: distribution resembles that of the Sacred Ibis (53) and the two mostly breed in the same colonies/established heronries, but the spoonbill is scarcer, shyer and in most places more local (present in 82 squares, as against the ibis' 115), and is restricted to the actual margins of water bodies of all types (Br; Ja; MPG).

Although recorded at many localities throughout the year (*e.g.* La; St), few birds are likely to be resident (Br). The low breeding:presence square ratio (13%) implies at least local movements and, like the Sacred Ibis, this spoonbill is a non-breeding visitor to Somalia (AM). Se recorded an Oct wanderer around 3000 m at Mau Narok 61D.

The African Spoonbill forms part of a species-group with the Eurasian Spoonbill (58)(Sn). Other reference: Kahl 1983.

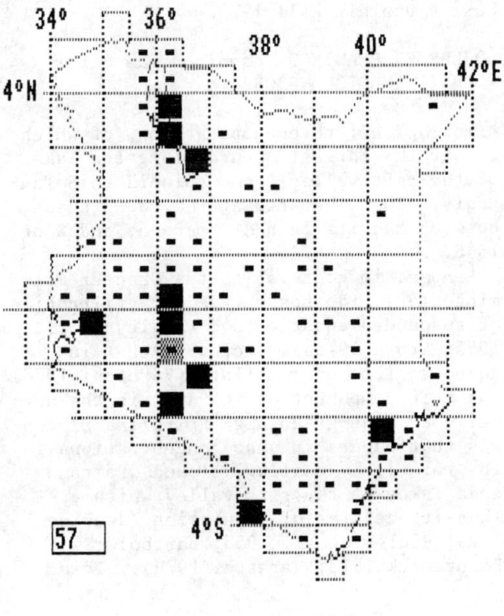

58 EURASIAN SPOONBILL *Platalea leucorodia*
B56 / MPG69 / D2C0
NO MAP: post presence 13B 62AC; pre presence 14C

Afrotropical: this predominantly Palearctic bird breeds around 20°N on the W coast of Africa, and also along the Ethiopian and Somali Red Sea coasts. As a migrant from the Palearctic, it also occurs in small numbers along the S fringe of the Sahara from Senegambia to Ethiopia, with wanderers S to Uganda and Kenya.

Kenyan: the six records are all from L Turkana, L Nakuru 62A and L Naivasha 62C Aug and Oct-May, and in all but one case ("several") they involved single birds (Ba; EABR). These are thought to have been Palearctic migrants that wandered down the rift valley lakes, S of their normal wintering range. Vagrancy of Red Sea breeding birds inland to Kenya and Uganda is considered unlikely (Ba). See African Spoonbill (57).

Family Phoenicopteridae: Flamingos

Four species, with representatives in all the tropical and warm temperate areas of the world except Australasia. Two are found in the Afrotropical Region, and both have bred in Kenya.

59 GREATER FLAMINGO *Phoenicopterus ruber*
B58 / MPG71 / D3C3

Afrotropical: mainly E and S Africa, from the Red Sea coast along the lakes of the E rift valley and adjacent areas, and thence over much of the continent S of 15°S. Range also includes W coasts. L Elmenteita 62A is one of its very few major African breeding sites. Palearctic migrants penetrate S at least to Senegambia, Sudan, Ethiopia and Somalia, and probably to Kenya (Br; Backhurst 1981b).

Kenyan: although also a filter feeder, this species differs significantly from the Lesser Flamingo (60) by taking mostly invertebrate prey from bottom mud (Ridley *et al*. 1955; Brown 1973a), but has the same preference for carbonate, saline and brackish waters. It has only about 1% of the Lesser's abundance (Brown 1973a; Bartholomew & Pennycuick 1973), but can still flock locally in thousands. Like the Lesser, it performs erratic and unpredictable movements, mainly at night (*e.g.* St).

Thus its map is similar to that of the Lesser Flamingo, with greatest concentrations on the rift soda lakes and much smaller, transitory numbers on the rift freshwater lakes (*e.g.* St; EABR 1984). Wandering away from the rift occurs more frequently than in the Lesser, but most records from the coastal lowlands inland to Tsavo E 101B occur in the N winter and probably refer to Palearctic migrants from the Middle East moving along the E African seaboard (Britton & Britton 1976b; Br; La; EABR). Coastal birds are mainly in sheltered estuaries and creeks *e.g.* regular flocks of up to 50 in Mida Creek 102B (Br). The record at Wajir 42A was in Apr (J E Miskell *in litt.*).

Like the Lesser, this species breeds colonially but only sporadically and at very few sites *i.e.* about 30 fledglings on Koro Lokuaria Island 4C, L Turkana, during Aug 1984 (J R Ness *in litt.*); L Elmenteita 62A (Brown 1958; EABR); L Nakuru 62A (Brown *et al*. 1973; Meinertzhagen 1958; Haas & Siebenrock 1983); and with the record Lesser Flamingo colony at L Magadi 75C in 1962; abortive breeding is again quite common (Brown 1955b, 1973a; Brown *et al*. 1973; EABR 1978). Other reference: Kear & Duplaix-Hall 1975.

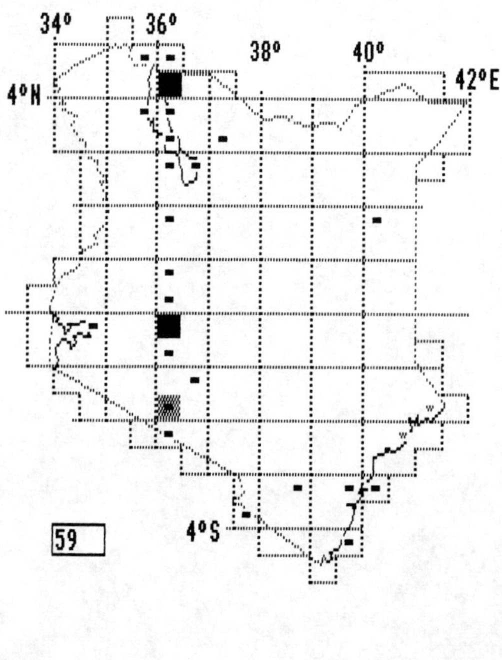

60 LESSER FLAMINGO *Phoeniconaias minor*
B57 / MPG72 / D3C3

Afrotropical: three populations, of which by far the largest occurs along the lakes of the rift valley from Ethiopia to S Tanzania. The other two are on the W coast between Mauritania and Cameroon, and S of 15°S.

Kenyan: in contrast to the Greater Flamingo (59), the Lesser feeds by filtration of suspended aquatic food (Ridley *et al*. 1955; Brown 1973a), including that in aqueous films on mud flats (Vareschi 1978). The main component of its diet is the unicellular blue-green alga *Spirulina platensis* that occurs in vast concentrations in the relatively shallow, closed drainage, soda lakes of the rift valley, which are thus its centres of population (Jenkin 1929; Ridley *et al*. 1955; Bartholomew & Pennycuick 1973; Vareschi 1978). This species is approximately five times as numerous as the combined total of all the world's other flamingos, and its occurrence in gatherings which exceed a million, particularly at L Bogoria 50C and L Nakuru 62A, means that these waters support the largest biomass of birds per unit area of any inland water body in the world (Brown 1959, 1973a; Bartholomew & Pennycuick 1973; Vareschi 1978).

The Lesser Flamingo does not undergo regular migrations, but is nomadic, performing movements which are not always in response to fluctuating food supplies, water levels or alkalinity, and which are generally poorly understood (Brown 1973a; Vareschi 1978). These movements can be very rapid *e.g.* an exodus of 400 000 individuals from L Bogoria during seven days in Aug 1956 (Brown 1959, see Vareschi 1978). Nestlings from Magadi 75C have been recovered at 8°N in Ethiopia and 3°S in Tanzania (Backhurst 1981b), and there is apparently frequent interchange between birds on the various rift lakes (Vareschi 1978). Tuite (1979) suggests the possibility of large scale movement from the Kenyan and NE Tanzanian rift lakes to S Africa. Possibly for reasons of energy (Vareschi 1978) or to avoid aerial predators, movements are mainly by night (*e.g.* Se; St), though very high altitude diurnal flights are also known (BUN).

The map reflects three types of occurrence: (a) on the rift soda lakes, where it is most numerous and can feed adequately *e.g.* L Turkana, L Bogoria 50C, L Nakuru 62A, L Elmenteita 62A and L Magadi 75C (Simbi Crater L 60B (Br) may also fit in here); (b) on the other freshwater rift lakes, where it occurs infrequently and in small numbers in transit between the soda lakes *e.g.* L Baringo 50A (St) and L Naivasha 62C; (c) with the exception of Simbi Crater L 60B, records outside the rift refer to rare and often moribund wanderers that have presumably become disorientated during nocturnal movement, to descend in habitats as diverse as pools on mountain streams 61D (Se), Aruba dam in Tsavo E 101B (La) and the coast 103A (Britton & Britton 1976b).

Like its movements, the breeding of the Lesser Flamingo is still not well understood, with no clear correlation with either food supply or climate (BUN). Its main E African breeding ground is L Natron (N tip = 87A), but even there breeding is not annual. In Kenya, early records of breeding at L Nakuru (van Someren 1922; Meinertzhagen 1958) lack details and are unsatisfactory. Nest building and even the production of a few eggs have been followed by desertion at L Suguta 26D (see Chambers 1960), L Bogoria 50C and L Nakuru 62A (Brown & Root 1971; Brown 1973a; Haas & Siebenrock 1983). The only confirmed successful breeding records for Kenya are a colony of 30 000 pairs on Central Island 14C in 1957, one of 1 200 000 pairs at L Magadi 75C in 1962 (the largest known anywhere, when heavy rain had rendered the nearby L Natron breeding ground unsuitable: Brown & Root 1971; Brown 1973a; BUN), and a relatively very small colony at L Bogoria 50C in 1980 (EABR 1980). Other references: Akeley 1912; Brown 1955b, 1971d; Jenkin 1957; Grzimek & Grzimek 1960; Pennycuick & Bartholomew 1973; Kear & Duplaix-Hall 1975; Tuite 1978.

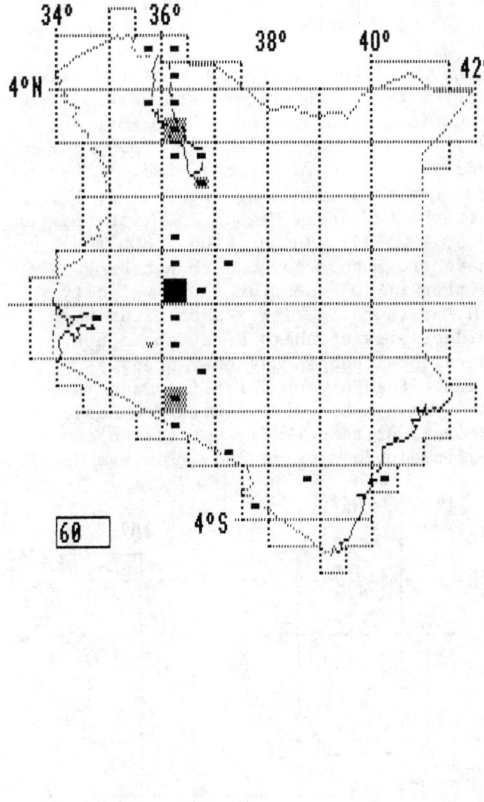

Family Anatidae: Ducks, Geese

Approximately 145 species, with representatives in all zoogeographical regions. Of the 31 (21%) found in the Afrotropical Region, 24 (77%) are known in Kenya. The Kenyan anatids comprise 14 (58%) which breed or have done so, 9 (38%) which are non-breeding visitors, and one vagrant. The Mallard has not certainly occurred in the wild state in Kenya.

61 FULVOUS WHISTLING DUCK *Dendrocygna bicolor*
B59 / MPG95 / D2C2

Afrotropical: widespread S of 15°N, apart from the arid areas of E Somalia and the SW, and the lowland equatorial forests of the W.

Kenyan: found on all types of relatively quiet water, including coastal creeks and estuaries, mainly below 1500 m (Br; Ja; MPG). Flocks may exceed 1000 (*e.g.* EABR 1983), but this is a local bird. There are only three breeding records: at L Naivasha 62C and L Magadi 75C (van Someren 1916, 1925-35), and at Kiunga 80D (Ja). These are mapped here, although Brown & Britton (1980) question their validity. G N Paterson (*in litt.*) shot birds with oviduct eggs at Ahero Rice Scheme 60B, where local people say they breed.

Hence the Fulvous Whistling Duck is mainly a non-breeding visitor to Kenya (Brown & Britton 1980), and the absence of confirmed breeding records from Uganda (Br) and Somalia (AM) suggests that movements are extensive. The few published dates appear to conform to a pattern. The movements that produce large Nov influxes to L Baringo 50A (St) may also be responsible for Nov-Dec stragglers at Mau Narok 61D (Se) and Tsavo E 101B (La). Apr dispersal of non-breeding migrants from Baringo agrees with Apr nesting dates in Tanzania (Brown & Britton 1980), Apr-May stragglers at Mau Narok 61D and Baringo 50A (*op. cit.*) and MPG's note of June-Aug dispersal from E African waters. Br suggests that this duck may be resident at L Turkana and L Jipe 100D, but there is no breeding record and, as at L Baringo 50A (St), the species may be present in all months yet still undergo marked fluctuations in abundance.

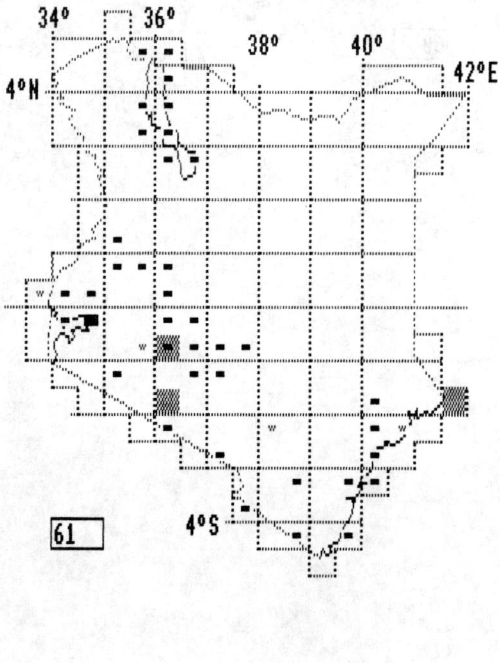

62 WHITE-FACED WHISTLING DUCK *Dendrocygna viduata*
B60 / MPG94 / D3C3

Afrotropical: widespread S of 15°N, excepting the driest areas of E Somalia and the SW interior, and interior areas of the Congo basin.

Kenyan: this species resembles the Fulvous Whistling Duck (61) in that it can occur on any type of relatively quiet water (Br). But the White-faced is decidedly the more widespread (66 squares as against 44) and numerous of the two and, particularly in lowland areas, is often the most abundant Afrotropical duck (Ja; Britton & Britton 1974b; Br). It has two breeding areas, in 500+ mm rainfall country. Nesting is most commonly reported from the coast (Brown & Britton 1980), but there are also scattered sites above 1000 m in the highlands.

Movements occur (*e.g.* St), but these would seem less necessarily extensive than those of the Fulvous Whistling Duck. There are few details, but peak counts and/or definite abundance increases occur during Oct-Apr at L Turkana, L Baringo 50A and other localities throughout the country (HH; St; EABR). These may partly represent off-season flocking, since coastal breeding peaks Jul-Aug and 19 of 29 (65%) E African breeding records are May-Sept (Brown & Britton 1980; EABR).

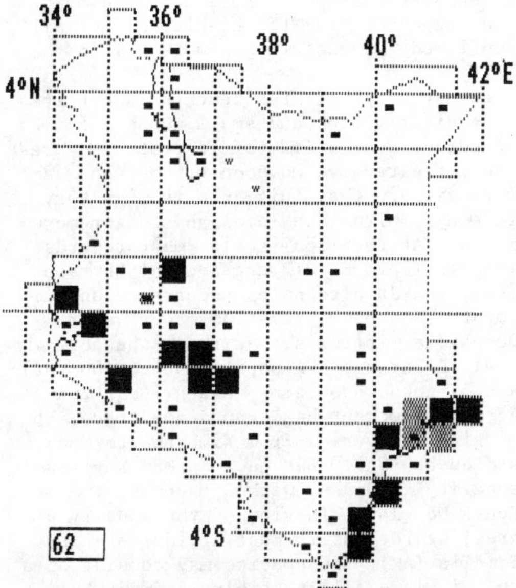

63 RUDDY SHELDUCK *Tadorna ferruginea*
B81 / MPG96 / D2C0
NO MAP: pre presence 13B

Afrotropical: a predominantly Palearctic duck that winters in the Sudan and Ethiopia. It has recently been discovered breeding 2½°N of the Kenyan border in S Ethiopia (Ash 1977), but there is only a single occurrence S to Kenya.

Kenyan: the single record is of one N of Ferguson's Gulf 13B, L Turkana, in the late 1950s (Ba).

64 EGYPTIAN GOOSE *Alopochen aegyptiacus*
B61 / MPG99 / D3C3

Afrotropical: widespread S of 15°N, apart from some areas in and around the Congo basin. One of the commonest and most widespread ducks in Africa (BUN).

Kenyan: common and widespread, up to 3000 m (Br; Se). Coastal records are mostly from the vicinity of the Tana 91 and Sabaki 103A estuaries (Br)(that in 102D is inland), but range elsewhere includes all types of waters, breeding occurring even on seasonal bodies *e.g.* La. It is largely resident, but local movements occur. A Nov-Mar influx has occurred at L Baringo 50A (St), while most of Se's records at Mau Narok 61D were Mar-Apr or Oct-Dec. Gatherings of hundreds (Ja;

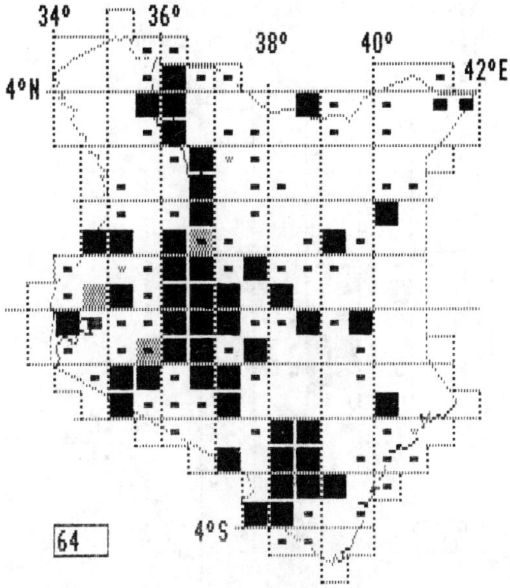

MacInnes 1933-34; Davey 1983b) are most probably undergoing wing moult (BUN).

65 SPUR-WINGED GOOSE *Plectopterus gambensis*
B79 / MPG100 / D3C2

Afrotropical: widespread S of 15°N, apart from the interior of the Congo basin and the dry areas of Somalia and the SW. Distribution in Kenya represents the E limit of the range.

Kenyan: can occur on most types of inland water up to 3000 m, but prefers freshwater, particularly floods near lakes and swamps (Br). It has been reported from 85% of the squares containing eutrophic lakes. Range is 70% above 1000 m, mainly in the highlands and their peripheral plateaux. Away from the highlands, it is uncommon at L Turkana (HH), and there are few records from Tsavo E 101B (La). Occurrence at the coast is restricted to squares containing the Tana 91 and Sabaki 103A estuaries (*cf.* Red-billed Teal (76) and Yellow-billed Duck (74)), and it extends inland along the Tana R 65D 79C. Seven of the eight breeding squares are above 1000 m. Congregations of hundreds at Ahero 60B and L Jipe 100D are presumably the off-season flocks mentioned by BUN, but much smaller numbers are usual (Ja; Br; St). BUN consider it migratory over most of its African range, but no details of movements are known in Kenya.

(map overleaf)

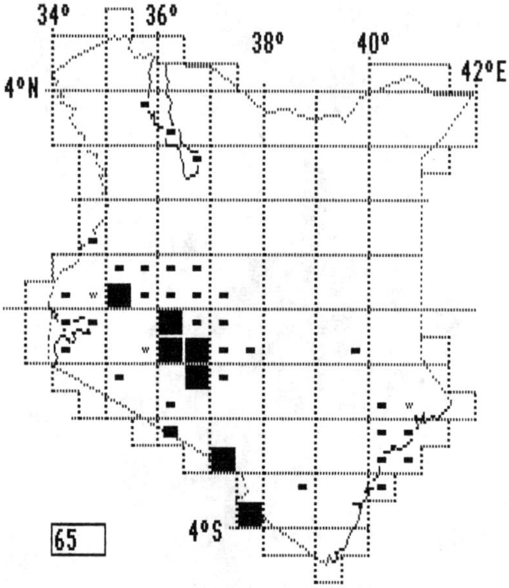

66 KNOB-BILLED DUCK *Sarkidiornis melanotos*
B80 / MPG98 / D3C2

Afrotropical: widespread S of 15°N with the exceptions of the E Somalia and SW dry areas. Extensive intra-African movements are known.

Kenyan: frequents all types of water body in relatively open country up to 3000 m (Br; Se), but is commoner on fresh water. It is found in most areas of Kenya, but is rare at L Turkana (HH). While usually in small flocks, gatherings up to 50 are unexceptional (Br), and the maximum count is of 700+ at Ahero Rice Scheme 60B in Feb (EABR 1983). Probable breeding has been reported from the highlands, but all confirmed records are below 1000 m in the SE.

It may be partly resident (Br), but its wide distribution and small number of breeding squares (12% of its total squares) suggest extensive movements. MPG and Ja were certain that this duck is migratory in Kenya, with both local and longer movements. At the local scale, most records in arid Tsavo E 101B are soon after rain (La), and Ja also notes occurrence on temporary rain pools far from other water. Occasional influxes that raise the normally small L Baringo 50A population to 100 (St) are of unknown origin. Extensive intra-African movement has been demonstrated by ringing recoveries from Zimbabwe to Chad and Sudan (Sn). BUN quote a May-Nov seasonality in Sudan, with S movement in Kenya Dec-Jan. In view of the paucity of breeding records in East Africa (Br) and Somalia (AM), the species may consist of a relatively small, possibly resident breeding population that is regularly augmented by trans-equatorial intra-African migrants.

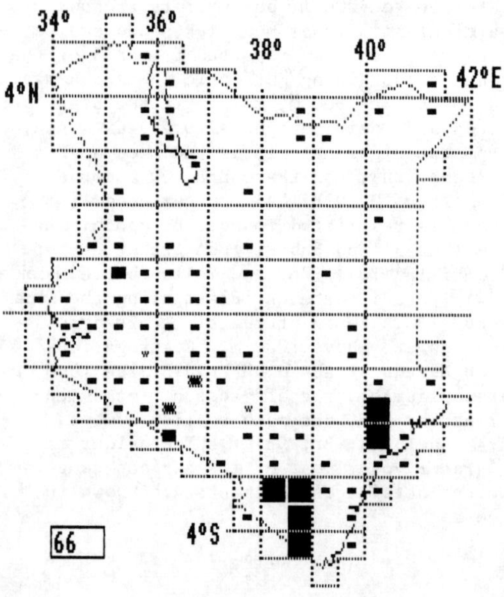

67 AFRICAN PYGMY GOOSE *Nettapus auritus*
B77 / MPG97 / D2C2

Afrotropical: widely distributed S of 15°N, apart from the dry areas of E Somalia and the SW.

Kenyan: rather local, preferring waters with rich emergent vegetation, especially water-lilies *Nymphaea*, mainly below 1600 m (Br; Ja; MPG). Like the Black Heron (37), this is mainly a bird of the coast and L Jipe 100D, where it can occur in small flocks and in most months (Br; EABR 1979). Both breeding squares are also at the coast. Br also mentions regular occurrence in Uasin Gishu 49AC (up to 2000 m) and in Nyanza 60, but a published request for records from these areas produced no data. P L Britton (*in litt.*) reports small numbers (usually 15-25) on dams in Siaya District 48C: these birds are apparently resident although incompletely grown young have yet to be seen. Elsewhere, like the Black Heron, the Pygmy Goose is only an uncommon, occasional wanderer *e.g.* a pair at L Turkana (HH), records at L Baringo 50A during May 1983 (EABR 1983), and a Jan pair around 3000 m at Mau Narok 61D (Britton 1979). At least local movements must occur, *e.g.* to produce the Turkana

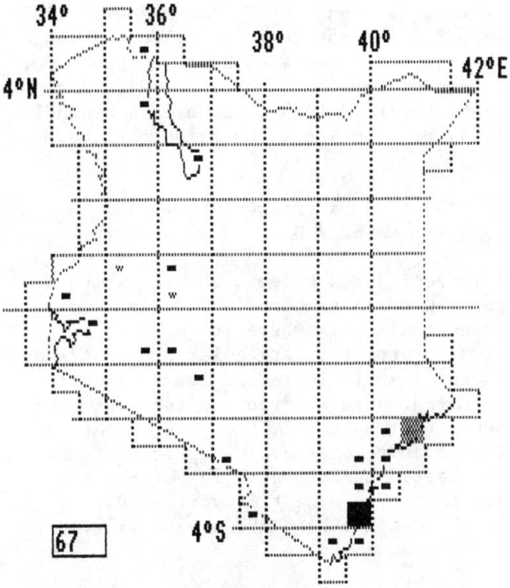

and highlands records, but no details are known. The species is absent Dec-Feb in Somalia (AM).

68 AFRICAN BLACK DUCK *Anas sparsa*
B70 / MPG84 / D2C2

Afrotropical: mainly the E and S, from the Ethiopian highlands through mainly highland areas of S Sudan, East Africa and the W rift to a more widespread distribution S of 15°S, excluding most of the dry SW.

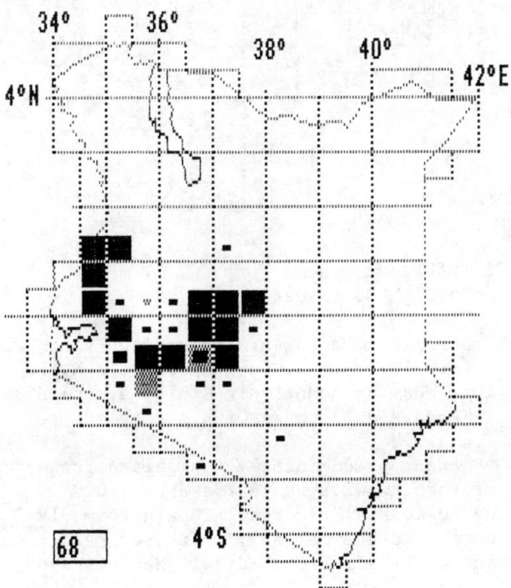

There are also isolated populations in Cameroon, Gabon and other W areas. Distribution in the highlands of Kenya reflects the E limit of the range.

Kenyan: local in pairs along well wooded streams in the highlands, up to 4250 m (Br; Ja; MPG; Se). Breeding has occurred up to 4200 m on Mt Elgon 36D (MPG), and 81% of the breeding squares are above 1500 m. Br considers it sedentary, and the high proportion of breeding squares (55% of its total) support this conclusion. Local wandering to dams occurs occasionally, however, and stragglers reach atypically low altitudes (Br). A striking example of the latter involved a pair at around 600 m on the Tiva R 89A in June (EABR 1983).

69 CAPE WIGEON *Anas capensis*
B63 / MPG89 / D2C2

Afrotropical: breeds on lakes along the rift valley from Ethiopia S through Kenya to Tanzania, and is widespread S of 15°S. It also occurs in Sudan and Chad, and rarely W to Ghana. Distribution in Kenya reflects total range at these latitudes.

Kenyan: all recent breeding records are from rift valley soda lakes, and the species is restricted to the rift (where hundreds may occur on alkaline lakes : Winterbottom 1974) and closely adjacent areas, with solitary wanderers to freshwater as far distant as L Kanyaboli 48C (Br) and Aruba dam, Tsavo E 101B (La). Local movements are demonstrated by ringing recoveries from L Nakuru 62A to L Bogoria 50C and Baringo 50A (Backhurst 1974, 1977),

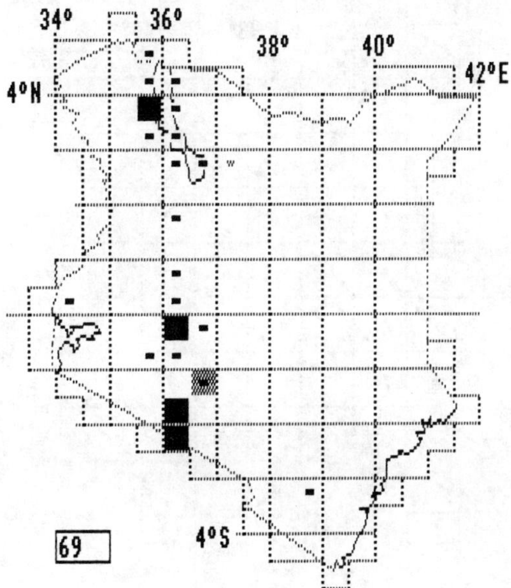

and by intermittent small numbers at L Baringo 50A that are interspersed with absences of months (St). Other references: McKay 1960; Britton 1979.

70 WIGEON *Anas penelope*
B68 / MPG86 / D2C2

Afrotropical: a Palearctic species that winters locally in small numbers across the N tropics, most abundantly in Ethiopia. Small numbers winter annually in Kenya. A few penetrate S to N Tanzania, and it is sporadic in Uganda.

Kenyan: though more interest in Palearctic wildfowl has subsequently shown both species to be commoner than originally thought (Meadows 1978, 1980), Ba were essentially correct in comparing the distribution of this duck to that of the Teal (72). Thus, with the notable exception of L Turkana, the Wigeon mainly winters Nov-Feb above 1600 m in the highlands of the C Kenya rift valley and adjacent areas, and occurs in flocks that are usually below, often far below, 20 (Br; EABR). Counts of 133 (Nov) and 60 (Jan-Feb) at L Naivasha 62C were exceptional (Meadows 1978; EABR 1979). The main difference from the Teal's distribution is that, while the Wigeon is rare elsewhere at lower altitudes (*e.g.* La), it is occasional in flocks of up to 20 on L Turkana (HH), with exceptional concentrations of 100 in El Molo Bay 26B in Feb and Dec 1976 (OS-C 1977; Steinbacher 1977). Our map's coverage of L Turkana is probably inadequate for this species. The Wigeon also tends to prefer larger, more open waters than the Teal. Meadows (1984) suggests the total wintering population of the Wigeon in Kenya to be less than 1000. Other references: Lynn-Allen 1942; MacInnes 1946; Leuthold 1973b; Leuthold & Leuthold 1974; Britton & Britton 1976b; Britton 1979.

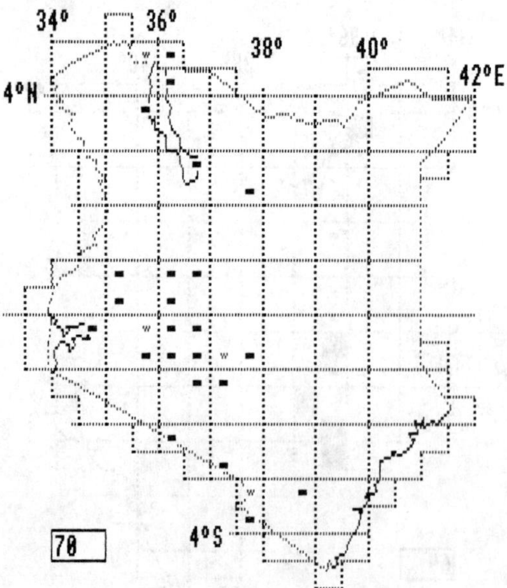

71 GADWALL *Anas strepera*
B71 / MPG85 / D1C0

Afrotropical: a Palearctic bird which winters in small numbers in N Sudan, and occasionally elsewhere in the N tropics. It is a very rare straggler S to Kenya and there is a single record for N Tanzania.

Kenyan: a rare, drab and inconspicuous duck (Ba)(hence D1), S of its normal range. The seven records are of up to four birds together, Dec-Feb, and five of the sites are in or W of the rift valley (Ba; Leuthold & Leuthold 1974; OS-C 1977; Br).

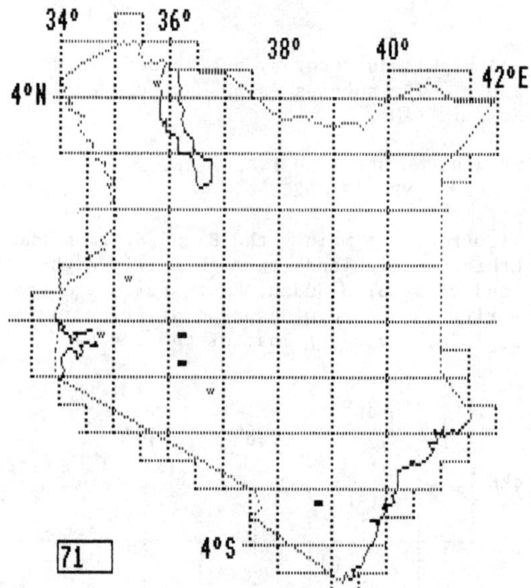

72 TEAL *Anas crecca*
B65 / MPG87 / D2C2

Afrotropical: a Palearctic species that winters in small numbers across the N tropics, most abundantly in Ethiopia, and S to Kenya and W Uganda with stragglers to N Tanzania.

Kenyan: predominantly a highland freshwater duck, with range 79% above 1000 m and no coastal records. Though formerly considered scarce (Ba), increased recording of Palearctic wildfowl (Meadows 1980)

Plate 5. A typically dense flock of Lesser Flamingos at Lake Nakuru (J.S. Karmali F.R.P.S.). *See page 74*.

Plate 6. The Secretary Bird, sole memeber of a family that is endemic to the Afrotropical Region (J.S. Karmali F.R.P.S.). *See page 87*.

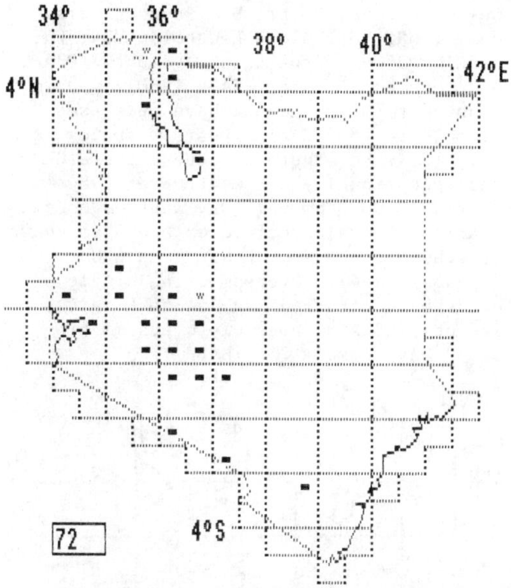

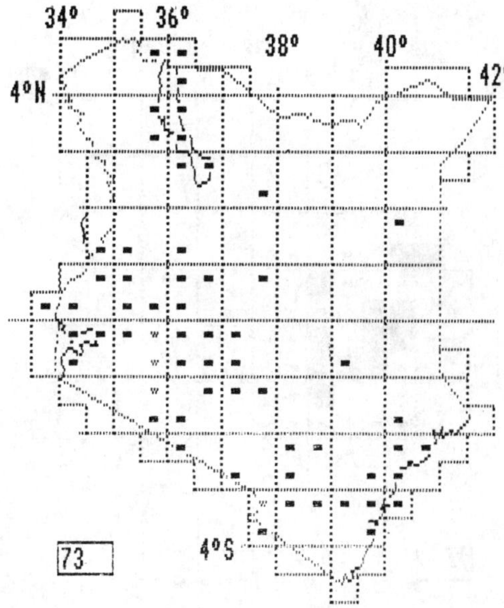

has shown the Teal to be of annual occurrence as a winter visitor Dec-Mar to the C Kenya rift valley and adjacent highlands, which constitute the S limit of its range, with only stragglers further S. Numbers are always small, with flocks up to 30 in the highlands (EABR)(exceptionally 60, Jan, Thika 76A: EABR 1981). It is rare in other areas, *e.g.* the far W 48C (Britton & Britton 1976b), and particularly so in the lowlands *e.g.* L Turkana (HH) and Tsavo E 101B (Leuthold & Leuthold 1974). Single birds at Limuru 75B in late May and Thika 76A in Aug were possibly oversummering (EABR 1982). See Wigeon (70). Other references: Lynn-Allen 1942; Britton 1979.

73 GARGANEY *Anas querquedula*
B69 / MPG88 / D2C2

Afrotropical: this Palearctic species winters commonly to very abundantly across the N tropics, S to 10°N in the W and, in the E, through East Africa to Zimbabwe, Botswana and South Africa.

Kenyan: though possibly scarcer at the start of this century (van Someren 1925-35; Ja), this is now the most widespread Palearctic duck in Kenya, occurring in flocks of up to thousands, mainly late Oct-Apr (Br; EABR), with an estimated annual wintering population of 20 000 (Meadows 1984). It is found up to 3000 m (Br; Se) and, in contrast to the Teal (72), Pintail (75) and Shoveler (78), forms substantial flocks at lower altitudes, *e.g.* L Turkana (HH) and the coast. Maximum counts at L Baringo 50A have increased from 40 in 'wet' years to 1200 in years with low rainfall (St).

- MALLARD *Anas platyrhynchos*
B- / MPG82 / D2C0
RECORDS UNSATISFACTORY

Afrotropical: a Palearctic duck that winters S along the Nile to Sudan and Ethiopia, with possible vagrants S to Kenya.

Kenyan: the uncertainty over the status of this species is summarised by Ba. Mallards are reported to have been shot at L Paradise, Marsabit 27D (Feb 1930) and also on Koronli Water 27A (undated)(van Someren 1925-35), and at L Naivasha 62C (Dec 1938: Lynn-Allen 1951), but none of the specimens survive. Furthermore, although stragglers could conceivably occur S from Sudan or Ethiopia, the possibility that the Kenya records refer to escaped feral birds from colonial settlers' farms cannot be excluded.

74 YELLOW-BILLED DUCK *Anas undulata*
B72 / MPG83 / D3C3

Afrotropical: E and S Africa, from Sudan and Ethiopia S through mainly highland areas of East Africa and the W rift, and more widely distributed over moister areas of the continent S of 10°S. Distribution in Kenya reflects the E limit of the range.

Kenyan: mainly on freshwater above 1600 m in the highlands, where it can be the most numerous Afrotropical duck (Br; Ja). Range is in 81% of the squares that contain eutrophic lakes. It is absent from L Turkana

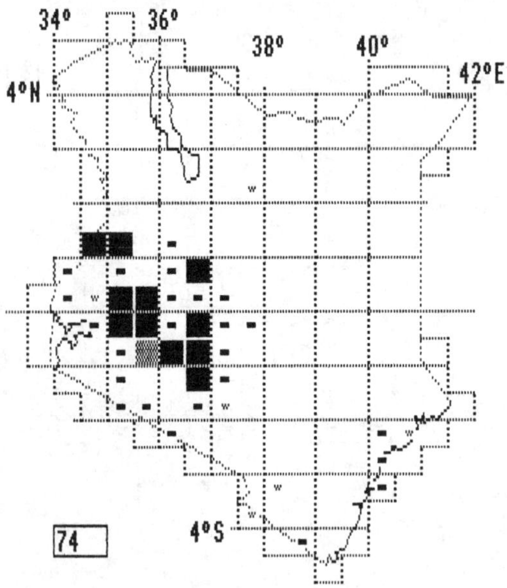

and, although Br describes it as locally common in papyrus swamps around the mainly oligotrophic L Victoria, we have records from only a single QSD (60B) on the lake. There are a few lower altitude squares and, as with the Red-billed Teal (76) and Southern Pochard (79), occurrence at the coast is restricted to squares containing estuaries and/or lakes.

Breeding is entirely above 1500 m, and the breeding range is more extensive than those of the Hottentot Teal (77) and Southern Pochard because, like the African Black Duck (68), the Yellow-billed Duck is a mainly sedentary species (Meadows 1984; Br; Ja). At least local movements must occur, however. Se observed small influxes at Mau Narok 61D, while occasional visitors to L Baringo 50A are mostly Oct-Feb (St). Ja thought this species a visitor to the coast near Lamu 91B, in Sept.

75 PINTAIL *Anas acuta*
B62 / MPG92 / D2C2

Afrotropical: a widespread and locally abundant Palearctic migrant, across the N tropics from Senegambia to Ethiopia and S Somalia, S to Uganda and N Tanzania. Stragglers reach S to Zambia and Zimbabwe.

Kenyan: mainly frequents permanent waters, the map quite closely resembling that of the Shoveler (78). The Pintail usually winters Nov-Mar in thousands on the rift valley lakes and in the W and C highlands (c 8000 on a Kinangop dam 62D in Jan: EABR 1981); flocks below 1400 m, E to the coast, seldom exceed 25 (Br; D J Pearson pers. comm.; HH; Leuthold & Leuthold 1974; La; EABR). 350+ at the Sabaki R mouth 103A in Dec were exceptional (EABR 1980). Maximum counts at L Baringo 50A have increased from 40 to 1800 during years with low rainfall (St), but there were none there during the exceptionally dry weather of 1984 (D J Pearson pers. comm.). Dry weather concentrations of Palearctic duck are also known at other waters *e.g.* L Naivasha 62C (Meadows 1984). Oversummering single birds have occurred May-Jul at L Naivasha, Nairobi 75B and (possibly oversummering, on 12 May) Limuru 75B (EABR).

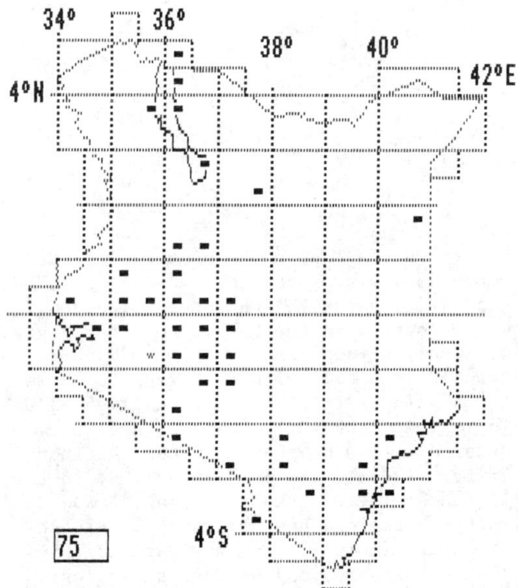

76 RED-BILLED TEAL *Anas erythrorhynchos*
B66 / MPG91 / D3C2

Afrotropical: the E and S, from Sudan and Ethiopia S through E Africa and S Zaire to most areas S of 10°S.

Kenyan: a common species that wanders extensively to all types of water, including small temporary pools, up to at least 3000 m (Br; Ja; MPG; Se; La). It is largely absent from L Turkana, and occurrence at the coast is confined to the vicinity of the Tana 91 and Sabaki 103A estuaries. The seven breeding squares are reasonably closely grouped in the highlands, with six above 1500 m. Movements to seasonal waters and possibly breeding areas presumably occur, but there are no data beyond a few local ringing recoveries (Backhurst 1977). In addition, intra-African migrants, possibly from the N or NE, augment the large resident population during the N winter (Meadows 1984).

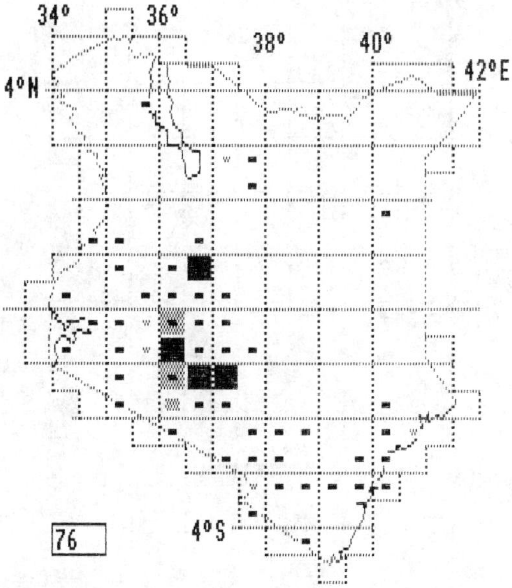

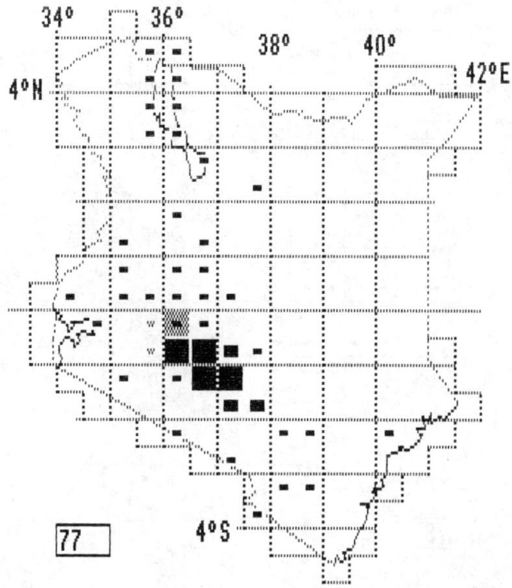

77 HOTTENTOT TEAL *Anas hottentota*
B67 / MPG90 / D2C2

Afrotropical: E and S Africa, from Sudan and Ethiopia S through East Africa to Mozambique, Zambia, S Zaire and all but arid areas S of 15°S. There is also an isolated population in Chad and N Nigeria. Distribution in Kenya represents the E limit of the range.

Kenyan: rather less widespread than the Red-billed Teal (76), except at L Turkana (HH). The Hottentot Teal wanders to most types of water, including seasonal pools, up to at least 3000 m, but is absent from the coast (Br; Ja; Se). Apart from L Turkana, range is mainly in the highlands and their peripheral plateaux. All breeding squares are above 1000 m, in a compact group in the C highlands and the adjacent rift valley. Movements to seasonal waters and breeding sites must occur, and ringed birds have moved between L Nakuru 62A and L Ol Bolossat 62A, L Nakuru and L Solai 50C, and L Naivasha 62C and Mwea Rice Scheme 63C (Backhurst 1971b, 1972, 1973b, 1977). Individuals from L Nakuru and L Naivasha have reached NE Tanzania (Smart 1967; Backhurst 1973b). There are no records from Tsavo E 101B during Apr-June and Sept (La) and, as in the Red-billed Teal, overall numbers in Kenya are greatest during the N winter, especially in years of low rainfall, when the local population is augmented by intra-African migrants that possibly originate from the N or NE (Meadows 1984). Other reference: Reynolds 1972b.

78 SHOVELER *Anas clypeata*
B64 / MPG80 / D2C2

Afrotropical: a locally common to abundant Palearctic winter visitor across the N tropics from Senegambia to Ethiopia, S to N Tanzania.

Kenyan: the map is similar to that of the Pintail (75). Shoveler occur late Oct-early Apr, mainly on permanent waters. They are practically confined to the rift valley lakes and W and C highlands, where they can be present in tens of thousands and are the commonest Palearctic duck, with large concentrations on L Solai 50C, L Ol Bolossat 62A, L Nakuru 62A and L Naivasha 62C (Br; Ja; EABR). Below 1500 m (Br), this duck is common only on L Turkana, where 10 000 have occurred in Feb-Mar at Ferguson's Gulf 13B (HH) and 50 000 have roosted in the Omo delta 4A in early Mar (Stevenson 1982b). Elsewhere in lowland Kenya it is occasional in small numbers, *e.g.* flocks of 15 in Tsavo E 101B (Leuthold & Leuthold 1974; La).

Meadows' (1984) minimum estimate for the Kenya population during a low rainfall year was 22 000, mainly centred on L Naivasha 62C and appreciably less in wetter years. A decrease in overall numbers may have occurred since the early 1970s (Meadows 1980). A bird ringed on the N of the Caspian Sea, USSR, was recovered at L Naivasha 62C (Backhurst 1969b).

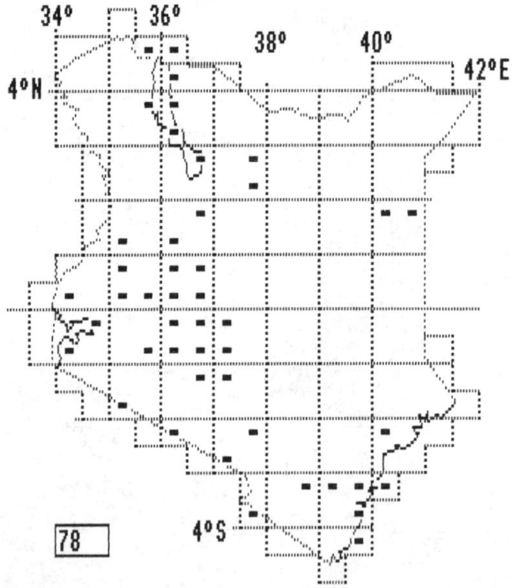

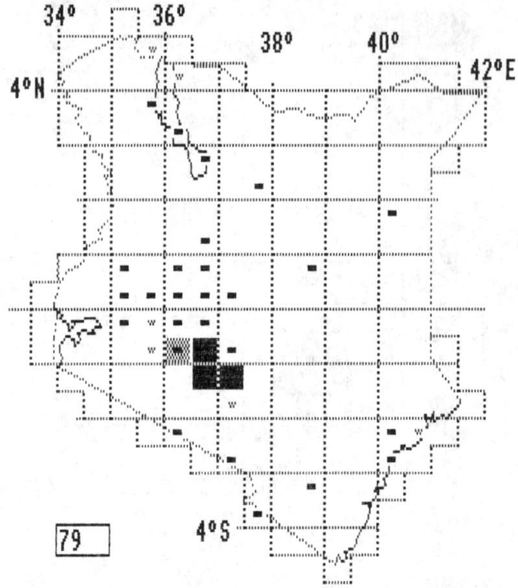

- MARBLED TEAL *Anas angustirostris*
B- / MPG- / D2CO
RECORDS UNSATISFACTORY

Afrotropical: mainly Palearctic N Africa, with small numbers S to Senegambia, Mali, Nigeria and Chad (BUN).

Kenyan: Ba listed two very indefinite records quoted by Lynn-Allen (1951) and did not admit the species to the Kenyan avifauna. These records are omitted (and by implication rejected) by Br.

79 SOUTHERN POCHARD *Netta erythrophthalma*
B76 / MPG77 / D2C2

Afrotropical: mainly E and S Africa, from Sudan, Ethiopia and S Somalia S through East Africa and E Zaire to become more widely distributed S of 10°S. Major intra-African movements are known.

Kenyan: 57% of this species' squares occur in a relatively compact area of the W and C highlands, and there are only scattered localities elsewhere. It is scarce at L Turkana (HH), absent from L Victoria and, like the Red-billed Teal (76), restricted on the coast to one degree square 91, where occurrence is most probably either in the Tana delta or on adjacent freshwater lakes. The Southern Pochard is mainly a bird of the larger, deeper fresh waters, and there are records from 73% of the squares that contain eutrophic lakes. It is often the most abundant diving duck, in flocks of hundreds or more (Br)(maximum 2000+ at L Nakuru 62A Oct: EABR 1983).

The four breeding squares are above 1500 m, in a very compact group in the SC highlands. Breeding was possibly more widespread formerly (Brown & Britton 1980). Numbers breeding in Kenya are usually very small (Meadows 1984), but increase in abnormally wet years (*e.g.* Cunningham-van Someren & Richards 1977b). This breeding and that in W Uganda are the most northerly known, and this bird is a non-breeding migrant to Ethiopia and S Somalia (Br; AM). It is probably also a mainly non-breeding, intra-African migrant from the S to Kenya, since a bird ringed in South Africa was recovered at L Naivasha 62C in Nov (Smart 1967), and influxes of migrants in Sept-Oct of most years produce peak numbers during the N winter, which decline after Feb (Britton 1979; Br; Meadows 1984; La; St; EABR). Nov records from L Turkana (HH) and Wajir 42A (J E Miskell *in litt.*), and the isolated occurrence in 52B, most probably represent migrants moving N or NE.

80 NORTHERN POCHARD *Aythya ferina*
B73 / MPG76 / D2CO
NO MAP: post presence 13B 60B 62A

Afrotropical: a Palearctic species, wintering in small numbers across the N tropics. Records in the E are S to Uganda, Kenya and N Tanzania. Numbers in both W and E Africa have declined in recent decades (Br; BUN).

Kenyan: a very rare straggler S of its normal range. The five records are: one

Ferguson's Gulf 13B Mar 1973 (HH); one L Ol Bolossat 62A 1975/6 (Davey & Harrison 1976) and two there 30-31 Dec 1976 (Duffus & Butterworth 1977); one Kisumu 60B 23 Dec 1977 (EABR 1977); two Ferguson's Gulf 13B 10 Feb 1981 (EABR 1981).

81 WHITE-EYED POCHARD *Aythya nyroca*
B75 / MPG75 / D1C0

Afrotropical: this Palearctic duck winters in the N tropics, sometimes in large numbers in the W, but in small numbers in Sudan and Ethiopia and rarely S to Uganda (one record) and Kenya. It has shown a decided numerical decrease in Africa in recent decades (BUN).

Kenyan: while possibly overlooked in flocks of the similar Southern Pochard (79)(hence D1), there is no doubt that this is only a rare straggler S to Kenya. Six of its eight squares are above 1000 m, and occurrence in 27D was on Mt Marsabit. All records concern single birds except those relating to a marked influx of probably less than 10 individuals to the C highlands at L Naivasha 62C, Nairobi 75B and Thika 76A in Jan-Mar 1981, when up to four were seen together (Meadows & Pearson 1981). With the marked exception of Ol Joro Orok 62A on 6 Sept (Bednall 1959b), all of the birds in the C highlands and Mt Elgon 36D were in the period 1 Jan-1 Mar (Ba; Meadows & Pearson 1981; Lewis 1983e; D J Pearson *in litt.*), whereas those at L Turkana 13B (HH) and Marsabit (Harper & Harper 1974c) were Mar-Apr and possibly on passage.

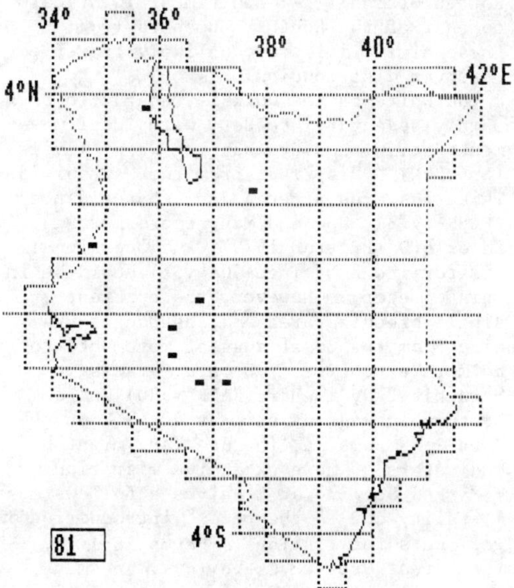

82 TUFTED DUCK *Aythya fuligula*
B74 / MPG78 / D2C1

Afrotropical: a Palearctic species that winters in small numbers across the N tropics, particularly in Sudan and Ethiopia, and S to Uganda, Kenya, N Tanzania and, rarely, Malawi. Distribution in Kenya represents the E limit of the range.

Kenyan: this duck's status as a relatively uncommon winter visitor, mostly Nov-Mar, does not appear to have altered significantly since the beginning of this century (Ba; Br; Ja; EABR 1978), the total population most probably never exceeding 1000 birds (Meadows 1984). It is predominantly a species of the C Kenya rift valley and adjacent highlands (range 75% above 1000 m), occurring almost annually and usually seen singly or in pairs (EABR), though flocks of up to 30+ are known (L Nakuru 62A Jan: OS-C 1977). Occurrences away from the C highlands are few, but they can involve small flocks *e.g.* 20 at Ferguson's Gulf 13B in Mar (Br), 10 on L Paradise, Marsabit 27D in Dec (DEP pers. obs.) and nine at Aruba dam 101B in Jan-Feb (La). There is a single coastal record, of a female at Bamburi 114B Jan 1975 (Britton & Britton 1976b). Other references: Steinbacher 1977; Lewis 1983e.

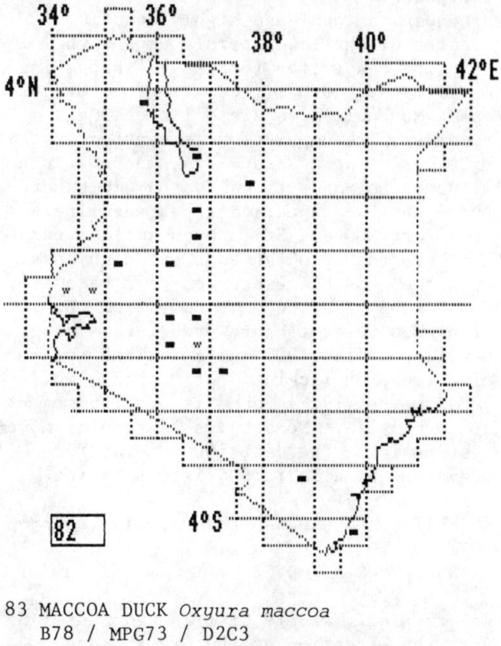

83 MACCOA DUCK *Oxyura maccoa*
B78 / MPG73 / D2C3

Afrotropical: in two distinct areas. In the NE, this is a predominantly inland bird from N Tanzania and the E Zaire

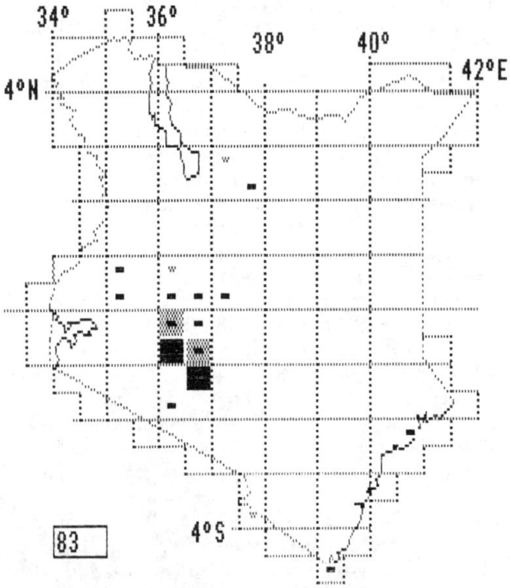

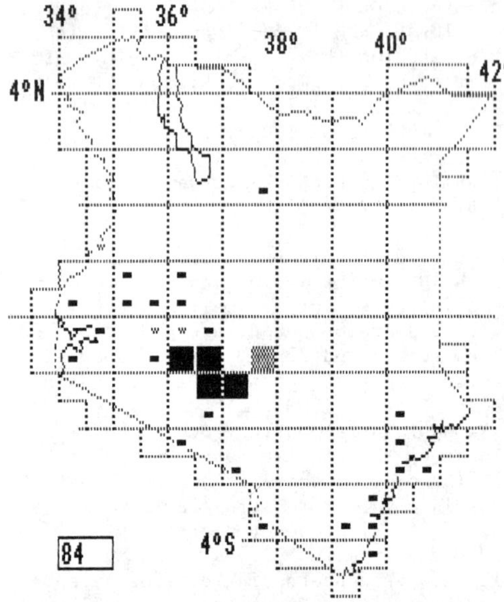

borders, NE through Uganda and mostly highland areas of Kenya to Ethiopia. It is also widely distributed in all but interior areas S of 20°S. Since there are no recent records from Uganda (Br), distribution in C and N Kenya describes the range at these latitudes.

Kenyan: a local and highly aquatic species of shallow waters/swamps with fringing vegetation (Br; MPG). Range is mainly above 1500 m (occurrence in 27D is on Mt Marsabit), as are all breeding records. The population is centred in the rift valley at L Nakuru 62A, where it can be numerous, and at L Naivasha 62C, where there has been a marked recent decline (contrast MPG and Br). Much smaller numbers are found in the adjacent highlands. With the possible exception of L Paradise at Marsabit 27D, records away from the highlands refer to rare wanderers. Ja mentions L Jipe 100D and L Turkana (precise location lacking: not mapped), while van Someren (1922) found it on floodwaters at Koroli 27A. No details of regular movements exist. The distribution map for Kenya and Somalia in BUN is misleading.

84 WHITE-BACKED DUCK *Thalassornis leuconotus*
B82 / MPG74 / D2C2

Afrotropical: widely distributed S of 5°S with the exception of the dry SW, and ranging further N through East Africa and the E Zaire borders to Sudan and Ethiopia. It also occurs along the S edge of the Sahara from Senegambia to Chad. Distribution in Kenya represents the E limit of the range, with the exception of an old record from S Somalia (AM).

Kenyan: a rather local and unobtrusive, highly aquatic species that prefers quiet, fresh waters, with both floating and fringing vegetation, up to 3000 m (Br; Ja; Britton 1979). Coastal records are not marine. Flocks of up to 50 can occur (EABR), and there is a count of 500 from L Naivasha 62C (Meadows 1984). Four of the five breeding squares are above 1000 m, in a compact breeding area in the C highlands that is similar to that of the Hottentot Teal (77) and almost identical to that of the Southern Pochard (79): these similarities are unlikely to reflect observer distribution.

The White-backed Duck is thought to be a mainly sedentary breeder, with local movements stimulated by changing water levels (Br; BUN). Dispersal from the Nairobi 75B-Thika 76A area after a Jan-Feb residence (EABR 1978), Nov and May records at Mau Narok 61D (Britton 1979), and occasional visitors to L Baringo 50A (St) could be in this category. However, the apparent single breeding area and absence of breeding from the coast suggest movements to be more extensive. 80 on L Paradise, Marsabit 27D, in Mar (EABR 1980) and a presumably vagrant pair at 42-43°E on the Juba R, S Somalia, in June (AM) might be indicators of more extensive migration.

Marked decline at L Naivasha follows mass mortality due to the recent introduction of gill-nets for fishing (J Hayes *in litt.*) (*cf.* Great Crested Grebe (4)), and may also reflect the decline of water-lilies there (Meadows 1984).

Family Sagittariidae: Secretary Bird

A monotypic family, endemic to the Afrotropical Region. It breeds in Kenya.

85 SECRETARY BIRD *Sagittarius serpentarius*
B83 / MPG103 / D3C3

Afrotropical: widely distributed S of 17°N, absent only from S Somalia and the equatorial lowland forest areas of the W. Distribution in Kenya represents the E limit of the range.

Kenyan: a terrestrial predator, usually in pairs and typical of habitats ranging from wooded grassland to grassy plains, mainly up to 3000 m (Br; Ja; MPG). Pitman (1956) found it at about 4000 m on Mt Kenya 63A. The scarcity of records in 64 65 77 78 resembles the distribution of the Ostrich (1: *q.v.*), and represents an area of thicker bush/woodland and, to a lesser extent, cultivation, that is unattractive to ground species. Scarcity of this species in W Kenya 48 60 73 (Br) and at the coast probably reflects a combination of high rainfall (which produces unsatisfactorily tall and dense vegetation (Ja)) and human pressure (see below). Range is only 29% arid, reflecting the fact that few of the driest areas support the high density of prey necessary for a predator that walks at only 3 km/hour, as opposed to one that covers vast areas by flight (Brown 1970a; BUN). Some arid areas may only be utilised during the rains (see below). In the arid NW, HH describe only occasional occurrence around L Turkana. The arid NE and E are mainly beyond the limits of this species' range. Breeding is likewise only 10% arid.

The number of areas adjacent to square 62 where breeding has apparently not occurred since 1969 is distinctive, and may be partly due to the proliferation of smallholder farms, which render nests especially vulnerable. Decrease due to this factor occurred in Embu district 63CD between 1950 and 1970 (Brown 1970a).

Although predominantly terrestrial, Secretary Birds can soar well (BUN) and they wander widely. The 48C and 49A records involved single visitors to Nyangweso on 9 Oct 1968 (Br) and to Soy in Sept 1969 (A G Johnston *in litt.*) respectively. These dates are of interest because Oct wanderers are also mentioned at L Turkana (HH 1973 edition) and in NW Tanzania (Br), while La describe a possible Nov-Apr influx to arid Tsavo E 101B. These wanderers may thus be a result of movement to arid areas at the start of the 'short' rains, to take advantage of temporary prey abundance. DEP (pers. obs.) found this bird resident in 74B 75D 88D 89C 100B 101A, which are all semiarid-dry subhumid. Other references: Henley 1943; Brown 1952-3, 1955a.

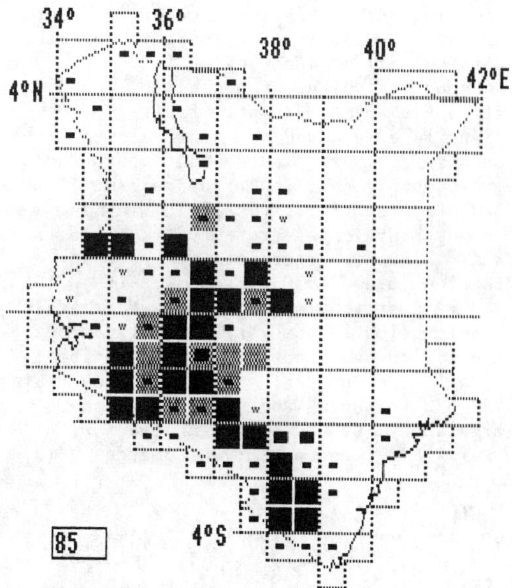

Family Accipitridae: Vultures, Eagles, Hawks, Buzzards, Kites etc.

Around 211 species, with representatives throughout the world. 72 (34%) occur in the Afrotropical Region, which is very rich in diurnal raptors and particularly so in the larger ones that inhabit savannah and other open habitats (Sn). 55 (76%) of the species that occur in the Afrotropical Region have been recorded in Kenya, where 41 (74%) breed or probably do so, 13 (24%) are non-breeding migrants or vagrants from the Palearctic and one is a non-breeding intra-African migrant.

86 PALM-NUT VULTURE *Gypohierax angolensis*
B84 / MPG161 / D2C2

Afrotropical: widespread N to Senegambia, Nigeria, Chad, S Sudan and Kenya, but avoiding the dry SW and much of interior Tanzania. Distribution in Kenya represents the NE limit of the range.

Kenyan: a vulture renowned for its diet of *Raphia* and *Elaeis* palm fruit, conventionally considered to occur only in areas supporting these plants. But other foods are probably more important in East Africa and, for example, birds living in mangroves on the Kenya coast probably take live fish and various invertebrates, as well as carrion (Brown 1970a; Brown & Britton 1980; BUN). Distribution is in two broad areas, mainly below 1400 m (Br). Firstly it extends along the coast (where it can be locally common: Br), and up the Tana R 91A 79C 65D to the NE and E edges of the C highlands 51BD 63D, with wanderers W to Mweiga 62B (Cunningham-van Someren 1971e) and Trans Nzoia (mapped as 37C)(Ja). This extension into the interior via the Tana is seen in other coastal species of S affinity *e.g.* Retz's Helmet Shrike (872) and the Black-breasted Glossy Starling (877). Secondly, this vulture has a centre of distribution in the lowlands N and E of Mt Kilimanjaro, NE to Kibwezi 88B and Tsavo W 89C. This species is thought to be predominantly sedentary (Br).

87 LAPPET-FACED VULTURE *Torgos tracheliotus*
B89 / MPG108 / D2C3

Afrotropical: a somewhat discontinuous distribution N of 10°N from Senegambia E to Ethiopia and Somalia, and thence S through East Africa and the E Zaire borders to the whole of the continent S of 15°S.

Kenyan: a bird of open country up to 3000 m (Br; Se), with 68% of its squares in the intermediate, 250-1000 mm rainfall regime. It forms a species pair with the White-headed Vulture (88), and differs

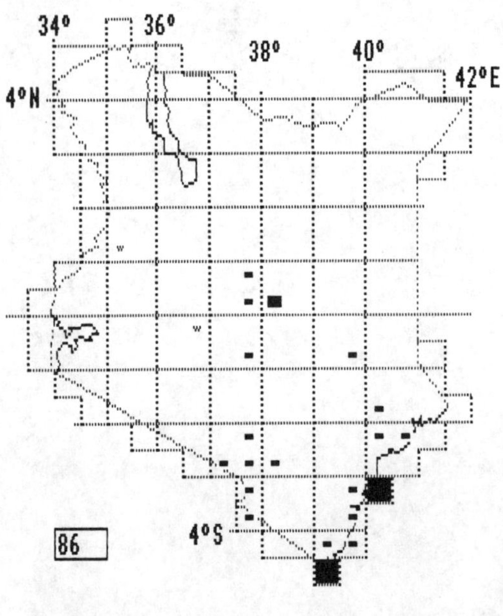

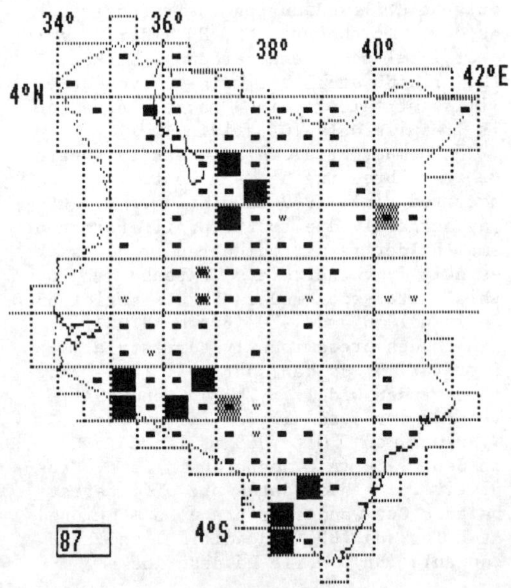

from the African White-backed (90) and Ruppell's (89) Vultures, the other large plains species, by occurring in pairs on relatively small territories (BUN). It has a more powerful bill than the others, which rely on it to open the larger carcasses (Kruuk 1967). It can also be kleptoparasitic on the other vultures, and kill small or weak animals (BUN; Sn).

The Lappet-faced's territorial nature produces the most distinctive of the vultures' maps. It is absent or rare in the high rainfall areas of the W 48 49 60 and the coastal strip because the few game animals in these areas provide insufficient carcasses, while the less open, well wooded or forested habitats may obscure this aerial scavenger's view of potential prey items on the ground. Also, although there are numerous settlements in these areas, this bird does not take human refuse (*cf.* the other large plains vultures). Rarity of this species and the other vultures in 49 is demonstrated by the fact that, despite pleas to observers resident in the area and a published request for records, we received only a single additional report, of two Hooded Vultures (92) in 49C during Mar (Siemens 1985).

There is also a paucity of records from 63 64 65 76 77 that has already been remarked upon for the Ostrich (1) and Secretary Bird (85) as probably being thick bush (and in 63, forest), which inhibits ground species. Such bush would again obscure aerial scavengers' visibility of carrion on the ground, thus making foraging less productive. This feature also appears in the other species' maps. Thus the White-backed Vulture, which is more a bird of wooded country, shows a few records in this area, while Ruppell's which prefers more open areas, shows fewer. Both of the small vultures avoid these squares, presumably because they scavenge in company with the larger species. The White-headed Vulture is also scarce here. Thus the open country ground species and vultures occur in the savannahs of SW Kenya from 74 to 101, in the open rift valley floor 62AC, and thence northwards out over the extensive plains of the N lowlands.

The Lappet-faced Vulture breeds solitarily in trees, and 86% of the breeding squares are in 250-1000 mm rainfall areas. There is no evidence of other than local foraging movements, and DEP (pers. obs.) found the species resident in 74B 88D 89C 100B 101A.

88 WHITE-HEADED VULTURE *Trigonoceps occipitalis*
B90 / MPG109 / D2C3

Afrotropical: widely distributed S of 15°N, apart from the dry areas of E Somalia and the SW, much of South Africa, and the equatorial lowland forest areas of the W.

Kenyan: like the Lappet-faced Vulture (87: *q.v.*), this species occurs in pairs on rather restricted territories (Sn; BUN). It eats carrion flesh and also kills medium-sized mammals and, because it often arrives first at a carcass, appears to fulfil the role of searcher in mixed vulture parties, after which it feeds on scraps dropped by the other species (Kruuk 1967; BUN). Range is in open and wooded game country up to 3000 m (Br; Se), and 68% of the squares are within the intermediate, 250-1000 mm rainfall regime. As with the other large vultures, it is scarce or absent on the coast and in W Kenya, and in 63 64 65 76 77. The White-headed Vulture is a solitary tree-nester, and no non-foraging movements are known.

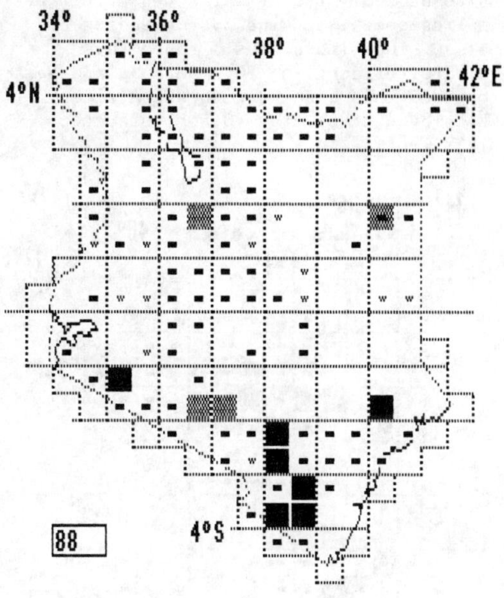

89 RUPPELL'S VULTURE *Gyps rueppellii*
B86 / MPG106 / D2C3

Afrotropical: the N tropics, from Senegambia E between 10°N and 20°N to Ethiopia, and thence S to Uganda, Kenya and N Tanzania. Distribution in Kenya represents the E limit of the range.

Kenyan: differs from the similar African

White-backed Vulture (90: *q.v.*) by roosting and breeding on cliffs, and by inhabiting generally more open, often treeless, arid country that entails more extensive foraging flights (BUN; Sn). This difference in habitat cannot be recognised on our maps since we do not illustrate abundance, but Ruppell's is by far the most numerous of the two species in arid N Kenya (Ja; Brown & Britton 1980), with particularly large concentrations (up to thousands) at such favoured roosting/breeding sites as Mt Nyiru 26D, Maralal 38D, Mt Ololokwe 51B and the Njorowa Gorge, Naivasha 62C (Br; BUN). Definition of this species' (and the White-backed's) distribution must also be diffused by its potential daily foraging range of 150 km (BUN), during which it could traverse three QSDs. This being the case, areas of non-occurrence assume more significance. They are in the W 48 49 60 61, the coastal strip, and C Kenya 63 64 77: see Lappet-faced Vulture (87).

Breeding requires suitable cliff sites and, although some colonies may remain undiscovered, their apparent scarcity (10% of the species' total squares: *cf.* African White-backed Vulture, which nests in trees) implies some extensive post-breeding dispersal. DEP (pers. obs.) found this vulture resident in 75C. Both Ja and Se confused this species with the African White-backed Vulture. Other reference: Kruuk 1967.

90 AFRICAN WHITE-BACKED VULTURE *Gyps africanus*
B85 / MPG107 / D2C3

Afrotropical: widespread S of 15°N, although absent from E Somalia, much of South Africa and Namibia, and the equatorial forest areas of the W.

Kenyan: a bird of open or wooded country up to 3000 m, where carcasses of large mammals provide a source of carrion (Br; Se). It is rare in the high rainfall areas of the W 48 49 60 61 and the S coastal strip, and uncommon around 64: see Lappet-faced Vulture (87). This is the most abundant vulture in Kenya (Br; MPG: La; Pomeroy 1977), though Ruppell's Vulture (89)(with which it forms a species pair: Kruuk 1967) can outnumber it locally, particularly in the N (Brown & Britton 1980). The African White-backed Vulture breeds and roosts on trees, and 77% of the breeding squares are in intermediate, 250-1000 mm rainfall areas. Breeding is far more widespread than in Ruppell's Vulture, which is restricted to cliffs (40 squares, as against Ruppell's 11). The White-backed is thought to be more sedentary than Ruppell's (Sn; BUN), but both undertake extensive daily searches for food, and follow the regular migrations of ungulates (Br). DEP (pers. obs.) found the White-backed resident in 74BD 75 88ABD 89C 100B 101A. See Egyptian Vulture (91).

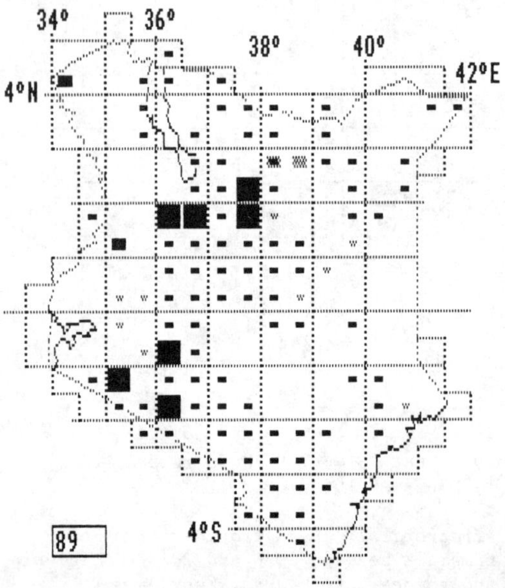

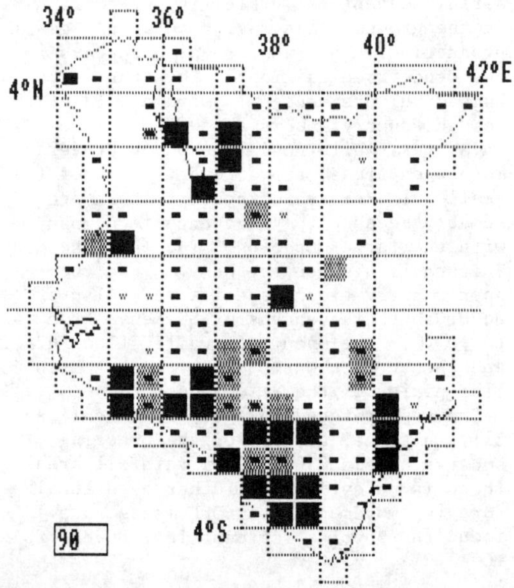

91 EGYPTIAN VULTURE *Neophron percnopterus*
B88 / MPG110 / D3C3

Afrotropical: although there are isolated and possibly extinct populations in S Africa, this is mainly a bird of the N tropics from Senegambia, N of about 10°N, E to Ethiopia and Somalia, and thence S to Kenya, Uganda and N Tanzania.

Kenyan: this small, versatile vulture forms a species pair with the Hooded Vulture (92)(Kruuk 1967). It is omnivorous, foraging widely for carrion with mixed vulture parties, or scavenging around small settlements for human refuse, including faeces (BUN; Ja; MPG). It is renowned for its habit of throwing stones to break Ostrich (1) eggs (van Lawick-Goodall and van Lawick-Goodall 1966). Occurrence is solitary or in pairs, uncommonly in small flocks, and it nests on rocky crags (BUN; Br).

Range is mainly below 2000 m (exceptionally 3000 m: Se) and, unlike the Hooded Vulture, is usually in open habitats including light acacia woodland (Br; BUN). 62% of the squares are within 250-1000 mm rainfall areas, so that BUN's estimate of 600 mm as the normal upper limit may be a little low.

Distribution shows two main features. Range is 77% arid-semiarid (*cf.* Br) and, in contrast to the Hooded Vulture, it is rare in the humid W 48 49 60 61 (*cf.* Ja) despite its omnivorous, commensal tendencies. Relative scarcity in the apparently suitable SE lowlands S of Wajir 42A and E of Kilimanjaro (*e.g.* La; *cf.* Ja) is comparable to distribution in Somalia, where it is only occasional at 8-1°N and absent further S (AM). This is believed to be due to the fact that these areas represent the SE extremities of its range.

Breeding squares are 87% above 500 m and 62% within the 500-1000 mm rainfall regime, and no breeding is known from the SE lowlands mentioned above. Equatorial populations are apparently sedentary (BUN), and there is no evidence for the presence of non-breeding migrants from the N tropics (Br). DEP (pers. obs.) found this species resident in 75C.

Thus the African White-backed (90) and Ruppell's (89) Vultures, and the Hooded Vulture (92) and this species, form two species pairs, with the former pair being larger, carrion-eating wanderers of open and wooded country, while the latter are smaller, more versatile, omnivorous forms that can wander widely or scavenge around settlements (Kruuk 1967). Each pair has a cliff-nester of generally more open country and a tree-nesting form. See Lappet-faced Vulture (87). Other reference: North 1947.

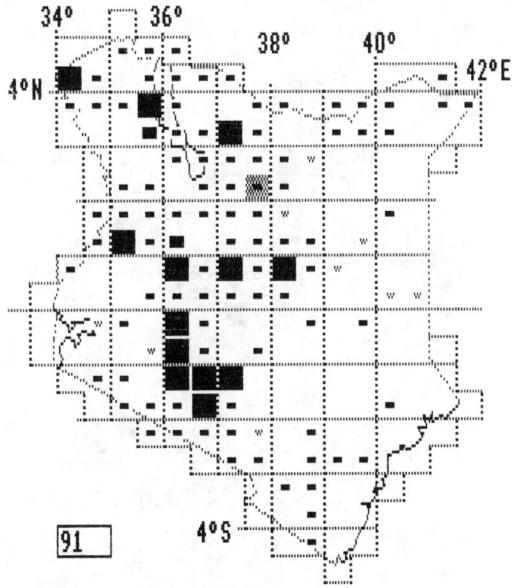

92 HOODED VULTURE *Neophron monachus*
B87 / MPG111 / D2C3

Afrotropical: widely distributed S of 17°N, with the exceptions of many of the equatorial lowland forest areas of the W, the E Somalia and SW arid areas, and indeed the S tip of the continent as a whole.

Kenyan: while the larger plains species are more numerous, this is the most widespread and common vulture in East Africa because it is omnivorous, can be commensal with man, and occurs in all types of habitat from deserts to wooded country and forest edge, up to 3000 m (van Someren 1929; Ja; North 1939, 1942a; Brown 1970a; Br; BUN). Thus, as well as scavenging for plains carrion with mixed vulture parties, it also takes garbage from human settlements. It is a well known urban scavenger in Uganda and North (1942a) noted this activity in Kakamega 48D, but in the rest of Kenya it is not seen around towns, being more a bird of smaller settlements N of the equator.

The Hooded Vulture is found over most of the country, but is absent from the 1000+ mm rainfall coastal strip S of Malindi 103A and apparently uncommon in the region around 77A: see Lappet-faced Vulture (87). In the W it is uncommon in 49 61 (again see Lappet-faced Vulture, but also Se) like the African White-backed (90) and Ruppell's (89) Vultures but, unlike them, is found in the far W 48 60, presu-

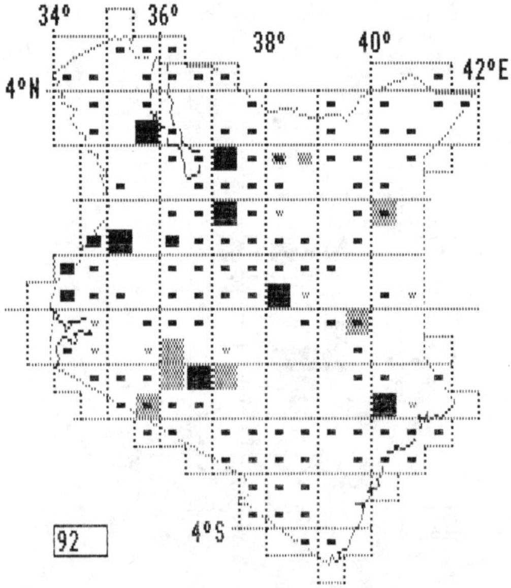

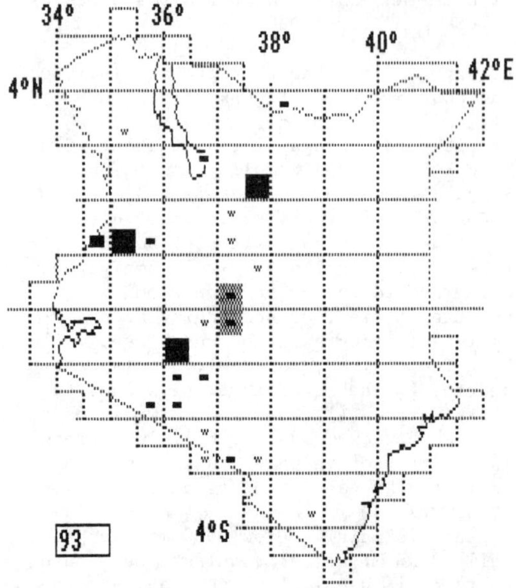

mably due to its omnivorous and commensal tendencies.

The Hooded Vulture nests in trees, and breeding squares are reasonably evenly distributed throughout Kenya N of 2°S. A concentration of pre-1970 records from the Nairobi region 61D 62C 63C 75A 76A may denote a local decline. With the exception of foraging flights this species is thought to be sedentary, and DEP (pers. obs.) found it so in 74BD 75C. See Egyptian Vulture (91). Other references: van Someren 1956; Kruuk 1967.

93 LAMMERGEYER *Gypaetus barbatus*
B91 / MPG162 / D3C2

Afrotropical: mainly the NE, from N Tanzania and NE Zaire through Uganda and Kenya to the Ethiopian highlands. There is also a population in S Africa. Distribution in Kenya represents the E limit of the range.

Kenyan: this vulture eats carrion flesh, skin and bones, and may associate with pastoralists to take advantage of the high mortality of their stock (Brown 1970a). It probably kills smaller mammals and also has the unique habit of dropping overlarge bones on to rocks to reduce them to ingestible size (*e.g.* BUN). In other countries *e.g.* Ethiopia, it is a numerous urban scavenger, but the only record of this type for Kenya is of one at a lodge rubbish dump in Amboseli 88C (Bleazard 1982). Ja could find only one record for Kenya (Blayney Percival 1914) and had not seen the species himself, but this was almost certainly due to the fact that it is widespread in only very small numbers, which mainly utilise remote montane nest sites (Br; Gerhart 1971): hence C2 even though unmistakeable. It frequents cliff faces on mountains and inselbergs and range is 73% above 1000 m.

Solitary nests are found on cliffs, and breeding squares are 83% above 1500 m. The map depicts a number of certain or probable breeding sites, with other records probably referring to wanderers and/or foragers to lower altitudes (Sn). From S to N, the breeding sites are: (a) the Kilimanjaro massif, with wanderers to 87BD 88CD (Bleazard 1982); (b) the Nguruman Escarpment 74D (Br); (c) the Njorowa Gorge, Naivasha 62C, with wanderers as far as Nairobi 75B (Blayney Percival 1914; Dewhurst 1971; Allen 1973c); (d) Mt Kenya 63A and the Loldaiga Hills 51C (Pollard 1944; Schuz 1973; Dyer 1974); (e) Mt Elgon 36D and the Cherangani Hills 37CD (Brown & Brown 1963; Gerhart 1971; Clifton 1976; Cheke 1978; Siemens 1985); (f) possibly the Matthews Range 39C or the Ndoto Mts 39A; (g) Mt Marsabit 27D; (h) Mt Kulal 26B (North 1944; Clifton 1976); (i) Mt Loima 13C (Brown 1970a); (j) squares 19B (van Someren 1929) and 16A may represent wanderers S from the large Ethiopian highlands population. With the exception of wanderers to lower altitude, this species is sedentary (Br; BUN; Sn). BUN quote daily 20 km trips to food, but some of the wanderers mentioned above *e.g.* to Nairobi 75B had certainly travelled further. Other references: North 1948; Brown 1977b.

94 HARRIER HAWK *Polyboroides radiatus*
B96 / MPG185 / D3C3

Afrotropical: widespread S of 15°N, apart from the dry areas of E Somalia, Ethiopia and the SW.

Kenyan: rather uncommon in woodland and forest edge, including wooded watercourses in dry country *e.g.* at L Turkana (HH), up to 3000 m (Br; Se). Range is 77% within 500+ mm rainfall areas (*cf.* Br) and only 14% arid. This bird has the unique habit of searching for prey in holes and crevices on tree trunks while clinging to them with its double-jointed legs (BUN), but also employs more normal methods of foraging.

Breeding squares are 91% above 1000 m and absent from the apparently suitable lowlands SE of 88, including the coastal strip. This species is however generally thought to be resident (Br) and is known to be secretive while nesting (Brown 1972a), so that breeding in these areas has presumably been overlooked. In this region, it is resident in semiarid 89A (DEP pers.obs.) but a sporadic, non-seasonal visitor to arid Tsavo E 101B (La). Brown (1966) noted a decline in the Eagle Hill area 63D due to human population pressure. Other references: Copley 1945; Brown 1952-53; 1955a; Forbes-Watson 1977b.

95 PALLID HARRIER *Circus macrourus*
B93 / MPG183 / D2C3

Afrotropical: a Palearctic species that winters across the N tropics S to 8°N, and thence S through E Africa to Namibia and E South Africa.

Kenyan: both this species and the Montagu's Harrier (96) are reasonably common passage migrants and winter visitors, up to at least 3000 m (Br; Se). The Pallid Harrier can be seen to be the more widespread of the two at the coast and along the Tana R 65C 78B 79AC. Although we map only one square for Montagu's at L Turkana, it presumably traverses this area on passage (see below), and HH describe it as occasional there. Neither species was recorded by observers resident in 19A 42A 65D, so that absence from arid NE Kenya may be real. This is supported by these birds' absence at 3-9°N in Somalia, where most S records are restricted to the vicinity of the coast (AM). Migration of both species would thus appear to be through NW Kenya, and BUN suggest the rift valley to be the Pallid's major migration route. Both of these harriers are very uncommon in the L Victoria basin (Br).

Montagu's Harrier is approximately twice as numerous as the Pallid (BUN; Br), and both of these grassland species greatly outnumber the more localised Eurasian (97) and African (98) Marsh Harriers.

The Pallid Harrier is a locally common winterer on the grasslands of the highlands and SE Kenya, late Oct-Mar. Br notes exceptional passage dates Jul-Sept, but there seems no reason why a bird in Maasai

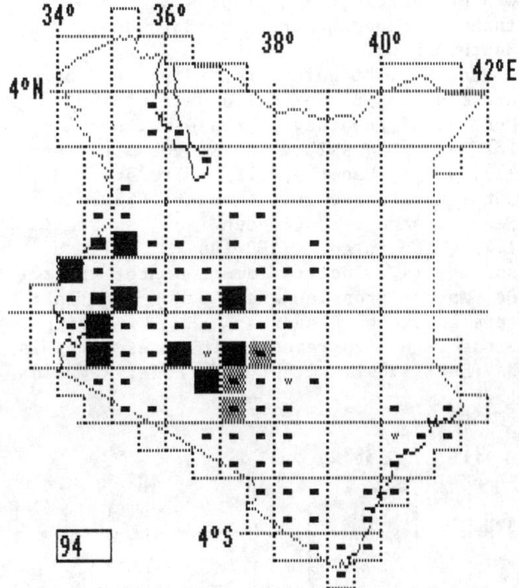

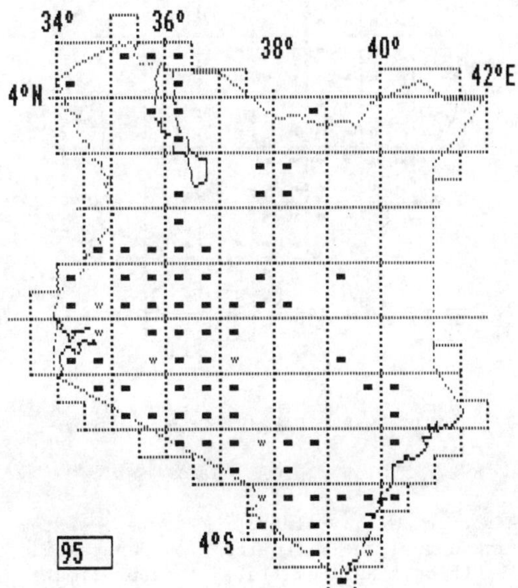

Mara 74A on 13 Jul was not oversummering (EABR 1978). Brown (1970a) notes severe declines of both this species and Montagu's during recent decades, possibly due to agricultural developments in Europe and/ or the Middle East. Other reference: Reynolds 1974.

96 MONTAGU'S HARRIER *Circus pygargus*
B94 / MPG182 / D2C3

Afrotropical: a Palearctic harrier that winters across the N tropics S to 8°N, and thence through E Africa to Botswana and E South Africa.

Kenyan: most points regarding this bird are made under the Pallid Harrier (95). Montagu's frequency increases with rainfall and it possibly hunts over more wetland areas than the Pallid Harrier (BUN), but it is sparse at the coast: 80CD = a Sept passage bird (Fogden 1963), while 114C = one in Apr (EABR 1984). This species is a locally common winter visitor Oct-Mar to grasslands and other open habitats in the highlands and SE. Wintering birds were also recently found at Kamathia 3A (EABR 1984). Other reference: Reynolds 1974.

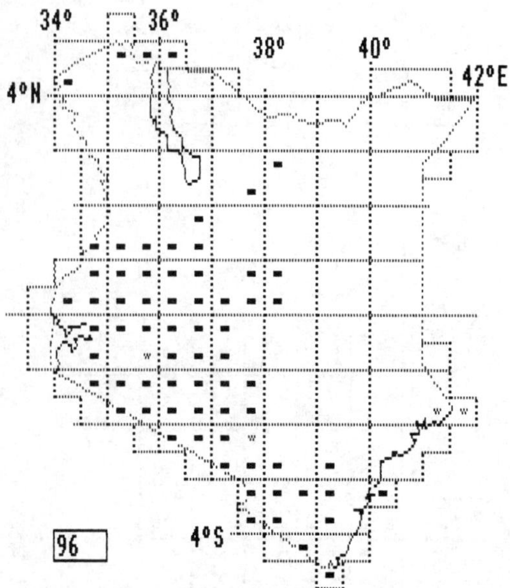

97 EURASIAN MARSH HARRIER *Circus aeruginosus*
B92 / MPG184 / D2C2

Afrotropical: a Palearctic species, wintering across the N tropics from Senegambia to Ethiopia and Somalia, S to approximately the equator in the W, and in the E through East Africa S to Zimbabwe and Mozambique.

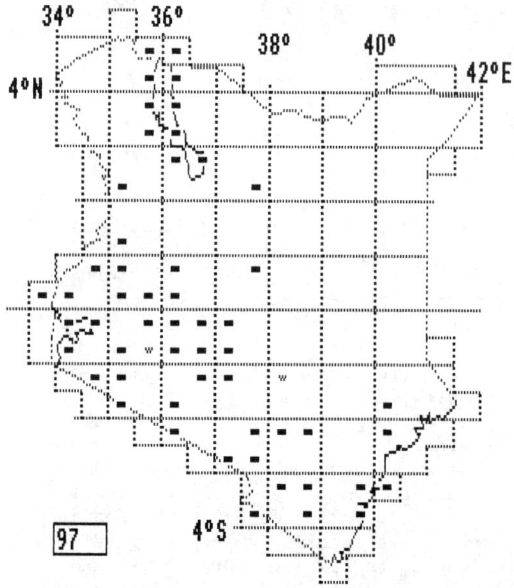

Kenyan: a common winter visitor to L Turkana (HH), the central rift valley and adjacent highlands and the SE, mainly Oct-Mar, but generally uncommon in the L Victoria basin (Br). Numbers are decidedly smaller than those of the other two Palearctic harriers (*e.g.* Se; La; Br), but Se recorded increased abundance during Oct and Mar-Apr passage around 3000 m at Mau Narok 61D. It occurs over the fringing vegetation of mainly still waters, and also over tall grassland. See African Marsh Harrier (98).

98 AFRICAN MARSH HARRIER *Circus ranivorus*
B95 / MPG184 / D2C2

Afrotropical: the S and E, with a patchy distribution S of the Congo basin, N to NE Zaire, S Sudan, Uganda, Kenya and Somalia (AM). Distribution in Kenya represents the N limit of the range.

Kenyan: an often local and never numerous bird of permanent swamps, fringing aquatic vegetation and adjacent open habitats, up to 3000 m (Br; Se; Ja). It can quarter grasslands, but this niche is largely unoccupied by any Afrotropical species and really only filled by the three migrant Palearctic harriers (Brown 1970a; Reynolds 1974).

It forms a superspecies with the more numerous Eurasian Marsh Harrier (97)(Sn). The two have similar distributions, but

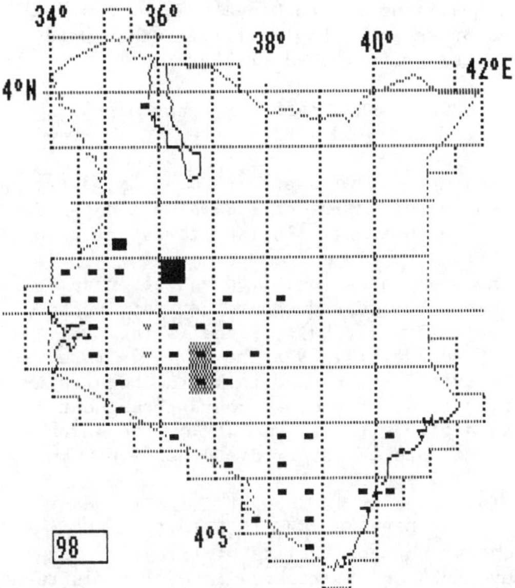

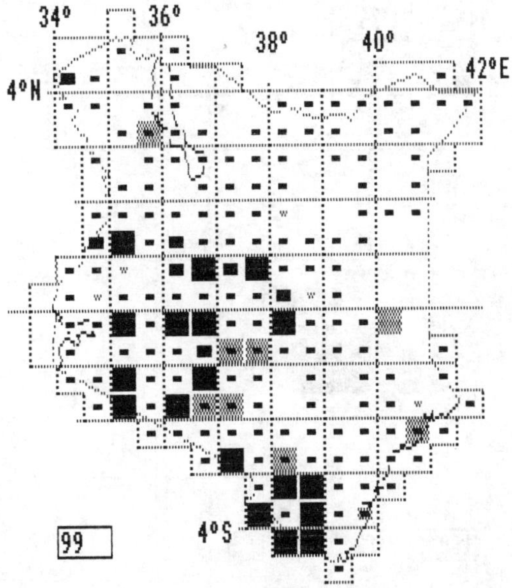

the African is less common at L Turkana and the coast, and possibly around L Victoria.

Although possibly mainly sedentary, the African Marsh Harrier can move to and breed on seasonal swamps (Br), and individuals at L Baringo 50A are mostly transient (St). The few dated records from SE Kenya and S Somalia are Nov-Apr: Nov Ngulia 101A (D J Pearson *in litt.*), Jan Tsavo E 101B (La), Feb Sokoke swamp 102B (N R Thomas *in litt.*), Apr Sabaki R 103A (EABR 1980) and Feb-Mar S Somalia (AM). Recent wetland drainage in some areas (*e.g.* Murang'a 63C and lower Embu 63D) has apparently caused a decrease in numbers (Br).

99 BATELEUR *Terathopius ecaudatus*
B101 / MPG159 / D3C3

Afrotropical: widespread S of 15°N, with the exceptions of the dry SW and the equatorial lowland forest areas of the W.

Kenyan: common in open habitats, including woodland, up to 3000 m, but scarce in some heavily settled areas (Br; Se; La; HH; St). There are records from 76% of Kenya's squares. This high coverage is due to this bird's unique appearance and easy detectability during its habitual low foraging flight: "It is probably the most characteristic bird of Africa" (MPG). 95% of the breeding squares receive over 250 mm annual rainfall, and it requires large, leafy trees for nesting. It is thus possible that the apparent absence of breeding records from the arid N and E is real, although the effects of low observer frequency in these remote areas cannot be excluded. The existence of at least local movements is shown by La's note of large, annual Jan-Mar influxes of non-breeding birds to arid Tsavo E 101B. DEP (pers. obs.) found the species resident in 75BCD 88ABD 89AC 100B 101A: these squares are with the exception of 75B entirely semi-arid. Other references: Brown 1952-3, 1955a; Brown & Cade 1972; Smeenk 1974; Short & Horne 1985a.

100 SHORT-TOED SNAKE EAGLE *Circaetus gallicus*
B100 / MPG152 / D2C2

Afrotropical: widely distributed S of 15°N, apart from the equatorial lowland forest areas of the W. Three distinctive races occur (Brown 1974d). Beaudouin's Snake Eagle *C.g. beaudouini* is resident across the N tropics from Senegambia to Uganda and W Kenya, while the Black-chested Snake Eagle *C.g. pectoralis* overlaps with it in W Kenya and occupies the remainder of the Afrotropical range quoted above. The Short-toed Snake Eagle, the nominate Palearctic race, winters across the N tropics from Senegambia to Ethiopia, with a single certain record in N Kenya.

Kenyan: the majority of sightings of this species are not racially determined, but will inevitably refer to *C.g. pectoralis*. This is a reasonably common bird of open country up to 3400 m (Br), including grass plains, wooded grassland and open bush, and it is more aerial than the

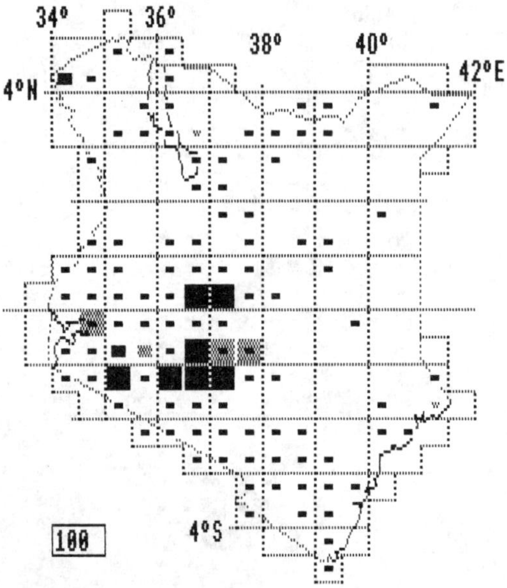

other snake eagles (*e.g.* Se), often hovering. It is commoner than its congeners, and more widespread than the other open country species, the Brown Snake Eagle (101)(107 squares as opposed to 75). Both of these birds are more widespread than the two woodland species, the Banded Snake Eagles (102, 103). Breeding is apparently concentrated in the highlands (83% above 1000 m and entirely within 500+ mm rainfall areas), with a single probable record in extreme NW Kenya 2C (at 500-1000 mm rainfall). The absence of breeding records from the apparently suitable S and coastal lowlands may imply at least local movements and, while this bird can occur in Tsavo E 101B in any month, it is commonest there Dec-Feb (La). Br however could find no evidence of a reputed wet season migration from Kenya to S Sudan (Sn). DEP (pers. obs.) considered this bird resident in 75C 88D 89C 100B 101A.

Confirmed records of the other two races are few. Beaudouin's is regular in small numbers at Ukwala and L Kanyaboli 48C in the far W, with wanderers SE to Tsavo E in Nov and Feb (La; EABR 1980), and to Mwea 63C (Br). One of this race nested with a Black-chested on Raboor Island 60B (Brown 1974d). All records of this race are from open country with trees, near water (Br).

The sole certain record of the Palearctic bird is one collected 48 km N of Loyengalani 14D on 27 Oct 1958 (Owre & Paulson 1968), while one with the characteristics of this race was seen in Tsavo E in Jan (EABR 1980). This bird could be regular in small numbers on Kenya's N borders (Ba). See Brown Snake Eagle (101). Other references: Brown 1952-3, 1955a.

101 BROWN SNAKE EAGLE *Circaetus cinereus*
B98 / MPG153 / D2C2

Afrotropical: widespread S of 15°N, apart from the dry areas of E Somalia and the SW, and the equatorial lowland forest areas of the W.

Kenyan: rather uncommon and never numerous in bushed and wooded habitats up to 2000 m (Br; Ja; La). Range is 68% semiarid-subhumid: this snake eagle is rare in the arid N and E, and in humid, high rainfall areas. Breeding occurs throughout its altitude range, and six of the seven squares are in intermediate, 250-1000 mm rainfall areas.

This species is thought to be resident (Br; La; DEP: 88D 89C 100B 101A). BUN suggest the possibility of irregular local movements and St reports irregular fluctuations in the semiarid Baringo area 50A. Pairs require as much as 200 km² of suitable habitat in arid Tsavo E 101B and semiarid Embu 63D (Br), and a decline in the Eagle Hill area of the latter square is probably due to human population pressure (Brown 1966).

The Brown Snake Eagle is found in more open, less wooded areas than the Banded Snake Eagles (102, 103) and can be broadly sympatric with them. It is also ecologically separated from the Short-toed Snake Eagle (100)(BUN; Sn).

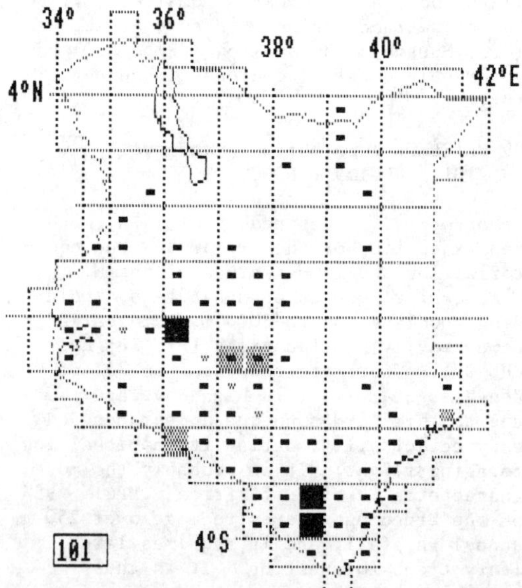

102 BANDED SNAKE EAGLE *Circaetus cinerascens*
B97 / MPG157 / D2C2

Afrotropical: a somewhat patchy distribution across the N tropics from Senegambia to Ethiopia, S to Angola and Zimbabwe but absent from most of the equatorial lowland forests of the W. Distribution in Kenya reflects the E limit of the range.

Kenyan: a rather uncommon, silent and secretive species of woodland and forest edge, including riverine growth (*e.g.* at L Turkana: HH), at 400-2000 m (Br; MPG). It is most common to the W of the rift valley (Br), where it may outnumber the Brown Snake Eagle (101: *q.v.*)(BUN). Further E, the latter is decidedly more abundant while the Banded, at the E limit of its range, is scarce and mainly confined to river valleys S of Mt Kenya 63A, especially those of the Tana, Thika and Thiba (Br; EABR 1977). Range is 80% within 500+ mm rainfall areas. Frequency generally increases with rainfall, but this eagle is replaced in the high rainfall areas of the coast by the allopatric Southern Banded Snake Eagle (103)(Sn). These two birds have been reported from adjacent squares of 101, but both are probably rare there (*e.g.* La).

There are no confirmed breeding records of the Banded Snake Eagle from East Africa (Brown & Britton 1980; EABR), probably due to observer coverage of a secretive species. This view is supported by the similar paucity of breeding data from Africa as a whole (BUN). This eagle is thought to be largely sedentary, although immatures probably wander (Br; BUN). BUN suggest that a possible Aug-Feb population peak in East Africa represents an influx of unknown origin.

103 SOUTHERN BANDED SNAKE EAGLE *Circaetus fasciolatus*
B99 / MPG156 / D2C2

Afrotropical: restricted to the E coastal strip, from S Somalia S to E South Africa. Distribution in Kenya thus reflects range at these latitudes.

Kenyan: a moderately common bird of the coastal lowlands, easily overlooked in its densely wooded and forested habitat (Br). It is most frequent near the coast, with inland occurrences only along the Tana R 79C 91A (Zimmerman & Mumford 1965; AGH; EABR) and (a single Mar record) in riverine forest near Voi 101B (La). All squares with the exception of 101B are within 500+ mm rainfall areas.

Few nests have ever been found (BUN), and squares 102BD represent the only two breeding records for East Africa (Brown 1969; Zimmerman 1970). Brown & Britton (1980) rightly reject Kreuger's (1970) breeding at Ngong 75B. Whereas Br regards regular movements as unlikely but sees seasonal absences at some localities as possibly reflecting local movements, BUN suggest that it may migrate N Jul-Oct, as it is relatively more numerous in N coastal Kenya at this time. See Brown Snake Eagle (101) and Banded Snake Eagle (102). Other reference: Short & Horne 1985a.

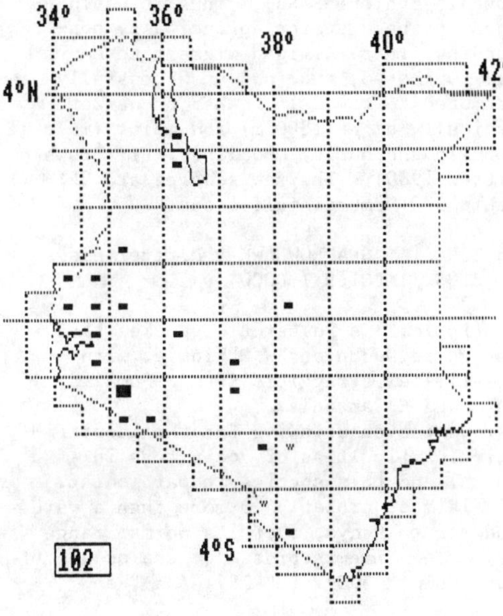

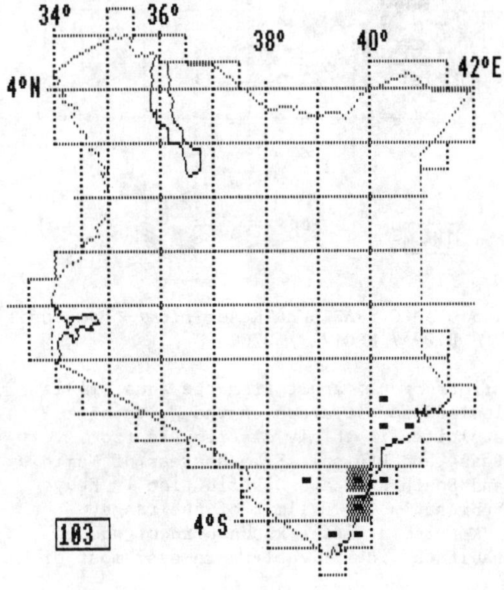

104 GREAT SPARROWHAWK *Accipiter melanoleucus*
B106 / MPG174 / D2C2

Afrotropical: from Ivory Coast E to Ethiopia and Kenya, and S throughout the rest of the continent except the dry SW. Apart from a single record at 1°S in Somalia (AM), distribution in Kenya represents the E limit of the range.

Kenyan: solitary and rather local in thick woodland and forest, including exotic plantations and mangroves, up to 3000 m (Br; Se; Ja; MPG). It mostly preys on birds, caught both in the open and in cover (BUN; Sn). Range is 90% within 500+ mm rainfall areas, and includes the forest islands of Mt Marsabit 27D 28C and the Taita Hills 101A. This sparrowhawk is considered resident (Br). The sole mention of movement is an extraordinary note of migration in some numbers through totally unsuitable bush country at Wajir 42A (Brown & Brown 1979). Br accepts this record, but we map it here with strong reservations. See African Goshawk (108).

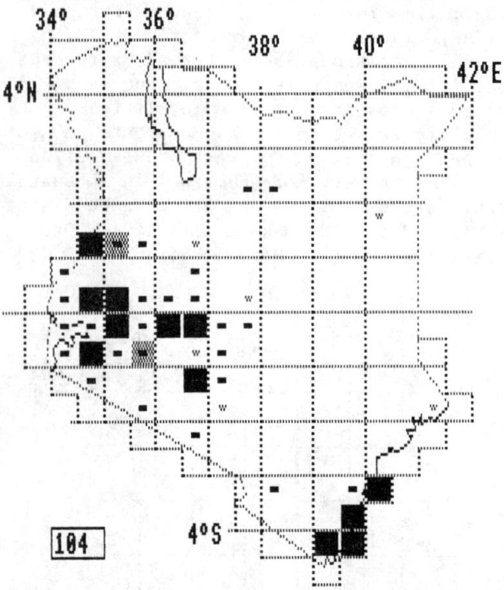

105 OVAMPO SPARROWHAWK *Accipiter ovampensis*
B109 / MPG172 / D2CO

Afrotropical: apart from the equatorial lowland forest areas of the W, this accipiter is thinly distributed from Ivory Coast and Ethiopia S to N areas of Namibia and South Africa. Distribution in Kenya represents the E limit of the range.

Kenyan: a little known bird of wooded habitats, inexplicably rare over most of

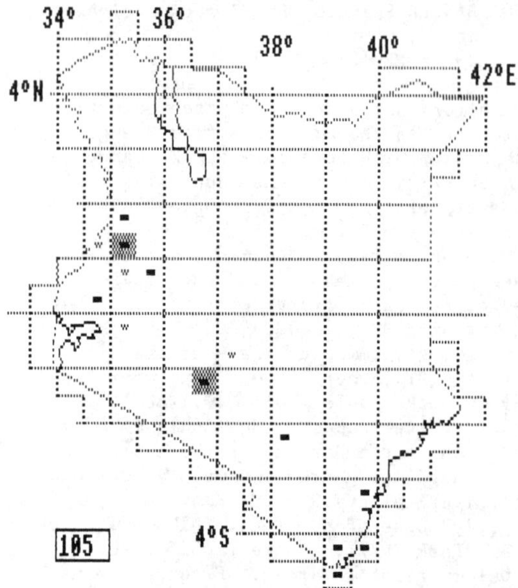

its Afrotropical range (Brown 1970a; BUN) (*cf.* Ayres' Hawk Eagle (126) and Taita Falcon (145)). Br could find less than 40 records for East Africa. The few recent reports from Kenya have been in Jan, Mar, Jul, Aug and Oct (EABR), and there have been unsuccessful breeding attempts at Kitale 37C in May 1954 (Br) and at Nairobi 75B in Sept 1968 (Brown 1970b)(eggs were produced in both cases, so that these records are mapped as confirmed breeding). This species is thought to be migratory in all but S Africa (BUN) and no breeding is known in Ethiopia (Sn). Thus it would appear to be a scarce, principally non-breeding, intra-African migrant to Kenya from further S, that has unsuccessfully attempted to breed in rather uncharacteristically high (1800 m) and moist (moist subhumid and humid) woodlands (Br; Brown & Britton 1980). The few squares are 92% within 500+ mm rainfall areas.

106 EURASIAN SPARROWHAWK *Accipiter nisus*
B108 / MPG169 / DOCO

Afrotropical: a Palearctic species that winters in Sudan and W Ethiopia, with extralimital stragglers S to S Somalia, Kenya and E Tanzania.

Kenyan: even allowing for the identification difficulties of accipiters in general and this species in particular, this is most probably no more than a rare wanderer to Kenya, S of its normal range (Br). The five records are: one on the lower Molo R 50C 2 Feb 1911 (Ja), an

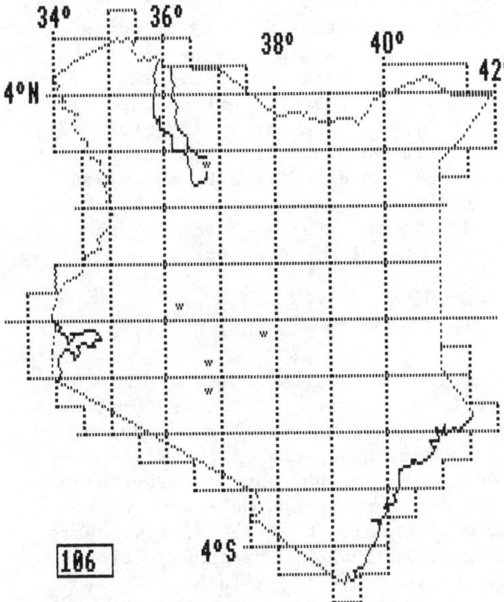

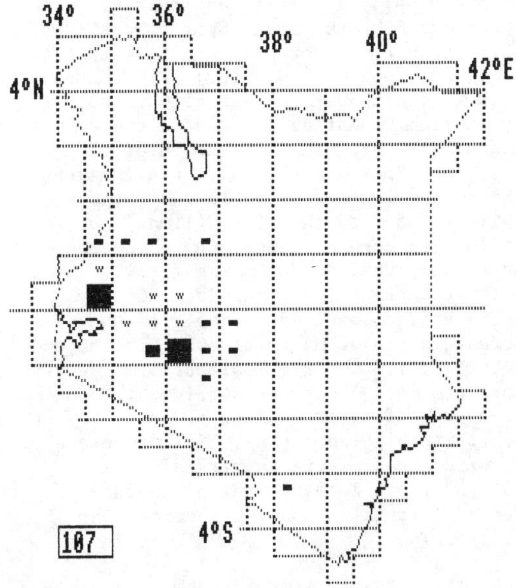

immature male collected on the Kinangop plateau 62D 4 Dec 1928 (Leakey 1943a), three together at Ishiara 63B 21/22 Jan 1952, one at Ruiru 75B 23 Jan 1953 (Ba), and a female collected near Loyengalani 26B (Owre & Paulson 1968). These last authors omit a date for their specimen, but other rare raptors procured by them at this site were in the Oct-Nov period, so that this bird was probably on S migration. If this is correct, then the Dec-Feb wintering records are all in or on the peripheries of the C highlands. This sparrowhawk forms a superspecies with the resident Rufous Sparrowhawk (107)(Sn).

107 RUFOUS SPARROWHAWK *Accipiter rufiventris*
B110 / MPG173 / D2C2

Afrotropical: mainly the E, from highland forested areas of Ethiopia, S Sudan, East Africa and the W rift, S through Malawi and Mozambique to the Cape. There is also an isolated population in Angola. Distribution in Kenya represents the E limit of the range.

Kenyan: sometimes reasonably common in highland forests, mainly at 2000-3000 m (Br; Se). It often hunts over adjacent open habitats. Presence at around 1500 m on the forest islands of the Taita Hills 101A (D A Turner *in litt.*) compares to a record at 1900 m in the Usambara Mts of adjacent NE Tanzania (EABR 1981). The concentration of pre-1970 records around 61B is more likely to reflect poor coverage of a forest accipiter (*e.g.* Se) in infrequently visited areas than a contraction of range. This sparrowhawk is apparently sedentary (Br; Se). See Eurasian Sparrowhawk (106).

108 AFRICAN GOSHAWK *Accipiter tachiro*
B111 / MPG177 / D2C2

Afrotropical: widely distributed S of 10°N (and further N along the W coast and in Ethiopia), but absent from the drier areas

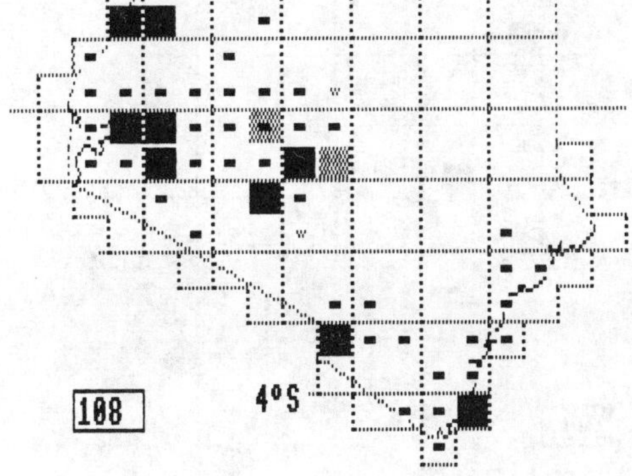

of E Ethiopia, E Somalia and the SW.

Kenyan: like the Great Sparrowhawk (104), this is an accipiter of thick forests and woodlands, up to 3000 m (Br; Se). Range is 90% within 500+ mm rainfall areas. Apart from Mt Marsabit 27D 28C, the African Goshawk is more widely distributed than the Great Sparrowhawk, particularly along the coast (including the Tana R forests 79C: AGH) and to the E of Kilimanjaro, and it is also more numerous, being the commonest accipiter in many areas (Br).

It is thought to be mainly sedentary (BUN) and, despite extending N to the forest island of Mt Kulal 26B (Diamond & Keith 1980), the arid lowlands of N Kenya have provided a barrier sufficient to produce a distinctive race isolated in the Ethiopian highlands (Sn). Six wanderers in arid and semiarid country at Baringo 50A and Tsavo E 101B were in the Oct-Apr period (St; La). Other reference: van Someren 1956.

109 LITTLE SPARROWHAWK *Accipiter minullus*
B107 / MPG170 / D1C2

Afrotropical: widespread to the S of the Congo basin apart from the dry SW, and further NE through East Africa to S Sudan, the Ethiopian highlands and S Somalia.

Kenyan: a somewhat uncommon species that captures prey within thick woodland canopies and other dense cover, and is found up to 3000 m (Br; Se; Ja; BUN). Range is 82% within 500+ mm rainfall areas. The map's coverage of its range contains gaps similar to those of the Shikra (110), probably due to the low detectability (D1) of a small, unobtrusive bird of dense cover. C2 is awarded because it is not difficult to identify, while being sufficiently rarely seen to warrant reporting. It is thought to be mainly sedentary (Br; BUN), although wanderers have reached dry country at Kapedo 38C (EABR 1983, 1984), and in 39B (J D Gerhart *in litt.*). Moyale 17A (Ja) represents the S edge of the Ethiopian highlands population.

110 SHIKRA *Accipiter badius*
B102 / MPG176 / D2C2

Afrotropical: occures widely S of 15°N, apart from the arid areas of E Somalia and the SW, and most of the equatorial lowland forest regions of the W.

Kenyan: a somewhat uncommon species of wooded and forest edge habitats, up to 3000 m (Br; Se). Only 15% of its squares receive less than 250 mm annual rainfall, and occurrence in the arid N and E is mostly near water (Br; *e.g.* La). However, as is evident from the gaps in its coverage on the map, distribution in most of the moister woodlands and wooded savannahs of the highlands and the SW is also patchy, and this absence from apparently suitable equatorial habitats remains enigmatic (Br; BUN). The Shikra is thought to be mainly resident (Br; St; DEP pers. obs.: 74D 89A). The only indication of regular movement comes from arid Tsavo E 101B where, apart from a single Jul record, it is a Nov-Apr visitor (La). Se also quotes a single Jul record from the

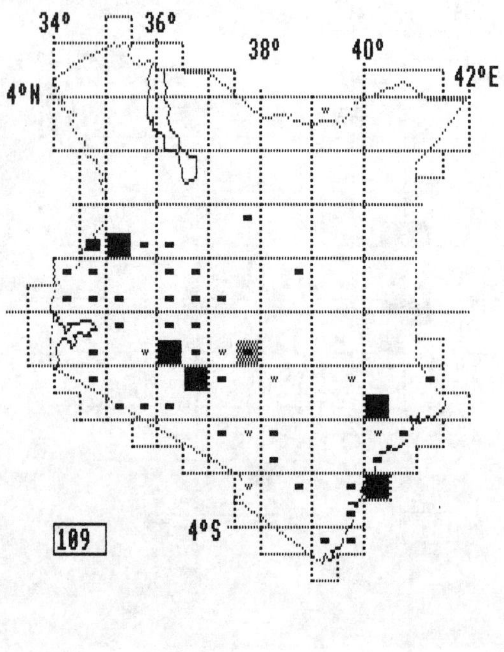

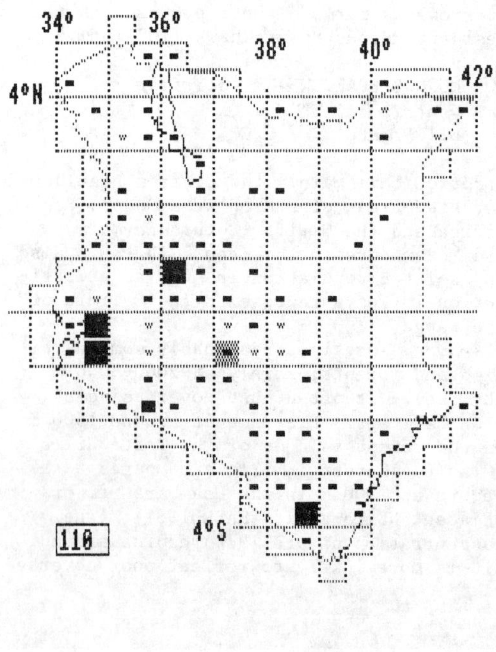

top of the altitude range at Mau Narok 61D. This accipiter forms part of a species-group with the Levant Sparrowhawk (111) (Sn).

111 LEVANT SPARROWHAWK *Accipiter brevipes*
B103 / MPG175 / D0C0
NO MAP: post presence 52C

Afrotropical: due to confusion with the very similar Shikra (110: *q.v.*), the range of this Palearctic species is very poorly known. It almost certainly winters in Ethiopia, and there are a few records from Sudan. Other occurrences are S to Niger, Zaire, NW Tanzania and Kenya.

Kenyan: the single record is of an adult male in Meru NP 52C on 8 Nov 1983 (EABR 1983). Other possible records have occurred, and the species is probably less rare than this single sighting suggests.

112 DARK CHANTING GOSHAWK *Melierax metabates*
B132 / MPG180 / D2C2

Afrotropical: widespread from 17°N to 16°S in the W and 25°S in the E, apart from the equatorial forest areas of the W. It is replaced by the Pale Chanting Goshawk (113) in E areas of Ethiopia, Kenya and Tanzania. Further S, it reaches the E coast.

Kenyan: a common bird of bushed and wooded country, up to 3000 m (Br; Se; Ja; MPG; St). It is found in W Kenya, mainly E to and within the rift valley, and it is virtually allopatric with the Pale Chanting Goshawk, which is a member of the same superspecies (Sn) and occurs further E. The separation of these two birds at QSD scale is incomplete, but there are only two squares in which both breed (37C 50A), and overlap elsewhere probably reflects non-breeding wanderers and/or squares containing diverse habitat (Sn; Br). The claimed allopatry of these two goshawks E and W of L Turkana and N and S of Mt Longonot 62C (Brown 1970a; BUN) is incorrect.

The Dark Chanting Goshawk appears to prefer moister habitats than the Pale (Br), since it is absent from the arid W shore of L Turkana (HH), while the Pale has been reported from 67% of the most arid squares (moisture index below 1.2) further E. The Dark is also less frequent in humid highlands W of the rift. Eight of the nine breeding squares are in 500+ mm rainfall country. No details of movements are known, beyond the wandering mentioned above.

113 PALE CHANTING GOSHAWK *Melierax poliopterus*
B133 / MPG179 / D3C3

Afrotropical: this species replaces the Dark Chanting Goshawk (112: *q.v.*) in the NE, from Somalia and E Ethiopia S to N Uganda (where both occur), Kenya and E Tanzania.

Kenyan: common in open bushed and wooded country (Br; MPG; HH; La). Range is 85% below 1500 m and 83% arid-semiarid. DEP (pers. obs.) finds this goshawk resident in 74B 75BCD 88AD 89C 100B 101A, and the

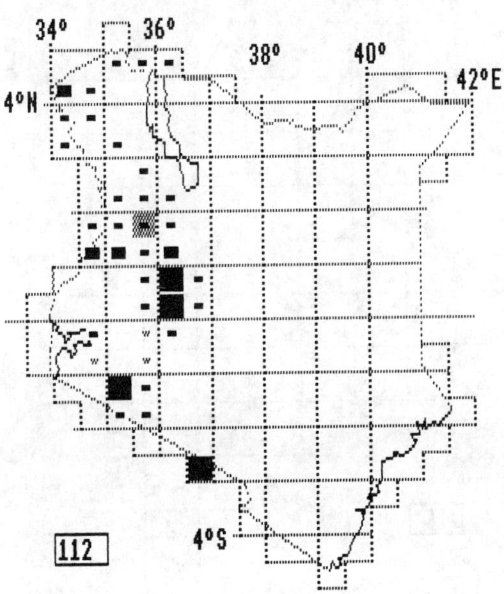

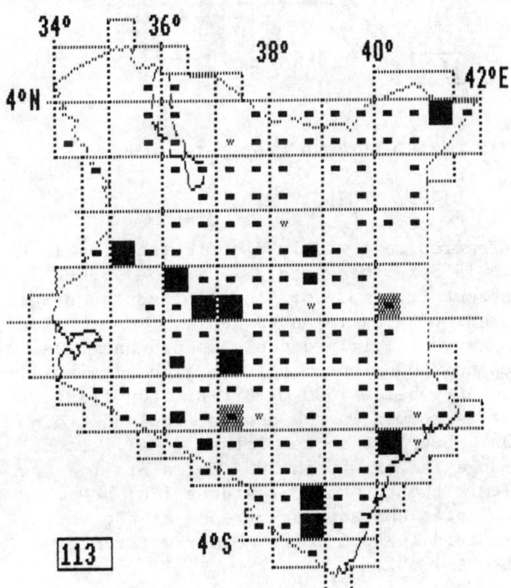

only evidence of movements comes from wanderers within the breeding range of the Dark Chanting Goshawk (Br). Other reference: Smeenk & Smeenk-Enserink 1975.

114 GABAR GOSHAWK *Melierax gabar*
B131 / MPG178 / D2C2

Afrotropical: widespread S of 16°N, apart from the equatorial lowland forests of the W.

Kenyan: locally common in a wide variety of fairly dense bushed and wooded habitats, up to 2000 m but mainly in lower rainfall areas (Br; Ja; La; St; HH). It is generally considered resident (Br; La: 101B; DEP pers. obs.: 74D 75C 88D 89AC 100B 101A). Other references: van Someren 1956; Smeenk & Smeenk-Enserink 1975.

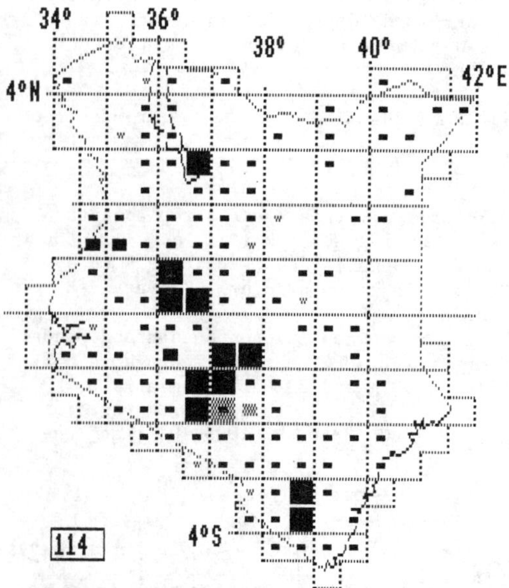

114

115 LIZARD BUZZARD *Kaupifalco monogrammicus*
B129 / MPG150 / D2C2

Afrotropical: widely distributed from 15°N to 15°S in the W and 30°S in the E, but absent from most of Ethiopia and the drier areas of Somalia and N Kenya.

Kenyan: fairly densely wooded habitats, up to 3000 m but mainly in higher rainfall country below 1500 m (Br; Britton 1979). It is common in the coastal lowlands (*e.g.* Ja), but less so and rather local inland. Range is 76% within 500+ mm rainfall areas. It is thought to be resident (Sn; BUN). La notes sporadic occurrence at the Voi R in arid Tsavo E 101B. Other reference: Lewis 1983f.

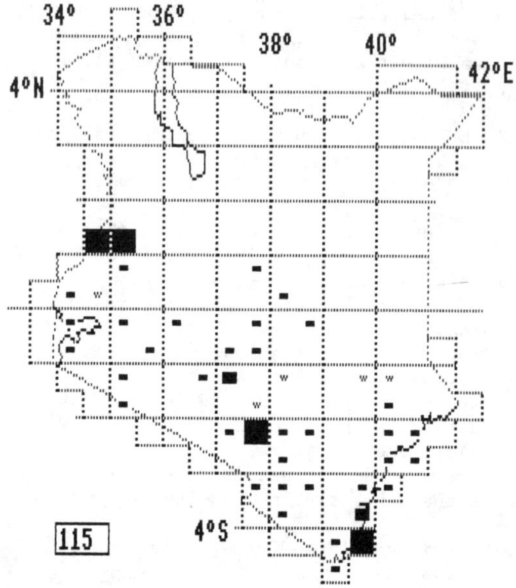

115

116 MOUNTAIN BUZZARD *Buteo tachardus*
B124 / MPG165 / D1C2

Afrotropical: in the NE and E, a bird of montane forest from Ethiopia and S Sudan to suitable areas of East Africa and the E Zaire borders. Further S, it is at lower altitudes in Malawi and South Africa. Distribution in Kenya reflects the E limit of the range.

Kenyan: seldom common, in pairs in highland forest, up to 3800 m (Br; Se; BUN).

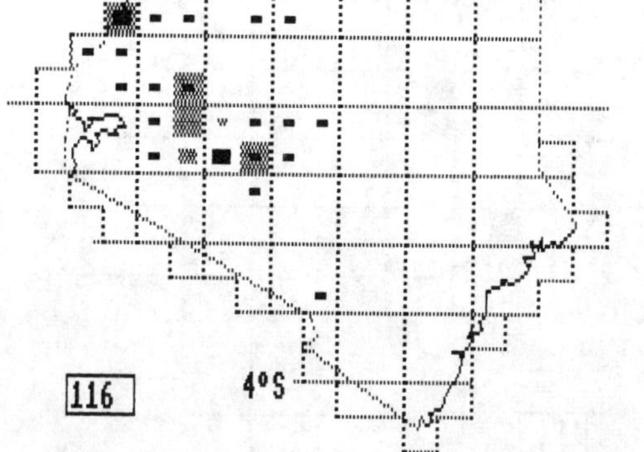

116

All squares contain extensive highland forests, or the forest islands of Mts Kulal 26B (Diamond & Keith 1980) and Marsabit 27D 28C, the Mathews Range 39C and the Chyulu Hills 88D. D1 reflects its dense habitats, and the possibility of confusion with other species *e.g.* Common Buzzard (117)(with which it forms a superspecies: Sn). C2 reflects its restricted habitat range. The Mountain Buzzard is apparently resident (Br; Be; Se), and is ecologically separated from the closely related Augur Buzzard (119) in denser habitats (BUN).

117 COMMON BUZZARD *Buteo buteo*
B122 / MPG163 / D2C2

Afrotropical: a Palearctic species, wintering in E and S Africa from Sudan and Ethiopia S through East Africa and E Zaire to Namibia and South Africa; and also along the W periphery of the continent S to Liberia.

Kenyan: mostly a passage migrant Sept-Nov and Feb-Mar (Br), when it can be quite numerous, far outnumbering the resident Augur Buzzard (119) in many areas (Brown 1963). Birds migrating to and from S Africa pass around L Victoria, and these movements follow a well known route between NE Tanzania, Wire Hill 60B, Ng'iya Hill 48C and Mt Elgon 36D, using the lift over the high points to gain height. S passage is most marked, in flocks of up to 50, but N passage also occurs, in flocks of up to 20 (Brown 1970a; Br; Ja; Britton & Britton 1976b; Lewis & Pearson 1981a). Br knew of no records from L Turkana, but small numbers of N passage migrants have recently been seen there (D K Richards *in litt.*), so that the rift valley may act as a secondary migrational pathway. Limited passage is also known from other areas *e.g.* Nairobi 75B (EABR 1978, 1984). Small numbers overwinter (Br).

Common Buzzards are found in bushed and wooded habitats (Br). Range is 72% above 1000 m and 68% within 500+ mm rainfall areas and, apart from migrants at Turkana and rare records at the coast (P L Britton, D J Pearson *in litt.*; and in coastal Somalia: AM), overall distribution is similar to that of the closely related Augur Buzzard. See Mountain Buzzard (116).

118 LONG-LEGGED BUZZARD *Buteo rufinus*
B123 / MPG166 / D1C0

Afrotropical: a Palearctic species that winters in small numbers across the N tropics from Senegambia to Ethiopia mainly N of 10°N, with very small numbers S to N Uganda and N Kenya, and a single record in NE Tanzania.

Kenyan: like the Palearctic *Aquila* eagles, this is not an easy species to identify, and the first record came as recently as 1958, when an adult male was collected during a period of strong S raptor passage at Loyengalani 26B on 6 Nov (Owre & Paulson 1968). Since then, it has been recognised as a regular winter visitor in very small numbers at L Turkana, with stragglers S to Taveta 100B and Tsavo E 101B (Ba; Turner & Forbes-Watson 1974; HH; La; Br; EABR).

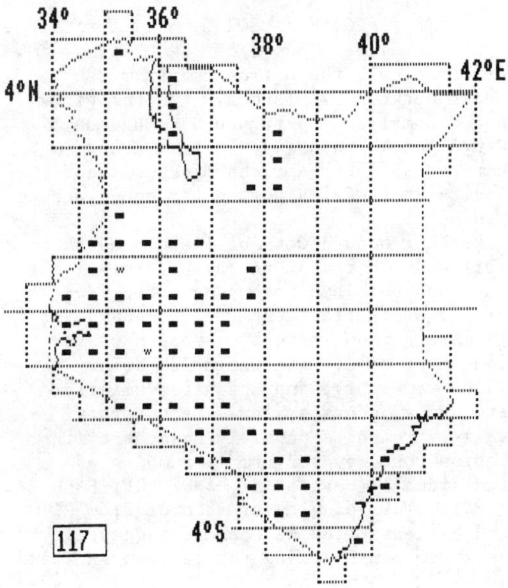

117

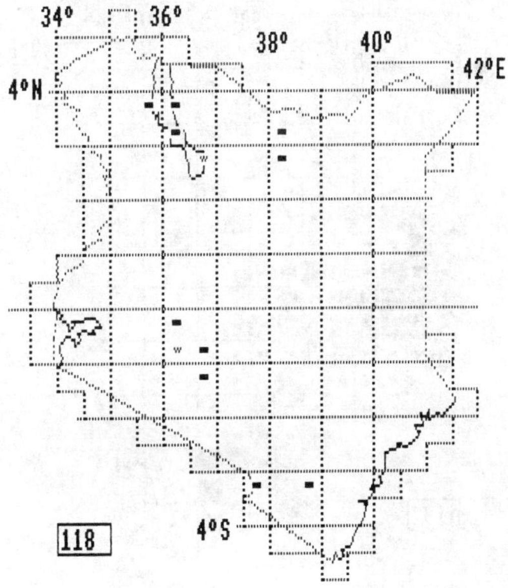

118

66% of its squares are arid-semiarid, and birds in higher altitude, subhumid-humid localities on the S of the C highlands 62CD 75B were all Oct, Nov or Mar and thus possibly on passage, though one remained at Tigoni 75B for at least 10 days (EABR 1978) and one wintered at L Nakuru 62A Oct 1976-Feb 1977 (EABR 1983).

119 AUGUR BUZZARD *Buteo augur*
B120 / MPG167 / D3C3

Afrotropical: the E and S from Sudan, Ethiopia and extreme NW Somalia S through mainly highland areas of East Africa and the E Zaire borders to scattered areas between 10°S and 25°S. Distribution in Kenya represents the E limit of the range.

Kenyan: common in open country, typically in pairs (Br; Ja; MPG; Se). While often thought of as a bird of the highlands, distribution is in fact most closely correlated with the presence of rocky outcrops and hills (Br), and it occurs in these habitats below 500 m in Tsavo E 101B (La) and the Shimba Hills 114A, and has been recorded at 150 m near Kwale 114A (EABR 1980), though nowhere else near the coast. But it is certainly more plentiful in the highlands (Ja) and, while being recorded in 88% of the squares above 1000 m, can be seen at over 5000 m around the peaks of Mt Kenya 63A. N of the highlands, it occurs around the forest islands of Mts Kulal 26B, Nyiru 26D and Marsabit 27D 28C. Squares 15AC 16AB represent the Huri Hills volcanoes and the S edge of the Ethiopian highlands population. Breeding has been

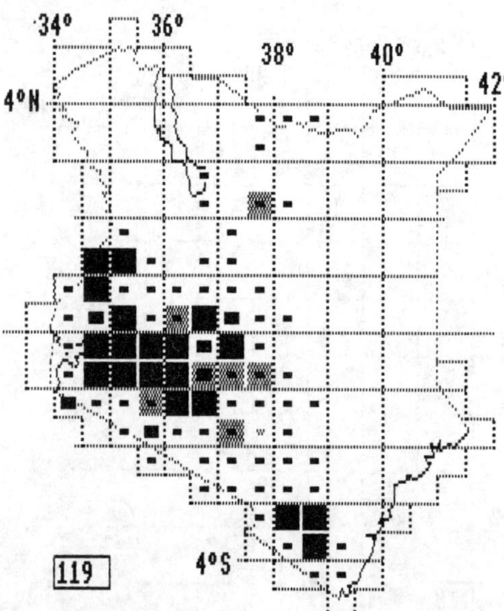

recorded up to 4000 m, in the Mt Elgon crater (Brown 1970a).

The Augur Buzzard is considered sedentary (Br; Se: 61D; La: arid 101B). DEP (pers. obs.) found similar status in 74BD 75B 88D 89C 100B 101A, but also noted a regular Nov-Jan influx to the Ruaraka-Kahawa-Ruiru area of 75B, which coincides with the species' non-breeding season (Brown & Britton 1980).

Melanistic birds, which constitute over 10% of the population, rise to 25% in some high rainfall areas *e.g.* Kisii 60D and Kericho 61A (Br; *cf.* Se). A decline in the Eagle Hill area 63D is attributed to human population pressure (Brown 1966). The Augur Buzzard has been treated as conspecific with the next species. Other references: Brown 1952-3, 1955a; Lewis & Pomeroy 1982a.

\- RED-NECKED BUZZARD *Buteo auguralis*
B121 / MPG168 / D2C0
RECORDS UNSATISFACTORY

Afrotropical: replaces the Augur Buzzard (119) in most of the N tropics and the W, occurring from Mauritania S to Angola, and E across the N tropics to C Ethiopia and NW Uganda.

Kenyan: a single sight record at Kaibibich 37C on 9 Jul 1969 (Cheke 1978) is supported by insufficient details for such a unique observation, particularly one made only four days after the observers had started their expedition, and Br rejected it.

120 GRASSHOPPER BUZZARD *Butastur rufipennis*
B119 / MPG158 / D3C2

Afrotropical: the N tropics, from the W coast between 17°N and 5°N to Ethiopia and W Somalia, and thence S to Uganda, Kenya and E Tanzania. It is a trans-equatorial intra-African migrant that visits East Africa during its non-breeding season.

Kenyan: occurs Oct-Apr (mainly Nov-Feb) (Br) and, in recent years, has in all but one instance (see below) been recorded singly or in flocks up to 13 (EABR). It is most common in bush or wooded grassland E of 38°E, with range 72% below 1000 m (*cf.* Br), and 85% arid-semiarid. It rarely occurs as far W as the rift valley, which is possibly used as a migration route by very small numbers of individuals (Haas *et al.* 1982; St; EABR 1984). This paucity of records from C, W and NW Kenya (and NE Uganda) suggests separate non-breeding populations in W and

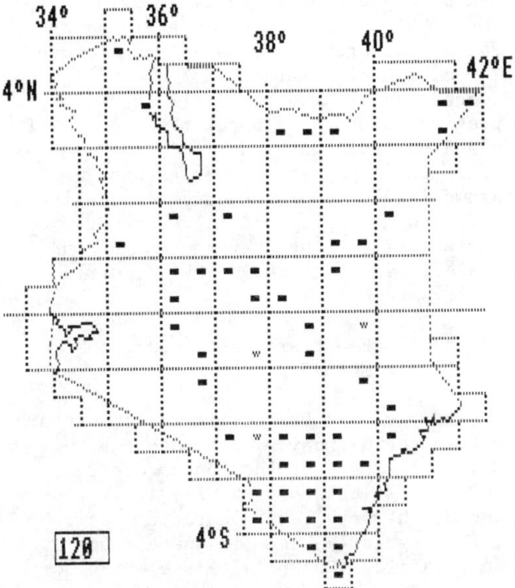

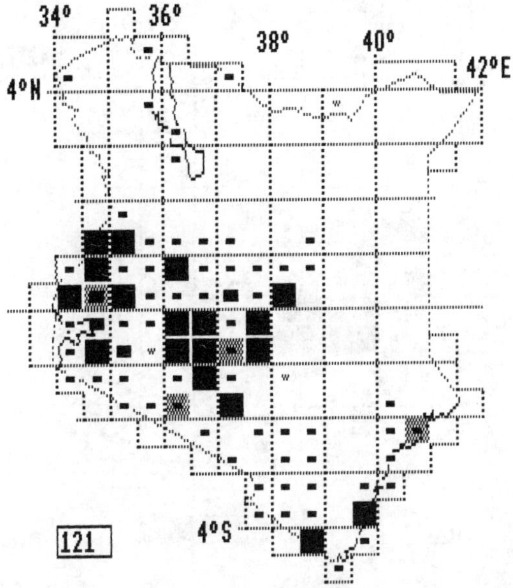

E East Africa (Br).

The abundance of records S of 2°S (with the notable exception of the coastal strip) is likely to reflect observer frequency, but also denotes an area where regular concentrations have occurred (particularly 101BD 102C)(Br). Despite an apparently considerable decline in numbers in SE Kenya during recent years (EABR 1977; Br), scores appeared during heavy rain in Tsavo W 89C 101A in Nov (EABR 1982). An exceptional record of a single bird near Moyale 17C on 21 June 1975 (D A Turner *in litt.*; Br) may be related to occurrence in all months S of 4°N in Somalia (AM).

121 LONG-CRESTED EAGLE *Lophaetus occipitalis*
B130 / MPG149 / D3C3

Afrotropical: widely distributed S of 15°N, apart from the E Somalia and SW arid areas.

Kenyan: a locally common, solitary bird of wooded and forest edge habitats, mainly below 2000 m but also reaching 3000 m (Br; Se). It is common in even heavily settled areas that have been partially cleared for agriculture (Br; Ja; Se). Range is 75% within 500+ mm rainfall areas. The records along the N borders 2C 3B 5C 17A represent the S edge of birds in moister country in Sudan and Ethiopia.

Breeding squares are 96% within the 500+ mm rainfall regime, and concentrated in the highlands and coastal lowlands. If breeding is indeed found only in these high rainfall areas, birds elsewhere are non-breeding wanderers. BUN suggest this origin for individuals associated with riverine trees in arid country (Brown 1970a; *e.g.* 40D 52B and at L Turkana). Furthermore, in arid Tsavo E 101B, an area which lies between these two breeding concentrations, this species is an irregular and sporadic visitor Nov-Apr (La). This seasonality compares with rare Sept and Dec-Mar wanderers in coastal S Somalia (AM). In the highlands breeding area, DEP (pers. obs.) found this bird resident in 75B. Other references: Brown 1952-3, 1955a.

122 CROWNED EAGLE *Stephanoaetus coronatus*
B135 / MPG147 / D2C2

Afrotropical: inhabits wooded and forested areas from Senegambia and Ethiopia S to Angola and South Africa. Distribution in Kenya represents the E limit of the range.

Kenyan: rather uncommon and local in forests and thick woodland, including riverine growth (*e.g.* the lower Tana R 79C: AGH), up to 3000 m (Br; Se; Ja). This is the most powerful African eagle, and the closed habitat equivalent of the similar but not closely related Martial Eagle (123) (Sn). Range is 84% within 500+ mm rainfall areas, and the 98% well wooded squares include the forest islands of Mt Marsabit 27D 28C, the Mathews Range 39C and Mt Endau 77B (Tennent 1964). It is resident (Br; Brown 1972b; Se). Human population pressure has possibly caused decline in the Embu 63D area (Brown 1966). Other references: Oulton 1938; Brown 1952-3, 1955a, 1971c, 1982; Se; Gregory 1971;

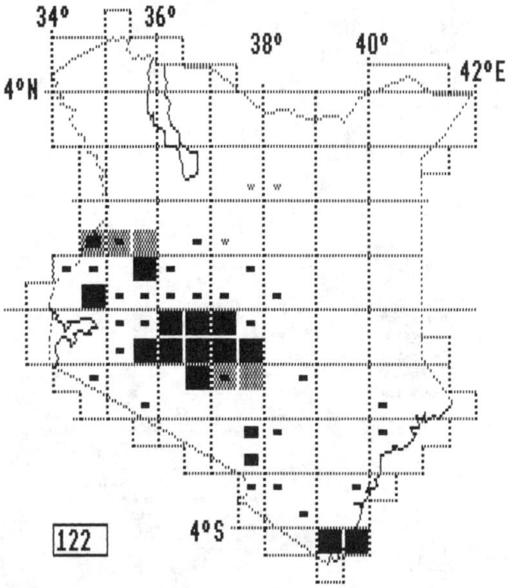

122

Leuthold 1971a; Lewis & Pomeroy 1982b.

123 MARTIAL EAGLE *Polemaetus bellicosus*
B134 / MPG146 / D2C2

Afrotropical: widely distributed S of 15°N, though rather sporadic in the W and absent from most of the lowland equatorial forest areas.

Kenyan: widespread in small numbers in open habitats including woodland, reaching 3000 m but mainly lower down (Br; Se). It makes very extensive daily foraging trips (Sn; BUN) and may occur in every square. 96% of the breeding squares receive over 250 mm annual rainfall, 88% are well wooded, and it requires tall trees for nesting (St; BUN). Hence the absence of breeding from the arid N and E may be real. Dense human population was thought by Brown (1966) to be the cause of decline in the Embu 63D area (*cf.* Br). See Crowned Eagle (122). Other references: Brown 1952-3, 1955a; Leuthold 1971a; Norris 1972; Smeenk 1974.

124 BOOTED EAGLE *Hieraaetus pennatus*
B127 / MPG145 / D2C2

Afrotropical: a Palearctic eagle, wintering across the N tropics, S to about 5°N in the W but passing S of the Congo forests at least to Zimbabwe in the E. It is also one of the very few Palearctic species which regularly breed in the extreme S of South Africa (Brown 1978):see White Stork (45).

Kenyan: as with the Palearctic *Aquila* eagles, problems in identification seem to have caused under-recording until about 1971, when only eight records were known (Ba). But now both phases are recognised as regular but rather uncommon winter visitors, Oct-Apr, to mainly bushed or wooded habitats (Br; EABR). Range is 69% within 500+ mm rainfall areas. With the exception of L Turkana, where it appears to be a passage migrant (*e.g.* HH), overall distribution is similar to that of the closely related African Hawk Eagle (125).

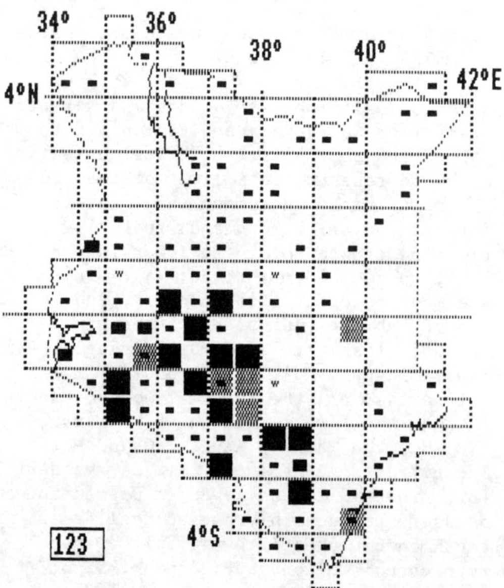

123

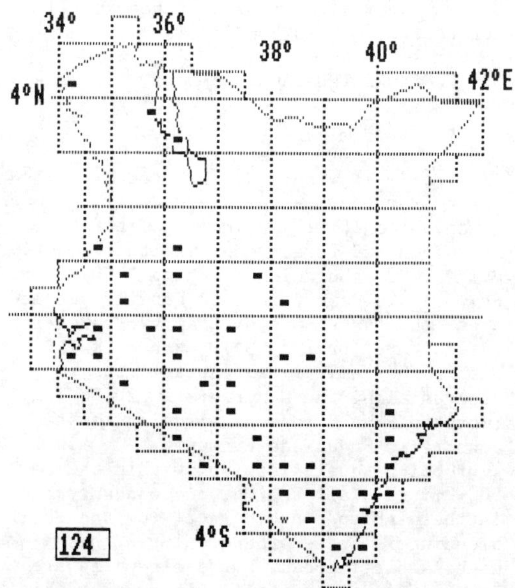

124

It has been observed using the Mt Elgon 36D-Wire Hill 60B migration route E of L Victoria with Common Buzzards (117) and, as in that species, S passage may be more pronounced than N (Br; *e.g.* Lewis & Pearson 1981a; EABR 1983). A bird at Nairobi 75B on 18 Aug may have been oversummering (EABR 1982). Other reference: Mann 1973b.

125 AFRICAN HAWK EAGLE *Hieraaetus spilogaster*
B128 / MPG143 / D2C2

Afrotropical: widespread S of 15°N, apart from the arid SW and the lowland equatorial forests of the W.

Kenyan: small numbers are found in open bushed and wooded country, up to 3000 m but mainly below 1500 m (Br; Se; MPG; La). They are often near hills and rocky outcrops. Occurrence in the arid N and E is either in squares adjacent to semiarid regions S of the equator *e.g.* 64B 77B 78C 89BD 101B, or in riverine trees (*cf.* Br) at L Turkana (HH), and along the Daua R 7C 18B 19A (E J Roberts, ADL pers. obs.) and the Tana R 65BD. 87% of its squares are well-timbered, but this eagle is uncommon in the high rainfall areas of the highlands and, particularly, the L Victoria basin (Br; *e.g.* Se). It is resident, in pairs (Br; La; St). Recent declines, in some cases perhaps due to human population pressure, are reported from the Embu 63D and Tsavo E 101B areas (Brown 1966; La). See Booted Eagle (124). Other references: Brown 1952-3, 1955a; Smeenk 1974.

126 AYRES' HAWK EAGLE *Hieraaetus dubius*
B126 / MPG144 / D2C2

Afrotropical: widely but thinly distributed throughout Africa S of Ivory Coast, Ethiopia and S Somalia, but absent from most of the SW and possibly much of the Congo basin forest.

Kenyan: as in the rest of its Afrotropical range, this species is inexplicably scarce and local despite being a very efficient predator with an apparently plentiful supply of food (Brown 1974b; Br; Ja)(*cf.* Ovampo Sparrowhawk (105) and Taita Falcon (145)). It inhabits thick woodland and all but the densest forest, up to 3000 m, but often hunts over adjacent open habitats (Br; Se). Range is 89% within 500+ mm rainfall areas, and includes the forest islands of Mt Marsabit 27D 28C and the Chyulu Hills 88D. Rather than any decline, the sprinkling of pre-1970 records 61D-37D-28C more probably represents observer coverage of a rather obscure species of dense habitats in infrequently visited areas (*e.g.* Se; BUN).

Ayres' Hawk Eagle is thought to be basically sedentary (Br; Sn; EABR 1977), though possibly wandering during the non-breeding season (BUN). It can only be presumed resident in the coastal lowlands since it appears to be exceptionally scarce and irregular there (Britton & Zimmerman 1979; Br; EABR) and there are no breeding records (*cf.* status along the Juba R of adjacent S Somalia: AM). It is conceivable that this eagle's nest is harder to find than those of some other,

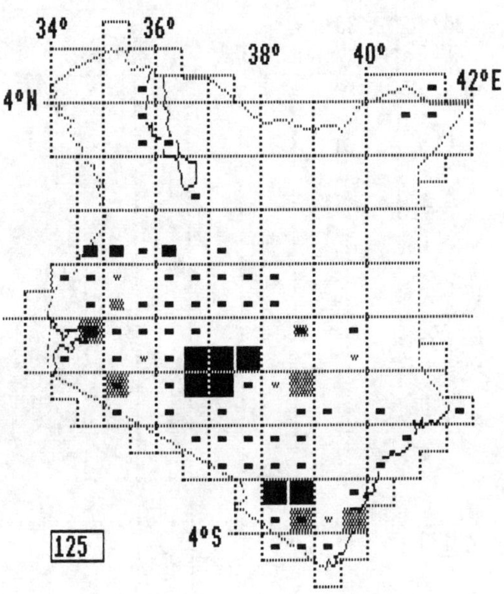

125

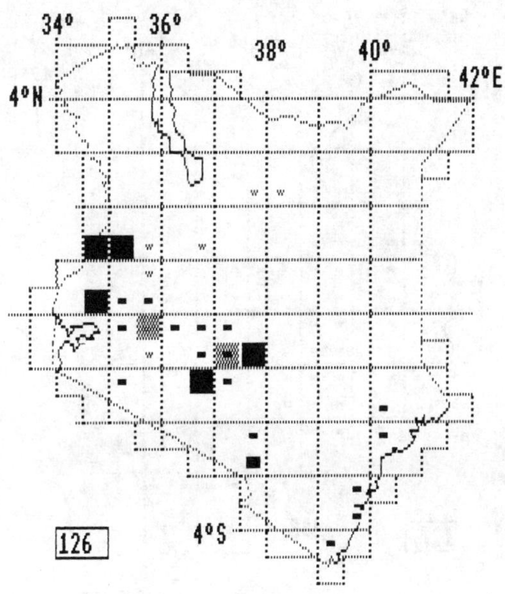

126

commoner species that may breed along the coast only rarely or not at all *e.g.* Hamerkop (44) and Hadada (54). Other references: Brown 1952-3, 1955a, 1966; Brown & Davey 1978; Wolf 1984.

127 LESSER SPOTTED EAGLE *Aquila pomarina*
B115 / MPG141 / D1C1

Afrotropical: a Palearctic eagle that migrates through E Africa, mainly wintering from Zambia S to Namibia and NE South Africa. Small numbers in Kenya in Jan-Mar probably represent the N edge of its wintering range, and distribution in Kenya represents the E limit of the total range.

Kenyan: the great difficulties in distinguishing between the Palearctic members of this genus in many of their plumages, combined with the modern disinclination to collect specimens, has probably resulted in under-recording of this species: in 1973, Ba could quote only one certain Kenya record, that of Mackworth-Praed (1917)(*cf.* Greater Spotted Eagle (128) and Imperial Eagle (131)). However, recent developments in field recognition (Porter *et al.* 1976) have shown this species to be more numerous in Kenya, mainly as a passage migrant that winters further S.

Occurrence can be split into three periods. It is not uncommon on S passage late Oct-early Dec, but apparently most of the population overflies Kenya on its way S, only descending to near ground (and observer) level during bad weather (Brown 1970a). An account of "many thousands" of *Aquila* sp., thought to be this species, at Sultan Hamud 88A in Nov (Bowles 1967) provides the only scanty record of the large scale S movements that presumably occur regularly at high altitude (see also Backhurst 1985). Small numbers overwinter Jan-Mar, mostly around the rift lakes and dams W of the rift 49 61 (*e.g.* EABR 1984). An influx in the winter of 1978/9, with flocks of up to 20 in mid-Jan, was possibly due to unusually heavy rainfall combined with drought further S (Pearson & Meadows 1979). Records of N passage are very few in Kenya, possibly because overflying is more direct and uninterrupted, or because the main movement occurs further W (Pearson & Meadows 1979; Br). N movement with other migrant raptors at Homa Bay 60C in Mar (also after heavy rain) supports the latter view (Lewis & Pearson 1981a).

This eagle's squares are 79% above 1000 m and 80% within 500+ mm rainfall areas. Most Jan-Feb wintering records are above 1500 m in the rift valley highlands, in the 500+ mm rainfall regime, although birds have occurred as low as L Jipe 100D and Aruba dam 101B (EABR). Occurrence to the W of the rift is less common overall (*e.g.* Lewis & Pearson 1981a). Other references: Campbell & Campbell 1975; Dowsett 1975.

128 GREATER SPOTTED EAGLE *Aquila clanga*
B112 / MPG140 / D1CO

Afrotropical: a Palearctic bird that winters in small numbers in Sudan and Ethiopia, with a few records S to Kenya.

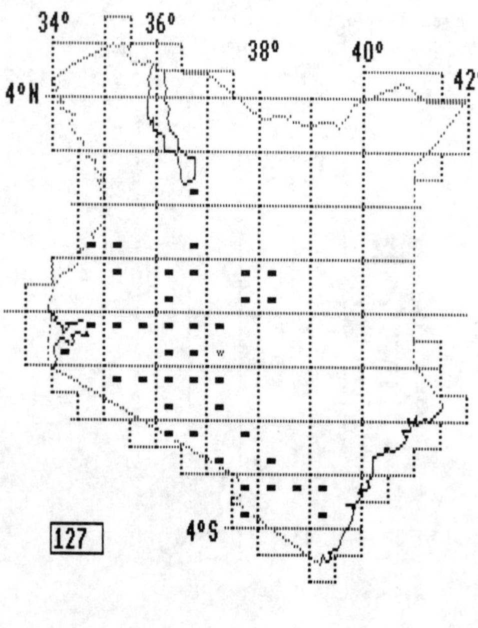

127

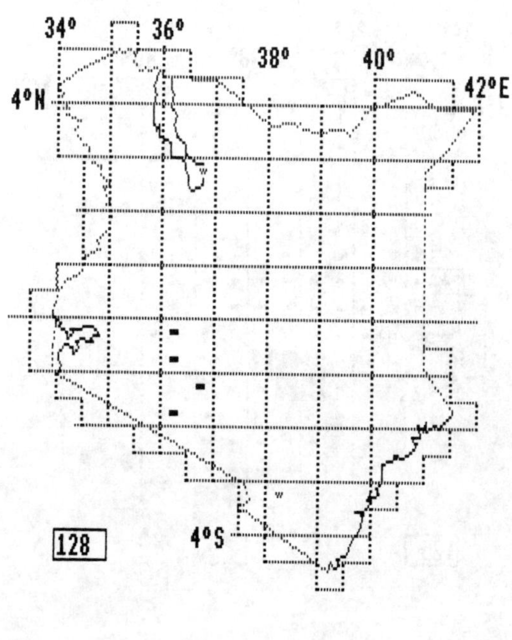

128

Kenyan: the identification problems mentioned under Lesser Spotted Eagle (127) apply to an even greater extent here, despite recent advances in field recognition. Further knowledge of field characters may necessitate a revision of all Greater Spotted Eagle records from East Africa (EABR 1980).

The first confirmed record of this species was as recent as 1958, when Owre & Paulson (1968) collected an immature female at El Molo Bay 26B on 19 Oct. Since then, there have been records of single birds at Ngulia 101A on 3 Dec 1969, L Nakuru 62A on 20 Nov, 5 & 7 Dec 1970, and 16 Oct and 2 Dec 1971, L Naivasha 62C on 16 Feb 1977, Nairobi NP 75B on 11 Dec 1979, 12 Jan 1980 and 26 Feb 1983, Olorgesaille 75C on 17 Dec and Magadi 75C on 23 Nov 1975 (Ba; OS-C 1977; EABR). In addition, Bednall (1959a) made a possible identification in Nairobi NP 75B on 10 Nov 1958, and he mentions (*in litt.*) two more in the conspicuous immature plumage, undated, at Oleolondo 62A.

Despite the identification problems, however, this species can be no more than a scarce winter visitor on the S extremities of its range (*cf.* Imperial Eagle (131)). Concentration of records in and around the rift valley may reflect observer frequency and/or a migration route/wintering area. The 26B Oct bird was presumably on passage since it was at 380 m in an arid environment, but all other records except the two in 75C have been above 1000 m in 500-1000 mm rainfall areas.

129 TAWNY EAGLE *Aquila rapax*
B116 / MPG139 / D2C2

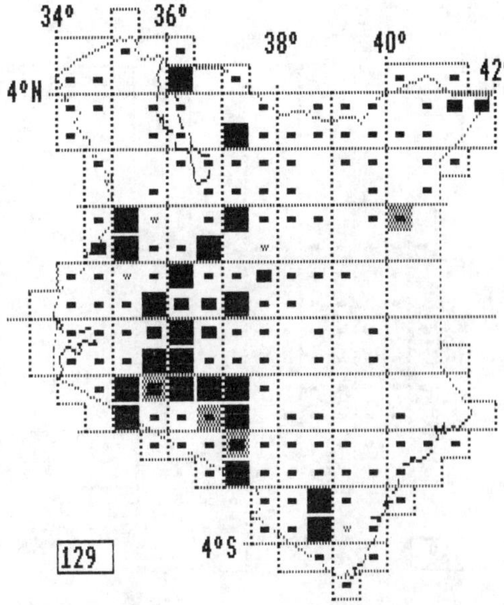

Afrotropical: widely distributed S of 17°N, apart from the arid SW and the equatorial lowland forest areas of the W.

Kenyan: a sometimes common bird, usually encountered singly or in pairs, in open habitats up to 3000 m (Br; Se; Ja; MPG; La; St). The identification problems of the genus have little effect on coverage in this species because it is reasonably well known, and the only large brown *Aquila* present in the N summer: hence D2C2. It is a fierce and strong predator, but its catholic feeding habits also include kleptoparasitism and scavenging, with limited commensalism at settlements in N Kenya (Sn; BUN).

Range is 70% arid-semiarid (*cf.* Br and BUN), but there are records throughout our rainfall and moisture categories. Breeding squares are 66% within intermediate, 250-1000 mm rainfall areas and the apparent paucity of nesting in the arid N and E may be real, although the effect of low observer frequencies in these areas cannot be excluded. Apparent absence of breeding from the high rainfall areas of the far W 48 60 and the S coast may be due to dense habitats and/or high human density (*cf.* Br), although lack of breeding records from other areas of the coastal lowlands is less easily explained.

The Tawny Eagle is thought to be resident (Br; Ja; Be; Se; La; St; DEP pers. obs.: 74BD 75 88ABD 89AC 100B 101A), but the possibility of local wandering is mentioned by several authors (*e.g.* Ja, Se). This, or movement of non-breeding birds into arid areas (see above), may account for this bird's ubiquity on our map, but no details of movements are known.

The Tawny Eagle has been considered conspecific with the Palearctic Steppe Eagle (130)(see Br). Their maps clearly contrast, with the Steppe Eagle absent from the coastal lowlands, and virtually so from the arid N and E. Other reference: Smeenk 1974.

130 STEPPE EAGLE *Aquila nipalensis*
B114 / MPG138 / D2C2

Afrotropical: a Palearctic species commonly wintering in E and S Africa, mainly from Kenya S and SW to South Africa and Namibia. Distribution in Kenya reflects the E limit of the range.

Kenyan: another species affected by the identification problems cited for the

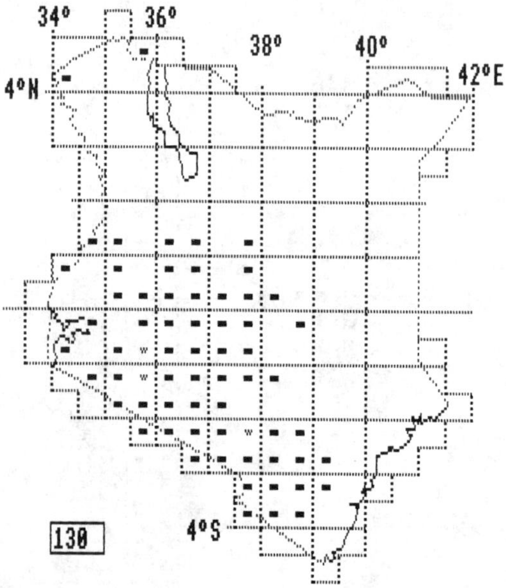

three other Palearctic members of the genus, but to a far lesser extent because most birds in East Africa are in the distinctive immature plumage (hence D2C2). Adults, which fluctuate in abundance in East Africa from year to year, mostly winter elsehwere (Br).

The Steppe Eagle is very numerous in E and NE Africa, outnumbering both the resident Tawny Eagle (129: *q.v.*) and the two Palearctic Spotted Eagles (127, 128)(BUN). It shows many similarities to the Lesser Spotted Eagle. Thus S passage Oct-Dec is most marked, in flocks of up to 250 and 350+ (Lewis & Pearson 1981b; EABR 1982). Because the Steppe Eagle is rare in Uganda, most birds bound for S Africa presumably overfly Kenya (Pearson & Meadows 1979), probably only descending in bad weather, as in the Lesser Spotted Eagle. It is a common winterer in C and S Kenya, W to Mara 74 and E to Tsavo 101B (Br), particularly on grasslands and open farmland, and usually in drier areas than wintering Lesser Spotted Eagles (Pearson & Meadows 1979). Flocks of up to 30 are not unusual, with 65 at Mtito Andei 89C in Feb (EABR 1979), hundreds in wet weather at Maasai Mara 74AC in Dec (EABR 1983), and an exceptional 500 over 1 km² near Voi 101B on 25 Feb (EABR 1978). As in the Lesser Spotted Eagle, N passage is obscure, presumably at high altitude (Reynolds 1974; BUN; *e.g.* Lewis & Pearson 1981a).

Overall, the Steppe Eagle is rare in the W (Britton & Britton 1976b) and, although common in Tsavo E 101B (La), is strikingly absent from the adjacent coastal lowlands. Range is absent from the driest, 0-250 mm rainfall areas, and 72% of the squares are above 1000 m. Other reference: Stoneham 1944c.

131 IMPERIAL EAGLE *Aquila heliaca*
B113 / MPG137 / D1C0

Afrotropical: this Palearctic eagle winters in Sudan and, less commonly, Ethiopia, with very few records S to Kenya and 3°S in NE Tanzania.

Kenyan: this species has undoubtedly been affected by the identification problems mentioned under Lesser Spotted Eagle (127). Although the first sighting came as late as 1962 (Brown 1963a), the Imperial Eagle is now recorded almost annually (Ba; Smeenk 1974; OS-C 1977; St; La; Lewis & Pearson 1981b; EABR) and we now regard it as another scarce winter visitor on the S extremities of its range (*cf.* Greater Spotted Eagle (128)). The 16 records are Nov-Mar, with eight in Nov. Apart from Aitong 74A (EABR 1983), they are in the rift valley and areas to the SE. The squares are 84% above 1000 m, and entirely within intermediate, 250-1000 mm rainfall areas.

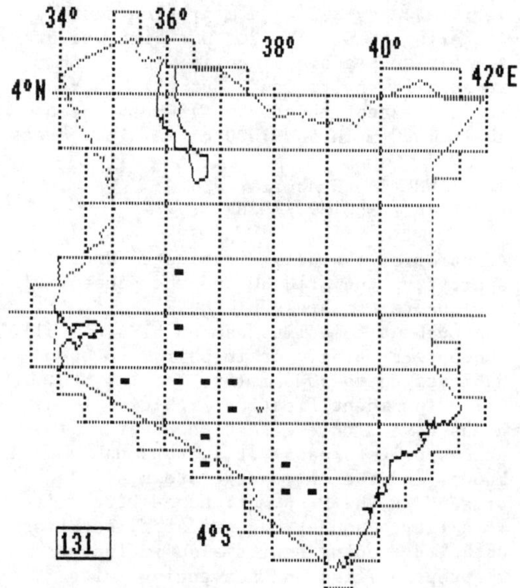

132 VERREAUX'S EAGLE *Aquila verreauxi*
B117 / MPG136 / D2C2

Afrotropical: mainly the E and S, from Ethiopia and NW Somalia S through East Africa and E Zaire to all but interior areas of the continent S of 10°S. There

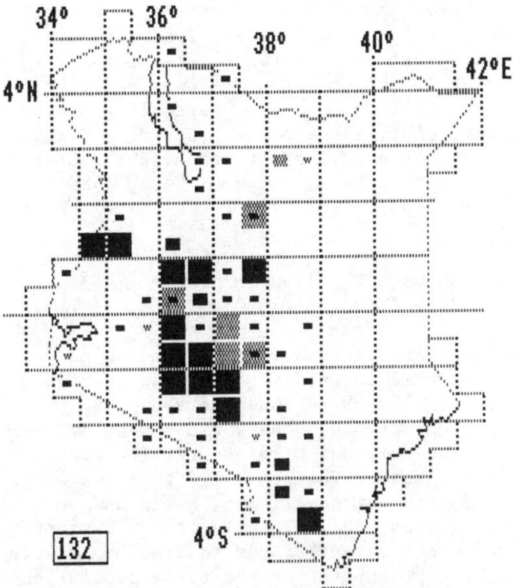

is also a population in W Sudan and Chad. Distribution in Kenya represents the E limits of the range.

Kenyan: Verreaux's Eagle is a local and uncommon bird usually seen in pairs, up to 3500 m (Br). Its basic ecological requirements are precipitous nesting cliffs and a plentiful supply of its main prey, the hyrax *Procavia* and *Heterohyrax*, so that it is found on rocky mountains, hills and gorges, and inselbergs in drier country (BUN; Ja; MPG; La). Cliffs in heavily wooded areas are avoided (BUN), and hence this bird is absent in the W 48 49 60 61 73. Many cliffs in arid areas presumably support insufficient prey. Occurrence is concentrated in and E of the rift valley highlands, and in scattered areas to the NW (Mt Elgon 36D, the Cherangani Mts 37C) and N (the Ndoto and adjacent mountains 39AB and inselbergs and volcanoes in the L Turkana basin).

65% of the breeding squares are concentrated in the rift and C highlands from 50-51 S to 75-76, with a periphery of presence records defining an apparent non-breeding area to the S and SE. This eagle is however thought to be mainly resident (Br; St; DEP pers. obs.: 75C), although immatures may wander (BUN; Sn), so that the absence of breeding records from this S and SE periphery more probably reflects the difficulties of detecting nests on extensive cliffs in remote areas. Similar factors undoubtedly apply to the more remote Turkana basin, particularly around Mts Kulal 26B and Nyiru 26D. A decline in the Eagle Hill area of 63D has been ascribed to human population pressure (Brown 1966). Other references: Firmin 1944; Pollard 1946c; Brown 1952-3, 1955a, 1974c; Frere 1982.

133 WAHLBERG'S EAGLE *Aquila wahlbergi*
B118 / MPG142 / D2C2

Afrotropical: widely distributed S of 15°N, apart from the lowland equatorial forests of the W and the dry areas of E Somalia and the SW.

Kenyan: mainly a trans-equatorial intra-African migrant that breeds in East Africa Aug-Apr, when it can be the most common eagle in suitable habitat, and then spends the May-Jul non-breeding season in the N tropics (Brown 1970a; Br; Brown & Britton 1980). Relatively small numbers are probably resident near the equator (Br; *e.g.* La), while others breed in the N tropics (Brown 1978). Concentrations, probably on N passage, occurred late Jan-early Mar 1971 in Tsavo E 101B (Br).

Wahlberg's Eagle inhabits bushed and wooded country, up to 3000 m but mainly below 1800 m (Br; Britton 1979). Its squares are only 6% within the driest, 0-250 mm rainfall areas. The few records from the arid N and E presumably refer to migrants. Nesting occurs throughout our altitude categories, and the 73% of breeding squares in areas with above 500 mm annual rainfall agrees with BUN's estimate of 600-1000 mm for breeding areas. Unlike the Tawny Eagle (129), Wahlberg's breeds in both the W 60 and the

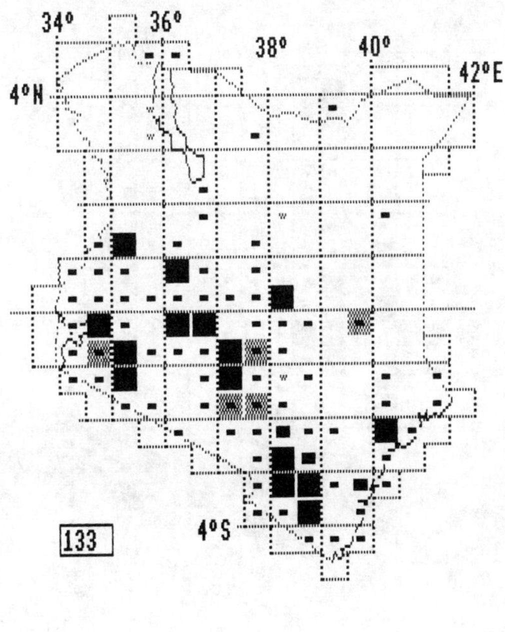

coastal lowlands, perhaps because it is a smaller bird of denser habitats. Other references: Brown 1952-3, 1955a; Davey 1978a.

134 FISH EAGLE *Haliaeetus vocifer*
B137 / MPG160 / D3C3

Afrotropical: widespread S of 16°N, apart from the dry areas of E Somalia and the SW.

Kenyan: this locally common, noisy and conspicuous species is one of the most familiar African birds of prey. It is found in pairs near waters of any type that contain fish, the main prey item, but it also takes birds and carrion and is a kleptoparasite (BUN; Sn). Distribution at the coast is mainly confined to calm creeks and estuaries while, at the other end of its altitudinal range, it extends up to artificially stocked dams and trout streams at 3000 m (Br; Se).

The Fish Eagle breeds by large eutrophic, oligotrophic and carbonate waters, especially lakes, but seldom at the coast (Brown & Britton 1980). Since both it and its nests are conspicuous, our coverage of breeding sites is probably good, so that the low proportion of breeding squares (24% of its total) denotes a high degree of post-breeding dispersal (Lewis & Pomeroy 1982b). This has previously been observed by Brown (1966) and Brown & Cade (1972) and, while some breeding pairs can be resident (*e.g.* at L Baringo 50A: St), others may leave their breeding areas altogether (*e.g.* in the Winam Gulf 60AB: Brown 1960). Individuals can then occupy even small seasonal waterholes temporarily (*e.g.* La). Immatures are forced to disperse, but most movements are probably under 150 km in extent (Brown 1960). Records at L Paradise, Mt Marsabit 27D, in particular, involve the crossing of large tracts of unsuitable terrain (BUN). Other references: Brown 1952-3, 1955a, 1980; Brown & Hopcraft 1973; Parsons 1977; Davey 1984.

135 BLACK KITE *Milvus migrans*
B138 / MPG132 / D3C3

Afrotropical: this species, including its Palearctic races, is widespread S of 18°N.

Kenyan: a gregarious and abundant species, probably the world's most successful and numerous bird of prey (BUN). As in the Hooded Vulture (92), this success is due to commensalism with man and an omnivorous diet. Thus, although migrants may occur in any of our squares on passage, a high proportion of the total population is found around settlements of any size, preferably but not necessarily near trees and water (BUN; Sn; Br). And kites are extremely versatile and agile scavengers of any refuse, carrion or offal, as well as being opportunistic predators of small creatures, and kleptoparasites.

Three races occur (Brown 1970a; Br). The nominate Palaearctic race is an abundant winter visitor and passage migrant Oct-Mar, mainly at medium altitudes but up to 3000 m. Britton & Britton (1976b) record flocks of up to 200 on S passage in the W 48 60. *M.m. parasiticus* is an intra-

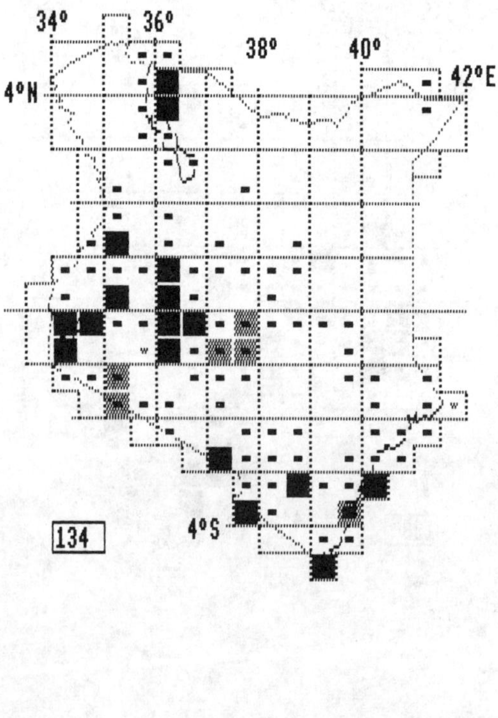

134

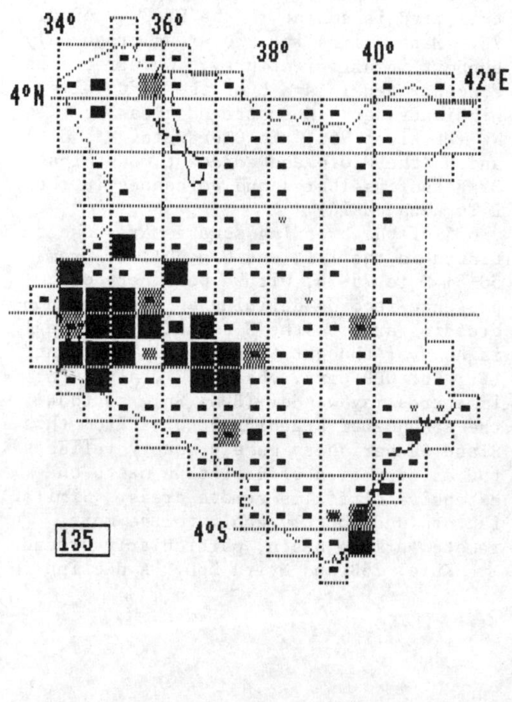

135

Plate 7. White-backed Vultures on a carcass: a typical plains scene (J.S. Karmali F.R.P.S.). *See page 89*.

Plate 8. Jackson's Francolin, a montane species virtually confined to kenya (D.K. Richards F.R.P.S.). *See page 128.*

African migrant that arrives in flocks of up to thousands Aug-Oct, breeds and then departs in Mar: it also occurs up to 3000 m, and some are resident. Finally, *M.m. aegyptius* is an uncommon, non-breeding Palearctic migrant Oct-Mar. Hence this species is far more abundant during the N winter, when three migrant races are present (North 1942a; *e.g.* Se; St; La).

The Black Kite is most common in moister areas. *M.m. parasiticus* nests in tall trees, and its breeding squares are 89% well timbered and 85% within 500+ mm rainfall areas: breeding is thus apparently concentrated in the high rainfall, densely populated highlands and coastal lowlands. Prior to commensalism, this kite was probably a bird of woodlands: its numbers have increased and no doubt continue to do so due to human population expansion (Brown 1970a; BUN). Other reference: van Someren 1956.

136 CUCKOO HAWK *Aviceda cuculoides*
B139 / MPG130 / D1C1

Afrotropical: in wooded and forested areas S of Senegambia, W Ethiopia and S Somalia, apart from the drier SW.

Kenyan: a little known, retiring and rather secretive inhabitant of dense woodland and forest edge habitats, up to 3000 m (Br; Britton 1979; Ja). There appear to be two discrete populations, in the highlands and at the coast, which both occur in areas with over 500 mm annual rainfall. Coastal squares are all below 500 m: this is the only area of Kenya where this species is at all regular or common, and all records are May-Nov, peaking Aug (Br; EABR). This seasonality also occurs in coastal Tanzania (Br) and extreme S Somalia (Ash 1983). These birds are presumably breeding intra-African migrants from further S.

The highland population's squares are 73% above 1500 m. Occurrence in all months and breeding have recently been reported from the Nairobi area 75B (EABR 1979, 1980), and this may be the case in other highland areas. A spate of sightings around Nairobi in late Jul-Aug 1985 (several obs.) were in the same period as peak migrant numbers at the coast (see above) and may represent a similar influx to the highlands.

Apart from a Mar date at Nairobi (EABR 1981), the few breeding indications from both of these populations, and from NE Tanzania and S Somalia, are all Aug-Nov (Brown & Bursell 1968; Ash 1983; EABR).

137 HONEY BUZZARD *Pernis apivorus*
B140 / MPG135 / D1C1

Afrotropical: a Palearctic species wintering S to South Africa, but apparently mainly in the lowland forests of the W. Distribution in Kenya represents the E limit of the range.

Kenyan: this rather secretive and obscure bird is doubtlessly under-recorded to some extent (hence D1C1) but, in view of the recently increasing expertise in the awareness of Palearctic raptors (*cf.* the *Aquila* eagles 127-131), it can be no

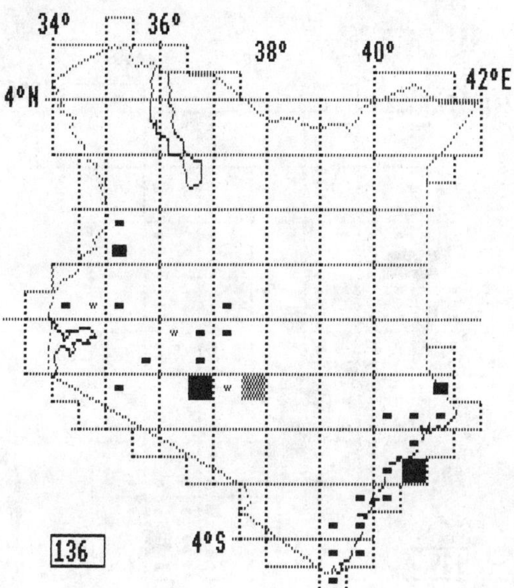

136

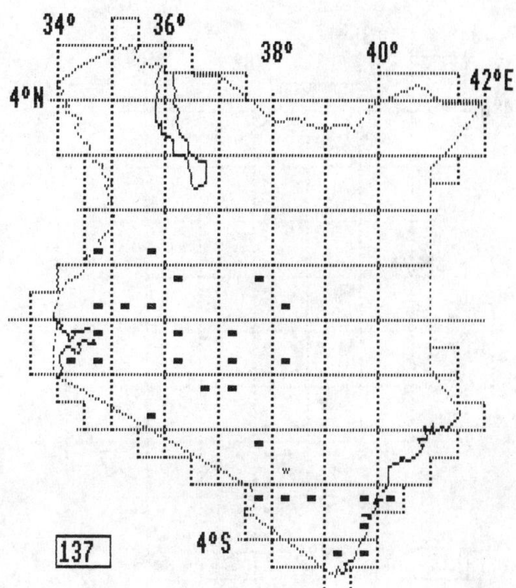

137

more than an uncommon passage migrant and winter visitor, mostly Oct-Apr (Br).

Range is 89% within 500+ mm rainfall areas. The Honey Buzzard winters in woodland and near forest edge, and most records are from the C highlands and the coast. Records of active passage appear to be confined to a group of four moving N up the rift valley, near the Ngong Hills 75B, on 19 Apr (EABR 1980), and a single bird on 17 Mar 1981 using the established migration route E of L Victoria with other Palearctic storks and raptors (see Common Buzzard (117) and Booted Eagle (124)) (Lewis & Pearson 1981a). Br notes seven oversummering records, mostly around Nairobi 75B. This relative abundance is all the more unusual considering the species' overall scarcity (Ba; Blom-Bjorner 1945d; Voous 1948; Cunningham-van Someren 1972d). Other references: Brown 1973d; Mann 1973b; EABR 1984.

138 BAT HAWK *Macheiramphus alcinus*
B143 / MPG134 / D1C1

Afrotropical: occurs in the Ethiopian highlands and, with the exception of the dry SW, throughout the rest of Africa S from Senegambia and S Somalia.

Kenyan: uncommon in wooded and forest edge habitats, up to 2000 m (Br). It is difficult to detect (D1) because it is a silent and crepuscular bird that is confined to a small hunting range and spends the day in thick cover (BUN; Sn; MPG). Its principal requirements are firstly a plentiful supply of small bats which, along with some opportunistically taken small birds, form the bulk of its diet. It also needs adequate open space for attacking bats at speed, and thickly canopied trees in which to roost when not engaged in its two or so hours of dusk hunting.

Range is 86% within 500+ mm rainfall areas, and 95% of the squares are well wooded (19A = woodland along the Daua R: EABR 1984). The Bat Hawk is often found in towns *e.g.* Nairobi 75B (EABR) and Malindi 103A, and is apparently most common at the coast (Brown & Britton 1980), and possibly in the W 48 60. The absence of confirmed breeding records results from its low detectability. Sn mentions possible post-breeding dispersal or migration, and AM describe this bird as a visitor to S Somalia. Nothing is known of any movements in Kenya. Br considers it resident, and St mentions sightings throughout the year at Baringo 50A. Other references: Percival 1914a; Thomsett 1981.

139 BLACK-SHOULDERED KITE *Elanus caeruleus*
B142 / MPG133 / D3C3

Afrotropical: widespread S of 16°N, apart from arid E Somalia.

Kenyan: usually encountered singly or in pairs in open country up to 3000 m, where it is often the commonest hawk (Br; Se; Ja). At Mau Narok 61D, it colonised areas as they were cleared for cultivation (Se). Breeding squares are 90% above 1000 m and 90% within 500+ mm rainfall areas, and the species may be a non-breeding visitor to the arid N and E, probably during the

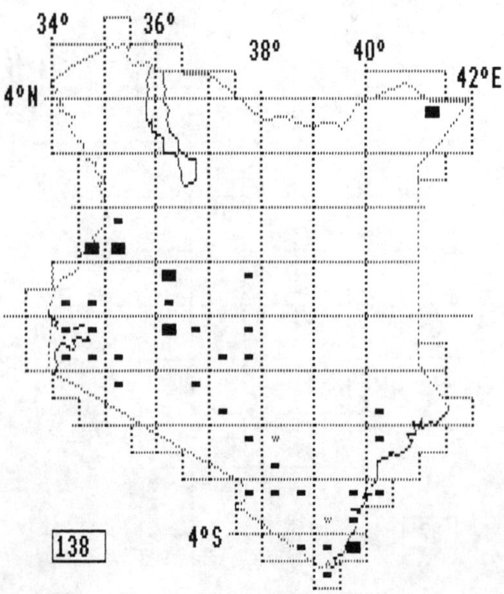

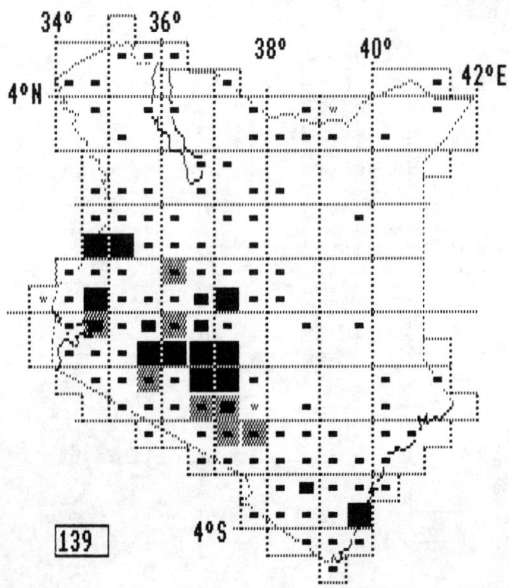

rains. BUN suggested that its migrations are linked to seasonal changes in semiarid grasslands, and most records from arid Tsavo E 101B are Nov-Mar (La).

Be found this bird resident in the Narok area 74, and Br suggested this type of occurrence to be general in higher rainfall areas. DEP (pers. obs.) confirmed these observations in semiarid-subhumid 75 88AB, but also noted an annual Jul-Oct influx to the Ruaraka-Kahawa-Ruiru area of moist-subhumid 75B. Occasional visitors to the semiarid Baringo area 50A stay for months at a time (St). Other references: van Someren 1956; Forbes-Watson 1977d.

140 SWALLOW-TAILED KITE *Chelictinia riocourii*
B141 / MPG131 / D2C2

Afrotropical: the N tropics, mainly between 8°N and 16°N, from the W coast to Ethiopia and N Somalia, and thence S to N Uganda, Kenya and S Somalia.

Kenyan: an elegant and often gregarious bird of open country, mainly at lower altitudes (Br; Ja; MPG). Range is 76% below 1500 m and 72% within intermediate, 250-1000 mm rainfall areas.

The Swallow-tailed Kite is a pronounced intra-African migrant, breeding during the N tropics' rains and then moving S to avoid the N dry season (Sn; Br). Distribution in Kenya appears to show three distinct features. Firstly, N of 2°N, the species shows a Mar-Oct seasonality, with Mar-June breeding which is simultaneous with that further N (Brown 1970a; HH; EABR 1978; Davey & Davey 1980; Brown & Britton 1980; C Dewhurst *in litt.*). These areas are the SE limits of the main, N tropics breeding range, and although they are shown as arid on our overlay, breeding occurs during their brief wet season.

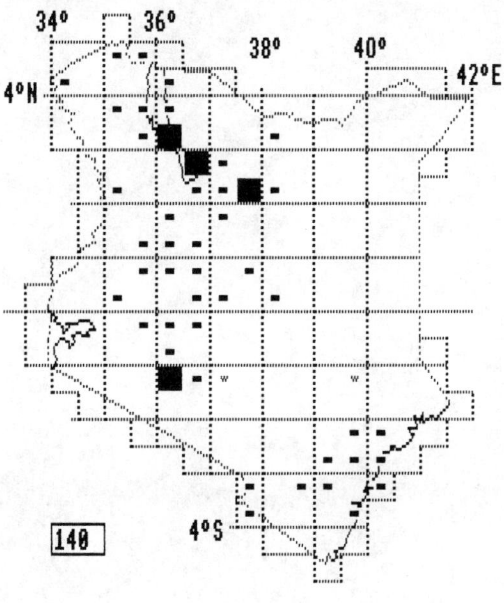

Secondly, there is in the Mt Suswa 75A area a small resident breeding population (largest observed flock 24 birds: EABR 1980)(Blankenship *et al.* 1971), with the single known laying date in Aug (Sutton *et al.* 1984; Dewhurst & Fishpool 1984). Small annual numbers at L Baringo 50A Aug-Jan (St) suggest limited post-breeding dispersal of the Turkana population to the S, whilst the records and rather confused seasonality S to the Suswa 75A area, particularly in the rift valley 62AC, support Br's view that the Suswa population is not disjunct.

The third area of occurrence is in the lowlands of SE Kenya, where records are mostly Nov-Mar; *i.e.* in the non-breeding season. All records in Somalia S of 4°N are also along the coastal strip and mostly Nov-Mar (J S Ash *in litt.*). These records represent birds that have migrated S from their breeding range. The extent of this movement is apparently dictated by prevailing weather conditions to the E and NE, since unusual numbers occurred at the coast during the severely dry period of early 1984 (EABR 1984), when other dry country species from the NE, *e.g.* the Lilac-throated Roller (455), also penetrated far SW of their usual range. Following these unusual numbers of Swallow-tailed Kites in the SE, an unprecedented roost of about 150 were photographed at Solio 62B on 31 Mar (D Fawcett *in litt.*; EABR 1984), presumably returning N to breed. The origin of these birds in SE Kenya and S Somalia is uncertain. The lack of records of such a distinctive species from the reasonably well-observed country between 87-88 and 64-65 suggests that they are not directly SSE-moving elements of the Turkana or Suswa breeding populations. It is possible that the Turkana population migrates SE into the poorly observed arid area around 41 and thence S to the coast, but the total lack of records, particularly from observers resident in 19A 42A 65B again makes this doubtful. Thus it seems equally possible that birds in the SE lowlands breed in Ethiopia and then move SE to coastal Somalia and thence SW down the coastal strip to Kenya (see Fox Kestrel (157)). Other references: Norris 1971a; Clifton 1976; Binks 1984.

Family Pandionidae: Osprey

A monotypic family: ospreys occur in all of the world's zoogeographical regions.

141 OSPREY *Pandion haliaetus*
B144 / MPG186 / D3C2

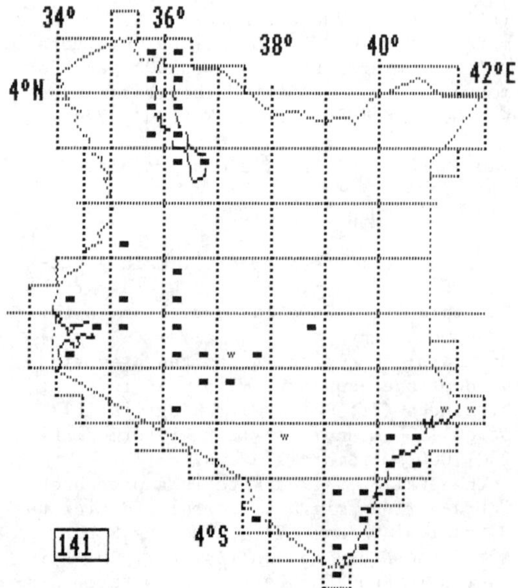

Afrotropical: widely distributed S of 17°N, apart from the arid SW. There are a few unconfirmed breeding records S to the Cape, but the Osprey is mainly a non-breeding migrant from the Palearctic. Its only regular breeding area in the Afrotropical Region is along the Red Sea coasts.

Kenyan: a reasonably common Palearctic passage migrant and winter visitor Sept-Apr that is found by fish-bearing and mainly large waters, usually singly (Br; Ja). These waters are mainly eutrophic and carbonate lakes, while it occurs in sheltered creeks and estuaries along the coast (Ja; Br), and shows little correlation with permanent rivers.

Brown & Britton (1980) convincingly reject Loveridge's (1922) record of breeding at L Naivasha 62C, and can find no substantial evidence of more recent breeding records at L Turkana (HH) or Naivasha. Single birds are recorded oversummering May-Aug almost annually (EABR), and several reported breeding records at Naivasha have involved birds using old Fish Eagle (134) nests as feeding platforms (Brown 1970a).

Light passage has been observed only in Aug at L Turkana (BUN) and in Oct-Dec at L Baringo 50A (St), so that N passage may be faster and at a high level and thus obscure, as discussed for the Lesser Spotted (127) and Steppe (130) Eagles. The above mentioned two passage records, together with the Osprey's rarity in Kenya away from the rift lakes and the coast, and its virtual restriction to the coast in Somalia (AM), imply that most migration occurs along these two routes.

Family Falconidae: Falcons, Hobbies, Kestrels

61 species, represented in all of the world's zoogeographical regions. 19 (31%) occur in the Afrotropical Region and all have been recorded in Kenya, where 8 (42%) are non-breeding migrants from the Palearctic, 2 breed but are augmented by Palearctic races during the N winter, 8 (42%) breed or are presumed to do so, and Dickinson's Kestrel (154) is a wanderer N of its usual range. The Afrotropical species are never common, and are often greatly outnumbered by their Palearctic counterparts, particularly the Eastern Red-footed Falcon (150), Hobby (152), Kestrel (159) and Lesser Kestrel (160). These migrant species, together with the similarly Palearctic Red-footed (149), Eleonora's (155) and Sooty (156) Falcons, mainly winter S of Kenya in areas where the rains produce abundant prey. Much of their migration may be obscure because it is performed at great height, beyond human vision, with the birds only approaching the ground if forced down by bad weather, or perhaps if coming to feed on the dense termite (Isoptera) swarms that the rains bring forth, or if descending to engage in some brief crepuscular hunting prior to overnight roosting (Brown 1970a). Ja comments on the apparently crepuscular habits of the Sooty and Eastern Red-footed Falcons, and the rarer falcons in general are often seen only in difficult light conditions, at dusk or dawn and in bad weather. Other reference: Leakey 1943a.

142 PYGMY FALCON *Polihierax semitorquatus*
B163 / MPG129 / D3C3

Afrotropical: occupies two distinct areas: in the NE, it is known from interior Tanzania N through Kenya and NE Uganda to S Sudan, Ethiopia and Somalia, while in the S it occurs from N South Africa to Namibia and Botswana.

Kenyan: a comparatively common bird, found singly or in pairs in dry, open bushed and wooded habitats below 1600 m (Br; Ja; HH; La; St). Range is 86% arid-semiarid and only 5% within 1000+ mm rainfall areas. Hence it is very scarce in the moist highlands 61 62 63 and L Victoria basin. It is also virtually absent from the coastal lowlands, including even semi-arid squares. The Pygmy Falcon roosts and breeds in unoccupied nests of the White-headed Buffalo Weaver (1004)(Br; BUN; Muringo 1980), and is absent from the only parts of the country that this similarly dry country weaver avoids 48 49 60 61 102CD 114.

This falcon is thought to be mainly sedentary (La; DEP pers. obs.: 75D 88AD 89AC 100B 101A), although substantial wandering occurs (Br). Examples of the latter are very rare birds in humid 61A, and a pair in moist-subhumid Nairobi 75B during Jan 1954 (D K Bednall *in litt.*).

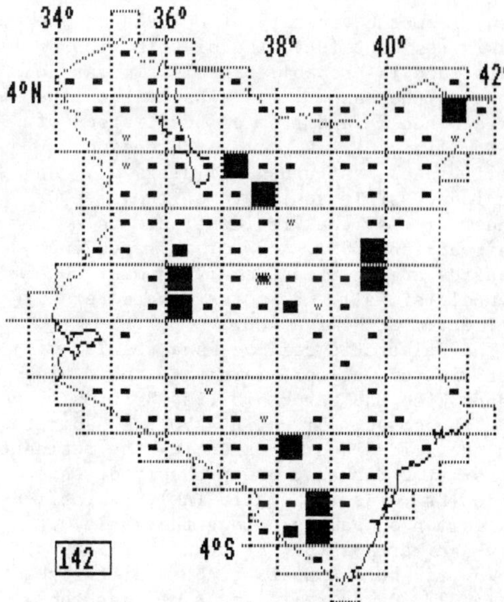

143 PEREGRINE FALCON *Falco peregrinus*
B158 / MPG112 / D2C2

Afrotropical: widely distributed S of 15°N, either as a resident or a non-breeding migrant from the Palearctic. It is generally absent from the lowland equatorial forests of the W.

Kenyan: the features of the Peregrine's map are blurred by the presence of the resident race *F.p. minor* and the Palearctic migrant *F.p. calidus*: the former is treated first. Resident Peregrines are uncommon, occurring around cliffs in open country unless excluded by the larger and more adaptable Lanner (146: *q.v.*)(Brown 1970a; Br; BUN; Britton *et al.* 1984).

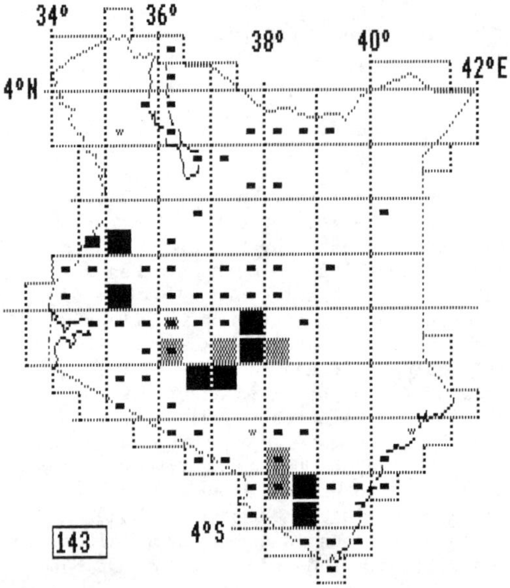

143

Away from the vicinity of L Turkana, where it is resident (HH), the distribution of this race is probably broadly similar to that of the Lanner, *i.e.* absent from the arid N and E (and from adjacent areas of W Somalia: AM). *F.p. minor* breeds on cliff faces, and urban buildings *e.g.* on an inaccessible sculpture on the side of the Office of the President, central Nairobi 75B (BUN). 79% of its breeding squares are in areas receiving over 500 mm annual rainfall. Br notes a wanderer collected at Ng'iya 48C.

The Palearctic *calidus* is a regular winter visitor Oct-Apr, mainly to the C highlands W to Mau Narok 61D (Br; Britton 1979) and the coast. Birds in atypical habitats in the arid N and E, excepting the resident *minor* at L Turkana, are probably of this race (Br). As in Somalia (AM), *F.p.calidus* is common on passage along the coast Oct-Nov and Mar-Apr (Br; Britton *et al.* 1984). However, there appears to be a difference inland in Kenya reminiscent of the Hobby (152), where sites around 36°E experience little N movement (L Baringo 50A: St; Mau Narok 61D: Se), while the more easterly Tsavo E 101B experiences the opposite (La). EABR records also appear to show heavier N passage at the coast.

The Peregrine Falcon may be conspecific with the Barbary Falcon (144), and forms part of a species-group with the Taita Falcon (145)(Sn).

144 BARBARY FALCON *Falco pelegrinoides*
B157 / MPG112 / DOCO
NO MAP: pre presence 26B

Afrotropical: this mainly Palearctic falcon has occurred in the N tropics S to about 10°N, with only a single record further S in N Kenya.

Kenyan: the only record of this obscure and extremely easily overlooked species is of one collected from two near Loyengalani 26B on 4 Nov 1958, during strong S raptor passage (Owre & Paulson 1968). See Peregrine Falcon (143).

145 TAITA FALCON *Falco fasciinucha*
B155 / MPG114 / D1C1

Afrotropical: a rare species that occupies two distinct areas: from SW Ethiopia S through Kenya to NE Tanzania; and in Malawi, Zambia, Zimbabwe and N Mozambique. Distribution in Kenya represents its range at these latitudes.

Kenyan: like the Ovampo Sparrowhawk (105) and Ayres' Hawk Eagle (126), this predator is rare throughout its range (Brown 1970a; Sn). Distribution is apparently governed by the occurrence of cliffs (on which it nests) in semiarid country at moderate to high altitude (Lack 1976c). Williams (1968, and *in* Dowsett 1977) described it as resident in the remote Lokitaung Gorge 3D, but most records are from the regularly observed country around the Taita Hills 101A, the type locality. However, it is only very infrequently seen even there (Ja; Lack 1976c; Pelchen 1977ab; La; EABR).

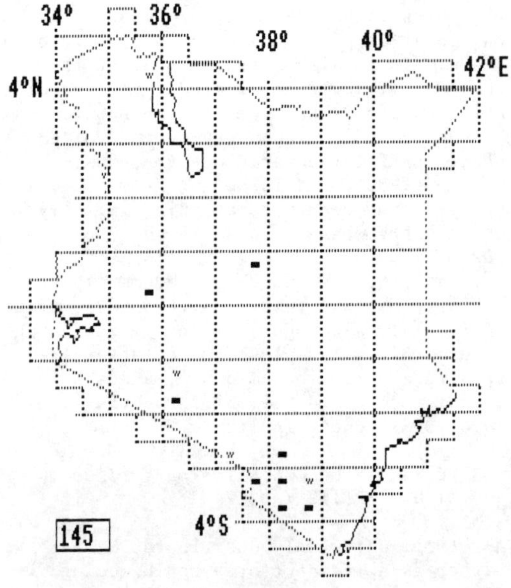

145

There is no breeding record, but it may be resident at Lokitaung 3D (Williams 1968), and records around Taita 101A show no seasonality. Br suggested that Williams' (1968) records in Amboseli 88C and Mann's (1973b) near L Magadi 75C represented wandering in search of concentrations of small birds near water, as in the Lanner (146) in Tsavo E 101B, but an old sight record as far distant as Malindi 103A (MPG) is not sufficiently substantiated (Dowsett 1977). Including possible wanderers, range is 75% at 500-1500 m, and 91% arid-semiarid. See Peregrine Falcon (143).

146 LANNER FALCON *Falco biarmicus*
B148 / MPG113 / D2C2

Afrotropical: widely distributed S of 17°N, apart from the equatorial lowland forest areas of the W. Distribution in Kenya represents the E limit of the range.

Kenyan: this fierce predator is a rather uncommon bird of open habitats up to 3200 m, although also found on cliffs in forest on Mt Elgon 36D (Br; BUN; Ja). It is generally more common than the Peregrine (143) and, while these two large falcons usually occur in different habitats, this species may exclude the Peregrine, perhaps explaining why in some areas only one occurs (Brown 1970a; Br; BUN). Range is 81% above 500 m and 71% within intermediate, 250-1000 mm rainfall areas. Hence it is scarce in the high rainfall areas of the highlands and the far W 48 60 73, as well as in the arid N and E (apart from L Turkana: HH). Away from Turkana, the Lanner is found in only 10 arid squares, most of which are adjacent to semiarid regions. Virtual absence from even moister areas of the coastal lowlands reflects general scarcity in Kenya E of 39°E, on the E limits of the range.

While breeding pairs may be resident (*e.g.* at Baringo 50A: St), movements in response to rainfall, probably mostly by immatures or non-paired adults, are known (BUN). Influxes into arid Tsavo E 101B during the Nov-Mar wet season feed on termite swarms and large Red-billed Quelea (986) colonies (La).

The Lanner forms part of a superspecies with the Saker Falcon (147)(Sn).

147 SAKER FALCON *Falco cherrug*
B149 / MPG121 / D1C0

Afrotropical: a Palearctic species that winters in very small numbers in Sudan and Ethiopia, with a few records S to Kenya, and one in Tanzania (Br).

Kenyan: the nine records are all within or close to the rift valley (Williams 1948a; Ba; EABR). The birds' predominant Nov and Mar-May seasonality is apparently real (Ba) and suggests passage migrants. While undoubtedly overlooked both in Kenya and probably also further S (hence D1C0), the Saker is unquestionably a scarce visitor to Kenya. See Lanner Falcon (146).

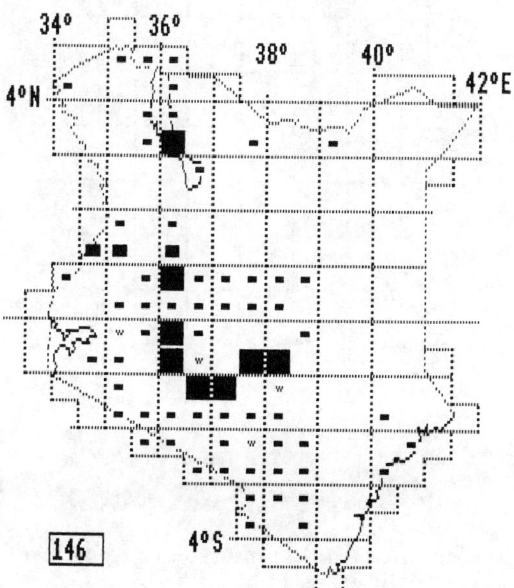

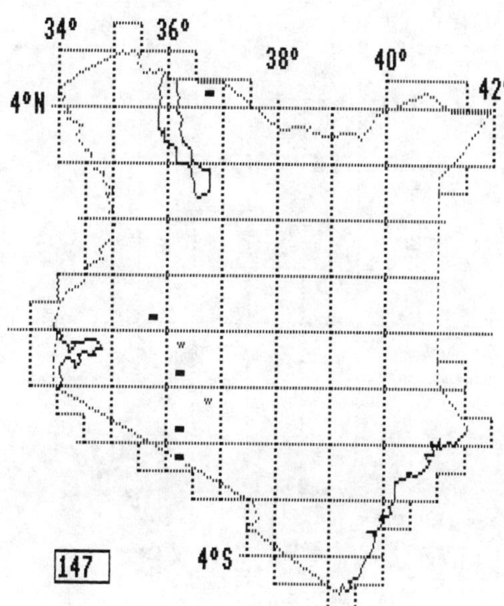

148 RED-NECKED FALCON *Falco chicquera*
B150 / MPG122 / D2C2

Afrotropical: widely distributed S of 15°N, with the exceptions of most of the equatorial lowland forest areas of the W. A vagrant in S Somalia (AM) represents the only record to the E of Kenya.

Kenyan: an uncommon falcon encountered singly or in pairs in low, open country with palm trees (*Borassus* or *Hyphaene*), in which it nests (Br; Mann 1973b). Distribution is 80% below 1000 m, in two very different areas.

It is widespread in the coastal lowlands (all above 500 mm rainfall), extending inland along the Tana R 79C 78B 64B to Meru 52C and along the Galana R 103A 102B to Tsavo E 101B (where there is a single June record: La)(Br). The three breeding squares range 250-1000 mm in rainfall, and at least two of these records were in *Borassus* palms (Brown & Britton 1980). Apart from an old record in 88D, the only other inland sighting S of the equator is of a single bird, presumably a wanderer, at Amboseli 88C in Nov (EABR 1977).

HH describe this falcon as frequent and nesting in tall *Acacia* trees in the arid L Turkana area, which is the other centre of population, with records of presumed wanderers S to Baringo 50A in Apr and Oct (EABR 1981) and E to a possible family party in the Dida Galgalla Desert 28A in Mar (EABR 1980).

BUN describe this species as exclusively sedentary (and Br agrees) but map a vagrant on the Angola coast, while Ash (1983) describes an immature vagrant at 2°N in coastal Somalia in Jan-Feb. All presumed wanderers except the one in Tsavo were Oct-Apr, while the two coastal lowlands breeding records are Jul and Aug (Brown & Britton 1980).

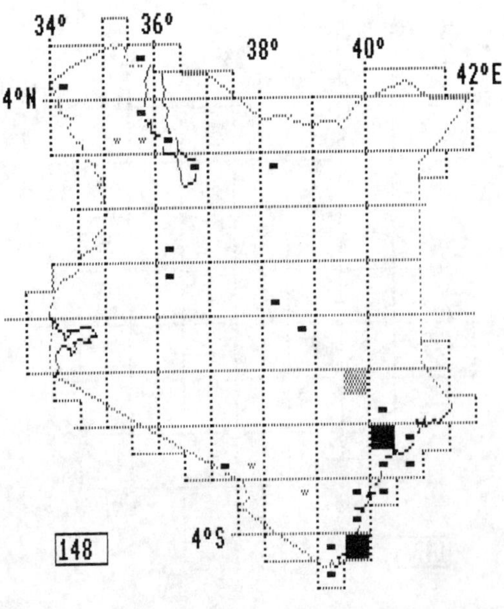

149 RED-FOOTED FALCON *Falco vespertinus*
B162 / MPG119 / D2C0

Afrotropical: migrates from the W and C Palearctic to winter in SW Africa, mainly from Angola and Zambia to Namibia and Botswana.

Kenyan: in contrast to the Eastern Red-footed Falcon (150), which may be conspecific, this species' migration occurs predominantly W of E Africa, with only very rare stragglers E of L Victoria (Ba; Br; BUN). The five records, all except the latest of single birds, were at Elmenteita 62A on 12 Oct 1968 (Ba), Ng'iya 48C on 15 Oct 1972 (Britton & Britton 1973a), Kamboyo 89C on 26 Apr 1980 (Pearson 1981b), Nairobi 75B on 3 May 1980 (Lewis 1981a), and 25+ moving N between Maktau and Taveta 101A on 20 Mar 1981 (Taylor 1981a). The last three of these records were in the company of other migrating Palearctic falcons.

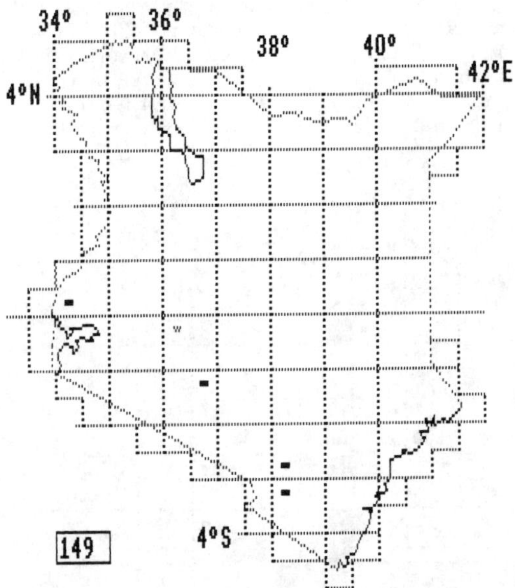

150 EASTERN RED-FOOTED FALCON *Falco amurensis*
B146 / MPG120 / D2C2

Afrotropical: a migrant from the E Palearctic that winters in SE Africa, mainly

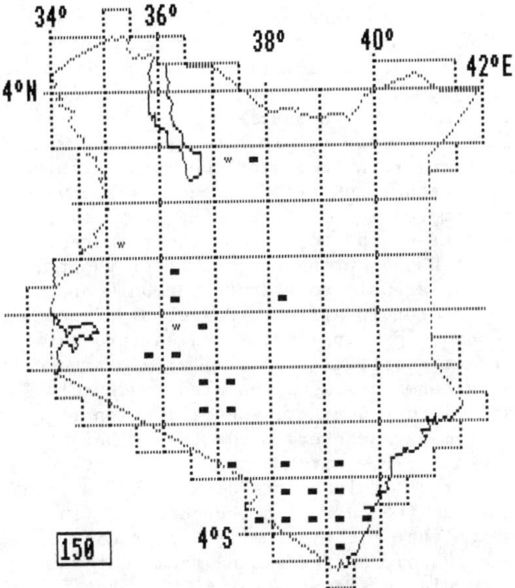

from Zambia and Mozambique to E South Africa.

Kenyan: BUN describe this species' migration as "certainly the most remarkable of any raptor known": it breeds in E Siberia, NE of Mongolia, and by a largely unknown route migrates SW across the Himalayas, India and the Indian Ocean to make a landfall in Kenya and possibly also further S, to winter in the areas described above (Ba; BUN). Thus this gregarious species of open country is predominantly a passage migrant in Kenya Nov-early Dec and Apr (Ba; EABR; Se; Sessions 1975; Britton 1979; St; La). There are only three late Dec-Jan records, all of single birds: at Lukenya 76A (EABR 1978), Konza 76C (EABR 1981), and in Nairobi NP 75B (20 Jan 1985: D J Pearson pers. comm.).

S migration enters extreme S Kenya from the E or NE, and there appears to be a strong passage across at least the lowlands of the far S each year: flocks of 500+ at Salt Lick Lodge 101C on 30 Nov, scores in the Maktau 101A-L Jipe 100D area on 1 Dec (EABR 1982), and flocks of up to 80 in the Ngulia area 89C 101A (EABR 1984), may well be usual (see also Backhurst 1985). Migrants apparently fly high (Brown 1970a), with flocks only nearing the ground when forced down by large storms (*cf.* Lesser Spotted Eagle (127)). Thus scores appeared in torrential rain near Ngulia Lodge 101A on 26 Nov (EABR 1982), just prior to the other large flocks in that area noted above. Stragglers have been recorded NW to 37C during S passage (Ba).

N passage is again largely obscure and probably also at height. Most may leave Africa S of Kenya, but the movement that is seen in Kenya is further W than the S migration, NW to Koroli 27A (BUN; Ba). More westerly N passage has also recently been discovered near the Somali coast at 2-3°N, where S passage is unrecorded (AM), and also occurs in the closely related Red-footed Falcon (149)(BUN). Palearctic falcons often migrate in mixed flocks, and some of this species roosted with thousands of Lesser Kestrels (160) during N migration at Nairobi 75B (Smalley 1983) and near Kajiado 75D (Taylor 1984a) in 1983.

The rather scattered movements of this species make overlay interpretations less useful, but 76% of its squares are in the 250-1000 mm rainfall regime.

151 AFRICAN HOBBY *Falco cuvieri*
B152 / MPG116 / D2C2

Afrotropical: the N tropics, mainly between 4°N and 14°N, from the W coast to Ethiopia, and thence S through East Africa and the E Zaire borders to S Zaire, Zimbabwe and E South Africa. Distribution in Kenya represents the E limit of the range.

Kenyan: uncommon, usually singly or in pairs, in woodland and forest edge up to 3000 m (Br; Se; Ja; MPG). Range is 85% within 500+ mm rainfall areas, and this falcon is most common in the highlands and the L Victoria basin (Br), including deforested, densely populated areas of the

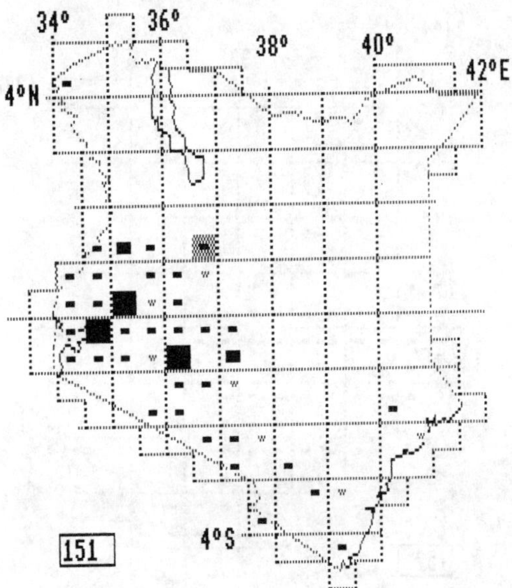

W (BUN). BUN's estimate of 1200 mm annual rainfall appears too high. Distribution is 81% above 1000 m, and virtually absent from the apparently suitable coastal lowlands (Britton *et al.* 1980), which are at the E limit of the range. Breeding takes place in old nests of corvids or raptors, especially those of the Pied Crow (577) and the Black Kite (135) in the W (BUN). All breeding squares are above 1000 m, in 500+ mm rainfall areas.

While this falcon is probably resident in higher rainfall areas *e.g.* Kisumu 60B (EABR 1981), Ja noted influxes at Nandi 49C in May and Eldama Ravine 49D in Jul and Sept, and the few records in the marginal habitat around L Baringo 50A mainly exhibit similar seasonality in Apr, May and Jul (St; EABR 1984).

The African Hobby forms part of a species-group with the Hobby (152)(Sn). Other reference: Forbes-Watson 1963.

152 HOBBY *Falco subbuteo*
B160 / MPG115 / D2C2

Afrotropical: this Palearctic bird winters in scattered areas of East Africa but the main wintering range is further S, from S Zaire and Zambia to Namibia and E South Africa.

Kenyan: a widespread and relatively numerous passage migrant Oct-Dec and Mar-Apr, when it sometimes outnumbers the resident African Hobby (151: *q.v.*) by hundreds to one (Brown 1963); it also overwinters, possibly irregularly, in small numbers in parts of the highlands (Br).

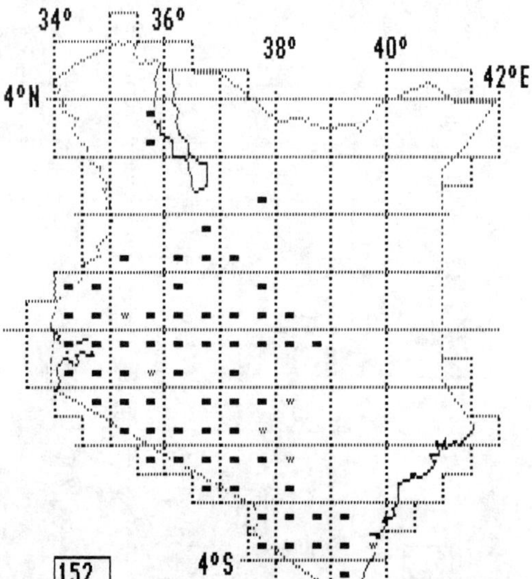

The points made under Eleonora's Falcon (155) regarding rainfall, termite hatches and high level migration apply to the Hobby too, and are better demonstrated in this more common species (Brown 1970a; BUN). Passage is clearly associated with S-moving rain fronts of the inter-tropical convergence zone (ITCZ), and birds give the impression of descending from migration at height to feed on the termites which hatch during or just after the rain and apparently constitute the bulk of their diet in Africa. Bowles (1967) recorded migrating *Aquila* eagles (probably Lesser Spotted (127)) similarly attracted to a termite swarm. Hobbies are rarely seen in dry weather, when migration is presumably uninterrupted at some height: Brown (1970a) watched this and other Palearctic falcons at Naivasha 62C, rising towards the base of a thundercloud (in which they might gain lift to greater altitudes), and estimated them lost to binocular vision at over 6000 m above ground level. N passage is thought to be similarly associated with rain around the N-moving ITCZ.

Br records flocks of up to hundreds at termite swarms and, like Eleonora's Falcon, the areas of Kenya where the Hobby occurs near the ground are at high altitude (73% of its squares over 1000 m) and in at least moderate rainfall regimes (73% over 500 mm). However, it is notably rare at the coast, which also receives higher rainfall.

Ja thought that its N and S migrations occur by different routes, and BUN describe the N passage as more widely dispersed, and in smaller flocks. Thus areas of Kenya around 36°E receive both N and S passage (Baringo 50A: St; Mau Narok 61D: Se), while in Tsavo E 101B at 38°30'E there are only a few S passage records but N passage is regular (La). This distinction continues to the NE of Tsavo, in Somalia, where N passage is regular and heavy but there are only four records of S movements (AM). A similar restriction to N passage in Somalia is also seen in Sooty (156) and Eleonora's Falcons.

153 GREY KESTREL *Falco ardosiaceus*
B147 / MPG127 / D2C2

Afrotropical: from the W coast between Senegambia and Angola, E to Ethiopia, W Kenya, N and W Tanzania and N Zambia. Distribution in Kenya represents the E limit of the range.

Kenyan: solitary and rather uncommon in open wooded areas below 1800 m (Br; MPG; BUN). This is a bird of the W, with rare

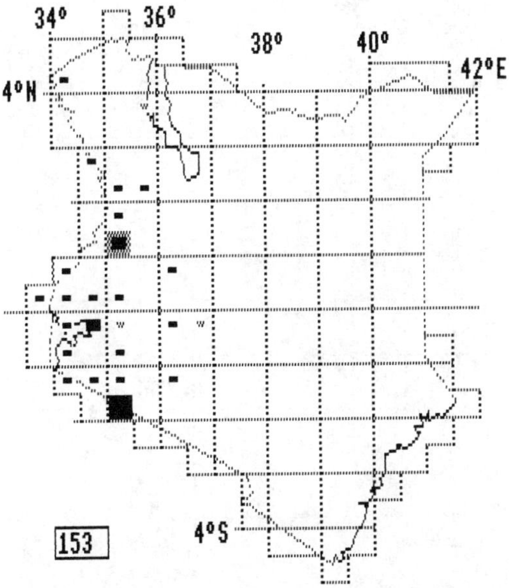

wanderers E to the rift valley 13B 50A 62A 75A and a single record on the E plateau near Nyeri 62B (Dowsett 1977; Br; St; EABR). Range is 88% above 1000 m and 96% within 500+ mm rainfall areas. Breeding takes place, possibly exclusively, in old Hamerkop (44) nests (BUN), and all three breeding squares are within that species' breeding range, in 500+ mm rainfall country. The Grey Kestrel is apparently sedentary (Br; BUN), and is in the vicinity of old Hamerkop nests in Maasai Mara 74AC throughout the year (P R A Davey *in litt.*). Wanderers E to the rift show no seasonality (Feb, June, Nov: *op. cit.*).

The Grey Kestrel is a member of the same species-group as the virtually allopatric Dickinson's Kestrel (154: *q.v.*)(Sn; Br), which is a rare wanderer to S Kenya. The two birds have yet to be recorded in the same QSD.

154 DICKINSON'S KESTREL *Falco dickinsoni*
B153 / MPG128 / D3C0
NO MAP: post presence 75B 88C 101C

Afrotropical: a bird of S Africa, from Angola and NE South Africa N to S Zaire, S and coastal Tanzania and, as a wanderer, S Kenya. It effectively replaces the Grey Kestrel (153: *q.v.*) to the SE of the latter's range (Sn).

Kenyan: found in open wooded country, usually near water and palm trees (Br; BUN). Although widespread on the Tanzanian coast N to 7°S, it is inexplicably absent from the apparently suitable Kenyan coast (*op. cit.*). The three records, all of single birds, were at Amboseli 88C on 30 Jul and 2 Aug 1975 (Horwitz & Anderson 1976), on 23 June and 14 Aug 1978 at Salt Lick Lodge 101C (Zimmerman 1978a; EABR 1978), and on 17-18 Aug 1983 in Nairobi NP (EABR 1983). These were evidently wanderers not far N of their normal range, and the 101C records suggest a temporary residence of nearly two months. Late June-late Aug coincides almost exactly with the species' Jul-Aug laying dates on Pemba (Brown & Britton 1980).

155 ELEONORA'S FALCON *Falco eleonorae*
B154 / MPG117 / D2C1

Afrotropical: this Palearctic falcon migrates from its Mediterranean breeding grounds to winter in S Tanzania, Madagascar and Mauritius. A few occur annually in Kenya.

Kenyan: in recent years, Eleonora's Falcon has been shown to be an annual passage migrant in very small numbers, usually late Oct-Nov and more occasionally Apr-early May (Br; EABR). A bird at Kapedo 38C on 27 Jan (EABR 1984) may have been wintering. While the possibility exists that our map reflects the distribution of experienced observers, this species seems to occur mainly in C and S Kenya, 81% above 1000 m and 81% within 500+ mm rainfall areas.

As with other Palearctic falcons, migration may normally be obscure at considerable heights. A bird that struck Ngulia Lodge 101A in Nov (EABR 1978) supports the

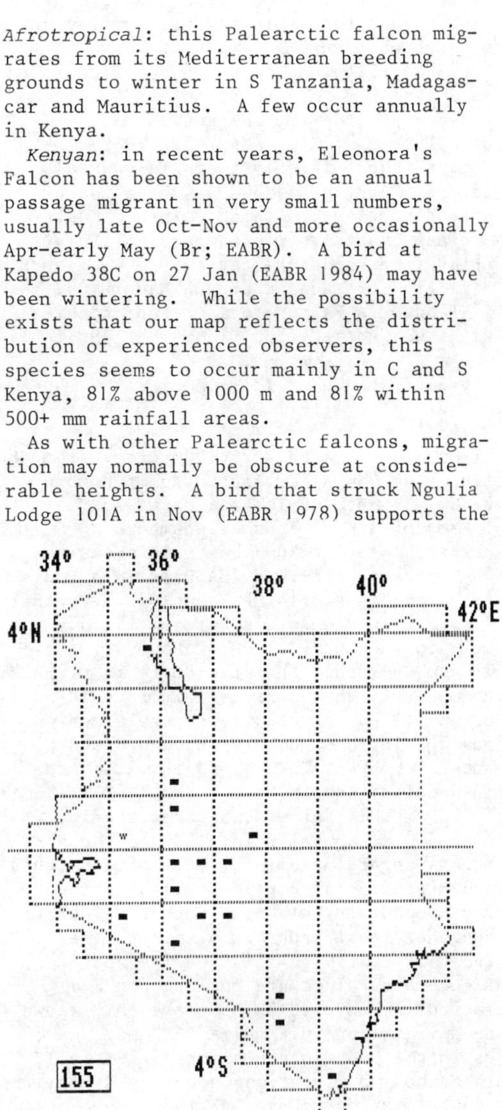

concept of descent in bad weather and raises the possibility that migration is usually at night. It is also possible that this species' increased frequency in higher rainfall areas reflects migrants either forced down by storms or frequenting rain fronts to feed at termite hatches (*e.g.* MPG), as occurs around Nairobi 75B. Numbers wintering in S Tanzania are proportional to rainfall and the availability of winged termites (Br). Based on localities and seasonality, Br argues that birds seen in Kenya are bound for wintering areas in S Tanzania, as opposed to Madagascar or Mauritius. Given that the Red Sea is this falcon's known regular migration route (Ba; BUN), migration S down the Ethiopian rift valley and on further S through areas of Kenya to the E of the rift would provide a direct route to S Tanzania. There is only a single S passage record from the Kenyan rift itself, at L Turkana 13B in Nov (EABR 1983). Because there is only a single coastal record (EABR 1979), BUN's suggestion of migration along the high rainfall East African coastal strip appears unlikely, unless the birds are far out to sea, bound directly for Madagascar and Mauritius.

Eleonora's Falcon forms a superspecies with the Sooty Falcon (156: *q.v.*)(Sn).

156 SOOTY FALCON *Falco concolor*
B151 / MPG118 / D1C1

Afrotropical: a Palearctic migrant from NE Africa and SW Asia, wintering in Mozambique, E South Africa and Madagascar.

Kenyan: annual in small numbers on S passage late Oct-Nov, but only occasional on N passage Apr-June (EABR). Inland distribution is similar to that of the rather less common Eleonora's Falcon (155:*q.v.*), in mainly C and SE areas. However, there is some evidence that the Sooty is slightly commoner at the coast, and it is said by Brown (1970a) and Ba to be regular on passage in some numbers through E and coastal Kenya. The record in 19C, actually in Somalia, was in May or June (AM).

72% of this falcon's squares receive over 500 mm annual rainfall, and the points regarding rainfall, termite hatches and migration at considerable height made for Eleonora's Falcon and other Palearctic congeners most probably also apply here. The fact that both this species and Eleonora's Falcon are commoner on S migration in Kenya (EABR) but are only known on N migration in Somalia (AM) may mean, as in the Red-footed Falcons (149, 150), that the two migrations occur by differing routes. Meinertzhagen's (1930) note of

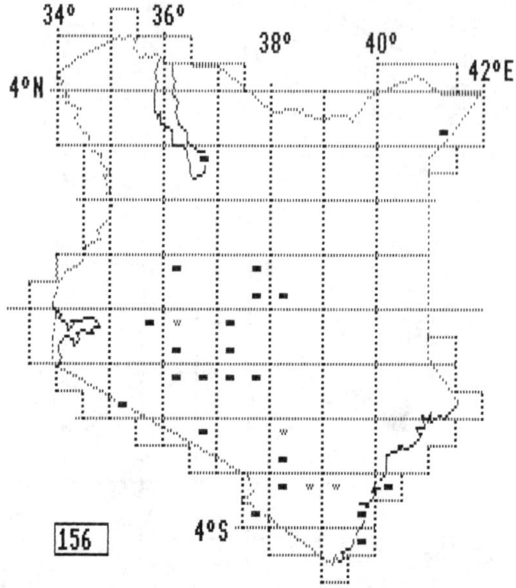

breeding at Mombasa 114B was convincingly rejected by Moreau (1969). Other reference: Dowsett 1977.

157 FOX KESTREL *Falco alopex*
B145 / MPG125 / D2C2

Afrotropical: the N tropics, from the W coast, mainly between 8°N and 15°N, to W Ethiopia, S Sudan, N Uganda and, on the SE edge of the range, mainly NW areas of Kenya. This distribution resembles that

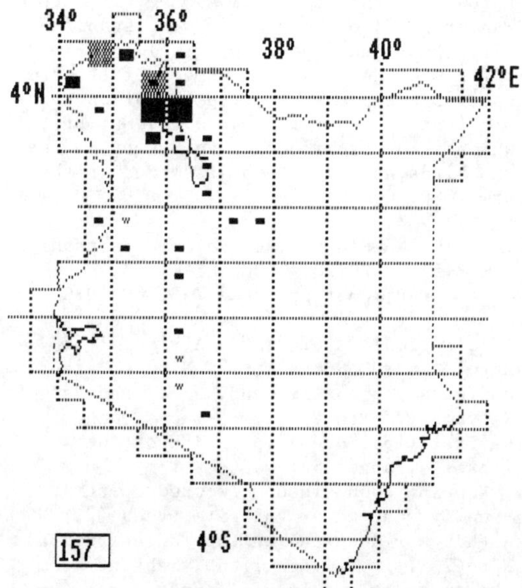

of the Swallow-tailed Kite (140).

Kenyan: resembles but is in fact not closely related to the White-eyed Kestrel (158)(Sn). It is locally common in open country around rocky hills, inselbergs and cliffs in the L Turkana basin (Br; HH) and further NW. Hence, range is mainly below 1000 m and 80% arid-semiarid, with only stragglers to different environments further S. This kestrel nests in inaccessible sites on cliffs and rocky slopes, and the breeding in the NW represents the SE extremity of the N tropics breeding range. Breeding occurs during the rains in Mar-June, with up to 25 pairs reported in a colony (Williams 1968; Brown & Britton 1980). 10-20 around suitable cliffs at Kamathia 3A in late Feb were apparently prospecting nest sites (EABR 1984).

Whether this species is sedentary is uncertain, but all records S of 1°N are Sept-Feb, and apparently represent single individuals wandering S during the non-breeding season, to the N edge of the highlands and thence further S down the rift valley to 75D and Serengeti (1°58'S in NE Tanzania)(St; Br; EABR). Only occasional individuals appear as far S as Baringo 50A (St), so that this S movement is certainly not commonplace.

158 WHITE-EYED KESTREL *Falco rupicoloides*
B159 / MPG124 / D2C2

Afrotropical: seems to occur in four distinct areas: N Ethiopia; E Ethiopia and Somalia S to 6°N (AM); Kenya, NE Uganda

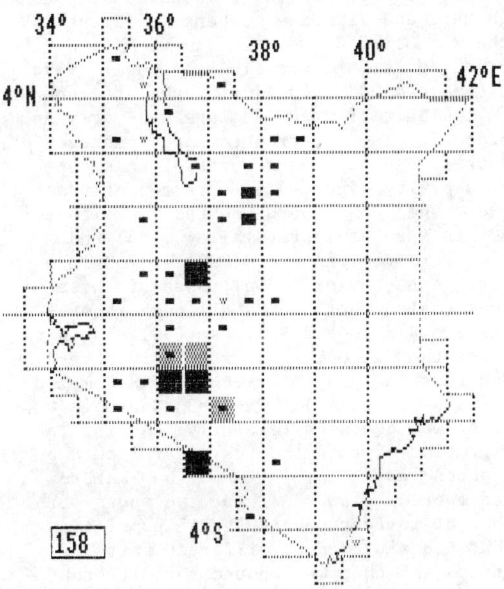

and N Tanzania; and over much of the continent S of 15°S.

Kenyan: generally uncommon and local on dry, open plains with scattered trees, below 1800 m (Br; Ja; MPG). Range is 65% within intermediate, 250-1000 mm rainfall areas, and only 7% below 500 m.

Three races possibly occur (Br). The type specimen, from Mombasa 114B, was at a uniquely low altitude and was presumably a vagrant N from the nominate S African population, which is known to wander in the dry season. Birds N of Eldoret 49C and Kabarnet 49B (Br) or Marsabit 27D 28C (BUN) may be non-breeding wanderers of the race *fieldi* which AM describe as a visitor S to 2°N on the Somali coast. Records in the C highlands and adjacent areas are the resident race *arthuri*, and its breeding records S of 1°N are all in areas with over 500 mm annual rainfall (87D is from Tanzania). The probable breeding 27D 39B derives from copulation in May, just after the rains (Lewis 1981e): these birds may have been *arthuri*, and the date agrees with breeding records of this race from the Suswa 75A area (EABR). This kestrel breeds exclusively in old nests of other species (*cf.* African Hobby (151) and Grey Kestrel (153)), especially those of the Cape Rook (580), which has a similar range in Kenya.

159 KESTREL *Falco tinnunculus*
B161 / MPG123 / D2C2

Afrotropical: widely distributed S of 15°N, apart from the equatorial forest areas of the W. Palearctic migrants penetrate S to Angola, Zambia and Malawi.

Kenyan: a bird of open country up to 4300 m (Br). Although three visually rather indistinct races occur and confuse our interpretation, the following points can be made. The nominate Palearctic race is mostly a winter visitor Nov-Mar: early Nov concentrations in the W are probably birds on S passage, as are Nov-Jan birds at L Baringo 50A (Br; St). These Palearctic birds are common, easily outnumbering resident races and also, in the lowlands, other kestrel species (Brown 1963; Br; Reynolds 1974).

The resident race *archeri* is known only from Lamu 91B (MPG): AM describe it as common only in N Somalia. The remainder of the resident birds are *F.t. rufescens*, which is usually sedentary, in pairs, and associated with rocky hills and cliffs (Br; MPG). Several authors comment on its inexplicable scarcity, both in terms of numerical abundance and numbers of potential breeding sites, in many apparently

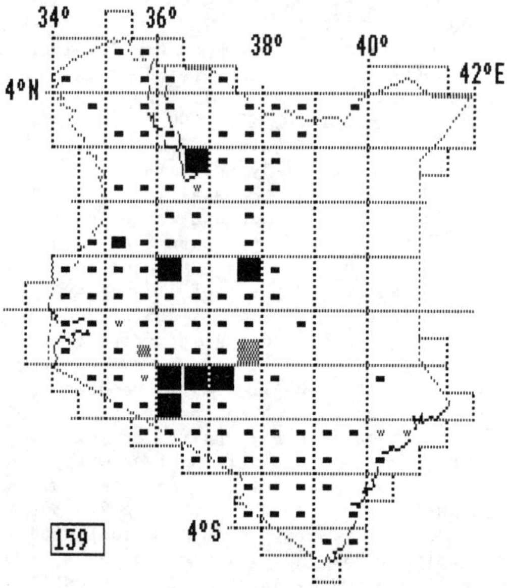

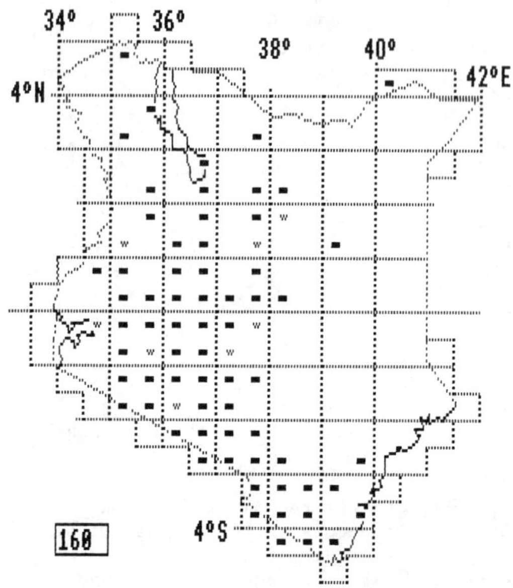

suitable areas of Kenya.

Only 17% of the Kestrel's squares are below 500 m. Nov-Apr seasonality in Tsavo E 101B suggests that, apart from *archeri* at Lamu, only Palearctic birds reach these low altitudes in SE Kenya (La). This suggestion is supported by the lack of breeding records in this region. In the NW, however, HH describe *rufescens* as a frequent to common breeding bird around L Turkana (400 m).

Absence of Kestrels from the far E of Kenya is apparently real in view of the lack of reports from observers resident in 19A 42A 65B, and absence in adjacent areas of S Somalia (AM). These low, arid areas are presumably unsuitable for either of the resident races, while Palearctic migrants may enter the country principally through the L Turkana basin and adjacent areas. See Lesser Kestrel (160). Other reference: Stevenson 1983a.

160 LESSER KESTREL *Falco naumanni*
B156 / MPG126 / D2C2

Afrotropical: a Palearctic species, wintering throughout sub-Saharan Africa and avoiding only arid E Somalia and the Congo basin forests.

Kenyan: a common to locally abundant, gregarious species of open country, late Oct-early May, that greatly outnumbers the resident White-eyed Kestrel (158) and the Afrotropical races of the Kestrel (159) (Br; Brown 1963). Range is 89% above 500 m and only 8% within the driest, 0-250 mm rainfall areas. It is rare in the coastal lowlands (EABR 1984) and the L Victoria basin 48 60 73 (Br).

S passage is very light with a flock of 60 at L Nakuru 62A in Nov being exceptional (EABR 1982), while in many areas (*e.g.* Tsavo E 101B: La) it is uncommon before Dec (Br). Most birds continue further S but small numbers winter, mainly in the W and C highlands and the less arid interior (Br).

N passage is much heavier than S, and Ja describes 2000+ moving through Nandi 49C in Mar, and "immense numbers" roosting in the Nairobi 75B area mid-Jan to Mar. Other notes (MacInnes 1948; Bednall 1959c) of up to 4500 moving during the afternoons in the Nairobi area 75B late Mar-early Apr probably also referred to large passage roosts, but further reports of this type of activity occurred only as recently as 1983, when up to 5000 roosted in Nairobi NP 75B Mar-late Apr (Smalley 1983) and 2500 were roosting near Kajiado 75D in early Apr (Taylor 1984a); both of these roosts also contained small numbers of Eastern Red-footed Falcons (150). Numbers using the Nairobi roost were only several hundreds in 1984 (D K Richards, H Gomez de Silva pers. comm.). The 1983 Nairobi roost was censused regularly. An initial migration of this species was followed by a second wave that included the Eastern Red-footed Falcons. These two waves were thought to represent separate populations that had wintered in different parts of Africa and that were bound for different breeding grounds (Smalley 1983).

Like the Kestrel and for the same

reasons, the Lesser Kestrel is virtually absent from E Kenya and adjacent W Somalia, in fact it is very scarce in Somalia overall (AM). In view of this, and of its regular N passage through E Tanzania but very uncommon status in much of Uganda and W Tanzania (Br), the bulk of N passage must occur through Kenya. N passage apparently follows the vicinity of the rift valley, while the more obscure S passage is either widely dispersed or occurs at considerable heights. This contrast between the two migrations is the opposite to that found in the Steppe (130) and Lesser Spotted (127) Eagles. Other reference: Reynolds 1974.

Family Phasianidae: Quails, Francolins, Stone Partridge

165 species, with representatives in all zoogeographical regions. 41 (25%) are found in the Afrotropical Region, and 16 of these (39%) reach Kenya. Apart from the Chestnut-naped Francolin (163), which is on the very edge of its range on the Ethiopian border, all of these species are known to have bred in Kenya at some time. The Quail (173) is the only phasianid with both Palearctic migrant and Afrotropical breeding races. Many of this family are difficult to identify and also rather skulking and retiring. Confirmed identifications, particularly of the more obscure species, are therefore less frequent than when the taking of specimens was routine. Hence concentrations of pre-1970 records may not necessarily indicate areas where species are no longer found. Other references: Jackson 1910, 1911, 1926.

161 SCALY FRANCOLIN *Francolinus squamatus*
B184 / MPG204 / D2C2

Afrotropical: C Africa, from Nigeria and Congo E to Ethiopia, Kenya, E Tanzania and Malawi. Distribution in Kenya reflects the E limit of the range.

Kenyan: often numerous in bamboo and highland forest, and adjacent dense habitats, up to 3000 m (Br; Ja; MPG; Se). Range is 88% within 500+ mm rainfall areas and entirely above 1000 m, and includes the forest islands of Mt Nyiru 26D and the Chyulu Hills 88D (Br). Se found it resident around 3000 m at Mau Narok 61D, and DEP (pers. obs.) concurs for 74D. Ja noted radical decline in the Nairobi area 75B consequent upon extensive hunting, but thought that the species had adapted to deforestation. But it is no longer common in this densely populated and cultivated area.

162 JACKSON'S FRANCOLIN *Francolinus jacksoni*
B173 / MPG206 / D2C3

Afrotropical: a montane francolin, endemic to the Cherangani Hills and C highlands of Kenya, and to Mt Elgon on the Kenya/Uganda border (hence C3).

Kenyan: a locally common but rather shy and secretive bird, found in the upper reaches of highland forest, in the bamboo zone and on montane moorland at 2300-3000 m on both sides of the rift valley (commoner on E), in the C highlands and on the Cherangani Hills 37CD and Mt Elgon 36D (Turner 1977; Ja; Ripley & Bond 1971; EABR 1983). Be and Se found it resident.

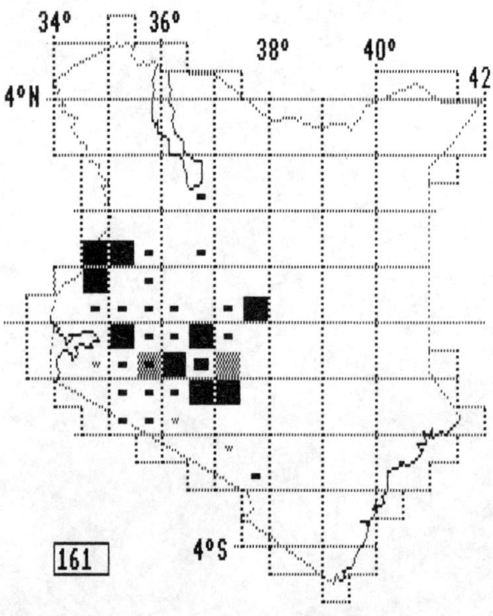

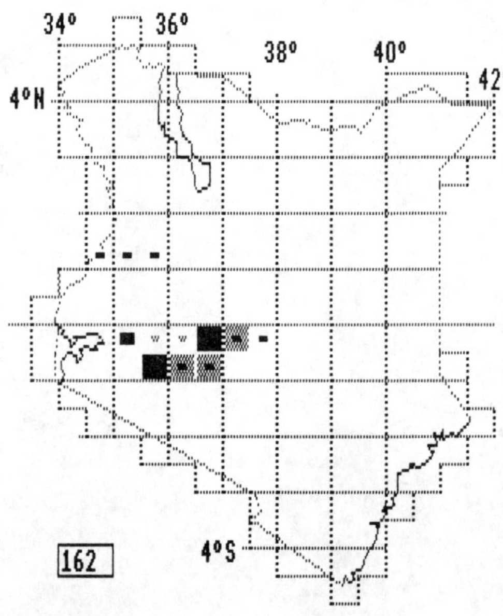

Together with the Chestnut-naped Francolin (163), it forms part of a superspecies predominantly associated with montane forests (Sn).

163 CHESTNUT-NAPED FRANCOLIN *Francolinus castaneicollis*
B168 / MPG197 / D2C0
NO MAP: post presence 17C

Afrotropical: a highland species, restricted to S Ethiopia and extreme NW Somalia, with a single record near Moyale on Kenya's N border.

Kenyan: the single sight record is one on a rocky hillside 6 km S of Moyale 17C on 21 June 1975: this was presumably the race *F.c. atrifrons*, which is known from the adjacent Mega area of Ethiopia (Turner & Forbes-Watson 1976b). See Jackson's Francolin (162).

164 YELLOW-NECKED SPURFOWL *Francolinus leucoscepus*
B175 / MPG210 / D3C3

Afrotropical: the NE, from E Tanzania N through Kenya and NE Uganda to Somalia, Ethiopia and S Sudan.

Kenyan: the Yellow-necked Spurfowl is common in a wide range of less open habitats including bushed grassland, scrub around watercourses in dry country and cultivation, up to 2400 m (Br; Ja; La; Lack 1985). It is allopatric with the Red-necked Spurfowl (165: *q.v.*), which is a member of the same species-group (Sn, Br).

The map shows three notable features. The Yellow-necked is absent from the S coastal lowlands 102CD 114 for reasons that are not clear: not all of this area is occupied by the Red-necked Spurfowl. The Yellow-necked occurs throughout our rainfall and moisture categories in other areas to the E of the rift, but is absent from much of the moist subhumid-humid country further W. Again, not all of these W areas contain Red-necked Spurfowls *e.g.* 49. In N Kenya, the Yellow-necked is found mainly on the E sides of both L Turkana (HH) and L Baringo 50A (St).

The Yellow-necked is resident in arid Tsavo E 101B (La; Lack 1985). DEP (pers. obs.) found similar status in 74BD 75 88AD 89AC 100B 101A, but also noted an annual Feb-Apr influx to the Ruaraka-Kahawa-Ruiru area of 75B. There are nesting records throughout our moisture categories, and the concentration of breeding squares around 37-38°E presumably reflects observer coverage. However, paucity of breeding records in moist areas W of the rift, especially in the well covered 74 and adjacent squares, might reflect allopatry with the Red-necked Spurfowl. Other reference: Swank 1977.

165 RED-NECKED SPURFOWL *Francolinus afer*
B167 / MPG208 / D3C2

Afrotropical: widely distributed S of the Congo forests, occurring SW around the periphery of the continent to the Cape. To the NE, it extends through E Zaire and W-C Tanzania to Uganda and W Kenya, and through E Tanzania to the coastal lowlands

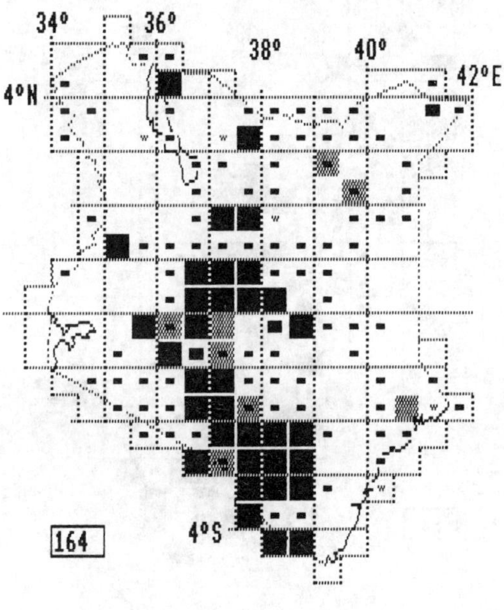

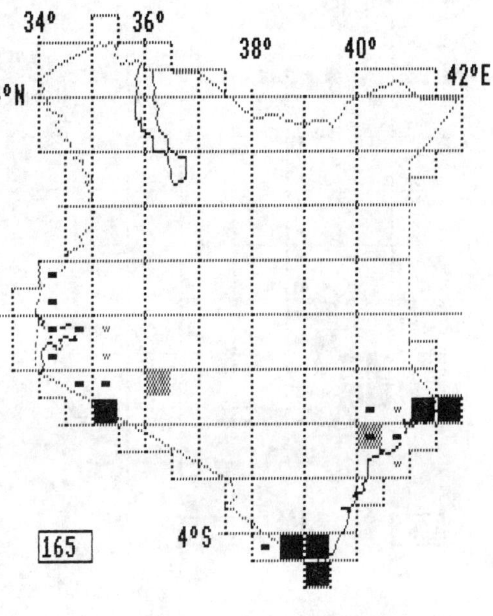

of Kenya. Distribution in Kenya thus represents the NE limit of the range.

Kenyan: in thickets and open bushed and wooded habitats, up to 1500 m (Br; Ja). *F. a. cranchii* is found E to Fort Ternan 61A and Maasai Mara GR 74AC (Br), in moist subhumid-humid areas above 1000 m. Jackson (1910) described it as "fairly plentiful" in the Kisumu 60B-Fort Ternan area but later thought it rare there due to hunting (Ja), and Br mentions it as local in the area. It is fairly common in Lambwe Valley 60C (EABR 1978).

F.a. leucoparaeus occurs in the coastal lowlands, mainly in semiarid areas. It is common N to the Shimba Hills 114A (Br), where the allopatric Yellow-necked Spurfowl (164: *q.v.*) is absent. Absence from the C coast can be only partly due to the Yellow-necked's presence there (Ja; Br), as neither species is reported from 102CD. These two birds are presumably separated by habitat preferences N of 3°S in the coastal lowlands, with the Red-necked in moister situations (Sn).

166 HILDEBRANDT'S FRANCOLIN *Francolinus hildebrandti*
B171 / MPG203 / D2C2

Afrotropical: the E, from N Mozambique, Malawi and Zambia through Tanzania to the N limit of the range in Kenya.

Kenyan: frequents rather densely bushed, thicketed or grassed habitats, often associated with rocky ground (Br; Ja; La). Range is 85% above 1000 m and 80% within intermediate, 250-1000 mm rainfall areas, and it includes the isolated highlands of Mt Marsabit 27D 28C, the Ndoto Mts 39A and Mt Ololokwe 51B (Ja; Br). La describe it as resident around its lower altitudinal limit in arid Tsavo E 101B, and DEP (pers. obs.) concurs for 74D. Relatively numerous pre-1970 squares at the N limits of the range may represent a contraction, but observer coverage is low in these areas and the species is easily overlooked in dense cover.

167 CRESTED FRANCOLIN *Francolinus sephaena*
B182 / MPG194, 195 / D2C2

Afrotropical: from S Sudan, Ethiopia and W Somalia, S through Uganda, Kenya and all but W Tanzania to Mozambique, and thence W to Angola and S to NE South Africa.

Kenyan: certainly the most widespread and probably the most plentiful francolin in Kenya, this bird frequents dense bushed and wooded habitats, up to 2100 m but mainly below 1500 m (Br; Ja; Lack 1985). It breeds throughout our moisture categories, but is present in only 22% of the humid squares. In the far W 48B 61A, some of this humid country is, or has been, occupied by the mainly allopatric Ring-necked Francolin (168)(Br), with which the Crested forms a species-group (Sn). DEP (pers. obs.) considers the Crested Francolin resident in 75CD 88AD 89AC 100B 101A and, as in the Yellow-necked Spurfowl (164), the relative concentration of breeding squares around 38°E probably reflects observer coverage. Ja mentions declines in the heavily populated Kisumu 60B and

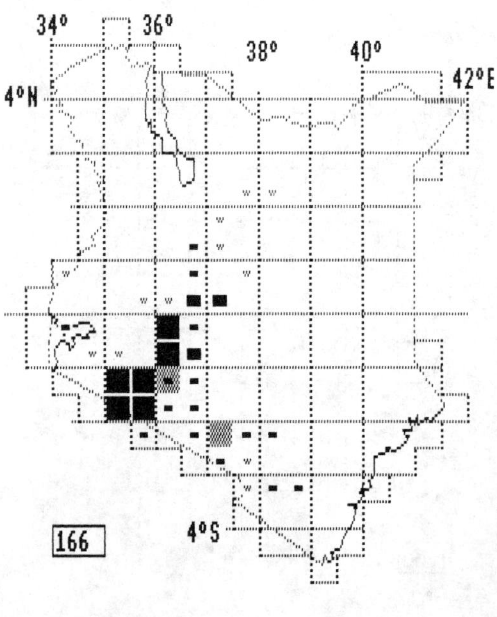

166

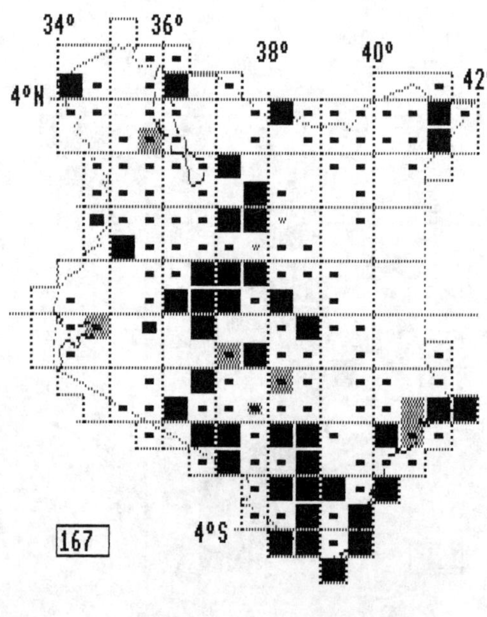

167

coastal areas, apparently due to habitat changes and hunting. Other reference: Swank 1977.

168 RING-NECKED FRANCOLIN *Francolinus streptophorus*
B185 / MPG193 / D2C0
NO MAP: pre presence 48C; pre confirmed 48B 61A

Afrotropical: an uncommon species, restricted to W Kenya, Uganda, NW Tanzania and Cameroon.

Kenyan: the type specimen of this skulking species was collected by Jackson at Mangiki 48B, in the S foothills of Mt Elgon. He did not encounter it again for a decade, when he found numbers in the Nyando Valley between Muhoroni 61A and Fort Ternan 61A. He noted breeding at both of these localities, and also mentioned presence in the Samia Hills 48C (Jackson 1910; Ja). Clearly this is a very uncommon and local bird, that may only occur in Kenya, the E extremity of its range, very sporadically. Ja describes his Kenyan localities as stony hillsides with short grass and scrub, and all three squares are humid. See Crested Francolin (167).

169 RED-WINGED FRANCOLIN *Francolinus levaillantii*
B176 / MPG199 / D2C1

Afrotropical: five distinct and scattered populations: W Kenya; Uganda, W Tanzania and E Zaire; from S Tanzania SW to Zambia; Angola; and S and E South Africa.

Kenyan: local and uncommon in grassland with patches of woodland or scrub, at 1800-3000 m in the W (Br; Ja). Range is entirely within 1000+ mm rainfall areas. The very few recent records suggest a substantial decline, and this isolated population may be near extinction. It has not been recorded in the well-watched Saiwa area 37C since Jan 1975 (T J Barnley *in litt.*).

This species forms part of a species-group with Shelley's Francolin (170) and the Moorland Francolin (171)(Sn). All three are allopatric (Br). Shelley's is in drier country (only 17% within the 1000+ mm rainfall regime) to the E of the Red-winged's range, and both of these birds are altitudinally separated from the Moorland Francolin (Sn).

170 SHELLEY'S FRANCOLIN *Francolinus shelleyi*
B183 / MPG198 / D2C2

Afrotropical: E Africa, from NE South Africa N through Mozambique and S Zaire to scattered areas of Tanzania and Uganda and, on the NE of the range, S and C Kenya.

Kenyan: locally common but rather unobtrusive in grassland, scrub and woodland clearings at 700-3000 m (Br; MPG). Range is 81% within intermediate, 250-1000 mm rainfall areas. Distribution is mainly in and E of the rift valley, and there are no recent records further W. See Red-winged Francolin (169).

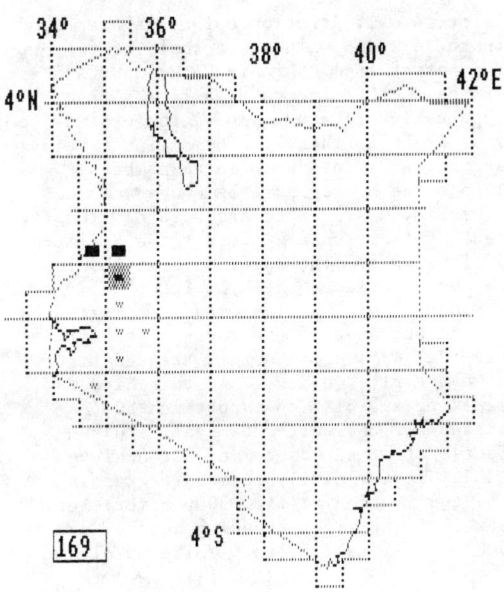

169

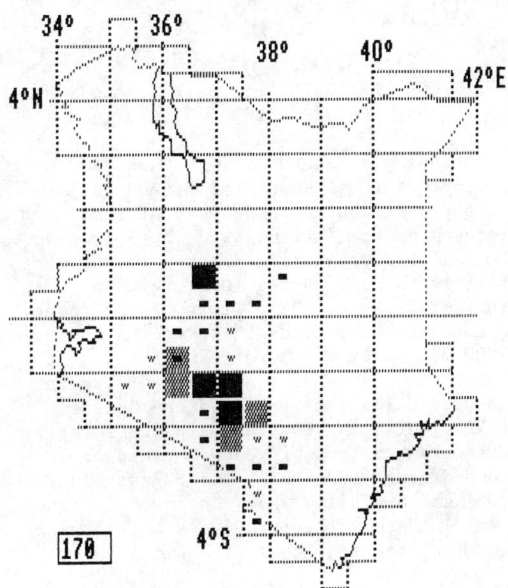

170

171 MOORLAND FRANCOLIN *Francolinus psilolaemus*
B180 / MPG198 / D2C3

Afrotropical: a montane species, endemic to the highlands of S Ethiopia and C Kenya, and to Mt Elgon on the Kenya/Uganda border (hence C3).

Kenyan: an inhabitant of montane heath and grassland at 2400-4000 m (Br). All squares receive over 1000 mm annual rainfall. This francolin is reasonably common W of the rift valley on the Mau 61D (Se) and Mt Elgon 36D (Britton & Sugg 1973), but there appears to be only one recent record E of the rift (EABR 1980; *cf.* Br). See Red-winged Francolin (169).

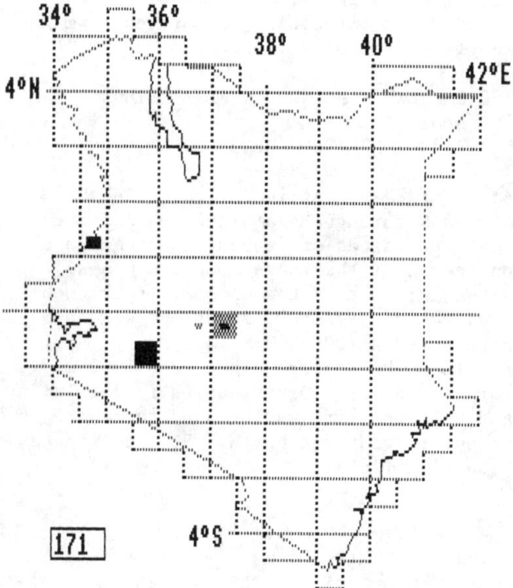

172 COQUI FRANCOLIN *Francolinus coqui*
B170 / MPG191 / D2C2

Afrotropical: extends from S Sudan and S Ethiopia through East Africa, and is widely distributed in areas S of the Congo forests except for the SW. There is also an isolated population in the W, from Nigeria to Mauritania. Distribution in Kenya represents the E limit of the range.

Kenyan: Br records this species up to 2200 m, and draws the distinction that the coastal lowlands nominate race occurs in woodland clearings (N to Sokoke Forest 102B), while the two races in the interior of Kenya are found in open, bushed or wooded grassland. Range is 82% within 500+ mm rainfall areas. Be found it resident in Narok district 74. Even allowing for the fact that this is a small and very

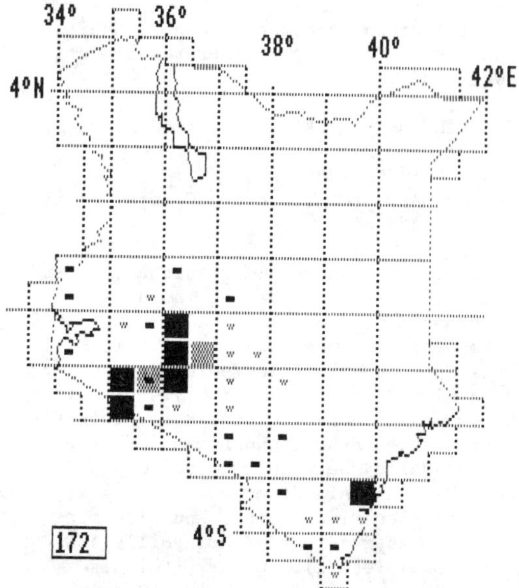

close-sitting bird (Ja), the concentration of pre-1970 squares in the relatively well-covered country around 63 and 76 probably reflects a decline in these densely populated and cultivated areas. van Someren (1925-35) noted marked decreases at the coast and in the Nakuru-Elmenteita area 62A, which he attributed to hunting by local peoples. Other reference: Woosnam 1912.

173 QUAIL *Coturnix coturnix*
B165 / MPG211 / D2C2

Afrotropical: Afrotropical birds are found in the E and S. They are resident in highland areas from Ethiopia S through East Africa and E Zaire to Malawi, while an apparently migratory population extends to most areas further S. The nominate Palearctic race winters in large numbers across the N tropics from Senegambia to Ethiopia, mostly S to about 10°N: the few records from Kenya represent the S limits of its range. The species' distribution in Kenya reflects the E limit of the range.

Kenyan: this species' map is complicated by the presence of two races. The Afrotropical *C.c. erlangeri* is treated first. Like the Blue Quail (175), this bird is skulking and difficult to flush but, because of the well-known call and the fact that it can be common, it achieves D2C2. It is locally common in grassland and cultivation at 1400-3000 m (Br; Se; Ja). Seven of the eight breeding squares are in 500+ mm rainfall areas. While mainly resident, it is suspected also to wander (Br; Se), but a bird at the Ngulia lights 101A in Dec (EABR 1983), presumably moving at night, is the only published evidence of this.

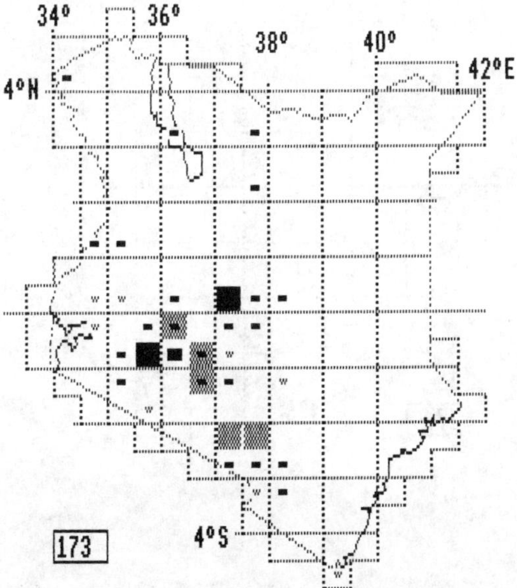

The nominate Palearctic race cannot be distinguished from *erlangeri* in the field (Ba), and so records are restricted to specimens, S to Kisumu 60B, the Loita Plains 74D, Mbuyuni 100B and Shimoni 114C (Br; Ja). Birds in apparently unsuitable habitat *e.g.* several Mar-Apr passage migrants on Central Island 14C, L Turkana (HH), were probably also of this race.

The Quail forms part of a species-group with the Harlequin Quail (174)(Sn). When the two bred alongside each other in NE Tanzania, the latter chose shorter, more degraded grassland (Br).

174 HARLEQUIN QUAIL *Coturnix delegorguei*
B166 / MPG212 / D2C2

Afrotropical: widely distributed S of Ivory Coast, Ethiopia and S Somalia, excepting arid E Somalia and the SW, and much of the Congo basin. It is, however, rare in the W, and a pronounced intra-African migrant.

Kenyan: an at least local migrant or wanderer, seldom if ever resident, that moves, often in very large numbers, into areas of lush grassland and cultivation during good rains, and breeds opportunistically (Br; Ja; Brown & Britton 1980; La; Lack 1985). It can move to the vicinity of even local rainfall, but is not always attracted to rain (*e.g.* St). Breeding is

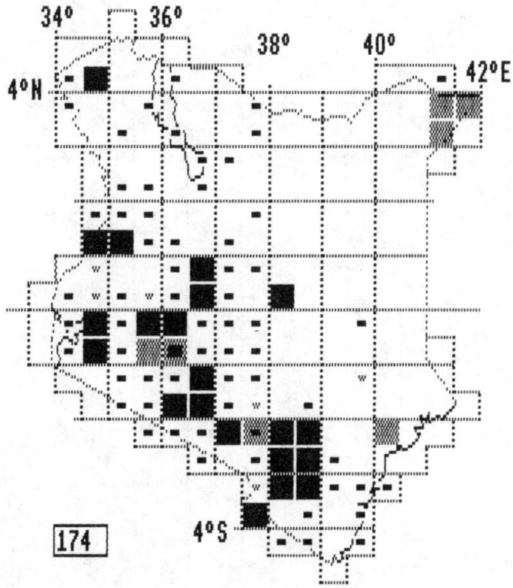

known up to 3000 m (Se; Br).

These movements complicate the map. However, with range only 5% within the most arid areas (moisture index below 1.2) and breeding squares only 10% within the driest, 0-250 mm rainfall regime, this quail is absent from much of the NE and E. There are no records from an observer resident in Wajir 42A, nor from adjacent S Somalia (AM). The breeding records from the arid 19 area were all in the May rainy season (Erlanger 1905, 1907). This bird also appears to be uncommon and to breed rarely in the lush coastal strip.

Most if not all movements probably occur at night, since birds have struck lighted windows at Mau Narok 61D (Se) and this is the most numerous Afrotropical species at the Ngulia lights 101A (Backhurst & Pearson 1977b). The Harlequin Quail is commonly trapped and eaten by local peoples (Ja; Ba; ADL pers. obs.), but is not endangered. See Quail (173) and Black-rumped Button Quail (182).

175 BLUE QUAIL *Coturnix chinensis*
B164 / MPG213 / D1C1

Afrotropical: widespread S of 10°N, apart from E Ethiopia, Somalia, the arid SW, and many of the lowland equatorial forest areas of the W. Distribution in Kenya represents the E limit of the range.

Kenyan: a scarce, local and very skulking bird of wetter grasslands than the Harlequin Quail (174), up to 1800 m (Br; Ja; Sn). Range is entirely in 500+ mm

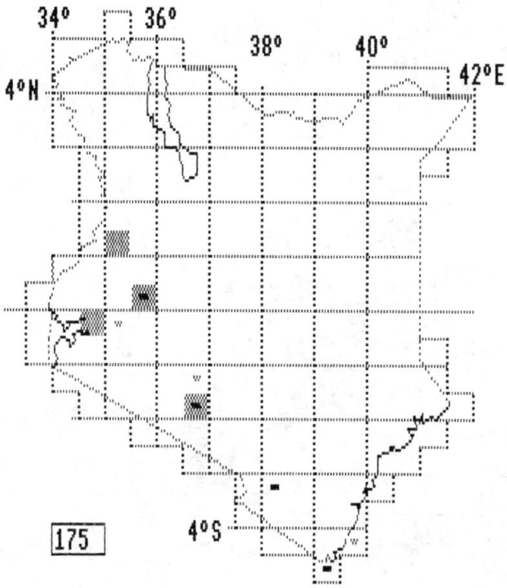

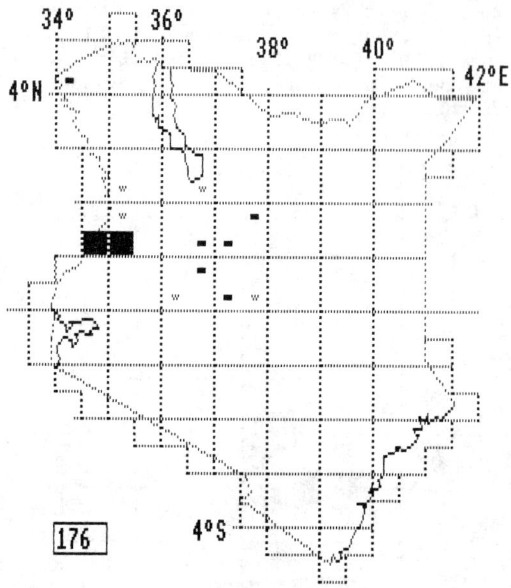

rainfall areas. The four breeding squares are above 1000 m, but van Someren (1925-35) reports nesting at the coast, without giving details.

It is at least partially migratory. Most records are apparently in the Apr-May 'long' rains, with breeding May-Aug (Br). However, a few pairs arrived and bred at Eldama Ravine 49D in Oct (Jackson 1911; Ja). Single birds, presumed nocturnal migrants, were attracted to lights at Kabarnet 49D on 10 Nov (Br) and at Ngulia Lodge 101A on 17 Dec (EABR 1982). See Black-rumped Button Quail (182). Other references: Bannerman 1910; EABR 1981.

176 STONE PARTRIDGE *Ptilopachus petrosus*
B186 / MPG214 / D1C1

Afrotropical: the N tropics, from the W coast, mainly between 6°N and 16°N, E to Ethiopia, N Uganda and, on the SE extremity of the range, NW Kenya.

Kenyan: a gregarious, shy and local species of rocky hills and cliffs in arid-semiarid, frequently remote country (hence D1C1), at 600-1500 m (Br; Ja). It is essentially a bird of the NW, extending SE to the N periphery of the W and C highlands.

Family Numididae: Guineafowls

Seven species, endemic to Africa. Four are resident in Kenya.

177 VULTURINE GUINEAFOWL *Acryllium vulturinum*
B187 / MPG219 / D3C2

Afrotropical: the NE, from NE Tanzania through mainly E areas of Kenya to S Sudan, SE Ethiopia and Somalia. Distribution in Kenya represents the W limit of the range.

Kenyan: often common in dry, open, bushed and wooded country, exceptionally up to 1900 m (Br; Ja; HH). Range is 86% below 1000 m, 70% within 0-500 mm rainfall areas, and 95% arid-semiarid. This guineafowl can exist without drinking (MPG; Sn), and there are reports from 56% of the most arid squares (moisture index below 1.2). It is confined to semiarid parts of the coastal strip 80D 91C. The rift valley and L Turkana (HH) form the W edge of the range N of the equator. N of L Turkana, it extends further W to 2C 3B and S Sudan (*cf.* Heuglin's Bustard (208)).

This species essentially occupies the arid N and E regions of the country, where the Helmeted Guineafowl (180: *q.v.*) is uncommon or absent: in their wide areas of overlap, the Helmeted occupies the less arid, more thickly vegetated habitats, and is usually near water (Sn; Br).

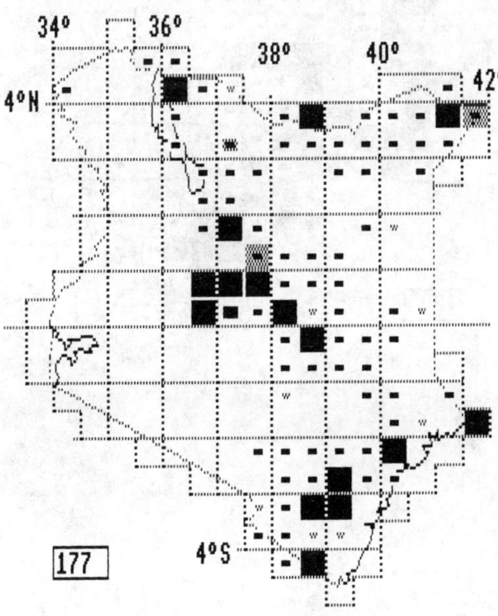

178 KENYA CRESTED GUINEAFOWL *Guttera pucherani*
B189 / MPG218 / D2C2

Afrotropical: the E, in S Somalia, and in areas of Kenya and Tanzania to the E of the E rift valley.

Kenyan: a rather retiring species of forests, woodland and thickets, including groundwater forests at Kibwezi 88B and Kitovu 100B, and riparian thickets at Ngulia 101A and Voi 101B, up to 1800 m (Br; Ja; MPG; La). Range is 76% within intermediate, 250-1000 mm rainfall areas. 74% of the squares are below 1000 m, and this is essentially a bird of the SE, inland to the E and SE edges of the C highlands. Occurrence inland along the Tana R 79C 78B 65ABD and other rivers to the periphery of the highlands (and in this case, N to Marsabit 27D) is known in other lowlands/coastal species *e.g.* Zanzibar Sombre Greenbul (619), Retz's Helmet Shrike (872) and Black-breasted Glossy Starling (877).

The Kenya Crested Guineafowl forms part of a superspecies with the Crested Guineafowl (179)(Sn). The two are allopatric (Br), approximately either side of the

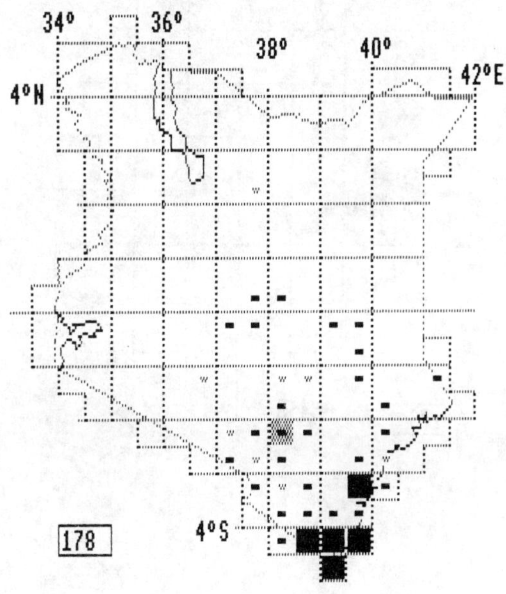

rift valley, with the Crested's range 87% moist subhumid-humid and the Kenya Crested's 79% arid-semiarid. 75B is the only square in which both have occurred, and the Kenya Crested is unknown there subsequent to Ja's reports from Kikuyu and Ngong.

179 CRESTED GUINEAFOWL *Guttera edouardi*
B188 / MPG217 / D1C2

Afrotropical: apart from the drier SW, in many areas S from Senegambia and S Sudan. Distribution in Kenya represents the E limit of the range.

Kenyan: rather uncommonly observed, shy and retiring in forests, thickets and woodland (hence D1), at 1700-3000 m (Br; Se; Ja; MPG). It mainly occurs to the W of the rift valley, but is found in the rift at Nakuru 62A (Br) and extends very locally onto the Kieni/Kikuyu escarpment 62D 75B (EABR 1982), *i.e.* on the W edge of the E plateau that is otherwise occupied by the allopatric Kenya Crested Guineafowl (178: *q.v.*). Range is 83% within 1000+ mm rainfall areas. The Crested Guineafowl is bound to suffer from habitat destruction and trapping in densely populated areas and, while formerly throughout the Trans Nzoia area 36D 37CD, is now confined to the Mt Elgon 36D and Cherangani Hills 37CD forests (T J Barnley *in litt.*).

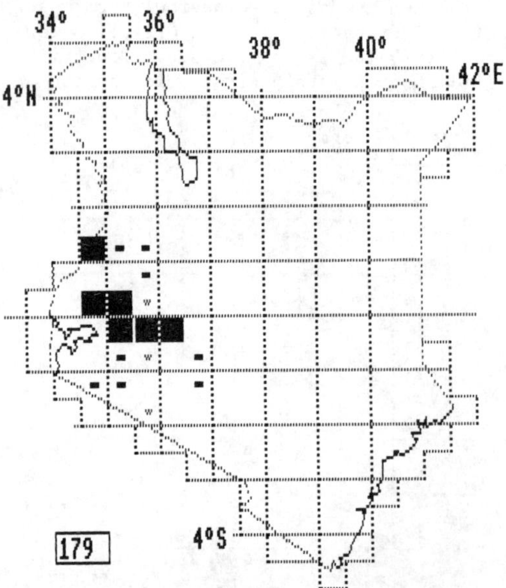

180 HELMETED GUINEAFOWL *Numida meleagris*
B190 / MPG215, 216 / D3C3

Afrotropical: widespread S of 15°N, apart from most of Somalia and the arid SW, and most of the equatorial lowland forest areas of the W. Distribution in Kenya represents the E limit of the range.

Kenyan: a well-known species, in a great variety of bushed and wooded habitats, including woodland and cultivation, usually near water and up to at least 2200 m (Br; Be; HH; La). It can be very numerous, Ja finding a flock of over 2000 near the upper Turkwell R 37. Range is only 32% within arid areas, as against 67% for the Vulturine Guineafowl (177: *q.v.*). The Helmeted Guineafowl is uncommon in the dry N, where it is restricted to well wooded localities around L Turkana (HH) and elsewhere. Absence from the far E appears real in view of the lack of records from recent observers resident in 42A 65B, and absence S of 3°30'N in adjacent S Somalia (AM). This guineafowl has been described as resident in 74 (Be), 101B (La), and 74BD 75CD 88AD 89AC 100B 101A (DEP pers. obs.). Other reference: Swank 1977.

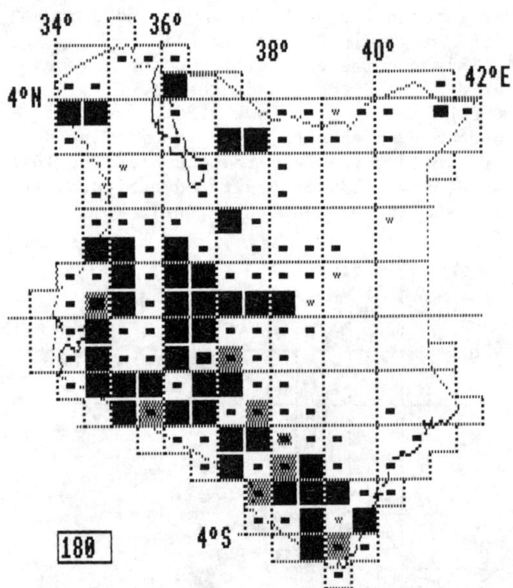

Family Turnicidae: Button Quails, Quail Plover

14 species, mainly in the Oriental and Australasian Regions, with only three (21%) in Africa (Sn). All of these three are also known from Kenya, but there are breeding records for only one. They are small, shy, retiring and very skulking birds, and the Black-rumped Button Quail (182) and Quail Plover (183) in particular are very little known here. As with the Francolins, modern disinclination to take specimens results in fewer confirmed identifications.

181 BUTTON QUAIL *Turnix sylvatica*
B193 / MPG365 / D1C1

Afrotropical: widely distributed S of 15°N, except for many of the equatorial lowland forest areas of the W, and the drier areas of E Somalia and the SW.

Kenyan: a locally common but skulking, secretive and infrequently seen species of open grassy habitats, up to 2000 m but mainly below 1500 m (Br; Ja; MPG; La). Range is only 4% in 0-250 mm rainfall areas and 16% arid, and virtually absent from the dry N and E. The records on the N borders in 2A 19AB represent the S limits of populations in higher rainfall areas to the N. Regular arrivals at the Ngulia lights 101A suggest nocturnal movements (Backhurst & Pearson 1977b; EABR), probably of the type that bring numbers into adjacent arid Tsavo E 101B during the rains (La). It is also intermittently abundant on the coast (MPG). See Black-rumped Button Quail (182). Other reference: van Someren 1925-35.

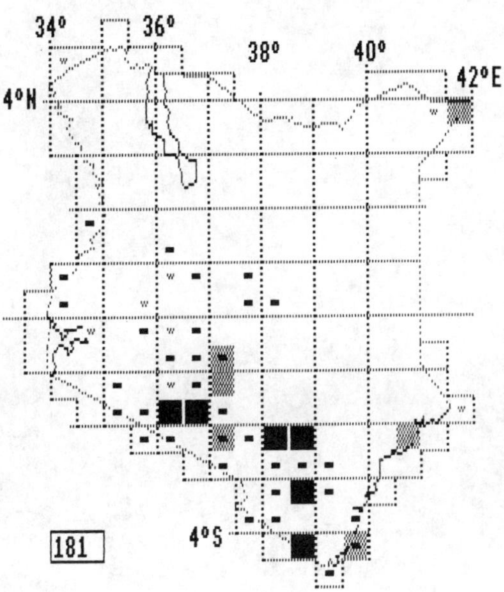

182 BLACK-RUMPED BUTTON QUAIL *Turnix hottentota*
B192 / MPG366 / D0C0
NO MAP: pre presence 37C

Afrotropical: rather scarce in scattered areas from Senegambia to Angola in the W, and from NW Kenya S to the Cape in the E.

Kenyan: a bird of open and often flooded, short grassland (Br; Ja). Its only minor distinctions from the more common and widespread but similarly very shy and skulking Button Quail (181) must render field identification very difficult (hence D0), and there are only two records. Stoneham collected the type specimen, and found several clutches of eggs during June and Jul, in Trans Nzoia (approximately 36D-37C: not mapped)(Ja), and one was caught near Kipsain 37C on 2 Feb 1958 (T J Barnley *in litt.*). Its continued presence in these areas, on the NE edge of its range, is an open question.

Sn points out the strikingly similar relationships between this species and the Button Quail, and between the Blue (175) and Harlequin (174) Quails: in each case, the first member of the pair prefers shorter, often wetter grassland, while the second is in longer, lusher growth.

183 QUAIL PLOVER *Ortyxelos meiffreni*
B191 / MPG367 / D1C0

Afrotropical: the N tropics, mainly between 12°N and 17°N, from the W coast to C Sudan. Further SE, it also extends from the S borders of Ethiopia to Kenya and NE Uganda.

Kenyan: on the ground this is a well camouflaged, secretive and very close-sitting bird (hence D1), but once in flight it has a unique appearance, so that the striking paucity of records confirms its local and uncommon status (Br). The Quail Plover frequents open grassy and bushed habitats in the dry N and E, with five of

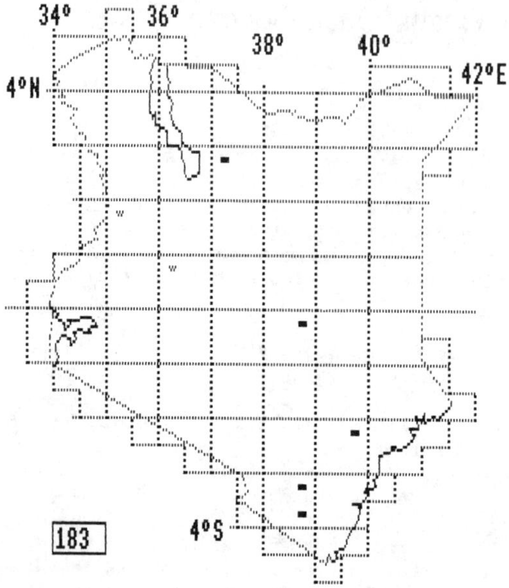

its seven squares below 1000 m and six arid-semiarid. The few records can be summarised.

Ja found two in Sept at Baringo 50A, while occasional individuals along the Turkwell R (37A 25 13D: not mapped) included a male in breeding condition on 30 Dec. This latter record is very likely the source of MPG's note of breeding condition in N Kenya Dec-Jan, and there are no other breeding indications. Before 1970, T J Barnley (*in litt.*) and J G Williams shot one at Nakait (Chepterr) 37A. It was found in 1967 in the Tsavo E 101BD area by A D Forbes-Watson (Br), and there have subsequently been records there in all months (Lack 1975b, 1985; La). More recent observations have come from Olturot 27A (DEP pers. obs.) and Kora 64B, while sightings on Galana Ranch 90B in Mar 1984 (EABR 1984) were repeated shortly afterwards (B R Heath pers. comm.).

Family Gruidae: Cranes

14 species which, while mainly Holarctic (Sn), have representatives in all zoogeographical regions except the Neotropical. The Crowned Crane is a common breeding resident in Kenya, while there is also an unsatisfactory old record of the Palearctic Crane *Grus grus* (van Someren 1919b).

184 CROWNED CRANE *Balearica pavonina*
B194 / MPG244, 245 / D3C3

Afrotropical: widely distributed S of 16°N but absent from the dry SW and from the equatorial forest areas of the W. Distribution in Kenya represents the E limit of the range.

Kenyan: two races (sometimes considered separate species *e.g.* MPG) occur. Most records are referrable to *B.p. gibbericeps* which is common in highland marshes and grasslands, up to 3000 m (Br; Se; Ja; Be). Burke (1965) counted over 800 in 757 mile^2 (1960 km^2) of the Kisii district 60D that had a human population of 650/mile^2 (250/km^2). Range is 93% above 500 m and 82% within 500+ mm rainfall areas. This race is considered basically sedentary by Br, Be, Se, St and DEP, and this conclusion is supported by the high proportion of its total squares in which breeding occurs (55%).

Movements do occur, however. This race breeds in solitary pairs, but non-breeding birds can flock to produce large concentrations. Ja mentions congregations of 200+ near rift lakes and, at L Baringo, the few resident pairs are occasionally augmented by influxes that produce flocks of up to 150 (St). In arid Tsavo E 101B, at around 500 m, this bird is a sporadic visitor to waterholes, mainly in Jan-May (La). More extensive wandering has recently been demonstrated by the first records from the coast, of two small groups in 91B 103A.

The S edge of the Ethiopian and Sudanese population of the race *B.p. ceciliae* is represented by records at low altitudes around N L Turkana (Owre 1966), where it is frequent to common (HH), and at 16D. This race may be only a seasonal visitor from further N (Br). Other reference: Dobbs 1914.

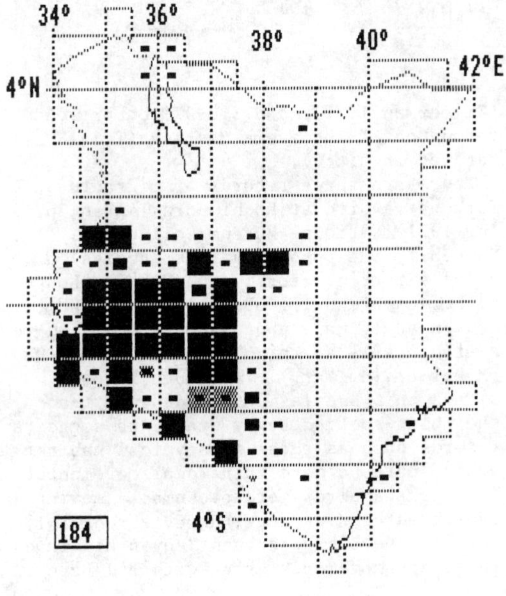

Family Rallidae: Rails, Crakes, Gallinules, Moorhens, Coots

124 species, with representatives in all zoogeographical regions. 24 (19%) occur in the Afrotropical Region and, of these, 17 (71%) are known in Kenya. Two of the Kenyan species (12%) are non-breeding migrants from the Palearctic, while the remainder either breed, at least occasionally, or are presumed to do so, or have done so in the past. The local populations of the Lesser Spotted Crake (195) are possibly augmented by Palearctic migrants. Many members of this family are extremely skulking and difficult to observe so that, as with the Francolins and the Button Quails, absence of collecting in recent years leads to fewer certain recent identifications and thus an often high proportion of pre-1970 squares, which may be misleading.

185 WHITE-SPOTTED PYGMY CRAKE *Sarothrura pulchra*
B213 / MPG235 / D1C2
NO MAP: post presence 48AD 49C; pre confirmed 48D

Afrotropical: a W African forest species, around the Gulf of Guinea from Senegambia to Angola, and E through the Congo basin to E Zaire, Uganda, NW Tanzania and, on the E extremity of the range, W Kenya.

Kenyan: a reasonably common and well-known (at least vocally) bird of swampy areas near streams and pools in the Kakamega, Yala R and Nandi Forests 48D 49C (Br; Ja; EABR)(hence D1C2). The record in 48A is from Uganda (J S Ash *in litt.*). All squares are above 1000 m and receive over 1000 mm annual rainfall. It is thought to be sedentary (Br), and breeding in 48D is based upon a juvenile (van Someren 1939).

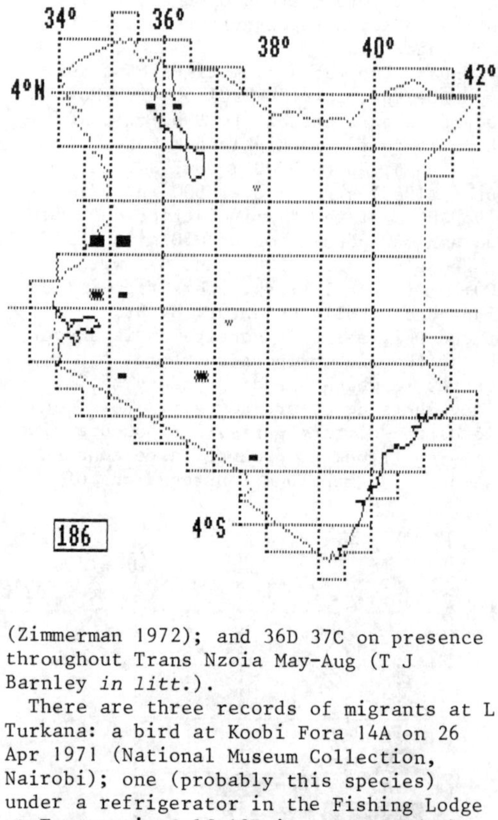

186 BUFF-SPOTTED PYGMY CRAKE *Sarothrura elegans*
B211 / MPG236 / D0C0

Afrotropical: records are thinly scattered over most areas S of 10°N, apart from the dry SW. Distribution in Kenya represents the E limit of the range.

Kenyan: uncommon in forest, bamboo and thicket (Br), in drier situations than the other forest-dwelling *Sarothrura*, the White-spotted Pygmy Crake (185)(Sn). It is known at 2600 m on Mt Kenya 63A (Ja). Apart from migrants at L Turkana (see below), and records from the forest islands of Mt Marsabit 27D and the Chyulu Hills 88D (Br)(which are highlands of too restricted extent to appear on our environmental overlays), all squares are above 1000 m in 500+ mm rainfall areas. There are no confirmed breeding records: 75B is based upon a juvenile in the National Museum Collection, Nairobi; 48D on enlarged ovaries and continual presence in Kakamega Forest 48D (Zimmerman 1972); and 36D 37C on presence throughout Trans Nzoia May-Aug (T J Barnley *in litt.*).

There are three records of migrants at L Turkana: a bird at Koobi Fora 14A on 26 Apr 1971 (National Museum Collection, Nairobi); one (probably this species) under a refrigerator in the Fishing Lodge at Ferguson's Gulf 13B in Apr 1975 (HH); and an adult male photographed clinging to a window sill at this lodge on 13 Apr 1985 (M Palmer *in litt.*). Although Sn considers this species less sedentary than the White-spotted Pygmy Crake, he suggests records such as these in atypical habitat to indicate no more than local movements. The Turkana birds were presumably moving over greater distances, however, possibly between the Ethiopian and Kenyan highlands. These Apr movements agree with T J Barnley's note (above) of May appearance in the Kitale area.

187 RED-CHESTED PYGMY CRAKE *Sarothrura rufa*
B214 / MPG233 / D1C1

Afrotropical: in many areas S of 10°N, but not Somalia or the arid SW. Distribution in Kenya represents the E limit of the range.

Kenyan: locally common and reasonably frequently reported in well-vegetated marshes and bogs up to 2700 m (Br; MPG; EABR). Range is 84% above 1000 m, and 92% within 500+ mm rainfall areas. It is most common and widespread in areas of the L Victoria basin to the N of the Winam Gulf (Britton 1978; Br; EABR 1984). E of the rift valley, it appears to occur on the periphery of the C highlands. A vocal record at sea level near Mombasa 114B (EABR 1978) represents the NE extremity of birds in NE and coastal Tanzania (Br). The sole confirmed breeding record is from van Someren (1939).

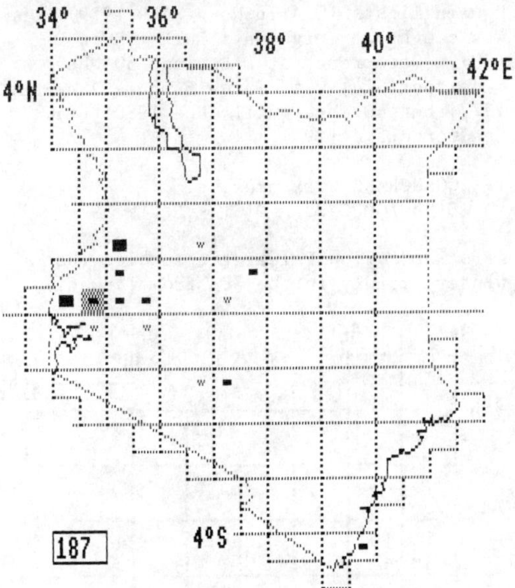

188 STREAKY-BREASTED PYGMY CRAKE *Sarothrura boehmi*
B210 / MPG231 / D0C0
NO MAP: post presence 37C 75B; pre presence 76C; pre probable 60B 75B; pre confirmed 37C

Afrotropical: from Cameroon to the highlands of Kenya, and S to Zimbabwe and S Mozambique.

Kenyan: a rarely recorded bird of open short grassland and river margins (Br; Ja). All four squares are above 1000 m in 500+ mm rainfall country. This pygmy crake may be an intra-African migrant that mostly breeds in the S tropics (Br). The few records are: an incompletely grown imm near Kitale 37C (Br) and frequent vocal records in the same area (T J Barnley *in litt.*); birds in breeding condition at Kisumu 60B (MPG) and Nairobi 75B (Ja); a female at Nairobi 75B in June (EABR 1980); Machakos 76C (Ja). MPG also mention breeding at Trans Nzoia 36D 37C.

189 CHESTNUT-TAILED PYGMY CRAKE *Sarothrura affinis*
B209 / MPG230 / D0C0

Afrotropical: a discontinuous distribution from South Africa N through Mozambique and Malawi to Tanzania, Kenya and S Sudan. Distribution in Kenya represents range at these latitudes.

Kenyan: another skulking and rarely recorded species, found on montane moorland and marsh, and known up to 3700 m on Mt Kenya 63A (Sn; Br). Ja's description of the type area Ndassekera appears to relate to Entesekera 74D, in which case Br's mention of this species in NE Tanzania is incorrect. It has recently been recorded in 74D by Fayad and Fayad (1980). Breeding is known also from Trans Nzoia (36D 37C (not mapped): Ja).

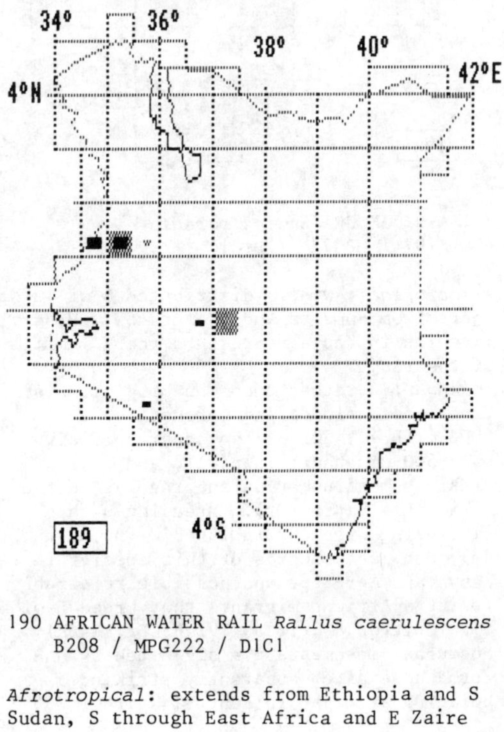

190 AFRICAN WATER RAIL *Rallus caerulescens*
B208 / MPG222 / D1C1

Afrotropical: extends from Ethiopia and S Sudan, S through East Africa and E Zaire to become more widespread S of 10°S where, as in others of this family, it appears commoner in the E. Distribution in Kenya represents the E limit of the range.

Kenyan: uncommon and skulking, predominantly in freshwater swamps, up to 3000 m (Br; Britton 1979; Ja). 88% of the squares receive over 500 mm rainfall. It is apparently sedentary (Br; Sn), and has been recorded throughout the year at Nairobi 75B and Thika 76A (EABR 1982). It may wander (Br), but there are no details of possible movements beyond sporadic occurrence at L Baringo 50A (St) and large numbers at L Naivasha 62C in Jan (EABR 1982).

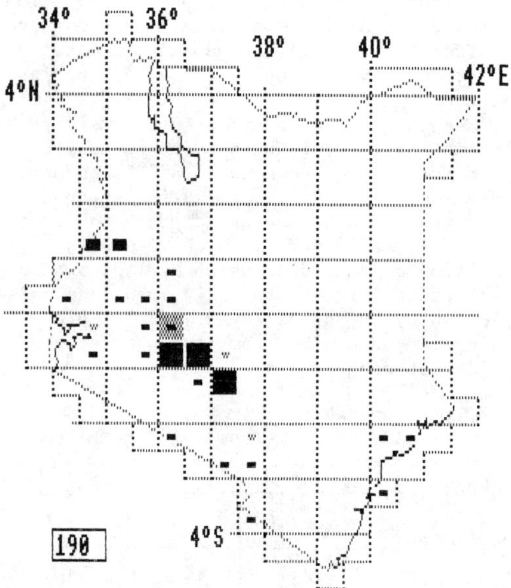

191 AFRICAN CRAKE *Crex egregia*
B197 / MPG224 / D1C1

Afrotropical: widely distributed S of 12°N, apart from Somalia and the dry SW. Distribution in Kenya represents the E limit of the range.

Kenyan: a rather uncommon bird of moist grasslands, floods and marshes, up to 2000 m (Br). 80% of the squares receive over 500 mm annual rainfall, and this crake is most numerous and regular in the L Victoria basin (Br). Breeding occurs (including Trans Nzoia 36D 37C (not mapped): Ja), but most records of this species in Kenya are Apr-Sept and possible referrable to intra-African migrants that breed S of 5°S (Britton & Britton 1976b; Br; EABR). Nocturnal movements are suggested by one found dead after apparently striking a building at night in central Nairobi 75B

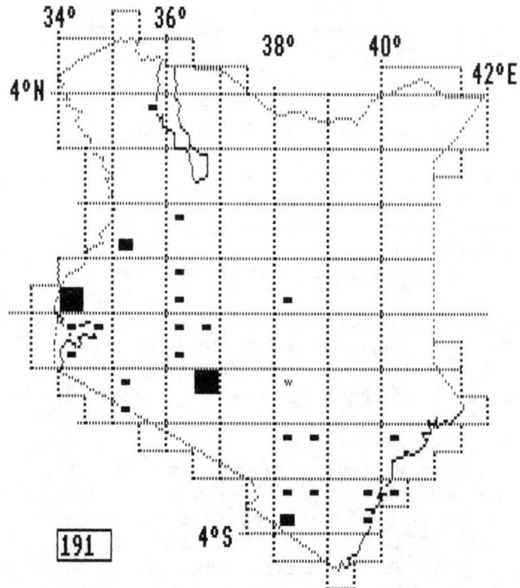

in June (Lee 1974), and by singles at the Ngulia lights 101A in Nov (EABR 1979). One was caught at Ferguson's Gulf 13B in Dec (HH). Occurrence at L Baringo 50A has been sporadic, with large numbers June-Aug 1977, but no more until June 1980 (St). See Corncrake (192).

192 CORNCRAKE *Crex crex*
B196 / MPG223 / D1C1

Afrotropical: a Palearctic species that winters mainly in the SE, from Tanzania S

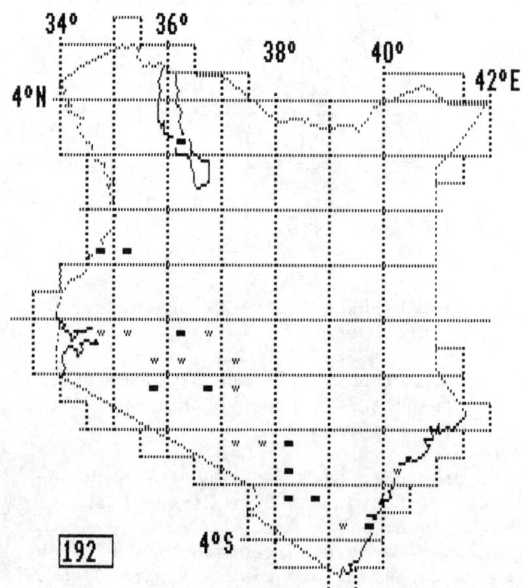

to Zambia and E South Africa.

Kenyan: a skulking and very infrequently observed passage migrant, mainly Oct-Dec and Mar-Apr, although there are records as late as May and 2 June (Ja; Hudson 1973; Ba; Br; EABR). It is known up to 3000 m (Britton 1979) and, like the closely related African Crake (191), is a bird of grassland. Range is 76% within 500+ mm rainfall areas. Most records are in the C highlands and the SE, perhaps reflecting observer distribution (Ba), whilst the high proportion of squares with only pre-1970 records (50%) probably reflects this bird's low detectability. N and S passages appear of similar intensity, and individuals are caught at the Ngulia lights 101A (Br; EABR).

193 BLACK CRAKE *Limnocorax flavirostra*
B201 / MPG225 / D2C2

Afrotropical: widely distributed S of 15°N but absent from the drier areas of Somalia and the SW.

Kenyan: the commonest and most widespread of Kenya's rails, capable of occurring on any patch of swamp that combines some open water with fringing vegetation, up to 3000 m (Br; Se; Ja; MPG; HH; St). Range is 70% within 500+ mm rainfall areas, and mostly absent from the dry N and E. The record in 17C is from a small dam near Moyale, while presence in 19A is along the Daua R. HH describe it as common at L Turkana, but there is no breeding record (*cf.* Red-knobbed Coot (197), Purple Gallinule (198), Common Moorhen (201)). Virtual absence from the Tana R is notable.

The Black Crake is thought to be mainly sedentary (Br) and absence from remote waters such as those at Marsabit 27D and Wajir 42A argues against extensive movements, but at least local mobility is indicated by a possible nocturnal migrant at the Ngulia lights 101A in Dec (Backhurst & Pearson 1977b), by colonisation of temporary waters and newly constructed dams (Se), and by appearance on the Voi R 101B when it is flowing (La).

194 SPOTTED CRAKE *Porzana porzana*
B206 / MPG227 / D1C1

Afrotropical: this Palearctic species winters mainly in the S, from Zambia to N Namibia and E South Africa. Distribution in Kenya represents the E limit of the range.

Kenyan: a silent, extremely secretive and rarely recorded species of moist grasslands and swamps (Br; MPG; Ba). Like the Corncrake (192), it appears to be most frequent in the C highlands and the SE but, more so than in that species, this distribution almost certainly reflects that of competent observers. There are particularly few records W of 36°E (Br): 37C = an individual collected at Kipsain, Apr 1961 (T J Barnley *in litt.*). Passage dates are Nov-Jan and late Apr-early May (distinctly later in both cases than the Corncrake), with the majority of records in spring (Br; Ba; EABR). Feb birds at Londiani 61B, Naivasha 62C and Athi River 75B were

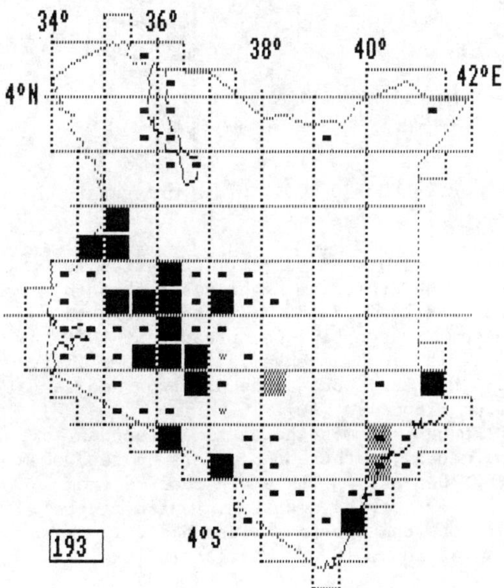

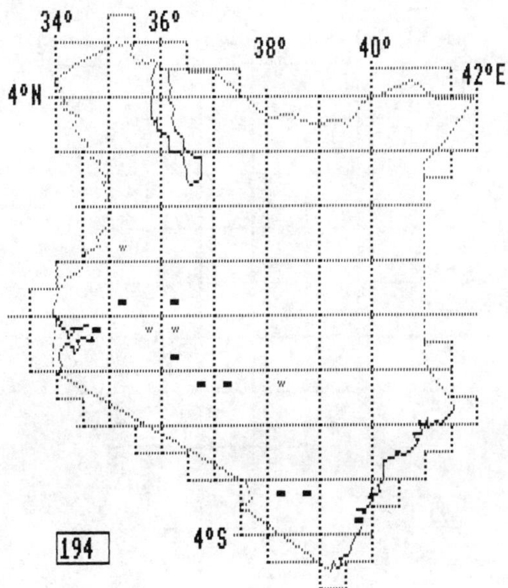

probably wintering (Ja; Ba; EABR 1982): these localities are all above 1500 m in the highlands, with annual rainfall over 500 mm.

195 LESSER SPOTTED CRAKE *Porzana pusilla*
B207 / MPG228 / D1C0

Afrotropical: another very secretive crake which appears to be fairly widely distributed in Africa S of the Congo basin forests, and further N through E Zaire and East Africa to Ethiopia. The extent to which Palearctic migrants penetrate this range is uncertain. Distribution in Kenya represents this species' E limit.

Kenyan: an extremely secretive and skulking inhabitant of well-vegetated swamps (Br; MPG): the map again reflects the frequency of competent observers. 80% of this crake's squares are above 1000 m, 90% receive over 500 mm rainfall, and there is only a single record from the coastal lowlands, in June (EABR 1981). It is assumed to be at least partly resident (Br) but, in contrast to most of the equally obscure *Sarothrura* crakes, there is no breeding record, and Sn knew of no confirmed breeding between 9°N and 15°S. A notable influx of nine or more at L Naivasha 62C in Jan and at least three at L Baringo 50A in Feb-Mar (EABR 1982), and a bird at Benane 52B in Nov (Tomlinson 1946), might have involved Palearctic migrants, but there is no other evidence of this race. See Jack Snipe (248).

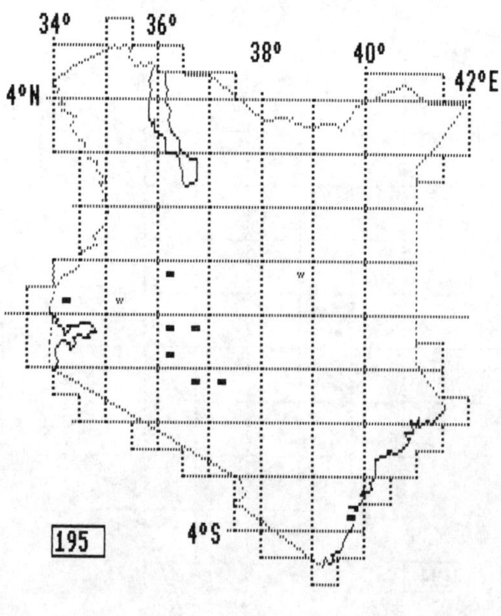

196 STRIPED CRAKE *Porzana marginalis*
B204 / MPG226 / D1C1

Afrotropical: most records are between 7°N and 20°S, with Kenya at the NE limit of the range.

Kenyan: a highly secretive or actually rare bird, in marshes and flooded grassland (Br; Ja; Sn). Seven of the nine squares receive over 500 mm annual rainfall. It is an intra-African migrant, and the few records are May-Sept and Nov (EABR; Br; St). Presumed nocturnal migrants in breeding condition have struck lighted windows in Nairobi 75B on 17 and 29 May (Br), but a June clutch at Kabete 75B (Pitman 1965a), and a female carrying nesting material at L Naivasha 62C on 27 Nov (Br), are the only breeding records between 6°N and 11°S (Sn; Br)(*cf*. Lesser Spotted Crake (195)).

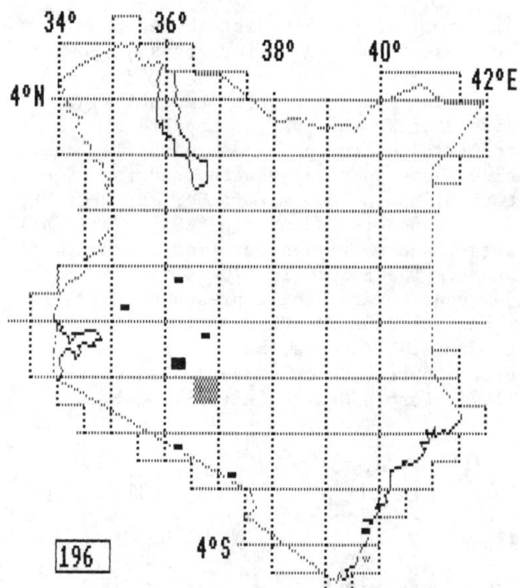

197 RED-KNOBBED COOT *Fulica cristata*
B215 / MPG242 / D3C3

Afrotropical: E and S Africa, extending from the Ethiopian highlands S through East Africa and the E Zaire borders to become reasonably widespread S of 10°S. Distribution in Kenya is near the E limit of the range, but there are two records from S Somalia (AM).

Kenyan: common and at times abundant on permanent, rather open waters up to 3000 m (Br; Se; MPG). There are records from only 19% of the squares in which rivers are the only surface water bodies. Range is 74% above 1000 m but, at lower alti-

Plate 9. The Helmeted Guineafowl, one of a small family found only in Africa (J.S. Karmali F.R.P.S.).
See page 136.

Plate 10. The Kori Bustard (J.S. Karmali F.R.P.S.). *See page 152*.

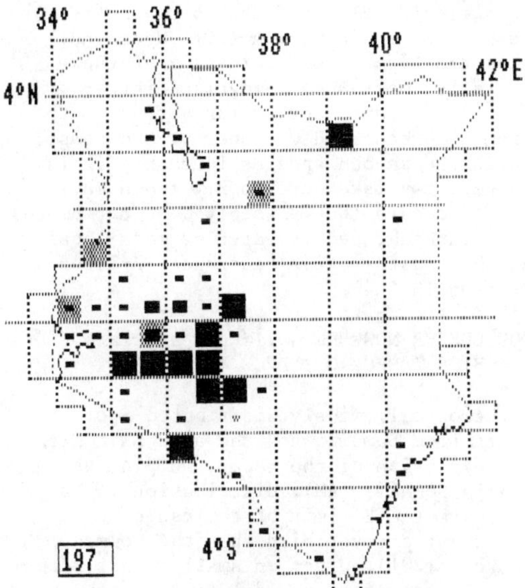

tudes, coots are common on L Turkana (HH), and frequent on pans at Wajir 42A (Ash 1983). The species is however decidedly rare at the coast, where records are confined to the vicinity of the Tana R 91. Breeding is concentrated above 1000 m in the W and C highlands. The only two other records are from L Paradise, on Mt Marsabit 27D, and from a dam near Moyale 17C (ADL pers. obs.). The latter probably represents the S edge of the large Ethiopian breeding population.

The Red-knobbed Coot moves extensively. Br mentions seasonal fluctuations at L Naivasha 62C and other sites, while Backhurst (1969b) records movement of ringed individuals from Ngorongoro crater (at 3°S in NE Tanzania) N to Eldoret 49C and L Naivasha. Breeding peaks are May-Jul and Sept-Oct (Brown & Britton 1980), and flocks seem to form during the non-breeding season Oct-May. Hence it is resident in small numbers at L Baringo 50A but increases to 300 Nov-May (St), while peak counts of 3000 at Lessos 49C and 2600 at Sergoit 49A were on 31 Jan and 1 Feb respectively (EABR 1981). Flocks of 500 have occurred at L Turkana Oct-Dec (HH), but the apparent absence of breeding there suggests that the species is an off-season visitor to this low altitude locality (*cf.* Purple Gallinule (198)). Non-breeding visitors are also found on dams near Nairobi 75B Sept-Mar (Pomeroy & Tengecho 1982a). Finally, two presumed vagrants at 2°N in Somalia were Dec and Mar, while the single dated record from Wajir 42A was in Nov (J E Miskell *in litt.*; AM; Ash 1983). Other reference: Plumb 1978c.

198 PURPLE GALLINULE *Porphyrio porphyrio*
B203 / MPG237 / D2C2

Afrotropical: widely distributed in the E from Ethiopia S through East Africa and E Zaire to South Africa. In the W, it avoids the equatorial lowland forests and, like the Common Moorhen (201), appears to be far more scattered and less common in other areas. Distribution in Kenya represents the E limit of the range.

Kenyan: inhabits thickly vegetated, mainly permanent swamps, up to 3000 m but usually below 2400 m (Br; Britton 1979; MPG; St). Distribution is centred in the C highlands and their peripheral plateaux, and the bird is frequent at L Turkana (HH). While possibly overlooked, it is rare at the coast (*e.g.* EABR 1981) and apparently so to the W of the rift valley. Breeding is more possibly overlooked and, while this species has occurred Jan-Nov at the coast (EABR 1981) and is considered resident at L Jipe 100D and Turkana (Br), all eight breeding squares are in or around the C highlands, seven of eight in 500+ mm rainfall areas. Lack of breeding records from L Turkana parallels the Red-knobbed Coot (197) and Common Moorhen (201) and, although the effects of low observer coverage in this remote area cannot be excluded, there remains the possibility that these species are non-breeding visitors from further N. The Purple Gallinule is thought to be resident,

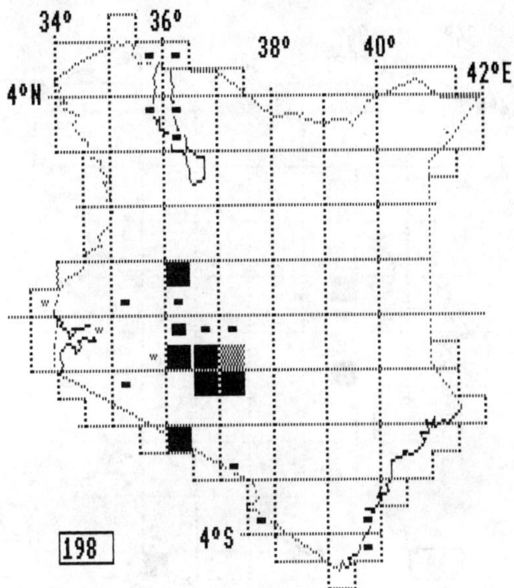

though there is a record of a May-June wanderer around 3000 m at Mau Narok 61D (Br; St; Britton 1979). Brown & Britton (1980) describe dispersal to and breeding on temporary waters in Embu district 63CD during unusually heavy rains. See Allen's Gallinule (199).

199 ALLEN'S GALLINULE *Porphyrio alleni*
B202 / MPG238 / D2C2

Afrotropical: widespread S of 15°N, apart from the dry areas of N and E Somalia and the SW.

Kenyan: like the closely related Purple Gallinule (198), this usually uncommon species is found in dense vegetation fringing permanent swamps, up to 1900 m (Br; Sn; MPG). Range is 93% within 500+ mm rainfall areas, and virtually restricted to the rift valley lakes and the coastal lowlands. It may be resident in the latter, and not uncommon (Br; EABR). Influxes of hundreds at L Baringo 50A occur in late May-early June when suitable habitat exists, and these breed (tiny chicks seen: T Stevenson *in litt.*); numbers then decrease rapidly Aug-Oct (St; EABR). A single wanderer in June at Ferguson's Gulf 13B, L Turkana (HH) may have been associated with the movements responsible for the Baringo influxes. Other dated occurrences include two separate Jan birds at Nairobi 75B (EABR), and Aug and Dec records at 1°-2°N in coastal Somalia (AM).

There are striking similarities in the relationship between this species and the Purple Gallinule, and that between the Lesser (200) and Common (201) Moorhens: thus the first species in each of these pairs is smaller, a recognised intra-African migrant, absent (or virtually so) from L Turkana and commoner at the coast, while the second species in each pair is larger, may make more poorly known movements, occurs (but possibly does not breed) at L Turkana, and is rarer or more local at the coast. Other reference: Britton & Britton 1976b.

200 LESSER MOORHEN *Gallinula angulata*
B198 / MPG240 / D2C2

Afrotropical: widely distributed S of 15°N apart from Somalia and the arid SW. With the exception of the record in 31A, which was in Somalia (AM), distribution in Kenya reflects the E limit of the range.

Kenyan: a shyer bird than the Common Moorhen (201), often on small waters with fringing vegetation, including temporary floods, up to 2000 m (Br; Ja; MPG). Range is 87% within 500+ mm rainfall areas, and the bird is most common in the L Victoria basin, in the highlands, and at the coast (Br). Unlike the Common Moorhen, it is absent from L Turkana.

The Kenyan breeding records (all four squares in areas with over 500 mm rainfall) are some of the very few between 8°N and 13°S (Sn) and, while the Lesser Moorhen can occur in most months, it is a pronounced intra-African migrant. The predominance of records during and after the Apr-Jul rains, when the species can be tempo-

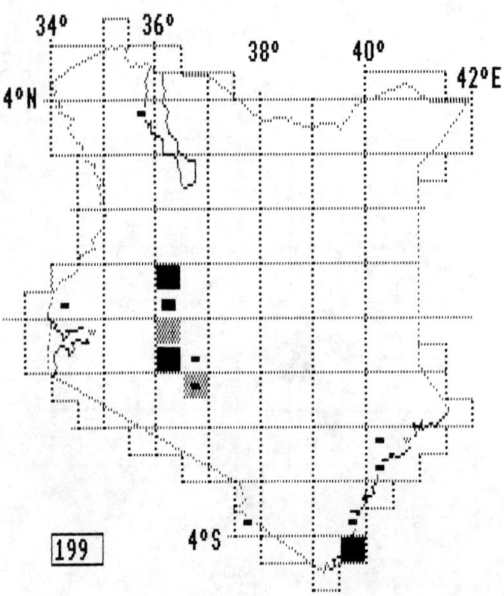

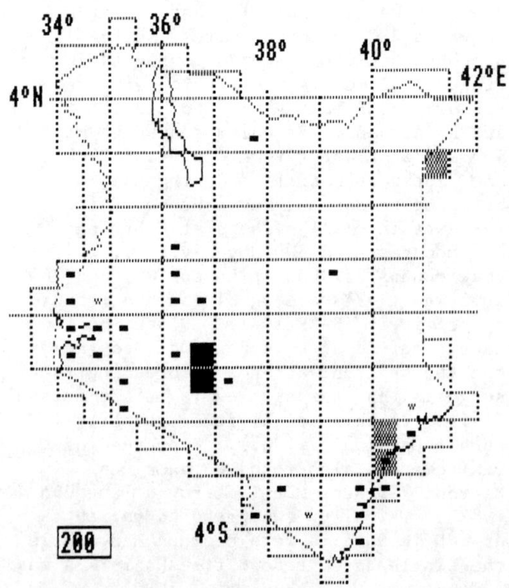

rarily abundant, is thought to indicate non-breeding migrants from the S tropics (Br; EABR). These movements account for occurrences on often small and temporary pools in hostile arid country *e.g.* an overgrown donga 15D (R G Allan pers. comm.), 53A (D A Turner *in litt.*) and the Taru bush 101D (Ja). Appearance at the Ngulia lights 101A in Dec and Jan indicates nocturnal migration (Backhurst & Pearson 1977b; EABR 1984). Ja described it as "very plentiful" and breeding along the flooded lower Tana R 91AC in June 1891, and these details have been reiterated by MPG and Br. This abundance is not a regular phenomenon, however, as it was not seen by the Jul-Sept Tana R Expedition (K L I Campbell *in litt.*), nor during the Jul-Aug fieldwork of AGH. Occurrence at L Baringo 50A has been sporadic (St). See Allen's Gallinule (199).

201 COMMON MOORHEN *Gallinula chloropus*
B199 / MPG239 / D3C2

Afrotropical: widely distributed S of 15°N, mostly on the E, from Ethiopia S to South Africa. It is absent from most of the lowland forest areas of the W, and from the arid areas of E Somalia and the SW. Palearctic migrants may penetrate as far S as Senegambia and N Sudan (Sn).

Kenyan: common but often local on permanent waters with thick fringing vegetation, up to 3000 m and mostly above 1200 m (Br; Se; MPG). This species is often in less open situations than the Red-knobbed Coot (197). It is common in the highlands (Br) but more local at the coast. Occurrence N of 2°N is restricted to L Turkana (where frequent in swampy edges: HH), L Paradise on Mt Marsabit 27D, and a small dam S of Moyale 17C (ADL pers. obs.). Breeding squares are 86% above 1000 m and 86% within 500+ mm rainfall areas, and there are no records from L Turkana (*cf.* Purple Gallinule (198)). Little is known of any movements but it presumably, for example, crossed extensive hostile terrain to reach Marsabit. Br mentions influxes to permanent waters and breeding on seasonal ones, and this moorhen is sporadic at L Baringo 50A (St). One at the Ngulia lights 101A in Dec suggests nocturnal movements (EABR 1982). See Allen's Gallinule (199).

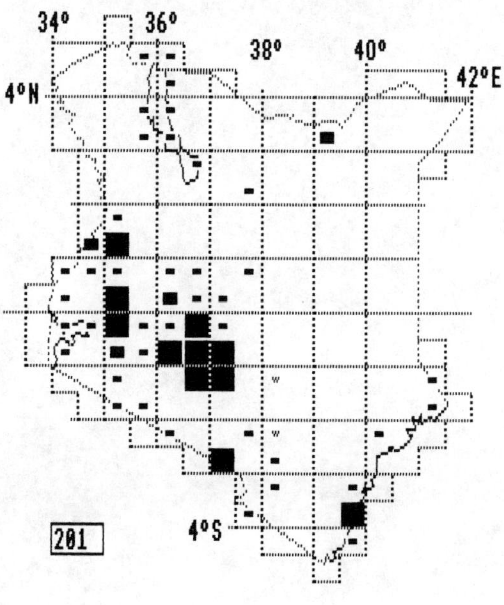

Family Heliornithidae: Finfoots

Three species, one each in the Neotropical, Afrotropical and Oriental Regions. The African Finfoot is a breeding resident in Kenya.

202 AFRICAN FINFOOT *Podica senegalensis*
B216 / MPG243 / D1C1

Afrotropical: widely distributed S of 10°N -15°N, apart from the dry areas of Somalia and the SW.

Kenyan: uncommon but also secretive on permanent rivers and streams that are bordered with thick and at least partly overhanging vegetation (Br; MPG; Ja). 89% of its squares contain permanent rivers but it is known from only 20% of the total squares that contain these waters, and it is almost certainly under-recorded (hence D1C1). Range is 67% below 1500 m (*cf*. Br) and 84% within 500+ mm rainfall areas.

It is known along the Galana R 102A and the Tana R 91A 79C 65BD, at Mzima Springs 89C and in the C highlands. W of the rift valley, there are reports from the Mara R 74AC (EABR 1981) and the Migori R 73B (A Lohding *in litt*.). Absence of recent records from the coast 114 may show a decline. Br considers this bird resident. Other references: Sn; Carver & Carver 1979; Gerhart 1980.

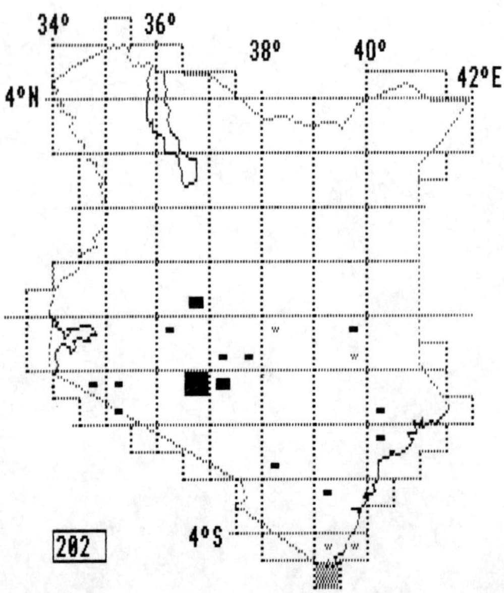

Family Otididae: Bustards

22 species with representatives in all zoogeographical regions except the Nearctic and Neotropical. 14 (64%) are endemic to the Afrotropical Region, while another, the Arabian Bustard (210) also extends marginally to the Palearctic. Eight of the 22 species (36%) occur in Kenya where all but one, again the Arabian Bustard, are breeding residents. All species make local movements. Distribution of breeding records tends to reflect observer frequency. These terrestrial birds are excluded by intensive small-scale agriculture, but this is practically restricted to the highlands, and while the larger Denham's Bustard (207) has declined in these areas, the smaller Black-bellied Bustard (205) is apparently less affected. There are also fears that Kenyan bustards will be exported to the Middle East for sporting purposes (see Fanshawe 1985). This family has a precarious ecology, and all members should be regarded as potentially endangered.

203 WHITE-BELLIED BUSTARD *Eupodotis senegalensis*
B220 / MPG256 / D2C2

Afrotropical: in much of Africa between 17°N and 17°S, and also in E South Africa, but naturally avoiding the major forested areas.

Kenyan: generally uncommon in open habitats, up to 2000 m (Br; Be; MPG; La). Range is 83% within intermediate, 250-1000 mm rainfall areas and 74% less arid-semiarid, and this bustard is scarce in the dry N and E, and in the high rainfall areas of the coastal strip, the highlands and the W. DEP (pers. obs.) found this species resident in 74B 75BD 88AB, and La note similar status in arid Tsavo E 101B. Br considers it generally so.

204 HARTLAUB'S BUSTARD *Eupodotis hartlaubii*
B217 / MPG259 / D2C2

Afrotropical: the NE, from E Tanzania N through Kenya and NE Uganda to S Sudan, Ethiopia and Somalia.

Kenyan: rather local and often uncommon on open grasslands, mainly E of the rift valley, up to 1600 m (Br; Ja; MPG; La). Hartlaub's Bustard forms a superspecies with the Black-bellied Bustard (205: *q.v.*) (Sn). The two are mainly allopatric, and separated by differing habitat preferences where both occur (Br). We agree with Sn's summary of their broad environmental differences: Hartlaub's tends to be 'lower and drier'. Thus Hartlaub's has 74% of its squares below 1500 m and is rare above that altitude (MPG; Br), whereas the Black-

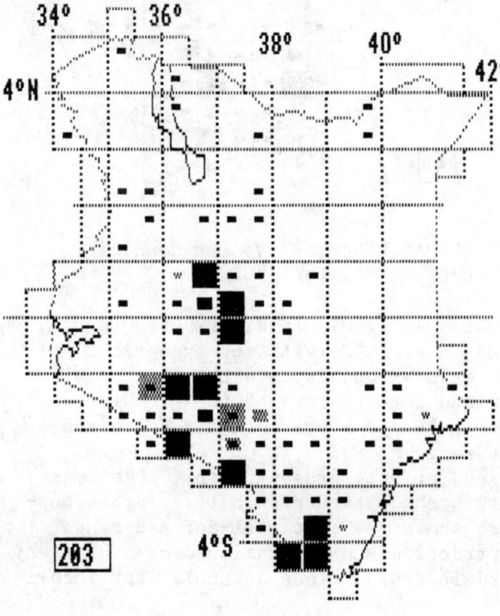

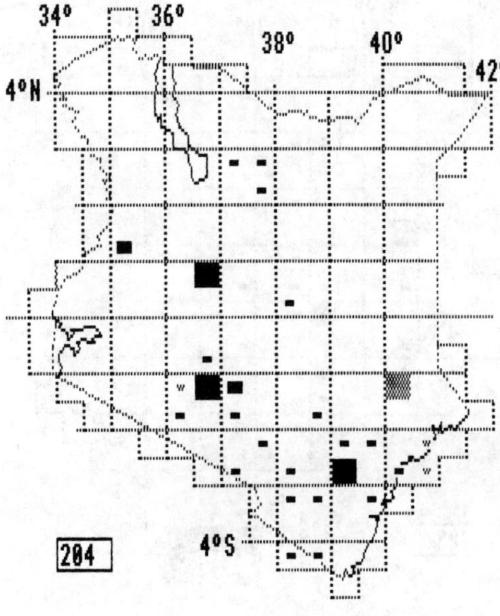

bellied occurs up to 2500 m (Br) and is present in 70% of the squares above 1500 m. Also, Hartlaub's has 81% of its squares arid-semiarid, whereas the Black-bellied is reported from 58% of the subhumid-humid squares.

Hartlaub's Bustard is thought to be resident (Br; La: 101B; DEP pers. obs.: 88D 89C 100B 101A). The population in NE Uganda has been considered isolated (Br), but this may not be the case in view of the probable breeding in 37C (T J Barnley *in litt.*).

205 BLACK-BELLIED BUSTARD *Eupodotis melanogaster*
B218 / MPG258 / D2C2

Afrotropical: widely distributed S of 15°N apart from the arid areas of E Somalia and the SW, and the equatorial lowland forests in the W.

Kenyan: the Black-bellied Bustard is a commoner bird than Hartlaub's Bustard (204: *q.v.*)(70 squares as against 28) and, in addition to grassland, also frequents areas with bush and trees, including open woodland (Br). Range is 84% above 500 m and only 25% within arid areas. This bustard is most common in the highlands and the L Victoria basin, and scarce at the coast and in the lowlands of the N and E (Br). Br regards this species as resident. DEP (pers. obs.) found similar status in moist subhumid 75B, but also recorded an annual Aug-Nov influx to the Ruaraka-Kahawa-Ruiru area of the square.

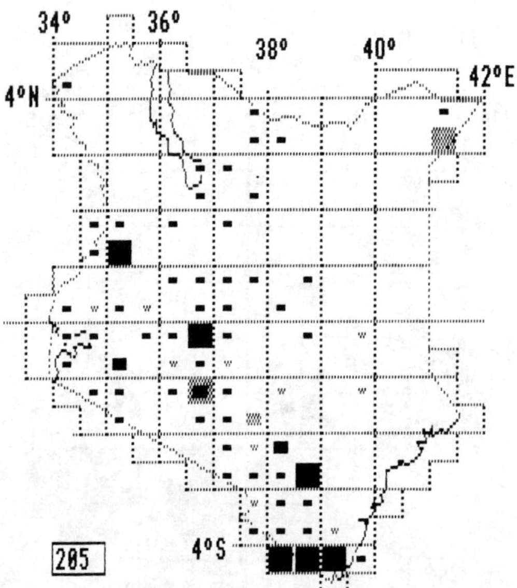

206 BUFF-CRESTED BUSTARD *Eupodotis ruficrista*
B219 / MPG257 / D2C2

Afrotropical: occupies four areas: in the S, between 15°S and 30°S; in the W in Senegambia and Mauritania; in the N tropics from Niger to W Sudan; and in the NE from E Tanzania N through Kenya and NE Uganda to S Sudan, Ethiopia and Somalia.

Kenyan: a common bird, more often heard than seen, in dry, thickly bushed habitats below 1400 m (Br; MPG; HH; La; Lack 1985). Range is 94% arid-semiarid (*cf.* Br), including 65% of the most arid squares (moisture index below 1.2). Thus this species is peripheral to the highlands, and the line 48-63-74 represents part of the W edge of its East African range. It is also absent from the subhumid country in the coastal strip. This bird is resident (Br; La: 101B; DEP pers. obs.: 75C 88AD 89C 100B 101A).

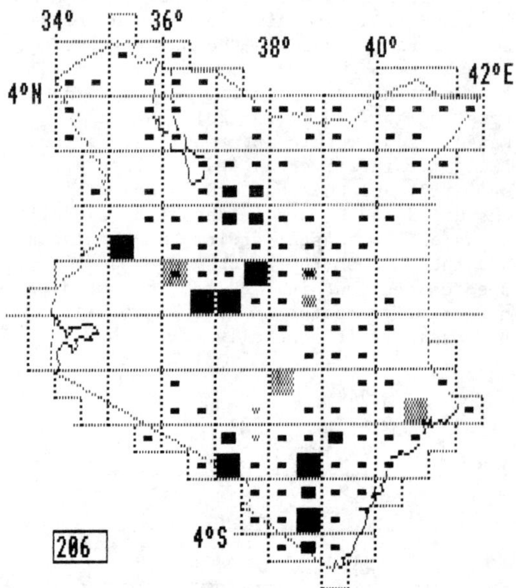

207 DENHAM'S BUSTARD *Neotis denhami*
B221 / MPG251 / D2C2

Afrotropical: widely distributed in many areas S of 16°N, with the exceptions of the arid NE and SW, and the equatorial lowland forest areas of the W. Distribution in Kenya represents the E limit of the range.

Kenyan: open grassy plains, sometimes with bush or tree growth (Br). This bustard shows the most well-defined range contraction and population decrease of any bird in Kenya. Thus Ja could list locali-

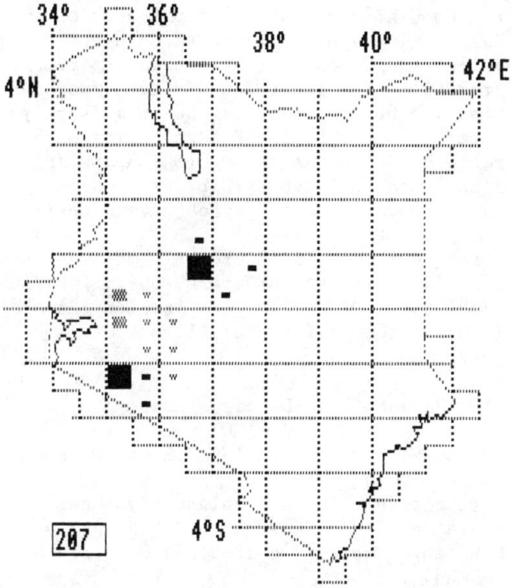

ties in 49CD 61AB 62AC 75A and, though he considered the bird rather scarce and local, could count over 50 in a day near Nakuru 62A. MPG still described it as "abundant" in some areas, but Se found it uncommon at Mau Narok 61D. Now this bustard has retreated from the rift and W highlands to their NE and SW peripheries, to produce a disjunct population in the Timau 51C-Archer's Post 51B-Maralal 38D area, and another centred in Mara-Narok 74. There has been a considerable decline in the latter area in recent years (EABR 1977). Breeding in 50B represents an adult with a chick (C Mortensen *in litt.*), while that in 74A is of a single egg laid in late 1982 (T Stevenson *in litt.*).

Both Ja and MPG describe the bird as shy and apt to leave its nest if at all disturbed. This steep decline (53% of its squares are pre-1970) is presumably due to increased human population and agriculture. Sn mentions possible competition with the larger Kori Bustard (209) in S Africa, but there are no records of that species in Denham's former squares W of the rift.

Br notes Denham's up to 3000 m, and all squares (excluding 51B: see below) are above 1500 m. All recent records (again excluding 51B) are in the 500-1000 mm rainfall regime, while the bird has disappeared from the 1000+ mm rainfall squares W of the rift valley, and the 500-1000 mm squares in the rift itself.

The record in 51B is of two adults photographed near Archer's Post on 4 Jan 1985 (A J Holcombe & party). The area was lush after recent rain, and these birds were presumably wanderers E or NE of their 50B-51C range, since there are no previous records from this well covered area. This is the only square below 1500 m, and within the 250-500 mm rainfall regime. The range of the disjunct N population of Denham's Bustard is now adjacent to that of the drier country Heuglin's Bustard (208), which is a member of the same species-group (Sn), and 51B constitutes their first area of overlap at QSD scale.

Although perhaps secure in the Maasai Mara GR, Denham's Bustard is the most endangered of its family in Kenya.

208 HEUGLIN'S BUSTARD *Neotis heuglini*
B222 / MPG252 / D2C1

Afrotropical: the NE, from N and E Kenya N to Ethiopia and Somalia.

Kenyan: a local but not uncommon species of low, dry, open country (Br; MPG; HH). Range is 87% below 1000 m, 87% arid and 93% within 0-500 mm rainfall areas. It may be reasonably regular N of about 1°N, but was not recorded by observers resident in 42A 65B, and is absent S of 3°30'N in adjacent S Somalia (AM). Kenyan records S of the equator thus apparently represent occasional wanderers (La; Powiss & Wacher 1982; H M H Braun *in litt.*). Wanderers also move SW into moist subhumid country on the NE periphery of the C highlands, and records in 39A 51BD are adjacent to the range of the closely related Denham's Bustard (207: *q.v.*). Restriction to the E of L Turkana (HH) but occurrence to the W

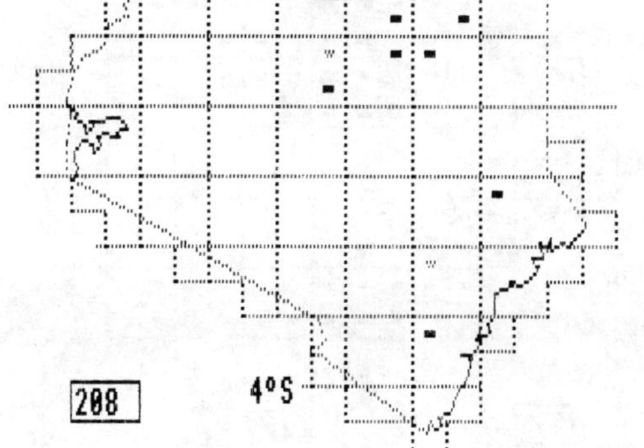

further N parallels the distribution of the Vulturine Guineafowl (177), which is another cursorial species at the SW limits of its range in these areas. Other reference: Williams 1968.

209 KORI BUSTARD *Otis kori*
B224 / MPG250 / D3C3

Afrotropical: occupies two distinct areas: in the S, it is widely distributed S of 15°S, while in the NE it extends from NE Tanzania N through Kenya and N Uganda to S Sudan and S Ethiopia. Absence of recent records from Somalia (AM) means that distribution in Kenya now reflects the E limit of the range.

Kenyan: found on grasslands up to 2000 m, and most common in the rift valley highlands (Br; Ja). Range is 82% arid-semiarid and only 3% within the wettest, 1000+ mm rainfall areas. Like the similarly terrestrial Ostrich (1), Secretary Bird (85) and Crowned Plover (220), this bustard is absent from the area 63-65 76-78, probably due to the combined effects of a high human population and agriculture (mostly in the W), and thick bush, which impairs ground species' range of vision. Similar factors, with woodland and forest instead of bush, probably account for its absence in the W 48 49 60 61 (*cf.* Br), while near the coast it is only present in the semiarid Shimba Hills 114A and Tana R 79C 91AC areas.

Since this species is absent W of L Victoria (Sn; Br) and apparently now so in Somalia (AM), the sole connection between the populations N and S of the equator in East Africa appears to be through the rift valley 62AC, and there is no recent record from 62A. The lower Tana 79C 91AC population may be disjunct, as the bird is only a rare visitor to the E edge of its main range on Galana Ranch 90C (B R Heath *in litt.*) and in Tsavo E 101B (La), while it has not been reported by observers resident in 65B.

The Kori Bustard is probably mainly resident (Br; St: 50A; DEP pers. obs.: 74B 75CD 88ABD 89C 100B 101A). See Arabian Bustard (210). Other reference: Allen & Clifton 1972.

210 ARABIAN BUSTARD *Otis arabs*
B223 / MPG249 / D1C0
NO MAP: ONE IMPRECISELY LOCATED RECORD

Afrotropical: the N tropics, from Mauritania and Ivory Coast E to N Ethiopia and S Sudan. There is a single old record S to Kenya.

Kenyan: the only accepted record concerns a pair collected by V G L van Someren during Jan 1932 in "Turkana, East Lake Rudolph area" (van Someren 1925-35). While noting presence in Sudan adjacent to 2ABC, Turner (1982) rejects all other Kenyan records as having been insufficiently distinguished from the Kori Bustard (209), which is a mainly allopatric member of the same superspecies (Sn).

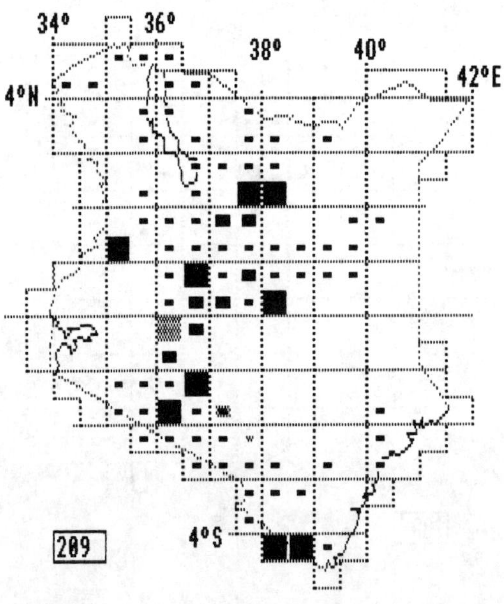

Family Jacanidae: Jacanas

Seven species, with representatives throughout the tropics. Two (29%) occur in the Afrotropical Region, and both breed in Kenya.

211 JACANA *Actophilornis africanus*
B225 / MPG264 / D3C2

Afrotropical: widespread S of 16°N.

Kenyan: common in well vegetated, swampy habitats up to 3000 m, foraging both along the shore and on floating aquatic plants *e.g.* water lily pads (Br; Se; Ja; MPG). It is confined to sheltered bays and river mouths at L Turkana (HH), and is not found on sea coast. Range is 73% within 500+ mm rainfall areas (with L Turkana a marked exception).

While it must obviously undertake movements if its habitat dries out, Br states that it does not wander to seasonal waters whereas AM note that in Somalia it often does so. Whichever is the case, it must be capable of crossing extensive arid areas to have reached Mt Marsabit 27D, and Se recorded two solitary, highly transient wanderers in Apr and June at Mau Narok 61D. It is an occasional visitor to Tsavo E 101B (La).

212 LESSER JACANA *Microparra capensis*
B226 / MPG265 / D1C1

Afrotropical: a localised and discontinuous distribution from Niger E to Ethiopia, and S through East Africa and E Zaire to Angola, Botswana and E South Africa. Distribution in Kenya represents the E limit of the range.

Kenyan: a small, little known and retiring species, easily overlooked and/or confused with the juvenile Jacana (211): hence D1C1. Like the Jacana, it requires waters with substantial floating vegetation but it is far more uncommon and local than that species (Br). 85% of the squares and all of the breeding squares receive over 500 mm annual rainfall. The Lesser Jacana is most frequently observed at L Jipe 100D, where it occurs throughout the year (Br; EABR), but is possibly most common in the W highlands. Br mentions scattered highland records at 1800-3000 m. Ja described this species as "plentiful" on L Naivasha 62C, but there has presumably been a decline, possibly due to the recent radical ecological changes there, since it is not so now (EABR). It has recently been found in the coastal lowlands (EABR). The only record of wandering comes from NE Tanzania (Br).

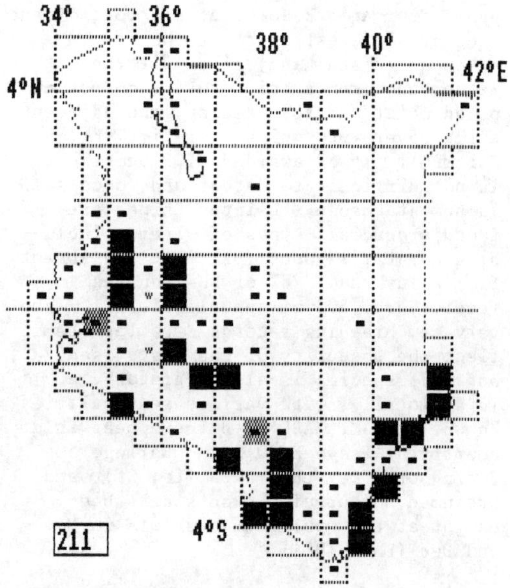

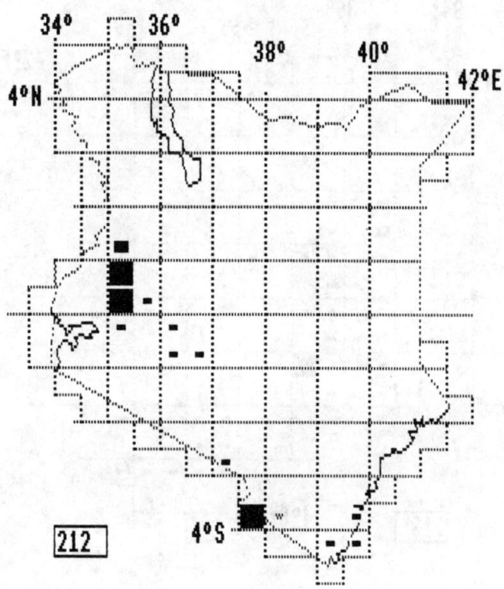

Family Rostratulidae: Painted Snipes

Two species, represented in all zoogeographical regions except the Nearctic. One occurs in the Afrotropical Region, and breeds in Kenya.

213 PAINTED SNIPE *Rostratula benghalensis*
B227 / MPG297 / D2C2

Afrotropical: widely distributed S of 15°N, apart from arid E Somalia, and most of the equatorial forest areas of the W.

Kenyan: a skulking, close-sitting, rather uncommon species that requires exposed moist mud for feeding, and adjacent thick cover into which to retire (Br; Sn; Ja; MPG). It is awarded D2C2 because it is not difficult to detect and, once seen, is unmistakeable. Painted Snipes are found around all types of waters, including coastal lagoons (Br), but are absent from L Turkana. 78% of the squares receive over 500 mm annual rainfall. The very few breeding records doubtless reflect the obscurity of the nests (see Ja), but this species is also migratory, at least locally, with various seasonalities known (St; AM; EABR). Birds appear in dry country at Tsavo E 101B and Baringo 50A during or after the rains (La; St), and presumed nocturnal migrants have been caught at the Ngulia lights 101A in Nov and Dec (EABR 1981).

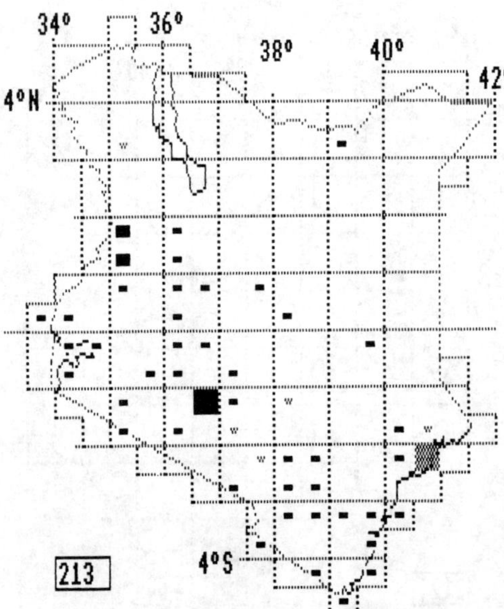

Family Haematopodidae: Oystercatchers

Seven species, with representatives in all zoogeographical regions. Two (29%) are found in the Afrotropical Region, and one is a non-breeding migrant to Kenya. In an important paper, Dowsett (1980) reviewed evidence for the migration of various Palearctic wading birds across the continent from Ethiopia to their wintering grounds in coastal South Africa. Whether this migration is on a broad front or follows the line of the rift valley lakes is uncertain, but the important point in relation to this atlas' maps is that this hypothesis explains the occasional occurrences inland of species which are otherwise almost exclusively coastal. Inland records are commoner during S passage, suggesting this movement to be more interruptable, with birds descending to feed around waters en route, whereas N passage may be more motivated and continuous: a similar distinction between the two migrations has also been mooted for some Palearctic raptors *e.g.* Steppe (130) and Lesser Spotted (127) Eagles.

214 OYSTERCATCHER *Haematopus ostralegus*
B228 / MPG294 / D3C3

Afrotropical: a Palearctic species, wintering on African coasts S to Senegambia and N Somalia, with only occasional records further S.

Kenyan: recorded annually on the coast in very small numbers since at least 1977 (EABR). Earlier records were mostly in the very early Jul-Sept period and on rocky offshore islands N of Malindi 103A, and the species was considered rare on the mainland (Smart 1949; Fogden 1963; Ba; Spindlow 1974; Britton & Britton 1976b; Br). Subsequent observations have involved many later dates in the Palearctic season, and many mainland and more southerly sites (EABR): these differences probably derive from better observer coverage. There are records for most months, and oversummering also occurs in Somalia (AM). Early (1932) descriptions of Oystercatchers as "common" on the coast no longer apply so that there may have been a decline (Ba; *e.g.* Pearson 1984d).

Old records (van Someren 1925-35) at L Naivasha 62C and L Victoria are ignored by Ba and treated as unconfirmed by Dowsett (1980), but there have been Feb, Mar and Jul records at L Turkana (Ba; EABR 1979), and two Sept birds at L Nakuru 62A (OS-C 1977; EABR 1982). In addition to Dowsett's (1980) overflying theory, this species' abundance on offshore islands (*e.g.* 18 on Manda Mtoto 91B in Aug: Britton & Britton 1976b) and an Apr bird on N passage 8 km offshore (Ba) have raised the possibility of regular offshore passage. Ash (1981b) describes limited coastal migration in S Somalia.

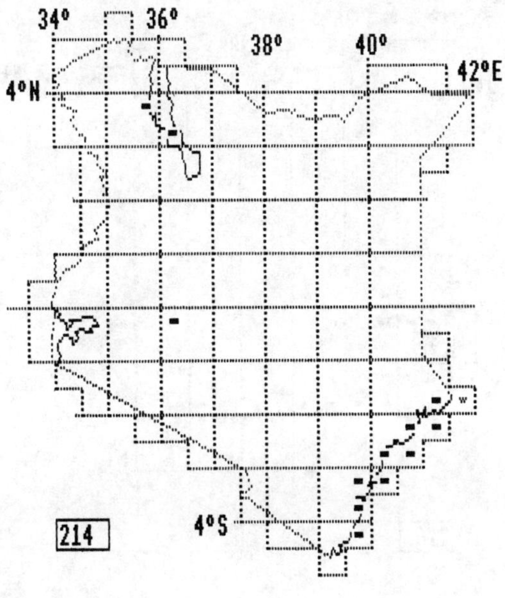

Family Charadriidae: Plovers

62 species, with representatives worldwide. 28 (45%) occur in the Afrotropical Region, and of these, 21 (75%) are known in Kenya. Of Kenya's species, 8 (38%) are non-breeding migrants from the Palearctic, 11 (52%) breed, the Wattled Plover (216) lacks breeding records, and the Brown-chested Wattled Plover (221) is a non-breeding intra-African migrant. Inland records of the Palearctic members of this family which predominantly winter in coastal Africa can be considered in the light of Dowsett's (1980) proposals regarding migration across continental Africa (see Haematopodidae).

215 LONG-TOED PLOVER *Vanellus crassirostris*
B245 / MPG293 / D2C2

Afrotropical: a swamp bird of the E, from S Sudan through East Africa and E Zaire to Angola, Botswana and E South Africa. There is also an isolated population at L Chad. Distribution in Kenya reflects the NE limit of the range.

Kenyan: locally common on permanent swampy areas up to 2200 m and, like the jacanas (211, 212), requiring substantial floating vegetation to walk and forage on (Br; MPG). Range is 85% within 500+ mm rainfall areas and, while present on both eutrophic and oligotrophic waters, there are no reports from carbonate lakes. The Long-toed Plover is most common W of the rift valley, in the L Victoria basin and peripheral highlands (Br; Britton 1978), while occurrences at the coast are confined to the vicinity of the lower Tana R 79C 91. The curiously fragmented distribution may be due to the fact that it is extending its range. Thus Ja, who spent much time hunting and egg collecting in and around lakes and swamps, could only quote a single Kenya locality (91D), at second hand. The species has colonised L Baringo 50A since 1977 (St; EABR), and appearance in further new localities may be expected. Aside from this range expansion, Br thinks this bird resident.

216 WATTLED PLOVER *Vanellus senegallus*
B248 / MPG291 / D2C2

Afrotropical: widely distributed S of 15°N but absent from the equatorial lowland forest areas of the W, the E periphery of the continent from Somalia to N Mozambique, and most of the SW. Distribution in Kenya reflects the E limit of the range.

Kenyan: a widespread but local bird of the moist W plateau that frequents damp, short grass near water, up to 2200 m (Br;

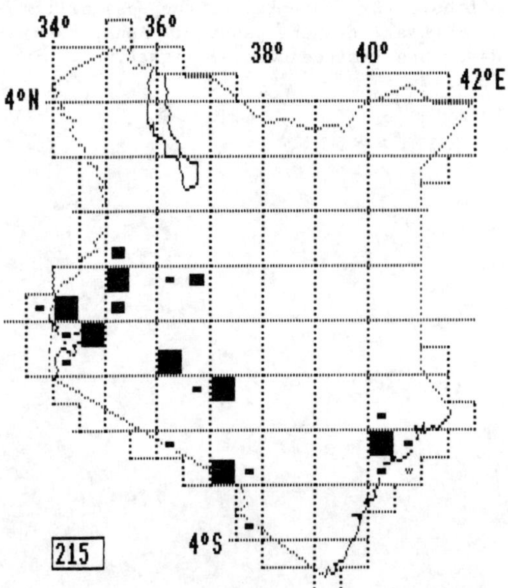

215

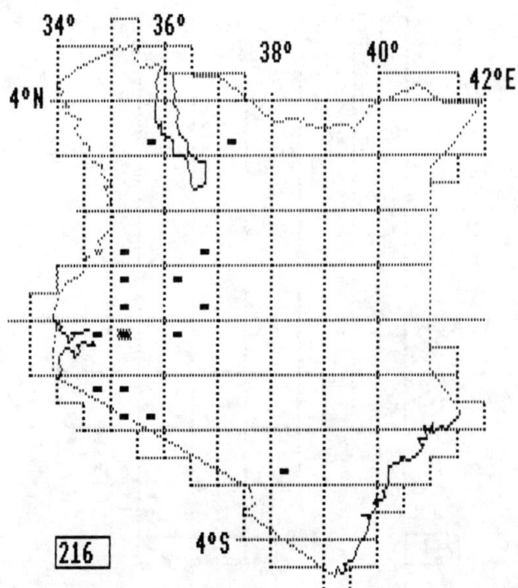

216

MPG). Excluding wanderers (see below), range is entirely above 1000 m, in 500+ mm rainfall areas. Single birds or small parties can wander E to the rift valley and marginally on to the E plateau, E to North Horr 15C and Tsavo W 89C. These birds reach areas below 1000 m, where annual rainfall is 0-250 mm, and have occurred in May (50A) and Jul (13D)(HH; St; Br).

Ja mentions breeding in the 61A area during May (mapped as pre probable), but admits that he could never find eggs or chicks. Br records occurrence throughout the year in the W highlands, but well defined seasonality at similar longitudes in adjacent NE Tanzania. The lack of confirmed breeding records is particularly striking when compared to those known for other areas of East Africa (Brown & Britton 1980), and those known for equally detectable congeners in even remote areas of Kenya *e.g.* Black-headed Plover (219). Hence we suggest that this species is a non-breeding visitor to the E edge of its range, with non-breeding, perhaps first year individuals, temporarily resident.

217 BLACKSMITH PLOVER *Vanellus armatus*
B243 / MPG287 / D3C2

Afrotropical: mainly the S, widely distributed S of 10°S and further N through S and E Tanzania to the N extremity of the range in Kenya. This species and the Spur-winged Plover (218) form a species-group and replace each other in the S and the N of the Region respectively: they are predominantly allopatric, and Kenya is noted as their principal area of marginal sympatry (Sn).

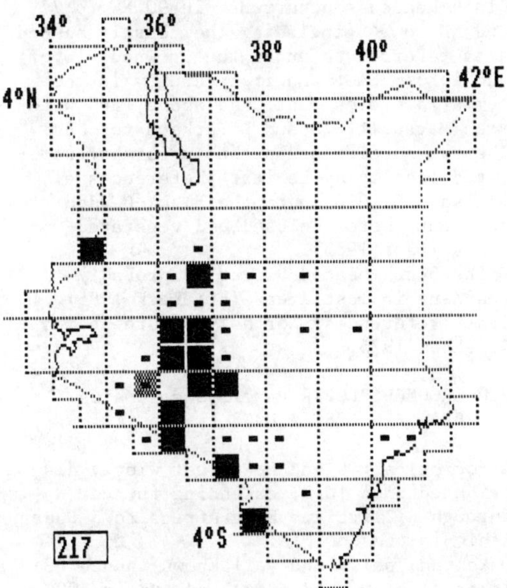

Kenyan: common, well known and noisy (hence D3), on muddy shores and short grass areas near water (Br; MPG)(*cf.* Spur-winged Plover). Here, where their ranges overlap, the Blacksmith is primarily a bird of the rift valley highlands and the adjacent E plateau, while the Spur-winged Plover, particularly as a breeding species, tends to occupy the N, E and coastal lowlands. Thus, ignoring evident wanderers (52B 64B 65B 91AB 101B), the Blacksmith has 72% of its squares and 92% of its breeding squares above 1000 m (36D: T J Barnley *in litt.*), while the Spur-winged has 72% of its squares and 72% of its breeding squares below 1000 m. Similarly, 45% of the Blacksmith's squares (excluding wanderers) are subhumid-humid, while 81% of the Spur-winged's are arid-semiarid.

Their maps show considerable overlap of range, but Reynolds (1980) suggests this to be mostly due to wandering of Spur-winged Plovers during their non-breeding season. The Blacksmith wanders much less (see below). In fact, even at QSD scale, the two species share only two breeding squares, 62C Naivasha and 88C Amboseli. They are only known actually to breed alongside each other, occasionally, at the latter (where they do not exhibit inter-specific aggression (Reynolds 1980)). A possibly hybrid individual was seen with Blacksmith Plovers at Ziwani 100B on 17 Dec 1982 (Pearson 1983a), about 70 km SE of Amboseli.

Br states that most Blacksmith Plovers are apparently resident at their breeding sites but that wandering can occur, and this is illustrated by records below 500 m in the E lowlands E to Garissa 65B (R & L M Greenham *in litt.*), Witu area 91AB (V Haas *in litt.*) and Tsavo E 101B (very sporadically, Apr-Nov: La), and probably by Dec, Feb and Jul birds around 3000 m at Mau Narok 61D (Britton 1979). Blacksmith Plovers were rare or absent on dams at Ruaraka 75B in May-June (Pomeroy & Tengecho 1982a). Other reference: Kutilek 1974.

218 SPUR-WINGED PLOVER *Vanellus spinosus*
B249 / MPG286 / D2C2

Afrotropical: the N tropics, from the W coast between 5°N and 16°N to Ethiopia and W Somalia, and S through Uganda and Kenya to the E Zaire borders and Tanzania. See Blacksmith Plover (217).

Kenyan: common on mud and sand flats and short grass areas near water (Br). These habitats are very similar to those of the

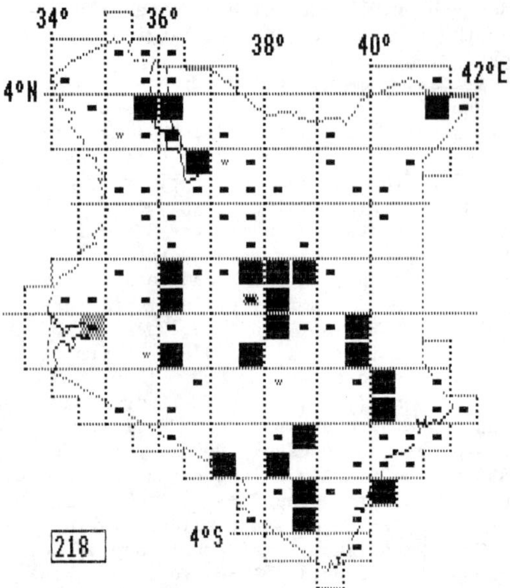

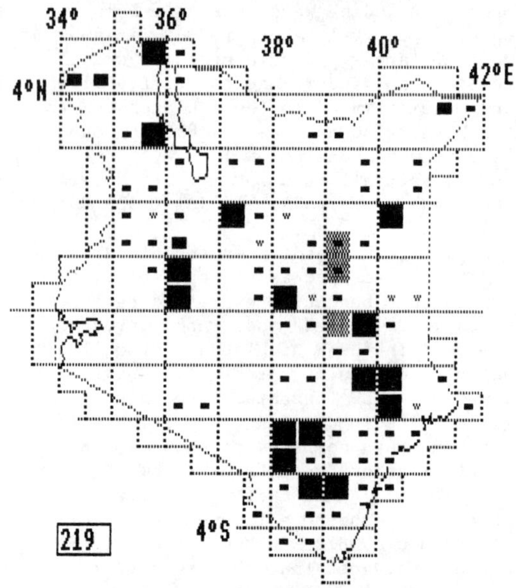

closely related Blacksmith Plover. The Spur-winged is common in the N and E lowlands and also occurs in small numbers in even littoral habitats N of Mombasa 114B, exceptionally even on the open shore (Br). It is most common on the rift valley lakes, especially L Turkana (HH), but scarce on L Victoria (Br). Although thought mainly resident (Br; La; St), Reynolds (1980) mentions wandering of Spur-winged Plovers during the non-breeding season. Examples of wanderers may include a bird on Diani Beach 114B in Feb (EABR 1981), and a late Nov individual around 3000 m at Mau Narok 61D (Britton 1979). This plover disperses miles inland from L Turkana after rain (HH), and may be a recent colonist in Tsavo E 101B (La).

219 BLACK-HEADED PLOVER *Vanellus tectus*
B251 / MPG292 / D2C2

Afrotropical: the N tropics, from the W coast between 10°N and 17°N to Somalia, and S to Kenya and N Uganda. Apart from a single record in NE Tanzania (Br), distribution in Kenya represents the SW limit of the range.

Kenyan: locally common in dry bush and grassland (Br; Ja; MPG; La; St). Range is 83% below 1000 m and 90% arid-semiarid (*cf.* Br), and its SW limits are defined by the N and E peripheries of the highlands. While there is only a single record known from Tanzania (Br), there are presumably more frequent Tanzanian occurrences in the 100D-113AB border area. This species is rare in the subhumid areas of the coast.

Although the Black-headed Plover is widespread in semiarid country elsewhere, it is noticeably sparse in the extensive semiarid plains to the S and SE of the C highlands 76BD 87 88. This may be because these areas are occupied by the Crowned Plover (220), which appears to be the Black-headed's ecological counterpart (Sn). Although there is overlap between these two species at our scale of mapping, some separation is still discernible *e.g.* the Black-headed's occurrence in 90 NW to 77 and NE to 54 (including the densely bushed area referred to under the Crowned Plover) and in 37-38AC; and its scarcity in the semiarid plains mentioned above, and between Marsabit 27D and L Turkana (*cf.* Br). The two species are probably further separated locally by habitat preferences *e.g.* in Tsavo E 101B where the Black-headed frequents more enclosed and vegetated areas (Lack 1985).

The Black-headed Plover is probably resident in most areas (Br; Lack 1985). Other references: North 1936; Greenham & Greenham 1975.

220 CROWNED PLOVER *Vanellus coronatus*
B244 / MPG281 / D3C3

Afrotropical: S and E Africa: widely distributed S of 10°S, extending further N through E Zaire and East Africa to S Sudan, Ethiopia and Somalia.

Kenyan: noisy and well known (hence D3) in open and bushed grassland, up to 3000 m

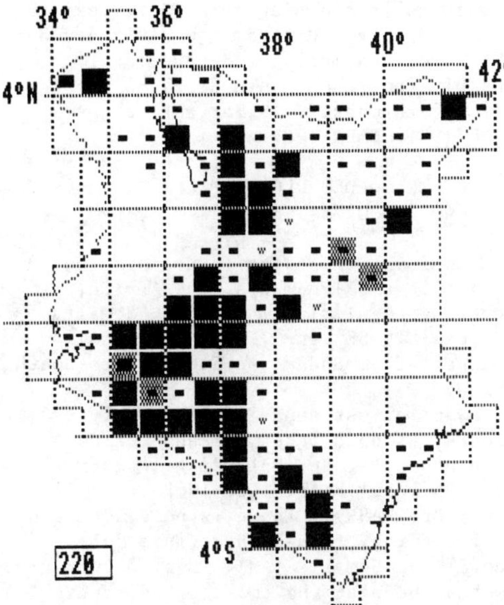

220

(Br; Se; MPG). Its virtual absence in 64 65 77 78 parallels that of other terrestrial species of open habitats: Ostrich (1), Secretary Bird (85), Kori Bustard (209) and Chestnut-bellied Sandgrouse (318). This is an area of thick bush that restricts the range of these species' vision. Interspecific competition with the Black-headed Plover (219: *q.v.*) may be a factor in the Crowned Plover's absence from other parts of the SE (with the notable exception of the vicinity of the lower Tana R 79C 91) and from parts of the NW 37 38. The Crowned is similarly absent S of 3°N in adjacent S Somalia, where the Black-headed Plover is very common (AM). The Crowned is furthermore not found in the high rainfall areas of the W 48 49 60, nor in adjacent areas of Uganda (Br).

Breeding seems to occur throughout the range, and this species is considered resident by Br, Be (74), La (101B) and DEP (pers. obs.: 74BD 75 88ABD 89C 100B 101A). Local movements probably occur, however, and this bird is often heard on the wing at night (Ja; La). Se recorded two Apr wanderers around 3000 m at Mau Narok 61D.

The Crowned Plover is closely related to the Senegal (222) and Black-winged (223) Plovers (Sn).

221 BROWN-CHESTED WATTLED PLOVER *Vanellus superciliosus*
B250 / MPG288 / D2C0
NO MAP: post presence 60B

Afrotropical: an intra-African migrant that breeds in Ghana and Nigeria, and then moves through Zaire to spend its non-breeding season in SE Zaire, mostly W areas of Tanzania and Uganda and, occasionally, on the E limit of its range, the far W of Kenya.

Kenyan: open grassland, particularly recently burnt ground (Br). East African occurrences are virtually confined to the Jul-Dec period (Br) and the only Kenyan records are five specimens from Kisumu 60B in Aug and Dec (van Someren 1925-35), and a bird at Kendu Bay 60B on 10 Sept 1977 (Br). van Someren's (*op. cit.*) mention of poorly mobile juveniles produced suggestions of breeding in Kenya, but these birds were in fact fledged immatures, quite capable of migration from their W African breeding grounds (Dowsett 1977).

222 SENEGAL PLOVER *Vanellus lugubris*
B246 / MPG282 / D2C2

Afrotropical: sporadically distributed around the Gulf of Guinea to Angola, extending E through Zaire to East Africa, and thence S through Malawi and Zambia to E South Africa. Distribution in Kenya represents the N limit of the range.

Kenyan: reasonably common in open grassy and bushed habitats, often near water, up to 3000 m but mostly below 1500 m (Br; Britton 1979; MPG). 87% of the squares and the five breeding squares are in 500+ mm rainfall areas, and this plover is most common in the coastal lowlands and L Victoria basin (Br).

It is a pronounced local migrant that shows diverse but apparently regular sea-

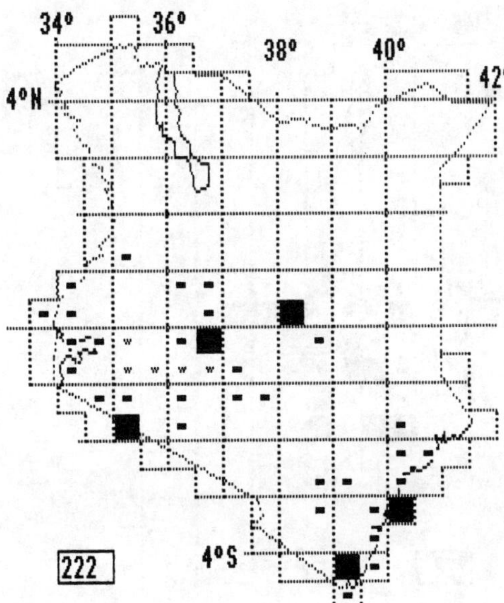

222

sonalities at most if not all localities: thus mainly June-Aug in Tsavo E 101B (La), mainly Apr-Jul at the coast (and on the adjacent SE Somalia coast: AM), and Nov-June in Siaya 48C (Br). St recorded flocks in June-Aug 1978 at L Baringo 50A, and movements are nocturnal (Ja).

The Senegal Plover forms a species-group with the Black-winged Plover (223)(Sn). The two frequent similar habitats, but the Black-winged is confined to the highlands, and less migratory. See Crowned Plover (220). Other reference: Donnelly 1976.

223 BLACK-WINGED PLOVER *Vanellus melanopterus*
B247 / MPG283 / D2C2

Afrotropical: in two distinct areas: in the NE, a highland bird from N Tanzania through Kenya to Ethiopia and NW Somalia; while in the S it occurs in E South Africa. Distribution in Kenya represents the E limit of its range.

Kenyan: a common and gregarious species of short grassland and recently burnt ground, mainly above 1500 m but breeding down to 1300 m on the Athi Plains 75BD (Br; Be; Se; Ja; MPG). Range is 92% within 500+ mm rainfall areas.

In contrast to the Senegal Plover (222: *q.v.*), this bird is resident in many areas (Br). But at Mau Narok 61D (2700-3000 m), it is a migrant which appears in Jan, breeds Mar-Apr or May-June, and then congregates into flocks of over 1000 that leave the area by Sept: they are heard migrating at night, presumably to lower elevations in the highlands (Se; Sessions 1975). Old records from low altitudes 88B 100B were presumably wanderers (Br), possibly from the Chyulu Hills 88D or the Kilimanjaro massif. Other references: North 1951; EABR 1984.

224 LESSER GOLDEN PLOVER *Pluvialis dominica*
B240 / MPG278 / D1C0

Afrotropical: the Asiatic race *P.d. fulva* winters in small numbers in E Africa, occurring regularly in Eritrea, Somalia and possibly SE Kenya, and sporadically elsewhere from Sudan and Ethiopia to South Africa.

Kenyan: almost annual in extremely small numbers, at open lake, dam and marine shore habitats, and Malindi golf course 103A, 26 Sept-18 Apr (Lee 1962a; Ba; Plumb 1978a; Br; EABR). There is no evidence of oversummering, though there is a Jul record from Somalia (AM). Most occurrences have been at the coast or adjacent areas, and the presence of flocks of up to five around L Bilisa 91A in four consecutive years may represent the species' most southerly regular wintering area (EABR; D J Pearson pers. comm.). Three records in and W of the rift valley have involved single birds at L Nakuru 62A on 15 Oct 1972, Sergoit dam 49A on 9 Oct 1976, and Ferguson's Gulf 13B on 25 Dec 1976 (Plumb 1978a; EABR 1977). Other reference: Ash 1980.

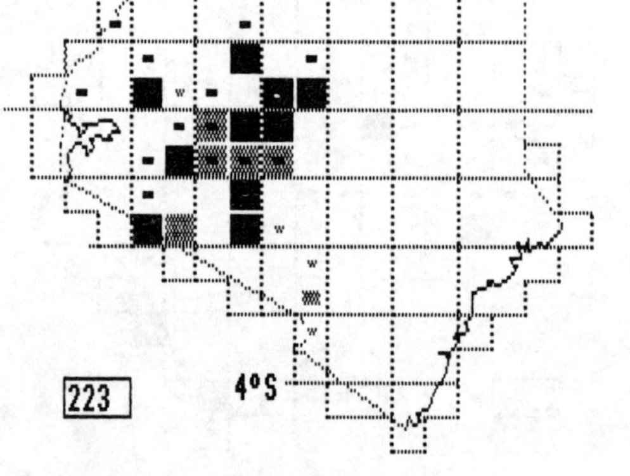

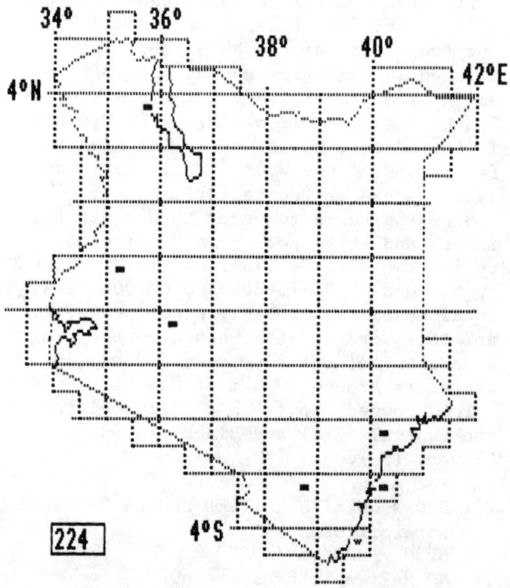

225 GREY PLOVER *Pluvialis squatarola*
B241 / MPG279 / D2C2

Afrotropical: this Palearctic species is an abundant winter visitor to the coasts of Africa, and occasionally appears inland, especially in the E and S.

Kenyan: a common passage migrant and winter visitor of open shores at the coast (Br), where Pearson & Britton (1980) estimated a wintering population of well over 1000. It mainly occurs Aug-Apr and, while many first year birds oversummer, all individuals in breeding plumage have migrated by May (Pearson & Britton 1980). Inland records are mostly on the rift valley lakes and, apart from L Turkana, mainly during Sept-Nov or Apr passage (Br; Pearson & Stevenson 1980). Dowsett (1980) mentions this species in his review of transcontinental migration by waders, and the fact that inland S passage records outnumber those during the N movement (St; La; EABR) supports his view that the former is more interrupted. Ash (1981b) noted N passage along the coast of Somalia. Other references: Blencowe 1962b; Pearson 1984d.

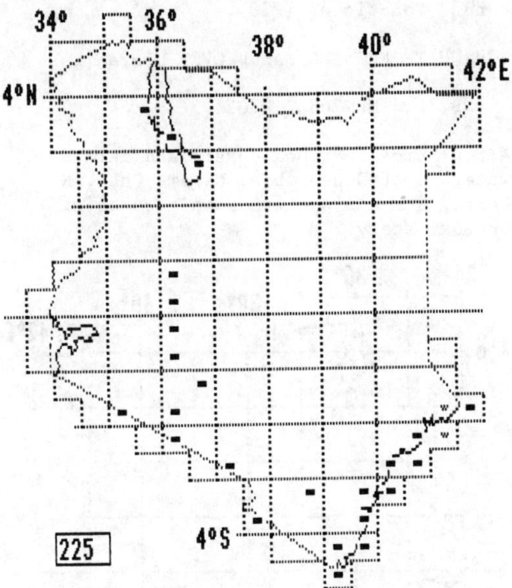

226 LITTLE RINGED PLOVER *Charadrius dubius*
B231 / MPG267 / D2C2

Afrotropical: a Palearctic plover that winters S approximately to the equator, reaching S to NE Tanzania in East Africa (Br).

Kenyan: scarce and unobtrusive, usually encountered singly or in flocks of up to

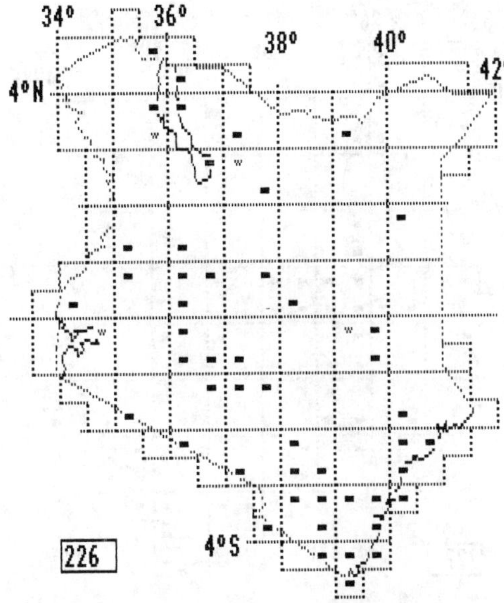

10 (maximum count 30 at L Nakuru 62A in late Dec), and far less common than the closely related Ringed Plover (227)(Br; Ja; Ba; HH; St; Pearson & Stevenson 1980; EABR). It is a winter visitor Oct-Mar, and there is no record of oversummering. Most records are inland on rivers, dams and lakes, where this plover is found on open shores and adjacent dry areas with short grass (Br). Unlike the Ringed Plover, it appears to be rare at L Victoria and possibly W of the rift valley generally, though this may reflect observer frequency. The Little Ringed Plover does not occur on the open seashore and the very small numbers at the coast are found around rainwater pools, pans, creeks and on short grass areas (Br). One ringed at L Nakuru 62A was recovered in USSR, on the NE coast of the Black Sea (Backhurst 1969b, 1971b).

227 RINGED PLOVER *Charadrius hiaticula*
B233 / MPG266 / D2C2

Afrotropical: a common to abundant Palearctic visitor to coasts and inland waters throughout Africa.

Kenyan: a common and widespread passage migrant and winter visitor, feeding on the open shores of all types of water body, but particularly at the coast and around the rift valley lakes (Br). It is mainly present Sept-early May, and few oversummer (Br). Small numbers winter on the rift valley lakes (Pearson & Stevenson 1980), and at Aruba dam in Tsavo E 101B (La), and

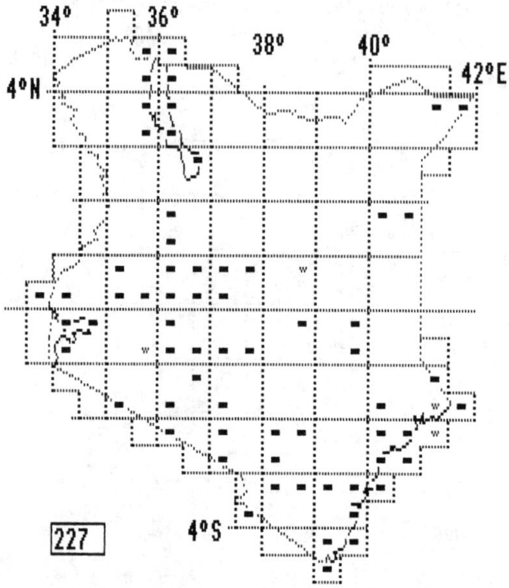

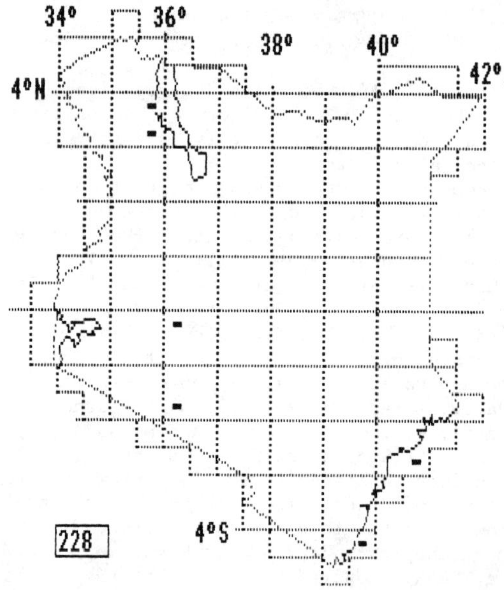

relatively small numbers on the coast (Pearson & Britton 1980). Greatest abundance at L Baringo 50A occurs Jan-Apr (St). Offshore N passage has been recorded in Somalia (Ash 1981b). Other reference: Pearson 1984d.

228 KENTISH PLOVER *Charadrius alexandrinus*
B229 / MPG269 / D1C0

Afrotropical: a predominantly Palearctic species which breeds on the coasts of Mauritania, the Red Sea and Socotra, while Palearctic migrants penetrate S to N Zaire, Ethiopia, Somalia and (apparently regularly) the far N of Kenya. The White-fronted Sandplover (229), a member of the same superspecies (Sn), breeds in areas of Africa to the S of the Kentish's breeding range.

Kenyan: this open shore species occurs Oct-Apr (Br) but, because earlier reports were not sufficiently substantiated (Ba), the first certain record came as recently as 19 Mar 1972, at Ferguson's Gulf 13B, L Turkana (Fry & Horne 1972b). Numerous subsequent records from this locality have shown occurrence to be probably annual there, in loose flocks of up to 10 or 20, so that this represents the S extremity of the main wintering range (Hopson 1974a; HH; Br; EABR; T Stevenson pers. comm.). Further S, there have been seven records of single birds on the carbonate lakes of the rift valley: Nakuru 62A, Elmenteita 62A and Magadi 75C (EABR). The only records outside of the rift valley are coastal, with two at Kipini 91D on 21-22 Jan 1984, and one near Likoni 114B on 21 Oct 1984 (EABR 1984). AM mention migration along the Somali coast. See Chestnut-banded Sandplover (230).

229 WHITE-FRONTED SANDPLOVER *Charadrius marginatus*
B235 / MPG268 / D2C2

Afrotropical: around coasts and inland waters S of Senegambia, Niger, Chad, N Zaire, N Kenya and the Somali coast at approximately 5°N.

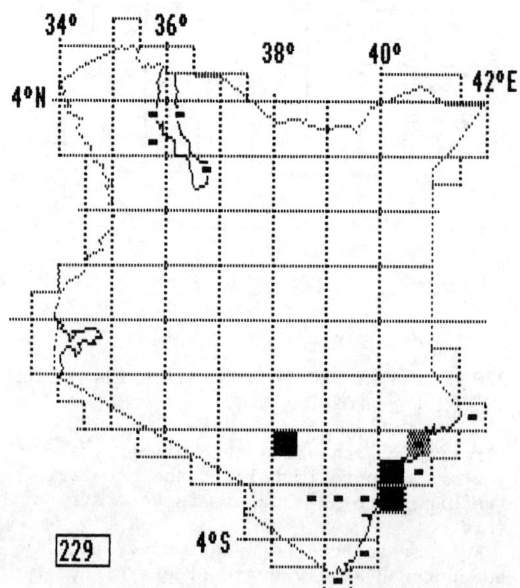

Kenyan: inhabits open, usually sandy, coastal, river and lake shores, and is most numerous at L Turkana and on the coast N of Malindi 103A (Br; HH; MPG; Ja). Small numbers are regular along the Galana -Athi R 103A 102AB 101B, inland to Bushwackers' Camp 89A (Br). The possibility of regular movements is suggested by small numbers of non-breeding birds that occur seasonally at Aruba dam 101B (3 of 4 records June-Jul: La) and at Mombasa 114B (Br). See Kentish Plover (228).

230 CHESTNUT-BANDED SANDPLOVER *Charadrius pallidus*
B237 / MPG270 / D2C3
NO MAP: post confirmed 75C 87A

Afrotropical: occupies two distinct areas: S of approximately 20°S it occurs on coasts and less commonly in the interior, while in East Africa it is confined to strongly alkaline lakes in SW Kenya and NE Tanzania.

Kenyan: entirely confined to the shores of alkaline L Magadi 75C 87A, at around 580 m altitude, where it is a common resident (Br)(hence C3). While Sn mooted the possibility of dispersal, Br doubted presence at 970 m altitude at L Baringo 50A (Ellis 1971) and breeding at 1757 m at L Nakuru 62A (Cunningham-van Someren 1971c). The former is supported by a photograph but the possibility of an escape from a collection adjacent to the lake cannot be excluded, while the latter remains unconfirmed due to the unlikely altitude, the lack of corroboration by the numerous other observers that have visited the area, and the possibility of confusion with other species (Dowsett 1977).

This sandplover is very closely related to the preceding superspecies (Sn).

231 GREAT SANDPLOVER *Charadrius leschenaultii*
B234 / MPG275 / D2C2

Afrotropical: a Palearctic species that winters on E African coasts from Egypt to South Africa, but is only occasional inland. It has possibly bred in extreme NW Somalia (AM).

Kenyan: a common passage migrant and winter visitor at the coast, mainly Aug-Apr, where it prefers open tidal flats and is one of the most abundant waders *e.g.* 1000+ on Galu beach 114B in Jan (EABR 1982), and 1100 at Shelley Beach 114B (Pearson 1984d)(Britton & Britton 1976b; Pearson & Britton 1980; Br). Many first year birds oversummer.

Apart from L Turkana, where it is fre-

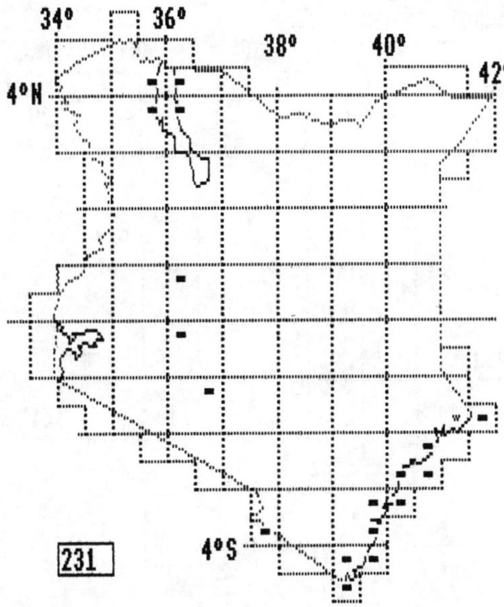

quent in all months except Jul (HH), Br could offer no inland records. This species is quoted by Dowsett (1980) in his review of transcontinental migration by waders, and six of seven recent inland occurrences in Oct-Nov (St; EABR) support his view of more interrupted S migration: three of these seven records have been away from rift valley lakes. This species is noticeably rarer inland than the closely related Mongolian Sandplover (232), except at L Turkana (HH).

232 MONGOLIAN SANDPLOVER *Charadrius mongolus*
B236 / MPG274 / D2C2

Afrotropical: a Palearctic species, wintering on E African coasts mainly from Eritrea S to Tanzania.

Kenyan: a common to abundant passage migrant and winter visitor of open shores at the coast, mainly late Aug-early May but with many first year birds oversummering (Pearson & Britton 1980; Br). Inland it is occasional at L Turkana (but probably commoner than is implied by our map)(HH: *cf.* Great Sandplover (231)), and erratic elsewhere. Dowsett (1980) mentions this species in his review of transcontinental migration by waders and the great majority of inland records are Aug-Nov (St; La; EABR), which supports his view of more interrupted S migration. These inland occurrences are confined to the rift valley lakes and E plateau, but this distribution may reflect observer

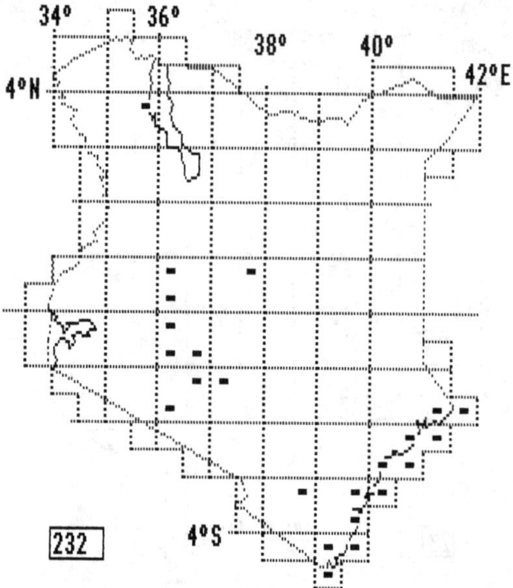

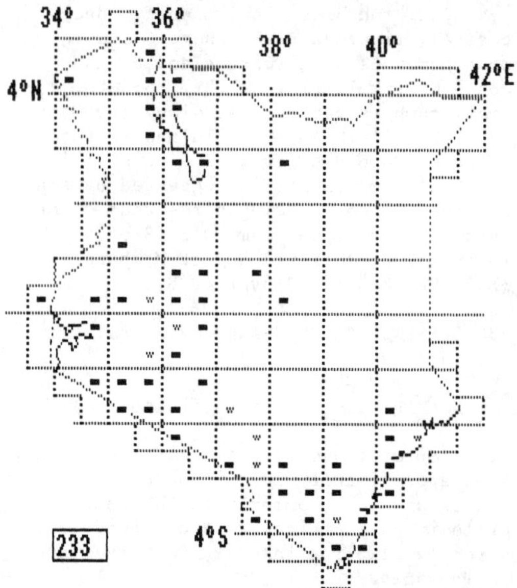

frequency. Old records from L Victoria (van Someren 1925-35) are insufficiently substantiated (Dowsett 1980). A bird ringed at Mida Creek 102B has been recovered in Pakistan (Anonymous 1985b).

233 CASPIAN PLOVER *Charadrius asiaticus*
B230 / MPG277 / D2C2

Afrotropical: this Palearctic species winters in E and S Africa, S to Angola and South Africa.

Kenyan: a locally common to very common passage migrant and winter visitor, mainly Oct-Mar, that usually prefers open, short grass plains (including recently burnt ground), but is also found in shore habitats (Br; Ja; MPG; Lack 1985).

This plover is a common passage migrant in Oct and Apr at L Turkana (HH), and wintering is also known there (D J Pearson *in litt.*). It is mainly a passage bird around 3000 m at Mau Narok 61D (Se) and at L Baringo, where N passage is less marked than the S movement (St). Records from the Nairobi area 75B are as early as Aug (Lee 1962b), but the main arrival is in Oct. It winters in very large numbers, in flocks up to 200, on the plains to the S of Nairobi 75B, including Maasai Mara 74AC, Amboseli 88C and Tsavo (mainly E 101B)(Br; Ja; La). It is uncommon in the coastal lowlands (*e.g.* Pearson 1984d) but large passage flocks can occur, even on the open shore (*e.g.* Campbell 1972a; EABR 1980). Oversummering is rare: van Someren (1925-35) quotes 22 June, and there is another June date from Somalia (AM).

Wintering birds at L Turkana are in 0-500 mm rainfall country, while those on the plains S of the C highlands are in the intermediate, 250-1000 mm regime.

234 KITTLITZ'S SANDPLOVER *Charadrius pecuarius*
B238 / MPG271 / D2C2

Afrotropical: widely distributed S of 16°N, apart from arid E Somalia and the equatorial lowland forest areas of the W.

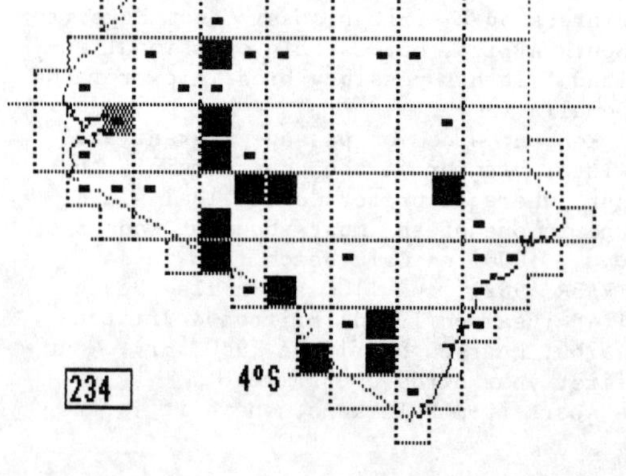

Kenyan: a locally common bird that prefers dry, short grass areas or bare mud adjacent to any type of water body except the open sea, up to 2300 m (Br; MPG). Occurrence at the coast is restricted to squares containing the Sabaki 103A and Tana R 91, and Br reports this plover from dry salt flats and pans at Ngomeni 91C 103A. Though widespread, Kittlitz's is considerably less so than the ecologically more catholic Three-banded Plover (235) (50 squares, as against 100).

This sandplover is probably mostly resident around large waters (Br), but movements occur. Ja mentioned local movements due to changing water levels, and there is occasional colonisation of seasonal waterholes in Tsavo E 101B (La). Backhurst (1970) refers to a L Nakuru to L Elmenteita ringing recovery (both 62A, approx. 30 km apart), while Br mentions the possibility of migrants in NE Tanzania. Other reference: Kutilek 1974.

235 THREE-BANDED PLOVER *Charadrius tricollaris*
B239 / MPG272 / D2C2

Afrotropical: a bird of the E and S, widespread to the S of the Congo basin forests and further N through E Africa and E Zaire to S Sudan, Ethiopia and Somalia. There are apparently isolated populations in Mali and Cameroon-Chad-Niger. Apart from two records in S Somalia (AM), distribution in Kenya represents the E limit of the range.

Kenyan: a common plover, singly or in pairs, which can be found at any type of water body except the open sea, up to 3000 m (Br; Se; Ja; MPG). It prefers the muddy edges of small pools and streams (*e.g.* HH; La), and birds at the coast are on lagoons, creeks and salt pans (Br). This plover is probably mostly sedentary (Br; Be: 74; La: 101B; DEP pers. obs.: 89A), so that the distribution of breeding records reflects observer frequency: for example, nesting almost certainly occurs around L Turkana (HH). The existence of at least local movements is demonstrated by a L Naivasha 62C to L Nakuru 62A ringing recovery (Backhurst 1971b). Transient migrants around 3000 m at Mau Narok 61D were mostly Nov-Jan or June-Aug (Se), and apparent absence at L Baringo 50A Sept-Dec (St) coincides with the first of these periods. On small dams near Nairobi 75B, it is most numerous after rains, *i.e.* Jul-Sept and Dec-Feb (Pomeroy & Tengecho 1982a). See Kittlitz's Sandplover (234).

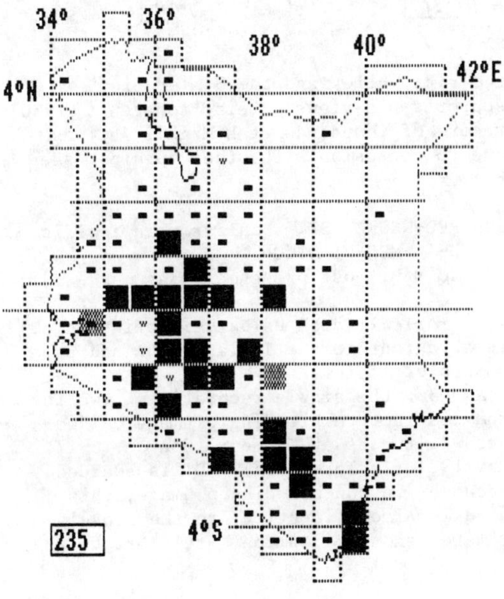

Family Scolopacidae: Sandpipers, Snipes, Stints, Godwits etc.

81 species, with representatives worldwide. 38 (47%) occur in the Afrotropical Region, and of these, 31 (82%) are known in Kenya. Of the Kenyan species, the African Snipe (250) is a breeding resident, while the rest (97%) are non-breeding migrants or vagrants from the Holarctic (mainly Palearctic). This extraordinary imbalance between a single resident breeding species and vast numbers of Holarctic migrants applies to this family both throughout the Afrotropical Region and, less strikingly, in the other S continents (Sn). Inland records of the Palearctic members of this family which predominantly winter in coastal Africa can be considered in the light of Dowsett's (1980) proposals regarding migration across continental Africa (see Haematopodidae).

236 KNOT *Calidris canutus*
B269 / MPG307 / D2C0
NO MAP: post presence 103A 114B; pre presence 80CD

Afrotropical: a Palearctic species, wintering abundantly on the NW and SW coasts of Africa, which has occurred in Kenya only as a coastal vagrant.

Kenyan: all early records of this species are insufficiently substantiated (Ba). The more recent sightings, all of single coastal birds, were at Kiunga 80CD on 16-17 Sept 1961 (Fogden 1963), Mombasa 114B on 2 Apr 1973, and near the mouth of the Sabaki R 103A on 21 Sept 1977 (Taylor 1978).

237 SANDERLING *Calidris alba*
B267 / MPG308 / D2C2

Afrotropical: a common to abundant Palearctic winter visitor to most African coasts, also occurring inland in small numbers in the E and S.

Kenyan: a common passage migrant and winter visitor Aug-Apr, preferring open sandy shores or extensive sand flats in creeks (Br). It is primarily a bird of the coast, where thousands overwinter and small parties of first year birds oversummer (Pearson & Britton 1980; Pearson 1984d). Inland, loose parties of up to 10 winter at L Turkana (HH), and there is a regular S passage of small numbers Aug-Nov on the rift valley lakes (Br; St; EABR). The Sanderling's restriction inland to the rift valley and adjacent areas, and its absence from L Victoria, are notable. There is little evidence of N passage (Br; EABR), though it is known along the Somali coast (Ash 1981b). Dowsett (1980) quotes this species in his review of transcontinental wader migration, and the virtual restriction of inland records to the rift valley and the S migration supports his views. Meinertzhagen's (1950) record of thousands at L Magadi 75C was due to confusion with other species (see Ba).

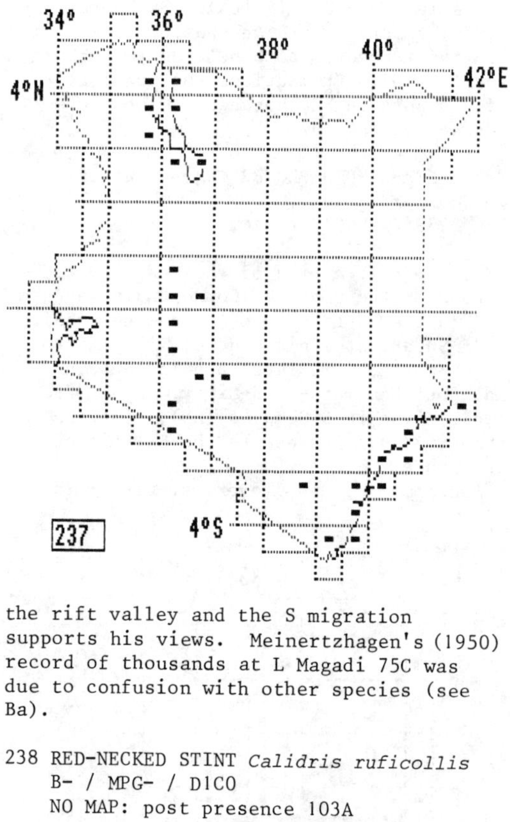

238 RED-NECKED STINT *Calidris ruficollis*
B- / MPG- / D1C0
NO MAP: post presence 103A

Afrotropical: an E Palearctic stint which is a vagrant to the Indian Ocean and Cape coasts of Africa.

Kenyan: the single record is of two in the distinctive breeding plumage on disused salt pans at Ngomeni 103A on 16-17 May 1981 (Taylor 1981b). D1 is awarded because, in non-breeding plumage, this bird is almost identical to the closely related and very common Little Stint (239).

239 LITTLE STINT *Calidris minuta*
B272 / MPG305 / D2C2

Afrotropical: this Palearctic species is a common to abundant winter visitor to coasts and wetlands throughout.

Kenyan: a common passage migrant and winter visitor, mainly Aug-May (Br). It is the most abundant Palearctic wader of inland mudflat habitats, particularly on the lakes of the rift valley, where tens of thousands are present at L Turkana and many thousands on the lakes of the S rift (Br; HH; Pearson & Stevenson 1980). Wintering birds at the coast are far less numerous than the Curlew Sandpiper (243), and frequent more sheltered mudflats, lagoons, estuaries, creeks and pans, rather than the open shore (Pearson & Britton 1980; Br; Pearson 1984d; Britton & Britton 1976b). Few oversummer (*op. cit.*). As in the Curlew Sandpiper, birds in the rift valley migrate N two to three weeks after those at the coast, so that these two populations must adopt very different northward migration strategies (Pearson & Britton 1980). Ringing recoveries have demonstrated movements between the various rift valley lakes, and between Kenya and Sudan, USSR, Zaire, Zambia and Zimbabwe (Backhurst: E A Bird Ringing Reports). Other reference: Pearson 1984a.

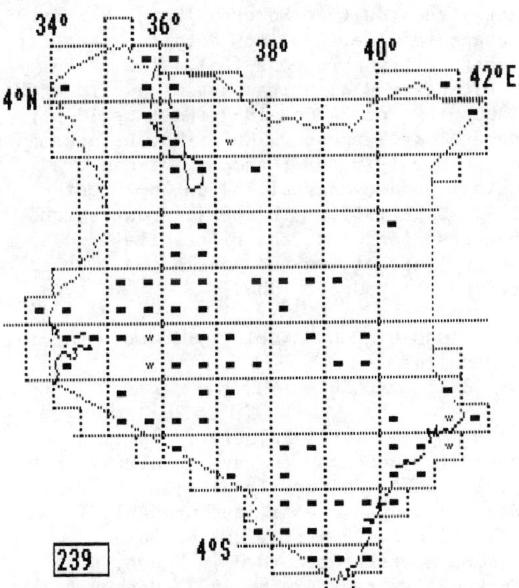

240 TEMMINCK'S STINT *Calidris temminckii*
B274 / MPG306 / D2C2

Afrotropical: this Palearctic species winters S to approximately 10°N in the W, and

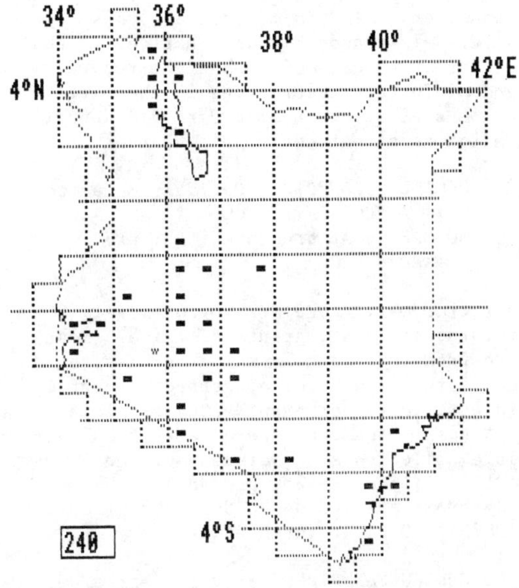

5°S in the E.

Kenyan: this rather unobtrusive species is a winter visitor, late Aug-early May, mainly on freshwater marshes, pools and lake edges (Br; Ba; HH). In contrast to the Little Stint (239), it occurs in small numbers and often singly, and is only locally distributed in the coastal lowlands, where the relatively few records are from lakes, dams and estuaries (Britton & Britton 1976b; Br; D J Pearson *in litt.*; MPG; EABR). It is regular at some of the lakes of the S rift valley and C highlands (Pearson & Stevenson 1980). There is no evidence of oversummering. Other references: Stoneham 1944d; van Someren 1945; Williams 1960a.

241 LONG-TOED STINT *Calidris subminuta*
B273 / MPG- / D1C0
NO MAP: post presence 62AC 75B 103A

Afrotropical: there are seven records from Africa of this vagrant from the C Palearctic: five of these are from Kenya.

Kenyan: the five records are: one collected at L Naivasha 62C on 27 Apr 1969 (Backhurst & Britton 1969) and one ringed there on 2 Feb 1974 (OS-C 1977); one ringed at L Nakuru 62A on 9 May 1970 (Ba); one seen at Dandora sewage ponds, Nairobi 75B on 20-22 Feb 1984 (EABR 1984); and one seen in Malindi Harbour 103A on 21 Nov 1984 (EABR 1984). Four of these five records are above 1500 m, and in or adjacent to the S rift valley. Assuming absence of the very similar Nearctic Least

Sandpiper (*Calidris minutilla*), this species is reasonably distinctive, so that the great paucity of both sight records and of further occurrences in the large numbers of stints that are ringed annually indicates an extremely rare vagrant.

242 PECTORAL SANDPIPER *Calidris melanotos*
B271 / MPG307A / D1C0
NO MAP: post presence 114B; pre presence 62C

Afrotropical: there are approximately 20 records of this vagrant to Africa. Most are from the S and E, which perhaps suggests that the birds concerned were of E Palearctic rather than Nearctic origin.

Kenyan: the two records are of an adult male collected at L Naivasha 62C on 11 May 1952 (Williams 1952), and an adult seen near Mombasa 114B on 28 Sept 1981 (Taylor 1982b). Other reference: Ba.

243 CURLEW SANDPIPER *Calidris ferruginea*
B270 / MPG303 / D2C2

Afrotropical: this Palearctic species winters abundantly on African coasts S of about 25°N, and more locally inland.

Kenyan: a common passage migrant and winter visitor, Aug-mid May, which frequents open shores and creeks, both coastal and inland (Br). It is probably the most abundant wader at the coast, wintering in thousands (Pearson & Britton 1980; Pearson 1984d; Ja). Inland, it is widespread on passage, but wintering is largely confined to the rift valley lakes, with particular abundance at L Turkana, L Naivasha 62C and L Magadi 75C (Br; HH; Pearson & Stevenson 1980). Many first year birds oversummer (Br). Both N and S migrations are marked (Br) but, as in the Little Stint (239), birds wintering in the rift valley migrate N two to three weeks later than those at the coast, so that these two populations must adopt very different northward migration strategies (Pearson & Britton 1980). N passage is the heavier in the rift valley at L Baringo 50A (St), L Nakuru 62A and (possibly) L Turkana (D J Pearson *in litt.*), whereas the opposite applies in Tsavo E 101B (La). Ringing recoveries have demonstrated movements between the rift valley lakes, and also to Iraq and Iran (Backhurst 1973b, 1974, 1977).

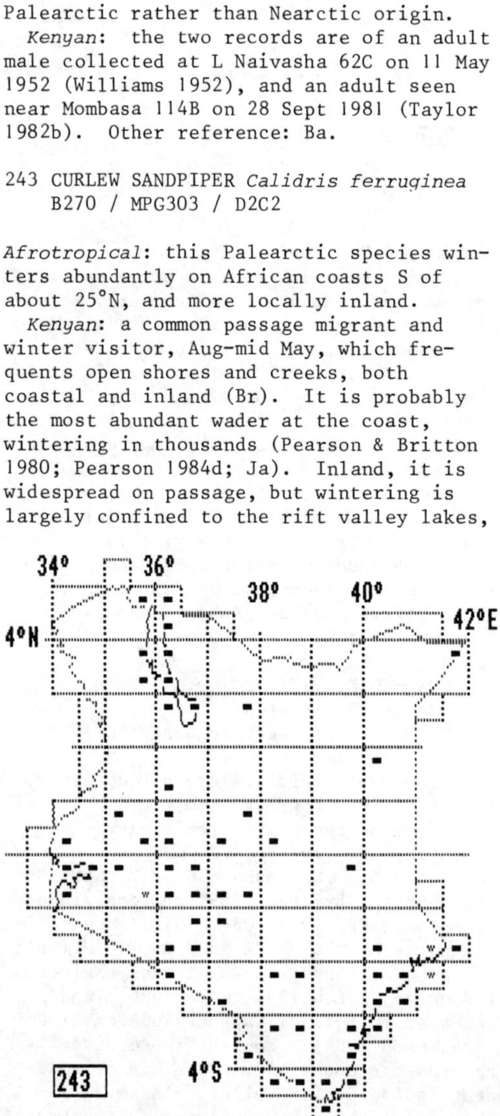

244 DUNLIN *Calidris alpina*
B268 / MPG304 / D1C0
NO MAP: post presence 13B 75B; pre presence 62A

Afrotropical: a Palearctic species that is a common to abundant winter visitor to coasts S to approximately 10°N in the W and 15°N in the E, and uncommon further S. It occurs in only small numbers inland in the N tropics.

Kenyan: a wader of open shores that is a vagrant S of its normal wintering range. Early records (van Someren 1918a; Meinertzhagen 1950) either lack detail or were at least partly erroneous (Ba). The three substantiated sightings are of a bird at L Nakuru 62A on 24 Oct 1953 (Wallace 1975); several at Ferguson's Gulf 13B, L Turkana on 28 Mar 1970 (HH); and one at Dandora sewage ponds, Nairobi 75B on 28-29 Oct 1983 (EABR 1983). The passage dates, and the restriction to the rift valley and nearby Nairobi, may be noteworthy. Other reference: Dowsett 1980.

245 BROAD-BILLED SANDPIPER *Limicola falcinellus*
B275 / MPG307B / D2C1

Afrotropical: this Palearctic species is generally uncommon to rare in Africa, but small numbers winter locally on the Red Sea and Kenyan coasts, and probably in coastal E South Africa and Namibia.

Kenyan: first recorded in Kenya, at L Turkana, as recently as 1972 (Hopson & Hopson 1972), this species has subsequently been found to winter annually early Aug-mid Apr, in numbers up to 60, on wet mud and sand in the estuary of the Sabaki R 103A (Britton 1974d; Br; Britton & Britton 1973B; OS-C 1977; EABR). There are also a

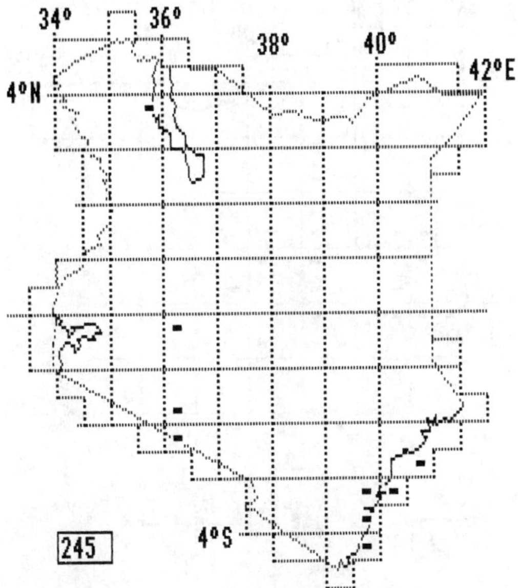

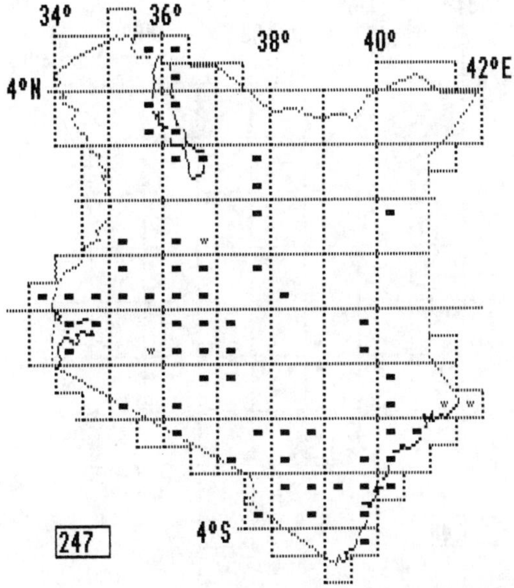

few records of very small numbers (up to seven) from other parts of the coast (EABR). Inland, it is known in numbers up to 10 at Ferguson's Gulf 13B, L Turkana (Hopson & Hopson 1972, 1973, 1974; HH), where the predominant S passage dates, together with those of single birds at L Nakuru 62A and L Magadi 75C 87A (OS-C 1977; EABR), agree with Dowsett's (1980) view of more interrupted transcontinental wader migration during the autumn (he quotes this species as a specific example of this type of migrant).

246 BUFF-BREASTED SANDPIPER *Tryngites subruficollis*
B280 / MPG- / D2C0
NO MAP: post presence 26A

Afrotropical: an occurrence of this Nearctic vagrant at L Turkana, Kenya, constitutes one of five records from the African continent.

Kenyan: the single record is of one seen just N of the Kerio R delta, L Turkana 26A, on 8 Dec 1973 (Hopson & Hopson 1974a).

247 RUFF *Philomachus pugnax*
B279 / MPG309 / D2C2

Afrotropical: this Palearctic species winters on inland wetlands throughout, apart from the lowland forest areas of W Africa.

Kenyan: a common to abundant passage migrant and winter visitor, Aug-May (Br). Inland, it is one of the most widespread waders, frequenting muddy and marshy edges of water bodies, as well as rice fields and open grassland near water; and it is particularly numerous on the rift valley lakes, and on dams and other waters in the W *e.g.* 2000 on Sergoit dam 49A Feb, 4000-5000 at Ahero Rice Scheme 60B Feb and 3000 at the Tana Bridge 63C late Jan (Reynolds 1974; Br; Pearson & Stevenson 1980; PH; EABR 1983, 1984). It is scarce at the coast, even on freshwater (Britton & Britton 1976b; Br). A few first year birds oversummer (Br; St). Ringing recoveries have demonstrated movements between the rift valley lakes, and migration to N India and the USSR (Backhurst: E A Bird Ringing Reports; Backhurst 1985a). Other reference: Pearson 1981a.

248 JACK SNIPE *Lymnocryptes minimus*
B266 / MPG302 / D1C1

Afrotropical: a Palearctic species that winters rather locally, mainly S to approximately 8°N in the W and 5°N in the E.

Kenyan: a scarce migrant, mainly Oct-Mar, occurring sporadically and in very small numbers to the S of its main wintering range (Ja; Br; EABR). It is a close-sitting and rather secretive bird of marshes and is not recorded annually, though it could be regular in very small numbers (Br; MPG). Records are mainly from the S rift valley and C highlands, with five of seven squares above 1500 m and six subhumid-humid. A bird at Loyengalani 26B in Nov 1971 (HH) was presumably on passage: this record is omitted by Br. Ja quotes the Sio R 48C (not mapped), but with reservations.

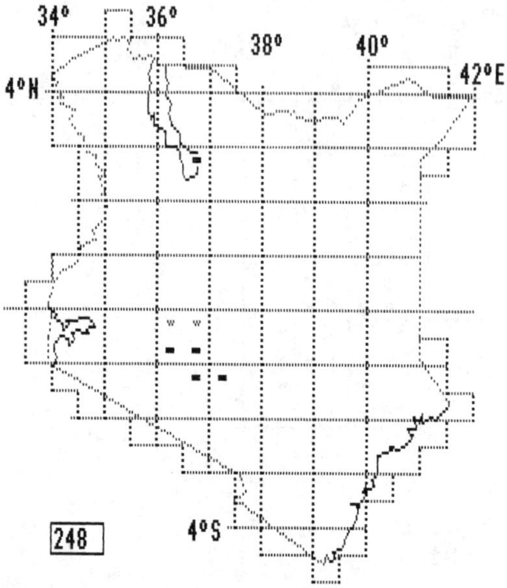

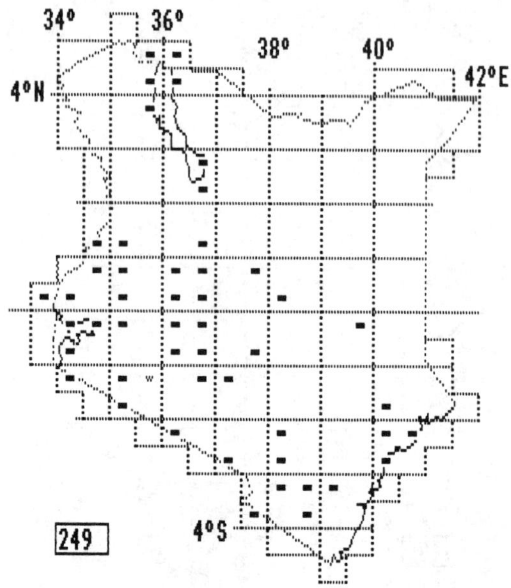

A notable influx of 16+ to L Naivasha 62C in Jan-early Mar 1982 occurred at the same time as unprecedented numbers of Lesser Spotted Crakes (195), which were possibly of Palearctic origin (EABR 1982): this is strongly reminiscent of the simultaneous arrival of this species and the Little Crake *Porzana parva* at Butiaba, Uganda, in early Dec 1906 (Ja). Other references: Meinertzhagen 1922; Ba.

249 COMMON SNIPE *Gallinago gallinago*
B262 / MPG298 / D2C2

Afrotropical: this Palearctic species winters S to N Zaire and N Tanzania.

Kenyan: a common passage migrant and winter visitor, late Sept-late Apr which, like the similar African Snipe (250), is found in swamps and marshy lake edges (Br; MPG). It is regularly seen in reasonable numbers on the S rift valley lakes (*e.g.* *c* 900 on the E and SE shores of L Naivasha 62C in Jan: EABR 1982; Pearson & Stevenson 1980) and in the L Victoria basin (Br), but is apparently less common and widespread than the African Snipe at L Turkana (HH). It is uncommon at the coast, mainly restricted to the swamps around the Tana R 91. Both N and S migrations are observed, especially N of the equator, but in many areas this species is only present Nov-Mar (Br; La). There is no evidence of oversummering. At L Baringo 50A, snipes are uncommon Dec-Mar during years of low rainfall (St). Other reference: Taylor 1983.

250 AFRICAN SNIPE *Gallinago nigripennis*
B264 / MPG300 / D2C2

Afrotropical: apart from the dry SW, widely distributed S of 10°S, and further N through East Africa and the E Zaire borders to S Sudan and Ethiopia. Distribution in Kenya represents the E limit of the range. This is the only member of the family that regularly breeds in this Region (Sn).

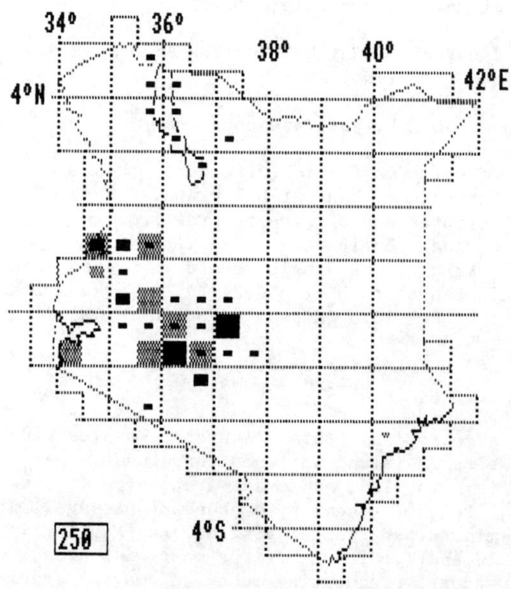

Kenyan: distribution can be split into two very different areas. This snipe is best known as a reasonably common breeding bird of marshes, bogs and lake edges at 1700-3000 m (and regularly to 4000 m) in the highlands (Br; Ja; MPG). Range here is entirely within the 500+ mm rainfall regime.

In striking contrast to the highland range, this species is also regular in marshy areas of L Turkana (HH; Br), at 400 m in 0-500 mm rainfall country, but there is no breeding record.

The African Snipe is certainly not sedentary, but the precise nature of its movements is uncertain. There is the suggestion that it is a non-breeding migrant or wanderer to low altitudes, but there are records from Turkana during all seasons (Sn; Br). Se noted influxes to the Mau Narok area 61D (3000 m) after prolonged rainfall had produced more extensive habitat. Single birds killed at night at a light at Kapsabet 49C in Nov (Br) and by a vehicle in the Chalbi Desert 15C on 6 Oct 1983 (F Alexander *in litt.*) were evidently nocturnal migrants. A bird collected near sea level at Witu 91A (Ja) may have been a disorientated migrant.

251 GREAT SNIPE *Gallinago media*
B263 / MPG299 / D1C1

Afrotropical: this Palearctic snipe spends the Sept-Oct period mainly in the N tropics and Dec-Apr mainly in the S, from Angola to Zambia and Malawi. In E and NE Africa, it is known mainly as a S passage migrant, but status varies locally. Distribution in Kenya represents the E limit of the range.

Kenyan: this species is awarded D1C1 because it resembles the closely related Common Snipe (249) and is undoubtedly overlooked, and is uncommon. Like the Common Snipe, it occurs in marshes throughout our altitude categories, but in far smaller numbers than that species. Virtual absence from localities where formerly plentiful *e.g.* Kisumu 60B and Nairobi 75B suggests a decline in recent decades (Ba). The few recent records are mostly Oct-Dec and Apr-May, and both the N and S migrations are regular in the W (Ja; Br). A bird was caught at the Ngulia lights 101A in Nov (EABR 1984). There are Jan-Feb records from Mau Narok 61D at 3000 m (Britton 1979), L Naivasha 62C, Limuru 75B and L Bilisa 91A (OS-C 1977; Pearson & Stevenson 1980; EABR). There is no evidence of oversummering.

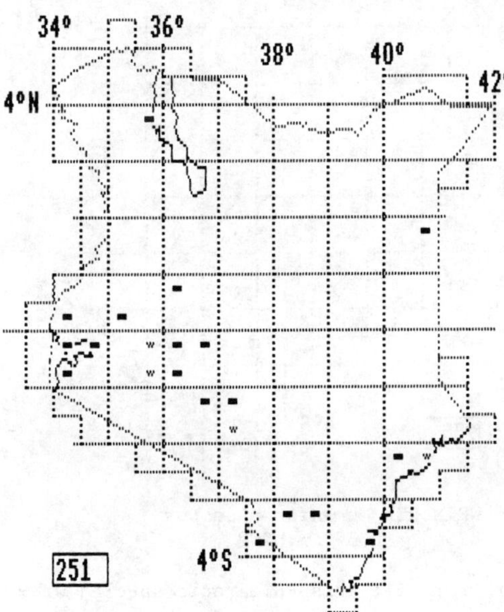

252 PINTAIL SNIPE *Gallinago stenura*
B265 / MPG301 / D0C0
NO MAP: post presence 62C 102D 114B

Afrotropical: a vagrant from the C Palearctic, recorded less than 10 times in Africa, all but two of these occurrences having been in C or coastal Kenya.

Kenyan: the extreme similarity of this species to the closely related Common Snipe (249), both in terms of appearance and habitat, means that it may very easily be overlooked (hence D0C0). Hence, while at present treated as a vagrant, the Pintail Snipe is possibly of more regular occurrence. It is best identified in the hand and, while possible old records from sportsmen have been lost (Ba), the first documented occurrence was one netted at L Naivasha 62C on 1 Jan 1969 (Backhurst 1969a). Subsequently, two were present at Bamburi 102D/114B from 22-31 Oct 1981, with at least one remaining until 31 Dec (EABR 1983), and another was seen at L Naivasha 62C on 24 Jan 1982 (Pearson 1984c).

253 ASIATIC DOWITCHER *Limnodromus semipalmatus*
B276 / MPG- / D1C0
NO MAP: pre presence 62A

Afrotropical: an occurrence at Nakuru, Kenya, constitutes the only African record of this vagrant from the C Palearctic.

Kenyan: the single record is of one seen at the mouth of the Njoro R, L Nakuru 62A, on 20-21 Nov 1966 (Smart & Forbes-Watson 1971).

254 BLACK-TAILED GODWIT *Limosa limosa*
B278 / MPG319 / D2C2

Afrotropical: this Palearctic species winters commonly in Africa S to the vicinity of the equator, mainly at inland rivers, marshes and irrigation schemes; it is decidedly uncommon further S.

Kenyan: uncommon Aug-Apr on the S extremity of its main wintering range (see above), with occasional oversummering June-Jul (Ba; Br; EABR; *cf.* AM). In contrast to the Bar-tailed Godwit (255), the Black-tailed mainly frequents inland waters. It is most frequent and numerous at L Turkana, where parties of 40 or 50 are regular in muddy inlets and lagoons, with a peak count of *c.* 80 at Ferguson's Gulf 13B in Jan (HH; Br; EABR 1980). Somewhat smaller numbers have occurred at L Nakuru 62A and (regularly) L Naivasha 62C (Meadows 1977b; Pearson & Stevenson 1980; EABR). With the notable exception of the Ahero Rice Scheme 60B, where up to 40 may be regular (EABR 1984; D J Pearson *in litt.*), other occurrences of this godwit in the interior of Kenya are usually in groups of less than 10, and it is infrequent W of the rift valley (Britton & Britton 1976b; Br; EABR).

Apart from the lower Tana flood lakes 91A (*e.g.* EABR 1984), this bird is also infrequent in the coastal lowlands (*e.g.* La). It has been recorded on salt pans at Ras Ngomeni 103A late May-early June (EABR 1981), but the only actual coastal records are of single birds at Sabaki estuary 103A in Jan and Oct (EABR 1984), Kiunga 80CD on 16-17 Sept 1961 (Fogden 1963), and near Malindi 103A on 12 Aug (EABR 1982).

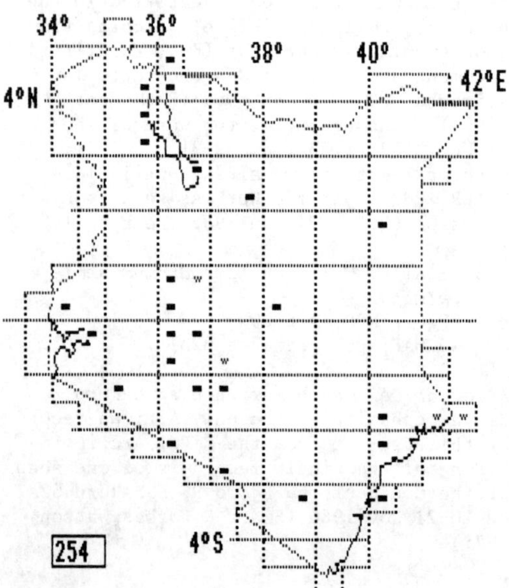

255 BAR-TAILED GODWIT *Limosa lapponica*
B277 / MPG320 / D2C2

Afrotropical: although wintering commonly to abundantly on the coasts of NW and SW Africa and the Red Sea, this Palearctic species occurs in only small numbers on the Indian Ocean seaboard, and is generally rare inland.

Kenyan: in contrast to the closely related Black-tailed Godwit (254), this bird is mainly an uncommon winter visitor to the coast, where parties of up to 10 are annual, Aug-Apr, on the extensive mudflats of Mida Creek 102B; other coastal occurrences are of very small numbers in the S migration period, and there is a single June oversummering record (Br; Britton & Britton 1976b; EABR). The Bar-tailed Godwit is very rare inland, with records of single birds from L Turkana, L Baringo 50A, L Nakuru 62A (including 2-3 there Sept 1969: OS-C 1977), L Naivasha 62C and Nairobi 75B (Ba; T Stevenson *in litt.*; HH; St; EABR). Dowsett (1980) includes this species in his review of transcontinental wader migration but, while S passage is known at the coast, the very few inland dates are not confined to this season.

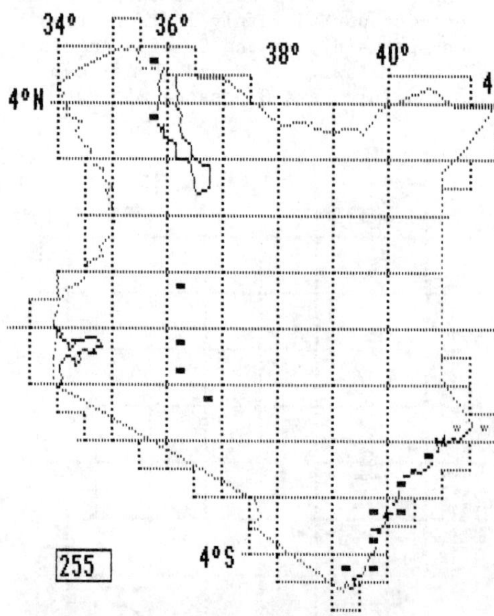

256 WHIMBREL *Numenius phaeopus*
B254 / MPG322 / D2C2

Afrotropical: this Palearctic species win-

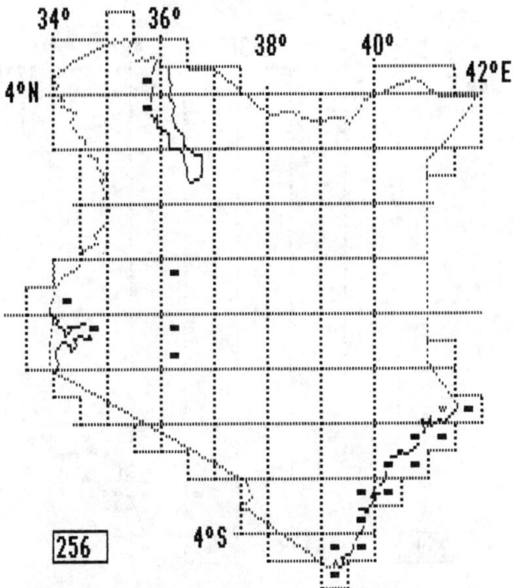

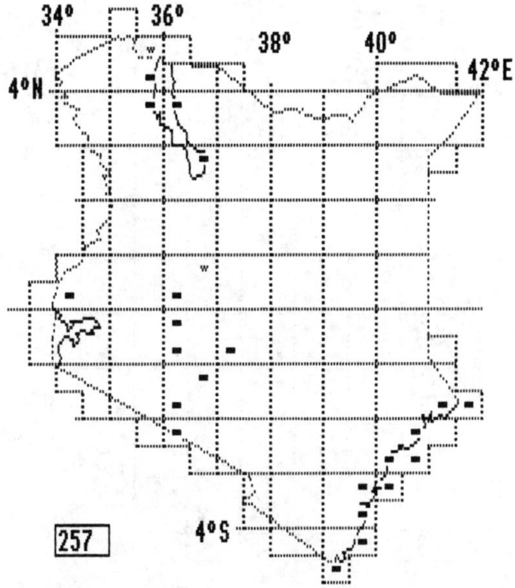

ters on all coasts, but is rare to uncommon inland.

Kenyan: a passage migrant and winter visitor, Aug-Apr, that is common at the coast, where hundreds winter on open shores and a few birds commonly oversummer (Britton & Britton 1976b; Pearson & Britton 1980; Br; Pearson 1984d). Inland, very small numbers occur on L Turkana and L Victoria, while the very few other records are from rift valley lakes (HH; St; Br; EABR). The Whimbrel is far more numerous at the coast than the closely related Curlew (257) but, with the possible exception of L Victoria, the rarer of the two inland.

Heavy N and S migrations occur along the Somali and Kenyan coasts respectively (Ash 1981b). Dowsett (1980) mentions the Whimbrel in his review of transcontinental migration by waders and, conforming with his views, the very few inland records are mostly during the S migration.

257 CURLEW *Numenius arquata*
B253 / MPG321 / D2C2

Afrotropical: a Palearctic species that winters on all African coasts, though rather locally and less abundantly on the Indian Ocean seaboard. Inland, it occurs locally and in small numbers on rivers and lakes.

Kenyan: a passage migrant and winter visitor, Aug-Apr, far less common than the closely related Whimbrel (256)(Br). Up to 60 are regular in Mida Creek 102B but, except possibly in the Lamu area 91B, it is generally uncommon and sporadic along the rest of the coast (Britton & Britton 1976b; Br). Loose parties of up to 10 are frequent throughout the year at L Turkana (HH), but it is very uncommon elsewhere inland, occurring mainly on the other rift valley lakes (Britton & Britton 1976b; Br). Small numbers oversummer, notably at L Turkana and in Mida Creek 102B (HH; Br). S migration, including nocturnal movement, has been recorded on the coast in Jul-Oct (Ja; EABR 1983, 1984), while N migration is known along the Somali coast (Ash 1981b).

258 SPOTTED REDSHANK *Tringa erythropus*
B255 / MPG316 / D2C2

Afrotropical: this Palearctic wader is a somewhat local winter visitor S to Cameroon, N Zaire and N Tanzania, frequenting inland marshes, lakes and irrigation schemes, and coastal salt pans.

Kenyan: first recorded only as recently as 1953 (Williams 1953), this species has extended its wintering range considerably further S during recent decades to become a regular winter visitor, mainly Oct-Apr (Ba; Meadows *et al*. 1974; Gardiner 1975; Pearson & Stevenson 1980). It is known up to 3000 m (EABR 1977), and is widespread inland on the margins of freshwater and carbonate lakes and marshy pools (Br). The Spotted Redshank appears particularly partial to irrigated areas, and the growing development of rice schemes, sewage ponds and other projects involving shallow flooding may attract greater num-

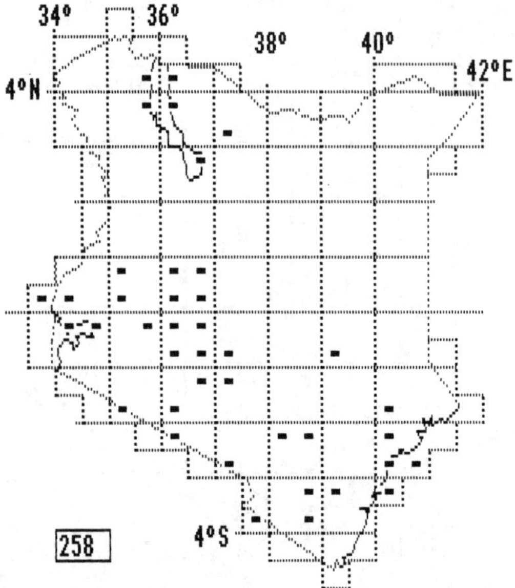

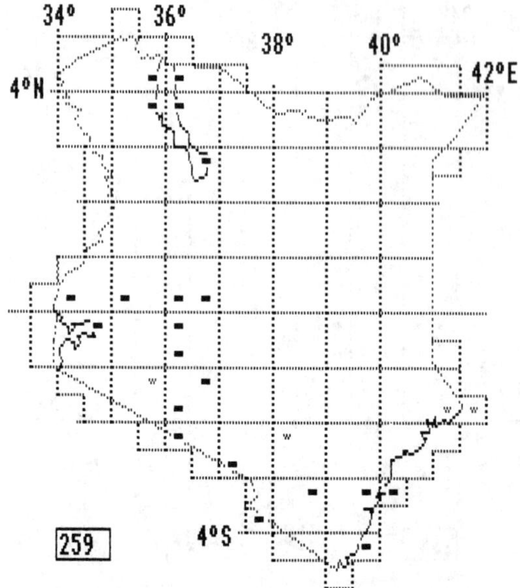

bers in the future (EABR 1983). Records at the coast are confined to the vicinity of the Sabaki 103A and Tana 91 R. This bird is usually solitary or in flocks up to 10, but larger concentrations have occasionally occurred, with a maximum of 165 on Ahero rice fields 60B in Feb (EABR 1983). At both L Baringo 50A and Tsavo E 101B, it is usually present only during Dec-Feb/Mar (St; La). There is no evidence of oversummering.

259 REDSHANK *Tringa totanus*
B260 / MPG315 / D2C2

Afrotropical: a Palearctic species that is an abundant winter visitor to coasts S to Guinea and Somalia, while on coasts further S, and inland generally, it is uncommon to rare.

Kenyan: an uncommon but regular passage migrant and winter visitor, late Jul-late May (Ba; Br). It is only at all common at Mida Creek 102B, where 5-20 winter annually on open shores, mainly Oct-Mar (Br; EABR). Up to seven have occurred in the nearby Sabaki estuary 103A in Dec (EABR 1980). Elsewhere it is distinctly less common than the closely related Spotted Redshank (258), being only at all regular at L Turkana (HH), and occurring in other areas as very irregular, usually solitary birds (OS-C 1977; EABR). A few individuals oversummer, particularly at Mida Creek (EABR). Other reference: Reynolds 1975c.

260 MARSH SANDPIPER *Tringa stagnatilis*
B259 / MPG317 / D2C2

Afrotropical: apart from the lowland forest areas of W Africa, this Palearctic species is a common to abundant winter visitor to coastal lagoons and estuaries and inland lakes, dams and streams throughout.

Kenyan: a common passage migrant and winter visitor, Sept-Apr, to coastal estu-

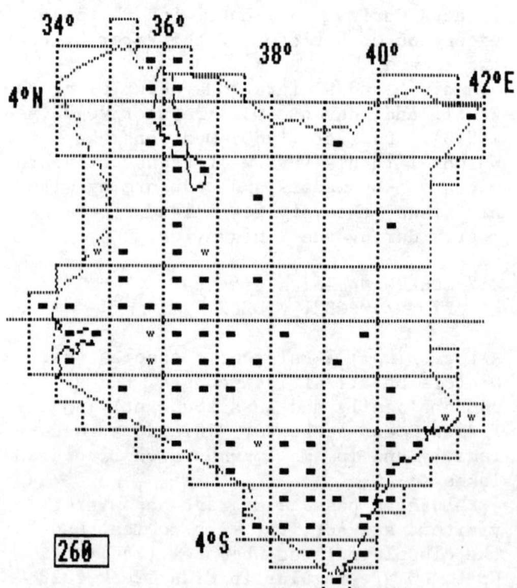

aries and open shores, and to all types of inland waters, especially the rift valley lakes (Br; Pearson & Stevenson 1980). Its map and general abundance are similar to those of the closely related Greenshank (261), though the Marsh Sandpiper is the commoner of the two at L Turkana, where over 1000 can occur in Ferguson's Gulf 13B (HH; Br; MPG). While occurring at L Baringo 50A throughout the winter, there is a marked Feb-Mar influx (St). Many first year birds oversummer (Br). Birds ringed at L Nakuru 62A have been recovered E of the Ural Mts in the USSR, and in E South Africa (Backhurst 1977). Other reference: Reynolds 1972b.

261 GREENSHANK *Tringa nebularia*
B257 / MPG318 / D2C2

Afrotropical: a widespread and common Palearctic species, wintering throughout on coasts, lakes and rivers.

Kenyan: a passage migrant and winter visitor, Aug-early May, which is much more common that the Spotted Redshank (258)(Br). Greenshanks are found around all types of water body, but are most abundant at the coast, where hundreds winter on mudflats and coral shores, and in creeks (Pearson & Britton 1980; Pearson 1984d). The species is widespread inland, but only in small numbers away from the rift valley lakes and L Victoria (Br; Pearson & Stevenson 1980). First year birds commonly oversummer (*op. cit.*; La). See Marsh Sandpiper (260).

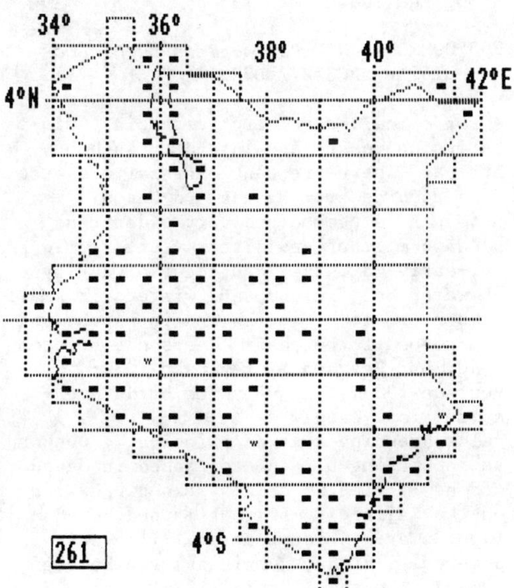

262 GREEN SANDPIPER *Tringa ochropus*
B258 / MPG313 / D2C2

Afrotropical: another common Palearctic wader, wintering on inland waters S to approximately 15°S.

Kenyan: a widespread passage migrant and winter visitor, mainly Sept-Apr, but far less numerous than the Common (265) and Wood (263) Sandpipers: the map shows only marginally fewer points than those species, but the Green is noticeably less widespread at L Turkana (*e.g.* HH), and absent from much of the Tana R 65ABD 78B 79A. The Green Sandpiper is shy, and occurs solitarily or in pairs on secluded pools, lake edges and small streams, even inside forest (Reynolds 1974; Br; Pearson & Stevenson 1980). Records at the coast refer to birds in secluded mangrove creeks and brackish lagoons (Br). There is no evidence of oversummering. As noted for other wader species by Dowsett (1980), S passage is more evident than N (St; La). Other reference: Britton & Britton 1976b.

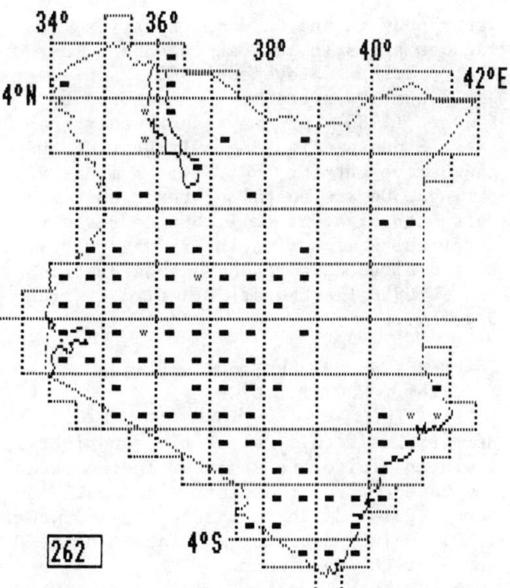

263 WOOD SANDPIPER *Tringa glareola*
B256 / MPG314 / D2C2

Afrotropical: a common to abundant Palearctic winter visitor to inland wetlands throughout.

Kenyan: a common passage migrant and winter visitor, late Jul-early May, and probably the most widespread wader in East Africa (Br). Its map closely resembles that of the Common Sandpiper (265). Wood Sandpipers are found around all types of

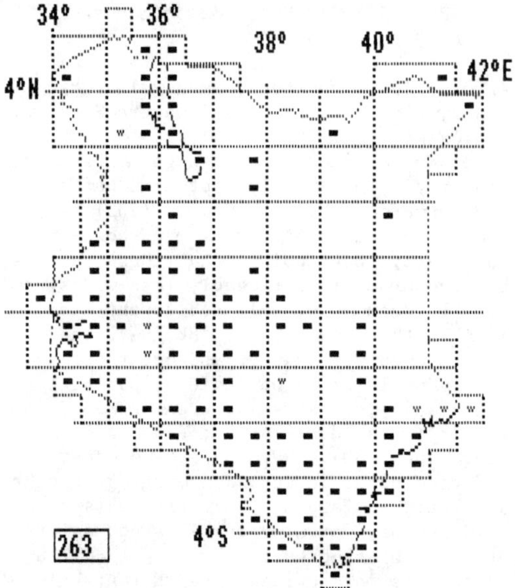

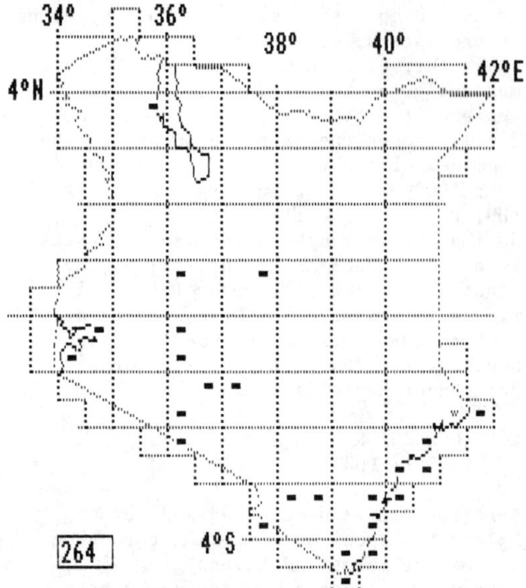

water body inland, and on estuaries, lagoons and salt pans at the coast (Br; MPG; Pearson & Stevenson 1980). Only very small numbers oversummer. As noted by Dowsett (1980) for many other wader species, S passage is generally heavier than the N movement (Br; HH; St). Numbers at L Baringo 50A are sustained throughout winters with high rainfall, but decrease after the S migration in dry years (St). Ringed birds have reached E South Africa and NW USSR (Backhurst: E A Bird Ringing Reports).

264 TEREK SANDPIPER *Xenus cinereus*
B261 / MPG311 / D2C2

Afrotropical: this Palearctic sandpiper is a winter visitor to Red Sea, Indian Ocean and Cape coasts of Africa, with small numbers regular during S passage (see Dowsett 1980) on inland rivers and lakes in the E and S of the continent.

Kenyan: a reasonably common passage migrant and winter visitor, late Aug-Apr. It is essentially a bird of the coast, where hundreds winter on salt pans and a variety of other shore habitats, including mangrove creeks (Pearson & Britton 1980; Br; Pearson 1984d; EABR 1981, 1982, 1984). Inland, the Terek Sandpiper is annual in very small numbers throughout the year at Ferguson's Gulf 13B on L Turkana (HH), but records elsewhere (mostly from the other rift valley lakes) mainly concern single birds on Sept-Nov S migration (Leuthold 1974; St; La; Br; Pearson & Stevenson 1980; EABR). Small numbers of first year birds oversummer at L Turkana and the coast (*op. cit.*), but 90+ at Mida Creek 102B on 30 May, thought to be in this category, were exceptional (EABR 1981). Dowsett (1980) quotes this species in his review of transcontinental wader migration, and the restriction of inland records to the S migration supports his views. Both migrations are known in Kenya (Br). The N migration is heavier in Somalia (AM), where at least some movement takes place along the coast (Ash 1981b).

265 COMMON SANDPIPER *Actitis hypoleucos*
B252 / MPG312 / D2C2

Afrotropical: a Palearctic species, wintering on coasts and inland throughout Africa. There are controversial old breeding records from Uganda and Kenya.

Kenyan: a common and very widespread passage migrant and winter visitor, mid Jul-early May, that can occur on any type of water body inland, and winters in small numbers on coral shores and in creeks and mangroves at the coast (Pearson & Britton 1980; Br; Pearson & Stevenson 1980; Pearson 1984d). A very few birds oversummer (*op. cit.*).

The breeding records principally concern an incubating bird photographed in Uganda, and newly fledged chicks seen during June on the Kajiado R 75D (not mapped here) (van Someren & van Someren 1911; van Someren 1916; Ja; Cunningham-van Someren 1973a). Benson & Irwin (1974) review

Plate 11. Migrant waders from the Palearctic mingle with Afrotropical waterbirds at Lake Magadi during the northern winter (D.K. Richards F.R.P.S.). *See page 166.*

Plate 12. The White-bellied Go-away Bird, a species of turaco, an Afrotropical endemic family (D.K. Richards F.R.P.S.). *See page 219*.

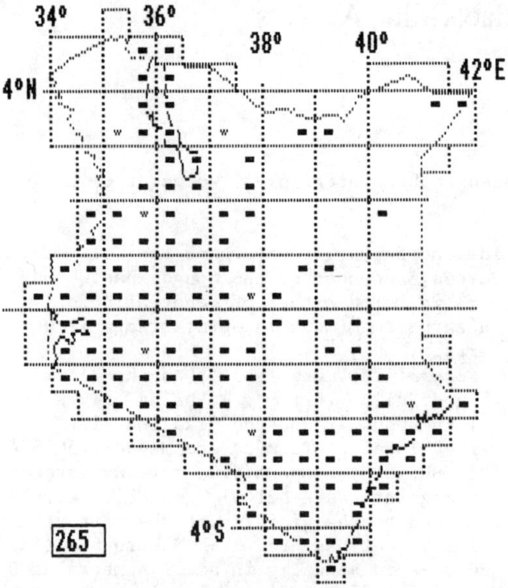

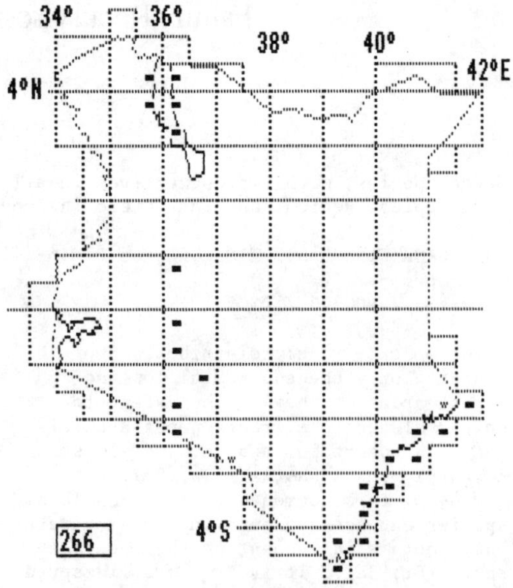

these records and, while accepting the Uganda one on the basis of the photograph, make the valid points that (a) breeding must have been a very rare occurrence because it has never again been observed by the large numbers of ornithologically interested visitors subsequently in these areas (or indeed anywhere else in Africa), so that this species cannot in any sense be regarded as a regular member of the Afrotropical breeding avifauna; and (b) that the very few Palearctic species that have commenced breeding in the Afrotropical Region (White Stork (45), Black Stork (46), Booted Eagle (124), Eurasian Bee-eater (448), House Martin (563)) have done so in the far S where variations in day length and season are more comparable to those in the Palearctic, rather than in the more uniform equatorial regimes: see also Brown & Britton (1980).

With the exception of the L Turkana area (HH), the Common Sandpiper is particularly common and widespread during S passage, while the N movement is much more obscure (Br): Dowsett (1980) notes this distinction in many wader species. One ringed at L Nakuru 62A reached 57°N 54°E, just W of the Ural Mts in USSR (Backhurst 1970, 1971b). Other reference: Pearson 1977b.

266 TURNSTONE *Arenaria interpres*
B281 / MPG310 / D2C2

Afrotropical: a common to abundant Palearctic winter visitor to African coasts, also occurring inland in small numbers in the E and S during S passage.

Kenyan: a common passage migrant and winter visitor, Aug-Apr. This is mainly a bird of the coast, where relatively small numbers winter on coral shores and on sandy beaches with accumulations of wrack; and a few oversummer (Pearson & Britton 1980; Pearson 1984d; Britton & Britton 1976b). Inland, Turnstones are annual in small numbers on open shores of the rift valley lakes Aug-Nov (Br; St; EABR), and particularly frequent at L Turkana, where they prefer shores with abundant decaying plant debris (HH): 25+ were counted at Ferguson's Gulf 13B in Jan (EABR 1980). The virtual restriction of inland records to the S movement agrees with Dowsett's (1980) views on transcontinental wader migration, and he quotes this species as an example of this type of migrant. N migration is known along the Somali coast (Ash 1981b).

Family Recurvirostridae: Stilts, Avocets

Seven species, with representatives in all zoogeographical regions. Two occur in the Afrotropical Region, and both breed in Kenya.

267 BLACK-WINGED STILT *Himantopus himantopus*
B282 / MPG296 / D3C2

Afrotropical: widely distributed S of 16°N, avoiding only the equatorial lowland forest areas of the W. The extent to which Palearctic migrants penetrate this range is uncertain, though some almost certainly reach East Africa (Br).

Kenyan: this common but sometimes local species can occur around any type of water body, but is infrequent on the open seashore (Br; Ja). It is far more widespread than the Avocet (268)(92 squares: 36). The cluster of old records around 61 reflects the lack of recent observer cover. Marked increases at various sites throughout the country during the Oct-Apr period (Ja; HH; St; La; Pomeroy & Tengecho 1982a) possibly represent Palearctic migrants, but there are no ringing recoveries.

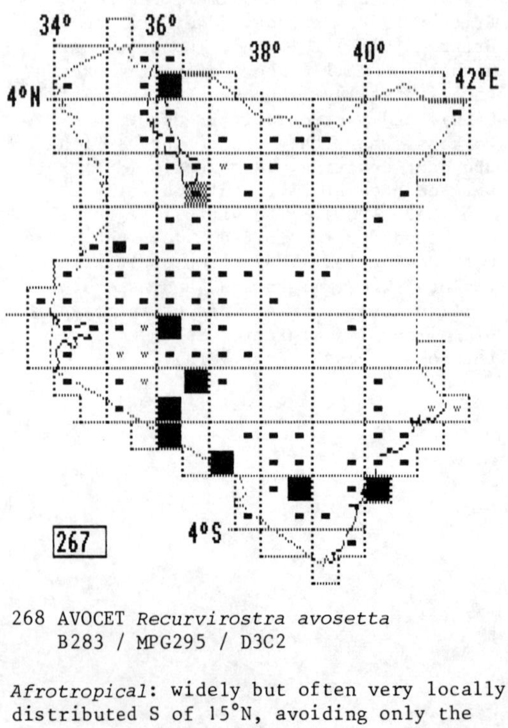

268 AVOCET *Recurvirostra avosetta*
B283 / MPG295 / D3C2

Afrotropical: widely but often very locally distributed S of 15°N, avoiding only the lowland forest areas around the Gulf of Guinea, and most of the Congo basin and Somalia. The extent of penetration of Palearctic migrants into this range is uncertain.

Kenyan: far more local than the Black-winged Stilt (267: *q.v.*), being only at all numerous around the alkaline rift valley lakes at Turkana and Magadi 75C 87A (Br; HH; MPG). Apart from rare wanderers (or migrants, see below), range is largely confined to the vicinity of the rift valley.

The Avocet has bred at L Nakuru 62A (Ja) and does so annually in small numbers at L Magadi 75C 87A. MPG mention large numbers breeding on Nyeri marshes 62B, but both the numbers involved and the habitat seem very unlikely, and this record is ignored by Brown & Britton (1980) and not mapped here.

The very small breeding population suggests the presence of non-breeding migrants. There are marked Oct-May increases at L Turkana (HH) and L Baringo 50A (St), rare Oct-Jan wanderers at Moyale 17C, L Kanyaboli 48C, Tsavo E 101B, the Sabaki estuary 103A and Likoni 114B (Britton & Britton 1976b; Br; La; D J Pearson *in litt.*;

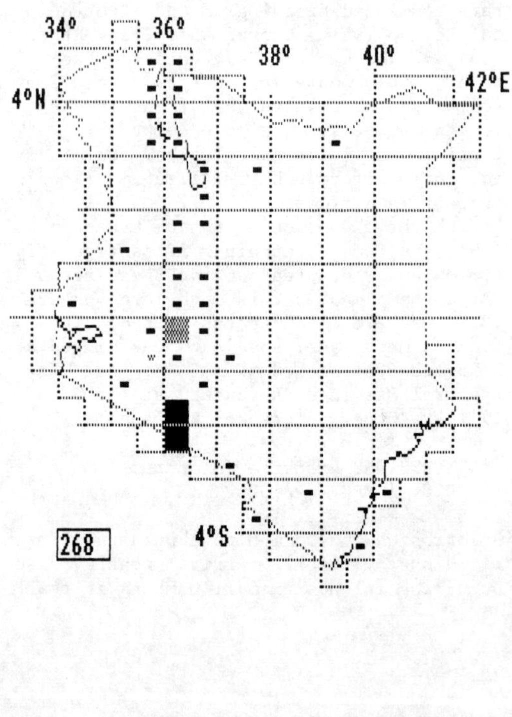

EABR 1984), and a record of 1000+ at L Ol Bolossat 62A in Feb (EABR 1981). Although the seasonality of these occurrences strongly suggests the presence of Palearctic birds, Br feels the evidence to be inconclusive. Flocks of 200 at L Nakuru 62A in June and 500+ at L Magadi 75C 87A in early Aug may well have been non-breeding migrants from the S tropics (Br).

Family Phalaropidae: Phalaropes

Three species, which breed in the Holarctic and then migrate S to the S oceans or the Neotropical Region. All three are known in the Afrotropical Region. One is a regular migrant and another a rare vagrant to Kenya.

269 RED-NECKED PHALAROPE *Phalaropus lobatus*
B285 / MPG323 / D2C1

Afrotropical: a Palearctic species that winters off the coasts of W, SW and NE Africa, and occasionally appears inland.

Kenyan: flocks of up to hundreds of phalaropes appear to be regular, mostly during Nov and Jan-Mar, off the coast, but their remoteness (usually beyond the reef), this species' close resemblance to the Grey Phalarope (270), and the fact that many of the observations are made by ornithologically inexperienced fishermen, have all combined to prevent identification to species level in most cases (Br). However, small numbers of this species have been certainly identified inshore and, in view of known ranges (*e.g.* Harrison 1983), it is generally agreed that the other records most probably belong here (Britton & Britton 1979; Br; Cunningham-van Someren 1974; EABR).

Inland, this phalarope is regular at L Turkana, where flocks build up to 30 during the Oct and Apr migration periods (HH), and it also seems to be annual or nearly so at L Nakuru 62A, where up to six have been seen (Ba; Backhurst 1973a; Br; EABR). Elsewhere inland, this species is erratic in very small numbers (La; OS-C 1977; EABR). Dowsett (1980) mentions this family in his review of transcontinental wader migration, but the inland records are by no means confined to the S migration.

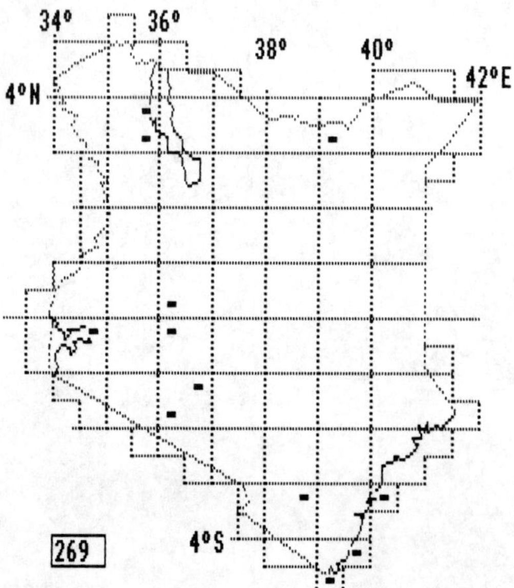

270 GREY PHALAROPE *Phalaropus fulicarius*
B284 / MPG324 / D2C0
NO MAP: post presence 13B 75B; pre presence 62A

Afrotropical: this Palearctic species winters off the coasts of W and SW Africa, and occurs occasionally in the Red Sea and the Gulf of Aden.

Kenyan: the five records of this vagrant concern one at L Elmenteita 62A 17 Feb 1953 (Ridley 1953), two on 15 Feb and one on 28 Mar 1959 at L Nakuru 62A (Lee 1959), one at Ferguson's Gulf 13B 12-13 Apr 1972 (Mann 1976), and one in Nairobi NP 75B 1 Sept 1979 (EABR 1980). Other reference: Dowsett (1980).

Family Dromadidae: Crab Plover

A monotypic family, found on the coasts of the Indian Ocean, the Persian Gulf and the Red Sea. The Crab Plover is a non-breeding migrant to Kenya.

271 CRAB PLOVER *Dromas ardeola*
B286 / MPG335 / D3C3

Afrotropical: breeds on the coasts of the Red Sea, and is a non-breeding visitor to the E coast of Africa, S to E South Africa.

Kenyan: a winter visitor, mainly Aug-Apr to the coast (Br; EABR). Flocks of up to 400 winter on extensive mud flats in the Lamu archipelago 91B and in Mida Creek 102B, with rather smaller numbers regular in many other areas (Britton & Britton 1976b; Br; EABR).

Small numbers oversummer annually (*op. cit.*) and MPG mention breeding in the Kilifi 102D and Sabaki 103A areas, but this and all subsequent records have involved juveniles which, although often dependent on their parents for food, are free-flying (including one "just able to fly": EABR 1978): until nest burrows or truly flightless young are found, it should be assumed that these juveniles have migrated from breeding grounds far to the N or NE (van Someren 1939; Britton & Britton 1976b; Brown & Britton 1980).

Very small numbers occur along the coast during the Oct-Nov migration (Br; EABR), but Ash (1981b) suggests that, for the Somali coast at least, the N movement appears to take place mostly far offshore, with at least some nocturnal flights. The sole inland record, of one at L Magadi 75C on 1 Sept 1984 (EABR 1984), was presumably a disorientated migrant.

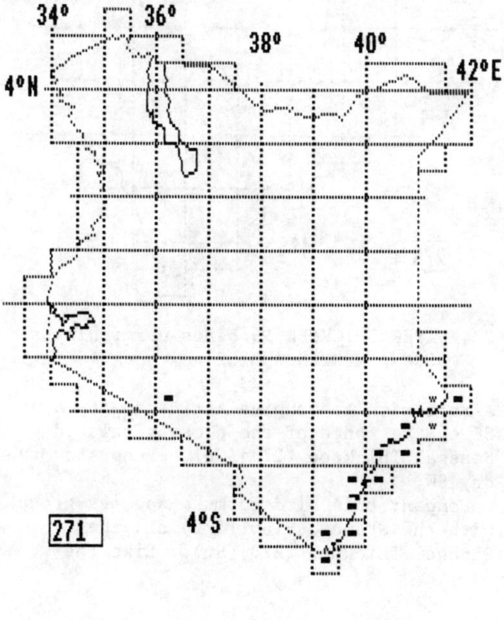

Family Burhinidae: Thicknees, Stone Curlews

Nine species, with representatives in all zoogeographical regions except the Nearctic. Four occur in the Afrotropical Region including Kenya, where three breed and one is a migrant from the Palearctic. They are most active at night, but are often visible while resting in sheltered situations during the day.

272 STONE CURLEW *Burhinus oedicnemus*
B288 / MPG260 / D2C1

Afrotropical: a Palearctic species, wintering in the S periphery of the Sahara, E to Ethiopia, and S to N Uganda, Kenya and the NE borders of Tanzania.

Kenyan: a scarce but probably regular winter visitor, Oct-Mar (Ba; Br), although some records may be lost due to confusion with Afrotropical congeners. It occurs in very small numbers S to the Kenya/Tanzania borders (Ja; Br; OS-C 1977; EABR 1978), but appears most frequent on the N side of the C highlands, including the rift valley (Ba; EABR). Range is 75% above 1000 m and 70% within 500+ mm rainfall areas. See Water Thicknee (274).

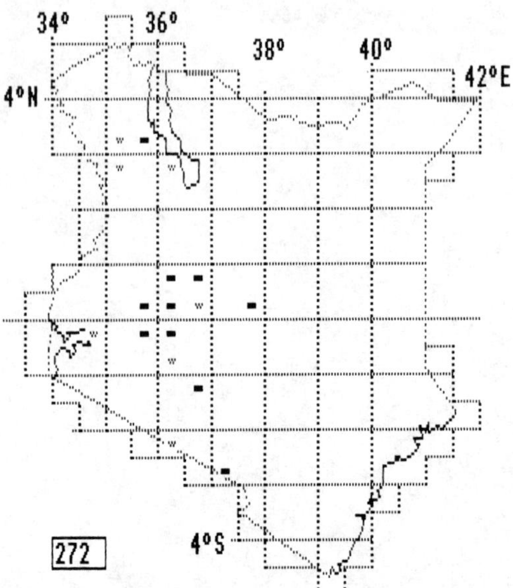

273 SENEGAL THICKNEE *Burhinus senegalensis*
B289 / MPG261 / D2C2

Afrotropical: the N tropics between 5°N and 16°N, from the W coast to Cameroon and L Chad, and from NE Zaire to Sudan, Ethiopia, N Uganda and, on the SE extremity of the range, NW Kenya.

Kenyan: reasonably common at L Turkana, mostly on the islands and E shores (HH; Br), at the SE limit of its N tropics breeding range. Records further S on the rift valley lakes, S to L Bogoria 50C, overlap the range of the closely related and otherwise allopatric Water Thicknee (274: *q.v.*), and most probably refer to non-breeding wanderers (Br). Suggestion of possible resident status at L Baringo 50A (St) has been retracted (T Stevenson *in litt.*). A definite wanderer, in exhausted condition, was seen at L Kanyaboli 48C on 8 Nov 1969 (Br).

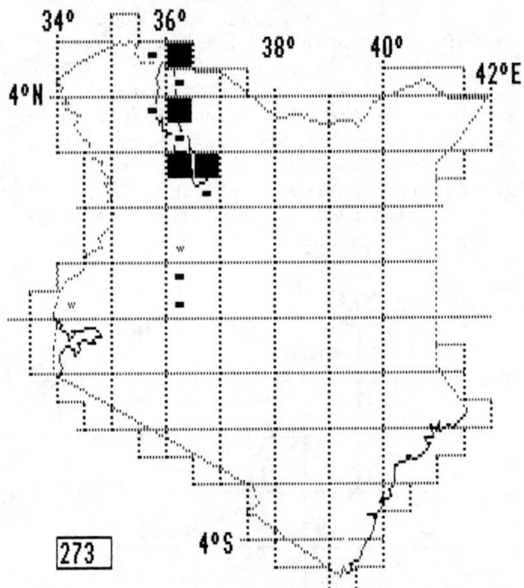

274 WATER THICKNEE *Burhinus vermiculatus*
B290 / MPG263 / D2C2

Afrotropical: occupies Africa to the S and SE of the range of the closely related Senegal Thicknee (273)(Sn), except for the dry SW.

Kenyan: this bird forms a species-group with the Stone Curlew (272) and the Senegal Thicknee (273)(Sn). Like the

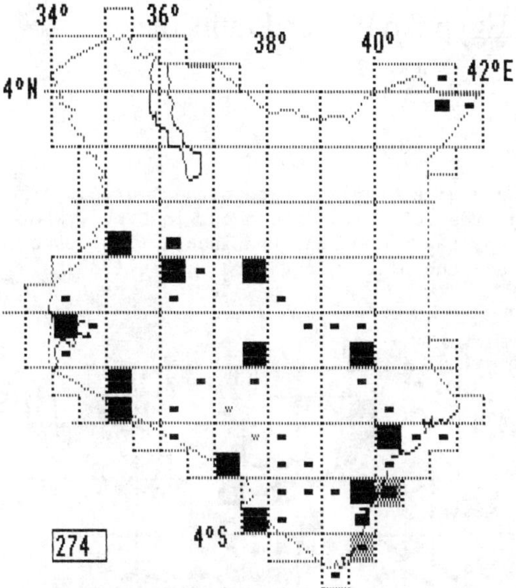

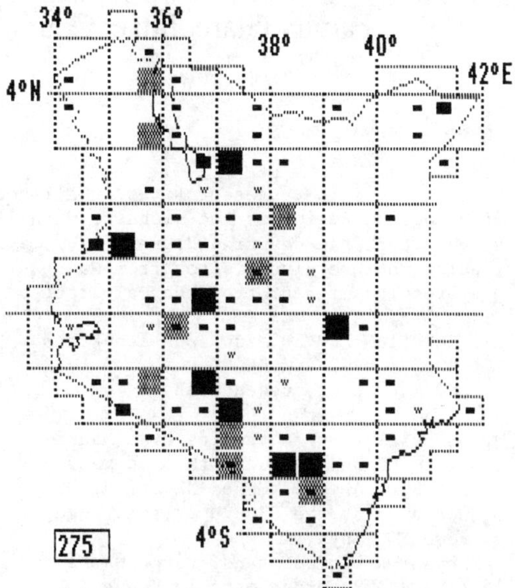

Senegal, with which it is allopatric, the Water Thicknee is resident on lake shores and river banks adjacent to woodland or bush, in which it shelters during the day (Sn; Br). It is widespread but local by these aquatic habitats, particularly those with sandy beaches or bars, to the S of L Turkana, and also along the Daua R 7C 19AB (ADL pers. obs.), while at the coast it is found in the vicinity of creeks, mangroves and islands (Br; Ja). 82% of its squares are below 1500 m.

275 SPOTTED THICKNEE *Burhinus capensis*
B287 / MPG262 / D2C2

Afrotropical: widely distributed S of 16°N, apart from most of the lowland forest areas of the W.

Kenyan: rather uncommon in open bushed and wooded habitats, including woodland edge, up to 2000 m (Br; MPG; Lack 1985). Range is 66% within intermediate, 250-1000 mm rainfall areas and 61% less arid-semiarid, but virtually absent from even semiarid squares on the coast. This species is mainly resident, but Ja noted local movements and Br mentions a wanderer in Uganda. The Spotted Thicknee is separated ecologically from the Senegal (273) and Water (274) Thicknees by not being associated with water (Ja; Sn).

Family Glareolidae: Coursers, Egyptian Plover, Pratincoles

17 species, with representatives in all zoogeographical regions except the Americas. The Afrotropical Region is the richest, with 11 species (65%)(Sn). 10 occur in Kenya, where seven breed, the Egyptian Plover (276) may do so, the Madagascar Pratincole (285) is a regular non-breeding visitor from Madagascar, and the Black-winged Pratincole (283) a non-breeding migrant from the Palearctic.

276 EGYPTIAN PLOVER *Pluvianus aegyptius*
B293 / MPG334 / D2CO
NO MAP: post presence 3B

Afrotropical: the N tropics from Mauritania to NE Sudan and N Ethiopia, and thence S to Angola, C and NE Zaire, NW Uganda and, at the SE tip of the range, extreme NW Kenya.

Kenyan: a bird of sandy river banks and lake edges, recorded once in Kenya, when eight were seen in Aug 1971 between Todenyang and Namuraputh 3B, on the W edge of the Omo Delta (Goddard 1972). It was common in June along the Omo R just inside Ethiopia, only 10 km N of Namuraputh (Hopson 1974b), and the paucity of records in Kenya most probably reflects lack of observers in a very remote area.

277 CREAM-COLOURED COURSER *Cursorius cursor*
B291 / MPG325 / D2C2

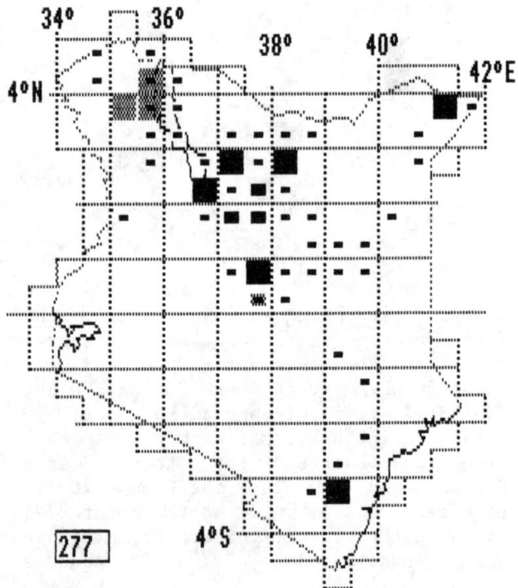

Afrotropical: the N tropics and NE Africa, E between 14°N and 20°N from the W coast to N Sudan and N Ethiopia, and S around the E edge of the Ethiopian highlands to Somalia and Kenya. Distribution in Kenya represents the SW limit of the range.

Kenyan: rather local in dry country with bare, grassy or open bushed habitats (Br; HH; MPG). It forms part of a species-group with Temminck's Courser (278) and, apart from presumed non-breeding wanderers, the two are allopatric (Sn; Br). The Cream-coloured is found in the N and E, where range is 86% below 1000 m, 87% within the 0-500 mm rainfall regime and 79% arid, while Temminck's replaces it in the W, C and S (including parts of the coastal strip), with range 69% above 1000 m, 72% within the 500+ mm rainfall regime and 80% non-arid. Buffalo Springs 51B is the sole locality where both species are known to breed in close proximity, with this species nesting on bare open plains and Temminck's on bare ground in nearby woodland (Br). Imperfect Sept-May seasonality on the edge of the Cream-coloured's range in the Tsavo E 101B-Galana Ranch 90C 102A area constitutes the only evidence of movements. Interestingly, most records of Temminck's Courser, which is also on the edge of its range, occur in Tsavo E in the other half of the year (Br; La; Lack 1985).

278 TEMMINCK'S COURSER *Cursorius temminckii*
B292 / MPG326 / D2C2

Afrotropical: widely distributed S of 15°N, apart from most of the lowland forest areas of the W, the arid SW, and those areas occupied by the Cream-coloured Courser (277: *q.v.*).

Kenyan: locally common in open habitats, including burnt ground (Br; Ja; Lack 1985). Breeding is known up to 2000 m (Br), while occasional migrants or wanderers reach 3000 m around Mau Narok 61D (Se). Sn notes erratic occurrence and a tendency to colonise recently burnt areas, and records from arid country well within the range of the Cream-coloured Courser, *e.g.*

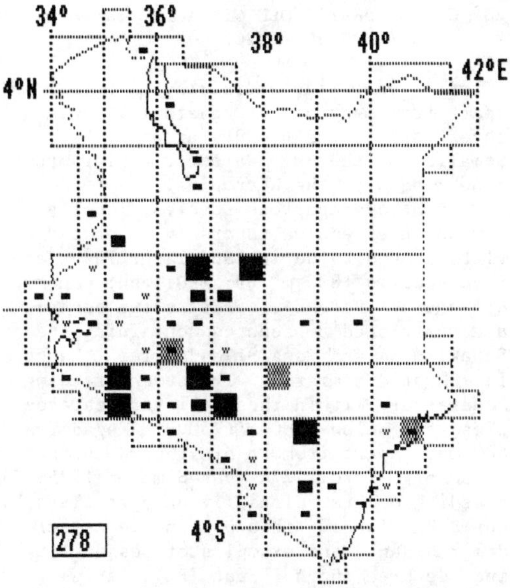

at L Turkana and Garissa 65B, presumably refer to non-breeding wanderers. Temminck's Courser appears after the rains on the edge of its breeding range at semiarid Baringo 50A (St).

279 TWO-BANDED COURSER *Rhinoptilus africanus*
B294 / MPG327 / D2C2

Afrotropical: in three distinct areas: in the NE, from E Ethiopia to N Somalia; in the E, from E Tanzania to SW Kenya; and in the dry SW of the continent.

Kenyan: rather uncommon and local on open or sparsely bushed plains, up to 1800 m (Br; Be; MPG). Range is 90% within intermediate, 250-1000 mm rainfall areas and 59% semiarid. The record in 7C J S Ash *in litt*.) was actually in Ethiopia, but is mapped because it represents the SW extremity of the E Ethiopia-N Somalia population. The other records define the N limits of the E Tanzania-Kenya population, with 79C (P L Britton *in litt*.) presumably representing a wanderer.

Br and Be describe this species as resident. See Heuglin's Courser (280).

280 HEUGLIN'S COURSER *Rhinoptilus cinctus*
B296 / MPG328 / D2C2

Afrotropical: an E species, from Zimbabwe and Zambia NE through areas of East Africa E of L Victoria, to S Sudan, E Ethiopia and Somalia. There is also a population in N Namibia.

Kenyan: a widespread but mainly nocturnal bird of dry, bushed habitats, readily overlooked by day (Br; Ja; MPG; Sn; Lack 1985): D2 reflects the number and widely distributed nature of its records. Range is 80% below 1500 m, 80% within intermediate 250-1000 mm rainfall areas and 76% less arid-semiarid. This courser is absent from even semiarid areas of the coastal strip, and ascends only to the peripheries of the highlands and the moist W plateau. It is segregated from the closely related Two-banded Courser (279) by habitat, being

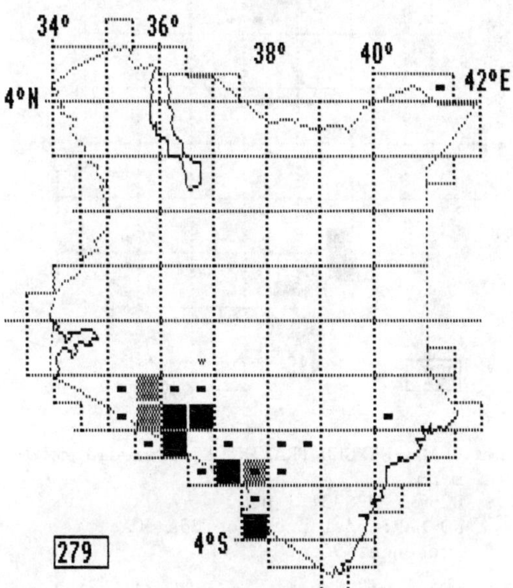

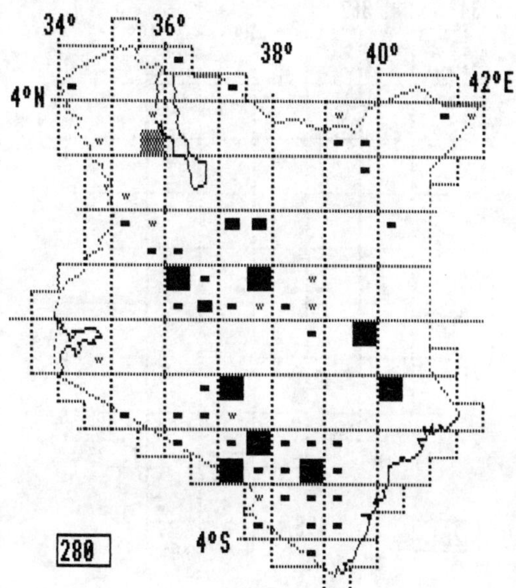

restricted to thorn scrub even in N Kenya, where that open plains species is absent (Sn). Br considers Heuglin's probably resident in many areas, and this view is shared by La (101B) and DEP (pers. obs.: 88A). However, unusual abundance in the Tsavo area 89 101 during Dec 1977 and Mar-Apr 1978 (EABR) presumably demonstrates at least local movement.

281 VIOLET-TIPPED COURSER *Rhinoptilus chalcopterus*
B295 / MPG329 / D1C1

Afrotropical: widely distributed S of 15°N, but absent from Somalia, the dry SW, and most of the equatorial lowland forest areas of the W. Distribution in Kenya represents the E limit of the range.

Kenyan: a rarely recorded, strictly crepuscular and nocturnal bird of woodland and bush up to 2200 m, easily overlooked during the day (Br; MPG; Sn): hence D1C1. It is found in the S lowlands and around the E and N peripheries of the highlands, with range 72% semiarid-dry subhumid. There is only a single confirmed breeding record, at Stony Athi 76C, and this species may be a predominantly non-breeding intra-African migrant from the S tropics (Br). Most Tsavo E 101B records are Dec-May (La) while, of the four Kenyan records since 1977, two at the coast were Apr and two in the highlands were June-Jul (EABR). A bird has recently been caught at the Ngulia lights 101A (not on map)(EABR 1985).

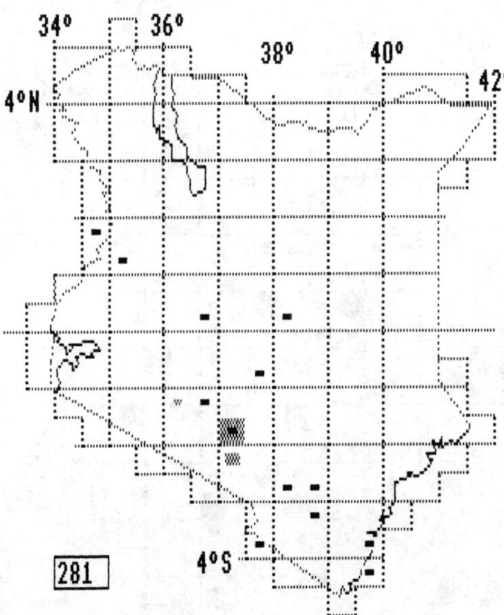

282 COMMON PRATINCOLE *Glareola pratincola*
B300 / MPG330 / D2C2

Afrotropical: widely distributed S of 17°N, apart from many of the equatorial lowland forest areas, and the dry areas of E Somalia and the SW. Palearctic migrants penetrate S to the N tropics.

Kenyan: a gregarious species, locally common on extensive, sparsely vegetated flats around estuaries and large lakes and rivers up to 1800 m (Br; MPG), but uncommon elsewhere. Recent records at the coast are restricted to squares containing the Sabaki 103A and Tana 91 estuaries. Range is 79% arid-semiarid. The seven breeding squares are within these moisture categories, and below 1500 m (75C = L Magadi, at 580 m). Apart from an old record on the seasonal L Koroli 27A (van Someren 1925-35), breeding occurs colonially on open flats by permanent lakes. This pratincole is seldom resident, and examples of seasonality include breeding Apr-Sept (Br), large flocks at L Baringo 50A Oct-Dec (St), and sporadic occurrence in Tsavo E 101B Dec-Apr (La).

The Common Pratincole forms part of a superspecies with the Black-winged Pratincole (283)(Sn), which is a vagrant to Kenya. Other reference: Reynolds 1977.

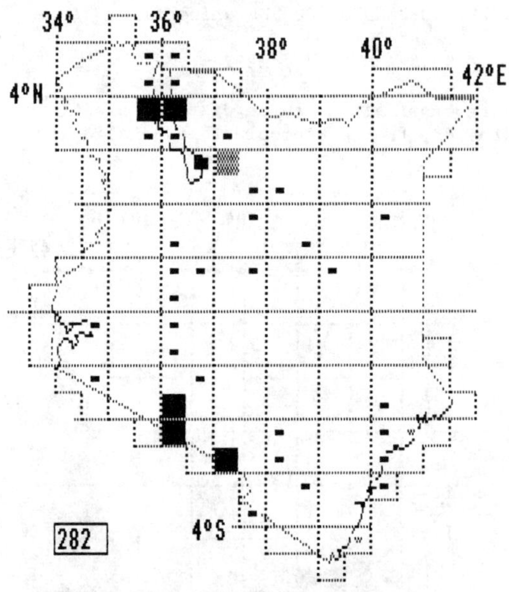

283 BLACK-WINGED PRATINCOLE *Glareola nordmanni*
B297 / MPG331 / D2C0
NO MAP: post presence 28A 50A; pre presence 49C 62C

Afrotropical: a Palearctic species that

migrates S through C Africa, Zambia and Angola, to winter in Namibia and South Africa.

Kenyan: a vagrant E of its normal migrational range (Ba). The four records concern three adults and an immature at L Naivasha 62C 12 Oct 1953 (Wallace 1975), one found dead near Lessos 49C 12 Oct 1969 (Mann 1971), one 30 km N of Marsabit 28A 25 Mar 1978 (EABR 1978), and six together at L Baringo 50A 2 Apr 1981 (EABR 1981). See Common Pratincole (282).

284 WHITE-COLLARED PRATINCOLE *Glareola nuchalis*
B298 / MPG333 / D2C2
NO MAP: post presence 91A; post confirmed 48C

Afrotropical: there are scattered records in the N tropics to 18°N from Niger to W Ethiopia, but this is mainly a bird of W and C areas that is associated with rocks in rivers from Sierra Leone S to Angola, and E to the extreme W of Kenya, E Zaire, SE Tanzania, Zambia and Zimbabwe.

Kenyan: the single known breeding colony is on bare rocks in the Nzoia R, within the grounds of a large sugar factory at Mumias 48C. Ja described the species as fairly plentiful on this river but, while other, undiscovered colonies may exist, there is also the possibility that its exposed nest sites have proved vulnerable in this densely populated area. A single bird with other pratincoles at L Bilisa 91A on 16 Feb 1985 (M A C Coverdale & D J Pearson *in litt.*) is the first record of a vagrant in East Africa (see Br).

285 MADAGASCAR PRATINCOLE *Glareola ocularis*
B299 / MPG332 / D2C2

Afrotropical: breeds on Madagascar and spends the Mar-Sept non-breeding season in coastal areas of Kenya and S Somalia (Britton 1977e; AM).

Kenyan: most frequent on estuarine flats and coastal dunes between the Sabaki R and Malindi 103A, where 800+ regularly winter Apr-Sept (Britton 1977e; Br). The S passage Aug-Sept is most marked: migrants appear in small numbers elsewhere along the coast, while the Sabaki flocks increase to 2500+, with a quite exceptional 9-10 000 there on 17 Aug 1978 (*op. cit.*; EABR). Small numbers are regular inland at sites near the coast at Garsen 91A and Malindi 103A (Br). The only other inland record is of hundreds at Kendu Bay 60B on 21-22 Aug 1920 (Granvik 1923), which were presumably disorientated migrants. Other reference: Meadows 1974.

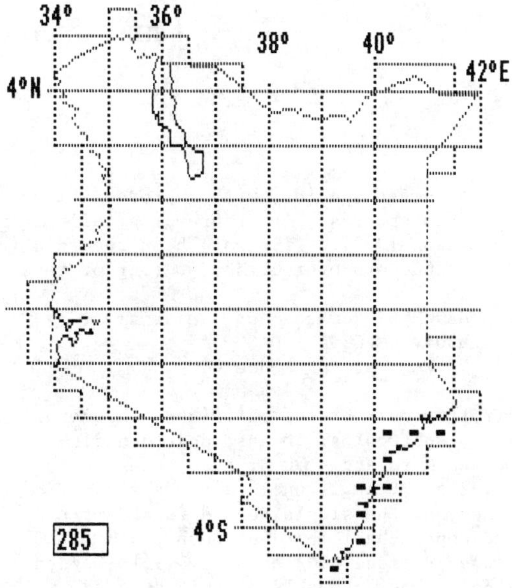

285

Family Stercorariidae: Skuas

Six or seven species, with representatives throughout the temperate and arctic oceans of both hemispheres. After breeding, all migrate towards, or to, the opposite hemisphere (Harrison 1983). Five have been recorded in the Afrotropical Region: there are extremely low numbers of records for four of these in Kenya.

286 BROWN SKUA *Catharacta antarctica*
B301 / MPG336 / D1C0
NO MAP: pre presence 80C

Afrotropical: breeds on islands throughout the S subtropical oceans, and then disperses N to about 30°S; its status in the N Indian Ocean is uncertain (Harrison 1983).

Kenyan: the single record is of a tired and approachable bird at Suthi, 5 km S of Kiunga village 80C, on 6 Aug 1961 (Fogden 1963). One, probably of this species, was seen off Shimoni 114C in Aug 1976, though confusion with the similar South Polar Skua *C. maccormicki* could not be excluded (Br).

287 POMARINE SKUA *Stercorarius pomarinus*
B304 / MPG338 / D1C0
NO MAP: post presence 13B 103A 114B

Afrotropical: this Palearctic species regularly winters off the W coast, but is rarely seen along the Red Sea and Indian Ocean seaboards.

Kenyan: the few records are a pale phase adult near Mombasa 114B 7 Dec 1973 (Britton & Britton 1974a), an immature near Mombasa 9 Jan 1974 (Mann 1974a), a pale phase adult at the Sabaki estuary 103A Feb-Mar 1976 (Britton & Britton 1976b), a pale phase bird at Ferguson's Gulf 13B, L Turkana, 31 Dec 1977-1 Jan 1978 (EABR 1977, 1978), a pale phase subadult at the Sabaki estuary 103A 18 Jan-14 Feb 1981 and possibly the same individual at Diani 114B 6 Mar 1981 (EABR 1981), and a pale phase adult at Ferguson's Gulf 13B 21-24 Oct 1983 (EABR 1983). Thus, apart from the most recent record, and in contrast to those of the Arctic Skua (288), the occurrences of this species have all been Dec-Mar.

288 ARCTIC SKUA *Stercorarius parasiticus*
B303 / MPG337 / D1C0
NO MAP: post presence 13B 103A

Afrotropical: this Palearctic species winters off the W and S coasts. It is a regular passage migrant off Somalia (Ash 1983; AM), but is otherwise rare off the E coast N of South Africa.

Kenyan: the three records are of a pale phase adult near Malindi 103A 9 Apr 1966 (Backhurst 1971a), a dark phase adult at Watamu 103A 13 Apr 1980 (EABR 1980), and a dark phase adult at Ferguson's Gulf 13B, L Turkana, 20 Oct 1983 (EABR 1983). In addition, two immatures at Ferguson's Gulf 13B Sept-Oct 1973, and another there Oct 1974 were probably this species (Robertson *et al.* 1974; HH), as were a dark phase adult at Shimoni 114C 14 Mar 1976 (Br), a dark phase bird off Ras Ngomeni 91C 19 Oct 1980 (EABR 1980) and another dark phase bird there on 23 Apr 1983 (EABR 1983). The restriction of all records to Sept-Oct and Mar-Apr passage dates agrees with the species' migrant status off Somalia (see above).

289 LONG-TAILED SKUA *Stercorarius longicaudus*
B302 / MPG- / D2C0
NO MAP: post presence 91C; pre presence 13B

Afrotropical: a Palearctic species, wintering off the Atlantic coasts.

Kenyan: the two records of this vagrant involve one at Ferguson's Gulf 13B, L Turkana, 25-26 Aug 1961 (Ba), and two moving S past Ras Ngomeni 91C 19 Oct 1980 (Moore 1981). These are the only records for East Africa; all were adults.

Family Laridae: Gulls, Terns, Noddies

About 87 species, with representatives on all oceans and continents. 37 (43%) have been recorded in the Afrotropical Region and, of these, 27 (73%) are known in Kenya. Of the Kenyan species, 10 (37%) breed, have done in the past or are thought to do so, four are non-breeding visitors or vagrants from other Indian Ocean or Red Sea coasts, and 12 (44%) are non-breeding migrants or vagrants from the Palearctic.

290 SOOTY GULL *Larus hemprichii*
B309 / MPG348 / D2C2

Afrotropical: breeds on the coasts of the Red Sea, Somalia and E Kenya, and extends S as a non-breeding visitor along the E coast to S Tanzania.

Kenyan: this scavenger can be found anywhere along the coast, and up to at least 7 km offshore, throughout the year (Britton & Brown 1974; Br). The breeding of under 50 pairs on islands N of Kiunga village 80CD Jul-Sept (Ja; Fogden 1964) represents the S extremity of its breeding range, and this gull is a markedly seasonal non-breeding visitor further S. Thus it is abundant Oct-May, with regular flocks of 500+ at Malindi 103A and Mombasa 114B and a peak count of 1500 on the Sabaki estuary 103A in Jan (EABR 1982), whereas seldom more than 30 (mainly young birds) are seen together during the rest of the year (Britton & Brown 1974; Britton & Britton 1976b; Br; EABR). The single inland record is of three immatures on Central Island 14C, L Turkana, on 25 Apr 1974 (Hopson & Hopson 1974b): this date is within this species' migration period, so that these were probably wandering passage birds.

Since Ja mentioned that it was nowhere plentiful and had only just started frequenting Mombasa harbour 114B, while MPG noted that it was becoming more numerous in East African harbours, this gull seems to have reached its present level of abundance only recently.

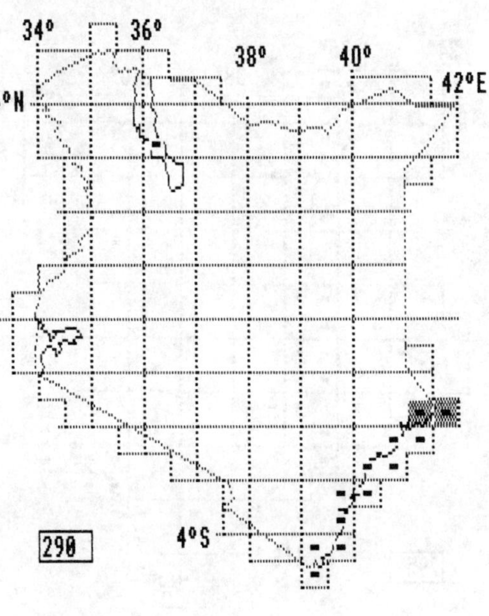

291 WHITE-EYED GULL *Larus leucophthalmus*
B311 / MPG347 / D1C0
NO MAP: post presence 13B

Afrotropical: breeds in the Red Sea, and mostly winters in the Gulf of Aden and along the adjacent Somali coast (Harrison 1983).

Kenyan: the single record is of an immature in a large gull flock at Ferguson's Gulf 13B, L Turkana, 4-14 Apr 1975 (Hopson & Hopson 1975b). Br suggests that it might occur intermittently at the coast.

292 GREAT BLACK-HEADED GULL *Larus ichthyaetus*
B310 / MPG343 / D1C0

Afrotropical: this Palearctic gull is a winter visitor in small numbers to the coast and rift valley lakes of NE Africa, occasionally S to Kenya and, more rarely, Uganda and Somalia.

Kenyan: since the first record, of an immature in the vicinity of the Sabaki estuary and Malindi 103A on 19 Jan 1974 (Britton & Duffus 1974), Great Black-headed Gulls have been found to occur annually Dec-Mar in this area, singly or with two together and usually in subadult plumage (Britton & Britton 1976b; Br; EABR). The very few inland records are concentrated at L Turkana, where there were three on the Ethiopian border 4AC on 18 Mar 1975 (Ash & Ashford 1977; J S Ash *in litt.*), one at Ferguson's Gulf 13B on 10-13 Apr 1975 (Hopson & Hopson 1975b) and three

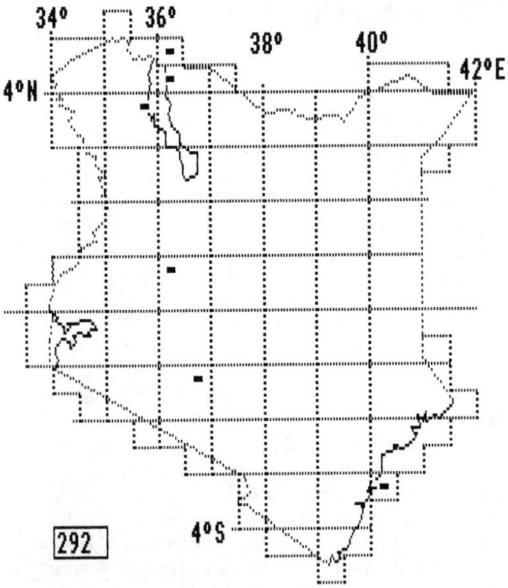

there 26 Dec 1976 (Pearson 1977a). An unprecedented 26 in this Gulf on 23-25 Jan 1984 (EABR 1984) coincided with two birds at Dandora sewage works, Nairobi 75B (EABR 1984), and were in the same period as an influx of Slender-billed Gulls (297) to L Nakuru 62A. The other inland record is of a bird at L Baringo 50A on 5-31 Jan 1980 (St). Other references: Britton 1974b; Mann 1976.

293 MEDITERRANEAN GULL *Larus melanocephalus*
B312 / MPG- / D1C0
NO MAP: post presence 13B 91B

Afrotropical: a Palearctic species that winters on the W coast, S to Mauritania and Senegambia. Two vagrants on the Sudan Red Sea coast and two in Kenya constitute the only records for the E.

Kenyan: the two records are of an immature at Ferguson's Gulf 13B, L Turkana, 12-15 Apr 1975 (Hopson *et al.* 1975), and an adult at Lamu 91B on 17 Aug 1978 (EABR 1978). The presence of a Red Sea species, the White-eyed Gull (291), at Ferguson's Gulf 4-14 Apr 1975 (Hopson & Hopson 1975b), and recent occurrences of the Mediterranean Gull on the Sudan Red Sea coast in late Mar and Apr (Nikolaus 1984), may be relevant to the origin of the first of the Kenyan birds.

294 LITTLE GULL *Larus minutus*
B313 / MPG344 / D2C0
NO MAP: post presence 26B

Afrotropical: this Palearctic species has only been recorded on the W coast (Sierra Leone and Nigeria), and in the rift valley in Ethiopia and Kenya.

Kenyan: the single record involved a flock of 60 adults and immatures just N of Loyengalani 26B, L Turkana, on 7 Jan 1979 (Groh 1979). The birds could not be found on the following day.

295 BLACK-HEADED GULL *Larus ridibundus*
B314 / MPG345 / D2C2

Afrotropical: a Palearctic gull that winters with increasing frequency on the coasts and inland lakes of Africa, S to Nigeria in the W and Tanzania in the E.

Kenyan: the Black-headed Gull constitutes an excellent example of sudden range expansion. Although unrecorded prior to 1949 (Smart 1952), and with only five records between 1949 and 1971 (Ba), this gull appeared in large numbers both inland and at the coast during the winter of 1971/72, and has continued to winter regularly in hundreds ever since. Early suggestions that the birds are Slender-billed Gulls (297) have been refuted by specimens, sightings of adults in breeding plumage and rare, later examples of that species (Britton & Brown 1974; Britton 1976; Br; Mann 1976; Pearson 1972; Brown 1973b; Bruce 1972; Fry & Horne 1972a; Standish King 1972a). Similar range extension has

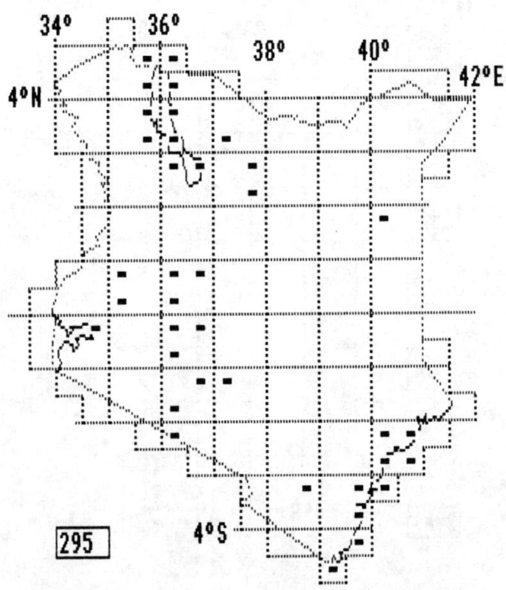

been noted in Somalia (AM), Sudan (Nikolaus & Hamed 1984) and Tanzania (Harvey 1972).

The main wintering sites are L Turkana, where flocks of 500+ are regular, peaking at *c.* 1000 in Ferguson's Gulf 13B during Mar-Apr (HH), and L Nakuru 62A, Malindi 103A and the Sabaki estuary 103A (Br; EABR). Small numbers are found at other waters, including those isolated in arid country, *e.g.* a desert pan 27B (Britton & Britton 1976b), waterholes at Wajir 42A (J E Squire *n.d.*), and Aruba dam 101B (La). Immatures predominate, and breeding plumaged adults form a minority of even the small numbers that regularly oversummer (Br; HH). Other references: Deane 1971; Isenmann 1978; Steinbacher 1977.

296 GREY-HEADED GULL *Larus cirrocephalus*
B306 / MPG342 / D2C2

Afrotropical: widely distributed in the E from South Africa N through East Africa to S Sudan and Ethiopia. In the W, it is widespread on coasts, and inland in the vicinity of the R Niger and L Chad. Distribution in Kenya represents the E limit of the range.

Kenyan: primarily a bird of the rift valley lakes, where it can be numerous (*e.g. c.* 8000 at L Nakuru 62A in Feb: EABR 1978, and 5000 in Jan: EABR 1984) and is the commonest gull (Br; Ja; MPG). Local movements between lakes most probably occur, but data exist only for L Baringo 50A, where small numbers occur Dec-Mar (just prior to its Mar-June breeding season: Brown & Britton 1980),

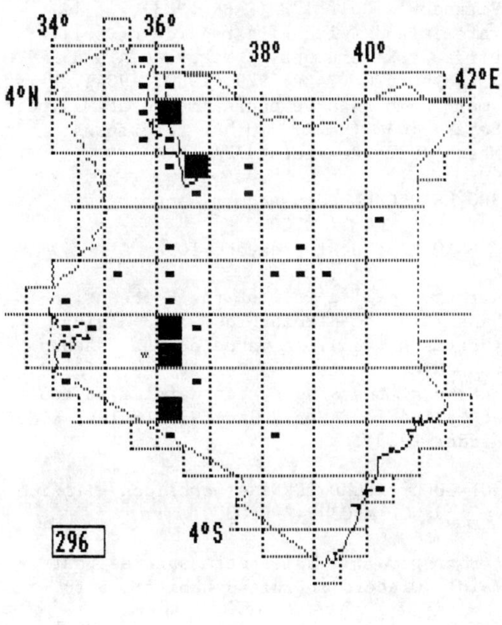

but it is otherwise only occasional (St). Small numbers wander widely to other inland waters (*e.g.* occasional transient visitors to dams at 3000 m at Mau Narok 61D: Se), but this gull is rare on the edge of its range in the E lowlands, with only a single record from the coast (Malindi 103A: Br), despite MPG's mention of occurrence there.

297 SLENDER-BILLED GULL *Larus genei*
B308 / MPG341 / D1C0
NO MAP: post presence 13B 26B 62A 91D 114B

Afrotropical: a Palearctic species that winters regularly on coasts S to Mauritania and Sudan, with recent records S to the Nigerian and Kenyan coasts, and Ethiopian and Kenyan rift valley lakes.

Kenyan: there has been considerable confusion between this species and the Black-headed Gull (295: *q.v.*), both during the latter's 1971/72 influx and subsequently. The few substantiated records of the Slender-billed Gull have occurred in three areas. At L Turkana, the first Kenya record was of two at Ferguson's Gulf 13B on 25-27 Dec 1976 (Pearson 1977a), followed by two at Loyengalani 26B on 20 Dec 1977 (EABR 1977), one at Ferguson's Gulf on 2 Jan 1978 (EABR 1978) and four there on 25 Nov 1981 (Stevenson 1982a): the last author suggested that this species might be annual at Turkana in small numbers. One was seen at L Nakuru 62A on 28 Feb 1978 (Berg-Schlosser 1979), and a marked influx of at least six birds occurred there in Nov 1983-Jan 1984 (EABR 1983, 1984). There have only been two coastal records: one at Ras Iwetine 114B on 10 Oct 1981 (Taylor 1982c) and two at Kipini 91D on 22 Dec 1984 (EABR 1984). See Nubian Shrike (859).

298 LESSER BLACK-BACKED GULL *Larus fuscus*
B307 / MPG340 / D2C2

Afrotropical: this Palearctic gull winters on African coasts, S to at least Nigeria in the W and South Africa in the E. Inland, it is common to abundant on large lakes and rivers S to Zimbabwe.

Kenyan: in contrast to the Herring (299) and Black-headed (295) Gulls, this species has been a regular winter visitor Oct-Apr for at least several decades (Br), although increase in numbers and/or range extension, as has recently also occurred in Somalia (AM), is indicated by Ja's note of absence at the coast, which contrasts with the present regular occurrences there (Britton & Brown 1974; Britton & Britton 1976b).

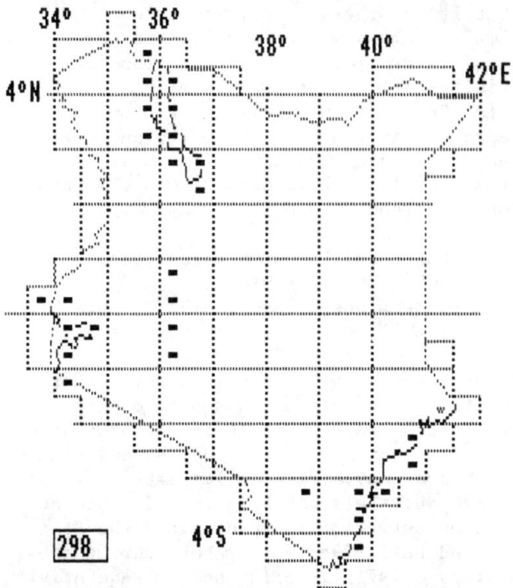

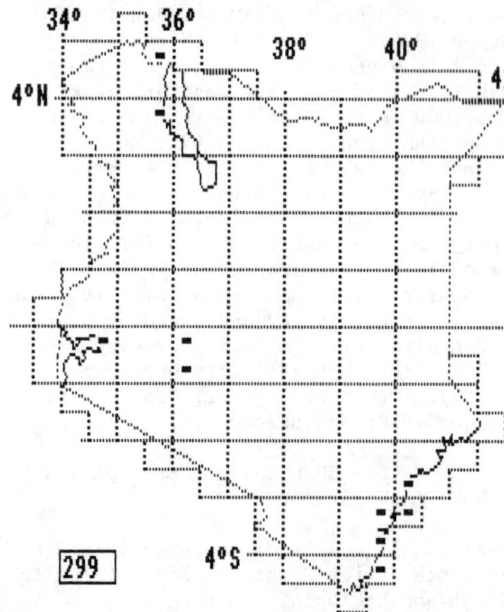

In contrast to the wandering of the Black-headed Gull, this species has a very well defined range on the coast, the large lakes of the rift valley, and L Victoria. The only record away from these squares is of two birds at Aruba dam 101B on 20 Nov 1975 (La). It is mainly an inland species, with hundreds on L Victoria, and large numbers on L Turkana peaking at 1000 in the Ferguson's Gulf area 13B Mar-Apr (HH). Further evidence of its increase are recent records of 50+ on L Nakuru 62A, and on the coast at Malindi 103A and Mombasa 114B, though numbers over 20 are unusual elsewhere (Br; EABR). A few birds over-summer. Birds ringed in Finland have been recovered at Ferguson's Gulf 13B (Smart 1967) and at Homa Bay 60C (Backhurst 1972). Other references: Smart 1952; Voous 1965.

299 HERRING GULL *Larus argentatus*
B305 / MPG339 / D2C2

Afrotropical: a Palearctic bird that winters in increasing numbers on African coasts, S to Mauritania in the W and Tanzania in the E, and regularly also in the Ethiopian rift valley.

Kenyan: like the Black-headed Gull (295), this species is extending its wintering range S, although less rapidly. Thus, while the first record, of one at Malindi 103A in Dec, was published only a decade ago (Britton & Brown 1974), numbers wintering at the most favoured localities of Malindi and the Sabaki estuary 103A have steadily increased from peak counts of five in 1973/4 to 300 in 1982: it is now a regular winter visitor there Nov-Apr, outnumbering the Lesser Black-backed Gull (298) (Britton 1974d; Br; Britton & Britton 1976b; EABR).

In contrast to the Lesser Black-backed Gull, the Herring is predominantly coastal but, although 20 were seen at Nyali 114B in Jan (EABR 1978), counts away from Malindi/Sabaki 103A have been small or very small (EABR). Apart from L Turkana, where numbers have reached 25 in Oct at Ferguson's Gulf 13B (EABR 1983), it is rare inland, with five records of solitary birds at Kisumu 60B, L Nakuru 62A and L Naivasha 62C (EABR 1979, 1980, 1984). Very small numbers have oversummered at Malindi/Sabaki 103A (EABR). Other references: Meeth 1977; Pearson 1977a.

300 KELP GULL *Larus dominicanus*
B- / MPG- / D2C0
NO MAP: post presence 103A

Afrotropical: a gull of the S oceans, N to N Namibia and S Mozambique in S Africa (Harrison 1983). A wanderer has reached N to Kenya.

Kenyan: the single record is of an adult at Malindi 103A on 2 Jan 1984 (Mulder & de Ridder 1985).

301 GULL-BILLED TERN *Gelochelidon nilotica*
B320 / MPG349 / D2C2

Afrotropical: a Palearctic species that mainly winters in inland habitats S to

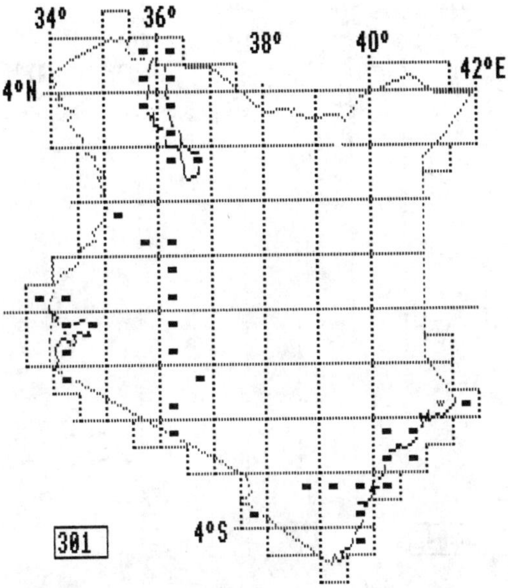

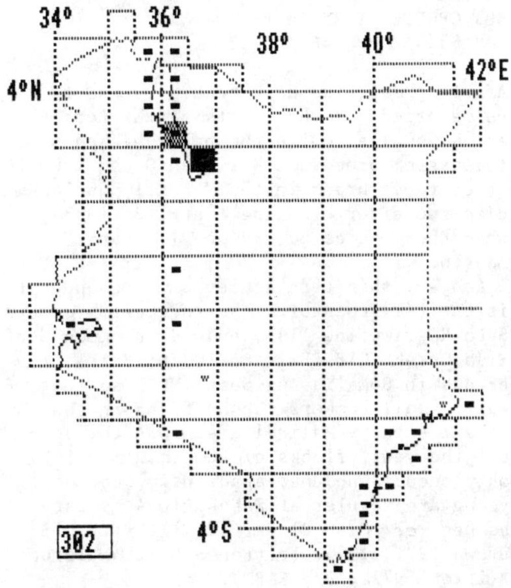

about 10°N in the W and 8°S in the E.

Kenyan: Britton & Brown (1974) noted the similar wintering distributions of this species and the Lesser Black-backed Gull (298) within Africa, and this comparison extends to their Kenyan ranges. Thus, while this tern is slightly more widespread, both are essentially birds of the rift valley lakes and L Victoria, and rather local on the coast.

The Gull-billed Tern is a reasonably common winter visitor Aug-Apr (Br). Inland, it is common in flocks of 30-40, particularly on the rift valley lakes, where roosts of up to 800 occur at L Nakuru 62A (Britton & Brown 1974; EABR 1979, 1983, 1984). A population of 100+ winters on the coast between Malindi 103A and Mida Creek 102B, but this tern is distinctly uncommon further S (Britton 1977b; Br). The possibility of further wintering birds in the Lamu archipelago (*op. cit.*) is upheld by 400 in late Jan at Kipini 91D (EABR 1984). Small numbers regularly oversummer, but suggestions that breeding might occur undetected at L Turkana are less convincing, on distributional grounds, than those for the Little Tern (311) (Britton & Brown 1974; Br). Passage occurs along the Kenyan coast (Britton & Brown 1974), as in Somalia (AM).

302 CASPIAN TERN *Sterna caspia*
B325 / MPG350 / D3C2

Afrotropical: occurs in winter as a Palearctic visitor to estuaries, large lakes and rivers, locally S to at least equatorial latitudes. Scattered breeding colonies also exist in the W and S of the Region, in Eritrea and possibly, very sporadically, at L Turkana, Kenya.

Kenyan: there is no evidence for non-breeding migrants from the S: both coastal and Turkana birds exhibit Palearctic seasonality, and the species is very rare on the East African coast S of Mida Creek 102B (Britton & Brown 1974; Br). Peak numbers off the Somali coast during Mar and Sept (AM) suggest Palearctic passage. Migrants are present in the Lamu archipelago 91B and Malindi 103A areas late Jul-Apr, and about 30 (occasionally up to 90) winter in the Sabaki estuary-Malindi area 103A Dec-Feb (Britton & Brown 1974; Britton & Britton 1976b; Br; Britton 1977b; EABR).

Status at L Turkana is unclear: regular flocks of 100 during the N winter (HH) are presumably of Palearctic origin, but records of Mar-Apr breeding on Central Island 14C (MPG) and of parents with begging young at El Molo Bay 26B Nov 1971 (HH)(both mapped as confirmed breeding) may represent occasional nomadic breeding birds from the N (Britton & Brown 1974). Of five records (all of single birds) from other inland sites, four during Oct-Nov in or adjacent to the rift valley may reflect wandering migrants (Britton & Brown 1974; St; EABR 1979, 1982); the fifth was on L Victoria 60A, in Feb (EABR 1977).

303 CRESTED TERN *Sterna bergii*
B324 / MPG356 / D2C3

Afrotropical: the dark-backed race *S.b. velox* breeds on the coasts of the Red Sea and N Somalia, while the paler-backed *S.b. thalassina* breeds on W Indian Ocean islands; other races breed in the S. All populations disperse after breeding (Harrison 1983), when the species occurs on all E and S coasts.

Kenyan: this tern achieves C3 because it is entirely coastal. The race *velox* occurs S to Ras Iwetine 114B, and is uncommon S of Mida Creek 102B (Britton 1977b; Br). It breeds in Somalia Aug-Sept (MPG) so that, while small numbers can be found in the Sabaki estuary/Malindi area 103A throughout the year, flocks of 20+ (maximum 180) only occur Jan-June; apparently dependent young are regular at Sabaki 103A as late as Dec (*cf.* Crab Plover (271))(Britton & Brown 1971, 1974; Britton & Britton 1976b; Britton 1977b; Br; EABR).

The first record of *thalassina* was as recently as 1979 (EABR 1979), but it is an abundant breeding species at 7°S off the coast of Tanzania, so that small numbers are probably regular N to Kenya (*e.g.* EABR 1979-81), and there are three records N to S Somalia (AM). Other reference: Britton & Osborne 1976.

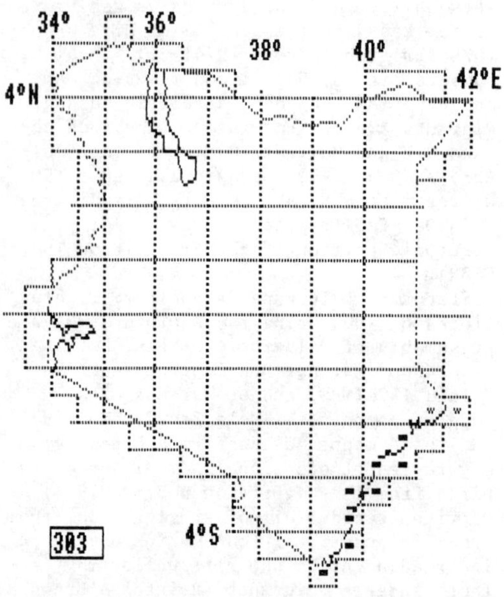

304 LESSER CRESTED TERN *Sterna bengalensis*
B323 / MPG355 / D2C3

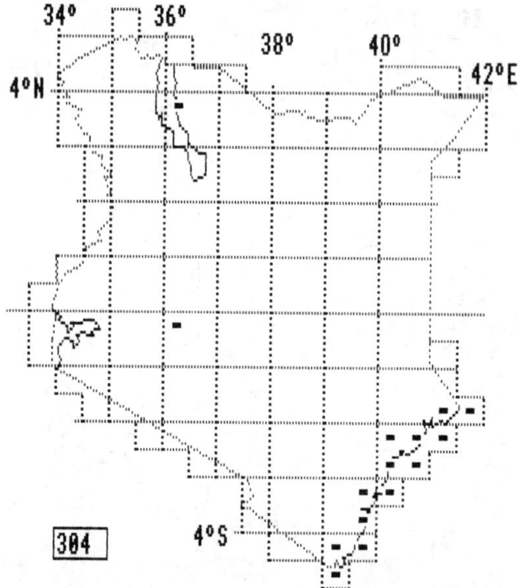

Afrotropical: breeds on the coasts of the Red Sea and N Somalia, and winters along the E African coast, S to E South Africa.

Kenyan: this distinctive species is awarded C3 because of its virtual restriction to the coast. It can be found there in small numbers throughout the year, but the Red Sea-N Somalia breeding season is June-Sept (MPG) so that it is more frequent in Kenya Sept-Mar (Ja), and most frequent Dec-Mar, when regular counts of 700 and peaks of 2000 are recorded in the Sabaki estuary-Malindi 103A area (Britton & Brown 1971, 1974; Britton & Britton 1976b; Britton 1977b; Br; EABR). It is very rare inland, the only records being of single birds at Entebbe (Uganda) Aug, L Nakuru 62A Aug, and three separate occurrences L Turkana Aug-Sept: this very restricted seasonality probably reflects S-bound migrants drifted inland (Britton & Brown 1974; HH).

305 SANDWICH TERN *Sterna sandvicensis*
B330 / MPG354 / D2C1
NO MAP: post presence 91BCD 103A 114B

Afrotropical: a Palearctic species that winters commonly on the NW and W coasts of Africa, but is only rarely recorded on the Indian Ocean seaboard.

Kenyan: first recorded in 1973 (Madge 1974), this tern has now been reported 14 times and, with the exception of one year, has been annual since 1976 (Britton 1974a, 1977c; EABR). Nine have been single birds and four more occasions involved groups of up to four, but a flock of 30 at Kipini 91D on 22 Jan (EABR 1984) is unprecedented in East Africa (*op. cit.*) and Somalia (Ash 1983). In contrast to the spate of recent records from Somalia, which are predominantly Mar-May (Ash 1983), the Kenyan birds have been almost evenly distributed throughout the Palearctic season Aug-Apr, with a single bird as late as 2 June (EABR 1984).

306 ROSEATE TERN *Sterna dougallii*
B326 / MPG352 / D2C2

Afrotropical: an abundant breeding species on the coasts of East Africa and Somalia. It also breeds in South Africa, and occurs along the whole W coast.

Kenyan: breeds annually in thousands on the Kiunga islands 80CD, and erratically at Whale Island 102B (up to 1500 pairs) and Kisite Island 114C (up to 1000 pairs), late Jul-early Sept (Britton & Brown 1974; Britton 1977b; Br; Donnelly 1978; EABR 1983). However, numbers using each site can fluctuate radically from year to year (Brown & Britton 1980). In view of the apparently extensive confusion with the Common Tern (307: *q.v.*), accounts of the Roseate's status and seasonality during its non-breeding season must be open to question (see EABR 1983 p. 127).

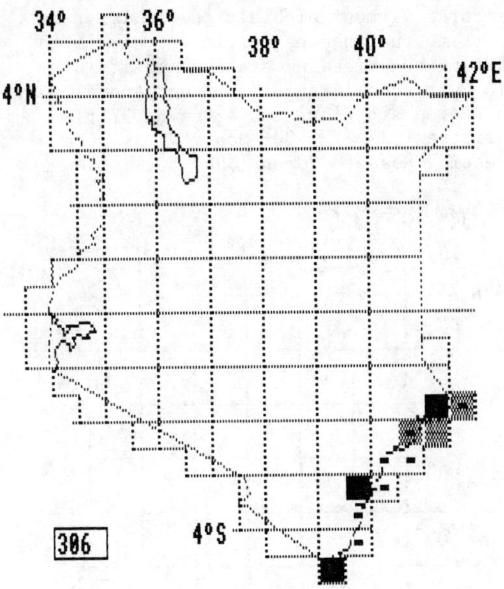

307 COMMON TERN *Sterna hirundo*
B328 / MPG351 / D2C2

Afrotropical: a common to abundant Palearctic visitor to all coasts, but rare inland.

Kenyan: the true status of this species is only just being realised, due mainly to

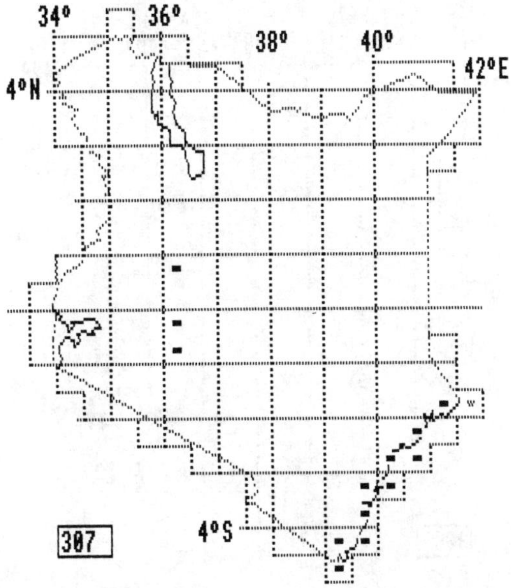

confusion with the closely related and similar Roseate Tern (306). Thus, ignoring all previous accounts, the Common Tern occurs mainly on S passage (maximum known concentrations 4000), with smaller numbers of N passage and oversummering birds, and usually very few in Jan-Feb (EABR 1983, 1984)(*cf.* AM). Contrary to Britton (1977b), during Aug-Nov at least, it is usually the commonest tern on the coast (*op. cit.*).

It is rare inland, with four single vagrants during the S migration, at L Nakuru 62A in Sept and Oct (Ba; Br), L Baringo 50A in Nov (St), and L Naivasha 62C in Sept (G C Backhurst pers. comm.). A bird ringed N of the Caspian Sea, USSR, was recovered at Kilifi 102D (Backhurst 1981b), while one from Poland was recovered at L Naivasha 62C (G C Backhurst pers. comm.).

308 WHITE-CHEEKED TERN *Sterna repressa*
B329 / MPG353 / D2C2

Afrotropical: breeds on the coasts of the Red Sea, Somalia and Kenya, non-breeding birds extending S to Tanzania.

Kenyan: up to 1000 pairs breeding annually Jul-Sept on the Kiunga Islands 80CD, and a colony on Tenewe Island 91B, represent the S tip of the breeding range (Britton & Brown 1974; Britton 1977b). Numbers breeding fluctuate from year to year, possibly inversely to the breeding abundance of the Roseate Tern (306)(Brown & Britton 1980). Further S along the coast, it is rather local in small numbers, throughout the year but mainly Apr-June

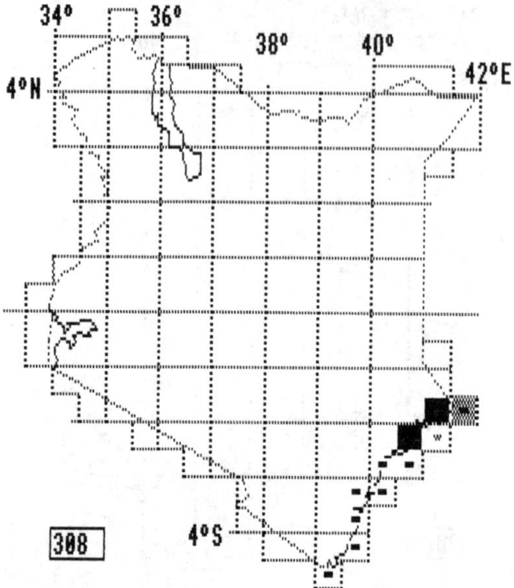

(Britton 1977b; D J Pearson *in litt.*; EABR). The largest count is 1400 at the Sabaki estuary 103A in June (Britton 1977b), while hundreds have occurred at Watamu 103A Apr (*per* D J Pearson), Ras Ngomeni 91C Apr, Galu 114B May (EABR 1983) and at Diani 114B in June (EABR 1984).

309 BRIDLED TERN *Sterna anaethetus*
B322 / MPG359 / D2C2

Afrotropical: this mainly pelagic tern

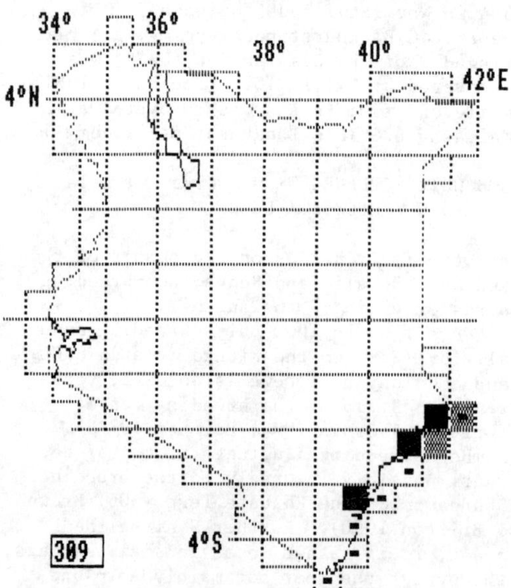

breeds on the coasts of the Red Sea, Somalia and Kenya, and on Indian Ocean islands, and can occur on E African coasts S to the S tropic. In the W, it breeds in the Gulf of Guinea, but is far less widespread.

Kenyan: about 750 pairs breed annually mid Jul-late Aug in the Lamu archipelago 80CD 91B 92A, and far smaller numbers, erratically, at Whale Island 102B (Ja; Britton & Brown 1974; Britton 1977b; Br). As in Somalia, where the breeding months are almost identical (AM; MPG), the few inshore records of small numbers away from the breeding sites are mostly Apr-Nov (Britton 1977b; EABR).

310 SOOTY TERN *Sterna fuscata*
B327 / MPG358 / D2C2

Afrotropical: a pelagic tern that breeds on islands along the tropical E African coast and in the Indian Ocean. In the W it breeds in the Gulf of Guinea, and extends to mainly S tropical coasts.

Kenyan: Smart (1965) found 5000 pairs breeding on Tenewe Island 91B in Aug, and 'clouds' of terns seen around this island from light aircraft in later years were thought to be this species (P L Britton *in litt.*). The lack of more recent breeding records is most probably due at least in part to the inaccessibility of the island during the breeding season, and it is thought to be a regular if not annual breeding site (Britton & Brown 1974; Britton 1977b). 100+ Sooty Terns with nesting Roseate Terns (306) on Kisite

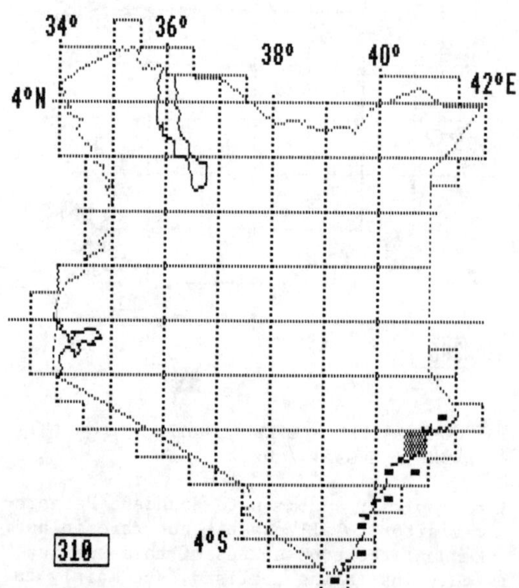

Island 114C in Aug 1976 may have been breeding, but direct evidence is lacking (EABR 1977).

Although possibly confused with the Bridled Tern (309), the Sooty Tern is a more pelagic species and, apart from an exceptional 100+ off Malindi 103A in Sept (EABR 1983), the very few inshore records away from Tenewe involve numbers under five: as in coastal Somalia (AM), these records are mostly Mar-Nov (Britton & Brown 1974; EABR).

311 LITTLE TERN *Sterna albifrons*
B321 / MPG357 / D2C3

Afrotropical: occurs off all coasts, the race *S.a. saundersi* breeding on those of the Red Sea and Somalia.

Kenyan: C3 is given because range is virtually confined to the coast and L Turkana. All birds seen in breeding plumage and all specimens have been of the race *saundersi*, and there is no evidence for migrants of the Palearctic nominate race (Ba; Britton & Brown 1974).

The Little Tern can be found throughout the year on the coast, but mainly Oct-Apr, when flocks of up to 5000 are present in the Sabaki-Malindi area 103A (Britton & Britton 1976b; Britton 1977b; EABR). Small numbers have recently been seen on L Bilisa and L Shakababo 91A, Jan-Feb (EABR 1984; D J Pearson *in litt.*). There is no indication of breeding on the coast, and occurrence there complements the Apr-Sept breeding season in S Somalia (AM).

The population at L Turkana is anomalous, but shows characteristics of *saundersi* (Britton & Brown 1974). Flocks of 500 occur at Turkana at all times of the year, but the paucity of records in inland Kenya and the species' absence from interior Ethiopia suggest that this is an isolated, resident breeding population: absence of actual breeding records has been attributed to unlocated colonies hidden in lakeside dunes (Britton & Brown 1974; Britton 1977b; Br; HH). The six records at L Baringo 50A and L Naivasha 62C were in Jan-May and Oct-Dec (Bednall 1959d; Ba; EABR 1981, 1982, 1984), and presumably represent wanderers S from L Turkana. Interestingly, all but one of these dates fall outside of the Apr-Sept S Somalia breeding season (AM).

312 WHISKERED TERN *Chlidonias hybridus*
B317 / MPG362 / D2C2

Afrotropical: the resident race *C.h. delalandii* ranges from the SE of the Region, locally N to Kenya. The nominate Palearctic race winters at inland sites in the N tropics, mainly N of 5°N, but some probably penetrate as far S as Kenya.

Kenyan: the Afrotropical race is a local bird that appears to be extending its range N. Thus Ja never saw it, while Britton & Brown (1974) knew of very few records away from L Naivasha 62C, and only two N of the equator. More recently, however, small numbers have become regular on lakes, dams and rice fields at Moiben 49A, Sergoit 49A, L Baringo 50A, Ahero 60B, L

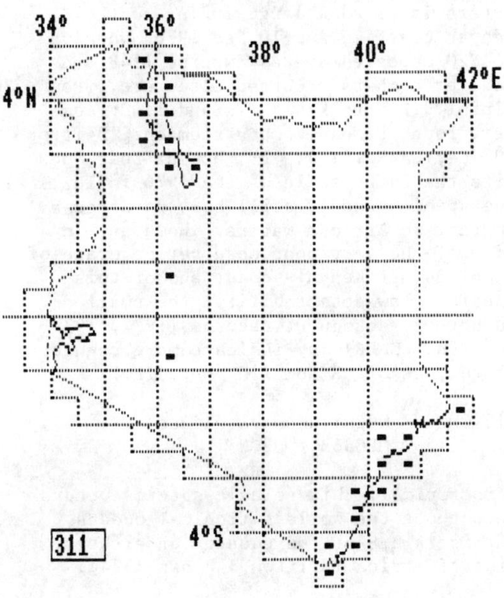

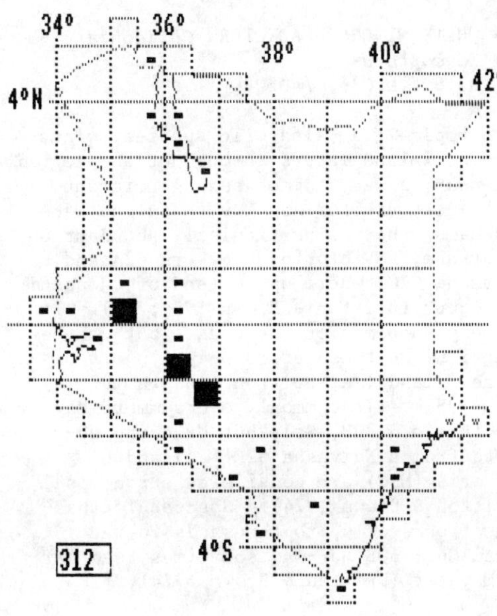

Nakuru 62A, L Naivasha 62C and L Jipe 100D (Br; St) and, in addition to possibly regular breeding at L Naivasha 62C (Britton & Brown 1974; Britton 1977b), other small colonies have been found at L Pirei 49C (Siemens 1984; EABR 1983) and Lari swamp 75B (EABR 1983). All of these breeding sites are above 1500 m.

Although there is still no specimen, Palearctic migrants probably occur regularly in small numbers on the rift valley lakes, and also on the coast, where the species is uncommon though possibly overlooked (Britton & Brown 1974; Britton 1977b).

313 BLACK TERN *Chlidonias niger*
B319 / MPG360 / D1C0
NO MAP: post presence 75B 76A; pre presence 60B 62A

Afrotropical: this Palearctic tern is a common to abundant winter visitor to coastal W Africa S to Namibia, but scarce elsewhere.

Kenyan: the four records are of a female in breeding plumage at Kisumu 60B on 30 Apr 1916 (Ja), and single winter plumage birds at L Nakuru 62A on 23 Sept 1953 (Wallace 1975), Nairobi 75B 12-18 Oct 1980 (Pearson & Lewis 1980), and Thika 76A 5 Feb 1983 (EABR 1984). These birds were on lakes or sewage oxygenation ponds. This vagrant is awarded D1 because, while reasonably distinctive even in winter plumage, it is almost certainly overlooked to some extent amongst flocks of the similar and far more abundant White-winged Black Tern (314)(Ba).

314 WHITE-WINGED BLACK TERN *Chlidonias leucopterus*
B318 / MPG361 / D2C2

Afrotropical: a Palearctic species, wintering at inland waters throughout the Region.

Kenyan: a very common passage migrant and winter visitor Aug-May to inland lakes and dams, that is particularly abundant on L Turkana, L Victoria, L Nakuru 62A and L Naivasha 62C, where up to tens of thousands may occur (Britton & Brown 1974; Br). It is very uncommon on the coast, but can be plentiful on freshwaters only a short distance inland (Ja; Britton & Brown 1974; EABR 1981). It commonly oversummers May-Jul (Br; *e.g.* HH, St), but breeding reports from L Naivasha 62C (Ja) and L Victoria (MPG) are considered erroneous (Britton & Brown 1974). Spectacular passage flocks of up to thousands in the Winam Gulf 60B Mar-May sometimes feed far from water (Britton & Brown 1974).

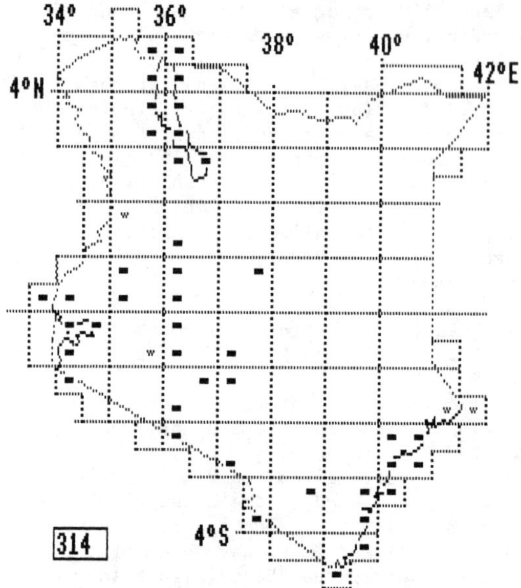

314

315 LESSER NODDY *Anous tenuirostris*
B316 / MPG- / D1C1
NO MAP: post presence 91C 102D 103A 114BC

Afrotropical: a normally sedentary species that breeds on islands in the Indian Ocean, and on the W coast of Australia. Formerly thought a vagrant to African waters, it has recently been found to occur regularly off Somalia (Ash 1983; AM).

Kenyan: the first records for African waters involved a large influx to the Kenyan coast 25 Aug-10 Dec 1976, when up to 220 appeared at Ras Iwetine 114B and smaller numbers occurred elsewhere, from Kipini 91D S to Kisite Island 114C: some were found to have empty stomachs (Britton 1977bd; EABR 1977). Br thought that, despite the large scale of the 1976 influx, the species should still be considered a vagrant to African waters. However, in view of the very poor observer coverage of seabirds off Kenya's coast and of this species' low detectability, the small number of subsequent records (1977, 1979-81, 1983: EABR) may indicate more regular status, as off S Somalia (*op. cit.*)

316 COMMON NODDY *Anous stolidus*
B315 / MPG363 / D2C2

Afrotropical: this pelagic species occurs throughout the world's tropical oceans, but is rare S of the equator on either side of Africa (Britton & Brown 1974).

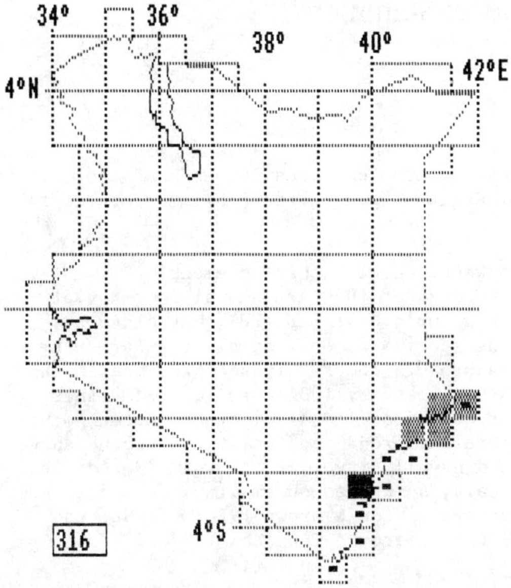

Kenyan: small numbers can be found offshore throughout the year, but it is less common inshore; exceptional inshore numbers Aug-Oct 1976, including *c.* 1000 at Tenewe Island 91B, coincided with an influx of the Lesser Noddy (315)(Britton 1977b; Br; Britton & Britton 1976b; EABR). Hundreds breed annually in the Lamu archipelago 91B June-Oct, and breeding has also occurred at Whale Island 102B, and probably at Kisite Island (114C: not mapped)(Britton & Brown 1974; Britton 1977b; EABR 1977). The predominance of pre-1970 breeding squares reflects lack of recent observer coverage during the rough weather of the breeding season.

Family Rhynchopidae: Skimmers

Three species, with single representatives in mainly tropical areas of each of the Nearctic-Neotropical, Afrotropical and Oriental Regions. The African Skimmer breeds in Kenya.

317 AFRICAN SKIMMER *Rhynchops flavirostris*
B331 / MPG364 / D3C2

Afrotropical: widely distributed from Senegambia E to Ethiopia and thence S, avoiding the equatorial lowland forests of the W, to Namibia, Botswana and Mozambique. Apart from a recent record of two in S Somalia (Ash 1983), distribution in Kenya reflects the E limit of the range.

Kenyan: breeds at L Turkana Mar-May: there is an old record from Ferguson's Gulf 13B (MacInnes 1933-34), and up to 50 pairs regularly bred on Central Island 14C until fishermen began to camp there in 1973, since when the birds have shifted to Allia Bay 14A and other less disturbed areas (HH; Modha & Coe 1969; Turner & Gerhart 1971; Britton & Brown 1974).

Most records from elsewhere in Kenya and from N Tanzania and E Uganda are in the Aug-Mar period, and are thought to represent post-breeding dispersal of both the Turkana population and probably others in the N tropics (Britton & Brown 1974). Two June-Aug birds at 1°57'N on the coast of Somalia (Ash 1983) might also have been associated with this movement.

Flocks of 1000 are annual Sept-Nov at Ferguson's Gulf 13B (HH), but elsewhere the species is only at all regular at L Nakuru 62A and L Naivasha 62C, and in the Sabaki estuary 103A: records from other areas conform to no distinct pattern other than the broad seasonality mentioned above, and usually involve less than 20 individuals, sometimes on small water bodies *e.g.* dams (Britton & Brown 1974; Br; St; La; EABR; Pomeroy & Tengecho 1982a).

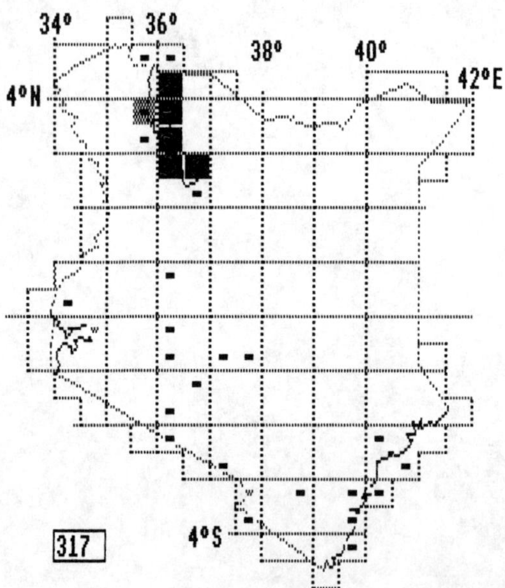

Family Pteroclidae: Sandgrouse

16 species, confined to the Palearctic, Afrotropical and Oriental Regions. Of the eight that are found in the Afrotropical Region, five breed in Kenya.

318 CHESTNUT-BELLIED SANDGROUSE *Pterocles exustus*
B333 / MPG369 / D2C2

Afrotropical: the N tropics, from the W coast between 10°N and 19°N, E to Ethiopia and Somalia, and S to Sudan, Kenya and NE Tanzania.

Kenyan: locally common in open country, mainly below 1500 m (Br; Ja) and 90% within arid-semiarid areas. This and the Black-faced (319) are the commonest of Kenya's sandgrouse, but their ranges show marked differences. Thus, being a bird of the N tropics, this species is abundant in and NW of the L Turkana basin, where the Black-faced, a bird of the E and NE, is virtually absent (HH). Secondly, because the Chestnut-bellied frequents open or at most lightly bushed habitats, it is virtually absent from the thickly bushed area 63-65, 76-78 already commented upon in this context for the Ostrich (1), Secretary Bird (85), the vultures and the Crowned Plover (220): the effectiveness of this densely vegetated barrier to gene flow in this species is shown by the fact that the populations to either side have slightly differing characteristics (Ja), although not sufficiently so to merit differing racial status (Br). This species is also absent from the dense habitats of even the semiarid areas of the coast, while the bush-loving Black-faced is found in both this and the 63-65, 76-78 area mentioned above.

Ja and Br found the Chestnut-bellied resident, and DEP (pers. obs.) concurs for 75C. A small June-Oct influx to Tsavo E 101B (La) is immediately subsequent to the May-June breeding peak given for this part of East Africa by Brown & Britton (1980).

319 BLACK-FACED SANDGROUSE *Pterocles decoratus*
B332 / MPG370 / D2C2

Afrotropical: the E, from N Mozambique and S Tanzania, N through areas of East Africa to the E of L Victoria, to S Ethiopia and Somalia.

Kenyan: locally common up to 1600 m in bushed habitats (Br; Ja; MPG; La). Range is 95% arid-semiarid, and absent from the subhumid squares in the coastal strip. N of 1°S, the only records W of the rift

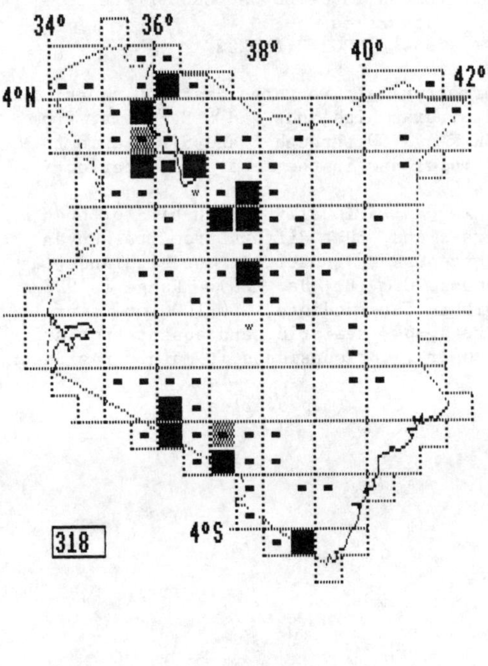

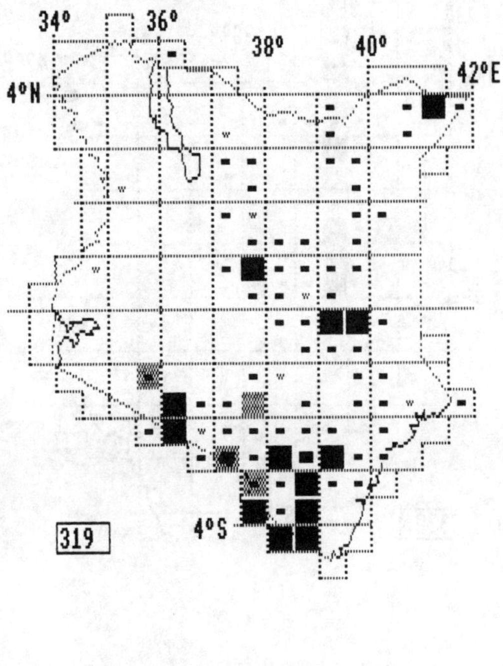

valley are from the Karasuk Hills 25C and S Elgon 48B (Ja), and from NE Uganda at 3°46'N (Br). There is no evidence for movements other than those undertaken daily to water (Br; Ja; La; DEP pers. obs.).

The Black-faced Sandgrouse forms part of a species-group with Lichtenstein's (320) and the Four-banded (321) Sandgrouse: in Kenya, which is a notable area of overlap for these three species, the Black-faced is ecologically separated from the other two in denser habitats (Sn; *e.g.* Lack 1985). See Chestnut-bellied Sandgrouse (318).

320 LICHTENSTEIN'S SANDGROUSE *Pterocles lichtensteinii*
B335 / MPG373 / D2C2

Afrotropical: the N tropics between 15°N and 20°N, from the W coast to Ethiopia and Somalia, and S to S Sudan, NE Uganda and Kenya. Distribution in Kenya represents the SW limit of the range.

Kenyan: an often uncommon species of low, dry, sparsely bushed country (Br; Ja; MPG). Range is 70% below 1000 m and 84% arid-semiarid. MacInnes (1933-34) found it common on the W side of L Turkana but HH describe it as common to the E and only occasional on the W. The lower Tana R population (79C 91A: AGH) may be isolated, because the species is only known N of 3°N in Somalia (AM), and was not recorded by observers resident at Garissa 65B. There is no evidence for movements other than daily flights to water, and small numbers occur at Baringo 50A throughout the year (St; EABR 1981). See Black-faced Sandgrouse (319).

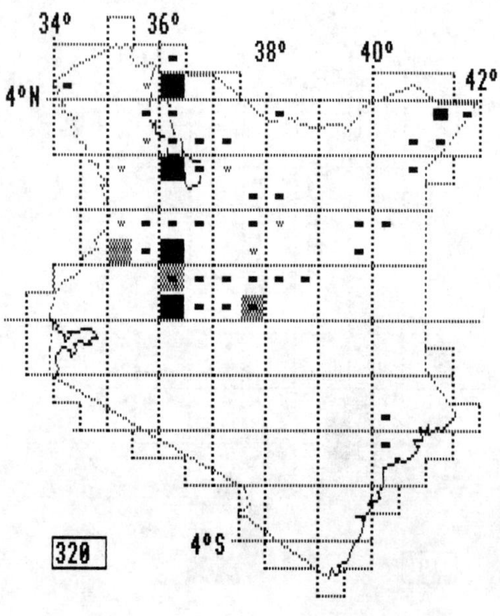

321 FOUR-BANDED SANDGROUSE *Pterocles quadricinctus*
B336 / MPG375 / D2C0
NO MAP: post presence 2C 3A 37A 50A; post confirmed 36B

Afrotropical: a bird of the N tropics which, replacing Lichtenstein's Sandgrouse (320) to the S of the latter's range, extends E from the W coast between 8°N and 17°N to S Sudan and Ethiopia, and thence S to N Uganda and NW Kenya.

Kenyan: this species replaces Lichtenstein's Sandgrouse in less arid habitats (Sn), but our map's coverage is insufficient to demonstrate this. There are few records. A specimen taken at L Rudolf (Turkana) on 5 Aug 1905 is in the National Museum Collection, Nairobi. Jackson (1911, Ja) found it plentiful on the upper Turkwell R (probably 25AC: not mapped), while Pitman (*in* Ja) described breeding along the Kerio R (49B-26A: not mapped) in 1923. Three more recent records from the Kodich area 36B 37A, the Lokichokio area 2C, and Kamathia 3A are detailed by Lewis *et al.* (1984), who suggest that it is regular in the NW, at least occasionally in considerable numbers, and that the few records reflect its crepuscular drinking habits and the very low observer frequency in its remote range. A pair on Laikipia 50A in May 1985 (Short & Horne 1985b) were at the SE extremity of the range. This species is apparently absent from the W side of L Turkana, where Lichtenstein's Sandgrouse is only occasional (HH). See Black-faced Sandgrouse (319).

322 YELLOW-THROATED SANDGROUSE *Pterocles gutturalis*
B334 / MPG372 / D2C2

Afrotropical: in three distinct areas: the Ethiopian highlands and S Ethiopia; from SW Kenya SW through E and S Tanzania to N Zambia; and in the less arid interior S of 14°S.

Kenyan: a locally common bird of open grassland, that differs from its Kenyan congeners by inhabiting the highlands, at 800-2000 m (Br; Ja; MPG). Range is 93% within intermediate, 250-1000 mm rainfall areas, 64% semiarid, and absent from arid country. Br considers it mainly resident.

(map overleaf)

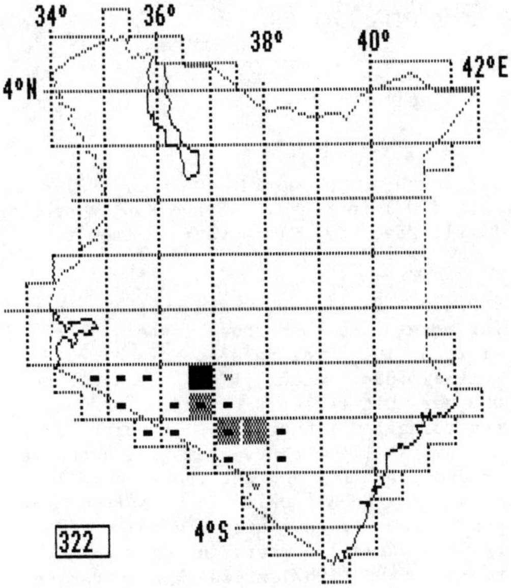
34°
36°
38°
40°
42°E
4°N
4°S
322

Family Columbidae: Pigeons, Doves

Approximately 290 species, with representatives throughout the world. Only 27 (9%) are found in the Afrotropical Region, with 19 of these (70%) in Kenya. Of the Kenyan species, 15 (79%) breed and three may do so, while the Turtle Dove (331) is a vagrant from the Palearctic.

323 FERAL PIGEON *Columba livia*
B342 / MPG376 / D2C2

Afrotropical: in the wild state, this mainly Palearctic pigeon is known in Africa S to 10°N, E to Sudan. Feral populations are associated with human habitation, but their Afrotropical distribution is imperfectly known since they are often left unrecorded by observers intent on truly wild birds. In the E and NE, feral populations are reasonably widespread in Somalia (AM), Kenya (this study), Uganda (DEP pers. obs.), and Tanzania (EABR 1982), though apparently absent from Ethiopia (J S Ash *in litt.*). They are also in most cities in the S (Maclean 1985). Occurrence of feral birds in W Africa is least known: there are small numbers in Dakar (Senegambia) and Bamako (Mali), but they may be scarce or absent elsewhere (G J Morel *in litt.*).

Kenyan: the status and history of this species are difficult to elucidate, due to the factors mentioned above. Neither Ja nor MPG mention this bird in Kenya, but whether this implies its absence is uncertain. Similarly, while our map shows far more sites than those listed by Br (Isiolo 51D, Karatina 63A, Nairobi 75B, Malindi 103A, Mombasa 114B), we are unsure whether this reflects expansion or underrecording. McVicker (1982) thought status at Mombasa 114B to have remained unchanged 1953-82, but L V J Grant (pers. comm.) considers the Nanyuki 51C and Nairobi 75B populations to have steadily increased from very small numbers during the last 15 years. EABR (1982) gives the impression of a recent expansion to towns and villages in many areas of Tanzania.

Range is 96% within 500+ mm rainfall areas and 76% subhumid-humid. Dark, 'Rock Dove' type and pale morphs are found in all populations, with dark birds predominant at Nairobi and Mombasa (Br).

324 SPECKLED PIGEON *Columba guinea*
B341 / MPG379 / D2C2

Afrotropical: widely distributed S of 16°N, apart from the equatorial lowland forests and areas S from the Congo basin to Bots-

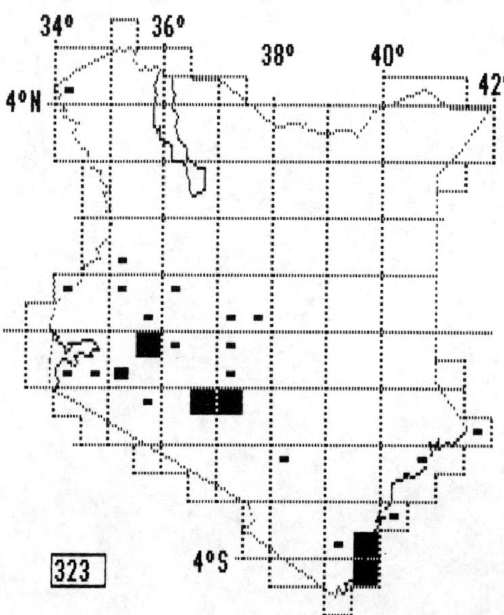

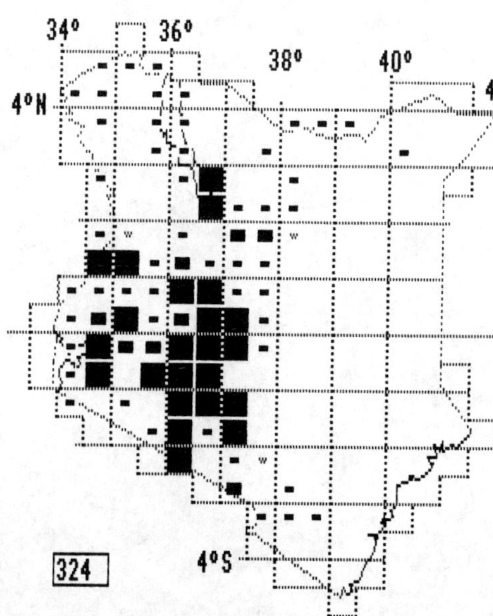

wana. Distribution S of 2°N in Kenya represents the E limit of the range at present, although the species is extending S into adjacent areas of S Somalia (AM).

Kenyan: a locally common, open country pigeon that requires rocks and cliff ledges on which to breed, and which, since early this century, has also shown an increasing tendency to colonise buildings (*e.g. cf.* Se with Brown & Britton 1980) (Br; Ja; MPG; St). This commensalism, together with deforestation, has produced definite increases in both the species' range and abundance (Sn), and thriving urban populations have become established in Eldoret 49AC, Kisumu 60B, Nakuru 62A, Nairobi 75B and Thika 76A, in some cases only during the last decade (Br; EABR 1981, 1983). Dramatic range extension via towns has reached S to 2°N in Somalia (AM), and birds have even nested inside small boats on L Naivasha 62C (Cunningham-van Someren 1973b).

Speckled Pigeons are known at 500-3000 m (Br; Britton 1979), and absence from the E lowlands is at present paralleled in adjacent S Somalia. Absence from the apparently suitable towns along the Kenyan coast contrasts with the situation in Somalia (AM), with the nearest record being an Apr group in Tsavo E 101B (La).

325 OLIVE PIGEON *Columba arquatrix*
B339 / MPG380 / D2C2

Afrotropical: a forest species, S from Ethiopia and S Sudan through mainly highland areas of East Africa and the E Zaire

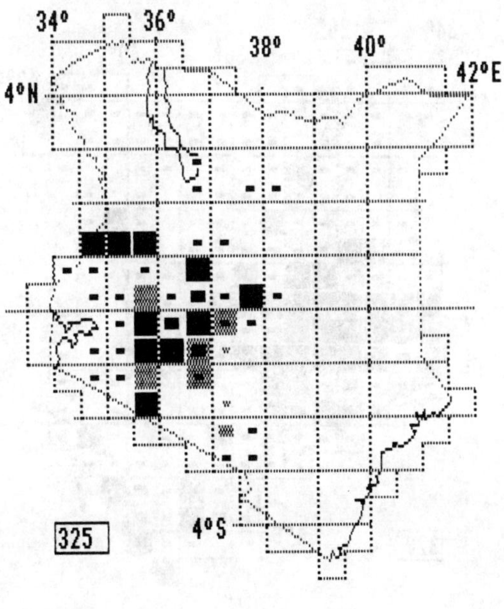

325

borders to Malawi. Further S, it is found W to Angola, and S around the E periphery of the continent to the Cape. Distribution in Kenya represents the E limit of the range.

Kenyan: locally common in forest canopy, mainly above 1500 m (Br; Ja; MPG; Be). Frequency increases with rainfall, and range includes the forest islands of Mts Kulal 26B, Nyiru 26D (Pelchen 1977d) and Marsabit 27D 28C, and the Chyulu Hills 88D (Br). Apparent absence from the extreme W 48C 60AC compares with range in Uganda (Br).

The Olive Pigeon is semi-nomadic (Brown & Britton 1980), making several types of movement to food sources (*e.g.* see Ja). Firstly, Ja and Brown & Britton (1980) record influxes to the relatively low altitude Nairobi area 75B in Apr and Apr-Jul respectively, while five wanderers in Meru NP 52C, the sole record below 1000 m, were in June (EABR 1980). This seasonality coincides with the 'long' rains and may reflect a regular altitudinal migration, either to exploit fruiting trees, and/or to escape from the wet, cold conditions at high altitudes during this season; similar altitudinal movements are also known in some hornbills and sunbirds (*cf.* Bronze-naped Pigeon (326)). Secondly, Se mentions fluctuations in the Mau Narok area 61D at 3000 m, where the species may be a breeding resident in some years, but only sporadic in others. This suggests other, more nomadic and possibly opportunistic movements, lacking regular seasonality. Finally, this pigeon is known to make spectacular daily flights between its high altitude roosting sites and favoured food sources far below (Sn). Thus, in Sept, Ja saw large numbers daily flying W from Escarpment 62D 75B at 0800h, to return at 1700h. The influxes to the Nairobi area, mentioned above, may be of this type, depending upon the location of their roosting site. Presence on forest islands may also suggest the ability for considerable movements.

326 BRONZE-NAPED PIGEON *Columba delegorguei*
B340 / MPG383 / D2C2

Afrotropical: in three distinct areas of the E: from S Sudan through NE Uganda to Kenya and E Tanzania; in S Malawi, Mozambique and Zimbabwe; and in E South Africa. Distribution in Kenya represents the E limit of the range.

Kenyan: a locally common, forest species, in two distinct areas. It is principally a highland forest bird, which extends to the forest islands of Mts Kulal 26B

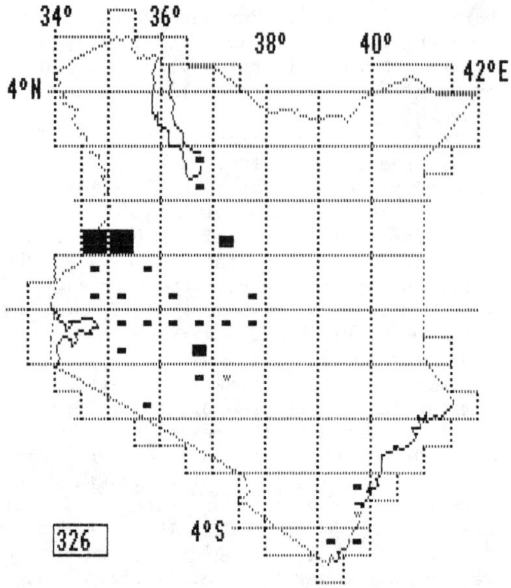

(H Gomez de Silva *in litt.*), Nyiru 26D (Pelchen 1977d) and the Matthews Range 39C (L Bennun *in litt.*). Excluding these localised highlands, which are too small to appear on our overlays, this population's squares are 77% above 1500 m. Records around 1650 m in Nairobi 75B are mainly during the cool Apr-Aug period (F Ng'weno, ADL pers. obs.), and may represent altitudinal migrants (*cf.* Olive Pigeon (325)).

There are also very small numbers of records from the coastal forests (Britton & Zimmerman 1979; EABR 1984), where van Someren (1925-35) described the bird as a non-breeding visitor, attracted by fruiting trees. MPG also mention local movements, and it appears most likely that these few coastal records represent occasional wanderers NE from the Usambara Mts of extreme NE Tanzania, where the species commonly descends as low as 300 m (EABR 1980), only 100 km SW of the Shimba Hills forests 114A (*cf.* Eastern Honeybird (508)). Presence on forest islands may also denote a degree of mobility.

327 WHITE-WINGED DOVE *Streptopelia reichenowi*
B349 / MPG389 / D3C3
NO MAP: post presence 7C 19AB; pre presence 6D

Afrotropical: restricted to the vicinity of large rivers in S Somalia, SE Ethiopia and extreme NE Kenya.

Kenyan: confined to the vicinity of the Daua R between Malka Mari 6D and Mandera 19B (Br), along the Ethiopian border, at 150-500 m (hence C3). It is common in the riparian trees and scrub, but also feeds in adjacent sparsely bushed grassland (ADL pers. obs.). There are no breeding records, but presumably it does so since it is present throughout the year (E J Roberts *in litt.*).

The White-winged Dove is closely related to the Ring-necked (328) and Mourning (329) Doves, and all three are sympatric (Sn), at least at QSD scale. Other references: Brown 1977a; Mann 1976.

328 RING-NECKED DOVE *Streptopelia capicola*
B346 / MPG388 / D3C3

Afrotropical: a bird of the S and E, widespread S of the Congo basin, and further N through East Africa to S Sudan, Ethiopia and Somalia.

Kenyan: very common in a great variety of reasonably open habitats, including towns and cultivation (Br; MPG), and likely to occur in every atlas square. It is mainly below 2000 m (Br; Be), but breeds in all but one of the 2000+ m squares and is colonising the Mau Narok area 61D at 3000 m, where it is displacing the Dusky Turtle Dove (332)(Se; Britton 1979; Br). Unlike the Laughing (333) and Namaqua (339) Doves, there are breeding records from the coastal lowlands.

DEP (pers. obs.) finds this dove resident in 74D 75 88AD 89AC 100B 101A and Be concurs for 74, but movements occur. Gichuki & Pomeroy (1983) relate fluctuations in numbers in the semiarid Bush-

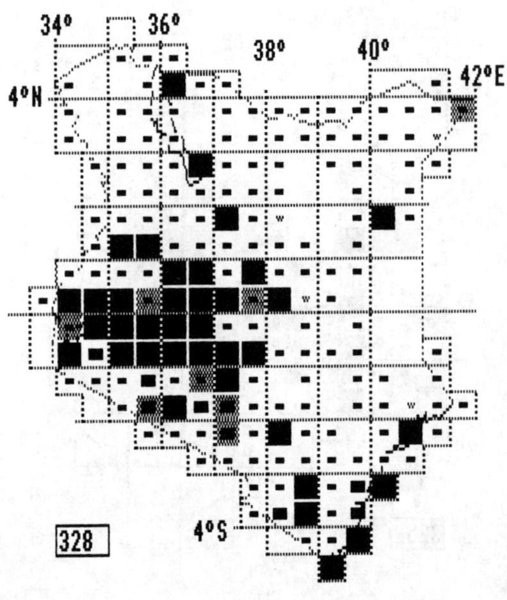

wackers area 89A to seasonal changes in food abundance and habitat: rapid increase after the Nov-Dec rains produces high numbers in Feb-Mar (*cf.* Lack 1985), while May-Jul breeding is followed by dispersal and low Jul-Nov numbers. 76% of the breeding squares receive over 500 mm annual rainfall, and the very few records from the driest areas probably reflect a real trend, despite the very low observer frequency (Pomeroy *et al.* 1982). Ja notes a concentration of many thousands feeding on recently burnt grassland near Simba 88B. See White-winged Dove (327). Other references : North 1959a; Short & Horne 1985a.

329 MOURNING DOVE *Streptopelia decipiens*
B347 / MPG387 / D2C2

Afrotropical: widespread S of 16°N, excepting the equatorial lowland forests and adjacent areas of the W, drier E Somalia and much of the SW.

Kenyan: common in dry areas that have some bush or trees and are near water, mainly below 1400 m (Br; Ja; MPG; La; St). Noticeably scarce in the squares containing the highlands 61-63, this is a bird of lower, hotter and drier country than the Ring-necked Dove (328). In further contrast to the latter species, there are reports from only two coastal squares, and no breeding records from the coastal lowlands. DEP (pers. obs.) found the Mourning Dove resident in 75C 88AD 89C 100B 101A. See White-winged Dove (327).

330 RED-EYED DOVE *Streptopelia semitorquata*
B350 / MPG386 / D3C3

Afrotropical: widespread S of 12°N, reaching 17°N in Ethiopia, and avoiding only the driest areas of E Somalia and the SW, and parts of the primary lowland equatorial forests of the W.

Kenyan: a common bird, found up to 3000 m (Br) although Se had only three records from that altitude. It contrasts with the similar Ring-necked (328) and Mourning (329) Doves by being far more restricted to wooded, higher rainfall areas, including forest edge and particularly near water (Br; Ja; MPG; Lack 1985). Hence range is 87% within well wooded areas, as opposed to 65% for the Ring-necked Dove and 67% for the Mourning Dove.

The Red-eyed Dove has only 21% of its squares arid, and occurrence in the dry N and E is mainly in locally moist situations. It is present on the forest islands of Mts Kulal 26B, Nyiru 26D and Marsabit 27D 28C, while records along the N border from 2C E to 17D represent the S extremities of Sudanese and Ethiopian birds, in moister areas to the N of the N Kenyan arid zone. This dove is noticeably absent from apparently suitable well wooded watercourses around L Turkana (HH) and along the Daua R 7C 19AB (E J Roberts *in litt.*; ADL pers. obs.). Further S, it is reported from the well defined strip of arid squares, 40D S to 101D, which are adjacent to semiarid country. Occurrence along the Tana R is apparently localised: 79C (AGH), 65B.

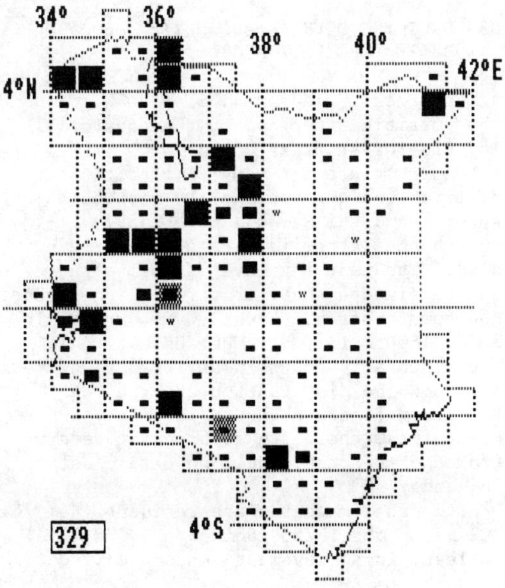

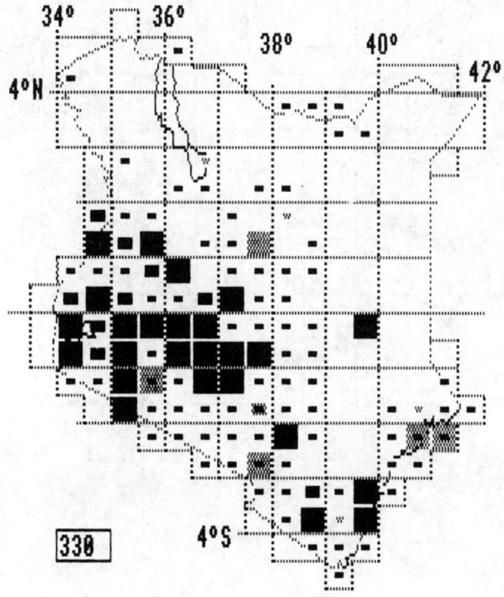

This species is considered resident (Be: 74; La: 101B; DEP pers. obs.: 74BD 75BCD 89A), but 85% of the breeding squares receive over 500 mm annual rainfall, and wandering outside the breeding season may take it to the edge of arid areas *e.g.* 2C-17D, 40D-101D (Pomeroy *et al.* 1982).

331 TURTLE DOVE *Streptopelia turtur*
B352 / MPG384 / D2C0
NO MAP: post presence 38D 51B 88C

Afrotropical: this Palearctic dove winters abundantly in the N savannas of Africa, S to about 10°N. There are very small numbers of records S to Uganda and Kenya.

Kenyan: three records of single individuals S of their normal wintering range: at Barsaloi 38D on 16-19 Oct 1976 (Horne & Short 1977), Samburu GR 51B on 3 Jan 1981 (Haas *et al.* 1981), and Amboseli GR 88C on 20 Oct 1983 (EABR 1983). These squares are semiarid (2) and dry subhumid. The first records for Uganda occurred in 1983, with at least eight birds Oct-Nov, S to the equator (EABR 1983). Virtual restriction of these East African records to Oct-Nov is marked. This species may well be overlooked.

The Turtle Dove forms part of a species-group with the Dusky Turtle Dove (332)(Sn).

332 DUSKY TURTLE DOVE *Streptopelia lugens*
B348 / MPG385 / D2C2

Afrotropical: a mainly highland species of the E and NE, from Ethiopia and S Sudan S through East Africa and the E Zaire borders to Malawi. Distribution in Kenya represents the E limit of the range.

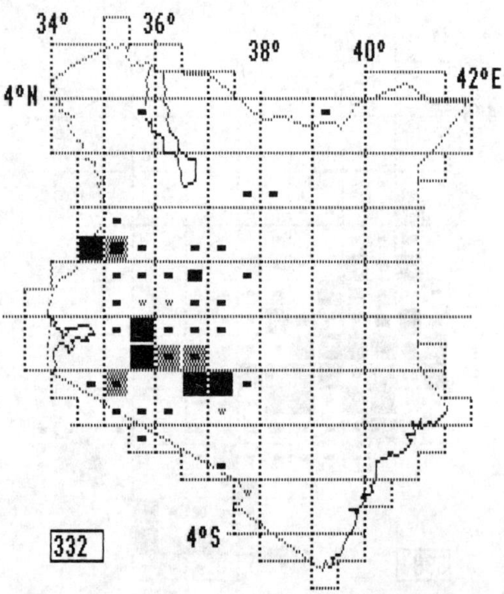

Kenyan: a locally common bird that differs markedly from the closely related Ring-necked (328), Mourning (329) and Red-eyed (330) Doves by being confined to high altitudes: 1800-3200 m (Br; Be; Se). It wanders and, like the Olive Pigeon (325), visits lower country to feed and drink. Thus it can occur in a variety of habitats, from sparsely wooded areas to cultivation, gardens and the montane bamboo and lower heather zones (MPG; Br). MPG, together with Ja and Be, also mention this dove in forest interiors, which would be a further marked contrast to the three other members of *Streptopelia* mentioned above, but Br considers it a forest edge species.

79% of this dove's squares receive over 500 mm annual rainfall, and it is present on the forest islands of Mt Marsabit 27D 28C and the Matthews Range 39C. Moyale 17A represents the S extremity of the Ethiopian highlands population, while 88C and 100B presumably represent wanderers from Kilimanjaro. Like the similarly high altitude Olive Pigeon, and probably for similar reasons, the Dusky Turtle Dove is more abundant in the relatively low altitude Nairobi 75B area Apr-Jul (Brown & Britton 1980). This seasonality was also shown by large numbers on recently burnt ground in the Njorowa Gorge 62C in Jul (Ja), by a single Apr vagrant to 400 m at Kalokol 13B, L Turkana (HH), by the Jul Moyale 17A record (J E Miskell *in litt.*), and by six individuals during late May 1985 at L Baringo 50A (T Stevenson *in litt.*). See Turtle Dove (331).

333 LAUGHING DOVE *Streptopelia senegalensis*
B351 / MPG392 / D3C3

Afrotropical: widespread S of 20°N, with the possible exception of the equatorial lowland forest areas of the W.

Kenyan: there are many similarities in distribution and seasonality between this species and the Namaqua Dove (339), although the latter is more of a dry country bird. The Laughing Dove is common in relatively open habitats, including gardens and town centres, up to 3000 m but mainly below 1800 m (Br; Se; MPG; HH; St). It could probably occur in any of our squares. Although the distribution of breeding squares reflects observer frequency in most areas, there are no breeding records from the reasonably well covered coastal lowlands.

Although some birds are resident (Be: 74; DEP pers. obs.: 75BC 88AD 89AC 100B 101A), at least local movements undoubtedly occur.

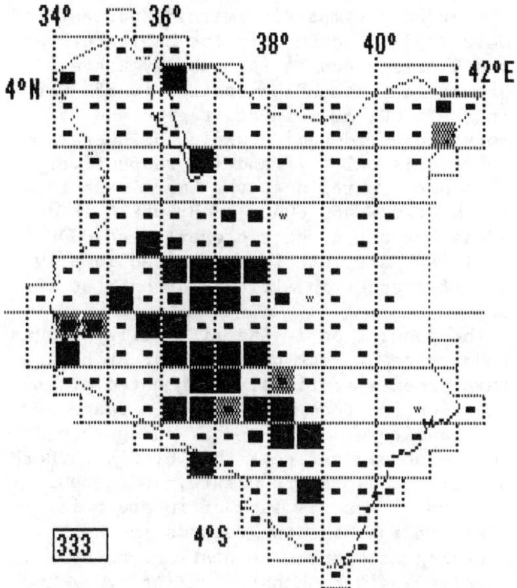

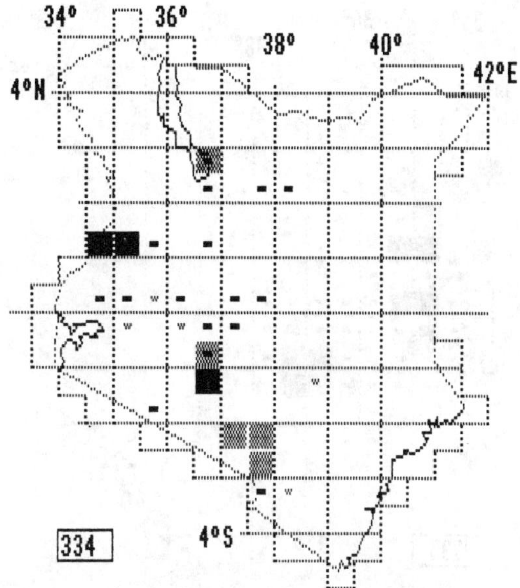

Thus, in arid Tsavo E 101B, small numbers present throughout the year (mainly near water) are augmented after the Nov-Dec rains by a large influx of breeding visitors (La; Lack 1985). A similar influx occurs in Tsavo W 89C 101A, peaking in Feb with many thousands of birds (D J Pearson *in litt.*). Just to the N, in the semiarid Bushwackers area 89A, similar population fluctuations are related to seasonal changes in food abundance and habitat: thus a rapid increase after the Nov-Dec rains produces high numbers Feb-Mar (*cf.* Tsavo), and May-Jul breeding is followed by local dispersal Jul-Nov (Gichuki & Pomeroy 1983). Birds apparently arriving through the mist at the Ngulia lights 101A in Nov-Dec (Backhurst & Pearson 1977b) may be involved in the influxes to 89A and 101B. Br mentions concentrations at post-rains seed crops.

334 LEMON DOVE *Aplopelia larvata*
B337 / MPG398, 399 / D1C1

Afrotropical: mainly the E, from Ethiopia and S Sudan through mainly highland habitats of East Africa and E Zaire, to S Zaire, Mozambique, and the SE periphery of the continent to the Cape. It is also present in the Cameroon-Gabon area of the W. Distribution in Kenya represents the E limit of the range.

Kenyan: a shy, rather uncommon inhabitant of highland forest up to 3000 m, feeding on the ground and usually obscure in dense, shady undergrowth (hence D1C1)(Br; MPG; Sn). Range includes the forest islands of Mts Kulal 26B, Nyiru 26D, Marsabit 27D 28C and Endau 77B, and the Chyulu 88D and Taita 101A Hills (Br; Tennent 1964; Diamond & Keith 1980). Apart from these localised highlands, which are too small to appear on our overlays, this species' squares are entirely above 1000 m, and 90% within 500+ mm rainfall areas. The concentration of pre-1970 records in 88 most probably represents this bird's low detectability, and infrequently visited habitats.

335 TAMBOURINE DOVE *Turtur tympanistria*
B357 / MPG394 / D2C2

Afrotropical: widely distributed between 9°N and 15°S, S on the E of the continent to the Cape.

Kenyan: a shy and locally common species, sometimes overlooked, that occurs up to 2500 m and is distinct from its three congeners in occupying denser and more shaded, bushed, wooded and forested habitats, including large urban gardens *e.g.* in Nairobi 75B (Br; Ja; MPG; Sn). Range is 85% within 500+ mm rainfall areas. Occurrence in the arid N and E is virtually restricted to the forest islands of Mt Kulal 26B (F Alexander *in litt.*) and Mt Marsabit 27D 28C (Br), and the vicinity of the lower Tana R 79C (AGH) and the Voi R (from which it has disappeared in recent years: La).

(map overleaf)

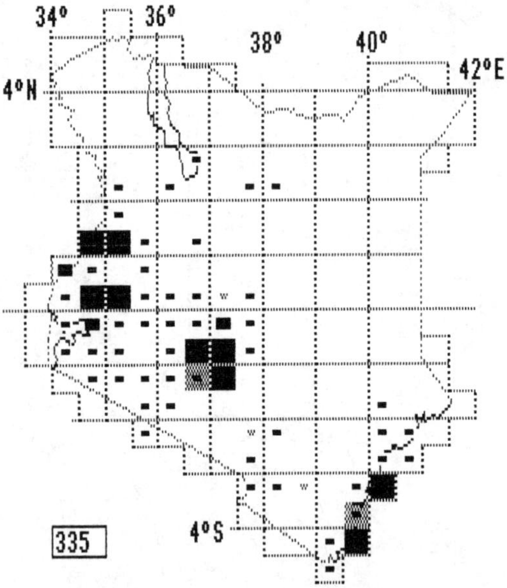

336 BLUE-SPOTTED WOOD DOVE *Turtur afer*
B355 / MPG395 / D2C2

Afrotropical: from Senegambia S to Angola in the W, and thence E to S Sudan, W Kenya, Malawi, Mozambique and Zimbabwe. Further E, it is also present in Ethiopia and parts of E Tanzania. Distribution in Kenya represents the E limit of the range.

Kenyan: common in well vegetated country up to 2000 m (Br; Ja). This species and the broadly sympatric Emerald-spotted Wood Dove (337) frequent similar habitats, but the Blue-spotted is in markedly wetter areas: hence the Blue-spotted is 85% within the 1000+ mm rainfall regime and 95% moist subhumid-humid, while the Emerald-spotted is only 17% and 15% respectively. This species replaces the Emerald-spotted in the far W 48C 60ACD 73AB, while in areas further E the two overlap at QSD scale but are largely segregated locally by differences in altitude or habitat (Br; Sn).

The population in the rift valley around L Naivasha 62C may be isolated: Br notes this species' curiously fragmented distribution in areas E of its main range. The paucity of confirmed breeding records most probably reflects this bird's rather skulking and obscure nature, and recent observers' general lack of interest in nest-finding: the two records in this breeding category were nests found by T J Barnley and D K Chepus (*in litt.*).

337 EMERALD-SPOTTED WOOD DOVE *Turtur chalcospilos*
B356 / MPG397 / D2C2

Afrotropical: S from S Sudan, Ethiopia and Somalia, through Kenya, E Uganda and most of Tanzania, to Zambia and Zimbabwe. Further S, it is present W to Angola, and S to E South Africa.

Kenyan: common in dry, bushed and wooded habitats with restricted undergrowth, mainly below 1600 m (Br; Ja; St; Lack 1985). It is generally in drier habitats

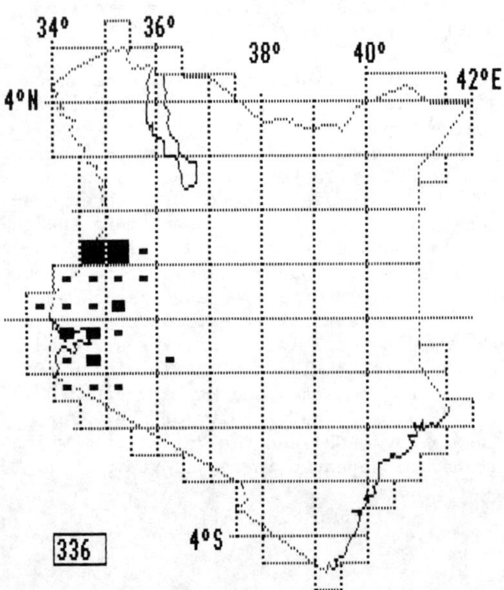

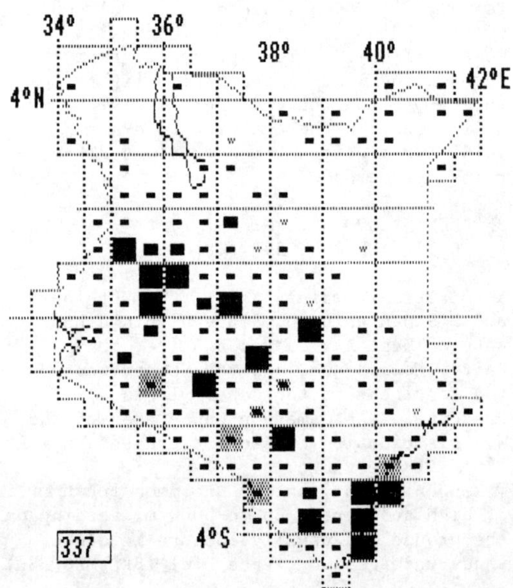

than the closely related Blue-spotted Wood Dove (336: *q.v.*) and, although Br mentions the Emerald-spotted's abundance in the forest interiors of coastal Kenya as an exception to this, these coastal squares are in fact distinctly drier than the Blue-spotted's habitats in the W, *i.e.* semiarid-dry subhumid as against moist subhumid-humid. There is no evidence of movements.

The Emerald-spotted Wood Dove forms a superspecies with the Black-billed Wood Dove (338)(Sn), which may replace it at L Turkana.

338 BLACK-BILLED WOOD DOVE *Turtur abyssinicus*
B354 / MPG396 / D1C1
NO MAP: post presence 13B 14C

Afrotropical: the N tropics, with a rather patchy distribution between 15°N and 5°N, from the W coast to W and N Ethiopia, S Sudan, N Uganda and possibly, on the SE tip of the range, the far NW of Kenya.

Kenyan: this species may replace the closely related and very similar Emerald-spotted Wood Dove (337: *q.v.*)(Sn) at L Turkana, although the situation there is by no means clear. There are very few records. MacInnes' (1933-34) mention of *Turtur afer sclateri* in the area between Eliye 13D and Todenyang 3B presumably refers to the Black-billed, because this area is unsuitable for the Blue-spotted Wood Dove (336), while MPG note that the distinction between the two species had only just been realised at the time.

The only recent records are from HH, who describe the Black-billed as frequent to common in wooded watercourses and thick bush around the lake: the 1973 edition of their paper mentions the Kalokol 13B and Turkwell 14C watercourses, and these squares are tentatively listed here.

However, the absence of further records either reflects a rather skulking species of habitats in which most visitors to the Turkana area spend little time, or mistaken identification. Although HH noted the Black-billed in watercourses, J R Ness (*in litt.*) collected an undoubted Emerald-spotted Wood Dove from the Lugha Ileret 4C, and he considers this the only wood dove species in the area. The question remains open.

Br also mentions the Black-billed at Amudat 36B, in Uganda.

339 NAMAQUA DOVE *Oena capensis*
B345 / MPG393 / D3C3

Afrotropical: widespread S of 20°N, avoiding only the equatorial lowland forest

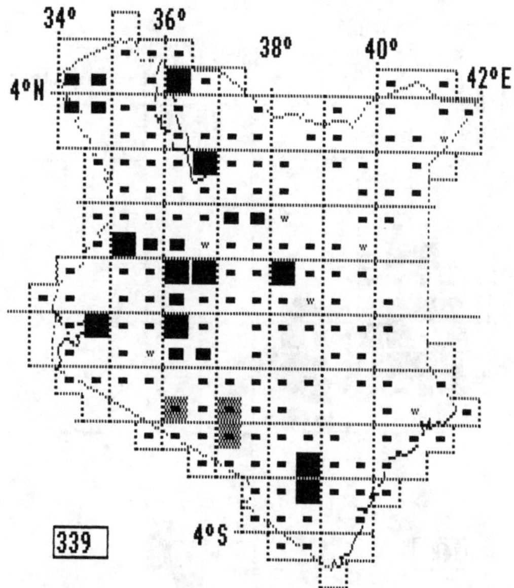

areas of the W.

Kenyan: a common bird of bushed habitats in drier country, mainly below 1600 m (Br; MPG; HH; St). Some may be resident (La: 101B; DEP pers. obs.: 75C 88D 89C 100B 101A), but the species performs frequent but poorly understood movements of unknown extent (*e.g.* see Ja), occurring occasionally in higher rainfall areas. Se recorded transient singletons as high as 3000 m at Mau Narok 61D, so there appears no reason why this dove should not occur in all of our squares, if only occasionally. The only possibly valid points to make about its map concern the absence of breeding records from the coastal lowlands (*cf.* Mourning Dove (329) and Laughing Dove (333)), and the Namaqua's apparent absence/scarcity in the extreme S, 114.

The Namaqua Dove's movements are probably at least partly in response to abundance of food (Ja; Sn), and concentrations at seed crops are known (Br). In arid Tsavo E 101B, this dove is most abundant for two-three months after the Nov-Dec 'short rains', but absent during the June-Nov dry season (La; Lack 1985). The possibility of nocturnal movement is raised by birds at the Ngulia lights 101A (Backhurst & Pearson 1977b).

340 GREEN PIGEON *Treron australis*
B358 / MPG401-403 / D2C2

Afrotropical: apart from the dry SW, widely distributed in Africa S of Senegambia, N Nigeria, Cameroon, N Zaire, S

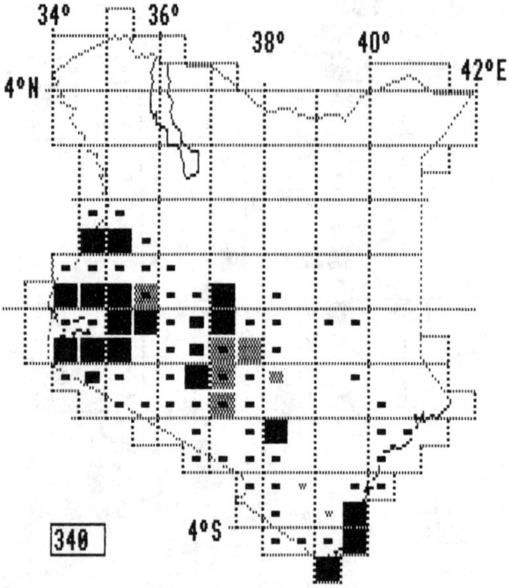

Sudan, W Ethiopia and SE Kenya. Distribution in Kenya represents the NE limit of the range.

Kenyan: a rather shy bird of wooded habitats, including forest edge, up to at least 2100 m (Br; MPG). It can be locally common in the vicinity of large, wild fruiting trees, particularly figs *Ficus*, which are its principal food source (*op. cit.*; Sn). Range is 84% within 500+ mm rainfall areas, while the few arid squares are either along the Tana R 65AB 78B 79C (AGH), or adjacent to semiarid country (the Voi R 101B: La; 101C; Duruma 102C: Ja). Despite its great mobility (see below), the Green Pigeon has yet to be recorded on any of the more remote forest islands of the N, *e.g.* Mts Kulal 26B and Marsabit 27D 28C. This is possibly because these are beyond its usual range, and within the range of the closely related Bruce's Green Pigeon (341: *q.v.*).

The Green Pigeon makes frequent movements in response to fluctuating food supplies. These are likely to be local in many cases, but some degree of regularity is suggested by Ja's note of Jan arrival in Trans Nzoia (approximately 36D-37C), followed by Apr breeding and virtual absence thereafter. In adjacent areas to this, seasonal abundance has been recorded on Mt Elgon 36D Mar-Apr (MPG), and at Eldama Ravine 49D Mar (Ja), while a single wanderer to the edge of the range at semiarid Baringo 50A was in Dec (St).

In view of this mobility, the correlation of breeding squares with areas receiving over 1000 mm annual rainfall (73%) and within the subhumid-humid moisture regimes (84%) may be real. If this is the case, then the three concentrations of breeding squares may represent the principal breeding areas of three races: *T.a. wakefieldii* at the coast, *T.a. brevicera* in the interior E of the rift valley, and *T.a. gibberifrons* W of the rift. Other areas, mostly semiarid, may be utilised only for their temporarily available food sources.

341 BRUCE'S GREEN PIGEON *Treron waalia*
B359 / MPG400 / D2C1

Afrotropical: the N tropics, from 15°N S to the N limit of the Green Pigeon (340: *q.v.*), and overlapping locally from the W coast E to S Sudan. Further E, it occurs S to N Kenya, SE Ethiopia and C Somalia.

Kenyan: this species forms part of a superspecies with the Green Pigeon, and similarly undertakes at least local movements to frequent the vicinity of large, wild, thickly foliaged fruiting trees (Ja; Sn). These two birds are mainly allopatric (Sn), and this one replaces the Green in the N where, in marked contrast to its congener, six of its seven squares are arid-semiarid. The sole overlap square is dry subhumid 37A where, at 1°30'-2°N on the N edge of the Green's range, T J Barnley (*in litt.*) recorded three of this species around fig trees at Nakuijit, on the Suam R, in the early 1970s. Overlap has been recorded in NW Uganda at 2-3°N (EABR 1983).

Other localities are: the Lokichokio-

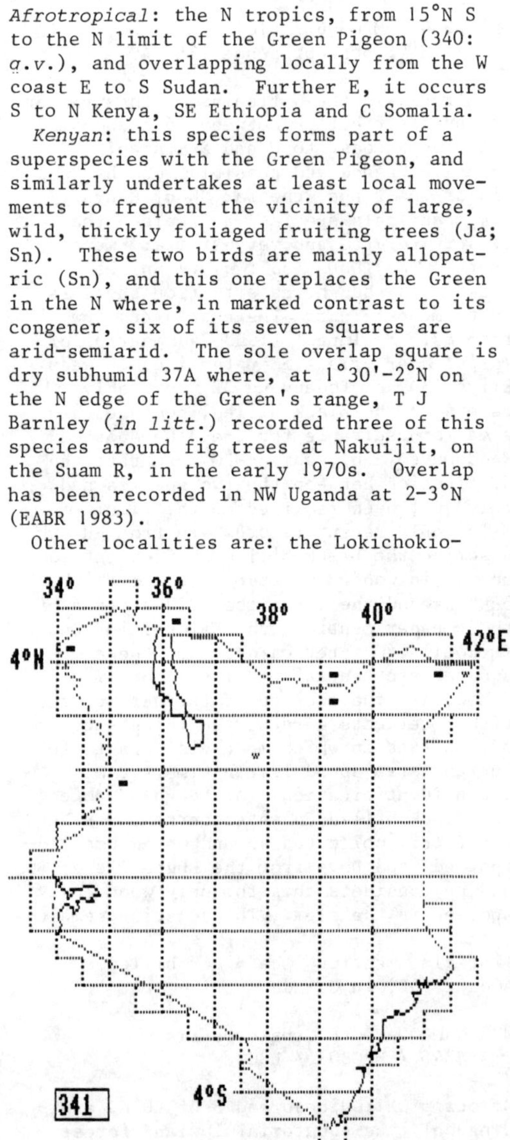

Sudan border area 2C Nov (G Nikolaus *in litt.*), Moyale 17AC Mar, June, Jul (J E Miskell, D A Turner *in litt.*; ADL pers. obs.), Rhamu 19A May, June, Nov (E J Roberts *in litt.*; EABR 1984), Mandera 19B Jul (van Someren 1929) and Marsabit 27D (Br). There are no Kenyan breeding records.

Other records, adjacent to Kenya's borders but not mapped here, are: 4B (J S Ash *in litt.*), 7D (AM), Karamoja 24B and Teso 35B (Br).

Family Psittacidae: Parrots, Lovebirds, Parrakeet

About 328 species, mostly in the Neotropical, Oriental and Australasian Regions. Only 19 (6%) are found in the Afrotropical Region and, of the nine in Kenya, three have been introduced.

342 RED-FRONTED PARROT *Poicephalus gulielmi*
B366 / MPG444 / D2C2

Afrotropical: inhabits lowland and medium altitude forests from Cameroon and Angola E to E Zaire and SW Uganda, with isolated populations in the primary highland to montane forests of N Tanzania and C and W Kenya.

Kenyan: locally common in forest at 1800-3250 m (Br; Ja). Range is entirely within subhumid-humid, 500+ mm rainfall areas. Absence from forest islands is perhaps surprising in view of its strong and swift flight: Se observed these birds embarking on flights to feeding grounds in company with the far-ranging Olive Pigeon (325), and Ja records daily flights from the Cherangani Hills 37CD to feed on Mt Elgon 36D. Se noted occasional absences at Mau Narok 61D during the middle months of the year.

Ja found this species particularly plentiful in the vicinity of the Kijabe and Escarpment railway stations 62D-75B, but this is certainly not the case now, and there is no recent record for 75B: this decline is presumably due to deforestation, possibly coupled with aversion to secondarily wooded habitats (Sn).

343 ORANGE-BELLIED PARROT *Poicephalus rufiventris*
B369 / MPG448 / D2C2

Afrotropical: the NE, from NE Tanzania through E Kenya to Somalia and SE Ethiopia.

Kenyan: this and the Brown (344) and Brown-headed (345) Parrots are allopatric members of a superspecies (Sn): their combined ranges cover most of Kenya. This species replaces the Brown at lower altitudes to the E of the rift valley, and is replaced by the Brown-headed in the coastal strip (the record in 102D was well inland).

The Orange-bellied Parrot is fairly common in dry, bushed and wooded country, below 1200 m (Br; MPG; La). Range is 91% arid-semiarid. Mostly in well wooded areas to the S of the equator, it extends into much less wooded country, *e.g.* along watercourses, further N. This parrot is considered resident (Br; Lack 1985: 101B;

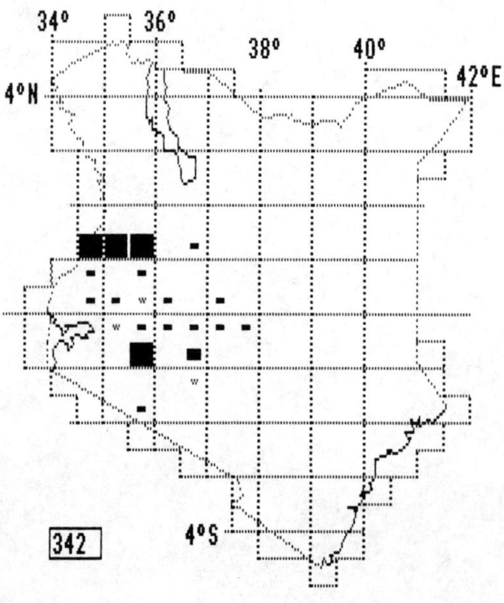

342

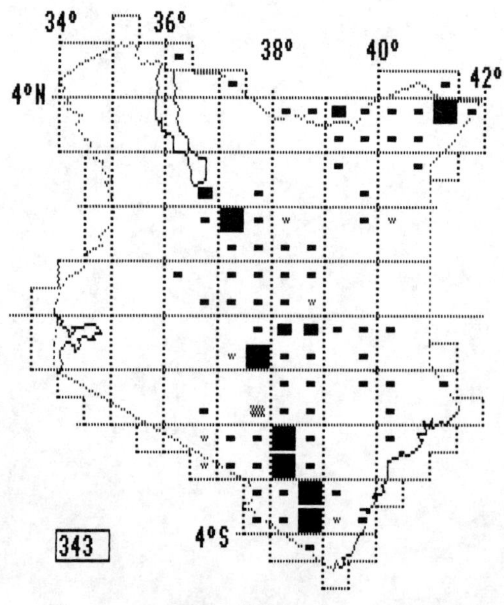

343

DEP pers. obs.: 88D 89AC 100B 101A).

344 BROWN PARROT *Poicephalus meyeri*
B367 / MPG449 / D2C2

Afrotropical: whilst absent from the lowland forests of the Congo basin, this parrot occupies interior Africa from NE Namibia and N South Africa, N through Angola, Zambia, Malawi, the E Zaire borders and East Africa, to Chad, S Sudan and W and N Ethiopia. Distribution in Kenya represents the E limit of the range.

Kenyan: locally common in wooded areas, and in more open country with occasional large trees, up to 2200 m (Br; Ja; MPG; HH). Range is 75% within the 500+ mm rainfall regime, but this parrot is also frequent in suitable habitats in the arid L Turkana area (HH). Records from the coast (Ja) lack confirmation (Br). Br considers this species resident, and Ja mentions only local movements. See Orange-bellied Parrot (343).

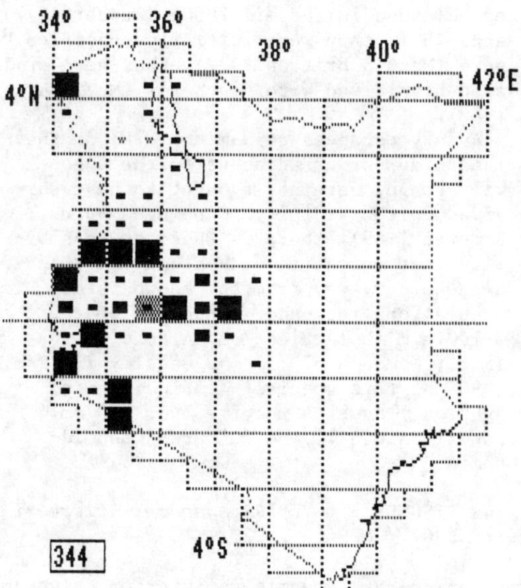

345 BROWN-HEADED PARROT *Poicephalus cryptoxanthus*
B365 / MPG445 / D2C2

Afrotropical: the SE periphery of Africa, from E South Africa N through Zimbabwe, Mozambique, Malawi and E Tanzania, to the N extremity of the range in the coastal lowlands of Kenya.

Kenyan: semiarid-dry subhumid bush and woodland, including coconut plantations and mangroves, in the coastal strip N to Boni 80C (Br; Ja). Common in 1920 on the coast around Mombasa 114B (Ja), it remains so at Shimoni 114C but is rather local and uncommon elsewhere (Br). It is considered resident (Br). See Orange-bellied Parrot (343).

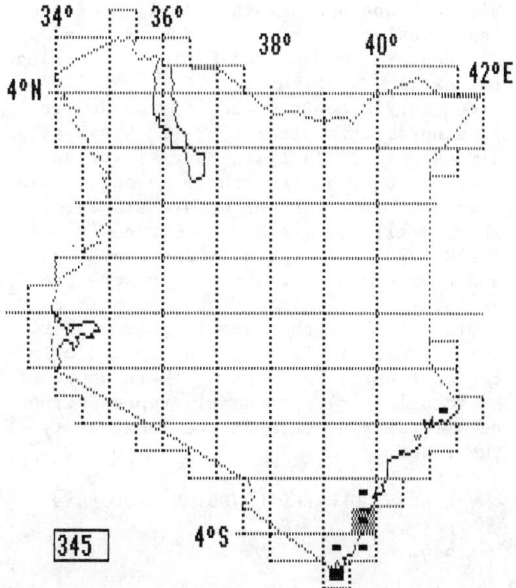

346 GREY PARROT *Psittacus erithacus*
B371 / MPG442 / D2C2
NO MAP: post presence 48AD; pre presence 48C 49C 61A

Afrotropical: a bird of lowland and medium altitude forests in W and C Africa, occurring around the Gulf of Guinea and in the Congo basin, and E to the E Zaire borders, NW Tanzania, W and S Uganda and, on the E extremity of the range, W Kenya.

Kenyan: this forest species' squares are all above 1000 m, and humid. There are old records from Mumias 48C (van Someren 1916), Nandi 49C, Lessos 49C and Nyarondo 61A (Ja), and the virtual restriction of modern sightings to the Kakamega Forest 48D, where it is locally common, presumably reflects a decline. A bird at Alupe 48A (R E M and T Scholtens *in litt.*) had probably wandered from the Bungwe Forest in adjacent Uganda. There is no breeding record, and nestlings in the possession of local people near Rusinga Island 60A (EANHS nest record card) were of unknown and quite possibly Ugandan provenance. Other reference: EABR 1983.

347 RED-HEADED LOVEBIRD *Agapornis pullaria*
B363 / MPG452 / D2C0
NO MAP: post presence 48A; pre presence 48B 61A

Afrotropical: E from the W coast between Sierra Leone and Angola, mainly N of the Congo basin forests, to S Sudan, Uganda, NW Tanzania and, on the E tip of the range, the extreme W of Kenya.

Kenyan: in contrast to its two Kenyan congeners, this bird occurs in Kenya as a wild species. Ja lists Mangiki 48B and Nyarondo 61A, but the three recent records have come from further W, at Malakisi 48A, where small numbers have occurred in Jul (EABR 1978) and a pair in Jan (EABR 1981), and from Alupe 48A, where four were present 24 Aug 1984 (R E M and T Scholtens *in litt.*). Together with a pair at Tororo 48A (Uganda) in 1967, these are the only recent records E of the northern shore of L Victoria (Br), presumably representing occasional wanderers at the E extremity of the range.

348 YELLOW-COLLARED LOVEBIRD *Agapornis personata*
B362 / MPG455 / D2C2

Afrotropical: in its wild state, this lovebird is endemic to interior areas of E and S Tanzania, with a single old record from the SW Kenyan border. It is only present in Kenya as an introduced feral species like Fischer's Lovebird (349), with which it frequently hybridises.

Kenyan: two birds at Taveta 100B (van Someren 1922) are thought to constitute the sole Kenyan record of this species in its truly wild state (Forbes-Watson 1972b), because of the proximity of this area to its N Tanzanian range, and the fact that the early date most probably precludes escaped aviary birds. This view has been questioned (Turner 1977) and, even if wild, the birds must in any case have been rare wanderers since Ja failed to relocate them during a lengthy stay in the area.

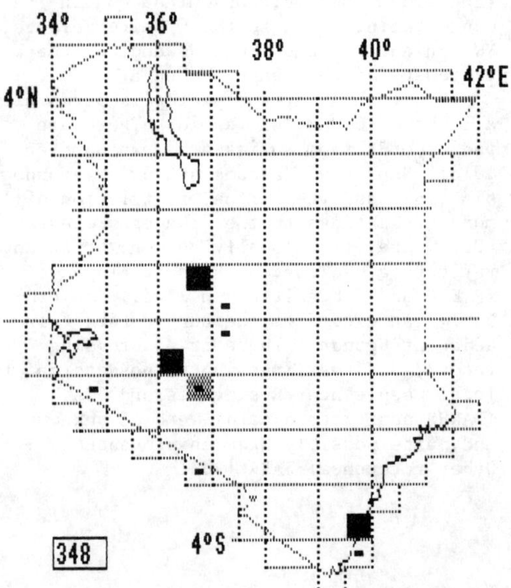

This species has been exported from Tanzania as a cage bird, and all subsequent Kenyan records have been considered to relate to aviary escapees (Cunningham-van Someren 1969). Substantial feral breeding populations have become established at L Naivasha 62C and in Nairobi 75B, with smaller numbers at a few other sites in the highlands, and these have freely interbred with similarly introduced Fischer's Lovebirds to form stable hybrids (Cunningham-van Someren 1969; Forbes-Watson 1972b). Similar feral populations of the Yellow-collared at the coast were established in the mid 1960s (Turner 1977) and, while they were initially specifically pure (Mann & Britton 1972), they have since also hybridised with Fischer's (McVicker 1982).

Aviary escapees at Namanga 87D, on the Tanzanian border adjacent to the species' wild range, were not thought to have survived in the vicinity (Cunningham-van Someren 1969): there is thus the possibility that recent records along the Tanzanian border away from the known feral populations represent wanderers of the wild form (Lolgorien 73B: A Lohding *in litt.*; Namanga 87D: J S S Beesley *in litt.*).

90% of this species' (including its hybrids') squares receive over 500 mm annual rainfall, and 70% are subhumid-humid.

349 FISCHER'S LOVEBIRD *Agapornis fischeri*
B361 / MPG454 / D2C2

Afrotropical: endemic to interior N Tanzania, from the S shores of L Victoria E to the Kenyan border and S to about 5°S. It occurs in S Kenya as an introduced feral species, that frequently hybridises with the Yellow-collared Lovebird (348).

Kenyan: the points made for the Yellow-collared Lovebird regarding introduction to Kenya, feral breeding, and subsequent extensive hybridisation also apply here, but there is no evidence that Fischer's ever occurred in Kenya as a wild species (*op. cit.* under Yellow-collared Lovebird).

Birds at Isiolo 51D were originally thought to be truly wild (Zimmerman 1967),

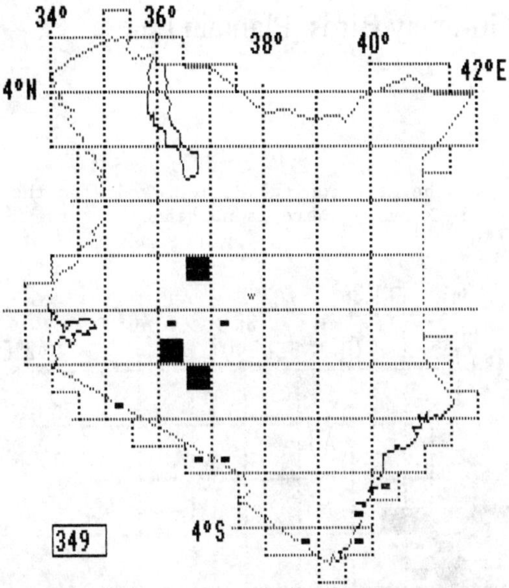

but Cunningham-van Someren (1975c) convincingly argued otherwise. Squares 74C (T Skoglund *in litt.*) and 87D 88C (J S S Beesley *in litt.*) represent apparently specifically pure birds seen on the Tanzanian side of the national border: as with the Yellow-collared Lovebird, these were away from known modern feral populations, so that they may have been truly wild stock on the edge of their natural range.

84% of this species' (and its hybrids') squares receive over 500 mm annual rainfall.

350 ROSE-RINGED PARRAKEET *Psittacula krameri*
B370 / MPG450 / D3C1
NO MAP: post presence 75B 102D 114B; pre confirmed 75B

Afrotropical: in the wild, a bird of the N tropics that extends from the W coast E between 7°N and 17°N to N Ethiopia, Sudan and N Uganda. Feral birds occur sporadically in Kenya, and formerly on Zanzibar.

Kenyan: this parrakeet occurs only infrequently as escapees from captivity around Nairobi 75B (Anonymous 1943; Cooper *et al.* 1971; Cunningham-van Someren 1969, 1971g; Norris 1971d; Forbes-Watson 1972b) and Mombasa 114B (Hamel 1971; McVicker 1982). A pair was seen at a nest hole in Nairobi 75B (Cunningham-van Someren 1969).

Family Musophagidae: Turacos, Go-away Birds, Plantain Eaters

This family is one of the few which are endemic to the Afrotropical Region, and also the largest, with 22 species (Sn). 11 members occur in Kenya. There is no tendency for other than short-distance, seasonal movements (Sn).

351 GREAT BLUE TURACO *Corythaeola cristata*
B372 / MPG437 / D3C3
NO MAP: post presence 48A 49C; post confirmed 48D

Afrotropical: W and C Africa, from the equatorial forests around the Gulf of Guinea, E through the Congo basin to S Sudan, W and S Uganda, NW Tanzania and at the E tip of the range, W Kenya.

Kenyan: this magnificent species, by far the largest of its family, is rated C3 in view of its very restricted range. It is locally common in the Kakamega 48D and S Nandi 48D 49C Forests, and also occurs in adjacent mainly cleared areas, including gardens set in tea and avocado plantations (Br; Candy 1984). There are also recent records from the Malakisi area 48A in Jan and Mar-May (F N Bruce-Miller *in litt.*). All three squares are W of the rift valley in humid, 1000+ mm rainfall areas.

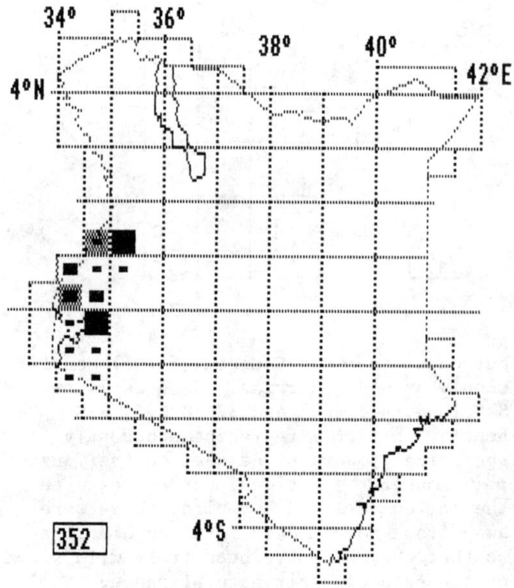

352 EASTERN GREY PLANTAIN-EATER *Crinifer zonurus*
B376 / MPG438 / D3C3

Afrotropical: from SE Zaire and the E Zaire borders, NE through NE Tanzania, Uganda, W Kenya and S Sudan to Ethiopia, and NW to Chad. Distribution in Kenya represents the E limit of the range.

Kenyan: C3 is given because of this species' prominent noisy character and restricted range. This is a common bird of open wooded habitats in the L Victoria basin and surrounding highlands, up to 1800 m (Br). Range is entirely above 1000 m, in moist subhumid-humid, 1000+ mm rainfall areas.

353 BARE-FACED GO-AWAY BIRD *Corythaixoides personata*
B375 / MPG441 / D2C2

Afrotropical: occupies two distinct areas: in the NE, it extends from the S Ethiopian border N through the Ethiopian rift valley highlands. In the E, range extends from Malawi, N Zambia and SE Zaire N through the E Zaire borders and W and C Tanzania to SW Uganda and W Kenya.

Kenyan: rather scarce and local in open wooded habitats up to 2200 m, in moist areas to the W of the White-bellied Go-away Bird (354)(Br; Ja; MPG). Sn considers these two species fairly closely related and, while mapping extensive overlaps of their ranges in other countries, suggests them to be probably ecologically separate locally. They are virtually geographically separate in Kenya, where the only overlap squares are 61B (old records) and 37C. This species sharply contrasts with the White-bellied by being 84% at 1000-2000 m, 84% moist subhumid-humid, and entirely within 500+ mm rainfall areas. There is no confirmed breeding record.

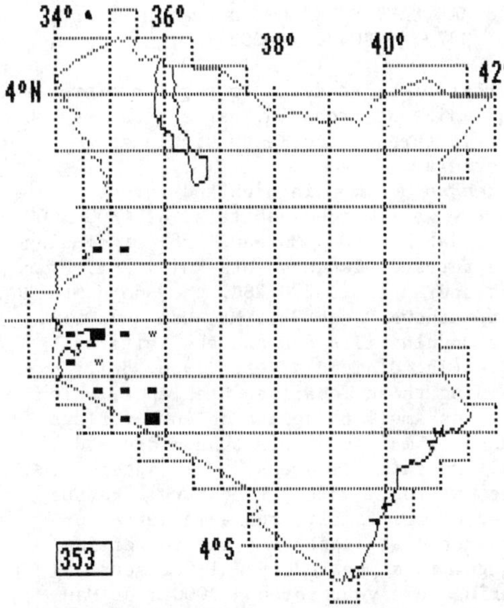

354 WHITE-BELLIED GO-AWAY BIRD *Corythai-xoides leucogaster*
B374 / MPG440 / D3C3

Afrotropical: the E and NE, from E Tanzania through Kenya and NE Uganda to S Sudan, SE Ethiopia and Somalia.

Kenyan: common in open wooded or (less frequently) bushed habitats, up to 2000 m but mainly below 1300 m (Br; Ja; MPG; HH; St; Lack 1985). Range is 87% arid-semiarid, and absent from the dry subhumid squares in the coastal strip. It is also absent from the moist highlands and their peripheral plateaux: the line 37-63-75 represents the W edge of the range, W of which some areas are occupied by the Bare-faced Go-away Bird (353: *q.v.*). Br described the White-bellied as absent from the most arid areas of the N and E, but it is in fact present in 68% of the most arid squares (moisture index below 1.2).

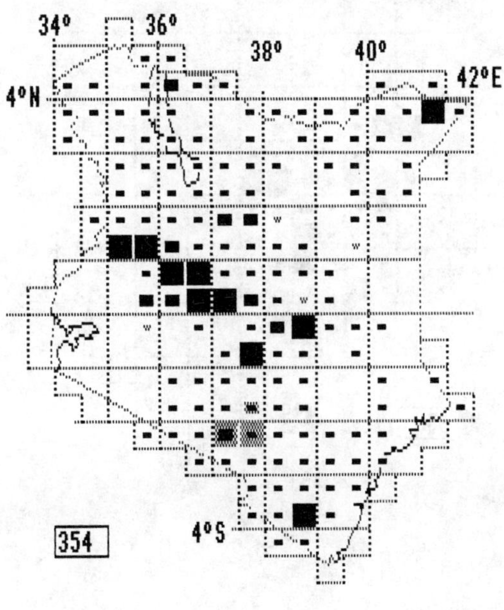

355 FISCHER'S TURACO *Tauraco fischeri*
B378 / MPG431 / D2C2

Afrotropical: the E coastal lowlands, from extreme NE Tanzania, through Kenya to the far S of Somalia.

Kenyan: this, Livingstone's (356) and the Black-billed Turaco (357) are allopatric members of a superspecies (Sn; Br): their ranges are completely separate, even at QSD scale.

Fischer's Turaco is reasonably common in forested, well wooded and thicketed parts of the coastal lowlands, and inland along the lower Tana R 79C 91A (Br; Ja; AGH; Britton & Zimmerman 1979). All squares are within the 500+ mm rainfall regime. Other reference: Britton & Britton 1976a.

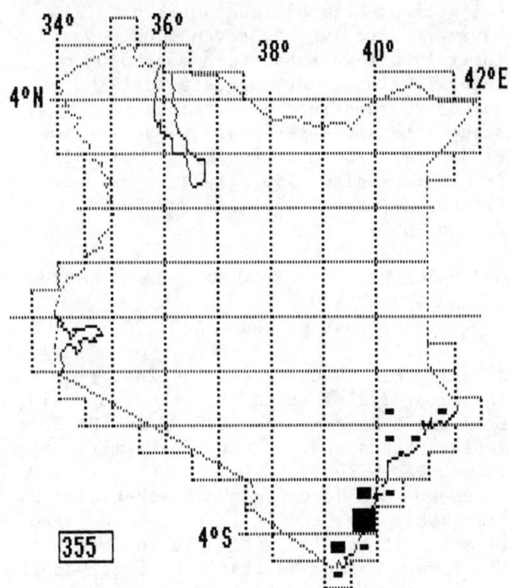

356 LIVINGSTONE'S TURACO *Tauraco living-stonii*
B382 / MPG427, 428 / D2C2

Afrotropical: from NE South Africa to Botswana and Angola, and N through Zambia, Zimbabwe and Mozambique to all but C areas of Tanzania. On the N extremity of the

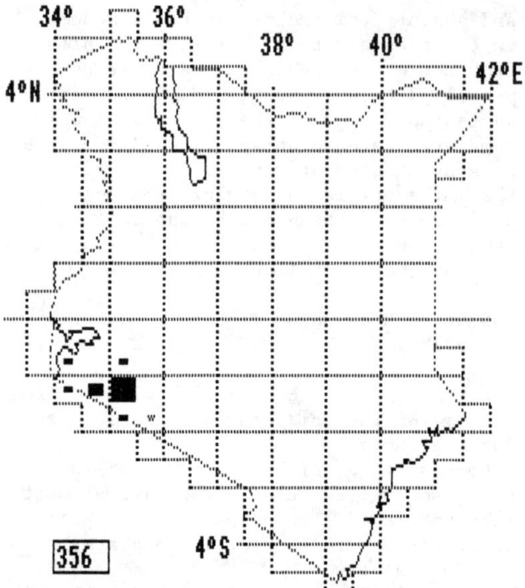

range, there is an isolated population in SW Kenya.

Kenyan: locally common in forested and thickly wooded areas of the extreme SW (Br; Ja), N to Rapogi 60C (National Museum Collection, Nairobi) and Sotik 61C (R A Davey *in litt.*). The seven squares are entirely within subhumid-humid, 500+ mm rainfall areas. Br suggests that there may be two distinct populations in Kenya, since extensive open plains separate the birds over 2200 m in the Loita Hills 74D from those below 1750 m further NW. See Fischer's Turaco (355) and Hartlaub's Turaco (358).

357 BLACK-BILLED TURACO *Tauraco schuetti*
B384 / MPG430 / D1C3
NO MAP: post presence 48D 49C

Afrotropical: a C African species, from the equatorial forests of the Congo basin, E to the E Zaire borders, S Sudan, NW Tanzania, Uganda and, at the E extremity of the range, W Kenya.

Kenyan: MPG describe the Black-billed as remarkably shy, and it is therefore rated D1 since it is also difficult to detect in its forest canopy habitat. C3 is given in view of the very restricted range: this turaco is found only in the Kakamega 48D and S Nandi 48D 49C Forests, and at Kaimosi 48D (Br). Both squares receive 1000+ mm annual rainfall and are humid. There is no breeding record. See Fischer's Turaco (355).

358 HARTLAUB'S TURACO *Tauraco hartlaubi*
B379 / MPG432 / D2C2

Afrotropical: endemic to highland forests and adjacent areas in East Africa, from E Uganda through the Kenya highlands to NE Tanzania.

Kenyan: common in highland forest and nearby well wooded habitats, at 1600-3000 m (Br; Ja; MPG; Se; Moreau 1958), including the forest islands of Mts Kulal 26B, Nyiru 26D and Marsabit 27D 28C, the Ndoto Mts 39A, the Mathews Range 39C, Mt Uraguess 51A, the Chyulu Hills 88D and the Taita Hills 101A (Ja; Diamond & Keith 1980; Br). Excluding these localised highlands, which are too small to appear on our overlays, range is mainly within subhumid-humid, 500+ mm rainfall areas. This species is thought to be mainly sedentary like the rest of the family, but some degree of movement, at least near the top of the range, is suggested by Jul-Oct scarcity during some years around 3000 m at Mau Narok 61D (Se).

Hartlaub's Turaco forms part of a superspecies with the White-crested Turaco (359) (Sn). Hartlaub's is mainly segregated from the other turacos by altitude (Br), but apparently coexists with Livingstone's (356) in the main Loita Hills forests 74D, while being separate from that species on small forest islands just to the SW (Sn).

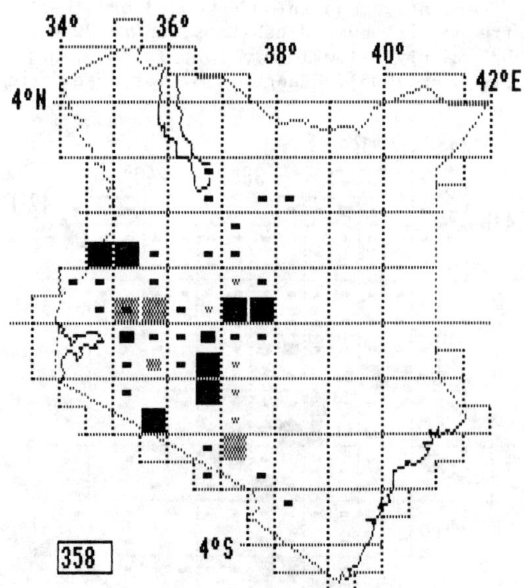

359 WHITE-CRESTED TURACO *Tauraco leucolophus*
B381 / MPG433 / D2C2

Afrotropical: the N tropics, from Cameroon and E Nigeria, N of the main Congo forests, to S Sudan, N Uganda and, on the SE extremity of the range, W Kenya.

Kenyan: locally fairly common in open bushed and wooded habitats, and forest edge, at 1000-2200 m (Br; Ja; MPG). Although Br describes its Kenyan and Ugandan distribution as mainly in low rainfall country, this species' squares are in fact entirely within 500+ mm rainfall areas, 76% within the 1000+ mm regime, and 75% moist subhumid-humid.

This turaco is most frequent N of the equator and W of the rift valley, but very small numbers reach the latter at L Baringo 50A (annually June-Jul and once in Oct: St), and there is also a single record of a Jan wanderer SE to L Nakuru 62A (EABR 1981). The only breeding records come from T J Barnley (*in litt.*), who has found it nesting from Saboti 48B NE to Kitale 37C and the base of the Cherangani Hills 37D. See Hartlaub's Turaco (358).

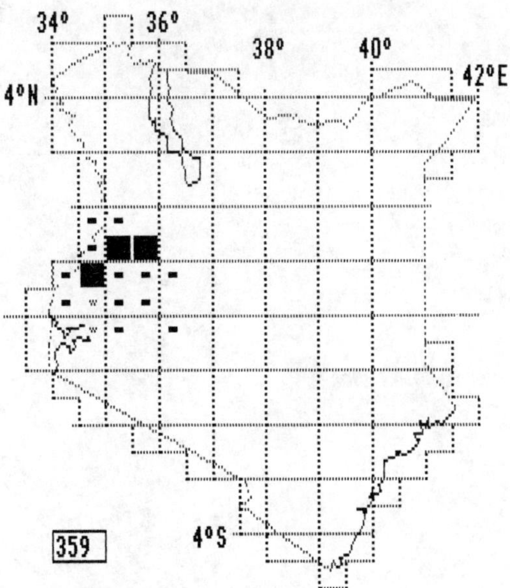

360 VIOLET-CRESTED TURACO *Tauraco porphyreolophus*
B383 / MPG434 / D2C2

Afrotropical: the E, from E South Africa N through Zimbabwe, Zambia and Mozambique to Tanzania and, on the NE tip of the range, S Kenya.

Kenyan: a locally common turaco of bushed

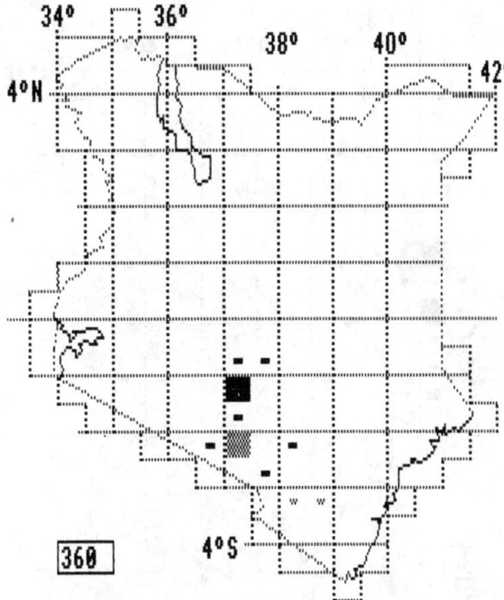

and wooded country, and forest edge and riverine forest, mainly below 1300 m but probably resident at nearly 2000 m on the Mua Hills 76A (Br; Ja; MPG; EABR 1982; ADL pers. obs.). Range is 90% within intermediate, 250-1000 mm rainfall areas, and 70% semiarid. Ja recorded this bird from the Duruma country (approximately 102C: not mapped) but there are no subsequent records; while it has more recently disappeared from riverine woodlands in the Tsavo area 101AB (La), probably due to radical habitat modification (Br).

361 ROSS'S TURACO *Musophaga rossae*
B377 / MPG436 / D3C2

Afrotropical: the periphery of the Congo basin forests, from Angola E to Zambia, N through the E Zaire borders to N Tanzania, Uganda, W Kenya and S Sudan, and thence NW to the Central African Republic and Cameroon. Distribution in Kenya represents the E extremity of the range.

Kenyan: locally common at 1400-2300 m in wooded and forested habitats to the W of the rift valley (Br). Range is 84% within 1000+ mm rainfall areas, and 94% moist subhumid-humid. Ja mentions local movements to food sources.

(map overleaf)

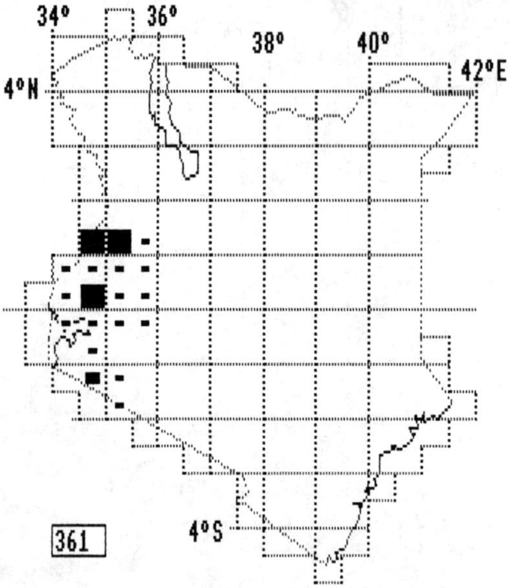

361

Family Cuculidae: Cuckoos, Coucals

Approximately 127 species, represented in all tropical and temperate areas of the world. 24 (19%) occur in the Afrotropical Region, 18 of them (75%) in Kenya. Of the Kenyan species, 14 (78%) have bred at some time or are thought to have done so, two are apparently non-breeding intra-African migrants, and two are non-breeding migrants from the Palearctic and/or Madagascar. The cuckoos are brood parasites whereas the coucals are not. The status and distribution of some of the cuckoos are obscured by the facts that most are migratory and difficult to detect unless calling or migrating, and that some have Afrotropical, Palearctic, Oriental and Malagasy breeding populations (Br).

362 GREAT SPOTTED CUCKOO *Clamator glandarius*
B392 / MPG413 / D2C2

Afrotropical: widely distributed S of 17°N, apart from the arid areas of E Somalia and the SW. Migrants from the Palearctic certainly penetrate the N tropics, and may reach East Africa (Br).

Kenyan: rather uncommon in open bushed and wooded country, up to 3000 m (Br; Se). 80% of its squares are above 500 m, and virtual absence from the coast and the E lowlands compares with status in adjacent S Somalia (AM).

This cuckoo's overall status is obscure due to the presence of locally breeding birds, possibly together with non-breeding migrants from the Palearctic and the S tropics. The Oct-Mar seasonality of most records, however, suggests the presence of N breeders, possibly including Palearctic birds (Br; Ja; La; Lack 1985; EABR). Presumed local breeders are occasional to frequent at L Turkana Apr-Jul (HH) and, further S in the rift valley, the species is present in all but very dry times at Baringo 50A (St). Locally breeding birds are very rare in Tsavo E 101B (La; Trevor & Lack 1976). Occasional birds in the Ngulia lodge netting area 101A at dawn Nov-Dec are of unknown origin (Backhurst & Pearson 1977b). See Black and White Cuckoo (363). Other references: Angwin 1977; Lewis 1981e.

363 BLACK AND WHITE CUCKOO *Clamator jacobinus*
B393 / MPG415 / D2C2

Afrotropical: widely distributed S of 16°N, except for the arid SW and the lowland equatorial forests of the W. Afrotropical birds may be augmented by migrants from

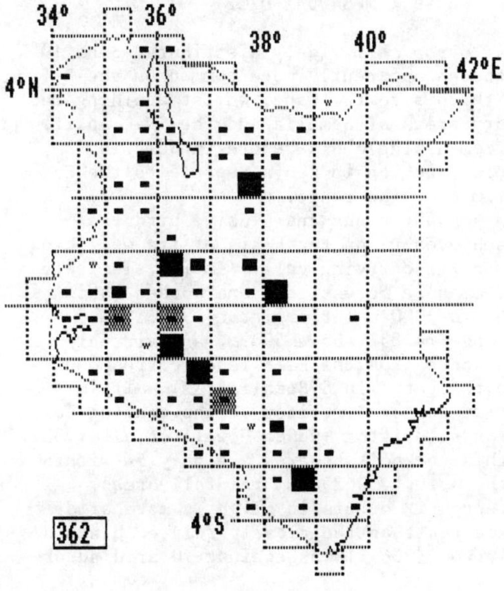

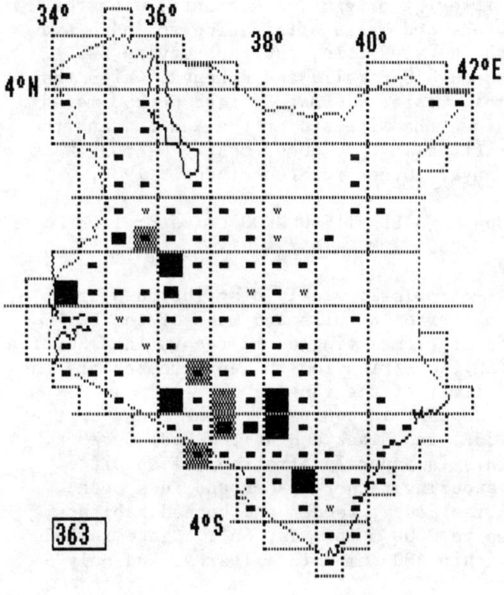

India and the E Palearctic.

Kenyan: an inhabitant of open bushed and wooded country, up to 3000 m but most frequent below 1500 m (Br; Se; D J Pearson *in litt*.; Ja; MPG; Lack 1985). Occurrence is only 10% within the driest, 0-250 mm rainfall areas, while the few breeding squares are above 500 m and absent from the coastal lowlands.

The Black and White Cuckoo appears to be a migrant in most if not all areas but, as with the Great Spotted Cuckoo (362), overall status is confused by the occurrence of locally breeding birds, along with non-breeding migrants from regions possibly as diverse as the N and S tropics, N and C India, and the E Palearctic (Br; MPG).

It shows at least two seasonalities in Kenya. W of, and including, the rift valley, this cuckoo is most numerous Mar-Aug, when breeding occurs (Br; St; EABR). In the Tsavo 89 101 area and elsewhere on the E plateau, however, it is mainly a Nov-Jan and (to a lesser extent) Apr-May passage migrant, which is often numerous around the lights at Ngulia Lodge 101A (Br; Backhurst & Pearson 1977b; D J Pearson *in litt*.; Ja; La; EABR). This cuckoo is however uncommon at the coast S of Malindi 103A, and only small numbers take part in the annual N migration of Palearctic cuckoos *Cuculus* through the coastal lowlands (D J Pearson *in litt*.; Br).

In contrast to the Great Spotted Cuckoo, the Black and White is uncommon at L Turkana (HH), but does occur in the E plateau and lowlands, and is widespread in Somalia S of 7°N (AM). The contrasting distributions of these two species may reflect differing origins of migrants, with the Black and White originating in India and the E Palearctic, and thus moving SW through S Somalia and E Kenya, while the more westerly Great Spotted Cuckoo moves S along the Nile and/or the Red Sea and the rift valley, to occur mostly W of 39°E in Kenya. Other reference: Huels 1982.

364 LEVAILLANT'S CUCKOO *Clamator levaillanti*
B394 / MPG414 / D2C2

Afrotropical: widely distributed S of 15°N, apart from Somalia and much of the SW. Apart from a single old record in S Somalia (AM), distribution in Kenya represents the E limit of the range.

Kenyan: in contrast to the Great Spotted (362) and Black and White (363) Cuckoos, this is a purely Afrotropical species, favouring rather lusher and less open, forest edge, wooded and bushed habitats, up to 2000 m (Br; Ja; Sn). Range is 85% within 500+ mm rainfall areas and only 9%

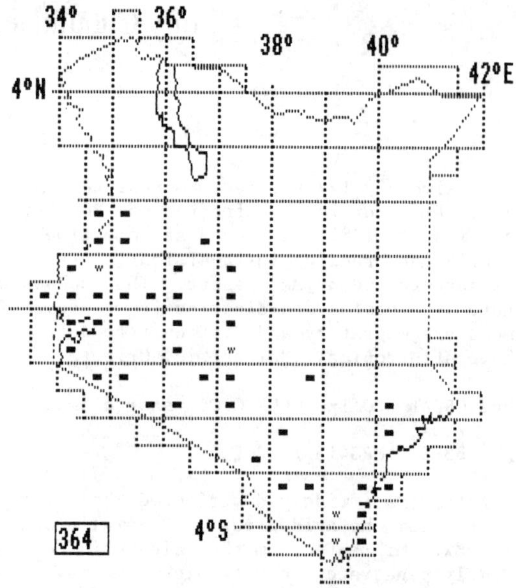

arid.

Levaillant's Cuckoo is only at all common W of the rift valley May-Sept, while small numbers of a small melanistic morph occur in the coastal lowlands, mainly Mar-Sept (Br; EABR 1978, 1979). It is a well known intra-African migrant, and the lack of breeding records is most probably real. Occasional presumed migrants are seen at dawn in the Ngulia area 101A Nov-Dec (Backhurst & Pearson 1977b).

365 RED-CHESTED CUCKOO *Cuculus solitarius*
B399 / MPG406 / D2C2

Afrotropical: widely distributed S of 10°N, though apparently more common on the E from Ethiopia to the Cape, whilst avoiding the dry areas of Somalia and the SW. Apart from a single old record in Somalia (AM), distribution in Kenya represents the E limit of the range.

Kenyan: a shy and elusive bird that achieves D2 on the basis of its well known and far-carrying call. It is relatively common in forest edge and wooded habitats, up to 3000 m (Br; Britton 1979; MPG). Range is 85% above 500 m, and virtual absence from the E lowlands corresponds with status in S Somalia (AM). It is a rare seasonal visitor to the coastal lowlands (Britton 1980c; Howell *et al*. 1980). This cuckoo's squares are only 5% within the driest, 0-250 mm rainfall areas. Although it occurs in slightly more arid squares than the closely related Black Cuckoo (366)(19 as against 10 arid squares),

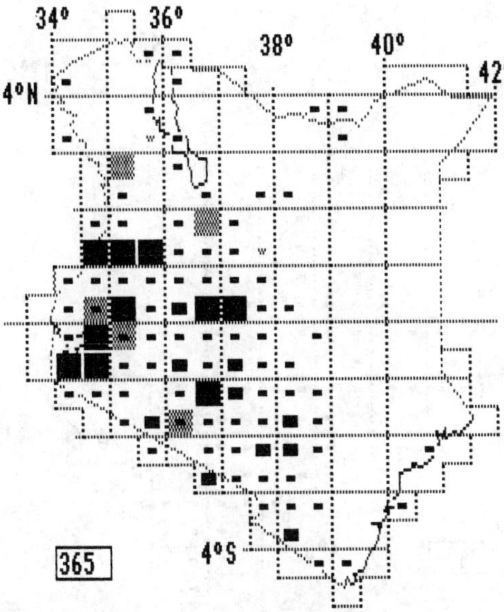

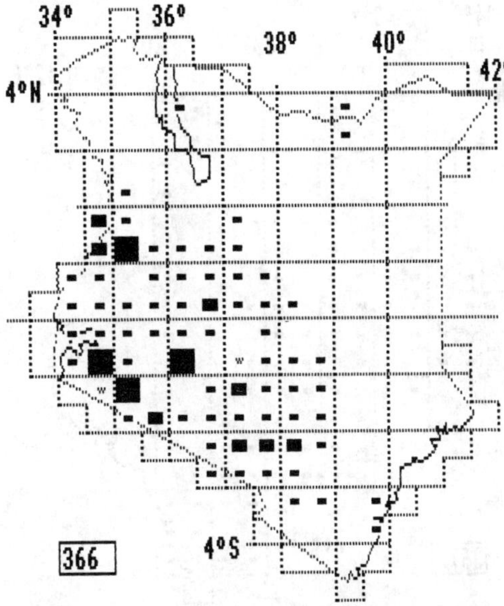

records of the Red-chested Cuckoo in dry N Kenya are largely confined to: (a) the S edge of the Sudanese and Ethiopian populations 2C 3B 16B 17AC; (b) the vicinity of highland islands *e.g.* Mts Nyiru 26D and Marsabit 27D 28C (Br; Ja); and (c) to riverine woods at L Turkana, where at least some records have been after heavy rain (HH).

The Red-chested's breeding squares are 84% above 1000 m and 80% within 500+ mm rainfall areas. This restriction of the breeding range may be real because this bird is a known migrant, although its movements in Kenya are not understood. In two arid-semiarid areas, at Baringo 50A and in Tsavo E 101B, it only appears in the wet seasons, while at the former locality it is more numerous during and after the Apr-May 'long' rains than during the other rainy season (St; La; Lack 1985). Occurrence after rain at arid Turkana has been mentioned above. Other references: Skinner 1978; Archer 1985.

366 BLACK CUCKOO *Cuculus clamosus*
B396 / MPG407, 408 / D2C2

Afrotropical: widely distributed S of 10°N, extending to 15°N in Ethiopia and other areas, and absent only from the dry areas of Somalia and the SW. Distribution in Kenya represents the E limit of the range.

Kenyan: a shy bird that is given D2 because of its distinctive and far-carrying call. It is found up to 2000 m (Br), but has 88% of its squares above 500 m and is notably scarce at the coast (*e.g.* Britton & Zimmerman 1979). Range is virtually absent from the arid N and E (*e.g.* St). There is a single Oct record from Koobi Fora 14A (EABR 1984), while 17AC represents the S extremity of birds from the Ethiopian highlands.

Two races occur. Nominate birds are common intra-African migrants, frequenting forest edge, bushed and wooded habitats and occurring in all months (Br; La; Lack 1985; EABR). The race *C. c. gabonensis* inhabits forest in the W, E to the Karissia Hills 38D and Kapsabet 49C (Br; EABR 1980). See Red-chested Cuckoo (365).

367 AFRICAN CUCKOO *Cuculus gularis*
B397 / MPG404 / D2C2

Afrotropical: widely distributed S of 12°N, though less common in the lowland equatorial forests of the W, and absent from the arid areas of Somalia and the SW. Distribution in Kenya represents the E limit of the range.

Kenyan: uncommon in bushed and wooded habitats up to 3000 m (Br; Se). This is an intra-African migrant that mainly occurs Oct-May (Br; La; EABR), and an immature ringed at Ngulia Lodge 101A (EABR 1979) suggests nocturnal movements. There are no definite breeding records, but juveniles, in some cases newly fledged, have occurred at Baringo 50A Aug, L Naivasha 62C Sept and Garissa 65B Nov (EABR).

The African Cuckoo forms a superspecies with the Eurasian Cuckoo (368), and the two are widely sympatric (Sn). Thus, rather

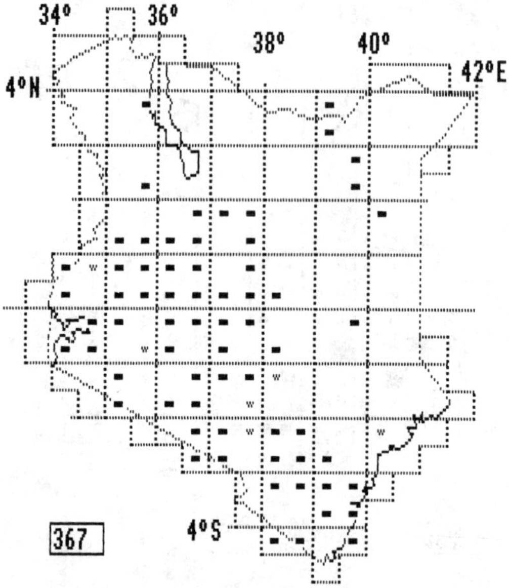

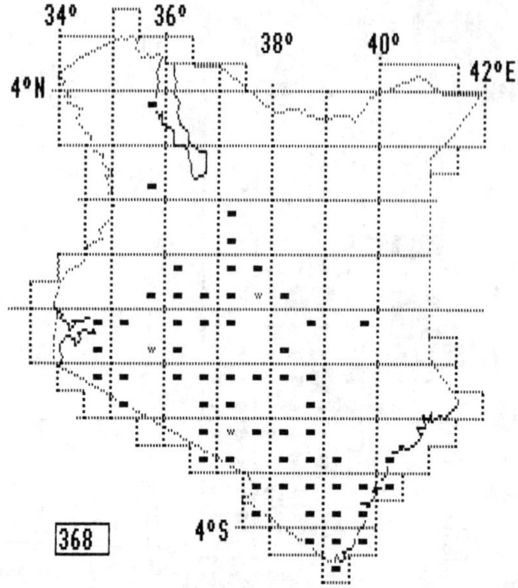

like those of the Eurasian, this species' squares are 92% in 250+ mm rainfall areas and 63% semiarid-subhumid. The differences in their ranges occur in the W, 37 48 60 and Uganda, where this species is more widespread; and in the NE 17 29 42 where the Eurasian is absent. In contrast to the rest of its range, the African's few squares in the NE are mostly arid, not wooded, and receive less than 250 mm rainfall, but all of these records occurred during rainy periods (J E Miskell *in litt.*; D A Turner pers. comm.; ADL pers. obs.). They apparently demonstrate temporary presence in normally inhospitable areas during clement conditions (possibly by migrants: see above). St notes absence from the semiarid but rather less marginal Baringo area 50A during very dry periods, together with marked influxes during the rains.

368 EURASIAN CUCKOO *Cuculus canorus*
B395 / MPG404 / D2C2

Afrotropical: a Palearctic species, wintering mainly from about 10°S to N Namibia and E South Africa.

Kenyan: a fairly common passage migrant in bushed and wooded country, up to 3000 m but mainly below 1700 m (Br; Se). Range is 97% within 250+ mm rainfall areas and 71% semiarid-subhumid. The concentration of squares in S and C areas is apparently real, since its areas of absence in the W 36D 37CD 48 49C 60AC have received reasonable observer coverage, while absence from the extreme N coast and from the E and NE lowlands complements its rarity in Somalia S of 8°N (AM).

The Oct-Dec S passage is light and mainly on the E edge of the plateau, but the Mar-Apr N movement is more marked and, occurring as far inland as L Turkana, L Baringo 50A and L Victoria, can produce marked abundance in the coastal lowlands (Br; Ja; HH; La; St). Given this species' rarity in S Somalia (*op. cit.*), migration N of Kenya presumably takes place through Ethiopia. There is a single oversummering record, on 6 June 1922 at Nairobi 75B (Britton 1982b). See African Cuckoo (367).

369 LESSER CUCKOO *Cuculus poliocephalus*
B398 / MPG405 / D1C1

Afrotropical: birds from the SE Palearctic migrate to the E, from Somalia S to Zimbabwe and E South Africa, Nov-Apr. Birds from Madagascar occupy a similar mainland African non-breeding range, Apr-Oct.

Kenyan: a usually uncommon visitor to forest edge, wooded and bushed habitats, 84% in areas with over 500 mm rainfall.

The nominate Palearctic race occurs in the E Nov-Apr, and is mostly seen on N passage in the coastal lowlands Mar-Apr, when it often moves with the Eurasian Cuckoo (368) and can be numerous (Ba; Br; EABR). Single birds in Sokoke Forest 102B in Nov (EABR 1983, 1984) and ringed at Ngulia 101A in Dec (Backhurst & Pearson 1977a) constitute the only evidence of the S migration, which may enter Africa S of Tanzania (*cf.* the migration of the Eastern Red-footed Falcon (150))(Ba; Br).

Ba knew of no certain record of the

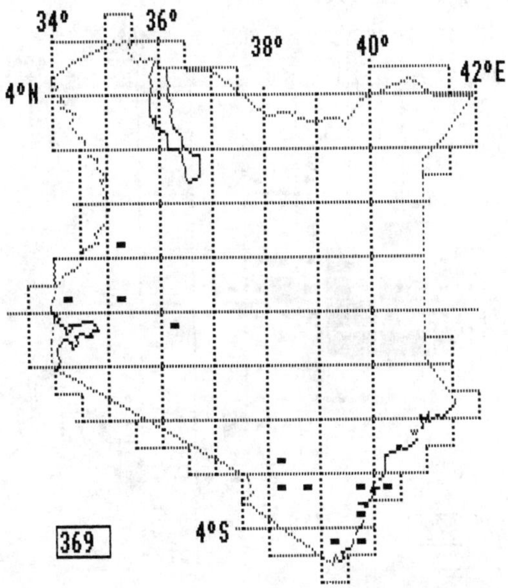

Malagasy race *C. p. rochii*, but it is now generally thought that May-Sept records belong here (Br): these comprise birds seen at Kapenguria 37C (collected 13 June 1968) and Saiwa Swamp 37C (seen and heard 18 Jul 1973)(T J Barnley *in litt.*); Ng'iya 48C and Kapsabet 49C (Br); Eldoret 49C (Ba); L Nakuru 62A (EABR 1983); Sokoke Forest 102B (EABR 1978).

370 THICK-BILLED CUCKOO *Pachycoccyx audeberti*
B400 / MPG412 / D1C1
NO MAP: post presence 79C 91AB 103A; post probable 102B

Afrotropical: this scarce and little known species extends from E South Africa, Botswana and Angola, N to Guinea, Burkina Faso, NE Zaire, W Uganda and SE Kenya. Distribution in Kenya represents the NE tip of the range.

Kenyan: a scarce and local bird of woodland and forest in the coastal lowlands (Br). The five squares are all in 500+ mm rainfall areas.

A female with an oviduct egg was collected at Muniuni (precise location unknown) on the lower Tana R in Oct 1879 (Ja). The few subsequent records have been from Sokoke Forest 102B (Britton & Zimmerman 1979; Short & Horne 1985a) N to Witu Forest 91B (EABR 1984), and along the Tana R to the Tana R Reserve 79C (AGH; EABR 1981). The female with an oviduct egg constitutes the only certain evidence of breeding, though this cuckoo is thought to be generally non-migratory (Sn; Br), while its range lies entirely within that of Retz's Helmet Shrike (872), which is its only known host (Br). Display flight (mapped as probable breeding) has been observed over Sokoke Forest 102B in Dec (Short & Horne 1985a).

371 BARRED LONG-TAILED CUCKOO *Cercococcyx montanus*
B386 / MPG411 / D1C0
NO MAP: post presence 63A 103A 114BC; pre probable 63A

Afrotropical: an E bird, from S Mozambique, N through Malawi and E Tanzania to the NE limit of its range in coastal and highland Kenya. It also occurs along the E Zaire borders.

Kenyan: the first record of this secretive forest species was on 6 Mar 1962, when a female with an oviduct egg was collected, and several birds were heard calling, at 2100 m in the Irangi Forest 63A (Keith 1968). None were heard during later visits to the area (*op. cit.*), but birds have again been located in this forest during late 1985-early 1986 (D A Turner pers. comm.).

The other three records are from the coastal lowlands. Single birds have been caught at Shimoni 114C on 3 Sept 1976 (Britton 1977a), found dead below a power line at Watamu 103A on 30 Oct 1978 (Britton & Zimmerman 1979), and seen at Tiwi 114B on 12 Nov 1980 (EABR 1980). The Watamu bird is considered indicative of nocturnal movement, and these very few coastal records probably refer to wanderers or migrants on the edge of their range (Britton 1977a; Br) (see also Britton 1980c).

372 EMERALD CUCKOO *Chrysococcyx cupreus*
B389 / MPG416 / D2C2

Afrotropical: widely distributed S of 10°N, and further N in Ethiopia and on the W coast, but absent from the drier areas of Somalia and the SW. Distribution in Kenya represents the E limit of the range.

Kenyan: a bird of wooded and forested country at 400-2000 m, infrequently observed when high in dense foliage, but achieving D2 because of its loud, distinctive and commonly heard call (Br; Ja; MPG). Range includes the forest islands of Mt Marsabit 27D 28C, the Ndoto Mts 39A, the Mathews Range 39C, Mt Endau 77B and the Taita Hills 101A (Br; Tennent 1964; G C Backhurst *in litt.*). Apart from these isolated highlands, which are too localised to appear on our overlays, this cuckoo's squares are 91% within 500+ mm rainfall

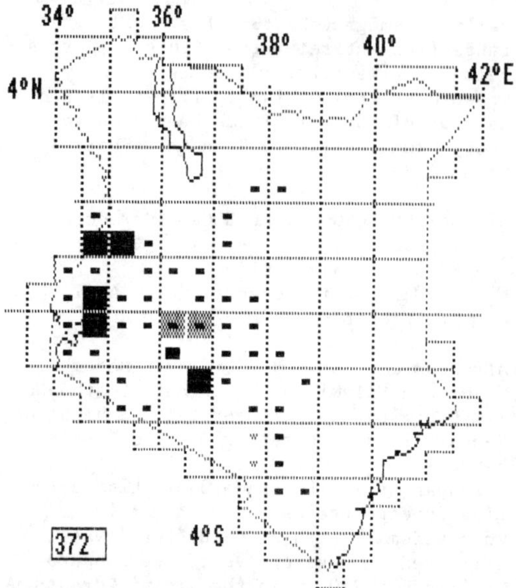

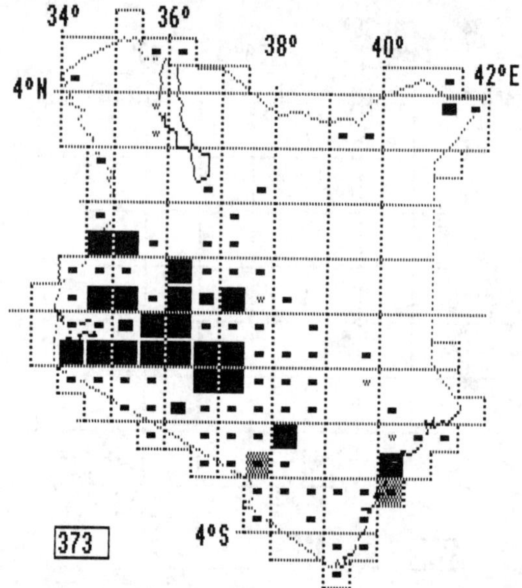

areas. It is absent from the coastal lowlands (Br), and the only records below 500 m are single Nov and Dec birds in Tsavo E 101B (La). Br suggests this species to be mainly resident, and a single Jul bird at L Baringo 50A (St) was presumably a wanderer. See Klaas' Cuckoo (373). Other references: Ryves 1959; Stagg 1973.

373 KLAAS' CUCKOO *Chrysococcyx klaas*
B391 / MPG418 / D2C2

Afrotropical: widespread S of 15°N, except for the drier areas of E Somalia and the SW.

Kenyan: a shy but sometimes common cuckoo of well wooded habitats, up to 3000 m (Br; Britton 1979; Ja; MPG). Like the Emerald Cuckoo (372), it is hard to see in high, dense foliage, but is rated D2 on the basis of its distinctive and far-carrying call. DEP (pers. obs.) found this species resident in 75C 89A, but at least local movements also occur (Ja). Thus, while present in arid Tsavo E 101B in all months, it is only common there during the rains (La). This type of occurrence in dry areas during wet seasons presumably serves to capitalise on the intense moisture-stimulated breeding activity of the dry country species. To the NE, this cuckoo is a presumed Feb-Nov migrant in S Somalia (AM).

Sn summarises the habitat of this species as being between those of the Emerald and Didric (374) Cuckoos, with some overlap: thus, where the Emerald would favour forest interior, Klaas' is on the forest edge; and where the Didric is in more open areas, Klaas' favours denser, more wooded terrain. The maps show these contrasts, to some extent. Hence, while the Emerald Cuckoo and Klaas' have almost the same proportions of their squares well wooded (94% for the Emerald, as against 86%), the former is 91% within lusher, 500+ mm rainfall areas, whereas Klaas' is 94% within 250+ mm areas. Unlike the Emerald, Klaas' extends along the Tana R 64B 65D 78B 79C, and is present and breeds in the coastal lowlands. Furthermore, the Sudanese and Ethiopian populations of Klaas' extend S to Kenya's arid N borders, whereas the Emerald Cuckoo is confined to moister areas, further N in those countries.

But while the Didric Cuckoo can certainly be found in drier and more open country than Klaas', this distinction is not clearly discernible on their maps. The Didric similarly occurs in areas with 250+ mm rainfall, and the best distinction between their very comparable maps lies in the distribution of the breeding squares, though even this quite possibly reflects observer frequency to some extent. Thus, where Klaas' breeding squares are 84% in areas receiving 500+ mm rainfall, the Didric's breeding extends into more 250-500 mm areas, particularly to the NE and SE of the highlands, in 51B 39B 27D and 89 90 101 respectively.

This partial ecological separation of these three reproductively parasitic species is reinforced by their varying host preferences (Sn).

374 DIDRIC CUCKOO *Chrysococcyx caprius*
B388 / MPG417 / D2C2

Afrotropical: widespread S of 16°N, avoiding only the most arid areas of Somalia and the SW.

Kenyan: sometimes common in open bushed and wooded country, up to 2000 m (Br; Ja; Lack 1985). Range is only 6% within the driest, 0-250 mm rainfall areas and 21% arid, and this cuckoo's few squares in the far N and E are mostly confined to the vicinity of permanent water *e.g.* L Turkana (where uncommon: HH), the Daua R 19AB, waterholes at Wajir 42A, and the Tana R 65ABD 79C: it is also scarce in adjacent W Somalia (AM).

While the Didric Cuckoo can occur throughout the year in many areas (Br; *e.g.* DEP pers. obs. in 75BC), it is also known to be an intra-African migrant, although its movements in East Africa are not well defined (Br; Ja; MPG). Common during both wet seasons at semiarid Baringo 50A, it is more so during the Apr-May 'long' rains (St; T Stevenson *in litt.*). In contrast, this species is common in arid Tsavo E 101B during the Nov-Jan 'short' rains, but sparse during Apr-May (La). There is an annual Jan-May influx to the Ruaraka-Kahawa-Ruiru area of moist subhumid 75B (DEP pers. obs.). See Klaas' Cuckoo (373).

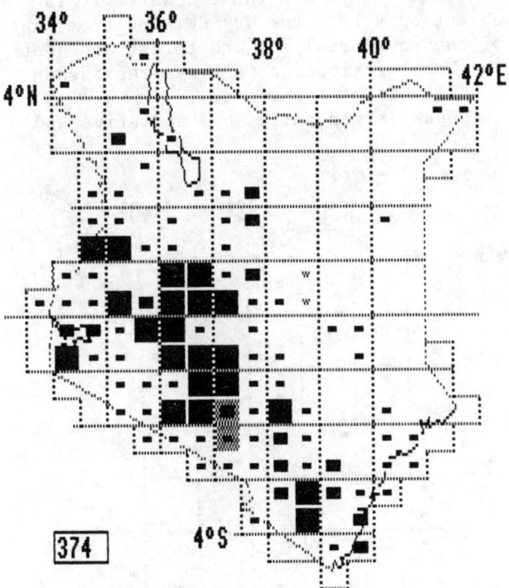

375 BLUE-HEADED COUCAL *Centropus monachus*
B404 / MPG421 / D2C2

Afrotropical: from W Ethiopia and S Sudan, through Kenya, Uganda and the Tanzania-E

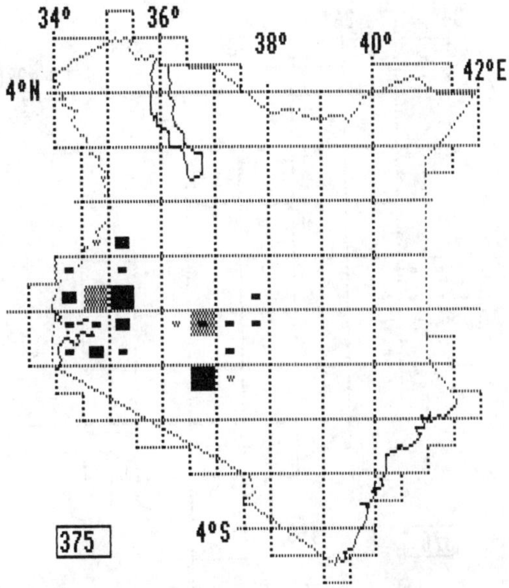

Zaire borders to SE Zaire. W of this, it occurs around the Gulf of Guinea from Senegambia to the Congo, and on the N of the Congo forests. Distribution in Kenya defines the E limit of the range.

Kenyan: a shy bird of thick vegetation near water, local in the W and C highlands up to 2000 m, but more common in the L Victoria basin (Br; Ja; MPG). It is also found in tea bushes at Nandi 49C and in the Nyambeni Hills 51D (Br). Range is 85% within the 1000+ mm rainfall regime, 95% moist subhumid-humid, and entirely within 1000+ m country. Br considers this coucal resident.

376 SENEGAL COUCAL *Centropus senegalensis*
B405 / MPG422 / D2C2

Afrotropical: widely distributed from 15°N to 5°S, E to Ethiopia, Kenya and S Zaire. Further SE, it is also found from Zambia and Malawi S to Botswana and Zimbabwe. Distribution in Kenya represents the E limit of the range.

Kenyan: a locally common but shy species of the W, in bush, thicket and forest edge (Br; Ja; MPG). It is the least associated of the coucals with wet habitats and undergrowth, and this preference may segregate it to some extent from the widely sympatric White-browed Coucal (377: *q.v.*), which is a possible ecological competitor (Sn; Br; MPG). Range is 91% at 1000-2000 m, 91% within 1000+ mm rainfall areas, and 75% humid. There is only a single, old breeding record, from Mt Elgon 36D, but this

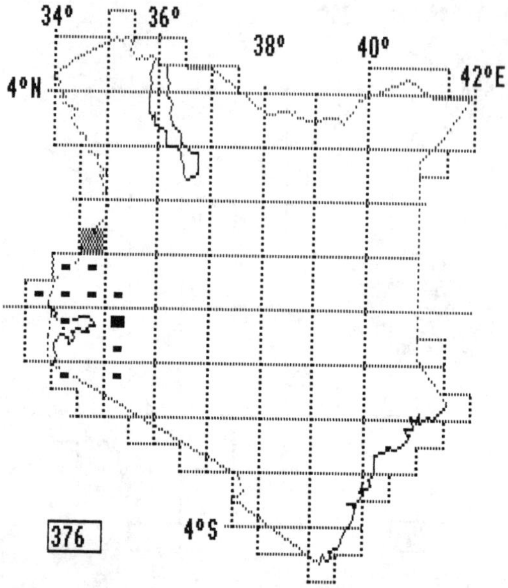

coucal is generally thought resident (Br; Ja).

377 WHITE-BROWED COUCAL *Centropus superciliosus*
B406 / MPG423 / D2C2

Afrotropical: widely distributed S of the Congo forests, and further N through East Africa to S Sudan, Ethiopia and W Somalia. It is absent from the dry SW, and from many areas occupied by the Senegal Coucal (376: *q.v.*).

Kenyan: the most widespread of the coucals, a common but rather retiring bird of dense grass, thicket, bush and reeds, up to 2200 m and often near water (Br; Ja; MPG; St). Range is only 13% within the driest, 0-250 mm rainfall areas, and most of this species' squares in the arid N and E are in the vicinity of water, *e.g.* L Turkana (where frequent: HH), the Daua R 19AB, Wajir 42A and the Tana R 65BD 78B 79C (AGH). It is similarly mostly around rivers in the adjacent interior of S Somalia (AM).

This coucal is generally resident (Br; DEP pers. obs.: 74D 75BC 88AD 89AC 100B 101A), but at least local movements also occur. Thus, in arid Tsavo E 101B, it is most common during the rains, while range contracts into the vicinity of the main rivers during the dry seasons (La; Lack 1985). And, although resident in the Ruaraka-Kahawa-Ruiru sector of moist subhumid 75B, there is also an annual influx, Sept-Nov (DEP pers. obs.). Other reference: Frere 1984.

378 BLACK COUCAL *Centropus grillii*
B403 / MPG420 / D1C1

Afrotropical: a generally scarce species, patchily distributed S of 15°N, E to W Ethiopia, Kenya and the E coast. It is absent only from the dry SW, and from many of the equatorial lowland forest areas of the W. Distribution in Kenya represents the E limit of the range.

Kenyan: a generally shy, secretive and

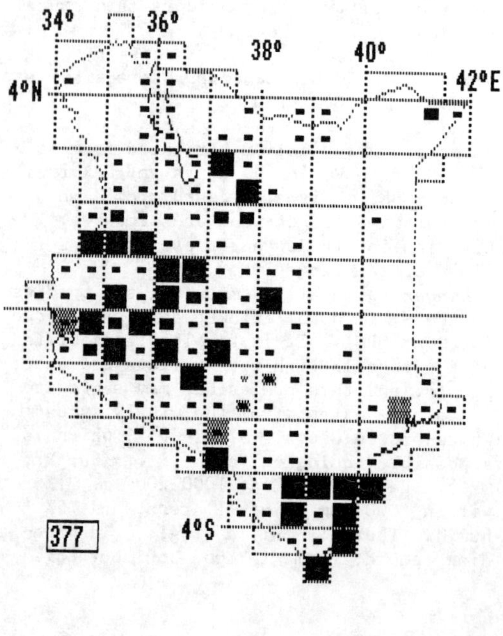

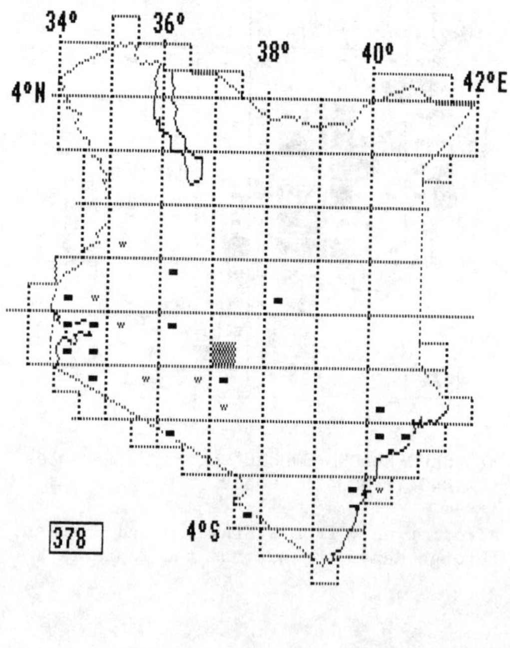

scarce bird, that may well be overlooked in its rather drab non-breeding plumage: hence D1C1 and, probably, the moderately high proportion of pre-1970 records. It inhabits marshes and seasonally flooded grasslands with only limited tree and shrub growth, and is more purely a swamp species than the Blue-headed Coucal (375) (Sn; Britton 1978; Br; Ja; MPG). 95% of its squares receive over 500 mm annual rainfall; and those above 1000 m are 83% subhumid-humid, whereas those in the SE lowlands, *i.e.* L Jipe 100D, the coastal strip and the lower Tana R 79C 91A, are five of six arid-semiarid. There is a single breeding record, from Embu 63C (Br).

This coucal is seasonal in the W 60, and present at L Kanyaboli 48C Jan-Mar and June-Jul (Britton 1978; Br). Three recent records from semiarid squares on the periphery of the C highlands (Baringo 50A, Meru 52C and Shombole 87A) have been late May-Aug (EABR 1980, 1982, 1983). Seasonal disappearance of temporary waters will be a factor in this species' movements (Sn).

379 YELLOWBILL *Ceuthmochares aereus*
B401 / MPG424 / D1C1

Afrotropical: the two areas of Kenya occupied by this species are parts of its two separate, though in places closely juxtaposed, ranges in Africa. On the E periphery of the continent, it extends N from coastal South Africa to Tanzania and, at the N extremity of the range, S areas of Kenya and Somalia. The other population occurs throughout the equatorial lowland forests of the W, E to W Tanzania, Uganda, S Sudan and, on the E limit of the range, SW Ethiopia and W Kenya.

Kenyan: a skulking, unobtrusive and usually silent bird of dense thicket and forest, up to 2000 m (Br; Ja; MPG; Sn): hence D1C1, despite its distinctive appearance. The nominate race occurs in the W, in humid, 1000+ mm rainfall areas above 1000 m. *C. a. australis* is found in the SE lowlands, where range is 63% below 1000 m, 74% semiarid-dry subhumid and 79% within the 500+ mm rainfall regime.

The absence of breeding records for this species may be real because, at least in the coastal lowlands, it is a locally numerous migrant May-Sept/Oct, but rare at other times (Br; EABR). An example of *australis* caught at Ngulia Lodge 101A on 29 Nov was presumably on migration (EABR 1978), and Ja mentions a wanderer of this race as far NW as Escarpment 75B.

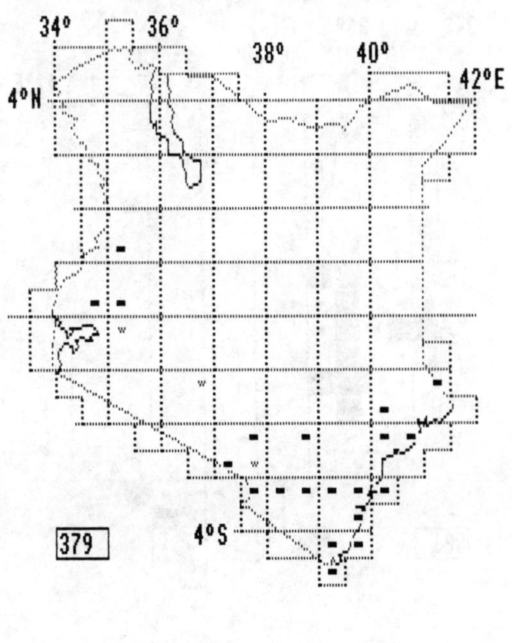

Family Tytonidae: Barn Owls

10 species, mainly in the Oriental and Australasian Regions. Three, including the cosmopolitan Barn Owl (381), are found in the Afrotropical Region and, of these, two breed in Kenya. These birds' crepuscular and nocturnal habits certainly lower their detectability and coverage.

380 CAPE GRASS OWL *Tyto capensis*

B408 / MPG529 / D1C1

Afrotropical: with the exception of most of the SW, fairly widely distributed in Africa to the S of the Congo forests, and further N to highland areas of E Zaire, East Africa and Ethiopia. There is also an isolated population in Cameroon. Distribution in Kenya represents the E limit of the range.

Kenyan: local and uncommon in long grasslands up to 3200 m, particularly in the vicinity of damper, marshy areas (Br; Ja; MPG; Brown 1970a). In contrast to the Barn Owl (381), this is purely a highland species, with range 76% moist subhumid-humid, and entirely within 500+ mm rainfall 1000+ m areas. The compact distribution shows no sign of wanderers, and this owl is apparently resident (Br). Brown (1970a) mentions temporary movements to concentrations of rodent prey.

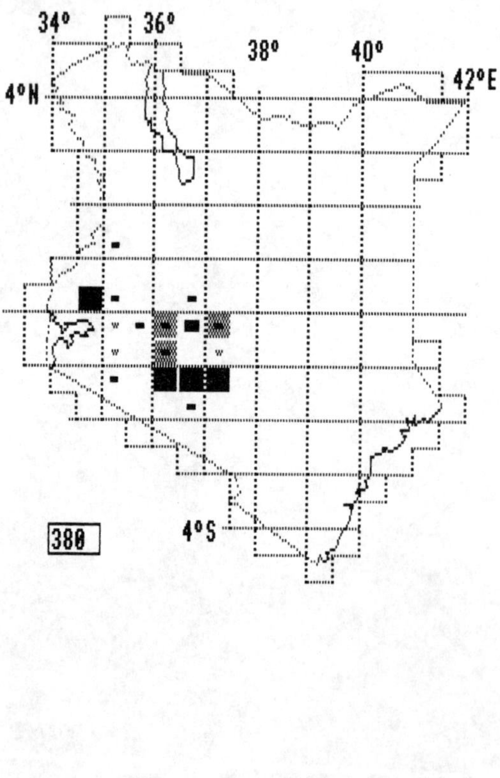

381 BARN OWL *Tyto alba*

B407 / MPG528 / D2C2

Afrotropical: widely distributed S of 17°N, avoiding only N Somalia and many of the equatorial lowland forest areas of the W.

Kenyan: while it can be found in the vicinity of rocks, caves, disused wells and Hamerkops' (44) nests in more remote areas, the Barn Owl is now largely a commensal of man that roosts and breeds on permanent buildings, even in urban centres: numbers may well have increased with the spread of suitable buildings during this century (Brown 1970a; Ja; MPG; Br; Sn). It has been recorded up to 3000 m (Se) and, with the notable exception of L Turkana (where occasional to frequent: HH), 72% of its squares are in 500+ mm rainfall areas. Thus, apart from Turkana, this owl is absent from the arid N and E, presumably due to insufficient prey abundance: it is similarly rare in adjacent S Somalia (AM). It is also scarce in the L Victoria basin (Br). There are no mentions of any movements (Br; Ja; St). Other reference: Cunningham-van Someren 1972c.

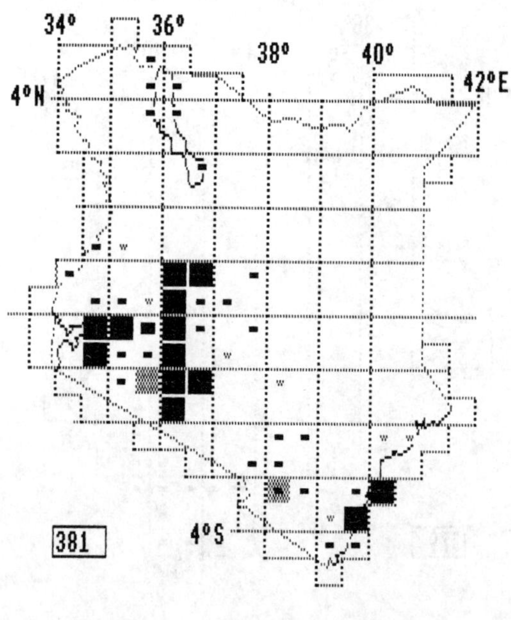

Family Strigidae: Owls

About 130 species, represented in all zoogeographical regions. 26 (20%) are found in the Afrotropical Region. Of these, 14 (54%) occur in Kenya, where the Short-eared Owl (395) is a Palearctic migrant, Scops Owl (382) is a resident that is regularly augmented by Palearctic visitors, and the rest (86%) are Afrotropical residents, including the endemic Sokoke Scops Owl (383). The crepuscular and nocturnal habits of this family undoubtedly lower the species' detectability and coverage, although to varying extents. High proportions of pre-1970 squares may reflect a reduction in collecting rather than declines.

382 SCOPS OWL *Otus scops*
B424 / MPG534 / D2C2

Afrotropical: widely distributed S of 14°N, avoiding only the drier areas of E Somalia and the SW, and most of the equatorial lowland forest areas of the W. Palearctic migrants winter S to about 6°N in the W, and 3°S in the E.

Kenyan: Afrotropical races, which constitute the bulk of this species' numbers in Kenya, are locally common but rarely seen, nocturnal birds that achieve D2 because of their frequently uttered and distinctive call (Br; Ja; Brown 1970a; La). They frequent a great variety of wooded habitats up to 2000 m (Br; MPG), and are found throughout our rainfall and moisture regimes. The clustered records of confirmed breeding all originate from T J Barnley (*in litt.*). These birds are thought resident (Br; DEP pers. obs.: 74D 75C 88A).

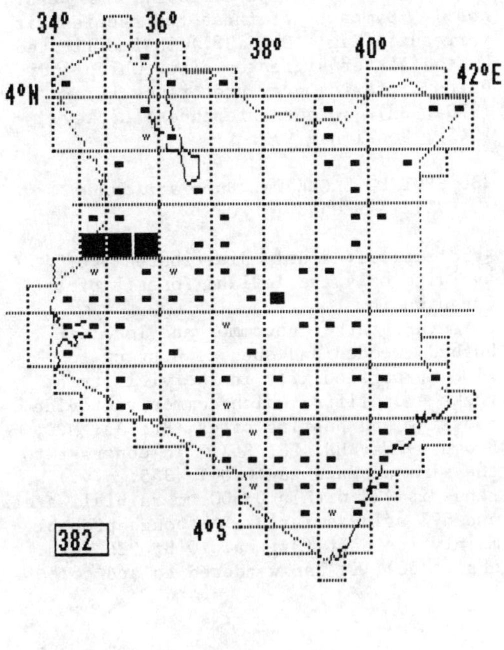

Palearctic migrants are probably annual Nov-Mar in small numbers in forest edge and woodland habitats, and Kenyan occurrences are amongst the most S known (Br; Ba; OS-C 1977; EABR). Records of Palearctic birds are predominantly in areas above 1500 m and with 500+ mm rainfall: exceptions are the only African record of the E race *O. s. turanicus* at Bura 78B (at 100 m: Friedmann & Keith 1968); and birds on migration in Tsavo W NP 89C 101A, some of which are attracted to the vicinity of Ngulia Lodge (at 900 m: EABR 1979, 1983).

383 SOKOKE SCOPS OWL *Otus ireneae*
B421 / MPG- / D2C3
NO MAP: post presence 102D; post probable 102B

Afrotropical: endemic to the Sokoke Forest in coastal Kenya.

Kenyan: the type specimen was obtained as recently as 9 Apr 1965, from *Brachystegia* woodland (Ripley 1966). All subsequent records have been in *Cynometra-Manilkara* forest, at 50-170 m, on red sands in the moister parts of Sokoke, where it is reasonably common: the total population is estimated at 1300-1500 pairs (Britton & Zimmerman 1979; Br; Britton *et al*. 1984; EABR 1977, 1983). This owl achieves D2 on the basis of its distinctive call, and C3 because of the extremely restricted range, though Br suggests that it may also occur on similar sands in the Mundane Range 80C.

384 WHITE-FACED SCOPS OWL *Otus leucotis*
B422 / MPG536 / D1C1

Afrotropical: widely distributed S of 15°N, but apparently rare in the Congo basin forests, and absent from S Somalia and the S tip of the continent. Distribution in Kenya represents the E limit of the range.

Kenyan: uncommon in bushed and wooded

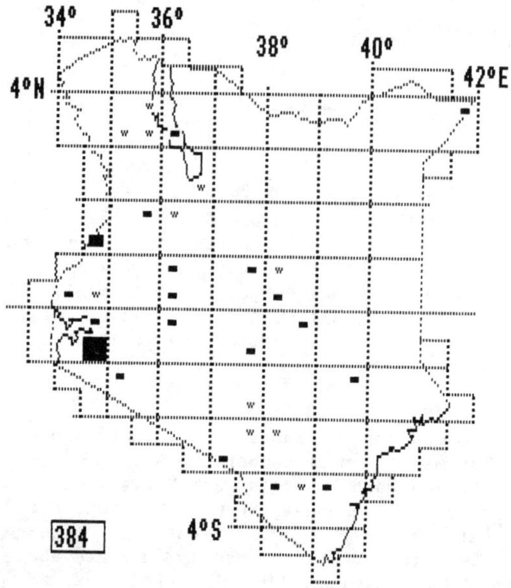

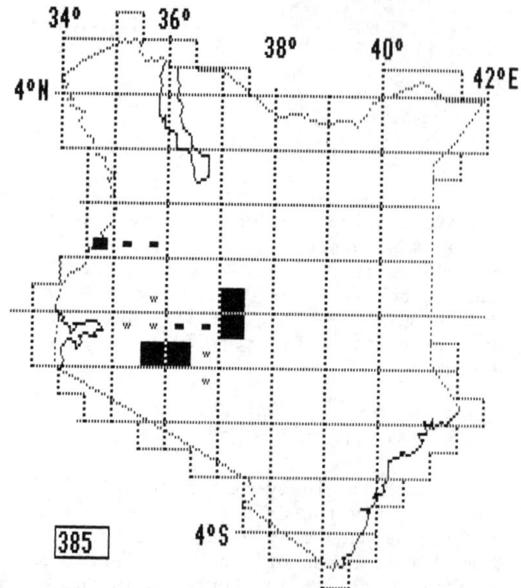

habitats up to 1700 m, but slightly more diurnal than its congeners (Br; Ja; MPG; Brown 1970a; Sn). Range is 73% within intermediate, 250-1000 mm rainfall areas and 69% less arid-semiarid. There are no records from the coastal strip, and virtual absence from the dry NE is probably real in view of the arid nature of the country and this species' absence from adjacent S Somalia (AM). The record in 19B is from locally moist habitat near the Daua R at Mandera (B S Meadows *in litt.*). This owl is generally considered resident (Br; St; EABR 1978, 1984), and the single confirmed breeding record (Maas 1983) doubtless reflects its unobtrusive and largely nocturnal character.

385 CAPE EAGLE OWL *Bubo capensis*
B413 / MPG541 / D2C2

Afrotropical: the E, with a discontinuous distribution in highland areas from Ethiopia S through Kenya, Tanzania, Malawi, Zimbabwe, and Botswana to E South Africa, and to lower altitudes at the Cape. Distribution in Kenya describes its range at these latitudes.

Kenyan: a locally common bird of rocky montane moorland and also more timbered country, mainly at 2400-4200 m, but also descending to 1800 m, particularly in the Nyeri, Nakuru, Gilgil, Naivasha, Kijabe and Limuru areas 62 75B (Br; MPG; Se; Brown 1970a; Lewis 1978). Range is entirely within subhumid-humid, 500+ mm rainfall country. The species is considered resident (Br; Se). The relatively high proportion of pre-1970 squares (all representing specimens) most probably reflects this bird's nocturnal habits, its preference for infrequently visited areas at high altitude, and possibly some loss of recent records due to confusion with the better known Spotted Eagle Owl (386).

The Cape Eagle Owl is mainly segregated from the similar Spotted Eagle Owl by altitude, although both occur in fairly close proximity around L Naivasha 62C, at 1900 m (D A Turner pers. comm., ADL pers. obs.). Sympatry of these two species with Verreaux's Eagle Owl (387) is facilitated by the latter's greater size, which indicates differing prey preference (Brown 1970a; Sn). Other references: Leakey 1943c; Sessions 1972.

386 SPOTTED EAGLE OWL *Bubo africanus*
B412 / MPG543 / D2C2

Afrotropical: widely distributed S of 15°N, avoiding only the lowland forests of the Congo basin.

Kenyan: rather uncommon and local in bushed, wooded and open country up to 2100 m, particularly in the vicinity of rocks and cliffs, which commonly provide roosting and nesting sites (Br; Ja; MPG; Brown 1970a; HH; La; St). In contrast to the similar Cape Eagle Owl (385: *q.v.*), range is 42% within 0-500 mm rainfall areas and 69% arid-semiarid. Although probably mainly resident (Br; La: 101B; DEP pers. obs.: 75C), it has wandered to and colo-

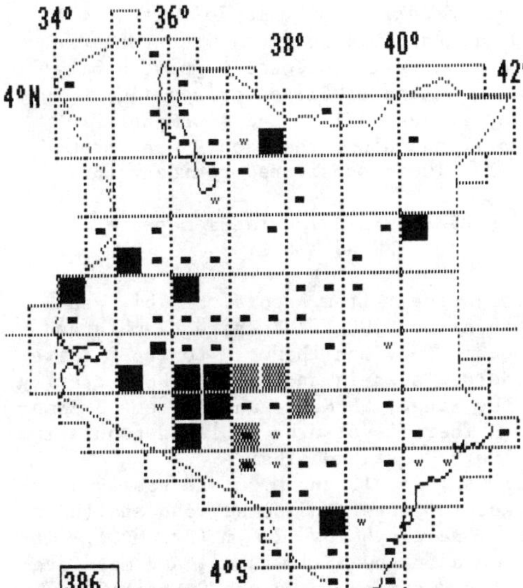

nised quarries near Mombasa 102D 114B, where it was formerly absent (Br).

387 VERREAUX'S EAGLE OWL *Bubo lacteus*
B414 / MPG544 / D2C2

Afrotropical: widely distributed S of 14°N, except for the equatorial lowland forests of the W, and the arid areas of E Somalia and the SW.

Kenyan: moderately common in a wide variety of wooded habitats, including occasional large trees in fairly open country, and especially in riverine and other trees near water (Ja; MPG; Br; Brown 1970a; Brown & Britton 1980; La). It has reached 3000 m (Se) but is more frequent below 2000 m, although relatively scarce at the coast (Br; Britton & Zimmerman 1979). Range is only 11% within the driest, 0-250 mm rainfall regime, and records from the arid N and E are in locally moist situations *e.g.* near trees and permanent water on the Mogilla Range 2C (T Romig *in litt.*), along a wooded lugga near Ileret 4C (J R Ness *in litt.*), in riverine woods at Kalokol 13B and along the Daua R 19AB, and in isolated trees near the single dam at Moyale 17AC (E J Roberts *in litt.*, ADL pers. obs.). There is no evidence of movements (Br; La: 101B; DEP pers. obs.: 75C 88D 89AC 100B 101A), but both records around the top of the altitude range at Mau Narok 61D were Mar-Apr (Se). See Cape Eagle Owl (385). Other references: Brown 1965b; Loefler 1984.

388 PEL'S FISHING OWL *Scotopelia peli*
B425 / MPG546 / D1C1

Afrotropical: this rarely observed species extends from Senegambia, Nigeria, Ethiopia and S Somalia, S to Congo, Botswana and E South Africa.

Kenyan: a rarely observed, highly unobtrusive species, inhabiting forest or woodland along large, permanent rivers, from which it takes fish (Br; MPG). All squares contain riverine woodland/forest, and all but one (see below) large, permanent rivers.

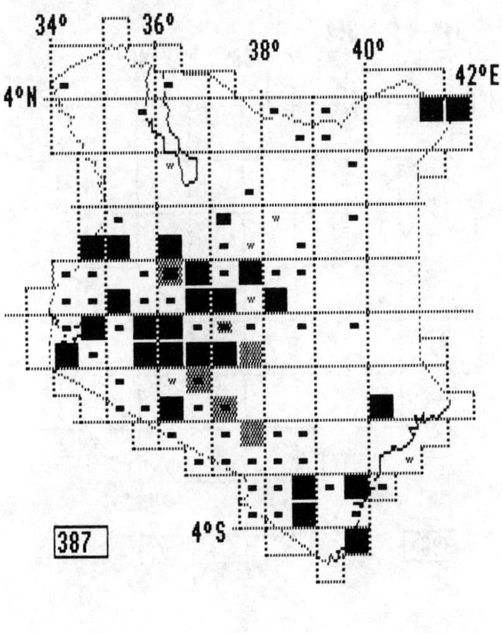

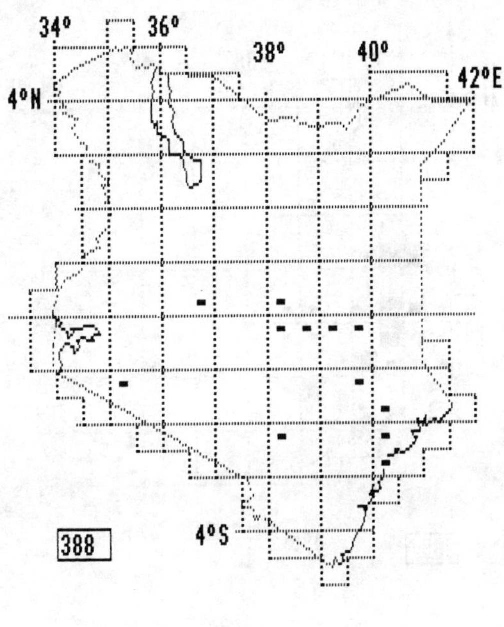

It has been recorded up to 1700 m (Br), but 83% of the squares are below 1000 m.

This owl has been most frequently observed along the Tana R 52C-91C, where Campbell (1977) described it as generally not rare, and suggested that the mangroves at the river's mouth might support a relatively large population. Fischer noted similar status near the Tana mouth in the 1870s (Ja). There are also recent records from the Ewaso Ngiro R 50D (L V J Grant *in litt.*), the Mara R 74A (EABR 1977) and the Athi R near Bushwackers Camp 89A (Schluter & Harcourt 1984; EABR 1983). While other authors (Brown 1970a; Sn) consider this species strictly nocturnal, MPG regularly observed a pair flying along the swamp edge at L Jipe 100D at 1600 h: this record is also unique in being from a non-riverine locality. This species is considered resident (Br), and the lack of breeding records no doubt reflects its obscure nature. Other references: Reynolds 1973b; Davey 1978b.

389 AFRICAN WOOD OWL *Ciccaba woodfordii*
B416 / MPG533 / D2C2

Afrotropical: widely distributed S of 7°N, and further N in Ethiopia and on the W coast, but avoiding the dry areas of N Somalia and the SW.

Kenyan: a locally quite common but nocturnal species that favours densely foliaged, wooded and forested habitats, including urban gardens, up to 2700 m (Br; Ja; MPG; Brown 1970a; Britton & Zimmerman 1979). It is rated D2 because of its distinctive call. Range is 76% within the 500+ mm rainfall regime. Occurrence in arid country is in locally moist habitats *e.g.* woodland at Eliye Springs 14C, L Turkana, in Aug 1971 (A L & T Campbell *in litt.*), and along the Tana R 65ABD 79C (AGH). It is considered resident (Br).

390 BARRED OWLET *Glaucidium capense*
B417 / MPG540 / D1C1

Afrotropical: from Angola, Namibia and E South Africa, NE to SE Zaire, Zambia and N Mozambique, and thence N to the E Zaire borders, Tanzania and, on the NE extremity of the range, SE Kenya and extreme S Somalia. There is also an isolated population in Ivory Coast and Liberia.

Kenyan: local, uncommon and easily overlooked in forest and dense woodland (hence D1C1), mainly below 1200 m (Br; MPG). It is found in the SE lowlands, and most frequently recorded in Sokoke Forest 102B (Britton & Zimmerman 1979; EABR 1977, 1981). Excluding an obvious wanderer around 1800 m in the rift valley at Gilgil 62A (Br), six of nine squares are in 500+ mm rainfall country, while another two are in the vicinity of the Tana R near Garissa 65BD (Br). This owlet is thought to be resident (Br; EABR 1977), and the lack of breeding records most probably reflects its obscure nature and dense habitats.

Away from the coastal strip, this species' squares are mostly also occupied by the Pearl-spotted Owlet (391), but interspecific competition is probably averted by the latter's smaller size (Sn), as well

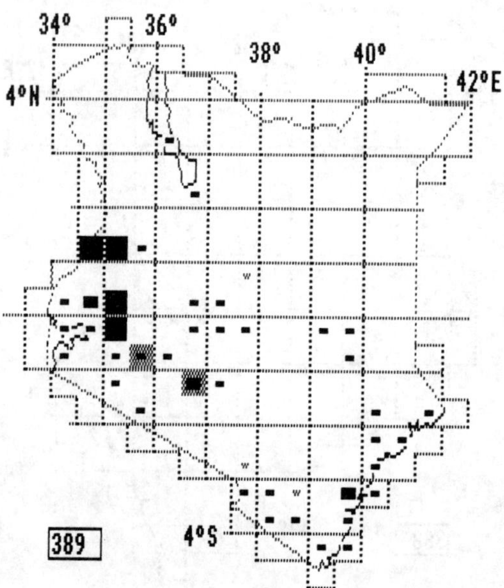

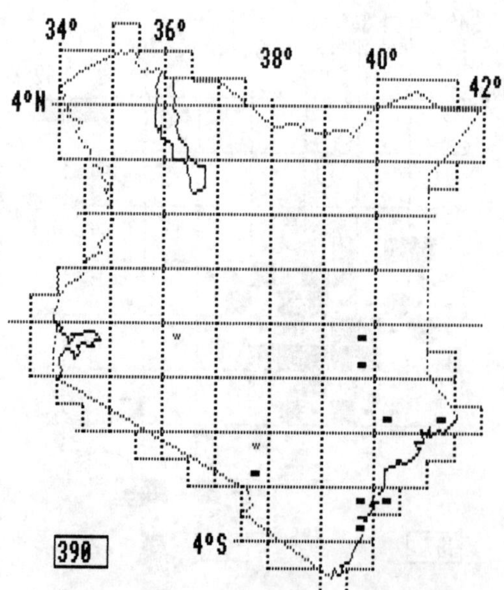

as these two birds' differing habitat preferences.

391 PEARL-SPOTTED OWLET *Glaucidium perlatum*
B419 / MPG538 / D2C2

Afrotropical: widely distributed S of 15°N, avoiding only the drier areas of E Somalia and the SW, and the equatorial lowland forest areas of the W.

Kenyan: by far the most widespread and locally common member of the genus in Kenya, the Pearl-spotted Owlet is found in open bushed and wooded environments, up to 2200 m (Br; Ja; MPG; Brown 1970a; Lack 1985). D2C2 reflects its partly diurnal nature and well known calls. Range is 74% within intermediate, 250-1000 mm rainfall areas (*cf.* Br). Only 26% of the squares are arid, and while this may reflect observer frequency to some extent, the bird is actually uncommon in the arid country around L Turkana (HH), and also in Somalia S of 4°30'N (AM). While predictably avoiding the high rainfall L Victoria basin 60, this owlet is also absent from even semi-arid areas of the coastal strip (Br). It is considered resident (Br; La: 101B; DEP pers. obs.: 75C 88A 89A).

The Pearl-spotted Owlet is ecologically segregated from the forest-dwelling Red-chested Owlet (392), which is a member of the same species-group (Sn). See Barred Owlet (390). Other reference: Brown 1963c.

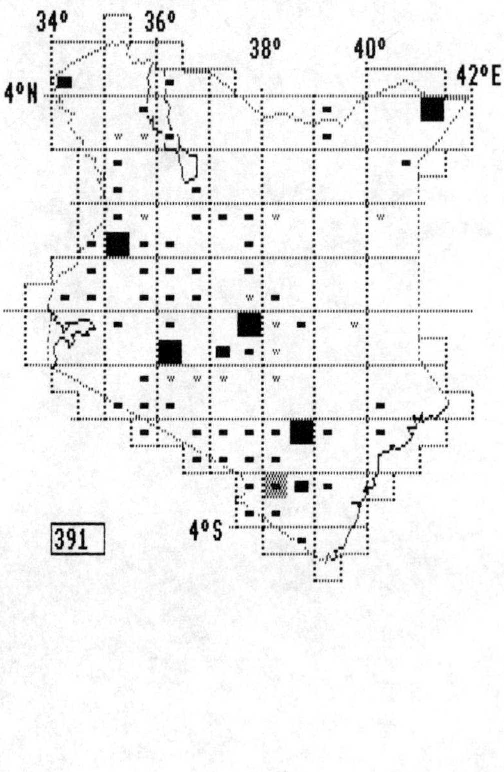

392 RED-CHESTED OWLET *Glaucidium tephronotum*
B420 / MPG539 / D2C2
NO MAP: post presence 48D 49C 61A; post confirmed 36D

Afrotropical: W and C Africa, from the lowland forests around the N of the Gulf of Guinea and in the Congo basin, E to the Zaire-Uganda borders and, at the E limit of the range, W Kenya.

Kenyan: an uncommon and little known forest species that achieves D2C2 on the basis of its distinctive call and restricted range. The type specimen of the race *elgonense* was taken on Mt Elgon 36D in 1926 (Ja), but it was not relocated until 1965, when a second specimen was obtained in the Kakamega Forest 48D (Ripley & Bond 1971). It has subsequently been recorded regularly at the latter locality (Mann 1976; EABR 1977), and has also been found in the adjacent N Nandi Forest 49C (Br) and in the NW Mau Forest 61A, where it is locally common (EABR 1977). The four squares are above 1000 m in humid, 1000+ mm rainfall areas. Br mentions it up to 2150 m. There is only a single breeding record (T J Barnley *in litt.*), but the species is thought to be resident (Br). See Pearl-spotted Owlet (391).

393 LONG-EARED OWL *Asio otus*
B411 / MPG530 / D0C3
NO MAP: pre presence 63A

Afrotropical: an extremely localised and little known owl, in montane forests in E Zaire, on the Rwenzori Mts and Mt Kenya, and in Ethiopia and S Sudan.

Kenyan: this apparently rare species achieves C3 because it is believed to be restricted to Mt Kenya 63A, where it inhabits the upper reaches of the montane forest and possibly the giant heath zone (Br). There are only two recent records. An adult female was collected by J R M Tennent in *Hagenia* woodland at 3350 m, above Naro Moru, on 10 Sept 1961 (specimen now in the National Museum, Nairobi: Lewis 1983e). Turner (1976) heard two owls, thought to have been this species, at 2800 m on the Sirimon track, on 7 Aug 1975.

394 AFRICAN MARSH OWL *Asio capensis*
B409 / MPG532 / D2C2

Afrotropical: with the exception of much of the dry SW, widely distributed in Africa to the S of the Congo basin forests, and further N through the E Zaire borders and scattered areas of East Africa, to Ethiopia and S Sudan. There is a separate population in the W, between Cameroon, Chad and Mali.

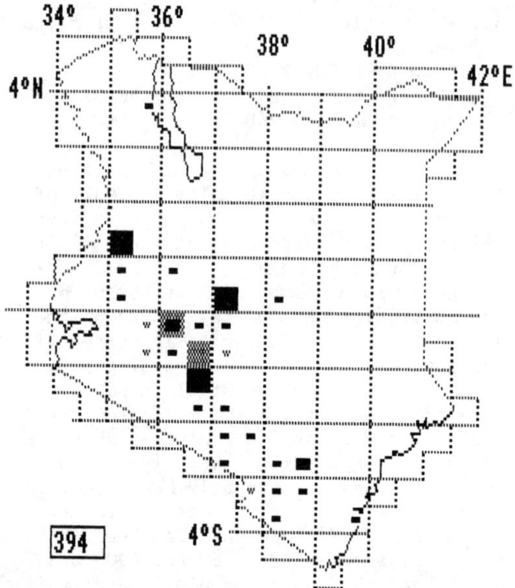

Distribution in Kenya represents the E limit of the range.

Kenyan: a largely gregarious species, locally fairly common, that frequents open marshy and grassland areas, up to 3000 m (Br; MPG; Se). Range is 74% above 1000 m, and this owl is most frequent in the rift valley and adjacent plateaux, especially those to the E (Ja; Br): only 18% of the squares are W of the rift, and there is but a single record from the coastal lowlands (EABR 1981). All squares receive over 250 mm annual rainfall, and this species has occurred near water in arid country at Kalokol 13B, L Turkana (where uncommon: HH), and at Aruba dam 101B, Tsavo E (Br). It is thought to be resident (Br), but Ja also gives examples of local influxes. The single record from 3000 m (at Mau Narok 61D) was in Apr, while that at Bamburi 102D was in May (EABR 1981).

The African Marsh Owl is the ecological equivalent of the Palearctic Short-eared Owl (395), which is a vagrant to Kenya, and the two form a species-group (Sn; Brown 1970a). The Marsh Owl is apparently the less diurnal of the two (Sn), but may hunt on overcast mornings and afternoons (EABR 1982; ADL pers. obs.).

395 SHORT-EARED OWL *Asio flammeus*
B410 / MPG531 / D1C0
NO MAP: post presence 4C 102B; pre presence 36D 80CD-92A

Afrotropical: a winter visitor from the Palearctic in small numbers across the N tropics, from Senegambia to Ethiopia.

Kenyan: this owl's rarity, together with the potential for confusion with the African Marsh Owl (394: *q.v.*), lead to D1. There are four records, which refer to wanderers S of their normal wintering range. A female was collected at Makuiwa, Mt Elgon 36D, on 27 Jan 1927 (Granvik 1934). Of two that came aboard a ship travelling S from the Red Sea, one left on 12 Nov 1936 at 2°S, *i.e.* off the coast 80CD-92A, just N of Lamu 91B (Moreau 1938). On 5 Jan 1971, one flew in from the sea at Mida Creek 102B (Brown 1971a), while another was seen at Koobi Fora 4C, L Turkana, in Dec 1973 (HH).

Family Caprimulgidae: Nightjars

A nocturnal and insectivorous family of approximately 72 species, with representatives in all tropical and temperate areas. 19 (26%) are found in the Afrotropical Region. Of these, 15 (79%) are known in Kenya, where one is a migrant from the Palearctic, at least three are non-breeding, intra-African migrants, and the remainder (73%) breed or are of uncertain status. Particularly now that specimens are infrequently taken, this is one of the most confusing families in Kenya. Few observers are familiar with the birds' calls and, apart from species which exhibit distinctive flight silhouettes, *e.g. Macrodipteryx* spp. and the Long-tailed Nightjar (407), most records mapped here have been identified from corpses (including many road-kills) with the aid of MPG and museum skins. Concentrations of pre-1970 squares are unlikely to signify declines or contractions, and absence of breeding records in some species is most probably due to lack of observation.

396 STAR-SPOTTED NIGHTJAR *Caprimulgus stellatus*
B438 / MPG555 / D1C1

Afrotropical: a little known bird of the NE, from N Kenya N to S Sudan and Ethiopia.

Kenyan: inhabits open, bushed country (Br), with 70% of its squares less arid-semiarid. Br describes it as only at all common in the Dida Galgalla Desert 28A, but recent records suggest that it may be so S to the N edges of the highlands 37C 50BD (Lewis 1984d; EABR 1983). See Donaldson-Smith's Nightjar (404).

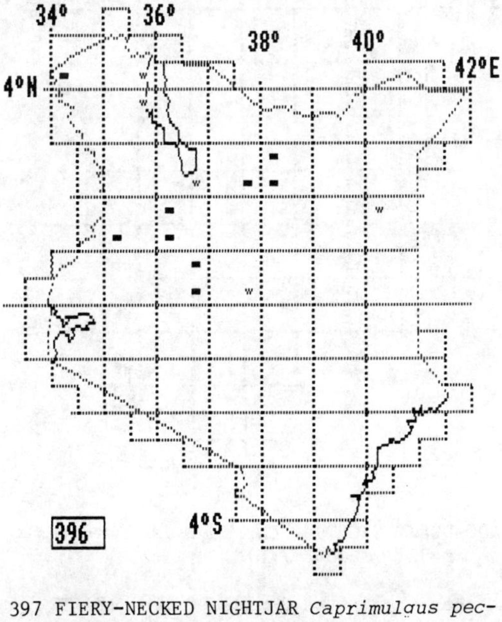

397 FIERY-NECKED NIGHTJAR *Caprimulgus pectoralis*
B436 / MPG551 / D1C1

Afrotropical: apart from the dry SW, widely distributed to the S of the Congo basin forests, extending further NE to E Tanzania and the coastal lowlands of Kenya; and further N through the E Zaire borders, W Tanzania and Uganda to extreme W Kenya and S Sudan. Other, isolated populations occur on the N of the Gulf of Guinea.

Kenyan: this account draws heavily upon that of Britton & Britton (1973c). The Fiery-necked Nightjar was first recorded on 26 Nov 1964, when a breeding condition female was collected in Sokoke Forest 102B (Ripley & Bond 1971): it is thought to be resident (Br), but this remains the only evidence of breeding.

The race *C.p. fervidus* has subsequently been found to be common in Sokoke (Britton & Zimmerman 1979), and to occur in forest, woodland and closed-canopy thicket through-

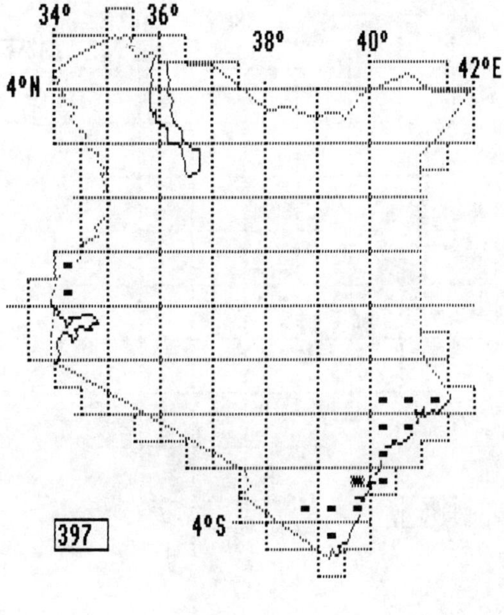

out the coastal lowlands. This race's squares are 75% semiarid-subhumid, and 92% within the 500+ mm rainfall regime.

The race *C.p. nigriscapularis* has been recorded around Malakisi 48A (N Bruce-Miller *in litt.*) and in extensive thickets in Siaya 48C. This race's two squares are above 1000 m, humid and within the 1000+ mm rainfall regime. Br mentions the species up to 1300 m. See Montane Nightjar (398).

398 MONTANE NIGHTJAR *Caprimulgus poliocephalus*
B437 / MPG549, 558, 559 / D2C2

Afrotropical: apart from a population isolated in Angola, a bird of the E and NE from Malawi, N through suitable highland habitats of East Africa and the E Zaire borders, to Ethiopia and S Sudan. Distribution in Kenya represents the E limit of the range.

Kenyan: this species achieves D2C2 because of its distinctive and widely known call, and restricted, highland range. It is locally common along forest edges and in other wooded habitats, including suburban gardens, at 1000-3000 m (Br). Range is 79% subhumid-humid and 91% within the 500+ mm rainfall regime. This nightjar is apparently resident (Ja; Br), and one ringed in Jan 1970 in Nairobi 75B was recovered in Nov 1973 only 4 km away (Backhurst 1974). At the highest altitudes, however, around 3000 m at Mau Narok 61D, Se described it as both resident and partially migrant.

The only montane African nightjar, this species is apparently closely allied to the Fiery-necked Nightjar (397)(Sn): the two are allopatric at QSD scale, and virtually so altitudinally.

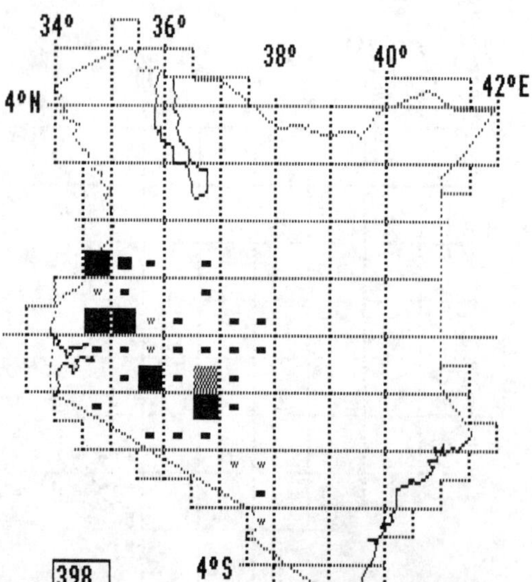

399 NUBIAN NIGHTJAR *Caprimulgus nubicus*
B435 / MPG553 / D1C1

Afrotropical: the NE, from Kenya N to Sudan, Ethiopia and Somalia.

Kenyan: local and uncommon in bushed habitats, mainly at 600-900 m (Br; MPG), although it has reached 1650 m at Nairobi 75B (National Museum Collection, Nairobi). Range is 79% less arid-semiarid, and 79% within intermediate, 250-1000 mm rainfall areas. With the exception of the Nairobi record, distribution in the lowlands of the N and E forms the SW limit of the range. Occurrence in 12D (Br; National Museum Collection, Nairobi), 37A (EABR 1983) and 113B (corpse: J Bartram *in litt.*) most probably means that this species reaches NE areas of Uganda and Tanzania, where it is at present unknown.

There are no breeding records, and this nightjar's status in Kenya is uncertain. It is occasionally recorded at the Ngulia lights 101A (Backhurst & Pearson 1977b; EABR).

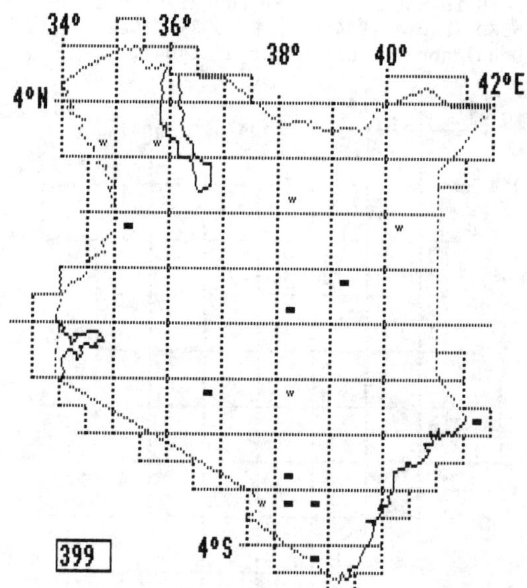

400 DUSKY NIGHTJAR *Caprimulgus fraenatus*
B432 / MPG548 / D1C1

Afrotropical: the E and NE, from the E Zaire borders and S Tanzania NE through Kenya to the far NE of Uganda, S Sudan, Ethiopia and W Somalia.

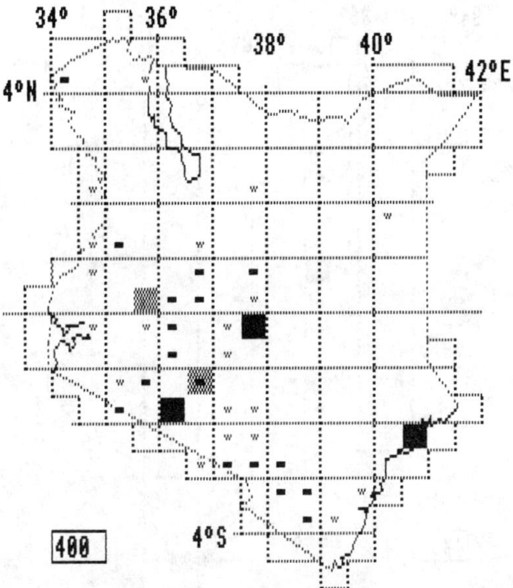

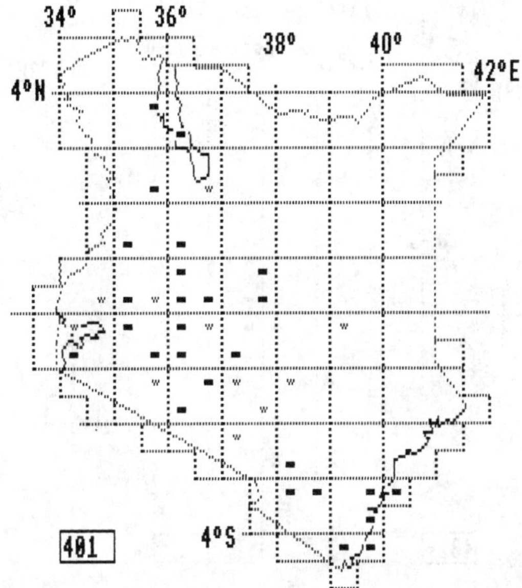

Kenyan: locally common in bush and grassland up to 3200 m (Br). Range is 89% above 500 m, and 72% semiarid-subhumid. These parameters imply virtual absence from the N, E and coastal lowlands and, while this may at least in part reflect observer frequency and a D1 species, it is similarly scarce further E, in S Somalia (AM). 71% of the squares are in or E of the rift valley, and the only records from Uganda are in 24A (not mapped)(Ash 1985c). The high proportion of squares with only pre-1970 records (51% of its total) most probably reflects its D1 characteristics.

The Dusky Nightjar makes considerable movements at least locally (Ja), and birds at the Ngulia lights 101A, where it is scarce in Nov but common in Dec, are presumably migrants (Backhurst & Pearson 1977b; D J Pearson *in litt.*; Br; EABR). This species may be closely allied with the Eurasian Nightjar (401)(Sn).

401 EURASIAN NIGHTJAR *Caprimulgus europaeus*
B430 / MPG547 / D2C2

Afrotropical: the main wintering areas of this Palearctic species appear to be in the S or SE. It is thus a passage migrant in the NE and E, but status in the W remains obscure.

Kenyan: a fairly common passage migrant late Oct-Nov and late Mar-Apr, in bushed and wooded country up to 3000 m (Br; EABR). D2 reflects its large size and pale coloration. Range is 75% within 500+ mm rainfall areas and 73% semiarid-subhumid. While the question of observer frequency must arise, virtual absence from the arid NE and E parallels status in S Somalia (AM): this may represent overflying of arid country while on migration, or migration through NW Kenya.

This nightjar is often common at the Ngulia lights 101A in Nov (Backhurst & Pearson 1977a; EABR). Most occurrences in Tsavo E 101B are in Nov (La), and seven of ten records from Somalia have also been during the S migration season. Four individuals handled at Mau Narok 61D, at around 3000 m, were Mar-early Apr however (Britton 1979). See Dusky Nightjar (400).

402 PLAIN NIGHTJAR *Caprimulgus inornatus*
B433 / MPG556 / D1C1

Afrotropical: an intra-African migrant, found in the N tropics at 5-18°N, from the W coast to Ethiopia and N Somalia; and SE to the E Zaire borders, Uganda, Kenya and E Tanzania. Distribution in Kenya represents the E limit of the range.

Kenyan: rather uncommon in bushed habitats up to 1800 m (Br; Ja). Distribution is 79% at 500-1500 m, and virtual restriction to medium altitudes W of 39°E represents the E edge of the range. 92% of the squares are within the 500+ mm rainfall regime.

The Plain Nightjar is mainly an Oct-Apr non-breeding migrant to Kenya (Br; La; EABR). The few breeding records are from the NW and W, at Turkana (Br: precise location uncertain), Mt Elgon 36D (Ja), and

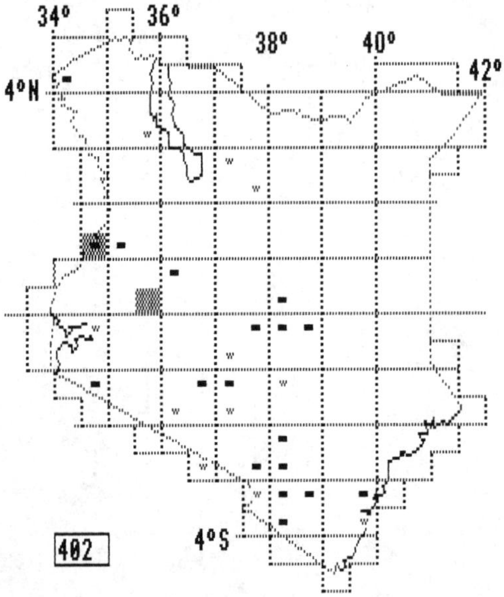

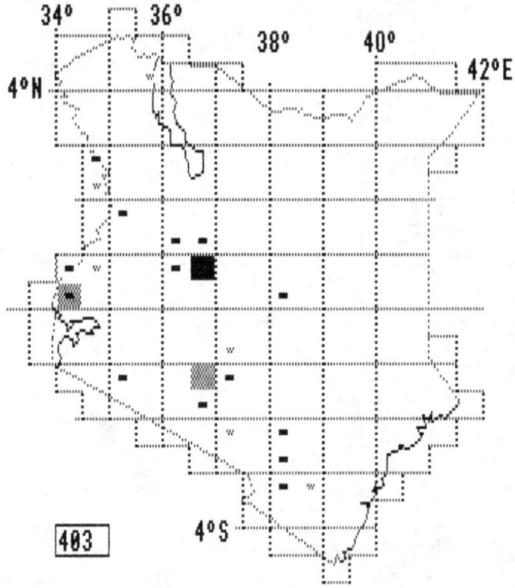

Eldama Ravine 49D (MPG): their May-Aug seasonality compares with that of Mar-Sept breeding immigrants in NW Somalia (AM). Birds at the Ngulia lights 101A Nov-Dec, where it can be common, are presumed to be migrants (Backhurst & Pearson 1977b; EABR), while a very fat Apr individual at Olorgesaille 75C was presumably on a long-distance movement (Br).

This species is possibly allied to the Dusky (400) and Eurasian (401) Nightjars (Sn).

403 FRECKLED NIGHTJAR *Caprimulgus tristigma*
B439 / MPG554 / D1C1

Afrotropical: widespread S of 15°S, and N through S Zaire and Zambia to the E Zaire borders, East Africa, S Sudan and Ethiopia. It is also found in the W, from Guinea to the Central African Republic. Distribution in Kenya represents the E limit of the range.

Kenyan: local and uncommon in rocky habitats, at 600-2000 m (Br; Ja; MPG). Range is 76% semiarid-subhumid and virtually absent from the arid N and E, and restriction to the peripheries of the W and C highlands, W of 38°30'E, represents its E edge. This nightjar is considered resident (Br; DEP pers. obs.: 89A).

404 DONALDSON-SMITH'S NIGHTJAR *Caprimulgus donaldsoni*
B429 / MPG552 / D2C2

Afrotropical: a NE species, from NE Tanzania through mainly E areas of Kenya to S Sudan, S Ethiopia and Somalia.

Kenyan: locally common in bush country up to 1050 m, and perhaps the commonest nightjar in the Tsavo region 89CD 101 and

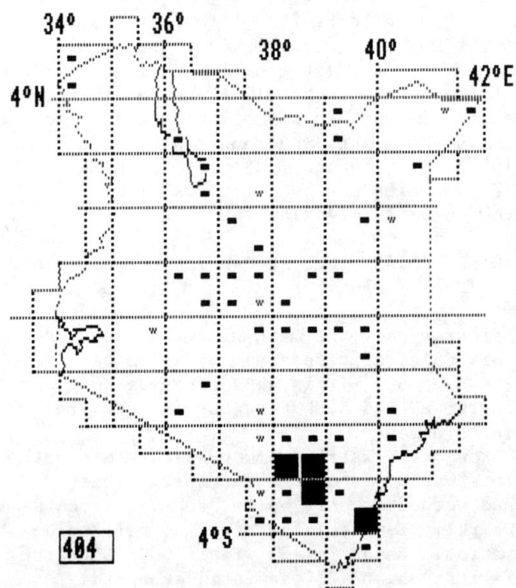

the NE (Backhurst & Pearson 1977b; D J Pearson *in litt.*; Br; MPG; La; EABR). D2C2 reflects its abundance and distinctively small size. Range is 71% within intermediate, 250-1000 mm rainfall areas and 76% less arid-semiarid, but this nightjar is sparse in even semiarid areas of the coastal strip. The cluster of breeding squares probably represents observer frequency, since the species is thought to be resident (Br). However, it is occasional during Sept-Oct each year at Baringo 50A (St). Possible allopatry with the Star-spotted Nightjar (396)(Sn) is not seen at QSD scale.

405 WHITE-TAILED NIGHTJAR *Caprimulgus natalensis*
B434 / MPG557 / D1C1

Afrotropical: principally W and C Africa, around the Gulf of Guinea from Liberia to Congo, E to Zambia, the E Zaire borders, Uganda and, on the E limit of the range, W Kenya. There are also outlying populations in Mali, Chad, S Sudan and E South Africa.

Kenyan: this locally common nightjar frequents grassland, particularly wet areas, and woodland and forest edge, up to 2200 m (Br; MPG). The eight squares are all above 1000 m in 500+ mm rainfall country, and seven of eight are subhumid-humid. Apart from semiarid Baringo 50A, where small numbers appeared in Sept-Nov 1978 (St; EABR 1978), range is W of the rift valley, and mostly N of the Winam Gulf 60AB. The single breeding record is from grass plains N of Chepterr 37A, in Apr-June 1976 (T J Barnley *in litt.*). The species is thought to be resident (Br), however, although Ja mentions local movements (of which the Baringo influx is presumably an example).

406 SLENDER-TAILED NIGHTJAR *Caprimulgus clarus*
B427 / MPG565 / D2C2

Afrotropical: the NE, from Uganda, N Tanzania and Kenya to S Sudan, Ethiopia and Somalia.

Kenyan: the taxonomy of this species and the Long-tailed (407) and Gabon (408: *q.v.*) Nightjars is complex, and has been interpreted in various ways (see Sn, Br). As with all other families, we follow Br's views throughout, such that the Slender-tailed is given full species status. Other authors have maintained otherwise, however, and this may well have led to confusion of some of the records collected by this atlas. In order to minimise these errors, two precautions have been taken. Only records of full-plumage adult male Long-tailed Nightjars, with their uniquely elongated central tail feathers, have been accepted: this species is in any case rare in Kenya. Secondly, the N limit of the more common Gabon Nightjar has been taken as that described by Br: this may in time prove to have been an incorrect procedure but, at the present state of knowledge, we feel it to be the most certain option.

The Slender-tailed Nightjar is a common to locally abundant bird of bushed habitats

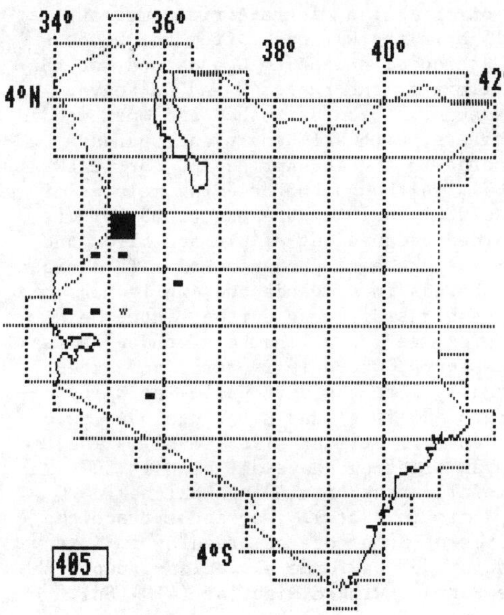

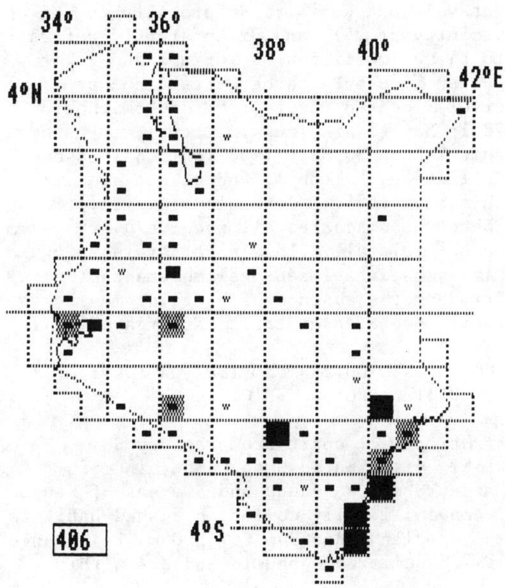

(Br; MPG; La; St). D2C2 reflects this abundance, and the fact that its slightly or moderately elongated central tail feathers are a good fieldmark. Br mentions it up to 2000 m and it is present in 2000+ m squares, but it is 80% below 1500 m and absent from much of the W and C highlands. Apart from L Turkana, where it is fairly common (HH), this nightjar is virtually unrecorded in the driest, 0-250 mm rainfall areas of the N and E: whether this absence is real or a product of low observer frequency is uncertain, but the species is widespread further E, in S Somalia (AM).

This nightjar appears at the Ngulia lights 101A, but is present throughout the year at Baringo 50A, Samburu 51B, Meru 52C and Magadi 75C (and possibly along the coast), and may be generally resident (Backhurst & Pearson 1977b; St; D J Pearson *in litt.*; DEP pers. obs.; EABR 1979, 1980, 1984).

407 LONG-TAILED NIGHTJAR *Caprimulgus climacurus*
B428 / MPG565 / D2C0
NO MAP: post presence 2C 37C; pre presence 25A

Afrotropical: the N tropics at 5-15°N, from the W coast to Sudan, N Uganda and, on the SE limit of the range, NW Kenya.

Kenyan: a scarce, possibly rare species that is at or beyond the SE limits of its normal range in bushed and wooded country of the NW. D2 reflects the male's greatly elongated central tail feathers.

Ja mentions this nightjar along the Turkwell and Kozibiri R (probably in the vicinity of 25), and Br apparently refers to these localities as Turkana and Pokot. We can find only three certain and precisely located records: NW of Lokichokio 2C in May (G Nikolaus *in litt.*), at Lorugumu 25A on 30 Sept 1958 (National Museum Collection, Nairobi), and at Kongelai 37C on 6 Sept 1975 (T J Barnley *in litt.*). These three squares all receive over 500 mm annual rainfall. This nightjar is known to make substantial seasonal movements (Ja; Br), and the absence of breeding may be real. See Slender-tailed Nightjar (406).

408 GABON NIGHTJAR *Caprimulgus fossii*
B431 / MPG560 / D1C1

Afrotropical: apart from the dry SW, widely distributed N to Gabon, interior Zaire, S and W Uganda and S areas of Kenya.

Kenyan: locally common in bushed habitats and woodland edge, up to 1800 m (Br). Range is 77% semiarid-subhumid and 82% within

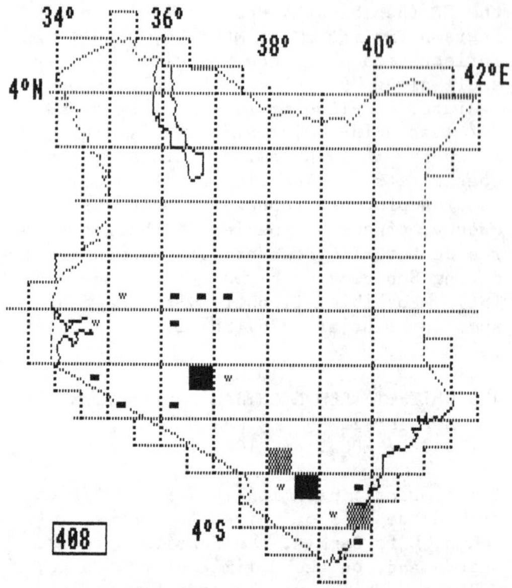

500+ mm rainfall areas. Br notes that there is no evidence for the calling of this species and the Slender-tailed Nightjar (406: *q.v.*) alongside each other, and that apparent sympatry could be the result of non-breeding migrants of this species from the S, penetrating the Slender-tailed's breeding range.

409 STANDARD-WINGED NIGHTJAR *Macrodipteryx longipennis*
B440 / MPG563 / D2C1

Afrotropical: an intra-African migrant, at 5-15°N in the N tropics from the W coast to Ethiopia, and SE to Uganda and, at the SE limit of the range, W and NW Kenya.

Kenyan: a bird of bushed and open wooded habitats, with all Kenyan and Ugandan records in the Aug-Apr period (Br; EABR 1983). Although the breeding male is of absolutely unique appearance, D3 is not awarded because out-of-plumage birds and females are easily overlooked. This nightjar breeds in N Uganda and, while most of these birds disperse further into the N tropics, very small numbers wander as far E as Kenya, where it is rare, and exceptionally E to the rift valley at Nginyang-Kapedo 38C (Aug) and L Baringo 50A (twice in Jan)(Br; St; EABR 1983, 1984). The few Kenyan sightings have all been in 500+ mm rainfall areas, including Lokichokio 2C (T Romig *in litt.*). We cannot trace the source of Sn's note of breeding in N Kenya.

This species forms a species-group with the Pennant-winged Nightjar (410)(Sn).

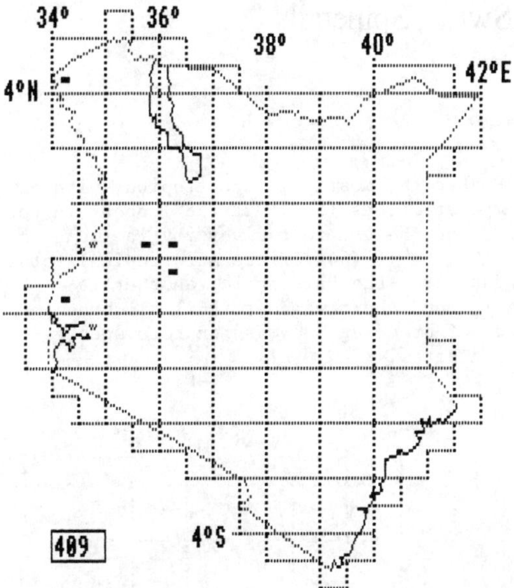

during Aug-Mar, and then migrates NW, when it appears in bushed and wooded habitats, Mar-Oct, in Nyanza 60 and the peripheral highlands (Br; Ja; J Hayes *in litt.*; EABR 1984). Passage birds wander further E, exceptionally to the rift valley (Kapedo 38C Sept, Makutani 50A Aug: EABR 1983; L Naivasha 62C Oct (twice): J Hayes *in litt.*) and the E plateau (Mugie 50B: P C Faull *in litt.*; Laikipia 50D Jul: L V J Grant *in litt.*; Nanyuki 51C: Br; Mweiga 62B: I W Hardy *in litt.*; Nairobi 75B Aug: Ja). A single Aug vagrant has also reached E to S Somalia (AM).

This species' squares are all above 1000 m in 500+ mm rainfall areas, mainly on the peripheries of the W and C highlands. It is given D2 for the same reasons as the Standard-winged Nightjar (409: *q.v.*).

410 PENNANT-WINGED NIGHTJAR *Macrodipteryx vexillarius*
B441 / MPG564 / D2C1

Afrotropical: this intra-African migrant is widespread S of the equator, excepting only the dry SW. Further N, it extends through Gabon, Zaire and Uganda to Ghana, Chad and S Sudan. Distribution in Kenya represents the E limit of the range.

Kenyan: this nightjar breeds S of 6°S

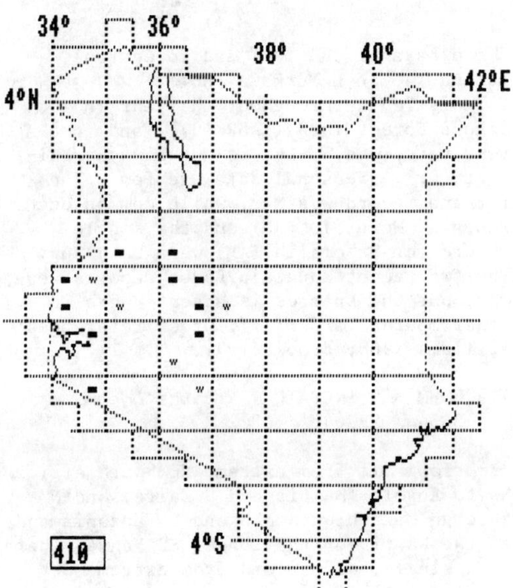

Family Apodidae: Swifts, Spinetails

An insectivorous and highly aerial family of 79 species, mostly in the tropics, with a few migratory members breeding in Holarctic temperate areas (Sn). 22 (28%) occur in the Afrotropical Region, where 18 (82%) are endemic, and the greatest diversity is found in equatorial regions from E Zaire to Kenya (Sn). Of the 14 (64% of the Afrotropical total) that occur in Kenya, one is a non-breeding migrant from the Palearctic, another is thought to be a non-breeding migrant from Somalia, and the remainder (86%) all breed, have bred or probably do so. Paucity or absence of breeding records for some species reflects their inaccessible and often remote nest sites.

411 SABINE'S SPINETAIL *Rhaphidura sabini*
B456 / MPG649 / D2C2
NO MAP: post presence 36D 48D

Afrotropical: a bird of W and C Africa, found in the equatorial forests of the W, E to Uganda and, on the E tip of the range, the far W of Kenya.

Kenyan: this forest species was first recorded in 1959 at Kakamega Forest 48D (Williams 1960b), and has subsequently been found to be not uncommon there in small numbers (Mann 1976; ADL pers. obs.). There is also a Jul record from Mt Elgon 36D (EABR 1977). Both squares are humid, with over 1000 mm rainfall. C2 is given in view of its distinctive appearance and this very restricted range. There are no breeding records, but specimens with enlarged gonads have been taken at Kakamega (Brown & Britton 1980).

412 MOTTLE-THROATED SPINETAIL *Telacanthura ussheri*
B457 / MPG647 / D2C2

Afrotropical: from the W coast between Senegambia and Angola, E through the Congo basin forests to S and W Uganda, mainly E areas of Tanzania and, on the NE edge of the range, Kenya. There is a single old record from extreme S Somalia (AM). It also occurs in the SE, in Zimbabwe and Mozambique.

Kenyan: a rather local or uncommon bird of dry woodland and forest edge (Br; Ja; EABR 1977). Br records it up to 1800 m and there are records from Mountain Lodge at 2200 m on Mt Kenya 63A (EABR 1981, 1982), but 74% of the squares are below 1000 m and it is essentially a lowland species. Range is 78% semiarid-subhumid. This spinetail is found at the coast, along the lower Tana R 79C 91A (AGH), and along the Galana-Athi and upper Tana R systems N around the E periphery of the C highlands to Meru 51D and Mt Kenya 63A (Br; La)(*cf.* Nicator (633)). Single individuals have wandered further

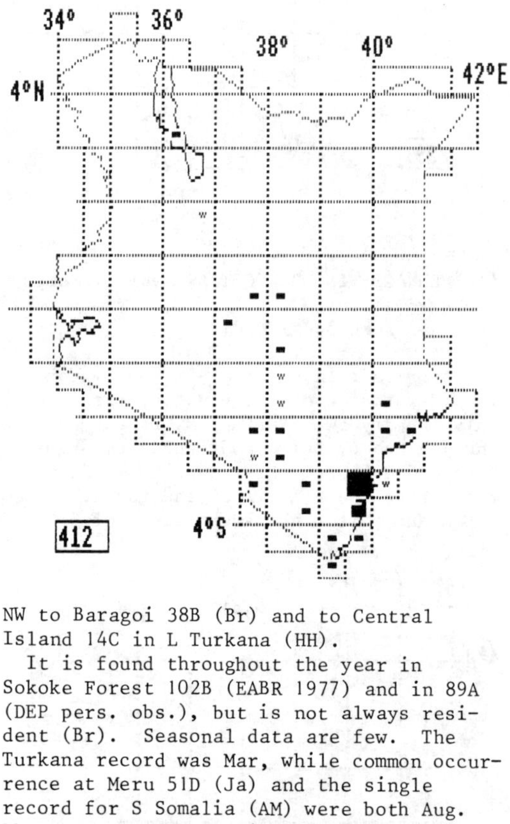

NW to Baragoi 38B (Br) and to Central Island 14C in L Turkana (HH).

It is found throughout the year in Sokoke Forest 102B (EABR 1977) and in 89A (DEP pers. obs.), but is not always resident (Br). Seasonal data are few. The Turkana record was Mar, while common occurrence at Meru 51D (Ja) and the single record for S Somalia (AM) were both Aug. The two recent Mountain Lodge reports were Oct, but the species is absent there in other months (*op. cit.*). Other references: Williams 1968; Brooke 1971.

413 BOHM'S SPINETAIL *Neafrapus boehmi*
B454 / MPG648 / D2C2

Afrotropical: from extreme NE South Africa, NW to Angola, Zambia and S Zaire; and N through Mozambique to S and E Tanzania and, at the NE tip of the range, SE Kenya. There is a single, old record from extreme S Somalia (AM).

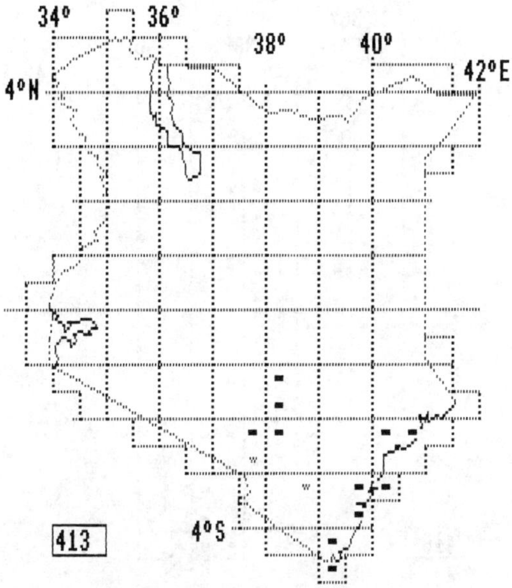

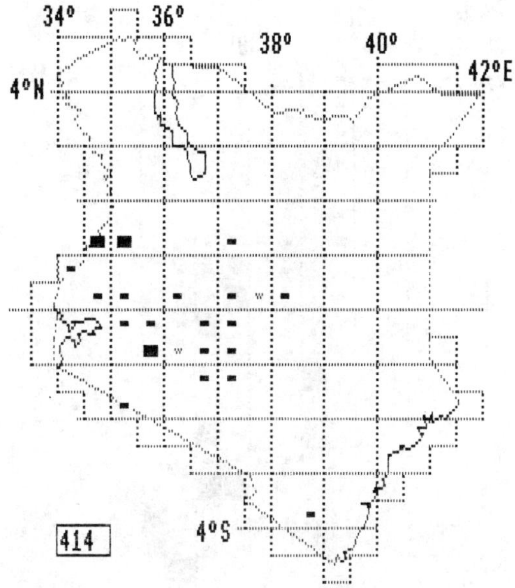

Kenyan: a bird of forest strips and woodland (Br), primarily in the coastal lowlands, where at least locally more common than the Mottle-throated Spinetail (412) (EABR 1977). Bohm's is also found along the Galana-Athi R to the SE edge of the C highlands, but here is less common than the Mottle-throated, being very local away from the coastal strip although not uncommon at Bushwackers' 89A (Br; La; DEP pers. obs.). Br mentions occurrence on the upper Tana R (possibly 64 or 65), but omits actual localities. Range is 83% below 1000 m and 92% semiarid-dry subhumid.

The single breeding record, of young in an underground chamber in Sokoke Forest 102B, is treated with caution by Brown & Britton (1980) and is not mapped here. Short & Horne (1985a) suggest that hollow trees are more likely nest sites. There are few seasonal data, but it is present throughout the year at Sokoke Forest 102B (EABR 1977) and in 89A (DEP pers. obs.). The record from S Somalia was in Aug (AM).

414 SCARCE SWIFT *Schoutedenapus myoptilus*
B453 / MPG639 / D2C2

Afrotropical: little known, principally in highland areas of the E Zaire borders, N Tanzania, Kenya, SW and NE Uganda and SE Sudan. Other, scattered records occur as far afield as Angola, the Gulf of Guinea, Zimbabwe, Mozambique and Ethiopia. Distribution in Kenya represents the E limit of the range.

Kenyan: local but not uncommon in the W and C highlands, nesting on rocky crags and foraging widely at lower altitudes over forest (Br; EABR). Frequency increases with altitude, with records in most of the 2000+ m squares, and the species has been observed at 3600 m on the Mt Kenya 63A moorlands (EABR 1982). 95% of the squares receive over 500 mm rainfall. Two Jan individuals at 600 m around the base of the Mt Kasigau forest island 101D (EABR 1981) were probably related to the species' presence on the Usambara Mts of adjacent NE Tanzania (EABR 1980, 1981).

As with other members of this family that nest on remote or inaccessible cliffs, there are no confirmed breeding records, but the three probable records are all above 1500 m in 1000+ mm rainfall areas. Williams' (1968) suggestion of breeding on the cliffs of Mt Kenya 63A may be correct, but direct evidence is lacking. Other references: Mann 1976; Dowsett & Dowsett-Lemaire 1978.

415 PALM SWIFT *Cypsiurus parvus*
B452 / MPG646 / D2C2

Afrotropical: widely distributed S of 15°N, except for N Somalia and the SW.

Kenyan: the range of this swift is principally governed by the distribution of tall palms, including exotic species (Sn), in which it roosts and nests (Br; MPG; Irvine & Irvine 1980b). Thus it is primarily a bird of lower altitudes, mostly reaching only the peripheries of the highlands, and with 91% of its squares below

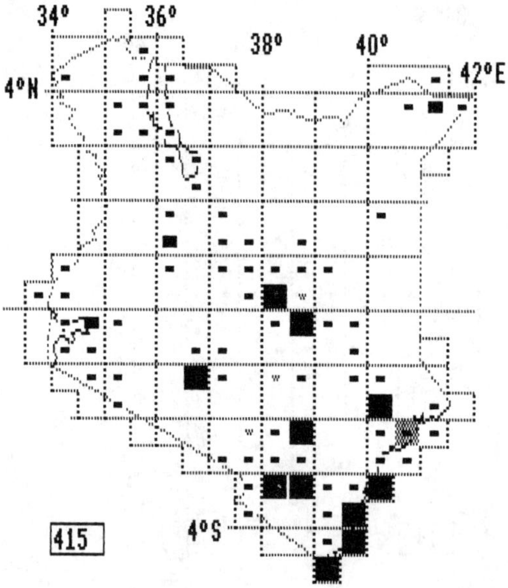

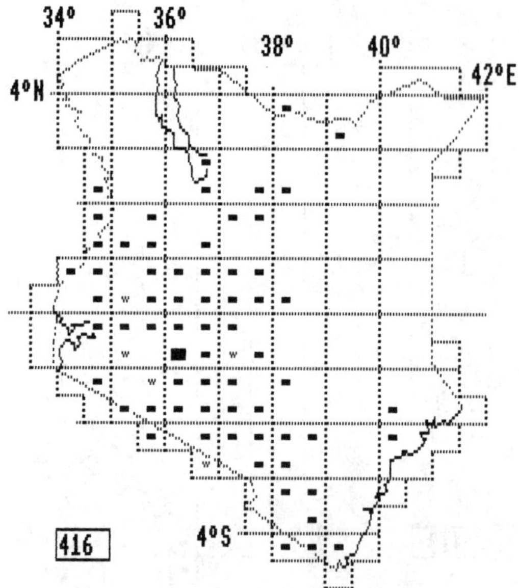

1500 m. Br describes it as mostly below 1400 m, but locally resident up to 1800 m. In arid country, this species is confined to the vicinity of the permanent water necessary for the palms' existence *e.g.* at L Turkana (HH), the Daua R 7C 19AB, water-holes at Wajir 42A, and the Tana R 64B-79C. This swift is mainly resident (Br; La: 101B; DEP pers. obs.: 88D 89C 100B 101A), but Br mentions wanderers up to 2000 m. St lists a single, Feb visitor to L Baringo 50A.

416 MOTTLED SWIFT *Apus aequatorialis*
B442 / MPG641 / D2C2

Afrotropical: from Zimbabwe N through Mozambique and Zambia to the E Zaire borders, S and E Uganda, E Tanzania, Kenya and Ethiopia. Scattered occurrences also extend W to Angola, Sudan and Guinea. Distribution in Kenya represents the E limit of the range.

Kenyan: roosts and apparently nests on rocky crags and cliffs in the highlands, up to 3000 m, but also disperses widely over lower, open country, particularly to wetland feeding areas (Br; MPG; DEP pers. obs.). 89% of its squares are above 500 m, and virtual absence from the E lowlands reflects the E limit of the range. It is rare on the W edge of these lowlands in Tsavo E 101B (La), but was recorded without comment from the lower Tana R 79C 91A by AGH: this latter record initially appeared dubious, but is probably correct in view of the lack of potential confusion species, the Jul-Aug dates, and this swift's affinity for lower altitude wetlands. Britton *et al.*(1980) omit these Tana records in their summary of the species in E Kenya, and suggest that two Jan birds at Mrima Hill (114A: altitude 300 m) were wanderers from the Usambara Mts of adjacent NE Tanzania. Several Feb birds at 500 m on Kuranze Ranch 113B were most probably also from these mountains, or the adjacent Pare Mts or Mt Kasigau 101D (M A C Coverdale, ADL pers. obs.).

This swift's frequency increases with rainfall, only 2% of its squares being in 0-250 mm areas, while 69% receive 500+ mm: the sole exception is the high rainfall L Victoria basin 60, where it is apparently scarce. Occurrence N of 2°N is virtually restricted to the vicinity of the highland islands of Mts Kulal 26B (Diamond & Keith 1980), Nyiru 26D and Marsabit 27D 28C (Br); records from 16A (J S Ash *in litt.*) and Moyale 17C (ADL pers. obs.) presumably represent the S edge of the Ethiopian highlands population.

The Mottled Swift is thought to breed in the crater of Longonot volcano 62C (Gurney 1909), on the cliffs of the Elgeyo scarp (approximately 37D 49B (not mapped): Ja), and in large numbers on the cliffs of the Njorowa Gorge 62C (Ja; Brown & Britton 1980), but all of these records apparently refer to birds visiting possible nest crevices, and there are no confirmed records of even used eggshells.

Kenya is one of the very few areas of Africa where both this species and the related Alpine Swift (417) breed. Their breeding areas are considered to be allopatric, with this species nesting at moderate altitudes and the Alpine Swift higher up (Sn): the few apparent breeding indications for the Mottled Swift, at approximately 2000-2800 m, are significantly lower than the two similar records for the Alpine Swift, at 3650 m on Mt Kenya 63A (Hale 1958), and at 3500 m on the Aberdare Mts 62B or D (not mapped: Brown & Britton 1980).

Apart from flights to feeding areas, and wanderers as mentioned above, this species is thought to be sedentary (Br). A Jan bird that entered Ngulia Lodge 101A during heavy night rain was probably not a migrant, since the species frequents the area during the day (Backhurst & Pearson 1977b), but a flock of 2500 moving S at L Baringo 50A on 3 Nov was exceptional (St). Other references: Bednall 1958, 1959f.

417 ALPINE SWIFT *Apus melba*
B449 / MPG640 / D3C2

Afrotropical: occupies two distinct areas, between which there may be regular migration (Sn). In the S, it occurs from Angola and E South Africa S to the Cape. In the NE, it extends from the E Zaire borders and N Tanzania N through Kenya and Uganda to S Sudan, Ethiopia and N Somalia. Palearctic migrants are thought to penetrate this range S to Uganda, but not to Kenya or Tanzania (Br). Distribution in Kenya represents the E limit of the range.

Kenyan: like the Mottled Swift (416: *q.v.*), this species apparently breeds on rocky crags in the highlands, but there are no completely certain records (Hale 1958;

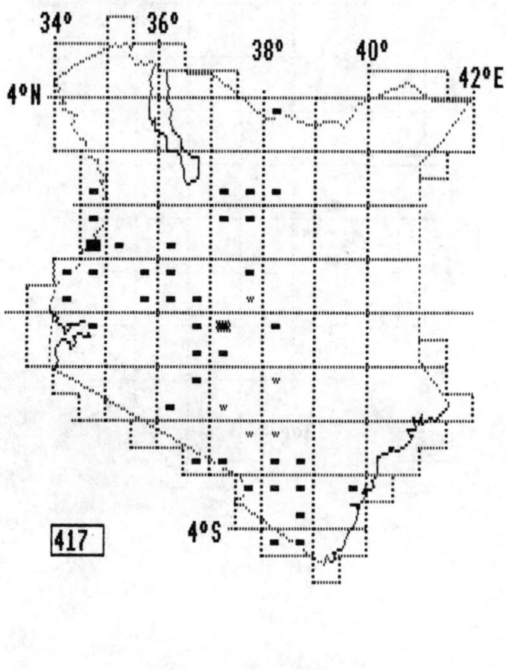

Brown & Britton 1980; Br). If this species' breeding areas are separated from those of the Mottled by greater altitude (Sn), then Br's suggestion of probable breeding in the Ngulia area 101A at 900 m is unlikely.

Br describes the Alpine Swift as typically above 1800 m, and its squares are 90% above 500 m and only 20% within arid areas. Hence virtual absence from the E lowlands, which are beyond the E limit of the range; and restriction N of the highlands to 16A (J S Ash *in litt.*), which represents the S extremity of the Ethiopian highlands population, and to wanderers from the Ndoto Mts 39A and Mt Marsabit 27D 28C (Lewis 1981e; EABR 1979, 1980). Only 18% of the squares are to the W of the rift valley, where the species is apparently centred on Mt Elgon 36D and the Cherangani Hills 37C.

Because of their relative remoteness from the highlands, small numbers of Dec-Mar birds over Sokoke Forest 102B were initially believed to be of Palearctic origin: but this species undertakes daily foraging flights of up to 100 km and these birds were more probably wandering residents (Br; Britton & Britton 1976b; EABR 1980). This seasonality compares with the predominance of Jan records in adjacent Tsavo E 101B (La), and with fairly common sightings in Feb at 500 m altitude on Kuranze Ranch 113B (M A C Coverdale, ADL pers. obs.).

The only evidence of movements, other than daily feeding flights or wanderers, are flocks of up to 180 at L Baringo 50A, which have been observed moving S in Apr, Jul-Aug and Oct (St).

418 EURASIAN SWIFT *Apus apus*
B444 / MPG636 / D1C1

Afrotropical: a Palearctic species that winters from approximately 10°N in the W and from the equator in the E, S to Namibia and E South Africa. Distribution in Kenya apparently represents the E limit of the range.

Kenyan: the Eurasian Swift is closely related to the Black Swift (419), which forms part of a superspecies with Forbes-Watson's Swift (420); this superspecies, in turn, forms part of a species-group with the Nyanza Swift (421)(Sn). These all dark swifts are readily confused in the field (*e.g.* La, St) and are thus all awarded D1 or D0: identification becomes far easier if they can be viewed from above, *e.g.* from a clifftop.

The status and movements of the Eurasian Swift are still largely unknown (EABR 1983), but the following points can be made. It

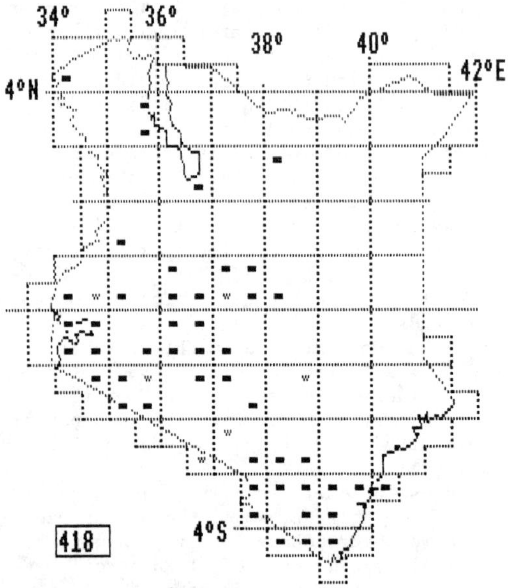

occurs Aug-Apr over open country, particularly where wetlands provide abundant aerial insect prey. Whereas Br quotes it at 500-2000 m, it has been recorded from most of the 2000+ m squares and Se lists possible records from 3000 m. There are recent records, including a count of 3000+, from the SE lowlands (EABR 1980, 1983), which may be its main centre in Kenya (D J Pearson *in litt.*).

Range is 73% within 500+ mm rainfall areas and only 14% arid, and hence virtually absent from the dry N and E. This swift is occasional near L Turkana (HH), and presumably overflies the NW on passage but, as in the rest of this family, arid habitats away from lakes or permanent rivers presumably do not provide sufficient insect prey, or the opportunity to drink whilst in flight.

There may be variations in status across the country. Most birds in the W are on Sept-Nov passage, although some winter (Br; Britton & Britton 1976b; EABR 1983, 1984). Further E, in the highlands, Br notes mainly Apr passage. While admitting their identification to be tentative, Se records heavy Apr passage in 61D, La note particular Dec-Feb occurrence in Tsavo E 101B, and St contrasts large Oct-Dec flocks with small Mar ones. Storms and rain tend to bring swifts nearer to the ground, and St quotes a striking record of hundreds of thousands, thought to be this species, moving S over L Baringo 50A after rain on 3 Nov.

419 BLACK SWIFT *Apus barbatus*
B445 / MPG636 / D1C1

Afrotropical: primarily the E, from the Cape and E South Africa through Zimbabwe and Malawi to the E Zaire borders, E Tanzania, NE Uganda and Kenya. Status in the W is unclear. Distribution in Kenya represents the NE limit of the range.

Kenyan: this species and Forbes-Watson's Swift (420) are certainly the least frequently recorded of the all-dark swifts (see Eurasian Swift (418)).

The Black Swift is a local bird of the highlands (most common in the W), where it is widely sympatric with the smaller Nyanza Swift (421)(Br; D J Pearson *in litt.*). In the S, it is also found around the isolated highlands of the Taita Hills 101A, and the Pare and Usambara Mts of adjacent NE Tanzania (Br; La). Range is 85% above 1000 m, and 90% within 500+ mm rainfall areas.

This swift nests in old cedar trees *Juniperus procera* around 2400 m at Molo 61B, but breeding is more typically in rocky crags at 1600-2000 m (Br). As with the Mottled Swift (416) however, these latter breeding records only relate to birds entering presumed nest sites in inaccessible cliffs, *e.g.* at the Njorowa Gorge 62C (where it is common), and at Thomson's Falls 50C (D A Turner pers. comm.). Other reference: Brooke 1969.

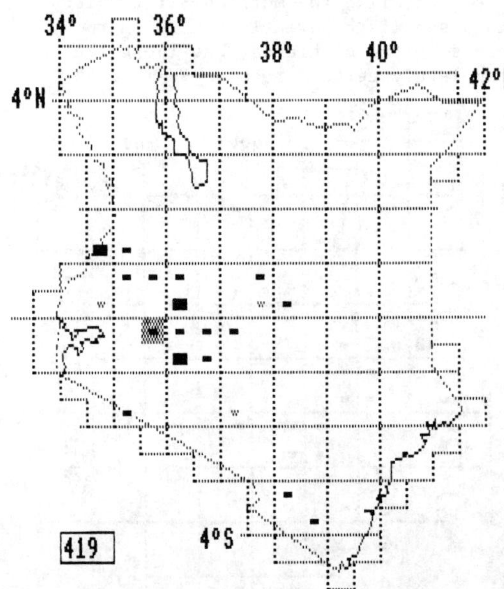

420 FORBES-WATSON'S SWIFT *Apus berliozi*
B446 / MPG- /D0C0
NO MAP: post presence 102BD 114AB

Afrotropical: a little known swift that breeds in E and probably N Somalia (AM), and is partially migratory as a non-breeding visitor S to the coastal lowlands of Kenya (Brooke 1972).

Kenyan: an apparently uncommon, but little known and quite possibly overlooked winter visitor, late Nov-Mar, to the S coastal lowlands (Br; Britton & Zimmerman 1979; EABR). The four squares are entirely within semiarid-dry subhumid, 500+ mm rainfall areas. See Eurasian Swift (418). Other reference: Brooke 1969.

421 NYANZA SWIFT *Apus niansae*
B450 /MPG637 / D1C1

Afrotropical: a NE species, from NE Zaire and N Tanzania through Kenya and NE Uganda to Ethiopia and N Somalia. Distribution in Kenya represents the E limit of the range.

Kenyan: the highlands and peripheral plateaux at 400-2800 m (Br). Range is 70% within 500+ mm rainfall areas and only 5% arid, so that it is absent from the dry N and E. The resident nominate race breeds on rocky crags and buildings (Schwab 1979), and particularly abundantly in the Njorowa Gorge 62C (Brown & Britton 1980). The four breeding squares are all above 1500 m, in 500+ mm rainfall areas. The race *A.n. somalicus* is a sporadic, Nov-Mar, non-breeding visitor from N Somalia, S to Nanyuki 51C (Br).

La list four certain Dec-Jan records of the Nyanza Swift at the lower limit of its altitudinal range in arid Tsavo E 101B. St tentatively states most records at L Baringo 50A to be June-Jul or after rain, the latter presumably reflecting birds descending from the neighbouring highlands (*cf*. Eurasian Swift (418)).

422 WHITE-RUMPED SWIFT *Apus caffer*
B447 / MPG644 / D2C2

Afrotropical: widely distributed S of Senegambia, Chad and Ethiopia, avoiding only the drier areas of E Somalia and the SW, the equatorial lowland forest areas of the W, and possibly much of the S interior.

Kenyan: the White-rumped Swift forms a species-group with the sympatric Horus (423) and Little (424) Swifts (Sn) but, unlike those congeners, mostly breeds in old nests of swallows *Hirundo* spp. and inhabits range largely governed by their presence (Br). Sn states preference for the nests of the Red-rumped Swallow (556) and, with the exception of the coastal lowlands, the two species' breeding ranges overlap extensively. Utilisation of the nests of the Striped Swallow (559) has been reported from Baringo 50A (where the Red-rumped Swallow is very uncommon: St), and from Mombasa 114B (Ja: where the Red-rumped is absent).

Br mentions this swift up to 3000 m, and its squares are 74% within 500+ mm rainfall areas and only 13% arid. Hence virtual absence from the arid N and E, where the

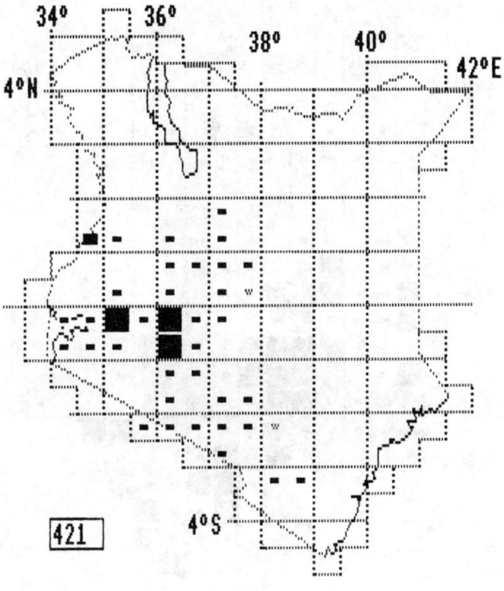

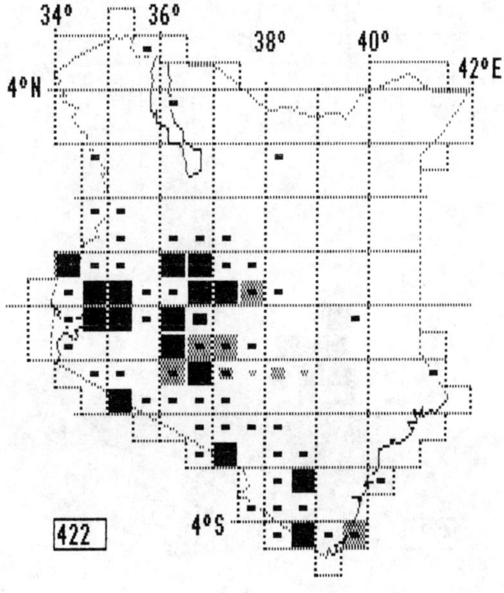

few records presumably represent wanderers, those in the far N probably from the Ethiopian highlands. This scarcity in the N and E matches status in S Somalia (AM), and clearly contrasts with the Little Swift at L Turkana and the coast.

Like the Little Swift, the White-rumped has benefitted from the spread of buildings in that it uses the swallows' nests that have spread to them (Sn). Apart from in towns, where the Little Swift is usually paramount, this is often the commonest breeding swift (Br). MPG mention local movements, and the White-rumped leaves the semiarid Baringo area 50A during very dry periods (St). Other reference: van Someren 1944.

423 HORUS SWIFT *Apus horus*
B448 / MPG645 / D2C2

Afrotropical: apparently mainly in the E, from Ethiopia and S Sudan through the E Zaire borders and East Africa to Zambia, Zimbabwe and E South Africa. There are also scattered records W to Angola, Congo and Chad. Distribution in Kenya represents the E limit of the range.

Kenyan: unlike the similar White-rumped (422: *q.v.*) and Little (424) Swifts, this is a bird of the highlands, with 92% of its squares and all breeding squares above 1000 m. Range (including eight of the nine breeding squares) is 81% within 500+ mm rainfall areas.

The Horus Swift is particularly common in the rift valley between Nakuru 62A and Narok-Kedong 74B-75A, typically nesting in abandoned river-bank holes of the White-fronted Bee-eater (441) at 1600-2000 m (Br). Birds appear at their breeding sites just prior to the Mar-May breeding season, and leave the area equally suddenly afterwards (Be; Brown & Britton 1980), and Br notes the paucity of records from other areas. Our map could suggest post-breeding dispersal throughout the highlands, but this swift is for example only an occasional visitor to L Baringo 50A (St). Paucity of records could reflect confusion with the White-rumped Swift, or post-breeding migration to the S. Regular small numbers end Apr-early Jul around 3000 m at Mau Narok 61D (Se) and large Mar flocks in Maasai Mara 74AC (EABR 1981) are both in or close to the breeding season, whereas two records of wanderers in lowland Tsavo E 101BD were Oct and Dec (La). A wanderer has also reached N to the Horr Valley 26D (Ja). Other reference: Collins 1980.

424 LITTLE SWIFT *Apus affinis*
B443 / MPG643 / D2C2

Afrotropical: widespread S of 16°N, possibly excepting the driest areas of E Somalia and the SW, and much of the S interior.

Kenyan: a common species, found up to 3000 m and very abundant in the vicinity of its breeding colonies (Br; Britton 1979; Brown & Britton 1980). Nesting sites were presumably originally confined to overhanging rocks on cliffs etc., and this swift is still commonly associated with this type of habitat in less developed areas such as Turkana (HH) and Baringo 50A

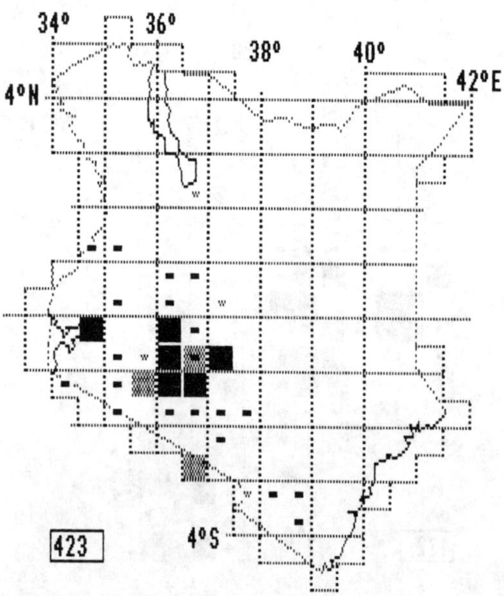

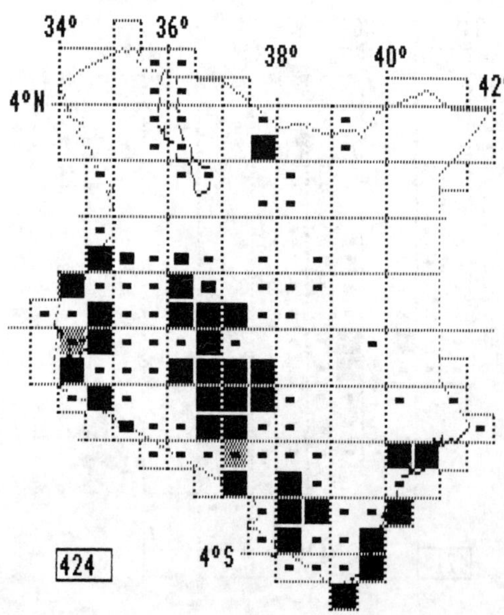

(St), but it has taken advantage of the comparatively recent spread of large buildings, bridges etc. to increase its population and to expand into areas where nest sites were formerly absent, so that it is now a familiar bird of urban and smaller settlement centres (Sn; Br; MPG).

Range is only 26% arid and 10% within the driest, 0-250 mm rainfall country so that, apart from the vicinity of L Turkana, where it is locally common (HH), the Little Swift is very scarce in the N and E lowlands. It is thought to be resident (La: 101B; DEP pers. obs.: 75BC 89C 100B 101A), but is absent from semiarid Baringo 50A during the Dec-Jan period (St) and has 83% of its breeding squares in 500+ mm rainfall areas. See White-rumped Swift (422). Other reference: Long 1963.

Family Coliidae: Mousebirds

Six species, endemic to the Afrotropical Region. Three breed in Kenya.

425 SPECKLED MOUSEBIRD *Colius striatus*
B459 / MPG566 / D3C3

Afrotropical: with the exception of most of the Congo basin forest, widely distributed from Nigeria, the Central African Republic, S Sudan, Ethiopia and S Somalia, to 10°S. Further S, it is confined to the E of the continent, S to E South Africa and the Cape.

Kenyan: common and conspicuous in small parties in moderately open bushed and wooded habitats, including forest edge and suburban gardens, up to 2600 m (Br; Ja; MPG). Range is 69% within 500+ mm rainfall areas and only 20% arid. This mousebird is thus scarce in the arid N and E, where it is largely confined to highland islands *e.g.* Mts Kulal 26B, Nyiru 26D and Marsabit 27D 28C (Br), to the S extremities of the Sudanese and Ethiopian highlands populations 3B 3ACD 17AC 19AB, and to the vicinity of the Tana R 64B 65ABD 78B 79C.

There are two well-defined breeding areas. That of the race *C.s. kikuyuensis* is in the interior, where breeding squares are entirely above 1000 m, and 97% within 500+ mm rainfall areas. *C.s. mombassicus* breeds at the coast, where four of the five squares receive over 1000 mm annual rainfall. The area between these two breeding zones, *i.e.* 87-89 and 101, is distinctly drier, with most squares in the 250-500 mm rainfall regime. Because this area receives regular coverage from birdwatchers, *e.g.* Amboseli GR 88C, Bushwackers' Camp 89A, Tsavo NP 89C 101ABC (La), this absence of breeding records may be real, suggesting this to be a marginal area, only utilised outside the breeding season. This view is supported by the data of La from arid Tsavo E 101B, where very small numbers of 'resident' birds (which were never found breeding) were regularly augmented by considerable Aug-Sept influxes that exploited abundance of ripe fruit.

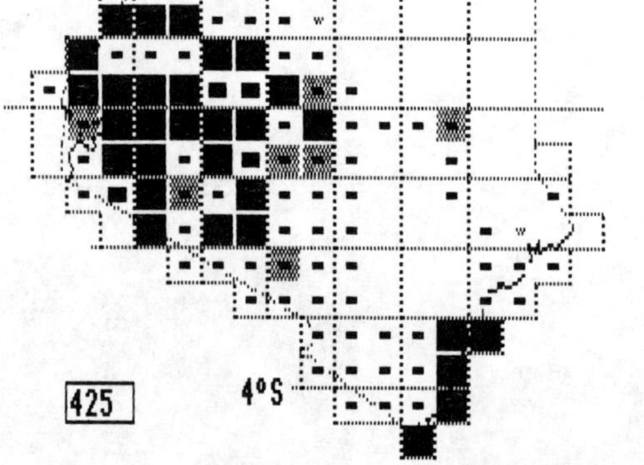

Hence, although the Speckled Mousebird is generally considered resident (Br; Be: 74; St: 50A; DEP pers. obs. 74D 75BCD), movements apparently occur, at least in dry country. Although finding birds present during all months in moist subhumid 75B, DEP (pers. obs.) also noted an annual May-Aug increase in the Ruaraka-Kahawa-Ruiru sector of the square: this is after the main breeding season (Brown & Britton 1980), and may partly be due to the addition of young birds to the population.

McVicker (1982) attributes the recent disappearance of this species from Mombasa 114B to either destruction of nests and young by Indian House Crows (575), or to the extensive use of insecticides since the 1950s. By comparison with equally urban and polluted Nairobi 75B, where this mousebird thrives but the crow is absent, the former explanation appears more likely.

The Speckled Mousebird forms part of a species-group with the broadly sympatric White-headed Mousebird (426)(Sn), which is a less common, drier country species. The White-headed is present and breeds in much of the Speckled's S non-breeding area 88 89 100 101.

426 WHITE-HEADED MOUSEBIRD *Colius leucocephalus*
B458 / MPG567 / D2C2

Afrotropical: in contrast to its two Kenyan congeners, restricted to four possibly distinct areas of the E and NE:

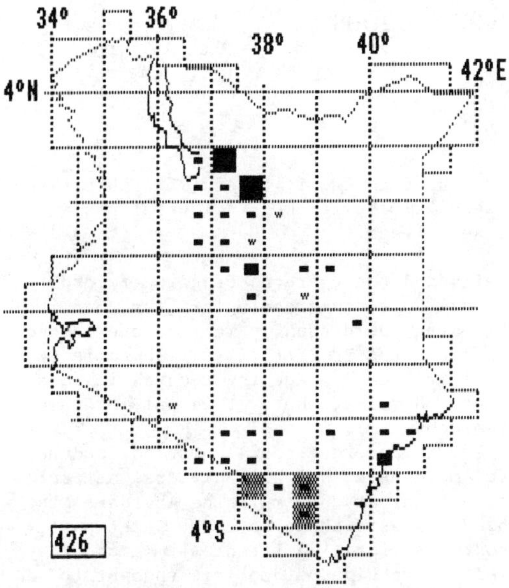

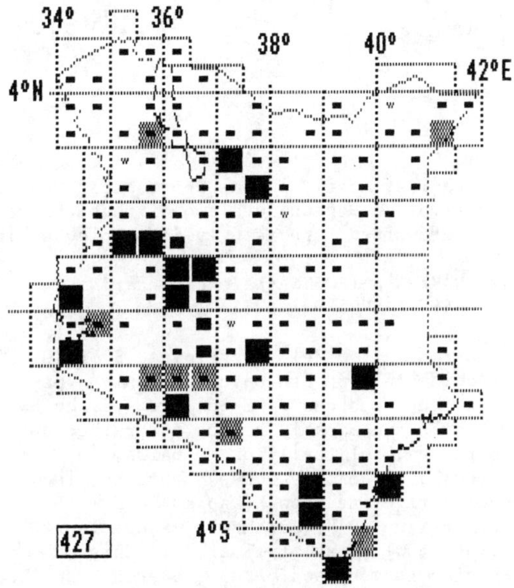

NE Tanzania-SE Kenya; central N Kenya; along the Tana R in Kenya; and in S Somalia and the SE Ethiopian borders.

Kenyan: a rather uncommon bird, found in small parties in dry bush country up to 1400 m (Br; MPG). Range is 79% within intermediate, 250-1000 mm rainfall areas, and 93% arid-semiarid.

The race *C.l. turneri* occurs in the N, from 52D NW to 26B, while the nominate race is S of 1°S. The population around the lower Tana R 79C 91 is supposedly isolated (Br), but a Mar flock near the dry Tiva R in the very poorly covered QSD 90A (EABR 1984) may suggest connection with the main S Kenya-NE Tanzania population. Nov-Jan birds at Garissa 65B (R & L M Greenham *in litt.*) could have been wanderers from either race (Lewis 1984d), and the species is certainly not resident at Selengei 88A (DEP pers. obs.). A recent paucity of records from the SE (Br) probably reflects observer frequency. See Speckled Mousebird (425).

427 BLUE-NAPED MOUSEBIRD *Urocolius macrourus*
B461 / MPG568 / D2C3

Afrotropical: a discontinuous distribution across the N tropics from the W coast, E between 10°N and 20°N, to Sudan, Ethiopia, and Somalia; and thence S to East Africa and the E Zaire borders.

Kenyan: common, usually in small parties in bushed and open wooded habitats typically below 1600 m (Br; MPG; Be; HH; La; St). In sharp contrast to the Speckled Mousebird (425), the Blue-naped occurs and breeds virtually throughout the country. The only constraint upon its range that can be derived from the map is that it is less widespread than the Speckled in the 1000+ mm rainfall areas of the W 48 49 60 61 73. Although at QSD scale it appears widely sympatric with the Speckled Mousebird, in reality these two are ecologically allopatric, with the Speckled frequenting decidedly lusher and moister vegetation (Sn; La; Lack 1985; ADL, DEP pers. obs.). Sympatry of the Blue-naped and the similarly dry country White-headed Mousebird (426) is much more extensive (Sn), but the latter species is more locally distributed.

The Blue-naped Mousebird is considered resident (Br; Be: 74; La: 101B; DEP pers. obs.: 74D 75C 88D 89AC 100B 101A), but local movements occur *e.g.* appearance in Nairobi NP 75B during dry weather (ADL pers. obs.), and annual Apr-May and Nov-Jan influxes to the Ruaraka-Kahawa-Ruiru area of 75B (DEP pers. obs.).

Family Trogonidae: Trogons

34 species, mostly in the Neotropical and Oriental Regions. Only three (9%) inhabit the Afrotropical Region, and two of these breed in Kenya. Despite their colourful plumage, they are often surprisingly inconspicuous in dense foliage.

428 NARINA'S TROGON *Apaloderma narina*
B462 / MPG570 / D2C2

Afrotropical: widely distributed S of 10°N, reaching 16°N in Ethiopia, and only avoiding the dry areas of E Somalia and the SW.

Kenyan: reasonably common but rather unobtrusive, solitarily or in pairs, in forested or densely wooded habitats, including riverine growth, up to 3000 m (Br; Se; Britton 1979; Ja; MPG). Range is 74% within 500+ mm rainfall areas. This species is very scarce in the arid N and E, where largely restricted to the S extremity of the Ethiopian highlands population, in the Omo riverine woods 4A (J S Ash *in litt.*), to highland forest islands *e.g.* Mts Kulal 26B (Diamond & Keith 1980), Marsabit 27D (Ja) and Endau 77B (Tennent 1964), and to the vicinity of the Tana R 65D 79C (AGH).

The nominate race occurs inland, where it mostly frequents highland forest in the W, but is usually below 1800 m in other areas; it is scarce in Nyanza 60 (Br). The race *A.n. littoralis* inhabits the coastal lowlands, where numbers fluctuate seasonally and there are no breeding records: it may be a non-breeding visitor, but its breeding could equally well be overlooked (P L Britton *in litt.*; Br). There is a similar lack of breeding records in adjacent S Somalia, where it is only presumed resident (AM).

The inland race shows a marked tendency for movements, and an individual attracted to a light at 2800 m in the Aberdare Mts 62BD suggests these to be at least partly nocturnal (Br). Altitudinal migration (*cf.* Olive Pigeon (325)) is thought to occur, with movement to lower levels during wet seasons (Brown 1970b). Thus it appears during the rains in rather marginal habitats at Ololua Forest, Nairobi 75B (Brown 1975b; Brown & Britton 1980), and records of wanderers at semiarid Baringo 50A and arid Tsavo E 101B have been in the Mar-Apr or Aug cold, wet periods (St; EABR 1982, 1983; La). Virtual restriction of this race's breeding squares to 1500+ m, 500+ mm rainfall areas may thus be real. A decline on the edge of the range in the Tiva R area (approximately 89AB: not mapped)(La) is probably also reflected in the lack of recent records from 89CD (P C Lack *in litt.*). See Bar-tailed Trogon (429). Other reference: Cunningham-van Someren 1973d.

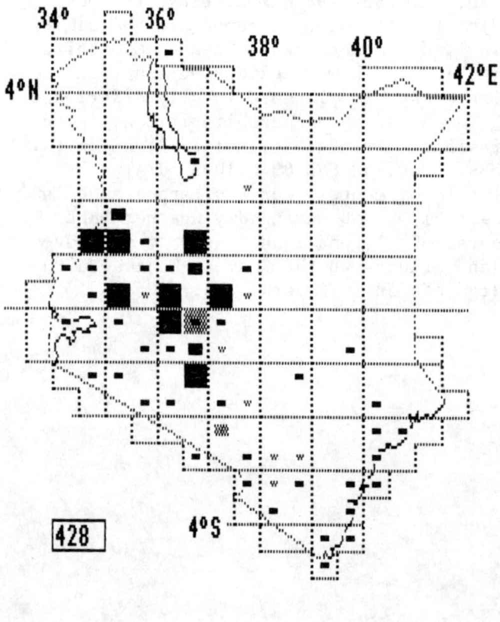

429 BAR-TAILED TROGON *Apaloderma vittatum*
B463 / MPG571 / D2C2

Afrotropical: a highland species of the E, from N Mozambique and Malawi NW to the E Zaire borders, and NE through E Tanzania to the NE limit of the range in the W and C highlands of Kenya. There are also isolated populations in Angola and Cameroon.

Kenyan: in contrast to Narina's Trogon (428), this rather uncommon bird is confined to highland forest (Br; MPG). Range is 91% moist subhumid-humid, 76% within the 1000+ mm rainfall regime, and entirely above 1500 m. The Bar-tailed Trogon is usually segregated from Narina's by greater altitude, but both occur around 1700 m in Kakamega Forest 48D (where Narina's is rare) and at 2000 m in the Nguruman Hills 74D (Br). It further con-

Plate 13. The Speckled Mousebird, one of an Afrotropical family and a common species of Kenya's moister areas. (J.S. Karmali F.R.P.S.). *See page 254*.

Plate 14. The exceptionally long-tailed white phase of the Paradise Fly-catcher, on its nest (J.S. Karmali F.R.P.S.). *See page 431*.

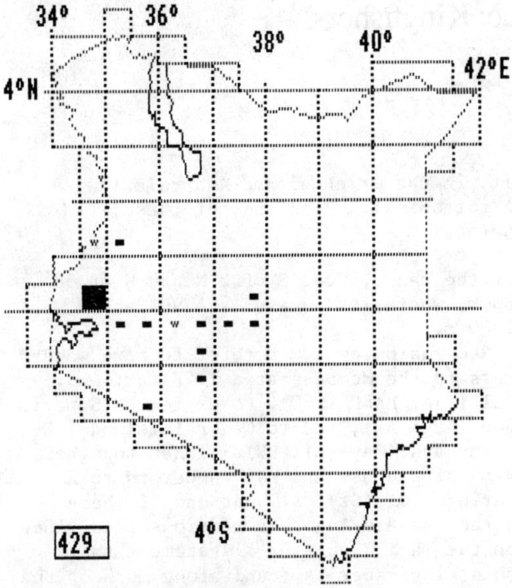

trasts with Narina's by being apparently resident (Br), and noticeably absent from highland forest islands.

Family Alcedinidae: Kingfishers

Approximately 90 species, with greatest diversity in the Oriental and Australasian Regions (Sn). Only 16 (18%) are found in the Afrotropical Region and, of these, 11 (69%) occur in Kenya, where one is an extralimital vagrant.

430 GIANT KINGFISHER *Ceryle maxima*
B464 / MPG466 / D2C2

Afrotropical: widely distributed S of 15°N, apart from Somalia, most of the dry SW, and some areas to the N of the Congo basin. With the exception of an old record in extreme S Somalia 80B (not mapped)(AM), distribution in Kenya represents the E limit of the range.

Kenyan: this unmistakeable, piscivorous species is awarded D2 because it is shy and rather uncommon, and often inconspicuous along the well-wooded banks of permanent rivers and lakes (Br; Ja; MPG). It is widely sympatric with the smaller Pied Kingfisher (431), but prefers more closed habitats (Sn). 92% of the squares are well wooded, 88% receive 500+ mm annual rainfall, and the Giant Kingfisher is thus absent from the arid N and E (but see below). Br states that this species is found up to 2700 m, typically as a bird of streams and dams in the highlands above 1500 m, while also being resident in some lower areas *e.g.* Kiboko 88B and Taveta 100B. This summary encompasses the bulk of the range, *i.e.* S of 2°N and W of 38° 30'E, where its squares are 90% above 1000 m.

Records below 500 m refer to rare wanderers to the Mombasa area 114B (Br; *e.g.* Jul 114A: EABR 1980), to extreme S Somalia 80B (Aug: AM), and to Central Island 14C, L Turkana (May-Jul: HH). Taken together with single Nov and May wanderers to L Baringo 50A (St), all but one of these dates are May-Aug. Ja mentions occurrence on the Tana R, but Br's statement that this kingfisher is found along much of the river's length appears erroneous because it was not recorded by the Tana R Expedition (K L Campbell *in litt.*). We can map only two Tana squares: 63C (D J Pearson *in litt.*) and 79C (P L Britton *in litt.*). Only the latter of these localities is away from the main highland range and, as noted above, this kingfisher is apparently only a rare wanderer to the E lowlands.

431 PIED KINGFISHER *Ceryle rudis*
B465 / MPG465 / D3C3

Afrotropical: widespread S of 15°N, avoi-

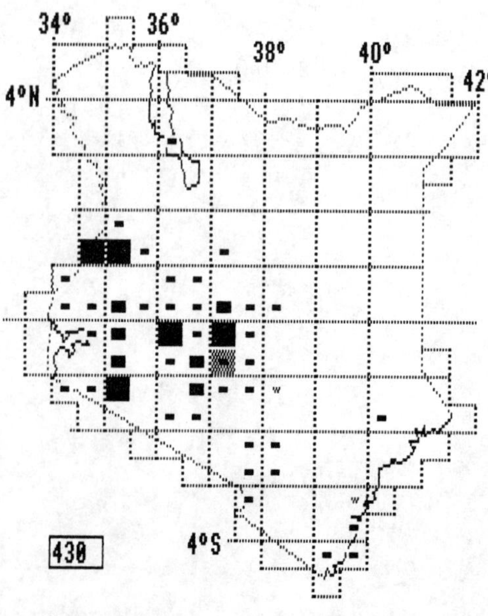

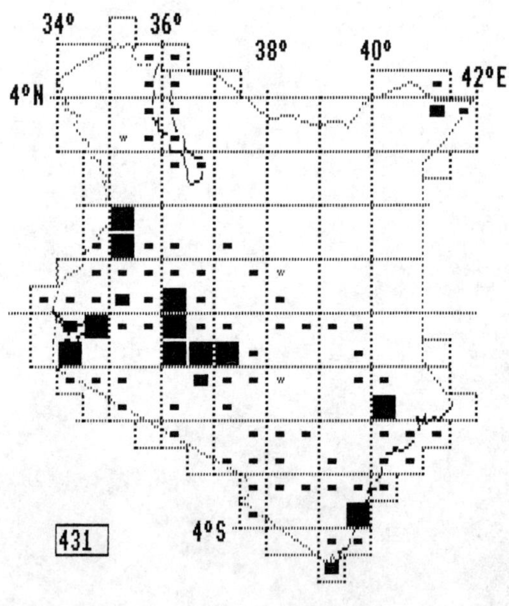

ding only the drier areas of E Somalia and most of the SW, and some areas to the N of the Congo basin.

Kenyan: this common, piscivorous, conspicuous and relatively approachable species frequents open waters, from coastal inlets up to 2300 m, and can be numerous (Br; Ja; MPG; HH; St; EABR 1979). In contrast to the Giant Kingfisher (430: *q.v.*), independence of extensive well-wooded fringing vegetation allows occurrence in the arid N and E, *e.g.* at L Turkana (HH), on the Daua R 7C 19AB (E J Roberts, ADL pers. obs.) and along the Tana R 64B-79C (AGH). This kingfisher is thought to be a resident and wanderer, although a ringed bird moved 760 km from W Ethiopia to N Uganda (Br; St; Backhurst 1974). Other references: Sugg 1974; Reyer 1980.

432 HALF-COLLARED KINGFISHER *Alcedo semitorquata*
B469 / MPG468 / D2C0
NO MAP: post presence 51B 62AC; post confirmed 100B; pre presence 101A

Afrotropical: apart from the dry SW, widely distributed S of 14°S, and further N through Zambia and Malawi to Tanzania, Kenya, SE Sudan and Ethiopia.

Kenyan: a local, scarce and easily overlooked piscivorous bird of well-wooded rivers and streams (Br). It is regular only in the Taveta area 100B, particularly in the Kitovu Forest, where it breeds and may be resident (Ja; Forbes-Watson & Turner 1973; Meadows 1976; Br; EABR 1980). Presence in this area is related to occurrence at a number of sites around the adjacent Kilimanjaro massif (Br). Vagrants have been seen at L Nakuru 62A and L Naivasha 62C, and on the Uaso Nyiro R near Archer's Post 51B (Meadows *et al.* 1975; Br; J Hayes *in litt.*).

It is allopatric with the Shining-blue Kingfisher (433)(Sn), which is a vagrant to Kenya.

433 SHINING-BLUE KINGFISHER *Alcedo quadribrachys*
B468 / MPG469 / D2C0
NO MAP: post presence 48D

Afrotropical: a forest species of W and C Africa, from Senegambia and Congo, E through the Congo basin to Zambia, the E Zaire borders and Uganda. A single wanderer has reached the Kakamega Forest 48D of W Kenya.

Kenyan: a piscivorous species of streams and lakes in forest that, at its closest to Kenya, is known from S and W Uganda (Br). The single Kenyan record is of an individual of the race *A.q. guentheri* at the Lugusida R, in Kakamega Forest 48D, 20 Nov 1974-27 Jan 1975 (Bengtsson 1975; Meadows *et al.* 1975). Failure to locate further individuals indicated this to be a solitary vagrant E of its normal range. See Half-collared Kingfisher (432).

434 MALACHITE KINGFISHER *Alcedo cristata*
B466 / MPG470 / D2C2

Afrotropical: widespread S of 15°N, but absent from the drier areas of E Somalia and the SW, and some areas to the N of the Congo basin.

Kenyan: a common, piscivorous and insectivorous species that frequents reasonably open fringing vegetation by all types of water, from coastal inlets up to 3000 m (Br; Ja; MPG). While inhabiting similar places to the Pied Kingfisher (431) and being similarly common at L Turkana (HH), the Malachite is distinctly less widespread than that species in the NE and E, *i.e.* along the apparently suitable Daua R 7C 19AB and Tana R 65ABD. The reasons for this absence are unknown, but it is paralleled in adjacent S Somalia, where absence on the more westerly Juba R (of which the Daua is a tributary) contrasts with presence on the more easterly Shebelle R (AM).

The Malachite Kingfisher's breeding squares are 72% above 1000 m, and there is only a single confirmed record from the coastal lowlands (91A: Ja). This pattern may well reflect observer frequency and a rather unobtrusive, hole-nesting species. Movements in response to fluctuating water

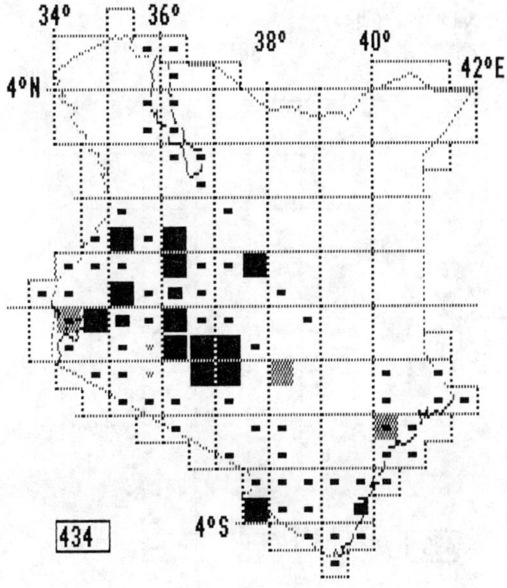

levels are known (Ja), and it is a sporadic visitor to arid Tsavo E 101B, mainly during and after the rains (La). Two single birds that entered Ngulia Lodge 101A at night during the Nov-Dec rainy period may have been related to this Tsavo E movement: they were away from suitable habitat, but were assumed to be wanderers because this species is not known to be migratory (Br; EABR 1978, 1982). Other reference: Meadows 1977a.

435 PYGMY KINGFISHER *Ispidina picta*
B478 / MPG471 / D2C2

Afrotropical: widely distributed S of 15°N, avoiding only E Somalia, an extensive area of the SW, and some areas to the N of the Congo basin.

Kenyan: a common species, feeding mainly on terrestrial insects but often found near water (Br; MPG). Range is 70% within 500+ mm rainfall areas and only 17% arid. Hence it is scarce in the arid N and E, where largely restricted to the vicinity of permanent water *e.g.* L Turkana (where confined to wooded habitats and uncommon: HH), the Daua R 19AB, and the Tana R 64B 65AB 79C.

Two races occur. The resident nominate race is mainly a bird of the W, in forested, wooded or bushed habitats below 2000 m, and mostly below 1500 m: there are a few records E of the rift valley, E to Taveta 100B and Voi 101B (see below)(Br). Breeding squares are all above 1000 m in 500+ mm rainfall areas, with 91% in or W of the rift valley. The race *I.p. natalensis* is an intra-African migrant that breeds in the S tropics and then moves N, to winter Apr-Sept, mostly in dense forest, in the coastal lowlands (Britton & Britton 1974c; Br; Britton & Rathbun 1978; EABR 1977, 1980). Two early Aug individuals that struck windows at Shimo la Tewa 102D were probably on S passage (Britton & Britton 1974c).

Sn mentions movements of this species during the rains, and two single birds in Tsavo E 101B 102A during the Nov-Dec wet season (La) were, on the basis of date, presumably members of the nominate race wandering beyond the E edge of their range.

436 CHESTNUT-BELLIED KINGFISHER *Halcyon leucocephala*
B473 / MPG477 / D2C2

Afrotropical: widely distributed between 15°N and 23°S, avoiding only dry E Somalia and most of the Congo basin forest areas.

Kenyan: a common, insectivorous species of bushed and wooded habitats up to 2200 m, often associated with water, partly because it uses river banks etc. as nest-hole sites (Br; MPG; Sn; Lack 1985). Range is only 8% within the 0-250 mm rainfall regime, and occurrence in these very dry areas is largely near permanent water *e.g.* L Turkana, the Daua R 7C 19AB, the Ewaso Nyiro R 40D and the Tana R 65AB.

That this is Kenya's most widespread kingfisher is most probably due to the fact that three migratory races occur (following information from Br except where otherwise indicated). The nominate

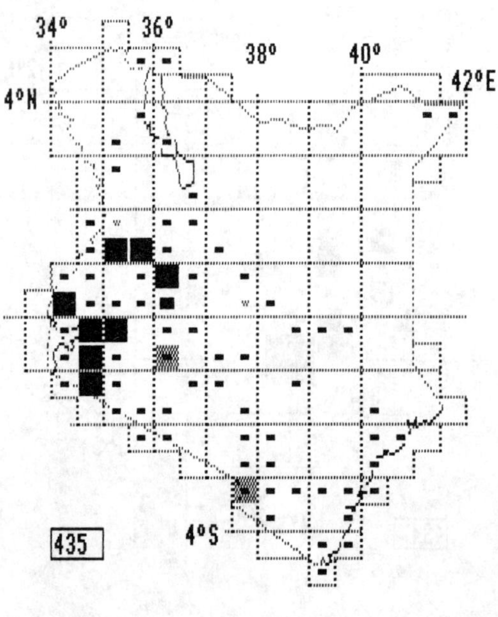

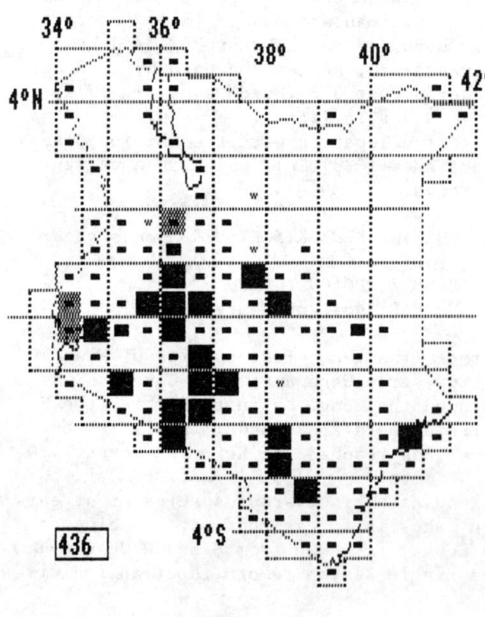

race occurs throughout, breeds widely, and is shown by ringing recoveries (Ethiopia to Kisumu 60B in Nov, and to Mombasa 114B in Mar: Backhurst 1977) to be augmented by migrants from the N. *H.l. hyacinthina* is endemic to the coastal lowlands, but scarce S of Malindi 103A. Finally, *H.l. pallidiventris* is an intra-African migrant that breeds in the S tropics, and then moves N Apr-Sept (*cf.* Pygmy Kingfisher (435)), at least as far as Nandi 49C (Ja). Presumed nocturnal migrants have appeared at the Ngulia lights 101A Nov-Dec (Backhurst & Pearson 1977b; EABR 1979).

Patterns of occurrence in the semiarid and arid environments of Baringo 50A and Tsavo E 101B are similar, with largest numbers Dec-Mar and Nov-May respectively, and absence/very small numbers Apr-Aug and June-Oct (St; La): on seasonality and geographical location, these maxima would appear to refer to nominate birds.

The Chestnut-bellied Kingfisher is broadly sympatric with the allied Brown-hooded (437) and Striped (438) Kingfishers (Sn; see Short & Horne 1985a).

437 BROWN-HOODED KINGFISHER *Halcyon albiventris*
B470 / MPG476 / D2C2

Afrotropical: a S species, widely distributed to the S of the Congo forests, and further N through E Tanzania to Kenya and S Somalia; but absent from the dry SW. Distribution in Kenya represents the N limit of the range.

Kenyan: locally common in wooded and forest edge habitats, often near water, up to 1800 m (Br; MPG). Range is 95% well wooded (as opposed to 80% for the Chestnut-bellied Kingfisher (436: *q.v.*)), and only 4% within the driest, 0-250 mm rainfall areas.

Like the Mottle-throated Spinetail (412), though more extensively, this kingfisher extends inland along the Athi R and the Tana R to the C highlands and their E periphery and, by striking coincidence, single wanderers of both species have reached NW to Central Island 14C, L Turkana (HH).

This kingfisher has also wandered W to the rift valley at L Baringo 50A (T Stevenson *in litt.*; EABR 1984) and L Nakuru 62A, and even as far W as the mouth of the Nzoia R 47D (Br). Status further S in the rift, in the Nguruman 74D-Magadi 75C area, is uncertain (P L Britton *in litt.*; DEP pers. obs.). The few records of wanderers show seasonality. Rare occurrences in the Voi R forest 101B have all been Oct-May (La), while single birds have appeared at L Turkana in Nov, and at Baringo 50A in Nov-Dec and Jan (*op. cit.*).

438 STRIPED KINGFISHER *Halcyon chelicuti*
B472 / MPG479 / D2C2

Afrotropical: widely distributed S of 15°N, apart from the drier areas of E Somalia and the SW, and most of the lowland forest regions in the Congo basin.

Kenyan: a usually common bird, up to 2300 m (Br; MPG). Its basic ecological requirement appears to be trees, from

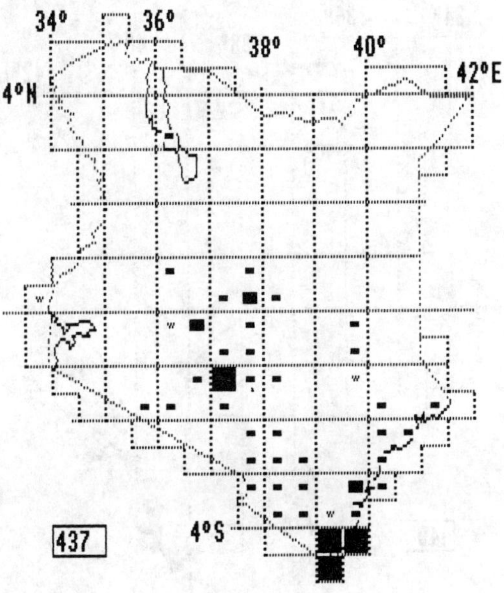

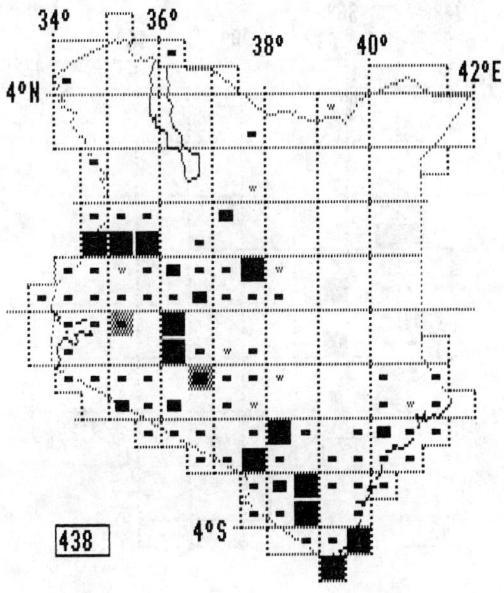

which it can detect its terrestrial insect prey (*e.g.* see La). Range is 75% within 500+ mm rainfall areas, and occurrence in the arid N and E is virtually restricted to the southernmost elements 2C 4A 15C 17A of populations in moister areas to the N, in Sudan and Ethiopia. As noted by Ja, trees are apparently more necessary to this species than water, and it is notably absent from wooded waterside habitats of arid areas *e.g.* L Turkana, the Daua R 19AB, and most of the Tana R 64B-79C (*cf.* Chestnut-bellied Kingfisher (436)).

This kingfisher is thought to be largely resident (Br; Lack 1985: 101B; DEP pers. obs.: 75C 88D 89AC 100B 101A) but, near the edge of the range at semiarid Baringo 50A, occasional single wanderers are mostly Mar-May and Oct-Nov (St): this seasonality broadly corresponds to the wet seasons in the highlands to the S.

439 WOODLAND KINGFISHER *Halcyon senegalensis*
B475 / MPG473 / D2C2

Afrotropical: widespread S of 15°N, E to Ethiopia, W Kenya, C Tanzania, Malawi and the SE coast, but avoiding the dry SW and the E periphery (see Mangrove Kingfisher (440)). Distribution in Kenya represents the E limit of the range.

Kenyan: a common, mainly insectivorous species of bushed and wooded habitats, often near water and mainly below 1500 m, but also known up to 2000 m (Br; MPG). Range is 84% within 500+ mm rainfall areas and 73% subhumid-humid.

This is a bird of the W that mainly occurs in Nyanza 60 and the peripheral highlands (Br), but there are records (including confirmed breeding) E to the rift valley which, in all but one instance, forms the E edge of the range. This kingfisher is possibly extending its range E into the rift at Baringo 50A where, formerly a very uncommon visitor June-Aug, up to three birds were present Jan-June 1982 (St; EABR 1979, 1982). There have subsequently been records at Baringo in nearly all months (T Stevenson *in litt.*). Br described the well known L Naivasha population 62CD as disjunct and, like the occurrences at Baringo and the breeding at Loboi 50C (H-U Reyer *in litt.*), this may be a further manifestation of the species' E extension into the rift. The sole record for the E plateau, of a pair displaying on 6 Jan 1984 at Ngurunet 39A (DEP pers. obs.), is most probably also connected with this eastward spread.

The race *H.s. cyanoleuca* is an intra-African migrant that breeds Oct-Apr in the S tropics, and then moves N to S Sudan: it is a passage migrant in W Kenya (Br).

The Woodland Kingfisher forms a superspecies with the Mangrove Kingfisher (440) (Sn), and the two are allopatric in Kenya.

440 MANGROVE KINGFISHER *Halcyon senegaloides*
B476 / MPG474 / D2C3

Afrotropical: the E periphery, in areas mostly not occupied by the Woodland Kingfisher (439: *q.v.*)(Sn). It extends up the

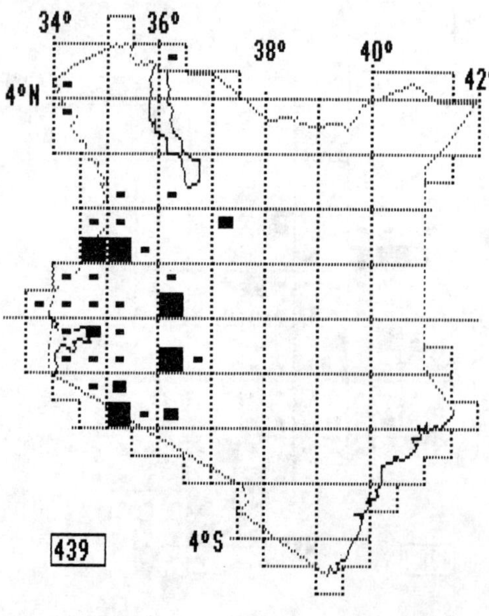

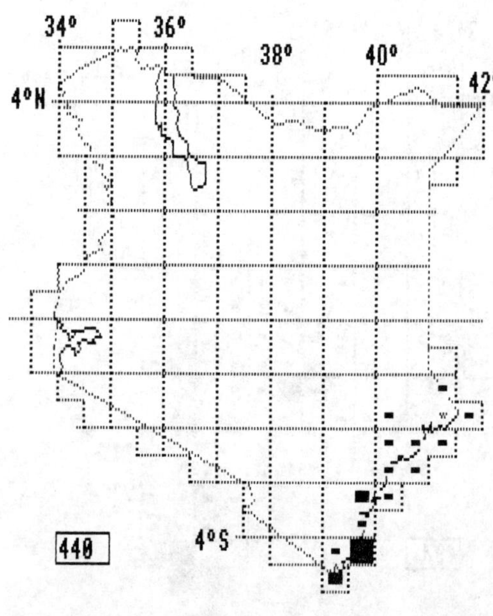

E coast and adjacent areas from E South Africa, through Mozambique and Tanzania, to the N extremities of the range in Kenya and S Somalia.

Kenyan: reasonably common throughout the coastal lowlands, NE to the lower Tana R 79C (AGH) and the Somali border (AM): hence C3. Range is entirely in 500+ mm rainfall areas. In addition to mangroves, this kingfisher also frequents wooded and forested areas, and extends quite far inland (Br).

Family Meropidae: Bee-eaters

24 species (Fry 1984), absent only from the Nearctic and Neotropical Regions. 18 (75%) are found in the Afrotropical Region, where 14 (78%) are endemic (Sn). 12 (67% of the Afrotropical total) occur in Kenya, where nine breed at least intermittently, two are non-breeding migrants from the Palearctic, and the Swallow-tailed Bee-eater (445) is an extralimital wanderer.

441 WHITE-FRONTED BEE-EATER *Merops bullockoides*
B482 / MPG493 / D2C2

Afrotropical: a bird of the S that, apart from the dry SW, is widely distributed N to 5°S. Further N, it extends along the W coast to Gabon; along the E Zaire borders; and through E Tanzania to the NE limits of the range, in Kenya.

Kenyan: a locally common, social and apparently resident bird of fairly open country, typically at 1400-2000 m, where earth banks of any type provide nest sites (Br; Ja; MPG; Be). Range is almost confined to the rift valley and adjacent plateaux, 91% above 1000 m and 77% within 500+ mm rainfall areas. Br also mentions breeding below 1000 m along the Kerio and Turkwell R (precise locations omitted: not mapped).

Br suggests post-breeding dispersal for a record at Kisumu 60B, and this may also be the origin of a single Jul flock at Baringo 50A (St), but a bird at Malindi 103A (Br) was a longer distance wanderer, over 200 km E of its normal range. Other references: Hegner *et al*. 1979; Emlen 1979, 1981.

442 LITTLE BEE-EATER *Merops pusillus*
B491 / MPG488 / D2C2

Afrotropical: widely distributed S of 15°N, except for the dry areas of E Somalia and the SW, most of the lowland forests of the Congo basin, and some areas to the N of this basin. It is probably the most abundant African bee-eater (Sn).

Kenyan: common up to over 2000 m in open country with scattered bushes or small trees, that provide perches from which it pursues prey (Br; Ja; MPG; St; La). Range is only 5% within the most arid areas, and this bee-eater is scarce in much of the NE, and around L Turkana, where it is mainly confined to the Omo 3B 4A and Kalokol 13B riverine habitats (HH). Occurrence in arid Tsavo E 101B is also in locally moist situations (Lack 1985).

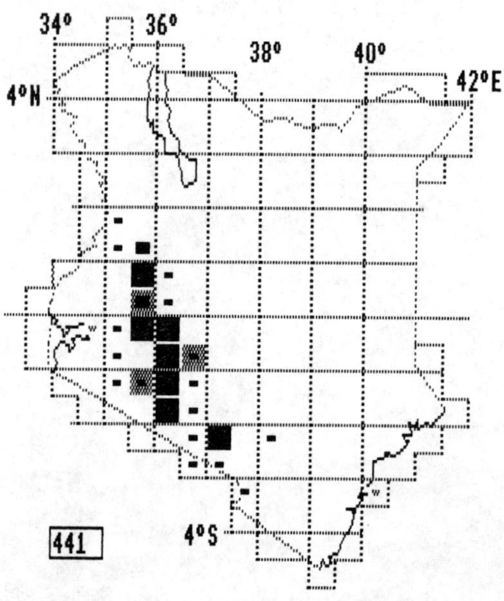

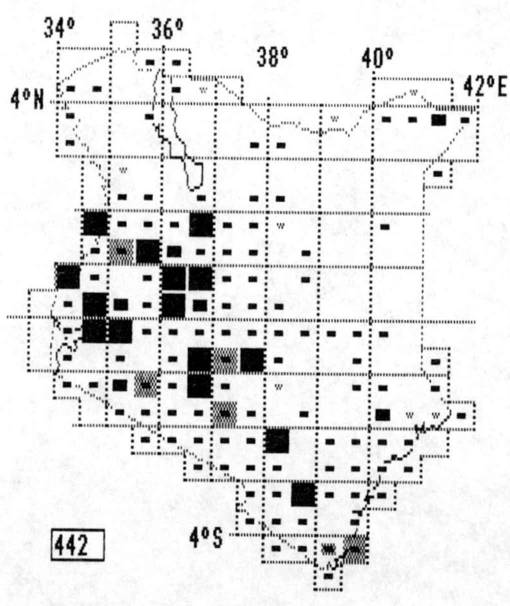

This species is thought to be resident (Br; Be: 74; La: 101B; DEP pers. obs.: 75C 88D 89AC 100B 101A). Breeding squares are 96% within 250+ mm rainfall areas and 87% semiarid-humid and, while this most probably reflects observer frequency, there is also the possibility that this bird is a non-breeding visitor to the most arid parts of its range.

The Little Bee-eater forms a species-group with the Blue-breasted (443), Cinnamon-chested (444) and Swallow-tailed (445) Bee-eaters (Sn). In Kenya, the last of these species is known from only a single record, while the other three are broadly sympatric but ecologically segregated: the Blue-breasted from the Little in longer, wetter grass, and the Cinnamon-chested from both in forest edge and woodland (Sn; Br).

443 BLUE-BREASTED BEE-EATER *Merops variegatus*
B494 / MPG490 / D1C3
NO MAP: post presence 48AC; post confirmed 60A; pre presence 60B

Afrotropical: from Cameroon and Congo E through the Congo basin to NW Tanzania, Uganda, S Sudan and, on the E limit of the range, the far W of Kenya. It is also found in S Zaire and Zambia, while a population in Ethiopia may be conspecific with the Cinnamon-chested Bee-eater (444)(Sn).

Kenyan: a local bird of moist grassland and swamp (Br; Britton 1978). D1 reflects the potential for confusion with the far better known and more widespread Little Bee-eater (442: *q.v.*), while C3 is given on the basis of the very restricted range: this bee-eater extends S from Alupe 48A (R E M & T Scholtens *in litt.*) to Yala Swamp 48C, L Sare and Usengi 60A (Britton & Harper 1969; Mann 1976; Britton 1978), and there is an old record from Kisumu 60B (Ja). The four squares are at 1000-1500 m, in moist subhumid-humid, 1000+ mm rainfall country. It is presumably resident (Br), but there is as yet only a single breeding record, of a pair feeding newly fledged young at Usengi 60A in May 1983 (T Stevenson *in litt.*).

444 CINNAMON-CHESTED BEE-EATER *Merops oreobates*
B488 / MPG489 / D2C2

Afrotropical: a highland species of the E, along the E Zaire borders and adjacent areas of Tanzania and Uganda; and N through E Tanzania to Kenya, E Uganda, S Sudan and extreme S Ethiopia. Apart from a single wanderer in S Somalia (AM), distribution

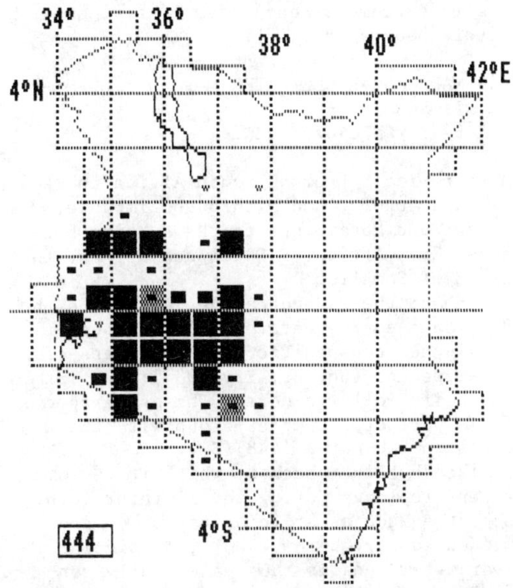

in Kenya reflects the E limit of the range.

Kenyan: common in forest edge and tall wooded habitats in the highlands, typically at 1600-2300 m (Br; Be; MPG). Range is 90% in 500+ mm rainfall areas and 81% subhumid-humid, and includes the forest islands of Mts Nyiru 26D (Ja) and Marsabit 27D (Br). The high proportion of breeding squares (59% of its total squares) suggests a reasonably sedentary species, and this view is concurred with by Br, Be (74), DEP (pers. obs.: 74D) and ADL (pers. obs.: 75B). This bee-eater is an Oct-Apr breeding visitor around 3000 m at Mau Narok 61D (Se) however, and AM quote a May vagrant to 2°N in S Somalia. See Little Bee-eater (442). Other reference: Reynolds 1960c.

445 SWALLOW-TAILED BEE-EATER *Merops hirundineus*
B485 / MPG496 / D3C0
NO MAP: pre presence 114C

Afrotropical: occupies two distinct areas, with differentiating populations N and S of a divide formed by the Congo basin forests, that continues further E through S Uganda and Kenya. Thus the N population occurs across the tropics from Senegambia to Uganda, while the S one is widely distributed S of 5°S: this situation is comparable to that of the two races of the Carmine Bee-eater (452)(Sn).

Kenyan: the single old record, from Vanga 114C (Ja), presumably involved a member of the S population wandering just N of its normal range. Further occurrences in extreme S Kenya seem likely. See Little Bee-eater (442).

446 MADAGASCAR BEE-EATER *Merops superciliosus*
B493 / MPG482 / D2C2

Afrotropical: from NE South Africa to Namibia and Angola, and N through Zimbabwe, Malawi and Mozambique to the E Zaire borders, East Africa, S Sudan and all but the dry E of Somalia.

Kenyan: the Madagascar Bee-eater and the Blue-cheeked Bee-eater (447) have been considered conspecific, or as members of a superspecies (Sn; Br; Fry 1984): the latter course is followed here. This superspecies forms part of a species-group with the Eurasian Bee-eater (448)(Sn).

Like the Blue-cheeked, which it to some extent replaces during the S winter (*e.g.* St), the Madagascar inhabits bushed and wooded country below 1500 m, particularly near water, and is thus essentially confined to the SE lowlands, the L Victoria basin, and areas peripheral to the W and C highlands (Br; Lack 1985). Hence, as in the Blue-cheeked, the squares are 82% below 1500 m and 62% semiarid-subhumid.

In W Kenya, the Madagascar Bee-eater is mainly a May-Sept migrant from SE Africa and possibly Madagascar, but it occurs erratically in all months at the coast, where it can be abundant in the Witu-Lamu area 91AB Aug-Sept (Ja; Britton & Britton 1976b; Br; Fry 1981). Jan-Mar flocks in Tsavo 89CD 101AB, Kitui 77 and other E bushlands are in drier habitats than the birds which appear in the L Victoria basin later in the year, and may be migrants en route to breeding sites in Somalia (Fry 1981; D J Pearson *in litt.*). Nov birds in Tsavo may be returning from Somalia (Lack 1985).

There are breeding records from Meru 52C (Bullock & Bullock 1976), the Athi R 89C and Sala 102A (La), Lamu 91B (Serle 1943; EABR 1978), and Malindi 103A (Fry 1981): these localities are all below 1000 m in the E or SE, in contrast to an old record, accepted with reservations by Brown & Britton (1980), from 1200 m on the Kano Plains 60B of the L Victoria basin (Ja).

447 BLUE-CHEEKED BEE-EATER *Merops persicus*
B490 / MPG482 / D2C2

Afrotropical: a Palearctic bee-eater that winters in the W from Senegambia S to Nigeria, and in the E from Sudan and Somalia S to Zambia, Botswana and E South Africa.

Kenyan: a locally common passage migrant and winter visitor Oct-early Apr, in bushed and wooded habitats, especially near water, below 1500 m (Reynolds 1974; Br; La). It is generally at lower altitudes than the Eurasian Bee-eater (448), mainly on the peripheries of the W and C highlands (*e.g.* 50A :St) and in the SE lowlands. It can be common in the L Victoria basin and around S swamps (Br). See Madagascar Bee-eater (446).

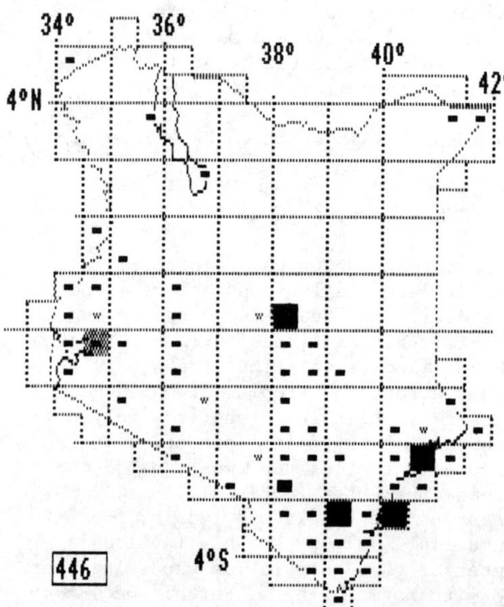

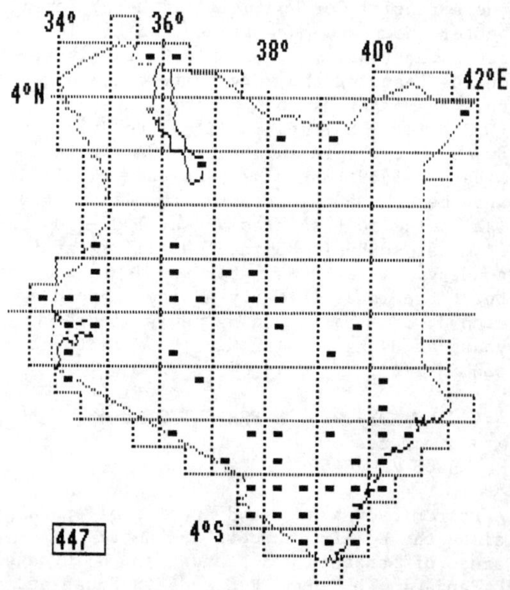

448 EURASIAN BEE-EATER *Merops apiaster*
B480 / MPG481 / D2C2

Afrotropical: this predominantly Palearctic species winters mainly in Africa, from Senegambia and Cameroon in the W, and from Kenya, Tanzania and S Zaire, S to the Cape. It is one of the very few mainly Palearctic species that also breed in the Afrotropical Region, N to 26°S (Sn). Rare in Somalia (AM), Kenyan distribution reflects the E limit of the main range.

Kenyan: frequents open habitats with at least some trees, and has range 72% within 500+ mm rainfall areas and only 16% arid. Virtual absence from the arid E is quite possibly real, since the species was not recorded by observers resident in 19A 42A 65B, and there are only four records from S Somalia (AM): these areas may be E of the normal migration route, or overflown (*cf.* La in arid Tsavo E 101B). The few records from the arid N, mostly near L Turkana, are of passage migrants (*e.g.* HH), most of which probably overfly the area. Further S and W, this bird is a common to abundant passage migrant late Sept-early Nov and late Mar-early Apr, particularly during the S movement (Br).

Smaller numbers wintering in open wooded habitats of the W and C highlands are at the N edge of the wintering range. Winterers occasionally occur down to 1000 m at Baringo 50A (St), but Se noted only migrants at 3000 m at Mau Narok 61D. Birds in the coastal lowlands are mostly on passage, although some winter in sisal plantations (D J Pearson *in litt.*). See Madagascar (446) and Blue-cheeked (447) Bee-eaters.

449 BLUE-HEADED BEE-EATER *Merops muelleri*
B486 / MPG494 / D2C3
NO MAP: post presence 49C; post confirmed 48D; pre presence 36D 61A

Afrotropical: a lowland forest bird of W and C Africa, from Guinea to Cameroon in the W, and E through the Congo basin to E Zaire. An isolated population E of the main range occurs in W Kenya.

Kenyan: a not uncommon inhabitant of forest edge and clearings (Br; MPG). Subsequent to records from Nyarondo 61A (van Someren 1919a), Lerundo 48D and Elgon 36D (Ja), range has contracted into the Kakamega 48D and S Nandi 48D 49C Forests (Br): all four squares are humid, with 1000+ mm annual rainfall. This retreat may be the latest manifestation of the much larger scale contraction that presumably isolated this relict W Kenya population from the main range in Zaire; deforestation may be a factor too. C3 is given because of the restricted range but, while this species is unmistakeable, D2 is used to indicate its forest environs.

450 WHITE-THROATED BEE-EATER *Merops albicollis*
B479 / MPG486 / D2C2

Afrotropical: this intra-African migrant is widely distributed between approximately 17°N and the equator, extending further S to Congo and, in the E, to S Kenya and NW

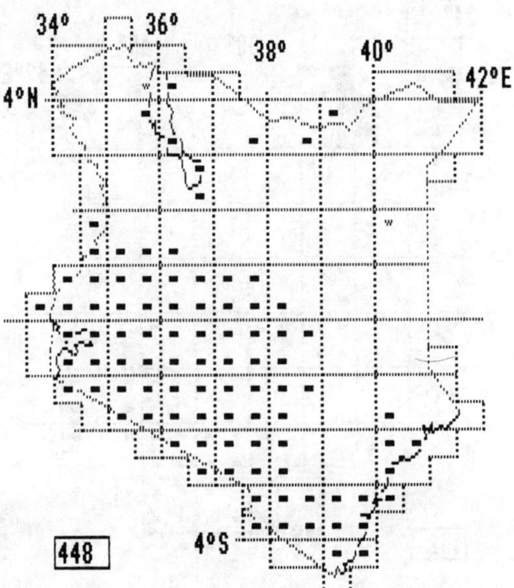

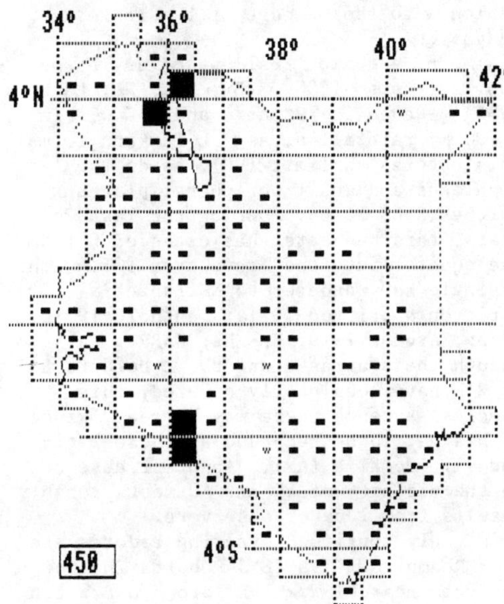

and E Tanzania, but avoiding arid E Somalia. It breeds along the S fringes of the Sahara and then migrates S: nesting in Kenya is thus exceptionally far S (Sn).

Kenyan: a seasonally quite common bird of bushed and wooded habitats, including forest edge, typically below 1400 m (Br; MPG), although a small Jul flock reached 3000 m at Mau Narok 61D (Se). Range is only 8% within the most arid areas, so that virtual absence from the NE may be real: there is a similar dearth of records from adjacent areas of S Somalia (AM).

Breeding at least occasionally takes place during the rains near Ileret 4C (May-June 1984: J R Ness *in litt.*), near Ferguson's Gulf 13B (HH; Hopson & Hopson 1977), in the Magadi-Olorgesaille area 75C 87A (where the species is present in all months: Cunningham-van Someren 1971f, 1972e; DEP pers. obs.), and possibly in other areas of the S rift valley (Br). The breeding squares are all less arid-semiarid, with three of four in 250-500 mm rainfall areas.

Otherwise, this bee-eater is a fairly common and widespread non-breeding migrant from the N tropics, mainly Sept-Apr, occurring only on passage in the W (Britton & Britton 1976b; Br). Jan-Sept seasonality occurs in the N at L Turkana (HH) and Baringo 50A (St), these birds presumably moving further S during the off-season. Other reference: Reynolds 1972c.

451 SOMALI BEE-EATER *Merops revoilii*
B492 / MPG491 / D2C2

Afrotropical: the NE, from Somalia and C Ethiopia to the SW edge of the range in Kenya.

Kenyan: a common resident of dry bushed habitats up to 1000 m (Br; MPG; HH; Lack 1985). Range is 79% arid and 85% within 0-500 mm rainfall areas. L Turkana forms this species' W limit (HH), as do the highlands and their peripheral plateaux further S 63 76 88. During the last 25 years, this bee-eater has extended S from the equator into the Tsavo area 101, with a single Nov wanderer to NE Tanzania (Britton & Britton 1971a; Leuthold 1971b, 1973a; Archer 1979; Br; La; EABR). Although the squares along the E borders S of 2°N have been poorly covered, this species' absence there may be real, since it has yet to be recorded at these latitudes in Somalia (AM). Apparent absence in the vicinity of 29 and 40 most probably results from lack of observers.

The only confirmed breeding records are from Rhamu town 19A (E J Roberts *in litt.*), and from near El Wak 18D (Brown & Britton

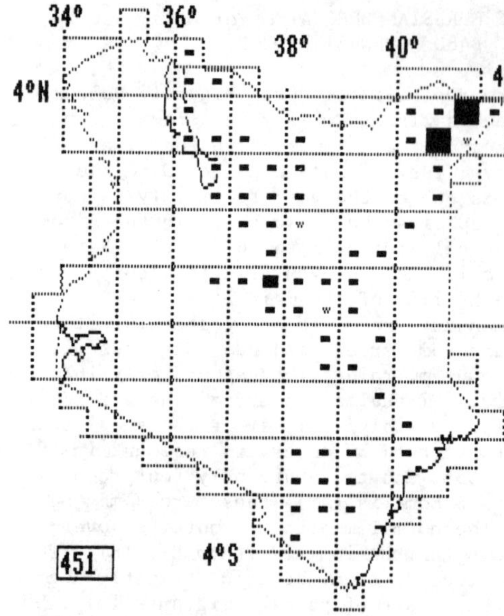

1980): these localities are adjacent to the single breeding record from S Somalia (immediately SE of 19D: AM). Probable breeding on the SW edge of the range at Shaba 51B represents a bird possibly incubating in a nest hole (EABR 1982).

452 CARMINE BEE-EATER *Merops nubicus*
B487 / MPG484, 485 / D3C2

Afrotropical: this intra-African migrant has two populations, both of which move

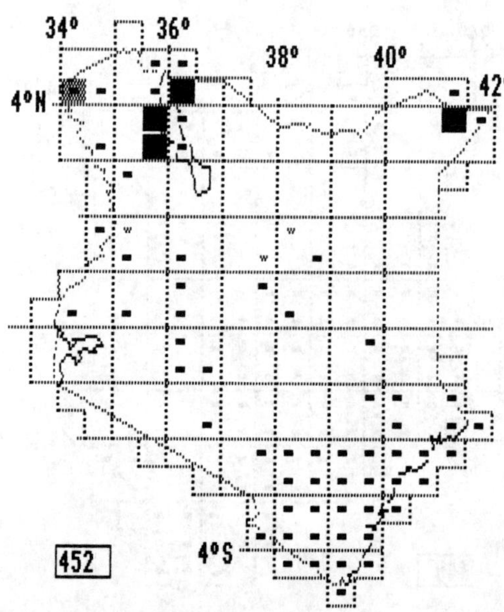

towards the equator after breeding. In the N tropics, the nominate race extends E between 15°N and the N edge of the equatorial forests from the W coast to Ethiopia, and SE to N Uganda, S Somalia, Kenya and E Tanzania. The S race *M.n. nubicoides* is widely distributed N from Angola and NE South Africa to the S edge of the equatorial forests, the E Zaire borders and W Tanzania, with wanderers NE to Kenya.

Kenyan: a species of bushed habitats, mainly below 1200 m (Br; MPG). All but three records concern the nominate race. 77% of its squares are below 1000 m and 78% arid-semiarid, but occurrence in the arid areas of the N, and of the E to the N of 2°S, is largely restricted to the vicinity of permanent water *e.g.* L Turkana (HH) and the Daua R 7C 19AB, the Ewaso Nyiro R 40D, and the Tana R 65B 78B 79AC.

The few breeding records, which mostly concern colonies of thousands of birds, represent the S edge of the N tropics breeding range. In the NW, these sporadic records are from Lokichokio 2C and various sites around L Turkana (Spoerry 1976; Hopson & Hopson 1977; Brown & Britton 1980; J R Ness *in litt.*). This breeding is Apr-June, and the species shows Apr-Sept seasonality in the Turkana area. A similar colony, of at least 1000 holes inaccessible in a wadi wall, has recently been discovered near Rhamu 19A, and was active in May 1984 and in previous years (E J Roberts *in litt.*). These breeding squares are all below 1000 m, with four of five arid.

Scattered records of small numbers of this bee-eater in interior Kenya are mainly in the Sept-Apr non-breeding season (Br; St; EABR). It is far commoner and more numerous in the SE, and particularly so at the coast, where it occurs in all months (though in small numbers during the Apr-Jul N breeding season) and forms huge roosts in the Sept-Mar off-season (Britton & Britton 1976b; Donelly & Donelly 1977; EABR 1979, 1984).

There are three records of the S race wandering N of its usual range to Kenya, during its non-breeding season: L Kanyaboli 48C Jul (Britton & Britton 1972b), the coast at 3°35'S 102D (Sn) and near Kijabe 62D Jul (EABR 1977).

Family Coraciidae: Rollers

10 species of tropical and temperate areas, absent from only the Nearctic and Neotropical Regions. Eight occur in the Afrotropical Region. Five are found in Kenya, where four breed and the other is a non-breeding migrant from the Palearctic.

453 ABYSSINIAN ROLLER *Coracias abyssinica*
B495 / MPG458 / D2C2

Afrotropical: the N tropics, from the W coast, between 17°N and the N edges of the equatorial forests, E to Ethiopia and SE to N Uganda and mainly NW areas of Kenya.

Kenyan: the Abyssinian Roller forms part of a species-group with the Eurasian (454) and Lilac-breasted (455) Rollers (Sn). While conspicuous, it is awarded D2 due to the potential for confusion, particularly of immatures, with the far commoner Palearctic species.

The Abyssinian Roller is principally a bird of Turkana and areas further NW, where it is at the SE limit of its main, N tropics range. Thus it is a locally common resident around L Turkana, in open, bushed or wooded habitats below 1000 m, breeding Apr-June (HH; Br; Brown & Britton 1980). There is also a breeding record from Lopading, near Lokichokio 2C, during the same months (T Romig *in litt.*). Ignoring all the squares cited below for wanderers, this NW range has its squares (including three of the four breeding squares) 77% arid.

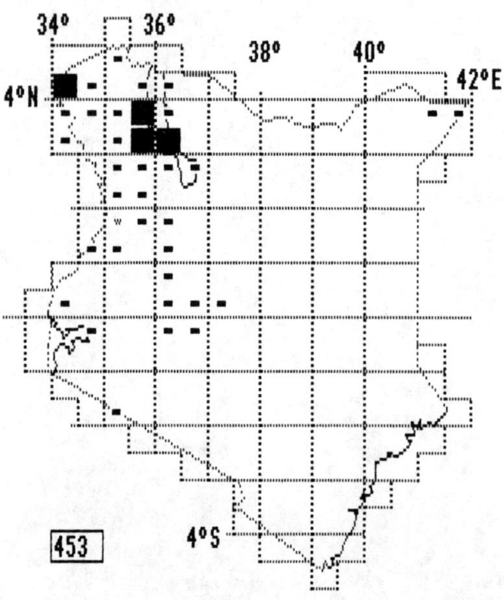

Records S of approximately 1-2°N refer to scarce, solitary but probably annual wanderers into the breeding range of the Lilac-breasted Roller (St; EABR), that reach humid, 2000+ m squares on the N edges of the W and C highlands. These occurrences are particularly frequent in the rift valley, which presumably provides an easy route S. Similar wanderers to the W and S of the W highlands 48C 60B 74C may originate in E Uganda, where the species is regular S to 1°30'N (Br; EABR). Similar scarce and sporadic status at Rhamu-Mandera 19AB (E J Roberts *in litt.*) presumably represents stragglers SE from the Ethiopian range. While low observer frequency is probably a factor, the apparent absence of birds wandering E or S into the area between Turkana and Rhamu is notable.

Wanderers S of Turkana are normally Nov-Jan (St; EABR), and a mid June-Aug influx of solitary birds to the N edges of the highlands during 1984 is unique, and presumably due to very dry weather further N. These unseasonal birds occurred at Kongelai 37C Aug (R E M & T Scholtens *in litt.*), Kapedo 38C Aug (T Stevenson, D A Turner *in litt.*), Timau 51C June (L V J Grant *in litt.*), L Nakuru 62A Jul and the Ark 62B (EABR 1984). A similar movement of Lilac-throated Rollers (455) had occurred earlier in the year, which was one of particular drought (*cf.* Rufous-crowned Roller (456)).

454 EURASIAN ROLLER *Coracias garrulus*
B497 / MPG457 / D3C3

Afrotropical: a Palearctic species wintering mainly in the E, from Kenya and S Somalia S to Zimbabwe and E South Africa. It is only local in the W.

Kenyan: a common to abundant passage migrant and winter visitor Nov-Apr, that often greatly outnumbers resident congeners, particularly in the E (*e.g.* Brown & Brown 1974). It is typically in open bushed or wooded habitats below 1500 m (Br), but rare migrants reach 3000 m at Mau Narok 61D (Se).

Heavy Nov-Dec and Apr passages occur in the E and SE, between the C highlands 63 76AB, Garsen 91A and the Mombasa hinter-

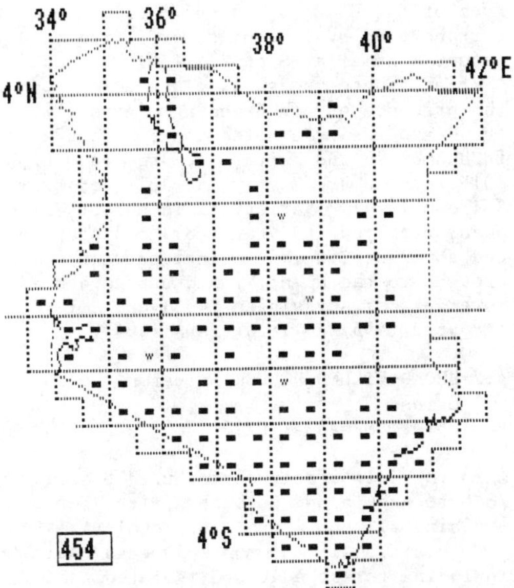

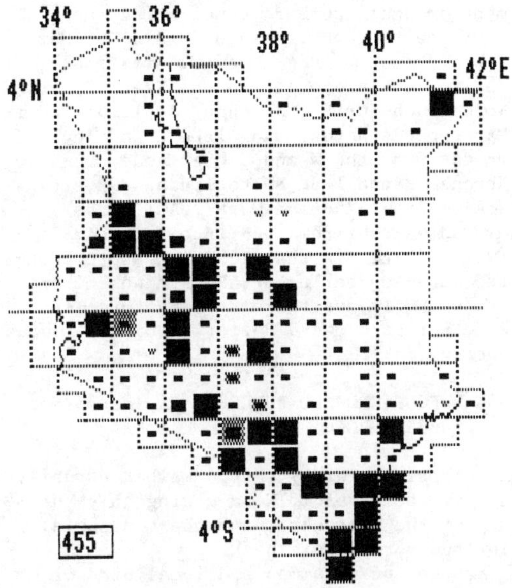

land 102C (Br; Lack 1985; *e.g.* Ja). 1000 passed Ngulia 101A during 30 minutes in Nov (EABR 1980), and concentrations are particularly associated with storms (La; EABR 1983)(*cf.* Eastern Red-footed Falcon (150)). Large numbers also winter in this SE area, and sporadically in the NE, where our coverage thus appears poor (Br; La; EABR 1982).

Further W, particularly in and W of the rift valley, smaller numbers are mainly on autumn passage, and the species is relatively scarce in Nyanza 60 and E Uganda (Britton & Britton 1976b; D J Pearson *in litt.*). In the rift, mention of only Oct passage at L Turkana (HH) corresponds to seasonality directly to the S at Baringo 50A, where regular Oct-Nov passage contrasts with only occasional birds in Apr (St). The contrast between small S passage numbers at Baringo and large numbers at Turkana (*op. cit.*) demonstrates prevalent movement SE from the latter, around the E periphery of the highlands, to the E and SE areas already mentioned. See Abyssinian Roller (453).

455 LILAC-BREASTED ROLLER *Coracias caudata*
B496 / MPG460 / D3C3

Afrotropical: widely distributed S of the Congo basin forests, apart from the dry SW. Further N, it extends through the E Zaire borders and Tanzania to E Uganda, Kenya, Ethiopia and W Somalia.

Kenyan: this is a common and conspicuous bird of open bushed or wooded habitats up to 2000 m, and two distinctive and hardly intergrading races occur (Br; MPG; Lack 1985). The nominate Lilac-breasted Roller is found SW and W of Marsabit 27D, Chanler's Falls 52A, and the Tana R 65B-91A (along which both races occur), while the Lilac-throated Roller *C.c. lorti* occurs further NE and E (Sn; Br). These two distributions cover the whole country apart from the Turkana basin, where the Abyssinian Roller (453: *q.v.*) is resident. Thus, excluding apparently non-breeding wanderers to this basin (HH; Br),the nominate's squares are approximately 65% above 1000 m, 75% in 500+ mm rainfall areas and 85% semiarid-humid, while *lorti's* are all below 1000 m, 65% in areas with below 500 mm rainfall, and 65% arid. Breeding records for the two races are as closely juxtaposed as Malindi 103A (nominate) and Garsen 91A (*lorti*), but breeding sympatry is unknown (Sn; Br).

Although this species is generally resident (Br; La: 101B; St: 50A; DEP pers. obs.: 75C 88ABD 89AC 100B 101A), both races undertake at least local movements. DEP's (pers. obs.) note of very small numbers of nominate birds annually Aug-Dec in the Ruaraka-Kahawa-Ruiru area of 75B complements Ja's mention of a Nov-May influx to the Nairobi 75B-Murang'a 63C-Machakos 76C area. This race is commoner after the rains in arid Tsavo E 101B (Brown & Brown 1974); and Se recorded two single Sept-Oct wanderers around 3000 m at Mau Narok 61D. Tomlinson (1946) observed a large Jan-Feb influx of *lorti* to Merille 39D 40A, and La mention small numbers penetrating the nominate's range in Tsavo E 101B Dec-Mar.

During early 1984 however, presumably due to severe drought to the N and E, *lorti* moved much further SW than usual, with Jan-Mar records across the plateau and lowlands to the S of the W and C highlands from Kuranze Ranch 113B NW to Ikutha 89A, Thika 76A, Nairobi 75B and Mara 74AC; birds around Nairobi remained into Jul (EABR 1984). Later in that year, an unprecedented June-Aug influx of Abyssinian Rollers was probably due to related factors (see also Rufous-crowned Roller (456) and Nubian Shrike (859)).

456 RUFOUS-CROWNED ROLLER *Coracias naevia*
B498 / MPG461 / D3C3

Afrotropical: widely but somewhat patchily distributed S of 15°N, avoiding only the S tip of the continent, and the equatorial lowland forests of the W.

Kenyan: an uncommon and local bird of open bushed and wooded areas, typically below 1200 m (Br; MPG). Though mainly in drier country, it occurs throughout the rainfall regimes and is regular in some areas of the highlands *e.g.* Kajiado 75D (Br). A Dec wanderer reached 3000 m at Mau Narok 61D (Se). This roller is scarce in Nyanza 60 (Br), while the first record from the coast, during a year of exceptional abundance (see below), involved an individual on the tip of Ras Ngomeni 91C on 26 Mar 1984 (EABR 1984). The single record in 102D was well inland, as was that in 91A (EABR 1984). Although these areas are poorly covered, apparent absence from much of the E 42 53-4 65-6 78-80 may be corroborated by lack of records from adjacent W Somalia, S of 2°N (AM).

Br attributes seasonal fluctuation to the presence of non-breeding migrants from the N tropics. Ja noted Jan-Feb abundance in Ukamba 76 and Taita 101, and Tsavo E 101B records are mostly Mar-Jul (La; Brown & Brown 1974). Variations in numbers also occur from year to year (La), and pronounced abundance in 1984 coincided with severe drought to the N and E, and unprecedented movements of Abyssinian (453) and Lilac-throated (455) Rollers (EABR 1984).

457 BROAD-BILLED ROLLER *Eurystomus glaucurus*
B500 / MPG463 / D2C2

Afrotropical: widely distributed S from 15°N to Angola and E South Africa, but avoiding all but S areas of Somalia.

Kenyan: locally common in wooded habitats, including forest edge and isolated tall trees near water, usually below 2000 m (Br). Range is 82% within 500+ mm rainfall areas, and thus virtually absent from the arid N and E. 4A represents the S limit of birds in Sudan and Ethiopia (J S Ash *in litt.*).

While this roller is present during all months in many areas, some records may refer to non-breeding migrants from Madagascar (Br; *e.g.* EABR 1979). It is a Nov-May migrant to arid Tsavo E 101B, but reappearance there after unseasonal rain in other months suggests movements to be only local (La; Lack 1985). A single late Sept-early Oct wanderer has reached semiarid Baringo 50A (EABR 1983).

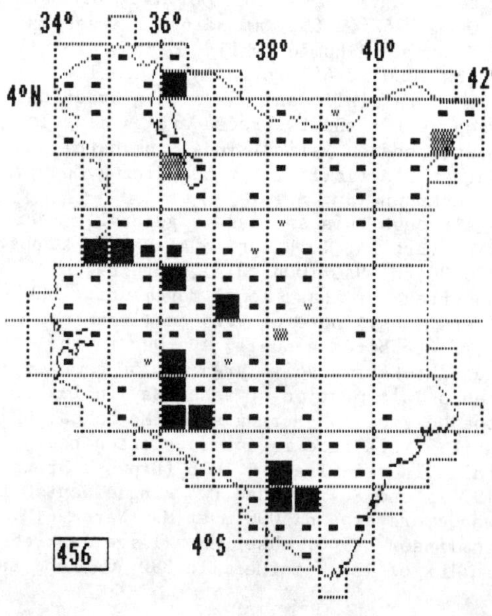

456

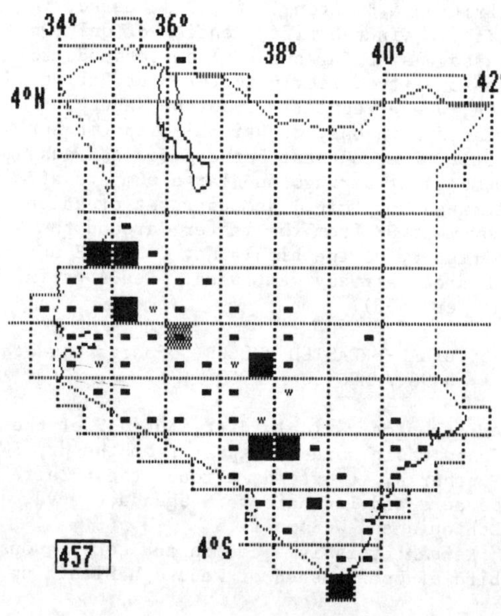

457

Family Upupidae: Hoopoe

A monotypic family, found in the Palearctic, Afrotropical and Oriental Regions. Both resident and migrant races occur in Kenya.

458 HOOPOE *Upupa epops*
B502 / MPG517, 518 / D3C2

Afrotropical: widely distributed S of 17°N, with the exception of most of the equatorial lowland forest areas of the W. Palearctic migrants winter S to about 10°N in the W, and to the equator in the E.

Kenyan: open bushed or wooded habitats, usually below 2000 m (Br; MPG). Only 14% of the squares receive less than 250 mm annual rainfall and, in these areas, it is largely confined to the vicinity of permanent water *e.g.* L Turkana, the Daua R 7C 19AB, the Ewaso Nyiro R 40D and the Tana R 65B, and waterholes at Wajir 42A. General paucity of records in the NE and E is paralleled in S Somalia, where occurrence is mostly N of 4°N (AM).

Three migratory races occur and, in certain months and regions, can by sympatric. The race *U.e. africana* ranges S of approximately 2°N and is reasonably common (Br). Although present throughout the year in the highlands (Br; St: 50A; DEP pers. obs.: 74BD 75BCD 88A), some degree of movement is indicated by influxes to the moist subhumid Nairobi area 75B: in Aug and Oct (Ja), and annually May-Oct (DEP pers. obs.). This bird's breeding squares are 80% above 1000 m and absent from 0-250 mm rainfall areas. Br notes the species' aversion to humid areas and, while this is not apparent from its total squares, 80% of this race's breeding squares are semiarid-subhumid. There are no breeding records from the lowlands SE of 89A, and regular Dec-June non-breeding birds in Tsavo 89CD 101AB may be migrants from S Tanzania, where it occurs Jul-Dec (Br; La; Lack 1985). La describe this race as rare June-Dec in Tsavo E 101B, and AGH note Jul-Aug occurrence on the lower Tana R 79C 91A. There are two Sept records from the far S of Somalia (AM).

The nominate Palearctic race typically occurs in dry bushland, S to Baringo 50A, Nairobi 75B, Kibwezi 88B, Ngulia 101A, and Tsavo E 101B (Br; EABR 1983, 1984). Generally uncommon, it is locally frequent in some N localities *e.g.* Baringo 50A (*op. cit.*). It occurs Oct-Mar, and Ja quotes a June oversummering record from the N Ewaso Nyiro R (possibly in the 51-52 area).

The race *U.e. waibeli* is a non-breeding Nov-Mar migrant from the N tropics: it has been recorded S to Nakuru 62A and Nairobi 75B (Ja), and may be resident in far N Kenya (Br). Other reference: Pollard 1946b.

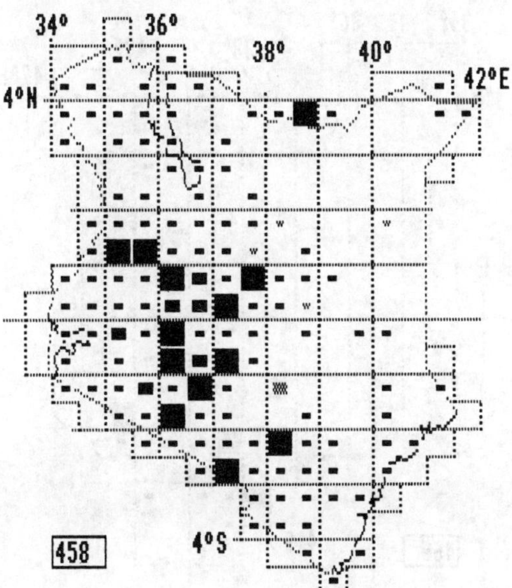

Family Phoeniculidae: Wood Hoopoes, Scimitarbills

Seven species, endemic to the Afrotropical Region. Five breed in Kenya, while the presence of a sixth requires confirmation.

459 GREEN WOOD HOOPOE *Phoeniculus purpureus*
B508 / MPG519 / D2C2

Afrotropical: widely distributed S of 15°N, avoiding only the drier areas of E Somalia and the SW, and the equatorial lowland forests of the W.

Kenyan: locally common in wooded habitats up to 2800 m (Br; St; Lack 1985). Range is only 13% within the driest, 0-250 mm rainfall areas, and scarcity in much of the NE and E, particularly away from permanent water, is apparently real. This species is similarly largely restricted to the vicinity of the Juba R in adjacent S Somalia (AM). The black-billed race *P.p. somaliensis*, thought to be an incipient species (Sn), is predominantly confined to the far NE 7C 16-19 29-30 (Br). This species is probably mostly resident (*e.g.* Br; La; St; DEP pers. obs.). Se lists single Oct and Nov transients around 3000 m at Mau Narok 61D.

The Green Wood Hoopoe forms part of a superspecies with the Violet (460: *q.v.*) and White-headed (461) Wood Hoopoes, and is mainly allopatric with the former (Sn; Br). Both have, however, been collected on the same day at Archer's Post 51B (Sn), and AGH mention the two in forest along the lower Tana R 79C 91A. Both also occur in the vicinity of Bushwackers' Camp 89A (many obs.). Other reference: Ligon & Ligon 1978.

460 VIOLET WOOD HOOPOE *Phoeniculus granti*
B506 / MPG520 / D1C2

Afrotropical: endemic to E Kenya.

Kenyan: a generally uncommon species of bushed and wooded habitats, usually near water and palm trees (MPG; Br). It is found in the E lowlands and plateau, with range 81% below 1000 m and 88% arid-semiarid (*cf.* the Green Wood Hoopoe's (459) 46% and 63% respectively). Br mentions the Violet Wood Hoopoe N to 2°N, but this range is extended by records from 17C and 29B (J E Miskell *in litt.*). The high proportion of pre-1970 squares is likely to reflect a combination of the modern disinclination for collecting specimens, confusion of sight records with the better known but similar Green Wood Hoopoe (hence D1) and, particularly N of the equator, low obser-

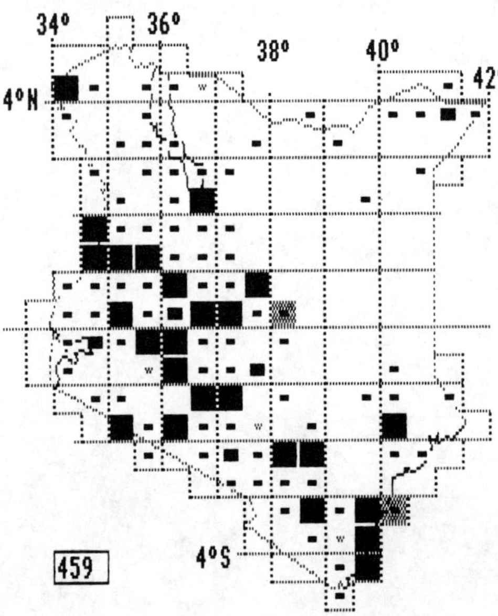

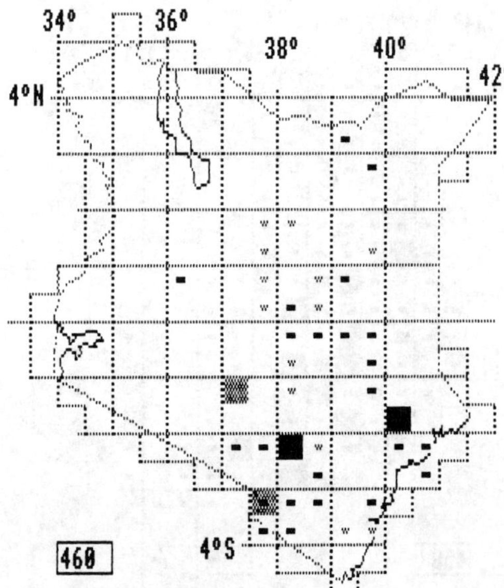

ver frequencies. Extralimital wanderers may also be a factor. These are thought to be the source of records within the Green's range, at the Amala R (approximately 61C: not mapped) and Takaungu 102D (Ja), S of Maralal (probably 50B: not mapped: Br), and at L Baringo 50A Feb-Apr 1978 (St). Other references: Brown 1963d; Kaestner & Kaestner 1978.

461 WHITE-HEADED WOOD HOOPOE *Phoeniculus bollei*
B503 / MPG522 / D3C2

Afrotropical: the forests around the Gulf of Guinea, from Ivory Coast to Cameroon, and E across N areas of the Congo basin forests to the E Zaire borders, S Sudan, Uganda, NE Tanzania and, at the E limit of the range, the highlands of Kenya.

Kenyan: a reasonably common resident of highland forest, mainly at 2000-3000 m but exceptionally down to 1500 m (Br; Ja; MPG; Se; Be). Range is 68% within 1000+ mm rainfall areas and 80% moist subhumid-humid. It does not extend to forest islands. See Green Wood Hoopoe (459).

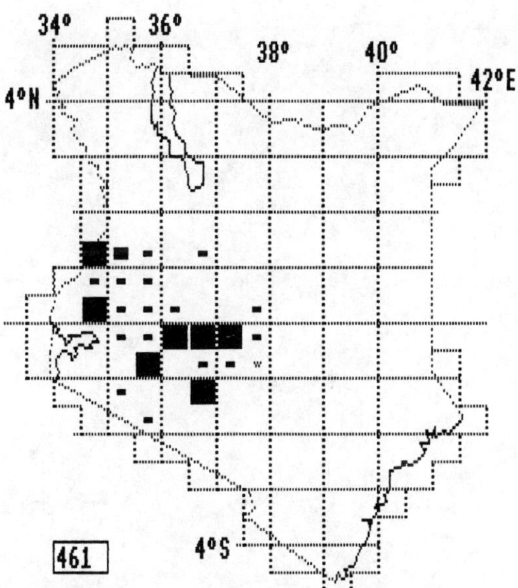

462 SCIMITARBILL *Phoeniculus cyanomelas*
B505 / MPG523, 526 / D2C2

Afrotropical: widely distributed S of 16°N, apart from E Somalia, areas S of 30°S, and the equatorial lowland forests of the W.

Kenyan: moderately common in bushed and wooded country up to 2400 m (Br; Ja). Range is 72% in 500+ mm rainfall areas and 76% semiarid-humid. Occurrence in the arid N

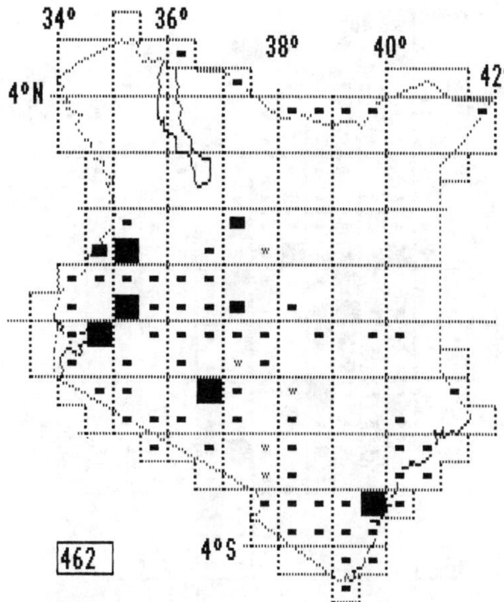

and E is virtually confined to locally moist situations *i.e.* the S edges of the population in the Ethiopian highlands 4A 5C 16AB 17AB (our environmental data from these border squares may not be entirely accurate, so that they may not actually be arid), the Daua R 19B and the Tana R 64B 65B 79C (AGH). Presence in arid Tsavo E 101B is restricted to two Mar records, which presumably represent wanderers beyond their usual range (La; Br).

The Scimitarbill forms a species-group with the smaller Abyssinian Scimitarbill (463)(Sn), which is typically in lower and drier country. Where apparently sympatric, these two are segregated by differing habitat preferences, with the Abyssinian on more sparsely vegetated ground (Br). The Scimitarbill is particularly common in the coastal strip (MPG; Br), where the Abyssinian is mostly absent.

463 ABYSSINIAN SCIMITARBILL *Phoeniculus minor*
B507 / MPG527 / D2C2

Afrotropical: the NE, from E Tanzania through Kenya and E Uganda to S Sudan, Ethiopia and Somalia.

Kenyan: a sometimes common bird of open bushed and wooded country, exceptionally reaching 2000 m but mainly below 1400 m (Br; Ja; MPG; HH; La; St). In contrast to the Scimitarbill (462: *q.v.*), range is 80% arid-semiarid. It is noticeably absent from the higher rainfall areas of the highlands 61 63, and from the coastal strip S

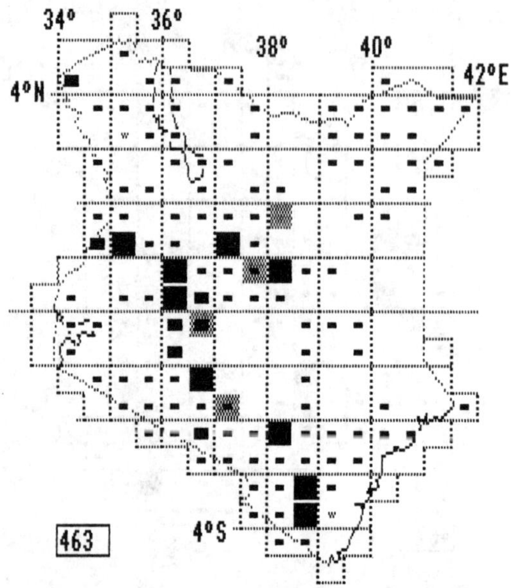

of 2°30'S. This species is considered resident (Br; La: 101B; DEP pers. obs.: 74BD 75CD 88A 89A). Other reference: Madge & Cunningham-van Someren 1975.

- FOREST WOOD HOOPOE *Phoeniculus castaneiceps*
  B504 / MPG524 / D1C0
  NO CONFIRMED RECORDS

Afrotropical: the W and C African equatorial lowland forests, E to the Kampala-Elgon area of E Uganda.

Kenyan: the sole evidence for this species' occurrence seems to be Ja's note of it on the Kenyan side of Mt Elgon 36D, and White's (1965) description of range from Elgon to Kavirondo 60 (Mann 1976). These statements require confirmation (Br).

Family Bucerotidae: Hornbills

46 species, confined to the Afrotropical and Oriental Regions. 13 of the 23 Afrotropical members (57%) are known in Kenya, where 11 (85%) breed regularly, one may do so, and one is an extralimital wanderer.

464 BLACK AND WHITE CASQUED HORNBILL *Bycanistes subcylindricus*
B513 / MPG500 / D2C2

Afrotropical: W and C Africa, through the N reaches and adjacent areas of the equatorial lowland forests, from Ivory Coast E to Nigeria, S Chad and S Sudan, and thence S to NW Tanzania and the E Zaire borders, and E through Uganda to the E limit of the range in W Kenya. There is also an isolated population in Angola.

Kenyan: a common and conspicuous species of wooded and forested habitats, including gardens and isolated stands of fruiting trees, at 1500-2600 m (Br; Ja; MPG; Be). Range is 80% within 1000+ mm rainfall areas and 66% humid. In striking contrast to the Silvery-cheeked Hornbill (465: *q.v.*), breeding squares constitute 38% of its total. MPG describe breeding as regular around 2400 m on Mt Elgon 36D. Be considered this hornbill resident along the Mara R (approximately 61CD). Other reference: North 1942b.

465 SILVERY-CHEEKED HORNBILL *Bycanistes brevis*
B509 / MPG501 / D2C2

Afrotropical: the E, from Ethiopia and S Sudan, S through Kenya and E Tanzania, to Malawi, Mozambique and NE South Africa.

Kenyan: locally common up to 2600 m in similar habitats to the allopatric Black and White Casqued Hornbill (464), with which it forms part of a species-group (Br; MPG; Sn). Range is 82% within 500+ mm rainfall areas. Apart from the Londiani-Molo area 61B (D A Turner pers. comm.) and the E edge of the Loita Hills 74D (Br; Fayad & Fayad 1980), the Silvery-cheeked's squares are all in or E of the rift valley. The Black and White Casqued Hornbill replaces it further W. Apparent sympatry with the Trumpeter Hornbill (466) may involve wanderers of either or both species (Br).

Although free-flying but dependent immatures are known from both the highlands and the coast, the restriction of actual nesting records to a single example is intriguing in that (a) this record was on

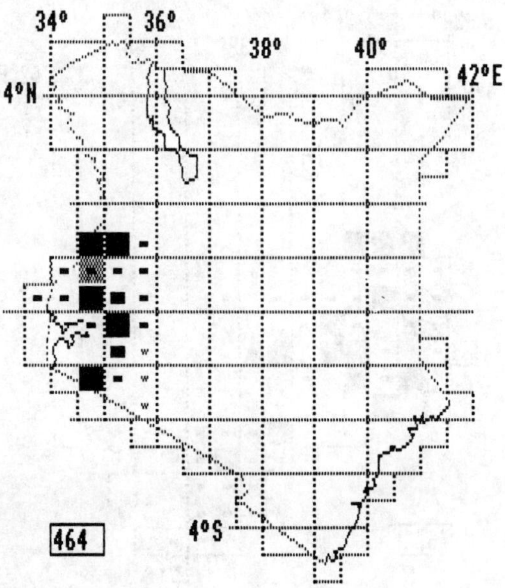

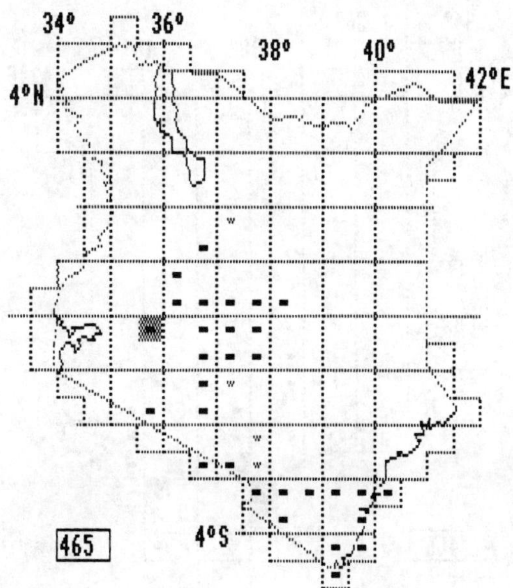

the W limit of the range, at Molo 61B (Start & Start 1978), (b) this hornbill's range includes well observed areas *e.g.* the C highlands and the coastal strip, (c) Moreau (1935) makes no mention of the Silvery-cheeked being particularly secretive while nesting in Tanzania, and (d) there are nesting records for the Black and White Casqued Hornbill, which inhabits more poorly known areas to the W of the rift. Hence the Silvery-cheeked can only be presumed to breed regularly in Kenya. The alternative, that it is mainly a non-breeding visitor from NE Tanzania, appears unlikely but cannot be totally precluded.

The Molo breeding began in Oct, and the family party left the area in Mar, as part of a large flock (*op. cit.*). In arid Tsavo E 101B, the three records subsequent to former occurrence in the Voi R Forest have been Oct, Feb and Mar (La). Other references: Baker & Allen 1978; Short & Horne 1985b.

466 TRUMPETER HORNBILL *Bycanistes bucinator*
B510 / MPG497 / D2C2

Afrotropical: from E South Africa N to Zimbabwe, Angola, S Zaire, W Tanzania and Malawi, and through E Tanzania to the NE tip of the range in Kenya.

Kenyan: locally common in wooded and forested areas of the SE lowlands and plateaux, and up to 1650 m on the SE periphery of the C highlands (Br; MPG; Lewis 1982e). Range is 73% within 500+ mm rainfall areas. Br notes occurrence in the coastal strip S to Sokoke Forest 102BD, with the Silvery-cheeked Hornbill (465: *q.v.*) extending from this forest S. These distributional limits are incorrect, however, since M A C Coverdale, who was resident in 114B, states (*in litt.*) that a few Trumpeters are usually present in the Shimba Hills 114A and in Gogoni Forest 114A, with rare wanderers to Diani Forest 114B. Br considers the Trumpeter Hornbill to be a resident and wanderer, and DEP (pers. obs.) found it resident in 88D 89C 100B. Other reference: Pomeroy & Tengecho 1980.

467 GROUND HORNBILL *Bucorvus cafer*
B528 / MPG515 / D3C2

Afrotropical: widely distributed to the S of the Congo basin forests, except for the dry SW. To the E of these forests, it extends N through the E Zaire borders and Tanzania to the N edge of the range in E Uganda and Kenya.

Kenyan: a locally common, predominantly terrestrial bird, found up to 3000 m in open habitats that serve to reduce the risk of predation (*cf.* Ostrich (1) and Kori Bustard (209))(Br; Se; Ja; MPG). Range is 79% within 500+ mm rainfall areas but of restricted extent in the moist coastal strip. Since records in 101B are confined to the W edge of the square (La) and the bird is very uncommon on Galana Ranch 90C (B R Heath *in litt.*), the population further E along the lower Tana R 79C 91AC may be isolated. This hornbill is considered resident (Br; Be: 74; Se: 61D;

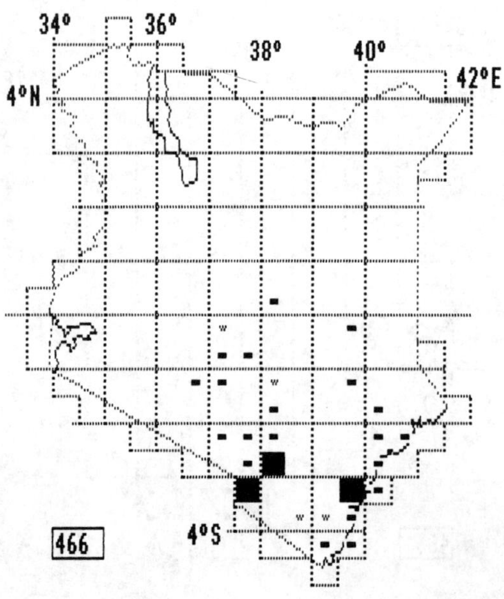

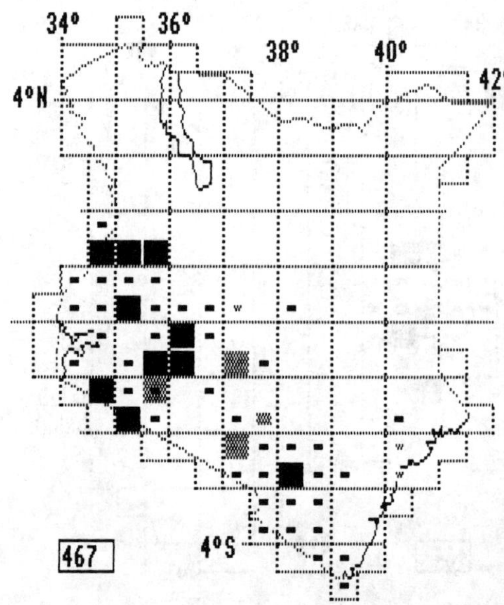

DEP pers. obs.: 88D 89C 100B 101A), although Br also mentions wanderers. The latter are indicated by occasional appearance in Tsavo E (La) and at Bushwackers 89A (DEP pers. obs.).

The Ground Hornbill forms a species-group with the very similar Abyssinian Ground Hornbill (468), which replaces it in the N tropics (Sn). Their ranges come into contact just to the N of the equator, in Kenya and Uganda (Br). In Kenya, their breeding is completely allopatric at QSD level, and the seven squares of overlap for presence records probably refer to occasional wanderers (see above) *e.g.* a Dec record of the Abyssinian from Ol Joro Orok 62A, at approximately 2500 m (D K Bednall *in litt.*). Other references: Stagg 1972; Hofer 1982.

468 ABYSSINIAN GROUND HORNBILL *Bucorvus abyssinicus*
B527 / MPG516 / D3C2

Afrotropical: the N tropics, from the W coast between 15°N and the N edge of the equatorial forests to Ethiopia, and SE to S Sudan, N Uganda and, on the SE edge of the range, NW Kenya.

Kenyan: typically a rather uncommon resident of the arid-semiarid NW, S to Baringo 50A (Br; HH; St; EABR 1983). Records up to 2500 m in the highlands (*e.g.* 49ABC: Br) are well within the breeding range of the Ground Hornbill (467: *q.v.*), and probably refer to wanderers. Even including this presumed wandering, this species clearly contrasts with the Ground Hornbill by occupying only 13% of the moist subhumid-humid squares, as against 72%.

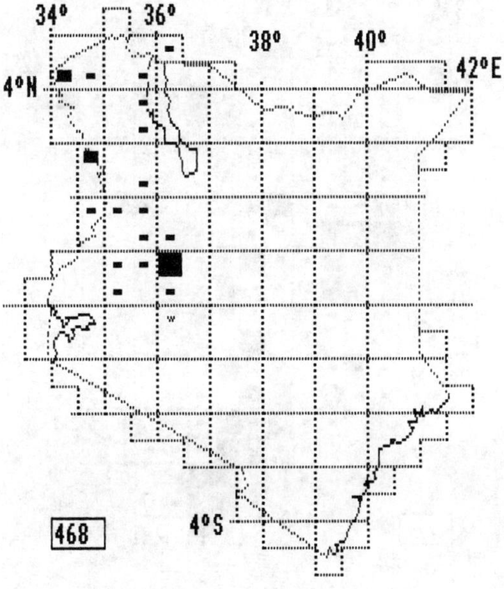

469 YELLOW-BILLED HORNBILL *Tockus flavirostris*
B520 / MPG506 / D3C2

Afrotropical: occupies two distinct areas. In the NE, it extends from NE areas of Uganda and Tanzania, through S Sudan and Kenya to Ethiopia and Somalia. It is also widespread between 15°S and 30°S.

Kenyan: generally rather uncommon in open bushed and wooded country, mainly below 1400 m (Br; Ja; HH; St). Range is 89% arid-semiarid, and absent from the moister parts of the highlands and the W. Absence from even semiarid parts of the coastal strip is paralleled in Somalia (AM). Br thinks this species resident, and the only mention of movement concerns a slight influx to arid Tsavo E 101B during the Apr rains (Lack 1985). This hornbill is widely sympatric with Von der Decken's (470: *q.v.*) and Jackson's (471) Hornbills (Sn). Other reference: North 1942b.

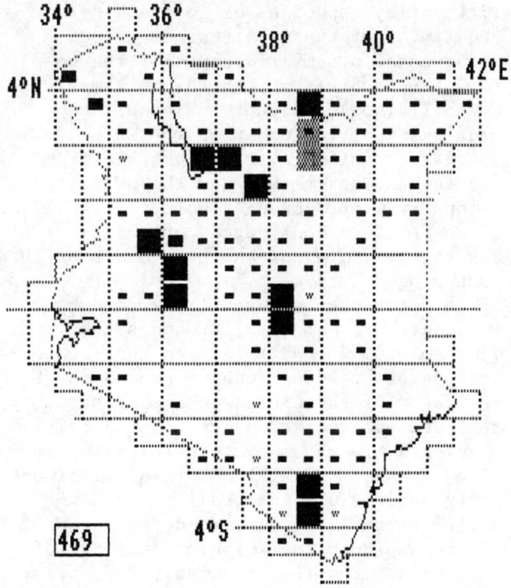

470 VON DER DECKEN'S HORNBILL *Tockus deckeni*
B517 / MPG507 / D3C3

Afrotropical: the E and NE, from C and E Tanzania through Kenya to SE Ethiopia and all but the arid E of Somalia.

Kenyan: common in open bushed and wooded habitats up to 1700 m (Br; MPG; Lack 1985). Range is 84% arid-semiarid, and mostly ab-

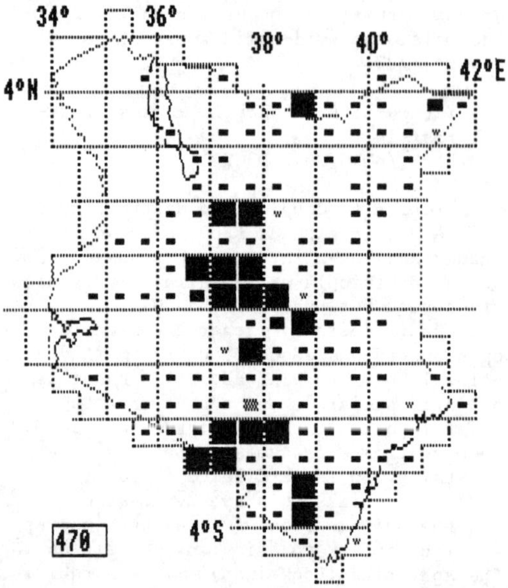

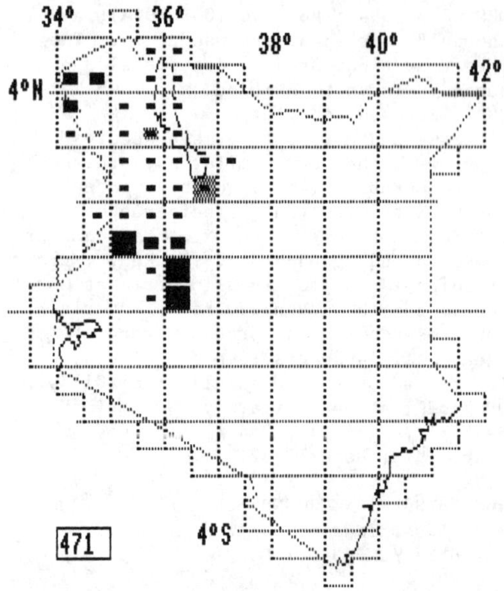

sent from the highlands and the moist W. This hornbill is mainly in or E of the rift valley, and squares to the W define the limits of the Afrotropical range. Occurrence in semiarid parts of the coastal strip contrasts with the Yellow-billed Hornbill (469), which is an otherwise widely sympatric member of the same species-group(Sn). Virtual absence from the Turkana basin (HH) and the NW reflects allopatry with Jackson's Hornbill (471), which is often considered conspecific (Sn; Ja): the two are not known to occur on the same ground (Br).

Von der Decken's Hornbill is thought resident (Br; La: 101B; DEP pers. obs.: 75CD 88AD 89AC 100B 101A), so that concentration of breeding squares around the E periphery of the C highlands most probably reflects observer frequency; but some movements are known *e.g.* influxes to fruiting trees in arid Tsavo E 101B, particularly during Jul-Sept (La)(*cf.* Yellow-billed Hornbill). Occurrence to the W of L Turkana (HH), well within the range of Jackson's Hornbill, presumably reflects wandering.

471 JACKSON'S HORNBILL *Tockus jacksoni*
B523 / MPG508 / D2C2

Afrotropical: the NE, from NW Kenya and NE Uganda to S Sudan and C Ethiopia.

Kenyan: relatively common in open bushed and wooded country of the NW, mainly at 350-1400 m but also reaching 2000 m (Br; Ja; HH; St). Range is 71% less arid-semiarid, and 65% within intermediate, 250-1000 mm rainfall areas. See Yellow-billed (469) and Von der Decken's (470) Hornbills. Other reference: Williams 1978a.

472 RED-BILLED HORNBILL *Tockus erythrorhynchus*
B518 / MPG505 / D3C3

Afrotropical: the N tropics between 9°N and 17°N from the W coast to Ethiopia and N Somalia, and thence S to NE Uganda,

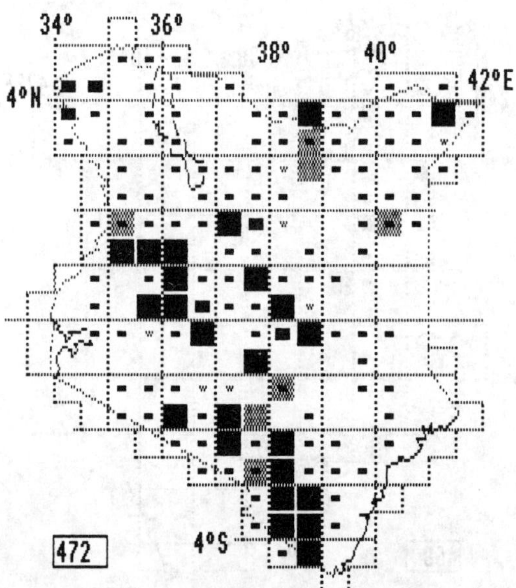

Kenya, Somalia and interior Tanzania. Further S, it extends W to Angola and Namibia, and S to NE South Africa.

Kenyan: common in open bushed and wooded country, mainly below 1400 m but also reaching 2000 m (Br; Ja; MPG; HH; St; Lack 1985), and widely sympatric with the Von der Decken's Hornbill (470) species-group. Range is 81% arid-semiarid (*cf.* Br), and the bird is scarce in the highlands and moist W. Absence from even semiarid areas of the coastal strip is paralleled by the Yellow-billed Hornbill (469). Absence from the E borders S of 1°30'N may be real in view of similar status in adjacent S Somalia (AM).

This hornbill is considered resident by La (101B) and DEP (pers. obs.: 75CD 88AD 89AC 100B 101A), although some movements occur. Possibly moving in and out of the arid Tsavo E area 101B, it is most common there during the wet season (Lack 1985; *cf.* Yellow-billed and Von der Decken's Hornbills). Birds on the E slopes of Mt Kenya 63B migrate to lower levels during the wet seasons (BUN)(*cf.* Olive Pigeon (325)). Other references: North 1942b; Williams 1978a.

473 HEMPRICH'S HORNBILL *Tockus hemprichii*
B522 / MPG512 / D2C1

Afrotropical: the NE, from Ethiopia and N Somalia SW to S Sudan, NE Uganda and, on the S limit of the range, Kenya.

Kenyan: uncommon and local, in the vicinity of rocky hillsides and cliffs (Br; MPG). Distribution is in two distinct areas. Records from near Lokichokio 2C (National Museum Collection, Nairobi; T Romig *in litt.*), Kamathia 3A Feb (D J Pearson, ADL pers. obs.) and Moyale 17A Oct (Ja) represent the S edge of the Sudan/Ethiopian population. These squares are arid-semiarid, at 500-1100 m. Further S, localities in the W and C highlands from Kongelai and the Marich Pass 37C SE to Nakuru town and lake 62A (Mann 1976; Hegner 1980; St; ADL pers. obs.; EABR 1978, 1983) are semiarid-humid, at approximately 1000-2000 m. Roughly midway between these two areas, the sole locality in Uganda is at Mt Moroto 24B (Br: not mapped).

This hornbill is a breeding resident at Baringo 50A (St; Williams 1978a). Range appears to be extending S into the Nakuru area 62A (Hegner 1980), and there is a recent breeding record from the nearby Menengai Crater (J P Dittami pers. comm.). Flocks of up to 14 at Nakuru 62A are thought to indicate post-breeding dispersal (Hegner 1980), and St noted a Jan flock of 13 at Baringo 50A.

Hemprich's Hornbill forms part of a species-group with the Crowned Hornbill (474)(Sn). Other reference: Short & Horne 1985b.

474 CROWNED HORNBILL *Tockus alboterminatus*
B515 / MPG509 / D2C2

Afrotropical: widely distributed to the S of the Congo basin forests, but absent from the dry SW. Further N, it extends through E Africa and the E Zaire borders to S Sudan, W Ethiopia and extreme S Somalia.

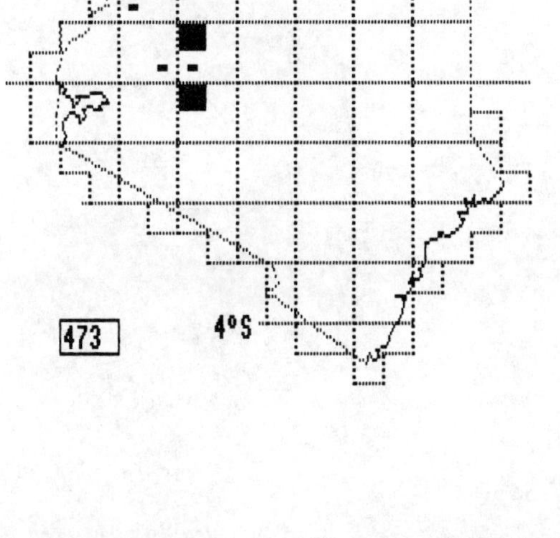

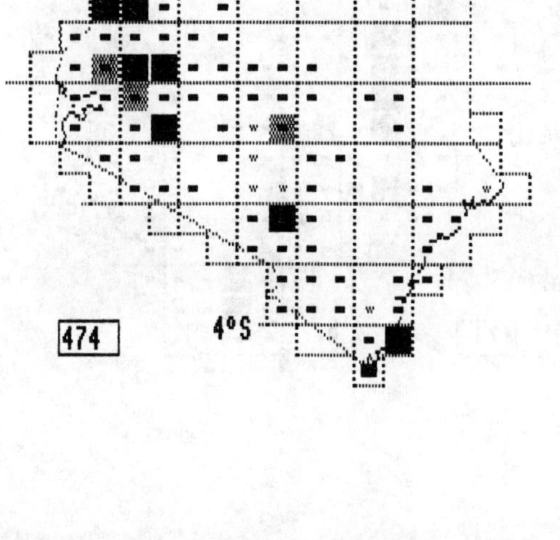

Kenyan: common in wooded habitats including forest edge up to 3000 m, although scarce in the L Victoria basin (Br; Se; MPG). Range is 81% within 500+ mm rainfall areas. Occurrence in arid country is in locally moist situations *e.g.* along the Tana R 65ABD 79C (AGH), on the Mt Endau forest island 77B (Tennent 1964; A W Diamond *in litt.*) and in the Voi R Forest 101B (La). This hornbill is thought to be resident (Br; Se). At the local limits of the range, occasional birds in the Voi R Forest are mostly Nov-Mar (La), while a single wanderer appeared at Baringo 50A in Jul (St). See Hemprich's Hornbill (473). Other reference: Siemens 1981.

475 GREY HORNBILL *Tockus nasutus*
B524 / MPG504 / D2C2

Afrotropical: the N tropics, between approximately 7°N and 17°N, from the W coast to Ethiopia, and thence S to East Africa and S Somalia. Further S, it extends SW to Angola and Namibia, and S to NE South Africa.

Kenyan: common in open wooded habitats up to 1700 m (Br; MPG). Range is 66% semiarid-subhumid and 70% within the intermediate, 250-1000 mm rainfall regime, and Br notes this bird's aversion to the driest and wettest areas. Exceptions occur, however, *e.g.* frequent sightings in the woods of the Turkwell Delta 14C at L Turkana (HH), and locally common occurrence in some humid areas to the W of the rift valley 60 (Ja; Br).

Although present throughout the year in many areas (La: 101B; DEP pers. obs.: 75C 88D 89AC 100B 101A), seasonal fluctuations certainly occur (Br; MPG). May-Jul absence or scarcity at Baringo 50A contrasts with common occurrence throughout the rest of the year (St), while birds on the E slopes of Mt Kenya 63B move to lower levels during the wet seasons (BUN)(*cf.* Olive Pigeon (325)). Furthermore, breeding records are entirely above 500 m and, while observer frequency may have contributed to this pattern, there is a notable absence of reports from the well covered coastal lowlands. This apparent absence of breeding continues to the NE in S Somalia, where AM can only presume the species resident. This raises the possibility of a migrant coastal lowlands population, although Lack (1985) could not detect seasonal fluctuations in nearby Tsavo E 101B.

The Grey Hornbill forms a species-group with the mainly allopatric Pale-billed Hornbill (476)(Sn; Br), which is an extralimital wanderer to Kenya. Other reference: Lewis 1984c.

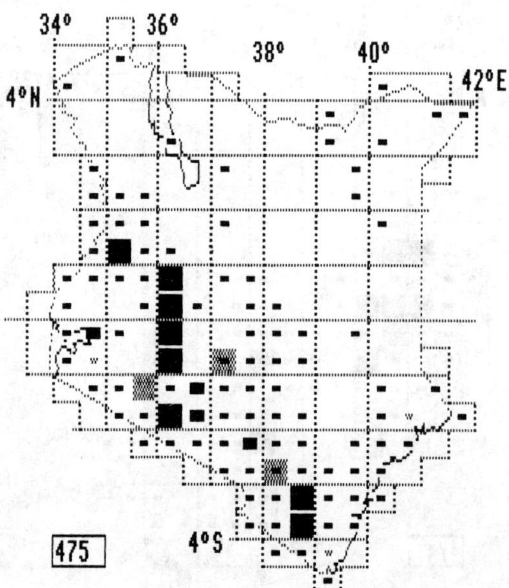

476 PALE-BILLED HORNBILL *Tockus pallidirostris*
B525 / MPG511 / D2C0
NO MAP: pre presence 100B

Afrotropical: widely distributed in the S tropics, approximately between 7°S and 16°S. Scattered occurrences in E Tanzania culminate northwards in a single old record on Kenya's S border.

Kenyan: the single record is from Taveta 100B (Ja). Lack of subsequent records, despite specific searches, suggests an extralimital wanderer. See Grey Hornbill (475).

Family Capitonidae: Barbets, Tinkerbirds

Approximately 79 species, mainly in the Afrotropical and Oriental Regions. Only 12 (15%) are Neotropical. 42 (53%) are found in the Afrotropical Region, with 23 (55%) known in Kenya. Migratory movements are not thought to occur (Br), and all of Kenya's species either breed locally or are presumed to do so.

477 DOUBLE-TOOTHED BARBET *Lybius bidentatus*
B534 / MPG573 / D3C2

Afrotropical: occurs in the lowland forests and adjacent habitats around the Gulf of Guinea, from Senegambia to Angola. Further E, it extends around the N and E peripheries of the Congo basin forests, E to W Ethiopia, S Sudan, W Kenya and NE Tanzania.

Kenyan: common in forest edge, wooded and thicketed habitats, up to 2300 m (Br). A record in 4B to the N of L Turkana (J S Ash *in litt.*) represents the S tip of the Sudan-Ethiopian population, but is not mapped here. Excluding this record, this barbet is confined to subhumid-humid country above 1000 m, to the W of the rift valley. Range is 80% in 1000+ mm rainfall areas. Ja mentions local movements to fruiting trees.

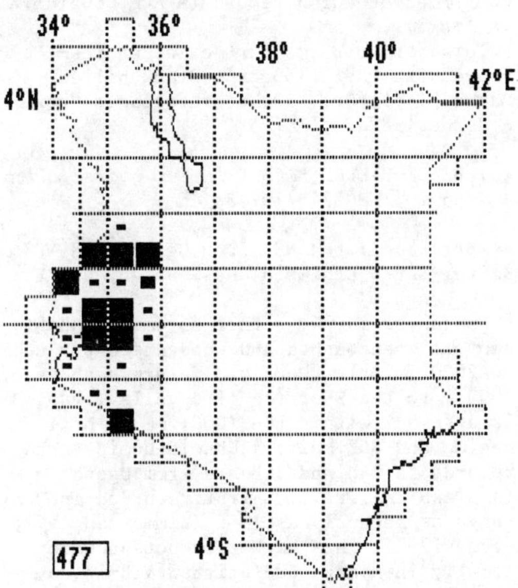

478 BROWN-BREASTED BARBET *Lybius melanopterus*
B542 / MPG580 / D2C2

Afrotropical: an E species, from Malawi and N Mozambique through E Tanzania to Kenya and S Somalia.

Kenyan: frequents wooded habitats, including forest edge and isolated fruiting trees (Br; Ja; MPG; Lack 1985), in the S and SE lowlands; range is 71% below 500 m. This barbet extends inland along the Tana R 65BD 79C, where it can be locally common (Br), and a Dec wanderer has reached NW to the Ewaso Nyiro R 51B (Lewis 1984d). Similar occurrence inland is observed along the Juba and Shebelle R of S Somalia (AM). Local movements to fruiting trees are known (Ja; La).

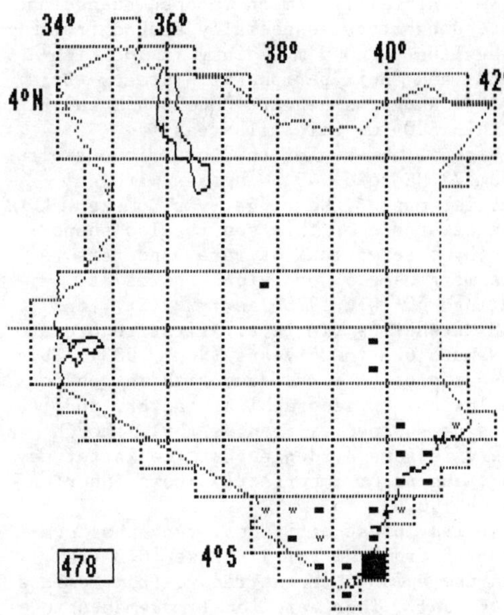

479 WHITE-HEADED BARBET *Lybius leucocephalus*
B540 / MPG576 / D3C2

Afrotropical: occupies three distinct areas. In the E, it is found from S Sudan and NE Zaire, SE through Uganda and N Tanzania, to the E limit of the range in Kenya and E Tanzania. There are other populations in Angola, and N from the Congo basin forests to N Nigeria and Chad.

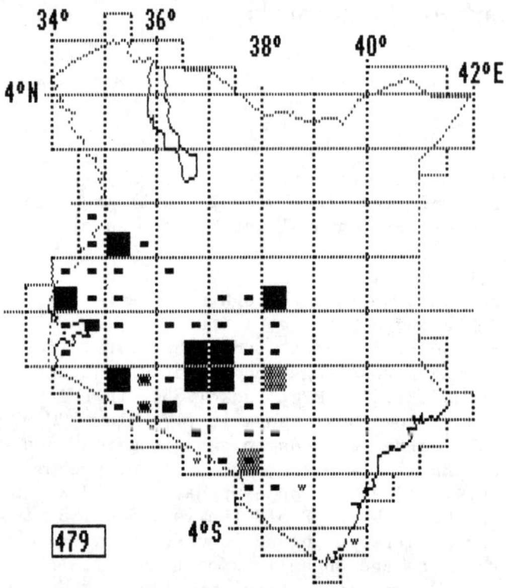

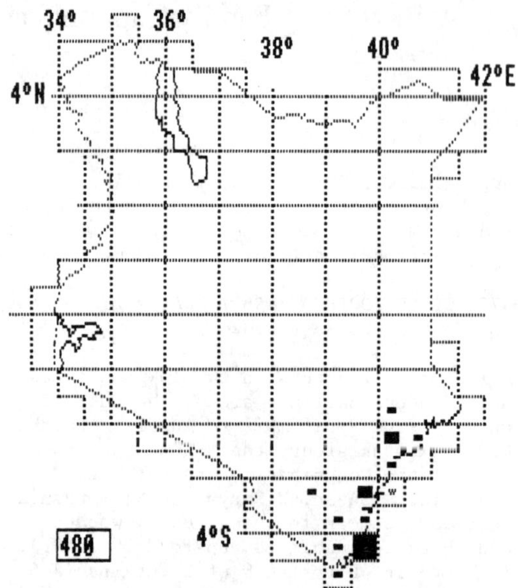

Kenyan: fairly common in open bushed and wooded habitats (especially around fruiting trees) up to 2000 m, mainly in the highlands and their peripheral plateaux (Br; MPG). Range is 66% at 1000-2000 m and 84% within 500+ mm rainfall areas.

Virtual absence of this conspicuous bird from 73 NE to 61 49 50 agrees with an earlier map of the range (van Someren 1939). The existence of this gap is also supported by observations of rare wanderers to its margins *e.g.* occasional birds at Baringo 50A Dec 1978-Jan 1979 (St), and I M Marshall's (*in litt.*) first record at L Nakuru 62A in 26 years (Sept 1983). This zone cannot be correlated with any obvious ecological or geographical factor, and it crosses subspecies' ranges (Br), though there is some evidence that the latter may be changing or imperfectly known (Short & Horne 1982b).

At its lowest altitudes, range has contracted from the Voi R Forest 101B (La), and there is no recent record from Mombasa 114B (Ja). There are local movements to fruiting trees (Ja). Other reference: Stagg 1973.

480 BLACK-COLLARED BARBET *Lybius torquatus*
B546 / MPG575 / D2C3

Afrotropical: a S bird, widely distributed S of 10°S, apart from the dry SW. Further N, it extends to SE Zaire and S Tanzania, and N through E Tanzania to the N tip of the range in the coastal lowlands of Kenya.

Kenyan: locally common in wooded habitats of the coastal lowlands and the lower Tana R 79C (Br; MPG; AGH; Britton & Zimmerman 1979). Range is 91% within the 500+ mm rainfall regime. The sole record outside this regime concerns a single Dec wanderer to the Voi R 101B (La). There is a single confirmed breeding record (M A C Coverdale *in litt.*).

This barbet forms part of a species-group with the allopatric Black-billed Barbet (481)(Sn). Other reference: Short & Horne 1982a.

481 BLACK-BILLED BARBET *Lybius guifsobalito*
B537 / MPG574 / D2C3

Afrotropical: the NE, from Uganda and NE Zaire, NE to Ethiopia and S Sudan, and E to W Kenya.

Kenyan: a mainly rather uncommon inhabitant of open bushed and wooded country up to 2000 m (Br). Confined to areas above 1000 m to the W of the rift valley, range is also 92% within the 1000+ mm rainfall regime and 85% moist subhumid-humid. The records in 24B and 36B are from Uganda (J S Ash *in litt.*). First recorded in Kenya as recently as 1965 (Zimmerman 1967; Payne 1967), this barbet is not uncommon locally in Nyanza 60 (Britton & Britton 1971b), and E range extension is apparently taking place (Br). See Black-collared Barbet (480). Other references: Mann 1976; EABR 1980.

(map overleaf)

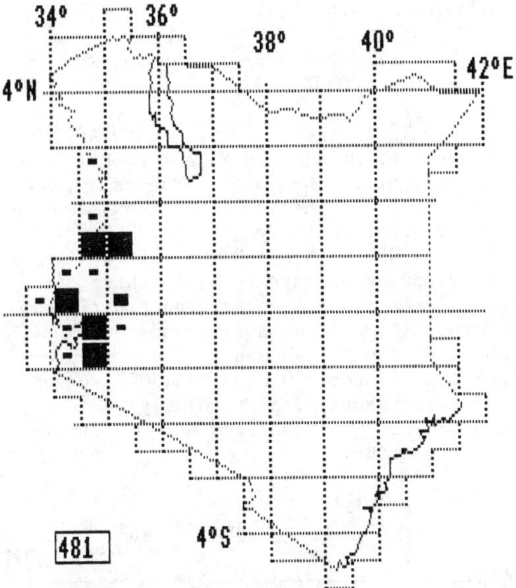

482 RED-FRONTED BARBET *Lybius diadematus*
B535 / MPG585 / D2C2

Afrotropical: the E and NE, from E Tanzania, through Kenya and N Uganda, to S Sudan and Ethiopia. Distribution in Kenya represents the E limit of the range.

Kenyan: a rather uncommon barbet of open bushed and wooded habitats at 600-2100 m (Br; St). Range is 86% above 1000 m, mainly in the highlands and their peripheral plateaux, and 81% within 500+ mm rainfall areas. The two arid squares are 77B, which contains the locally moist Mt Endau forest island, and Tsavo E 101B, where the bird was formerly rare and is now absent (La). See Spotted-flanked Barbet (484).

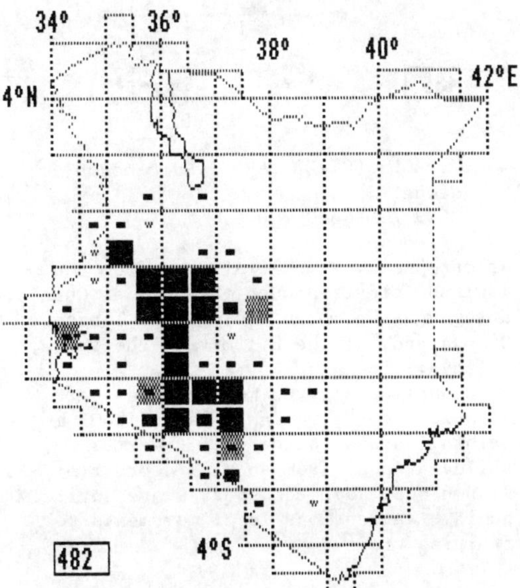

483 BLACK-THROATED BARBET *Lybius melanocephalus*
B541 / MPG583 / D2C3

Afrotropical: a NE species, from E Tanzania through Kenya to S Sudan, Ethiopia and Somalia.

Kenyan: sometimes common in dry bushed and wooded country, up to 1500 m (Br; MPG; HH; La; St). Range is 69% within 0-500 mm rainfall areas and 89% arid-semiarid: this barbet is virtually absent from the highlands and moist W 48 49 60-63, and also from even semiarid areas of the coastal strip. A single Sept record from 1550 m in Nairobi NP 75B (EABR 1980) presumably represents a local wanderer. Presence in 2C 25C 36B presumably means that this barbet reaches the extreme NE of Uganda, although there are as yet no records from that country. See Spotted-flanked Barbet (484).

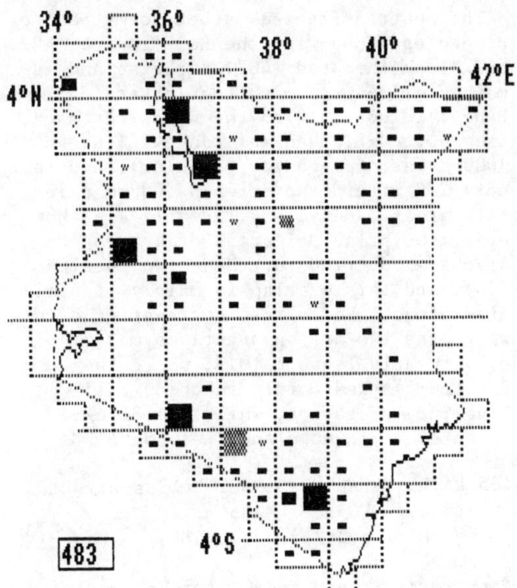

484 SPOTTED-FLANKED BARBET *Lybius lacrymosus*
B539 / MPG584 / D2C3

Afrotropical: the E, from NE Zambia, the E Zaire borders, Uganda and S Sudan to the E limit of the range in Kenya and E Tanzania.

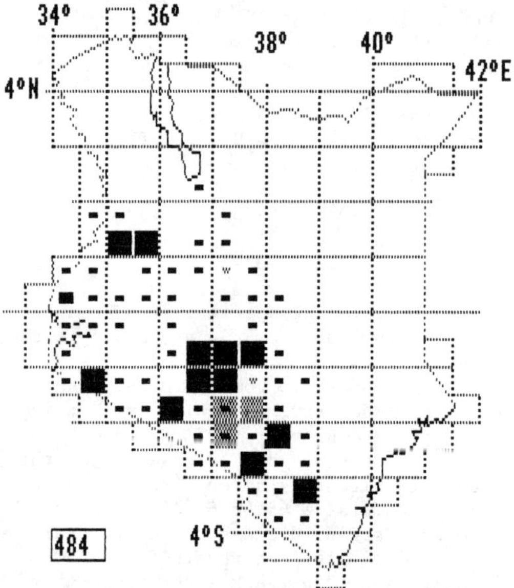

Kenyan: locally common in bushed and wooded habitats at 500-1800 m, mainly in the highlands and their peripheral plateaux (Br; MPG). Ja mentions local movements to fruiting trees.

The Spotted-flanked Barbet forms part of a species-group with the Red-fronted (482) and Black-throated (483) Barbets, and is mainly allopatric with the latter (Sn; Br). Hence, in contrast to the Black-throated, the Spotted-flanked has range 77% above 1000 m (as opposed to 70% below), and is only 27% within the drier 0-500 mm rainfall areas. Kenya and Tanzania are, however, the principal areas of broad geographical overlap for these two species (Sn), and this overlap is mainly in semi-arid areas. Ecological segregation occurs where the two are apparently sympatric *e.g.* in arid Tsavo E 101B, where the Spotted-flanked mainly inhabits tall riverine vegetation, while the Black-throated is in woodland (La; Lack 1985).

485 HAIRY-BREASTED BARBET *Lybius hirsutus*
B538 / MPG583 / D2C3
NO MAP: post probable 48D

Afrotropical: a W and C African species, from the equatorial lowland forests of the W, E to S Sudan, Uganda, NW Tanzania and, at the E extremity of the range, W Kenya.

Kenyan: a forest bird, known only from the Kakamega Forest 48D (hence C3), where it was first recorded by Ripley & Bond (1971). There have been fairly regular subsequent reports, but it appears to be rare (Mann 1976; Br).

486 WHITE-EARED BARBET *Buccanodon leucotis*
B530 / MPG588 / D2C2

Afrotropical: the E, from E South Africa N through Mozambique and Malawi to E Tanzania and, at the N limit of the range, Kenya.

Kenyan: a rather uncommon inhabitant of forest and forest edge, up to 2000 m (Br). Range is 85% within 500+ mm rainfall areas. Distribution appears to be in three discrete areas: the S coast, the E periphery of the Kilimanjaro massif, and the E periphery of the C highlands. The high proportion of pre-1970 squares (50%) suggests range contraction, particularly in the highlands. Congregations occur at fruiting trees (Br).

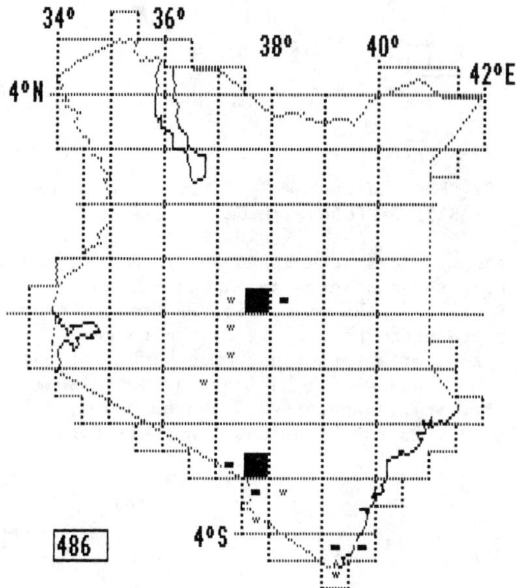

487 YELLOW-SPOTTED BARBET *Buccanodon duchaillui*
B529 / MPG590 / D2C2

Afrotropical: W and C Africa, from the lowland forests around the Gulf of Guinea, E through the Congo basin to NW Tanzania, Uganda and, at the E limit of the range, W Kenya.

Kenyan: locally common in forest at 1150-2000 m (Br). Range is in 1000+ mm rainfall areas to the W of the rift valley. Apart from an old record from Kisumu 60B (Ja), the squares are entirely humid. Ja mentions local movements to fruiting trees.

(map overleaf)

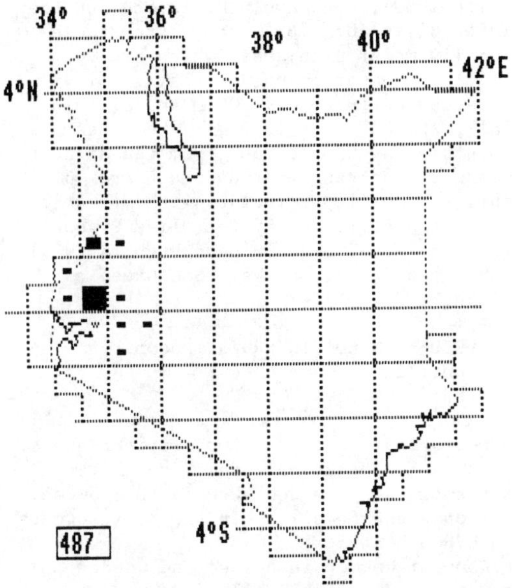

(Br; Britton & Zimmerman 1979). Range is in 500+ mm rainfall areas of the coastal lowlands. Sn mentions local movement to fruiting trees. Other reference: Short & Horne 1980.

489 GREY-THROATED BARBET *Gymnobucco bonapartei*
B533 / MPG586 / D2C2

Afrotropical: W and C Africa, from Cameroon and Congo E through the lowland forests of the Congo basin to the E Zaire borders, S Sudan, NW Tanzania, Uganda and, at the E limits of the range, W Kenya.

Kenyan: locally common in forest and forest edge, up to 2600 m (Br). Range is in moist subhumid-humid areas, above 1000 m, to the W of the rift valley. It is 92% within the 1000+ mm rainfall regime.

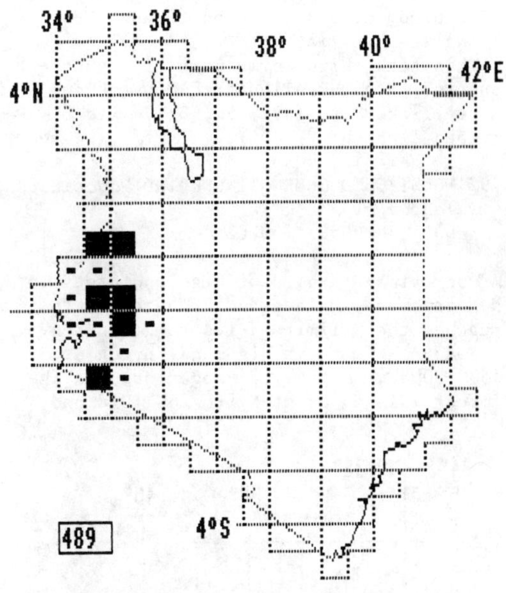

488 GREEN BARBET *Buccanodon olivaceum*
B531 / MPG589 / D2C3

Afrotropical: a highly disjunct distribution on the E of the continent, in E South Africa, Malawi and N Mozambique, S and E Tanzania and, at the N limit of the range, the coastal lowlands of Kenya. See Green Tinkerbird (491).

Kenyan: a common forest species, also frequent in woodland in Sokoke Forest 102B

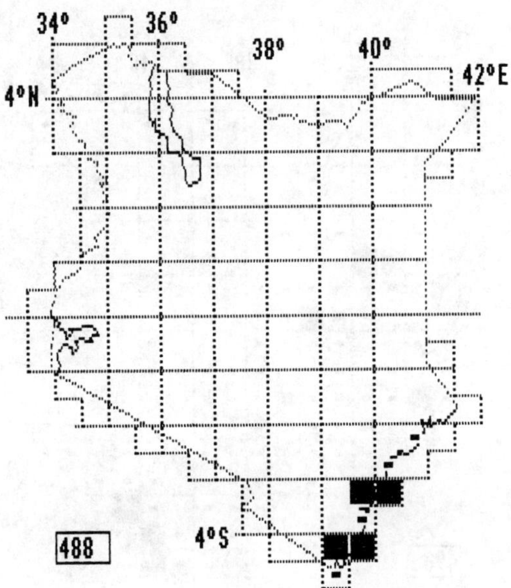

490 SPECKLED TINKERBIRD *Pogoniulus scolopaceus*
B553 / MPG599 / D2C0
NO MAP: post presence 48D; pre presence 36D 49A

Afrotropical: W and C Africa, from the equatorial lowland forests E to NE Zaire, Uganda and, on the E tip of the range, W Kenya.

Kenyan: a rare forest species. There are records from Mt Elgon 36D (the Uganda side: Ja), Kakamega Forest 48D and 25 km SE of Kitale 49A (Br). Ja also mentions possible occurrence at Naivasha 62C: this locality is not listed by Br, is most improbable, and is not accepted here. The three squares are all in 1000-2000 m, humid, 1000+ mm rainfall areas. There is no breeding record.

491 GREEN TINKERBIRD *Pogoniulus simplex*
B554 / MPG591 / D2C3
NO MAP: post presence 102BD 114ABC; post confirmed 103A

Afrotropical: in three distinct areas of the E: S Mozambique; Malawi; and from E Tanzania to the N limit of the range in the coastal lowlands of Kenya. This distribution is very similar to that of the Green Barbet (488), which may be a close relative (Sn).

Kenyan: uncommon in forest and rich woodland in the coastal lowlands, S from Sokoke Forest 102B (Br; Britton & Zimmerman 1979). All six squares receive over 500 mm annual rainfall. The single breeding record is of a brood at Gedi 103A on 20 Aug 1984 (F Ng'weno *in litt.*).

This tinkerbird forms part of a superspecies with the allopatric Moustached Green Tinkerbird (492)(Sn; Br). See Yellow-rumped Tinkerbird (495).

492 MOUSTACHED GREEN TINKERBIRD *Pogoniulus leucomystax*
B551 / MPG592 / D2C2

Afrotropical: a bird of highland areas in N Mozambique, Malawi and E Tanzania, that reaches the N limit of its range in Kenya.

Kenyan: common in highland forest at 1500-3000 m (Br; Se). Range includes the forest islands of Mt Nyiru 26D (Diamond & Keith 1980), the Chyulu Hills 88D and the Taita Hills 101A (Br): apart from these locally moist highlands, which are too small to appear on our overlays, all squares receive over 500 mm annual rainfall, and 65% over 1000 mm. Se notes a possible May-Sept exodus from the highest parts of the range, around Mau Narok 61D: this seasonality coincides with migration from high altitudes by the Olive Pigeon (325) and the Golden-winged Sunbird (926) during cold, wet months. See Green Tinkerbird (491).

493 RED-FRONTED TINKERBIRD *Pogoniulus pusillus*
B552 / MPG594 / D2C2

Afrotropical: occupies two distinct areas. In the NE, it extends from C and E Tanzania, N through Kenya and NE Uganda to S Sudan, Ethiopia and Somalia. In the S, it occurs in E South Africa.

Kenyan: common in bushed and wooded country, and occasionally forest edge, up to 2200 m but mainly below 1800 m (Br; MPG; La; St). Range is 68% within intermediate, 250-1000 mm rainfall areas, and only 27% arid. Hence, apart from squares along the S periphery of the Ethiopian highlands, this tinkerbird is largely absent from the arid N and NE including the L Turkana basin, where it is very local (HH). Restriction in the E and SE to the coastal lowlands and to squares containing the Tana R 65BD 79C (AGH) is comparable to virtual restriction to the coast and the Juba and Shebelle R in S Somalia (AM).

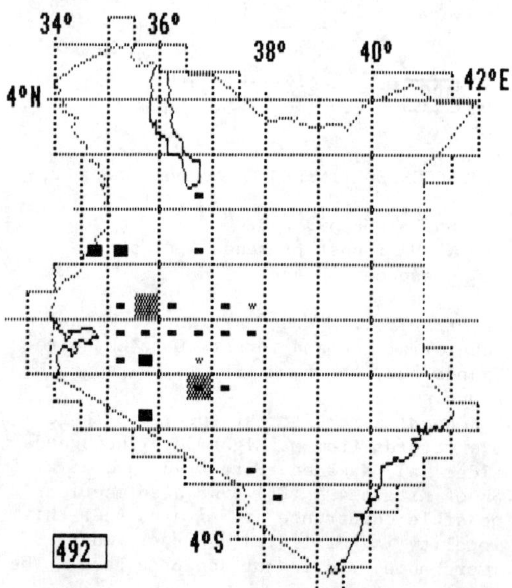

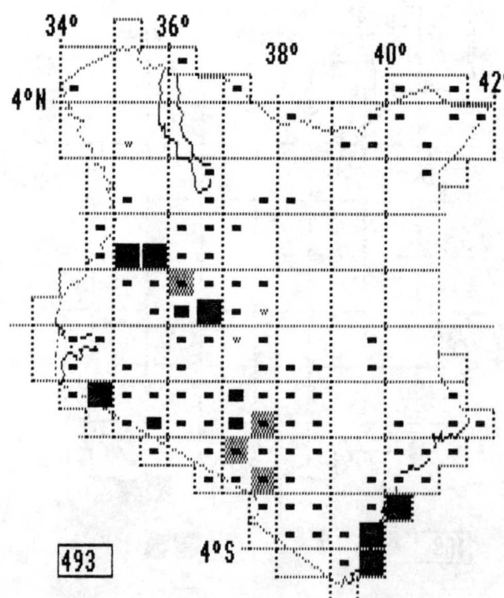

The Red-fronted Tinkerbird forms a superspecies with the allopatric Yellow-fronted Tinkerbird (494)(Sn; Br). At QSD scale, the two overlap only in the Winam Gulf area 60AB, where this species replaces the Yellow-fronted to the S of Kisumu 60B (Mann 1976). Whereas the Red-fronted's range is only 50% above 1000 m, 18% moist subhumid-humid and 19% within 1000+ mm rainfall areas, the Yellow-fronted's seven squares are entirely within these categories. See Yellow-rumped Tinkerbird (495).

494 YELLOW-FRONTED TINKERBIRD *Pogoniulus chrysoconus*
B549 / MPG595 / D2C3

Afrotropical: widely distributed S of 15°N, E to W Ethiopia, S Sudan, Uganda, W Kenya, C Tanzania and the E coast, and avoiding only the dry SW and the equatorial lowland forest areas of the W.

Kenyan: locally common in thicketed and wooded habitats, up to 1500 m (Br). See Red-fronted Tinkerbird (493). Other references: Britton & Harper 1969; Mann 1976; Short *et al*. 1983.

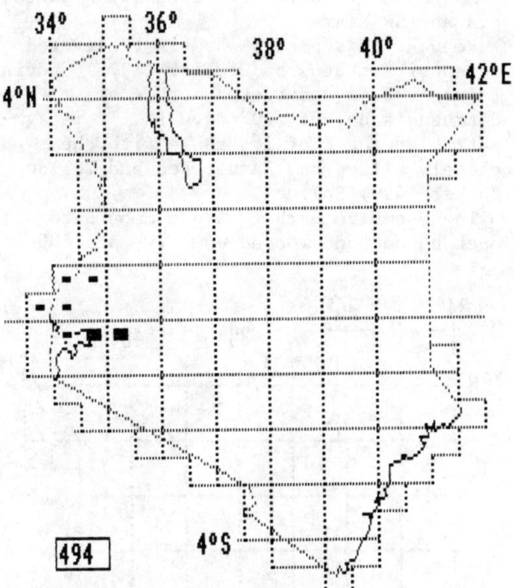

495 YELLOW-RUMPED TINKERBIRD *Pogoniulus bilineatus*
B548 / MPG596, 597 / D2C2

Afrotropical: the golden-rumped form of this species is a bird of the E, in two distinct areas: from E South Africa and Mozambique N through Malawi and E Tanzania to the N limit of the range in Kenya and E Uganda; and as an isolated population in W

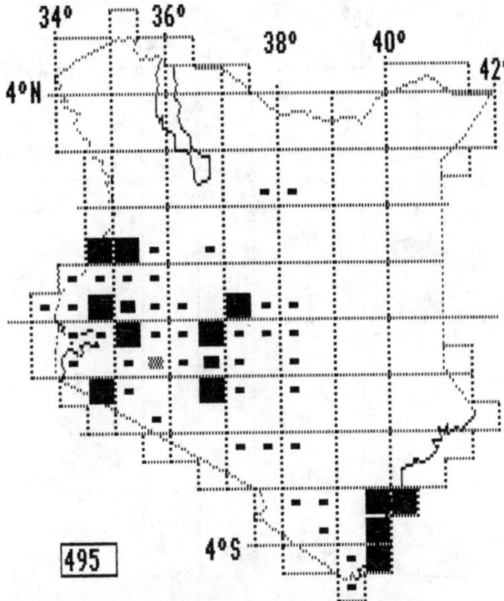

Tanzania and the E Zaire borders. The lemon-rumped form occurs further W, mainly in the equatorial lowland forests.

Kenyan: common in moist thicket, woodland and forest, up to 3000 m (Br; Se; MPG). Range is 87% within 500+ mm rainfall areas and only 7% arid (including locally moist riverine growth in Tsavo E 101B: La). It includes the forest island of Mt Marsabit 27D 28C.

This tinkerbird is usually segregated from its congeners by differing habitat preferences, but it breeds sympatrically with the Red-fronted Tinkerbird (493) near Mombasa 114B (Br), and is regular alongside the Green Tinkerbird (491) in Sokoke Forest 102B (Britton & Zimmerman 1979).

496 YELLOW-BILLED BARBET *Trachylaemus purpuratus*
B556 / MPG604 / D2C3

Afrotropical: W and C Africa, from the equatorial lowland forests E to S Sudan, Uganda and, at the E limit of the range, W Kenya.

Kenyan: common in forest, including clearings, at 1700-2800 m (Br; MPG). It is found in moist subhumid-humid, 1000+ mm rainfall areas to the W of the rift valley.

(map overleaf)

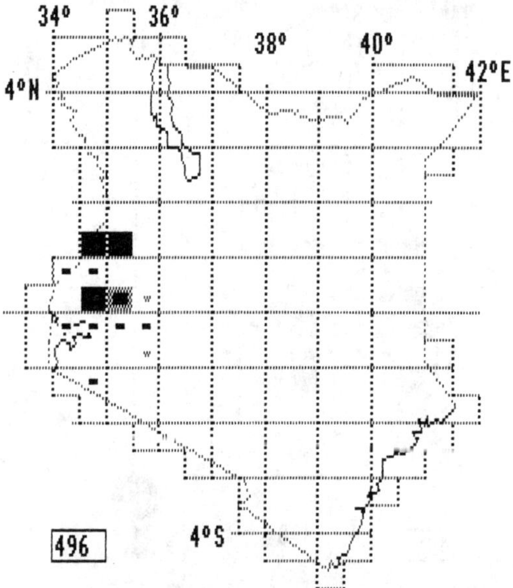

497 D'ARNAUD'S BARBET *Trachyphonus darnaudii*
B557 / MPG603 / D3C3

Afrotropical: the NE, from E Tanzania through Kenya and NE Uganda to S areas of Sudan, Ethiopia and Somalia.

Kenyan: d'Arnaud's Barbet forms a superspecies with the allopatric Usambiro Barbet (498: *q.v.*) and is widely sympatric with the closely allied but larger Red and Yellow Barbet (499: *q.v.*)(Br; Sn).

d'Arnaud's Barbet is common in bushed and dwarf wooded habitats, up to 1900 m (Br; Ja; La; St). Range is 76% within intermediate, 250-1000 mm rainfall areas, and this species is scarce in the very dry, 0-250 mm country of the N and, more especially, the NE. Absence from this latter area is supported by lack of records from observers resident in 19A and 42A, and by similar status in adjacent S Somalia (AM). At L Turkana, also mainly within this lowest rainfall category, d'Arnaud's is less common than the Red and Yellow Barbet (HH), which is generally more widespread in the arid N and NE.

d'Arnaud's is virtually absent from the 1000+ mm rainfall areas of the highlands and the equatorial W, and is replaced in S parts of the latter by the Usambiro Barbet. It is also virtually absent from the coastal strip. Other reference: Short & Horne 1984.

498 USAMBIRO BARBET *Trachyphonus usambiro*
B559 / MPG603 / D2C2

Afrotropical: endemic to interior N Tanzania and SW Kenya.

Kenyan: this barbet was first elevated to species status by Wickler (1973), having formerly been considered a race of d'Arnaud's Barbet (497: *q.v.*). It is sympatric on the E of its range with the closely allied but larger Red and Yellow Barbet (499)(Sn).

The Usambiro Barbet is a common bird of open bushed and wooded country, at 1100-

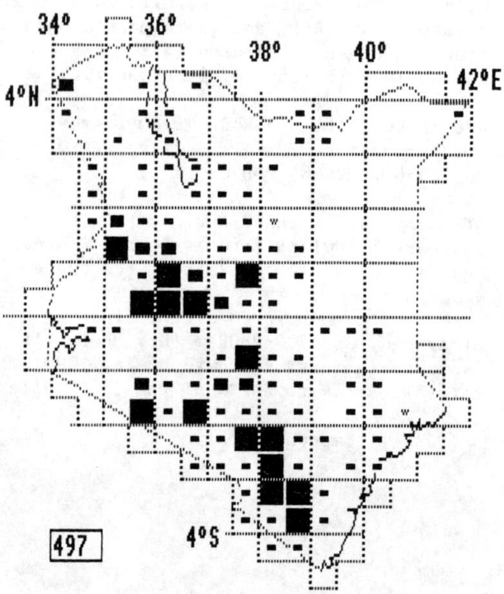

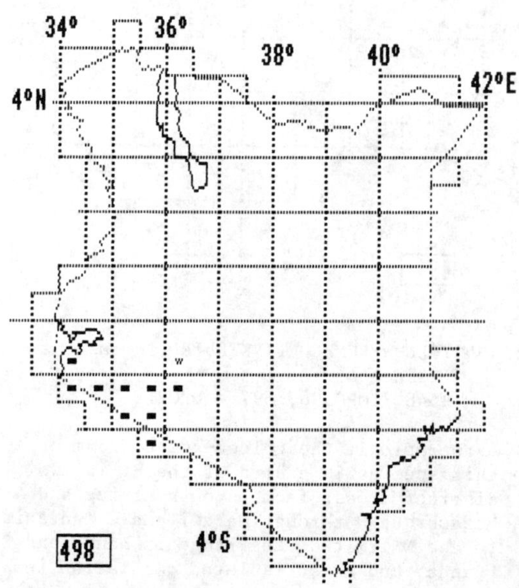

2100 m (Br). Br describes its Kenyan and Tanzanian range as semiarid but, of the 10 squares in Kenya, eight are subhumid-humid and nine are within the 500+ mm rainfall regime: in contrast, d'Arnaud's Barbet is 79% arid-semiarid. There is no breeding record. Other references: Turner 1977; Short & Horne 1984.

499 RED AND YELLOW BARBET *Trachyphonus erythrocephalus*
B558 / MPG601 / D3C3

Afrotropical: the NE, from NE Tanzania through Kenya and NE Uganda to S Sudan, Ethiopia and Somalia.

Kenyan: sometimes common, up to 1500 m. This species' habitat requirements broadly parallel those of the widely sympatric d'Arnaud's Barbet (497: *q.v.*), but the Red and Yellow is often in more open areas (Br; *e.g.* La, Lack 1985). Range is 84% arid-semiarid, and only 6% within the wettest, 1000+ mm rainfall regime. Absence from squares containing the highlands and the moist W is expected, but less explainable is the species' very local distribution in the SE and the coastal strip, where it is confined to the vicinity of the Tana R 65B 78B 79AC 91A (AGH). Absence from the Kenya-Somalia borders S of 2°30'N is apparently real, since 31A represents the S and SW limit of the species in Somalia (AM). Other reference: Short & Horne 1984.

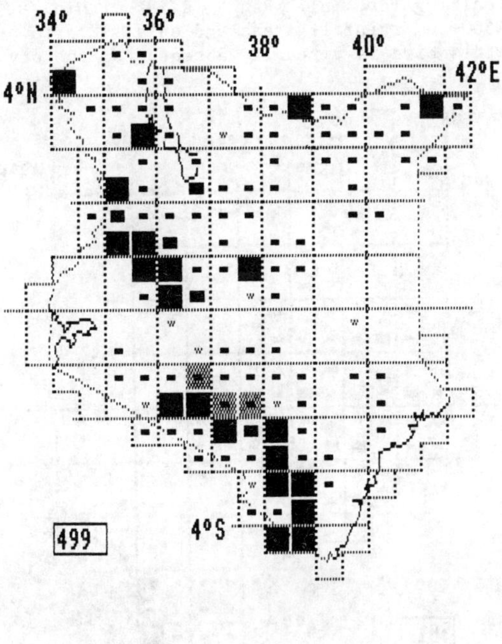

Family Indicatoridae: Honeyguides, Honeybirds

This family is predominantly Afrotropical, with 16 sub-Saharan species (89%) and only two in the Oriental Region. 10 (63% of the Afrotropical total) occur in Kenya, but there are breeding records for only five. This deficiency is typical of this poorly known, highly inconspicuous and taxonomically obscure family (see Forbes-Watson 1977a and Sn). Thus, particularly for the extremely similar smaller species, and especially in view of the modern disinclination for specimens, total numbers of atlas squares are inadequate as indicators of relative status, while concentrations of pre-1970 squares may not reflect retractions of range. The 'guiding' habit (see *e.g.* Ja) is well developed only in the Black-throated Honeyguide (500), but may also occur in some of its congeners (Sn). As far as is known, all species are brood parasites (Sn).

500 BLACK-THROATED HONEYGUIDE *Indicator indicator*
B563 / MPG605 / D2C2

Afrotropical: widely distributed S of 15°N, avoiding only the equatorial lowland forests of the W, and the dry areas of E Somalia and the SW.

Kenyan: fairly common in bushed and wooded habitats up to 3000 m (Br; Se; Ja). Range is only 13% arid and 2% within the 0-250 mm rainfall regime, so that this bird is scarce in the dry N and E, including the environs of L Turkana. Squares 4A 7C 16B 17A represent the S limit of birds in the Ethiopian highlands. Br considers this honeyguide resident, and DEP (pers. obs.) concurs for 75CD 88A 89A. Breeding is 81% above 1000 m and so far unrecorded from the coastal strip, but this may well reflect observer coverage of a comparatively obscure species (*cf.* Scaly-throated Honeyguide (503)). This honeyguide is only presumed resident in S Somalia (AM). Other reference: Madge & Cunningham-van Someren 1975.

501 LESSER HONEYGUIDE *Indicator minor*
B566 / MPG608 / D2C2

Afrotropical: widely distributed S of 15°N, but absent from the equatorial lowland forests of the W, and the dry areas of E Somalia and the SW.

Kenyan: locally common in a wide variety of bushed, wooded and forested habitats up to 3000 m, but inconspicuous when not calling (Br; Se). Range is 78% within 500+ mm rainfall areas and only 13% arid: this bird is virtually absent from the dry

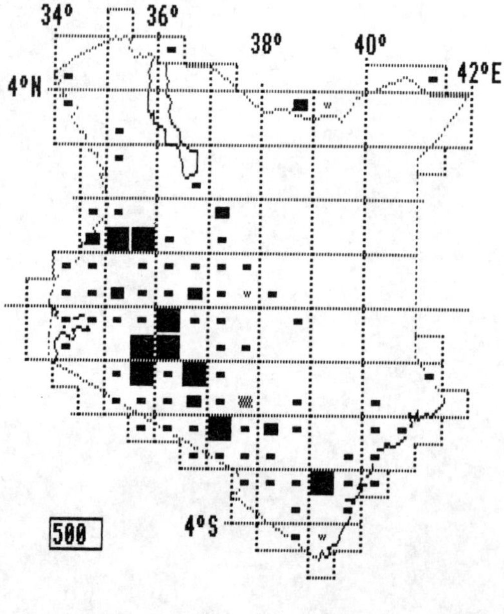

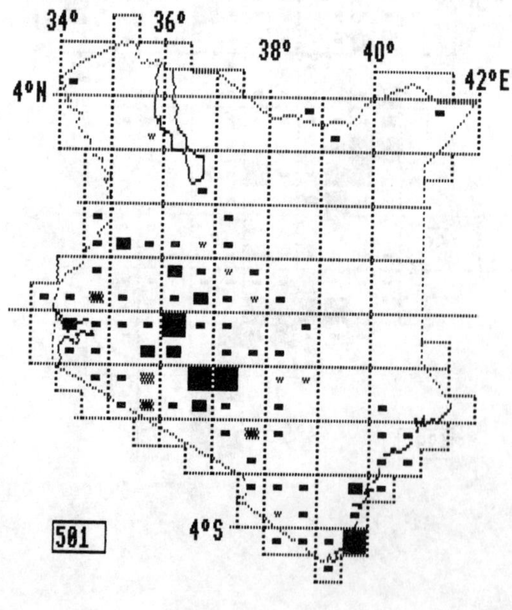

N and E. Squares 2C 16B 17C 19A represent the S edge of populations centred in moister country in Sudan and Ethiopia: AM also record presence in 7D (not mapped). This species is thought to be resident (Br; Se: 61D; DEP pers. obs.: 75CD 89A).

The Lesser Honeyguide forms a superspecies with the Thick-billed Honeyguide (502) (Sn) and, while their ranges closely interdigitate, allopatry is preserved by minor habitat preferences. Thus, in the W, the Thick-billed inhabits forest while the Lesser is in forest edge and woodland; further E, where only the Lesser occurs, it penetrates forest too (Forbes-Watson 1977a). Other references: Stagg 1973; Short & Horne 1983.

502 THICK-BILLED HONEYGUIDE *Indicator conirostris*
B561 / MPG609 / D1C2
NO MAP: post presence 48D 49C; post probable 36D 37C; pre confirmed 48D

Afrotropical: W and C Africa, from the equatorial lowland forests E to S Sudan, Uganda and, at the E limit of the range, W Kenya.

Kenyan: an uncommon species of forest interior (Br). Range is W of the rift valley and N of the Winam Gulf 60AB, in 1000+ m, humid areas within the 1000+ mm rainfall regime. See Lesser Honeyguide (501).

503 SCALY-THROATED HONEYGUIDE *Indicator variegatus*
B569 / MPG606 / D2C2

Afrotropical: mainly the E, from S Sudan, Ethiopia and S Somalia through East Africa, Zambia and Mozambique to E South Africa. S of the Congo basin forests, range also extends W to Angola.

Kenyan: uncommon to locally quite common in thicketed, wooded and forest edge habitats, up to 3350 m (Br; Sn; MPG; St). Range is 82% within the 500+ mm rainfall regime and 10% arid, including the forest islands of Mt Loima 13C and Mt Nyiru 26D (Diamond & Keith 1980). Br regards this bird as resident. The seven breeding squares are all above 1000 m but, while this pattern most probably reflects observer coverage of a rather obscure, canopy species, the absence of breeding records from the SE lowlands is paralleled in adjacent S Somalia, where this honeyguide is only presumed resident (AM)(*cf.* Black-throated Honeyguide (500)). There is only a single recent record from arid Tsavo E 101B, in Mar (La). Other reference: Britton & Britton 1971c.

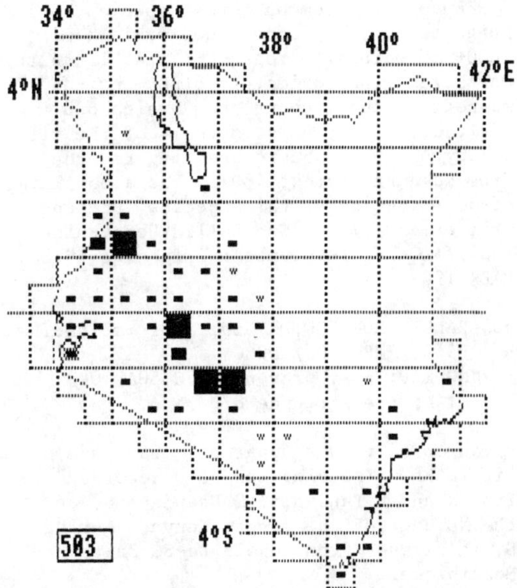

504 LEAST HONEYGUIDE *Indicator exilis*
B562 / MPG610 / D0C0

Afrotropical: a W and C African forest species, ranging between Senegambia and Angola in the W, E to S Sudan, NW Tanzania, Uganda and, on the E limit of the range, Kenya.

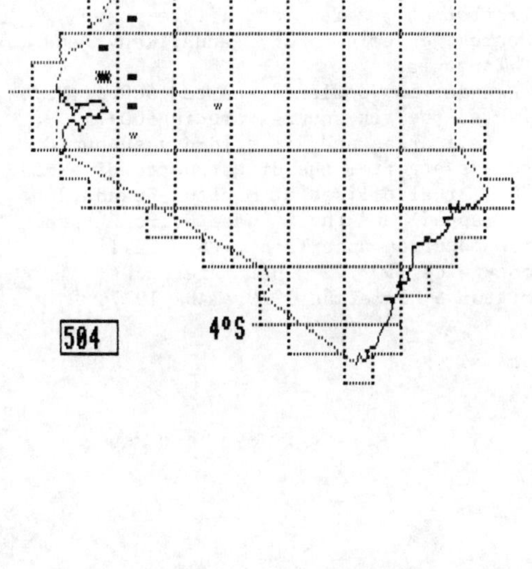

Kenyan: an uncommon forest species (Br). Range is entirely above 1000 m, in humid, 1000+ mm rainfall areas. Br also lists the Mau Forest (approximately 61BD, but not mapped) so that, apart from Mweiga 62B (Br), distribution is to the W of the rift valley. The probable breeding record from Kakamega Forest 48D is based on alarm of host species at this species' presence (Zimmerman 1972). See Pallid Honeyguide (505). Other references: Prigogine 1978a; EABR 1982.

505 PALLID HONEYGUIDE *Indicator meliphilus*
B565 / MPG610 / D0C0
NO MAP: post presence 37C 50A 100B 114A; pre presence 87D

Afrotropical: this apparently rare and little known species has been recorded from N and E Tanzania, E Uganda and, on the NE edge of the range, Kenya. Further S, it extends from Angola to SE Zaire and Mozambique.

Kenyan: the Pallid Honeyguide forms a superspecies with the Least Honeyguide (504)(Sn): while usually regarded as replacing that forest bird in woodlands, it too can occur in forest remnants (Br). Broad sympatry with the Kilimanjaro Honeyguide (506) is reported, but that species may prefer drier areas (Br).

Four of the Pallid's five squares are semiarid, and records are from the driest parts of humid 37C. Br also lists two localities on the Uganda borders: Mt Moroto 24B and the Siroko R 36C. Recent records have come from Kongelai 37C (May 1970: specimen, T J Barnley *in litt.*), the Shimba Hills 114A Nov (EABR 1977), Kitovu Forest 100B Aug (EABR 1978) and Laikipia 50A (Short & Horne 1985b).

506 KILIMANJARO HONEYGUIDE *Indicator narokensis*
B567 / MPG611 / D0C0

Afrotropical: a rare and little known species, endemic to NE Uganda, Kenya and NE Tanzania.

Kenyan: in woodlands, up to 1500 m (Br). Five of the six squares are in 500+ mm rainfall areas and semiarid-dry subhumid. Records are from the driest parts of humid 37C. Br also lists Mt Moroto (Uganda 24B: not mapped) and the S Ewaso Nyiro R (precise location uncertain). See Pallid Honeyguide (505). Other references: Britton & Zimmerman 1979; EABR 1977.

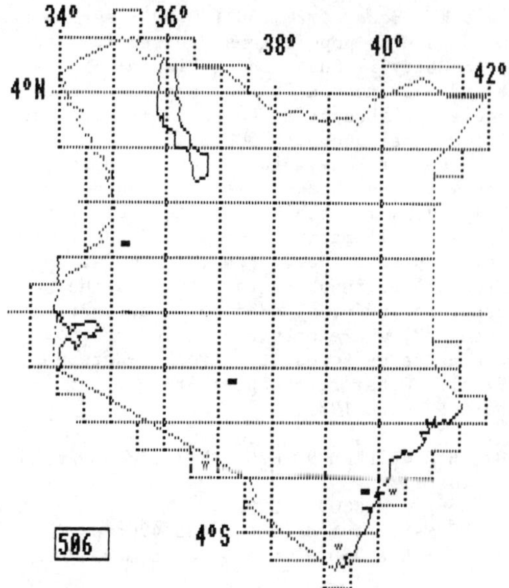

- DWARF HONEYGUIDE *Indicator pumilio*
B568 / MPG- / D0C0
RECORDS DELETED

Afrotropical: a forest honeyguide confined to SW Uganda and the E Zaire borders.

Kenyan: this bird forms a superspecies with the Kilimanjaro Honeyguide (506)(Sn). A supposed population isolated in Kakamega Forest 48D was known only from a single specimen (Williams & Friedmann 1965; Sn), but Prigogine (1978a) considered this bird to be an aberrant Least Honeyguide (504). Thus the Dwarf is not listed for Kenya by Br.

507 CASSIN'S HONEYBIRD *Prodotiscus insignis*
B572 / MPG613 / D1C1

Afrotropical: W and C Africa, from the equatorial lowland forests of the W, E to the E Zaire borders, S Sudan, Uganda and, at the E limit of the range, W Kenya.

Kenyan: a generally scarce bird of forests (Br; Ja). Range is above 1000 m, to the W of the rift valley. Three of four squares are within the 1000+ mm rainfall regime and humid, while the other (74D) is 500-1000 mm and dry subhumid. Records come from the Yala R and Kakamega Forest 48D (Br), Kericho 61A (EABR 1981), the Tinderet Forest 61AB (Squire 1976de: June 1976), and the Nguruman Forest 74D (Fayad & Fayad 1980). Br also lists Mt Moroto (Uganda 24B: not mapped).

This honeybird forms a superspecies with

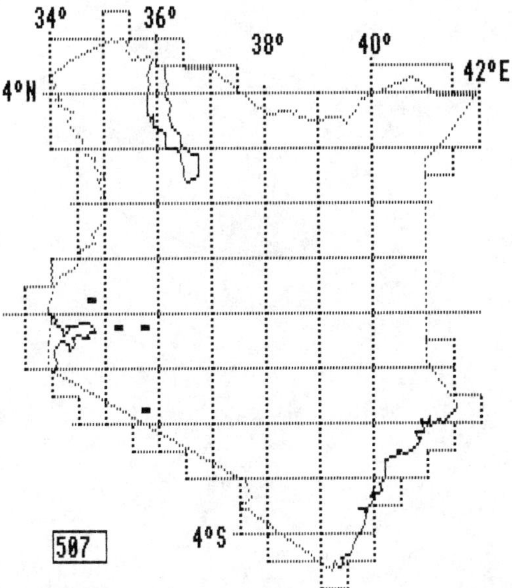

the allopatric Eastern Honeybird (508), which is found to the E of the rift (Sn; Br).

508 EASTERN HONEYBIRD *Prodotiscus zambesiae*
B574 / MPG613 / D2C2

Afrotropical: E from Angola to Mozambique and S Tanzania. Further N, apparently disjunct populations also occur in NE Tanzania-Kenya, and S Ethiopia.

Kenyan: entirely in 500+ mm rainfall areas to the E of the rift valley, in two distinct regions. In the C highlands, this is a locally common bird of forest edge and woodland up to 1850 m (Br), which can occur alongside Wahlberg's Honeybird (509) (Forbes-Watson 1977a). This population is entirely above 1000 m in subhumid-humid areas. At the coast, occasional birds in the semiarid woodland of Sokoke Forest 102B (Britton & Zimmerman 1979; Br) are presumably related to the species' presence in extreme NE Tanzania (*cf.* Bronze-naped Pigeon (326)). See Cassin's Honeybird (507). Other references: Cunningham-van Someren 1970; Ng'weno 1973.

509 WAHLBERG'S HONEYBIRD *Prodotiscus regulus*
B573 / MPG612 / D2C2

Afrotropical: widely distributed S of 9°S but absent from the dry SW. Further N, it extends through the E Zaire borders and Tanzania to Kenya, SW Uganda, S Sudan and Ethiopia. There are also scattered records in W Africa, N of 5°N. Distribution in Kenya represents the E limit of the range.

Kenyan: generally uncommon in bushed, thicketed and wooded habitats, up to 2000 m (Br). The squares are 90% in or E of the rift valley, and 85% above 500 m. This honeybird is apparently scarce in the coastal lowlands, where the only recent record is from Kipini 91D, on 6 Aug 1981 (D A Turner *in litt.*). Range is 85% within intermediate, 250-1000 mm rainfall

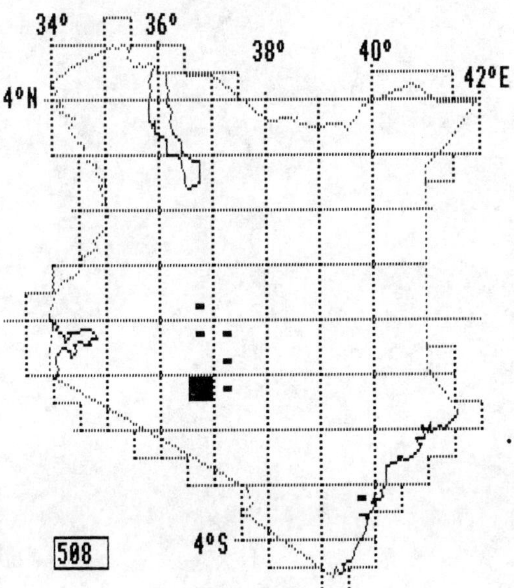

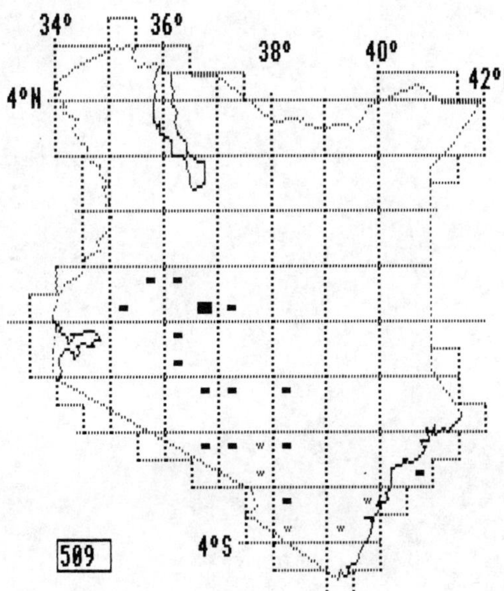

areas: the only squares with above 1000 mm are to the W of the rift 49BC (T Stevenson, P L Britton *in litt.*), and at the coast 91D. See Eastern Honeybird (508).

Family Picidae: Woodpeckers, Wrynecks

A nearly cosmopolitan family of approximately 200 species, with greatest diversity in the Neotropical and Oriental Regions (Sn). The Afrotropical Region has only 27 species (14%), and 14 of these (52%) occur in Kenya. Of the Kenyan species, 13 (93%) breed or are thought to do so, while the Eurasian Wryneck (511) is a non-breeding migrant from the Palearctic. The latter species is the only one known to undertake migratory movements (Br)(but see Grey Woodpecker (522)), and all are closely associated with trees or substantial bush.

510 RED-THROATED WRYNECK *Jynx ruficollis*
B575 / MPG635 / D2C2

Afrotropical: often localised, with a fragmented distribution from Ethiopia S through Uganda, Kenya and S and NW Tanzania to Malawi and E South Africa. It also extends around the broad peripheries of the Congo basin forests. Distribution in the Kenya highlands represents the E limit of the range.

Kenyan: locally common in wooded habitats, including forest edge, mainly at 1700-3000 m (Br; Se; MPG). Range is 96% within the 500+ mm rainfall regime and 89% subhumid-humid. Breeding records are clustered in a compact central area, entirely above 1500 m and subhumid-humid, but the significance of this is uncertain. A record from Simba 88B (Ja) may reflect a wanderer.

This species forms a superspecies with the Eurasian Wryneck (511)(Sn). Other reference: EABR 1980.

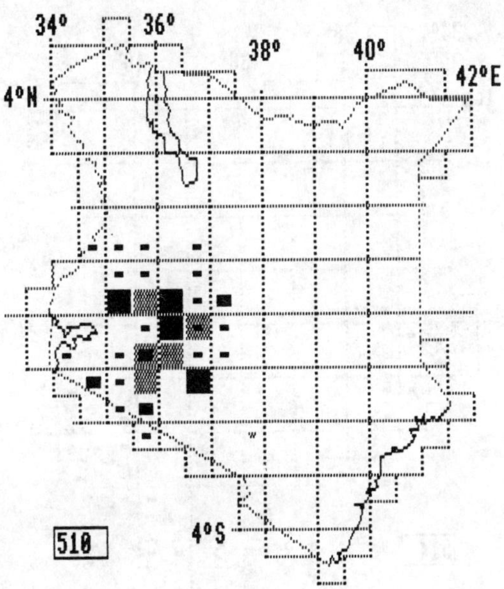

511 EURASIAN WRYNECK *Jynx torquilla*
B576 / MPG634 / D2C0
NO MAP: post presence 2C 27D 52C 62A 76C; pre presence 48C

Afrotropical: this Palearctic species winters in wooded savannas of the N tropics, S to about 5°N, from Senegambia to Ethiopia. The few records from Kenya represent the S limits of the range.

Kenyan: a very scarce winter visitor Oct-Apr, to open bushed and wooded habitats (Br). There are six records of single birds: 22 Feb 1969 Ng'iya 48C (Britton & Harper 1969), 12 Feb and 10 Apr 1972 L Nakuru 62A (probably the same individual twice), 11 Feb 1976 Gof Redo, Marsabit (mapped in 27D), 19 Jan 1977 Kyanguli, Kalama, Machakos District (mapped in 76C) (Cunningham-van Someren 1977d), 17 Dec 1978 NW of Lokichokio 2C (EABR 1978), 25 Oct 1982 Meru NP 52C (EABR 1982). A record from near Njoro 61B on 20 Dec 1976 (EABR 1977) lacks supporting initials and is ignored by Br and this study.

These records are all above 500 m. Five of six squares are in 500+ mm rainfall areas, and the sixth refers to the moist, isolated, highland of Mt Marsabit 27D. See Red-throated Wryneck (510).

512 NUBIAN WOODPECKER *Campethera nubica*
B583 / MPG619 / D2C2

Afrotropical: the NE, from N and E Tanzania, through Uganda and Kenya, to Sudan, Ethiopia and Somalia.

Kenyan: common in a wide variety of open bushed and wooded habitats, up to 2300 m (Br; Ja; MPG; St; Lack 1985). Apart from L Turkana, where it is common (HH), range is in only 35% of the most arid squares (moisture index below 1.2), which presumably do not support sufficient vegetation and insects. Scarcity in parts of the NE, *i.e.* around 29, is thus apparently real.

Breeding squares are 72% above 1000 m and there are none in the coastal lowlands, but this presumably reflects observer coverage (*cf.* Somalia: AM). Occurrence in the coastal strip S of 3°30'S is restricted to old records from Mombasa 114B and Shimoni 114C (Ja).

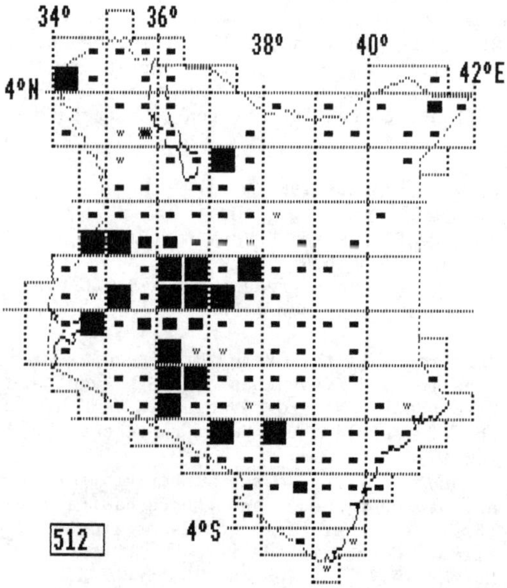

512

513 GOLDEN-TAILED WOODPECKER *Campethera abingoni*
B578 / MPG622 / D2C2

Afrotropical: widely distributed to the S of the Congo basin forests, except for the dry SW. Further N, it extends through the E Zaire borders to W Uganda, S Sudan and the Central African Republic; and also N through E Tanzania to the N limit of its range in S areas of Kenya and Somalia.

Kenyan: in forested, wooded and thicketed habitats (Br). Only at all common in the coastal lowlands (*e.g.* Britton & Zimmerman 1979), there are few inland records (*e.g.* EABR 1980). Range is 75% below 500 m and 81% within 500+ mm rainfall areas (88D = the high rainfall forest island of the Chyulu Hills: L L Short *in litt.*). Extension along the fringing forests of the Tana R 65AB 79C (National Museum Collection, Nairobi; AGH) is paralleled on the Juba R of adjacent S Somalia (AM).

514 LITTLE SPOTTED WOODPECKER *Campethera cailliautii*
B580 / MPG614, 620 / D2C2

Afrotropical: E from Ghana to W Ethiopia and extreme S Somalia, S to Angola and NE South Africa.

Kenyan: locally common in wooded and forested habitats, up to 2100 m (Br) in 500+ mm rainfall areas. The nominate race extends throughout the coastal lowlands, with range 83% semiarid-dry subhumid and entirely below 500 m. Granvik described it as very common in palm groves around Mombasa 114B (Ja). The race *C.c. nyansae* is in moist subhumid-humid, 1000+ m areas to the W of the rift valley: it is sparse in the W highlands 49CD (Br). There is no confirmed breeding record for the species.

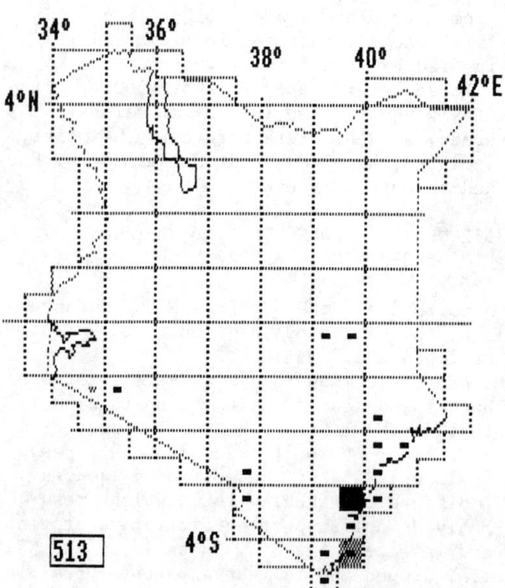

513

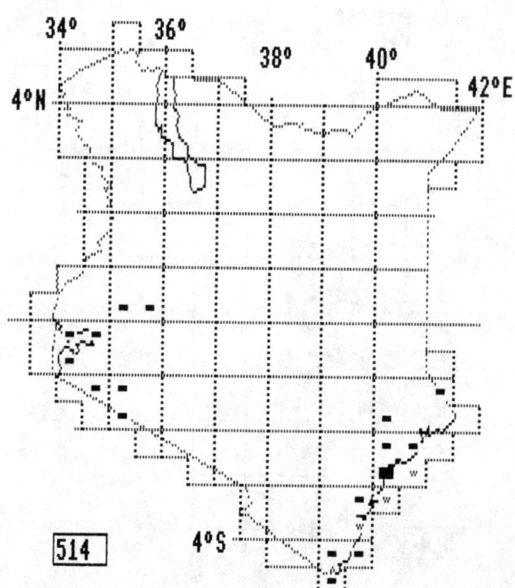

514

515 FINE-BANDED WOODPECKER *Campethera tullbergi*
B584 / MPG617 / D2C2

Afrotropical: in highland forests of three distinct areas: W and C Kenya; the E Zaire-Tanzania-Uganda borders; and Cameroon, SE Nigeria and adjacent offshore islands.

Kenyan: locally common in highland forest, up to 3000 m (Br; Se; Ja). MPG describe it as common on Mt Elgon 36D and around the rift valley, but rare on Mt Kenya 63A. Apart from a record from Kitui 77A (Br: see below), range is 73% above 1500 m, 82% moist subhumid-humid, and entirely within 500+ mm rainfall areas (74D = the Nguruman Forest: Fayad & Fayad 1980). The Kitui 77A record is also in 500+ mm rainfall country, but is otherwise unique in being below 1000 m in a semiarid area: this possibly reflects a wanderer.

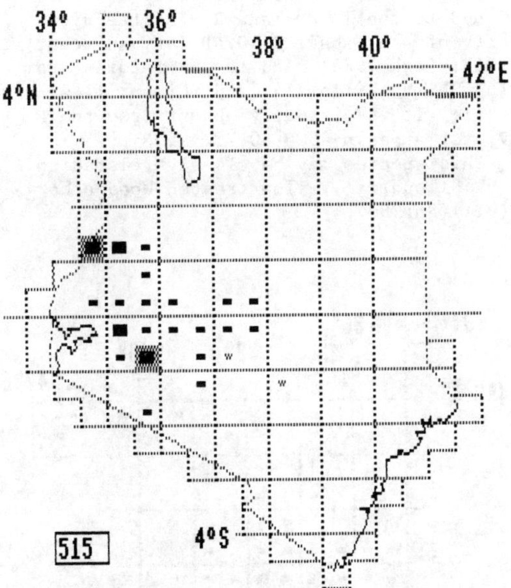

516 BUFF-SPOTTED WOODPECKER *Campethera nivosa*
B582 / MPG616 / D2C2
NO MAP: post presence 37C 48D; pre presence 36D 49C

Afrotropical: W and C Africa, between Senegambia and Angola in the W, and E through the Congo basin to NW Tanzania, S Sudan, Uganda and, at the E limit of the range, W Kenya.

Kenyan: locally common in forest and dense secondary growth in the Elgon 36D-Kitale 37C and Kakamega 48D-Nandi 49C areas (Ja; Br; T J Barnley *in litt.*). Range is above 1000 m, to the W of the rift valley and the N of the Winam Gulf 60AB. The squares are entirely humid and within the 1000+ mm rainfall regime. There is no breeding record.

517 BROWN-EARED WOODPECKER *Campethera caroli*
B581 / MPG615 / D2C3
NO MAP: post presence 37C; post probable 36D; post confirmed 48D

Afrotropical: a W and C African species, from Cameroon and Angola, E through the Congo basin to NW Tanzania, S Sudan, Uganda and, at the E extremity of the range, W Kenya. There is also an isolated population in Liberia and Sierra Leone.

Kenyan: a rather uncommon bird of forest and dense secondary growth, up to 1700 m (Br; MPG; Sn; ADL pers. obs.). Range is entirely above 1000 m in humid, 1000+ mm rainfall areas, to the W of the rift valley and N of the Winam Gulf 60AB. Unlike the allied Buff-spotted Woodpecker (516), it is not known from the Nandi forests 49C (Br).

518 UGANDA SPOTTED WOODPECKER *Dendropicos poecilolaemus*
B587 / MPG626 / D2C2

Afrotropical: to the N of the equatorial lowland forests, from Cameroon E to the Central African Republic; and also from S Sudan and NE Zaire E to Uganda and, on the E extremity of the range, W Kenya.

Kenyan: generally scarce in open wooded habitats (Br). Range is in 1000+ m areas,

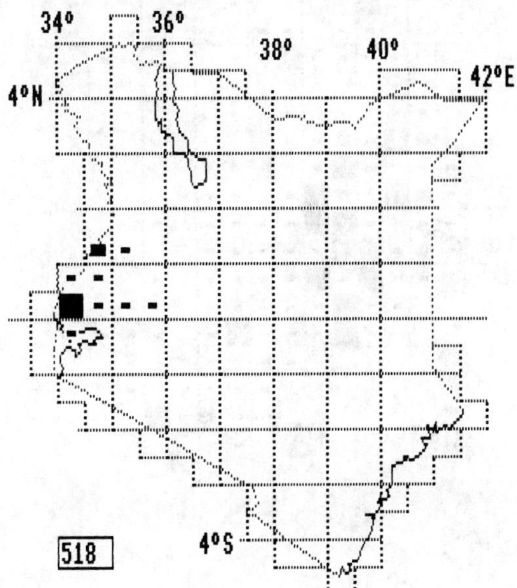

W of the rift valley and N or NE of the Winam Gulf 60AB. All nine squares are in moist subhumid-humid, 1000+ mm rainfall country.

519 CARDINAL WOODPECKER *Dendropicos fuscescens*
B585 / MPG623 / D2C2

Afrotropical: widespread S of 10°N, extending to 16°N in Ethiopia, and avoiding only the equatorial lowland forest areas of the W. This is one of the most numerous of Afrotropical woodpeckers (Sn).

Kenyan: common in a wide variety of open bushed and wooded habitats, up to 2600-3000 m (Br; Se; MPG; La; St). Apart from the L Turkana basin, where it is relatively common (HH), range is in only 35% of the most arid squares (moisture index below 1.2), which presumably support insufficient vegetation and insects. Sparse occurrence in the NE, *i.e.* around 29, is thus real (*cf.* Nubian Woodpecker (512), the only other member of this family which extends to dry country). Apparent absence on the E border S of 3°N may also be real as status in adjacent W Somalia is similar (AM), but these areas have received little observer coverage. Breeding is 69% above 1000 m but this again probably reflects observer coverage (*cf.* Somalia: AM). Other reference: Sn.

520 BEARDED WOODPECKER *Thripias namaquus*
B594 / MPG629 / D2C2

Afrotropical: widely distributed S of 10°S, avoiding only the dry SW. Further N, it extends through East Africa to S Sudan, Ethiopia and W Somalia.

Kenyan: this local and never numerous species frequents bushed and wooded habitats, including forest edge, up to 3000 m (Br; Se; La; St). Distribution is only 24% arid, and very sparse in the arid N and E. Squares 2C 7C 15C 17A 18A 19A represent the S extremities of the Sudan-Ethiopian population, and this woodpecker is also found on the forest islands of Mts Loima 13C, Nyiru 26D and Marsabit 27D (Br). Range in the arid E is mainly along the Tana R 64B 65ABD 79C (AGH), which corresponds to restriction to the vicinity of the Juba R in adjacent S Somalia (AM). Occurrence in the coastal strip is confined to the lower Tana R 91A, the vicinity of the Sabaki R 102AB (*cf.* adjacent 101B: Lack 1985), and to old records from the Shimba Hills 114A and Shimoni 114C (Ja). It is virtually absent from the L Victoria basin 48CD 60 73.

This species may be closely related to the allopatric Yellow-crested Woodpecker (521)(Sn; Br).

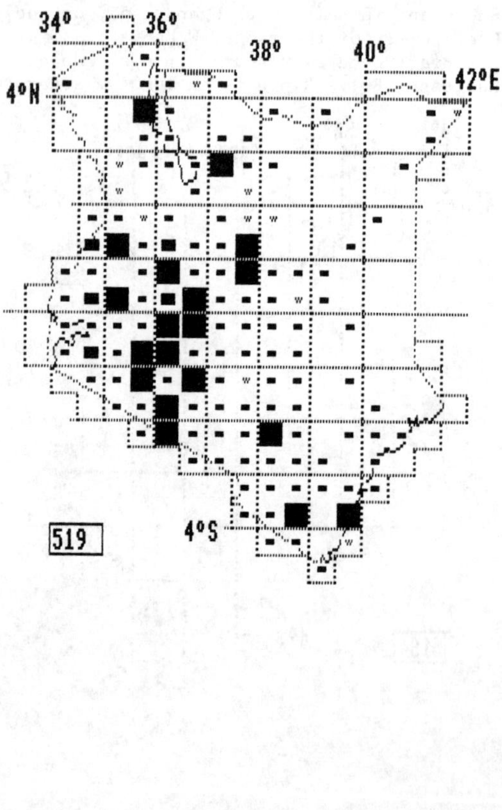

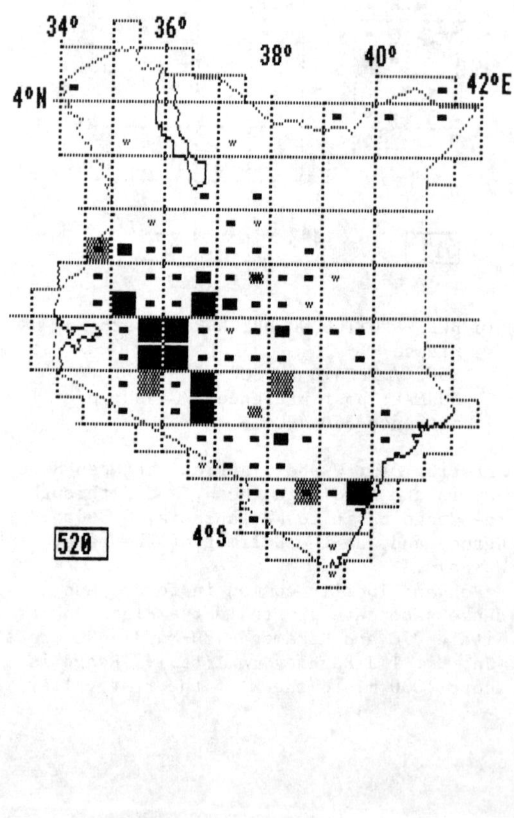

521 YELLOW-CRESTED WOODPECKER *Mesopicos xantholophus*
B592 / MPG632 / D2C3
NO MAP: post presence 49C; post probable 36D; post confirmed 48D

Afrotropical: W and C African forests, from Cameroon to Angola, E to the Congo basin, S Sudan, Uganda and, at the E tip of the range, W Kenya.

Kenyan: locally quite common in forest and dense secondary growth (Br), in humid, 1000+ m areas, W of the rift valley and N of the Winam Gulf 60AB. The squares are all within the 1000+ mm rainfall regime. There are records from Mt Elgon 36D, and the Kakamega and N Nandi Forests 48D 49C (T J Barnley *in litt.*; Br). See Bearded Woodpecker (520).

522 GREY WOODPECKER *Mesopicos goertae*
B590 / MPG630 / D2C2

Afrotropical: apart from Somalia, widely distributed between 17°N and 5°N, and further S through S Sudan to Uganda, Kenya and N Tanzania. There are also records on the S periphery of the Congo basin forests. Distribution in Kenya represents the E edge of the range.

Kenyan: a sometimes common bird, frequenting a wide variety of wooded habitats including forest edge, mainly at 700-2000 m in the highlands and their peripheral plateaux (Br; Ja; MPG). Sn mentions it up to 3400 m in Kenya, and Se lists wanderers around 3000 m at Mau Narok 61D. Range is 80% within the 500+ mm rainfall regime, and includes the forest island of Mt Nyiru 26D (Y Malcolm-Coe *in litt.*).

Breeding squares are 92% above 1500 m, entirely within 500+ mm rainfall areas, and loosely clustered in and around the W and C highlands. The significance of this distribution is uncertain, but some kind of movement may occur. DEP (pers. obs.) notes an annual May-Aug influx to the Ruaraka-Kahawa-Ruiru area of moist subhumid 75B. Three wanderers to semiarid Baringo 50A have all been in Mar (St), while single wanderers to 3000 m around Mau Narok 61D were in Mar and Nov (Se).

523 BROWN-BACKED WOODPECKER *Picoides obsoletus*
B593 / MPG627 / D2C2

Afrotropical: the N tropics, between 15°N and the N edge of the equatorial forests, E from the W coast to W Ethiopia and S Sudan, and SE to Uganda, Kenya and N Tanzania. Distribution in Kenya represents the E limit of the range.

Kenyan: uncommon and local in bushed and wooded habitats, up to 2300 m (Br; Sn; EABR). Excluding old records from Kitui 77A and Mazeras 102D (see below), range is 73% above 1500 m, and entirely within 500+ mm rainfall, subhumid-humid areas. The 77A and 102D records (Ja) were at 500-1000 m and near sea level respectively: the latter at least presumably represents a wanderer, since the species does not occur in adjacent NE Tanzania (Br). There are records for most months at Nairobi 75B (EABR 1983).

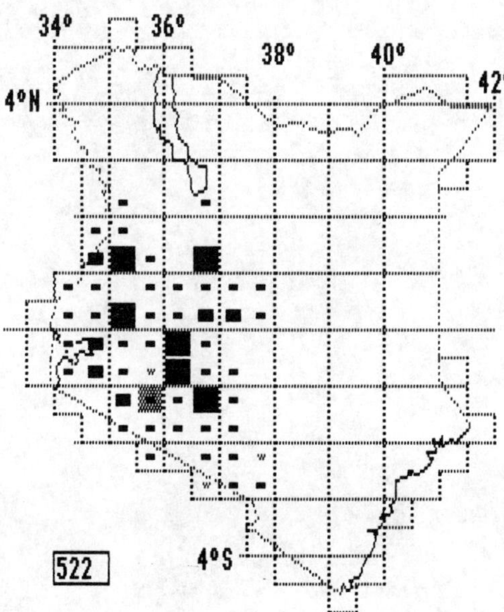

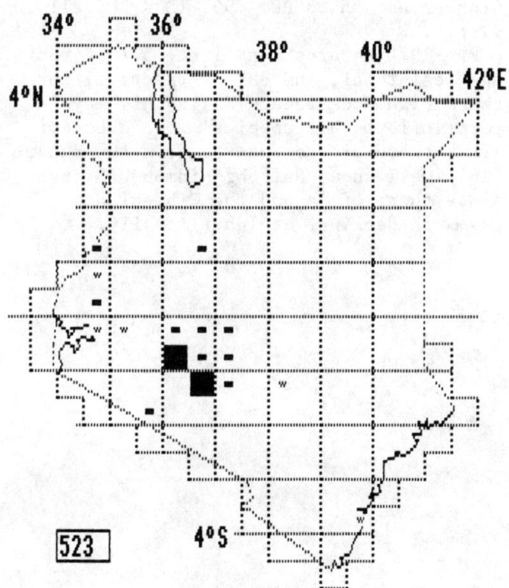

Family Eurylaimidae: Broadbills

14 species, 10 being Oriental and four Afrotropical. The African Broadbill (524) breeds in Kenya.

524 AFRICAN BROADBILL *Smithornis capensis*
B596 / MPG650 / D1C1

Afrotropical: widely distributed S of 5°S, but not in the dry SW. Further to the NE, it extends N along the E Zaire borders, and through E Tanzania to Kenya. Further NW, it occurs around the Gulf of Guinea from Angola to Sierra Leone. Distribution in Kenya represents the NE limit of the range.

Kenyan: a shy, inconspicuous and usually quiet bird of dense forest and thicket interiors (Ja; MPG; Br), only at all easily detected when calling: hence D1. Separate races occur in the highlands on either side of the rift valley, and at the coast (Br). Range is entirely in 500+ mm rainfall areas (88D = the Chyulu Hills forests). The coastal lowlands race is in semiarid-dry subhumid forest, and an individual was recaptured at the same site in Sokoke Forest 102B six months after ringing (Britton & Zimmerman 1979; EABR 1983). The highland races extend up to 1800+ m (Br), with range 92% above 1000 m and 83% moist subhumid-humid (EABR 1977, 1978, 1983). The record in 47D is from Uganda (J S Ash *in litt.*). Probable breeding in 62A is based on display flight in the Ol Joro Orok Forest on 23 Dec 1953 (D K Bednall *in litt.*).

Pre-1970 squares constitute 50% of the species' total, and 83% of the total for the C highlands race. While this may in part reflect this species' low detectability, there are no records from the reasonably well known Nairobi-Kiambu 75B area since those of Ja and the C highlands race may be in decline, at least locally.

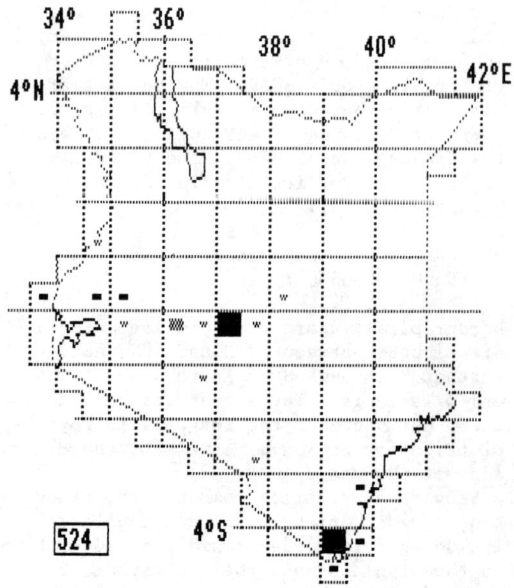

Family Pittidae: Pittas

26 species, which are mainly Oriental. Two (8%) occur in the Afrotropical Region, and one is a non-breeding intra-African migrant to Kenya.

525 AFRICAN PITTA *Pitta angolensis*
B598 / MPG652 / D1C1

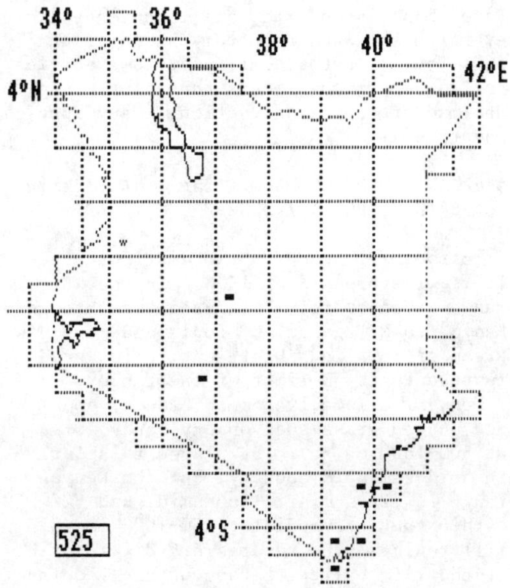

Afrotropical: occupies two distinct areas. In the W, range extends from Guinea to Angola. In the E, it is believed to breed from Zambia and SE Tanzania S to Botswana and Zimbabwe, and to disperse as a non-breeding migrant to South Africa, S Zaire, the E Zaire borders and coastal Kenya (HM). In addition to these regular non-breeding areas, exhausted or defunct migrants have occurred in several other parts of East Africa. Distribution in Kenya represents the NE limit of the range.

Kenyan: this very inconspicuous bird is an uncommon but regular non-breeding migrant, mainly Apr-Oct, to coastal forests N to Sokoke Forest 102B and Gedi Ruins 103A (Rathbun 1978; Britton & Zimmerman 1979; Br; EABR). This coastal range is entirely within semiarid-dry subhumid, 500+ mm rainfall areas. Although two Jul females with enlarged ovaries were collected in Sokoke 102B (Ripley & Bond 1971), all birds observed at Gedi 103A have been solitary and silent, and it is most unlikely that breeding occurs in Kenya (Britton & Rathbun 1978).

Records away from the coast concern exhausted or dead nocturnal migrants, in the highlands. These have principally occurred at Nairobi 75B (van Someren 1918c; Ja; Brown 1970c; Clifton 1976; Britton & Rathbun 1978), but one was netted at Kongelai 37C in Jul 1966 (T J Barnley *in litt.*), and one seen at Timau 51C on 4 Aug 1984 (E Allsop *in litt.*). These three highland squares are above 1500 m, in moist subhumid-humid, 500+ mm rainfall areas. These inland records have mostly been during the Nov-Dec and Apr-June passage periods (Br), but Jul and Aug dates are quoted above, and there is an Aug date from Nairobi 75B (Brown 1970c).

Family Alaudidae: Larks

72 species, with representatives in all zoogeographical regions. 52 (72%) occur in the Afrotropical Region, where 44 (85%) are endemic (HM). 22 species (30% of the world's total and 42% of the Afrotropical's) are recorded for Kenya. Of these, Williams' Bush Lark (535) is endemic, one is an extralimital migrant from the Palearctic, one is an extralimital wanderer from Tanzania, and 19 (86%) breed, have bred, or are assumed to do so. The low detectability of some species, coupled with the modern disinclination to collect specimens, means that clusters of pre-1970 squares need not necessarily denote decline or range contraction. Many species are rated C1 or C0 because they are hard to identify.

526 FAWN-COLOURED LARK *Mirafra africanoides*
B613 / MPG661 / D2C2

Afrotropical: occupies two distinct areas. It is widespread S of 15°S. In the NE, range extends from NE Tanzania, N through Kenya, to NE Uganda, Ethiopia and mainly N areas of Somalia. Distribution in Kenya represents the E limit of the range.

Kenyan: a locally common bird of open and bushed grasslands on dry sandy soils, at 500-1800 m (Br). Br places this lark in arid-semiarid country, but the Kenyan range is 72% semiarid-subhumid, and 80% within the intermediate, 250-1000 mm rainfall regime. The squares are 81% in or E of the rift valley, although the E lowlands are beyond this species' E limit. Hence range is on the plateaux to the N and S of the highlands, and in the intervening rift valley 62AC. Records 12C and 24D are from Uganda (J E Ash *in litt.*). There are recent records from the squares immediately NE and E of 19B (AM). Br considers this lark resident, and DEP (pers. obs.) concurs for 75D.

527 SINGING BUSH LARK *Mirafra cantillans*
B615 / MPG654 / D2C2

Afrotropical: a fragmented distribution across the N tropics N of 10°N from Senegambia to Ethiopia, and thence S through Kenya and coastal Somalia to NE areas of Uganda and Tanzania.

Kenyan: locally common on open and bushed grasslands, mainly at 400-1500 m but also reaching 1800 m (Br; MPG). Range is 88% within the intermediate, 250-1000 mm rainfall regime and 75% less arid-semiarid: this lark is absent from most of the arid N and E except (a) the W side of L Turkana

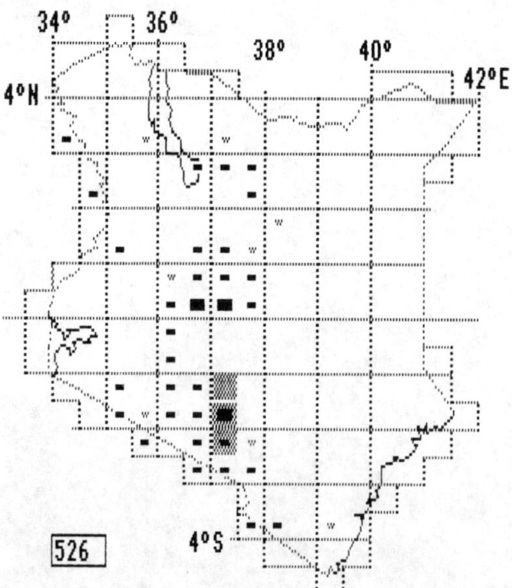

526

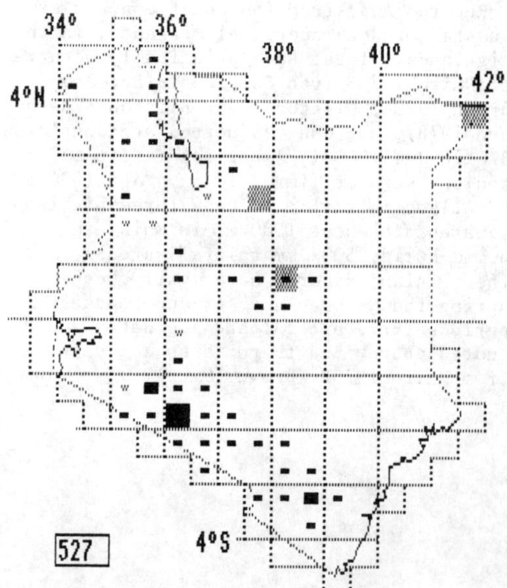

527

(and Central Island 14C)(where common: HH), and (b) arid squares adjacent to semiarid regions *e.g.* Tsavo E 89D 101BD (La). The 19B record in the extreme NE is from Erlanger (1905, 1907). Distribution is 83% in or E of the rift valley and, apart from L Turkana, it resembles that of the Fawn-coloured Lark (526) in being mainly on the plateaux N and S of the highlands, and in the intervening rift 62AC.

This lark has two modes of occurrence (Br). It can be resident (DEP pers. obs. in semiarid-subhumid 74B 75BCD), but it is also a partial migrant that often appears during or after substantial rains, *e.g.* in arid Turkana (HH), in arid Tsavo E 101B (La), and at semiarid Baringo 50A, where small numbers are usually Apr-Jul (St). Influxes to the semiarid Emali Plains 88AB and Tsavo W 89C 101A areas after good rains in Jan 1977 were unusual, and were not repeated after similar rains in Nov-Dec of that year (EABR 1977). Presumed nocturnal migrants have appeared at the Ngulia lights 101A Nov-Jan (Backhurst & Pearson 1977b; EABR 1982, 1983). See Northern White-tailed Bush Lark (528).

528 NORTHERN WHITE-TAILED BUSH LARK *Mirafra albicauda*
B614 / MPG655 / D1C1

Afrotropical: primarily the E, from Uganda and C Tanzania E to W and C Kenya. There is also a population in Chad.

Kenyan: this bird forms part of a superspecies with the Singing Bush Lark (527) (HM), and similarly prefers open and bushed grasslands (Br). Range is in slightly higher and moister areas than the Singing's, however: at 1200-2000 m (Br), 89% semiarid-subhumid and 80% within the 500+mm rainfall regime. In further contrast to that species, the Northern's squares are 52% to the W of the rift valley, while occurrence further E is virtually restricted to the vicinity of the high rainfall C highlands: this is thus a bird of the higher rainfall areas of W and C Kenya, which represent the E limit of its range. The record in 2C was NW of Lokichokio (G Nikolaus *in litt.*), while 12AC and 36B are from Uganda (J S Ash *in litt.*).

Br considers this lark a generally uncommon resident, but sometimes common in Kenya. This latter provision may derive from Ja, who described it as very plentiful on the Athi and Kapiti Plains 75BD 76C and not uncommon near Nairobi 75B. We have never observed such abundance, however, and are uncertain whether the high proportion of pre-1970 squares in and E of the rift valley reflects low detectability or a recent decline.

529 FLAPPET LARK *Mirafra rufocinnamomea*
B621 / MPG660 / D2C2

Afrotropical: widespread S of 15°N, except for the equatorial lowland forests and the dry areas of Somalia and the SW.

Kenyan: common in open bushed and wooded habitats, mainly below 1500 m but also up to 1800 m (Br; MPG). Range is 81% within 500+ mm rainfall areas (27D 28A = the vi-

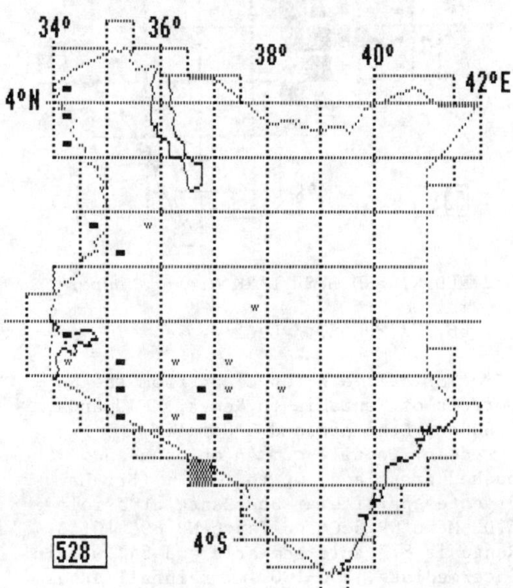

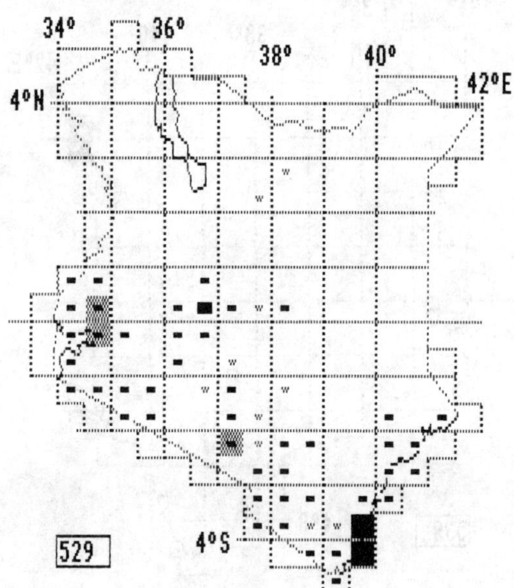

cinity of the locally moist Mt Marsabit: Ja) and only 13% arid (*cf.* Collared Lark (530)). Extension into arid areas along the Tana R 79C (AGH) parallels restriction to the Juba R of adjacent S Somalia (AM). DEP (pers. obs.) found this lark resident in 88AB.

530 COLLARED LARK *Mirafra collaris*
B616 / MPG663 / D2C0

Afrotropical: the NE, from E and NE Kenya to Somalia and S Ethiopia.

Kenyan: a very local species of bushed grassland, usually on red soils, at 100-1350 m (Br; MPG). It is generally uncommon (Br), although Serle (1943) reported presence in some numbers around Garissa 65B. Range is confined to the E and NE, 72% below 500 m and 81% within 0-250 mm rainfall areas: Br considers it arid-semiarid, but it is in fact entirely arid. Old records in 19 are from Erlanger (1905, 1907). Recent records in 19C and 31A are in Somalia (AM). Occurrence N of Marsabit (Br) is mapped in 28A.

The Collared Lark forms part of a superspecies with the allopatric Flappet Lark (529)(HM). AT QSD scale, the two only overlap in 28A (see above) and 79C. In the latter, the Collared is only known from the square's N boundary, at Hola (P L Britton *in litt.*; Br). It may be an occasional visitor to this S extremity of its range, since only the Flappet Lark was recorded in 79C by AGH.

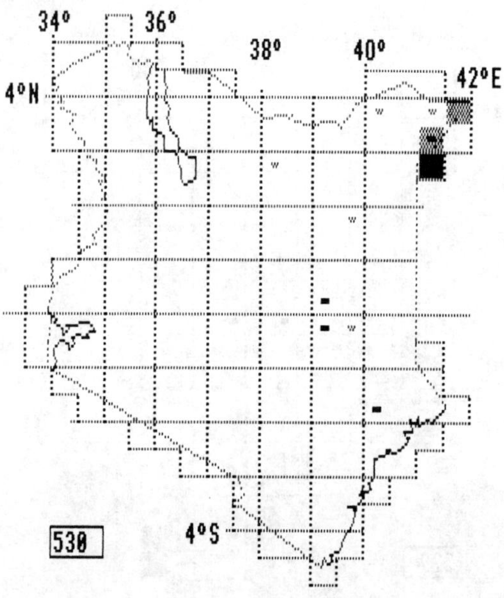

531 RUFOUS-NAPED LARK *Mirafra africana*
B612 / MPG659 / D2C2

Afrotropical: widely distributed S of 10°S, avoiding only the dry SW. Further N, it is mainly a bird of the E, from the E Zaire borders, through Uganda and N Tanzania, to the E limit of the range in W and C Kenya. There are other populations in N Somalia and W Africa.

Kenyan: locally common on open and bushed grasslands, typically at 1200-3000 m (Br; MPG). Range is 66% subhumid-humid and 80% within 500+ mm rainfall areas. It is considered resident (Br; Be: 74; DEP pers. obs.: 74B 75 88AB).

The Rufous-naped Lark forms part of a superspecies with the allopatric Red-winged Bush Lark (532)(HM), which is generally in lower, drier country. At QSD scale, the two overlap in semiarid-moist subhumid areas 50 51 88 89 101, but do not occur together (ADL pers. obs.). Other reference: White 1960.

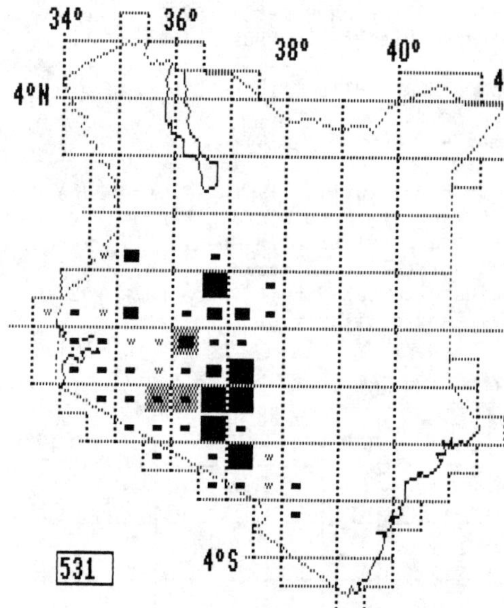

532 RED-WINGED BUSH LARK *Mirafra hypermetra*
B617 / MPG657 / D2C2

Afrotropical: a NE species, from the NE borders of Tanzania to Kenya, NE Uganda, S Sudan, Ethiopia and S Somalia.

Kenyan: locally common on open and bushed grassland, up to 1350 m (Br; La). Br notes particular abundance at Isiolo 51D, Meru NP 52C and Tsavo NP 89C 101AB. Range is 87% arid-semiarid and 84% within intermediate, 250-1000 mm rainfall areas (*cf*. Rufous-naped Lark (531)). Records in 12C and 36B are from Uganda (J S Ash *in litt*.), and presumably refer to *M.h. kidepoensis* (Br). Kenyan records are mainly from the plateaux to the NE and SE of the C highlands. An occurrence at Naivasha 62C (Ja) is the only Kenyan record in or W of the rift valley, and may reflect a wanderer. The species is otherwise thought to be resident (Br; La: 101B; DEP pers. obs.:88D 89C 100B 101A). There is no confirmed breeding record. Other reference: Dowsett-Lemaire & Dowsett 1978a.

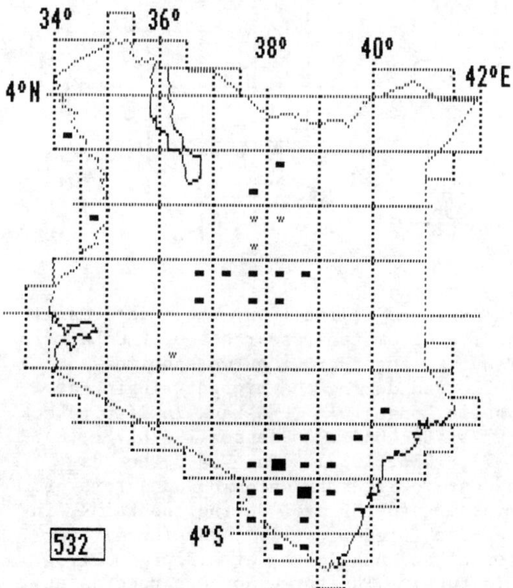

533 PINK-BREASTED LARK *Mirafra poecilosterna*
B619 / MPG665 / D2C2

Afrotropical: the E, from the NE borders of Tanzania, through Kenya to NE Uganda, S Sudan, Ethiopia and S Somalia.

Kenyan: locally common in open bushed and wooded habitats, on sandy soils with typically little or no grass cover, at 150-1800 m (Br; Ja; Lack 1985). Range is 92% arid-semiarid, only 2% within 1000+ mm rainfall areas and virtually absent from the coastal strip. 93% of the squares are in or E of the rift valley, and records further W represent the W limit of the range (excepting Moroto 24B, Uganda: Br). Occasional absences of several months' duration occur on this W range edge at Baringo 50A (St). This lark is otherwise considered resident (Br; La: 101B; DEP pers. obs.: 88D 89C 100B 101A).

It forms part of a superspecies with Gillett's Lark (534)(HM), which is con-

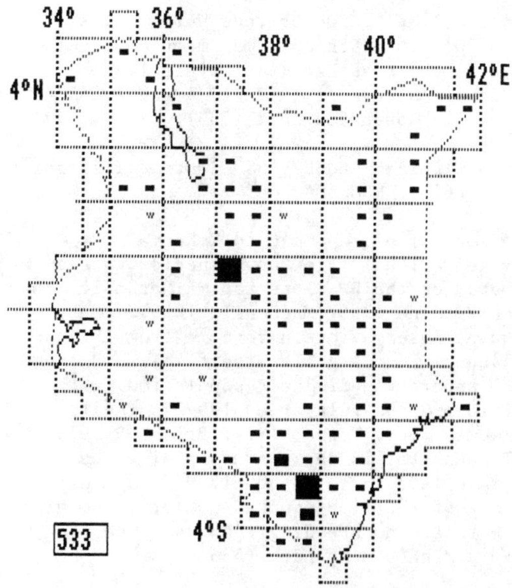

fined to the far NE. Other reference: Lack 1976a.

534 GILLETT'S LARK *Mirafra gilletti*
B- / MPG664 / D1C0
NO MAP: pre presence 19AB; pre confirmed 19C 31A

Afrotropical: the far NE, from W Ethiopia and N Somalia to S Somalia and the extreme NE of Kenya.

Kenyan: this species has recently been added to the Kenyan avifauna by Miskell & Ash (1985), who give details of at least 14 specimens taken by Erlanger in the extreme NE during May 1901. These records were published soon afterwards, but have been overlooked by subsequent authors. There are presence records from Handotu 19B, Karo-Lola 19B, Sarigo 19AB, Gordoba-Dschira 19A, Damaso 19C and Djeroko 19C. Breeding in 19C and 31A derives from Erlanger's (1907) records from Damaso and Wante: the former is on the Kenya-Somalia border, while the latter is probably so (J S Ash *in litt*.).

Of the four squares, three are below 500 m, three receive less than 250 mm annual rainfall, and all are arid. Occurrence in Somalia is in open bushland S to 2°N, and these Kenyan records are on the E edge of the range (AM). The very low observer frequency in NE Kenya, together with the species' low detectability and recent records from adjacent S Somalia (where common: Miskell & Ash 1985), suggest continued presence in these border areas to be likely. Occurrence W to at least 39°30'E in S Ethiopia may mean that this lark occurs as far W as Moyale 17A on Kenya's N borders (Miskell & Ash 1985). See Pink-breasted Lark (533).

535 WILLIAMS' BUSH LARK *Mirafra williamsi*
B622 / MPG660A / D2C2

Afrotropical: endemic to N Kenya.

Kenyan: a little known species, discovered on the SW slopes of Mt Marsabit 27D on 20 June 1955 (MacDonald 1956). Subsequent observations have come from the adjacent Dida Galgalla Desert 28A, and from an apparently disjunct population in the Isiolo-Garba Tula area 51BD 52A (Keith & Twomey 1968; Turner 1977; Br; EABR 1978). The habitat is short grass plains, on black lava soil, at 600-1350 m (*op. cit.*). Four of the five squares are arid-semiarid. There are no breeding records. See Friedmann's Bush Lark (536).

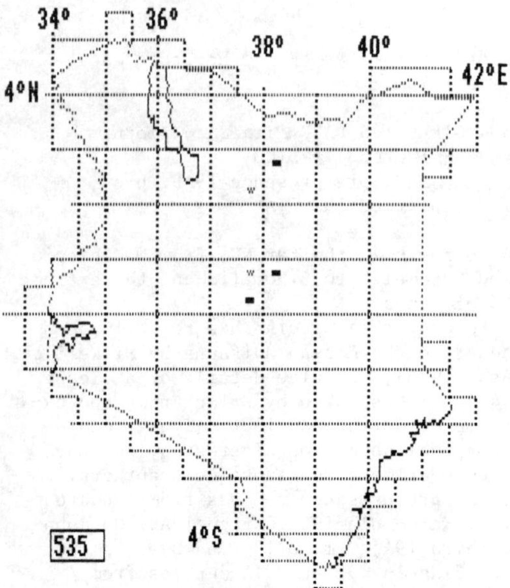

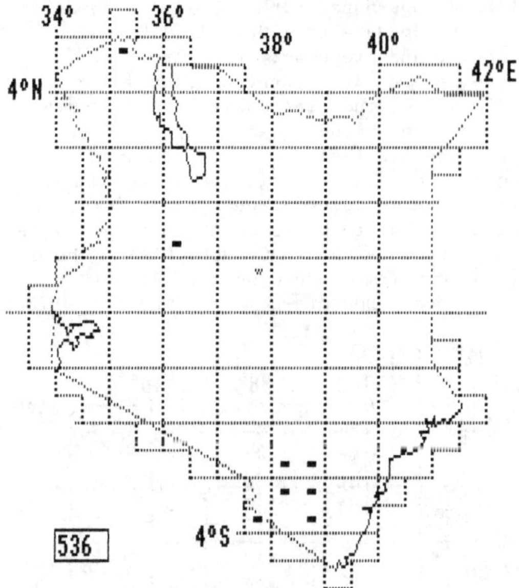

536 FRIEDMANN'S BUSH LARK *Mirafra pulpa*
B620 / MPG- / DOCO

Afrotropical: this very little known lark has been recorded from S Ethiopia S to NW, C and SE Kenya.

Kenyan: data for this very obscure species of bushed grasslands can be split into four discrete areas of occurrence.

An adult and two young birds were collected near Archer's Post 51B in 1912 (Friedmann 1930).

Until 1984, all recent records of the species were from the Tsavo area 89CD 101ABD, and L Jipe 100D: these data mainly derived from the researches of P C Lack in Tsavo E, together with two specimens collected during the migrant ringing at Ngulia Lodge 101A (P C Lack *in litt.*; Lack 1977a; Backhurst & Pearson 1977b; Br; EABR 1977, 1978, 1982). The Tsavo records appear to refer to uncommon wanderers or migrants to the area during the rains, in Dec-Jan (predominantly), Mar (once), Apr (once) and Aug (once, at L Jipe 100D)(*op. cit.*; La). The three Nov-Dec records at Ngulia Lodge (*op. cit.*) indicate nocturnal movements.

On 26 Feb 1984, six individuals were found near the Sudanese border at Kamathia 3A (EABR 1984). Shortly afterwards, tens located 5-30 km S of Kapedo 38C on 5 Mar remained in the area until mid Apr (EABR 1984).

Two points of interest emerge from these 1984 sightings. Firstly, the birds from Kamathia and Kapedo appear distinctly darker and less brown than specimens of the Tsavo population (D J Pearson pers. comm.). Plumage colour is affected by habitat soil colour in many lark species (see HM) and the browner Tsavo birds may reflect the well known red-brown soils of that area. The second question concerns these larks' status in the NW. Are they residents or regular migrants, or do the 1984 records represent an exceptional influx? The latter is a possibility in view of apparently drought-stimulated influxes by other species at that time (see Abyssinian (453) and Lilac-throated (455) Rollers), but the relocation of this lark at Kapedo in late Mar 1986 (D J Pearson, D A Turner, ADL pers. obs.), during more normal weather conditions, suggests otherwise.

This species' nine squares are at 550-900 m, in less arid-semiarid, 250-1000 mm rainfall areas. There is no breeding record. According to HM, this bird forms part of a superspecies with Williams' Bush Lark (535), but see Lack (1977a) for further taxonomic discussion.

537 SPIKE-HEELED LARK *Chersomanes albofasciata*
B604 / MPG- / D1C0
NO MAP: post presence 88C

Afrotropical: primarily the S, widespread S of 22°S but also extending N to Namibia and Angola. An isolated race occurs in NE Tanzania, and a vagrant from this population provides Kenya's only record.

Kenyan: the single record is of a bird near Amboseli Lodge 88C, on 21 Oct 1979 (Moore 1979). Although this individual was only 70 km NE of the NE Tanzanian range, the occurrence is nevertheless remarkable in view of the species' extremely restricted and highly sedentary nature (see references listed by Moore 1979). The lack of further records from this area, despite high observer frequency and specific searches, indicates a single extralimital vagrant, although the validity of the record has been questioned (Turner 1985).

538 MASKED LARK *Calandrella personata*
B602 / MPG689 / D2C2

Afrotropical: the NE, from C Kenya to Ethiopia.

Kenyan: a locally common lark of dwarf shrub grassland, particularly on sparsely grassed, black cotton and black lava soils, at 900-1350 m (Br)(*cf.* the range of Williams' Bush Lark (535), another bird of black lava areas). Range is 91% arid-semiarid and 84% in 0-500 mm rainfall areas. Three races occur (Br). *C.p. intensa* is endemic to the Isiolo 51D-Chanler's Falls 52A-Garba Tula 52B area, and *C.p. mcchesneyi* is endemic to the Marsabit plateau 27D 28C. *C.p. yavelloensis* extends from the Dida Galgalla Desert 28A N to S Ethiopia. Occurrence in 26D, an extension of known range, is based on a small flock in lava terrain approximately 30 km S of Loyangalani 26B, on 16 Aug 1983 (A J Holcombe *in litt.*).

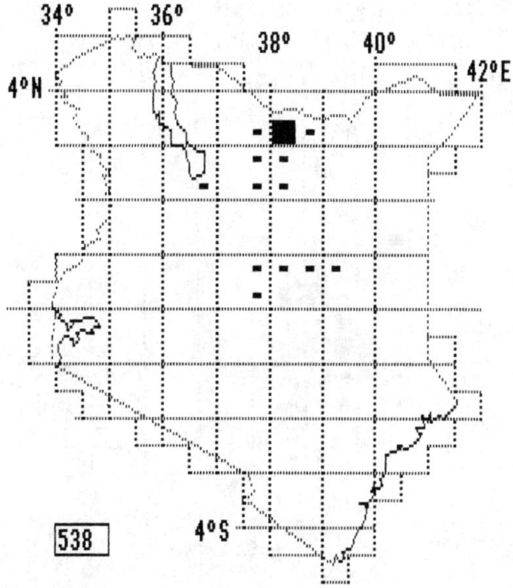

539 SHORT-TOED LARK *Calandrella brachydactyla*
B600 / MPG684 / D0C0
NO MAP: pre presence 114B

Afrotropical: this Palearctic species winters S to the S edge of the Sahara, and to S Somalia (HM; AM). Extralimital wanderers are known S to Kenya.

Kenyan: the two records are of single birds collected on the Athi R (precise location uncertain) on 14 Nov 1899 (Ja), and at Ukunda 114B on 19 Dec 1964 (Ripley & Bond 1971). These dates concur with very recent Nov-Dec records from 2°N on the coast of Somalia (AM).

This species forms part of a superspecies with the Red-capped Lark (540)(HM).

540 RED-CAPPED LARK *Calandrella cinerea*
B601 / MPG686 / D2C2

Afrotropical: widespread S of 5°S, and further N through mainly highland areas of East Africa to Ethiopia and N Somalia. Distribution in Kenya represents the E limits of the range.

Kenyan: a locally common bird of short grasslands, typically at 1200-3000 m but also down to 950 m (Br; Se; Ja). Range is mainly in the rift valley and surrounding highlands, 83% in 500+ mm rainfall areas and 64% subhumid-humid. This lark is considered resident by Be (74) and Se (61D), although the latter author also notes local movements of birds around 3000 m at

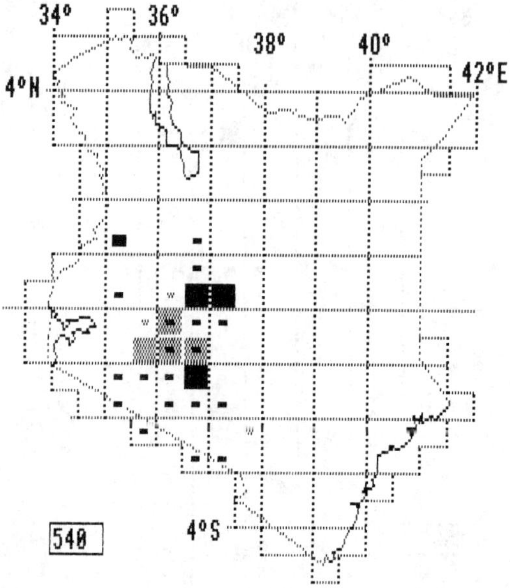

Mau Narok. See Short-toed Lark (539).

541 RUFOUS SHORT-TOED LARK *Calandrella somalica*
B603 / MPG685 / D1C1

Afrotropical: the NE, in NE Tanzania, Kenya, S Ethiopia and N Somalia.

Kenyan: rather uncommon on open, short grass plains at 1200-1650 m (Br). Range is 90% semiarid-subhumid and entirely within intermediate, 250-1000 mm rainfall areas. This lark reaches the rift valley at Naivasha 62C, but all other records are to the E, mainly on the plateau to the S of the C highlands. Two Mar birds to the N of the C highlands, at Samburu GR 51B, may have been wanderers from the S Ethiopian race *C.s. megaensis* (Br; EABR 1978). An individual was caught at the Ngulia lights 101A in Dec (EABR 1978).

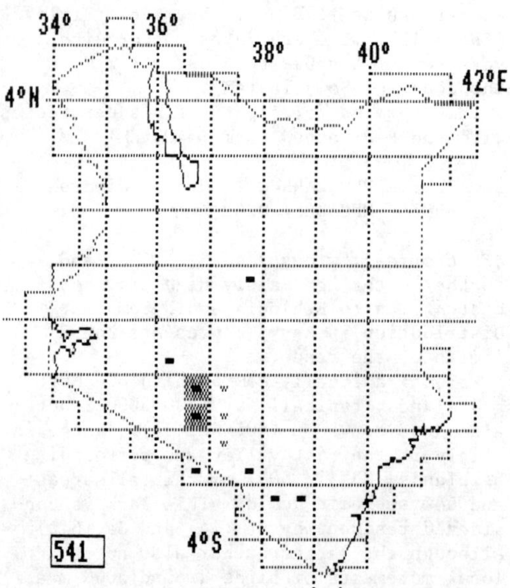

542 SHORT-TAILED LARK *Galerida fremantlii*
B609 / MPG678 / D2C1

Afrotropical: the NE, with a fragmentary distribution from the NE Tanzanian borders to S and N Kenya, S Ethiopia, and N and C Somalia.

Kenyan: locally common on short grass plains, particularly those with bare, open areas, at 1000-1700 m (Br). These habitat requirements are comparable to those of the Rufous Short-toed Lark (541), and these species' Afrotropical and Kenyan distributions are similar.

G.f. delamerei inhabits the plateau to the S of the C highlands, with range entirely above 1000 m, semiarid-subhumid, and within intermediate, 250-1000 mm rainfall areas. It is thought to be resident (Br; DEP pers. obs.: 75). *G.f. megaensis* extends from S Ethiopia to the vicinity of Marsabit 28A (Br; National Museum Collection, Nairobi): the single square is at 500-1000 m, arid, and within the 0-250 mm rainfall regime.

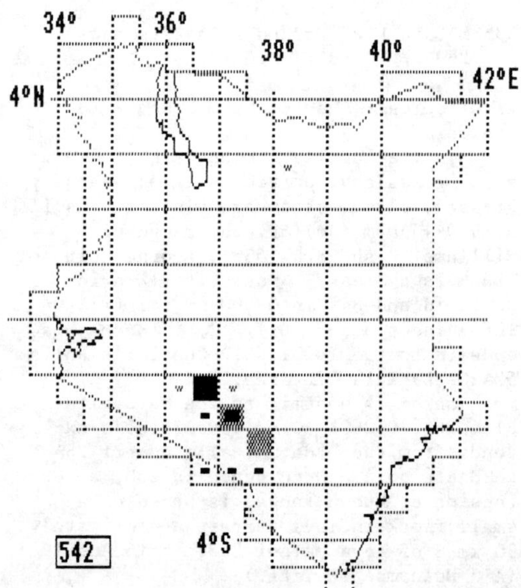

543 CRESTED LARK *Galerida cristata*
B608 / MPG675 / D2C2

Afrotropical: the N tropics, widely distributed between 10°N and 20°N, and extending further S through Ethiopia to N Kenya.

Kenyan: common in sandy semidesert with little or no grass, at 400-900 m (Br; Ja; HH). Range is 82% within 0-500 mm rainfall areas, and entirely arid-semiarid. The cluster of breeding squares 27CD 39AB reflects the records of a single observer in the Kaisut Desert just after the Apr-May rains (Lewis 1981e), and illustrates this atlas' poor coverage of breeding in remote areas (*cf*. Superb Starling (890)). See Short-crested Lark (544). Other reference: Skinner 1971.

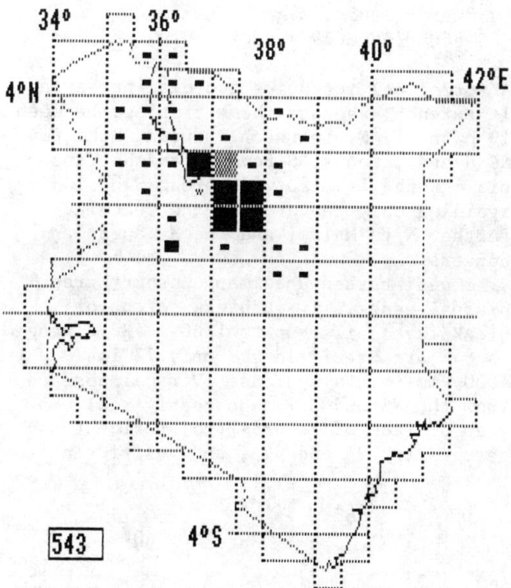

544 SHORT-CRESTED LARK *Galerida malabarica*
B610 / MPG676 / D2C1

Afrotropical: the NE, from N Kenya to Ethiopia and all but the extreme S of Somalia.

Kenyan: at times common in dwarf shrub grassland, on bare rocky ground, at 400-1300 m (Br). In contrast to the Crested Lark (543), which is a sibling species (HM), the range of the Short-crested (including 26A) is confined to the E of L Turkana (HH). These squares are entirely arid, and 72% within the 0-250 mm rainfall regime. This lark is especially numerous in much of the Dida Galgalla Desert 28A (Br).

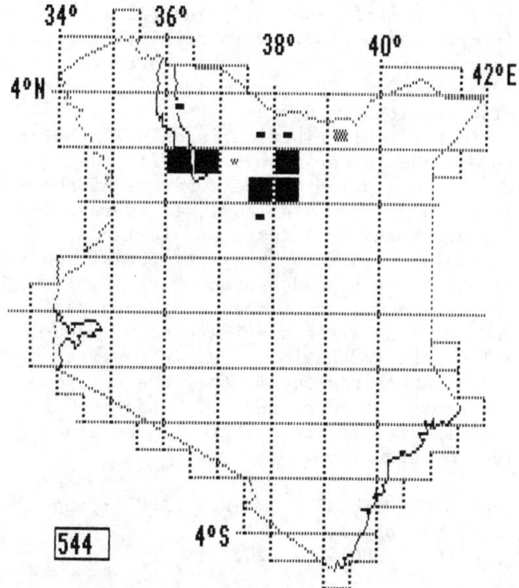

545 FISCHER'S SPARROW LARK *Eremopterix leucopareia*
B605 / MPG682 / D2C2

Afrotropical: an E species, from Malawi and N Mozambique to C Tanzania, and further N to NE Tanzania, Kenya and NE Uganda. Distribution in Kenya describes the N and NE limits of the range.

Kenyan: locally common on open grass plains, up to 1800 m (Br; MPG; St). Range

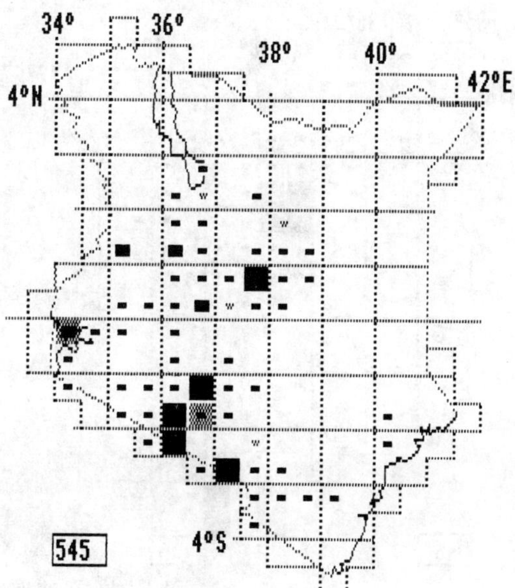

is 76% semiarid-subhumid, and 77% within intermediate, 250-1000 mm rainfall areas. This sparrow lark is considered resident (Br; St: 50A; DEP pers. obs.: 75), and Br also mentions wandering.

Fischer's distribution is only 27% below 1000 m and 16% arid so that, with the notable exception of the environs of the lower Tana R 79C 91A (AGH), it is rare in the dry N and E plateaux and lowlands. It is replaced in these areas by the allopatric Chestnut-headed Sparrow Lark (546: *q.v.*), which is a member of the same superspecies (HM). This latter species prefers generally drier situations and, in squares where both birds occur, they are not found together (ADL pers. obs.). Wanderers of both species may contribute to the apparent overlap (Br).

546 CHESTNUT-HEADED SPARROW LARK *Eremopterix signata*
B607 / MPG681 / D2C2

Afrotropical: the NE, from N and E Kenya to S Sudan, Ethiopia and Somalia. Distribution in Kenya represents the SW limit of the range.

Kenyan: locally common on bare desert or short grass areas, up to 1500 m (Br; HH; Lack 1985). Range is 92% arid-semiarid and 76% within the 0-500 mm rainfall regime, and its SW limit is defined by the higher rainfall areas of the highlands and their peripheral plateaux. Despite low observer frequency, apparent absence from the E border areas S of 2°N may be real, since in Somalia this bird is confined to the coastal strip at these latitudes (AM). The cluster of breeding squares 27CD 39AB reflect unusual observer frequency (see Crested Lark (543)).

Range in the SE lowlands has recently extended SW to Tsavo E 101B (where this species has virtually replaced Fischer's Sparrow Lark (545: *q.v.*), which was common there in the early 1960s: La) and Tsavo W 89C 101A (EABR 1979), and S to littoral grasslands and dunes between the Tana R 91AC and the Sabaki R 103A (Br; EABR 1978). A small June group at Diani Beach 114B (EABR 1980) may have been wanderers, or a continuation of this S extension. This sparrow lark is a partial migrant, commoner in the rains, in Tsavo E 101B (La).

547 CHESTNUT-BACKED SPARROW LARK *Eremopterix leucotis*
B606 / MPG679 / D2C2

Afrotropical: occupies two distinct areas. It extends W across the N tropics, between 10°N and 17°N, to reach S Sudan, Ethiopia, NE Uganda, Kenya, S Somalia and NE Tanzania. Range is also widespread S of 15°S, avoiding only the dry SW, and reaching further N to Malawi and the S Tanzanian borders.

Kenyan: rather uncommon on short grass plains, especially on black cotton or black lava soils, up to 1800 m (Br). Range is 84% arid-semiarid and only 7% in 1000+ mm rainfall areas. Virtual absence from the NE and E is apparently real, in view of lack of records from resident observers in 19A and 42A, and restriction to

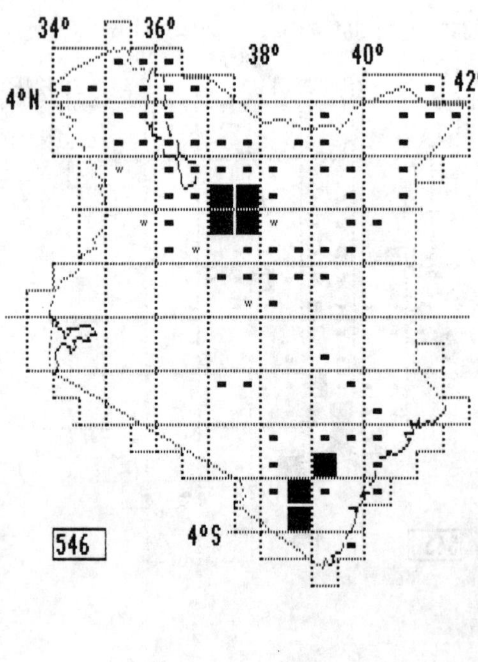

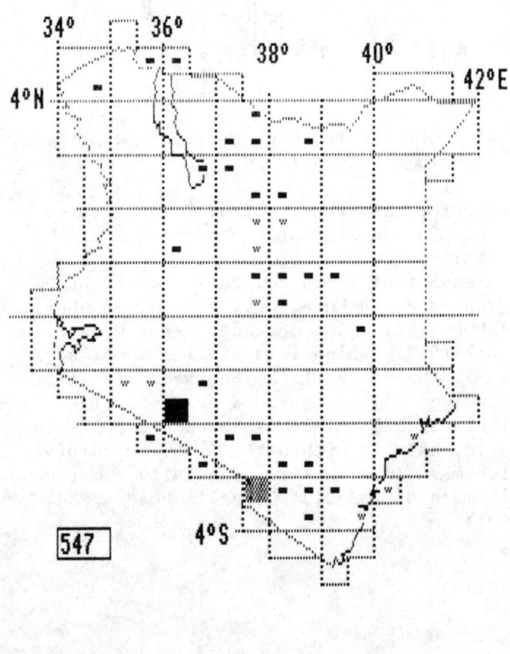

the coastal strip in adjacent S Somalia (AM). Similarly, HH never encountered it in the Turkana basin. Presence between these two areas, *i.e.* N of Marsabit 27D 28C, is thus presumably due to some specialised ecological requirement.

Br describes this sparrow lark as nomadic, and rather sporadic in many areas of Kenya. It is an uncommon visitor to Tsavo E 101B soon after rain (La), and a dawn individual at Ngulia Lodge 101A possibly indicates nocturnal movement (Backhurst & Pearson 1977b). There are only old records from the lower Tana R (not mapped) through the coastal strip to Malindi 103A (Ja)(date of Kilifi Creek 102D record (Br) unknown). Flocks with the Chestnut-headed Sparrow Lark (546) are common in N areas (Br).

Family Hirundinidae: Swallows, Martins, Rough-wings

An almost cosmopolitan family of aerial insectivores containing 74 species. Of the 35 (47%) recorded in the Afrotropical Region, 28 (80%) are endemic (HM). 18 species (51% of the Afrotropical total) have occurred in Kenya, where 13 (72%) breed, three (17%) are non-breeding migrants from the Palearctic, one is a vagrant from Madagascar, and one a scarce intra-African migrant on the E edge of its non-breeding range. The potential Afrotropical breeding ranges of many species have been immensely increased by their nesting on man-made structures (HM).

548 BANDED MARTIN *Riparia cincta*
B641 / MPG1070 / D2C2

Afrotropical: widely distributed to the S of Ghana and Ethiopia, avoiding only the arid areas of Somalia and the SW. Apart from two recent records at 1°N in coastal Somalia (AM), distribution in Kenya reflects the E limit of the range.

Kenyan: a rather local species of open or lightly bushed grassland, not necessarily near water, up to 3000 m (Br; MPG; Se). Range is 74% above 1000 m, and there is only a single record from the coastal lowlands (the Sabaki estuary 103A: Br). It is 68% in 500+ mm rainfall areas, and the only records from arid country are three Mar specimens from Bura 78B (National Museum Collection, Nairobi), and occasional birds at L Turkana (HH).

The seven breeding squares are six above 1000 m and six within 500+ mm rainfall areas. Although no doubt affected by observer coverage, these figures may point to a real pattern since Ja suspected the species to be mainly migrant in Kenya, and Br mentions poorly known movements. Se described it as a predominantly Feb-Jul visitor to 3000 m at Mau Narok 61D, breeding in June. The few dated records of presumed wanderers are in all but one case (Bura: see above) outside of this season. On the edges of its main Kenya range, a few birds were at semiarid Samburu GR 51B in late Sept (EABR 1980), while wanderers have reached semiarid Baringo 50A in Oct and Nov (St) and arid Aruba 101B in Nov (La). DEP (pers. obs.) thinks the species absent from the Emali area 88AB during this Sept-Nov period.

The occasional sightings at Turkana are probably *R.c. erlangeri*, the race in NE Uganda, S Sudan and Ethiopia (Br). The only dated records from this area are a flock of 30 at Todenyang 3B in Nov (HH), and a few at Ferguson's Gulf 13B in late Jan (EABR 1980). The two records from Somalia were Jan-Feb (AM)(*cf.* Bura, above).

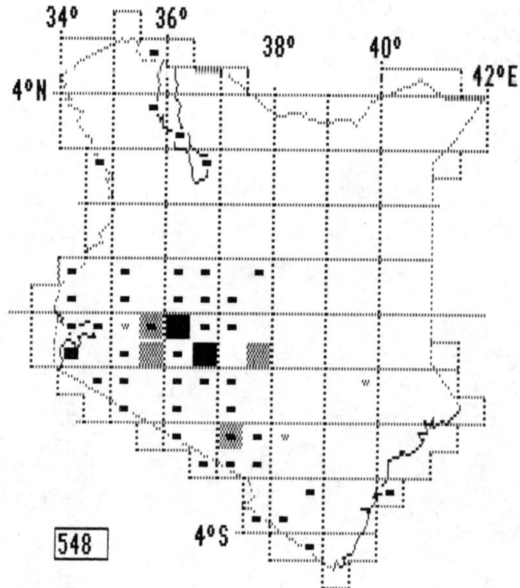

549 AFRICAN SAND MARTIN *Riparia paludicola*
B642 / MPG1069 / D2C2

Afrotropical: widely distributed S of 15°N, extending further N along the Nile, but absent from the equatorial forest areas of the W. Distribution in Kenya reflects the E limit of the range.

Kenyan: this open country species is typically near inland waters (Br), and 66% of its squares contain lakes. There are two distinct areas of occurrence. In the highlands and peripheral plateau, it is common at 1200-3000 m, but more sporadic lower down (Br; Se). Range here is 98% above 1000 m and 87% in 500+ mm rainfall areas. Breeding is confined to this region, being 72% above 1500 m, entirely within 500+ mm rainfall areas, and 87% subhumid-humid. This species is also widespread but occasional in arid country around L Turkana (HH). On analogy with the Banded

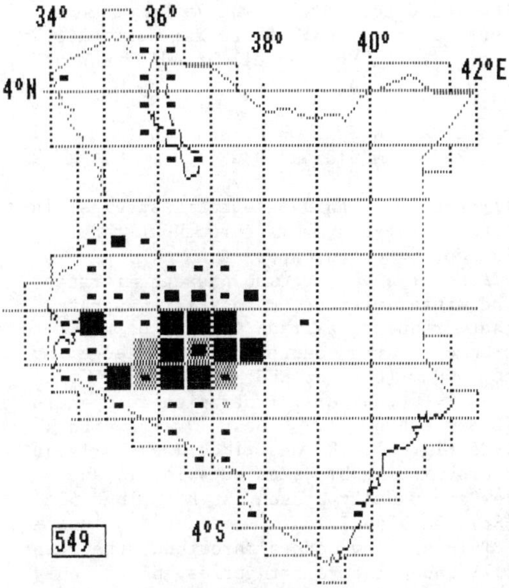

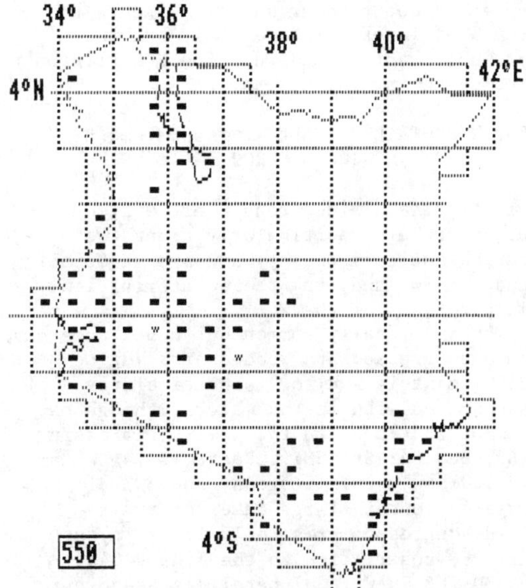

Martin (548), these birds may originate further N. The only other record in arid country is from the vicinity of the Tana R at Kora 64B. The sole record from the SE lowlands is of a single Oct bird at Bamburi 102D (EABR 1981).

Br suggests that this martin may be resident in some areas, but notes that its movements are poorly known. Se recorded widely fluctuating numbers around 3000 m at Mau Narok 61D. Most records on the range edge at semiarid Baringo 50A are June-Aug (St). DEP (pers. obs.) found the species in all months in the Ruaraka-Kahawa-Ruiru sector of moist subhumid 75B, but also noted an annual Sept-Dec influx there.

The African Sand Martin forms part of a superspecies with the widely sympatric Sand Martin (550)(HM), which is a Palearctic migrant.

550 SAND MARTIN *Riparia riparia*
B643 / MPG1068 / D2C2

Afrotropical: a Palearctic species that winters across the N tropics from Senegambia to Ethiopia and S Somalia, and S through C and East Africa to Botswana and E South Africa.

Kenyan: a locally common to abundant passage migrant and winter visitor Sept-May, mainly found over open grassland and swamp near large inland waters, below 1200 m (Br). Range is 71% within 500 + mm rainfall areas and only 18% arid. Only 6% of the squares lack large water bodies of some type (*cf.* 7% for the African Sand Martin (549: *q.v.*)).

The Sand Martin is a local but regular passage migrant Mar-Apr and Sept-Oct at L Turkana (HH). Wintering birds are most abundant in the L Victoria basin, but rather less so around the lakes of the W highlands and the rift valley (Br). It is most numerous at L Baringo 50A Jan-Mar, when 5000 can be present (St). 200 000 at nearby L Bogoria 50C in Feb was an unusual concentration for C Kenya (EABR 1982). In direct contrast to its two congeners (548, 549) however, the Sand Martin is regular if uncommon in the SE and coastal lowlands, and fairly common S of 4°N in adjacent S Somalia (AM). Flocks reach the Sabaki estuary 103A (Br), and small Sept-Jan numbers in arid Tsavo E 101B are mostly at Aruba Dam (La).

551 BLUE SWALLOW *Hirundo atrocaerulea*
B628 / MPG1060 / D2C0
NO MAP: pre presence 48BC

Afrotropical: breeds in the SE, from E South Africa N to S Tanzania. It migrates N to spend the non-breeding season in NE Zaire, Uganda and, rarely, on the E edge of the range, in W Kenya.

Kenyan: an intra-African migrant that frequents open grassland, and has occurred at Bungoma 48B, Mumias 48C and Busia 48C Apr-Sept (Ja; Br). These localities are W of the rift valley and N of the Winam Gulf 60AB, entirely above 1000 m in humid, 1000+ mm rainfall areas. Williams (1968) noted frequent occurrence in W Kenya to the E of Tororo (Uganda 48A), but there are no recent records despite P L Britton's residence in 48C.

552 WIRE-TAILED SWALLOW *Hirundo smithii*
B637 / MPG1061 / D2C2

Afrotropical: widespread S of 10°N, extending further N particularly along the Nile, but absent from the dry areas of E Somalia and the SW, and from the equatorial lowland forests of the W.

Kenyan: locally common but never numerous, in open grassed and wooded areas up to 2100+ m: this swallow is particularly associated with inland waters, though uncommon and local in the L Victoria basin (Br; Be; La; St; HM). Range is 74% within 500+ mm rainfall areas, and 80% of the squares contain large lakes or rivers. Arid squares comprise only 17%. In the arid E, restriction to the Tana R 64B 65ABD 78B 79AC (AGH) parallels association with the Juba and Shebelle R of adjacent S Somalia (AM). Small resident numbers in arid Tsavo E 101B are in the vicinity of water (La). Occurrences in the arid N are confined to shore and inland areas of 26B at Turkana Oct-Dec 1971 (HH), and to the locally moist, isolated highland of Mt Marsabit 27D 28C.

The Wire-tailed Swallow is probably mainly resident (Br; Be: 74; La: 101B; DEP pers. obs.: 88D 89AC 101A; HM), and two of three ringing recoveries at Nairobi 75B had moved less than 15 km (Backhurst 1971b). The third recovery showed 75 km of movement, from Nairobi 75B to Naivasha 62C (*op. cit.*). Most or all breed on man-made structures (Ja; Br).

553 EURASIAN SWALLOW *Hirundo rustica*
B634 / MPG1054 / D3C3

Afrotropical: this Palearctic species winters in large numbers throughout the Region, S to the Cape.

Kenyan: a very common passage migrant and winter visitor, mainly late Aug-Apr, that frequents a great variety of habitats, from open arid country to forest edge, up to 3000 m (Br; Se; MPG; HH; La; St). It greatly outnumbers Afrotropical congeners (Brown & Brown 1974; Best 1977; see Lack 1985 Fig. 6). Br describes June-early Jul oversummering birds as occasional, and a few are usually present at L Baringo 50A (St)(see also Ja).

This swallow occurs throughout the rainfall and moisture categories, but is only recorded from 41% of the 0-250 mm rainfall squares and 35% of the most arid (moisture index below 1.2). This implies apparently sparse distribution in the NE, which might reflect paucity of insect prey in these harsh environments. But Eurasian Swallows probably migrate through these areas, and this distribution is more likely to reflect low observer frequencies. This species is only mentioned as a passage migrant at L Turkana (HH), and this probably applies to Kenya's other arid areas too. There is also the possibility

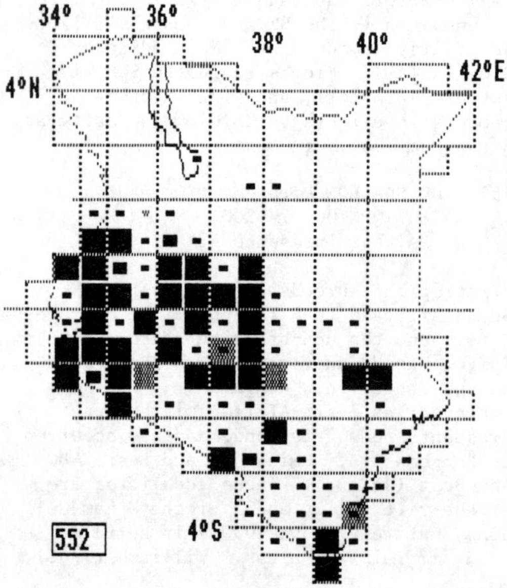

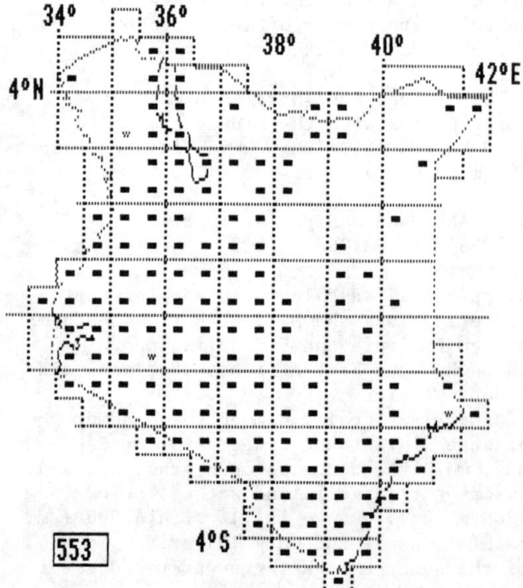

that this swallow only visits these areas after rains have stimulated insect activity, as happens in arid Tsavo E 101B (Brown & Brown 1974; La). These remote N and E areas are at their most inaccessible during wet periods (see Striped Swallow (559)).

Ringing recoveries have demonstrated movement between Kenya and South Africa, Uganda, Czechoslovakia, Poland, Hungary, Finland, the USSR, Greece, Lebanon and Iraq (E A Bird Ringing Reports). See Ethiopian Swallow (554). Other reference: Lewis & Pomeroy 1982a.

554 ETHIOPIAN SWALLOW *Hirundo aethiopica*
B625 / MPG1057 / D2C2

Afrotropical: mainly the N tropics at 5°N -15°N, from Benin and Niger to Ethiopia and N Somalia, and thence S to Uganda, Kenya, S Somalia and NE Tanzania.

Kenyan: a common bird of open country that is virtually confined to the coastal strip, and to areas N of the equator (Br). The coastal population is long-established, but range has apparently extended S into N Kenya during this century (Ja; Br; Grant & Lewis 1984). This spread may be associated with the recent proliferation of large, permanent buildings, which provide secure nest sites (*op. cit.*). In W and C Kenya, this spread has reached S to the N periphery of the highlands, which are occupied by the closely related Angola Swallow (555: see below), and has penetrated marginally further S in the rift valley 62AC.

Absence of records from the well known areas E to 63 and 76 is real. The situation further E is less certain due to far fewer observers, but there are no records from observers resident in 42A and 65B. Thus, since this species is apparently resident (see below), these coastal and N populations appear to be discrete, while merging further to the NE in Somalia (AM).

The total Kenya range is 75% below 1000 m, 73% arid-semiarid, and virtually absent from 1000+ mm rainfall areas inland (*cf.* Angola Swallow). Breeding squares are arid-dry subhumid (27D 28C = the moist, isolated highland of Mt Marsabit)(*cf.* Angola Swallow) with those at the coast below 500 m and those inland above.

The species is generally considered resident (MPG; Br; AM), and there is no evidence of post-reproductive dispersal or migration at El Karama Ranch 50D (Grant & Lewis 1984). On the local SW range edge, uncommon birds at L Baringo 50A are mostly in June and Dec (St).

The Ethiopian Swallow forms part of a superspecies with the Eurasian (553) and Angola (555) Swallows (HM). The two Afrotropical species show little overlap at QSD scale, and their breeding ranges are completely allopatric (*cf.* HM). The Palearctic Eurasian Swallow is widely sympatric with the other two. Other references: Dyson 1971; Lee 1971.

555 ANGOLA SWALLOW *Hirundo angolensis*
B627 / MPG1055 / D2C2

Afrotropical: from Gabon S to Angola, E to the S of the Congo basin forests to S and

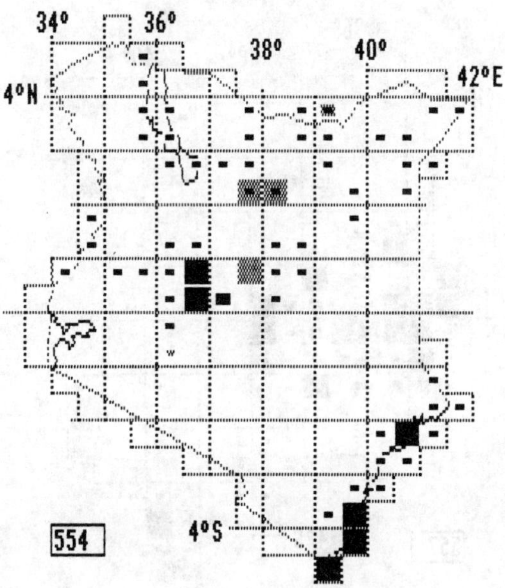

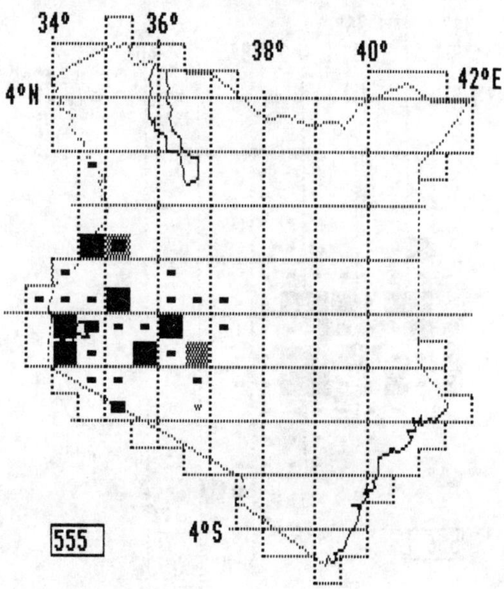

C Tanzania, the E Zaire borders, Uganda and, on the NE edge of the range, Kenya.

Kenyan: open and bushed habitats, and forest edge, up to 3000+ m (Br; Se). It is mainly found in the W, with 79% of the squares in or W of the rift valley, and it is the commonest breeding swallow in Nyanza 60 (Br). Wanderers are known E to Timau 51C, Naro Moru 63A (ADL pers. obs.) and Nairobi 75B (EABR 1980). In direct contrast to the Ethiopian Swallow (554: *q.v.*), range is 78% moist subhumid-humid, and entirely above 1000 m in 500+ mm rainfall areas. Breeding range is 90% in or W of the rift, and 80% in 1000+ mm rainfall areas.

This swallow is a breeding visitor to Mau Narok 61D (Se). Br mentions substantial movements, as evidenced by 500 on the tiny Magogo Islands 60A on 29 Jul. Wanderers are known at Mutara 50D in June (Olson *et al.* unpubl.), in 51C and 63A during May, and at Nairobi 75B Jul-Aug (*op. cit.*).

556 RED-RUMPED SWALLOW *Hirundo daurica*
B630 / MPG1062 / D2C2

Afrotropical: the N tropics and the E, from Senegambia E to Ethiopia, and S through East Africa and the E Zaire borders to Zambia and Malawi. Apart from a single wanderer in S Somalia (AM), distribution in Kenya represents the E limit of the range.

Kenyan: locally common in open habitats, frequently associated with rocky hillsides and buildings (Br; Brown & Britton 1980; Ja; La). It is mainly at 1200-3000 m in the highlands, and regular occurrence down to 500-600 m on their E periphery (52 S to 101) describes the E limit of the range (Br; La). Br considers this species absent below 1600 m in W Kenya, but there are records from all but one of the 1000-1500 m squares W of 35°E.

Range is 72% in 500+ mm rainfall areas and only 9% arid. Occurrence in the dry country N of 2°N is restricted to the locally moist, isolated highlands of Mts Kulal 26B, Nyiru 26D and Marsabit 27D (Br), and to Moyale 17A (J E Miskell *in litt.*), which represents the S limit of the Ethiopian highlands population.

While possibly mostly resident, this swallow is seasonal in some areas (Br; La; HM) Se noted apparent absences of one or two months around 3000 m at Mau Narok 61D. Very uncommon wanderers to the local range edge at semiarid Baringo 50A have been in May and Jul (St), while the single S Somalia record was in Feb (AM). Of four ringing recoveries, three at Nairobi 75B showed no significant movement, while the fourth had moved *c.* 240 km from Nairobi 75B to Kisii 60D (Backhurst 1971b, 1973b, 1977).

557 MOSQUE SWALLOW *Hirundo senegalensis*
B636 / MPG1063 / D2C2

Afrotropical: widespread S of 15°N, except for the equatorial lowland forests of the W, and the dry areas of Somalia and the SW. Apart from a single wanderer in S Somalia (AM), distribution in Kenya represents the

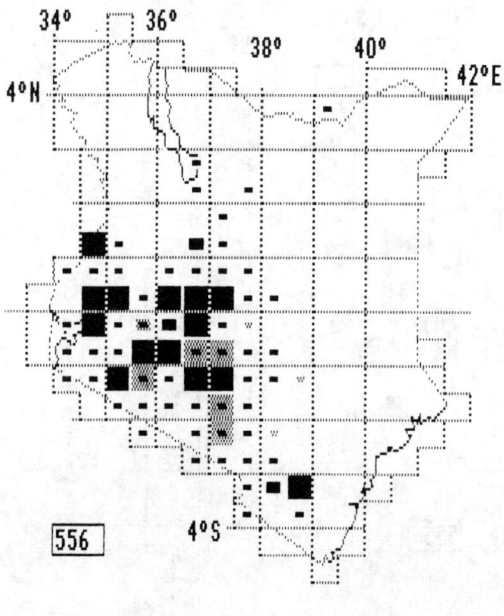

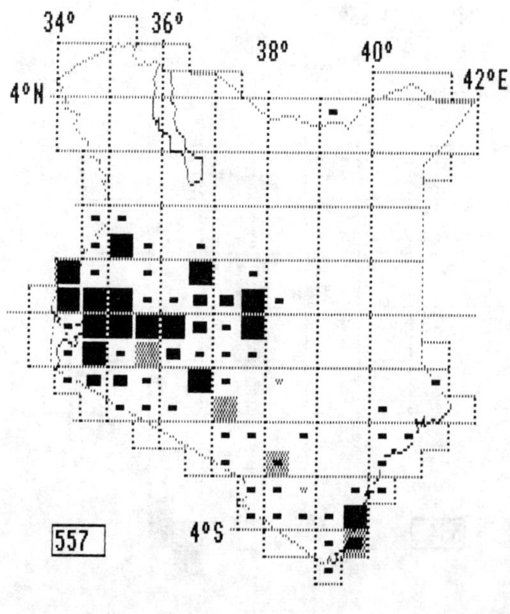

E limit of the range.

Kenyan: locally common in open wooded habitats, including forest edge, up to 3000 m (Br; Se; MPG). Range is 85% within the 500+ mm rainfall regime and only 11% in arid areas. Moyale 17A (EABR 1984) represents the S limit of the Ethiopian highlands population. Presence in the coastal lowlands contrasts with the Red-rumped Swallow (556). Breeding is 88% subhumid-humid, and apparently virtually restricted to the moist highlands and coastal strip: this may be a real pattern.

Se noted Aug-Dec absence near the top of the altitudinal range around Mau Narok 61D. The single wanderer to S Somalia was in Sept (AM): while these authors list this bird in 80B, Ash's (1983) description of the location appears to refer to 80A, and this square is mapped here.

The Mosque Swallow forms a species-group with the Rufous-chested Swallow (558), and the two are widely sympatric (HM).

558 RUFOUS-CHESTED SWALLOW *Hirundo semirufa*
B635 / MPG1064 / D2C2

Afrotropical: widespread S of 15°N, E to Sudan, Kenya, Malawi and S Mozambique, but absent from the dry SW and much of the low-land equatorial forest. Distribution in Kenya represents the E limit of the range.

Kenyan: rather uncommon in open country up to 1700 m (Br). Range is predominantly W of the rift valley, above 1000 m in subhumid-humid, 500+ mm rainfall areas. Rare wanderers E to the rift have occurred at Hell's Gate 62C Aug (EABR 1980), Mt Suswa 75A Sept (Standish King 1970), and the Ewaso Ngiro marshes 75C (DEP pers. obs.). The sole record E of the rift is of a Jul vagrant to arid Tsavo E 101B (EABR 1981). Rare nesting records reflect the E extremity of the equatorial breeding range (Lewis 1982f). See Mosque Swallow (557).

559 STRIPED SWALLOW *Hirundo abyssinica*
B624 / MPG1065 / D3C2

Afrotropical: widespread S of 15°N, avoiding only the drier areas of the SW and Somalia, and some of the equatorial low-land forest regions.

Kenyan: common in a wide variety of habitats, including forest edge, up to 2200+ m (Br; La; MPG). Range is only 6% in 0-250 mm rainfall areas and 20% arid, so that it is mainly absent from the arid N and E, including L Turkana (HH)(but see below). 5C and 17 represent the S edge of the Ethiopian highlands breeding range (J S Ash *in litt.*; F J McCartney and ADL pers. obs.).

Br considers this bird probably resident in many areas, and this view is endorsed by Be (74), La (101B) and DEP (pers. obs.: 75C 88AD 89AC 100B). Nine ringing recoveries from moist subhumid squares (eight from Nairobi 75B: E A Bird Ringing Reports) show no significant movement. However, movements certainly occur in some areas. This swallow is a breeding visitor Jan-early Oct to semiarid Baringo 50A (St), and Oct-Apr to the semiarid Ngulia area 101A (D J Pearson *in litt.*), while DEP

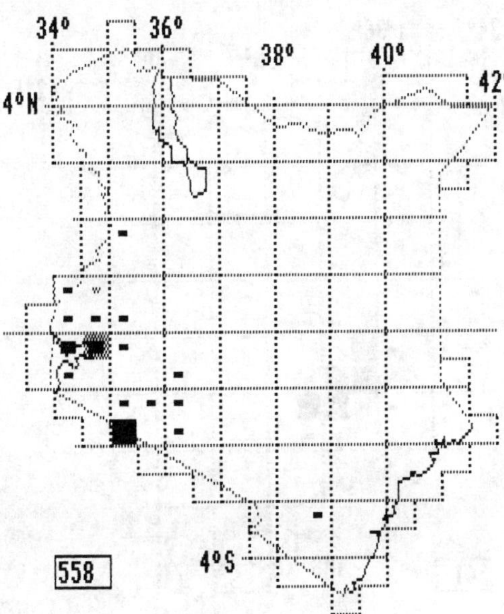

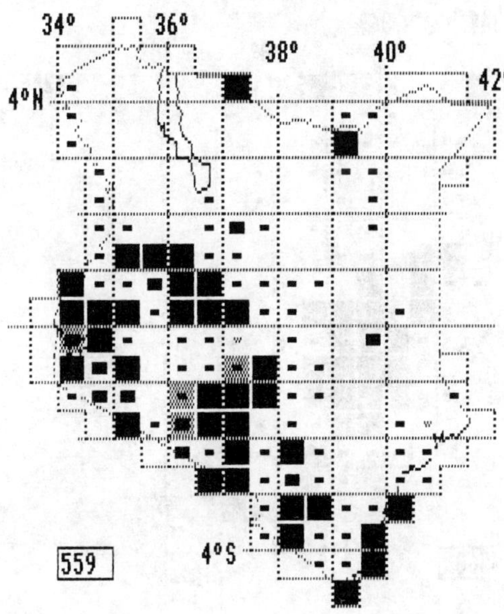

(pers. obs.) notes Oct disappearance from the Kahawa area of moist subhumid 75B.

Br notes movement into drier areas during the rains, and this is strikingly demonstrated by the records in arid 29 and 41B, which occurred in wet conditions (D A Turner *in litt.*). This type of rains-induced movement may be common in the arid areas of the N and E, although there are no records from observers resident at L Turkana (HH), and in 19A and 42A. In any case, the map primarily depicts this species' distribution in arid areas during the dry seasons. As with rains-induced breeding of arid country species (see Crested Lark (543)), these movements would occur when these remote areas are at their most inaccessible. Breeding is largely on man-made structures (Br; AM; La). Other reference: Reynolds 1975b.

560 AFRICAN ROCK MARTIN *Hirundo fuligula*
B632 / MPG1073 / D2C2

Afrotropical: local in many areas S of 20°N, but absent from the equatorial lowland forests and most of Somalia. Distribution in Kenya represents the E limit of the range.

Kenyan: relatively common birds around cliffs, rocky hills and buildings, on which they roost and nest (Br; MPG; HH; La; St). They are particularly associated with human dwellings, including Nairobi city centre 75B (Ja; Brown & Britton 1980). Range extends throughout the rainfall and moisture regimes, but the E and coastal lowlands are beyond its E limit. This bird occurs as low as 400 m at Turkana, but is more typical of 1500-3000 m areas of the highlands (Br; Se). There is a single record from L Victoria, at Kisumu 60B (Ja; Br).

Breeding is 84% above 1000 m and 85% within 500+ mm rainfall areas. This distribution presumably reflects observer coverage, since this martin is thought to be resident (Br; Be: 74; Se; La: 101B; St: 50A; DEP pers. obs.: 75BC 88D 89C 100B 101A). There is an annual May-Sept influx to the Ruaraka-Kahawa-Ruiru area of 75B (DEP pers. obs.). Other references: Hull 1944; Ussher 1944.

561 GREY-RUMPED SWALLOW *Hirundo griseopyga*
B633 / MPG1066 / D2C2

Afrotropical: widely distributed S of 10°N, E to W Ethiopia, Kenya, W Mozambique, and the coast S of 25°S, and avoiding only the dry SW. Distribution in Kenya represents the E limit of the range.

Kenyan: a local and often uncommon swallow of sheltered grassy flats and hillsides at 900-2200 m (Br; Ja). Range is 75% subhumid-humid and 86% within 500+ mm rainfall areas. Although present throughout the year at Nakuru 62A (EABR 1977), there are marked seasonal fluctuations at most localities that are only partly attributable to this swallow's affinity for burnt grassland (Br), *e.g.* Apr-Sept breeding in the Ruaraka-Kahawa-Ruiru sector of 75B, but complete disappearance from the area during Jan-Feb (DEP pers. obs.).

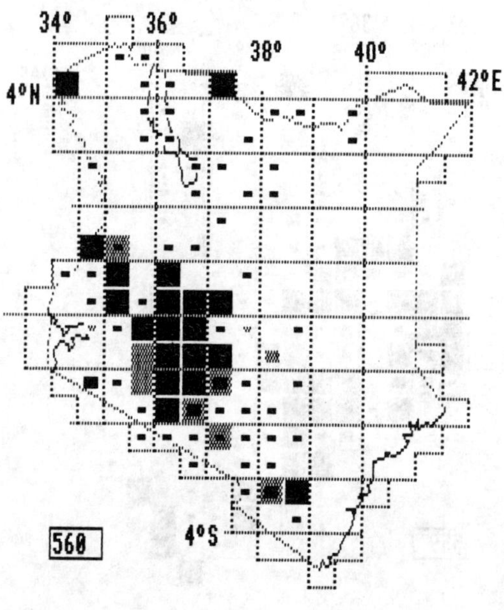

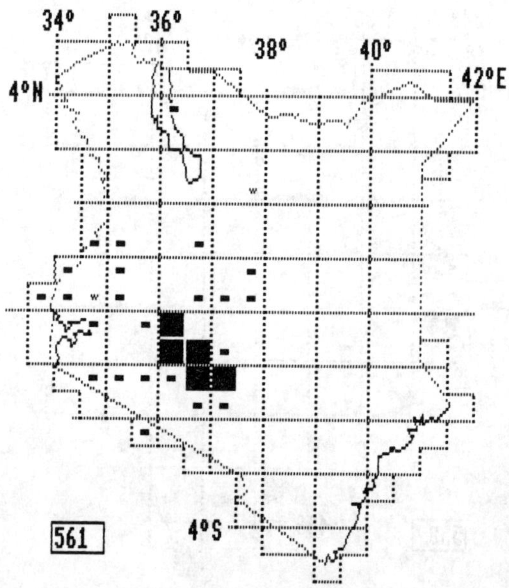

Two points arise from this pattern of behaviour. Firstly, the apparent restriction of breeding to a compact area need not necessarily reflect observer frequency. Of the five breeding squares, four are above 1500 m, and all receive 500+ mm annual rainfall. Br only mentions regular breeding in the rift valley between Nakuru 62A and Naivasha-Kedong 62C. Secondly, HH describe the species as uncommon at Allia Bay 14A, L Turkana, but this locality is omitted (and, by context, rejected) by Br. This report is admittedly from an atypical habitat, but we feel that it cannot be certainly rejected in view of the above-mentioned mobility, the species' presence in Ethiopia, and Tomlinson's (1946) record from Marsabit 27D.

Williams (1966) described Andrew's Swallow *H. andrewi* from a specimen taken from amongst large numbers of migrating hirundines at Naivasha 62C on 4 Apr 1965. In the absence of further evidence, this bird is currently regarded as a race of the Grey-rumped Swallow, possibly originating in Ethiopia (HM; Dowsett 1972a; Br).

562 MASCARENE MARTIN *Phedina borbonica*
B638 / MPG1067 / D3C0
NO MAP: post presence 100D 103A

Afrotropical: breeds on Madagascar, Reunion and Mauritius, and has occurred as a rare non-breeding vagrant to Africa, in Malawi, Pemba Island and Kenya.

Kenyan: the two records, on strikingly coincident dates, are of a flock of 16 at L Jipe 100D on 24 June 1978 (Zimmerman 1978b), and one with Striped Swallows (559) at Watamu 103A on 23 June 1980 (EABR 1980). There are Apr and late June dates from Malawi, but old records describe Sept-Mar seasonality on Pemba (at 5°S off Tanzania) (Br; Zimmerman 1978b).

563 HOUSE MARTIN *Delichon urbica*
B623 / MPG1074 / D2C2

Afrotropical: a Palearctic species, wintering in highland areas S to the Cape. It is one of the very few Palearctic birds to have established breeding populations within their wintering range in S Africa (MPG; Sn)(see White Stork (45)).

Kenyan: a passage migrant and winter visitor, mainly late Sept-Apr (Br). Range is 74% above 1000 m, 76% within 500+ mm rainfall areas and only 10% arid: this martin is virtually absent from the arid N and E (27D 28C = the locally moist Mt Marsabit), and also from the moist coastal strip. Birds in the arid Turkana area are only on passage (HH), but small numbers occur throughout the winter in arid Tsavo E 101B (La).

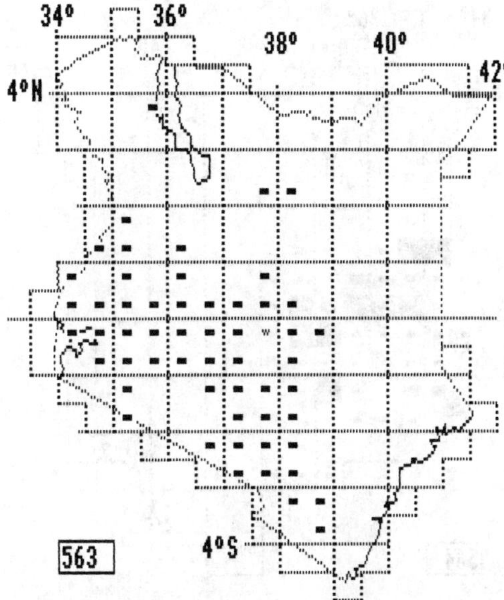

Wintering birds are common and often abundant in the W and C highlands, typically around cliffs and valleys at 1600-3000 m, but also out over adjacent open country at lower altitudes (Br). Passage birds are more widespread (Br). In areas around 36°E, S passage is the only one seen at L Turkana (HH) and Baringo 50A (St), and is the more marked at Mau Narok 61D (Se) and Nakuru-Elmenteita 62A (D J Pearson *in litt.*).

The only instances of oversummering are from Samburu GR 51B (four on 29 June: EABR 1978) and Nakuru 62A (one presumed oversummerer on 17 Jul: EABR 1982). A bird ringed in Nairobi 75B was recovered in the USSR, to the NE of the Black Sea (Backhurst 1971b). This martin commonly flies high above the ground, and concentrations lower down are usually associated with low overcast or heavy storm clouds (Ja; Harper & Harper 1974a; Sessions 1974a).

564 BLACK ROUGH-WING *Psalidoprocne pristoptera*
B640 / MPG1075, 1077 / D3C2

Afrotropical: widespread S of Cameroon and Ethiopia, except for the dry areas of the SW and Somalia. Distribution in Kenya represents the E limit of the range.

Kenyan: frequents open, wooded and forested habitats, in two distinct, non-arid areas. It is mainly a bird of the highlands, where the race *P.p. massaica*

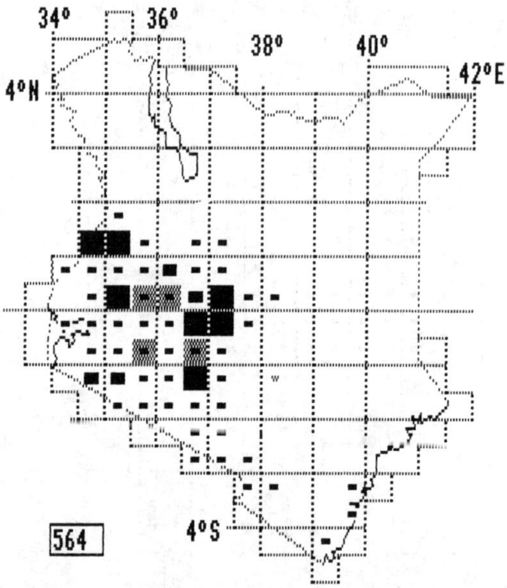

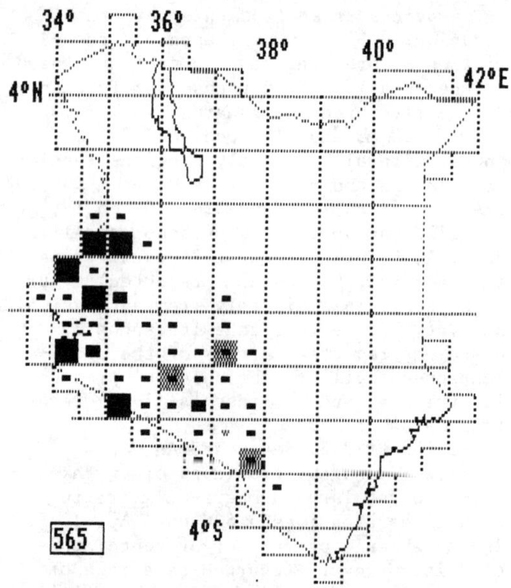

is a common resident at 1600-3200 m (Br; Be; Se). This race's range is 96% above 1000 m and 87% in 500+ mm rainfall areas. Restriction of breeding to areas that are 93% above 1500 m, 60% within the 1000+ mm rainfall regime and 93% subhumid-humid possibly reflects observer frequency.

Two other races have been recorded in 500+ mm rainfall areas of the coastal strip. *P.p. orientalis* apparently breeds on forested hills at Ribe 102D (breeding not mapped)(Br). Dec specimens from Sokoke Forest 102B were *P.p. holomelaena* (Ripley & Bond 1971). After two further Nov-Dec sight records from this forest, Britton & Zimmerman (1979) stated that this bird could be no more than a wanderer or migrant to the area, and that the dates do not suggest a S origin. There have been subsequent sight records of the species from the Shimba Hills 114A on 29 Apr (EABR 1979), and from Marafa 102B on 21 June (EABR 1983), and overall status at the coast is still unknown.

565 WHITE-HEADED ROUGH-WING *Psalidoprocne albiceps*
B639 / MPG1080 / D3C2

Afrotropical: primarily the E, from Zambia and Malawi N through W Tanzania and the E Zaire borders to Uganda, S Sudan and, on the E limit of the equatorial range, Kenya. There are also records from Angola and S Zaire.

Kenyan: in bushed, wooded and forest edge habitats (Br). This rough-wing is common in the L Victoria basin up to 1500 m, with much smaller numbers up to 2000 m in the W and C highlands (Br). Range is 96% above 1000 m and 87% within 500+ mm rainfall areas. DEP (pers. obs.) found this species resident in 74D 75D. Movements occur, but they are imperfectly understood (Br).

Family Dicruridae: Drongos

20 species, of which seven (35%) occur in the Afrotropical Region and the rest are Oriental (HM). Only three of the Afrotropical species are found on the mainland, the remainder being on Madagascar and adjacent islands. Two members of this family breed in Kenya.

566 DRONGO *Dicrurus adsimilis*
B644 / MPG1087, 1088 / D3C3

Afrotropical: widespread over the whole continent S of 15°N. The Velvet-mantled Drongo *D.a. coracinus* inhabits the equatorial forests from Nigeria and Angola, E through the Congo basin to the E Zaire borders, Uganda and the far W of Kenya.

Kenyan: races of this species show two contrasting modes of occurrence. The Velvet-mantled Drongo is an uncommon and unobtrusive bird of forest glades and clearings, up to 1700 m (Br; MPG). It is known only from Elgon 36CD (Ja) and the Kakamega Forest 48D. This range is above 1000 m, in humid, 1000+ mm rainfall areas. Formerly considered a species, it apparently intergrades with another race in W Africa (HM).

In distinct contrast, the Drongo is a common, conspicuous and noisy bird of forest edge and open wooded habitats, up to 2200 m (Br; Ja; MPG; HH; La; St). Range is throughout the rainfall and moisture regimes, and most probably includes every square. Breeding is 70% within 500+ mm rainfall areas and 87% in squares that are well wooded. Since we can find no mention of movements in this species, this distribution presumably reflects observer frequency. The Drongo often leads mixed species foraging parties (Start 1971). Other references: Forbes-Watson 1977c; Short & Horne 1985a.

567 SQUARE-TAILED DRONGO *Dicrurus ludwigii*
B645 / MPG1089 / D1C1

Afrotropical: locally distributed S of 5°S, avoiding the dry SW. Further N, it extends through E Tanzania to SE Kenya and S Somalia. In the W, this drongo occurs around the Gulf of Guinea, and E through the Congo basin to Uganda and W Kenya.

Kenyan: a local and rather unobtrusive resident of forest habitats (Br; MPG). The race *D.l. sharpei* occurs in the Kakamega and S Nandi Forests 48D 49C, where it is rather uncommon (Br). Presence in other W and C areas (Ja) remains unconfirmed (Br). Range is entirely above 1000 m, in humid,

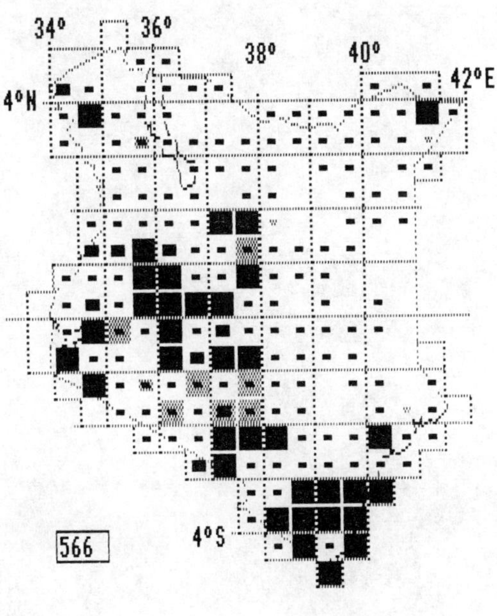

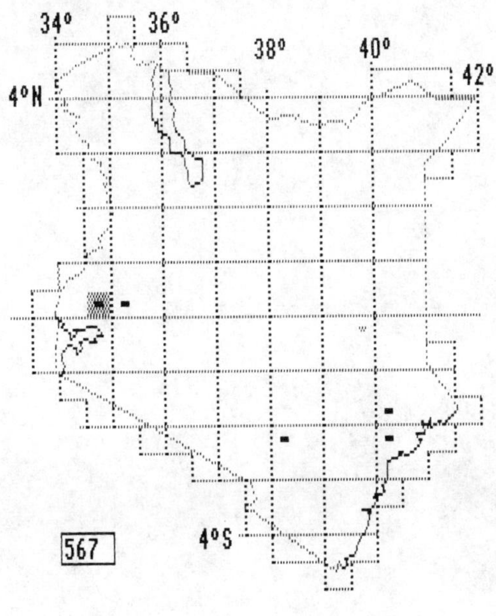

1000+ mm rainfall areas.

D.l. muenzneri is locally common (*e.g.* AGH) in riverine forest along the Tana N to Sankuri 65B (Ja; AGH; P L Britton *in litt.*), and along the Athi in 89A (DEP pers. obs.). These riverine habitats are in arid-semiarid country, with three of four squares below 500 m. Penetration inland along these rivers compares with range in adjacent S Somalia, which is largely associated with the Juba R, and which extends S to 80B (not mapped here) (AM).

Family Oriolidae: Orioles

Approximately 25 species, with representatives in all major zoogeographical regions except the Nearctic and Neotropical. Nine (36%) are known in the Afrotropical Region, and six of these occur in Kenya. Of the Kenyan species, four breed or are thought to do so, while the two others are non-breeding intra-African and Palearctic migrants respectively. Orioles are almost always associated with trees.

568 BLACK-HEADED ORIOLE *Oriolus larvatus*
B649 / MPG1167 / D3C3

Afrotropical: widespread S of 5°S, avoiding only the dry SW. Further N, it extends through East Africa to S Sudan, Ethiopia and all but the drier areas of Somalia.

Kenyan: the commonest of the orioles, typically a bird of woodland although also known in mangroves and forest (Br; Ja; MPG). It forms part of a superspecies with the Western Black-headed Oriole (569) and is closely related to the Montane Oriole (570) (HM), and where either of these forest birds is present, the Black-headed is restricted to forest edge (Br). Hybridisation with the Montane Oriole (*q.v.*) is known.

Br quotes the Black-headed up to 2300+ m, and Se mentions irregular visitors to 3000 m at Mau Narok 61D. Range is only 6% within the driest, 0-250 mm rainfall areas. Scarce occurrence in the arid N and E reflects paucity of trees, and is mainly restricted to the S edge of the Ethiopian highlands 4A 16B 17ACD, the Daua R 7C 19AB, Mt Marsabit 27D 28C and the Tana R 64B 65ABD 78B 79C. Virtual absence from the E borders N of 1°S compares with status in adjacent S Somalia (AM). This oriole is generally considered to be resident (Br; La: 101B; St: 50A; DEP pers. obs.: 74D 75D 88AD 89AC 100B 101A), but some movements occur. Thus, while finding the species in all months in moist subhumid 75B, DEP (pers. obs.) also noted an annual Jul-Nov influx to the Ruaraka-Kahawa-Ruiru sector of the square: this seasonality compares with that of the June-Dec visitors to 3000 m around Mau Narok 61D (Se). Ja mentions movements to fruiting trees.

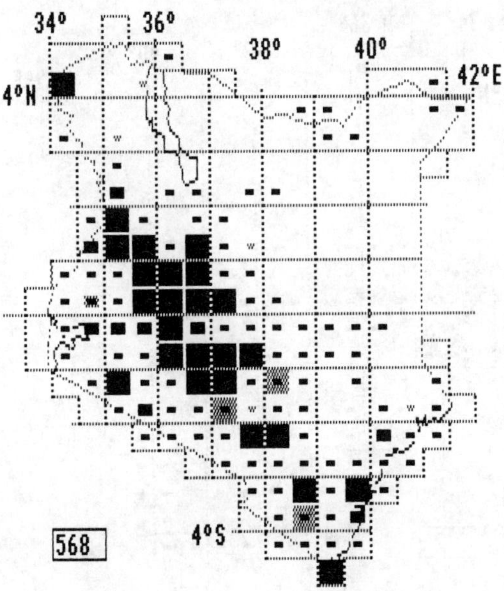

569 WESTERN BLACK-HEADED ORIOLE *Oriolus brachyrhynchus*
B647 / MPG1168 / D2C2
NO MAP: post presence 48D; pre probable 48D

Afrotropical: W and C Africa, from the forests around the Gulf of Guinea between Guinea and Angola, E through the Congo basin to Uganda and, on the E extremity of the range, W Kenya.

Kenyan: local in forest edge and clearings, at 1600-1800 m (Br). The sole locality is Kakamega Forest 48D, where there have been numerous observations in recent years (Mann 1976). Pre-1970 probable breeding derives from Zimmerman's (1972) mention of territory and nest-building there. This square is in a humid, 1000+ mm rainfall area. Br also mentions occurrence in Uganda at Moroto 24B and Elgon 36CD (not mapped here). See Black-headed Oriole (568).

570 MONTANE ORIOLE *Oriolus percivali*
B652 / MPG1169 / D1C1

Afrotropical: a highland species of the E, along the E Zaire-Tanzania borders E to Uganda and, on the E edge of the range, Kenya.

Kenyan: reasonably common in forest at 1530-2600 m in the W and C highlands (Br). Range is 88% within the 1000+ mm rainfall

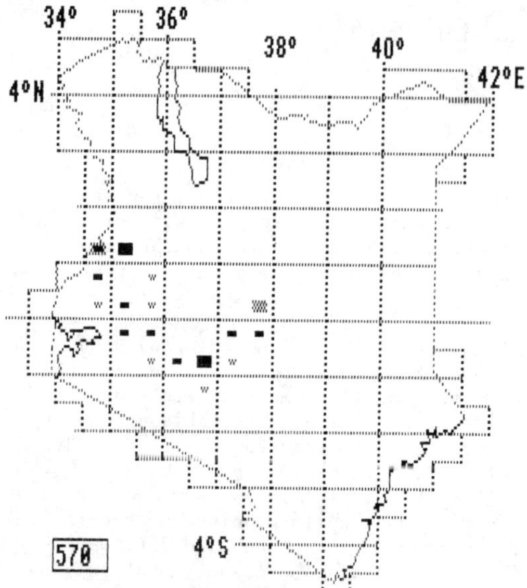

regime, and entirely within moist subhumid-humid areas. There is no confirmed breeding record, but Br considers this oriole resident.

The relatively high proportion of pre-1970 squares (41%) probably represents observer coverage of a little known species that has been considered conspecific with the Black-headed Oriole (568)(*e.g.* MPG). These two birds are generally segregated ecologically, and significant hybridisation is known only from the Nairobi-Kikuyu area 75B, where they coexist in relict forest strips (Prigogine 1978b).

571 GREEN-HEADED ORIOLE *Oriolus chlorocephalus*
B648 / MPG1170 / D2C3
NO MAP: post presence 102B; post probable 114AB

Afrotropical: an E species, in Malawi, Mozambique, E Tanzania and coastal Kenya.

Kenyan: Kenyan occurrences of this forest and rich woodland species represent the N extremities of its Afrotropical range, approximately 100 km NE of the Usambara Mts, the nearest Tanzanian locality. First recorded at around 400 m on the Shimba Hills 114A on 1 Jan 1964 (Keith 1968), it has subsequently been found to be fairly common in forest patches there (Br). The 114A probable breeding record derives from Britton's (1977a) tentative opinion of resident breeding status. Squire (1976a) recorded it near sea level in the Jadini Forest 114B, and probable breeding there is based on his mention of occasional juveniles. This oriole cannot be common in this forest, however, since there is but a single subsequent record despite reasonable observer coverage (M A C Coverdale *in litt.*). A single late Jan bird near sea level in Sokoke Forest 102B (Britton & Zimmerman 1979) was presumably a wanderer N of the normal range. The three squares are all below 500 m, in semiarid-dry subhumid, 500+ mm rainfall areas.

572 AFRICAN GOLDEN ORIOLE *Oriolus auratus*
B646 / MPG1165 / D2C2

Afrotropical: this intra-African migrant is widespread between 15°N and 22°S, except for the equatorial forests, N Mozambique, much of Ethiopia and all but S Somalia. Disjunct populations in the N and S tropics migrate towards the equator during the non-breeding season (Br).

Kenyan: a rather shy, non-breeding migrant which requires tall, densely canopied trees, and is found in sparsely wooded, wooded and forest edge habitats up to 1800 m (Br; Ja; MPG). Range is 79% within 500+ mm rainfall areas. Most of the 10% arid squares are along the Tana R 64B 65B 78B 79C. Association with riverine habitat in dry inland areas compares with occurrence near the Juba R in adjacent S Somalia (AM). La could list only four records from arid Tsavo E 101B. Small numbers on the local N range edge at semiarid Baringo 50A are mostly during or after rains (St).

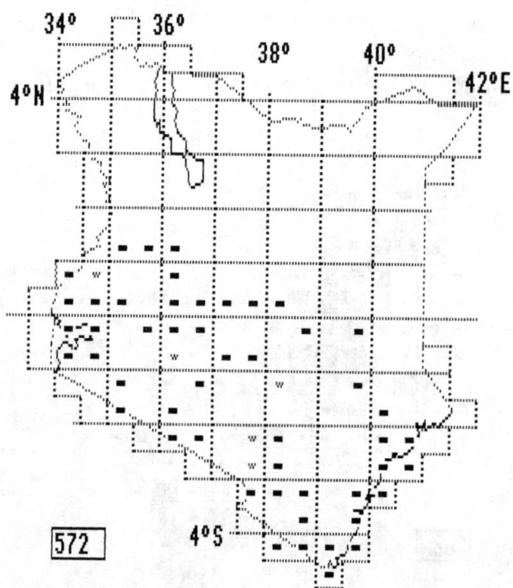

The nominate N tropics race occurs N of the equator in Uganda (Br), with a single record E to Mangiki 48B (Ja). The S tropics *O.a. notatus* is numerous at the coast (extreme dates Feb-Nov), but less common inland (Br; EABR). See Golden Oriole (573).

573 GOLDEN ORIOLE *Oriolus oriolus*
B651 / MPG1164 / D2C2

Afrotropical: this Palearctic oriole winters from approximately 8°N in the W and the equator in the E, S to the Cape.

Kenyan: this oriole is predominantly an Oct-Dec and Mar-early May passage migrant (Br; HH; St). Migration is heaviest to the E of the rift valley, and particularly so at the coast in spring, where concentrations are attracted to fruiting trees (Br; Ja). This abundance at the coast during N passage strikingly contrasts with occurrence in adjacent S Somalia, where the only three records are for Oct (AM). Small numbers at a few localities in SE Kenya represent the N edge of the Afrotropical wintering range (Br; *e.g.* La). A 1 June oversummering record from Nairobi 75B was doubted by Ja and is omitted by Br.

The Golden Oriole forms part of a superspecies with the African Golden Oriole (572)(HM): these two non-breeding migrants frequent similar habitats and can be widely sympatric, although the Palaearctic bird is generally the commoner and more widely ranging inland.

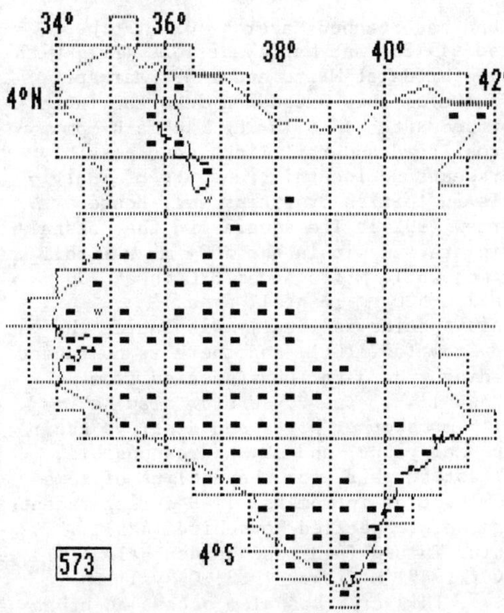

Family Corvidae: Crows, Ravens, Piapiac

Approximately 100 species, with representatives in all zoogeographical regions. Only nine are native to the Afrotropical Region, while a tenth has been introduced from Asia. Seven corvids occur in Kenya, where six (including the asiatic species) breed, and one is a wanderer E of its normal range.

574 PIAPIAC *Ptilostomus afer*
B659 / MPG1179 / D3C0
NO MAP: post presence 48AC 60B

Afrotropical: the N tropics, from Senegambia to S Sudan, Uganda and the W Kenya borders.

Kenyan: the Piapiac frequents sparsely wooded grasslands, particularly near *Borassus* palms, and is often associated with large mammals, both domestic and wild (Br). It extends E through Uganda to the Kenya border at Busia 48C (Br), and is frequently seen just within Kenya at Alupe 48AC (photographed: R E M & T Scholtens *in litt.*). J S Ash (*in litt.*) also has records from the Ugandan sector of 48A. There are two instances of wanderers SE to 60B: a single bird at Maseno on 16 Apr 1971 (Simpson 1971a), and a flock of 26 near Kisumu on 11 Aug 1983 (EABR 1983). All three squares are at 1000-1500 m, in moist subhumid-humid, 1000+ mm rainfall areas.

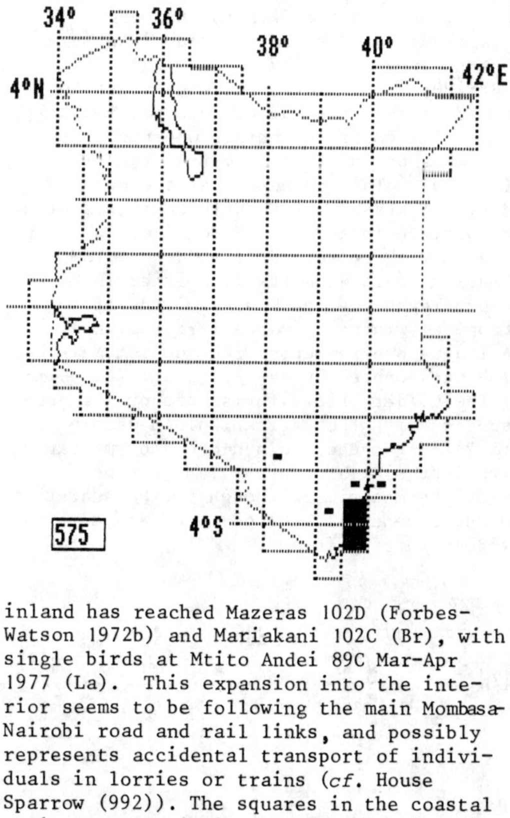

575 INDIAN HOUSE CROW *Corvus splendens*
B658 / MPG1174 / D3C3

Afrotropical: an asiatic crow, introduced to E coasts. It was brought to Zanzibar by the colonial authorities in the 1890s to act as an urban scavenger, and continues to thrive there (MPG; Forbes-Watson 1972b; Br). There are subsequent records from Port Sudan, Massawa (Ethiopia), Djibouti, Cape Guardafui (Somalia), coastal Kenya, Dar es Salaam (Tanzania), Inhaca (Mozambique) and Durban, East London and the Cape (South Africa)(J S Ash *in litt.*, Ash 1985ab).

Kenyan: first recorded in Mombasa 114B in 1947, this omnivorous scavenger is now uniquitous in the city and island, and has spread along the coast to Diani 114B and to areas N of Kilifi 102B (Br; McVicker 1982; EABR; *cf.* Ash 1985a). Occurrences at Malindi and Watamu 103A were considered to represent disjunct populations derived from local releases of birds (*op. cit.*), but a flock flying N 15 km to the N of Kilifi 102B (EABR 1981) suggests these N birds to be no longer isolated. Extension inland has reached Mazeras 102D (Forbes-Watson 1972b) and Mariakani 102C (Br), with single birds at Mtito Andei 89C Mar-Apr 1977 (La). This expansion into the interior seems to be following the main Mombasa-Nairobi road and rail links, and possibly represents accidental transport of individuals in lorries or trains (*cf.* House Sparrow (992)). The squares in the coastal strip are all within the 500+ mm rainfall regime, while Mtito Andei 89C is at 700 m in a 250-500 mm rainfall area.

This species is locally displacing the Pied Crow (577)(Br), and there is no recent breeding record for that species from Mombasa 114B. The House Crow also habitually mobs other birds and destroys their nests and young, and may be responsible, at least in part, for the decline of some other species in Mombasa 114B during recent years, *e.g.* Speckled Mousebird (425), Morning Thrush (660) and Golden Palm Weaver (945)(McVicker 1982; C Ryall *in litt.*). This crow has also become an urban pest in Mombasa, and control experiments were instituted in 1985 (Anonymous 1985a;

C Ryall *in litt.*). Other reference: Angwin 1977.

576 BROWN-NECKED RAVEN *Corvus ruficollis*
B657 / MPG1171 / D2C2

Afrotropical: the S edge of the Sahara from Mauritania E to Ethiopia and Somalia, and thence S to S Sudan, N Kenya and S Somalia.

Kenyan: a locally common scavenger that affects settlements (*e.g.* Tomlinson 1946) and arid open habitats in the N (Br; MPG). Range is 87% below 1000 m, 83% arid, and 88% within 0-500 mm rainfall areas. Absence from E border areas S of 1°N corresponds to status in adjacent S Somalia (AM). Occurrence in 2C and 25C suggests that this raven reaches the far NE of Uganda, although there are as yet no records from that country. See Pied Crow (577).

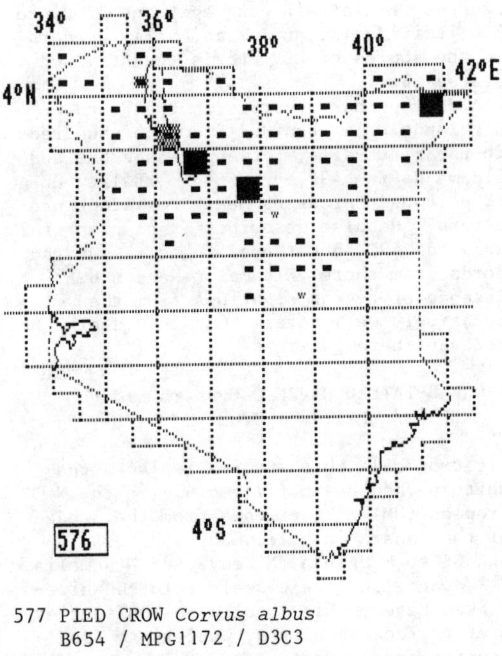

577 PIED CROW *Corvus albus*
B654 / MPG1172 / D3C3

Afrotropical: widespread S of 20°N, avoiding only the drier areas of Somalia and the SW interior. Where sympatric with the Brown-necked Raven (576), the Pied Crow is restricted to the vicinity of towns (HM).

Kenyan: an omnivorous scavenger of open country up to 3000 m, especially common around settlements below 2000 m (Br; Se; MPG). Range is 74% within 500+ mm rainfall areas and only 21% arid.

In the arid and semiarid N, this crow's range coincides with that of the Brown-necked Raven, which is a member of the

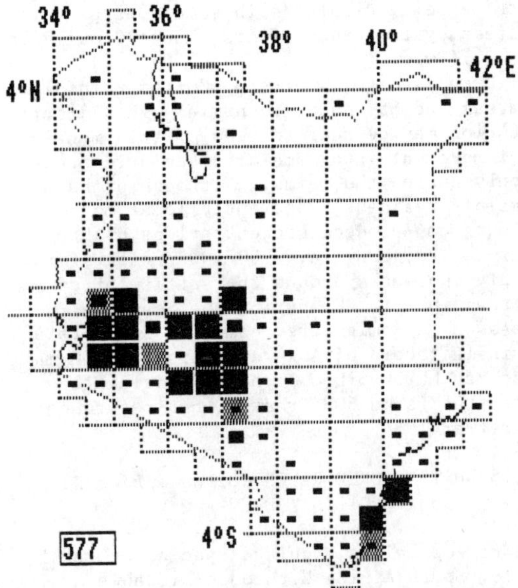

same superspecies. Where the two are sympatric, the Pied Crow is usually confined to settlements (HM; *e.g.* North 1942a). Occurrence in the arid N is restricted to squares that contain either L Turkana or the settlements of Nanam 2C (see below), North Horr 15C, Moyale 17A, Rhamu 19A, Mandera 19B, Marsabit 27D, Laisamis 39B and Wajir 42A, and uncommon status persists S to at least semiarid Baringo 50A (St). Occurrence around L Turkana is virtually restricted to occasional birds at Lodwar 13D (HH) and Loyengalani 26B: the record in 14A, for example, represents a single Oct individual at Koobi Fora (F Alexander *in litt.*). A single bird in Moyale town 17A contrasted with hundreds of Brown-necked Ravens (F J McCartney, ADL pers. obs.). Hybrids with this latter species are known from approximately 7°N in Ethiopia (HM) and from 1°N on the Somali coast (Ash 1983; AM), but have yet to be reported from Kenya.

Range in the E lowlands to the S of the equator, *i.e.* to the S of the Brown-necked Raven, is largely restricted to the coastal strip. Distribution in adjacent S Somalia is similar (AM). Occurrences in arid squares in these E areas of Kenya include Garissa town 65B, and Tsavo E 101B. This species is rare in the uninhabited areas of the latter, but resident at Voi and along the main Mombasa-Nairobi road (La; Lack 1985). Originally rare in the semiarid Makueni area 76D, this crow is now common around the markets that have arisen there during recent decades (Parsons 1959).

It is being displaced in some coastal areas by the exotic Indian House Crow (575: *q.v.*)(Br).

Breeding is 95% within 500+ mm rainfall areas and 85% subhumid-humid. This distribution may be real in view of Br's mention of movement into some arid-semiarid areas only during the rains. Although probably mainly resident, apparently local movements can produce concentrations of hundreds (Br; MPG), *e.g.* 50-100 present for only two weeks around the settlement of Nanam in arid 2C, during the Sept dry season (T Romig pers. comm.). Se lists a single record of Aug wanderers to 3000 m at Mau Narok 61D. Ja notes a Jul influx to the Kisumu 60B area. Other reference: Loefler 1984.

578 WHITE-NECKED RAVEN *Corvus albicollis*
B653 / MPG1175 / D2C2

Afrotropical: a highland species of the E, from South Africa N through Zimbabwe, Zambia and Mozambique to East Africa and the E Zaire borders. Distribution in Kenya describes the N and E limits of the range.

Kenyan: a locally common bird of cliffs and rocky slopes, which are used for roosting and nesting, that also makes long foraging flights over adjacent country and frequently scavenges around settlements (Ja; MPG; Br). Range is 70% semiarid-subhumid and 66% within 500+ mm rainfall areas. Typically at 1000-3000 m, it descends to 400 m on the E range edge, *e.g.* 102C, and has been recorded at 4100 m on Mt Elgon 36D (Br; Se; Britton & Sugg 1973). This raven is absent from the coastal strip, and virtually so from the L Victoria basin. A common resident at Voi 101B, it extends less than 10 km into the adjacent Tsavo E NP (La).

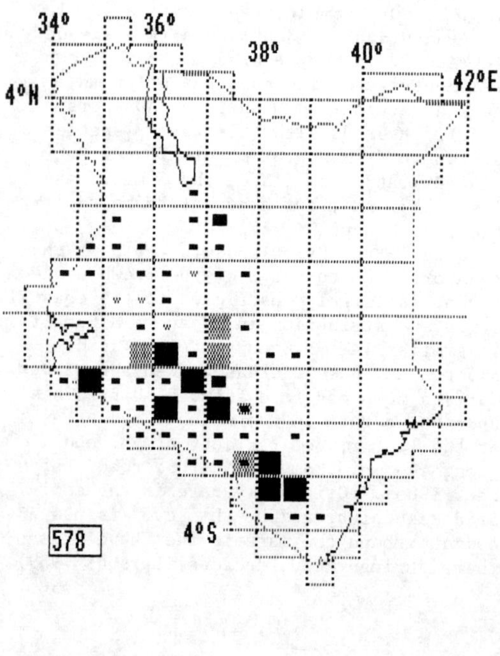

Virtual restriction of breeding records to areas S of the equator probably reflects observer coverage, although areas further to the N are within the breeding range of the Fan-tailed Raven (579: *q.v.*), which is a member of the same superspecies (HM). These two ravens are mainly allopatric, but have been seen together at 2600 m on the summit of Mt Nyiru 26D (Br). At semiarid Baringo 50A, where the Fan-tailed is a common breeding resident, the White-necked is very uncommon and often absent for months at a time (St).

Although typically resident (Br; La: 101B; DEP pers. obs.: 74D 88D 89C 101A), the White-necked Raven makes movements to food sources, can attain congregations of up to 800 individuals, and occasionally wanders to the shores of L Victoria (country not stated)(Br).

A decline has occurred in the W and C highlands. Ja recorded flocks of hundreds in the rift valley between Kijabe 62D and Eldama Ravine 49D, prior to mortality due to poisoned carcasses meant for stock predators. He also describes it as plentiful at Nandi 49C, but there are no recent records from there, Eldama or Nakuru 62A. Absence of recent sightings from the relatively well covered 63AC could also be relevant here.

579 FAN-TAILED RAVEN *Corvus rhipidurus*
B656 / MPG1177 / D2C2

Afrotropical: this species replaces the White-necked Raven (578: *q.v.*) in the N tropics (HM). It extends from the S edge of the Sahara in Mali, Chad and Sudan S and SE to E Uganda, N Kenya and N Somalia.

Kenyan: like the closely related White-necked Raven, this locally common species prefers rocky habitats, wanders to food sources and scavenges at settlements (Br; MPG; HH). It frequents lower and drier country to the N of the White-necked's range, being 63% at 500-1500 m and 79% arid-semiarid. Range extends S to the N edge of the highlands, and further S down their E margin (*e.g.* EABR 1981). Absence from the E lowlands is real, as there are no reports from observers resident in 42A and 65B, and occurrence is N of 5°N in adjacent Somalia (AM).

(map overleaf)

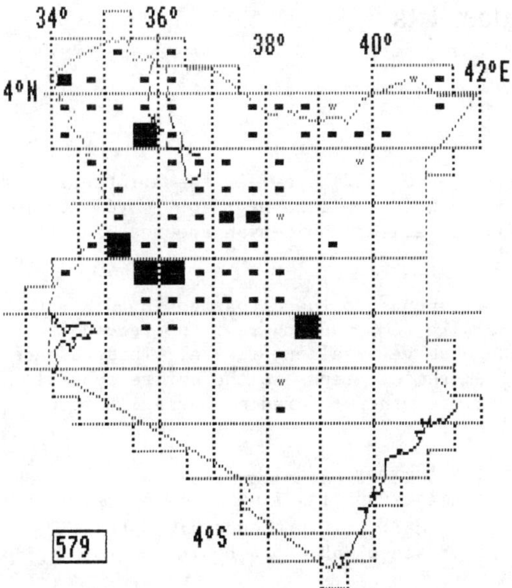

(Br; *e.g.* HH). Range extends throughout the rainfall and moisture regimes. Absence of recent records from the E compares with restriction to the coast in S Somalia (AM).

Increases during this century are apparently related to the advent of large scale farming (Ja; Brown & Britton 1980). This rook is now particularly common in the rift valley highlands, but feeding flocks of thousands are possibly exceptional (MPG; HM; Br). Br mentions S range extension into NE Tanzania, but the crescent of old records from 75D 88B NE to 40A 41D suggests contraction of the SE range edge. Very uncommon wanderers to this range edge in 75B and 76A have been Sept-Dec (EABR). Possible competition with the Pied Crow (577)(HM; *e.g.* AM) is not evident.

580 CAPE ROOK *Corvus capensis*
B655 / MPG1173 / D2C2

Afrotropical: occupies two distinct areas. In the NE, it extends from the N borders of Tanzania, through Kenya, to SE Uganda, S Sudan, Ethiopia and N Somalia. It is also widespread S of 10°S.

Kenyan: a locally common to rather uncommon bird of open habitats, typically at 1200-2500 m but in lower country in the N

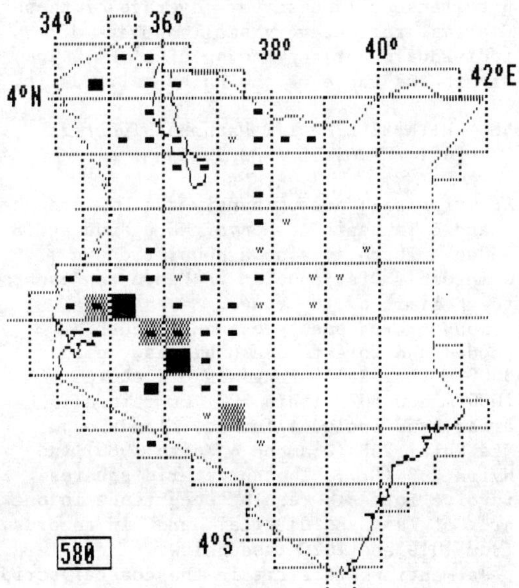

Family Paridae: Tits

46 species, present in all zoogeographical regions. 10 (22%) occur in the Afrotropical Region, and five in Kenya. None of this family are known to make migratory movements in East Africa (Br), and there are breeding records for all of the Kenyan species except the rare Black Tit (582).

581 GREY TIT *Parus afer*
B660 / MPG1151 / D2C2

Afrotropical: occupies two distinct areas. In the NE, it extends from the NE borders of Tanzania, through Kenya, to E Uganda, Ethiopia and Somalia. It is also widespread S of 15°S.

Kenyan: a rather uncommon tit of open bushed and wooded habitats (Br; Lack 1985; St), generally preferring lower and drier areas to the N and E of those of the White-bellied Tit (583). Br quotes the Grey up to 1600 m, but range is in fact 70% below 1000 m, with higher altitudes only being attained in N areas of the W and C highlands 49-51. Distribution is 86% arid-semiarid, but the L Turkana basin lies mostly beyond its W edge (HH). There is a single, old record to the W of the lake 3D (MacInnes 1933-34), and the species is not found in S Sudan (HM). There are reports from only 29% of the most arid squares (moisture index below 1.2), so that absence from the NE around 29 may be real. Absence from E border areas S of 2°30'N corresponds to the situation in adjacent S Somalia (AM), and there is no record from an observer resident in 42A. This absence from the E extends to the entire coastal strip. Other reference: Lewis & Pomeroy 1982a.

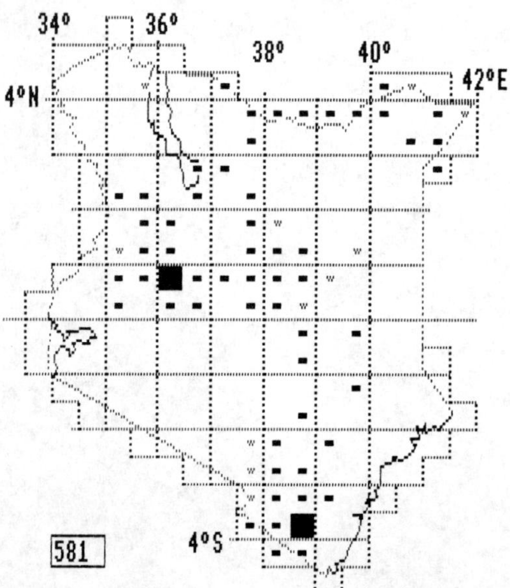

582 BLACK TIT *Parus leucomelas*
B666 / MPG1154 / D2C0
NO MAP: post presence 37C; pre presence 36D 48B 60B 61A

Afrotropical: widespread between 15°N and 15°S, E to Ethiopia, W Kenya, Tanzania and Malawi, but absent from the equatorial lowland forests.

Kenyan: frequents bushed and wooded habitats, and forms part of a superspecies with the allopatric White-bellied Tit (583) (HM; Br). There are records from Mt Elgon 36D, Kapenguria and Saiwa Swamp 37C, Bungoma 48B, Kibingei in 60B and the Nyando Valley 61A (Ja; Br; T J Barnley *in litt.*). Range is above 1000 m, to the W of the rift valley. All five squares are moist subhumid-humid and within the 1000+ mm rainfall regime. The few sightings, despite reasonable observer coverage of these W areas, reflect very small numbers of individuals (Br) on the species' E limits, within the range of the White-bellied Tit.

583 WHITE-BELLIED TIT *Parus albiventris*
B661 / MPG1155 / D3C3

Afrotropical: predominantly the E, through S and E Tanzania to Kenya, NE Uganda and S Sudan. There is also a population in Cameroon. Distribution in Kenya represents the E limit of the range.

Kenyan: reasonably common in bushed, wooded and forest edge habitats, up to 3400 m (Br; Se). Range is 87% above 1000 m and 80% within 500+ mm rainfall areas, and includes forest islands *e.g.* Mts Kulal 26B (Diamond & Keith 1980) and Nyiru 26D (Br). The three arid squares involve formerly regular occurrence in one area of Tsavo E 101B (La), and old records from 101D and 102C (see below).

Ja mentions this tit in the coastal strip

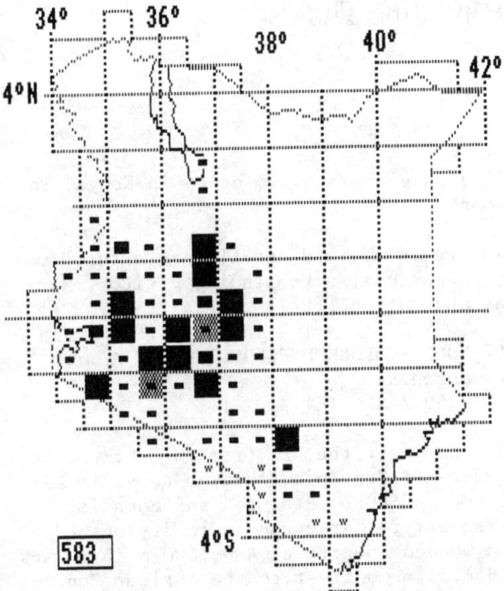

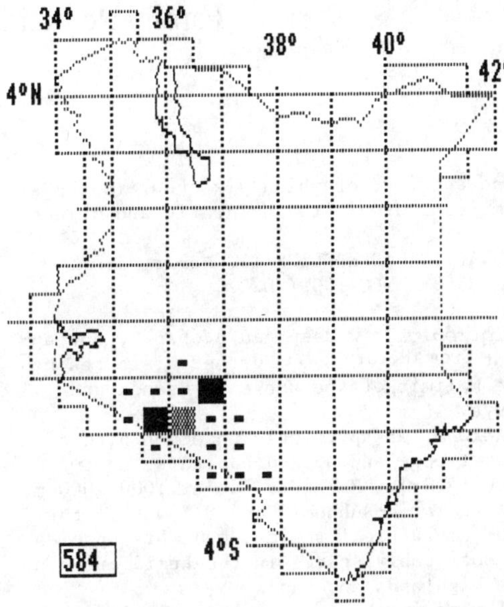

from Lamu 91B to Vanga 114C, but only quotes Sagala 101D and Samburu 102C as actual localities. Occurrence at the coast is reiterated by MPG, but doubted by Br. We share Br's opinion, particularly in view of Ja's lack of details. Aside from actual error, the only explanation for these statements would seem to be a retraction from the lowest altitudes of the range, which was well underway in Ja's day and which is still evident today (*e.g.* La)(*cf.* African Penduline Tit (586)). See Grey Tit (581) and Black Tit (582). Other reference: Lewis & Pomeroy 1982a.

584 RED-THROATED TIT *Parus fringillinus*
B663 / MPG1159 / D2C2

Afrotropical: endemic to interior areas of SW Kenya and NE Tanzania.

Kenyan: rather local and usually uncommon in bushed and wooded habitats, typically at 1000-1600 m (Turner 1977; Br). Range is entirely within semiarid-subhumid, 250-1000 mm rainfall areas. This species is sympatric with the White-bellied Tit (583) (see HM).

585 DUSKY TIT *Parus funereus*
B664 / MPG1157 / D2C2
NO MAP: post presence 49C; post confirmed 48D

Afrotropical: a bird of W and C Africa, with a very fragmented distribution around the Gulf of Guinea from Guinea to Angola, and E through parts of the Congo basin to S Sudan, Uganda and, on the E extremity of the range, W Kenya.

Kenyan: occurs as an isolated population in the Kakamega and Nandi Forests 48D 49C, and is reasonably common (HM; Diamond & Fayad 1979; ADL pers. obs.). These two squares are above 1000 m and to the W of the rift valley, in humid, 1000+ mm rainfall areas. This tit inhabits forest canopy, and is thus ecologically segregated from other local congeners (HM).

Family Remizidae: Penduline Tits

Nine species, of which six occur in the Afrotropical Region, and two breed in Kenya. The remainder are in the Holarctic and Oriental Regions.

586 AFRICAN PENDULINE TIT *Remiz caroli*
B668 / MPG1160 / D2C2

Afrotropical: widespread S of 5°S, but absent from the dry SW. Further N, it reaches the N limit of the range in Uganda and Kenya.

Kenyan: an apparently uncommon bird of forest edge and wooded habitats, up to 2200 m (Br). Range is 72% at 1000-2000 m, 84% semiarid-subhumid and 81% within the 500+ mm rainfall regime, and this species is noticeably scarce in the humid areas of the highlands and the W.

A race in the coastal lowlands is recorded from Taru 102C, Samburu 102C and Mombasa 114B by Ja, MPG and Br, and the last two of these publications also mention it N to the lower Tana R 79 91. In the absence of further localities and subsequent records, this situation is directly analogous to that seen in the White-bellied Tit (583)(including a record from Voi 101B: Br), and similar comments apply.

This species forms part of a superspecies with the Mouse-coloured Penduline Tit (587: *q.v.*), which replaces it in drier country to the N and E, with only marginal overlap (HM). Both occur at semiarid Baringo 50A, where the Mouse-coloured is a common breeding resident, and the African an uncommon and occasional non-breeding visitor, Mar-May and Oct-Nov (St).

587 MOUSE-COLOURED PENDULINE TIT *Remiz musculus*
B669 / MPG1161 / D2C1

Afrotropical: the NE, from the extreme NE borders of Tanzania, through Kenya to NE Uganda, S Sudan, Ethiopia and Somalia.

Kenyan: rather uncommon in dry bushed and wooded country at 400-1600 m (Br; Lack 1985). In contrast to the African Penduline Tit (586: *q.v.*), range is 86% arid-semiarid and 71% within intermediate, 250-1000 mm rainfall areas.

Both of the penduline tits are quite easily overlooked, but the African achieves C2 due to its range in the wetter, more observed areas. Apparent absence of the Mouse-coloured from much of the arid E and NE could reflect low observer frequencies. Alternatively, lack of records from the NE could represent aversion to the driest areas, with presence in only 15% of the 0-250 mm rainfall squares. Apparent absence from much of the E extends to Somalia S of 2°N (AM), but may reflect scarcity of observers.

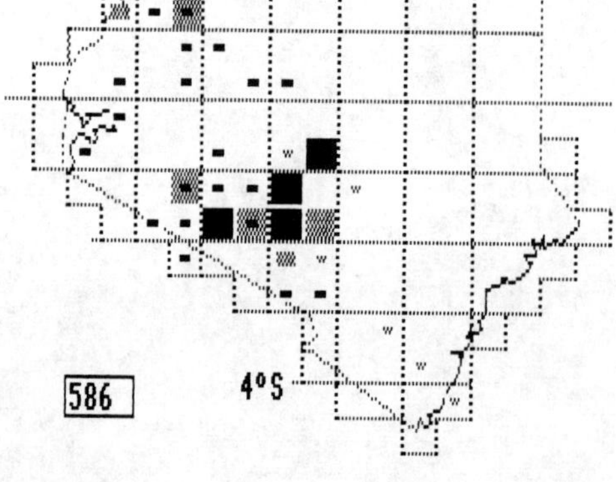

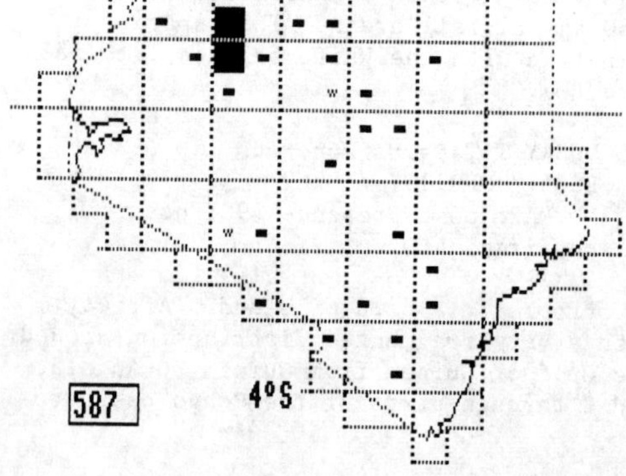

Family Salpornithidae: Spotted Creeper

A monotypic family, found in the Afrotropical and Oriental Regions. It breeds in Kenya.

588 SPOTTED CREEPER *Salpornis spilonota*
B670 / MPG1283 / D2C2

Afrotropical: the N tropics, from Senegambia through Nigeria and the Central African Republic to S Sudan and, at the E limit of the range, W Kenya and Ethiopia. It also occurs in the S tropics between Angola, S Tanzania and Mozambique.

Kenyan: distinctly uncommon and local in open wooded habitats, especially those with acacias, up to 2300 m (Br; Ja; HM). All six squares are above 1000 m, in humid 1000+ mm rainfall areas to the W of the rift valley. The record in 36D comes from Uganda (Ja).

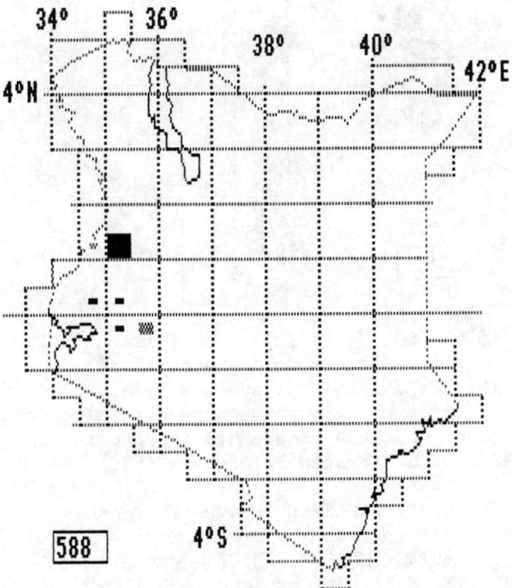

Family Timaliidae: Illadopsis, Babblers, Chatterers

252 species, mostly Oriental, but also extending to other zoogeographical regions. Of only 32 (13%) in the Afrotropical Region, 14 (44%) occur in Kenya. Hinde's Pied Babbler (600) is one of the very few Kenyan endemics. No migratory movements are known (Br). Many species can be at least temporarily obscure in dense vegetation, and concentrations of pre-1970 squares need not imply range contraction. Relatively high coverage estimates (C2) for the illadopsis species reflect their sedentary nature (Br) and restriction to reasonably well known country.

589 BROWN ILLADOPSIS *Trichastoma fulvescens*
B675 / MPG735 / DOC2
NO MAP: post confirmed 48D

Afrotropical: W and C Africa, from Guinea to Angola E through the Congo basin to W Tanzania, S Sudan, Uganda and, on the E extremity of the range, W Kenya.

Kenyan: a shy and skulking, forest undergrowth species, known only from Kabras 48D (Ja) and the Kakamega Forest 48D (Br). This square is in a humid, 1000+ mm rainfall area.

590 PALE-BREASTED ILLADOPSIS *Trichastoma rufipennis*
B677 / MPG736 / DOC2
NO MAP: post presence 37C; post confirmed 48D

Afrotropical: mainly W and C Africa, from the lowland forests around the Gulf of Guinea, E through the Congo basin to NW Tanzania, W Uganda and, at its equatorial E limit, W Kenya. There is also an endemic race in E Tanzania.

Kenyan: a skulking species of forest undergrowth, up to 1700 m (Br; Mann 1980). Range is entirely above 1000 m in humid, 1000+ mm rainfall areas to the W of the rift valley. It is known only from the Cherangani Hills 37C (a single record at 2300 m) and the Kakamega and S Nandi Forests 48D (Br).

It forms a superspecies with the Mountain Illadopsis (591) which, in areas of apparent sympatry, is typically at greater altitudes (HM; Br; but see Diamond & Fayad 1979).

591 MOUNTAIN ILLADOPSIS *Trichastoma pyrrhopterum*
B676 / MPG738 / D1C2

Afrotropical: a highland species of the E, from Malawi and S Tanzania, through the E Zaire-Tanzania-Uganda borders, to Kenya.

Kenyan: common in forest undergrowth at 1550-2800 m (Br; Ja). Range is entirely within humid, 1000+ mm rainfall areas. This is the only illadopsis to extend E of the rift valley, where Br mentions occurrence on the Aberdare Mts (mapped in 62B). See Pale-breasted Illadopsis (590).

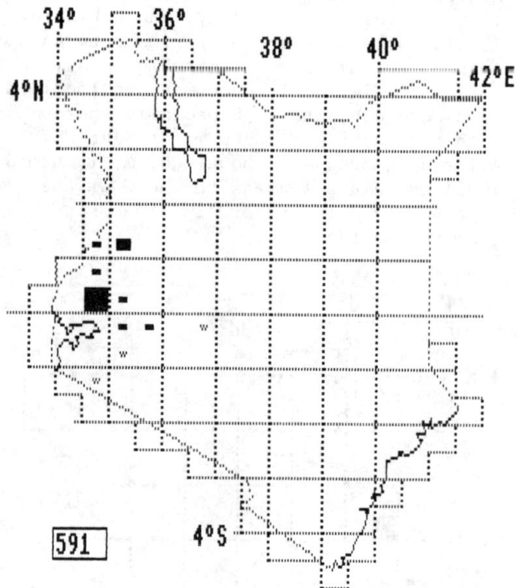

592 SCALY-BREASTED ILLADOPSIS *Trichastoma albipectus*
B674 / MPG737 / D1C2
NO MAP: post presence 48D 49C; pre presence 36D; pre probable 48D

Afrotropical: C Africa, NE from Angola, through mostly E areas of the Congo basin, to S Sudan, Uganda and, on the E limit of the range, W Kenya.

Kenyan: this skulking, forest undergrowth bird occurs at 1700-2100 m and can be locally common (Br; Diamond & Fayad 1979). It is known from Mt Elgon 36D (Ja), the Kakamega Forest 48D and Nandi 49C (Br; P L Britton *in litt.*). The three squares are all in humid, 1000+ mm rainfall areas to the W of the rift valley.

593 GREY-CHESTED ILLADOPSIS *Kakamega poliothorax*
B672 / MPG739 / D0C2
NO MAP: post presence 49C; post probable 48D; pre presence 36D

Afrotropical: apart from a population in Cameroon, a bird of the E that extends along the E Zaire-Uganda borders and also occurs in W Kenya.

Kenyan: uncommon, local and skulking in forest undergrowth at 1550-*c*. 2000 m (Mann *et al*. 1978; Mann 1980; Br), known from Mt Elgon 36D, Lerundo 48D and the Kakamega, N and S Nandi Forests 48D 49C (Ripley & Bond 1971; Mann 1976; Diamond & Fayad 1979; Br). The three squares are in humid, 1000+ mm rainfall areas to the W of the rift valley. Probable breeding in 48D is based upon immatures at Kakamega Forest (Mann *et al*. 1978).

594 AFRICAN HILL BABBLER *Alcippe abyssinica*
B671 / MPG740 / D2C2

Afrotropical: mainly the E highlands, from Malawi and S Tanzania NW along the E Zaire-Tanzania-Uganda borders, and NE through E Tanzania and Kenya, to S Sudan and Ethiopia. There are also isolated populations in Cameroon and Angola.

Kenyan: fairly common in forest undergrowth at 1500-3000 m (Br; Mann 1980; Se; Diamond & Fayad 1979), including the forest islands of Mts Loima 13C, Kulal 26B (Diamond & Keith 1980) and Nyiru 26D, and the Ndoto Mts 39A, Maralal 38D and the Chyulu Hills 88D (Br). Excluding these isolated highlands, which are too localised to appear on our overlays, range is 91% moist subhumid-humid and 82% within 1000+ mm rainfall areas. MPG mention seasonal movements, but Se noted presence throughout the year around 3000 m at Mau Narok 61D.

595 RUFOUS CHATTERER *Turdoides rubiginosus*
B685 / MPG732 / D2C2

Afrotropical: the E and NE, from C and E Tanzania, N through Kenya and N Uganda, to S Sudan, Ethiopia and S Somalia.

Kenyan: a rather shy and retiring bird of thickets and other low cover up to 1500 m, that is awarded D2 because it is reasonably common and often noisy (Br; MPG; La; St). It is found throughout the rainfall and moisture categories, but is less frequent in the highlands and the moist W. Apparent absence from much of the NE may reflect low observer frequencies or an aversion to the driest habitats: there are reports from only 32% of the most arid squares (moisture index below 1.2). Status is possibly similar in adjacent S Somalia (AM). Despite McVicker's (1982) note of a decline in the densely populated Mombasa area 114B, the high proportion of old breeding records in the coastal strip is probably not significant.

This species forms part of a species-group with the Scaly Chatterer (596: *q.v.*)

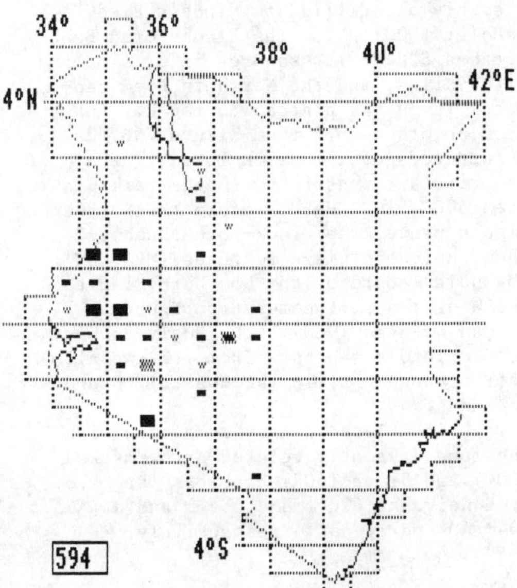

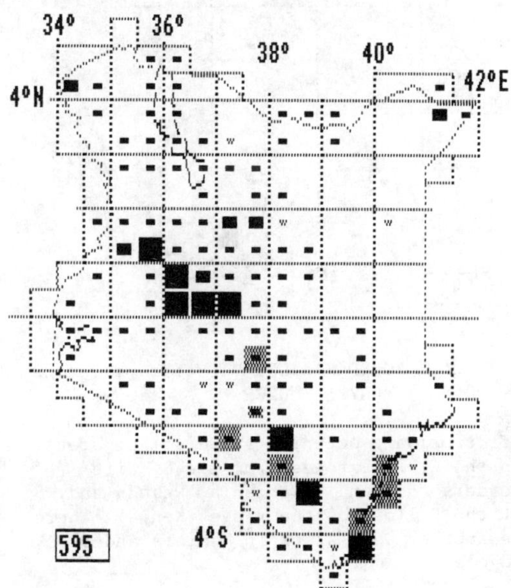

(HM). Where sympatric, these two birds are thought to be ecologically segregated, with the Rufous in rather denser vegetation (HM), but Br notes this distinction to be seldom clearly defined. In arid Tsavo E 101B, this species mostly inhabits riverine bush, while the Scaly is mostly in bush within woodland (La; Lack 1985). Both of these chatterers occupy generally drier country than the congeneric babblers (HM). Other reference: Huels 1982.

596 SCALY CHATTERER *Turdoides aylmeri*
B678 / MPG733 / D1C1

Afrotropical: the E and NE, from E Tanzania through mainly E areas of Kenya, to Somalia and SE Ethiopia. Distribution in Kenya describes the SW edge of the range.

Kenyan: local and uncommon in thickets and woodland, up to 1500 m (Br;Ja; La). Occurrence is mainly on the plateau N, E and S of the C highlands (19B: Erlanger 1905, 1907; 75C: T Stevenson *in litt.*; 86B: EABR 1981). In contrast to the Rufous Chatterer (595: *q.v.*), the Scaly's range is 96% arid-semiarid, and absent from even semiarid areas of the coastal strip.

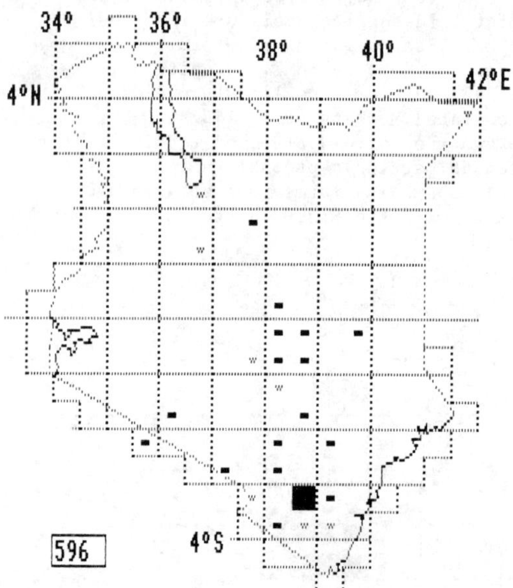

597 BLACK-LORED BABBLER *Turdoides melanops*
B683 / MPG726 / D2C2

Afrotropical: occupies two distinct areas. In the E, it extends from the E Zaire borders to W and C Tanzania, Uganda and, at the E limit of the range, Kenya. There is another population in Namibia and S Angola.

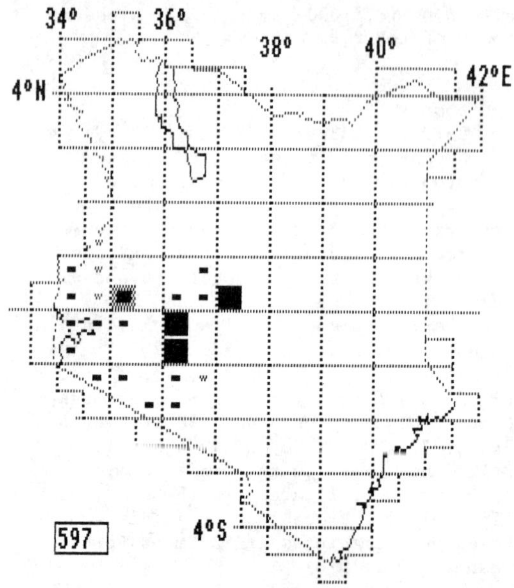

Kenyan: a locally common babbler of bushed and wooded habitats at moderate altitude, exceptionally up to 2200 m (Br). Range is 72% above 1500 m but also includes the 1000-1500 m country in and adjacent to the L Victoria basin. Br's note of apparent absence from Siaya 48C 60A and S Nyanza 60C is incorrect (ADL pers. obs.). Range is 85% subhumid-humid, and 95% within the 500+ mm rainfall regime.

This species forms part of a superspecies with the allopatric Northern Pied (598) and Scaly (599) Babblers, and is possibly also allied to Hinde's Pied Babbler (600)(HM). The Black-lored Babbler has 82% of its squares in or W of the rift valley, and the Northern Pied replaces it on the plateau to the E. Apart from an old record from Kikuyu 75B (Ja), occurrences of the Black-lored to the E of the rift are restricted to the Laikipia area 50BD 51C. Here, within the Northern Pied's range, the Black-lored Babbler shows unique plumage features (MPG; HM). The sole record of the Northern Pied to the W of the rift comes from Nguruman 74D (Fayad & Fayad 1980). The Scaly Babbler is completely separate from its two allopatric congeners, in the coastal, E and NE lowlands.

The Black-lored can by sympatric with the more distantly related Arrow-marked (601) and Brown (602) Babblers (Br) *e.g.* at L Naivasha 62C (many obs.) and at Akala 60A (ADL pers. obs.) respectively.

598 NORTHERN PIED BABBLER *Turdoides hypoleucus*
B680 / MPG729 / D3C3

Afrotropical: endemic to E and NE Tanzania, N to C Kenya.

Kenyan: locally common in bush, forest edge and suburban (Nairobi 75B) gardens, up to 1800 m (Turner 1977; Br). Range is 81% above 1000 m and, in clear contrast to that of the closely related Black-lored Babbler (597: *q.v.*), 95% to the E of the rift valley, 90% semiarid-subhumid, and 81% within intermediate, 250-1000 mm rainfall areas.

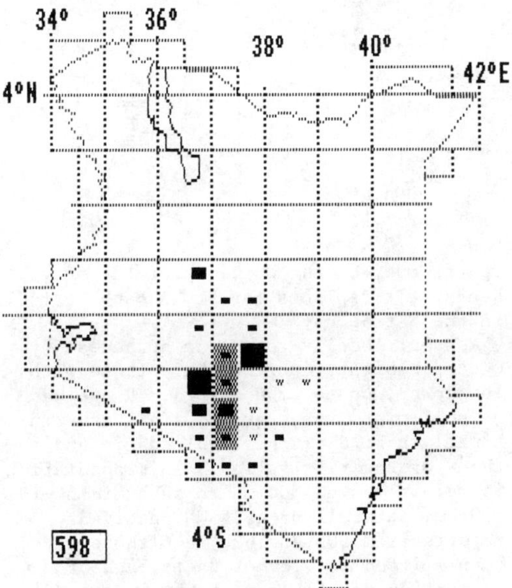

599 SCALY BABBLER *Turdoides squamulatus*
B686 / MPG724 / D1C1

Afrotropical: the E, in coastal areas of Kenya and S Somalia, and inland along the Tana, Juba and Shebelle R.

Kenyan: the distribution of this shy and retiring species of dense bush and thicket habitats is certainly imperfectly known, hence D1C1. Occurrence can be split into three areas. The nominate race is principally known from the coastal strip between Lamu 91B and Vanga 114C (Ja; Br). Absence of modern records in 91B and 114BC may well reflect its retiring, obscure nature. The record in 80A derives from Somalia (AM). This race also extends inland along the Tana R, at least as far as Garissa 65B, and possibly into the river's upper reaches in 63-64 (Ja; Br).

One or more other races occur in the extreme NE. J S Ash (*in litt.*) noted the

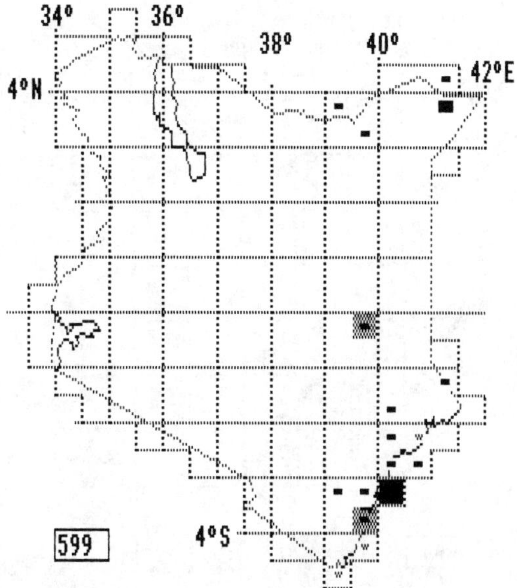

species in the Ethiopian sectors of 17AD, and it is resident on the Kenyan side of the border in thickets along the Daua R near Rhamu Dimtu 7C and Rhamu 19A (E J Roberts *in litt.*; EABR 1984). Ash (1981c) also mentions old specimens of this species, collected along this river immediately NE of 19B. However, while the Daua R is a tributary of the Juba, two independent field sketches of the Rhamu birds show more white on the head than Ash's (1981c) drawing of *T.s. jubaensis*, and more resemble *T.s. carolinae* from the more distant Webi Shebelle.

Apart from these far NE records, and in contrast to other members of this superspecies (see Black-lored Babbler (597)), this species' range is entirely below 500 m. The coastal population is within 500+ mm rainfall areas, but the riverine populations of the interior penetrate 0-250 mm rainfall country. Total range is 69% arid-semiarid.

600 HINDE'S PIED BABBLER *Turdoides hindei*
B679 / MPG730 / D2C2

Afrotropical: endemic to C Kenya.

Kenyan: an extremely local species, endemic to the S and E peripheries of the C highlands. This babbler frequents bushed and open wooded habitats, particularly where lantana thickets are well developed (Plumb 1979; EABR 1977). It favours small valleys at around 1500 m, but is also found in drier, more open, bushed and wooded country at 1300-1500 m

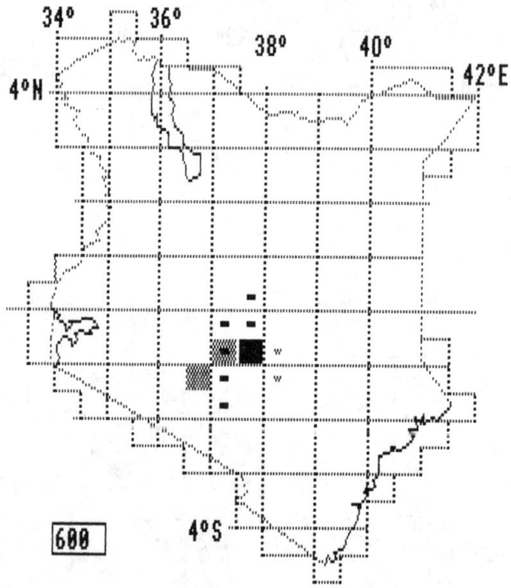

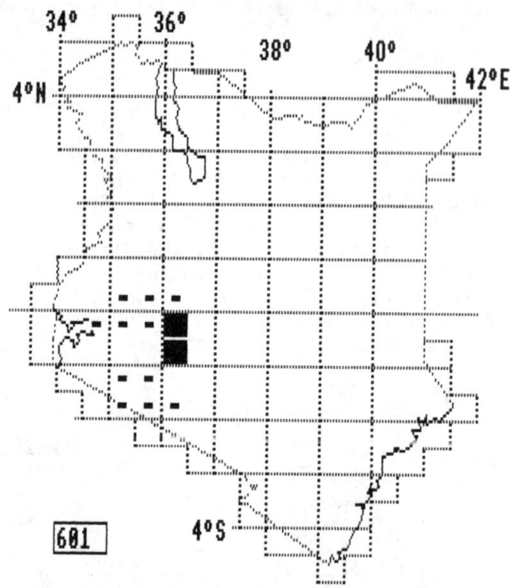

(*op. cit.*; Br). Occurrence is entirely within the 500+ mm rainfall regime, but the squares vary from semiarid to humid.

Plumb (1979) estimated total known range to be below 17 500 km², but considered recent contraction to have reduced this to only 1050 km², in the Embu-Saba Saba-Tana R area 63C. Recent records or revision of status from near Maua 51D (EABR 1984), S of Machakos 76C (Lewis 1984e) and Kitui 77A (Lewis 1983e) suggest this contraction to be less serious than supposed, possibly at least partly reflecting coverage of an often skulking species in rarely visited areas.

The suggestion that this bird is in fact a hybrid between the Northern Pied (598) and Arrow-marked (601) Babblers (HM; Turner 1977) is refuted by Plumb (1979). We support Plumb's view, particularly since Hinde's is sympatric with the former of these supposed parent species but allopatric with the other, whereas a true hybrid might be expected to be parapatric with both. Hinde's occurs on the same ground as the smaller Northern Pied, but the two have not been observed together (Blencowe 1960a; ADL pers. obs.). HM and Br quote Hinde's as forming a superspecies with the allopatric Arrow-marked and Brown (602) Babblers.

601 ARROW-MARKED BABBLER *Turdoides jardineii*
B681 / MPG725 / D2C2

Afrotropical: widespread S of 5°S, avoiding only the dry SW. Further N, it extends to Uganda and, on the NE limit of the range, Kenya. It replaces the Brown Babbler (602) in the S tropics.

Kenyan: locally common in bushed and wooded habitats, particularly those with low cover, up to 2200 m (Br). Range can be split into three areas. This is principally a bird of the highlands in and to the W of the rift valley. This population is mainly above 1500 m, in subhumid-humid, 500+ mm rainfall areas. We received reports from 50D and 62B, but these have been omitted in view of their lack of supporting detail, and of Br's note that occurrence in the E rift highlands requires confirmation. Secondly, there is an old record from Taveta 100B (Ja). Finally, presence on the coast at Lamu 91B is listed by Ja, but regarded as unconfirmed by Br: Ja's record is not mapped here.

The Arrow-marked Babbler forms part of a superspecies with the allopatric Hinde's Pied (600) and Brown Babblers (HM; Br), and occupies the area between their ranges. There appears to be overlap at QSD scale with the Brown Babbler, and both are recorded from the Kisumu area 60B (Ja; HM). These two birds are very similar in the field, however, and the possibility of misidentification, particularly where their ranges converge, cannot be totally precluded. See Black-lored Babbler (597).

602 BROWN BABBLER *Turdoides plebejus*
B684 / MPG723 / D2C2

Afrotropical: the N tropics, E between 15°N and 5°N to W Ethiopia, and SE through S Sudan and Uganda to the SE limit of the range in W and NW Kenya.

Kenyan: locally common in bushed and wooded habitats at 600-1500 m (Br). Range is 74% subhumid-humid and 91% within 500+ mm rainfall areas. However, as noted by Ja for Uganda, this babbler also descends to distinctly lower and drier country in the N. Ja's mention of occurrence at Naivasha 62C is apparently incorrect (HM). See Black-lored Babbler (597) and Arrow-marked Babbler (601).

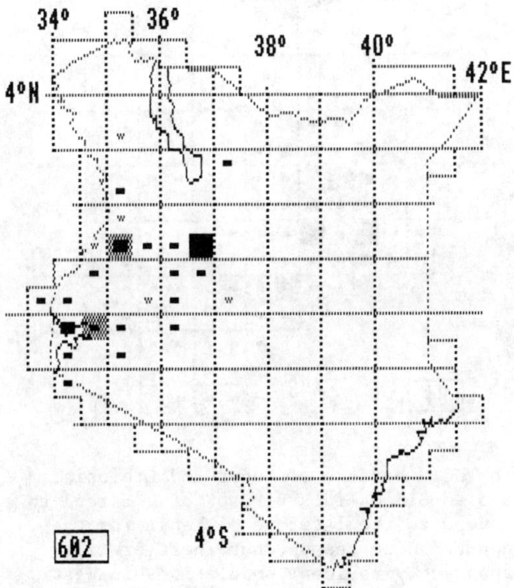

Family Campephagidae: Cuckoo Shrikes

70 species, predominantly Oriental and Australasian, and absent from the Americas. Only 10 (14%) are Afrotropical and, of these, six occur in Kenya. All of the Kenyan birds may have resident populations, but at least two are also regularly augmented by non-breeding intra-African migrants.

603 RED-SHOULDERED CUCKOO SHRIKE *Campephaga pheonicea*
B690 / MPG1083 / D2C1
NO MAP: post presence 48AC; pre presence 60B; pre probable 48D

Afrotropical: the N tropics, widespread between 15°N and the equator E to Ethiopia and W Kenya, but mostly avoiding the lowland equatorial forests.

Kenyan: a distinctly uncommon and little known bird of bushed and wooded habitats and forest edge (Britton 1973; Br). The E limit of its range is in the far W, in and N of the Winam Gulf. The four squares are entirely moist subhumid-humid, at 1000-1500 m, and within the 1000+ mm rainfall regime. The record in 48A derives from Uganda (J S Ash *in litt.*). The sole breeding record concerns nest building at Kakamega 48D (Tennent 1965). There is possibly a resident population, and Jan-Apr increases are thought to reflect non-breeding migrants from the N tropics (Br).

HM place this, the Black (604) and Petit's (605) Cuckoo Shrikes within a superspecies, with the following qualifications. The Red-shouldered and Petit's are considered distinct species. Interbreeding between Petit's and the possibly more closely related Black is thought to be mainly prevented by Petit's allopatry in forest interiors. The proximity of the Red-shouldered and the Black is uncertain, with possible intergradation of females in C and E Kenya, and an apparent lack of ecological segregation. Br, however, notes the Red-shouldered in generally denser habitats than the Black, and considers the latter to be mainly or entirely a non-breeding visitor to the Red-shouldered's range in W Kenya. Other reference: Cunningham-van Someren 1976.

604 BLACK CUCKOO SHRIKE *Campephaga flava*
B688 / MPG1081 / D2C2

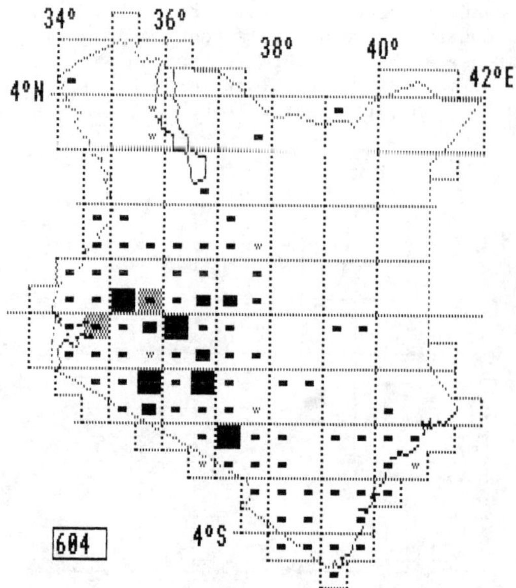

Afrotropical: this S tropics species is widespread S of 5°S, avoiding only the dry SW (*cf.* Red-shouldered Cuckoo Shrike (603)). Further N, it extends through East Africa to the S borders of Sudan and Ethiopia, and to S Somalia. Birds S of 7°S tend to move N to the latitude of Kenya for the non-breeding season, but there are also apparently resident populations in East Africa (Britton 1973; Br).

Kenyan: seasonally common in bushed, wooded and forest edge habitats, up to 3000 m (Britton 1973; Br; Se). Range is 74% within 500+ mm rainfall areas and only 15% arid. Breeding squares are entirely above 1000 m in the rift valley highlands and adjacent plateaux, and 91% within the 500+ mm rainfall regime. Br quotes breeding at 900-1700 m, and suggests its absence from the far W 48 to reflect allopatry with the Red-shouldered Cuckoo Shrike (603: *q.v.*).

Some East African populations are probably resident (Br). Tentative suggestion of this status in Tsavo E 101B (La) appears unlikely when this area's aridity and altitude are compared with those of the apparent breeding range (see above), and Lack (1985) thinks birds there mainly transient. Locally breeding birds are undoubtedly augmented by large numbers of Apr-Sept non-breeding migrants from the S tropics (Britton 1973; Br). These migrants exceptionally reach the arid N, but are rare even at semiarid Baringo 50A (St). Presumed nocturnal migrants have been recorded at Ngulia Lodge 101A at dawn (Nov, Jan: Backhurst & Pearson 1977b), and actually at night (Nov: EABR 1980). The proportion of yellow-shouldered morphs varies to some extent geographically (Br; *e.g.* EABR 1979).

605 PETIT'S CUCKOO SHRIKE *Campephaga petiti*
B689 / MPG1082 / D2C3
NO MAP: post presence 48D; pre presence 49C 61A; pre probable 48D

Afrotropical: appears to occupy three distinct areas: from Cameroon S to Angola; from NE Zaire to SW Uganda; and an isolated population in W Kenya.

Kenyan: regularly recorded in the forest interior at Kakamega 48D, there are also old records from Nandi 49C and Nyarondo 61A (Ja). Range is entirely W of the rift valley, in humid, 1000+ mm rainfall country. Br mentions it at 1400-1800 m. Although resident (Br), the only indications of breeding have been courtship display and individuals with enlarged gonads at Kakamega Forest 48D (Zimmerman 1972). See Red-shouldered Cuckoo Shrike (603) and Purple-throated Cuckoo Shrike (606).

606 PURPLE-THROATED CUCKOO SHRIKE *Campephaga quiscalina*
B691 / MPG1084 / D2C2

Afrotropical: from the E edge of the Congo basin to NW Tanzania, S Sudan, Uganda and, on the E limit of the range, the highlands of W and C Kenya and Oldeani in N Tanzania. Further W, it is found from Sierra Leone S to Angola, E to S Zaire and Zambia. There is also an endemic race in SE Tanzania.

Kenyan: a generally rather uncommon bird of highland forests at 1700-2500 m (Br; HM), although MPG mention local abundance on Mt Elgon 36D. Range is 84% moist subhumid-humid and 71% within 1000+ mm rainfall areas. It does not extend to forest islands. Presence in the coastal Sokoke Forest 102B was reported by Brown (1973c), and quoted by Mann (1976). In view of the observer's uncertainty and the lack of subsequent records (*e.g.* in Britton & Zimmerman 1979), we concur with Br's omission of this record.

This species is sympatric with the very similar Petit's Cuckoo Shrike (605) in Kakamega Forest 48D, and no ecological distinctions are known (HM; Br). Other reference: Cunningham-van Someren 1976.

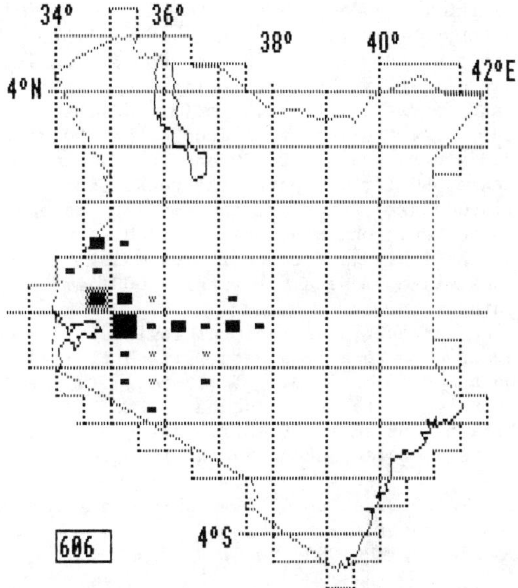

607 GREY CUCKOO SHRIKE *Coracina caesia*
B693 / MPG1086 / D2C2

Afrotropical: mainly the E, from S Sudan and Ethiopia S through East Africa to Mozambique and the Cape. There is also an isolated population in Cameroon. Distribution in Kenya represents the E limit

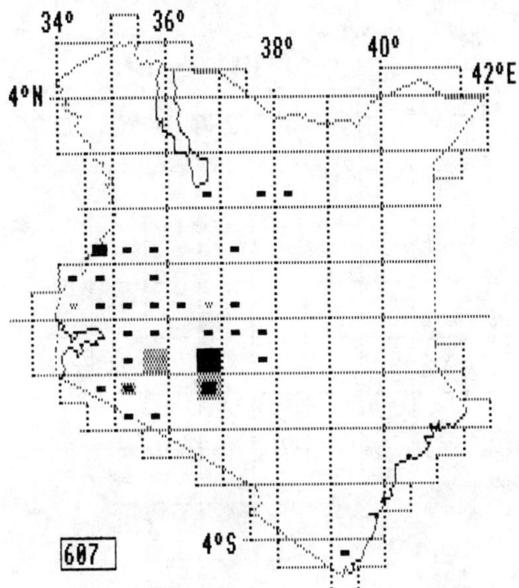

of the range.

Kenyan: distribution is divisible into two distinct areas. This cuckoo shrike is predominantly a common resident of highland forest at 1600-3000 m (Br; Ja; Se), including the forest islands of Mts Nyiru 26D and Marsabit 27D 28C and the Mathews Range 39C (Br). Apart from these localised peaks, which are too small to appear on our environmental overlays, this cuckoo shrike's highland range is 78% moist subhumid-humid and 67% within 1000+ mm rainfall areas.

Occurrence in the coastal lowlands is restricted to a female on Mrima Hill 114A on 6 Nov (EABR 1983). This record is presumably related to presence on the Usambara Mts of adjacent NE Tanzania (Br). See White-breasted Cuckoo Shrike (608).

608 WHITE-BREASTED CUCKOO SHRIKE *Coracina pectoralis*
B695 / MPG1085 / D2C1

Afrotropical: widespread between 15°N and 20°S, E to Ethiopia, W Kenya, coastal Tanzania and Mozambique, and avoiding only the equatorial lowland forests of the W.

Kenyan: a rare species of woodlands. Occurrence at Elgon 36D, Mangiki 48B, Busia 48C and Kakamega 48D (Ja; Br) defines the E limit of its normal range. This distribution is entirely W of the rift valley and N of the Winam Gulf, in humid, 1000+ mm rainfall areas. Ja considered it rare in this region, and there are no breeding records. There are, furthermore, no recent records, despite frequent coverage of the Kakamega-Nandi area 48D 49C, and P L Britton's several years' residence in 48C. These points suggest a W recession of the range. This cuckoo shrike is probably mainly resident (Br), but a wanderer reached E to the Tsavo R 89C (22 Aug 1914: Mackworth-Praed 1917): this extralimital record was at 500-1000 m, in a semiarid, 250-500 mm rainfall region.

This species forms a superspecies with the allopatric Grey Cuckoo Shrike (607) (HM). Where their ranges overlap in the far W 36D 48, they may be segregated by habitat and altitude, with this species seldom above 1400 m and the Grey typically above 1600 m (Br).

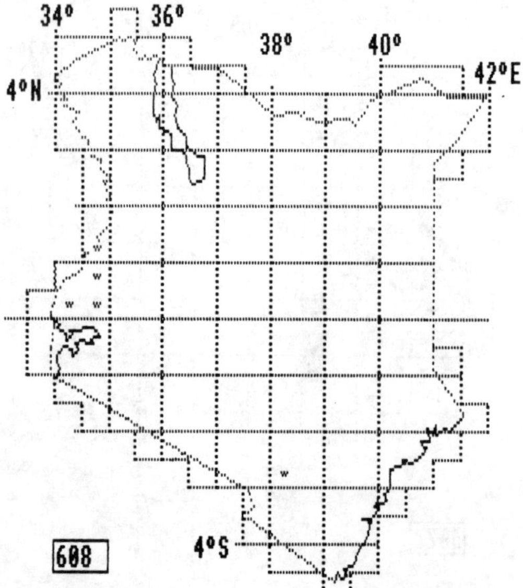

Family Pycnonotidae: Greenbuls, Brownbuls, Bulbul etc.

118 species, principally in the Afrotropical and Oriental Regions, and absent from the Americas. 50 (42%) occur in the Afrotropical Region, and half of these are recorded from Kenya. No migratory movements are known, but the abundant and ecologically catholic Common Bulbul (609) probably wanders (Br). Most members of this family are at least superficially similar inhabitants of forest. Problems in their field identification (Turner & Zimmerman 1979), together with the modern disinclination to collect specimens, means that high proportions or local clusters of pre-1970 squares do not necessarily signify declines or retractions. Paucity or absence of breeding records is certainly a result of under-recording.

609 COMMON BULBUL *Pycnonotus barbatus*
B732 / MPG741-744 / D3C3

Afrotropical: widespread S of 20°N, avoiding only the Cape and the dry SW. It is the most ubiquitous non-forest species in the Region.

Kenyan: one of the commonest and most widespread of birds, found up to at least 3000 m in any habitat except forest interior and open, bushless and treeless terrain (Ja; Be; Se; Br; Lack 1985). It is approachable, and common around human habitation. Range is throughout the rainfall and moisture regimes, but there are records from only 44% of the most arid squares (moisture index below 1.2). While this probably reflects low observer frequencies, there is a notable lack of reports from the main road through the Dida Galgalla Desert 28A, which is relatively often visited by birdwatchers. Observer frequency has more certainly biassed the location of breeding records, 82% of the breeding squares being within the 500+ mm rainfall regime. Se mentions post-reproductive flocking at Mau Narok 61D. Other references: Hudson 1945; Britton 1969, 1972; Lewis 1983g.

610 LITTLE GREENBUL *Andropadus virens*
B705 / MPG775 / D1C2

Afrotropical: mainly W and C Africa, from Senegambia to Angola, but also E through the Congo basin to W Kenya, S Tanzania, Malawi and N Mozambique. Range also extends through E Tanzania N to the coastal lowlands and adjacent areas of Kenya.

Kenyan: a shy and retiring but vocal species, found in forests, particularly along their edges, and in secondary growth

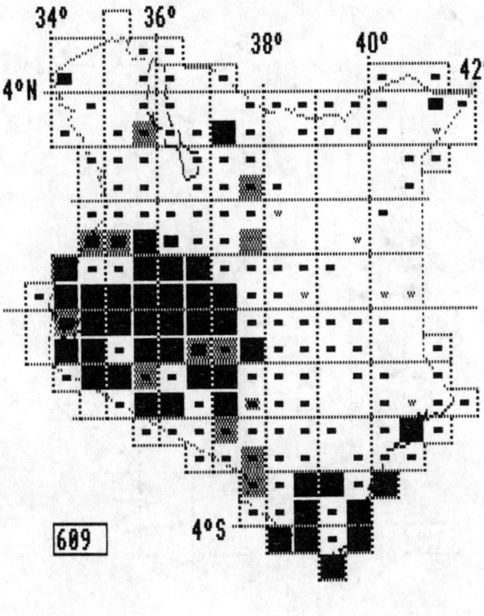

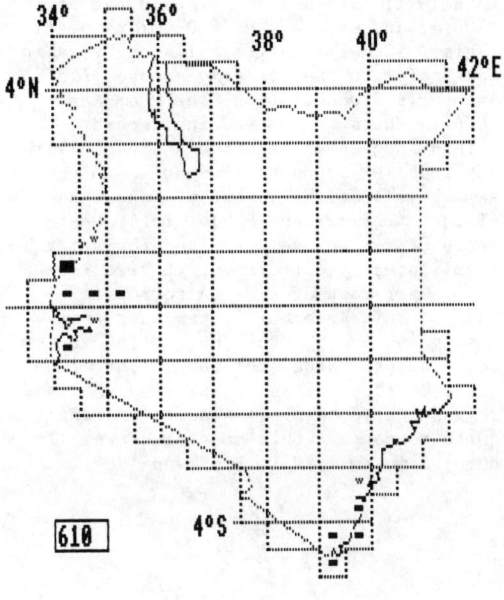

and thickets (Ja; HM; Turner & Zimmerman 1979; Br). Br quotes it exceptionally up to 1800 m, while Mann (1980) gives 1700 m as its altitudinal limit in the S Nandi Forest 48D 49C.

The race *A.v. zombensis* is common in the forests S of Mombasa 114B, but apparently in decline further N (Turner & Zimmerman 1979; Britton & Zimmerman 1979). Its range is within 500+ mm rainfall areas of the coastal lowlands, and entirely semiarid-dry subhumid.

The race *A.v. holochlorus* inhabits 1000+ m areas to the W of the rift valley: range here is entirely moist subhumid-humid and within the 1000+ mm rainfall regime. This race is best known from the Kakamega Forest 48D, where common, but is apparently excluded altitudinally from the adjacent and higher N Nandi Forest 49C (Zimmerman 1972; Turner & Zimmerman 1979; Mann 1980). Other localities are Elgon 36D, Kisumu and Kendu Bay 60B (Ja), and Rapogi 60C (Sugg & Sugg 1972). The record in 48C is from Uganda (J S Ash *in litt.*). The sole indication of breeding is nest building near Bungoma 48A (EABR 1978).

611 LITTLE GREY GREENBUL *Andropadus gracilis*
B699 / MPG772 / D1C3
NO MAP: post presence 48D

Afrotropical: W and C Africa, from the lowland forests around the Gulf of Guinea, E through the Congo basin forests to Uganda and, on the E tip of the range, W Kenya.

Kenyan: known only from the Kakamega and nearby Malaba Forests 48D, hence C3 (Br; EABR 1977). This area is humid, at around 1700 m, and within the 1000+ mm rainfall regime. Other localities listed by Ja in fact refer to the formerly conspecific Ansorge's Greenbul (612)(see MPG; Mann 1976). There is no breeding record.

This species and Ansorge's Greenbul are siblings (HM), and are sympatric in the same levels of the forest at Kakamega (Turner & Zimmerman 1979). This coexistence of very closely related forms may be facilitated by differences in diet (see HM). Over most of their Afrotropical ranges, this greenbul is commoner than Ansorge's, and HM thus suggest the latter to be a less successful competitor. The disjunct populations at Kakamega show the reverse trend however, Ansorge's being quite common and this species scarce (Turner & Zimmerman 1979; Br; Mann 1980).

612 ANSORGE'S GREENBUL *Andropadus ansorgei*
B696 / MPG772, 772A / D1C2
NO MAP: post presence 48D; pre presence 48B 61A; pre probable 48D

Afrotropical: W and C Africa, with a fragmentary distribution through the lowland forests around the Gulf of Guinea, and in E and NE Zaire. A race in W Kenya is disjunct and endemic.

Kenyan: the race *A.a. kavirondensis* is regularly recorded in Kakamega Forest 48D (Turner & Zimmerman 1979; Br), and there are old records from S Elgon 48B and Nyarondo 61A (Ja). This range is at 1000-2000 m to the W of the rift valley, in humid, 1000+ mm rainfall areas. Probable breeding in 48D is based upon Zimmerman's (1972) mention of birds in breeding condition at Kakamega Forest. See Little Grey Greenbul (611).

613 SLENDER-BILLED GREENBUL *Andropadus gracilirostris*
B698 / MPG771 / D2C2

Afrotropical: W and C Africa, from the lowland forests around the Gulf of Guinea, through the Congo basin to W Tanzania, S Sudan, Uganda and, on the E extremities of the range, the highlands of W and C Kenya.

Kenyan: fairly common in forest canopy and open edge areas, including suburban gardens in Nairobi 75B, up to 2400 m (Britton & Sugg 1973; Turner & Zimmerman 1979; Br; ADL pers. obs.). Range is 90% moist subhumid-humid and 80% within 1000+ mm rainfall areas, and includes the

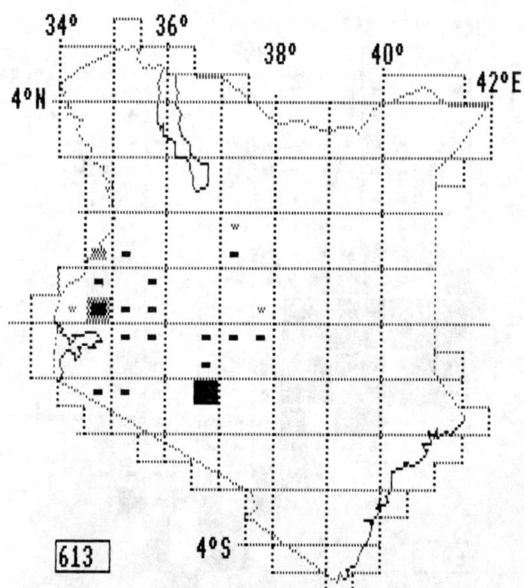

Ndoto Mts 39A and Mathews Range 39C forest islands. The confirmed breeding records are based upon a bird with a mature oviduct egg at Kakamega Forest 48D (Zimmerman 1972), and adults feeding fledglings at Nairobi 75B (EABR 1981; ADL pers. obs.).

614 CAMEROON SOMBRE GREENBUL *Andropadus curvirostris*
B697 / MPG774 / D1C2
NO MAP: post presence 48D 49C; pre presence 36D 48C

Afrotropical: a W and C African greenbul, from the lowland forests around the Gulf of Guinea, E through the Congo basin to S Sudan and Uganda, reaching the E extremity of the range in W Kenya.

Kenyan: locally common in the mid-levels and undergrowth of forests up to 2300 m (Turner & Zimmerman 1979; Br). All recent records are from the Kakamega and Nandi Forests 48D 49C, and there are old records from Elgon 36D and Mumias 48C (Br; Ja). This range is in 1000+ m areas to the W of the rift valley. All squares are humid and within the 1000+ mm rainfall regime. There is no breeding record.

615 MOUNTAIN GREENBUL *Andropadus tephrolaemus*
B704 / MPG765, 766 / D2C2

Afrotropical: apart from a population in Cameroon, a bird of E Africa from C and N Mozambique to Malawi and S Tanzania, and thence NW along the E Zaire borders to SW Uganda, and NE through E Tanzania to the highlands of Kenya. Distribution in Kenya thus reflects the NE limit of the range.

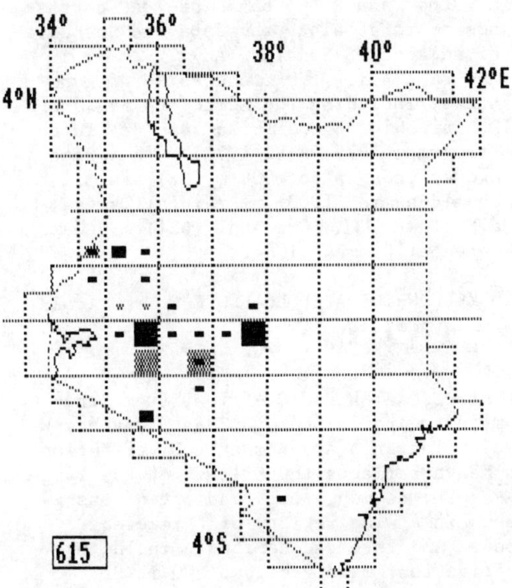

Kenyan: common in highland forest, including the bamboo zone, at 1850-3000 m (Turner & Zimmerman 1979; Br; Se). Range is 80% moist subhumid-humid and 70% within 1000+ mm rainfall areas, and includes the Taita Hills 101A forest island. The Taita record was of an individual of the distinctive race *A.t. usambarae*, observed on 14 Aug 1978 (Turner 1979). This race is common on the W Usambara and S Pare Mts of adjacent NE Tanzania (*op. cit.*; Br).

The Mountain Greenbul forms part of a species-group with the Stripe-cheeked (616) and Shelley's (617) Greenbuls (HM). The ranges of this species and the Stripe-cheeked are allopatric except on the Taita Hills 101A, where the individual *A.t. usambarae* mentioned above was actually in company with two Stripe-cheeked Greenbuls (Turner 1979).

616 STRIPE-CHEEKED GREENBUL *Andropadus milanjensis*
B703 / MPG767 / D2C3
NO MAP: post presence 88D; post probable 101A; pre confirmed 88D

Afrotropical: a highland species of the E, from Zimbabwe and Mozambique, N through Malawi and S and E Tanzania, to the N extremity of the range in S Kenya.

Kenyan: this rather shy, forest edge greenbul is not uncommon on the Chyulu 88D and Taita 101A Hills, at 1200-1850 m (Turner 1979; Turner & Zimmerman 1979; Br). The confirmed breeding from the Chyulu Hills 88D derives from van Someren (1939), while probable breeding in 101A reflects residence in the isolated forests of the Taita Hills. See Mountain Greenbul (615).

617 SHELLEY'S GREENBUL *Andropadus masukuensis*
B702 / MPG768 / D2C2
NO MAP: post presence 48D 49C 61A; pre presence 36D; pre probable 48D

Afrotropical: the E, from N Mozambique, Malawi and S Tanzania, N through the highlands of E Tanzania to 5°S, and NW along the E Zaire borders to Uganda. There is also an isolated population in W Kenya.

Kenyan: a fairly common and often conspicuous highland forest species, at 1500-2300 m (Turner & Zimmerman 1979; Br). There is an old record from Elgon 36D (Ja), but recent records are with a single exception from the Kakamega and Nandi Forests 48D 49C. A pair in forest above Kericho 61A in June 1979 may represent an E extension of range (EABR 1979). Probable breeding in 48D is based upon birds in breeding condition in Kakamega Forest (Zimmerman 1972). Range is entirely above 1000 m to the W of the rift valley, in humid, 1000+ mm rainfall areas. See Mountain Greenbul (615).

618 YELLOW-WHISKERED GREENBUL *Andropadus latirostris*
B710 / MPG776 / D2C2

Afrotropical: a lowland forest species found around the Gulf of Guinea, and E through the Congo basin to W Tanzania, S Sudan, Uganda and, on the E limit of the range, the highlands of W and C Kenya.

Kenyan: a rather shy bird of forests, bamboo and gardens at 1500-3000 m, with separate races to either side of the rift valley (Britton 1979; Br; Turner & Zimmerman 1979). It is the commonest greenbul in most areas (*op. cit.*; Diamond & Fayad 1979), and present on the forest islands of Mt Nyiru 26D, the Ndoto Mts 39A and Mt Uraguess 51A (Ja; Br). Excluding these isolated highlands, which are too localised to appear on our overlays, range is 83% moist subhumid-humid and 76% within 1000+ mm rainfall areas.

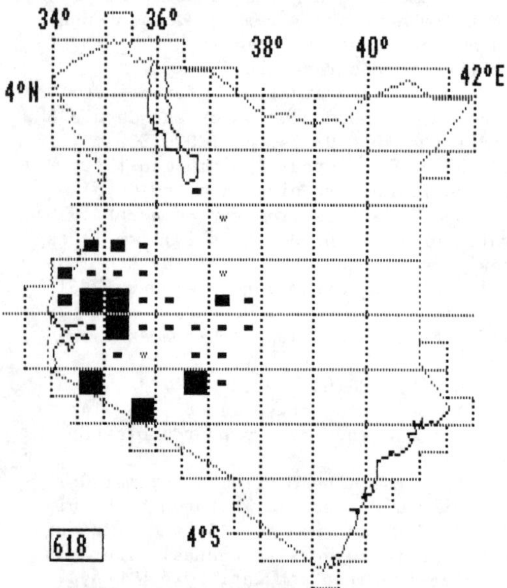

619 ZANZIBAR SOMBRE GREENBUL *Andropadus importunus*
B700 / MPG773 / D2C2

Afrotropical: the E periphery of Africa, from the Cape N through Mozambique, Zimbabwe and Malawi, to E Tanzania, Kenya and S Somalia.

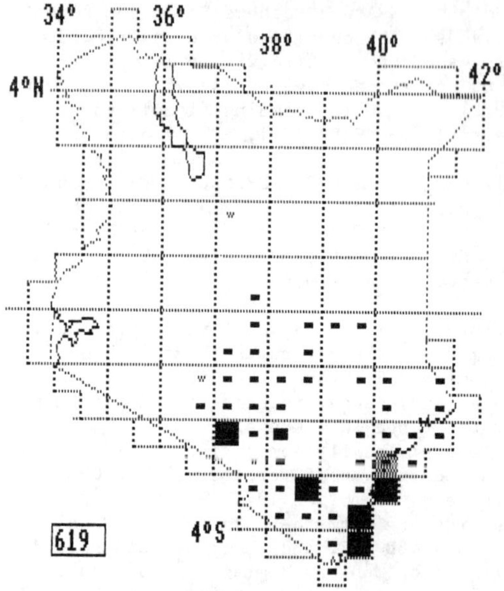

Kenyan: a common bird of bushed and wooded habitats, mostly at low to medium altitudes (Br; Lack 1985). The race *A.i. hypoxanthus* inhabits the periphery of the C highlands (approximately SE of 76A and 51D), and the E and coastal lowlands, being particularly common in coastal scrub and thicket (Br; Turner & Zimmerman 1979; La). Its range is approximately 82% below 1000 m, 85% arid-semiarid and 82% within the intermediate, 250-1000 mm rainfall regime. Penetration of the arid interior along the Tana R 64B 65AB 78B 79AC corresponds to range along the Juba and Shebelle R of adjacent S Somalia (AM).

The race *A.i. fricki* occurs NW of *hypoxanthus*, including the Ndoto Mts 39A, Meru 51D, Nairobi 75B and Thika 76A (Br; Turner & Zimmerman 1979; EABR 1977). This race's range extends up to 2000 m (Br), in moist subhumid-humid, 1000+ mm rainfall areas. See Yellow-bellied Greenbul (621). Other reference: Siemens 1984.

620 YELLOW-THROATED LEAFLOVE *Chlorocichla flavicollis*
B709 / MPG749 / D2C2

Afrotropical: W and C Africa, from Senegambia to Angola, E to S Sudan, Uganda, W Kenya, Tanzania and Zambia. Distribution in Kenya reflects the E limit of the range.

Kenyan: common, noisy and often conspicuous in a wide variety of thicketed, wooded and forested habitats, including cultivation, mainly below 1500 m (Br; Turner & Zimmerman 1979; Sugg & Sugg 1972).

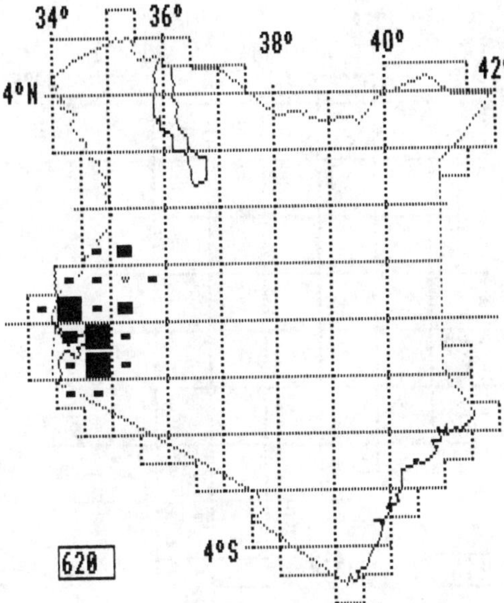

Range is to the W of the rift valley, in areas that are 94% at 1000-2000 m and entirely moist subhumid-humid and within the 1000+ mm rainfall regime. Other reference: Britton & Britton 1971c.

621 YELLOW-BELLIED GREENBUL *Chlorocichla flaviventris*
B710 / MPG769 / D2C2

Afrotropical: mainly the E, from E South Africa, through Zimbabwe, Zambia, Mozambique and SE Zaire, to S and E Tanzania, S Kenya, and Somalia S of 1°30'N (AM). It reaches the W coast around 15°S.

Kenyan: common but rather shy and retiring, typically in the undergrowth and thicket of wooded and forested localities (Ja; Britton & Zimmerman 1979; Turner & Zimmerman 1979; Br; Brown & Britton 1980). It is found up to 1700+ m on the E edge of the C highlands, and in the SE and coastal lowlands (*cf.* the Afrotropical and Kenyan ranges of the Zanzibar Sombre Greenbul (619)). Van Someren (1929) mentions occurrence on the Mathews Range 39C (not mapped) but this is omitted (and thus implicitly rejected) by Ja and Br.

Range is 70% below 1000 m and 73% semiarid-subhumid. Extension into arid areas along the Voi R 101B (La) and the Tana R 78B 79C corresponds to virtual restriction to the Juba R in adjacent S Somalia (AM). The records in 80AD are from the Kenya-Somalia border (J S Ash and J E Miskell *in litt.*).

622 JOYFUL GREENBUL *Chlorocichla laetissima*
B711 / MPG770 / D2C2

Afrotropical: the E Zaire borders, E to Uganda, S Sudan and, on the E extremity of the range, W Kenya.

Kenyan: a noisy and conspicuous bird, typically in forest edge canopy, at 1400-2300 m (Br; Turner & Zimmerman 1979). It is common in the Kakamega Forest 48D, but less so elsewhere (*op. cit.*; Mann 1980). The confirmed breeding in 48D is based upon a female in laying condition at Kaka-

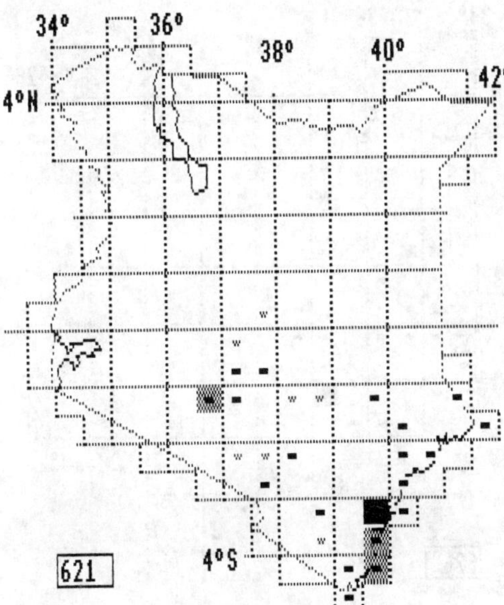

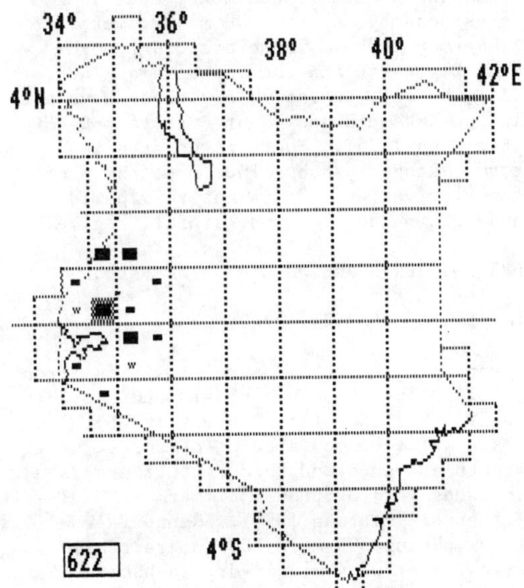

mega (Zimmerman 1972). Range is above 1000 m to the W of the rift valley: 83% humid, and entirely within the 1000+ mm rainfall regime. Other reference: Sugg & Sugg 1972.

623 BRISTLEBILL *Bleda syndactyla*
B708 / MPG746 / D2C2
NO MAP: post presence 37C; post probable 48D 49C; pre confirmed 36D 48D

Afrotropical: a W and C African species that has a fragmented distribution around the Gulf of Guinea, and E through the Congo basin to S Sudan, Uganda, NW Tanzania and, on the E extremity of the range, W Kenya.
Kenyan: common but shy in dense thicket and forest undergrowth, up to 2150 m (Br; Turner & Zimmerman 1979). There are records from Elgon 36D (Ja) E to the Kitale area 37C (T J Barnley *in litt.*), and from the Kakamega, North Nandi and Iruru Forests 48D 49C (Br). The confirmed breeding records are from Ja. All four squares are above 1000 m to the W of the rift valley. This distribution is entirely humid, and within the 1000+ mm rainfall regime.

624 HONEYGUIDE GREENBUL *Baeopogon indicator*
B706 / MPG751 / D2C2
NO MAP: post presence 49C; post probable 48D; pre presence 36D

Afrotropical: a W and C African species, from the lowland forests around the Gulf of Guinea, E through the Congo basin to S Sudan, Uganda and, on the E extremity of the range, W Kenya.
Kenyan: a rather uncommon greenbul of forest canopy, up to 2000 m (Br; Turner & Zimmerman 1979). There are old records from Elgon 36D and the upper Yala R 48D (Ja), but all recent observations have been in the Kakamega and Nandi Forests 48D 49C (Br; Mann 1976). There is no breeding record. Range is above 1000 m to the W of the rift valley. It is entirely humid, and within the 1000+ mm rainfall regime.

625 FISCHER'S GREENBUL *Phyllastrephus fischeri*
B723 / MPG758 / D2C2

Afrotropical: the E coastal lowlands, from N Mozambique through Tanzania and Kenya to the far S of Somalia (Dowsett 1972b).
Kenyan: a common species of forest undergrowth and woodland thicket in the coastal lowlands (Br; Britton & Zimmerman 1979; Turner & Zimmerman 1979). Range is entirely below 500 m in 500+ mm rainfall areas, and 90% semiarid-dry subhumid. See Olive Mountain Greenbul (626).

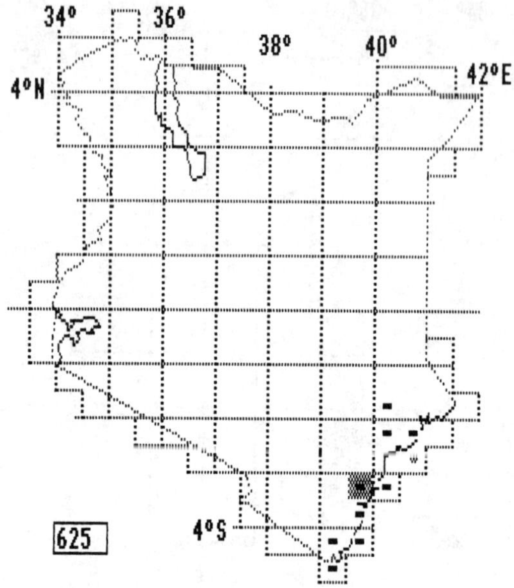

626 OLIVE MOUNTAIN GREENBUL *Phyllastrephus placidus*
B727 / MPG758 / D2C2

Afrotropical: an E species, from Malawi and N Mozambique N through Tanzania to the E Zaire borders, Uganda, S Sudan and, at the NE limit of the range, Kenya (Dowsett 1972b).
Kenyan: common in forest undergrowth at 1000-2700 m (Br; MPG; Sugg & Sugg 1972; Britton & Sugg 1973; Turner 1979; Diamond

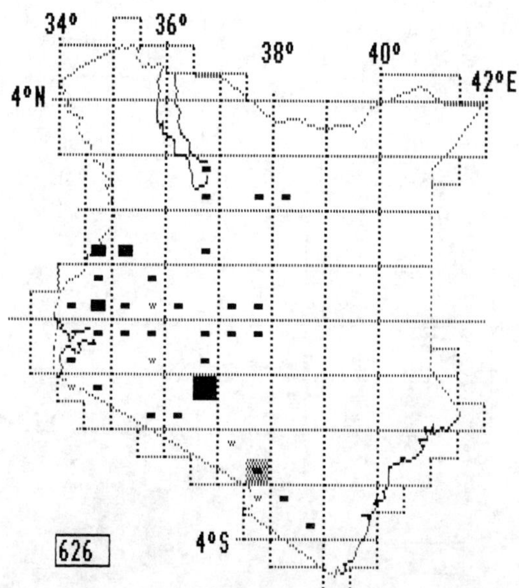

& Fayad 1979; Turner & Zimmerman 1979). Range includes the forest islands of Mts Kulal 26B (Diamond & Keith 1980), Nyiru 26D, Marsabit 27D 28C and Kasigau 101D, and the Chyulu 88D and Taita 101A Hills (Br). Apart from these isolated highlands, which are too localised to appear on our overlays, distribution is 75% moist subhumid-humid and 71% within 1000+ mm rainfall areas.

Dowsett (1972b) split this and another species from Fischer's Greenbul (625), and the three are now considered to be members of a superspecies. The Olive Mountain is allopatric with Fischer's, which replaces it in the coastal lowlands.

627 BROWNBUL *Phyllastrephus terrestris*
B730 / MPG753 / D1C2

Afrotropical: a predominantly E species, from around the E periphery of the dry SW of the continent, N through Zambia, Malawi and N Mozambique, to E Tanzania, Kenya and Somalia S of 2°N (AM).

Kenyan: a rather uncommon bird of forest and thicket (Br). Like the Zanzibar Sombre Greenbul (619), there are lowland and highland races. *P.t. suahelicus* is found up to 450 m in the coastal lowlands. Range is within 500+ mm rainfall areas, and 90% semiarid-dry subhumid. Extension into arid country along the Tana R 79C corresponds with restriction to the Juba R of adjacent S Somalia (AM).

P.t. bensoni occurs on the E edge of the C highlands, at 1200-1400 m (Br). It is known from the Meru 51D and Chuka 63B Forests (MPG), and there is also a breeding record from the Maraqua R 63C on 12 Feb 1970 (EANHS Nest Record Card). Range here is in moist subhumid-humid, 1000+ mm rainfall areas. In addition to these two regions, A W Diamond (*in litt.*) caught the species on Mt Endau 77B.

This bird forms part of a species-group with the similarly terrestrial Northern Brownbul (628) and Grey-olive Greenbul (629)(HM). The Brownbul is locally sympatric with the former, but seems to prefer richer, moister habitats (Br). HM mention partial sympatry between the Brownbul and the Grey-olive Greenbul (*q.v.*).

628 NORTHERN BROWNBUL *Phyllastrephus strepitans*
B729 / MPG754 / D2C2

Afrotropical: an E and NE species, from E Tanzania to Kenya, N Uganda, S Sudan, Ethiopia and S Somalia.

Kenyan: the Northern Brownbul is the most widespread of this family, except for the Common Bulbul (609). It is a common but shy bird of a wide diversity of scrub, thicket, wooded and occasionally forested habitats, mainly below 1000 m but also reaching 1600 m (Turner & Zimmerman 1979; Br; Lack 1985). Range is 66% within the intermediate, 250-1000 mm rainfall regime and 80% arid-semiarid (*cf.* Brownbul (627)). Presence in only 18% of the most arid squares (moisture index below 1.2) may be real, so that the bird is scarce in the driest areas of the NE, or a product of low observer frequencies and a shy species.

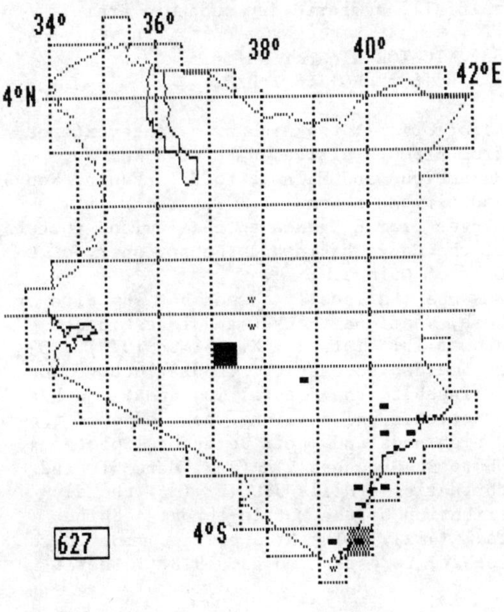

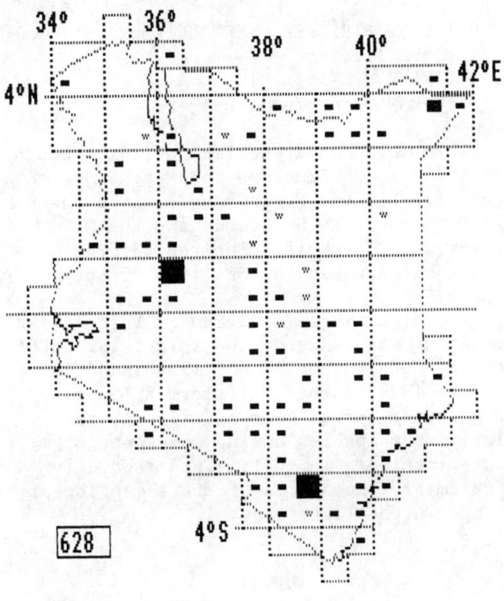

The cluster of old records 27 39 40 42 52 more certainly reflects paucity of recent observers.

629 GREY-OLIVE GREENBUL *Phyllastrephus cerviniventris*
B721 / MPG759 / D1C1
NO MAP: post presence 88D 100B; post confirmed 76A; pre presence 64A 101C

Afrotropical: a fragmented distribution that includes Malawi, Zambia and N Mozambique. Further N, it occurs in E Tanzania and, on the N limits of the range, Kenya.

Kenyan: a local, uncommon, secretive and little known species of forest undergrowth and streamside thicket, up to 1500 m (Turner & Zimmerman 1979; Br). There are records from Tharaka 64A (Ja: presumably the source of Br's mention of Meru), Thika 76A (EABR 1977), Lolterish 88D (Br), Kitovu Forest 100B (many obs. *e.g.* EABR 1978), and Bura 101C (Br). These localities are on the E periphery of the C highlands, and on the adjacent SE plateau. They are all at 500-1500 m, within the intermediate, 250-1000 mm rainfall regime. Three of these five squares are semiarid. Confirmed breeding in 76A is based upon a pair feeding fledged young in bush by the Thika R on 27 Dec 1976 (W J Plumb *in litt.*).

Br mentions probable occurrence on the coast at Rabai 102D and Shimoni 114C, and there is an EANHS nest record card from 102D: these localities require confirmation. Presence in these areas might imply some degree of sympatry with the Brownbul (627: *q.v.*), but their ranges are otherwise separate at QSD scale.

630 YELLOW-STREAKED GREENBUL *Phyllastrephus flavostriatus*
B724 / MPG755 / D1C0
NO MAP: pre presence 101D

Afrotropical: an E species, N from E South Africa through Zimbabwe and Mozambique to Malawi. Further N, it ranges along the E Zaire borders to SW Uganda and the NE borders of Tanzania. The record from S Kenya was thus at the NE limit of the range.

Kenyan: a single specimen of this forest greenbul was taken on the forest island of Mt Kasigau 101D on 18 Nov 1938 (Rand 1958; Turner 1979; Turner & Zimmerman 1979). D A Turner (pers. comm.) failed to relocate the species during an ascent of this inselberg in 1981. This record is presumably the source of MPG's mention of range in the Taita area.

631 TORO OLIVE GREENBUL *Phyllastrephus baumanni*
B719 / MPG760 / D1C2
NO MAP: post presence 48D; pre presence 36D; pre probable 48D

Afrotropical: one race extends from NE Zaire to S Sudan, Uganda and, on the E edge of the range, W Kenya; another occurs in the W, in the lowland forests to the N of the Gulf of Guinea.

Kenyan: a shy and uncommon bird of forest undergrowth (Turner & Zimmerman 1979). There are old records from Elgon 36D and Lerundo 48D (Ja), but all recent observations come from the Kakamega Forest 48D (Mann 1976, 1980; Br). Probable breeding in 48D is based upon juveniles and breeding condition females at Kakamega (Zimmerman 1972). The two squares are in 1000-2000 m areas to the W of the rift valley. They are both humid and within the 1000+ mm rainfall regime.

632 TINY GREENBUL *Phyllastrephus debilis*
B722 / MPG757 / D2C2
NO MAP: post presence 102D 114AB; post confirmed 102B; pre presence 91D

Afrotropical: the E periphery, in the coastal lowlands and adjacent areas of S Mozambique, and of Tanzania and Kenya.

Kenyan: a fairly common undergrowth greenbul of the coastal forests, extending up to 450 m in the Shimba Hills 114A (Britton 1972; Britton & Zimmerman 1979; Turner & Zimmerman 1979; Br). There is an old record from the mouth of the Tana R 91D (Ja). Range is entirely within 500+ mm rainfall, semiarid-dry subhumid areas.

633 NICATOR *Nicator chloris*
B716 / MPG1148 / D2C2

Afrotropical: the Nicator's range extends from E South Africa, N through Zambia, Mozambique and Malawi, to E Tanzania, Kenya and extreme S Somalia. Further W, it ranges from E Uganda to the lowland forests of the Congo basin and the periphery of the Gulf of Guinea.

Kenyan: a locally common but shy bird of bush, woodland thicket and forest undergrowth (Br; Turner & Zimmerman 1979). D2 is awarded because of its distinctive song. It inhabits the coastal lowlands, and has penetrated inland along the E edge of the C highlands and their peripheral plateaux, NW to Mt Uraguess 51A (Ja), Meru 51D and the Karissia Hills 38D (Br)(*cf*; the distribution of the Mottle-throated Spinetail (412)). The Nicator is common in the coastal forests N to Lamu 91B (Turner &

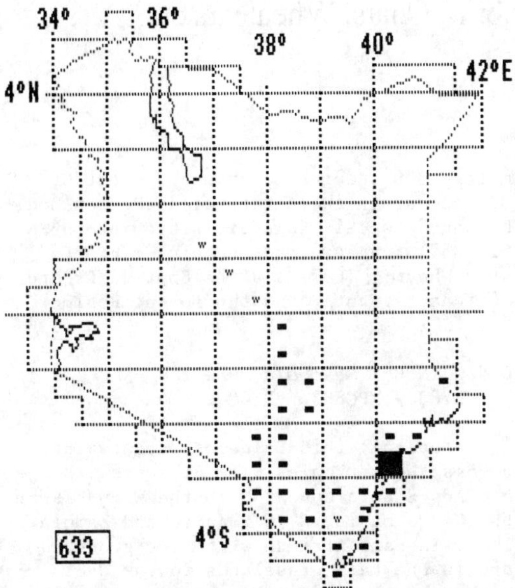

Zimmerman 1979), but rare in Tsavo E 101B (La). The record in 80A derives from Somalia (AM). There may have been a decline or retraction on the NW range tip 38D 51AD.

Distribution is mainly below 1300 m, but reaches 1900 m in the Karissia Hills 38D (Br). Range is to the E of the rift valley, in areas that are 76% below 1000 m, 96% less arid-dry subhumid and 83% within the intermediate, 250-1000 mm rainfall regime. The W race approaches Kenya as closely as W Elgon (approximately 36C) (Ja; Br).

Family Turdidae: Thrushes, Robins, Chats, Wheatears etc.

HM consider this cosmopolitan group of approximately 300 species to show the greatest habitat diversity of any family. Around 111 (37%) occur in the Afrotropical Region and, of these, 48 (43%) are recorded from Kenya. The Kenyan species are divisible into 33 (69%) that breed or are thought to do so, and 12 (25%) that are non-breeding migrants from the Palearctic. This leaves the Red-breasted Wheatear (637) and the Spotted Ground Thrush (681), which may be non-breeding intra-African migrants, and the Forest Robin (664), which is apparently a vagrant.

634 STONECHAT *Saxicola torquata*
B784 / MPG882 / D2C2

Afrotropical: widespread S of 5°S, avoiding only the dry SW. Further N, this chat extends through East Africa to S Sudan and Ethiopia. It is also found in the lower Congo basin, Cameroon and scattered areas further W. Distribution in Kenya describes the E limit of the range.

Kenyan: common in open habitats, mainly at 1800-3200 m, including the Chyulu 88D and Taita 101A Hills and the E periphery of the Kilimanjaro massif 88C 100B (Br; Se; Britton & Sugg 1973; St). Excluding these localised highlands, range is 97% within 500+ mm rainfall areas and 91% subhumid-humid. This chat is generally considered resident (Ja; Se; Br), and its map shows a good example of the distributional features of a resident species, *i.e.* relatively few non-breeding squares, all close to breeding areas.

635 WHINCHAT *Saxicola rubetra*
B783 / MPG883 / D2C2

Afrotropical: this Palearctic chat winters across the N tropics E to Ethiopia. Range extends S to around 7°N in the W and, E of the Congo basin, S to S Zaire and Zambia. The main East African wintering grounds are in the tall grasslands to the W of Kenya.

Kenyan: open habitats (including woodland edge) up to 3000 m, late Sept-early Apr (Br; Britton 1979). Br notes this species in medium rainfall areas, but Kenyan range is 81% within the 500+ mm regime and 48% within the highest, 1000+ mm category. Distribution is 65% subhumid-humid, and occurrence in the arid L Turkana basin is restricted to uncommon Oct-Nov and Mar-Apr passage migrants (Br; HH). Further S, at semiarid Baringo 50A, St records small numbers during Sept-Nov and Mar-May passage.

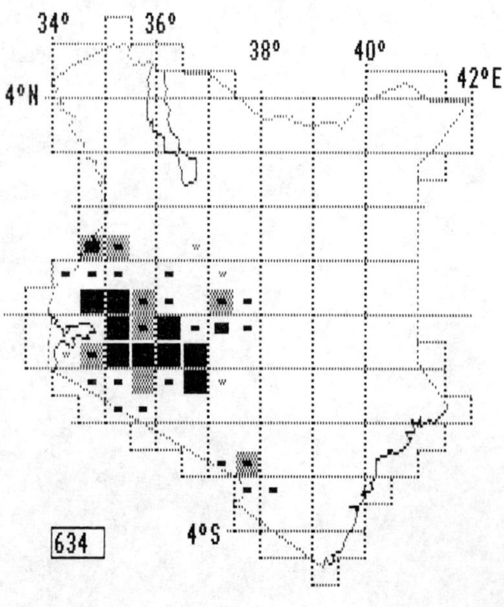

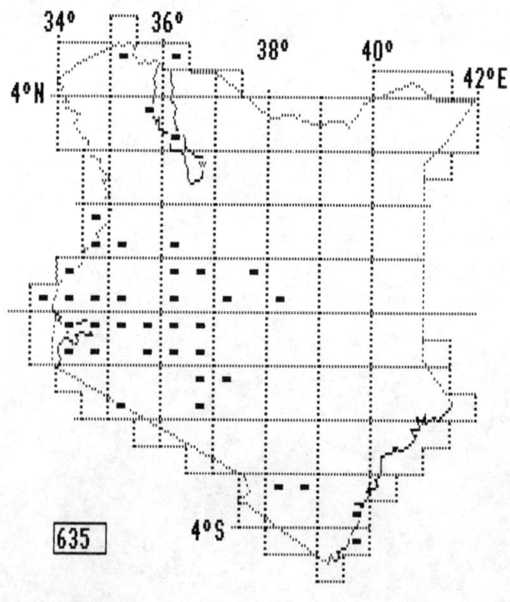

Most frequent to the W of the rift valley, this chat is common only in Nyanza 60, where small numbers overwinter and passage can be heavy (Br; EABR 1984). Small numbers are regular in the rift valley, particularly at L Nakuru 62A and L Naivasha 62C, but status is distinctly less common further E (Br; EABR). In the S, four single birds have been reported on Oct-Nov passage at Ngulia 101A (EABR 1978, 1982) and Bamburi 102D 114B (EABR 1981: the sole coastal record). The single record for arid Tsavo E 101B was on 10 Dec (La).

636 CAPPED WHEATEAR *Oenanthe pileata*
B778 / MPG868 / D2C2

Afrotropical: mainly a S species, widespread S of 5°S. Further N, it extends through Tanzania to the N limit of the range in Kenya. There is a single old Jul record from extreme S Somalia (AM).

Kenyan: this wheatear is typical of bare ground in open country, but its status is incompletely understood. It may be partly or mainly resident in the highlands and adjacent areas, from Amboseli 88C N to Nandi 49C, Laikipia 50 and Mt Kenya 63A (Br; Ja; Be; Se). Range here is mainly above 1500 m and within the 500+ mm rainfall regime. This area also contains all breeding records (Apr-June and Dec-Jan: *op. cit.*), the only recent one of which (87D) actually derives from Tanzania (J S S Beesley *in litt.*). Br notes all East African breeding records to be above 1400 m, and suggests birds at lower altitudes to be non-breeding migrants or wanderers.

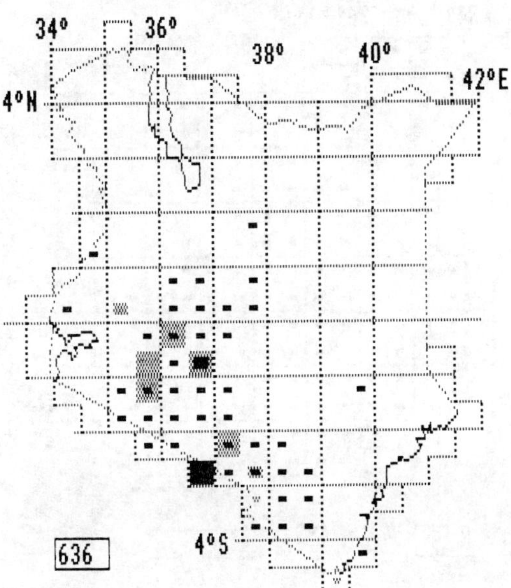

636

Thus virtually all dated records from the Tsavo 101-Mombasa 114 area are Apr-Sept, with the notable exception of presumed nocturnal migrants at the Ngulia lights 101A in Dec (Br; La; EABR 1978, 1982). Similarly, small annual numbers on the N range edge at semiarid Baringo 50A are May-Aug (St), while a lone bird at Korr 39B on 24 May (Lewis 1981e) was probably also part of this movement. This overall Apr-Sept migrant seasonality fits with the single Jul date from S Somalia (AM), and with the 4 June record at Bura 78B (F Alexander *in litt.*). Br mentions seasonal abundance in S Nyanza 60C, without details. There are only three coastal records (Ja; P L Britton *in litt.*).

In contrast to the presumed resident range, these migrants or wandering birds descend to sea level, and move widely into 250-500 mm rainfall areas. Influx into these more arid areas occurs during the Apr-May 'long' rains and presence there continues through the succeeding cooler, lusher months. The single bird at Korr 39B (moisture index below 1.2) occurred when the area was abnormally green after heavy rain (Lewis 1981e).

This wheatear forms a superspecies with the Red-breasted (637) and Isabelline (638) Wheatears (HM). It is allopatric with the former, which is confined to the far NW (N of 2°N), where it may only be a non-breeding visitor (Br). The maps show the Capped and the Isabelline Wheatears to be widely sympatric, but this interspecific contact is to some extent precluded in three ways. Firstly the Capped's apparently resident range is mainly above 1500 m, while the Isabelline is typically below 1800 m (Br). Secondly, migrant Capped Wheatears only enter the sub-1500 m, drier country so favoured by the Isabelline during Apr-Sept (see above), when Palearctic migrants are absent (*cf.* Lack 1985). Thirdly, research in Serengeti (in Tanzania, SW of 74CD) suggests overlap between this bird, the Isabelline and the Northern (639) Wheatears to only occur where the Capped is at medium to low population densities, although even here it still outnumbers the Palearctic birds (Br). Reynolds (1974) suggests competition to occur between this species and its Palearctic congeners in Serengeti. Increase of cultivation in the Mau Narok area 61D has apparently caused a decline of the Capped Wheatear, possibly due to ploughing up of nest holes (Se). Other reference: North 1951.

637 RED-BREASTED WHEATEAR *Oenanthe bottae*
B774 / MPG867 / D2C1

Afrotropical: the N tropics, from Mali E to Ethiopia, and SE through S Sudan and NE Uganda to the SE extremity of the range in NW Kenya.

Kenyan: like the Capped Wheatear (636:*q.v.*), this species frequents bare, open habitats, being particularly attracted to recently burnt ground (Ja; MPG; Br). It is poorly known due to its remote NW range. There are no breeding records, and it may be only a non-breeding visitor from further N (Br).

The very few Kenyan records fall in the period May-Dec. A wanderer S of the normal range was collected at Kisumu 60B on 24 Jul 1917 (Ja; MPG; Keith 1968). A bird on the Loiya Escarpment 25A is undated (Mann 1976). HH found it frequent in June 1974 at Kalom 3B (Ethiopia), and noted occasional birds at Ferguson's Gulf 13B after heavy rain in June-Jul 1974-5. G Nikolaus (*in litt.*) observed it NW of Lokichokio 2C during May and Dec, and later records from there were Jul and Sept (EABR 1983). The records in 12AC and 24B derive from Uganda (J S Ash *in litt.*), where it can be seasonally numerous at Kidepo 11B, and has been recorded Jan, Mar, June, Aug and Dec (Br; EABR 1983). The range mapped here is mainly within intermediate, 250-1000 mm rainfall areas, at 500-1500 m altitude. Excluding the Kisumu 60B wanderer, range is entirely arid-semiarid. See Isabelline Wheatear (638).

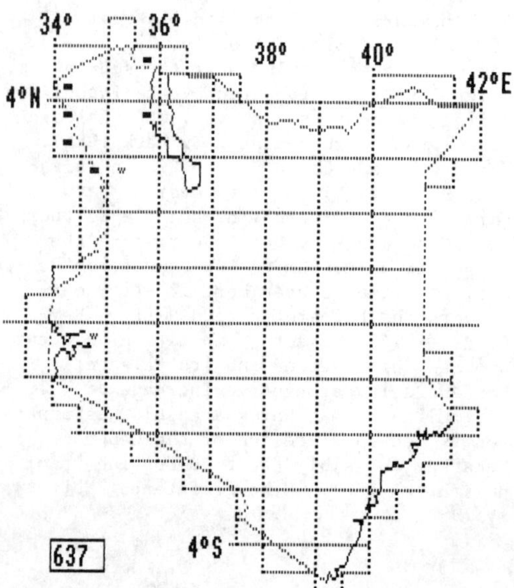

638 ISABELLINE WHEATEAR *Oenanthe isabellina*
B775 / MPG855 / D2C2

Afrotropical: this Palearctic wheatear winters in the N tropics, S to the S periphery of the Sahara in the W, and to Uganda and N Tanzania in the E.

Kenyan: common on bare ground or short grass habitats in open or thinly bushed country, mostly Oct-Mar (Br; La). Although typically in arid-semiarid areas below 1800 m (Br), there are records in all the altitude (see Britton 1979), rainfall and moisture categories. The restriction of records to only 40% of the 0-250 mm rainfall squares and to only 35% of those most arid (moisture index below 1.2) may be real, but exaggerated by low observer frequencies. Wintering birds are common in suitable habitat over much of the country, but noticeable passage is confined to Oct-Nov movement through the L Victoria basin and the coastal lowlands (Br). As in its congeners, local movements are stimulated by changes in the nature of the grass cover, caused by fires or rains (Br).

This species forms part of a superspecies with the Red-breasted Wheatear (637)(HM), and their ranges overlap. Actual sympatry may not be extensive, however, since most dated records for the Red-breasted, and notes of its abundance, are outside the Palearctic season (*cf.* Capped Wheatear (636)). Ecological separation between the Isabelline and the Northern Wheatear (639: *q.v.*) is imperfect, but the latter is more regular at greater altitude in hillier, more wooded and humid country (Br; MPG;

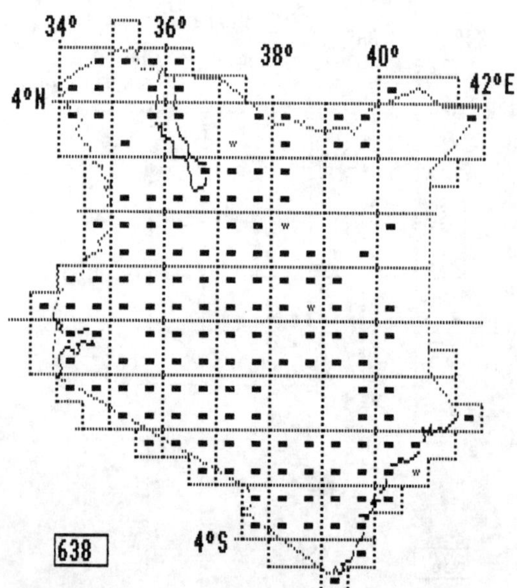

Reynolds 1974; Lack 1985). Abundance at semiarid Baringo 50A exceeds that of the Northern during years of lower rainfall (St). Other references: Morrison 1945; Dittami 1981; Leisler *et al.* 1983.

639 NORTHERN WHEATEAR *Oenanthe oenanthe*
B777 / MPG854 / D2C2

Afrotropical: a Palearctic species which winters S to approximately 8°N in the W whereas, to the E of the Congo basin, it penetrates S to N Zambia. Migrants from the W Nearctic possibly also occur (MPG; Br).

Kenyan: a common winter visitor and passage migrant, mainly Sept-Mar (Br). It favours short grass areas in open, bushed or wooded habitats, up to 3000 m (*op. cit.*; Se; Reynolds 1974). While most widespread during the pronounced Oct migration, a few overwinter in arid and semiarid areas of the N and E, and also in the rift valley highlands (Br; HH; La). At semiarid Baringo 50A, abundance is greatest in wetter years, when it is present throughout the N winter, whereas it is absent Jan-Feb during years of lower rainfall (*cf.* Isabelline Wheatear (638))(St). Single oversummering birds have been recorded in arid Tsavo E 101B Jul 1975 and May-June 1976 (La), and in semiarid Meru NP 52C June 1980 (EABR 1980).

In view of this wheatear's aversion to the driest areas, presence in only 29% of the most arid squares (moisture index below 1.2) and in only 32% of the 0-250 mm rainfall areas could be real. Apparent absence along the E border S of 2°N parallels status in adjacent W Somalia (AM), but may also reflect observer frequency. See Capped Wheatear (636). Other references: Morrison 1945; Dittami 1981; Leisler *et al.* 1983.

640 PIED WHEATEAR *Oenanthe pleschanka*
B779 / MPG859 / D2C2

Afrotropical: a Palearctic species, wintering in the NE from Sudan, Ethiopia and Somalia S to E Uganda and N Tanzania.

Kenyan: an often common bird of open, bushed and lightly wooded country, especially in drier and rocky or stony areas (Br). It is most frequent at moderate, 400-1500 m elevations (*op. cit.*), but wanderers reach 3000 m (Se). It is scarce in the L Victoria basin 60, and particularly so at the coast (Br). Records are mainly Oct-Mar, and oversummering is unknown (EABR). At semiarid Baringo 50A and in arid Tsavo E 101B, this wheatear is distinctly less common during Jan-Feb and Dec-Mar respectively, than in the passage months (St; La).

There are a few records of the white-throated morph (Pearson & Backhurst 1979; Lewis 1983h; EABR 1982, 1984). The Pied Wheatear more frequently perches on vegetation and other low prominences than its Palearctic congeners (Ja). In arid Tsavo E 101B, it is found in denser habitats than these congeners, though not in woodland (La; Lack 1985). Other references: Morrison 1945; Dittami 1981; Leisler *et al.* 1983.

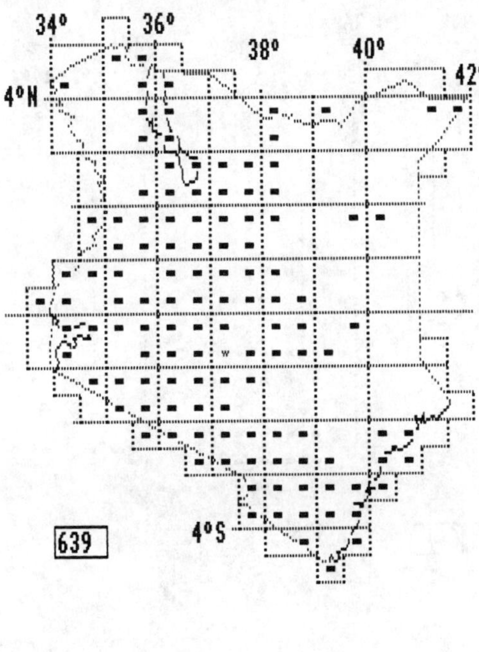

639

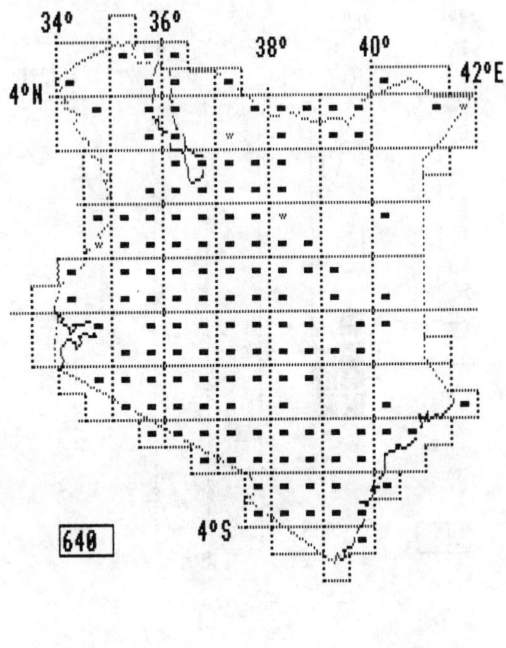

640

641 BLACK-EARED WHEATEAR *Oenanthe hispanica*
B- / MPG858 / D2C0
NO MAP: post presence 75B

Afrotropical: a Palearctic species that winters across the N tropics, S to about 14°N in the W and 12°N in the E. The single record for Kenya was a vagrant well S of the normal range.

Kenyan: a sight record from Nanyuki 51C on 5 Apr 1953 (Wallace 1975) was not acceptable to Br. The first accepted East African record involved an adult male of the white-throated morph of *O.h. melanoleuca*, at Athi River 75B 23 Mar-5 Apr 1984 (Backhurst *et al.* 1984).

642 DESERT WHEATEAR *Oenanthe deserti*
B- / MPG857 / D2C0
NO MAP: post presence 80C

Afrotropical: a Palearctic species that winters in the N tropics; it is found in SW Mauritania, and E from L Chad to C Sudan, Ethiopia and 1°N in Somalia. There is a single record S to Kenya.

Kenyan: the single acceptable record concerns two adult males at Kiunga 80C on 17 Feb 1984 (Backhurst 1985b). This is immediately SW of the species' range in Somalia (Ash 1981a). An earlier record from L Turkana has not been properly documented (Br).

643 MOURNING WHEATEAR *Oenanthe lugens*
B776 / MPG865 / D2C3

Afrotropical: the distinctive race *O.l. schalowi* (Schalow's Wheatear) is endemic to the rift valley and adjacent highlands of S Kenya and N Tanzania. Other races occur in N areas of Ethiopia and Somalia.

Kenyan: locally common, largely confined to dry water courses or boulder-strewn slopes, in bushed country at 1200-3000 m (Br; Ja). Range is 85% semiarid-subhumid and 85% within the intermediate 250-1000 mm rainfall regime. The records in 86B and 88C are from Tanzania (J S S Beesley *in litt.*). The only evidence of any movement or wandering are two records beyond the normal N range edge at Baringo 50A in Aug 1974 (St). Other reference: Haas 1982.

644 SPROSSER *Luscinia luscinia*
B763 / MPG922 / D1C2

Afrotropical: this Palearctic thrush migrates into E Sudan and Ethiopia during Nov-Dec, and passes S through Kenya to winter mainly from S Tanzania and E Zambia to Botswana and E South Africa.

Kenyan: both the Afrotropical and Kenyan accounts of this species are based mainly upon the data of Pearson (1984b, 1973). The Sprosser is a retiring and unobtrusive bird of dense thickets in bushed and wooded country (Br). Range is mainly along the E edge of the C highlands and plateau, approximately between 37° and 38°30'E. It is less common E to Garissa 65B and W to the rift valley (*e.g.* St), and there are also records from the far W at Siaya 48C and other sites (Br). This total range is 82% above 500 m, 78% semiarid-subhumid, and 81% within 250-1000 mm rainfall areas.

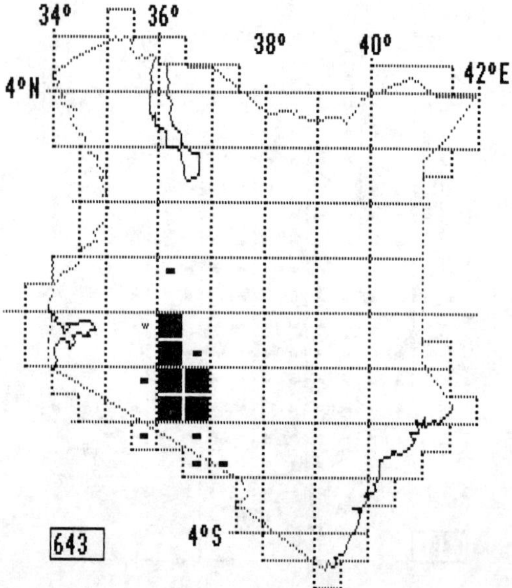

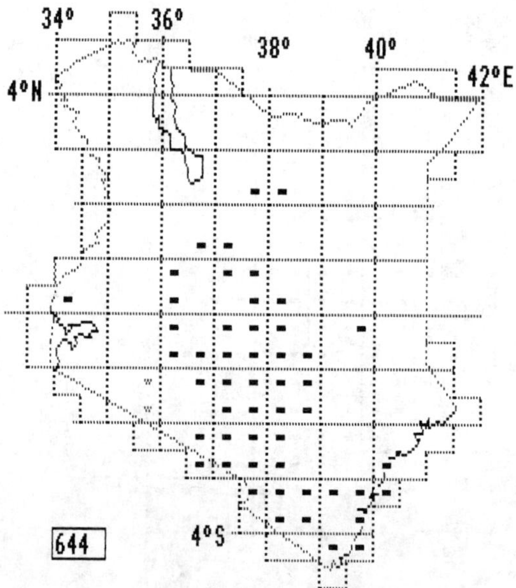

The two annual migrations, and the limited wintering area, show distinctive features. While the first S migrants reach Kenya in Oct, the main arrival is in Nov-Dec. Passage areas are to the E of L Turkana, Laikipia 50 and the Aberdare Mts 62BD, and the species is widespread and common, mainly below 1500 m (Br; OS-C 1977; La; EABR 1983). The E edge of this S movement is well defined, with only occasional records E to Garissa 65B, Kalalu 102A and Maungu 101D, and none from the coast, the lower Tana R or the NE. Very small numbers extend W to the rift valley, *e.g.* three Nov records at L Baringo 50A (St; EABR 1980).

The Sprosser winters in far smaller numbers than the closely related Nightingale (645: *q.v.*), and inhabits moister thickets, hollows and river courses. The main wintering area is at 500-1500 m, S from Meru 52C to Nairobi 75B and Namanga 87D in the W, and Kitui 77A, Machakos 76C, Kibwezi 88B, Voi 101B and the Taita Hills 101A in the E (Pearson 1984b; EABR 1980). This region is mostly semiarid. Kenya represents the N extremity of the Afrotropical wintering range.

N passage is less marked and shorter than the S migration, and more birds presumably overfly Kenya. N migration is regular along the E edge of the highlands, and in some years at the coast (EABR 1983). Occasional 'falls' of migrants during this movement have been recorded at Mtito Andei 89C (Pearson 1981d) and Ngulia 101A (Britton & Britton 1977). This N migration is further E than the S one, as in several of the falcons (see Hobby (152)). Birds ringed at Ngulia Lodge 101A have been recovered in Syria, Lebanon and the USSR (Backhurst 1977, 1981b, 1985a).

645 NIGHTINGALE *Luscinia megarhynchos*
B764 / MPG921 / D1C2

Afrotropical: a Palearctic species, wintering S to the N edge of the Congo basin in the W, and from S Sudan, Ethiopia and S Somalia S to Uganda and N Tanzania in the E.

Kenyan: both the Afrotropical and Kenyan accounts of this species are based mainly on the data of Pearson (1984b). The Nightingale is a reasonably common but unobtrusive bird of dense thicket and other cover in bushed and wooded localities, up to 1800 m. Most records are Nov-Mar and, as in the closely related Sprosser (644: *q.v.*), S passage is the more marked. In contrast to the Sprosser, however, the Nightingale is primarily a winter visitor near the S limits of its Afrotropical

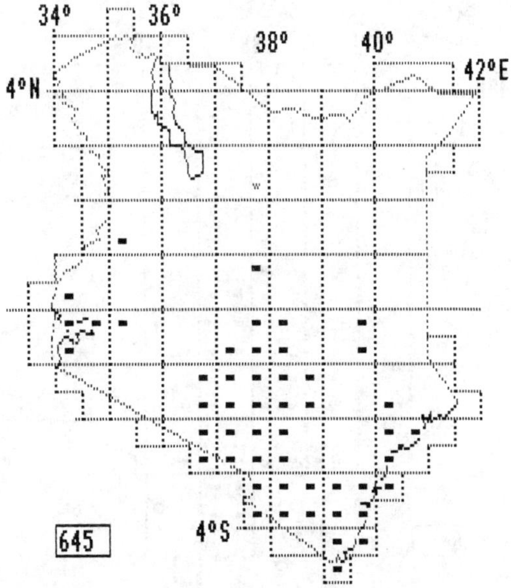

range, with apparently discrete populations W and E of the highlands.

The Nightingale is most common on the E and SE edges of the C highlands, SE to the coast. Range here is 93% below 1500 m, 75% semiarid-dry subhumid and 84% within 250-1000 mm, intermediate rainfall areas. Penetration of arid country along the Tana R 65BD 79C parallels restriction to the vicinity of the Juba and Shebelle R of adjacent S Somalia (AM). These birds in SE Kenya are thought to originate in Asia.

The other wintering area (known from only a few records) is in the L Victoria basin at 1000-2000 m, in moist subhumid-humid, 1000+ mm rainfall areas. These birds, together with those in adjacent areas of Uganda, possibly represent the S wintering limit of populations from SE Europe and/or Asia Minor. The record in 37C was a passage bird caught at Chepereria in Mar 1972 (T J Barnley *in litt.*), while occurrence on Marsabit 27D is mentioned by van Someren (1929).

646 IRANIA *Irania gutturalis*
B762 / MPG920 / D2C2

Afrotropical: this Palearctic thrush migrates from SW Asia to winter quarters that are apparently confined to Kenya and NE Tanzania. It is unknown, even on passage, in Uganda and S Somalia (Br; AM).

Kenyan: both the Afrotropical and Kenyan accounts of this species are based mainly on Pearson (1984b). The Irania is a reasonably common but retiring bird of low

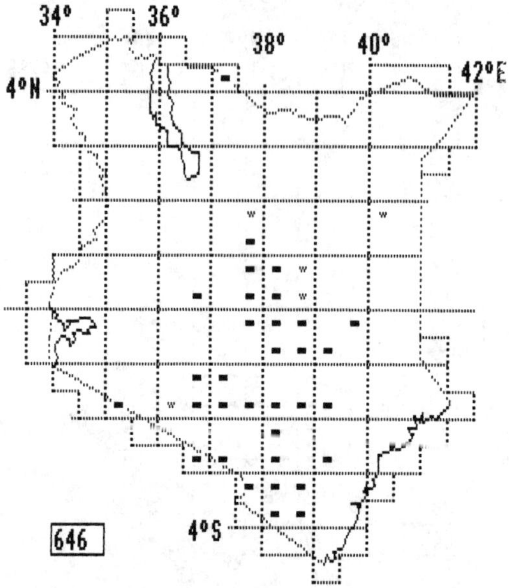

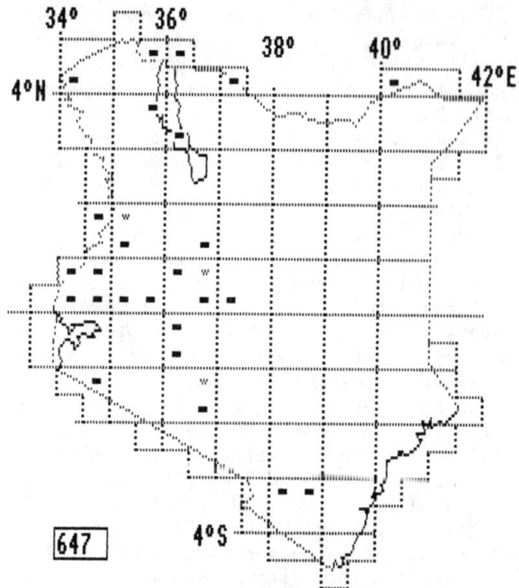

thickets in bushed country, often along gullies and dry watercourses (Ja; Ba). It seems to prefer drier country than the Sprosser (644) and Nightingale (645), but does not extend out into the most arid lowlands.

S passage is Nov-early Jan, and many are caught at Ngulia Lodge 101A (EABR). Wintering is principally on the plateau N and E of the C highlands, from 39D and 42A S to 87D and 101, at 300-1500 m. This range is 81% within the intermediate, 250-1000 mm rainfall regime. N passage is late Mar-Apr. There are no records from the L Turkana basin, or the coastal strip. The record in 5C represents passage in Ethiopia (J S Ash *in litt.*), and 52BD are from Tomlinson (1946). Other reference: OS-C 1977.

647 REDSTART *Phoenicurus phoenicurus*
B781 / MPG917 / D2C2

Afrotropical: a Palearctic species that winters across the N tropics from Senegambia to Ethiopia, S to about 9°N in the W, and to S Sudan, N Uganda and NW Kenya in the E. Distribution in Kenya represents the SE limit of the range, apart from two old records in S Somalia (AM).

Kenyan: distribution can be divided into two regions. Locally common in NE Uganda (Br), the Redstart is principally a bird of the NW and W of Kenya. It can be frequent during N passage through the L Turkana basin (HH), and is annual in small numbers at Maralal 38D (EABR 1981; D J Pearson pers. comm.) and S to the equator in the far W 48 (Ba; Britton & Britton 1973a). Records further to the SE involve scarce stragglers, SE to Ngulia Lodge 101A (EABR 1977, 1981) and Voi 101B (Br). Excluding the passage migrants in the Turkana basin, the Kenyan range is 76% at 1000-2000 m (*cf.* Br), 76% subhumid-humid and 95% within 500+ mm rainfall areas. Br describes the wintering habitat as dry woodland, thicket and bushland.

The second area of occurrence concerns the records along the N border 3B 4A 5C 6C, which derive from Ethiopia (J S Ash *in litt.*) and represent the S edge of the Ethiopian highlands population. Those in 5C and 6C were the race *P.p. samamisicus* (the White-winged Redstart) of Asia Minor, while all records for East Africa proper have involved the nominate race, which originates further to the N and NW in the Palearctic.

648 RED-TAILED CHAT *Cercomela familiaris*
B738 / MPG871 / D2C1

Afrotropical: widely distributed in all but the most interior areas S of 5°S. Further N, it extends through East Africa and the E Zaire borders to Sudan and Ethiopia. Another population occurs in the W, from the Central African Republic to Burkina Faso. Distribution in Kenya represents the E limit of the range.

Kenyan: rocky hills and scarps in open wooded and bushed country, mostly at 400-1600 m (Br). It is recorded throughout the rainfall and moisture categories, but

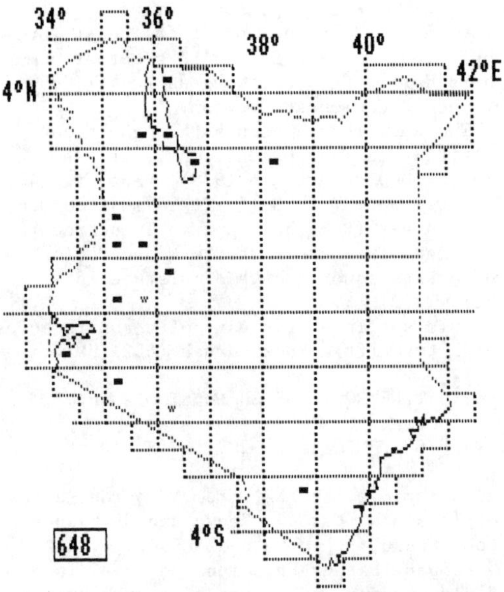

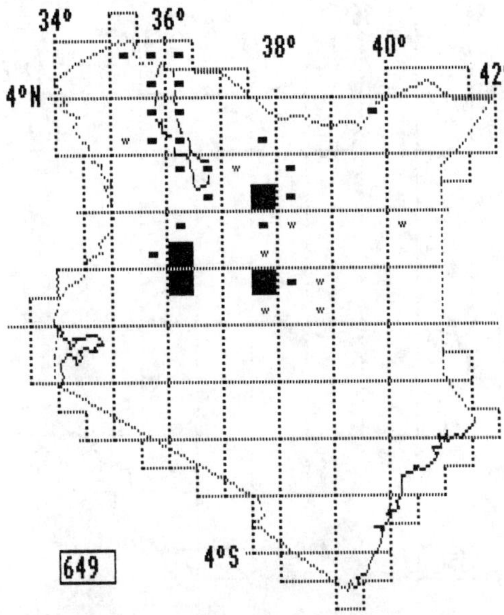

distribution is patchy. The few records along Kenya's SW border 60C 74A 75C 101B apparently represent the NE range edge of the race *C.f. falkensteini* (Br). There have been several sightings on the Ololololo Scarp 74A (EABR 1981, 1983), and MPG mention breeding in SW Kenya but omit details. There is no other Kenyan breeding record for the species. This chat is also locally common in the L Turkana basin, particularly the NE, which probably represents the S range edge of *C.f. omoensis* (HH; Br).

Br considers this species resident, but suggests records between these N and S areas to represent wanderers, including one at 3000 m at Timboroa 49D. It is curious that this chat should be uncommon in S Kenya, where it lacks obvious competitors (Br), yet more common at Turkana where sympatric with the Brown-tailed Rock Chat (649). This latter species forms a superspecies with the Red-tailed, and replaces it further to the NE (HM). These two chats are possibly in the same species-group as the Alpine Chat (650), which is segregated from them altitudinally (HM).

649 BROWN-TAILED ROCK CHAT *Cercomela scotocerca*
B739 / MPG870 / D1C1

Afrotropical: a bird of the NE, with a fragmented distribution in NE Uganda, W and NE Sudan, Ethiopia, N Somalia and, at the S limit of the range, N Kenya.

Kenyan: locally common in open, rocky, and usually bushed habitats, at 400-1200 m (HH; Br). It occurs S to the N edges of the highlands, with range 81% below 1000 m, 92% arid-semiarid and 80% within 0-500 mm rainfall areas. The record in 17B derives from Ethiopia (J S Ash *in litt.*). The cluster of old records around 39 40 42 52 probably reflects a D1 species in infrequently observed areas. It is considered resident (St; Br; EABR 1979, 1983, 1984). See Red-tailed Chat (648).

650 ALPINE CHAT *Cercomela sordida*
B740 / MPG873 / D2C3

Afrotropical: in alpine habitats of N Tanzania, E Uganda, Kenya and Ethiopia.

Kenyan: a relatively common resident on Mt Elgon 36D, the Cherangani Hills 37CD, the Aberdare Mts 62BD and Mt Kenya 63A (Br). It is characteristic of alpine moorland, from 4400 m down to the tree line, but has also been encountered in forest glades as low as 2300 m (Ja; Br). Five of six squares are humid, and receive over 1000 mm rainfall. Presence on the Kinangop Plateau 62D (Ja) requires confirmation (Br). See Red-tailed Chat (648). Other references: Pollard 1946a; Pitman 1956.

(map overleaf)

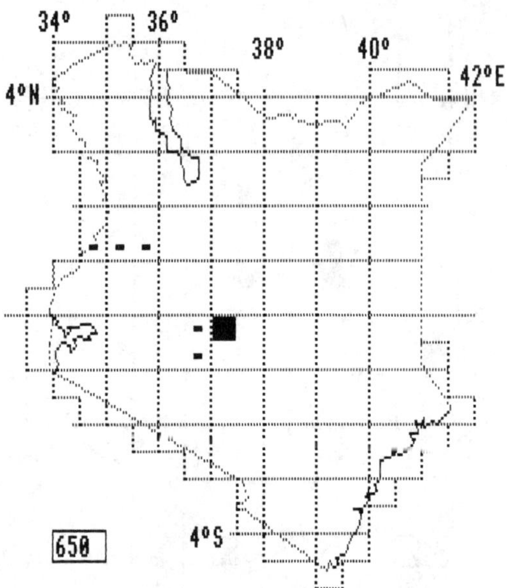

651 ROCK THRUSH *Monticola saxatilis*
B769 / MPG850 / D3C3

Afrotropical: a Palearctic species which winters in the N tropics, S to about 8°N in the W and 10°S in the E.

Kenyan: a locally common to common passage migrant and winter visitor, usually in open or sparsely bushed grasslands where trees, rocks or buildings provide elevated perches (Br). Occurrence is mainly Nov-Apr, with regular passage during Oct at L Turkana, and in Mar-Apr at the coast (Br; HH). Both Ja and St record greatest numbers during Feb-Mar, whereas Lack (1985) found it commonest Dec-Jan.

The Rock Thrush occurs throughout our altitude categories but, while it can reach 3000 m (Britton 1979), range is 74% below 1500 m (*cf*. Br). It is most common in E areas (Br), but apparently scarce in the most arid areas of the N and NE (only 6% of its squares have a moisture index below 1.2).

This species forms part of a superspecies with the Little Rock Thrush (652)(HM).

652 LITTLE ROCK THRUSH *Monticola rufocinerea*
B768 / MPG853 / D2C2

Afrotropical: the NE, from NE border areas of Tanzania, through Kenya and NE Uganda to S Sudan, Ethiopia and N Somalia.

Kenyan: rather uncommon and local in diverse rocky, open, wooded and thicketed habitats, at 1400-2500 m (Br). Breeding occurs in buildings at Nanyuki 51C (ADL pers. obs.). Range is 76% subhumid-humid and 84% within 500+ mm rainfall areas. The relatively high proportion of pre-1970 squares (27%) probably reflects its inconspicuous character: D2 is awarded because of its distinctive appearance. Be considered it resident near Narok 74B, but both Ja and MPG believed it to migrate. This latter view may well be correct but no details of movements are known (Br). The few published records show no obvious sea-

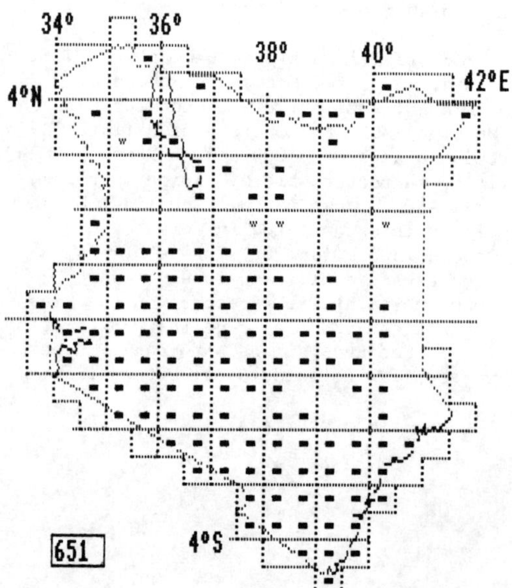

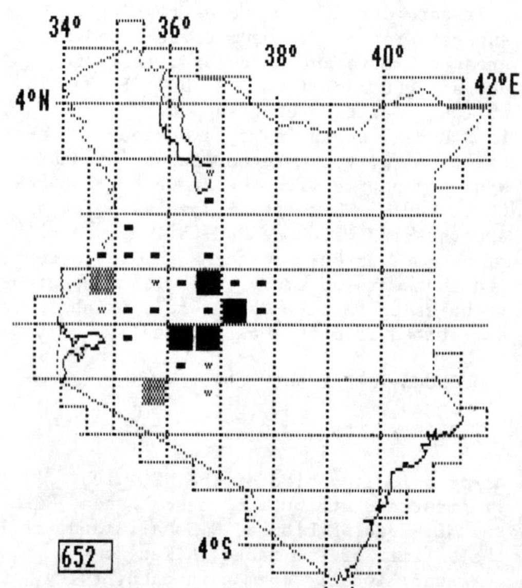

sonality (EABR). HM's suggestion that it is a non-breeding migrant in Kenya is clearly incorrect.

653 ANTEATER CHAT *Myrmecocichla aethiops*
B770 / MPG881 / D2C2

Afrotropical: principally the N tropics, N of 10°N from Senegambia E to S Sudan. The population in East Africa, from the highlands of W and C Kenya S to NE border areas of Tanzania, is isolated.

Kenyan: a common resident of open, grassy, bushed or wooded habitats, often associated with termite mounds (Br; Se; Be). It is a highland species at 1500-3000 m, but mainly above 1700 m (*op. cit.*). Range is 83% subhumid-humid and 94% within 500+ mm rainfall areas.

The Anteater Chat forms part of a species-group with the Sooty Chat (654), which replaces it in the W (HM). The two are allopatric but their ranges meet and possibly overlap at Lemek 74A (Br). The Anteater is very similar ecologically to the Sooty, but apparently prefers rather drier country (Br). Other references: Haas 1982, 1985.

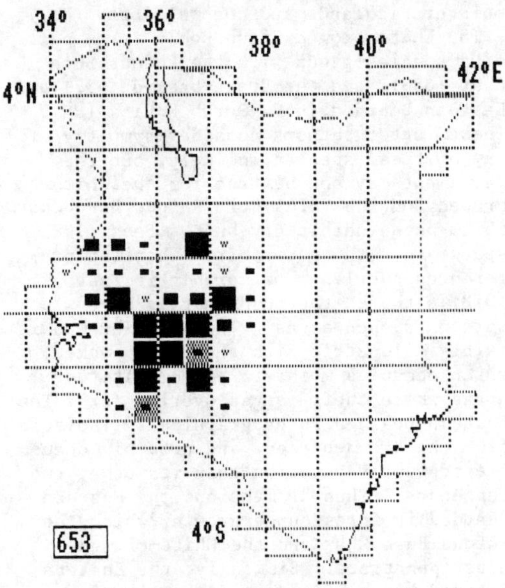

654 SOOTY CHAT *Myrmecocichla nigra*
B771 / MPG880 / D2C2

Afrotropical: widespread S of the Congo basin to 15°S. Further N, it extends through Tanzania, W Kenya and Uganda, NW to S Sudan and Cameroon. Distribution in Kenya represents the E limit of the range.

Kenyan: locally common in open, short

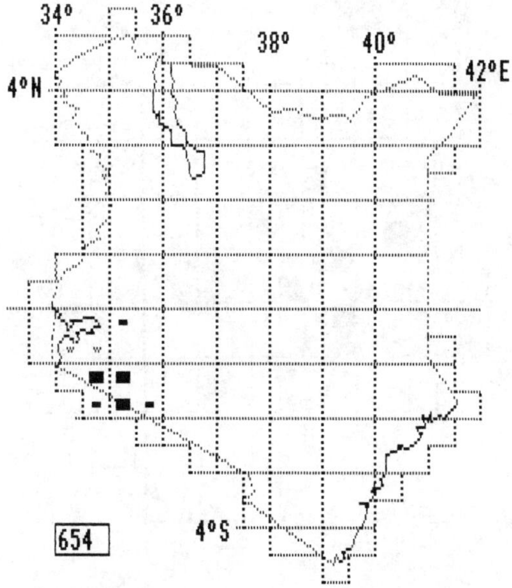

grass country, with some bush, at 1200-1700 m (Ja; Br; Brown & Britton 1980). Like the Anteater Chat (653: *q.v.*), it is frequently associated with termite hills (*op. cit.*). Range is entirely within subhumid-humid, 500+ mm rainfall areas. There is no confirmed breeding record. Ja notes occurrence on the N range edge at Fort Ternan, the Nyando Valley and Kericho (all 61A), but there is only a single recent record from this area (EABR 1981), despite the presence of resident observers. Nearly all records derive from the Lolgorien-Mara area of 73BD 74AC. Br notes the absence of records to the SW of the Kenyan population, but thinks it unlikely that the latter is disjunct.

655 CLIFF CHAT *Thamnolaea cinnamomeiventris*
B792 / MPG876 / D2C2

Afrotropical: the N tropics from Mali and N Nigeria to Ethiopia, and thence S through East Africa, Malawi and Zimbabwe to E South Africa. Distribution in Kenya represents the E limit of the range.

Kenyan: a local and uncommon species of slopes and scarps with sheer rock faces (Br; Ja). Br records this chat at 600-2200 m, but range is in fact 76% above 1000 m and 69% in 500+ mm rainfall areas, *i.e.* most squares are in the high rainfall highlands and their peripheral plateaux. Absence from most of the N and E, despite abundant rocky habitat and lack of obvious competitors (Br), presumably reflects

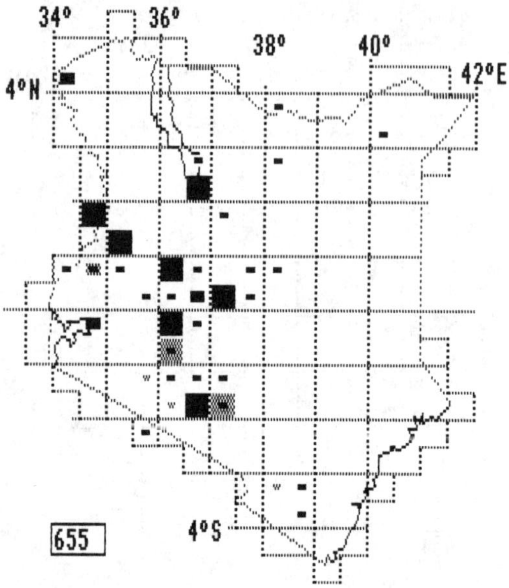

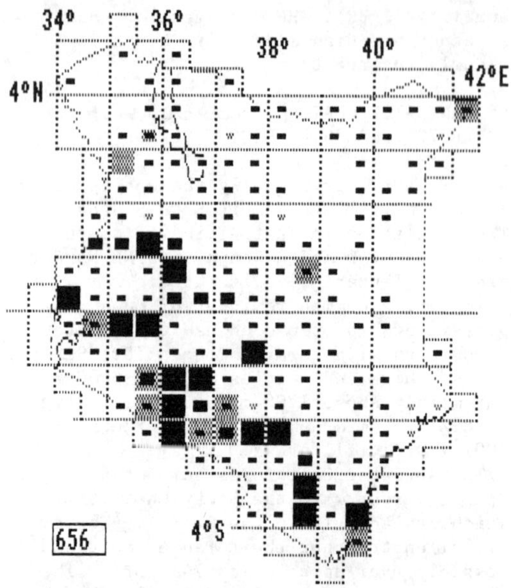

aversion to arid conditions. Range is only 18% arid, and the four arid squares in the N are all near or at locally lush areas, *i.e.* 16A (J S Ash *in litt.*) on the S edge of the Ethiopian highlands; 18C near dams at Takaba (E J Roberts *in litt.*); 26B on Mt Kulal; and 28A adjacent to Mt Marsabit 27D 28C. The E lowlands, including the coastal strip, are beyond the E range edge. The Cliff Chat is considered resident (Br; St).

656 WHITE-BROWED SCRUB ROBIN *Cercotrichas leucophrys*
B744 / MPG910, 911 / D2C3

Afrotropical: widespread S of 5°S, except for the dry SW. Further N, it extends to areas of the Congo basin, and through East Africa to S Sudan, Ethiopia and Somalia.

Kenyan: this common but rather retiring species frequents a very wide variety of bushed and wooded habitats with some ground cover, but avoids forests (Br; Lack 1985). It is found up to 2200 m, but mainly below 1500 m (*op. cit.*). Range extends throughout our altitude, rainfall and moisture categories, and it could probably occur in any square. There is no evidence of movements (Br; La; St; DEP pers. obs.), and virtual absence of breeding records from the N and E most probably reflects lower observer frequencies.

This species has red-backed and white-winged forms that display little intergradation and have been considered full species *e.g.* by MPG. Their taxonomic status is still uncertain. The White-winged Scrub Robin is found in the drier N and E, while the Red-backed is in the moister highlands and coastal strip (Br).

The White-browed Scrub Robin forms part of a species-group with the Rufous Bush Chat (658), and the Brown-backed (657) and Eastern Bearded (659) Scrub Robins (HM). These latter authors consider sympatry of any of these species unlikely, but the statement may not be true for the White-browed and the Palearctic Rufous Bush Chat, which both inhabit dry bush. Lack (1985) states that these two birds' habitat preferences overlap considerably in Tsavo E 101B, although the Bush Chat is more widespread in open areas. The White-browed is mainly allopatric with the Brown-backed, which occupies moister, higher altitude areas where their ranges overlap (Br). The Eastern Bearded is geographically isolated from the Brown-backed, and probably occupies denser vegetation than its other two congeners. Thus Br mentions the Eastern Bearded in forest undergrowth, which the Rufous Bush Chat and the White-browed never penetrate. Similarly, the Eastern Bearded was confined to dense riverine growth in Tsavo E 101B, with the White-browed in other dense habitats (La).

657 BROWN-BACKED SCRUB ROBIN *Cercotrichas hartlaubi*
B743 / MPG914 / D2C2

Afrotropical: from E Zaire to Uganda, NW Tanzania and, at the E limit of the range,

Kenya. There is another population in Cameroon.

Kenyan: a rather scarce resident of scrub, garden, tall grass and forest edge habitats, at 900-2200 m in the highlands (Br; Ja). Range is 83% moist subhumid-humid and 66% within 1000+ mm rainfall areas. There has been apparent decline in the Nairobi area 75B (Br), and the small cluster of old records 62AB 63C 74B suggest that this may be more general. See White-browed Scrub Robin (656).

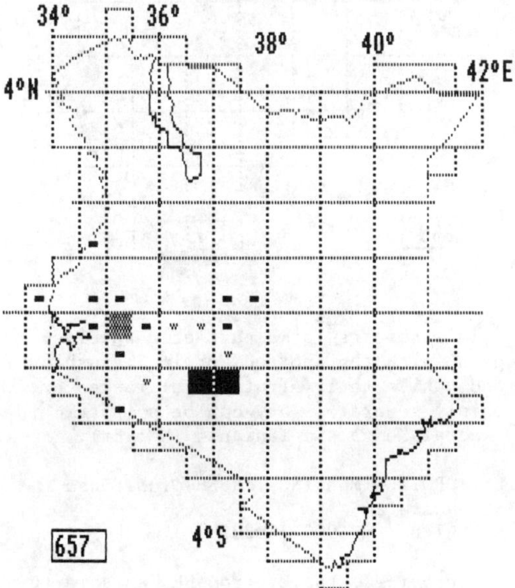

658 RUFOUS BUSH CHAT *Cercotrichas galactotes*
B742 / MPG934 / D2C2

Afrotropical: resident races extend across the N tropics, mostly N of 10°N, but further S in Ethiopia, Somalia and possibly N Kenya. Races from the Palearctic winter in the Sahel from Senegambia to Nigeria, and also in the E from Sudan, Ethiopia and Somalia S to Kenya and extreme NE Tanzania.

Kenyan: this species has possibly two modes of occurrence in Kenya. HH described the Afrotropical race *C.g. minor* as a common resident in woodland and thick scrub in the Ferguson's Gulf 13B and Turkwell Delta 14C areas of L Turkana. These areas would represent the S limit of this race's range. There have been no subsequent sightings however, and we agree with Br that these records require confirmation.

The Rufous Bush Chat is thus primarily a Palearctic passage migrant and winter visitor, near the S limit of its Afrotropical range. It is a locally common to common but retiring bird of dry, bushed habitats, typically below 1000 m, mainly Nov-Apr (Br; La). Range is 86% arid-semiarid and 74% within intermediate, 250-1000 mm rainfall areas.

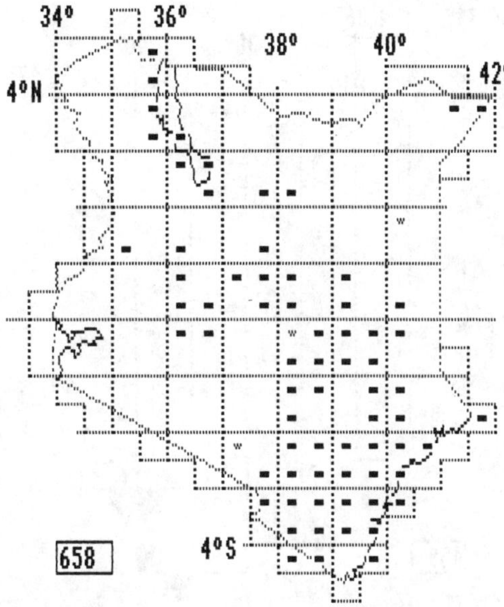

It is most frequent in the E and SE lowlands where it winters W to Marsabit 27D 28C, Baringo 50A and Samburu 51B, and SE from Garissa 65B, Kitui 77A, Mtito Andei 89C and Taveta 100B to the coast (Br; La; EABR). Nov passage is regular at Samburu 51B and in Tsavo 89 101 (Br). This chat could well be widespread in the NE (*i.e.* around 29), but these areas have received scant coverage, particularly during the Palearctic season (Ba; EABR 1977). It occurs as a common migrant Mar-Apr at L Turkana (HH). Virtual absence W of 35°30'E represents the W range edge, there being only a single record from Uganda (Br). See White-browed Scrub Robin (656).

659 EASTERN BEARDED SCRUB ROBIN *Cercotrichas quadrivirgata*
B746 / MPG913 / D2C2

Afrotropical: the E, from S Zambia and Mozambique to Malawi, E Tanzania, Kenya and S Somalia.

Kenyan: a locally common (*e.g.* see Britton & Zimmerman 1979) but retiring resident of dense thickets and undergrowth in bushed, wooded and forested areas below 1000 m (Br). Range is 68% below 500 m and 78% within 500+ mm rainfall areas, and

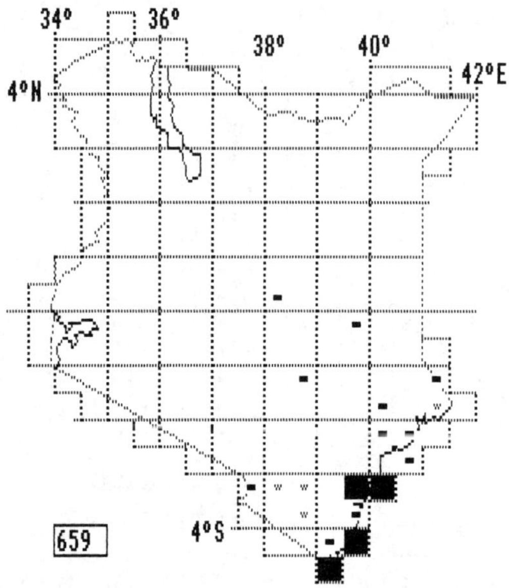

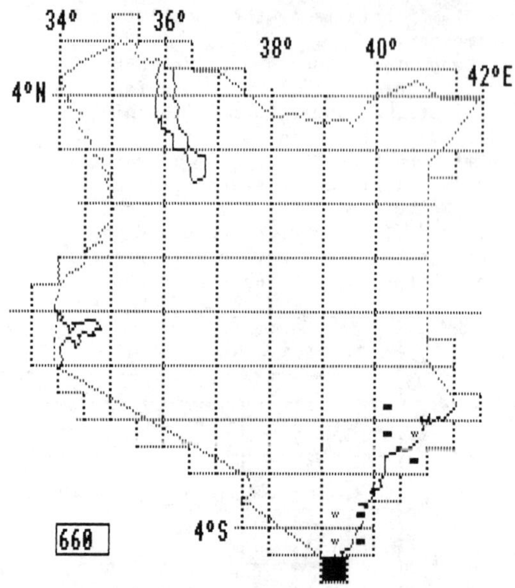

includes the Mt Endau forest island 77B (A W Diamond, D J Pearson *in litt.*).

Although typically a bird of the coastal strip, this scrub robin has penetrated inland along the Tana R 65B 79C, to reach the E edge of the C highlands 52C (*cf.* Brown-breasted Barbet (478) and Black-breasted Glossy Starling (877)). Range in adjacent S Somalia is similarly confined to the Juba R and coastal areas (including the record in 80A: AM). Occurrence at Garissa 65B was reported by Serle (1943), while R & L M Greenham (*in litt.*) recorded the species there in Feb-Mar 1974. 52C represents birds caught in Ngaia Forest (Br). The old records in 101 may reflect a decline, since La note recent disappearance from riverine habitats in Tsavo E 101B. See White-browed Scrub Robin (656). Other reference: Short & Horne 1985a.

660 MORNING THRUSH *Cichladusa arquata*
B747 / MPG908 / D2C2

Afrotropical: the E, from S Mozambique, Zimbabwe and Zambia to S Zaire, SW Uganda, Tanzania and, on the NE extremity of the range, coastal Kenya.

Kenyan: a generally scarce resident of the coastal lowlands and the lower Tana R 79C, frequenting thickets in the vicinity of *Borassus*, *Hyphaene* or coconut palms (Br). Range is 77% semiarid-dry subhumid, and entirely below 500 m in 500+ mm rainfall areas. There is no obvious reason for the lack of records from the reasonably well known Sokoke 102B and Malindi 103A areas.

This species forms part of a species-group with the Spotted Morning Thrush (661) and, while their Afrotropical ranges are mainly separate, they can be sympatric in coastal Kenya and Tanzania (HM; Br).

661 SPOTTED MORNING THRUSH *Cichladusa guttata*
B748 / MPG909 / D2C2

Afrotropical: the E, from SE Tanzania to Kenya, N Uganda and S areas of Sudan,

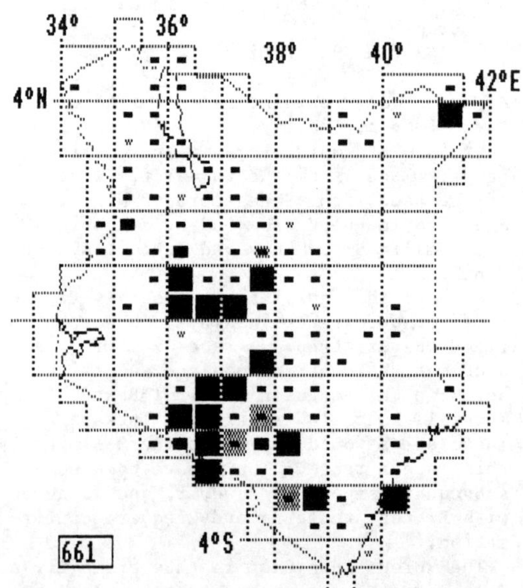

Ethiopia and Somalia.

Kenyan: a common but somewhat retiring resident of dense thicket and scrub in bushed country, up to 1600 m (Br; HH; St; Lack 1985). In contrast to the Morning Thrush (660: *q.v.*), range is 72% within the intermediate, 250-1000 mm rainfall regime. Hence it is scarce in the highlands and humid W. There are records from only 18% of the most arid squares (moisture index below 1.2) and only 26% of those with 0-250 mm rainfall, and this absence is probably real. This is supported by lack of records from an observer resident at Wajir 42A, and by apparently similar status in adjacent areas of Somalia (AM).

662 WHITE-STARRED FOREST ROBIN *Pogonocichla stellata*
B782 / MPG 915 / D1C2

Afrotropical: the E, from the Cape through Mozambique and Malawi to S Tanzania. Further N, it extends NW along the E Zaire borders to SW Uganda, and NE through E Tanzania to Kenya and S Sudan. Distribution in Kenya represents the NE limit of the range.

Kenyan: a locally common but retiring species of low cover in forest, forest edge and bamboo (Br). Range includes the forest islands of Mts Kulal 26B, Nyiru 26D (Diamond & Keith 1980), Uraguess 51A and Kasigau 101D, and the Chyulu 88D and Taita 101A Hills (Br).

Br quoted range at 1600-3300 m, but this estimate has been markedly revised by the recent discovery of a population on Mrima Hill 114A, at 300 m (Britton *et al.* 1980; EABR 1980). This compares with the same race's altitudinal range on the East Usambara Mts of adjacent NE Tanzania (EABR 1981). Confirmed breeding in 114A is based on very young immatures on Mrima (*op. cit.*). Excluding the forest islands, which each have their own localised moist climate, the species' range is entirely subhumid-humid and within 500+ mm rainfall areas: 67% of the squares receive over 1000 mm.

The large number of races (eight) in East Africa suggest this to be a sedentary species (Br), *i.e.* each montane population is evolving in isolation from its neighbours. Endemic races on Mt Elgon 36D and the Chyulu Hills 88D (Br) particularly illustrate this point. Se considered the species resident at Mau Narok 61D.

663 BROWN-CHESTED ALETHE *Alethe poliocephala*
B736 / MPG903 / D1C2

Afrotropical: a bird of W and C Africa, with a fragmentary distribution around the Gulf of Guinea, and along the E Zaire borders, E to W Tanzania, Uganda, S Sudan and, at the E extremity of the range, Kenya.

Kenyan: common but retiring in highland forest undergrowth, up to 2800 m (Br). Range is 74% above 1500 m, 91% moist subhumid-humid, and 83% within 1000+ mm rainfall areas. It does not extend to forest islands. *A.p. carruthersi* occurs to the W of the rift valley, while *A.p.*

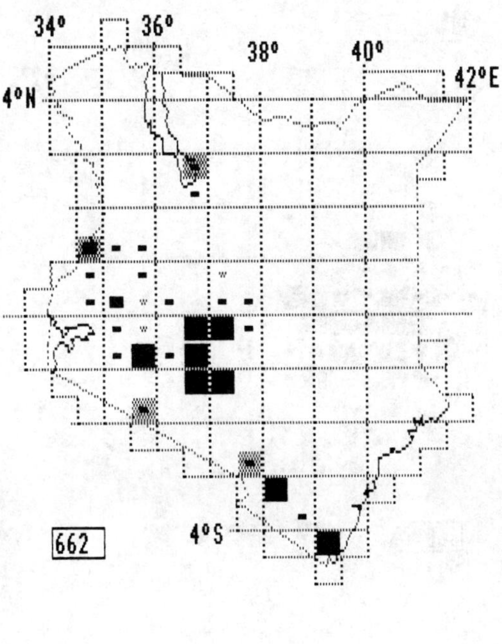

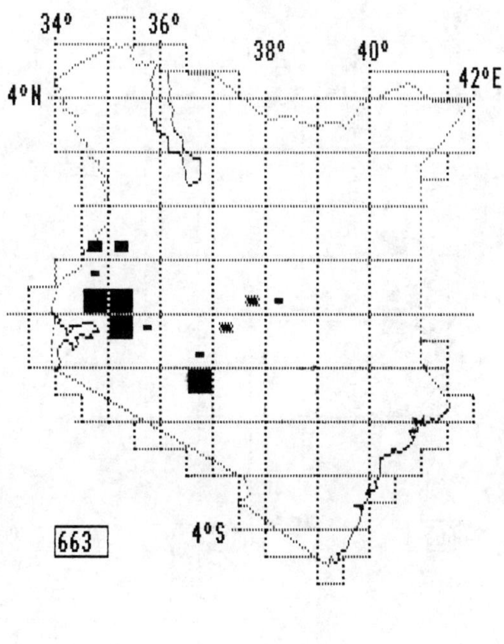

akeleyae is to the E (Br). Other references: Mann 1974d; Cunningham-van Someren 1977a; EABR 1978.

664 FOREST ROBIN *Stiphrornis erythrothorax*
B789 / MPG900 / D1C0
NO MAP: pre presence 49C

Afrotropical: W and C Africa, from the lowland forests around the Gulf of Guinea, E through the Congo basin to S Sudan and C Uganda.

Kenyan: this species is resident in forest undergrowth in Uganda (Br). The single Kenyan record, a female collected near Kipkabus 49C on 16 Apr 1966 (Zimmerman 1967), represents a vagrant over 200 km further E, 700 m above its Ugandan altitudinal limit (Br).

665 SNOWY-HEADED ROBIN CHAT *Cossypha niveicapilla*
B753 / MPG892 / D2C2

Afrotropical: W and C Africa, from Senegambia and Congo on the W coast, through the Congo basin, to the E limits of the range in W Ethiopia and W Kenya.

Kenyan: a locally common but retiring species of low cover in a wide variety of wooded and forested habitats, including gardens (Br; Ja; EABR 1981). It is mainly at 1500-2000 m (Br), and entirely to the W of the rift valley. The squares are 87% humid, and entirely within 1000+ mm rainfall areas. See White-browed Robin Chat (666). Other references: Sugg & Sugg 1971, 1972.

666 WHITE-BROWED ROBIN CHAT *Cossypha heuglini*
B751 / MPG884 / D2C2

Afrotropical: widespread to the S of the Congo forests, avoiding only the dry SW. Further N, it extends through East Africa to S areas of Sudan, Ethiopia and Somalia, and NW to Chad.

Kenyan: this common but retiring bird is easily the most widespread robin chat in Kenya (Br; Ja). It favours low cover in a wide variety of well vegetated habitats up to 2200 m, but is rarely in forest (*op. cit.*). Range is 81% within 500+ mm rainfall areas, and includes the isolated highlands of Mts Loima 13C, Nyiru 26D and Marsabit 27D 28C, the Ndoto Mts 39A, the Mathews Range 39C and the Taita Hills 101A. Range is only 10% arid, so that occurrence in the N and E is virtually restricted to these highland islands, and to the vicinity of the Tana R 64B 65B 79C. Distribution is similarly confined to the Juba and Shebelle R of S Somalia (AM). It is considered resident (Br; Be: 74; DEP pers. obs.: 74D), but individuals occasionally appear in the drier parts of semiarid Baringo 50A after rains (St). Range has recently contracted from arid Tsavo E 101B (La).

This, Ruppell's (667) and the Blue-shouldered (668) Robin Chats form a superspecies, and are part of a species-group with the Snowy-headed Robin Chat (665)(HM). The ranges of these birds overlap, but actual sympatry is largely avoided by differing ecological preferences (HM; Br).

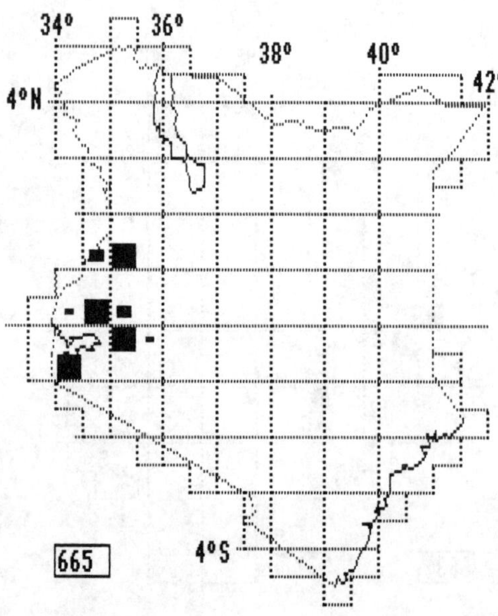

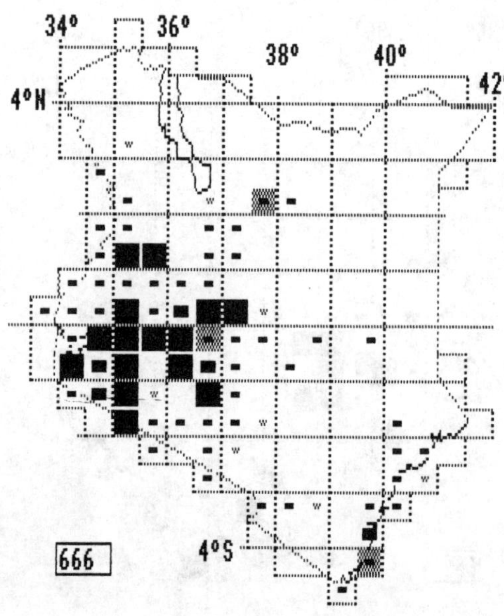

Thus, of the species W of the rift valley, the Blue-shouldered is confined to forest interior, while the Snowy-headed prefers forest edge. Ruppell's is also a forest bird, but is found only E of the rift. The White-browed occurs on both sides of the rift but, while it can occur on forest edge, it generally prefers other bushed and wooded habitats. However, while clearance of dense habitat at Nairobi 75B ought to encourage colonisation by the White-browed, Ruppell's seems to be displacing it locally, and individuals with the physical characteristics of Ruppell's have songs reminiscent of the White-browed (Br). See Robin Chat (670).

667 RUPPELL'S ROBIN CHAT *Cossypha semirufa*
B755 / MPG885 / D2C2

Afrotropical: a highland species of the NE, from N Tanzania through Kenya, to S Sudan and Ethiopia.

Kenyan: another common but skulking resident of dense cover and thicket (Br). Unlike the very closely related White-browed Robin Chat (666: *q.v.*) however, Ruppell's prefers more forested habitats, and is confined to the highlands at 1400-2300 m (Br). Range is entirely in or E of the rift valley, in 500+ mm rainfall areas. It includes the forest islands of Marsabit 27D 28C, the Chyulu Hills 88D, and the Taita Hills 101A. The records at Moyale 17A and Marsabit 27D 28C represent the S edge of the nominate race, which mainly inhabits Ethiopia and S Sudan, while all records further S are *C.s. intercedens* (Br).

668 BLUE-SHOULDERED ROBIN CHAT *Cossypha cyanocampter*
B750 / MPG889 / D1C2
NO MAP: post probable 48D 49C;
pre presence 36D

Afrotropical: the lowland forests of W and C Africa, along the N of the Gulf of Guinea from Sierra Leone to Cameroon. Further E, it is found from NE Zaire to Uganda, S Sudan and, on the E extremity of the range, W Kenya.

Kenyan: a locally common resident of forest undergrowth, particularly near streams, at Elgon 36D, Kakamega 48D, Kaimosi 48D, and Nandi 49C (Br). It is extremely shy and retiring, but is awarded C2 because of the restricted distribution. Range is W of the rift valley, in 1000-2000 m areas to the N of the Winam Gulf 60AB. The three squares all receive over 1000 mm rainfall. There is no confirmed breeding record. Probable breeding in 48D and 49C reflects singing birds on territories. See White-browed Robin Chat (666).

669 RED-CAPPED ROBIN CHAT *Cossypha natalensis*
B752 / MPG890 / D2C2

Afrotropical: widespread to the S of the Congo basin, avoiding only the dry SW. Further N, it occurs through East Africa to S areas of Sudan, Ethiopia and Somalia. There are also scattered records to the N of the Congo basin.

Kenyan: a locally common but retiring bird of thicket and undergrowth in bushed,

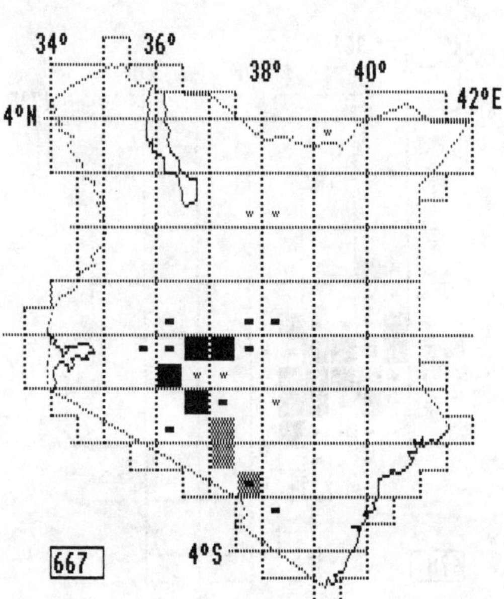

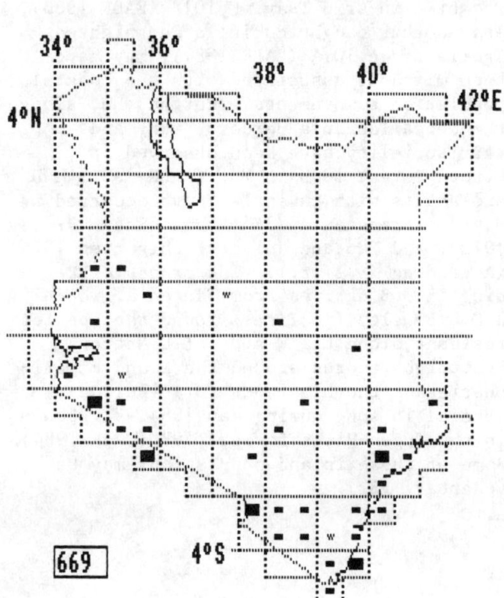

wooded or forested localities (Br). Range is 79% within 500+ mm rainfall areas. Br mentions it up to 2200 m, but it is mostly (78%) below 1500 m, mainly in the coastal lowlands, the Kenya/Tanzania borders, and at scattered sites around the N and E peripheries of the W and C highlands.

Status in Kenya is incompletely understood, and there are no confirmed breeding records. Most is known of its occurrence at the coast (Britton & Rathbun 1978), where it is common Apr-Dec, during and after the 'long' rains, but virtually absent Dec-Mar. This seasonality coincides with that of other migrants from the S (*e.g.* Pygmy Kingfisher (435), African Golden Oriole (572) and Black Cuckoo Shrike (604)), but this robin chat's migration cannot exactly parallel theirs, since it is still numerous at the coast in Sept-Nov, when peak numbers are also recorded in Zambia (Britton 1971b). Thus the location of these coastal birds during Dec-Mar remains uncertain, though they most likely move S to coastal Tanzania and Mozambique, where there are rains in these months. Britton's (1971b) suggestion of migration inland from the coast during Dec-Mar is now discounted. AM presume the species resident S of 3°N in Somalia, and the record in 80A is theirs.

Breeding has never been proved at the coast, despite the occurrence of song and territoriality there in May-June, and of immatures being fed by adults in later months (both types of activity mapped here as probable breeding).

Far less is known of status inland. A Dec bird in arid Tsavo E 101B (EABR 1983), and another caught during a Dec night at Ngulia Lodge 101A (EABR 1981), may have been migrants associated with the coastal population's movements. Further NW, along the Kenya/Tanzania borders, song and territoriality have been observed in Kitovu Forest 100B in Mar (EABR 1981, 1982), and adults with juveniles have occurred in May at Tarang'anya 73B (Sugg & Sugg 1971, 1972), and Feb and Jul near Lolgorien 73B (A Lohding *in litt.*). The probable breeding in 86B derives from Tanzania, where J S S Beesley (*in litt.*) found the species resident along the W side of L Natron. Scattered records around the N and E peripheries of the highlands (Br) include several in song during Jan 1984, at 2500 m in the Tugen Hills forests 49B (EABR 1984). Some of these inland populations may be sedentary (Br).

670 ROBIN CHAT *Cossypha caffra*
B749 / MPG893 / D2C2

Afrotropical: the E and S, throughout South Africa, and N through Zimbabwe and Malawi to S Tanzania. Further N, it extends NW along the E Zaire borders to W areas of Tanzania and Uganda, and NE through E Tanzania to E Uganda, Kenya and S Sudan. Distribution in Kenya represents the E limit of the range.

Kenyan: a common but rather skulking resident of the highlands, typically at 1800-3400 m (Br; Be; Se). It frequents a wide variety of bushed, wooded and forested habitats, including giant heather on alpine moorland (*op. cit.*; Britton & Sugg 1973). Range includes the forest islands of Mts Kulal 26B (Diamond & Keith 1980), Nyiru 26D and Marsabit 27D 28C (Br), the Mathews Range 39C, and the Chyulu 88D and Taita 101A Hills. Apart from these localised peaks, which are too small to appear on our overlays, range is 82% subhumid-humid and 87% within 500+ mm rainfall areas.

The Robin Chat is closely related to the White-browed Robin Chat (666) superspecies (HM). In areas like Kenya, where both the Robin Chat and the White-browed occur, the former is confined to highland areas whereas, in South Africa, where the White-browed is absent, the Robin Chat extends down to sea level (HM). Where these two species overlap, HM consider the Robin Chat to prefer more open habitats, but this distinction is hardly apparent in Kenya (ADL pers. obs.).

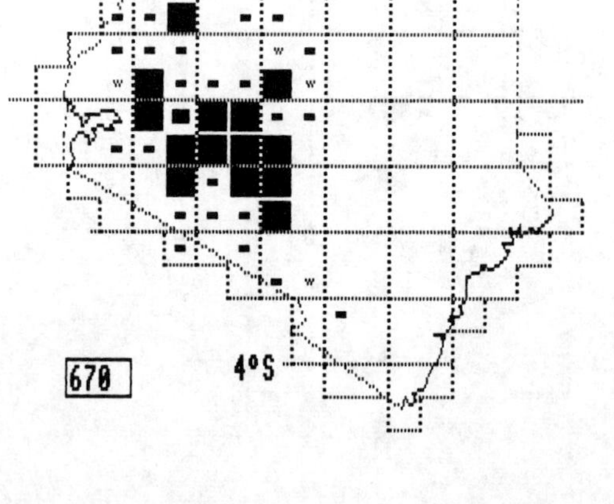

671 GREY-WINGED GROUND ROBIN *Dryocichloides poliopterus*
B761 / MPG887 / D1C1

Afrotropical: an apparently fragmented distribution from Guinea, Cameroon, Angola and Zambia, E to NW Tanzania, S Sudan, Uganda and, on the E extremity of the range, W Kenya.

Kenyan: extremely retiring and skulking in dense forest undergrowth at 1110-2000 m (Br). Sugg & Sugg (1972) found it common along forested streams at Rapogi 60C, but it is typically scarce elsewhere (Br). Range is in subhumid-humid country to the W of the rift valley, 90% within 1000+ mm rainfall areas. The cluster of recent breeding records derive from T J Barnley (*in litt.*), who has observed nesting on Mt Elgon 36D and the Cherangani Hills 37CD. Other reference: EABR 1978.

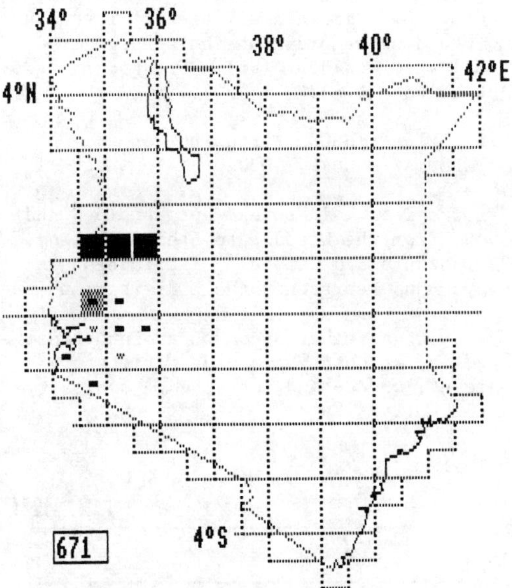

672 EQUATORIAL AKALAT *Sheppardia aequatorialis*
B785 / MPG898 / D1C2

Afrotropical: an E highlands species, from the E Zaire borders to SW Uganda, S Sudan and W Kenya.

Kenyan: this resident thrush can be common in dense forest undergrowth at 1600-2200 m (Br; Ja; Diamond & Fayad 1979). Range is entirely to the W of the rift valley, in humid, 1000+ mm rainfall areas.

It forms part of a species-group with the East Coast Akalat (673), which is a bird of the coastal lowlands (HM). Other reference: Mann 1974d.

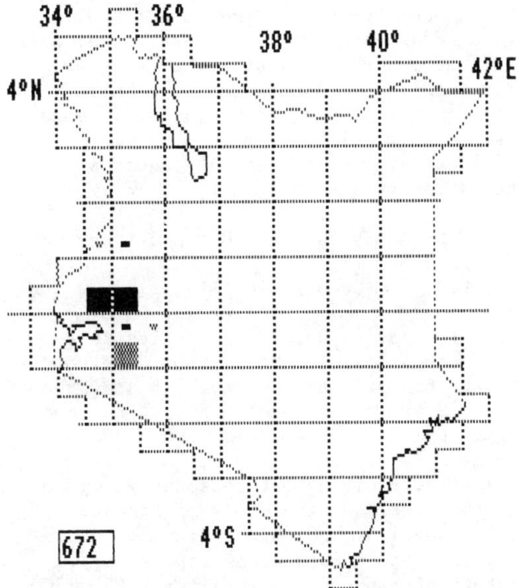

673 EAST COAST AKALAT *Sheppardia gunningi*
B787 / MPG896 / D1C2

Afrotropical: a localised species of the E, in S Mozambique, Malawi, and the coastal lowlands of NE Tanzania and Kenya.

Kenyan: forest undergrowth in the coastal lowlands and along the lower Tana R 91A 79C, up to 300 m altitude (Br; Keith 1968; AGH; EABR 1983). Most records come from the Sokoke Forest 102B where it can

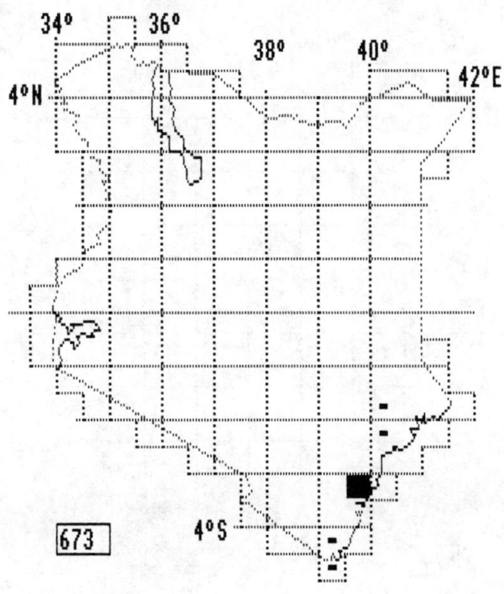

be common: Britton & Zimmerman (1979) regard it as apparently resident there, but Br mentions the possibility of seasonal fluctuations in abundance. In contrast to the Equatorial Akalat (672: *q.v.*), range is entirely below 300 m in 500+ mm rainfall areas, with five of six squares semiarid-dry subhumid.

674 RED-TAILED ANT THRUSH *Neocossyphus rufus*
B773 / MPG848 / D1C2

Afrotropical: apparently occupies three discrete areas: in the W, from Cameroon to the Central African Republic and Gabon; in NE Zaire and W Uganda; and in the coastal lowlands of Tanzania, Kenya and extreme S Somalia.

Kenyan: retiring but reasonably common in the lower to mid-levels of forest, moist woodland and thicket in the coastal lowlands and inland along rivers (Br). Extent of range along the Tana R 65B 79C is uncertain (Br), but Serle (1943) recorded it at Garissa 65B. There are no recent records from the Voi R 101B (La). Range is entirely below 500 m and, apart from the 65B 79C 101B riverine records, entirely within semiarid-dry subhumid, 500+ mm rainfall areas. Br considers this thrush to be probably mainly sedentary, but suggests that coastal birds might wander. The single record from Somalia was an Aug individual in 80B (not mapped here) (AM).

The Red-tailed forms a species-group with the White-tailed Ant Thrush (675), which is a bird of the far W; as their name implies, these species frequent ant columns (HM). Other reference: Brown 1970b.

675 WHITE-TAILED ANT THRUSH *Neocossyphus poensis*
B772 / MPG849 / D1C3
NO MAP: post presence 48D 49C

Afrotropical: a bird of W and C Africa, with a fragmented distribution through the lowland forests bordering the Gulf of Guinea, E to the Congo basin and W Uganda. A population in W Kenya is disjunct.

Kenyan: reasonably common but retiring in the lower and mid-levels of forest, at 1700-1900 m (Br). It is confined to the Kakamega, South and North Nandi Forests 48D 49C (Zimmerman & Mumford 1965; Simpson 1971c; Mann 1976; Diamond & Fayad 1979), hence C3. Range is to the W of the rift valley and N of the Winam Gulf 60AB, in humid, 1000+ mm rainfall areas. There is no breeding record, but the species is considered resident (Br). See Red-tailed Ant Thrush (674).

676 AFRICAN THRUSH *Turdus pelios*
B801 / MPG840 / D2C2

Afrotropical: widespread between 15°N and 10°S, from the W coast to Ethiopia, Kenya, SW Tanzania and Zambia. Distribution in Kenya thus represents the E limit of the range.

Kenyan: a rather uncommon and unobtrusive species, frequenting a wide variety of thicket, wooded and forest edge habitats,

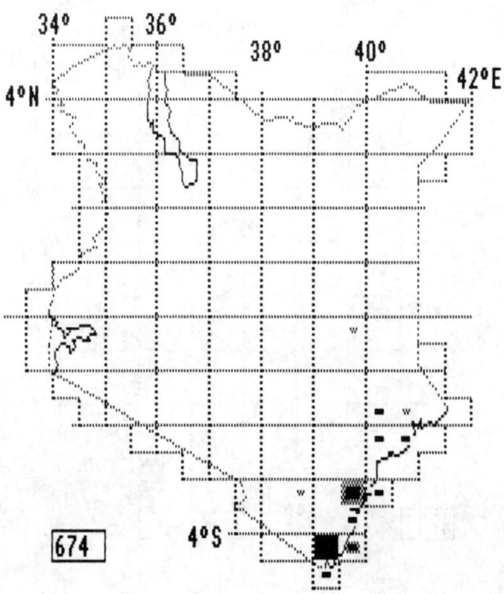

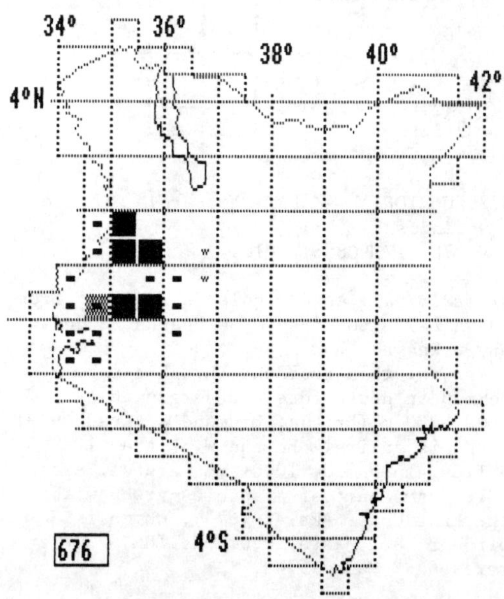

including cultivation, at 900-2000 m (Br; Ja). Range is in or to the W of the rift valley, apart from Br's note of occurrence on the E plateau in the Karissia Hills 38D and at Lorogi 50B. While status in these E plateau areas is unknown, single birds have been recorded in the rift valley floor near Mogotio 49D (May: EABR 1983), at Baringo 50A (Jul: St), and at L Nakuru 62A (Sept: Clifton 1976; Oct: EABR 1980). Including these records in and E of the rift, which may represent wanderers, range is entirely in 500+ mm rainfall areas, 65% within the 1000+ mm rainfall regime and 90% subhumid-humid.

The African Thrush forms part of a superspecies with the Bare-eyed Thrush (677), and this superspecies forms part of a species-group with the Northern Olive Thrush (678)(HM). The African and Northern Olive Thrushes are sympatric in parts of 48D, and on the E plateau at Lorogi 50B (Br). The Bare-eyed Thrush is apparently at least mainly allopatric with both the African and the Northern Olive, being mainly restricted to lower and drier areas further E, and the coastal lowlands.

677 BARE-EYED THRUSH *Turdus tephronotus*
B804 / MPG843 / D2C2

Afrotropical: the E and NE, from E Tanzania, through Kenya, to S areas of Ethiopia and Somalia. Distribution in Kenya represents the W limit of the range.

Kenyan: locally common in bush, thicket and woodland, up to 1600 m (Br; Ja; La). Range is entirely E of the rift valley, and apparently in three broad regions, namely the coastal lowlands, the E edges of the C highlands and their adjacent plateaux, and the S periphery of the Ethiopian highlands.

Virtual absence in the lowlands of the E interior could reflect low observer frequencies and a rather unobtrusive species, but several other factors suggest otherwise. There have been no records from observers resident in Garissa 65B, and this thrush appears similarly sparse and local in adjacent W Somalia (AM: the records in 67C and 80A are from their study). Range is 86% arid-semiarid, but most of the arid squares are adjacent to semiarid areas, and only 6% of the squares are in the most arid areas (moisture index below 1.2). Similarly, range is 79% within intermediate 250-1000 mm rainfall areas, but only in 11% of the driest, 0-250 mm country. Absence from even the vicinity of permanent water in arid country is also notable *e.g.* L Turkana, and the Tana R at Garissa 65B. If absence from these E lowlands is real, then range away from the coastal lowlands is 75% at 500-1500 m. This thrush is considered to be both resident and a wanderer (Br; La; M A C Coverdale *in litt.*). See African Thrush (676).

678 NORTHERN OLIVE THRUSH *Turdus abyssinicus*
B793 / MPG841, 842 / D3C2

Afrotropical: there is much disagreement on the taxonomy of the Olive Thrushes. The distribution given here is that of HM's

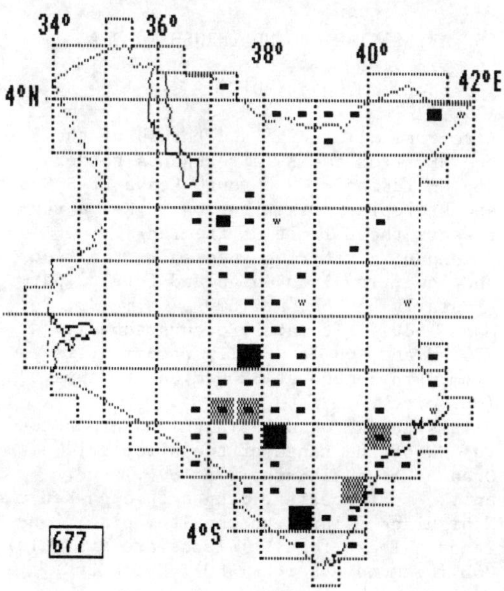

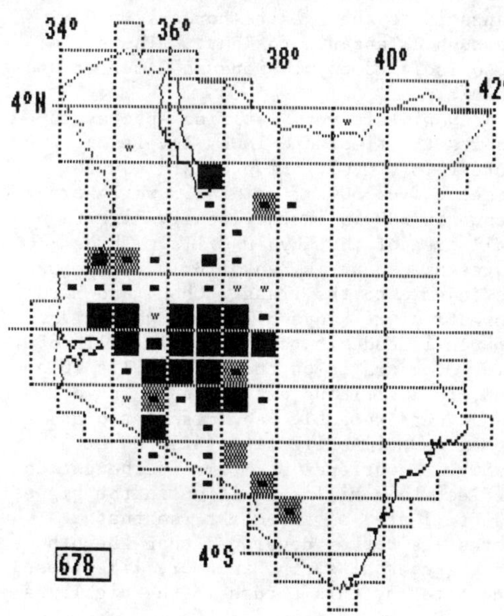

T.o. abyssinicus group, together with *T.a. helleri*, which are in the E and NE, from Malawi to the E Zaire-Uganda borders and S Sudan, and NE through E Tanzania to Kenya and Ethiopia. Distribution in Kenya represents the E limit of the range.

Kenyan: there are two races in Kenya. The distinctive *T.a. helleri* (the Taita Olive Thrush) is endemic to the Taita Hills 101A, where it is an uncommon inhabitant of the floor and lower levels of forest (Br).

The remainder of this account concerns the nominate race, which occupies the rest of the species' Kenyan range. This is a common and well known resident in a wide variety of wooded and forest edge habitats, including urban gardens, bamboo and alpine giant heath zone, mainly at 1700-3400 m (Br; Ja; Be; Se; Britton & Sugg 1973).

Range includes the forest islands of Mts Loima 13C, Kulal 26B, Nyiru 26D, Marsabit 27D 28C and Uraguess 51A, and the Ndoto Mts 39A and the Mathews Range 39C (Ja; Br; Diamond & Keith 1980). Apart from these localised highlands, which are too small to be shown on our overlays, range is 72% subhumid-humid and 81% within 500+ mm rainfall areas. The record at Moyale 17A (Ja) represents the S edge of the Ethiopian highlands population. Se records flocks of up to 30 individuals during the non-breeding season at 3000 m around Mau Narok 61D. See African Thrush (676).

679 ORANGE GROUND THRUSH *Turdus gurneyi*
B796 / MPG844 / D1C2

Afrotropical: mainly the E, from E South Africa, through Mozambique, Malawi and S Tanzania to the E Zaire borders, and NE through E Tanzania to Kenya. There are also isolated populations in Cameroon and Angola.

Kenyan: this very retiring species inhabits the floor and lower levels of forest (Br; MPG). Five of six squares receive over 500 mm rainfall, while presence in 88D is in the locally high rainfall area of the Chyulu Hills. Three races are known in Kenya (Br). *T.g. chuka* is endemic to the Irangi, Chuka and Meru Forests of Mt Kenya 51D 63AB, while *T.g. chyulu* is endemic to the Chyulu Hills 88D. *T.g. otomitra* is on the Taita Hills 101A and, in a striking extension of known range, has recently been discovered at 300 m on Mrima Hill 114A (Britton *et al.* 1980). Occurrence on Mrima is thought to reflect this hill's position in the highest rainfall area of the coast, so that its forest is better developed than the others in the coastal strip, and more like those inhabited by this thrush in the highlands

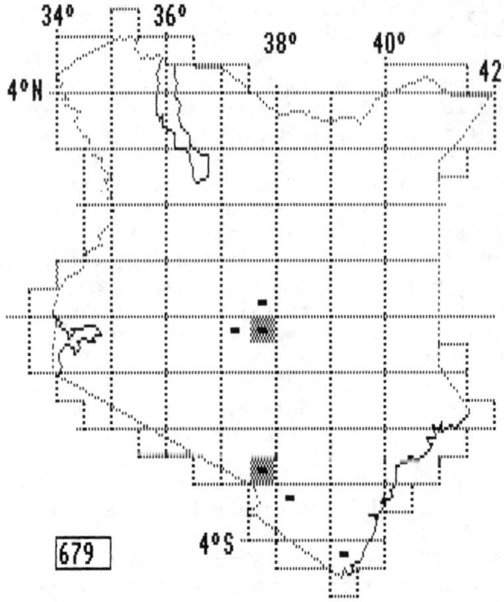

(*op. cit.*). This species has also recently been found at 450 m in the East Usambara Mts of adjacent NE Tanzania (EABR 1981).

This thrush forms part of a species-group with the allopatric Abyssinian Ground Thrush (680: *q.v.*), and Mt Kenya 51D 63AB is the only known locality where both occur (HM). They are ecologically segregated on this mountain, where the Orange is mainly below 2000 m on the wetter E slopes, and the Abyssinian mainly above 2300 m in drier areas (Br; HM). Other reference: van Someren 1930.

680 ABYSSINIAN GROUND THRUSH *Turdus piaggiae*
B802 / MPG845 / D1C2

Afrotropical: a highland thrush of the E and NE, from the E Zaire-Uganda borders and NE Tanzania, N through Kenya to S Sudan and Ethiopia. Distribution in Kenya represents the E limit of the range.

Kenyan: a retiring species of the floor and lower levels of highland forest, particularly bamboo, at 1800-3300 m (Br; Se; Mann 1980). It mainly occurs above 2300 m and, while usually uncommon, is common in bamboo at 3200 m on Mt Kenya 63A (*op. cit.*).

This thrush occupies highland areas to the NW of the range of the allopatric Orange Ground Thrush (679: *q.v.*) and, unlike that species, ranges through Kenya's N highland islands to the Ethiopian highlands. These forest islands are Mts Kulal 26B (Diamond & Keith 1980), Nyiru 26D,

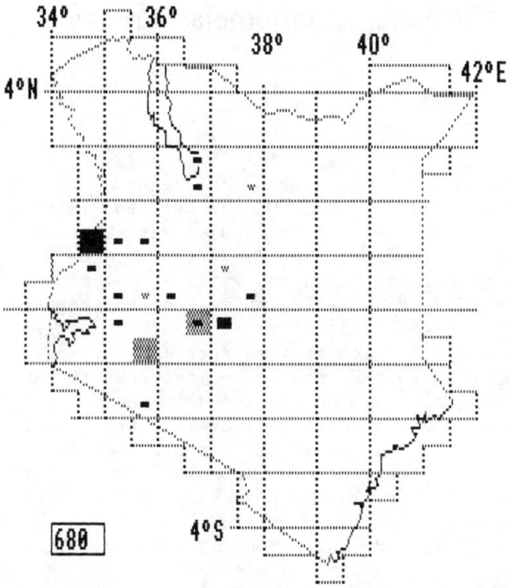

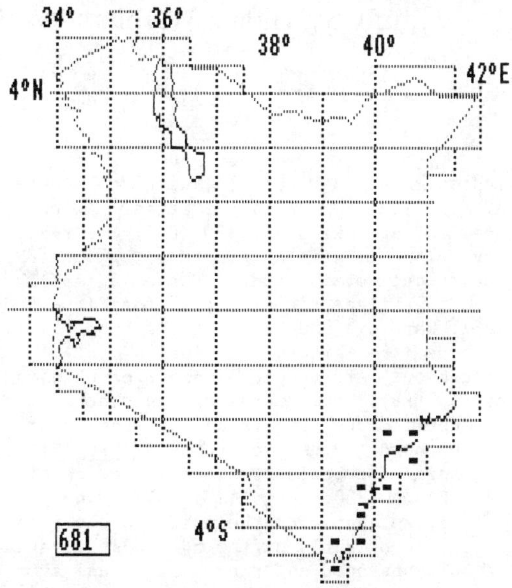

Marsabit 27D and Uraguess 51A (Ja; Br). Excluding these localised peaks, which are too small to appear on our overlays, range is entirely subhumid-humid, and 77% within 1000+ mm rainfall areas. Occurrence in the Nguruman forests 74D, a new locality, was reported by Fayad & Fayad (1980).

681 SPOTTED GROUND THRUSH *Turdus fischeri*
B795 / MPG846 / D1C2

Afrotropical: a very local species of the E periphery of Africa, found only in coastal South Africa, S Malawi, and the coastal lowlands of NE Tanzania and Kenya. It has recently been discovered in SE Sudan (Nikolaus 1982).

Kenyan: this account draws heavily upon that of Britton & Rathbun (1978). The nominate race is endemic to the coastal strip from Lamu 91B S to the Pugu Hills, at 6°53'S in NE Tanzania. The type was collected in 1885 but, due to lack of subsequent records, this race was considered probably extinct by MPG. However, it was rediscovered in Sokoke Forest 102B in 1964 (Keith & Twomey 1968) and there have been numerous subsequent observations in the coastal strip, mainly from Sokoke and the Gedi ruins 103A (Mann 1976).

It has been considered more common in the ruined town of Gedi than in Sokoke, and a recent survey estimated the Gedi population at 60-80 birds (about 1.5-2 birds/ha), while only three were seen in Sokoke during the same period (EABR 1983; *cf*. Britton & Zimmerman 1979). Bennun's (1985) estimate of the Gedi population is slightly higher still. This apparent preference for Gedi over Sokoke may reflect the former's higher rainfall (1000+ mm, as against Sokoke's 500-1000 mm), or the possibility that centuries of human occupation at Gedi have created improved feeding opportunities. This forest floor species may also prefer the uneven ground of forests on coral rag, as at Gedi. Bennun (1985) suggests that the species prefers coral rag forests *per se*, possibly for their particular vegetation characteristics.

Range is entirely below 60 m (Britton & Rathbun 1978), in semiarid-dry subhumid, 500+ mm rainfall areas. This thrush appears to be a migrant visitor, since there is no breeding record and all dates are Mar-Oct (EABR). Occasional birds in atypical bush or woodland thicket habitats may be migrants (Br), but the location of the presumed Dec-Apr breeding range is unknown. HM record regular migration of the South African race. Other reference: Burrell & Abel 1976.

Family Sylviidae: Warblers, Apalis, Cisticolas, Eremomelas, Prinias, Crombecs etc.

Approximately 340 species, with representatives in every zoogeographical region. The Afrotropical Region is particularly rich, with about 175 (51%). Of the Afrotropical species, approximately 150 (86%) are resident, and only six of these also extend to, or have close relatives in, other regions (HM). 98 species occur in Kenya (56% of the Afrotropical total), and of these 76 (78%) breed or apparently do so, 17 (17%) are non-breeding Palearctic migrants, four (the Red-winged (748), Savi's (689), and Grasshopper (687) Warblers and Kretschmer's Longbill (776)) are extralimital vagrants or wanderers, and one (the White-winged Apalis (752)) may be extinct. The African Reed Warbler (691) is the only Afrotropical member of this family known to make migratory movements in East Africa (Br): hence notes of resident status are in the main not included in these accounts. Many members of this family are skulking and/or difficult to identify if not in the hand, so that clusters of pre-1970 squares need not necessarily reflect declines or range contractions. The 27 species of cisticolas are a particular case in point here (see Lynes 1930; Lewis 1981-82). Many of the cisticolas also have very specialised ecologies, so that even the most closely related species can breed within view of each other, on slightly differing types of vegetation (HM). Because of this, range overlaps on our maps need not represent actual sympatry.

682 LITTLE RUSH WARBLER *Bradypterus baboecala*
B832 / MPG948 / D1C1

Afrotropical: widely distributed S of 5°S, excepting only the dry SW. Further N, it is found through East Africa and the E Zaire borders to Ethiopia and S Sudan. There are also populations in Chad, Nigeria, Cameroon, and probably other areas of the N tropics. Distribution in Kenya represents the E limit of the range.

Kenyan: locally common but easily overlooked in dense fringing vegetation around inland waters (Br; HM). Range is entirely within 500+ mm rainfall areas. The race *B.b. moreaui* is in the coastal lowlands, mostly around the lower Tana R and adjacent areas 91, but also inland to L Jipe 100D (Br; AGH). Five of its six squares are below 500 m, and all are semiarid-dry subhumid. There is no breeding record. *B.b. elgonensis* occurs in the highlands up to 2300 m (Br). Range is entirely above 1000 m, and 89% subhumid-humid. Br interprets appearance at man-made dams *etc.* as evidence of wandering, and there was a small Feb-Apr influx to L Baringo 50A (EABR 1982).

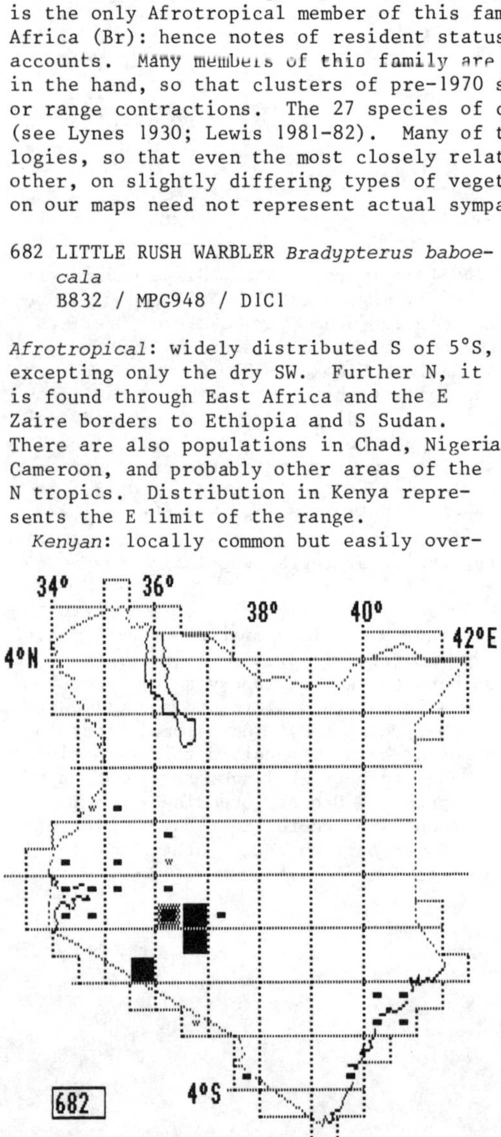

Britton (1978) recorded sympatry of this species and the larger White-winged Warbler (683) at L Kanyaboli 48C, and could find no clear cut zonation of their (and other warblers') foraging levels within the papyrus stands. He considered differences in body size to be the most likely factor separating species in this type of situation (this species has only 55-60% of the body weight of the White-winged).

683 WHITE-WINGED WARBLER *Bradypterus carpalis*
B834 / MPG950 / D1C2
NO MAP: post presence 60AB; post probable 48C

Afrotropical: a very restricted distribution, from NE Zaire to S and SW Uganda, NW Tanzania and, on the E extremity of the range, W Kenya.

Kenyan: a locally common but skulking and inconspicuous bird of the interior of papyrus swamps in the L Victoria basin (Br). First recorded in Kenya by Britton & Harper (1969), there are subsequent records from the Yala Swamp 48C, L Kanyaboli 48C, Ukwala 48C, Waturi 60A and Kisumu 60B (*op. cit.*; Britton 1978; Br). C2 reflects this restricted range. These three squares are all at 1100-1300 m, in moist subhumid-humid, 1000+ mm rainfall areas. Probable breeding in 48C is based upon a collected immature (Britton 1970a) and a male with enlarged testes (*op. cit.*), but there is no confirmed breeding record. See Little Rush Warbler (682).

684 CINNAMON BRACKEN WARBLER *Bradypterus cinnamomeus*
B835 / MPG952 / D2C2

Afrotropical: mainly the E and NE, from S Zaire, Zambia and N Mozambique, N through highland areas of East Africa and the E Zaire borders, to S Sudan and Ethiopia. There is also a population in Cameroon. Distribution in Kenya represents the E limit of the range.

Kenyan: a common but skulking species, resident in dense, low vegetation, including forest and bamboo undergrowth, mainly at 2000-4000 m (Br; MPG; Be; Se; Mann 1980; Britton & Sugg 1973). Range is 90% subhumid-humid and 90% within 500+ mm rainfall areas, and includes the forest islands to Mt Nyiru 26D (Diamond & Keith 1980) and the Chyulu Hills 88D. The Chyulus are inhabited by *B.c. rufoflavidus*, which is widespread in adjacent NE Tanzania, while the remainder of the Kenyan range is occupied by the nominate race (Br).

Where sympatric with the similar Evergreen Forest Warbler (685), that species is more typically in forest interior, while the Cinnamon Bracken Warbler favours forest edge and other non-forest habitats (HM; Br).

685 EVERGREEN FOREST WARBLER *Bradypterus barratti*
B833 / MPG953, 954 / D1C1

Afrotropical: mainly the E, N from E South Africa to Mozambique and Malawi. Further N, it ranges NW along the E Zaire-Tanzania-Uganda borders, and NE through E Tanzania to the NE limits of the range in the highlands of Kenya. There are also populations in Cameroon and Angola.

Kenyan: a locally common but very retiring species of dense undergrowth in highland forest, mainly at 1500-3100 m (Br; MPG), including the forest islands of the Chyulu Hills 88D and the Taita Hills 101A (EABR 1978). Excluding these two localised highlands, the other eight squares of the range are seven moist subhumid-humid and six within the 1000+ mm rainfall regime. The Taita Hills 101A are inhabited by *B.b. usambarae*, which is also found in the adjacent highlands of NE Tanzania, while the remainder of the Kenyan range is occupied by *B.b. mariae* (Br). There is no breeding record, but the species is probably resident. See Cinnamon Bracken Warbler (684).

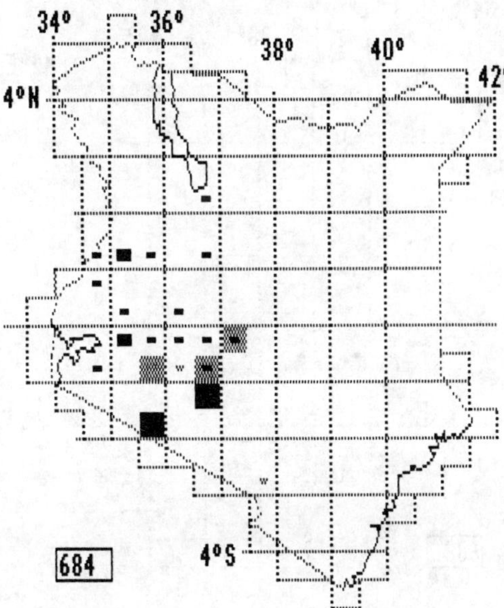

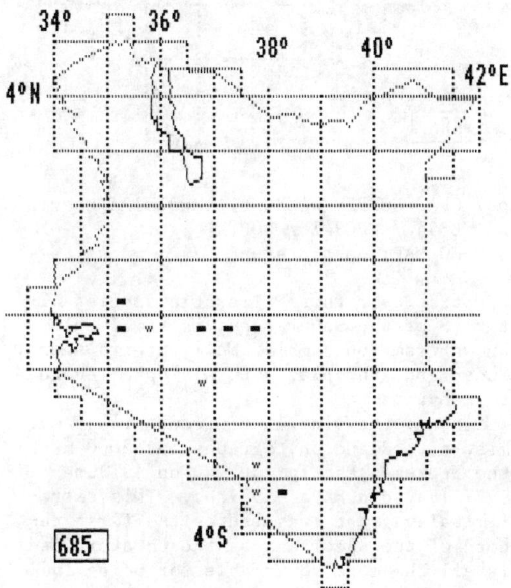

686 FAN-TAILED WARBLER *Schoenicola platyura*
B914 / MPG969 / D2C2

Afrotropical: widespread between 5°N and 15°S, though avoiding most of the Congo basin forests. Further N, it extends to S Sudan and Ethiopia, and there are isolated populations in Cameroon and Sierra Leone. Further S, it extends through S Mozambique to E South Africa. Distribution in the highlands of Kenya describes the E limit of the range.

Kenyan: locally common but skulking in scrub, tall grass, reeds and other rank vegetation, often near water, at 800-2000 m (Br; Ja; MPG). Range is 93% above 1000 m, including the isolated highland of the Chyulu Hills 88D. Excluding these hills, range is entirely within 500+ mm and 63% within 1000+ mm rainfall areas, and 94% subhumid-humid.

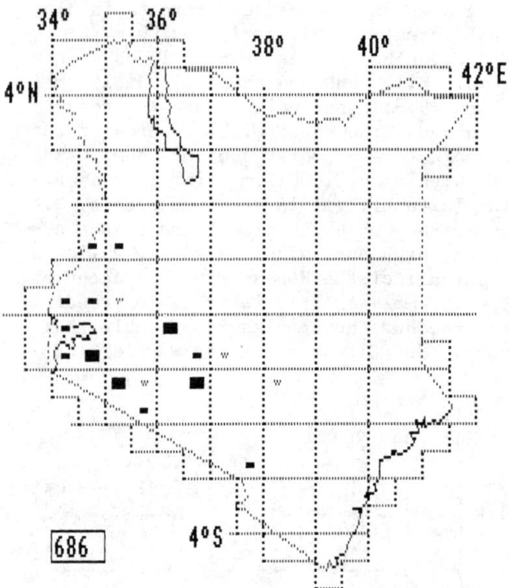

687 GRASSHOPPER WARBLER *Locustella naevia*
B894 / MPG941 / D0C0
NO MAP: post presence 74D

Afrotropical: this Palearctic species winters S of the Sahara, but is still only known from Senegambia, Mali, Sierra Leone, Sudan and Ethiopia, with a single record S to Kenya.

Kenyan: the single occurrence is of one mist-netted and photographed at 2000 m on the Nguruman Escarpment 74D on 19 June 1977 (Fayad & Fayad 1977ab). This extralimital vagrant constitutes the first record of the species S of the equator, and is all the more remarkable for being in mid-summer.

688 RIVER WARBLER *Locustella fluviatilis*
B892 / MPG940 / D0C1

Afrotropical: a Palearctic species that migrates through E Sudan, Ethiopia and Kenya to winter in the SE, from E Zambia and Mozambique to Zimbabwe, E Botswana and NE South Africa. See Pearson & Backhurst (1983).

Kenyan: an extremely retiring, inconspicuous and skulking bird of dense undergrowth, thickets and scrub (MPG; Ba). Presumably due to its elusive nature, there were few early records and both Ja and MPG considered it rare. A truer representation of its status in Kenya became apparent in 1969, when nocturnal migrants began to be attracted to the game-viewing lights of the newly erected Ngulia Lodge 101A, where nearly 2500 have now been ringed (Ba; Backhurst & Pearson 1984). A few were also attracted to car lights during a misty night on the nearby Chyulu Hills 88D (Nikolaus 1980). There are still relatively few records from other areas, but this almost certainly reflects its unobtrusive nature (Pearson & Jobson 1979; Pearson & Backhurst 1983).

It has been found to be a local and generally uncommon passage migrant Nov-Dec and Apr on the E and SE peripheries of the highlands, from Nairobi 75B to Tsavo 101 (Pearson 1980; *op. cit.*; OS-C 1977; EABR). This passage range is 83% at 500-1500 m and 94% in intermediate, 250-1000 mm rainfall areas. The isolated record in 37C

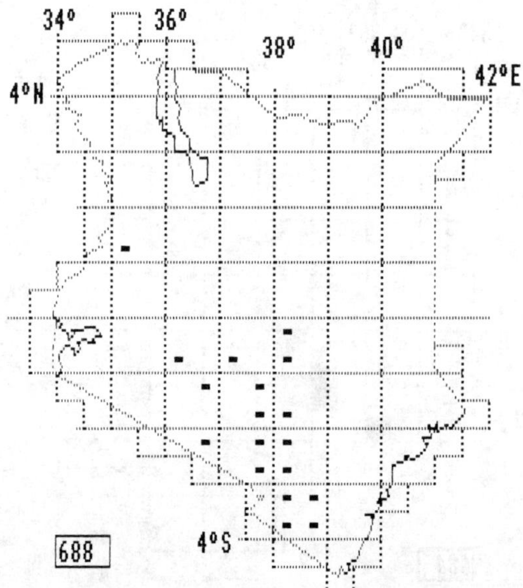

represents one caught on 23 Oct 1972 on the Kabolet R (T J Barnley *in litt.*). Further to the W, there is but a single record from Uganda (Br; EABR), and it is unrecorded in Somalia (AM). Thus range in Kenya appears to define its migration route at these latitudes.

689 SAVI'S WARBLER *Locustella luscinioides*
B893 / MPG939 / D0C0
NO MAP: post presence 101A

Afrotropical: a Palearctic species, wintering S of the Sahara, apparently in swampy areas N of 5°N, though its Afrotropical distribution in general is poorly understood. It occurs in S Sudan and Ethiopia, but there is only a single record S to Kenya.

Kenyan: the sole record involved a member of the E European race *L.l. fusca*, mist-netted at Ngulia Lodge 101A on 6 Dec 1975 (Backhurst & Pearson 1976). This was over 1000 km S of the southernmost Ethiopian locality (*op. cit.*).

690 SEDGE WARBLER *Acrocephalus schoenobaenus*
B811 / MPG947 / D2C2

Afrotropical: this Palearctic species winters S from the N tropics to N Namibia and E South Africa.

Kenyan: except where otherwise stated, this account draws upon the data and distributional map of Pearson (1982). The Sedge Warbler is a locally common to very common passage migrant and winter visitor,

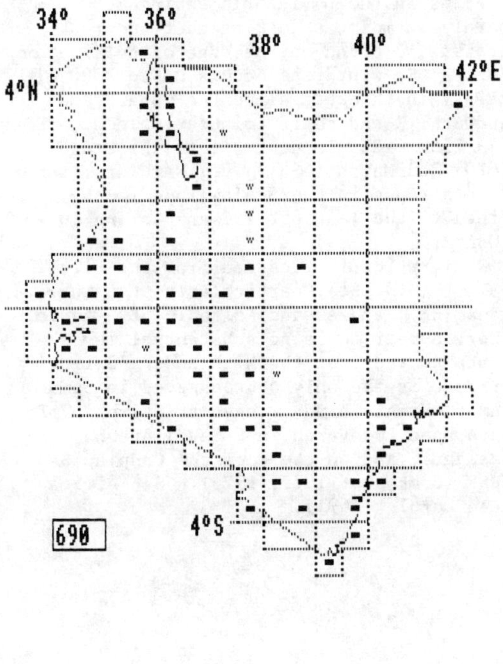

mainly Dec-May, and typically in dense waterside vegetation. Total range is 67% within 500+ mm rainfall areas, and 81% of the squares contain lakes or permanent rivers.

Autumn passage is in Nov-Dec, and only small numbers appear. This S movement may pass largely to the W of East Africa, *i.e.* through E Zaire, or most birds may overfly Kenya. Large numbers of Sedge Warblers in an unusual 'fall' of migrants caused by bad weather at Nairobi 75B in Dec supports the latter view.

Wintering Sedge Warblers appear in Dec-Jan, and reside in large numbers in dense waterside vegetation around L Victoria and the higher altitude rift valley lakes, and up to 2400 m on dams in the W and C highlands. Wintering in the SE lowlands is mainly restricted to the vicinity of the lower Tana R 79C 91 and inland to L Jipe 100D, though small numbers have recently been reported from Bamburi 102D 114B (EABR 1981). These wintering areas are all within the 500+ mm rainfall regime.

N passage during Apr-May is far heavier and more widespread than the S movement (Pearson 1980; St; La; EABR 1984), particularly in the area between the rift lakes and the coast. An Apr migrant reached 3000 m at Mau Narok 61D (Se). Birds on N migration are again mostly near water, but some occur in dry scrub and bush. This contrast between the N and S migrations is even more pronounced in Somalia, where there is a heavy spring movement but only a single autumn record (AM). This distinction is also known for other species, *e.g.* some falcons, and the Lesser Grey Shrike (863). A Sedge Warbler ringed at L Nakuru 62A was recovered to the N of the Black Sea, USSR (Backhurst 1973b). Other reference: Pearson *et al.* 1979.

691 AFRICAN REED WARBLER *Acrocephalus baeticatus*
B806 / MPG946 / D1C1

Afrotropical: most common S of 15°S. Further N, it extends through East Africa to the equator, with a very few records NW to Cameroon and Senegambia, and N to Chad, Sudan and Somalia.

Kenyan: inconspicuous, local and seldom common in dense, moist, low vegetation including tall grass, thicket and swamp edge, up to 1900 m (Br; Ja; MPG). Britton (1978) tentatively ascribed local scarcity around L Kanyaboli 48C to competition with other small Afrotropical warblers, though the African Reed is in fact the smallest species present, and ought thus to be ecologically segregated. The African Reed

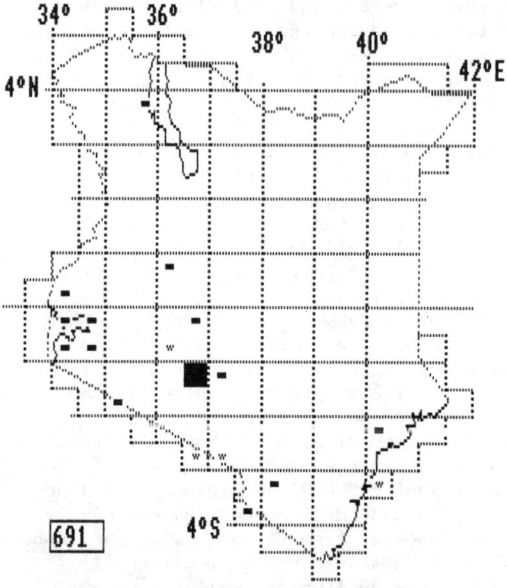

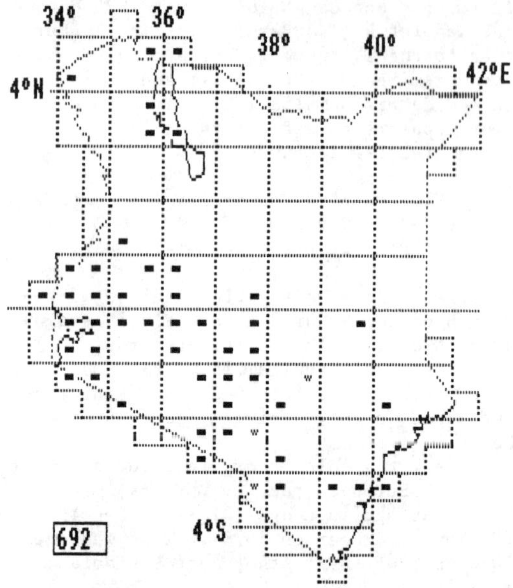

Warbler forms a superspecies with the Palearctic Reed Warbler (692)(HM). The total Kenyan ranges of the two are similar, though the Afrotropical bird is recorded from markedly fewer squares (18, as against 50).

There are records from the coastal lowlands (Br), and from Ferguson's Gulf 13B at L Turkana (HH; Mann 1976), but range is 77% above 1000 m. Further, 77% of the squares are semiarid-subhumid, and 82% within 500+ mm rainfall areas.

This warbler has been recorded throughout the year at some sites and may be mainly resident, but there are also indications of migration (Br). Four presumed nocturnal migrants have been caught at the Ngulia lights 101A in Dec (Backhurst & Pearson 1977b; Br; EABR 1980, 1982), and this bird is apparently a Mar-Apr migrant at 2°N on the coast of Somalia (AM). It is also a known migrant in other areas of the N and S tropics (HM).

692 REED WARBLER *Acrocephalus scirpaceus*
B812 / MPG944 / D2C2

Afrotropical: a Palearctic species that winters S to the N edge of the Congo basin in the W, and to Zambia and Malawi in the E.

Kenyan: this account draws mainly upon that of Pearson (1982). The Reed Warbler is a common passage migrant and winter visitor, mainly Nov-Apr. Total range is 70% above 1000 m and 78% within 500+ mm rainfall areas, and there are records from 66% of the 1000+ mm rainfall squares. Occurrence in arid areas is virtually restricted to the vicinity of permanent water, *e.g.* L Turkana and the Tana 65B 79C R.

S passage begins in Oct. This S migration must either pass through or overfly Uganda and W Kenya, since the few birds caught at Ngulia Lodge 101A indicate that little S movement takes place to the E of Mt Kilimanjaro.

The Reed Warbler is a common wintering species in the L Victoria basin 48CD 60ABC, mainly in moist scrub, thicket and rank grass, and often on the edge of cultivation. Further E, wintering occurs up to 2000 m, mainly in thickets along the upper Tana and Athi R and their tributaries *e.g.* 63BD 65B 75B 76AB. In contrast to the Sedge (690) and Marsh (693: *q.v.*) Warblers, the Reed's range in the SE lowlands is confined to the Tana 65B 79C and Galana 102AB 103A R.

N migration is recorded from the rift valley (*e.g.* St), particularly at L Turkana (HH), and around Nairobi 75B. Again there appear to be no substantial movements to the E of the highlands. Spring passage is the only one observed in Somalia (AM). Single birds ringed at Nairobi 75B have been recovered in C Saudi Arabia (on passage), and on the N of the Caspian Sea, USSR (Backhurst 1972, 1974). See African Reed Warbler (691).

693 MARSH WARBLER *Acrocephalus palustris*
B809 / MPG945 / D1C2

Afrotropical: a Palearctic species that migrates into Ethiopia during Aug-Sept, and later moves further S. The virtual absence of records from Uganda (Br) and S Somalia (AM) implies that movement between Ethiopia and the more S latitudes is mainly through Kenya. Further S, the species moves on through Tanzania, to winter mainly from E Zambia, Malawi and Mozambique S to E South Africa.

Kenyan: this account is based mainly upon those of Pearson (1973, 1982). The Marsh Warbler is mainly a Nov-Dec and Apr passage migrant, although small numbers winter in the E and SE. It is found in similar habitats to the closely related Reed Warbler (692:*q.v.*) and the two are largely allopatric (Br). This geographical separation can be seen to some extent in these species' climatic percentages: the Reed's total range is 78% within 500+ mm rainfall areas and 36% semiarid-dry subhumid, while the Marsh is 65% and 60% respectively.

Their migration routes and wintering areas are also largely separate. Thus the Marsh Warbler's Nov-Dec S migration is mainly to the E of the Reed's, being virtually restricted to areas E of L Turkana, Laikipia 50 and Mt Kenya 63A. This movement continues S across Tsavo 101 (where over 21 000 have been ringed at Ngulia Lodge 101A: EABR 1983), and onwards to the E of Mt. Kilimanjaro. At this time, the Marsh Warbler can be the most abundant Palearctic passerine in the more verdant parts of this area. The E edge of this S migration is unknown, but there are no records from Garissa 65B or the lower Tana R 78B 79AC 91A, though Ja quotes a Nov specimen from Wajir 42A (see below). Our map's representation of this E range edge may be reasonably accurate.

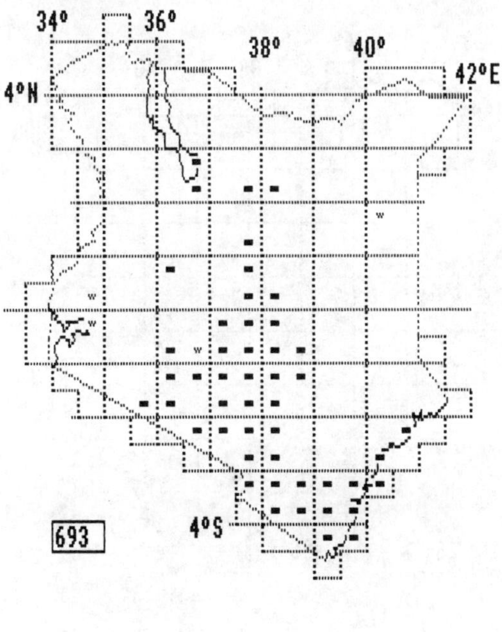

To the W, the species is regular at Nairobi 75B and Thika 76A, but virtually absent in the rift valley at L Turkana, Baringo 50A, Nakuru 62A and Naivasha 62C (EABR). The single record at Baringo 50A was of a migrant attracted to a light on 30 Nov 1983 (T Stevenson pers. comm.). Further W, Pearson (1982) mentions late Nov-Dec records from Nyanza 60, but maps only rare to uncommon occurrence on N passage in 48C.

Small overwintering numbers have recently been found in the region between Isiolo 51D, Bissell 87B, Endau 77B and Taveta-Taita 100B-101A. As with the autumn migrants, these overwintering birds favour moist scrub, thickets, herbaceous undergrowth and overgrown ditches. Wintering birds could not be found in the very dry conditions of Jan-Mar 1984 (EABR 1984). This wintering area is E of the Reed Warbler's principal winter quarters in the L Victoria basin and is in semiarid-dry subhumid habitats, whereas the lake basin is moist subhumid. Ja quotes Jan-Feb Marsh Warbler records from the Maragoli Hills and the Kakamega Road 48D 60B, but his specimens would need to be checked for confusion with the Reed Warbler before any conclusions re wintering in the lake basin could be drawn. The more E wintering areas of the Reed Warbler partly overlap those of the Marsh, but again are rather moister.

The Apr-May N migration of the Marsh Warbler is less marked than the S movement and, while similarly restricted to areas E of the Reed Warbler's movements, also extends further E to the coastal lowlands. Even fewer birds than on the S passage reach W to Nairobi 75B, and the sharp W limits of both of these migrations suggest that they pass around the mountain blocks of Ethiopia and Kenya. The fact that only small numbers of Marsh Warblers are found in Tsavo 101 and Ukambani 76 77 during this N migration suggests that most birds overfly the area. This view is supported by the very few Apr migrant 'falls' at Ngulia 101A, which have included hundreds of birds of these species (Pearson 1980).

Ringing recoveries have demonstrated movement of individuals between Kenya and Malawi, Saudi Arabia, the Sultanate of Oman, the USSR, Czechoslovakia, East

Germany, West Germany and Switzerland (Backhurst: E A Bird Ringing Reports; Backhurst 1985a). The occurrence of birds from the last two of these countries is of particular interest, since they show this species to be the only *Acrocephalus* that migrates to Kenya from both W and C European and Russian breeding grounds. The Marsh Warbler has provided the first extensive evidence of vocal imitation of Afrotropical species by a Palearctic migrant on its European breeding grounds. In a classic study, Dowsett-Lemaire (1979) detected imitations of 113 Afrotropical and 99 Palearctic species by Marsh Warblers in Belgium. Other reference: OS-C 1977.

694 GREAT REED WARBLER *Acrocephalus arundinaceus*
B805 / MPG942 / D2C2

Afrotropical: this Palearctic species winters across the N tropics from Senegambia to Ethiopia and Kenya, and S to Namibia and E South Africa.

Kenyan: this account is based mainly upon that of Pearson (1982). The Great Reed Warbler is a passage migrant and winter visitor, mainly Dec-Apr. It favours lush, moist thickets, tall grass and other dense vegetation and is most often near lakes or rivers. During passage, it can be found in drier and more open habitats.

This warbler is very scarce on S passage Oct-Nov (*e.g.* EABR 1979, 1982). As with the Sedge Warbler (690), most of this movement may occur further to the W, *i.e.* in Zaire, or migrants may directly overfly Uganda and W Kenya on their way S. The small numbers at Ngulia 101A show that a significant S migration around the E edges of the C highlands and Mt Kilimanjaro is unlikely (EABR).

This species winters commonly along the Tana R 79C 91A (EABR 1983). It also winters locally in Nyanza 48CD 60AC, and up to 1600 m in the C highlands and their peripheral lowlands, N to Isiolo 51D and E to Sultan Hamud 88A. These wintering areas are entirely within the 500+ mm rainfall regime.

Again as with the Sedge Warbler, the N migration is much heavier and more widespread than the S movement, and the species can be locally common during Apr in the rift valley *e.g.* between L Naivasha 62C and L Turkana, and further E, *e.g.* in the Nairobi 75B-Thika 76A area (HH; Br; La; St; EABR 1984). This N migration can also extend E to Tsavo 101 and the coast; significant movement E of Kilimanjaro is demonstrated by the species' presence in Apr migrant 'falls' at Ngulia 101A (Pearson 1980). The only four records from Somalia are also during this N migration (AM).

695 BASRA REED WARBLER *Acrocephalus griseldis*
B808 / MPG943 / D1C1

Afrotropical: a Palearctic species that breeds in Iraq, and winters in E Africa from Somalia S to Mozambique.

Kenyan: this account is based largely

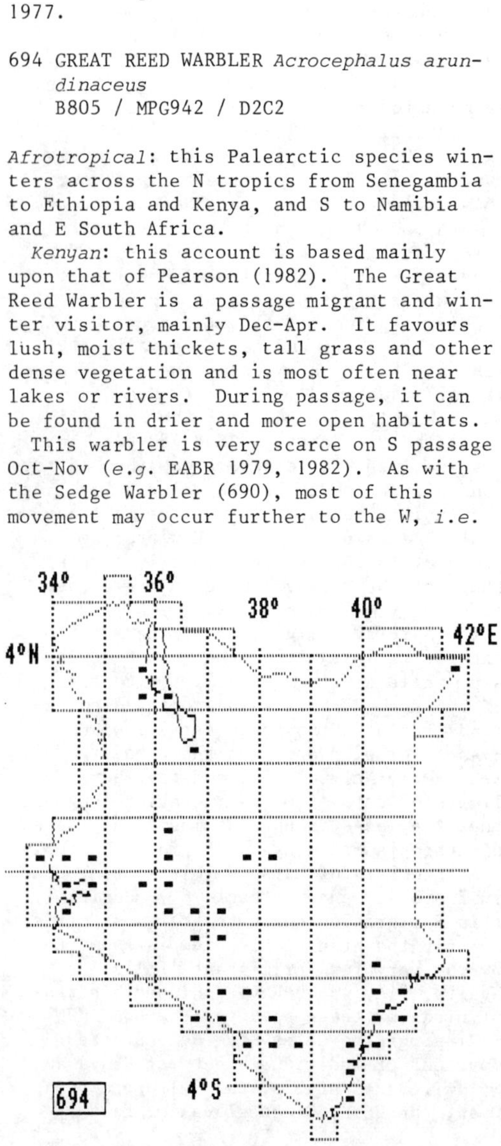

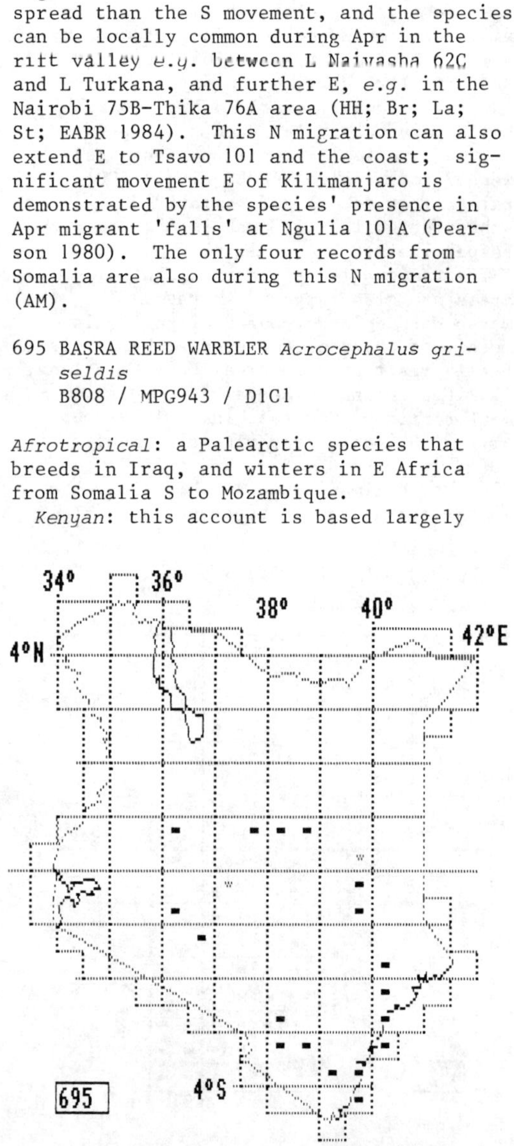

upon those of Pearson *et al.* (1978) and Pearson (1982). The Basra Reed Warbler is a locally common passage migrant and winter visitor, mainly Nov-Apr. Range is 75% below 1000 m, and entirely in or E of the rift valley. Most records are from the SE lowlands, the coastal strip, and the Tana R. Distribution is 70% within the intermediate, 250-1000 mm rainfall regime and, while 70% of the squares are arid-semiarid, this is actually a bird of waterside situations.

There is a regular Nov-Dec passage, mainly observed at Ngulia Lodge 101A, where over 500 have been ringed (EABR). Birds on S passage are also known from near Habaswein 52B, Mtito Andei 89C, Tsavo 101 and Bamburi 102D 114B.

Prior to 1977, there were only 23 dated records for this species in East Africa away from Ngulia Lodge (Ba; OS-C 1977) but, in that year, large Dec-Mar wintering populations were found along the lower Tana R 79C 91AB, and as far upstream as Garissa 65B (Pearson *et al.* 1978; EABR). Wintering birds have subsequently also been located at Mtito Andei 89C, Maji ya Chumvi 102C and Bamburi 102D 114B (EABR). Wintering habitats include moist green thicket with tall grass, the saltbush *Sueda monoica,* and occasionally woodland undergrowth. Like the Sedge Warbler (690), this species' habitats are more closely associated with water than those of the Reed (692), Marsh (693) or Great Reed (694) Warblers.

The very few Apr records that can be attributed to N passage have been mainly in the coastal and adjacent lowlands, at Ngulia 101A, Voi 101B, Bamburi 102D 114B, and Watamu 103A. Three individuals at L Baringo 50A during Apr 1984 were a marked exception to this pattern (EABR 1984).

696 LESSER SWAMP WARBLER *Acrocephalus gracilirostris*
B807 / MPG955, 957 / D2C2

Afrotropical: the S and E, widespread S of 15°S. Further N, it extends through East Africa to Chad, S Sudan, Ethiopia and S Somalia (Ash 1983).

Kenyan: locally common but somewhat retiring in dense vegetation near water, up to 2400 m (Br). Range is 93% within 500+ mm rainfall areas, 69% subhumid-humid, and only 3% arid. Hence it is absent from the N and E, including L Turkana (*contra* HH) and the Tana R. Range has apparently recently receded from Tsavo E 101B (La), the sole locality in arid country.

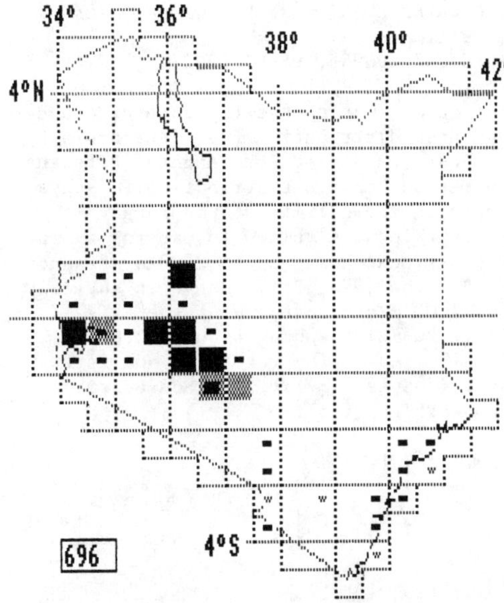

This species forms part of a species-group with the Greater Swamp Warbler (697) (HM). The two are sympatric in the L Victoria basin, but are segregated locally by various factors. Firstly, in these W overlap areas, the Lessers are a particularly small form of their species (Britton 1978; HM), so that they presumably take different food items to the distinctly larger Greaters. The two species also show colour differences in the gape (HM). The Greater frequents the interiors of papyrus swamps, where the Lesser is restricted to the swamp edges. The Lesser is uncommon in these edge habitats, and is in fact only really common in the L Victoria basin on small, rocky, shrub-covered islands where the Greater does not occur (*op. cit.*; Ja; Brown & Britton 1980). The arrival of Palearctic *Acrocephalus* species in these L Victoria habitats has minimal effect on these two resident members of the genus (Britton 1974f, 1978). Further E, where the potentially competing Greater Swamp Warbler is absent, the Lesser is a plentiful breeding species within *Typha* and papyrus swamps (*op. cit.*).

The Lesser's breeding records are all above 1000 m, but R A M McVicker (*in litt.*) considers breeding to occur in the coastal lowlands. There is a similar absence of breeding indications for the newly discovered population in S Somalia, which is at present presumed resident (AM). St records the Lesser throughout the year on the N range edge at L Baringo 50A.

697 GREATER SWAMP WARBLER *Acrocephalus rufescens*
B810 / MPG955, 956 / D2C2

Afrotropical: an apparently scattered and fragmented distribution from Senegambia E to S Sudan, and also S to Angola, Botswana, E Zaire and Kenya. Distribution in Kenya represents the E limit of the range.

Kenyan: locally common in papyrus swamp interiors, extending up to 2000 m at Nandi 49C (Britton 1978; Br). Range is entirely above 1000 m and to the W of the rift valley, in moist subhumid-humid, 1000+ mm rainfall areas. The record in 48A is from Uganda (J S Ash *in litt.*). See Lesser Swamp Warbler (696).

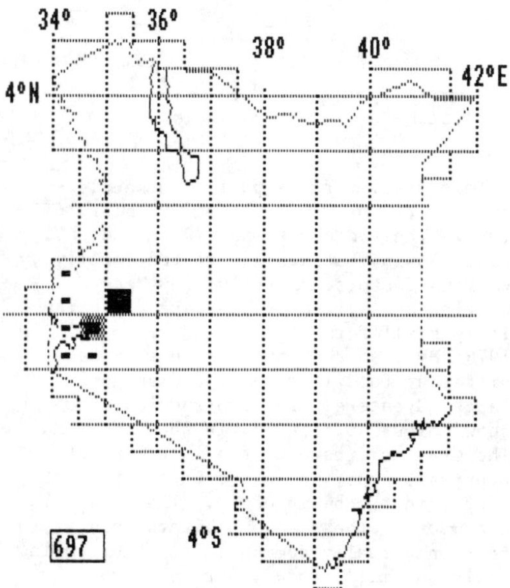

698 OLIVACEOUS WARBLER *Hippolais pallida*
B888 / MPG938 / D2C2

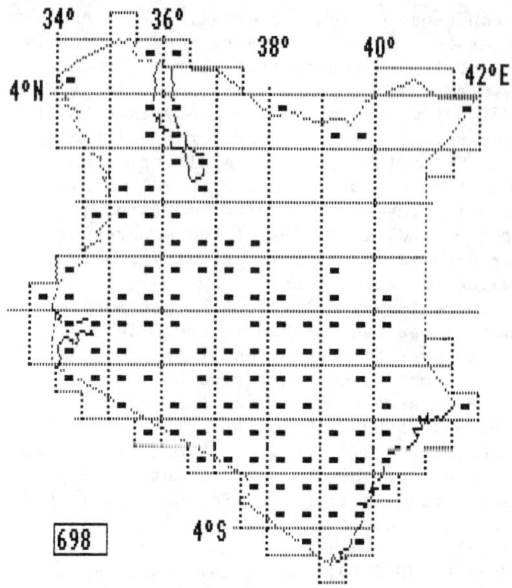

Afrotropical: mainly a Palearctic migrant. Birds from the Balkans and SW Asia winter in the NE, S to Uganda and N Tanzania, whilst those from the SW Palearctic migrate to W Africa. The breeding range also extends into the N edge of the Afrotropical, with resident races in Niger, Chad, Sudan and possibly NE Somalia (HM; AM).

Kenyan: a common winter visitor to dry bushed and wooded habitats, particularly those with tall acacias, mainly Nov-Apr (Br; Ja; HH; La; St). There is little evidence of passage movements (Br). While extending up to 1900 m, it prefers arid-semiarid country at low-medium elevations (Br; D J Pearson *in litt.*). These preferences are not evident from our map as we do not plot abundance.

Virtual absence from the NE may reflect low observer frequencies, but there are no reports from the relatively better known localities of Marsabit 27D 28C and Wajir 42A. Alternatively, this absence may reflect the species' aversion to the most arid environments: there are records from only 19% of the 0-250 mm rainfall squares and 12% of those with moisture index below 1.2. That range is 77% well timbered would also imply an aversion to harsh, barely vegetated areas. See Upcher's Warbler (699).

699 UPCHER'S WARBLER *Hippolais languida*
B886 / MPG937 / D1C1

Afrotropical: this Palearctic species winters mainly in Somalia and Kenya.

Kenyan: a common winter visitor to dry bushed and wooded habitats, mainly Dec-Apr (Br; La; EABR). Br notes a subdued Nov-Jan passage, but makes no mention of a N movement. Pearson (1981d) found small numbers of migrants grounded by heavy rain at Mtito Andei 89C in early Apr.

The distribution of this warbler shows several clear distinctions from that of the closely related Olivaceous Warbler (698). Firstly, while the Olivaceous frequents W Kenya and is widespread in Uganda (Br), Upcher's is a bird of the E, with 97% of its squares in or E of the rift valley. Upcher's is really only common E of approximately 37°30'E (Br; EABR). Further W, it is distinctly uncommon at

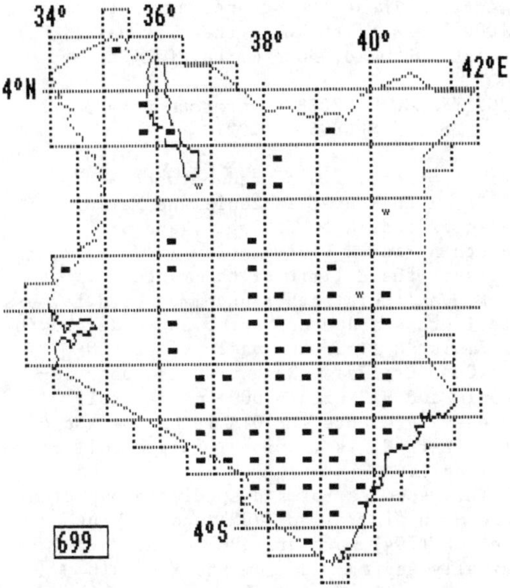

699

Nairobi 75B and in the rift valley (Ba; OS-C 1977), although apparently occasional to frequent at L Turkana (HH). There is a record from the extreme NW at Kamathia 3A, in Feb (EABR 1984). Further W still, in extreme NE Uganda, there is a report from 48A (D J Pearson pers. comm.), and a few sightings N and W of Mt Elgon 36D (Rolfe & Pearson 1973; Mann 1976; Br).

Upcher's range is 88% arid-semiarid, and the bird apparently endures arid conditions for longer than other Palearctic migrants *e.g.* Irania (646), Whitethroat (706) and Barred Warbler (705)(OS-C 1977). Thus, in direct contrast to the Olivaceous Warbler, Upcher's may well be widespread in the arid NE (*i.e.* around 29), but observers competent to identify this rather difficult species have yet to survey that region during the Palearctic season (Ba; OS-C 1977). In further contrast to the Olivaceous Warbler, although Upcher's is a bird of the E, with 92% of its squares below 1500 m, it is apparently absent from even semiarid areas of the coastal strip. Lack (1985) notes habitat differences between these two species in arid Tsavo E 101B. The Kenyan distribution of Upcher's Warbler is also close to its S range edge. Birds caught at Ngulia Lodge 101A are presumably en route to N Tanzania, but there are few records from that country (Br; EABR).

700 OLIVE-TREE WARBLER *Hippolais olivetorum*
B887 / MPG936 / DOCO

Afrotropical: a Palearctic species, wintering in dry country of the SE, from Kenya and Tanzania S to Zimbabwe and E South Africa. There are no records from either Uganda (Br; EABR) or S of 11°N in Somalia (AM), so that migration may pass wholly through Kenya.

Kenyan: an uncommon and little known species of dry, open, wooded habitats, including bush with scattered trees, up to 2300 m (Br). Range is 86% below 1500 m, 69% arid-semiarid and 82% within intermediate, 250-1000 mm rainfall areas. This species mainly occurs as a Nov passage migrant on the N and E peripheries of the C highlands (EABR). Over 280 have been ringed at Ngulia Lodge 101A, where it is one of the earliest migrants (EABR). The sole record W of the rift valley involved one ringed at Ukwala 48C on 21 Nov 1969 (Ba).

Br knew of no convincing overwintering records, but D J Pearson (*in litt.*) suggests Feb-Mar presence at various sites to indicate the N limit of the species' Afrotropical winter quarters. These wintering sites are at Baringo 50A (EABR 1981), the Archer's Post-Isiolo area 51BD (EABR 1980), and Tsavo E-Galana Ranch 101B 102A (EABR 1982). Four of these five wintering squares are arid-semiarid.

There are few records of N passage, but the single coastal record is of an Apr bird at Watamu 103A (OS-C 1977), and C F

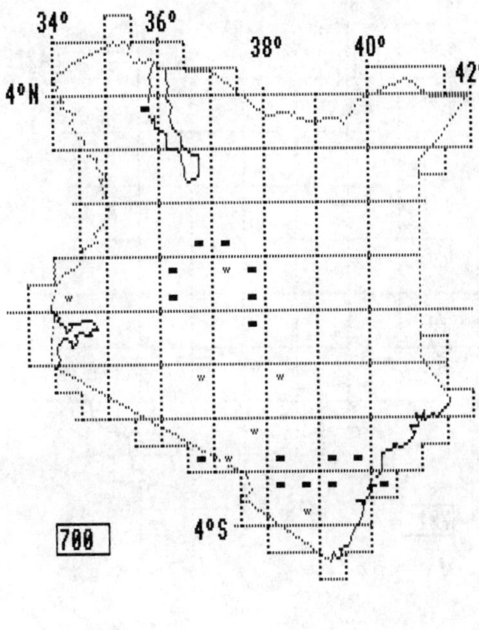

700

Mann found the species frequent in the Kalokol riverine woodland 13B at L Turkana, during Apr 1972 (HH). Pearson (1981d) found a few migrants grounded by heavy rain at Mtito Andei 89C in early Apr.

701 ICTERINE WARBLER *Hippolais icterina*
B885 / MPG935 / D2C1

Afrotropical: this Palearctic warbler winters in S woodlands, mainly from Rwanda and SE Zaire S to Namibia and South Africa. Its main migration routes are W of Kenya; the few Kenyan records mostly represent stragglers E of their normal range.

Kenyan: very scarce in leafy bush and open wooded habitats, up to 3000 m (Br). Range is 83% above 1000 m, 73% subhumid-humid and 82% within 500+ mm rainfall areas. 12 of the 14 records are Oct-Nov or late Mar-Apr, presumably referring to passage migrants (Ba; Meadows & Weston 1974; HH; EABR). These were all single birds, with the striking exception of five at L Naivasha 62C on 13 Apr 1970 (Ba). Seven of the 11 passage records refer to the N movement, and this migration may take a more E route than the S one (Ba). The Apr group at Naivasha (see above), and Williams' mention of not infrequent occurrence at this lake in spring (*in* Ba), might provide further support for this proposal, but there have been no subsequent Naivasha records.

Single birds in song on Rusinga Island 60A on 2 Mar (EABR 1982), and NE of Keekorok 74A on 21 Feb (EABR 1983) were considered to be wintering (D J Pearson *in litt.*). These two squares are both above 1000 m and to the W of the rift valley, in moist subhumid, 500+ mm rainfall areas.

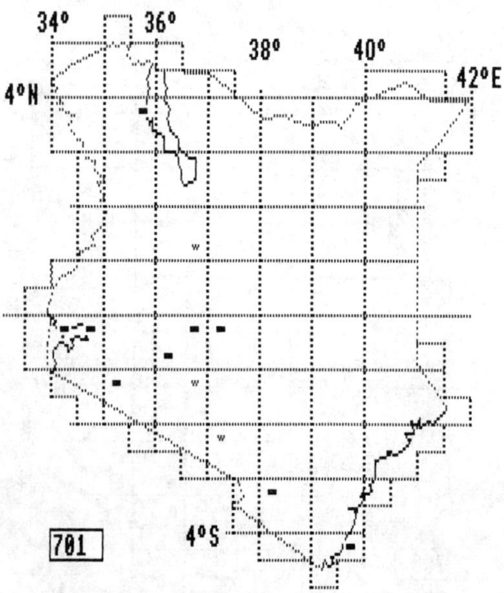

702 YELLOW WARBLER *Chloropeta natalensis*
B843 / MPG805 / D2C2

Afrotropical: widespread between 5°N and 15°S, E to Kenya, Tanzania and Malawi. It also extends N to Ethiopia, and S to NE South Africa. Distribution in Kenya represents the E limit of the range.

Kenyan: this rather uncommon bird favours tall grass and dense scrub and undergrowth in moist hollows and valleys, at 800-2300 m (Br; Ja). Range is 82% subhumid-humid and 91% within 500+ mm rainfall areas. Presence in 101A is on the top of the Taita Hills (National Museum Collection, Nairobi).

This species forms a species-group with the Mountain Yellow (703) and Papyrus Yellow (704) Warblers (HM). The Yellow is usually segregated from the Mountain Yellow by altitude and probably habitat preferences, although the two can occur together (HM; Br). The Papyrus Yellow is distinct from its congeners by being endemic to papyrus swamps on the shores of, or closely adjacent to, L Victoria.

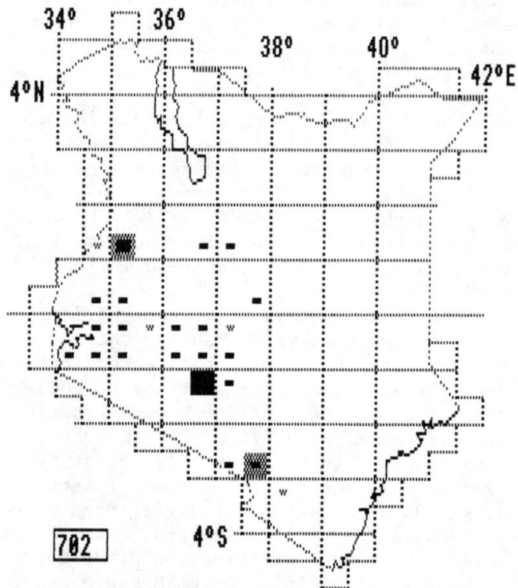

703 MOUNTAIN YELLOW WARBLER *Chloropeta similis*
B844 / MPG806 / D2C2

Afrotropical: a highland species of the E, from Malawi to East Africa and the E Zaire borders, N to S Sudan. Distribution in

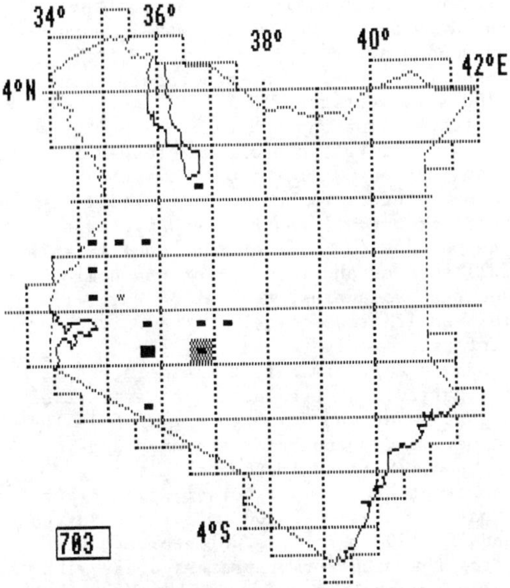

Kenya represents the NE limit of the range.

Kenyan: locally common in sheltered areas of bracken, bamboo and forest edge, at 1800-3400 m (Br; Ja; Be; Britton & Sugg 1973). Presence in 26D represents the forest island of Mt Nyiru (Diamond & Keith 1980). Excluding the localised habitat of Nyiru, range is entirely subhumid-humid, and 83% within 1000+ mm rainfall areas. The only breeding record is from the Kinangop plateau 62D (MPG). See Yellow Warbler (702).

704 PAPYRUS YELLOW WARBLER *Chloropeta gracilirostris*
B842 / MPG 958 / D1C2
NO MAP: post presence 48C 60B; pre confirmed 48C

Afrotropical: an endemic of papyrus swamps, with a very fragmented distribution in E Zaire, W Uganda and, at the E extremity of the range, W Kenya.

Kenyan: a scarce and retiring inhabitant of papyrus swamps, particularly those over standing water (Britton 1978; Br; Ja). First recorded in Kenya at L Kanyaboli 48C by Britton & Harper (1969), it has subsequently also been found on the S shores of the Winam Gulf at Kendu Bay 60B (EABR 1979). These two localities are on or closely adjacent to L Victoria. Their squares are at 1000-1500 m, in moist subhumid-humid, 1000+ mm rainfall areas. The confirmed breeding in 48C is based upon Britton & Harper's (1969) mention of a female at L Kanyaboli that had recently laid. See Yellow Warbler (702).

705 BARRED WARBLER *Sylvia nisoria*
B920 / MPG 933 / D2C2

Afrotropical: a Palearctic species which winters in the NE, mainly from E Sudan and Ethiopia S to Kenya and N Tanzania.

Kenyan: this account is based primarily on that of Pearson (1978). The Barred Warbler is a passage migrant and winter visitor, mainly Nov-Apr, that frequents scrub, thicket and undergrowth in dry, bushed or wooded country (Br). Total range is 89% below 1500 m (although it is absent from the coastal strip), 84% less arid-semiarid, and 81% within intermediate 250-1000 mm rainfall areas.

Both migrations are widespread, but the only four records from Somalia are during the N movement (AM). The S movement Oct-Nov is most evident in the Tsavo 101 area, where over 470 have been ringed at Ngulia Lodge 101A (EABR 1983). These birds are presumably en route to N Tanzania, but there are few records of wintering from that country (Br; EABR 1978, 1982). The N migration is most noticeable around Nairobi 75B.

Overwintering in Kenya is more localised. The two main areas are to the E and N of the C highlands. The former extends S from Meru 64D and Kitui area 64BD 77BD to Tsavo 89CD and Voi-Taita-Taveta 101 (*e.g.* La). Small numbers extend W to Nairobi 75B, Kajiado 75D and Amboseli 88C. These winter quarters are mostly at 500-1500 m, in semiarid country. The wintering area

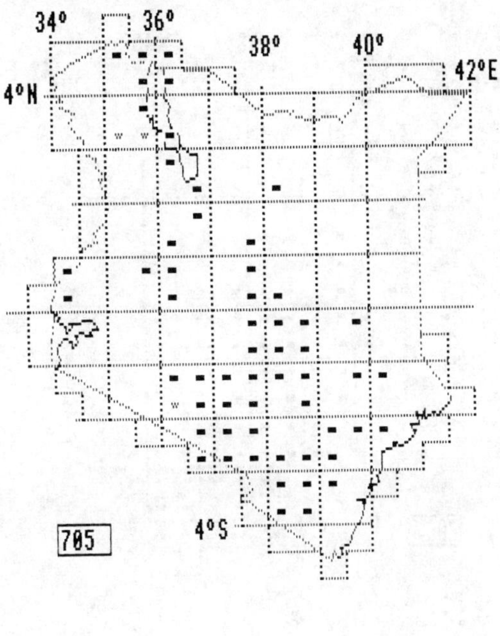

to the N of the highlands extends from Baringo 50A (St; EABR 1984) and Isiolo 51D N along the Turkwell R to Ferguson's Gulf 13B (HH) and South Horr 26D. These winter quarters are at similar elevations to those in the E, but extend N into more arid areas.

This warbler may also winter in the arid NE but, as noted under Upcher's Warbler (699), this area has yet to receive adequate coverage at the relevant season. Another, less important, wintering area is to the N of the Winam Gulf of L Victoria 48, in humid country. The Barred Warbler can be locally common or even abundant in the N and E wintering areas, but is usually less common than the other *Sylvia* species in Kenya (Br). A bird ringed at Ngulia Lodge 101A was recovered in Saudi Arabia (Backhurst 1974). See Whitethroat (706).

706 WHITETHROAT *Sylvia communis*
B919 / MPG924 / D2C2

Afrotropical: this Palearctic species winters S to approximately 9°N in the W whereas, in the E, the wintering range extends through Kenya, Tanzania and the far E of Uganda S to Botswana and NE South Africa. Since there is but a single record S of 8°N in Somalia (AM), distribution in Kenya effectively represents the E limit of the range.

Kenyan: this account is based primarily on that of Pearson (1978). The Whitethroat is a common passage migrant and

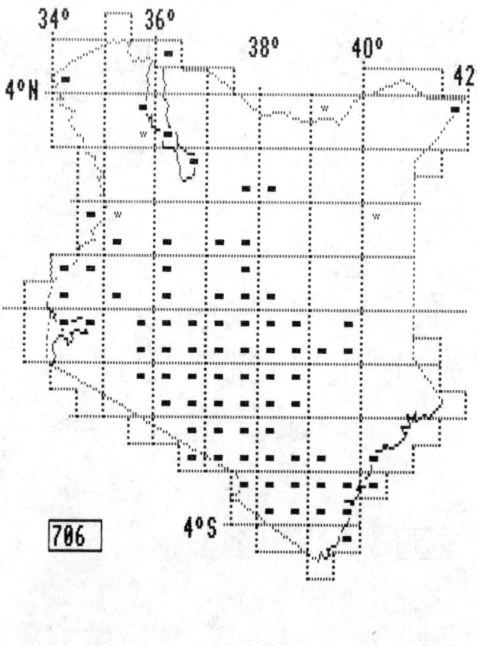

winter visitor, mainly Nov-Apr. It is found in scrub and undergrowth in bushed or wooded areas, typically in semiarid country at 800-1200 m, and rarely above 1600 m (Br; Britton 1979).

Distribution and habitat requirements are similar to those of the closely related Barred Warbler (705), but the Whitethroat prefers rather moister areas. In terms of overall range, the Whitethroat has records from 56% of the 1000+ mm rainfall squares and from 56% of the humid squares, as against the Barred Warbler's 16% and 11% respectively. The Whitethroat winters commonly to the E of the C highlands in approximately the same areas as the Barred, but the latter ranges further E into lower and drier country, while the former is commoner near the highlands. Pearson (1978) estimates the altitude ranges of these two species' E wintering areas as mainly 600-1000 m for the Barred and 700-1200 m for the Whitethroat, and notes that the former prefers areas with taller vegetation cover. Like the Barred, the Whitethroat also winters in the humid W 48 49, though more extensively and commonly. In complete contrast to the Barred Warbler, however, the Whitethroat does not winter in arid N Kenya, where the few records mainly refer to passage migrants *e.g.* at L Turkana (HH), at Moyale 17A and Wajir 42A (Ja), and at L Baringo 50A (St). Given the Whitethroat's preference for moister areas, wintering in the little known areas of the very arid NE would seem improbable (see Upcher's Warbler (699)).

Autumn passage Oct-Jan is particularly marked in Tsavo E 101B (*e.g.* La), where the ringing of over 18 500 individuals at Ngulia Lodge 101A has yielded only a single recovery in W Russia (Br; Backhurst 1981b, 1985a; EABR 1983). N passage in Apr is regular around Nairobi 75B, and rare 'falls' at Ngulia demonstrate direct overflying of Tsavo and the SE on a large scale (Britton & Britton 1977; Pearson 1980, 1981d). See Garden Warbler (707).

707 GARDEN WARBLER *Sylvia borin*
B918 / MPG925 / D2C2

Afrotropical: a Palearctic species that winters in equatorial and S Africa, from about 8°N in the W and 3°N in the E, S to N Namibia and E South Africa. Distribution in Kenya represents the E limit of the range.

Kenyan: this account is based primarily on that of Pearson (1978). The Garden Warbler is a common passage migrant and winter visitor late Oct-Apr. It frequents forest edge, and dense thicket and under-

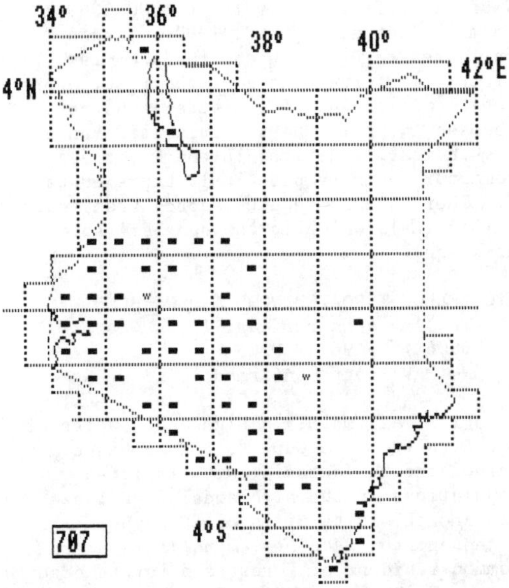

707

growth in bushed and wooded country, mainly at 800-1800 m (Br; Ja). Range is 75% above 1000 m, but there are also a few records from the subhumid areas of the coastal lowlands. Its squares are 75% within 500+ mm rainfall areas and only 8% arid, so that the bird is virtually absent from the N and E. Rare occurrences at L Turkana have been on Mar-Apr passage (HH), while a bird near the Tana R at Garissa 65B was in Nov (D J Pearson *in litt.*).

This warbler is widespread during the Oct-Dec and Apr migrations, sometimes reaching altitudes well over 2000 m (*e.g.* Britton 1979). The S migration is most evident in the W 48 60, while the N movement appears further E, being most noticeable in C Kenya.

Two areas of Kenya represent the N limit of the species' Afrotropical wintering range. The first of these areas is in Nyanza 48, at 1000-2000 m, within the 1000+ mm rainfall regime and humid. The other is along the E edge of the C highlands, from Murang'a 63C and Nairobi 75B, S to Kibwezi 88B and Ngulia 101A. This area is also at 1000-2000 m, but with a semiarid-moist subhumid, 500+ mm rainfall climate. There is a single oversummering record, at Eldama Ravine 49D on 2 Jul (Ja).

This species and the closely related Blackcap (708) are almost completely segregated by altitude in their wintering areas. Sympatry would mainly be at around 1600-1800 m, as for example on the E edge of the C highlands. When the two species occur together, the Blackcap tends to favour higher cover. The Garden Warbler is usually in moister, more verdant country than the Whitethroat (706)(Br). A wintering Garden Warbler ringed at Ngulia 101A was recovered in Jordan (Backhurst 1977).

708 BLACKCAP *Sylvia atricapilla*
B917 / MPG926 / D2C2

Afrotropical: many birds migrate from the Palearctic to winter in W Africa S to the equatorial forest belt, and in highland areas of the E from Ethiopia and S Sudan S to Malawi. Distribution in Kenya represents the E limit of the range.

Kenyan: this account is based primarily upon that of Pearson (1978). The Blackcap is mainly a common winter visitor, Nov-Mar, in all storeys and edges of highland forests and woodlands (Br; Se; Britton & Sugg 1973). Se and Sessions (1975) estimated 1000 individuals along 1.6 km of forest edge at 3000 m near Mau Narok 61D, and a possible total population of 1000-2000 on his farm there during the winter of 1974-5. Range is mainly at 1600-3000 m, but there are records up to 3600 m on Mt Kenya 63A. Occurrences below 1600 m are few, even during passage periods (Br), although small numbers are annual during Nov and Feb-Apr at 1000 m around L Baringo 50A (St).

The Blackcap is present on the forest islands of Mt Nyiru 26D, Marsabit 28C, the Mathews Range 39C, Mt Endau 77B, and the Chyulu Hills 88D. Excluding the local environments of these isolated highlands,

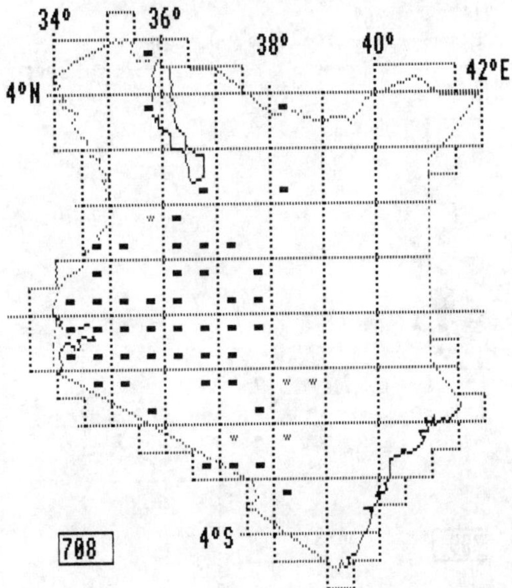

708

range is 87% within 500+ mm and 47% within 1000+ mm rainfall areas, and only 6% arid. Absence from the coastal lowlands contrasts with the closely related Garden Warbler (707: *q.v.*).

Neither N nor S migration is marked. Evidence for the Nov-Jan S movement includes Nov concentrations, birds with sufficient fat reserves to move further into the S tropics, and regular Nov occurrence at Ngulia Lodge 101A. Spring passage is even less marked, with an abrupt departure in late Mar. HH record a Mar male at Kalokol 13B, Turkana. A bird ringed at Ng'iya 48C was recovered in S Iran (Backhurst 1972).

709 MOUSTACHED WARBLER *Sphenoeacus mentalis*
B915 / MPG1051 / D2C2

Afrotropical: widespread between 10°N and 15°S, E to W Ethiopia, but apparently avoiding much of the Congo basin. It reaches the E limit of the range in Kenya, where there are two disjunct populations (Br).

Kenyan: locally common but rather retiring in dense, rank grass and bush, usually in moist situations, up to 2000 m (Br; Ja). Although two disjunct populations occur, they are both of the nominate race (Br). The population to the E of the rift valley is on the E and SE peripheries of the C highlands. Range here is 94% at 500-2500 m, 81% semiarid-dry subhumid, and 94% within intermediate 250-1000 mm rainfall areas. In contrast, the population

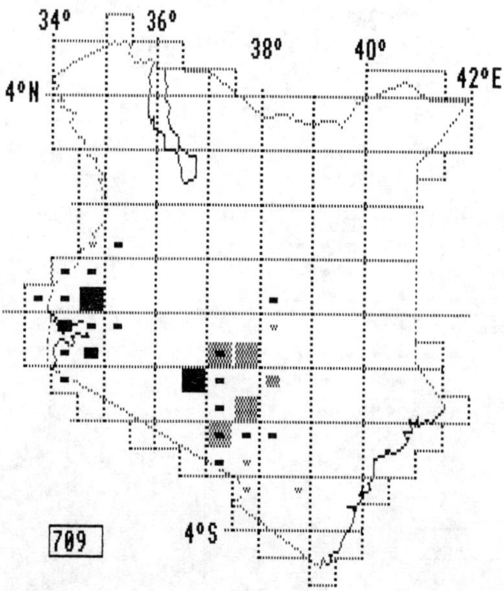

709

W of the rift is entirely at 1000-2000 m in moist subhumid-humid, 1000+ mm rainfall areas. Pre-1970 squares are distinctly more frequent E of the rift, where they comprise 44% of the total squares as opposed to 8% in the W. This difference may indicate a decline in the E, or commoner and more plentiful status in the W. There are no recent records from Tsavo E 101B (La), which is the species' sole arid square.

710 YELLOW-THROATED WOODLAND WARBLER
Phylloscopus ruficapilla
B906 / MPG963 / D2C3
NO MAP: post confirmed 101A

Afrotropical: mainly a highland species of the E, N from the Cape to S Mozambique, Malawi and SW Tanzania, with isolated populations in the highlands of NE Tanzania and the Taita Hills of S Kenya.

Kenyan: the race *P.r. minulla* is a common resident of forest and forest edge on the Taita Hills 101A (Br; ADL pers. obs.). This race is also found on the Usambara and South Pare Mts of adjacent NE Tanzania (Br).

This species forms part of a species-group with the Uganda Woodland Warbler (711)(HM), which is confined to areas W of the rift valley.

711 UGANDA WOODLAND WARBLER *Phylloscopus budongoensis*
B903 / MPG965 / D2C2
NO MAP: post presence 49C; post confirmed 48C; pre presence 61A

Afrotropical: a localised distribution from the E Zaire borders, through Uganda, to the E extremity of the range in Kenya.

Kenyan: forest interior at Kakamega 48D, in the North and South Nandi Forests 48D 49C, and at Nyarondo 61A (Br; Ja). Br also mentions occurrence on the Ugandan side of Mt Elgon 36CD (not mapped here). It is common at Kakamega, but sparse elsewhere (Br). Range is in 1000+ m areas to the W of the rift valley, with all three squares humid and within the 1000+ mm rainfall regime. See Yellow-throated (710) and Brown Woodland (712) Warblers.

712 BROWN WOODLAND WARBLER *Phylloscopus umbrovirens*
B909 / MPG964 / D2C2

Afrotropical: highlands of the E and NE, S from S Sudan, Ethiopia and N Somalia to the E Zaire-Uganda-Tanzania borders, Kenya and E Tanzania. Distribution in Kenya represents the E limit of the range.

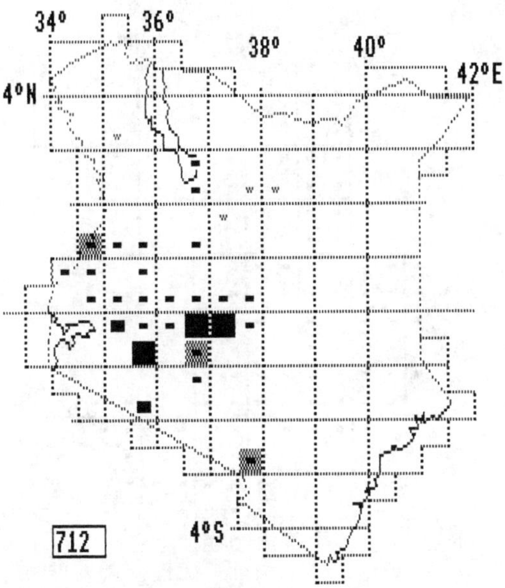

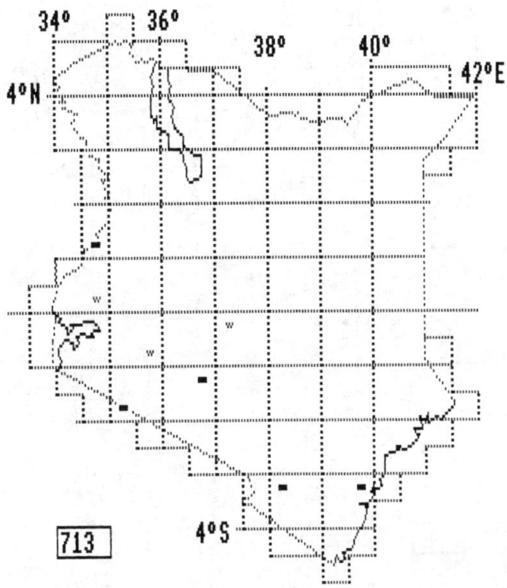

Kenyan: a common resident of forest, bamboo, woodland and giant heather at 1500-3300 m, but mainly above 1800 m (Br; MPG; Se; Britton & Sugg 1973). It is present on the forest islands of Mts Loima 13C, Kulal 26B, Nyiru 26D and Marsabit 27D 28C, the Ndoto Mts 39A, and the Chyulu Hills 88D (National Museum Collection, Nairobi; Br; Diamond & Keith 1980). Excluding the localised climates of these isolated highlands, range is entirely within subhumid-humid, 500+ mm rainfall areas. Its distribution overlaps that of the Uganda Woodland Warbler (711) in the W, but the Brown is at greater altitudes (Br).

713 WOOD WARBLER *Phylloscopus sibilatrix*
B907 / MPG961 / D2C1

Afrotropical: this W Palearctic species winters across equatorial Africa from Guinea to Somalia, and S to the forests of Gabon and Zaire. It is apparently scarce in the E, and there are few Kenyan records.

Kenyan: rare in forest and well developed woodland, up to 3000 m (Br; Se). The 10 very scattered records in all but one instance involve single birds: Naro Moru 63A Feb 1953 (Wallace 1975); Mau Narok 61D Oct 1963 (Se); Kakamega Forest 48D 1965 (Ripley & Bond 1971); Nairobi 75B Apr (Ba); Mt Elgon 36D Feb (P L Britton *in litt.*); Ngulia 101A Dec 1976 (EABR 1977); Mida Creek 102B (singing in mangroves) Mar (EABR 1979); Mara 74C Nov (EABR 1979); Ngulia 101A Dec (EABR 1982); and two at Mara 74C Mar (EABR 1983). These eight squares are in 500+ mm rainfall areas. Seven of the records are in subhumid-humid, 1000+ m areas of the W and C highlands. The two caught at Ngulia Lodge 101A were presumably en route to winter quarters S of Kenya. Br suggests that small numbers winter annually.

714 CHIFFCHAFF *Phylloscopus collybita*
B904 / MPG960 / D2C2

Afrotropical: many W Palearctic individuals of this species winter in the N tropics, S in East Africa to the Kenya highlands and Mt Kilimanjaro. Distribution in Kenya represents the E limit of the range.

Kenyan: this species achieves C2 because of its distinctive song, which immediately distinguishes it from the very similar but far more widespread Willow Warbler (715). The Chiffchaff is a Dec-Mar winter visitor to highland forest and bamboo, where it can be locally common in small numbers (Br; Ba). It is most common on Mts Marsabit 27D 28C, Elgon 36D and Kenya 63A, and in the Mau Forest 61ABD (*op. cit.*; Se; Sessions 1974b, 1975; Benson 1968; Oliver *et al.* 1971; Harper & Harper 1974b; Meadows & Weston 1974; OS-C 1977; Britton 1979; EABR). Wintering is mostly at 2200-3700 m, but there are also records at 1800-2000 m in the W and C highlands, and down to 1400 m in N Kenya (Br). The few records at lower altitudes are of birds on Dec or Mar-Apr passage (Br; Ba).

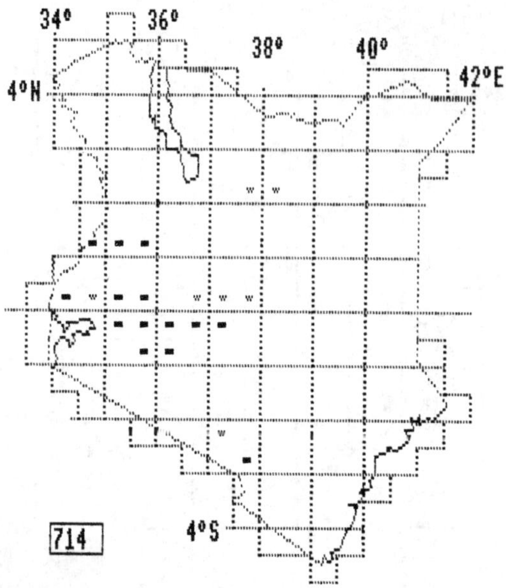

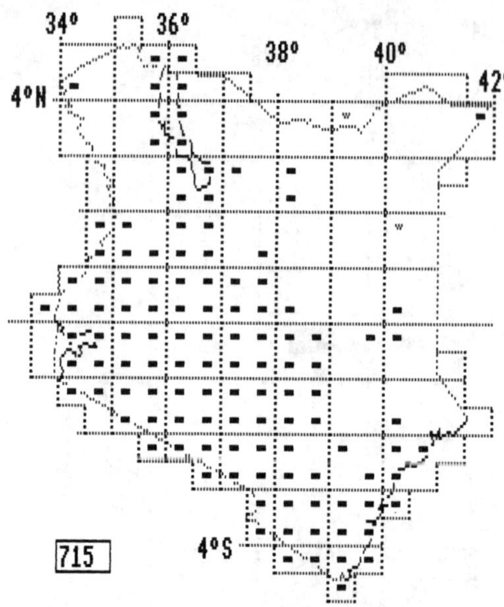

Excluding the localised climates of the forest islands of Mt Marsabit 27D 28C (Br) and the Chyulu Hills 88D (L L Short *in litt.*), range is 95% subhumid-humid, and 63% within 1000+ mm rainfall areas (*cf.* Willow Warbler).

The wintering ranges of the Chiffchaff and the closely related Willow Warbler are mainly segregated altitudinally. In addition, the former generally affects forest, whilst the latter prefers woodland and forest edge (Br).

715 WILLOW WARBLER *Phylloscopus trochilus*
B908 / MPG959 / D2C2

Afrotropical: a Palearctic species that winters from about 8°N in the W and about 12°N in the E, S to Namibia and E South Africa. There is but a single record from Somalia S of 5°N (AM), so that distribution in Kenya effectively describes the E edge of the range.

Kenyan: a common and sometimes abundant passage migrant and winter visitor to wooded areas and forest edge up to 3000 m, but mainly below 2200 m (Br; Ja; Se). Range is 80% well timbered and only 9% within the 0-250 mm rainfall regime, so that virtual absence from the arid NE may well be real. This is supported by similar status in S Somalia (see above).

Oct-Nov S passage is most marked in the W (Br), but the species is still common during this period in Tsavo E 101B (La). Dec-Feb overwintering is widespread in the moister areas of W and C Kenya, particularly at intermediate altitudes (Br; St). Very few remain during this period in arid Tsavo E 101B (La), and birds of this genus (probably mostly this species) at arid L Turkana are similarly restricted to passage dates (HH).

N passage during Apr is particularly heavy and widespread across the whole S of the country, from Uganda to the coast, with exceptional numbers sometimes occurring in C Kenya (Br; Ja; Se; La; *cf.* AM; EABR 1984). This heavy N movement is not annual around Baringo 50A (St). There is a single oversummering record, involving a bird at Kianyaga 63A on 1 Jul (EABR 1978). A bird ringed in Nairobi 75B was recovered in S Finland (Backhurst 1974). See Chiffchaff (714).

716 PECTORAL-PATCH CISTICOLA *Cisticola brunnescens*
B851 / MPG1020 / D1C2

Afrotropical: a rather fragmentary distribution in the E, from Ethiopia through Kenya and E Tanzania to C areas of Africa around 12°S. Further S, it reaches Zimbabwe and NE South Africa. There are also populations in Cameroon, the lower Congo basin and Angola, which may be isolated. Distribution in Kenya represents the E limit of the range.

Kenyan: locally common on mainly dry, short grass and moorland, at 1400-2500 m (Br). In contrast to the Wing-snapping Cisticola (717), range is 76% semiarid-subhumid and 68% within 500-1000 mm rain-

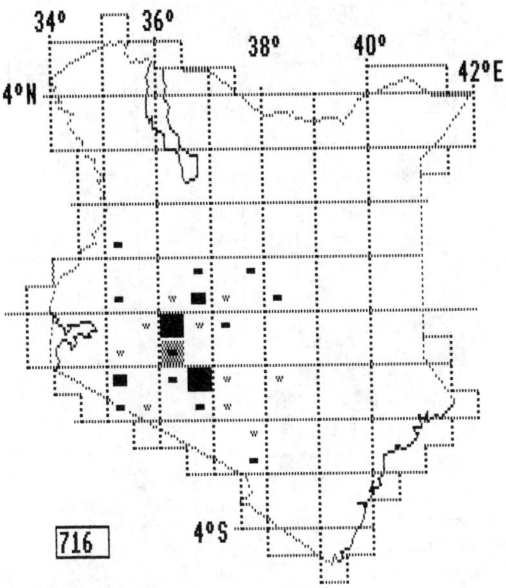

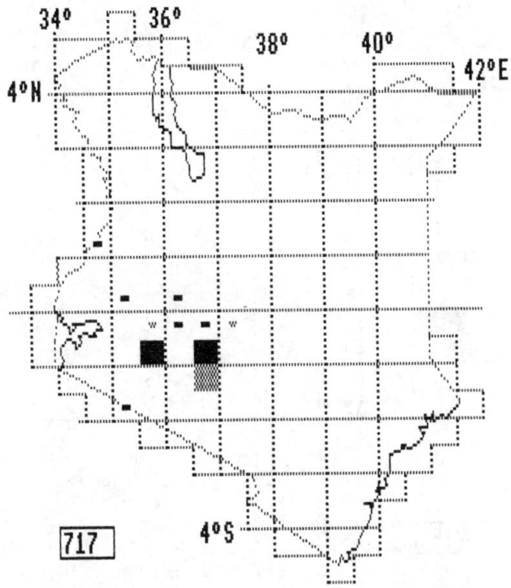

fall areas.

Together with the Wing-snapping and Black-backed (718) Cisticolas, this bird forms part of a species-group known as the Cloudscrapers because of their very high level song flights (often beyond unaided human vision)(HM; Lynes 1930). In the W and C highlands, the Wing-snapping is usually segregated from the Pectoral-patch on damper ground (see moisture indices of each) and at greater altitudes (Br). However, the two can occur and occasionally breed on the same ground, mainly at altitudes around 1700-1900 m (Br). The Black-backed would appear to be virtually completely segregated from both of the other species at lower altitudes (Br).

717 WING-SNAPPING CISTICOLA *Cisticola ayresii*
B848 / MPG1019 / D1C2

Afrotropical: mainly the E, with a fragmented distribution from the Cape N through Zimbabwe and Malawi to S Tanzania, the E Zaire borders, NW Tanzania, Uganda, S Sudan and, at the NE limit of the range, Kenya. There are also populations in Gabon, the lower Congo basin, S Angola and the Zaire-Zambia borders.

Kenyan: often common on short grassland, in both dry and damper situations at 1700-3700 m (Br; Ja; Se). In contrast to the closely related Pectoral-patch Cisticola (716: *q.v.*), range is 90% moist subhumid-humid and 63% within 1000+ mm rainfall areas.

718 BLACK-BACKED CISTICOLA *Cisticola eximia*
B858 / MPG1021 / D0C0
NO MAP: pre presence 48CD

Afrotropical: the N tropics, patchily distributed from Senegambia E to W Ethiopia and SE to S Sudan, Uganda and, on the SE extremity of the range, W Kenya.

Kenyan: short, dry grassland, at 900-1500 m (Br). The sole records are from Mumias and the Yala R 48CD (Ja). These two squares are to the W of the rift valley and N of the Winam Gulf 60AB, in humid, 1000+ mm rainfall areas. Br suggested this W Kenya population to be disjunct, since the Wing-snapping Cisticola (717) occupies areas to the E. Continued presence of the Black-backed in Kenya remains uncertain in the absence of recent records. See Pectoral-patch Cisticola (716).

719 ZITTING CISTICOLA *Cisticola juncidis*
B862 / MPG1016 / D2C2

Afrotropical: widespread S of 15°N, except for the equatorial lowland forests, and the drier areas of Somalia and the SW.

Kenyan: locally common on short to medium grassland in dry or seasonally flooded areas (Br; St). It is found at low to medium elevations (Br), with range 73% below 1500 m. Br considers its distribution apparently fragmented. The squares are 68% within 500+ mm rainfall areas and 90% semiarid-subhumid. The only

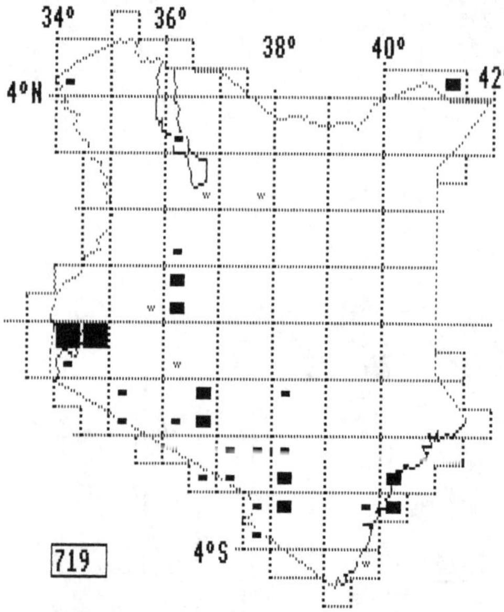

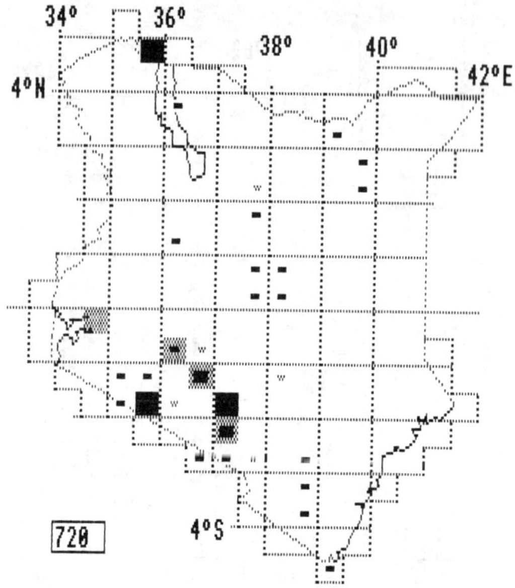

three reports from arid country are from locally moist situations: the mouth of the Kerio R 14C at L Turkana (HH), Mt Marsabit 27D (Ja), and lush grassland around the Daua R 7C (EABR 1984).

This cisticola forms part of a species-group with the Desert Cisticola (720), which favours drier habitats (HM). D2 reflects the Zitting's unique song flight.

720 DESERT CISTICOLA *Cisticola aridula*
B847 / MPG1018 / D1C1

Afrotropical: the N tropics, N of 10°N, from Senegambia E to Ethiopia and NW Somalia, and thence S to Kenya and E Tanzania. Further S, it is widespread S of 10°S, avoiding only the driest areas of the SW. Distribution in Kenya represents the E limit of the range.

Kenyan: locally common to abundant on dry, short grasslands, up to 1800 m (Br). There is a single record from the coastal strip 114C, during Apr (D J Pearson *in litt.*). Ja mentions occurrence on the Tana and Br lists the lower reaches of that river, but precise localities are omitted. In contrast to the closely related Zitting Cisticola (719: *q.v.*) range is 66% arid-semiarid and 73% within intermediate, 250-1000 mm rainfall areas. Most records in arid Tsavo E 101B are during the rains, but this may simply reflect greater activity while breeding (La; Brown & Britton 1980).

721 CROAKING CISTICOLA *Cisticola natalensis*
B866 / MPG1036 / D1C1

Afrotropical: widespread S of 10°N, except for most of the Congo basin, and the drier areas of Somalia and the SW.

Kenyan: common in a wide variety of bushed and wooded habitats with plenty of grass, up to 2200 m (Br). Range is 86% at 500-2000 m and 86% within 500+ mm rainfall

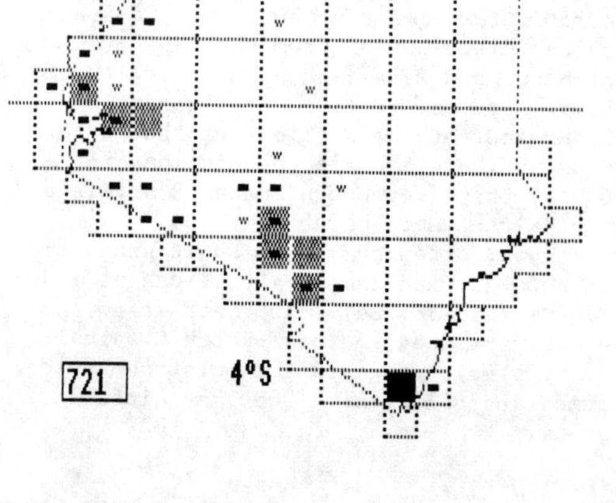

areas. The single locality in arid country is the locally moist highland of Mt Marsabit 27D (Br). The relatively high proportion of pre-1970 squares (33%) probably reflects a rather infrequently identified species, plus the modern disinclination for specimens and nest-finding.

This, the Stout (722) and the Aberdare (723) Cisticolas form a species-group (HM). The Aberdare inhabits grassland, and is mostly segregated from the other two species altitudinally (Br), though Turner (1977) notes it sympatric with the Stout in some areas. The Croaking and the Stout Cisticolas frequent similar, more vegetated habitats, and have been known to breed alongside each other (HM). However, where both occur, the Stout is usually at greater altitudes (HM). This is the case in Kenya, where the Stout is 66% above 1500 m and a bird of the highlands and their adjacent plateaux, whereas the Croaking is only 36% above this altitude, and mainly peripheral to the highest country 49 50 61 62 63.

722 STOUT CISTICOLA *Cisticola robusta*
B869 / MPG1035 / D2C2

Afrotropical: from Angola to S areas of Zaire and Tanzania, and thence N along the E Zaire-Tanzania borders to Uganda and the highlands of Kenya and NE Tanzania. There are also populations in Ethiopia and Cameroon. Distribution in Kenya represents the E limit of the range.

Kenyan: locally common in moist bushed or wooded grassland, at 1400-2500 m (Br; Ja). Range is 81% subhumid-humid and 96% within 500+ mm rainfall areas. The unusually high proportion (50%) of pre-1970 squares may reflect a decline. See Croaking Cisticola (721).

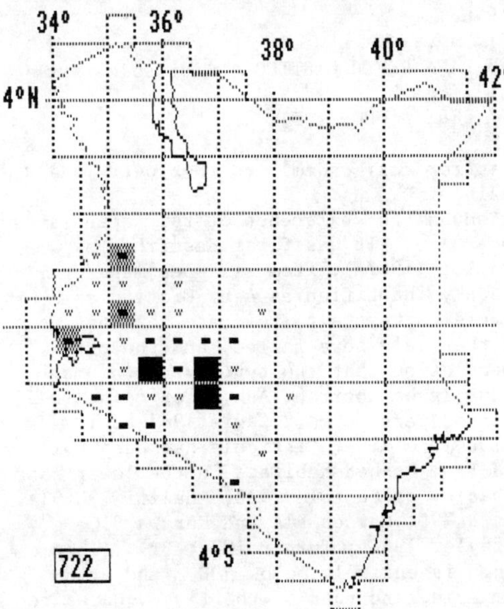

723 ABERDARE CISTICOLA *Cisticola aberdare*
B845 / MPG1035 / D2C3

Afrotropical: endemic to highland grasslands adjacent to the rift valley in C Kenya.

Kenyan: formerly considered a race of the Stout Cisticola (722), this bird was raised to species status by Traylor (1967b). It inhabits highland grasslands at 2300-3700 m, from Molo 61B and Mau Narok 61D E to the Aberdare Mts 62B (Turner 1977; Se). It is a common resident, often the commonest cisticola around 2400 m (*op. cit.*). Of the four squares, three are within the 1000+ mm rainfall regime, and all are humid. C3 is awarded because of the very restricted range. See Croaking Cisticola (721).

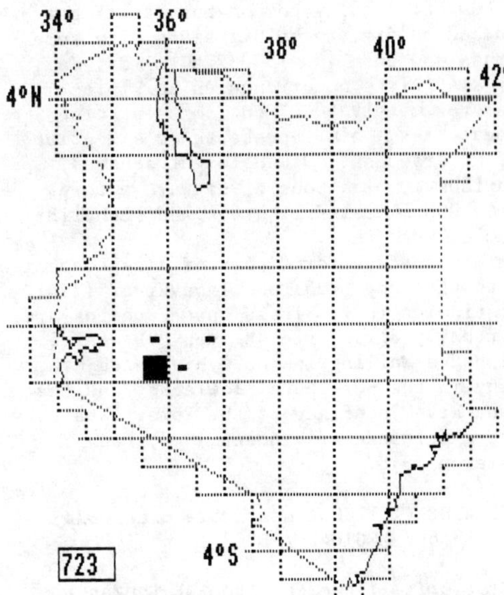

724 WAILING CISTICOLA *Cisticola lais*
B863 / MPG1022 / D1C2

Afrotropical: mainly the E, patchily distributed from the Cape N through Mozambique and Malawi to S Tanzania. There is another population in Angola. Further N, the race *C.l. distincta* (often considered a separate species) is endemic to Kenya and to adjacent areas of Uganda (Mt Moroto 24B: not mapped) and Tanzania (the Gol Mts 86A: not mapped)(Br; EABR 1981).

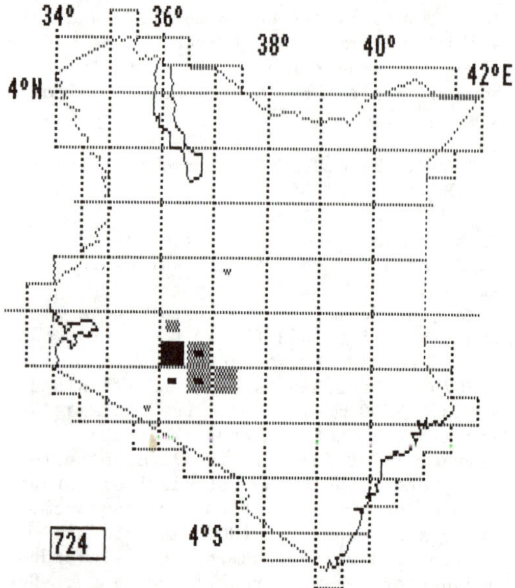

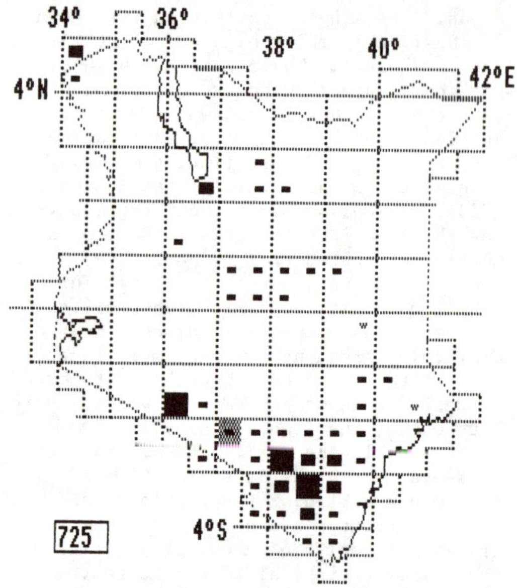

Kenyan: locally common but easily overlooked on dry, rocky hillsides with some grass and bush, at 1500-2750 m (Br; Ja). Br also lists occurrence on Laikipia (approximately 50CD) but omits precise localities. 51A represents Ja's mention of Mt Uraguess. Of the eight squares, various combinations of six are within 500-1000 mm rainfall areas and subhumid-humid.

This species forms part of a superspecies with the Ashy (725) and Tana River (726) Cisticolas (HM). These three species are virtually allopatric at even QSD scale, with the Wailing in the C highlands, the Ashy in the peripheral lowlands, and the Tana River confined to the lower Tana R. Other references: Vincent 1947; van Someren 1948.

725 ASHY CISTICOLA *Cisticola cinereola*
B856 / MPG1042 / D2C2

Afrotropical: the NE, from NE Tanzania, through Kenya to S Sudan, Ethiopia and Somalia.

Kenyan: sometimes common in dry, bushed habitats at 300-1300 m (Br). Range is 90% within intermediate, 250-1000 mm rainfall areas. This bird is mainly to the NE and SE of the C highlands, which constitute its local SW range edge, and virtually absent from the driest areas of the NE. This absence is supported by observers resident in 19A and 42A. Range is 94% arid but, with the notable exception of L Turkana (HH), occurrence in arid areas N of 2°S tends to be adjacent to locally moist situations *e.g.* Mt Marsabit 27D 28C, the Lorian swamp 52B 53A, and the Tana R 65B 78B 79A. This association with moist habitats could reflect observer frequency, but occurrence in adjacent S Somalia is similarly in the vicinity of the Juba and Shebelle R (AM). Apparent absence from the lower reaches of the Tana R 79C 91AC may reflect presence there of the Tana River Cisticola (726). See Wailing Cisticola (724) and Boran Cisticola (728). Other reference: Dowsett-Lemaire & Dowsett (1978b).

726 TANA RIVER CISTICOLA *Cisticola restricta*
B868 / MPG- / D1C1

Afrotropical: endemic to the lower Tana R basin of SE Kenya.

Kenyan: the existence of this species is uncertain. It was first described by Traylor (1967a), from six specimens previously identified as Ashy Cisticolas (725). Several attempts to relocate the species in the field have failed, and there is speculation that the type specimens may actually be aberrant Ashy Cisticolas (D J Pearson pers. comm.; Lewis 1981-82).

The Tana River Cisticola has been collected from bushed habitats in the lower Tana R basin, at Sangole 79BD, Mnazini 79C 91A, Ijara 79D, Garsen 91A and Karawa 91C (Traylor 1967a; Turner 1977; Br). This range is entirely below 500 m and arid-semiarid, and four of the five squares re-

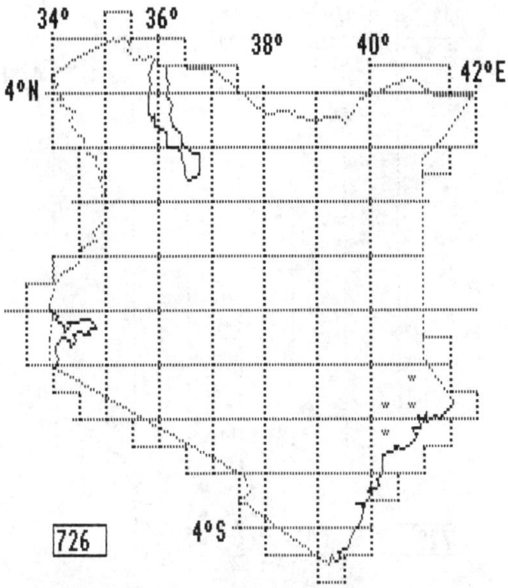

ceive 500-1000 mm annual rainfall. There is no breeding record. See Wailing Cisticola (724).

727 RATTLING CISTICOLA *Cisticola chiniana*
B854 / MPG1024 / D2C2

Afrotropical: widely distributed S of 5°S, except for the dry SW. Further N, it is found through mainly C and E areas of East Africa to S Sudan, Ethiopia and S Somalia. There may have been some confusion in N Kenya, Ethiopia and S Somalia between this species and the relatively recently described Boran Cisticola (728: *q.v.*)(HM; Br; AM).

Kenyan: a common and noisy resident of bushed and open wooded habitats, up to 2000 m (Br; Ja; Be; St). Br describes it in moderate to low rainfall areas, but range is in fact only 6% within the driest, 0-250 mm regime. Occurrence N of 2°N is thus restricted to the isolated highland of Mt Marsabit 27D 28C, and the S edges of populations in the moister areas of S Sudan and Ethiopia 2C 3B 4A 17AC 7C. With the exception of 4A, there are no records from even the relatively well vegetated river delta areas of L Turkana (HH). Scarcity in the NE corresponds to status in S Somalia, where the species is known only from 80B (not mapped) and from the coast at 3°N (AM). Status in N Kenya is incompletely understood, due to possible confusion with the probably closely related and very similar Boran Cisticola. There are no actual nesting records for the Rattling Cisticola from the Kenya coast, but breeding certainly occurs there (R A McVicker, C Ryall *in litt.*). Other references: Traylor 1967a; Carlson 1986b.

728 BORAN CISTICOLA *Cisticola bodessa*
B849 / MPG1024 / D1C1

Afrotropical: the total range of this recently recognised and little known species is uncertain, due to possible confusion with the very similar Rattling Cisticola (727). At present, there are records

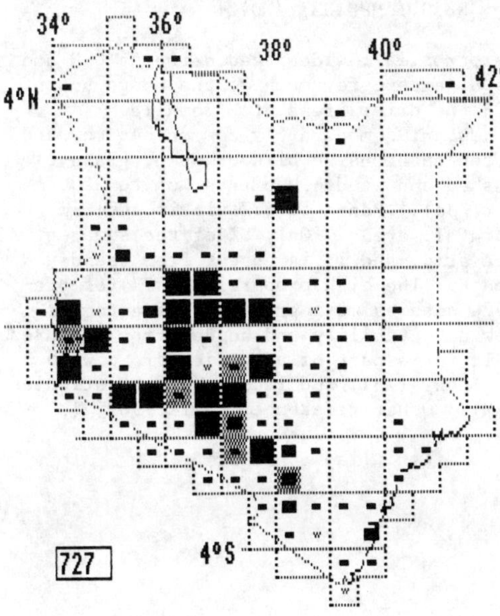

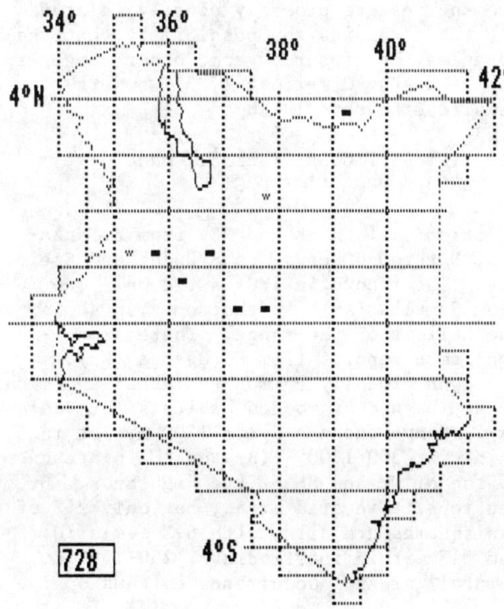

from SE Sudan and S Ethiopia, S to the N edge of the highlands of Kenya (HM; Dowsett-Lemaire & Dowsett 1978b), while AM mention possible occurrence in S Somalia.

Kenyan: formerly classified as a race of the Rattling Cisticola, this bird was raised to full species status by Ash (1974) and Erard (1974). It is apparently a very local species, frequenting rocky hillsides with scrub and some grass in N Kenya, up to 2100 m (Dowsett-Lemaire & Dowsett 1978b; Br). Range can effectively be summarised as the N and NE edges of the Kenya highlands, the S edge of the Ethiopian highlands, and the intervening isolated highland of Mt Marsabit 27D. Localities on the periphery of the Kenya highlands are Kapenguria-Kacheliba 37C (North & McChesney 1964), the Kito Pass 37D, Maralal 38D and NE of L Baringo 50A (all EABR 1983); and the Isiolo-Timau area 51CD, which is the best known and most frequently visited site (EABR 1977, 1978; Dowsett-Lemaire & Dowsett 1978b). Ja reported it from Marsabit 27D and Moyale 17A, and D A Turner (*in litt.*) recently relocated it at the latter site. Further localities for this easily overlooked bird will almost certainly be found.

Of the present eight squares, six are above 1000 m, six receive over 500 mm annual rainfall, and five are subhumid-humid. These figures are biassed below their true values by the moist, local highland of Marsabit, and by the scanty climatic data from Moyale. There is no breeding record.

The Boran Cisticola occurs alongside the Rattling on the Timau scarp 51C. These two species are probably closely related, but there remains the possibility that the Boran may be closer to some other congener, *e.g.* the Ashy Cisticola (725)(Dowsett-Lemaire & Dowsett 1978b).

729 TINY CISTICOLA *Cisticola nana*
B865 / MPG1041 / D2C1

Afrotropical: the E and NE, from E Tanzania, N through Kenya, to S Sudan and Ethiopia. The record in 19C is the only one from Somalia (AM), so that our map shows the E limit of the range. There are unconfirmed reports from NE Uganda (Br).

Kenyan: locally common to rather uncommon in dry bushed or wooded habitats (Br; Ja; La). Range extends up to 1500 m, but is mainly at 300-1300 m (Br) on the plateaux to the N, NE and SE of the highlands. Br mentions it in arid areas, but only 25% of its squares are arid, with 61% semiarid; and 91% within intermediate, 250-1000 mm rainfall areas. Occurrence is thus sparse

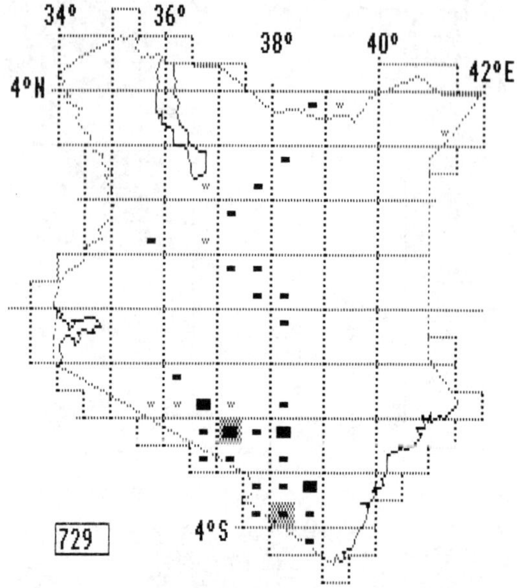

in the arid N, with notable absence from the better vegetated riverine habitats around L Turkana (HH). There are scattered records, including the locally moist highlands of Mts Nyiru 26D and Marsabit 27D, N to the S edge of the Ethiopian population at 16A (J S Ash *in litt.*) and Moyale 17A (Ja). The coastal and arid E lowlands, including the lusher habitats along the Tana R, are beyond the E range edge.

730 SIFFLING CISTICOLA *Cisticola brachyptera*
B850 / MPG1037 / D1C1

Afrotropical: widespread between 10°N and 15°S, except for much of the Congo basin, and the drier areas of E Somalia.

Kenyan: a moderately common but shy and often unobtrusive bird of open, grassy, bushed and wooded habitats, up to 2200 m (Br; Ja). Range is 88% within 500+ mm rainfall areas. Only 8% of the squares are arid, and it is absent from the dry N and E. The high proportion (47%) of pre-1970 squares most probably reflects its low detectability and nondescript features.

It forms part of a species-group with the Foxy (731) and Tabora (732) Cisticolas (HM). Other reference: Lewis 1982j.

(map overleaf)

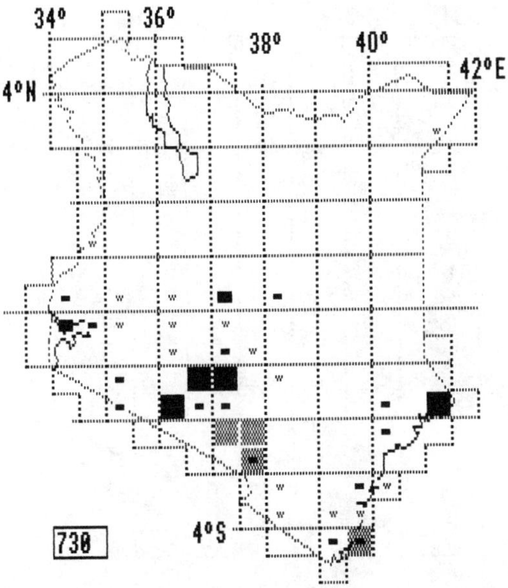

731 FOXY CISTICOLA *Cisticola troglodytes*
B872 / MPG1038 / D1C0
NO MAP: pre presence 37C 48B

Afrotropical: the N tropics, from Chad and the Central African Republic to W Ethiopia, and SE through S Sudan to N Uganda and, on the SE extremity of the range, NW Kenya.

Kenyan: a very unobtrusive bird of wooded grassland, up to 2200 m (Br). The only two precise localities are the Cherangani Hills 37C and Mangiki 48B (Ja). These two squares are above 1500 m, to the W of the rift valley. Both are humid, and within the 1000+ mm rainfall regime. Br's mention of occurrence in Pokot (approximately 37: not mapped) and up to 2200 m on Mt Elgon presumably derive from Ja's notes of Suk and Mangiki respectively. Br also lists W Turkana, but omits actual localities. There are no breeding or recent records, so that continued presence of the range edge in NW Kenya is uncertain. While of reasonably distinctive appearance, D1 reflects its easily overlooked nature. See Siffling Cisticola (730).

732 TABORA CISTICOLA *Cisticola fulvicapilla*
B859 / MPG1039, 1040 / D2C2
NO MAP: post presence 60C 61A 73B 74A; pre presence 60B

Afrotropical: widespread S of 10°S, avoiding only the dry SW. Further N, it extends to SE Tanzania, and through W Tanzania to the NE limit of the range in W Kenya.

Kenyan: local and often uncommon in open bushed or wooded habitats, up to 1400 m (Br). There are records from Kendu Bay 60B and Muhoroni 61A (Br), Gogo Falls 60C (National Museum Collection, Nairobi), Lolgorien 73B (photographed: A Lohding *in litt.*), and Mara GR 74A (Mann 1974c). These five squares are all moist subhumid-humid and above 1000 m, to the W of the rift valley. Four of the five are within 1000+ mm rainfall areas. There is no breeding record. See Siffling Cisticola (730).

733 ROCK-LOVING CISTICOLA *Cisticola aberrans*
B846 / MPG1027 / D0C0
NO MAP: post presence 89C; pre presence 75B 101ABC

Afrotropical: a discontinuous distribution across the N tropics, from Guinea and Mali to S Sudan. Further S, it occurs through East Africa, again rather discontinuously, to Malawi, Zimbabwe and E South Africa. There is also an isolated population in Angola. Distribution in Kenya represents the E limit of the range.

Kenyan: a local and little known species of bare, rocky places with some grass and stunted bush, at 600-2200 m (Ja; Br). It is possibly often overlooked, since the habitat is so atypical of the genus (*op. cit.*). There are records from Limuru 75B (MPG), Tsavo W 89C (D A Turner *in litt.*), the Taita Hills 101A and Voi 101B (Br), and Sagala 101BD (MPG). Four of these five squares are arid-semiarid, and all are within the intermediate, 250-1000 mm rainfall regime. There is no breeding record, and the only recent report is that from 89C.

This, the Whistling (734) and the Trilling (735) Cisticolas may form part of a species-group (HM). The Rock-loving is ecologically separate from the other two and, at QSD scale, allopatric with them.

734 WHISTLING CISTICOLA *Cisticola lateralis*
B864 / MPG1025 / D2C2

Afrotropical: W and C Africa, from Senegambia and Angola E to S Zaire. It also extends E from Cameroon, across the N periphery of the Congo basin, to S Sudan, Uganda and, on the E extremity of the range, W Kenya.

Kenyan: a wide variety of undergrowth, bushed, wooded and forest edge habitats, at 1100-1900 m to the W of the rift valley (Br; Ja). It forms part of a superspecies with the mainly allopatric Trilling Cisti-

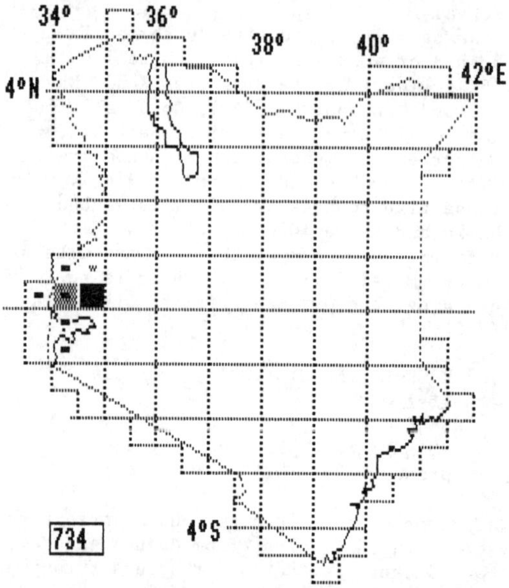

734

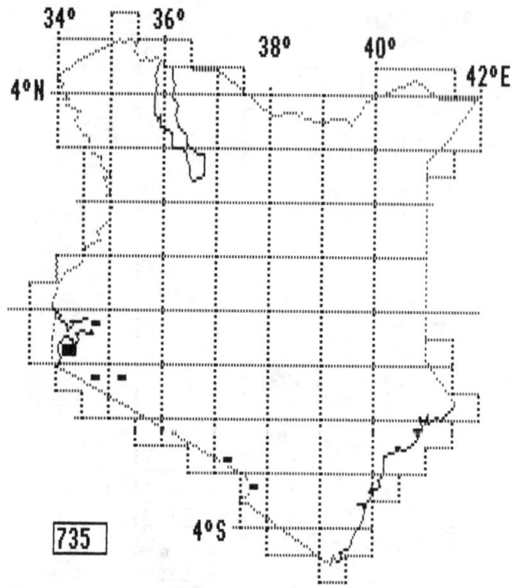

735

cola (735) (HM; Br). Both favour well-wooded country, but the Whistling's range is entirely moist subhumid-humid and within 1000+ mm rainfall areas, while the Trilling is mainly semiarid-subhumid and within the 500+ mm regime. In areas of overlap, the Trilling is in drier, more open and less densely vegetated country (*op. cit.*). See Rock-loving Cisticola (733). Other references: Mann 1973a, 1976.

735 TRILLING CISTICOLA *Cisticola woosnami*
B873 / MPG1026 / D2C1

Afrotropical: C and E Africa, from S Zaire, Zambia and Malawi to N Tanzania, Uganda and, at the NE limit of the range, W Kenya.

Kenyan: this locally common species has been reported from open, rocky, bush-clad hillsides in the W at Rapogi 60C, Wire Hill 60B, Lolgorien 73B and Mara River 74A (Sugg & Sugg 1972; Mann 1974c). Ja mentions it W of Mt Elgon 36D, but there is no record between Elgon and the Winam Gulf 60AB, despite good observer coverage (Br). The records in 88C and 100B are from Tanzania (J S S Beesley *in litt.*). All of the six squares are at 1000-2000 m and, in contrast to the Whistling Cisticola (734: *q.v.*), various combinations of five are semiarid-subhumid and in 500+ mm rainfall areas. The Trilling Cisticola is apparently resident near Rapogi 60C, and probable breeding in that square is based upon courtship display there in June (Sugg & Sugg 1972). See Rock-loving Cisticola (733).

736 RED-FACED CISTICOLA *Cisticola erythrops*
B857 / MPG1032 / D1C2

Afrotropical: widespread S of 10°N, apart from most of the Congo basin, and the dry areas of Somalia and the SW. Distribution in Kenya reflects the E limit of the range.

Kenyan: locally common but retiring in low, dense vegetation near water, at approximately 1000-2300 m (Ja; MPG; Br). It is found in the highlands, with the notable exception of the rift valley floor

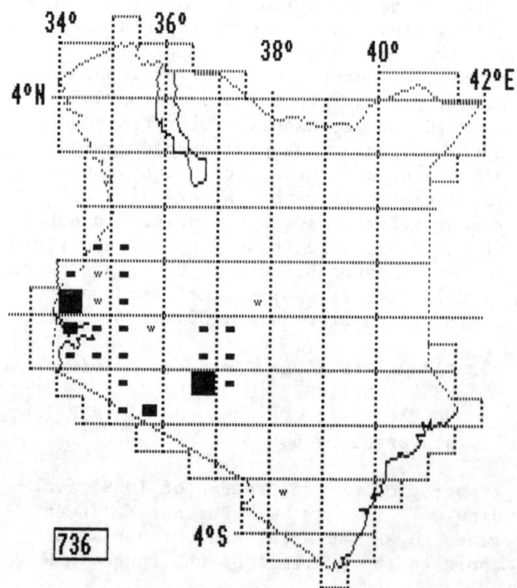

736

50AC 62AC 75AC, and is particularly common in the W (Br). There are specimens from the Taita Hills 101A (National Museum Collection, Nairobi). Excluding this isolated highland, range is entirely subhumid-humid, and 80% within 1000+ mm rainfall areas.

HM place this and the Singing Cisticola (737: *q.v.*) in a superspecies, but Br notes sympatry in Siaya 48C 60A; and very close coexistence around Nairobi 75B, where the Red-faced is characteristically in wetter, streamside sites, while the Singing can inhabit immediately adjacent but drier habitats (*e.g.* Lewis 1981-82). Other reference: Lewis 1982j.

737 SINGING CISTICOLA *Cisticola cantans*
B852 / MPG1031 / D2C2

Afrotropical: N from Mozambique, through Malawi and mainly highland areas of the E Zaire borders and East Africa, to Ethiopia. It also has a rather fragmentary distribution across the N tropics W to Senegambia. Distribution in Kenya represents the E limit of the range.

Kenyan: common but rather retiring in dense low vegetation, not necessarily near water, up to 2500 m (*cf.* Red-faced Cisticola (736))(Br). Range is 85% above 1000 m, and mainly in the highlands and adjacent plateaux. A specimen from the Shimba Hills 114A of the coastal lowlands (National Museum Collection, Nairobi) may be related to presence on the nearby Usambara Mts of adjacent NE Tanzania (Br). The Singing Cisticola extends into drier country than the Red-faced, being present on the rift valley floor, and 82% within 500+ mm rainfall areas and 35% semiarid. The record in 15B is of a pair with a nest on grassy, lightly bushed lava upland at 750 m altitude in the Huri Hills during Jul 1982 (F Alexander & H F Lamprey *in litt.*). This presumably represents the S edge of the species' range in the Ethiopian highlands.

738 HUNTER'S CISTICOLA *Cisticola hunteri*
B861 / MPG1030 / D2C2

Afrotropical: endemic to the highlands of NE Tanzania and W and C Kenya, just reaching Uganda on the W side of Mt Elgon 36D.

Kenyan: a locally common and conspicuous resident of highland forest glades and edge, bamboo and giant heather (Ja; Be; Se; Britton & Sugg 1973; Br). It occurs at 1550-4400 m, but is mainly above 1700 m (Br). It is present on the forest island of Mt Uraguess 51A (Ja). Excluding this isolated peak, range is entirely subhumid-humid and within 500+ mm rainfall areas.

This species forms a superspecies with the allopatric Chubb's Cisticola (739: *q.v.*) (HM). The only locality where both are known to occur is Mt Elgon 36D, and allopatry there is virtually sustained by altitudinal preferences, with Hunter's above 2500 m and Chubb's below (Britton & Sugg 1973). Very limited sympatry of apparently resident birds of these two species occurs on this mountain at 2500-2550 m in the bamboo zone, but no inter-

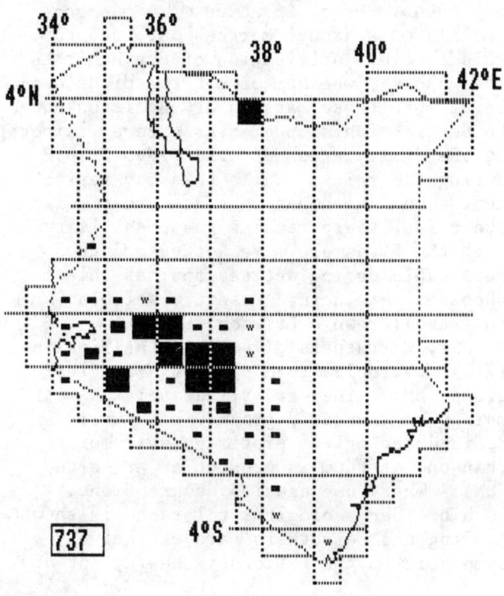

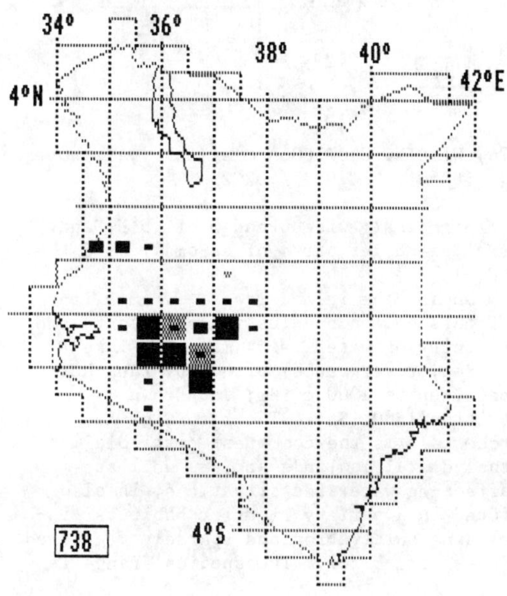

mediate individuals have been found (*op. cit.*).

739 CHUBB'S CISTICOLA *Cisticola chubbi*
B855 / MPG1028, 1029 / D2C2

Afrotropical: mainly the E, from Malawi and S Tanzania to the E Zaire borders, NW Tanzania, Uganda and, at the NE limit of the range, W Kenya. There is also a population in Cameroon.

Kenyan: a common and conspicuous cisticola of dense bush and scrub along forest edge and wooded streams to the W of the rift valley, usually at 1700-2200 m (Ja; Br). In contrast to the closely-related Hunter's Cisticola (738: *q.v.*), the squares are entirely moist subhumid-humid, and within the 1000+ mm rainfall regime.

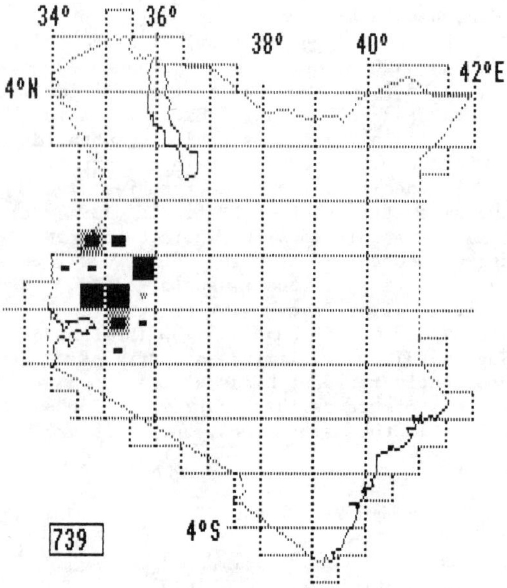

739

740 WINDING CISTICOLA *Cisticola galactotes*
B860 / MPG1033 / D2C2

Afrotropical: widespread S of 15°N, apart from the drier parts of E Somalia and the SW.

Kenyan: the inland race *C.g. amphilecta* of this common species is confined to the vicinity of water (*e.g.* marshes, lake shores, river banks) or nearby rank herbage, up to 2300 m (Br; Ja; Brown & Britton 1980; St). The race *C.g. haematocephala* is the commonest cisticola in the coastal lowlands and, as well as affecting waterside situations, is also found in a variety of drier habitats, including cultivation and sparsely vegetated dunes (*op. cit.*). The species' range is

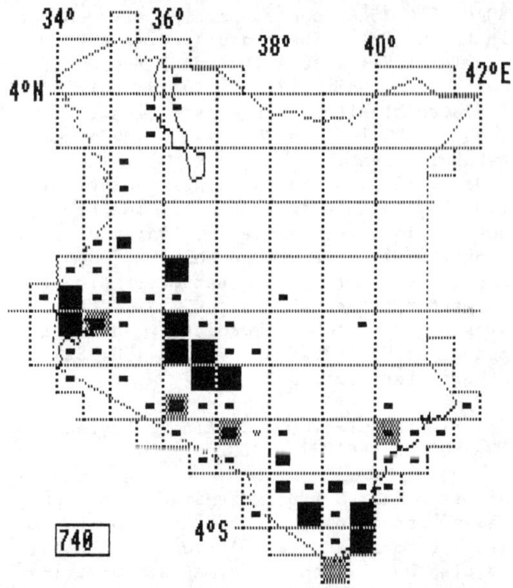

740

75% within 500+ mm rainfall areas, 89% within squares containing extensive permanent water of some type, and only 13% arid. Hence virtual absence from the arid N and E, with the notable exceptions of marshes at L Turkana (HH), and the locally lush habitats along the Tana R 65B (D J Pearson *in litt.*) and 79C. This riverine range parallels restriction to the vicinities of the Juba and Shebelle R in adjacent S Somalia (AM).

St has records throughout the year at L Baringo 50A, but at least local movements may occur. Hence Br notes *C.g. haematocephala* on seasonal marshes but, due to this species' utilisation of dry habitats (see above), whether or not the birds disperse when these marshes dry up is unknown. La describe this species as a rare visitor to the damper areas of arid Tsavo E 101B during the rains. While this may represent actual movement into dry areas during the rains, there remains the possibility that the birds are more active and thus detectable during wet seasons, as this is when they breed (*cf.* Desert Cisticola (720) in Tsavo)(Brown & Britton 1980).

This, Carruthers' (741) and the Tinkling (742) Cisticolas form part of a species-group (HM). They are all associated with marshy or inundated ground, with well defined ecological preferences if more than one of them is present at any site (HM). Where one species occurs alone, much broader ecological tolerance is shown. Winding and Carruthers' are extensively sympatric in the L Victoria basin, but are ecologically segregated, with Carruthers' endemic to the interiors of the papyrus swamps, while the Winding prefers swamp edge (Britton 1978). The Tinkling Cisticola is absent from Carruthers' range in the marshes around L Victoria, and elsewhere is mainly segregated from the Winding at greater altitudes.

741 CARRUTHERS' CISTICOLA *Cisticola carruthersi*
B853 / MPG1034 / D2C3
NO MAP: post presence 60AC; post probable 60B; post confirmed 48C; pre confirmed 60B

Afrotropical: a papyrus endemic of the E, from the E Zaire borders to NW Tanzania, Uganda and, on the E extremity of the range, W Kenya.

Kenyan: common in the interior of papyrus swamps in the L Victoria basin (Britton 1978; Br; EABR 1979, 1981). Range is entirely at 1000-1500 m in moist subhumid-humid, 1000+ mm rainfall areas. See Winding Cisticola (740). Other reference: Britton 1970a.

742 TINKLING CISTICOLA *Cisticola tinniens*
B871 / MPG1044 / D2C2

Afrotropical: mainly a S species, widespread S of 10°S but avoiding the dry SW and N Mozambique. It also reaches the E Zaire borders and, at the N limit of the range, *C.t. oreophila* is endemic to the W and C highlands of Kenya.

Kenyan: a reasonably common bird of permanent streams, swamps and marshes at 2000-3200 m in the highlands (Br; Ja). Range is 72% within 1000+ mm rainfall areas and 90% moist subhumid-humid. See Winding Cisticola (740).

743 TAWNY-FLANKED PRINIA *Prinia subflava*
B913 / MPG1045 / D2C2

Afrotropical: widespread S of 15°N, apart from the drier areas of the SW and most of Somalia.

Kenyan: a locally common warbler of dense grass and other low growth in bushed, wooded, forest edge and swamp habitats, up to 2300 m (Br; Ja; Be). Range is 79% within 500+ mm rainfall areas, and only 10% arid. Hence it is virtually absent from the dry N and E, including locally moist lacustrine, upland and riverine habitats. The records in 3B and 4A are from outside Kenya, in the vicinity of the Omo R delta (J S Ash *in litt.*) and represent the S edge of range in Sudan and Ethiopia. The record in 80A derives from Somalia, where the species occurs, uncommonly, S of 1°N (AM).

This species forms part of a superspecies with the allopatric Pale Prinia (744: *q.v.*), which replaces it in N and E Kenya, SE Ethiopia and most of Somalia (HM; AM). In areas of apparent sympatry, the Tawny-flanked is found in lusher, moister situations (Br). This segregation is evident in arid-semiarid Tsavo E 89ABD 90C 101B, where the Tawny-flanked has been recorded in riverine habitats, while the Pale is known from the drier area of 90C (La).

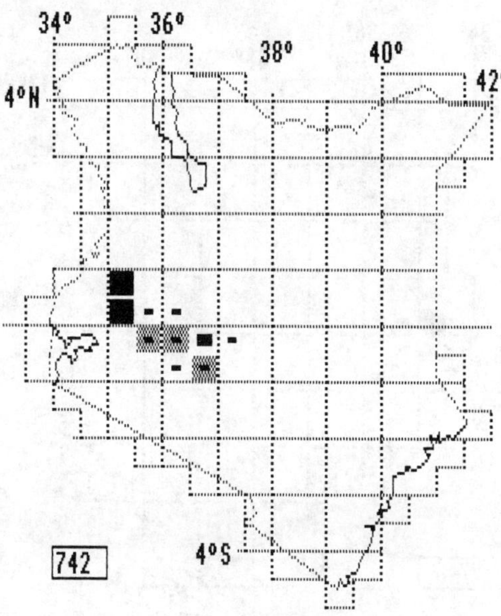

742

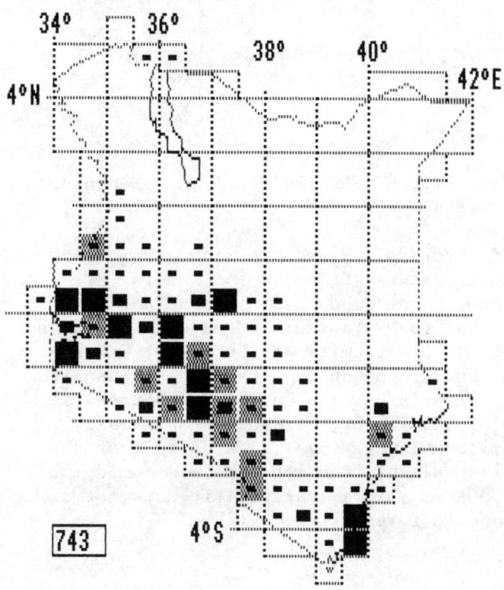

743

744 PALE PRINIA *Prinia somalica*
B912 / MPG1046 / D2C2

Afrotropical: replaces the Tawny-flanked Prinia (743: *q.v.*) in the NE, N from Kenya to NE Uganda, S Sudan, SE Ethiopia and all but the far S of Somalia.

Kenyan: rather local and uncommon in open, dry, bushed country, at 150-1300 m (Br; St). Range is in the lowlands of the N and E, 79% below 1000 m. In contrast to the Tawny-flanked Prinia, the squares are 93% arid-semiarid and 78% in 0-500 mm rainfall areas. MPG note breeding condition in N Kenya, but the single confirmed breeding record 19C derives from Somalia (AM).

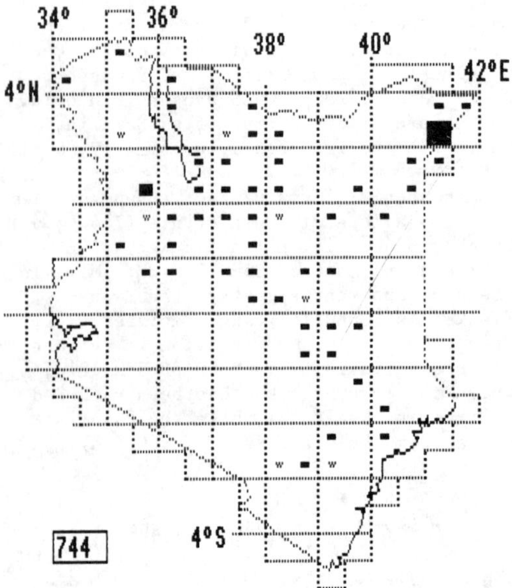

745 WHITE-CHINNED PRINIA *Prinia leucopogon*
B911 / MPG1048 / D2C2

Afrotropical: W and C Africa, from the W coast between Cameroon and Angola, E through the Congo basin and adjacent areas, to NW Tanzania, Uganda, S Sudan and, at the E extremity of the range, W Kenya.

Kenyan: common in forest edge, bush and other moist, well-vegetated habitats, at 1100-2400 m (Br; Ja; Sugg & Sugg 1972; Britton & Sugg 1973). Range is W of the rift valley, in moist subhumid-humid, 1000+ mm rainfall areas. It does not extend to forest islands.

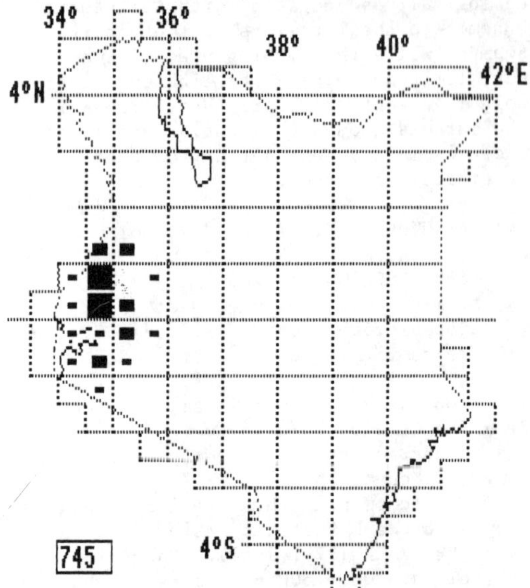

746 BANDED PRINIA *Prinia bairdii*
B910 / MPG1049 / D1C1

Afrotropical: a W and C African species, found in three distinct areas: in the far W, from Cameroon to Angola; in E Zaire and the E Zaire-Uganda-Tanzania borders; and an isolated population in W Kenya.

Kenyan: local, somewhat retiring and often uncommon in forest undergrowth and bush, mainly in edge areas, at 1700-3000 m

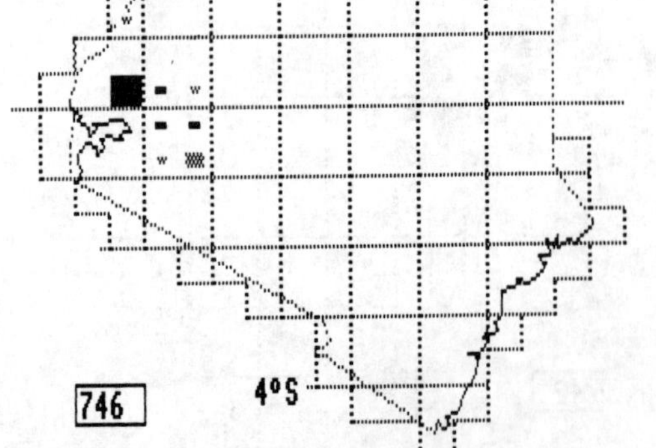

(Br; Ja; Se). Range is entirely to the W of the rift valley, in 1000+ mm rainfall areas. Six of seven squares are humid. The significance of the high proportion of pre-1970 squares (50%) is uncertain in view of the low detectability.

747 RED-FRONTED WARBLER *Spiloptila rufifrons*
B916 / MPG987 / D2C2

Afrotropical: the NE, from an apparently disjunct population in N Tanzania and S Kenya (Br) to N and E Kenya, Somalia, Ethiopia and W Sudan.

Kenyan: locally common in dry bush and scrub at 300-1300 m (Br; HH; La; St). It is found in the N and E lowlands, with an apparently disjunct population to the S of the C highlands, from Olorgesaille 75C S to L Natron 87A (Br) and N Tanzania (*e.g.* 87D: J S S Beesley *in litt.*). Range is 92% arid-semiarid and 74% within 0-500 mm rainfall areas. Absence from the E border S of 31A and from even semiarid areas of the coastal strip compares with absence S of 1°N in adjacent S Somalia (AM).

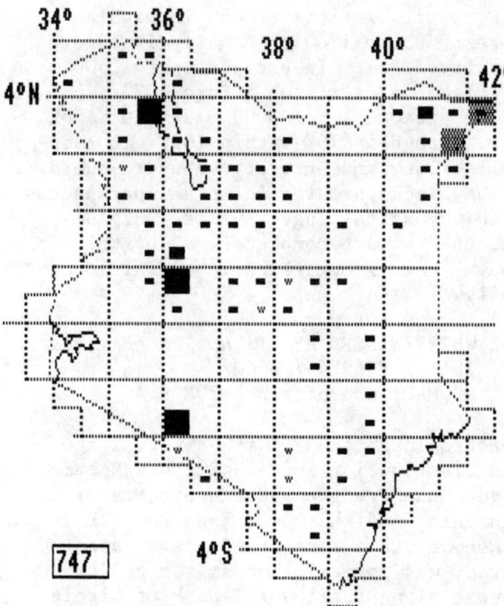

748 RED-WINGED WARBLER *Heliolais erythroptera*
B883 / MPG1050 / D2C0
NO MAP: pre presence 48C 61A

Afrotropical: a patchy and localised distribution from Mozambique N through Zambia, Malawi and E Tanzania, to W Kenya, NW Uganda and Ethiopia. From the latter country, it also extends W across the N tropics to Ghana.

Kenyan: grassy areas in bushed or wooded country (Br). The two records for Kenya, both from the L Victoria basin, were at Fort Ternan 61A on 27 Aug 1918, and at Ng'iya 48C on 7 May 1969 (Br; Ja). These birds were presumably wanderers E or N of their normal range.

749 BUFF-BELLIED WARBLER *Phyllolais pulchella*
B902 / MPG995 / D2C2

Afrotropical: from N Tanzania and NE Zaire to Kenya, Uganda, Ethiopia and Sudan. In the N tropics, there are scattered records W to Chad. Distribution in Kenya represents the E limit of the range.

Kenyan: a locally common species, characteristic of acacia canopies in bushed, wooded and forest edge environments, at 600-2000 m (Br; MPG; La; St). Range is 82% above 1000 m and 75% within 500+ mm rainfall areas. There are records from only three arid squares, at Lodwar 13D (Br), Kora 64B, and formerly in Tsavo E 101B (La).

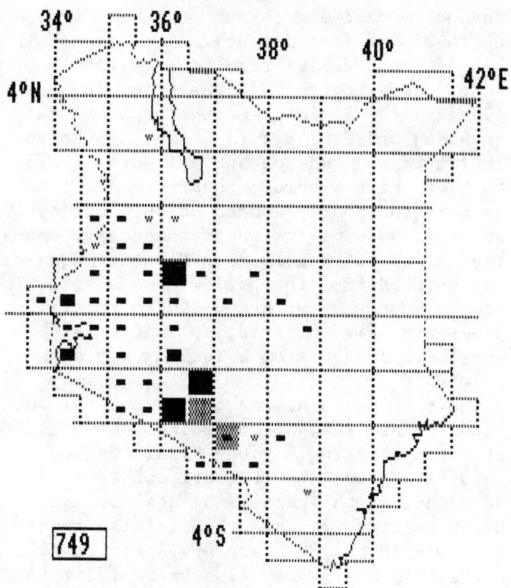

750 YELLOW-BREASTED APALIS *Apalis flavida*
B818 / MPG979, 980 / D2C2

Afrotropical: widespread S of 5°N, except for the dry SW and much of the Congo basin. Further N, it extends to Nigeria, Chad, S Sudan, Ethiopia and N Somalia.

Kenyan: a sometimes common species which is by far the most widespread apalis in Kenya, being less restricted than its con-

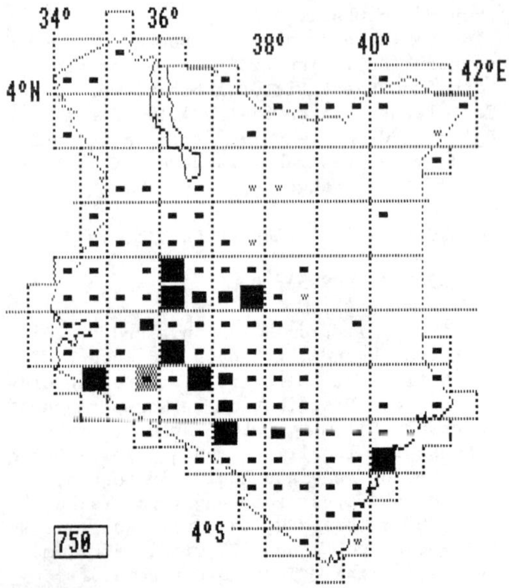

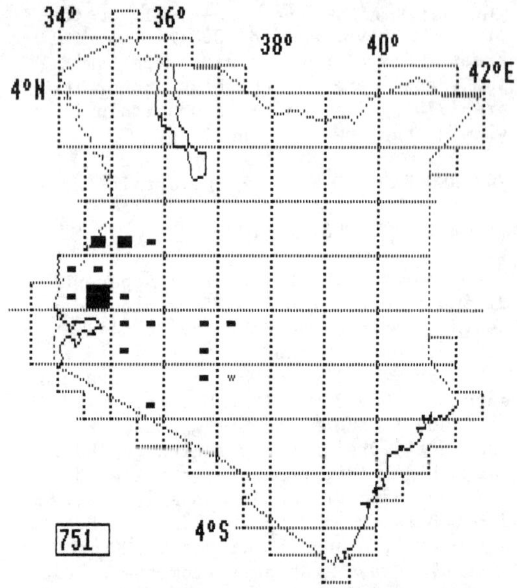

geners to forest areas (HM). Instead, its range includes a wide variety of thicket, bushed, wooded and forest edge habitats, up to 2200 m (Br; Ja; St). The records in 31A 67C 80AC derive from Somalia (AM).

Records along the N border from 2C E to 19B mark the S limits of the populations in Sudan and Ethiopia (J S Ash, G Nikolaus *in litt*.; D J Pearson and ADL pers. obs.). Most of these N border squares are shown as arid and poorly wooded on our overlays, but this possibly reflects inadequate ecological data. Hence these N border squares are omitted from the following analyses. Range is only 3% within 0-250 mm rainfall areas and only 18% arid, so that this apalis is absent from L Turkana and much of the NE.

The species is scarce in the coastal lowlands, where the sole breeding record is of a pair feeding a young Klaas' Cuckoo (373) at Karawa 91C (P L Britton *in litt*.). Be described this apalis as resident in Narok district 74, and this is also true in 74D 75BD 88A 89A (authors' pers. obs.). While this apalis may also be resident in arid Tsavo E 101B, all La's records there were May-Oct. Other references: Lewis 1982cd, 1983a.

751 BLACK-THROATED APALIS *Apalis jacksoni*
B819 / MPG982 / D2C2

Afrotropical: occupies three distinct areas. In the E, it ranges from the E Zaire borders, E to NW Tanzania, S Sudan, Uganda and, at the E limit of the range, the highlands of W and C Kenya. There are also populations in Cameroon and Angola.

Kenyan: sometimes common in highland forests above 1700 m, usually in edge habitats (Br). It extends up to 2400 m on Mt Elgon 36D (Britton & Sugg 1973), but Se had no records from 3000 m at Mau Narok 61D. Range is 76% within 1000+ mm rainfall areas, and entirely subhumid-humid.

The Black-throated forms a superspecies with the White-winged Apalis (752)(HM), and the two are completely allopatric in Kenya, both geographically and altitudinally.

752 WHITE-WINGED APALIS *Apalis chariessa*
B816 / MPG984 / D2CO
NO MAP: pre presence 79C 91A

Afrotropical: the E, with a very fragmented distribution in Malawi and N Mozambique, the Uzungwa and Uluguru Mts of E Tanzania, and the lower Tana R of SE Kenya.

Kenyan: this, one of Africa's rarest birds, was formerly known from gallery forest along the lower Tana R at Mitole 91A, and 37 km N of Garsen 79C (Br; Ja). Ja considered it very scarce, mentioning only the type and just one other specimen procured at Mitole. Despite numerous searches, there have been no records from the Tana since 1961 (Br), and the species may well be extinct in Kenya. The two squares are both below 100 m, arid-semiarid, and within the 500-1000 mm rainfall regime. See Black-throated Apalis (751). Other reference: Keith 1968.

753 BAR-THROATED APALIS *Apalis thoracica*
B828 / MPG970, 971 / D2C3
NO MAP: post presence 101A; pre presence 88D

Afrotropical: the E, from the Cape to Zimbabwe, E Zambia and Malawi. Further N, it extends through E Tanzania to the N limits of the range in S Kenya.

Kenyan: reasonably common in forest undergrowth and mid-levels at 1300-2200 m on the Chyulu Hills 88D and the Taita Hills 101A (Br; van Someren 1939). *A.t. griseiceps* on the Chyulu Hills is also widespread in adjacent NE Tanzania (Br), but *A.t. fuscigularis* is endemic to the Taita Hills. This species' restriction to isolated highlands renders overlay analysis impracticable. There are no breeding records.

The Bar-throated forms part of a species-group with the Grey Apalis (754), which replaces it in the W and C highlands (HM).

754 GREY APALIS *Apalis cinerea*
B817 / MPG973 / D2C2

Afrotropical: occupies three distinct areas. In the E, through the E Zaire-Tanzania-Uganda borders NE to S Sudan and, on the E limit of the range, Kenya. There are also populations in Cameroon and Angola.

Kenyan: locally common in highland forest and forest edge at 1700-3000 m, from undergrowth to treetop levels (Br; Ja; Se; Britton & Sugg 1973). It is present on the forest islands of Mt Marsabit 27D 28C (Mann 1972), the Ndoto Mts 39A and the Mathews Range 39C (Br). Excluding these isolated highlands, range is 68% within 1000+ mm rainfall areas and entirely subhumid-humid. See Bar-throated Apalis (753).

755 BUFF-THROATED APALIS *Apalis rufogularis*
B826 / MPG972, 976 / D2C2
NO MAP: post presence 36D 48BD 49C; pre presence 48C; pre probable 48D

Afrotropical: W and C Africa, from Nigeria to Angola, and E to E Zaire, NW Tanzania, S Sudan, Uganda and, on the E extremity of the range, W Kenya.

Kenyan: a locally common forest or more usually forest edge species, at 1700-2400 m (Br; Britton & Sugg 1973). There are records from Mt Elgon 36D (Britton & Sugg 1973), 48B (P L Britton *in litt.*), Mumias 48C (Ja), Kakamega Forest 48D (many obs.; Br), and the North Nandi Forest 48D 49C (Diamond & Fayad 1979). Range is entirely to the W of the rift valley and N of the Winam Gulf 60AB. All five squares are humid and within the 1000+ mm rainfall regime. Probable breeding in 48D is based upon song and birds in breeding condition at Kakamega Forest (Zimmerman 1972).

756 CHESTNUT-THROATED APALIS *Apalis porphyrolaema*
B824 / MPG985, 986 / D2C2

Afrotropical: highlands of the E, from Mozambique and Malawi to many highland areas of East Africa and the E Zaire bor-

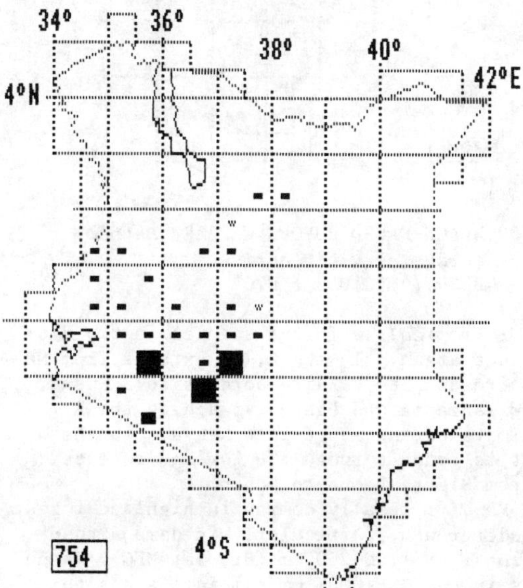

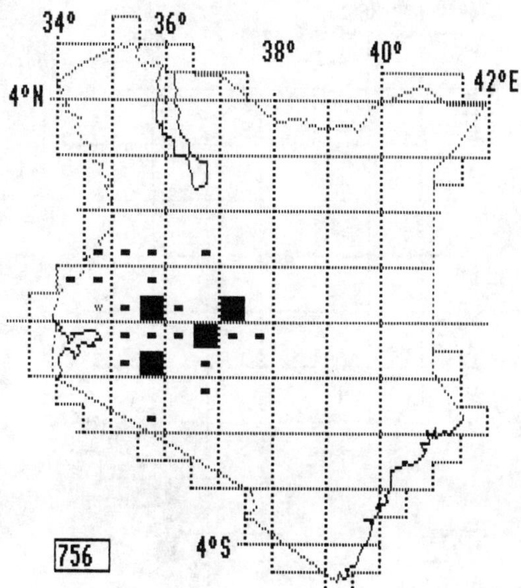

ders. Distribution in Kenya represents the NE limit of the range.

Kenyan: common in highland forest and forest edge, mainly at 1700-3400 m (Br; Ja; Se; Britton & Sugg 1973). Range is 69% within 1000+ mm rainfall areas and entirely subhumid-humid. This warbler has not been recorded from isolated forest islands. MPG mention possible movements to lower altitudes on Mt Elgon 36D during Nov-Dec, but Se considered the bird resident around 3000 m at Mau Narok 61D.

This species forms part of a superspecies with the Black-headed Apalis (757), which replaces the Chestnut-throated on the E of the C highlands (HM). Where the two are sympatric, *e.g.* around Nairobi 75B, the Chestnut-throated is mostly above 1500 m while the Black-headed is mostly lower (*op. cit.*).

757 BLACK-HEADED APALIS *Apalis melanocephala*
B821 / MPG975 / D2C2

Afrotropical: the E, from Mozambique to E Tanzania, Kenya and extreme S Somalia. Distribution in Kenya represents the N limits of the range.

Kenyan: locally common in forest and dense woodland, up to 2000 m (Br). It occurs throughout the coastal lowlands N to the lower Tana R 79C (AGH; Britton & Zimmerman 1979). The record in 80A is from Somalia (AM). Inland, it is found in the C highlands from Nairobi 75B NE to Meru 51D, and on isolated highlands to the SE. The latter are Mt Endau 77B (Br; A W Diamond *in litt.*), the Taita Hills 101A (Br) and Mt Kasigau 101D (EABR 1981). Excluding these isolated highlands, range is entirely within 500+ mm rainfall areas. The race at the coast is 82% semiarid-dry subhumid, while that in the highlands is moist subhumid-humid. See Chestnut-throated Apalis (756).

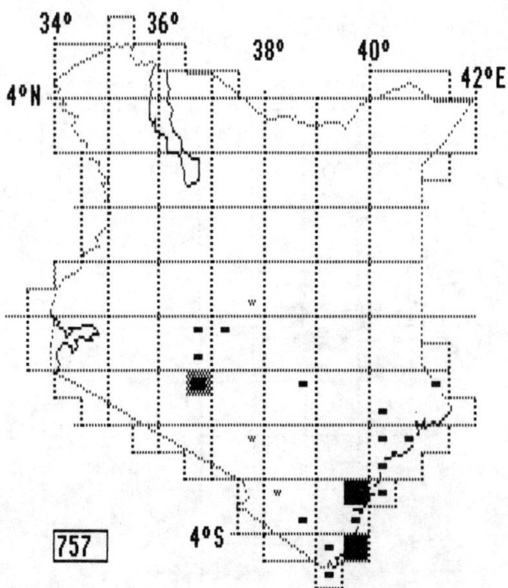

758 BLACK-COLLARED APALIS *Apalis pulchra*
B825 / MPG977 / D2C2

Afrotropical: a highland species of the NE and SE Zaire borders, S Sudan and, at the E limit of the range, the highlands of Kenya. There is also a population in Cameroon.

Kenyan: common in highland forest undergrowth at 1550-2400 m (Br; Ja; Britton & Sugg 1973). Range is 83% within 1000+ mm rainfall areas, and entirely moist subhumid-humid.

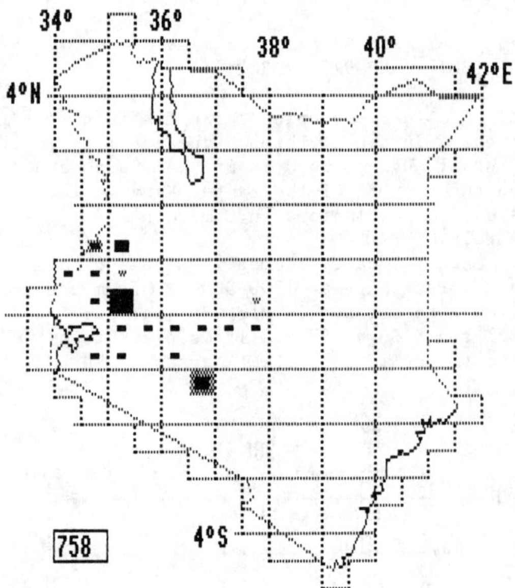

759 BLACK-FACED RUFOUS WARBLER *Bathmocercus cerviniventris*
B829 / MPG1053 / D2C2

Afrotropical: a forest species in two distinct areas. In the E, it extends from NE Zaire and the E Zaire borders to S Sudan, NW Tanzania and Uganda, reaching the E limits of the range in W Kenya. In the W, it is found through the lowland forests from Sierra Leone to Gabon.

Kenyan: locally common in highland forest undergrowth, particularly in damp situations, at 1700-2800 m (Br; Ja; MPG). Range is in humid, 1000+ mm rainfall areas to

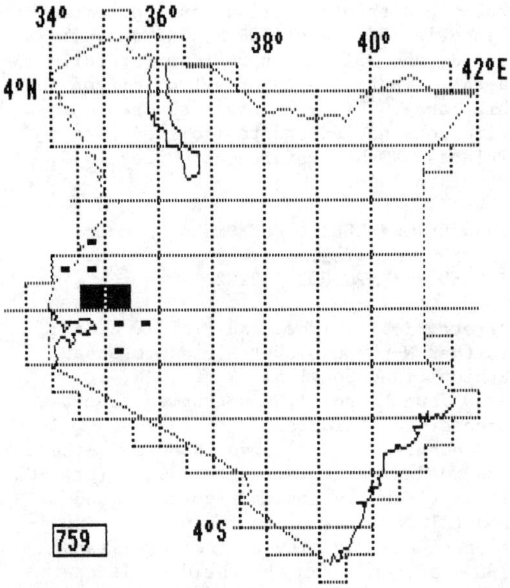

the W of the rift valley.

760 GREY-BACKED CAMAROPTERA *Camaroptera brachyura*
B837 / MPG1009, 1011 / D2C2

Afrotropical: widespread S of 15°N, avoiding only the drier areas of Somalia and the SW.

Kenyan: this common but retiring warbler frequents dense, low vegetation in a wide variety of habitats up to 2200 m (Br; St). D2 reflects its well known calls. Normally only in the edge of forest clearings, it commonly penetrates forest interior at Sokoke 102B (Br; Britton & Zimmerman 1979). Range is only 9% within the lowest, 0-250 mm rainfall regime, and 25% arid. Occurrence is thus rather scarce in the N, with the notable exception of L Turkana, where the bird can be frequent in suitable cover (HH). Absence from the arid E, including the riverine habitats of the Tana, is more pronounced, and does not correlate with distribution in adjacent S Somalia (AM). Occurrence in arid Tsavo E 101B is occasional, in wooded and riverine habitats (La). Green- and grey-backed members of this species are often given full species status (see HM; Br).

761 GREY WREN WARBLER *Camaroptera simplex*
B839 / MPG968 / D2C2

Afrotropical: occurs in the lower Congo basin and Angola, E to E areas of Zambia and Zaire. Further NE, range is through S and E Tanzania to Kenya, NE Uganda, S Sudan, Ethiopia and Somalia. There may be more than one species involved (HM).

Kenyan: a sometimes common bird of bushed and wooded habitats in dry country, up to 1300 m (Br; Ja; HH; La). While most frequent in dense undergrowth, the well known song is delivered from high, exposed perches (hence D2). Range is 88% arid-semiarid, and throughout the country with the exceptions of the highlands, the moist W, and the dry subhumid, 1000+ mm rainfall areas of the coastal strip. Distribution

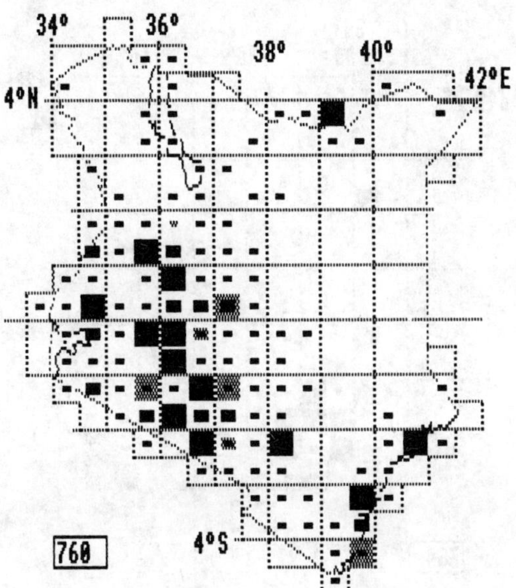

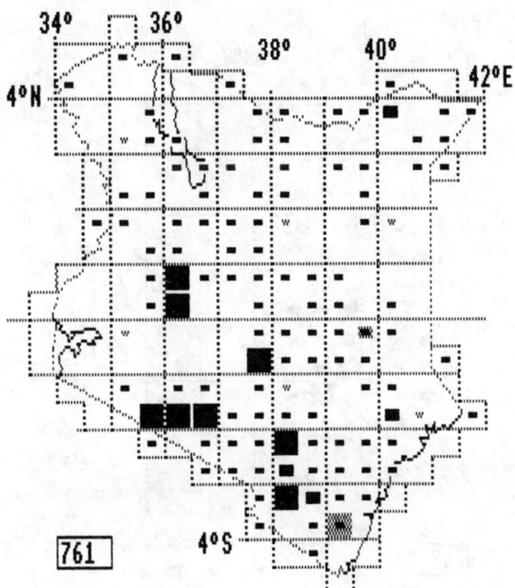

of breeding squares reflects observer frequency.

762 OLIVE-GREEN CAMAROPTERA *Camaroptera chloronota*
B838 / MPG1010 / D2C3
NO MAP: post presence 49C; post confirmed 48D

Afrotropical: W and C Africa, from the lowland forests that border the Gulf of Guinea through scattered areas of the Congo basin to Uganda, S Sudan and, at the E limit of the range, W Kenya.

Kenyan: locally common in forest undergrowth at 1700-2000 m in the Kakamega 48D and North Nandi 48D 49C Forests, and at Kaimosi 48D (Ja; Diamond & Fayad 1979; Br). The two squares are in humid, 1000+ mm rainfall areas to the W of the rift valley. Ja mentions occurrence on Mt Elgon, and Br refers to the lower slopes, but precise localities are omitted. D3 reflects this restricted range.

763 GREY-CAPPED WARBLER *Eminia lepida*
B875 / MPG993 / D2C2

Afrotropical: the E, from the E Zaire borders and S Sudan, through Uganda and N Tanzania, to the E limits of the range in Kenya.

Kenyan: reasonably common but retiring in dense, low and often moist vegetation at 1200-2500 m (Br; Ja; Britton & Sugg 1973). It can be found in a variety of situations, including suburban gardens, the edges of swamps and forests, and in the vicinity of watercourses (*op. cit.*). Range is 73% above 1500 m, 88% subhumid-humid, and 93% within 500+ mm rainfall areas. There are records from 81% of the highland 1000+ mm rainfall squares. This bird does not extend to isolated forest islands. Other references: Wiley 1945; Morrison 1946.

764 YELLOW-BELLIED EREMOMELA *Eremomela icteropygialis*
B878 / MPG1003 / D2C2

Afrotropical: widespread S of 10°S, and further N through East Africa to Sudan, Ethiopia and Somalia. N of 12°N, it is also found across the N tropics from the W coast to Ethiopia.

Kenyan: a seldom common bird of bushed and wooded habitats, up to 1900 m (Br; St). It is virtually absent from the E and coastal lowlands, with the notable exception of the vicinity of the lower Tana R 79C 91AC, and from the humid W plateau.

The Yellow-bellied Eremomela forms part of a superspecies with the Yellow-vented (765), and the varying amount of yellow on the abdomen of the former can cause confusion between the two (HM). They are mainly allopatric in Kenya, where the Yellow-bellied favours the moister areas of the W and SW, whereas the Yellow-vented inhabits the dry N and E. Their ranges are, respectively, 80% above 500 m and 82% below 1000 m, 77% at 250-1000 mm rainfall and 80% at 0-500 mm, and 28% arid as against 74%. However, Ja mentioned possible sympatry along the Turkwell R

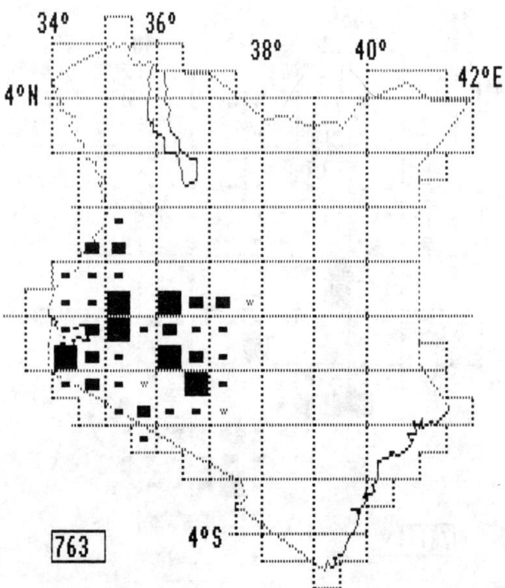

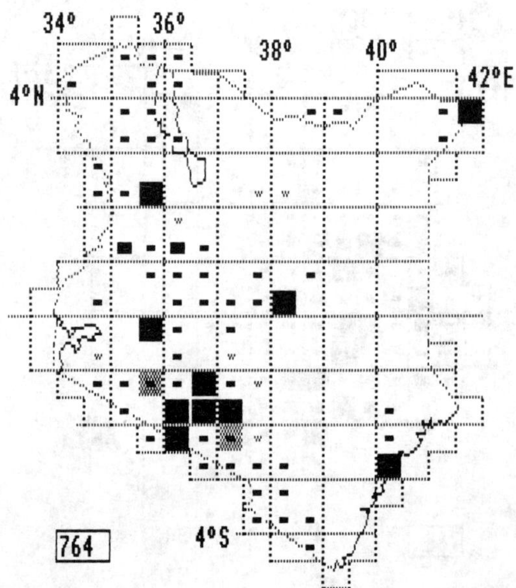

(37A 25AC 13D: not mapped) and in NE Uganda, and HM mapped both at the same sites in NW and SE Kenya. Br noted that apparent sympatry in these areas requires clarification.

The main areas of overlap at QSD scale are W of L Turkana. While these areas are mostly arid and hence apparently unsuitable for the Yellow-bellied, they are also mostly within the 500+ m altitude and 250-1000 mm rainfall categories that this species frequents. These parameters are also within the ecological tolerances of the Yellow-vented (see above). Hence both species could inhabit these areas, and perhaps be locally segregated by minor habitat preferences. This latter explanation is supported by observations near Kamathia 3A in Feb 1984: the two birds were seen within about 2 km of each other, with the Yellow-bellied in lusher, greener bush, and the Yellow-vented in drier, browner bush (D J Pearson, ADL pers. obs.).

Range overlap at QSD level in S Kenya is confined to Simba 88B and Maungu 101D (Ja). Ecological parameters in these areas are not as suitable for both species as they are in the NW, but are still within both birds' overall environmental tolerances, as suggested by our overlays. Apparent range overlap in the extreme NE 19 is consistent with AM's report of wide sympatry and coexistence in S Somalia. Confirmed breeding in 19B derives from Somalia (*op. cit.*).

765 YELLOW-VENTED EREMOMELA *Eremomela flavicrissalis*
B877 / MPG1004 / D2C1

Afrotropical: the NE, from Kenya and NE Uganda to Somalia and SE Ethiopia. Apart from Moroto 24B in NE Uganda (not mapped), distribution in Kenya represents the SW limit of the range.

Kenyan: uncommon in open bushed or wooded country, up to 1200 m (Br; La). Distribution is 80% within 0-500 mm rainfall areas. The zone of old records along the SW range edge may reflect range contraction, particularly in view of the absence of recent records from the very frequently covered country around L Baringo 50A (St). See Yellow-bellied Eremomela (764).

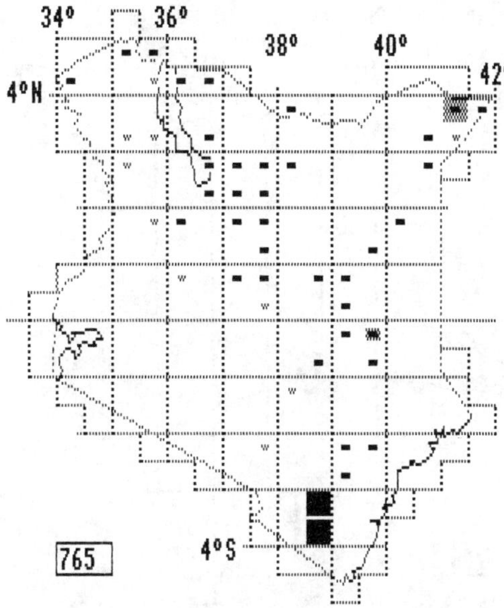

766 GREEN-CAPPED EREMOMELA *Eremomela scotops*
B880 / MPG1006 / D2C1

Afrotropical: mainly the S tropics, widespread S of 5°S apart from the dry SW. Further N, it reaches SW Uganda, N Tanzania and, at the N limits of the range, S Kenya. It is replaced in the N tropics by the Green-backed Eremomela (767).

Kenyan: uncommon and local in woodland and a wide variety of forest edge, garden and moist bushed habitats, up to 1800 m (Br; Ja). Range is entirely within 500+ mm rainfall areas.

Three geographically separate races occur (Br). *E.s. citriniceps* is in the W and S peripheries of the W highlands, from Kendu Bay 60B to Mara GR 74A and Loita District (precise location uncertain: mapped in 74B)(Ja; Br). These areas are subhumid-humid. *E.s. kikuyuensis* is found on the SE periphery of the C highlands, from Nairobi 75B to E of Embu 63D (Br; EABR). This region is semiarid-subhumid. Finally, *E.s. occipitalis* is known from very few records (all Apr-June) from the Mongeya Hill-Sokoke Forest area 102B of the coastal lowlands (Britton & Zimmerman 1979; EABR). This square is semiarid.

The Green-capped Eremomela forms part of a superspecies with the allopatric Green-backed Eremomela, which replaces it in more N areas of Kenya (HM; Br). The only overlap square is 60B, where both species have been recorded on the S side of the

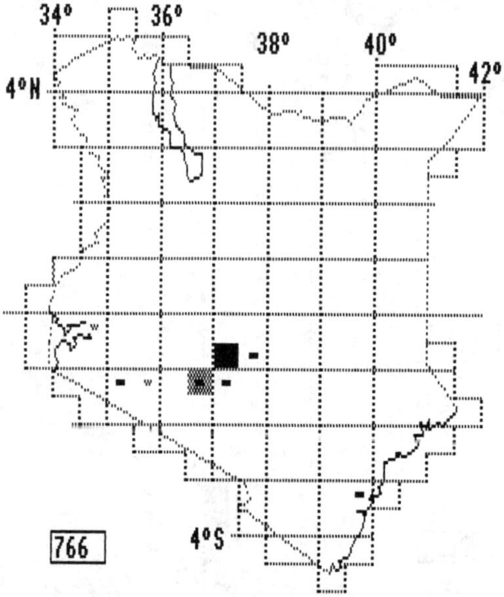

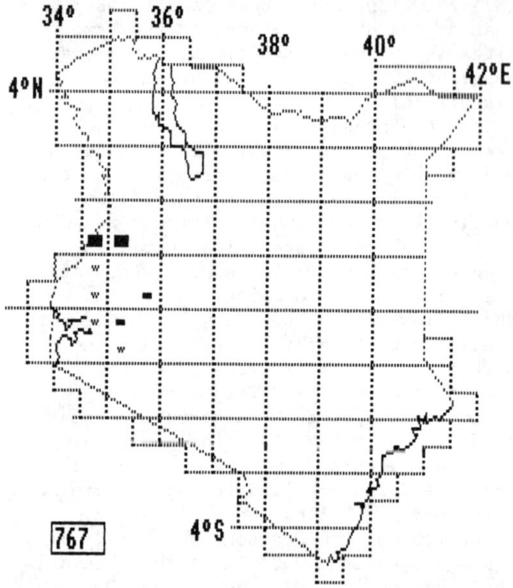

Winam Gulf at Kendu Bay (Ja; National Museum Collection, Nairobi). HM note apparent lack of sympatry where these two birds meet in Kenya, and suggest that some ecological segregation may occur. From the descriptions of habitat given by Br, the Green-capped seems to prefer rather more thickly vegetated situations while, as well as woodland, the Green-backed is found in more open bushed and wooded grassland. In further contrast to the Green-capped, the range of the Green-backed is entirely within 1000+ m areas to the W of the rift valley, with all squares moist subhumid-humid, and within the 1000+ mm rainfall regime. Other reference: van Someren 1930.

767 GREEN-BACKED EREMOMELA *Eremomela pusilla*
B879 / MPG1005 / D2C1

Afrotropical: the N tropics between 15°N and 5°N, from the W coast to Ethiopia, and SE to S Sudan, N Uganda and, at the SE limit of the range, W Kenya.

Kenyan: locally common up to 2200 m (Br; Ja). Details of range and habitats are given under the Green-capped Eremomela (766). Br lists occurrence in Pokot (approximately 37: not mapped). MPG mention breeding in W Kenya but omit details, and there is no confirmed breeding record. The significance of the high proportion of pre-1970 squares (50%) is uncertain.

768 TURNER'S EREMOMELA *Eremomela turneri*
B881 / MPG1007 / D2C3
NO MAP: post presence 48D; pre probable 48D

Afrotropical: from E Zaire to SW Uganda and the Kakamega Forest area 48D of W Kenya.

Kenyan: all recent records have come from Kakamega Forest 48D, where this warbler is locally common in edge habitats (Br; ADL pers. obs.). This forest is around 1700 m, in a humid, 1000+ mm rainfall area. There are old records from the nearby Yala R 48D (Ja). Probable breeding in 48D is based upon individuals at Kakamega with enlarged gonads (Zimmerman 1972).

769 WHITE-BROWED CROMBEC *Sylvietta leucophrys*
B923 / MPG1002 / D2C2

Afrotropical: a highland species along the E Zaire-Tanzania-Uganda borders and, as an isolated population to the E, in W and C Kenya.

Kenyan: locally common in undergrowth in highland forest and bamboo, up to 2600 m (Br; Ja; Diamond & Fayad 1979). Range is 80% within 1000+ mm rainfall areas, and entirely subhumid-humid. While common at 2000 m in the North Nandi Forest 48D 49C (Diamond & Fayad 1979), the White-browed is distinctly uncommon in the adjacent but lower altitude Kakamega Forest 48D

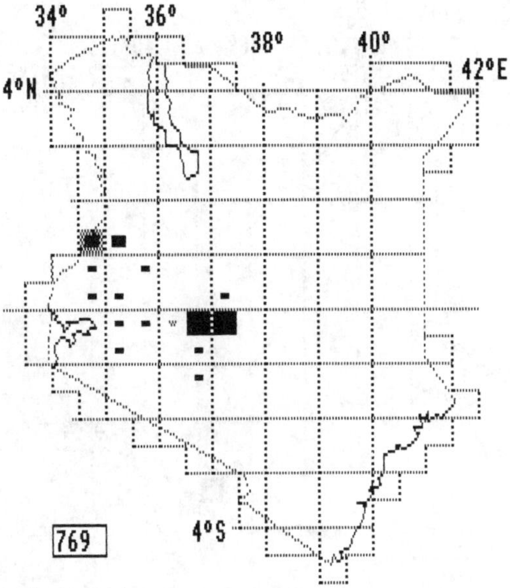

(Zimmerman 1972), and its lower altitudinal limit may be about 1540 m (Mann 1980).

The White-browed forms part of a species-group with the Green Crombec (770)(HM), and both are known from Kakamega Forest 48D. They are both rare there, however, and near the lower (White-browed) and upper (Green) limits of their altitudinal ranges (*op. cit.*; Br; Gerhart & Paxton 1980).

770 GREEN CROMBEC *Sylvietta virens*
B924 / MPG1001 / D2C2
NO MAP: post presence 48D

Afrotropical: forests of W and C Africa, from the Gulf of Guinea through the Congo basin to NW Tanzania, S Sudan, Uganda and, on the E extremity of the range, the Kakamega Forest 48D of W Kenya.

Kenyan: in the undergrowth and mid-levels of forests, often in edge habitats (Br). Unpublished reports and Williams' (1967) mention of the bird in the Kakamega Forest 48D lacked confirmation (Br). The first fully substantiated records for Kenya were those of Gerhart & Paxton (1980) at Kakamega on 1-2 June 1980. There have been no subsequent records, and there is no breeding record (EABR). Kakamega Forest is at 1700 m, in a humid, 1000+ mm rainfall area. Ja mentions occurrence on Mt Elgon 36CD, but omits precise localities. See White-browed Crombec (769).

771 RED-FACED CROMBEC *Sylvietta whytii*
B925 / MPG997 / D2C2

Afrotropical: the E, from Zimbabwe and S Mozambique, through East Africa, to S Sudan and S Ethiopia. Distribution in Kenya represents the E limit of the range.

Kenyan: locally common in a wide variety of habitats, from forest edge to dry wooded or bushed areas, up to at least 2000 m (Br; HM). While also extending down to sea level, range is 75% above 1000 m. The squares are 79% within 500+ mm rainfall areas. The species is virtually absent from even locally moist areas of the arid N and E. The predominance of old records (mostly from Ja) in the reasonably well covered coastal strip may indicate a contraction of range.

The Red-faced forms part of a superspecies with the Northern Crombec (772: *q.v.*)(HM). The two are mainly allopatric and, in areas of overlap, the Red-faced is usually in lusher, moister vegetation (Br; HM). While both can be common near the Red-faced's N range edge at semiarid Baringo 50A, the Red-faced is commonest there during wet periods, but can be absent or very uncommon for intervals of several months during dry weather (St; T Stevenson *in litt.*). See Somali Long-billed Crombec (773).

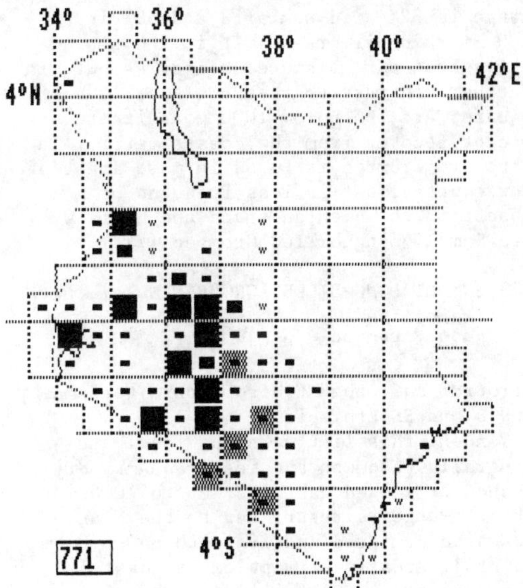

772 NORTHERN CROMBEC *Sylvietta brachyura*
B921 / MPG996 / D2C2

Afrotropical: a N tropics species, widely distributed N of the equatorial lowland

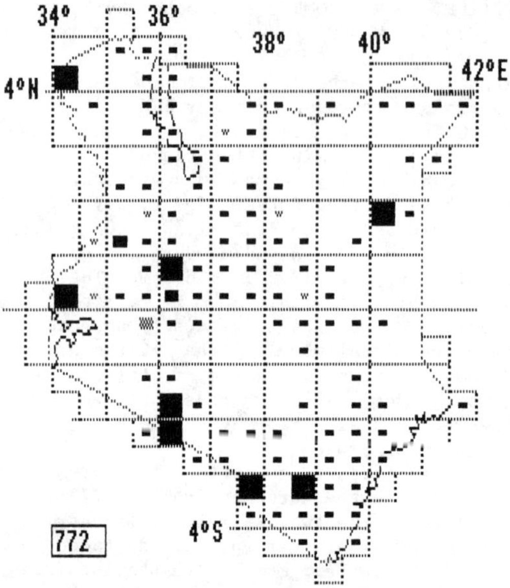

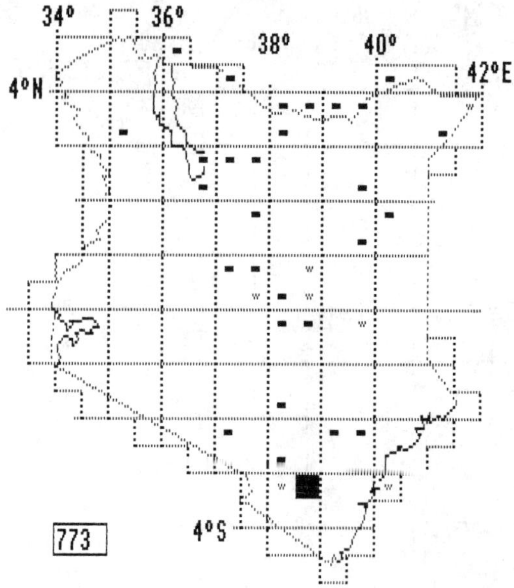

forests from Senegambia E to Ethiopia and Somalia, and SE to Uganda, Kenya and NE Tanzania.

Kenyan: locally common in a wide range of bushed and wooded habitats up to at least 2000 m (Br; HH; St; La). In contrast to the Red-faced Crombec (771: *q.v.*), range is 82% arid-semiarid and 58% in 0-500 mm rainfall areas. It is known throughout our moisture categories but, in further contrast to the Red-faced, the squares are 66% below 1000 m and there are recent records from the coastal strip. The Northern Crombec is found in most areas of the country, but is less frequent or absent in the highlands and the humid W. See Somali Long-billed Crombec (773).

773 SOMALI LONG-BILLED CROMBEC *Sylvietta isabellina*
B922 / MPG999 / D1C1

Afrotropical: the NE, from Kenya to Somalia and S and SE Ethiopia.

Kenyan: this little known, local and generally uncommon species frequents dry bushed or wooded habitats, up to 1000 m (Br). Range is restricted to the N and E, 93% arid-semiarid and 79% within 0-500 mm rainfall areas. Br mentions a coastal sight record at Sabaki 103A. The single breeding record is of a pair with eggs in Tsavo E 101B in June 1972 (La).

This species is in the same species-group as the Red-faced (771) and Northern (772) Crombecs (HM). The Somali Long-billed is widely sympatric with the Northern, although there may be some ecological segregation (HM). The Somali is virtually allopatric with the Red-faced at QSD scale, occupying the dry areas to the N and E of the latter's range.

774 BROWN PARISOMA *Parisoma lugens*
B900 / MPG789 / D1C2

Afrotropical: a highland species with a fragmentary distribution in the E and NE. It extends NW from Malawi to SE Zaire, and NE from NE Tanzania to Kenya, extreme NE Uganda, S Sudan and Ethiopia. Distribution in Kenya represents the Afrotropical range at these latitudes.

Kenyan: locally common but rather unobtrusive in forest edge and wooded habitats, especially those with *Acacia abyssinica*, at 1400-2500 m (Br; Ja; HM). Range is 79% subhumid-humid and entirely within 500+ mm rainfall areas: the apparently suitable areas of the moist W 48 60 are at insufficient altitudes. Records of single birds during Aug and Nov-Dec at 1000 m in semi-arid country near the range edge at Baringo 50A indicate some degree of wandering (St; EABR 1983).

The Brown Parisoma forms part of a species-group with the Banded Parisoma (775)(HM), which typically inhabits lower and drier country.

(map overleaf)

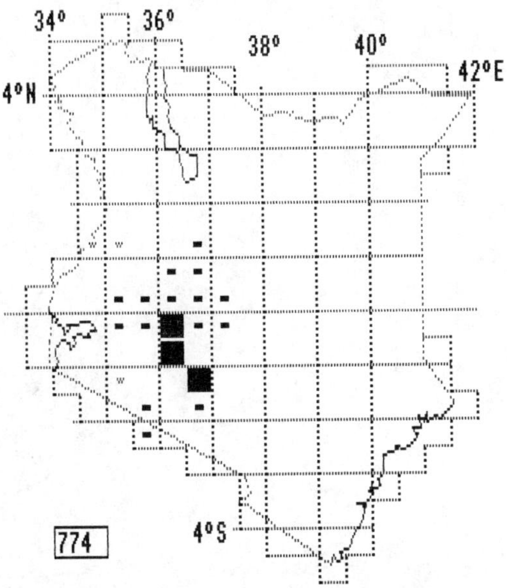

775 BANDED PARISOMA *Parisoma boehmi*
B899 / MPG787 / D2C2

Afrotropical: the E and NE, through E Tanzania to Kenya, Ethiopia and Somalia N of 8°N (AM).

Kenyan: reasonably common in dry bushed or wooded country, especially with acacias, up to 1700 m (Br; MPG). Range is 79% arid-semiarid (*cf.* Br) and 93% within intermediate, 250-1000 mm rainfall

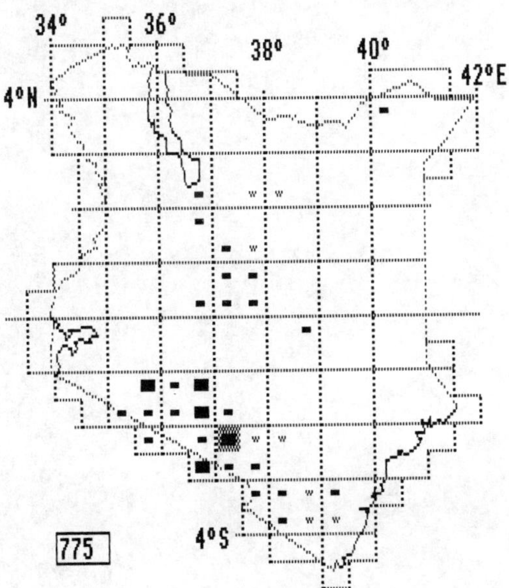

areas. The squares are 90% at 500-2000 m, and distribution is mainly on the plateaux to the NE and SE of the C highlands: these two areas hold *P.b. marsabit* and the nominate race respectively (Br). 27D 28C reflects the environs of the locally moist Mt Marsabit. The sight record in 18A (E J Roberts *in litt.*) represents the S edge of the S Ethiopian population, and is of uncertain race. In Kajiado district 75D 87B, it is mainly resident but also wanders to higher altitudes (DEP pers. obs.). See Brown Parisoma (774).

776 KRETSCHMER'S LONGBILL *Macrosphenus kretschmeri*
B897 / MPG764 / D1C0
NO MAP: pre presence 100B

Afrotropical: the E, with a fragmentary distribution from NE Mozambique to NE Tanzania, and a single record NE to the S borders of Kenya.

Kenyan: the single record of this forest undergrowth and edge species is from Kitovu Forest, at Taveta 100B (Ja; Br). Several subsequent searches in this area have failed to relocate the species, so that this record may refer to a wanderer some 100 km to the NE of its normal range in NE Tanzania (Br).

777 GREEN HYLIA *Hylia prasina*
B889 /MPG1281 / D1C3
NO MAP: post presence 48D 49C; pre probable 48D

Afrotropical: W and C Africa, from the lowland forests around the Gulf of Guinea, E through the Congo basin to NW Tanzania, Uganda, S Sudan and, at the E extremity of the range, the Kakamega-Nandi area of W Kenya.

Kenyan: in undergrowth and mid-levels of forest at Kakamega and Nandi 48D 49C, up to 1800 m (Br; P L Britton *in litt.*). Range is in 1000+ m areas to the W of the rift valley. Both squares are humid and within the 1000+ mm rainfall regime. Probable breeding in 48D is based upon a bird with a brood patch at Kakamega (Zimmerman 1972).

778 SOUTHERN HYLIOTA *Hyliota australis*
B890 / MPG814 / D1C1
NO MAP: post confirmed 48D; pre presence 49C

Afrotropical: mainly S-C Africa, across the S tropics from Angola to Malawi and S Mozambique. Further N, there are isolated populations in Cameroon, the Uganda-Zaire borders, NE Tanzania and the Kakamega-

Nandi area 48D 49C of W Kenya.

Kenyan: a scarce, unobtrusive and easily overlooked inhabitant of forest canopy, known only from Kakamega Forest 48D and Nandi 49C (Br; MPG; EABR). The two squares are W of the rift valley at 1500+ m. They are both humid and within the 1000+ mm rainfall regime.

The Southern Hyliota forms part of a species-group with the Yellow-bellied Hyliota (779)(HM). The two are allopatric in Kenya, where the isolated population of the Southern Hyliota exists within the Yellow-bellied's overall range.

779 YELLOW-BELLIED HYLIOTA *Hyliota flavigaster*
B891 / MPG813 / D1C2

Afrotropical: widespread between 10°N and 15°S, E to W areas of Ethiopia, Kenya and Mozambique, but avoiding the equatorial lowland forests.

Kenyan: an uncommon canopy species of open forested or well wooded habitats, up to 1800 m (Br; Ja). Range is in 1000+ m areas to the W of the rift valley. All seven squares are moist subhumid-humid and within 1000+ mm rainfall areas. There is no breeding record and, while this species is presumably resident, the few published dates are all June-Aug (Ja; EABR 1978, 1981). See Southern Hyliota (778).

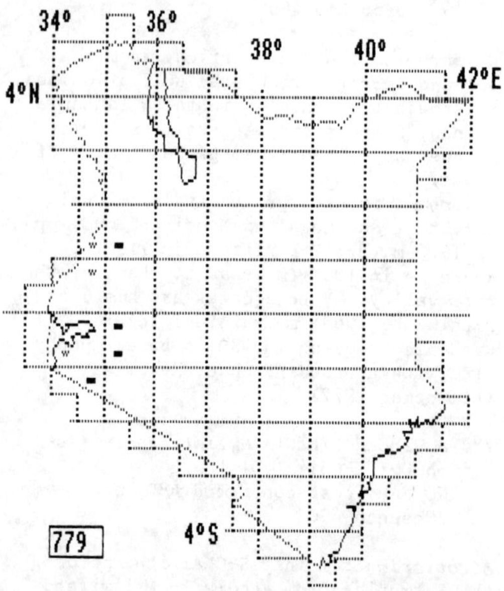

Family Muscicapidae: Flycatchers, Batis, Wattle-eyes etc.

Approximately 350 species, lacking representatives only in the Americas. 69 (20%) have been recorded in the Afrotropical Region and 31 of these (45%) are known in Kenya. 29 (94%) of the Kenyan birds are Afrotropical, while the Spotted (780) and Collared (786) Flycatchers are non-breeding migrants from the Palearctic.

780 SPOTTED FLYCATCHER *Muscicapa striata*
B945 / MPG778 / D2C2

Afrotropical: a Palearctic species that winters in sub-Saharan Africa, being widespread from about 7°N in the W and 3°N in the E, S to the Cape.

Kenyan: a locally common passage migrant and winter visitor, Oct-Apr, in forest edge and a wide range of open bushed or wooded habitats (Br; Ja; La; St). Se reported a single Apr migrant at 3000 m near Mau Narok 61D, but Br notes occurrence mainly below 2200 m.

Kenya is at or very near the N limit of this flycatcher's Afrotropical wintering range (see above). Wintering birds are rather sparsely distributed, except in the coastal strip (*cf*. Somalia: AM) and restricted areas of the E plateau, where they can be reasonably common (Br; *e.g.* Baringo 50A: St, and Tsavo E 101B: La). MPG mention frequent local movements of wintering birds. Migrants (Oct-Nov and Mar-Apr) are of much more widespread occurrence (Br).

The Spotted Flycatcher forms part of a superspecies with the Gambaga Flycatcher (781), and part of a species-group with the Dusky (782), Chapin's (783) and Swamp (784) Flycatchers (HM). The first three of these Afrotropical species appear to be segregated at even local level by differing combinations of geographical range, altitudinal range and habitat preference. The little known Chapin's is quite separate from the Gambaga and Swamp Flycatchers, but presumably exists in closer vicinity to the Dusky. The Palearctic Spotted Flycatcher is most probably temporarily sympatric with the Afrotropical species during its widespread migrations but, while the Spotted winters in some numbers in the coastal strip (see above), none of the African species are found there.

781 GAMBAGA FLYCATCHER *Muscicapa gambagae*
B941 / MPG778 / D0C1

Afrotropical: the N tropics, from Ghana to extreme NW Somalia, and SE to S Sudan, N Uganda and, on the SE limit of the range, Kenya.

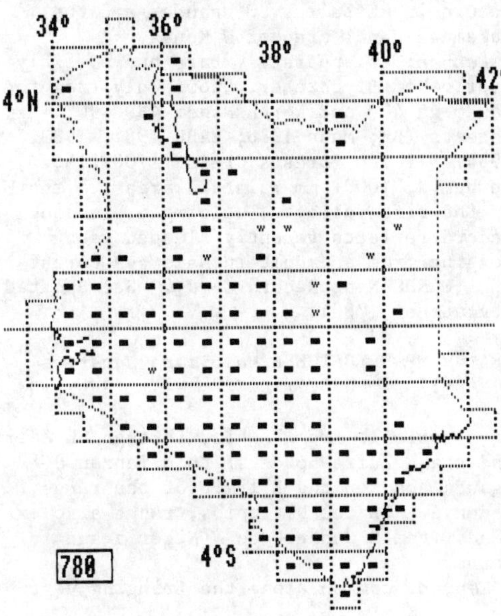

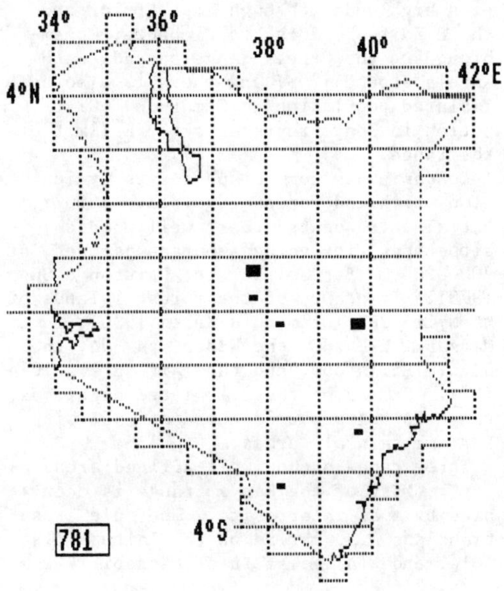

Kenyan: this little known and possibly much overlooked species frequents dry bushed or wooded country (Br; Pearson *et al*. 1980). There are records from Mt Elgon 36D, the Mt Ololokwe-Archer's Post area 51B and Isiolo 51D (Ja; Pearson *et al*. 1980; EABR 1980), the Ura Gate of Meru NP 64A (Pearson *et al*. 1980; EABR 1980), near Garissa 65B (EABR 1983), and Galana Ranch 90B Mar 1984 (EABR 1984). Three birds have been netted at night at the Ngulia lights 101A (EABR 1978, 1980).

Of the seven squares, six are below 1500 m and to the E of the C highlands and their adjacent plateaux, five are arid-semiarid, and five are in 500+ mm rainfall areas. Br noted that all records are Nov-Apr, such that the bird might be a non-breeding winter visitor from further N. The presumed nocturnal migrants during Nov at Ngulia might support this view.

Pearson *et al*. (1980) considered a young bird at Lerata 51B to indicate breeding in the Mar-Sept period, and suggested that the species may be a local and possibly not uncommon resident to the N and NE of Mt Kenya 63A. Another spotted juvenile with an adult near Garissa 65B in Nov also suggested breeding in that area (EABR 1983; D J Pearson pers. comm.). Both of these records are mapped as probable breeding: they are both below 1000 m, in 0-500 mm rainfall, arid-semiarid areas. See Spotted Flycatcher (780).

782 DUSKY FLYCATCHER *Muscicapa adusta*
B936 / MPG781 / D2C2

Afrotropical: mainly the E, from the Ethiopian highlands, through East Africa and the E Zaire borders, to Zimbabwe, Mozambique and the Cape. Range extends W to Angola around 12°S, and there is also an isolated population in Cameroon. Distribution in Kenya represents the E limit of the range.

Kenyan: this common species is typically along highland forest edge above 1500 m, but also frequents other, well wooded situations, including suburban gardens, at 900-3200 m (Br; Ja; Be; Se; Britton & Sugg 1973). It occurs on the forest islands of Mt Nyiru 26D (Diamond & Keith 1980), Marsabit 27D 28C, the Ndoto Mts 39A, the Chyulu Hills 88D, and the Taita Hills 101A (Br). Excluding these isolated highlands, range is 90% subhumid-humid and 95% within 500+ mm rainfall areas.

Three races occupy well-defined areas (Br). Most of the Kenyan range is occupied by *M.a. interposita*. The sole exceptions are *M.a. murina* on the Taita Hills 101A, and *M.a. marsabit* at Marsabit 27D

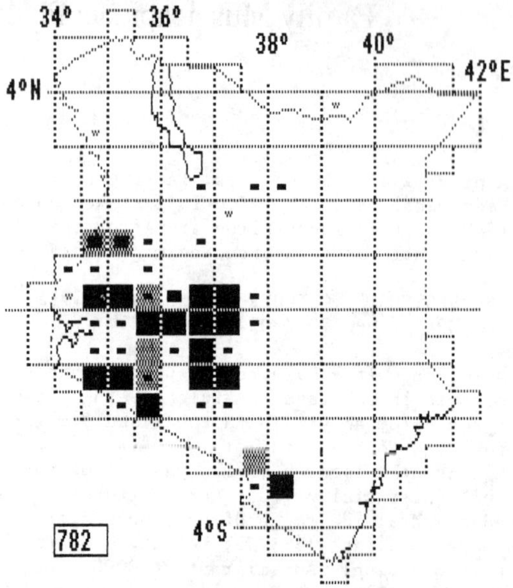

28C and Moyale 17A. This last locality (Ja) represents the S limit of the Ethiopian highlands population. Se considered this species resident around 3000 m at Mau Narok 61D, and DEP (pers. obs.) concurs for 74D. See Spotted Flycatcher (780).

783 CHAPIN'S FLYCATCHER *Muscicapa lendu*
B943 / MPG- / DOC1
NO MAP: post confirmed 48D; pre presence 49C

Afrotropical: this poorly known species occurs in NE Zaire, SW Uganda, and the Kakamega-Nandi area of W Kenya.

Kenyan: an apparently rare but possibly overlooked flycatcher, known only from the Kakamega 48D and North Nandi 48D 49C Forests (Br; Mann 1976; EABR 1981, 1983, 1984). These forests are at 1700-2150 m, in humid, 1000+ mm rainfall areas to the W of the rift valley. The single breeding record reflects recently fledged young begging from an adult in Kakamega Forest 48D (EANHS Nest Record Card). See Spotted Flycatcher (780).

784 SWAMP FLYCATCHER *Muscicapa aquatica*
B937 / MPG782 / D2C2

Afrotropical: from Zambia, through SE Zaire and the E Zaire borders, to N Tanzania, Uganda and, on the E limit of the range, W Kenya. In the N tropics, range also extends from S Sudan W to N Nigeria and Ghana.

Kenyan: common along the fringing vege-

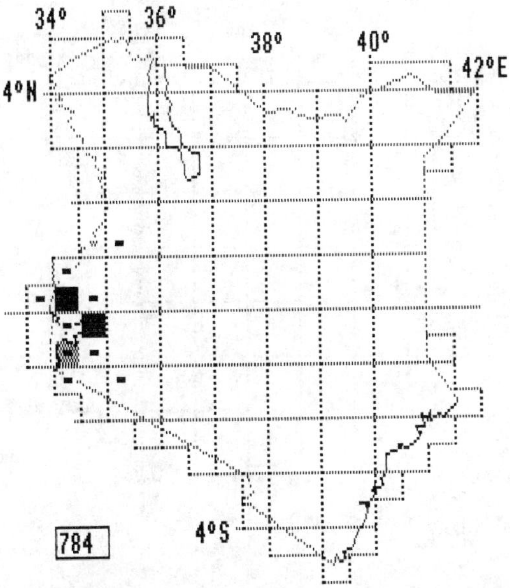

tation (including papyrus) of lakes, streams and swamps in the W, mainly at 1130-1400 m (Ja; MPG; Br; Britton 1978). Britton (1978) describes range as confined to the L Victoria basin, but notes that HM map presence on higher ground to the E. Br includes Mt Elgon 36D. There are also reports from the Kitale-Kapenguria area 37C (T J Barnley *in litt.*), and from near Keekorok, Mara 74A (D A Turner *in litt.*). Range is entirely moist subhumid-humid, and 91% within 1000+ mm rainfall areas. See Spotted Flycatcher (780).

785 ASHY FLYCATCHER *Muscicapa caerulescens*
B938 / MPG785 / D2C2

Afrotropical: widespread S of 5°N, apart from the drier areas of N Kenya, Somalia and the SW. In the W, there are also records to the N of the Gulf of Guinea. Distribution in Kenya represents the N limit of the range.

Kenyan: rather uncommon in forest edge, well developed woodland and similar riverine habitats, up to 1800 m (Br; Ja; MPG). Range is 76% semiarid-subhumid and 84% within 500+ mm rainfall areas. Restriction of range in the far SE to the lower Tana R 79C and the coastal strip parallels distribution in adjacent S Somalia (which includes presence in 80B (not mapped): AM). The Ashy Flycatcher is probably mainly resident, but there is evidence of nocturnal movement at Mombasa 114B (Br). Ja records it from Nairobi and Kiambu 75B, but there are no recent records from this well

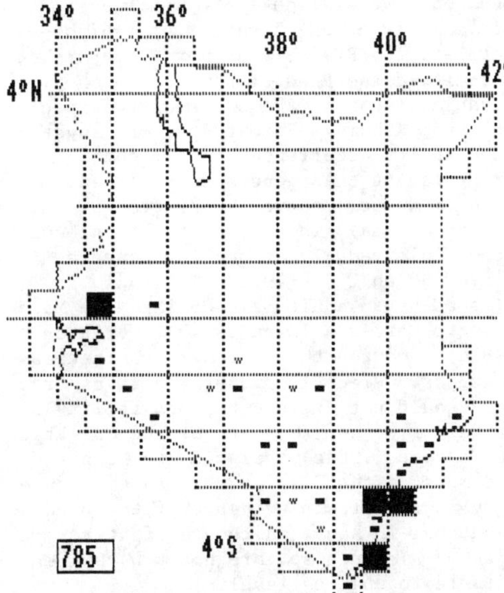

known area.

786 COLLARED FLYCATCHER *Ficedula albicollis*
B930 / MPG779, 780 / D2C1

Afrotropical: the migration of this Palearctic flycatcher passes mostly to the W of Kenya, and the main wintering range extends from S Tanzania and E Zambia S to Zimbabwe, NE South Africa and Namibia.

Kenyan: the occurrence of this forest and woodland canopy species in Kenya has

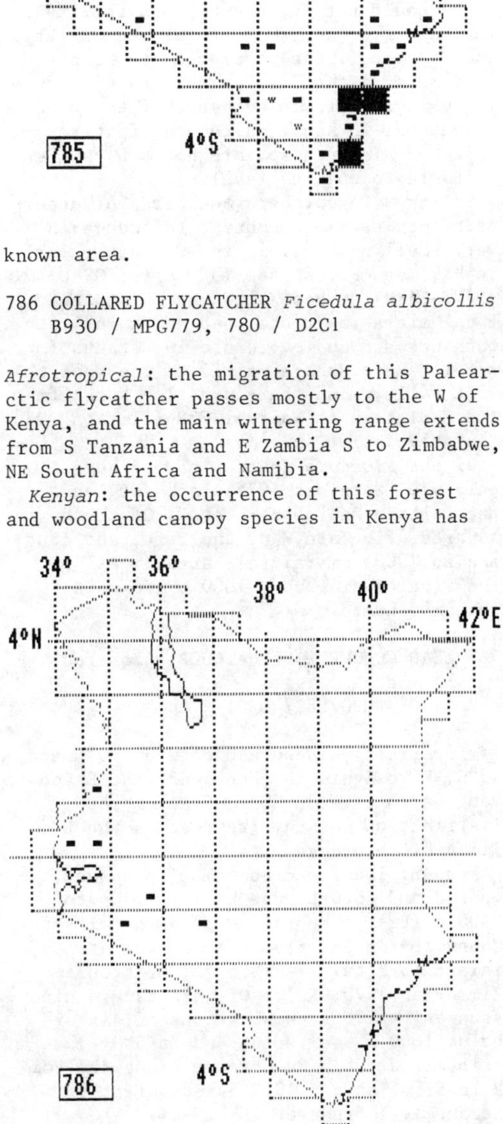

been controversial due to resemblance, particularly of the E race *F.a. semitorquata*, to the Pied Flycatcher *F. hypoleuca*. Ba admitted the Pied Flycatcher to the Kenyan avifauna on the basis of two specimens from Kakamega Forest 48D, and suggested that its occurrence there might be more frequent than generally realised. These views were shared by Britton & Britton (1973a), who caught the first confirmed Collared Flycatcher for Kenya at Ng'iya 48C on 2 Oct 1972. This was followed by an adult male Collared in Mara 74A on 3 Mar 1974 (Mann 1974b). Britton (1980b) reported that further examination of supposed Pied Flycatcher specimens from Kenya could not exclude the possibility of the race *F.a. semitorquata* of the Collared, and Pearson (1981c) assigned two supposed specimens of the Pied to this race. Hence there is no certain evidence of the Pied Flycatcher in Kenya, although sight records of *Ficedula* sp. are not always identifiable to species level.

Collared Flycatchers and birds of uncertain species are commonest (although in very small numbers) in the W, particularly in Kakamega Forest 48D (*op. cit.*; OS-C 1977; EABR). These birds are E of their usual migration route, and dates indicate both passage and overwintering status (*op. cit.*). Records are mainly to the W of the rift valley, but there are also a few from the E plateau at Nairobi 75B (Clifton 1976; Lewis 1981b; *op. cit.*).

Of the *Ficedula* squares mapped here, only 61D (Sessions 1975) lacks at least one certain Collared record. Of these six squares, five are W of the rift, and four are in 1000+ mm rainfall areas. All of the squares are above 1000 m and moist subhumid-humid.

787 LEAD-COLOURED FLYCATCHER *Myioparus plumbeus*
B946 / MPG788 / D2C1

Afrotropical: widespread between 10°N and 15°S, E to Ethiopia, Kenya and the E coast. Range also extends SE to S Mozambique. Distribution in Kenya represents the E limit of the range.

Kenyan: local and uncommon in bushed, wooded and forest edge habitats, up to 2000 m (Br). The race *M.p. orientalis* is found in the SE, from the coastal strip inland to Kitui 77A (Br; Ja; Britton & Zimmerman 1979; La). Of this race's nine squares, various combinations of six are below 1000 m, semiarid, and in 500+ mm rainfall areas. Occurrence along the Voi R in arid Tsavo E 101B has apparently become less frequent (La).

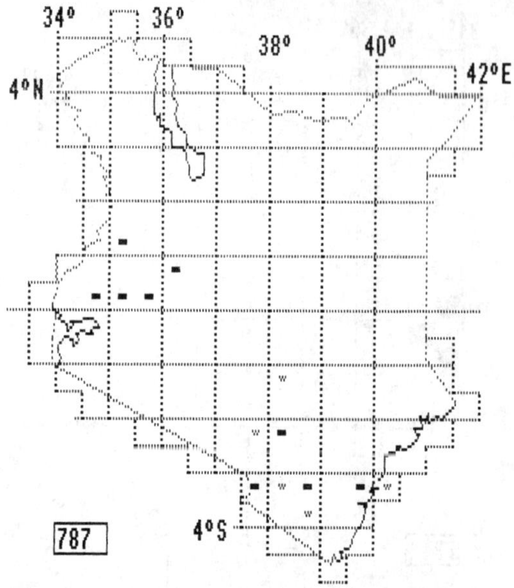

Records to the W of the rift valley are thought to involve the nominate race (Br), and this presumably also applies to reports from Baringo 50A (St; EABR). These highland squares are four of five subhumid-humid, and entirely above 1000 m in areas that receive 500+ mm rainfall. Nevertheless, the species has been resident throughout the year at semiarid Baringo 50A (EABR 1981). The very few other published records show no obvious seasonality.

788 SOUTHERN BLACK FLYCATCHER *Melaenornis pammelaina*
B935 / MPG799 / D1C2

Afrotropical: replaces the Black Flycatcher (789) in the S tropics. It is widespread S of 5°S, apart from the dry SW. Further N, range extends around the E of the Congo basin to Tanzania and, at its N limits, to Kenya and extreme S Somalia.

Kenyan: the accounts of this species and the extremely similar Black Flycatcher, which are allopatric members of a superspecies (HM), may well be influenced by observers' confusion of the two, and also by confusion with the similar Drongo (566) (MPG; HM). These flycatchers' maps may therefore contain errors, particularly where the species' ranges abut.

Br's mention of the Southern Black W to Rumuruti 50D and Namanga 87D is presumably based on Ja's specimens from these localities. Together with Nairobi 75B (many obs.), this would confine the Southern Black to areas E of the rift valley. We

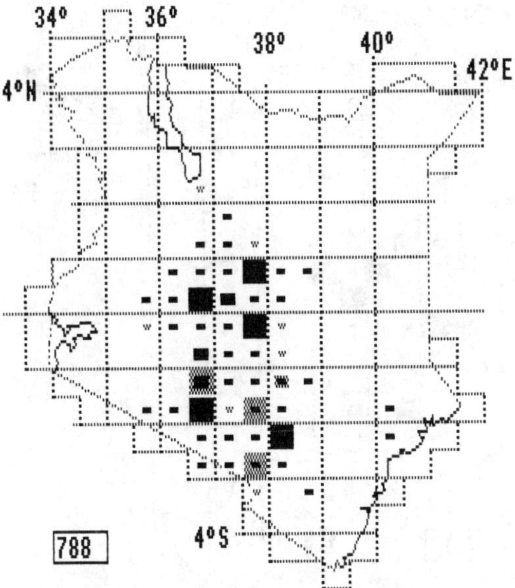

map records both in and (just) W of the rift valley, while stressing the need for further, careful fieldwork on the birds' ranges. HM note that the ranges meet in East Africa without apparent overlap, and the similarity of the species' habitats (Br) would probably preclude local ecological segregation. In fact, at QSD scale, overlap squares are four in number. In 26D, the Southern is on Mt Nyiru (Br), while the Black is in woodland around L Turkana (HH). In 39A, the Southern is again at high altitudes, on the Ndoto Mts (Br), while DEP observed the Black in distinctly lower country at Nguronet. Br also mentions the Black on Mt Uraguess 51A and at Nakuru 62A. These two localities (sources unknown) may well be correct but, on our map, they appear rather E of the rest of the range. Wanderers may be involved (see below), or our maps incorrect.

The Southern Black Flycatcher is a usually rather uncommon bird of bushed and wooded habitats up to 1800 m (Br; Ja). Subject to the above-mentioned uncertainties, range is 68% above 1000 m and 84% within 250-1000 mm rainfall areas. 80% of the squares are semiarid-subhumid, so that it is virtually absent from the arid N (where the Black occurs) and E. Presence along the lower Tana R 79C 91A (AGH) parallels occurrence along the lower Juba R in adjacent S Somalia (AM). The existence of at least local movements is demonstrated by a few birds in arid Tsavo E 101B during Aug 1970 (La).

789 BLACK FLYCATCHER *Melaenornis edolioides*
B934 / MPG798 / D1C2

Afrotropical: the N tropics, from Senegambia to Ethiopia and S Sudan, and SE to Uganda, N Tanzania and Kenya. Distribution in Kenya represents the SE limit of the range.

Kenyan: a common but sometimes crepuscular species of bushed and wooded habitats at 400-1800 m (Br; Ja; HH). In contrast to the closely related Southern Black Flycatcher (788: *q.v.*), the Black extends from humid, 1000+ mm rainfall areas to the arid, 0-500 mm environments of Turkana (HH) and the Kenya-Sudan-Ethiopian borders. The record in 16A is from Ethiopia (J S Ash *in litt.*), but those from Moyale 17AC are in Kenya (EABR 1984). MPG mention apparent seasonal movements of the race *M.e. lugubris*. The race *M.e. schistacea* was absent from the vicinity of Moyale 17AC during Mar 1984 but, after rains in the intervening period, was found feeding young there in the following Jul (F J McCartney, ADL pers. obs.; EABR 1984; *cf.* Brown & Britton 1980).

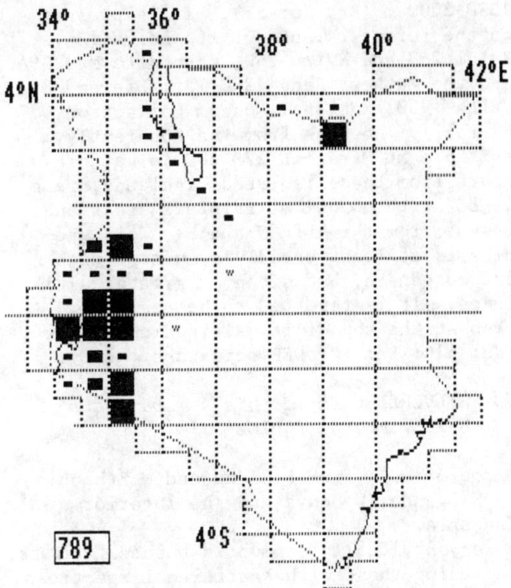

790 WHITE-EYED SLATY FLYCATCHER *Melaenornis chocolatina*
B933 / MPG796 / D3C3

Afrotropical: mainly the highlands of the E and NE, from Ethiopia, through East Africa and the E Zaire borders, to Malawi. Another population occurs in Angola. Distribution in Kenya represents the E limit

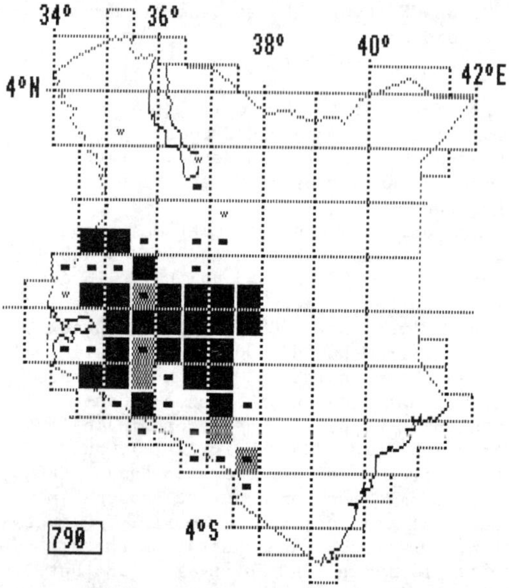

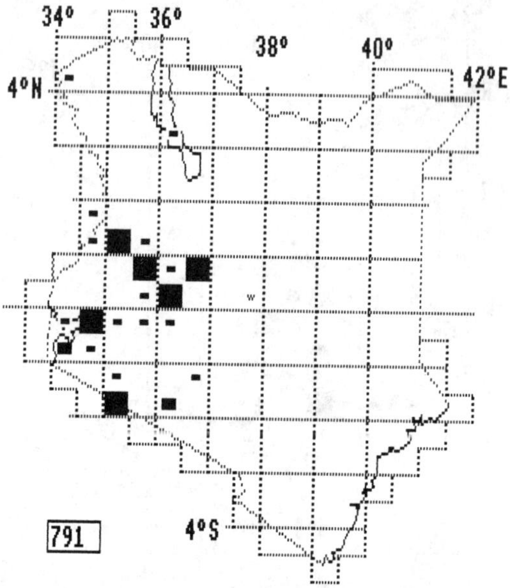

of the range.

Kenyan: common in forest edge and wooded habitats, including suburban gardens, at 1350-3000 m (Br; Ja; Se). It is present on the forest islands of Mts Loima 13C, Kulal 26B and Nyiru 26D, the Ndoto Mts 39A and the Mathews Range 39C (Br; Diamond & Keith 1980; National Museum Collection, Nairobi). Absence from the apparently suitable Mt Marsabit 27D 28C is notable. Apart from these isolated highlands, range is 88% within 500+ mm rainfall areas and absent from the arid N and E. This species is at least broadly sympatric with its congeners, but often at greater altitudes. It is reported to be resident, even at the top of its altitudinal range (Se: 61D; Be: 74; DEP pers. obs.: 74D).

791 SILVERBIRD *Empidornis semipartitus*
B929 / MPG801 / D2C2

Afrotropical: from S Sudan and W Ethiopia, to N Uganda, W Kenya and the interior of N Tanzania.

Kenyan: locally common in bushed habitats, including those with scattered large trees, at 400-1500 m (Br; St). Excluding presumed wanderers in 14C 51D 75B (see below), range is 95% in or to the W of the rift valley and 95% above 1000 m. Br mentions this species in moderately arid areas, but distribution is in fact 95% within the 500+ mm rainfall regime and 80% subhumid-humid. Occurrence in 2C was between Lokichokio and the Sudan border in Nov (G Nikolaus *in litt.*).

Br considers the species probably mainly resident, and St reports this status in the semiarid country around L Baringo 50A. Presumed wanderers have however occurred E of the normal range at Central Island 14C, L Turkana (Feb 1974: HH), at Isiolo 51D (Br), and up to 1800 m in the highlands including Nairobi 75B (Mar 1973: Edwards 1973).

Br considered absence from 48 and 49AC apparently real, in direct contrast to common occurrence in the adjacent areas of Nyanza 60. On this basis, he suggested the population in 60 and Mara 74AC to be separate from that further N in the Elgon 36-Bogoria 50C area. Birds at Rongai 61B and Nakuru 62A were considered further examples of wanderers. These 61B and 62A records fall between those supposedly separate populations and, rather than wanderers, may represent some degree of movement between them. In this context, it should be noted that observer frequencies in many parts of 61AB are low. Absence from other apparently suitable areas of the Afrotropical Region is also noted by HM.

792 PALE FLYCATCHER *Bradornis pallidus*
B928 / MPG792, 795 / D2C2

Afrotropical: widespread S of 15°N, avoiding only the dry SW and the equatorial lowland forests.

Kenyan: this flycatcher forms part of a species-group with the similar Grey Flycatcher (793)(HM), and the potential for

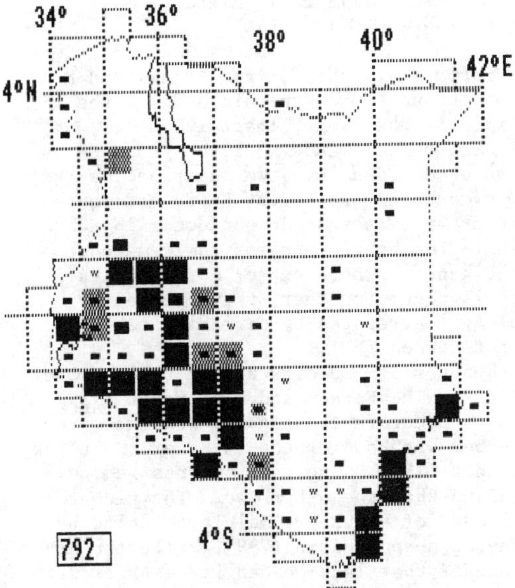

confusion of the two in the field is similar but possibly slightly less than that already noted for the Southern Black (788) and Black (789) Flycatchers. Thus our maps should be treated with caution. The situation is particularly difficult in the NW, and only records from very experienced observers are mapped there. HH's description of the Pale Flycatcher as widespread and frequent around L Turkana is implicitly rejected by Br, who follows Traylor's (1970b) authoritative account of the genus in East Africa. This view is supported by D A Turner (*in litt.*). An idea of the potential for misidentification of these two species is reflected in La's entry for the Pale.

The Grey and Pale Flycatchers are both common birds of bushed and wooded habitats and, at QSD scale, their ranges overlap extensively up to 2000 m (Br; Traylor 1970b). They are segregated by ecological preference at more local levels, however, with the Pale in moister habitats, and often confined in dry areas to the edges of wooded or forested riverine growth (HM; *op. cit.*). The Pale is mainly a bird of the moister areas, with range 72% within the 500+ mm rainfall regime and only 16% arid. Several of these arid squares are occupied by the race *B.p. bafirawari* (the Wajheir Grey Flycatcher, Number 795 of MPG), which ranges NE from Garissa 65B to Wajir 42A and Djiroko 19C. In direct contrast to the Pale's other races in Kenya, this bird's range is entirely below 500 m, in 0-250 mm rainfall country.

Elsewhere in the dry N and E, the Pale Flycatcher is restricted to the 250-1000 mm rainfall areas along the borders with Uganda 12AC 24B, Sudan 2C and Ethiopia 16A (J S Ash, G Nikolaus *in litt.*), and to the lusher vegetation along the lower Tana R 78B 79C (Traylor 1970b). Restriction to the vicinity of large rivers and/or coast also occurs in adjacent S Somalia (AM). The Pale Flycatcher is considered resident (Be: 74; DEP pers. obs.: 74BD 75BD).

793 GREY FLYCATCHER *Bradornis microrhynchus*
B927 / MPG793, 794 / D2C2

Afrotropical: a bird of the E and NE, which extends from S Tanzania, to the E of L Victoria, N to Kenya, NE Uganda, S Sudan, Ethiopia and Somalia.

Kenyan: most points regarding this species are mentioned under Pale Flycatcher (792). In contrast to the Pale, the Grey's range is 78% within the intermediate 250-1000 mm rainfall regime, and 43% arid. Apparent absence from much of the NE 28 29 40 41 53 may reflect very low observer frequencies, or very dry environments: the Grey is recorded from only 26% of the most arid squares (moisture index below 1.2), and 40% of the squares with 0-250 mm rainfall. It is also less frequent in the highlands and moist W, in only 39% of the humid squares and 56% of those with 1000+ mm rainfall. Similarly, it is absent from the dry subhumid areas of the coastal strip. The Grey is consi-

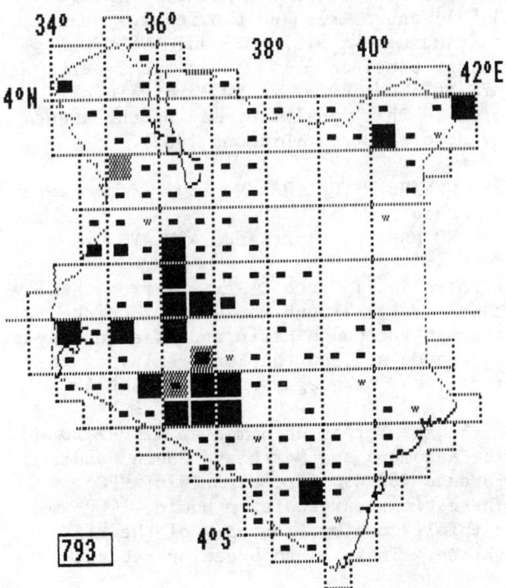

dered resident (St: 50A; La: 101B; DEP pers. obs.: 75C).

794 BLACK AND WHITE FLYCATCHER *Bias musicus*
B955 / MPG812 / D2C0
NO MAP: post presence 51D 52C; pre presence 114B; pre confirmed 102D

Afrotropical: from Sierra Leone to Angola, and E through the Congo basin to NW Tanzania and Uganda. Further E there is a disjunct population in C Kenya, and range also extends in a rather fragmented manner from coastal Kenya S to S Mozambique.

Kenyan: tall trees in bushed, wooded and forest (edge and canopy) habitats, up to 1700 m (Br). There are records from the coast at Rabai 102D, Takaungu 102D, Changamwe 114B and Mombasa 114B, but none of these has been during the last 50 years, and the bird was considered rare there even in the 1930s (Ja; Britton & Zimmerman 1979). However, it is still plentiful on the Usambara Mts of adjacent NE Tanzania (EABR), so that apparent absence from the Kenya coast may represent a small, S retraction of range. Both coastal squares are dry subhumid and within the 1000+ mm rainfall regime. This coastal range is unique in being confined to the divide between the differing forests N and S of Mombasa 114B-Kilifi 102D (Britton & Zimmerman 1979).

A disjunct population on the NE edge of the C highlands is known from a sight record from the Luazomela R 51D (Ja), a 1944 specimen from Maua 51D (National Museum Collection, Nairobi), a female seen in the Ngaia Forest 52C on 26 Nov 1978 (Diamond 1983), and a male and two females seen near Meru town 51D on 29 Jul 1985 (D A Turner pers. comm.). These two squares are at 500-1500 m, in 500+ mm rainfall areas. This population may be threatened by deforestation (Diamond 1983).

795 SHRIKE FLYCATCHER *Megabyas flammulata*
B956 / MPG811 / D2C3
NO MAP: post presence 48D 49C

Afrotropical: W and C Africa, from the lowland forests around the Gulf of Guinea, E through the Congo basin and adjacent areas to Uganda and, on the E extremity of the range, the Kakamega-Nandi area 48D 49C of W Kenya.

Kenyan: rather uncommon at 1700-2150 m in the Kakamega and North and South Nandi Forests 48D 49C (Br; Mann 1976; MPG). These two squares are in humid, 1000+ mm rainfall country, to the W of the rift valley. There is no breeding record.

796 FOREST BATIS *Batis mixta*
B950 / MPG815 / D1C2

Afrotropical: the E, from Malawi and N Mozambique to NE Tanzania and, on the NE limit of the range, coastal Kenya.

Kenyan: the members of the genus *Batis* can be divided into two superspecies (HM). One contains the Forest Batis and four other species that do not occur in Kenya, while the other contains the remainder of the genus. Where two or more species of batis occur together, they are nearly always segregated by differing habitat preferences (HM).

Kenya's six species are in three broad areas. The Forest and East Coast (797) Batis occur in the coastal lowlands, together with one of the races of the Black-headed (798). The last extends further inland than the other two. The higher altitudes of Kenya are mainly occupied by the Chin-spot Batis (799), while the other race of the Black-headed is restricted to areas W of the rift valley. The remaining two species, the Pygmy (800) and the Grey-headed (801), are found at lower altitudes in the drier country of Kenya's N and E.

The Forest Batis is common in forest undergrowth and mid-levels (Br). The race *B.m. ultima* is found in the coastal lowlands, entirely in semiarid - dry subhumid, 500+ mm rainfall areas. Ja quotes occurrence at Taveta 100B, but this is ignored and presumably rejected by Br. In view of the nominate race's range in the Kilimanjaro-Arusha area of adjacent NE Tanzania, occurrence at Taveta appears possible, and

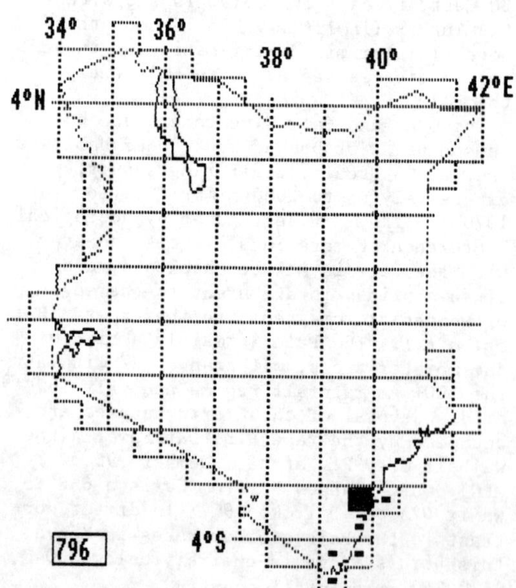

we map it here. With the obvious exception of altitude, the ecological characteristics of 100B (*i.e.* rainfall *etc.*) are identical to those of this species' coastal squares.

There is ecological segregation of the three coastal lowlands species (Br). While the Forest is in forest, the East Coast Batis favours woodland, and the Black-headed frequents more open habitats, only penetrating woodland edge. Segregation of the Forest and the East Coast was demonstrated by Britton & Zimmerman (1979) in Sokoke Forest 102B. The Forest was characteristic of forest and only recorded in woodland once or twice, while the East Coast was *vice versa*.

797 EAST COAST BATIS *Batis soror*
B954 / MPG817 / D2C3

Afrotropical: the E, from S Mozambique to Zimbabwe, Malawi, S and coastal Tanzania and, at the N extremity of the range, coastal Kenya.

Kenyan: a common, woodland species of the coastal lowlands and the lower Tana R 79C (Br). Range is 90% semiarid-dry subhumid, and entirely within 500+ mm rainfall areas. HM note overlap of this species and the Pygmy Batis (800) in SE Kenya, where the latter occupies drier country. Our maps do not show this overlap, even at QSD scale, and thus may be incomplete. HM also note that the East Coast Batis replaces the Chin-spot (799) geographically, without apparent differences in habitat, habits or voice. This is confirmed by the maps, where the two species approach each other most closely in 113B and 114A (M A C Coverdale & ADL, and D A Turner, pers. obs. respectively). See Forest Batis (796).

798 BLACK-HEADED BATIS *Batis minor*
B949 / MPG820 / D2C2

Afrotropical: from the peripheries of the Congo basin forests E to Ethiopia, S Sudan and W Kenya. Another race inhabits the coastal lowlands and adjacent areas of S Somalia, Kenya and Tanzania.

Kenyan: reasonably common in bushed and wooded habitats, including woodland edge, up to 1600 m (Br). The nominate race is found in the SE, where range is 69% below 500 m, 62% semiarid-dry subhumid and 77% within 500+ mm rainfall areas. Presence in 77B is on the forest island of Mt Endau (Tennent 1964; National Museum Collection, Nairobi). Restriction E of 40°E to the vicinity of the coast and/or the lower Tana R (AGH) parallels distribution in adjacent S Somalia (AM). The Black-headed is also largely confined to riverine vegetation in arid Tsavo E 101B, where it is an uncommon resident (La).

The race *B.m. erlangeri* occurs in 1000+ m areas to the W of the rift valley. Range here is entirely moist subhumid-humid and within the 1000+ mm rainfall regime. The record in 2C was on the Kenya-Sudan border in Nov (G Nikolaus *in litt.*). This was presumably also *erlangeri* (*i.e.* on the basis of ranges given by MPG): this square receives 500-1000 mm rainfall and

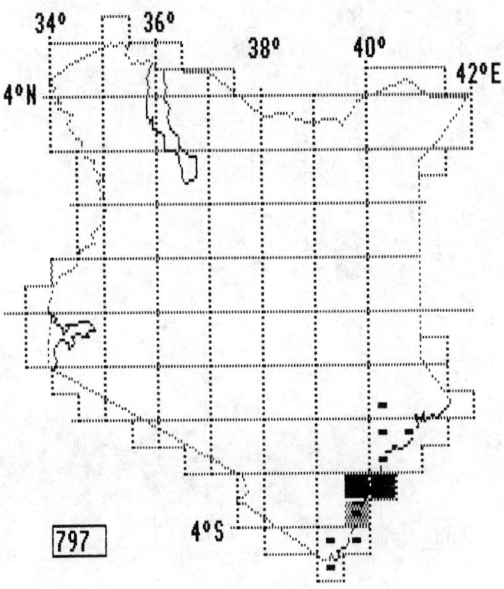

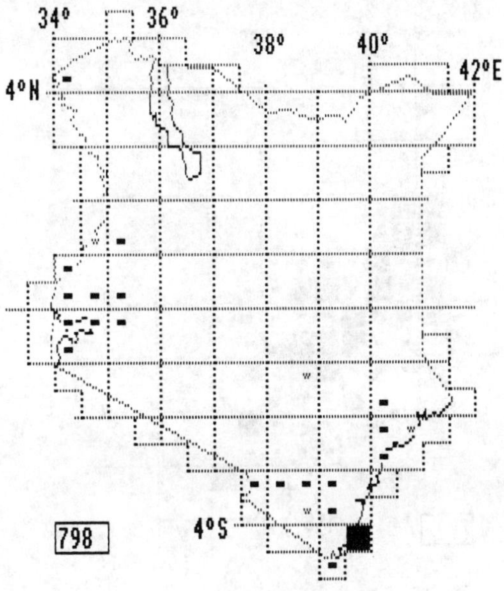

is semiarid.

Two birds were caught at dawn on 1 Dec 1976 at Ngulia Lodge 101A during a large fall of Palearctic migrants, but there is no other evidence of this species being a migrant (Backhurst & Pearson 1977b). The Black-headed and Chin-spot (799) Batis occupy similar habitats, but extensive sympatry in the W is probably not real at local level (Br; see La). See Forest Batis (796).

799 CHIN-SPOT BATIS *Batis molitor*
B951 / MPG817 / D2C2

Afrotropical: widespread S of 5°S, excepting the dry SW and those areas occupied by the East Coast Batis (797). Further N, it extends E of the Congo basin to East Africa and S Sudan. Distribution in Kenya represents the E limit of the range.

Kenyan: common in wooded habitats, including suburban gardens, at 500-3000 m (Br; Britton 1979; Ja). Range is 80% above 1000 m and 78% within 500+ mm rainfall areas. It has only been recorded from 2% of the arid squares, and is thus virtually absent from the arid N and E, which are occupied by the Pygmy (800) and Grey-headed (801) Batis. Presence in 26D was at 1700 m on Mt Nyiru (Pelchen 1977d). Absence from the moist coastal lowlands reflects allopatry with the East Coast Batis. Presumed allopatry with the Black-headed Batis (798)(Br) is demonstrated in arid Tsavo E 101B, where that species is in riverine habitats while the Chin-spot inhabits woodland (La).

The Chin-spot is generally considered resident (Be: 74; DEP pers. obs.: 74BD 75BD 88D 89AC 101A), even in arid Tsavo E 101B (La). However, St notes increases in numbers near the N range edge at semiarid Baringo 50A during the rains. See Forest Batis (796).

800 PYGMY BATIS *Batis perkeo*
B953 / MPG819 / D2C1

Afrotropical: the NE, from NE Tanzania, through N and E Kenya, to NE Uganda, SE Sudan, S Ethiopia and Somalia.

Kenyan: a reasonably common species of bushed and wooded habitats in dry country below 1200 m (Br): the W and C highlands and their adjacent plateaux form the local SW range edge. Absence from the coastal strip continues to 2°N in Somalia, where there is only a single square further S (AM). Range is 64% within the 0-500 mm rainfall regime and 88% arid-semiarid. Presence in only 18% of the most arid squares (moisture index below 1.2) and in only 21% of the lowest rainfall squares (0-250 mm) produces apparently scarce status from Turkana E to 18 30 41. This paucity may be real, since there are very few records from the relatively well covered L Turkana area (HH; many subsequent obs.).

The Pygmy Batis is considered resident (St: 50A; La: 101B). Mentioning a single bird caught at dawn at Ngulia Lodge 101A, Backhurst & Pearson (1977b) note that the species is not known to be migratory. HM state that, in overlap areas between the

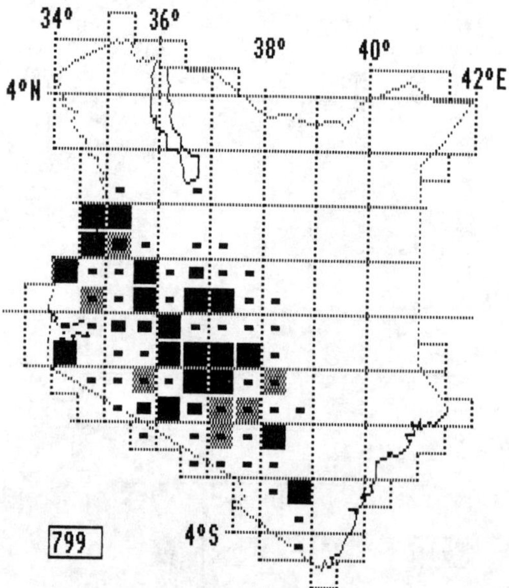

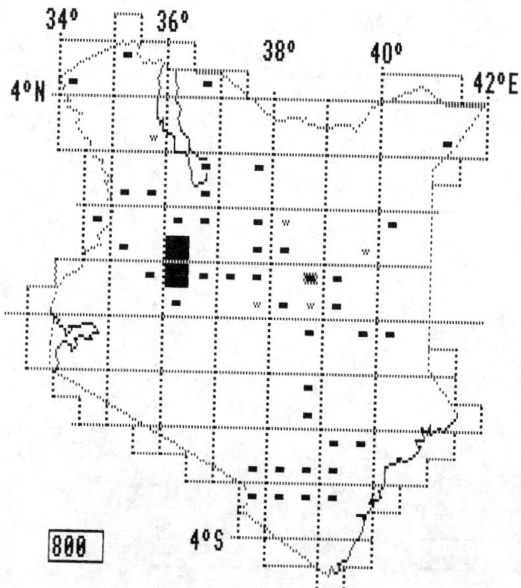

Pygmy and other congeners in SE Kenya, the Pygmy tends to prefer drier country. On our maps, these overlapping congeners are the Black-headed (798) and Chin-spot (799) Batis. See Forest Batis (796) and Grey-headed Batis (801).

801 GREY-HEADED BATIS *Batis orientalis*
B952 / MPG818 / D1C1

Afrotropical: the N tropics, from Chad and the Central African Republic to S Sudan, Ethiopia and Somalia, and SE to NE Uganda and Kenya. Distribution in Kenya represents the S limit of the range.

Kenyan: an infrequently recorded species of dry bushed and wooded habitats. Br mentions occurrence around 1200 m on the S edge of the Ethiopian highlands at Moyale 17A, and this batis descends to around 400 m at L Turkana (HH). There is also a record near sea level on the Somali border 80C (AM). Range is predominantly in N Kenya, from the borders with Sudan (2C Dec: G Nikolaus *in litt.*) and Ethiopia (4A 15B 16AB 17B: J S Ash *in litt.*), S to the N periphery of the highlands 50A 51AB 52C (St; EABR 1978-1980). This distribution is 77% below 1000 m, and entirely arid-semiarid. There is no breeding record, but AM regard this batis as resident in Somalia, and the few published dates from the N periphery of the Kenya highlands (EABR) show no clear seasonality.

Habitat distinctions between this species and the Pygmy Batis (800) are not well understood, but the Pygmy appears to tolerate more arid environments (Br; HM). Presence of the Grey-headed around L Turkana contrasts directly with the Pygmy (HH), and the Grey-headed is mainly restricted to the vicinity of the coast and/or large rivers in Somalia, whereas the Pygmy is not (AM). Also, whereas the Grey-headed is reported from only four of the 0-250 mm rainfall squares away from L Turkana, the Pygmy is in 10. Similarly, for the most arid squares (moisture index below 1.2), the totals are two and six respectively. In terms of total rainfall range, however, the Grey-headed is 78% below 500 mm, while the Pygmy is only 64%. The influence of observer frequency on all of these figures is difficult to assess. See Forest Batis (796).

802 WATTLE-EYE *Platysteira cyanea*
B960 / MPG822 / D2C2

Afrotropical: W and C Africa, from Senegambia and Congo, E through the Congo basin to the E limits of the range in Ethiopia, W Kenya and N Tanzania.

Kenyan: relatively common in a variety of bushed and wooded habitats up to 2200 m (Br; Ja). Range is in 1000+ m areas to the W of the rift valley. The squares are 78% within the 1000+ mm rainfall regime and 88% moist subhumid-humid.

The Wattle-eye forms part of a superspecies with the Black-throated Wattle-Eye (803)(HM). In areas of apparent sympatry in East Africa, the two are usually altitudinally segregated, with the Wattle-eye at lower elevations (*op. cit.*). The Wattle-eye is much commoner than the Black-

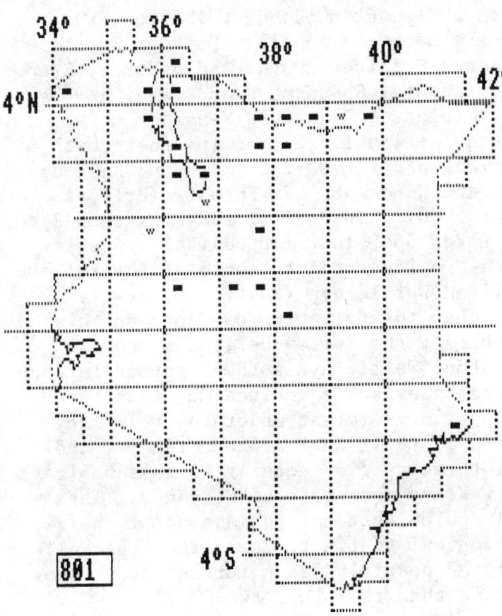

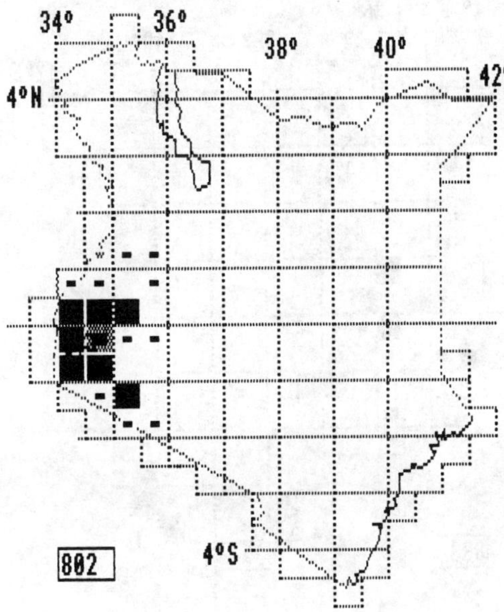

throated at 1700 m in the Kakamega Forest 48D, but this relationship is reversed (or the Wattle-eye is actually absent) at *c.* 2000 m in the adjacent North Nandi Forest 48D 49C (Diamond & Fayad 1979; Mann 1980; ADL pers. obs.). Other reference: Mann 1973a.

803 BLACK-THROATED WATTLE-EYE *Platysteira peltata*
B961 / MPG823 / D2C2

Afrotropical: replaces the Wattle-eye (802: *q.v.*) in areas further to the S and E. Hence widespread in the S tropics between 5°S and 15°S, and further S through Mozambique to NE South Africa. Further N, it reaches East Africa and S Somalia. There is an isolated population in Cameroon. Distribution in Kenya represents the N limit of the range.

Kenyan: rather local and uncommon in thicket, bushed and wooded habitats, and forest edge and patches, up to 3000 m (Br; Ja; Se). Two races occur (Br). *P.p. mentalis* inhabits the rift valley and country further W, with range 76% above 1500 m, 76% within 1000+ mm rainfall areas and 88% moist subhumid-humid. A single Sept record from around 3000 m at Mau Narok 61D (Se) might reflect a wanderer. Range widely overlaps that of the Wattle-eye to the W of the rift, but allopatry is apparently at least mostly preserved by the Black-throated's being at greater altitudes (see their altitude % data). To the E of the rift, where the Wattle-eye is absent, the Black-throated extends down to sea level.

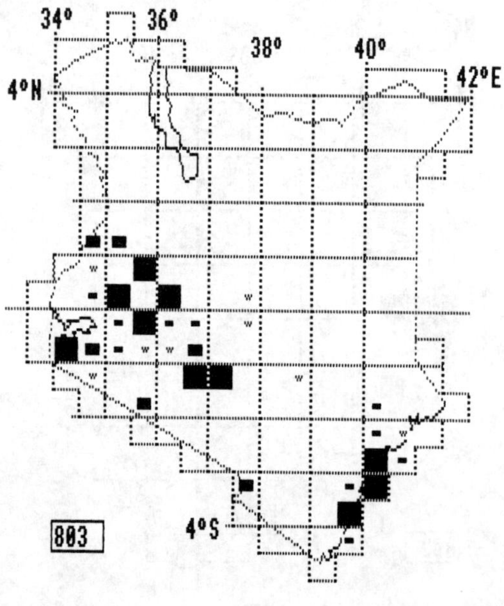

803

The nominate race occurs E of the rift, and is 94% within 500+ mm rainfall areas and 88% semiarid-subhumid. Ja mentions presence in Tsavo 89CD 101ABC (not mapped) and this is reiterated by Br, but precise localities are omitted. Tennent (1964) reported presence on the Mt Endau forest island 77B. Penetration of arid country along the Tana R 79C (AGH) parallels restriction to the Juba R in adjacent S Somalia (AM).

804 YELLOW-BELLIED WATTLE-EYE *Platysteira concreta*
B959 / MPG826 / D1C3
NO MAP: post confirmed 48D 49C

Afrotropical: W and C Africa, with a very discontinuous distribution between Sierra Leone and Angola and, on the E limits of the Congo basin, along the Zaire-Tanzania-Uganda borders. The E limit of the range is represented by an isolated population in the Kakamega-Nandi area 48D 49C of W Kenya.

Kenyan: uncommon in undergrowth at approximately 1700-2000 m in the Kakamega 48D and North and South Nandi Forests 48D 49C (Br; Ja; HM; Diamond & Fayad 1979; Mann 1980). These two squares are W of the rift valley, in humid, 1000+ mm rainfall areas. See Jameson's Wattle-eye (805).

805 JAMESON'S WATTLE-EYE *Platysteira blissetti*
B957 / MPG825 / D1C3
NO MAP: post confirmed 48D 49C

Afrotropical: a W and C African species with a discontinuous distribution from Sierra Leone to Angola. Further E, it reaches E Zaire, Rwanda and Burundi, E to NW Tanzania, S Sudan, Uganda and, at the E limit of the range, W Kenya.

Kenyan: locally common in undergrowth at approximately 1700-2300 m in the Kakamega 48D and Nandi 48D 49C Forests (Br; Ja; Mann 1980). In Uganda, range extends E to Mt Elgon 36CD (not mapped)(Br). Overlay data are identical to those of the Yellow-bellied Wattle-eye (804).

Br has the following comments on this species', the Yellow-bellied's and the Chestnut Wattle-eye's (806) interrelationships. Jameson's and the Chestnut are sympatric in forest undergrowth in many parts of their East African ranges, and all three species coexist in this habitat in W Kenya and parts of W Uganda. The only point that can be made about their ecological distinctions is that the Yellow-bellied possibly feeds further above the ground than the other two.

806 CHESTNUT WATTLE-EYE *Platysteira castanea*
B958 / MPG824 / D1C3
NO MAP: post presence 48D 49C; pre presence 36D 48C; pre probable 48D

Afrotropical: W and C Africa, throughout the lowland forests around the Gulf of Guinea and E through the Congo basin to NW Tanzania, S Sudan, Uganda and, at the E limit of the range, W Kenya.

Kenyan: reasonably common in forest undergrowth, up to 1800 m (Br). There are old records from Mt Elgon 36D and Mumias 48C (Ja), but all recent reports are from the Kakamega 48D and South Nandi 48D 49C Forests (Mann 1980; *op. cit.*). Overlay data are identical to those of the Yellow-bellied Wattle-eye (804). Probable breeding in 48D is based on courtship display and the presence of pairs and immatures at Kakamega (Zimmerman 1972). See Jameson's Wattle-eye (805).

807 LITTLE YELLOW FLYCATCHER *Erythrocercus holochlorus*
B964 / MPG807 / D2C2

Afrotropical: the E, in the coastal lowlands of Tanzania, Kenya and S Somalia.

Kenyan: a common species of thicket, woodland and forest, up to approximately 450 m, in the coastal lowlands and along the lower Tana R 79C 91A (Br; Ja; AGH; Britton & Zimmerman 1979). Range is 80% semiarid-dry subhumid, and entirely within 500+ mm rainfall areas. The record in 80A is from Somalia (AM). HM reject an old record of the closely related and allopatric Livingstone's Flycatcher *E. livingstonei* from Vanga 114C.

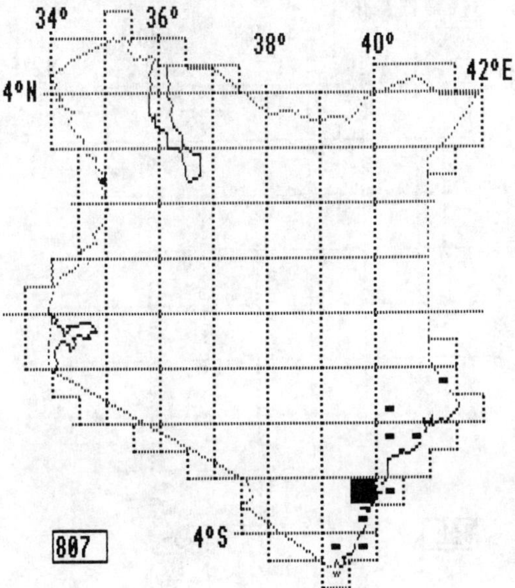

808 BLUE FLYCATCHER *Erannornis longicauda*
B963 / MPG827 / D3C2

Afrotropical: primarily a W and C African species, from Senegambia to Angola. It also occurs in the NE of the Congo basin, N to S Sudan, and E to NW Tanzania, Uganda and, on the E extremity of the range, W Kenya.

Kenyan: locally common in forest edge and a variety of bushed and wooded habitats, up to 2400 m (Br; MPG; Britton & Sugg 1973; Diamond & Fayad 1979). Range is in subhumid-humid country to the W of the rift valley. The squares are 90% at 1000-2000 m and 85% within the 1000+ mm rainfall regime.

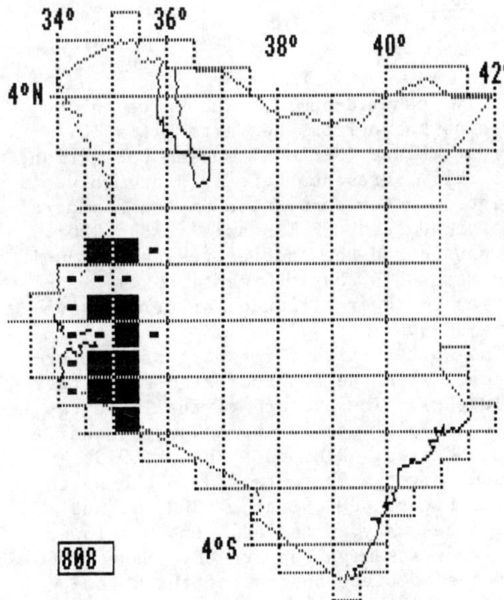

809 WHITE-TAILED CRESTED FLYCATCHER *Trochocercus albonotatus*
B970 / MPG829 / D2C2

Afrotropical: a highland species of the E, N from Zimbabwe, Mozambique and Malawi to most suitable areas of East Africa and the E Zaire borders. Distribution in Kenya represents the N and NE limits of the range.

Kenyan: reasonably common in forest undergrowth and mid-levels, up to 2700 m (Br; Ja; HM). Distribution is in two distinct areas. This is mainly a bird of the W and C highlands, where range is 79% within 1000+ mm rainfall areas, and en-

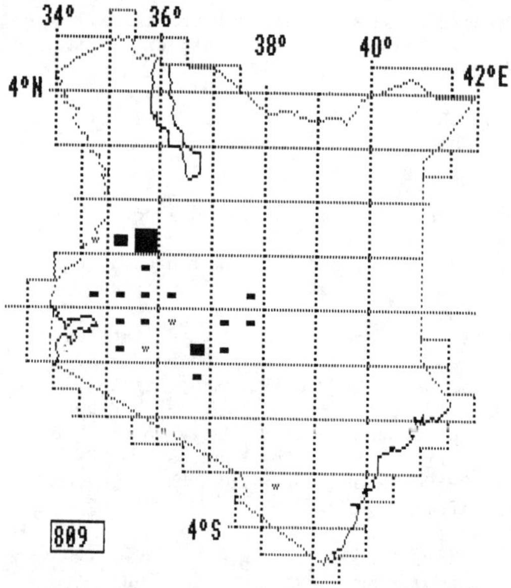

tirely subhumid-humid. The White-tailed is also known from the Taita Hills 101A forest island (National Museum Collection, Nairobi). Presence here is presumably related to the population on the Usambara Mts of adjacent NE Tanzania (EABR 1980). Se considered the species resident around Mau Narok 61D, but MPG mention some wandering to lower altitudes after the breeding season.

The White-tailed forms part of a superspecies with the Dusky Crested Flycatcher (810)(HM). The two are virtually segregated by altitude in the Kakamega-Nandi 48D 49C area. Diamond & Fayad (1979) found only the White-tailed in the North Nandi Forest 48D 49C at *c*. 2000 m, and only the Dusky in the adjacent but lower altitude Kakamega Forest 48D. Mann (1980) refutes complete absence of the White-tailed at Kakamega, but agrees that the Dusky is more common there. He suggests that, at 1650 m in this forest, the White-tailed and Dusky may be at their lower and upper altitudinal limits respectively.

810 DUSKY CRESTED FLYCATCHER *Trochocercus nigromitratus*
B972 / MPG831 / D1C2
NO MAP: post presence 48D 49C; pre presence 36D

Afrotropical: a W and C African species, from Nigeria S to Gabon. On the E of the Congo basin, it extends from Zaire E to NW Tanzania, S Sudan, Uganda and, at the E limit of the range, W Kenya.

Kenyan: a local bird of forest undergrowth (Br). There is an old record from Elgon 36D (Ja) and a recent sighting in Chemoni Forest 49C (EABR 1980), but most reports are from the Kakamega Forest 48D, where it is regularly seen. Range is above 1000 m in humid, 1000+ mm rainfall areas to the W of the rift valley, and N/NE of the Winam Gulf 60AB. There is no breeding record. D1 reflects its dense habitat. See White-tailed Crested Flycatcher (809).

811 CRESTED FLYCATCHER *Trochocercus cyanomelas*
B971 / MPG828 / D1C2

Afrotropical: the E, from the Cape, through E South Africa and Mozambique, to S Zaire, Zambia, Malawi, East Africa and extreme S Somalia. Distribution in Kenya represents the N limit of the range.

Kenyan: locally common but rather retiring in the dense undergrowth and mid-levels of forest, up to 1800 m (Br; MPG). It is present on the forest islands of Mts Uraguess 51A and Endau 77B (Br; Tennent 1964). Excluding the localised climates of these isolated highlands, range is 88% within 500+ mm rainfall areas, and 81% semiarid-subhumid.

The Crested Flycatcher is most common in the SE and coastal lowlands (*e.g.* Britton & Zimmerman 1979). The sole breeding record is reported by van Someren (1939) but, in the absence of any evidence of movement, the species is presumed resident. Similar status is reported from S Somalia, where

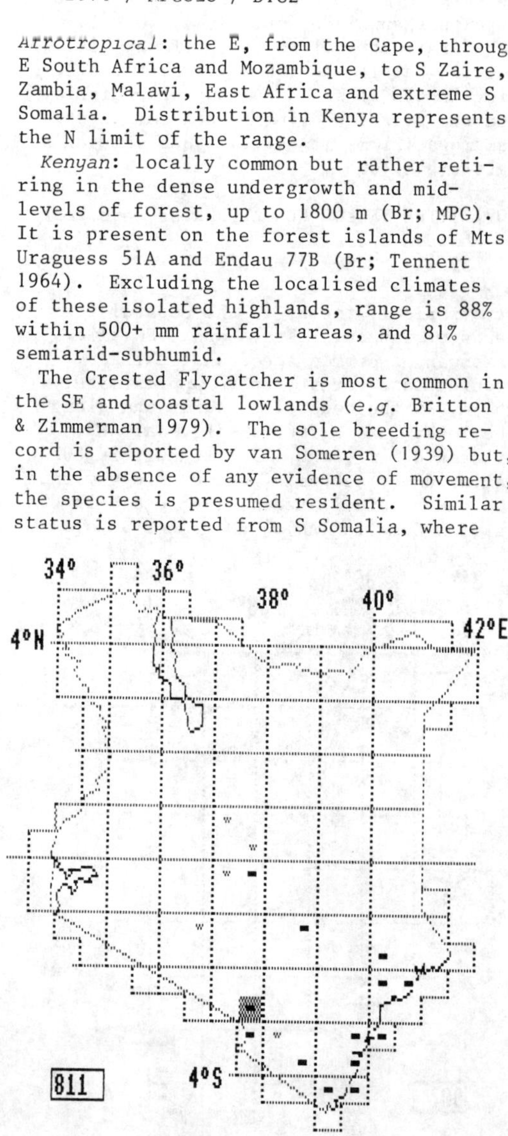

range is restricted to 80B (not mapped here) and the Juba R S of 1°N (AM).

This flycatcher is certainly uncommon in the highlands, where the preponderance of pre-1970 records (four of five squares) suggests a contraction of range. Despite this bird's low detectability, the absence of recent records from such well known localities as Nairobi, Ngong and Kiambu (all 75B)(Ja) and Mt Kenya 63A (Br) provides further evidence of this decline.

812 PARADISE FLYCATCHER *Terpsiphone viridis*
B968 / MPG832-834 / D3C3

Afrotropical: widely distributed S of 15°N, except for the dry SW, and lowland forest areas in the vicinity of Liberia and Ivory Coast.

Kenyan: locally common in bushed, wooded and forested areas, up to 2500-3000 m (Br; Se; Britton & Sugg 1973). Range is only 23% within the 0-250 mm rainfall regime and in 15% of the most arid squares (moisture index below 1.2), so that scarcity in the NE (*i.e.* around 29) is most probably real. This is supported by lack of records from an observer resident in 42A. That these dry, harsh environments lack sufficient vegetation and prey for this flycatcher is shown by its restriction in arid and semiarid country to the locally moist and well vegetated habitats around lakes and rivers, *e.g.* at L Turkana (HH) and L Baringo 50A (St), along the lower Tana R 79C 91A (AGH), and in Tsavo E 101B (La).

The race *T.v. ferreti* occurs throughout

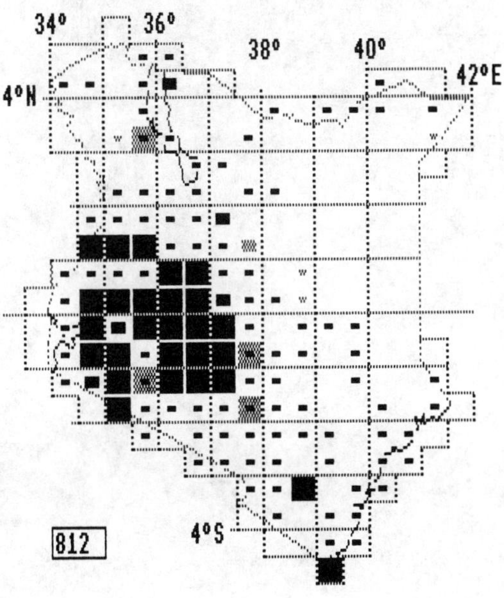

Kenya (Br) but with apparently varying status. Its breeding is seemingly largely restricted to the highlands and their peripheral plateaux: breeding squares are 82% above 1000 m, 86% in 500+ mm rainfall areas and 71% subhumid-humid. Be described *ferreti* as resident at 1500-2100 m in Narok district 74, but Se noted uncommon Jul-Nov visitors or migrants around 3000 m at Mau Narok 61D. Occurrence in arid Tsavo E 101B is mostly during the rains (La). Appearance in adjacent semiarid Tsavo W 89C 101A is similarly in the Nov wet period, and regular individuals at the Ngulia lights 101A during this period are probably nocturnal migrants associated with this movement (Backhurst & Pearson 1977b; Br). Movement of off-season birds into semiarid areas around the highlands during the rains may be regular, and might explain the absence of breeding records in areas such as 87-89, which include sites fairly regularly covered by birdwatchers.

The S tropics race *T.v. plumbeiceps* occurs within the range of *ferreti* in the coastal lowlands N to Sokoke Forest 102B (Br). This race is highly migratory in other parts of Africa, and may be a non-breeding, intra-African migrant to SE Kenya (Br). Virtual absence of breeding records from the coastal lowlands may be related to these migrants, and to AM's lack of breeding records from Somalia where the sole race *ferreti* is only presumed resident. P L Britton (*in litt.*) states that the Paradise Flycatcher is certainly not completely resident on the Kenyan coast. The sole evidence of coastal breeding that he knows are old nests on Mpunguti Island 114C in Aug, where the species was particularly common.

Polymorphism in this flycatcher affects males only and produces individuals that are largely white. This phenomenon is mostly known in *ferreti*, but birds with white tails are virtually absent from the C highlands (Ja; Br; Se).

The Paradise Flycatcher and the Red-bellied Paradise Flycatcher (813) are members of a species-group whose taxonomy is made particularly complex by frequent hybridization (HM). On the E limits of the Red-bellied's range, its race *T.r. somereni* has apparently hybridized extensively with the Paradise Flycatcher to produce *T.r. emini* (HM). This interbreeding continues in the Kakamega Forest-Kaimosi area 48D, and is probably facilitated by sympatry artificially induced by forest clearance (Br; Zimmerman 1972). Other references: Meneghetti 1943, 1944.

813 RED-BELLIED PARADISE FLYCATCHER *Terpsiphone rufiventer*
B967 / MPG835 / D2C2
NO MAP: post confirmed 48D

Afrotropical: W and C Africa, from Senegambia and Gabon E through the Congo basin to N Tanzania, Uganda and, at the E extremity of the range, the Kakamega area of W Kenya.

Kenyan: a reasonably common species of forest interior in the Kakamega Forest-Kaimosi area 48D (Br), at approximately 1700 m. This square is W of the rift valley in humid, 1000+ mm rainfall country. See Paradise Flycatcher (812).

Family Motacillidae: Pipits, Longclaws, Wagtails

Approximately 54 species, most diverse in the Afrotropical, Palearctic and Oriental Regions. Of the 30 (55%) recorded from the Afrotropical, the longclaws *Macronyx* are endemic, as are 11 of the 13 breeding pipits *Anthus* and *Tmetothylacus* (HM). 21 motacillids (70% of the Afrotropical total) occur in Kenya, where 15 (71%) breed or are assumed to do so and the remaining six are non-breeding migrants from the Palearctic. Some of the pipits *Anthus* are difficult to identify in the field, and clusters or high proportions of pre-1970 squares may not signify declines. Other references: Jackson 1905; Tyler & Ormerod 1986.

814 RICHARD'S PIPIT *Anthus novaeseelandiae*
B981 / MPG706 / D2C2

Afrotropical: the S and E, widespread S of 5°S and N through East Africa to S Sudan, Ethiopia and Somalia. There is also a population in Cameroon.

Kenyan: often common in a wide variety of open habitats, up to 3400 m (Br; Ja; Se; Britton & Sugg 1973). Range is 78% within 500+ mm rainfall areas and only 10% arid, and this pipit is virtually absent from the N and E (27D 28C = Mt Marsabit).

Breeding records occur in two distinct areas, in the coastal strip and in 1000+ m country mostly in or around the C highlands. These squares are 73% subhumid-humid, and 88% within 500+ mm rainfall areas. The validity of these apparent breeding areas is uncertain but there are, for example, no reports from experienced observers resident in the apparently suitable country of 37C and 48C. Br considers this pipit probably mainly resident, but the only mentions of this status from specific areas come from Se (61D), Be (74B) and DEP (pers. obs.: 75 88A), whose squares are in any case within the highlands' breeding cluster.

Movements are certainly known elsewhere. HM mention considerable local mobility, and Br notes rapid colonisation of newly cleared areas, together with some wandering. St reports absence from the grassy shores of L Baringo 50A during very dry periods. The only mention of regular seasonal movements is from arid Tsavo E 101B, where this pipit is an uncommon Nov-Apr visitor to short grass areas, particularly near Aruba Dam (La).

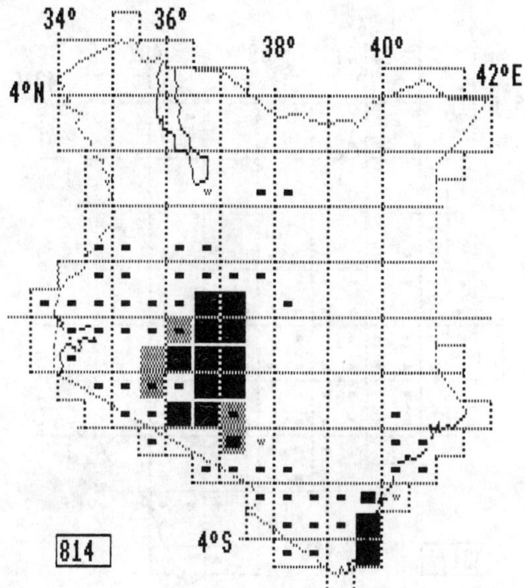

815 TAWNY PIPIT *Anthus campestris*
B976 / MPG702 / D1C0
NO MAP: post presence 50A 52C 65B 91B; pre presence 50B 62A 75B

Afrotropical: a Palearctic species that winters in the N tropics, S to about 12°N in the W and 7°N in Ethiopia. The few Kenyan records represent birds S of their normal wintering range, and there are reports from similar latitudes in Uganda (once: Br) and Somalia (occasionally: AM).

Kenyan: the eight records are of two males on 26 Mar 1918 in Tsavo (probably 89CD 101AB: not mapped; Ja); near Nairobi 75B on 11 Oct 1959, one at L Kisima 50B on 26 Mar 1967, one at L Nakuru 62A on 20 Oct 1968 (all in Ba); one near Garissa 65B on 11 Nov 1974 (OS-C 1977); one at Lamu 91B on 25 Oct 1982 (EABR 1982); one in Meru NP 52C on 5 Nov 1984 and one at L Baringo 50A on 13 Jan 1984 (EABR 1984). The Baringo bird is the only one outside the Mar or Oct-Nov migration periods. A Jul record from Lokitaung 3D is listed by Ba but omitted (and thus implicitly rejected) by Br. Particularly in view of the unusual date and of the possibility of confusion with the Long-billed Pipit (819) in this area, we also reject it here.

Of the seven squares, six are semiarid-subhumid, six are in 500-1000 mm rainfall areas, and all contain permanent water. This species is probably overlooked to some extent.

816 RED-THROATED PIPIT *Anthus cervinus*
B977 / MPG713 / D2C2

Afrotropical: this Palearctic pipit winters S to about 10°N in the W, but to Kenya and NE Tanzania in the E.

Kenyan: a fairly common winter visitor, mainly Nov-early Apr, typically in damp habitats, usually around permanent or temporary waters, up to 3300 m (Br; Se; Reynolds 1974; EABR 1983). Only 6% of its squares lack a large water body of some kind. This pipit has occurred on the coast in the Sabaki estuary 103A, and is particularly common around the lakes in both the high and low altitude sectors of the rift valley (Br). Occurrence in the arid N and E is particularly confined to the vicinity of permanent water, at L Turkana (where common: HH), at the single dam at Moyale 17C (late Mar: ADL pers. obs.), on Mt Marsabit 27D (Nov: G C Backhurst *in litt.*), near ponds at Wajir 42A (Apr: J E Miskell *in litt.*), near the Tana R in 64B and 65B (Garissa, Nov: D J Pearson *in litt.*), and around Aruba Dam in Tsavo E 101B (La).

Dec-Mar occurrence in arid Tsavo E 101B (*op. cit.*) is in the same period as scattered wintering records in S Somalia (AM). Br notes little evidence of passage. Ja reports large numbers at some highland sites in Mar and Apr. Reynolds (1974) mentions apparent fluctuations in overall abundance from year to year.

817 TREE PIPIT *Anthus trivialis*
B984 / MPG708 / D2C2

Afrotropical: a Palearctic species that winters across the N tropics from Guinea to Ethiopia, S of about 10°N. To the E of the Congo basin, it extends S to N Namibia, Botswana and NE South Africa. Distribution in Kenya represents the E limit of the range.

Kenyan: a locally common winter visitor and passage migrant, mainly late Oct-early Apr, in forest edge and wooded habitats (Br; Ja; Se). It extends up to 3000 m but is scarce below 700 m (*op. cit.*). Range is 71% within 500+ mm rainfall areas and only 19% arid. Virtual absence from the arid E represents the species' E limit. La quote only three records (Mar, Nov) from Tsavo E 101B, while the single record in the coastal strip also involved one on passage in late Mar at Mida Creek 102B (EABR 1984).

Occurrences in the arid N are mostly confined to the S periphery of the Ethiopian highlands 3B 4A (J S Ash *in litt.*) and 17A (Moyale, late Mar: ADL pers. obs.), and to birds on S passage at L Turkana (once: HH; Diamond & Keith 1980) and on Mt Marsabit 27D 28C (Ja; G C Backhurst pers. comm.). Restriction to autumn dates also occurs at Baringo 50A, where small

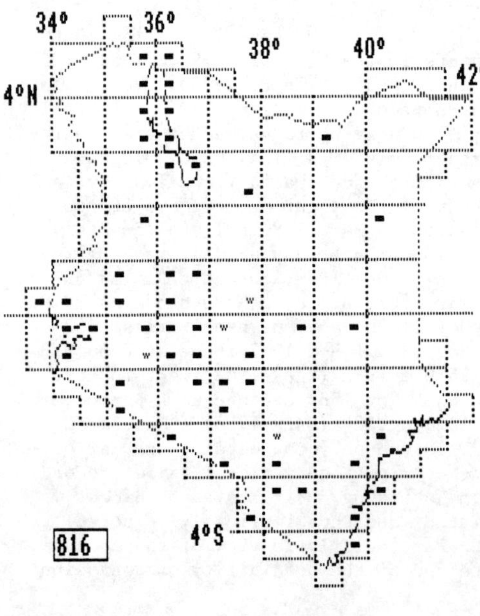

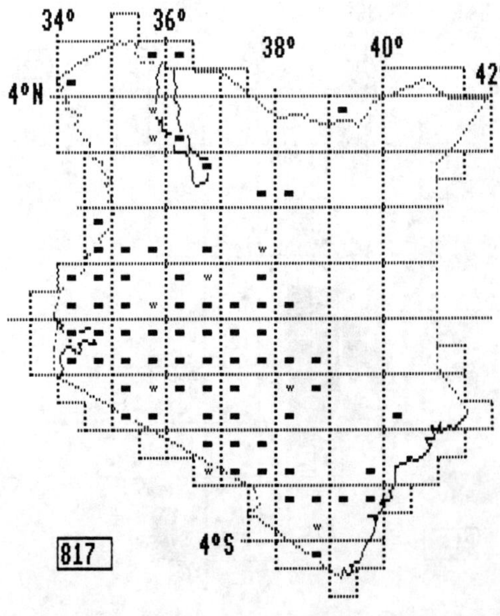

numbers are annual in Nov (St). This pipit is mainly an Oct-Nov and Mar-Apr migrant in the L Victoria basin (Br). Most wintering records are above 1500 m in the W and C highlands, with smaller numbers on the wooded grasslands of the moist plateaux (Br). The record in semiarid 2C was in Dec (G Nikolaus *in litt.*).

818 PLAIN-BACKED PIPIT *Anthus leucophrys*
B978 / MPG704, 705 / D2C2

Afrotropical: widespread S of 15°N, apart from most of the Congo basin and much of the coastal periphery of the E and SW. Distribution in Kenya represents the E limit of the range.

Kenyan: locally common in open country, including bare ground and bushed and wooded grassland, mainly at 700-2200 m (Br; Ja). It is primarily a bird of the highlands and their adjacent plateaux, where range (including all breeding records) is 88% above 1000 m and 82% in 500+ mm rainfall areas. The four arid squares (11%) are along the S periphery of the Ethiopian highlands (16A: J S Ash *in litt.*; 17AC: Ja, ADL pers. obs.), where the climatic data may be inaccurate, and in Tsavo E 101B (La).

Br considers the species resident, but single Jan and Mar records and former unspecified occurrence at approximately 500 m in the atypical country of Tsavo (*op. cit.*) may reflect wanderers or some more regular movement. Ja mentions nesting near Nairobi 75B, but there are no recent records from this well known area.

The races *A.l. zenkeri* (W Kenya) and *A.l. goodsoni* (in more E areas) intergrade, but occurrence (though not breeding) of the two at Nakuru 62A and Mara R (precise locality uncertain) has led to the suggestion that they are separate species (HM; Br).

819 LONG-BILLED PIPIT *Anthus similis*
B982 / MPG703 / D1C1

Afrotropical: a wide but sporadic distribution in much of Africa S of 5°S, and further N through East Africa to S Sudan, Ethiopia and mainly N Somalia. There are also isolated populations across the N tropics to Niger, Guinea and Cameroon. The one or two records on the Somali coast between 2°N and 1°S (AM), together with those in Kenya, represent the E limits of the range.

Kenyan: a local and generally uncommon pipit, in a variety of open and often rocky habitats, mainly at 1200-2000 m (Br; Ja; HH). It is little known, and possibly overlooked. Range is 83% above 500 m, and virtually absent from the E and coastal lowlands. Occurrence at Lamu 91B (Br; DEP pers. obs.) compares with apparently very scattered and uncommon distribution on the coast of adjacent S Somalia (see above). This pipit is found throughout the rainfall and moisture categories, but 70% of its squares are within 250-1000 mm, intermediate rainfall areas and it is scarce in the 1000+ mm highlands and W. The records in 12C and 24B are from Uganda (J S Ash *in litt.*).

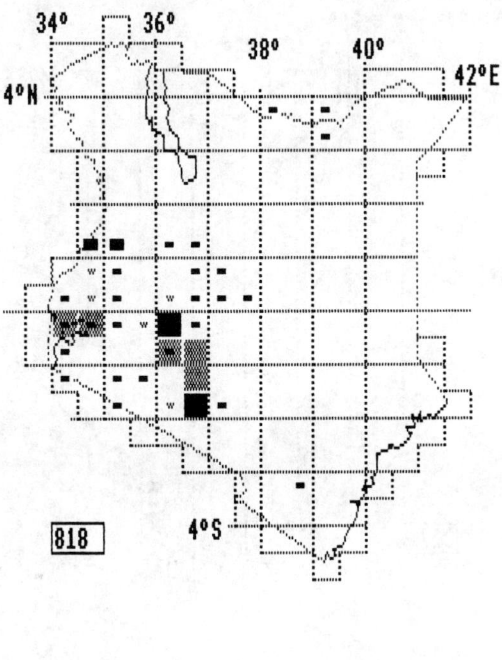

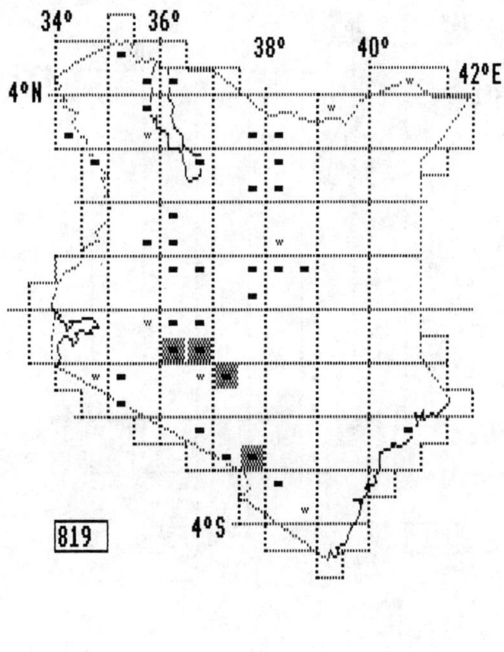

820 STRIPED PIPIT *Anthus lineiventris*
B979 / MPG714 / D2C2
NO MAP: post presence 101A; pre presence 88D

Afrotropical: mainly the E and SE, from E South Africa N through Zimbabwe, Mozambique and Malawi to NE Tanzania and, at the NE limit of the range, S Kenya. It also occurs in Angola.

Kenyan: a local and scarce resident on rocky, wooded slopes on the isolated highlands of the Chyulu 88D and Taita 101A Hills (Br; MPG). There is no breeding record.

821 LITTLE TAWNY PIPIT *Anthus caffer*
B975 / MPG712 / D0C1

Afrotropical: an apparently broken distribution in the E, from E South Africa, through Zambia, N Tanzania and S Kenya, to the S borders of Ethiopia. It also occurs in Angola.

Kenyan: local and scarce, favouring bushed grassland and sandy patches in open woodland, up to 2200 m but mainly at 1000-1800 m (Br; Ja; MPG). It occurs in the SW, on the plateau to the S of the highlands. Of the nine squares, six are within 500-1000 mm rainfall areas, and seven are semiarid-dry subhumid.

The only recent records are of one at Ngong 75B in May (EABR 1978), and of presence in the Tanzanian sector of 87D (J S S Beesley *in litt.*: Ja lists a record from the Kenyan side of this square). Br lists Mara R (approximately 61CD 74A: not mapped). The high proportion of pre-1970 squares (78%) most probably reflects its DO characteristics. Movements of unknown type occur S of 5°S in Tanzania (Br).

This species forms part of a species-group with the Sokoke Pipit (822)(HM), which is restricted to coastal forest and woodland.

822 SOKOKE PIPIT *Anthus sokokensis*
B983 / MPG710 / D1C2

Afrotropical: endemic to relict patches of forest in coastal areas of Kenya and NE Tanzania.

Kenyan: an uncommon resident in the Sokoke 102B and Gedi 103A Forests. Earlier authors described it in forest glade and edge habitats (Ja; MPG; HM; Turner 1977), but it is typically in the ground-stratum of uncleared evergreen forest, and only occasionally in open areas in *Brachystegia* woodland (Britton & Zimmerman 1979; Br; Britton *et al.* 1984). A population density of up to one pair per two hectares in this uncleared *Afzelia* forest produces a possible total population of 2500 pairs in this habitat alone: the total Sokoke population might reach 3000-5000 pairs (*op. cit.*). A more recent survey agreed with this density in *Afzelia* forest, and estimated the total Sokoke population to be over 2000 pairs (EABR 1983).

The old record in 102D derives from Ja's mention of Takaungu. Presence in 114B is noted by P L Britton (*in litt.*), though Squire's (1976a) note of one on forest floor S of Mombasa is omitted (and presu-

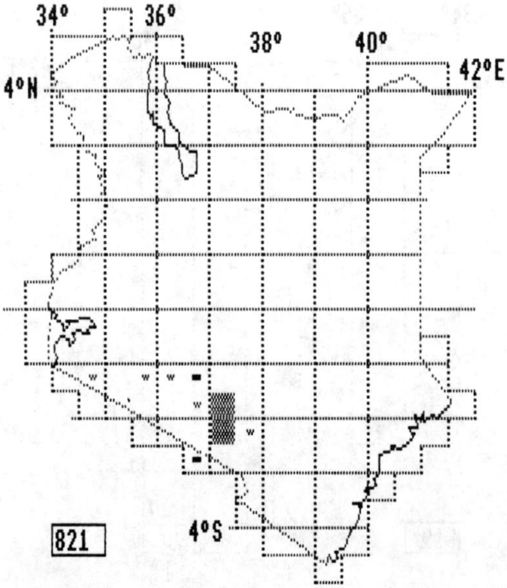

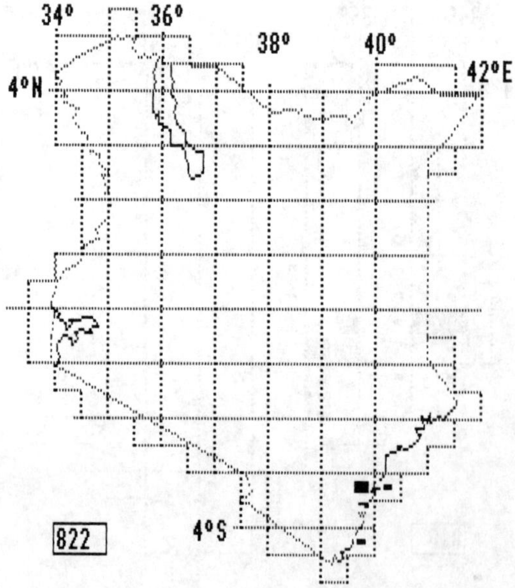

mably rejected) by Br. This pipit has never been found breeding in Kenya or Tanzania, and probable breeding in 102B is based on resident status and pairs in Sokoke Forest (*op. cit.*).

The four squares are all semiarid-dry subhumid and within 500+ mm rainfall areas. The species is threatened by deforestation (Turner 1977; Britton & Zimmerman 1979). See Little Tawny Pipit (821).

823 MALINDI PIPIT *Anthus melindae*
B980 / MPG709 / D2C2

Afrotropical: a very localised species, confined to the E coast and closely adjacent areas, from 4°N in Somalia (AM) S to Maji ya Chumvi 102C and Mombasa 114B in Kenya.

Kenyan: this distinctive bird is fairly numerous in coastal grasslands between Ngomeni 91C 103A and Karawa 91C, and widely distributed but less numerous on the riparian grasslands of the lower Tana R 79C 91A (Br; AGH). This very limited range is discussed by Britton & Britton (1978a), who point out that the coast in 91C 103A is unique in East Africa in lacking a fringing coral reef. This is due to the fact that coral cannot grow in the sediment-laden water discharged in this area by Kenya's two major rivers, the Tana and the Athi-Galana-Sabaki. This coast has sand dunes which, unprotected by a reef, are breached by the sea to form mangrove-covered flats. The pipit is most numerous on grasslands peripheral to these flats, whilst also extending to the contiguous grasslands of the Tana R. Distribution is similarly confined to the vicinity of the coast and the Juba and Shebelle R of S Somalia (AM).

Br suggests the Malindi Pipit to be probably resident, with records of occasional single birds in the well watched areas to the S of Ngomeni probably referring to wanderers. These records are from Mida 102B, Maji ya Chumvi 102c, Samburu 102C, Changamwe 114B and Mombasa 114B (Ja; *op. cit.*). One record from Mombasa (MPG) has been withdrawn by the observer (Britton & Britton 1978a). Similar records to the NE of the main range (Mokowe, Lamu 91B: Ja) may also represent wanderers (*op. cit.*) but, in this more poorly known area, there may be the possibility of interchange with birds in Somalia, which are known S to 80D (AM).

This pipit's squares are all in 500+ mm rainfall areas and, including those of presumed wanderers, seven of nine are semiarid-dry subhumid. In its main range (79C and 91AC), two of three squares are semiarid.

824 GOLDEN PIPIT *Tmetothylacus tenellus*
B997 / MPG715 / D3C2

Afrotropical: the E and NE, from mainly E areas of Tanzania and Kenya to S Sudan, W Somalia and SE Ethiopia. Distribution in Kenya represents the SW limit of the range, apart from records in extreme NE Uganda (SW of 2C: Br). A single wanderer of this highly mobile species has reached S to E South Africa (HM).

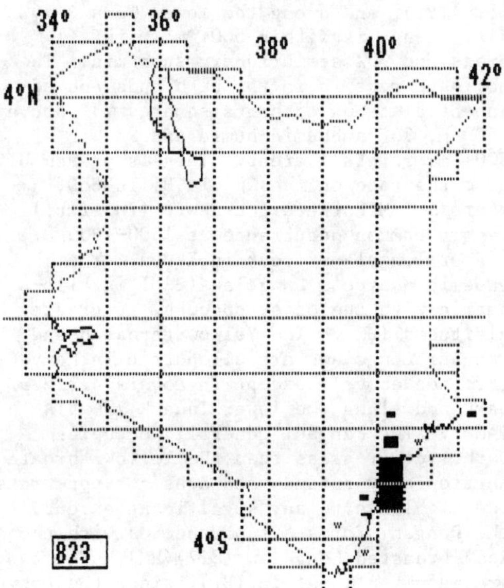

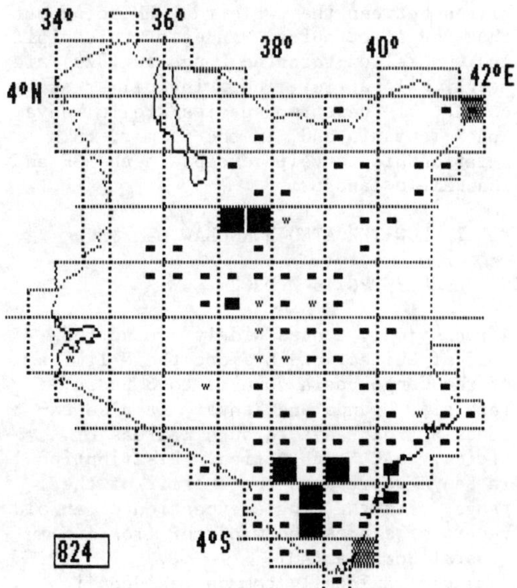

Kenyan: sometimes numerous in the bushed and wooded grasslands of the E lowlands (Ja; Br). Br describes distribution as mainly E of 37°E and, with 90% of the squares in these areas, this longitude approximately defines the W edge of the normal range. Scattered records further W appear to be mostly across the N periphery of the W and C highlands from 50ABD to 37D (J Harjula *in litt.*) and 49B (D J Pearson *in litt.*), although this may reflect observer frequencies to some extent. Other W occurrences have been at Macalder Mine 60C (Sugg & Sugg 1972), Nairobi 75B (Ja), and Kidepo NP (Uganda SW of 2C: Mann 1976). Jackson (1905) quotes a record from the Nyando Valley 61A (*cf.* 60C above), but omits it in Ja. The 77% of the Golden Pipit's squares that are below 1000 m agrees with Br's estimate but, whereas he describes the range as semiarid, it is in fact 46% arid and 40% semiarid.

This pipit regularly makes at least local movements, but these are poorly understood. It can be very common during and soon after the rains in arid Tsavo E 101B (La), and there are small numbers at the nearby Ngulia lights 101A in the Nov-Jan wet period (Backhurst & Pearson 1977b; EABR 1980). While it can completely disappear from Tsavo E during June-Oct (*op. cit.*), the very few records near the W range edge at Baringo 50A are in the succeeding Oct-Feb period (St; EABR 1978). This Oct-Feb seasonality agrees with the dates of a Feb migrant on the S slopes of Kilimanjaro (HM), and the late Sept record in 60C (*op. cit.*).

Remarking on the broad physical similarities between the two birds, HM point out that the Afrotropical ranges of this pipit and the Yellow-throated Longclaw (825) are nearly allopatric. The principal areas of contact of these two species are in Kenya and E Tanzania and, in the former, this relationship is valid except in the SE and coastal lowlands.

825 YELLOW-THROATED LONGCLAW *Macronyx croceus*
B988 / MPG716 / D2C2

Afrotropical: ranges widely around the Gulf of Guinea, and E along the N limits of the Congo basin forests to S Sudan, Tanzania, Uganda and Kenya. It also extends S from Kenya through E areas of Africa, to E South Africa. Distribution in Kenya represents the E limit of the range, with the single exception of an old record of a presumed wanderer from 1°N in coastal Somalia (AM).

Kenyan: a locally common resident in

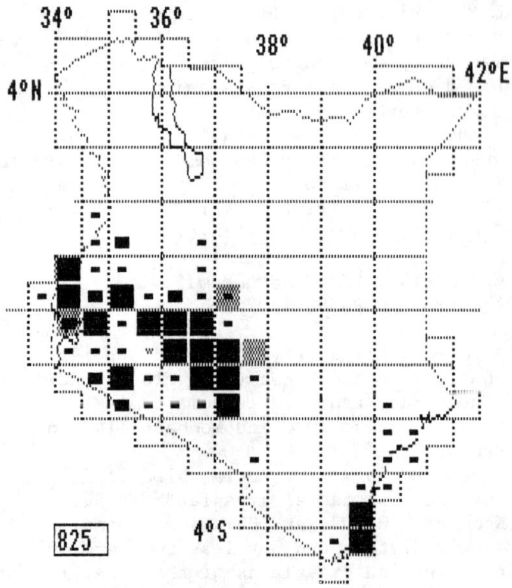

open and bushed grassland up to 2200 m (Br; Ja; Be). Longclaws breed on the ground during the rains, and are mostly in moderate-high rainfall areas where long grass hides their nests (Brown & Britton 1980; *op. cit.*). Birds of this genus are thus virtually absent from the arid N and E. The Yellow-throated's single arid square is 79C, where range is confined to the vicinity of the Tana R (see below).

Two races occur (Br). *M.c. tertius* is found in those areas of the coastal lowlands not occupied by the Pangani Longclaw (826)(Br), and along the lower Tana R 79C 91A. Range is within 500+ mm rainfall areas and 90% semiarid-dry subhumid. The nominate race is in the highlands and adjacent plateaux with its squares 73% above 1500 m, 86% subhumid-humid and 98% in 500+ mm rainfall areas. This is presumably also the race on the Chyulu Hills 88D, where J F M Horne & L L Short (*in litt.*) report common occurrence at 1500-1800 m.

Four longclaws occur in Kenya. The endemic Sharpe's Longclaw (828) is allopatric with the other species, at greater altitudes (HM). The Yellow-throated and Pangani Longclaws are allopatric in Kenya (Br: see above), except in *Dobera/Albizia* parkland along the lower Tana R 79C 91A (AGH). The Pangani generally occupies rather drier areas than the Yellow-throated, particularly inland, and does not penetrate the highlands to any significant extent. The Pangani forms a superspecies with the Rosy-breasted Longclaw (827)(HM). The two are mainly allopatric (Br), since the Rosy-breasted is generally at greater altitudes, where it is broadly sympatric with the Yellow-throated (*sensu* HM). The members of this genus generally display greater ecological tolerance when no congener is present (HM). The Yellow-throated shows striking evolutionary convergence with the Eastern Meadowlark *Sturnella magna* (Icteridae) of the New World. See Golden Pipit (824).

826 PANGANI LONGCLAW *Macronyx aurantiigula*
B987 / MPG720 / D2C2

Afrotropical: the E, from NE Tanzania through Kenya to SE Somalia.

Kenyan: a locally common resident of bushed and, locally, open grassland, mainly below 1000 m (Br; La). It is a bird of the coastal and SE lowlands, extending NW to the E and S peripheries of the W and C highlands at Meru NP 52C (Br), Tharaka 64B (Ja), the Athi Plains 75BD (not mapped: Br), Nairobi NP 75B (18 Nov 1974: D A Turner *in litt.*), Mara 74A (EABR 1981) and 86B (in Tanzania: J S S Beesley *in litt.*). In contrast to the Yellow-throated (825: *q.v.*) and Rosy-breasted (827: *q.v.*) Longclaws, this range is 64% semiarid and 94% within 250-1000 mm, intermediate rainfall areas (*cf.* Br's estimates).

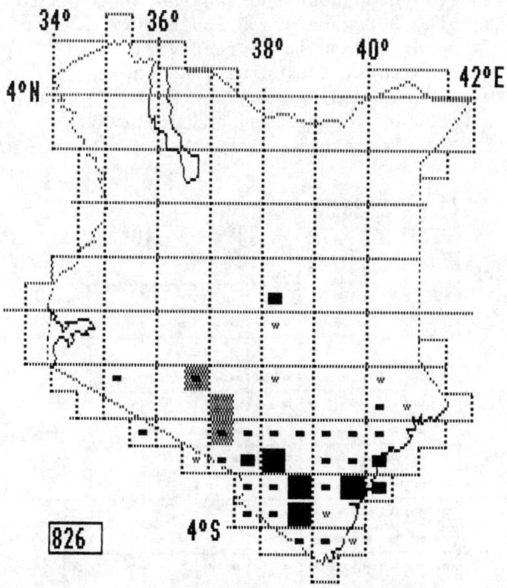

827 ROSY-BREASTED LONGCLAW *Macronyx ameliae*
B986 / MPG721 / D2C2

Afrotropical: occupies two distinct areas. In the S, it extends from NE South Africa N to Angola, S Zaire, Malawi and S Tanza-

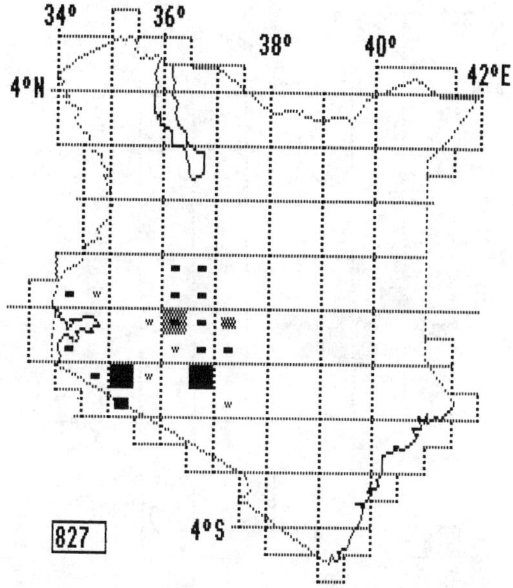

nia. A separate N population occurs in Kenya and the NE borders of Tanzania.

Kenyan: local in open grassland, often near water or in damp areas, up to 2200 m (Br; Ja; MPG). In contrast to the Pangani Longclaw (826), its squares are entirely above 1000 m in 500+ mm rainfall areas, and 85% subhumid-humid. Where apparently sympatric with the Pangani in NE Tanzania, that species is segregated in areas with trees and longer grass (HM): the Rosy-breasted does not apparently perch in trees (MPG).

There may have been a decline in some areas to the W of the rift valley. Ja notes occurrence at Elgeyu (approximately 37D 49AB), Uasin Gishu (approx. 49AC) and Kitotos (60B or 61A)(none of these localities mapped), but there are no recent records. Br mentions regular and numerous status in the L Victoria basin S and E of the Winam Gulf 60AB but, N of the Gulf, he found the bird only as an erratic wanderer, W to Nyangweso 48C. Br considered this longclaw probably mainly resident, but Ja noted local movements. A single wanderer to the N limit of the range in the semiarid country around L Baringo 50A was in Jul (St). See Yellow-throated Longclaw (825).

828 SHARPE'S LONGCLAW *Macronyx sharpei*
B990 / MPG718 / D2C3

Afrotropical: endemic to highland grasslands in W and C Kenya, although it may well reach Uganda on the W side of Mt Elgon 36D (Turner 1977).

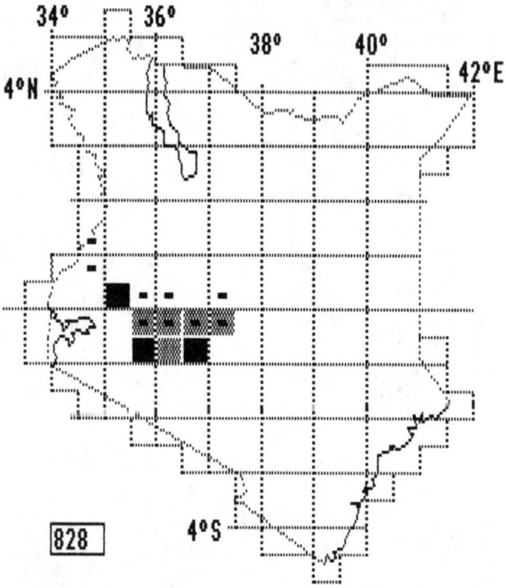

828

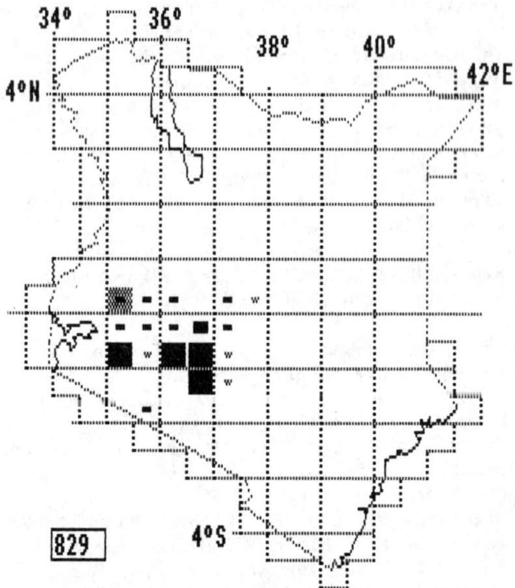

829

Kenyan: a resident species, locally common on short, open, montane grasslands at 2000-3400 m (Br; Ja; Be; Se; Britton & Sugg 1973; Brown & Britton 1980; EABR 1983). Range is 84% moist subhumid-humid and 61% within 1000+ mm rainfall areas. See Yellow-throated Longclaw (825).

829 CAPE WAGTAIL *Motacilla capensis*
B993 / MPG693 / D2C2

Afrotropical: the S, widespread S of 10°S and W of 32°E. Further N, it is found along the E Zaire borders to NW Tanzania, Uganda and, at the E limit of the range, Kenya.

Kenyan: local and usually uncommon in a variety of lakeside, swamp edge, grassland and forest clearing habitats, up to 3000 m in the W and C highlands (Br; Ja). Range is 82% above 1500 m, and entirely within subhumid-humid, 500+ mm rainfall areas. Presence in 74D represents birds around swamps in the Nguruman Forest (Fayad & Fayad 1980). Br considers the species resident but notes that wanderers occur. The latter status presumably accounts for single Sept and Dec individuals near the upper limit of the altitudinal range around Mau Narok 61D (Se).

This wagtail forms part of a superspecies with the Palearctic Yellow Wagtail (830)(HM), which is a much more abundant and widespread bird in Kenya during the N winter.

830 YELLOW WAGTAIL *Motacilla flava*
B996 / MPG695, 697-701 / D3C3

Afrotropical: a Palearctic species that winters throughout sub-Saharan Africa S to Namibia, Botswana and E South Africa.

Kenyan: common in a great variety of open habitats, including lake shores, swamp edges and, in moister W and C areas up to 3000 m, grassland and cultivation (Br; D J Pearson *in litt.*; Se; HH; St).

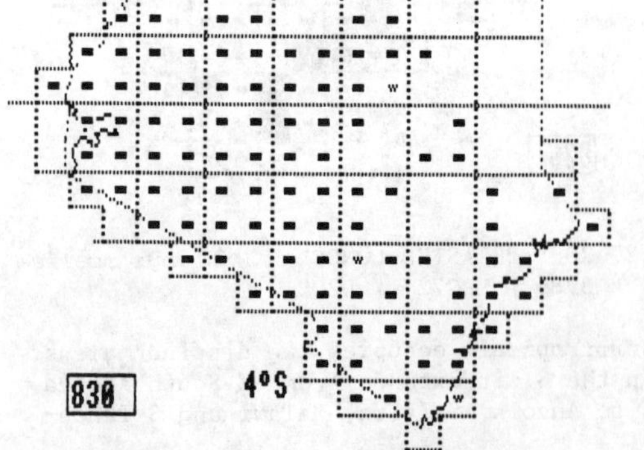

830

Records are throughout the altitude, rainfall and moisture categories illustrated on our overlays, and there seems no reason why this wagtail should not occur in every square, at least on passage. Apparent absence from the driest areas of the NE (around 29) may be real, but is more likely to reflect lack of observers.

It is particularly common, both as a passage migrant and winter visitor, in the L Victoria basin and the rift valley, and local wintering concentrations also occur in drier areas further E (Br). Thousands can congregate at sewage works, *e.g.* at Kariobangi in Nairobi 75B (Br).

Frequent association with large domestic or wild animals fills an ecological niche unoccupied by small Afrotropical passerines (Reynolds 1974; Ja; Se). The Yellow Wagtail species often greatly outnumbers Afrotropical congeners *e.g.* when heavy rains stimulate insect activity in Tsavo 89CD 101 (Brown & Brown 1974).

Occurrence is mainly late Sept-early Apr (*op. cit.*), but scarce stragglers persist into mid/late May (latest 20th)(Se; EABR 1982). An individual near Kapedc 38C on 10 June is the only oversummering record (EABR 1984).

At least seven races occur, and distinctions in range, habitat and seasonality can often be drawn (Br; Ja; Wallace 1955; Se; Reynolds 1974; HH; St). The most numerous races in Kenya are the Blue-headed Wagtail *M.f. flava*, which is particularly common in the rift and adjacent highlands, and the Kirghiz Steppe Wagtail *M.f. lutea*, in drier plains areas further E. Wallace (1955) recorded the W European Yellow Wagtail *M.f. flavissima* at Meru 52C, but Pearson & Backhurst (1973) showed that birds with plumage characteristics of this race are clearly identifiable as the E European *lutea* when they moult into breeding plumage.

Over 50 000 birds ringed at Nairobi 75B have produced recoveries in the USSR (41°-59°E, apart from one at 81°E), Saudi Arabia, Iran, Qatar, Tanzania and Uganda (Br; E A Bird Ringing Reports (including map of recoveries in Backhurst 1971b); G C Backhurst *in litt.* and 1985a). See Cape Wagtail (829).

831 AFRICAN PIED WAGTAIL *Motacilla aguimp*
B991 / MPG691 / D3C3

Afrotropical: widespread S of 10°N, extending to 15°N in Ethiopia and Niger, and only absent from the dry areas of Somalia and the SW. In most areas, this is the commonest Afrotropical wagtail and the most associated with man (HM).

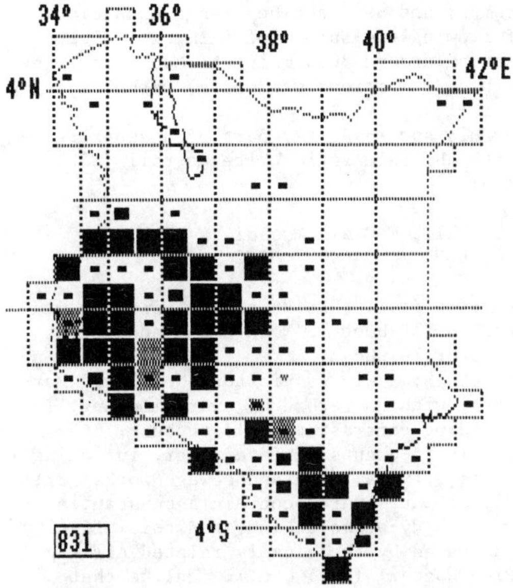

Kenyan: this fairly common species is found in the vicinity of water, whether as discrete bodies *e.g.* on lake shores and river banks, or as that available around human habitation. It is approachable, and particularly associated with settlements, ranging from Nairobi city centre 75B to villages and farms (Ja; MPG). Br mentions occurrence in forest clearings and up to 3000 m.

Apparent absence from all but 14% of the driest, 0-250 mm rainfall areas and 6% of the most arid squares (moisture index below 1.2) is most probably real. The 21% of squares in the arid N and E mainly contain settlements or permanent water, *e.g.* presence around habitation at L Turkana (HH), along the Daua R 19AB, on the isolated, inhabited highland of Marsabit 27D 28C, in the vicinity of the Ewaso Ng'iro R 40D 52B, along the Tana R 64B-91A, and along the Galana R in Tsavo E 101B (La). Restriction to the Tana (*e.g.* AGH) parallels virtual confinement to the Juba R of adjacent S Somalia (AM).

This wagtail is mainly resident (Br; Be; La; St; Pomeroy & Tengecho 1982a), and presence of 82% of breeding squares within 500+ mm rainfall areas most likely represents observer coverage of a species that frequently breeds in buildings (Brown & Britton 1980). The high number of breeding squares is probably of similar origin. Four ringing recoveries, all in the Nairobi area 75B, showed a maximum movement of only 30 km (Smart 1967; Backhurst 1971b). Ja mentions colonisation of newly erected camps, and Se describes range expansion following the spread of farming and habitation around 3000 m in the Mau Narok area 61D. Br mentions wanderers, without giving details.

This species forms part of a superspecies with the Palearctic White Wagtail (832) (HM).

832 WHITE WAGTAIL *Motacilla alba*
B992 / MPG690 / D2C2

Afrotropical: this Palearctic species winters S to about 12°N in the W and to the equator in the E.

Kenyan: a winter visitor, Nov-early Mar (Br), with no record of oversummering. In striking contrast to Ja's account, it mainly frequents surface water, including lakes, rivers, dams and sewage works (Br). 88% of the squares contain a substantial water body of some kind, and this affinity is shared by the closely related African Pied Wagtail (831: *q.v.*). Unlike that congener, however, occurrence around human habitation is hardly mentioned for Kenya, although AM note it in Somalia.

The White Wagtail is only at all common at L Turkana (HH; D J Pearson *in litt.*). Examples of the smaller numbers that occur further S include concentrations of up to 30 in the Dec-Feb period at Kariobangi sewage works, Nairobi 75B, and single birds at L Baringo 50A (Dec: St), Garissa 65B (Nov), and along the coast (Nov-Dec) (EABR). There are only two records S to Tanzania, in the E and NE (Br).

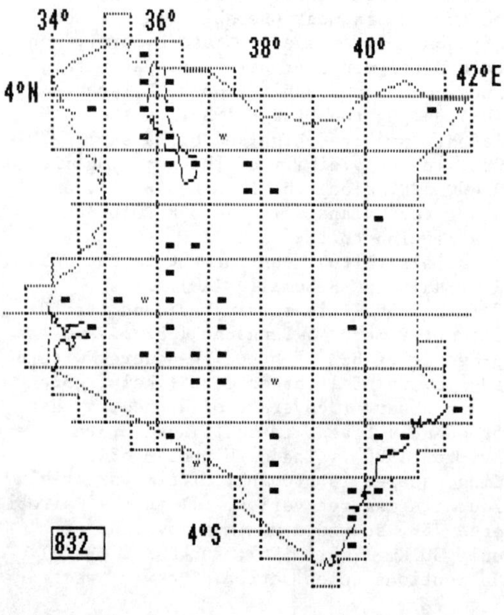

833 MOUNTAIN WAGTAIL *Motacilla clara*
B995 / MPG692 / D2C2

Afrotropical: widely distributed in many areas between 5°N and 15°S. Further N, it occurs in the Ethiopian highlands, in Cameroon-SE Nigeria, and in Liberia-Guinea. Further S, it extends to Zimbabwe, S Mozambique and E South Africa. Distribution in Kenya represents the E limit of the range.

Kenyan: local and rather uncommon along fast-flowing, often rocky streams, and on forest tracks (Br; Ja; Be). It is a highland bird, up to 3000 m (Br; Se), with range entirely above 1000 m, 87% subhumid-humid and 91% within 500+ mm rainfall areas. Occurrences at slightly lower altitudes in Kitovu Forest 100B (Br) and in the Tanzanian sector of 88C (J S S Beesley *in litt.*) represent the periphery of the population on the Kilimanjaro massif.

This wagtail is apparently mainly resident (Br; Be), but Se mentions regular, apparently non-breeding visitors near the species' upper altitudinal limit around Mau Narok 61D, and Br notes wandering.

It forms a superspecies with the Palearctic Grey Wagtail (834)(HM). The two are often found along the same stream, even though they seem to share very similar ecological requirements (Br).

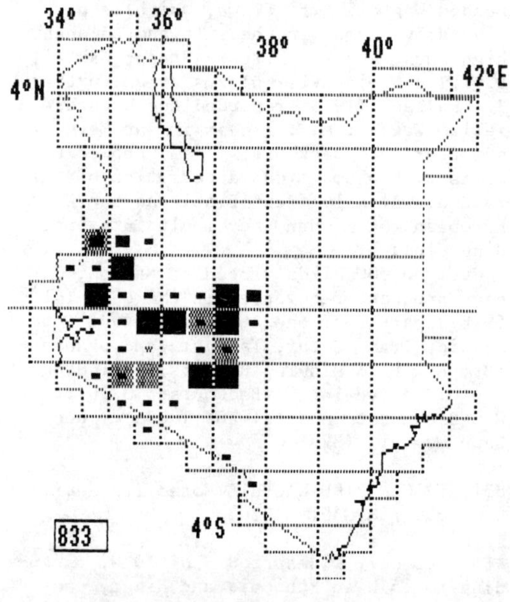

834 GREY WAGTAIL *Motacilla cinerea*
B994 / MPG694 / D2C2

Afrotropical: a Palearctic species that is

scarce and sporadic in the W, but more frequent in E highlands, from S Sudan and Ethiopia S to Malawi.

Kenyan: mainly a winter visitor, Oct-Mar (Br), with no record of oversummering. Like the closely related Mountain Wagtail (833: *q.v.*), it frequents fast-flowing streams, usually within forest, at up to 3000 m in the highlands (*op. cit.*; Se). It winters in small numbers on Mt Elgon 36D and in the W and C highlands (Br), although Se considered all individuals near the species' upper altitidinal limit around Mau Narok 61D to be transient.

The Grey Wagtail is very scarce away from the highlands. The record in 16A is from Ethiopia (J S Ash *in litt.*) and may represent the S extremity of the Ethiopian highlands' wintering population. Small numbers on Mt Marsabit 27D have been along streams (Sept & Nov: G C Backhurst, J H Fanshawe *in litt.*). One in Shaba GR 51B at approximately 800 m was on S passage on 29 Sept (EABR 1983). Of six single birds in the SE and coastal lowlands, four have been on S passage (Boni Forest 80C 6 Oct: Fogden 1963; Sokoke Forest 102B 1 Oct: Br, Britton & Zimmerman 1979; Tiwi 114B 7 Nov: EABR 1983; Mida Creek 102B 20 Oct: EABR 1984). One at Aruba Dam 101B on 19 Feb (La) is assumed to have been on passage (Br), but a bird on 20 Dec at Diani 114B could have been wintering (EABR 1980).

Excluding (a) 16A due to unreliable overlay data, (b) the passage records in 51B 80C 101B and 102B, and (c) the records from Mt Marsabit 27D due to the overlays' inability to show that highland's localised moist climate, range is entirely subhumid-humid, and 70% within 1000+ mm rainfall areas. With the striking exception of the Dec bird near sea level at Diani 114B, the squares are all above 1000 m, and 74% above 1500 m.

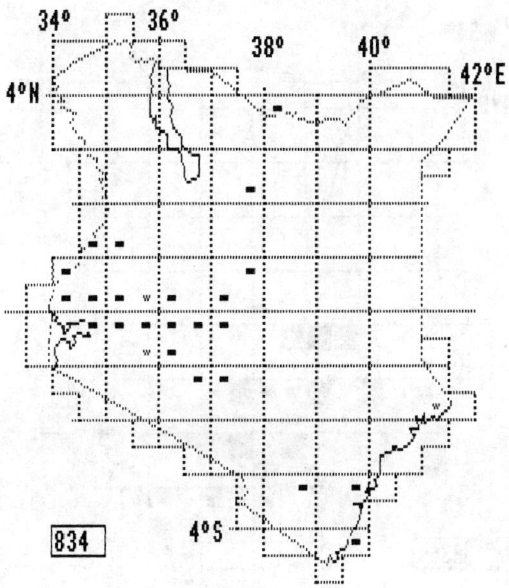

Family Malaconotidae: Bush Shrikes, Puffbacks, Boubous, Gonoleks, Tchagras etc.

These 45 species are endemic to the Afrotropical Region with the sole exception of the Black-headed Tchagra (842), which reaches the Palearctic in N Africa and SW Arabia (HM). 22 (49% of the Afrotropical total) occur in Kenya, while Kenyan records of two further birds remain unconfirmed. This family shows no sign of migratory movements (Br), so that paucity or absence of breeding records apparently reflects the secretive and retiring nature of most species. Some of the forest-dwelling members of *Malaconotus* exhibit the most pronounced colour polymorphism known in birds (HM).

835 BRUBRU *Nilaus afer*
B1020 / MPG1098, 1099 / D2C2

Afrotropical: widespread S of 17°N, with the exception of the S tip of the continent, and much of the Congo basin.

Kenyan: moderately common in bushed and wooded habitats, particularly the canopy of Acacia woodland, up to 2000+ m (Br; Ja; Lack 1985). Range is throughout our altitude, rainfall and moisture categories. Apparent absence from the well-observed coastal strip contrasts with distribution in S Somalia (AM) and part of E Tanzania (Br). The concentration of pre-1970 squares around 52B and the lack of reports from the areas around 29 and 54C may reflect paucity of observers.

The species is considered resident (Be; La; St), and the fact that 80% of the relatively few breeding squares are in the 250-1000 mm rainfall regime probably also reflects distribution of observers. There is no breeding record from Somalia (AM).

836 BLACK-BACKED PUFFBACK *Dryoscopus cubla*
B999 / MPG1128, 1129 / D2C2

Afrotropical: this species and the Northern Puffback (837) replace each other in the S and N tropics respectively. The Black-backed is widespread S of S Somalia, Kenya, Tanzania, S Zaire and Angola, except in the dry SW. Distribution in Kenya thus represents the N limit of the range.

Kenyan: often common in wooded and forest edge habitats, and sometimes forest interior, up to 2200 m (Br; Be; Britton & Zimmerman 1979). Different races occur E and W of the rift valley, and in the coastal lowlands (Br).

The Black-backed forms part of a superspecies with the Northern and Pringle's (838) Puffbacks (HM). These three species are similar in appearance and have on occasion been confused by more than a few

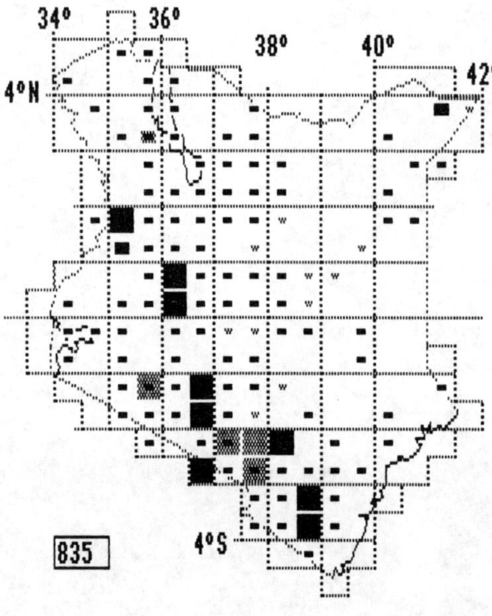

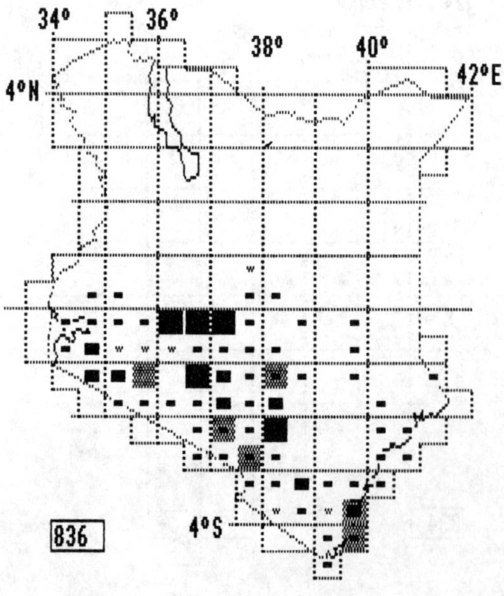

observers, both before and during the fieldwork for this atlas (*e.g.* the presumed confusion in Be, and HM's note of the similarity of Black-backed and Northern males). With this in mind, we have omitted records which disagree markedly with the distribution of the three species given by Br: none of these records are published, and none have supporting details. Even with this safeguard, our maps should still be viewed with caution.

These three species are at least mainly allopatric (HM; Br). The Black-backed and the Northern frequent similar habitats (Br), with the latter mainly N of the equator and the former mainly S (the Black-backed in 51B = Archer's Post: Br)(*cf.* their Afrotropical distributions). These two species occupy the moister areas of Kenya, and have ranges 74% and 74% within the 500+ mm rainfall regime and only 16% and 12% arid, respectively. The major difference in their ranges is that the Black-backed descends to the coast, while the Northern is in the interior, 83% above 1000 m. These two puffbacks are apparently sympatric in some W and C areas, and definitely occupy the same ground at Nandi 49C where, however, the Northern is far more widespread and numerous (Br).

In contrast to these two species, Pringle's Puffback is a bird of drier, bushed and wooded country at lower altitudes in the N and E. Its range is 79% below 1000 m, 87% arid-semiarid and 73% in 250-1000 mm, intermediate rainfall areas.

HM note that the Black-backed and Pringle's are always ecologically segregated where their ranges overlap. This is demonstrated in arid Tsavo E 101B, where the Black-backed is resident in lush riverine growth, while Pringle's is restricted to other, drier, bushed and wooded habitats (La). The single, major area of uncertainty with regard to these two species is the Tana R 65B-91A, where we map the Black-backed throughout. Ja mentions the latter on this river but omits localities, while Br, presumably out of caution, omits any mention of this river for either bird. Occurrence of the Black-backed along the Tana would compare with distribution in adjacent S Somalia, which is virtually restricted to the Juba and Shebelle R (AM). AGH list both the Black-backed and Pringle's in the lower Tana riverine forest 79C 91A without comment. While occurrence of the Black-backed there is conceivable (see above), we feel that presence of Pringle's should be treated with caution, and we omit it from the map for the following reasons. Firstly, the occurrence of Pringle's in riverine forest, and in the presence of a congener, is unlikely (*cf.* Tsavo E above). Secondly, in S Somalia, the two do not occur along the Juba R, and Pringle's has a rather scattered and local distribution (AM).

837 NORTHERN PUFFBACK *Dryoscopus gambensis*
B1000 / MPG1131 / D2C2

Afrotropical: the N tropics between 5°N and 15°N from the W coast to Ethiopia, and thence S to Kenya, Uganda and the E Zaire borders. It also occurs in the lower Congo basin. Distribution in Kenya represents the S limit of the range.

Kenyan: an often common bird, at 900-2200 m (Br). Records in the semiarid country around L Baringo 50A during Nov-Dec 1979 and Apr 1980 (St) represent birds just N of their normal range. All other details for this species are given under the Black-backed Puffback (836).

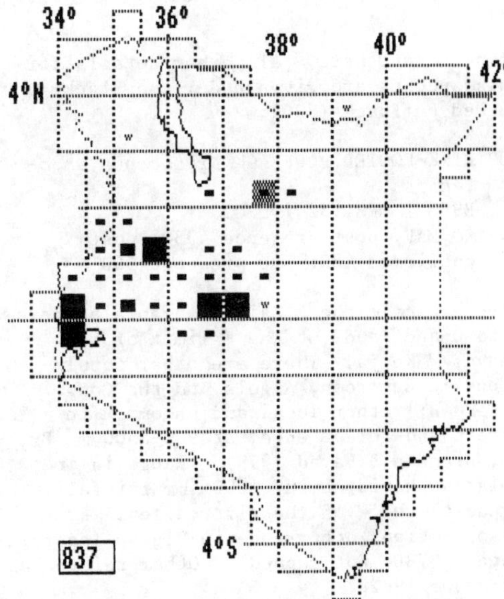

838 PRINGLE'S PUFFBACK *Dryoscopus pringlii*
B1001 / MPG1130 / D1C1

Afrotropical: the NE, from NE Tanzania through Kenya to S areas of Ethiopia and Somalia.

Kenyan: this uncommon or locally common bird occurs at 200-1000 m (Br; La). It inhabits dry country, but reports from only 14% of the 0-250 mm rainfall squares and 12% of those most arid (moisture index below 1.2) may imply an aversion to the driest habitats. Lack of records from an observer resident in 42A

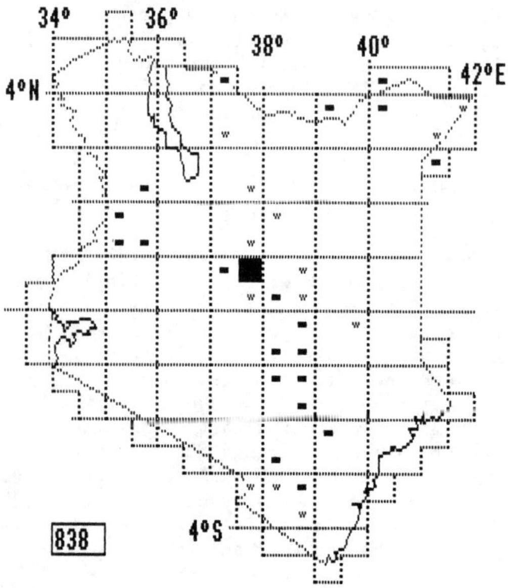

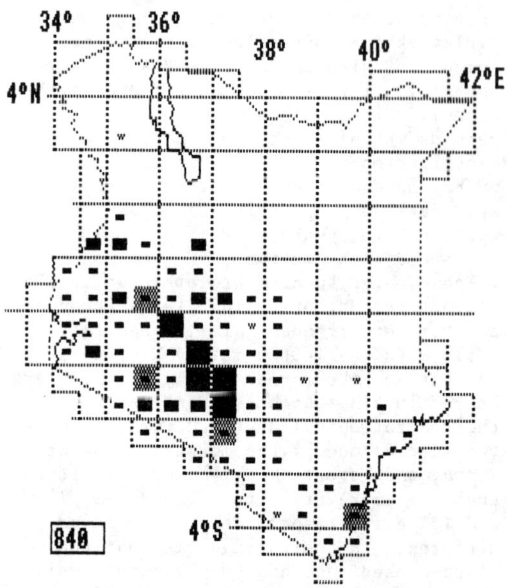

is pertinent here. All other details for this species are given under the Black-backed Puffback (836).

839 PINK-FOOTED PUFFBACK *Dryoscopus angolensis*
B998 / MPG1132 / D2C2
NO MAP: post presence 49BC; post confirmed 48D; pre presence 36D

Afrotropical: the E Zaire-Tanzania borders, E to Uganda and, at the E limit of the range, W Kenya. There are other populations in Cameroon, Angola and the Congo.

Kenyan: rather local and uncommon in forest and forest edge, at 900-2500 m (Br; Ja; Diamond & Fayad 1979). Range is in moist subhumid-humid, 1000+ mm rainfall areas to the W of the rift valley. Br also mentions occurrence in Elgeyu (approximately 37BD: not mapped). Other reference: Blencowe 1962c.

840 BROWN-HEADED TCHAGRA *Tchagra australis*
B1022 / MPG1134 / D2C2

Afrotropical: apart from much of the Congo basin and the dry SW, widespread S of approximately 7°N, E to S Sudan and to the coast from Kenya S. Distribution in Kenya represents the NE limit of the range.

Kenyan: an often common but shy and retiring bird of low bushy cover in bushed and wooded country up to 2000 m (Br; Ja; MPG). Its squares are 87% above 500 m, but it extends to the coast S of Sokoke Forest 102B (Br). Records further NE in the SE lowlands are of birds near the coast at Lamu 91B, along the Tana R in 79C (P L Britton *in litt.*), and at Bura 78B (National Museum Collection, Nairobi). Their status is uncertain. The species' range is 78% within 500+ mm rainfall areas and only 9% arid.

The Brown-headed Tchagra forms part of a superspecies with the mainly allopatric Three-streaked Tchagra (841), which replaces it in the dry N and E (HM; Br). A specimen from near Lodwar 13C is apparently a Brown-headed x Three-streaked hybrid (Keith & Twomey 1968). The Brown-headed is extensively sympatric with the slightly larger Black-headed Tchagra (842)(HM; Br). Both species occupy similar habitats, though HM suggest that the former frequents moister situations where sympatry occurs. Britton & Zimmerman (1979) recorded both during most visits to woodland in the Sokoke Forest 102B.

HM mention partial sympatry of the Black-headed and the smaller Three-streaked Tchagras in coastal Kenya, and Br confirms this for semiarid, E areas of the former's range. La recorded all three species in Tsavo E 89BD 101BD. The Three-streaked is the most numerous species there, being a fairly common resident in dense habitats. There was only a single Dec record of the Brown-headed from riverine habitat, and the Black-headed was a rare resident of very dense woodland, again near rivers (*cf.* the relationship of the Black-backed (836) and Pringle's (838) Puffbacks in this area). The sole record of the Brown-headed from the semiarid country around L Baringo 50A was also in Dec, in the artificially lush garden of L Baringo Club (T Stevenson pers. comm.).

841 THREE-STREAKED TCHAGRA *Tchagra jamesi*
B1023 / MPG1135 / D2C2

Afrotropical: the NE, from S Kenya and the far NE of Uganda to Somalia and S Ethiopia.

Kenyan: locally common but retiring in the dry bush country of the N and E, up to 1100 m (Br; Ja; HH; La; St). It is the commonest tchagra in most of Tsavo 89 101 (Br; La). The N and E edges of the highlands and their peripheral plateaux define the SW limit of the range, which is 85% arid-semiarid, and only 7% within 1000+ mm rainfall areas.

This species is apparently absent from even semiarid areas of the coastal strip S of 91A, while the mainly allopatric Brown-headed Tchagra (840: *q.v.*) is mostly S of 102B. The race *T.j. mandana* is endemic to the Witu-Lamu-Manda 91AB area of the coast (Br). The Three-streaked is considered resident by La and St, and one seen in atypically moist, high rainfall country at Yimbo Kangera 60A (Br) was presumably a wanderer.

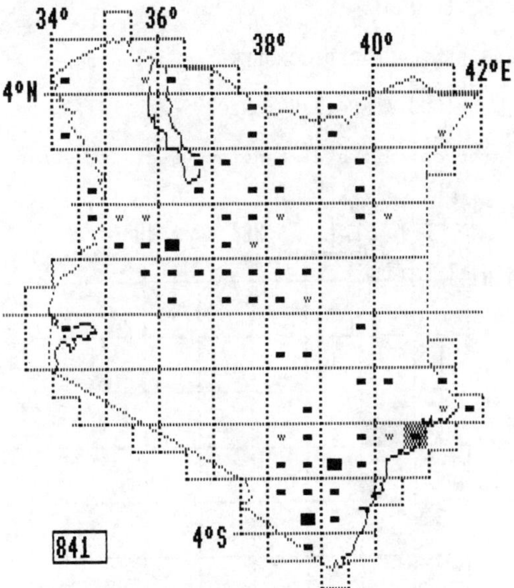

842 BLACK-HEADED TCHAGRA *Tchagra senegala*
B1025 / MPG1133 / D2C2

Afrotropical: widespread S of 15°N, apart from the Congo basin and the dry SW.

Kenyan: this common but rather retiring bird is found in very similar habitats to

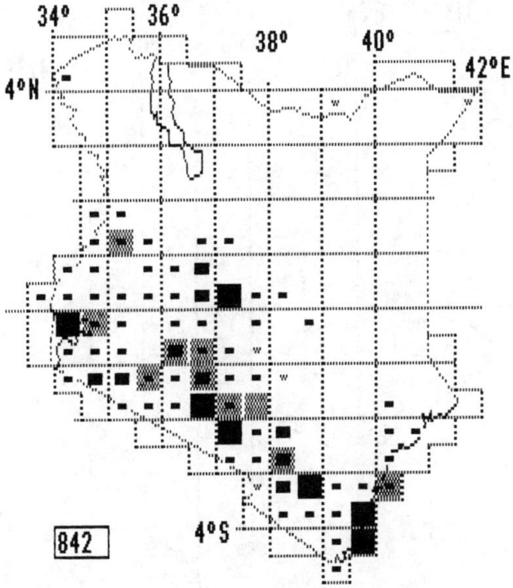

the Brown-headed Tchagra (840: *q.v.*), up to 1900+ m (Br; Ja; MPG). It occurs throughout our altitude categories, and is rather more widespread at the coast than the Brown-headed. Presence along the coast and the lower Tana R 79C 91A (AGH) largely parallels distribution in adjacent S Somalia (AM). Range is 84% within 500+ mm rainfall areas, and only 10% arid, the latter including occurrence near rivers in 79C (see above) and 101B (La). This tchagra is absent from the dry country N of 2°N, apart from records in 2C (NW of Lokichokio: G Nikolaus *in litt.*), Moyale 17A (Ja), and Karo Lola 19B (Erlanger 1905, 1907), which represent the S limit of populations in moister habitats in Sudan and Ethiopia.

843 MARSH TCHAGRA *Tchagra minuta*
B1024 / MPG1136 / D2C2

Afrotropical: widespread between 10°N and 15°S, E to W Ethiopia, Kenya, coastal Tanzania, and W Mozambique, while avoiding the Congo basin and some adjacent areas.

Kenyan: a locally common tchagra of tall rank grass and other dense vegetation in damp hollows, stream valleys, *etc.* up to 2000 m (Br; Ja; MPG). Although typically near permanent water, it does not extend into the interior of large swamps (*op. cit.*). Total range is 87% above 1000 m, 93% subhumid-humid, and entirely within 500+ mm rainfall areas, but this species has undergone striking retractions of range.

The race *T.m. reichenowi* is mentioned at

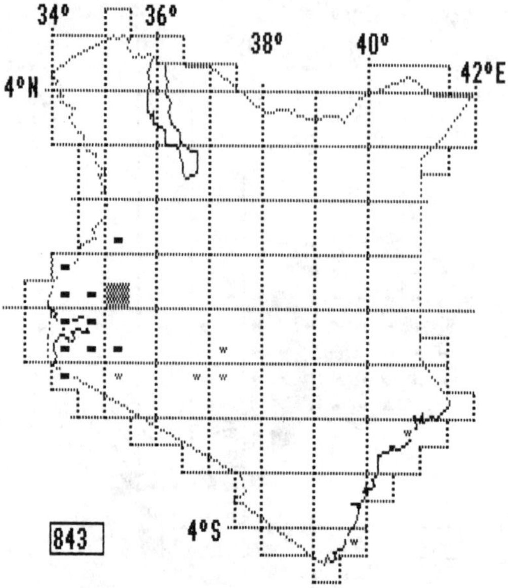

Lamu 91B and Mombasa 114B by Ja. There are no recent records from the coast and, in view of the frequency of observers in this area, this apparently reflects a genuine disappearance. Since this race persists in the Usambara Mts of adjacent NE Tanzania (Br), absence from the Kenyan coast represents a S range contraction. Similar points apply to the former occurrence (Ja) of the nominate race on the SE edge of the C highlands in the Nairobi 75B-Thika 76A-Murang'a 63C area, and possibly also to the lack of recent records from the moderately well known Nandi area 49C.

There are thus no recent records from E of the rift valley, or from below 1000 m. Range is now entirely in moist subhumid-humid, 1000+ mm rainfall areas, and only the nominate race is involved. The sole breeding record also comes from Ja, but this paucity probably reflects the species' secretive nature. This tchagra is partly sympatric with its congeners, but is largely confined to wetter habitats (HM).

844 ROSY-PATCHED SHRIKE *Rhodophoneus cruentus*
B1021 / MPG1147 / D2C2

Afrotropical: the NE, from NE Tanzania to Kenya, S Sudan, Ethiopia, Somalia and the extreme SE of Egypt. Distribution in Kenya represents the W limit of the range.

Kenyan: this locally common bird frequents bushed habitats in dry country, mainly below 1300 m but in places up to

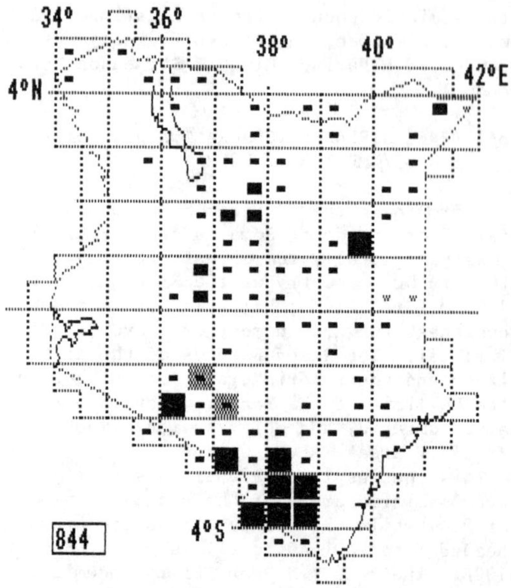

1600 m (Br; Ja; La). Range is 69% below 1000 m and 92% arid-semiarid. Absence from the coastal lowlands, and virtual absence from the E borders S of 2°N, corresponds with status in adjacent S Somalia (AM).

845 BLACK-HEADED GONOLEK *Laniarius barbarus*
B1003 / MPG1119 / D2C2

Afrotropical: a discontinuous distribution

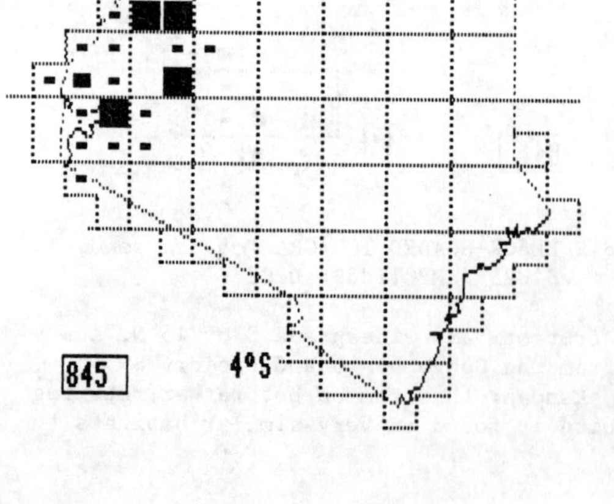

at 15°N to 5°N between the W coast and Ethiopia, and S through W Kenya, Uganda and W Tanzania to E Zaire. There is another population in the S, from Angola and Zimbabwe to C South Africa. Distribution in Kenya represents the E limit of the range.

Kenyan: this common species of thickets in bushed and wooded country is normally shy and retiring, but becomes less so when these habitats are replaced by more open gardens (Br; Ja; MPG). Distribution is in two distinct areas. The records in 2C (G Nikolaus *in litt.*) and along the Omo R 3B 4A (HH) represent the S limits of the populations in Sudan and Ethiopia. These three squares are in or W of the rift valley, and below 1000 m. They are arid-semiarid, but the 3B 4A records at least were around a permanent river.

This gonolek is better known as a bird of the moist W plateau, where range is 76% within 1000+ mm rainfall areas, subhumid-humid, and entirely above 1000 m. Br notes this species up to 1600 m. A bird in artificially lush habitat around L Baringo Club 50A during Aug 1983 was joined by another in Oct (EABR 1983). This represented an E extension of range into the semiarid rift floor, at roughly 1000 m altitude.

HM place this gonolek in a superspecies with the Papyrus Gonolek (846), Lühder's Bush Shrike (847), the Red-naped Bush Shrike (848), the Tropical Boubou (849) and other species. They state that the members of this superspecies are largely allopatric, and ecologically segregated where ranges overlap. This is certainly true of some of these birds. The Papyrus Gonolek is separated from congeners by its habitat, while the Red-naped Bush Shrike occupies dry bush country to the E of the rift valley. The forest habitat of Lühder's Bush Shrike segregates it from these two species and the Black-headed Gonolek. The final species here, the Tropical Boubou, is found in a wide variety of well vegetated habitats, such that its complete segregation from Lühder's Bush Shrike and the Black-headed Gonolek appears dubious. Br's implicit omission of the Tropical Boubou from this superspecies may reflect this uncertainty.

846 PAPYRUS GONOLEK *Laniarius mufumbiri*
B1009 / MPG1120 / D1C3
NO MAP: post presence 48A 60AB; post probable 48C

Afrotropical: endemic to the interior of papyrus swamps from E Zaire to NW Tanzania, Uganda and, on the E extremity of the range, W Kenya.

Kenyan: a locally common but usually shy and retiring bird, first reported from Kenya by Britton (1970b). It is found along the N shore of the Winam Gulf 60AB, and at Yala Swamp, L Kanyaboli 48C and associated waters (*op. cit.*; Britton 1978; Br; Mann 1976; EABR 1984). The record in 48A was in Uganda (J S Ash *in litt.*). The four squares are all at 1000-1500 m, in moist subhumid-humid, 1000+ mm rainfall areas. There is no confirmed breeding record. Probable breeding in 48C represents netted immatures (Britton 1978). See Black-headed Gonolek (845).

847 LÜHDER'S BUSH SHRIKE *Laniarius luehderi*
B1008 / MPG1127 / D2C2

Afrotropical: occupies two distinct areas. It is found on the E periphery of the Congo basin in E Zaire and W Tanzania, and E to S Sudan, Uganda and, on the E tip of the range, W Kenya. Another population occurs between Cameroon and Angola.

Kenyan: a locally common but retiring bird of low, dense cover in forest and forest edge, up to 2400 m (Br; Ja; Britton & Sugg 1973). Where forest has been cleared, this bush shrike sometimes persists in dense riverine vegetation (Br). Range is W of the rift valley, in 1000+ m, humid, 1000+ mm rainfall areas. See Black-headed Gonolek (845).

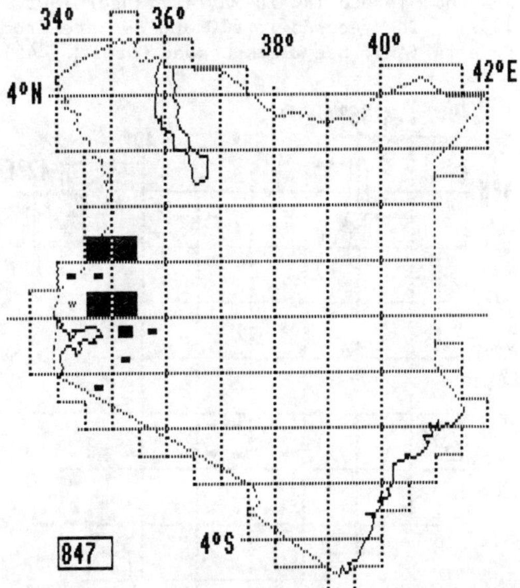

848 RED-NAPED BUSH SHRIKE *Laniarius ruficeps*
B1010 / MPG1126 / D2C1

Afrotropical: the NE, from SE Kenya to Somalia and SE Ethiopia.

Kenyan: this little known species inhabits bush country below about 750 m in the E lowlands and plateaux. Range is entirely arid-semiarid, but there are no reports from semiarid squares in the coastal strip S of 2°S. There were comparatively few records of this species (Br) before D J Pearson commenced regular visits to the SE during his studies of Palearctic migrants. He has regularly recorded this bush shrike in the area from 64AB S to 89C and 90D (EABR).

Serle (1943) described it as not uncommon around Garissa 65B in Oct, and there is a report (supported by a specimen identified by M E W North) of "many" in coastal bush near Kiunga 80C during Oct 1956 (Lewis 1983e). Pearson's reports have involved considerable numbers (EABR), with 10s or 20s per day seen in 64D during Nov 1983 and in 77D during Mar 1985 (pers. comm.). This species is thus far less uncommon, at least locally, than has been thought. Common occurrence around Ukasi 64D during Nov, but fewer there in moister, greener conditions in the following month (EABR 1983), may indicate some kind of movement.

The only breeding records are from Sarigo 19A and Djeroko 19C in May (Erlanger 1905, 1907). The records in 67C and 80A are from Somalia (AM). See Black-headed Gonolek (845).

849 TROPICAL BOUBOU *Laniarius ferrugineus*
B1004 / MPG1125 / D3C2

Afrotropical: widespread S of 10°N, extending to 16°N in Ethiopia, but absent from much of the Congo basin and the dry areas of E Somalia and the SW.

Kenyan: common but surprisingly inconspicuous in a wide range of densely vegetated bushed, wooded and forested habitats, up to 3000+ m (Br; Ja; Se). D3 reflects its loud, beautiful and well-known calls. Range is 76% within 500+ mm rainfall areas.

Only 15% of the squares are arid, and these are mostly in one of three categories. Firstly, there are those containing the locally moist forests of Mts Kulal 26B, Nyiru 26D and Marsabit 27D 28C (Diamond & Keith 1980; Ja). Secondly, the records in 2C (in riverine thickets: J Harjula *in litt.*) and at Moyale 17A (Ja) represent the S limits of populations in moister areas to the N. Thirdly, this boubou is found in the lush habitats along the Tana R 64B-79C (AGH). Restriction in the SE to this river and the coastal strip parallels distribution in S Somalia (AM). Similar confinement to riverine habitats occurs in arid Tsavo E 101B (La).

Separate races occur in Kenya to the E and W of the rift valley, in the coastal lowlands, and to the N of the N arid country at Moyale 17A (Br). Se mentions loose flocks that congregate to feed in favoured trees. In Tsavo E 101B, La mention a pair in Aug, and note that this species' rare occurrences are mainly

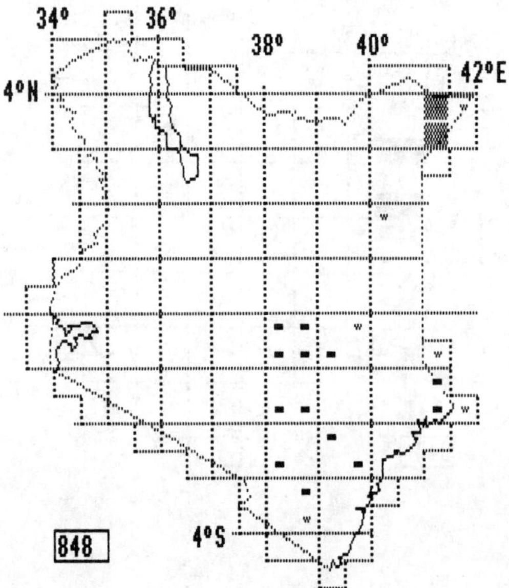

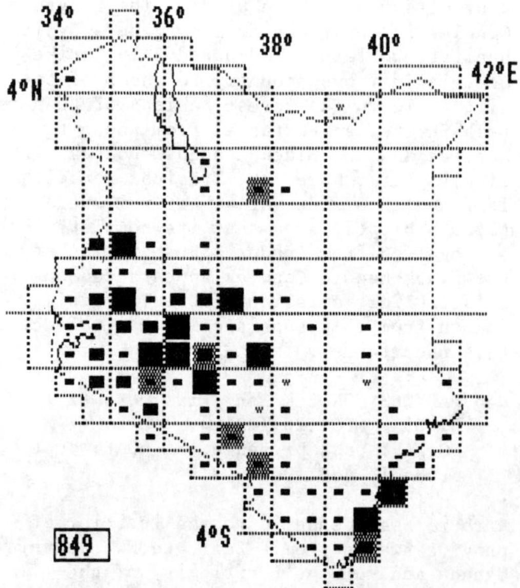

during the rains. The latter may indicate some movement into temporarily verdant habitats.

This species' range overlaps that of the Slate-coloured Boubou (850) but the two are ecologically segregated, with the Tropical in moister habitats and/or at greater altitudes (Br; see AM). In Tsavo E 101B, this interrelationship may be illustrated in the lush vegetation along the Voi R, where the Tropical Boubou occurs (rarely), but the Slate-coloured is apparently absent (La). See Black-headed Gonolek (845).

850 SLATE-COLOURED BOUBOU *Laniarius funebris*
B1006 / MPG1121 / D3C2

Afrotropical: the E and NE, from Tanzania, Kenya and parts of Uganda, N to S Sudan, E Ethiopia and Somalia.

Kenyan: a common but retiring bird, usually found in denser, drier, bushed habitats up to 1500 m or, occasionally, 2200 m (Br; Ja; Be; La; St). D3 reflects its frequent and well known calls, which are one of the characteristic sounds of Kenya's dry country. Range is 72% arid-semiarid, but presence in only 24% of the most arid squares (moisture index below 1.2) and 30% of those with least rainfall (0-250 mm) results in scarcity in the NE, possibly W to Turkana (HH).

Virtual absence from the far E of Kenya parallels distribution in adjacent W Somalia, where the sole record within 55 km of the national border is in 80A (AM). This absence in the E is supported by lack of records from the country around the lower Tana R 79C 91AC (AGH and many other observers), and from observers resident in 19A 42A 65B. There are only old records (Ja) from dry subhumid areas of the coastal strip (91C 114B), and this boubou is absent from all of this region S of Malindi 103A (Br). It is also absent from much of the humid W, *e.g.* one degree square 48, between Kongelai 37C and the Winam Gulf 60AB (Br). See Tropical Boubou (849) and the following species.

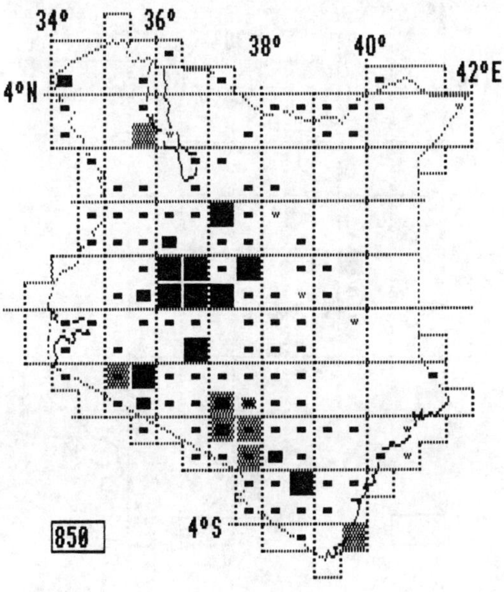

\- SOOTY BOUBOU *Laniarius leucorhynchus*
B1007 / MPG1122 / D1C0
NO CONFIRMED RECORD

Afrotropical: a W and C African species, from the lowland equatorial forests of the W, E to S and SW Uganda, and possibly W Kenya.

Kenyan: a shy and furtive inhabitant of dense forest undergrowth (MPG; Br). White (1962) noted presence in W Kenya, but Mann (1976) could trace neither the source nor the precise locality of this record, and Br considered it unconfirmed. We agree with this conclusion. This species forms part of a superspecies with the Slate-coloured Boubou (850) (HM).

851 GREY-GREEN BUSH SHRIKE *Malaconotus bocagei*
B1013 / MPG1141 / D2C2
NO MAP: post presence 49C; post confirmed 48D; pre presence 61A

Afrotropical: W and C Africa, from Cameroon and Angola E through mainly peripheral areas of the Congo basin to Uganda and, on the E extremity of the range, W Kenya.

Kenyan: rather scarce in thickets and forest edge, at 1100-2200 m (Br; Ja). It is regularly reported from the Kakamega Forest 48D and there are also records from Nandi 49C, Nyarondo 61A and Kericho 61A (Br). All three squares are in humid, 1000+ mm rainfall areas to the W of the rift valley. See Sulphur-breasted Bush Shrike (852).

852 SULPHUR-BREASTED BUSH SHRIKE *Malaconotus sulfureopectus*
B1019 / MPG1138 / D2C2

Afrotropical: widespread S of 10°N and locally to 14°N, but absent from the Congo basin and the drier areas of Somalia and the SW.

Kenyan: locally common in a diversity of bushed, wooded and forest edge habitats, up to 3000 m but usually below 1500 m (Br;

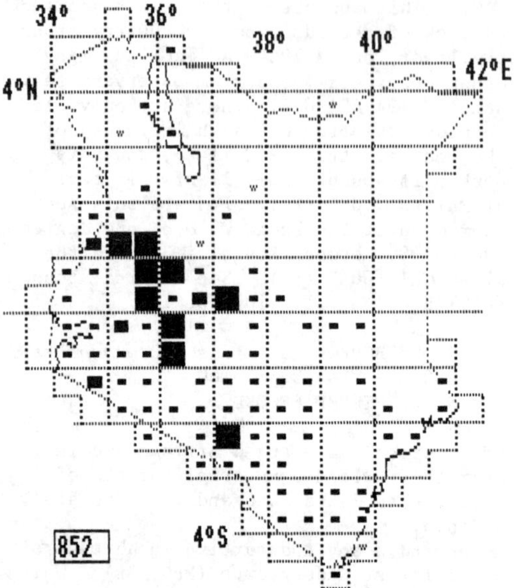

Britton 1979; Ja). Br states that it avoids many semiarid areas, but there are reports from 52% of Kenya's semiarid squares. Range is in only 16% of the arid squares and 7% of those with 0-250 mm rainfall.

This bush shrike is thus very scarce in the N and E, in locally moist situations. These include woodland around L Turkana (HH; 4A = Ethiopia: J S Ash *in litt.*), the S edge of the Ethiopian highlands' population at Moyale 17A, the isolated highland of Marsabit 27D (Ja), and the moist habitats around the Tana R 64B-79C (*e.g.* AGH).

St considers the species resident in the semiarid country around L Baringo 50A, but there are no Kenyan breeding records below 1000 m. This distribution probably reflects the bird's moderately retiring behaviour, but the following features of its status in the SE lowlands may suggest otherwise. La describe it as rare but probably resident in mainly woodland and riverine vegetation in arid Tsavo E 101B, but note that most records are of birds heard during the rains. Whether this indicates actual movement into arid areas during wet periods, or merely increased activity (possibly including breeding) of resident birds stimulated by the rains, is uncertain (*cf.* Tropical Boubou (849)). However, two individuals caught at dawn near the Ngulia lights 101A during the Nov wet season (Backhurst & Pearson 1977b) may have been moving at night into the Tsavo area, and Ja described local abundance during Nov on Manda Island 91B. There are no breeding records from Somalia and the bird is only presumed resident there (AM).

The Sulphur-breasted Bush Shrike forms part of a superspecies with the Many-coloured Bush Shrike (853)(HM), which is ecologically segregated in forest interiors (Br). The two species can occur close together in mixed habitats. This superspecies is extremely similar to that containing the distinctly larger and heavier billed Grey-headed Bush Shrike (854) and other congeners, both in ecological distribution and plumage colours (HM). Both groups contain a widespread species of more open habitats (the Sulphur-breasted and Grey-headed), that is yellow below and green above with some grey on the head, while the other members are all more localised inhabitants of forest, and in many cases more colourful. The Grey-green Bush Shrike (851) is closely related to these superspecies (HM).

853 MANY-COLOURED BUSH SHRIKE *Malaconotus multicolor*
B1017 / MPG1137, 1140 / D1C2

Afrotropical: from mainly highland areas of East Africa and the E Zaire borders SW to Zambia, and S to NE South Africa. It is also discontinuously distributed in the lowland forests around the Gulf of Guinea. Distribution in Kenya represents the NE limits of the range.

Kenyan: reasonably common but unobtrusive in the mid-levels and canopy of highland forest at 1500-2300 m (Br). Range is

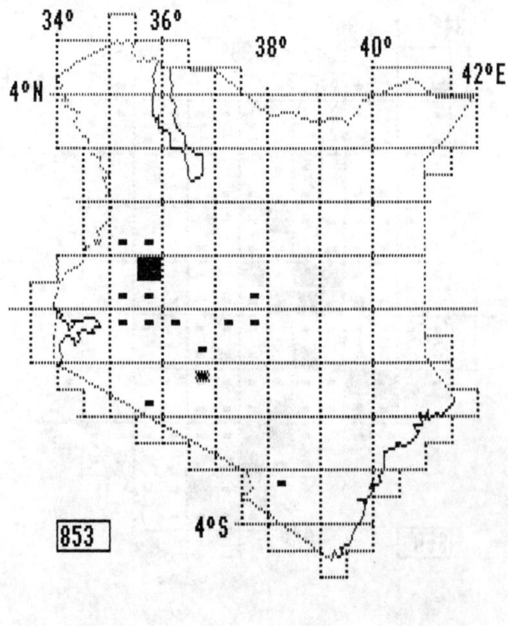

entirely within 500+ mm rainfall areas, and includes the Taita Hills forest island 101A (Br). Occurrence on the latter parallels presence on the Usambara Mts of adjacent NE Tanzania (EABR 1978). Excluding the Taita Hills, whose climate is too localised to be shown on our overlays, range is entirely subhumid-humid.

The sole confirmed breeding record represents an adult feeding a newly fledged, dependent juvenile in Kabarnet Forest 49B *c.* 1979 (T Stevenson *in litt.*). There is no evidence of movements, but the only two published records near the species' lower altitudinal limit at Nairobi 75B have been in Nov-Dec (EABR 1980; Ja). This bush shrike exhibits exceptional colour polymorphism (HM). The race in Kenya, *M.m. nigrifrons* (the Black-fronted Bush Shrike) is often considered a distinct species (HM; Br). See Sulphur-breasted Bush Shrike (852).

854 GREY-HEADED BUSH SHRIKE *Malaconotus blanchoti*
B1012 / MPG1144 / D2C2

Afrotropical: widespread S of 10°N and locally to 15°N, but absent from the lowland equatorial forests, and from the drier areas of E Somalia and the SW.

Kenyan: generally uncommon and somewhat unobtrusive in bushed and wooded country up to 1600 m, though extending locally up to 3000 m in W Kenya (Br; Britton 1979; Ja; La; St). Its squares are only 24% arid and 6% within 0-250 mm rainfall areas. Occurrence in the N and E is thus largely restricted to the S edge of the Ethiopian highlands' population 4AC 17AC 19AB, the isolated highland of Mt Marsabit 27D 28C, and the vicinity of the Tana R 64B 65ABD 79C (*e.g.* AGH). Br's mention of affinity for watercourses in semiarid country thus also extends to arid areas.

In the drier parts of Kenya, La and St consider this bush shrike resident, although MPG mention Jul movements in Uganda. See Sulphur-breasted Bush Shrike (852) and the following species.

\- FIERY-BREASTED BUSH SHRIKE *Malaconotus cruentus*
B1014 / MPG1146A / D1C0
NO CONFIRMED RECORD

Afrotropical: a W and C African species, from the lowland forests around the Gulf of Guinea E to W Uganda and, possibly, W Kenya.

Kenyan: inhabits the mid-levels and canopy of forest, and forms part of a superspecies with the Grey-headed Bush Shrike (854)(HM; Br). An old specimen from Kakamega Forest 48D has been lost. This bird was identified as this species by some workers, but HM treat it as another species, *M. monteiri*, which is otherwise known only from a population in Angola, and a single specimen from Cameroon (Br).

855 FOUR-COLOURED BUSH SHRIKE *Malaconotus quadricolor*
B1018 / MPG1142 / D1C2

Afrotropical: occupies two distinct areas.

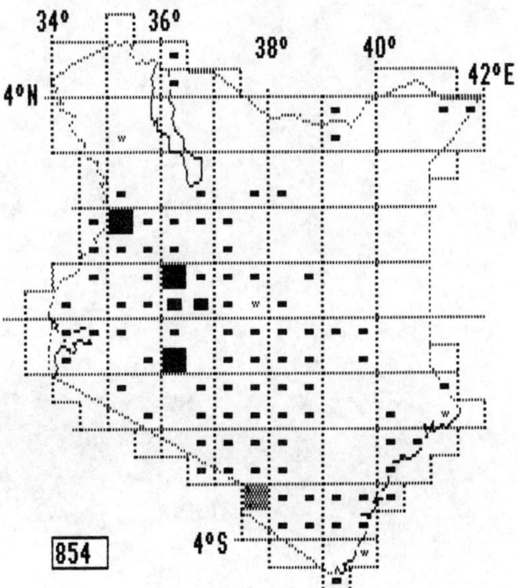

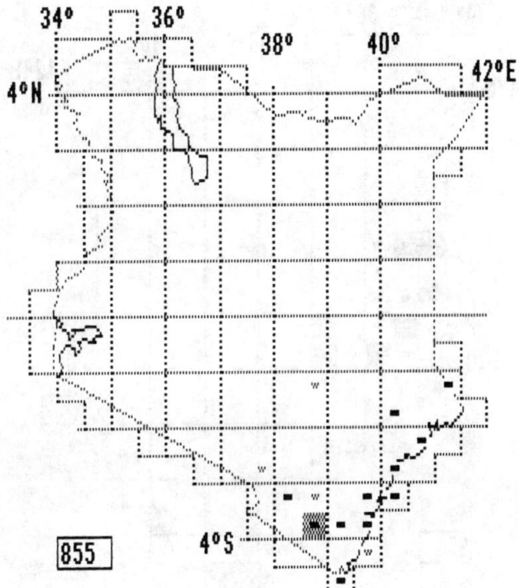

In the E, it occurs in the coastal lowlands and adjacent areas of Tanzania, Kenya and the extreme S of Somalia. In the SE, range extends from Malawi S to the coast at around 30°S.

Kenyan: a shy and very retiring bird of thicket and the undergrowth of woodlands and forest, mainly below 600 m in the coastal lowlands (Br; Ja; Britton & Zimmerman 1979). Range is 72% within 500+ mm rainfall areas. Occurrence in 77B is at 1200 m in mist forest on the Mt Endau forest island (Tennent 1964; Br). P L Britton (*in litt.*) notes presence along the lower Tana R 79C. It was formerly rare along the Voi R 101B, but there are no recent records (La).

Although not a known migrant, an immature caught at dawn at Ngulia Lodge 101A (altitude 900 m) in Dec had probably arrived in misty conditions during the previous night (Backhurst & Pearson 1977b). The only records for Somalia have come recently from 80A, in Sept (AM).

The Four-coloured Bush Shrike forms part of a superspecies with Doherty's Bush Shrike (856), which is completely allopatric at greater altitudes (HM; Br).

856 DOHERTY'S BUSH SHRIKE *Malaconotus dohertyi*
B1015 / MPG1143 / D1C2

Afrotropical: a highland bird of the E, found only along the E Zaire borders, in SW Uganda and in Kenya.

Kenyan: a locally common but shy and retiring inhabitant of undergrowth in forest and bamboo, at 1600-3350 m (Br; Ja; MPG; Se). It is most frequent in edge habitats, dense secondary growth and moist thickets (*op. cit.*). Range is in the W and C highlands, 68% humid and 81% within 1000+ mm rainfall areas. It does not extend to forest islands. See Four-coloured Bush Shrike (855).

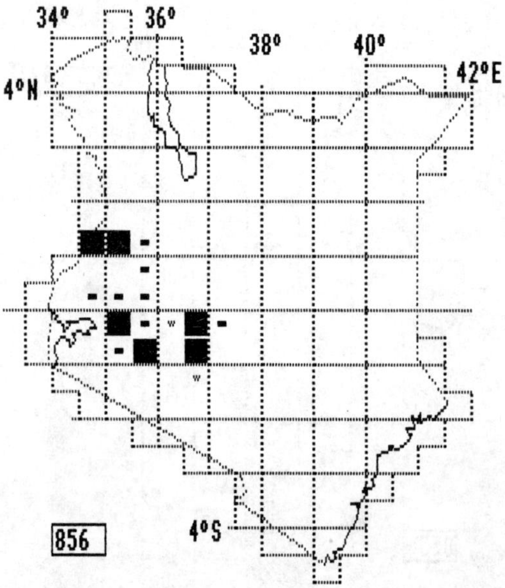

Family Laniidae: Shrikes, Fiscals

Some 26 species, which are predominantly Afrotropical, Palearctic and Oriental. 15 (58%) occur in the Afrotropical Region and, of these, 13 (87%) have been recorded in Kenya. Although some wandering may occur, no Afrotropical shrikes are certainly known to undertake migratory movements (Br; HM), and all eight (62%) of Kenya's African species either breed or are thought to do so. The remaining five shrikes recorded in Kenya are all Palearctic migrants. In contrast to most of the Bush Shrikes, most members of this family are prominently perching, conspicuous and often noisy and aggressive birds, with higher D ratings.

857 RED-BACKED SHRIKE *Lanius collurio*
B1030 / MPG1112 / D2C2

Afrotropical: this Palearctic shrike migrates through NE Africa to winter from S Kenya and Tanzania S to Namibia and E South Africa.

Kenyan: seasonally common in open country up to 3000 m, where bushes, small trees and other minor prominences provide low perches (Br; Se; La). It is mainly a passage migrant in Kenya, and apparent absence from much of the E and NE may reflect absence of observers in these areas during the migration periods. Records from Turkana (HH), Moyale 17A (ADL pers. obs.) and Wajir 42A (J E Miskell *in litt.*) have all been on Mar-Apr passage dates. Alternatively, this distribution may reflect the bird's aversion to the driest areas: there are reports from only 6% of the most arid squares (moisture index below 1.2)

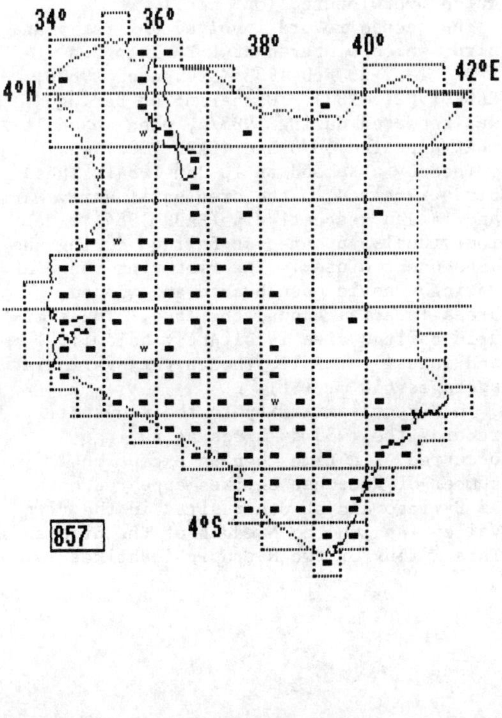

857

and 16% of those with the lowest 0-250 mm rainfall. Furthermore, apparent absence from E Kenya parallels distribution in adjacent S Somalia (AM).

S passage is most marked W of 33°E, and the mainly Nov movements are small in Kenya (Br; Se; La; St; *cf.* AM for S Somalia). N passage (mostly Apr) is further E than the S migration, being mainly E of L Victoria and particularly in and to the E of the Kenya-Tanzania rift valley (Br). This heavy and widespread N movement is mentioned by Ja, by HH at L Turkana, by Se for the country around 3000 m at Mau Narok 61D, and by La for Tsavo E 101B. An exception is Baringo 50A, where St records only small numbers. N passage can be heavy at the coast *e.g.* 250+ around the Sabaki estuary 103A in Apr (EABR 1981), but does not continue NE into Somalia, where this migration is mainly seen in the N of the country (AM).

Small numbers winter annually Dec-Mar in the area between Kitui 77A, Tsavo 89CD 101 and mainly S areas of the coastal strip 114BC, and also more occasionally in the W, *e.g.* 48C (Br; La; EABR 1981). These two areas have very different altitude, rainfall and moisture characteristics. St mentions occasional birds during this season around Baringo 50A. The sole East African record of this shrike (presumably) oversummering is one on 8 June in Meru NP 52C (EABR 1980).

The Red-backed Shrike has been considered conspecific with the Red-tailed Shrike (858: *q.v.*) (see Voous 1977), and very small numbers of hybrids between the two are regular in Kenya (Pearson 1979a; HH; EABR 1980, 1982, 1983).

858 RED-TAILED SHRIKE *Lanius isabellinus*
B1034 / MPG1114, 1116 / D2C2

Afrotropical: a Palearctic species that winters in the NE, W to Sudan and N and E Uganda, and S to NE Tanzania. This winte-

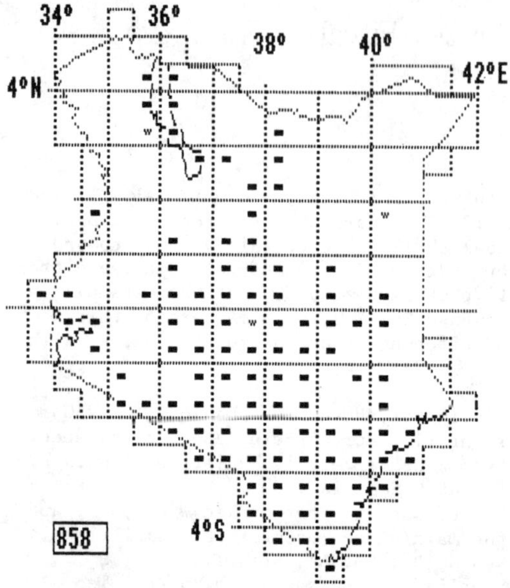

ring range is almost entirely NE of that of the closely related Red-backed Shrike (857: *q.v.*), the zone of overlap passing through Uganda, Kenya, NE Tanzania and S Somalia (Br; AM).

Kenyan: common in a variety of dry, open, bushed and wooded habitats, up to approximately 2000 m (Br; Ja; La). In contrast to the Red-backed Shrike, the Red-tailed is mainly a Nov-early Apr winter visitor, near the S limit of its Afrotropical range (Br; Ja; HH; La; St). Br mentions wintering as widespread in Kenya with the exception of the coastal strip (*cf.* Red-backed Shrike), and D J Pearson (*in litt.*) describes it as most common in bushland to the NE and E of the highlands. Migration is evident in the Tsavo area 89 101 during Nov, and in E areas between Nairobi 75B and Tsavo in early Apr (Br; D J Pearson pers. comm.).

The Red-tailed Shrike is a bird of rather lower and drier country than the Red-backed. This distinction is possibly detectable in the overlay analyses, even though our maps depict total ranges that, particularly for the Red-backed Shrike, include passage migrants in atypical (*e.g.* arid) environments. Quoting the Red-tailed's figures first in each case, the ranges of these two species are: 82% and 72% below 1500 m; 71% and 61% arid-semiarid; 14% and 9% within the 0-250 mm rainfall regime; and 7% and 2% within the most arid squares (moisture index below 1.2). The Red-tailed's aversion to overmoist conditions is demonstrated in the semiarid country around L Baringo 50A, where it is absent or very uncommon during years with above average rainfall, particularly if the rains are in Jan-Feb (St).

The Red-tailed's squares are 90% within or to the E of the rift valley, whereas the Red-backed's are 77%. In the far E, apparent absence of the Red-tailed from the Kenya/Somalia borders S of 1°N is similar to its distribution in S Somalia (AM). Two birds ringed at Ngulia Lodge 101A have been recovered in Kuwait (Backhurst 1977, 1981b). Other references: Pearson 1979a; Lack 1985.

859 NUBIAN SHRIKE *Lanius nubicus*
B1037 / MPG1109 / D3C0
NO MAP: post presence 50A; pre presence 48C

Afrotropical: a Palearctic species that winters in the N tropics S to approximately 10°N, mainly from Chad E to Ethiopia. There are records of four birds S to Kenya.

Kenyan: the occurrence of this shrike in an unspecified country or countries near L Turkana has been quoted by MPG and two other standard works of reference (Britton 1970b), but a British Museum specimen from the vicinity of this lake cannot be traced (Ba). The first adequately documented record was an immature collected at L Kanyaboli 48C on 9 Nov 1969 (Britton 1970b). This bird was described as a vagrant S of its normal range, and its appearance coincided with an exceptional influx of Palearctic migrants to the area, including several Woodchat Shrikes (860), which overwintered (*op. cit.*).

The second record involved a first year bird, which wintered at L Baringo 50A 6 Dec 1982-28 Feb 1983 (Stevenson 1983b; EABR 1982, 1983). Wintering at Baringo was repeated during 1983/4, when an adult that appeared on 8 Dec (EABR 1983) was joined by a second on 10 Jan 1984. These birds remained in the area until early Mar and 14 Jan respectively (EABR 1984). No individuals appeared at Baringo during the subsequent winter. The wintering range at Baringo was in open bushed and wooded areas (acacia) around the lake. The semiarid Baringo area is slightly below 1000 m, and annual rainfall, though very variable, averages 650 mm (St).

In direct similarity to the Kanyaboli record, the 1983/84 birds at Baringo occurred at a time when an exceptional number of Woodchat Shrikes appeared both at Baringo and at other sites in the rift valley, and on the N edges of the highlands. This influx of two N tropics' shrikes coincided with influxes of Abyssinian (453), Lilac-throated (455) and Rufous-crowned (456) Rollers, which were attributed to unusually dry conditions N of Kenya, and which were variously notable in terms of range, numbers of individuals and seasonality. Finally, this same winter also produced unprecedented influxes of Great Black-headed (292) and Slender-billed (297) Gulls, Palearctic species which also normally winter further N. HM suggest that the Nubian Shrike may be related to the Red-backed (857) and Red-tailed (858) Shrikes.

860 WOODCHAT SHRIKE *Lanius senator*
B1038 / MPG1115 / D3C1

Afrotropical: a Palearctic species wintering in the N tropics between Senegambia and W Ethiopia, S to approximately 8°N in the W and to N Uganda in the E. The few records from Kenya represent the SE limit of the wintering range.

Kenyan: a bird of open bush country. The few records are from three distinct areas, which may reflect distribution of observers. All involve single birds unless otherwise stated. In the W, there are reports from Mt Elgon 36D (Ja), the Maragoli Hills 48D (10 Jan: *op. cit.*; Ba), Bungoma 48B (28 Jan-4 Feb 1973: OS-C 1977), and several at L Kanyaboli 48C (8 Nov 1969-20 Mar 1970: Ba). These W localities are closest to the species' normal range in N Uganda (Br).

Secondly, there have been reports from the vicinity of the rift valley lakes at Baringo 50A (21 Dec 1979-23 Jan 1980: St; EABR 1979, 1980; and up to seven there Nov 1983-Mar 1984: EABR 1983, 1984), Bogoria 50C (two 22-29 Dec 1983: EABR 1983), Nakuru 62A (24 Jan 1984: K H Siebenrock *in litt.*) and Hell's Gate 62C (12 Feb 1978: EABR 1979). Finally, three Feb records from the NE periphery of the C highlands were at Samburu GR 51B (23rd, 1970: OS-C 1977; 7th, 1978: EABR 1979), and at Isiolo 51D (18th, 1984: EABR 1984).

This species' 10 squares are eight subhumid-humid and, apart from 51B, entirely above 1000 m in 500+ mm rainfall areas. The dates of these records are divisible into two groups. Those away from L Kanyaboli 48C and Baringo 50A are all Dec-Feb. The minor influxes of birds that wintered at Kanyaboli Nov-Mar and Baringo Nov-Jan were both accompanied by the considerably rarer Nubian Shrike (859: *q.v.*).

861 GREY-BACKED FISCAL *Lanius excubitorius*
B1032 / MPG1102 / D3C2

Afrotropical: mainly the E and NE, from N Mozambique and SW Tanzania to Uganda, W Kenya, Ethiopia and S Sudan, W to Chad. Distribution in Kenya represents the E limit of the range.

Kenyan: common and conspicuous in bushed and wooded habitats, including open woodland, at 1000-1900 m (Br; Ja; St). With the exception of 12AC (in Uganda: J S Ash *in litt.*), range is entirely within 500+ mm rainfall areas, with 83% of the squares subhumid-humid.

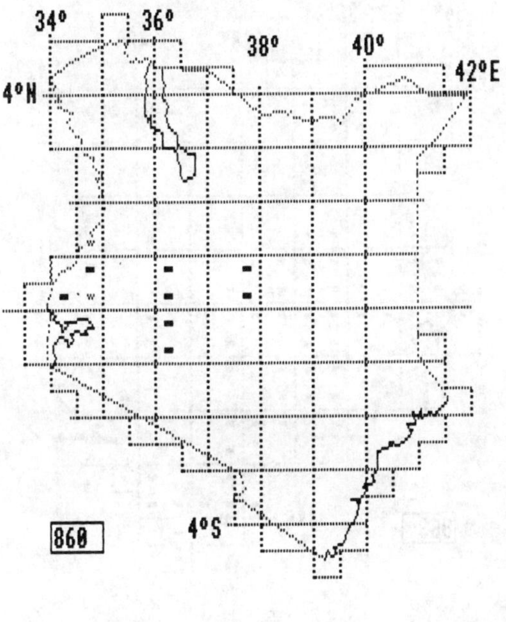

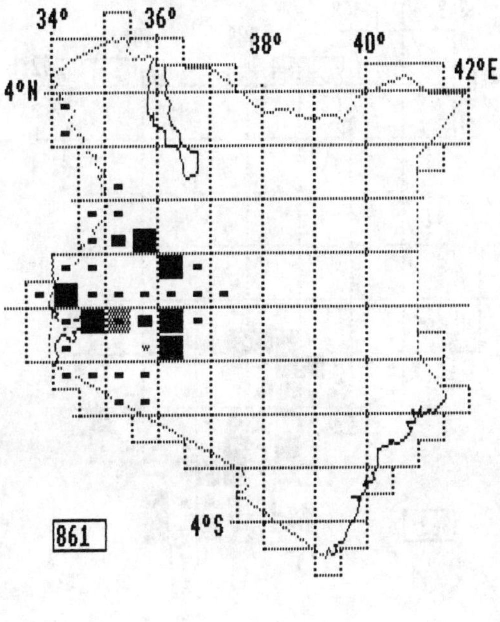

Reports of migration in W Kenya (MPG) and in the rift valley (White 1962) are rejected by Br who, in company with St, considers the species resident. A single Apr bird around 3000 m at Mau Narok 61D is presumed to have been a wanderer (Britton 1979).

The Grey-backed Fiscal forms a superspecies with the allopatric Long-tailed Fiscal (862: *q.v.*)(HM; Br), and the two occupy Kenya's non-arid areas (Lewis & Pomeroy 1982b; Pomeroy *et al.* 1982). The areas of overlap of these two shrikes at QSD scale are predominantly along the S periphery of the W highlands 73 74. The single other overlap square is 51C, where the Long-tailed is known only from an old record. Br mentions the Grey-backed E to the rift valley, but we map records on the E plateau in 50BD 51C and 62B. Similarly, the range of the Long-tailed on our map extends W of Br's summary. Since there is the slight possibility of confusion between immatures of these species, we have left our maps intact to await further, specific searches in these areas. See Fiscal (864).

862 LONG-TAILED FISCAL *Lanius cabanisi*
B1028 / MPG1108 / D2C2

Afrotropical: the E, from NE Tanzania to Kenya and S Somalia.

Kenyan: this common and conspicuous shrike frequents similar habitats to the Grey-backed Fiscal (861:*q.v.*)(Br; Ja), and replaces that species in drier but still mainly non-arid country (Pomeroy *et al.* 1982), from 1600 m on the S and E peripheries of the highlands to the coast (Br). In contrast to the Grey-backed, range is 81% semiarid-subhumid and 70% within intermediate, 250-1000 mm rainfall areas. It is resident in locally moist areas of arid Tsavo E 101B (La; Lack 1985). See Fiscal (864) and Taita Fiscal (865).

863 LESSER GREY SHRIKE *Lanius minor*
B1036 / MPG1103 / D2C2

Afrotropical: the S migration route of this Palearctic shrike is to the W of Kenya, passing from the Middle East, through Sudan and Zaire, to the wintering range in Namibia, Botswana and NE South Africa. The N migration is further to the E, through East Africa, Ethiopia and mainly N areas of Somalia.

Kenyan: because of the differing migration routes mentioned above, this species is almost entirely a N passage migrant in Kenya. There are no records of overwintering. Single birds at Eldama Ravine 49D on 8 June and at Larogi 50B on 20 Jul (Ja) were presumably oversummering, but there have been no subsequent records during this season. These two oversummering squares are above 1500 m in semiarid - subhumid, 500+ mm rainfall country.

S passage birds are only occasional even as far E as Uganda and W Tanzania (Br; EABR 1983). The only S passage records from Kenya, all of single birds, are from near Aruba 101B on 15 Oct 1971 (La), in Samburu GR 51B on 9 Nov 1980, at L Bogoria

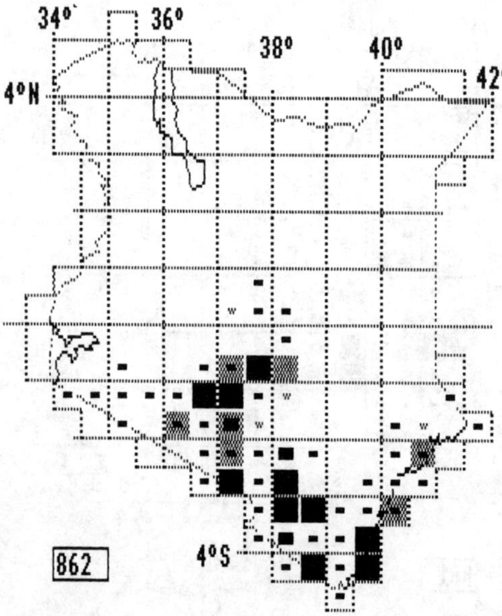

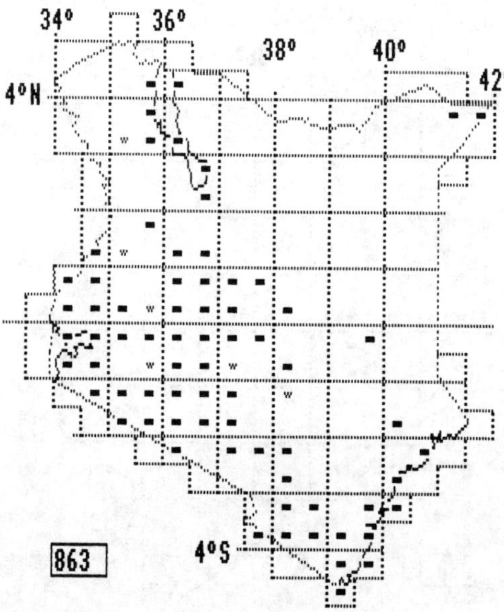

50C on 13 Nov 1980, near Ngulia 89C on 29 Nov 1980 (all EABR 1980), and in Meru NP 52C on 26 Oct 1981 (EABR 1981) and 3-4 Nov 1984 (EABR 1984). These five localities are all on the N, NE or SE margins of the C highlands and their peripheral plateaux. There is a single Sept record from Somalia (AM).

By contrast, the N migration (mainly in Apr) is widespread and often heavy, in open bushed country and cultivation up to at least 3000 m (Br; Ja; Se; La; St; EABR 1981). Because of the paucity of records from other seasons, our map mainly depicts this N passage's range, which is 72% within 500+ mm rainfall areas, and only 18% arid. Occurrence in the arid N and E is virtually restricted to the vicinity of permanent water, at L Turkana (HH), the Daua R 19AB (E J Roberts *in litt.*), and the Tana R in 65B (R & L M Greenham *in litt.*, and an older specimen) and 79C (P L Britton *in litt.*).

As with the Red-backed Shrike (857), apparent absence from other parts of the N and E could reflect lack of observers during the migration periods, but the Lesser Grey's marked affinity for the moister areas of Kenya, or for permanent water elsewhere, belies this, as does lack of records from experienced observers resident in 15D and 42A. Furthermore, this apparent scarcity in the NE and E compares with distribution during this migration in Somalia, where only 7% of the species' records are S of 6°N, and the sole QSD S of 2°N is over 100 km E of the Kenya border (AM). A bird ringed at Maranda 60A was recovered on the island of Chios, Greece (Backhurst 1974).

864 FISCAL *Lanius collaris*
B1029 / MPG1104, 1105 / D3C3

Afrotropical: widespread but often localised S of 10°N, E to the Ethiopian highlands (where it reaches 16°N) and Kenya, and to the E coast in areas further S. Distribution in Kenya thus represents the E limit of the range.

Kenyan: well known in open habitats, including cultivation and urban gardens, where bushes, trees or other prominences provide perches (Br; Ja). This shrike occurs at 500-3350 m, and is common in the highlands but less so elsewhere (Br; Se). Distribution in N Kenya is virtually confined to the isolated highlands of Mts Kulal 26B (Diamond & Keith 1980), Nyiru 26D (Br) and Marsabit 27D 28C (Ja), and to Moyale 17A (Ja), which represents the S limit of birds in the Ethiopian highlands. Excluding (a) the moist climates of these

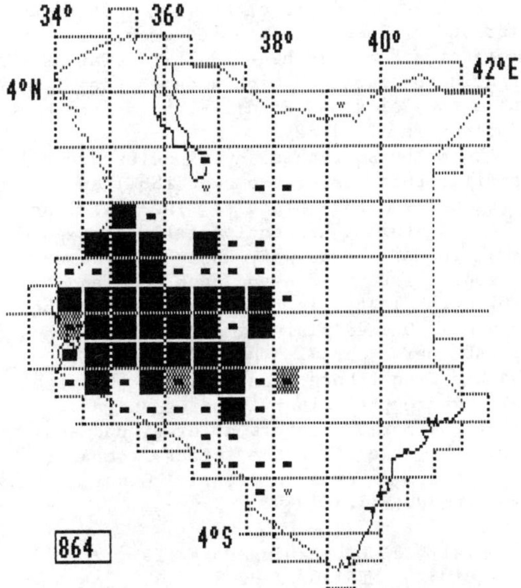

isolated highlands, which are too localised to appear on our overlays, and (b) the record at Moyale, where our overlay data may be incorrect, range is 84% within 500+ mm rainfall areas, while the only arid squares are 39D and 102C (see Pomeroy *et al.* 1982).

The Fiscal's breeding squares are 96% above 1000 m, 90% subhumid-humid, and entirely within 500+ mm rainfall areas. This distribution may reflect observer frequencies, or some degree of movement of non-breeding birds into semiarid areas during the rains. This latter alternative is demonstrated near the local N limit of the species' range at Baringo 50A. There, in semiarid country at approximately 1000 m altitude, the Fiscal is only an occasional visitor, usually after rains (St). MPG also mention seasonal movements, but omit details. At greater altitudes and in subhumid-humid regimes, this shrike is considered resident at Mau Narok 61D (3000 m: Se), in Narok district 74 (Be), and in Nairobi 75B (1650 m: D J Brass, M Sinclair, DEP pers. obs.).

The Fiscal forms part of a superspecies with Mackinnon's Shrike (867), and the Taita (865: *q.v.*) and Somali (866) Fiscals (HM; Br). The latter two closely resemble the Fiscal but, in areas of overlap, all behave as valid species, and have different calls (HM). Mackinnon's Shrike occupies high rainfall areas to the W of the rift valley. It is broadly sympatric with the Fiscal but, where both occur together, the latter frequents more open habitats while Mackinnon's is often in forest edge (Br; HM). These two species occur in moister and mainly higher altitude areas (including isolated highlands in the Fiscal) than the Taita and Somali Fiscals (see Pomeroy *et al.* 1982).

The members of this superspecies are all smaller than the Grey-backed (861) and Long-tailed (862) Fiscals. The Fiscal and the Long-tailed are ecologically segregated *e.g.* in Nairobi NP 75B, where the former occupies forest edge and open woodland, while the latter is found on the more open, sparsely bushed plains immediately to the S (ADL pers. obs.). The Fiscal and Grey-backed Fiscal both occur around L Naivasha 62C, where preliminary studies suggest that the former prefers drier ground away from the lake while the latter is nearer the shore (J Hayes *in litt.*). Other reference: Emlen 1979.

865 TAITA FISCAL *Lanius dorsalis*
B1031 / MPG1107 / D2C2

Afrotropical: the NE, from NE Uganda, Kenya and NE Tanzania to extreme SE Sudan, and S areas of Ethiopia and Somalia.

Kenyan: common in drier, open bushed and wooded habitats, including open woodland, up to 1500+ m (Br; Ja; Be; La). In direct contrast to the allopatric Fiscal (864: *q. v.*), the Taita's range is 69% below 1000 m and 89% arid-semiarid, with only 1% of its squares in 1000+ mm rainfall country. Virtual absence of records from the highlands and adjacent moist subhumid-humid areas (which are occupied by the Fiscal) defines the W limit of the Taita's range 36D-37CD-49-63-73. Apart from these moist areas, the only part of Kenya where this shrike is absent is the coastal strip S of 91C (Br), where there are no records from even arid squares. Absence of records from one degree square 77 is due to no obvious reason, and may represent a chance lack of coverage.

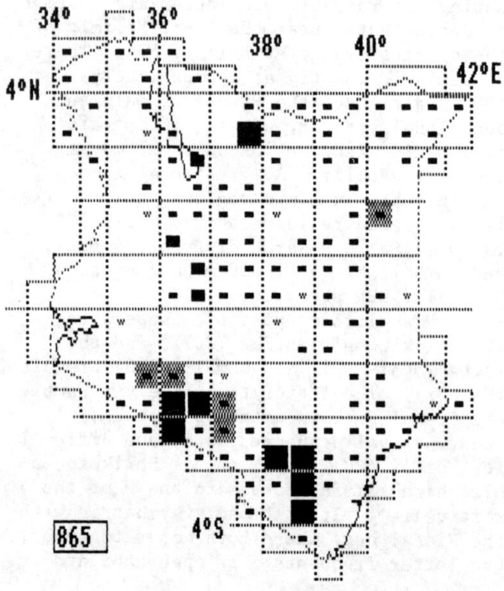

Be and La describe the Taita Fiscal as resident, but MPG mention apparent local movements, omitting details. A record from Kericho 61A (Ja) is the sole occurrence in humid country, and presumably represents a wanderer W of its normal range.

Br states that the Taita is sympatric with the less common Somali Fiscal (866) in many areas N of Marsabit 27D 28C, but that the Somali is usually in more arid and sparsely bushed habitats. This relationship was observed at Kamathia 3A during dry conditions in Feb 1984, where Taita Fiscals were confined to more thickly bushed plains, while one or two Somali Fiscals were in sparse bush atop a nearby rocky hill (D J Pearson, ADL pers. obs.). This distinction in habitat density also exists in Somalia (AM). The Somali Fiscal's somewhat greater tolerance of more arid conditions can be seen on our maps, where 35% of its squares are within the 0-250 mm rainfall regime and (excluding wanderers) 81% are arid, while the Taita's figures are 28% and 55% respectively (see Pomeroy *et al.* 1982).

HM state the Taita to be sympatric with the larger Long-tailed Fiscal (862) in the lowland scrub of SE Kenya, but Br notes this relationship to be more apparent than real because, locally, the Long-tailed is usually in moister situations. Thus in the arid country of Tsavo E 101B, the Long-tailed replaces the Taita in more verdant, riverine growth (La). Around the lower Tana R 79C 91A, AGH found both species in wooded grassland, but only the Taita extended out into the surrounding dry bush.

866 SOMALI FISCAL *Lanius somalicus*
B1039 / MPG1106 / D2C2

Afrotropical: a NE species, from extreme SE Sudan, Ethiopia and mainly N areas of Somalia, S to Kenya.

Kenyan: a thinly distributed bird of open country, often in areas with little or no bush, at 400-1000 m (Br). Resident status is known as far S as Kapedo 38C (T Stevenson *in litt.*; EABR 1984). Presumed wanderers have extended further S to the N periphery of the highlands, at Baringo 50A (early Dec-mid Feb: St; EABR

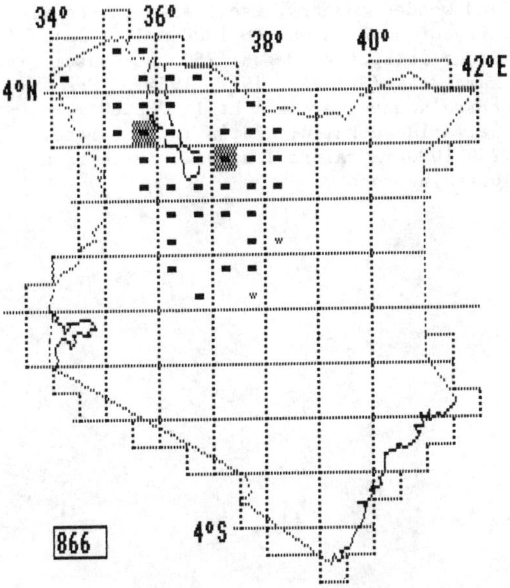

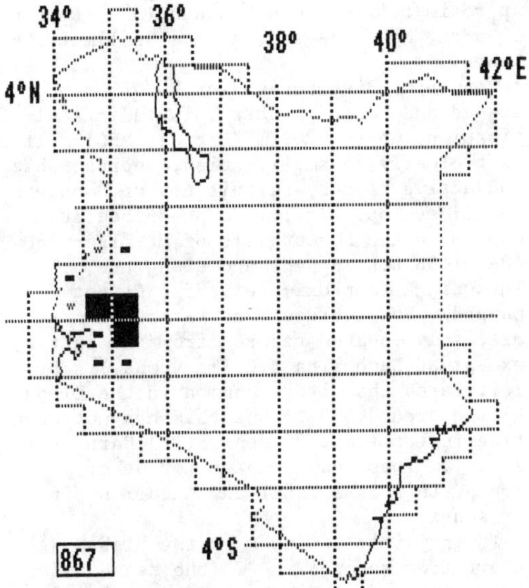

1978, 1979), El Karama Ranch 50D (L V J Grant *in litt.*), Wamba 51A and Isiolo 51D (Br). These wanderers reached atypically high and moist country *e.g.* *c.*1500 m and dry subhumid in 50D and *c.*1200 m and moist subhumid in 51D. Apart from these four squares, range is 89% within 0-500 mm rainfall areas and 78% arid.

There have been no reports from various observers in the Moyale 17AC area, nor from birdwatchers resident in 19A and 42A, and this Fiscal seems to be generally absent from the apparently suitable NE. This compares with distribution in Somalia where, at 1°N to 5°N, the bird is confined to the coastal strip (AM). See Fiscal (864) and Taita Fiscal (865).

867 MACKINNON'S SHRIKE *Lanius mackinnoni*
B1035 / MPG1110 / D2C2

Afrotropical: a W and C African bird with a discontinuous distribution from Cameroon, Congo and Angola, E through apparently S areas of the Congo basin to Rwanda, NW Tanzania, Uganda and, on the E limit of the range, W Kenya.

Kenyan: rather uncommon, local and unobtrusive, favouring bushed and wooded country, and forest edge, up to 2200 m (Br; Ja). Range is entirely within 1000+ m country to the W of the rift valley. All squares are moist subhumid-humid and within 1000+ mm rainfall areas (*cf.* the closely related Fiscal (864)). Ja mentions occurrence at Elgeyu (approximately 37CD 49B: not mapped), and Br lists the Mara R (61CD 74AC: not mapped). Other references: Britton & Britton 1971c; EABR 1979.

868 YELLOW-BILLED SHRIKE *Corvinella corvina*
B1026 / MPG1117 / D3C2

Afrotropical: this endemic genus contains two species, which replace each other either side of the equator. Kenya is the only country where both occur. The Yellow-billed Shrike inhabits the N tropics,

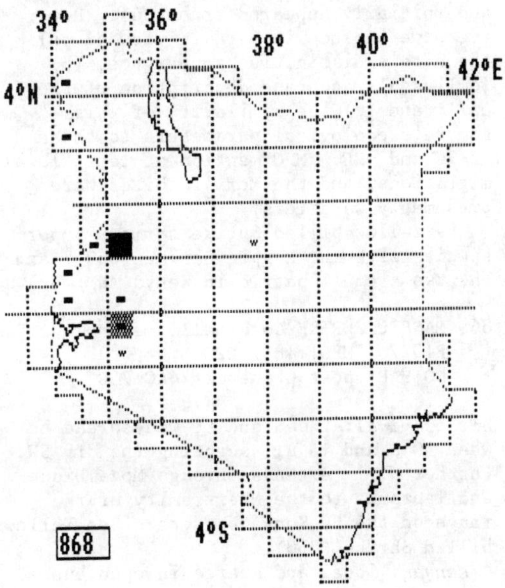

approximately between 5°N and 15°N, E to S Sudan, N Uganda and, at its SE limit, NW Kenya.

Kenyan: a distinctive shrike of open bushed and wooded country, including cultivation, up to 2200 m (Br; Ja; MPG). It is typically in small, noisy, approachable and active flocks, reminiscent of babblers *Turdoides* (*op. cit.*). Ja described it as common in the S foothills of Mt Elgon 36D 48AB (48B not mapped) and along the Turkwell R (approximately 25 13D: not mapped), but there are no recent or more precisely located records from the latter, except at Kacheliba 37C (many obs.). Br reiterates that it is common in the Elgon-Kitale area 36D 37C, but this has not been true for at least 20 years (T J Barnley *in litt.*). Thus, as Br notes for other areas, this is a local and uncommon bird in Kenya.

It is mainly confined to the highlands, apart from records in 2C (the Kenya/Sudan border in Dec: G Nikolaus *in litt.*), 12C (Uganda: J S Ash *in litt.*), and 39D (Tomlinson 1946: see below). The seven highlands squares are all above 1000 m, in humid, 1000+ mm rainfall areas (although Kacheliba is in the lowest and driest part of 37C). The other three squares descend to 500-1000 m, in arid-semiarid country. The record in 39D appears to have been overlooked by previous authors, and concerns a flock of 20 at Merille on 16 and 25 Jan 1941 (Tomlinson 1946). These birds were in company with large numbers of Wattled Starlings (899), at a swarm of young locusts and grasshoppers. This record is 2°E of the species' known range and initially appeared erroneous. However, we include it here in view of (a) the bird's distinctive appearance, (b) Tomlinson's accurate description of the call, and (c) the similarity of this square's ecological parameters to those of 2C and 12C. MPG mention at least local migrations, and the Merille birds were presumably wanderers.

The Yellow-billed Shrike forms a superspecies with the Magpie Shrike (869), and the two are allopatric in Kenya (HM; Br).

869 MAGPIE SHRIKE *Corvinella melanoleuca*
B1027 / MPG1118 / D2C2
NO MAP: post presence 74AC 75C

Afrotropical: the S and E, widespread between 30°S and 15°S, excepting the dry SW. Further N, it extends through Mozambique and Tanzania to the N extremity of the range on the SW Kenya borders. See Yellow-billed Shrike (868).

Kenyan: local and scarce in open bushed and wooded country, especially in the vicinity of acacias, up to 1800 m (Br; Ja). Range is between Magadi 75C and Maasai Mara 74AC (Br; EABR 1977, 1984). The three squares are all at 1500-2000 m, semiarid-subhumid, and in intermediate, 250-1000 mm rainfall areas. There is no breeding record.

Family Prionopidae: Helmet Shrikes

Nine species, endemic to the Afrotropical Region, of which five breed in Kenya. All probably wander to some degree, and members of the genus *Prionops* form small, noisy, active and often aggressive flocks which forage over extensive areas (Br; Ja).

870 HELMET SHRIKE *Prionops plumata*
B1043 / MPG1090, 1091 / D3C3

Afrotropical: widely distributed S of 15°N, apart from the dry SW and the lowland forest areas of the equatorial W.

Kenyan: locally common in generally dry, bushed and wooded country, up to 1800 m (Br; La). It occurs in small flocks which make extensive and apparently erratic foraging movements over large areas (Ja). This type of occurrence is well described by St, who notes irregular absences of many months' duration in the semiarid country around Baringo 50A, interspersed with appearances of several weeks or months. It tends to avoid the Baringo area during wetter periods (*op. cit.*).

Although there are records throughout our altitude categories, distribution is only 15% above 1500 m, and it is absent from much of the highlands, parts of which are occupied by the Grey-crested Helmet Shrike (871: *q.v.*)(see below). Range is 79% arid-semiarid, and absent from the L Victoria basin (Br), and from 1000+ mm rainfall areas of the coast.

The Helmet Shrike forms part of a species-group with the Grey-crested Helmet Shrike (HM), which replaces it in parts of the highlands and their peripheral plateaux. The two are predominantly allopatric, the single note of their association describing mixed flocks in the Loliondo district 86B of NE Tanzania (Williams 1963: not mapped). In the arid country of Tsavo E 101B, the Helmet Shrike occupies well vegetated habitats with the exception of riverine growth, where Retz's Helmet Shrike (872) occurs (La).

871 GREY-CRESTED HELMET SHRIKE *Prionops poliolopha*
B1044 / MPG1092 / D2C1

Afrotropical: an East African endemic species, confined to NE Tanzania and SW Kenya.

Kenyan: apart from Williams (1947), Lewis (1981c) summarised all published (and some unpublished) records of this species, up to the end of 1980. All unreferenced material in this account derives from this source. This uncommon

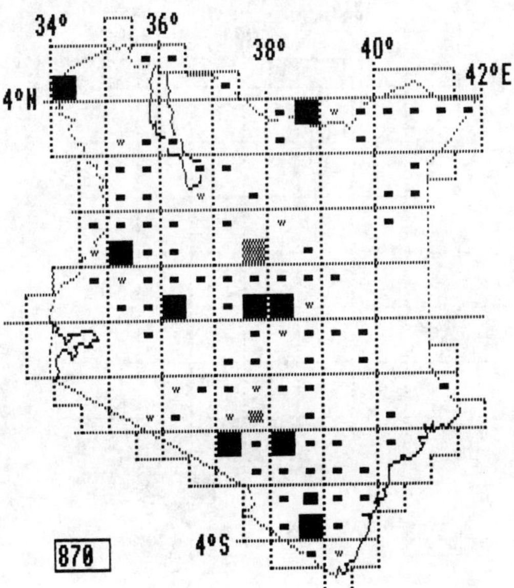

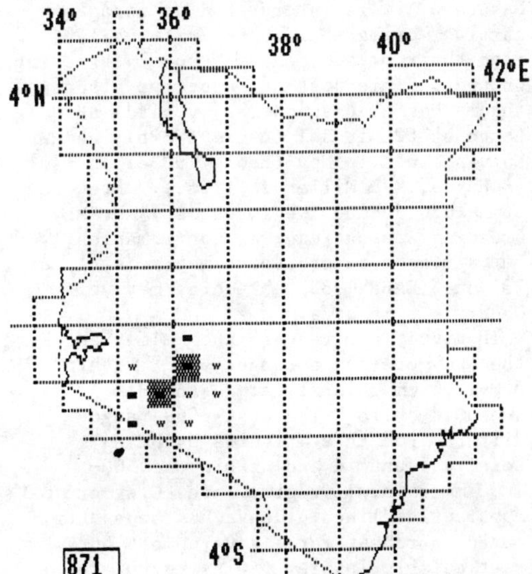

species frequents open woodland, particularly *Acacia drepanolobium* and *Tarchonanthus*, at 1200-2000 m (Br; Be). An apparent contraction of range during the last 100 years is now, at least in the rift valley, reversing.

The type specimen was shot on its nest at Naivasha 62C on 11 May 1884. Sporadic reports from the Naivasha area continued until 1926, but then ceased until 1982 (Lewis 1982b). 55% of this helmet shrike's squares contain only pre-1970 records and, for all areas except Naivasha, it is uncertain whether this reflects a genuine decline, lack of observer coverage of a sparse species, or loss of records due to confusion with the far commoner and better known Helmet Shrike (870: *q.v.*). The absence of even old records of the latter species from the very well known Naivasha 62C area indicates this 1926-1982 absence to be real.

Until 1978, range appeared to have contracted into the Mara GR 74AC-Narok 74B area, where it is frequently seen in the vicinity of Keekorok Lodge 74C. Similar reports come from the Serengeti region of adjacent NE Tanzania. On 25 Oct 1978, I G Marshall found a flock in L Nakuru NP 62A, and there followed a series of Oct-Feb reports from this park and the nearby Menengai Crater (EABR 1980). The only three breeding indications for this species have been Apr-May: the type specimen on 11 May, a juvenile in the Kedong Valley (possibly also 62C) on 17 Apr, and newly fledged young near Narok 74B on 27 May. In view of this breeding seasonality, the Oct-Feb reports from Nakuru 62A were interpreted as foraging parties moving NE into the rift valley from their presumed Mara-Narok 74 breeding grounds. This post-reproductive dispersal theory has been supported by birds at Longonot 62C in Jul (Lewis 1982b), and at Naivasha 62C in Nov (Lewis 1983d) and Jul (J Hayes, M H Miller *in litt.*). Presence at Nakuru during the (presumed) Apr-May breeding season requires confirmation, but a recent record there was as early as 23 June (EABR 1983): breeding may well occur in this area.

In direct contrast to the Helmet Shrike, the Grey-crested's range is 72% within 500-1000 mm rainfall areas and 81% subhumid-humid. The post-1969 range differs from these figures by being totally subhumid and within the 500-1000 mm rainfall regime. The Grey-crested's apparent decline in Kenya has been attributed to recent rural development but, particularly in view of its recent reappearance in the rift valley, this seems doubtful. The bird's history in Kenya can probably be better explained by range fluctuations in a rather uncommon species.

872 RETZ'S HELMET SHRIKE *Prionops retzii*
B1045 / MPG1095 / D3C2

Afrotropical: a S species, from Angola E to SE Zaire, Malawi and Mozambique, and NE to E Tanzania, Kenya and S Somalia.

Kenyan: a locally common helmet shrike of woodland and forest, particularly in riverine environments and mainly below 1500 m (Br; Ja; Britton & Zimmerman 1979; La). It is confined to the SE, from the coastal lowlands to the Tana R 52C-91A (ACH), the Athi-Galana R and the SE and E periphery of the C highlands (*cf.* Zanzibar Sombre Greenbul (619)). This range is 79% below 1000 m, 80% within intermediate, 250-1000 mm rainfall areas and 79% arid-semiarid.

There are old reports from 75B at Nairobi (frequently), Kiambu and Ngong (at approximately 1900 m)(van Someren 1916, 1939; Ja), but there are no recent records from these very well known areas despite continued presence around the Athi R SE of Thika 76A. Recent records from riverine areas in Meru NP 52C (D K Richards *in litt.*) are the northernmost for Kenya. In the E, records from the vicinity of the Tana R and the coast parallel distribution near the Juba R and the coast in S Somalia, where this shrike reaches 1°30'N (AM). See Helmet Shrike (870) and Chestnut-fronted Helmet Shrike (873). Other reference: Short & Horne 1985a.

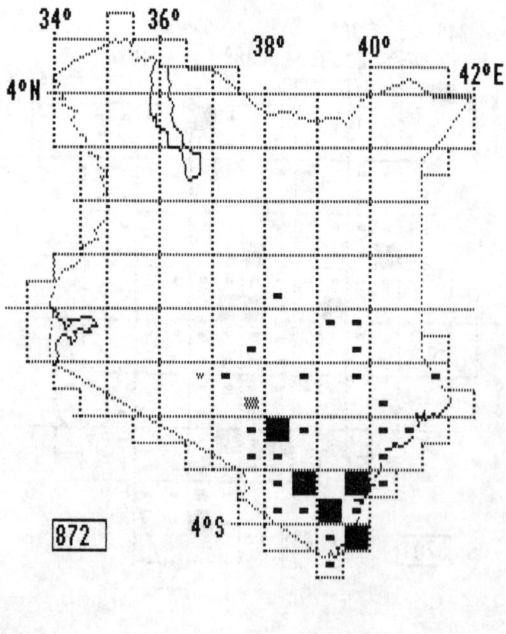

Plate 15. The Helmet Shrike, one of a family endemic to Afrotropical Africa (R. S. Daniell A.R.P.S.). *See page 462.*

Plate 16. Yellow-billed Oxpecker in characteristic position. The Oxpeckers are endemic to the Afrotropical Region (D.K. Richards F.R.P.S.). *See p. 480*.

873 CHESTNUT-FRONTED HELMET SHRIKE *Prionops scopifrons*
B1046 / MPG1096 / D3C2

Afrotropical: the E, with a discontinuous distribution from Mozambique N to E Tanzania, Kenya and the extreme S of Somalia. HM suggest this fragmented range to be largely a result of historically recent deforestation.

Kenyan: in contrast to Retz's Helmet Shrike (872), this species is confined to the coastal strip and the deep interior of the country, but absent from the intervening SE lowlands. The race *P.s. kirki* is locally common in coastal woodland and forest, including mangroves (Br; Britton & Zimmerman 1979). Br reports it NE to Lamu 91B, but the first records from Somalia have recently extended range to 80AB (AM). This race's distribution is entirely below 500 m and within the 500+ mm rainfall regime, in the coastal lowlands. 88% of its squares are semiarid-dry subhumid.

The race *P.s. keniensis* is confined to interior Kenya, being known only from Karoli 27A, Marsabit 27D, Meru 51D (Ja; National Museum Collection, Nairobi) and Embu 63D (Br). Of these four squares, 51D and 63D are on the E periphery of the C highlands, while the other two are on or near the forest islands of Mts Kulal 26B and Marsabit 27D. Particularly in view of this species' distinctive appearance, the lack of recent records of this interior race suggest that it is at or near to extinction. This decline is unlikely to have been due to the deforestation mentioned by HM, as substantial forests and/or woodlands persist in 27D 51D 63D.

The Chestnut-fronted is probably closely related to Retz's Helmet Shrike, which is generally in rather drier vegetation (HM), and which is less widely distributed and frequent than the Chestnut-fronted in Sokoke Forest 102B (Britton & Zimmerman 1979). Other references: Britton & Britton 1971b; Short & Horne 1985a.

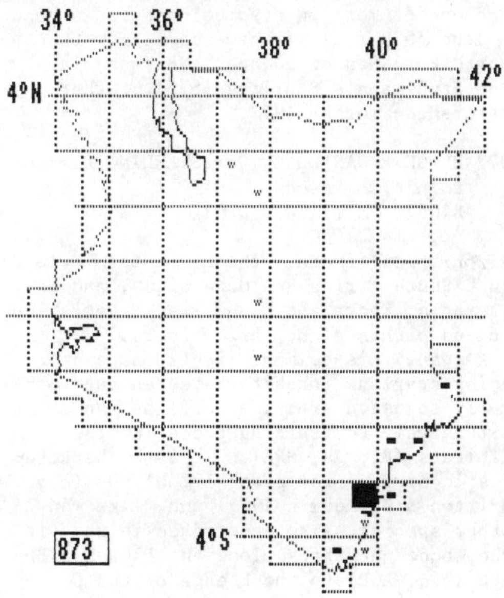

874 WHITE-CROWNED SHRIKE *Eurocephalus rueppelli*
B1041 / MPG1097 / D3C3

Afrotropical: the E and NE, from S Tanzania through East Africa to the E of L Victoria, to S Sudan, Ethiopia and Somalia.

Kenyan: a common and distinctive species that favours high, conspicuous perches in open bushed and wooded country, up to 2200 m but mainly below 1400 m (Br; Ja; MPG; Be; HH; La; St). Range is 83% below 1500 m and 79% arid-semiarid. It avoids many of the humid areas of the highlands and the moist W *e.g.* 48 61BD 62B 63A, and also most of the dry subhumid squares of the coastal strip *i.e.* 91D 102D 114BC (*cf.* Br). Other reference: Plumb 1978d.

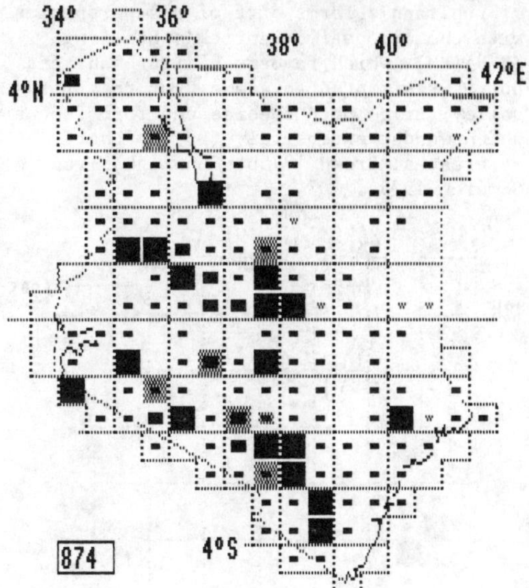

Family Sturnidae: Starlings, Oxpeckers

111 species that are mainly Afrotropical, Australasian and Oriental. 45 (41%) are found on the Afrotropical mainland, where 43 (96%) belong to endemic genera. The closely related genera *Spreo* and *Cosmopsarus* are concentrated in the NE of the Afrotropical Region, where five members regularly occur in the Taveta 100B-L Jipe 100D area. Congeners of some open country taxa *e.g. Lamprotornis* and *Spreo* can exhibit wide sympatry, whereas forest genera tend to be more allopatric. The two oxpeckers *Buphagus* are endemic to the Afrotropical Region and unique both in ecology and adaptive structure (all information from HM). 27 starlings (60% of the Afrotropical total) are known from Kenya, where 25 (92%) breed or are presumed to do so; Shelley's Starling (888) is an irregular, non-breeding, intra-African visitor, and the Ashy Starling (894) an extralimital wanderer.

875 STUHLMANN'S STARLING *Poeoptera stuhlmanni*
B1070 / MPG1209 / D2C2

Afrotropical: a highland species, from the E Zaire-Tanzania-Uganda borders NE to S Sudan and SW Ethiopia, and E to W Kenya.

Kenyan: locally common in highland forest canopy and nearby tall trees, at 1500-2600 m (Br; Ja; MPG). Range is in 1000+ mm rainfall areas to the W of the rift valley: 88% of the squares are humid.

Stuhlmann's forms part of a superspecies with the allopatric Kenrick's Starling (876)(HM), which favours similar habitats but is restricted to areas E of the rift valley (Br). Both species can form flocks which wander extensively (Br). Other references: Brown 1975b; Cunningham-van Someren 1975a.

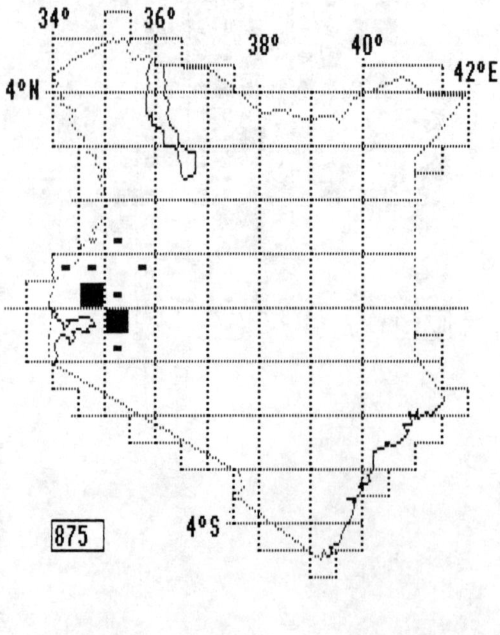

876 KENRICK'S STARLING *Poeoptera kenricki*
B1068 / MPG1210 / D2C2
NO MAP: post presence 63AB; pre presence 51D 100D

Afrotropical: endemic to the highlands of E Tanzania and C Kenya.

Kenyan: locally common in highland forest on Mt Kenya 63AB and the Nyambeni Hills 51D, up to 2500 m (Br). The three squares are in moist subhumid-humid, 1000+ mm rainfall areas of the E plateau. During the rains, Brown (1965a) noted these birds as low as 1400 m on the periphery of Mt Kenya near Embu 63C (not mapped because actual locality uncertain). Brown (1965a, 1975b) found birds prospecting a nest hole in Njukiini Forest 63A, but breeding apparently did not occur. There are no other breeding indications from Kenya. This starling can wander (Br), and one or more in atypical habitat around L Jipe 100D (Ja) may have come from the North Pare Mts of adjacent NE Tanzania. See Stuhlmann's Starling (875). Other reference: Turner 1977.

877 BLACK-BREASTED GLOSSY STARLING *Lamprotornis corruscus*
B1057 / MPG1194 / D2C2

Afrotropical: mainly the E coastal lowlands, in E South Africa and Mozambique, and in Tanzania, Kenya and S Somalia. Penetration inland occurs along large rivers.

Kenyan: this common, gregarious and noisy starling inhabits forested and formerly forested areas, as well as wandering extensively to fruiting trees in other habitats (Br; Ja; MPG). It is a characteristic bird of the coastal lowlands (*e.g.* Britton & Zimmerman 1979) and, like many other species, also penetrates inland via the wooded habitats along the Tana R 51BD-79C (*e.g.* AGH) to the E edge of the C

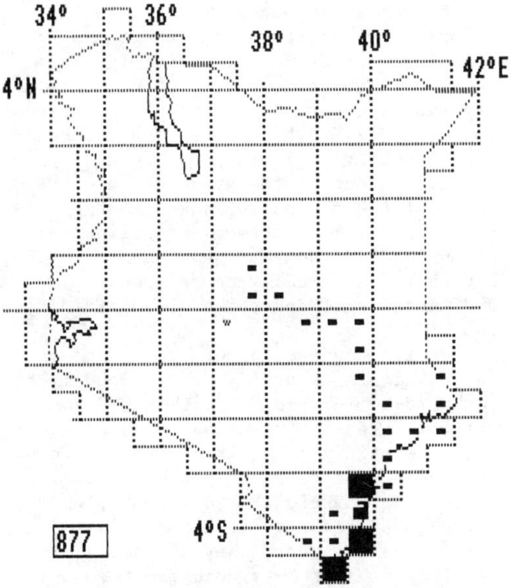

highlands. Distribution is similarly restricted to the vicinity of the coast and the Juba R in adjacent S Somalia (AM). Range is 83% below 500 m, but greater altitudes are attained near the highlands, and Br mentions occurrence up to at least 1400 m. Range in the coastal lowlands is entirely within 500+ mm rainfall areas, and 80% semiarid-dry subhumid. Similar environments are reached in the highlands. Overlay data from the intervening arid lowlands are misleading in that the bird is confined to the locally moist habitats around the Tana R.

The Black-breasted forms part of a superspecies with the Bronze-tailed Glossy Starling (878)(HM), which occupies open bushed and wooded country in Kenya's W (Br). Other reference: Short & Horne 1985a.

878 BRONZE-TAILED GLOSSY STARLING *Lamprotornis chalcurus*
B1054 / MPG1189 / D1C1
NO MAP: post presence 36B 49A; post confirmed 37C; pre presence 48B 49B 61A

Afrotropical: the N tropics, E from Senegambia to Nigeria, S Sudan, Uganda and, on the SE extremity of the range, W Kenya.

Kenyan: this local, apparently uncommon and very easily overlooked species is little known in Kenya. It frequents open bushed and wooded country, and there are reports from Kodich 36B S to Nyarondo 61A (Br; Ja; MPG; T J Barnley *in litt.*; EABR 1983). Ja also mentions occurrence on the Kerio R but gives no precise location. Range is in subhumid-humid areas, W of the rift valley. Various combinations of five of the six squares are at 1500-2000 m, and within the 1000+ mm rainfall regime. These figures contrast with Br's mention of the bird in medium rainfall areas up to 1200 m, which probably applies to its Ugandan range, but agree more with D Chepus and T J Barnley's summary (*in litt.*) of range in the Kitale-Kapenguria-Kongelai area 37C.

Although these latter observers do not have dated breeding records, they have extensive experience of this starling and its nest holes in the Kitale-Kapenguria area 37C, where breeding is annual during the 'long' rains. A juvenile found dead on 31 May 1985 was approximately 10-14 days out of the nest. There appears to be some post-reproductive dispersal down to around 1400 m at M'temberr 37C and nearby country (D Chepus, T J Barnley *in litt.*).

The Bronze-tailed often flocks with the more abundant Lesser Blue-eared Glossy Starling (880)(Br), *e.g.* at the base of the Kongelai escarpment 37C in Jul (EABR 1983). See Black-breasted Glossy Starling (877).

879 BLUE-EARED GLOSSY STARLING *Lamprotornis chalybaeus*
B1055 / MPG1188 / D3C3

Afrotropical: the N tropics at 10°N-15°N, from Senegambia to Ethiopia and NW Somalia. Further S, it extends through East Africa to Zambia, Angola and NE South Africa.

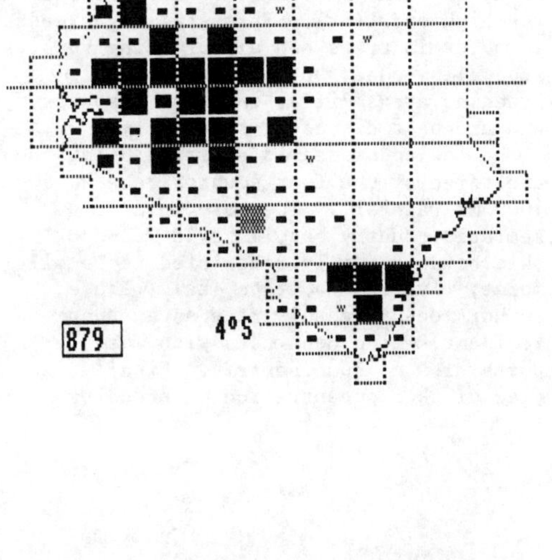

Distribution in Kenya represents the E limit of the range, with the exception of a single record in the QSD immediately E of 7D (AM).

Kenyan: a noisy and often common bird of open bushed and wooded habitats, including cultivation and the vicinity of settlements, up to 3000 m (Br; Ja; MPG; Se). Range is only 19% arid and 6% within 0-250 mm rainfall areas. Virtual absence from the NE and E is verified by the lack of recent records from the moderately well known highland of Marsabit 27D, and by the lack of reports from observers resident in 19A 42A 65B. Records in the N have either been from L Turkana, where the bird is occasional in wooded localities (1973 edition of HH), or from the peripheries of populations in moister areas in Uganda 12C 24BD, Sudan 2C and Ethiopia 4A 16B 17AC 19B (J S Ash, G C Backhurst, G Nikolaus *in litt.*; *cf.* 19B and AM's locality near 7D (above)).

Be considered this starling resident, and Br thought it probably mainly so, but these generalisations appear incorrect, at least on a national scale. Breeding records are concentrated in the highlands, and in the SE lowlands 101B 102ABC. There is the possibility that this pattern reflects frequency of observers, but there are no breeding reports from such relatively well known areas as Machakos 76C, Amboseli GR 88C, Bushwackers' Camp 89A, Tsavo W NP 89C 101AC, the Kilifi area 102D and Mombasa 114B. In view of the well known propensity of this species to flock and wander or migrate during the non-breeding season (Ja; MPG; HM; Dittami 1983), these apparently well defined breeding areas may be real, while intervening regions (particularly S of 2°N) may represent off-season wandering or regular non-breeding range.

The two clusters of breeding records apparently represent different races, *i.e.* *L.c. sycobius* near the coast (and possibly including 88A), and *L.c. cyaniventris* in the highlands (Br). This starling breeds in holes in trees and all breeding squares are well wooded, but the two apparent breeding areas differ in other respects, and are best discussed separately.

The lowlands area is entirely below 500 m, and three of the four squares receive 500-1000 mm rainfall. They are in arid-semiarid country but, in arid Tsavo E 101B, this bird is mainly restricted to locally moist, riverine habitats (La). These authors describe this race as an uncommon resident, which contrasts with most reports of *L.c. cyaniventris*. Finally, in view of the concentration of breeding in the wettest areas of the highlands (see below), it is strange that there are no breeding records for *L.c. sycobius* from the apparently suitable, 1000+ mm rainfall, dry subhumid parts of the coastal strip *i.e.* 102D 114B, which have both had resident observers. It is possible that this absence is due to the presence in these areas of the abundant Black-breasted Glossy Starling (877). These two starlings are not placed in the same superspecies by HM, but they specifically comment on the difficulties of detecting close affinities in this genus.

The breeding squares of *L.c. cyaniventris* in the highlands are 79% above 1500 m, 88% subhumid-humid, and 96% within 500+ mm rainfall areas. Extensive off-season movements certainly occur in this race and have been best demonstrated by Dittami (1983) in L Nakuru NP 62A. There, only 12 pairs remained on their territories throughout the year, whereas the majority of birds present only occupied them while breeding. After breeding, wandering flocks that began to form in June reached peak numbers of 300 in Aug (*cf.* Lack (1985) for seasonal peaks of *Spreo* spp. in Tsavo E). This flocking was followed by apparent dispersal from the area en masse, and only very low numbers were present from late Nov to early Jan. Specific searches for concentrations of this species during this period at Bogoria 50C, the Mau 61BD, Naivasha 62C and the Kinangop-Kikuyu area 62D 75B were unsuccessful, and Dittami concluded that the birds had left the area.

Similar phenomena to those observed by Dittami (1983) apparently also occur at Mau Narok 61D where, while describing the species as resident, Se noted fluctuations in abundance. Apr-May breeding was followed by wandering flocks of up to 200 individuals from Jul onwards. The seasonality of these wandering flocks at Nakuru 62A and Mau Narok 61D can be compared to the appearance of small, non-breeding flocks during Jul-Sept and Nov-Feb in the semiarid country around Baringo 50A (St), which is on the edge of the proposed highlands breeding area. Oct-Nov dispersal is also shown by birds which, on the basis of their location (Br), appear to be the lowland *L.c. sycobius*. Thus Nov-Apr absence occurs at Kitui 77A (Tennent 1962), and two individuals near the E range edge at Mida Creek 102B were on 20 Oct (EABR 1980). Fluctuations during Jan-Mar around Eldama Ravine 49D may have been partly due to the occurrence of fruiting trees (Ja).

The Blue-eared forms part of a species-group with the very similar but smaller Lesser Blue-eared Glossy Starling (880)

(HM), and the two are broadly sympatric in Tanzania (Br). Other references: Davey 1982c; Short & Horne 1985a.

880 LESSER BLUE-EARED GLOSSY STARLING *Lamprotornis chloropterus*
B1056 / MPG1190 / D1C1

Afrotropical: widespread in the N tropics between 5°N and 15°N, from the W coast E to Ethiopia, and SE to NW Kenya. Another population in the S tropics extends N from Zimbabwe to Zambia, Mozambique and E Tanzania, reaching its N limit in coastal Kenya.

Kenyan: locally common and usually gregarious in open bushed and wooded country, including open woodland, up to 1600 m (Br). It is very similar in appearance to the Blue-eared Glossy (879) and Bronze-tailed Glossy (878) Starlings (MPG; HM) and, like the latter, is probably widely overlooked (hence D1C1).

The Lesser Blue-eared is mainly a bird of the NW, where the nominate N tropics race is common throughout the year in the Tugen Hills 49B and the Kerio Valley (also approximately 49B)(Br), and also occurs at other sites (EABR 1983). Flocks irregularly move E to the rift valley, where St reports reasonably common occurrence around fig trees *Ficus* spp. from Nov 1978 to Feb 1979 at Baringo 50A. The only record of breeding is T J Barnley's (*in litt.*) mention of annual nesting below 2000 m on the Kongelai escarpment 37C and in West Pokot 37 during the 'long' rains (latter square not mapped). A small but apparently resident population of this race occurs around the base of Ol Doinyo Sapuk 76A, well on to the E plateau (Br). This race's range is entirely above 1000 m, in 500+ mm rainfall areas. Apart from Baringo 50A, all squares are subhumid-humid.

The S tropics race *L.c. elisabethi* reaches N to 7°S in NE Tanzania (Br), and old records from Mazeras 102D, Mombasa 114B and Vanga 114C (Ja) presumably represent wanderers just N of their normal range (Br). The tendency of some members of this genus to wander extensively is stressed by HM, and Br notes this species as both a resident and wanderer. The three squares of presumed *L.c. elisabethi* are below 500 m in the coastal lowlands, and entirely within 1000+ mm rainfall, dry subhumid areas.

881 RUPPELL'S LONG-TAILED GLOSSY STARLING
Lamprotornis purpuropterus
B1060 / MPG1196 / D3C2

Afrotropical: the E and NE, from N Tanzania, through East Africa and the E Zaire borders, to Sudan, Ethiopia and S Somalia.

Kenyan: this distinctive starling frequents open bushed and wooded habitats, including cultivation and the vicinity of settlements, up to 1800 m (Br; Ja; MPG). It is most common in semiarid-humid habitats to the W of the rift valley, but rather less common and more local further E, mainly near water in arid-semiarid

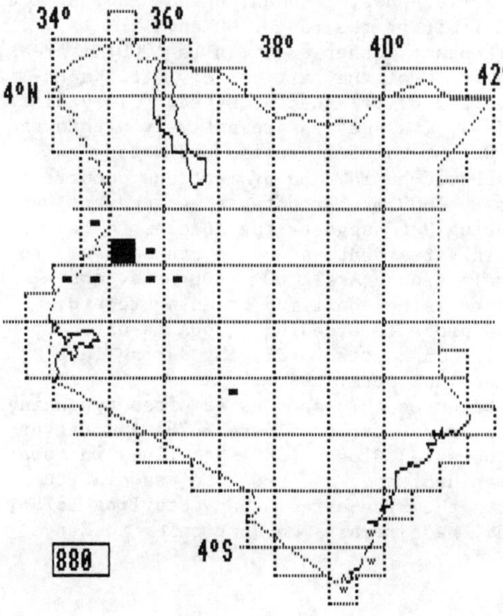

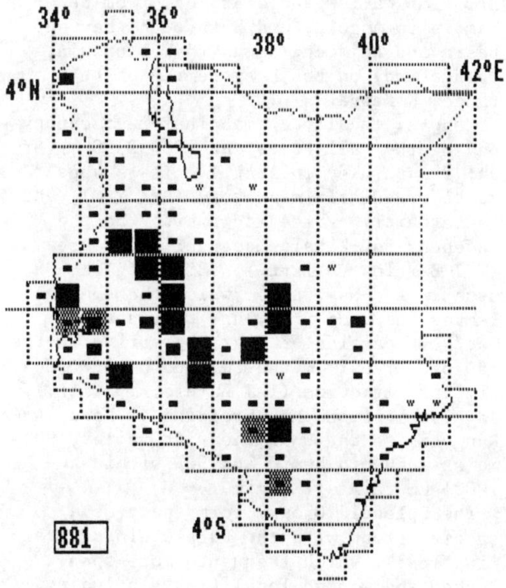

country (Br).

Range is only 10% within 0-250 mm rainfall areas. Absence from most of the dry areas of the N and E, and particularly the NE, is apparently real, being corroborated by lack of records from various observers at Moyale 17AC, from resident observers in 19A and 42A, and from recent observers at Marsabit 27D. Absence from the E borders compares with the data of AM, who list no Somali records between 4°N and 0°30'S within 100 km of Kenya. Records in these arid N and E regions are virtually confined to the vicinity of permanent water at L Turkana (HH), the Ewaso Ng'iro R 52B 53A, the Tana R 64B-79C (AGH), and the Tiva and Galana R in Tsavo E 89B 101B (La). Restriction E of 39°E to riverine and coastal areas parallels range S of 1°N in Somalia (AM). Whereas this starling is locally common in the wettest parts of the highlands and their adjacent plateaux on both sides of the rift, there are records from only semarid areas of the coastal strip.

81% of the breeding squares are above 500 m, and there are no breeding records from the coastal lowlands or Somalia (AM). The implication of this distribution is unclear. Br, St (50A) and DEP (pers. obs.: 75C 89A) consider this starling resident, although Br also mentions wandering. Other reference: Dyer 1971.

882 SPLENDID GLOSSY STARLING *Lamprotornis splendidus*
B1061 / MPG1192 / D2C2

Afrotropical: W and C Africa, from Senegambia to Angola, and E through the Congo basin and adjacent areas to W Tanzania, Uganda and, on the E extremity of the range, W Kenya.

Kenyan: locally common in forest canopy and nearby tall trees, at 1700-2300 m (Br; Ja; T J Barnley *in litt.*). It is apparently regularly migratory (Br; MPG), and is certainly a breeding visitor to the Endebess 36D-Kitale-Kapenguria 37C area (T J Barnley *in litt.*). Breeding is usually Apr-May, after which the species leaves the area from June to (usually) Oct/Nov: parties of migrants arrive in the 36D 37C area from the SW, and depart in a similar direction (T J Barnley *in litt.*). Ja also mentions May breeding at Nandi 49C. Range is to the W of the rift valley, 90% moist subhumid-humid and 90% within the 1000+ mm rainfall regime.

The Splendid Glossy forms part of a species-group with the Purple Glossy Starling (883), which frequents more open, bushed and wooded localities, including

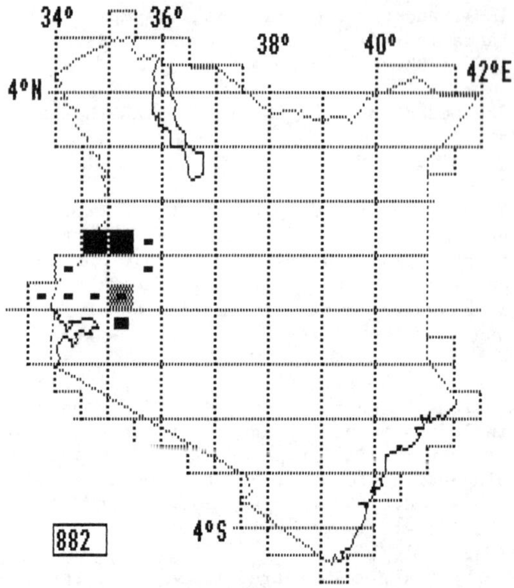

cultivation (HM; Br; Ja).

883 PURPLE GLOSSY STARLING *Lamprotornis purpureus*
B1059 / MPG1191 / D2C2
NO MAP: post presence 36D 37C 48AC; post probable 60A; pre presence 48B 61A

Afrotropical: the N tropics, at rather higher latitudes than the Splendid Glossy Starling (882: *q.v.*), E from Senegambia to Nigeria, Chad, S Sudan, Uganda and, at the SE limit of the range, W Kenya.

Kenyan: rather uncommon in 1000+ m areas to the W of the rift valley, E to Kapenguria 37C (Br) and the Nyando Valley 61A (Ja). Its squares are entirely within the 1000+ mm rainfall regime and moist subhumid-humid, and Br mentions occurrence up to 1800 m. Br notes some apparent wandering but suggests the species to be mainly resident, and this status seems to apply around Akala 60A (ADL pers. obs.). There is no confirmed breeding record, and probable breeding in 60A is based upon pairs in courtship display during late June (ADL pers. obs.).

Br notes this species as often replacing the Bronze-tailed Glossy (878) and Lesser Blue-eared Glossy (880) Starlings in more open habitats, and MPG note association with the Blue-eared Glossy Starling (879), but these remarks may not refer to Kenya.

884 BRISTLE-CROWNED STARLING *Onychognathus salvadorii*
B1065 / MPG1206 / D2C2

Afrotropical: the NE, from W Somalia, Ethiopia and extreme SE Sudan to NE Uganda and N Kenya.

Kenyan: a locally common starling of cliffs and gorges, often near water, at 400-1300 m (Br; MPG; Brown 1965a; HM; HH; St). Brown (*op. cit.*) mentions it in Marsabit town 27D 28C, and it also frequents towns in Somalia (AM; *cf.* Red-winged Starling (885)). Mainly a bird of the dry lowlands of N Kenya, the Bristle-crowned also extends S to higher and moister country on the N edges of the highlands, *i.e.* to the NE of the Tugen Hills 49A (D J Pearson *in litt.*), and at Bogoria 50C (Brown 1965a), Timau 51C (J Hayes *in litt.*) and Isiolo 51D (Br).

Range is 65% within 0-500 mm rainfall areas and 77% arid-semiarid (*cf.* Br). Squares with only old records form a distinct fringe on the S edge of the range. This pattern may be insignificant in view of the proximity of recent records in some cases *e.g.* 50AC 51BCD, and of the poorly covered nature of the areas in others *e.g.* 25C 37A 53A. Only the lack of records from an observer recently resident in Wajir 42A seems to indicate a real disappearance of the species from the area. Range reaches S to 2°30'N in Somalia (AM). MPG mention breeding at Turkana, but omit localities. Both Br and St consider this bird resident.

The Bristle-crowned Starling is placed in a species-group with the rest of its genus (HM). There are three congeners in Kenya. The Slender-billed Chestnut-winged Starling (886) is segregated from the Bristle-crowned at greater altitudes in the highlands, while Waller's Chestnut-winged Starling (887) is largely confined to the canopy of highland forest. The Red-winged Starling frequents similar cliffs and rocky situations to the Bristle-crowned, but is mainly S of the latter's range, in moister country.

885 RED-WINGED STARLING *Onychognathus morio*
B1064 / MPG1203 / D2C2

Afrotropical: N from the Cape through E areas to Zambia, Malawi, East Africa and Ethiopia. From the latter, it extends sporadically W across the N tropics to the W coast. Distribution in Kenya represents the E limit of the range.

Kenyan: an often common species that roosts and nests on cliffs and other rocky prominences in non-arid environments, at 600-4100 m (Br; MPG; Brown 1965a). Flocks undertake extensive foraging flights to other bushed, wooded and forested areas (*op. cit.*; Ja). An increasing tendency to breed on buildings, including those of Nairobi city centre 75B, has allowed the species to colonise areas without natural cliffs (Brown 1965a; Br; La).

This starling is most common to the E of the rift valley but, with the exception of Mt Elgon 36D, less so in the W (Brown 1965a). 74% of its squares are in or E of the rift. Range is 71% within the 500+ mm

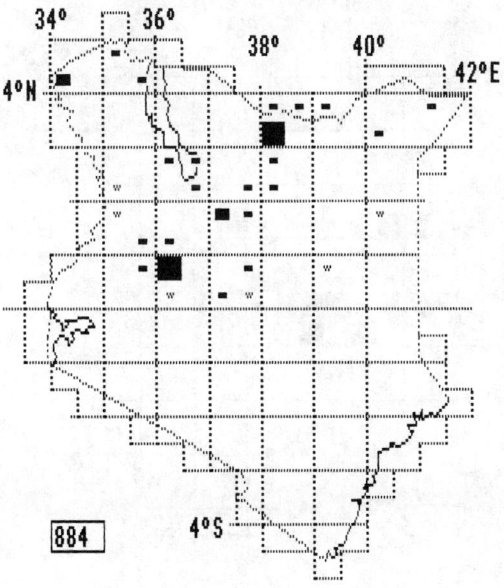

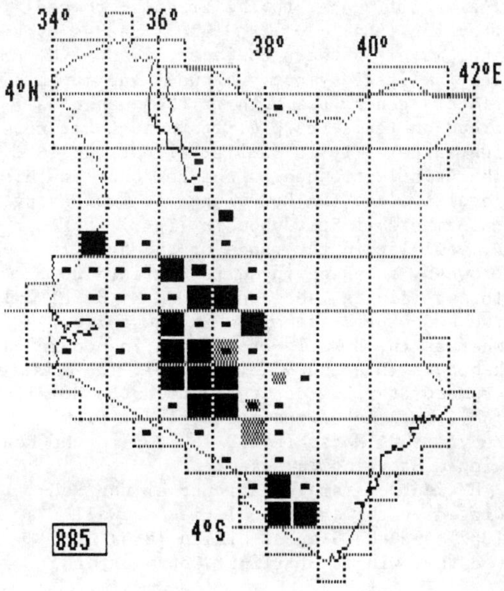

rainfall regime, which compares well with Brown's (1965a) 635+ mm estimate. Distribution is 77% above 1000 m, and the only square below 500 m is 101B Tsavo E (La). Range is only 12% arid, so that the Red-winged is virtually absent from the N and E.

Occurrence in arid squares E of 38°30'E is virtually restricted to the locally moist Mt Endau forest island 77B (A W Diamond *in litt.*), and to a resident pair on Voi Safari Lodge 101B (La). The apparently suitable moist coastal strip is beyond the species' E limit, and the coast is similarly avoided in Somalia (AM). Occurence in the arid and semiarid country N of 1°N in Kenya is confined to the locally moist forest islands of Mt Kulal 26B (Diamond & Keith 1980), Mt Nyiru 26D (source of record unknown, but included here on the basis of presence on forested peaks to the N and S), the Ndoto Mts 39A (Br) and the Mathews Range 39C (Brown 1965a). Absence from the nearby and apparently suitable highland of Marsabit 27D 28C is notable (see below). The record in 16A is from Ethiopia (J S Ash *in litt.*), and represents the S edge of the Ethiopian highlands population.

This starling is considered resident by MPG, La (101B), St (50A) and DEP (pers. obs.: 75B), but Br records seasonal congregations at Narok 74B. These, and an Aug-Sept influx to Tsavo E 101B (La), suggest regular movements, as opposed to foraging flights. Lee (1963) and Angwin (1971c) observed this starling perching on Klipspringers *Oreotragus oreotragus* on or near the Ngong Hills 75B.

The Red-winged Starling utilises similar rocky cliffs *etc.* to the Bristle-crowned Starling (884: *q.v.*) and their ranges overlap, mainly between the equator and 1°N. HM report little sympatry between any members of this genus, but both of these species breed on fault scarp cliffs near L Baringo 50A, apparently in similar situations (St). The Bristle-crowned is the commoner at this locality, but the Red-winged is increasing in numbers (T Stevenson *in litt.*). Further N, well within the range of the Bristle-crowned, the Red-winged is confined to forest islands (see above). On Mt Kulal 26B, the Red-winged is common in the forest, whereas the Bristle-crowned is in non-forest habitats (Diamond & Keith 1980). The Bristle-crowned Starling occurs in towns in Somalia (AM), but the only similar record from Kenya is at Marsabit 27D 28C, where the Red-winged Starling is absent.

HM mention rare occurrence of the Red-winged in the same localities as Waller's (887) and the Slender-billed (886: *q.v.*) Chestnut-winged Starlings, but note that sympatry is very restricted.

886 SLENDER-BILLED CHESTNUT-WINGED STARLING
Onychognathus tenuirostris
B1066 / MPG1205 / D2C2

Afrotropical: high altitudes of the E and NE, from the Ethiopian highlands S through East Africa and the E Zaire borders to Malawi. Distribution in Kenya represents the E limit of the range.

Kenyan: this locally common highland starling exhibits two habits apparently unique in African birds (HM). Firstly, its colonial roost and nest sites are typically behind waterfalls. This implies critical timing of breeding with regard to the volume of water flowing (Brown 1965a; Brown & Britton 1980). During incubation and the fledgling stage, drought would render the nest more vulnerable to predators, while flood conditions could make it completely inaccessible to the parents. However, low water levels are essential later, to enable the inexperienced young birds to leave the nest, and fledging is delayed if high water levels persist at this time.

The second feature of this species is its great altitudinal range while foraging. Flocks occur mostly in open or edge areas of highland forest, but can also move down to cultivation at 1400 m, or up to open alpine moorland at 4500 m (Br; Brown 1965a).

The Slender-billed's Kenyan range is uncertain, possibly due to some confusion with the similar Red-winged Starling (885). Ja lists the Slender-billed from only E of

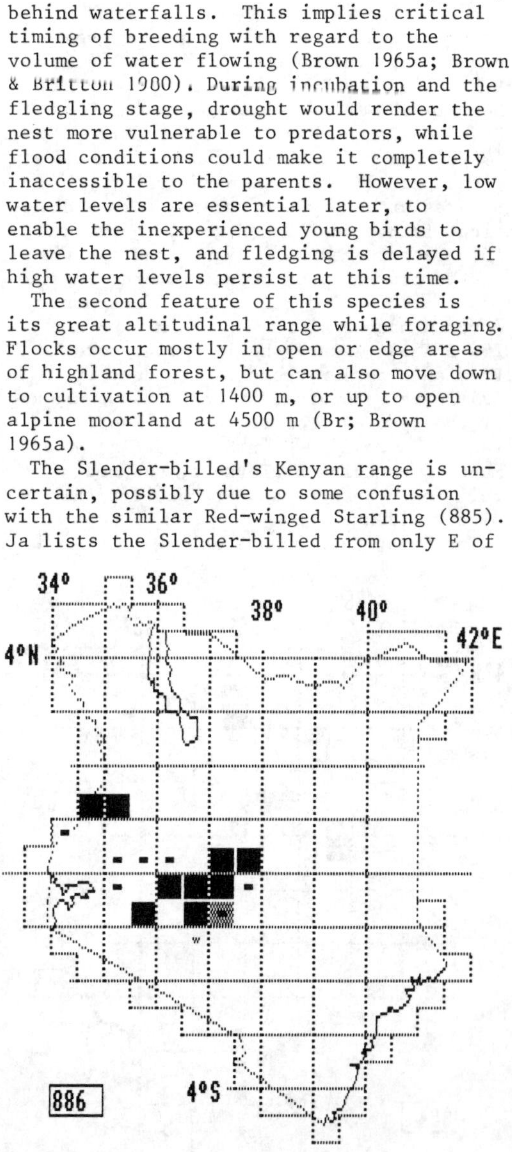

the rift valley in Kenya, and remarks on the notable lack of records from apparently suitable habitats of Mt Elgon 36D. He also mentions the recent proposal of a slender-billed race of the Red-winged Starling at high altitudes on Elgon, but was unable to verify it. MPG record the Slender-billed from the Aberdare Mts 62B and Mt Kenya 63A, and state the slender-billed race of the Red-winged Starling, *O.m. montanus*, to be confined to 2700+ m areas on Elgon.

Brown (1965a) makes the first mention of the Slender-billed itself on Elgon. He describes this species as most common E of the rift but less so on the Mau 61BD and Elgon, possibly due to lower numbers of suitable cave and waterfall sites in the W. HM map no Kenyan specimens of the Slender-billed from W of the rift, and mention partial sympatry of the Slender-billed and the Red-winged Starlings only in the E (*i.e.* C) highlands.

Br ignores Brown (1965a) and follows the other authorities in restricting the Slender-billed to the C highlands, W to Mau Narok 61D (from Se). Br reports the Red-winged Starling from Elgon but, possibly in view of the foregoing confusion, totally omits racial nomenclature. EABR (1981), however, describes breeding of the Slender-billed on Elgon 36D without comment, and we map this record. Finally, T J Barnley (*in litt.*) reports nesting of the Slender-billed behind waterfalls on the Morun R below Kaibibich 37C, and this record is also mapped here. The Red-winged is not known to utilise this type of nest site.

Hence we map records on and around Elgon, while noting the need for further field study of these species in W Kenya. Including Mt Elgon 36D, range is 76% within 1000+ mm rainfall areas and 87% moist subhumid-humid. Unlike the Red-winged Starling, the Slender-billed does not extend to forest islands.

If the Slender-billed exists on Elgon, then it is at least broadly sympatric with the very similar *O.m. montanus* (if that race is still recognised). For the C highlands, HM note partial sympatry of these two species, with local ecological and vocal segregation. Around Mau Narok 61D, apparently referring to this species, Se reports only a few Dec-May records (*cf.* Sharpe's Starling (897)), but notes it to be resident and numerous only a few miles away. See Bristle-crowned Starling (884).

887 WALLER'S CHESTNUT-WINGED STARLING *Onychognathus walleri*
B1067 / MPG1201 / D2C2

Afrotropical: mainly the highlands of the E, from Malawi through East Africa and the E Zaire borders, to S Sudan. There is also a population in Cameroon. Distribution in Kenya represents the NE limits of the range.

Kenyan: a locally common starling of highland forest canopy, up to 3000 m (Br; Ja), including the Mt Marsabit forest island 27D 28C. Excluding this isolated highland, with its very localised moist climate, range is 72% above 1500 m, 89% moist subhumid-humid, and 89% within 1000+ mm rainfall areas. The nominate race occurs E of the rift valley, while *O.w. elgonensis* is to the W (Br). The species is resident, but local movements and some wandering can occur during the off-season (Br; Ja; MPG). See Bristle-crowned Starling (884). Other reference: Brown 1965a.

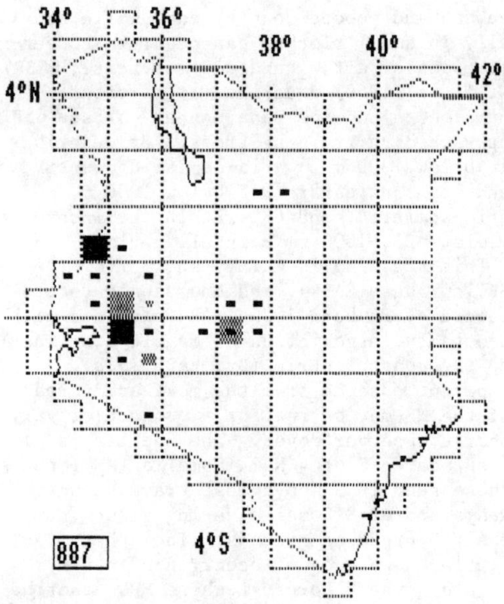

888 SHELLEY'S STARLING *Spreo shelleyi*
B1075 / MPG1215 / D2C1

Afrotropical: the NE, from Kenya to S Sudan, S and SE Ethiopia and Somalia. Distribution in Kenya represents the SW limit of the range.

Kenyan: a generally uncommon and erratic non-breeding visitor from the N, which often flocks with the similarly nomadic Magpie Starling (895)(Br; Williams 1968; La; EABR 1978, 1982). It frequents

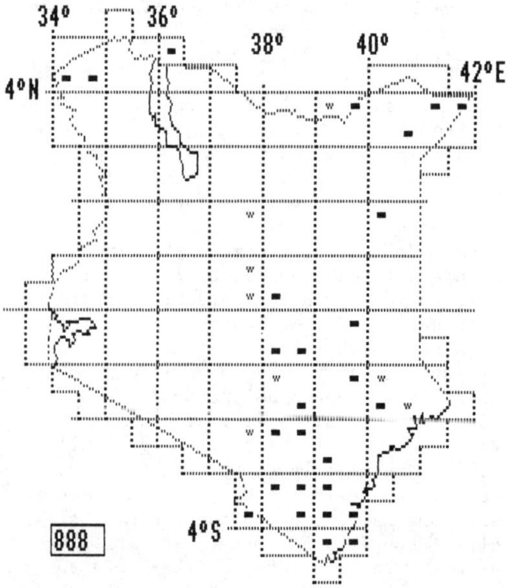

bushed and wooded country and, while usually in small flocks, can occur in concentrations of a few hundreds (*op. cit.*; EABR). Williams (1968) describes it as often numerous along the Tana R near Garissa 65B, attracted by fruiting trees. Br describes it below 1300 m in arid-semiarid country, and its squares are 91% below 1000 m, 90% arid-semiarid, and 68% within the intermediate 250-1000 mm rainfall regime.

This starling is mainly known from the SE lowlands (EABR), and the single coastal record at Bamburi 102D 114B coincided with a notably large influx of Magpie Starlings to the coastal strip (Br; EABR 1978). Apparent absence from the most arid areas of the NE may be real or a product of very sparse observer cover. The species is also scarce S of 4°N in Somalia (AM), but those seen in S Kenya must traverse NE Kenya and/or S Somalia during return to their N breeding grounds. The other, little known area of occurrence in Kenya is along the N borders, where MPG describe it as very local but not uncommon. The records in these areas are from the Kenya/Sudan borders 2C (Dec: G Nikolaus *in litt.*), 2D (T Romig *in litt.*), Moyale 17a (Ja), and the Ethiopian sectors of 4A and 17B (J S Ash *in litt.*).

Shelley's Starling is an Apr-Sept intra-African visitor to N Somalia (AM), breeding in Apr-June, and the four breeding records from Ethiopia are also in Apr, at 4°30'-5°30'N (J S Ash *in litt.*). Nearly all dated Kenyan records are during the Oct-Mar off-season (EABR; *op. cit.*). An 'early' flock on 17 Aug at Rukanga 101D contained juveniles (EABR 1981). Three 'late' dates were all in the NE, in Meru NP 52C (May: EABR 1983), between Rhamu 19A and El Wak 30B (June: D A Turner *in litt.*), and at Wajir 42A (Jul: J E Miskell *in litt.*).

Shelley's forms part of a superspecies with the allopatric Hildebrandt's Starling (889: *q.v.*)(HM; Br). HM s mention of certain breeding of both species in the Taveta 100B-L Jipe 100D area is most unlikely in the case of Shelley's (see Brown & Britton 1980), and Br describes only marginal overlap in SE Kenya, where one or both birds are wanderers. HM's suggestion that Shelley's occupies more arid country than Hildebrandt's is certainly correct, with the former's squares 53% arid, compared to the latter's 12%. See Wattled Starling (899).

889 HILDEBRANDT'S STARLING *Spreo hildebrandti*
B1074 / MPG1214 / D3C2

Afrotropical: endemic to Kenya and E Tanzania.

Kenyan: locally fairly common in open bushed and wooded country, mainly at 600-1700 m, but occasionally up to 2200 m (Br; Ja; Turner 1977). Br remarks upon its peculiarly restricted distribution. In contrast to the closely related Shelley's Starling (888: *q.v.*), range is 82% at 500-2000 m, 80% semiarid-subhumid and 89% within the intermediate, 250-1000 mm rainfall regime.

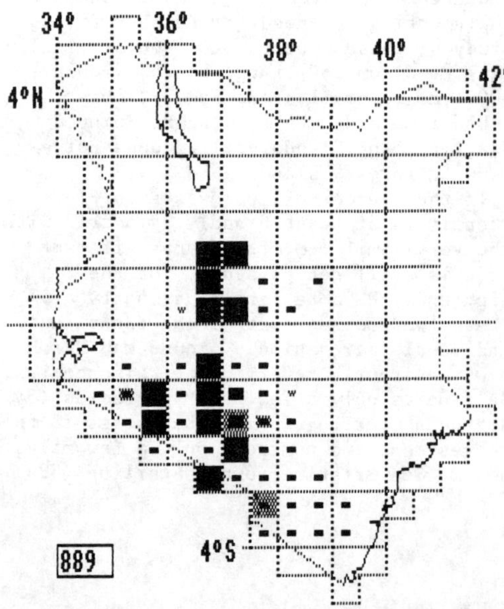

Br considers it mostly resident and DEP (pers. obs;) concurs for 74BD 75BD 88A 89A. Blencowe (1963) mentions apparent wandering into the Nyahururu area 50C, however, and a definite influx occurs in the Ruaraka-Kahawa-Ruiru area 75B during the June-Sept off-season (DEP pers. obs.). All breeding squares are above 1000 m, and this starling is a rare visitor to its lower altitudinal limit in arid Tsavo E 101B (La). In view of the lack of breeding records from such well-watched localities as Samburu GR 51B, Meru NP 52C, Bushwackers' Camp 89A and Tsavo 89C 101ABD, birds in these lower altitude areas may be only non-breeding wanderers/migrants (*cf*. Lack (1985) for Tsavo E 101B).

Hildebrandt's Starling often feeds and flocks with the similar Superb Starling (890)(Br; Ja). They are broadly sympatric but, while Hildebrandt's can be in drier and less wooded habitats (a distinction that is completely contrary to their overall ranges on the maps), their modes of ecological segregation are not understood (Br).

890 SUPERB STARLING *Spreo superbus*
B1076 / MPG1216 / D3C3

Afrotropical: an E and NE species, from E Tanzania to Kenya, N and E Uganda, S Sudan, Ethiopia and Somalia.

Kenyan: this common and gregarious starling is the most widespread *Spreo* in Kenya (Br; Ja; MPG; HH; La; St). It is found in a great diversity of open bushed and wooded habitats up to 2200 m (occasionally 3000 m: Britton 1979), from arid bushed grassland to open woodland, cultivation and urban gardens and centres *e.g.* Nairobi 75B (*op. cit.*; Be). There appears no reason why it should not be recorded for every square, with the exceptions of those in the moister areas of the W and the coast. Br notes absence from the coast S of Malindi 103A, and we have records from only one of the five most moist, 1000+ mm rainfall coastal squares. Complete absence of breeding records from even the well known areas of the coastal strip is notable. This bird is also scarce or absent in the humid, 1000+ mm rainfall areas of the W 36CD 48 60AD (*cf*. Br). Virtual absence of records from the E borders from 42CD S to 79AB probably represents lack of observers.

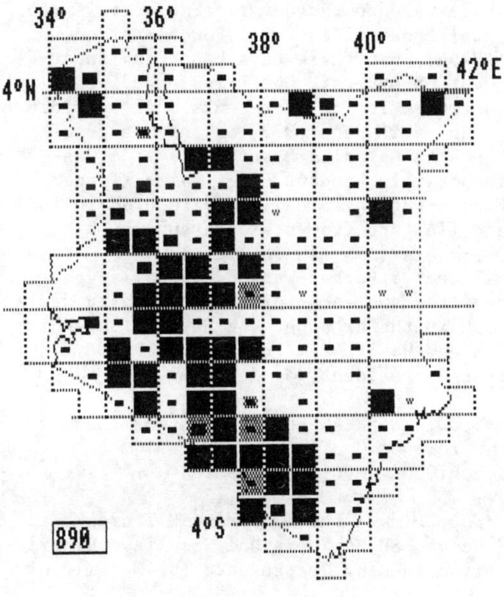

This species is generally considered resident (Br; Be: 74; T Stevenson *in litt.*: 50A; DEP pers. obs.: 74BD 75 88ABD 89A 100B 101A). In the arid country of Tsavo E 101B, La propose resident status, but Lack (1985) records an apparent June-Aug increase in numbers (*cf*. Blue-eared Glossy Starling (879)). J E Miskell's (*in litt.*) records from the Wajir area 42A are throughout the year (although not in every month), and E J Roberts (*in litt.*) recorded the bird at Rhamu 19A during every month of her May-Dec study.

In view of these comments, the restriction of breeding to squares that are 78% at 500-2000 m and 75% within 250-1000 mm rainfall areas probably reflects presence of observers, except in montane areas. This proposal is supported by St's data from the semiarid Baringo area 50A where, although most breeding is during the main Apr-Jul rains, it can also take place during other months if rains occur (*cf*. Brown & Britton 1980). This latter, opportunistic mode of breeding probably occurs in many areas of the arid N and E, when their low and irregular periods of rainfall render them temporarily verdant (and mostly inaccessible to observers: *cf*. Crested Lark (543)). Absence of breeding from the apparently suitable coastal strip, and from 91% of the harshest, most arid squares (moisture index below 1.2), may be real.

The Superb Starling forms a species-group with Fischer's Starling (891)(HM), and the two are widely sympatric (including breeding squares) in S and E Kenya. HM draw attention to this pair and the Golden-breasted (893)/Ashy (894) Starling superspecies, which both contain a very colourful member, and a drab one. The Superb Starling is the commonest *Spreo*, and is sympatric with four congeners at QSD scale in Kenya (HM; see Hildebrandt's Starling (889)). Other reference: Trevor & Lack 1976.

891 FISCHER'S STARLING *Spreo fischeri*
B1073 / MPG1211 / D2C2

Afrotropical: a NE species, from NE Tanzania and mainly E areas of Kenya to SE Ethiopia and S Somalia.

Kenyan: an often common and usually gregarious starling of dry, bushed country at 50-1200 m (Br; La). In contrast to the Superb Starling (890: *q.v.*), range is 80% below 1000 m, 94% arid-semiarid (*cf.* Br), and 98% to the E of the rift valley. It is probably mainly resident (Br; La: 101B; DEP pers. obs.: 88AD 89AC 100B 101A), and J E Miskell's (*in litt.*) records from the Wajir area 42A are throughout the year, although not in every month.

Wanderers reach atypical areas. Ja lists occurrences in the moist subhumid highlands at Naivasha 62C (1900 m), and near Nairobi 75B (Jul, at approximately 1500 m). The single record from the wettest, 1000+ mm rainfall parts of the coast involved a flock feeding on fruiting trees at Ukunda 114B, possibly in the Jul-Aug period (M A C Coverdale *in litt.*). Other reference: Miskell 1977.

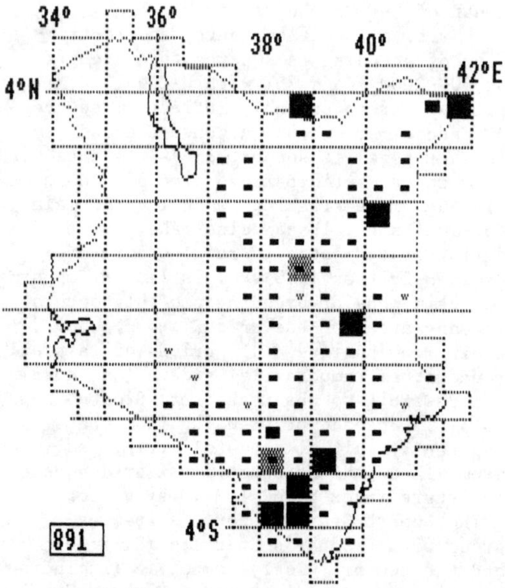

892 WHITE-CROWNED STARLING *Spreo albicapillus*
B1072 / MPG1212 / D3C2

Afrotropical: the far NE, from N Somalia to SE Ethiopia and N Kenya.

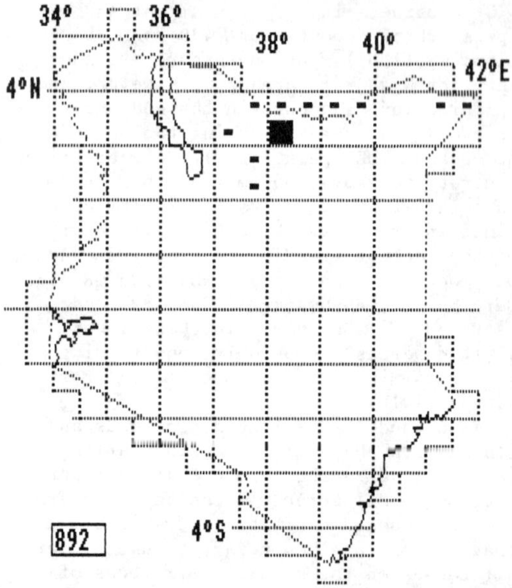

Kenyan: this extremely local starling frequents dry bush country, and the close vicinity of settlements and cultivation, at 300-1000 m in the N (Turner & Forbes-Watson 1976a; Squire 1977; Br; EABR 1982). Range is entirely in arid, 0-500 mm rainfall areas. The nominate race is found in the Mandera area 19AB (Turner & Forbes-Watson 1976a; E J Roberts *in litt.*), and extends S to similar latitudes in immediately adjacent parts of Somalia (AM).

The smaller *S.a. horrensis* occurs further W. First described by Keith (1964), there were no further records of this race until those of Turner & Forbes-Watson (1976a), but there have been many subsequent observers. Confirmed breeding of *horrensis* in 16C represents adults feeding young on 21 June 1975, on the N edge of the Dida Galgalla Desert: the area was temporarily verdant after rains (Turner & Forbes-Watson 1976a). The records in 16B and 17AB are from Ethiopia and may be *horrensis* (J S Ash *in litt.*).

Turner & Forbes-Watson (1976a) suggest that the two races are segregated by dense bush in the area between Moyale 17A, Rhamu 19A and Wajir 42A. Br considers the species resident, as do AM.

893 GOLDEN-BREASTED STARLING *Cosmopsarus regius*
B1050 / MPG1198 / D3C2

Afrotropical: the NE, from NE Tanzania to E Kenya, SE Ethiopia and Somalia. Distribution in Kenya represents the W limit of

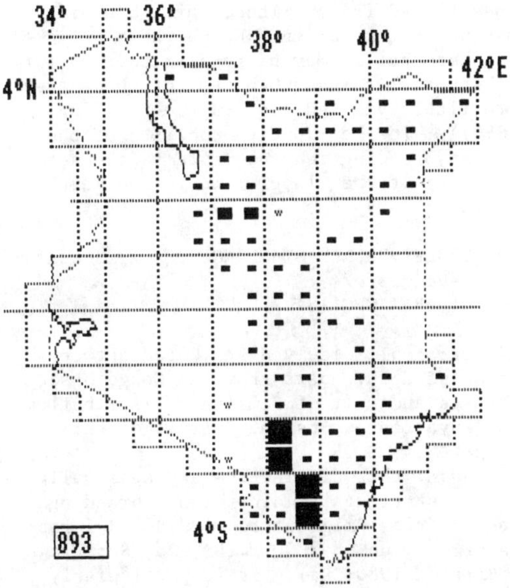

893

the range. Further to the S and SE in Tanzania, it is replaced by the closely related Ashy Starling (894: *q.v.*).

Kenyan: gregarious and locally common in dry, open, bushed and wooded country at 50-1200 m (Br; Ja; MPG; La). Range is confined to the lowlands to the E of the rift valley, apart from records in 4C (see below). Distribution is 84% below 1000 m and 94% arid-semiarid (*cf.* Br). The species is absent from all the dry subhumid, 1000+ mm rainfall areas of the coastal strip. Absence of records from the E borders between 2°N and 1°S agrees with AM's data from Somalia, but may partly reflect lack of observers, at least in the Kenyan areas. MPG mention breeding in N Kenya, but omit localities.

Br, La (101B) and DEP (pers. obs.: 89AC 100B 101A) consider the species resident, but Lack (1985) notes peak numbers during June-Aug in Tsavo E 101B. J R Ness (*in litt.*) reports rare occurrences at Ileret 4C during the rains.

The Golden-breasted Starling forms a superspecies with the allopatric Ashy Starling (HM; Br). Both have been reported from L Jipe 100D, but the records were in different years (HM), and that of the Ashy involved an extralimital wanderer. See Superb Starling (890). Other reference: Huels 1981.

894 ASHY STARLING *Cosmopsarus unicolor*
B1051 / MPG 1199 / D2C0
NO MAP: pre presence 100D

Afrotropical: endemic to Tanzania, with the single exception of an old record from the S Kenya border.

Kenyan: inhabits open bushed and wooded country, but in less arid areas than its close relative, the Golden-breasted Starling (893: *q.v.*)(Br; HM). The Ashy is locally common NE to approximately 3°30'S 36°E in NE Tanzania (Br), so that the single Kenyan record near L Jipe 100D (Ja) was only marginally E or NE of the normal range. The Jipe area is visited quite frequently and the lack of subsequent records of this distinctive species indicates that the early record was of an extralimital wanderer.

895 MAGPIE STARLING *Speculipastor bicolor*
B1071 / MPG1187 / D3C2

Afrotropical: a NE species, from the NE borders of Uganda and Tanzania to Kenya, S Ethiopia and Somalia.

Kenyan: gregarious and often numerous in the dry, bushed and wooded country of the N and E lowlands, mainly below 1200 m (Br; HH). Apparent absence from the area around 28 and 39 presumably represents a chance lack of observers. Distribution is 83% arid-semiarid (*cf.* Br). Apart from rare wanderers, the highlands and their peripheral moist plateaux form part of the SW limit of the species' Afrotropical range. Wanderers SW into the highlands have

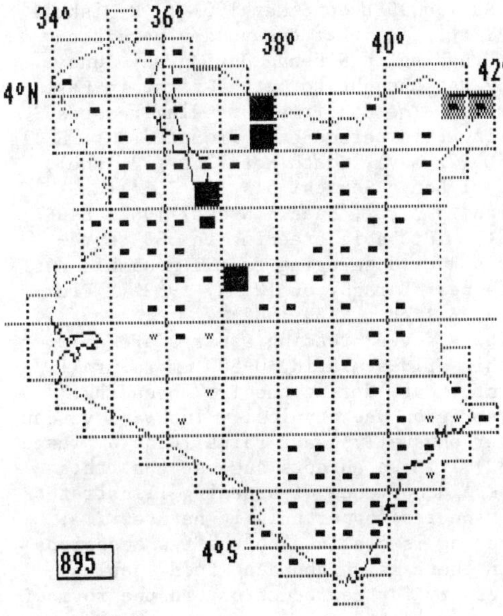

895

reached much higher and moister country *e.g.* near Nakuru 62A (Dec: EABR 1980) and Nairobi 75B, and as far SW as Magadi 75C (Ja).

This is a nomadic species, *i.e.* apparently having no regular migratory pattern (*cf.* AM, but also see below), that wanders erratically in search of food sources, notably ripe fruit (Ja; MPG; La; Haas & Nickel 1982; EABR 1979, 1980). However, despite this tendency for erratic and irregular occurrence, some pattern can be seen in the few dated records (HH (1973 edition); Br; St; EABR; J E Miskell, E J Roberts, D A Turner *in litt.*).

If Kenya is divided arbitrarily into regions N and S of the equator, then to the N there are (a few) breeding records and the species extends across the whole country. However, to the S, there are no breeding records and range is restricted to areas E of the highlands. To the N, St records varying numbers during all months around Baringo 50A, and there are also records throughout the year (although not in all months) from the region as a whole. To the S, there is a single record on 5 Apr (EABR 1981), but none for the remainder of the apparent Apr-June breeding season (see below). This periodicity suggests some S movement after breeding (*cf.* Archer & Godman's (1937, 1961) view of movements in Somalia, and the movements of Shelley's Starling (888)), and notable influxes S to the coastal lowlands have occurred in this off-season, during Sept-Oct (Br; EABR 1978) and Aug-Oct (EABR 1981).

Br and Brown & Britton (1980) overlooked Erlanger's (1905, 1907) record of breeding at Sarigo 19AB on 8 May 1901 (J E Miskell *in litt.*), so that the sudden occurrence of breeding in N Kenya during May-June 1982 was thought to be the first in East Africa. The confirmed breeding records during this period came from the Huri Hills 15BD (22 May), South Horr 26D (13 June), and Ol Donyo Sabachi 51A (11 June)(R G Allan *in litt.*; Haas & Nickel 1982; EABR 1981). Probable breeding in 38B represents a bird emerging from a possible nest hole near Baragoi on 12 May 1982 (M Pickford *in litt.*; EABR 1982).

These seven breeding squares are all within arid-semiarid, 0-500 mm rainfall country but, during the 1982 breeding, this region was atypically lush and verdant after unusually heavy rains (Haas & Nickel 1982). These authors suggest that this sudden occurrence of breeding illustrates the species' opportunistic nature, *i.e.* breeding is confined to the few occasions when there is an abundant food supply necessary for reproduction, in the form of insects and fruit. Thus, while breeding may not be annual in N Kenya, Haas & Nickel suggest that it may be regular there after substantial rains (when many of these areas are little visited, *cf.* Superb Starling (890)). The May-June seasonality of the Kenyan breeding records reflects the seasonality of the 'long' rains. See Wattled Starling (899).

896 VIOLET-BACKED STARLING *Cinnyricinclus leucogaster*

B1048 / MPG1184 / D3C2

Afrotropical: widely distributed between 15°N and 25°S, apart from the Congo basin forests and most of Somalia. Distribution in Kenya represents the E limit of the range.

Kenyan: this distinctive and seasonally common starling is mainly a non-breeding intra-African migrant to wooded and forested areas, up to 3000 m (Br; Ja; Se; Brown & Britton 1980; La). It is particularly attracted to fruiting trees, and small flocks scattered over an area can total hundreds of birds (*op. cit.*; EABR 1979). Range is 76% within 500+ mm rainfall areas and only 14% arid. Occurrence in arid country is mainly restricted to the vicinity of L Turkana (occasional: HH) and the Tana R 65AB 79C (AGH; K L Campbell, R & L M Greenham, D J Pearson *in litt.*). The bird is sporadic in arid Tsavo E 101BD (La; P C Lack *in litt.*).

Two races occur in Kenya, and one breeds at least intermittently. The N tropics nominate race is a non-breeding visitor to

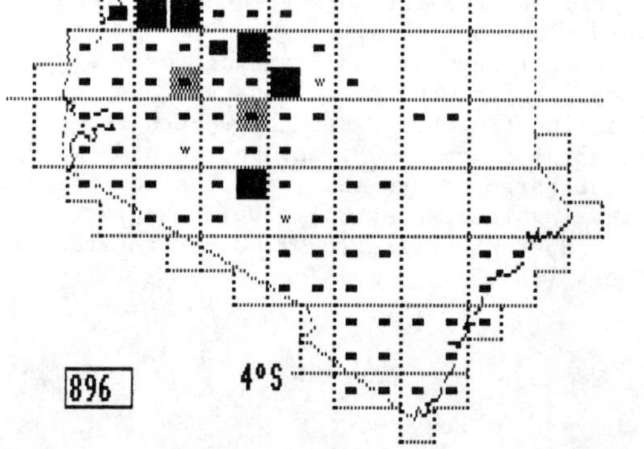

NW Kenya, E to Turkana and S at least to Nandi-Eldama Ravine 49CD (Br; Ja). It can appear during most months of the year (Br). The S tropics *C.l. verreauxi* extends to approximately 2°N, and is mainly a non-breeding migrant (Br). Breeding S of 8°S in interior Tanzania and adjacent countries is Sept-Mar, and large numbers in the coastal and adjacent lowlands of Kenya, mainly during Mar-Sept, may be from these S breeding populations (Br; MPG; La; AGH; K L Campbell *in litt.*; EABR 1979, 1980). Small flocks at Baringo 50A during May-Aug (annually) and Nov (occasionally)(St) have not been racially identified.

The few breeding records of *C.l. verreauxi* are all in the highlands and their peripheral plateaux. The breeding range is entirely above 1000 m, in 500+ mm rainfall areas. Breeding has taken place during Mar at Saiwa 37C (Br; Brown & Britton 1980) and Eldama Ravine 49D (Ja). Apparent breeding on W Laikipia 50A was in Apr-May (Short & Horne 1985b). A dependent juvenile at Nairobi 75B on 25 Sept had only limited powers of flight (H Gomez de Silva *in litt.*).

HM place Sharpe's (897) and Abbott's (898) Starlings, and possibly the Violet-backed, in a superspecies. Sharpe's and Abbott's both inhabit highland forest canopy and are considered allopatric (Br). Range overlap, including a record of the two flocking together in NE Tanzania (*cf.* HM), probably reflects wanderers (Br). At localities where both Sharpe's and the Violet-backed occur, HM consider the latter to be confined to woodland.

897 SHARPE'S STARLING *Cinnyricinclus sharpii*
B1049 / MPG1185 / D2C2

Afrotropical: a highland species of the E and NE, from S Tanzania through East Africa and the E Zaire borders, to S Sudan and S Ethiopia. Distribution in Kenya represents the E limit of the range.

Kenyan: gregarious but local and usually uncommon, in forest canopy at 1700-3000 m (Br; Ja; Se; Britton & Sugg 1973). Wanderers reach Kakamega Forest 48D (1700 m; EABR 1983) and Nairobi 75B (1650 m; EABR 1981)(Br). It is present on the forest islands of the Mathews Range 39C (EABR 1979), the Chyulu Hills 88D and the Taita Hills 101A (Br). Excluding the locally moist peaks of these isolated highlands, which are too small to feature on our overlays, range is 70% within 1000+ mm rainfall areas and entirely subhumid-humid.

Ja mentions local movements to fruiting trees but, at 3000 m around Mau Narok 61D,

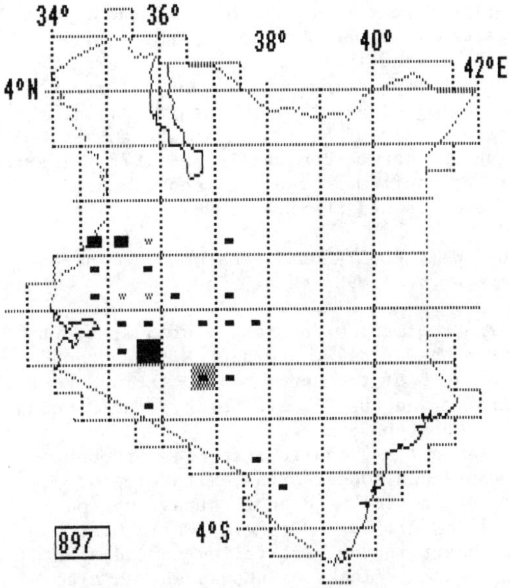

Se found it only a spasmodic visitor. Using Se's summary of this area's climate, this starling would arrive around Dec, in the second month of the Nov-Dec rains, remain through the Jan-Mar dry season, and then leave during the Apr-May rains (*cf.* Slender-billed Chestnut-winged Starling (886)). This may suggest movement to the top of the altitudinal range only for the longer of the dry seasons (Se mentions the heaviest rains in Jul-Aug). In some years, the birds did not appear at Mau Narok at all. See Violet-backed Starling (896).

898 ABBOTT'S STARLING *Cinnyricinclus femoralis*
B1047 / MPG1186 / D2C2
NO MAP: post presence 62B; post probable 63A; pre presence 62D 75B 88D

Afrotropical: endemic to highland areas of NE Tanzania, and of Kenya to the E of the rift valley.

Kenyan: local and little known, recorded from forest canopy at 1800-2500 m in the C highlands and the Chyulu Hills 88D (Br; Ja; van Someren 1939; Turner 1977; EABR 1978, 1980, 1981). It can be common only on Mt Kenya 63A, particularly on the S slopes (*op. cit.*). Excluding the Chyulu Hills forest island 88D, three of the other four squares are in humid, 1000+ mm rainfall areas. Records from 75B are in the higher, wetter parts of the square, around Limuru (Br).

van Someren (1939) found flocks of a hundred on fruiting trees in the Chyulu

Hills forest 88D, but there are no recent records. Since the tops of these hills are visited infrequently, van Someren's report may reflect occasional movement of foraging flocks from the nearby Kilimanjaro massif of NE Tanzania, where the species can be common (Turner 1977). There is no confirmed breeding record. See Violet-backed Starling (896).

899 WATTLED STARLING *Creatophora cinerea*
B1052 / MPG1182 / D3C3

Afrotropical: widespread S of 5°S, apart from the arid SW. Further N, it extends to the E of the Congo basin, through East Africa and the E Zaire borders, to S Sudan, Ethiopia and Somalia.

Kenyan: a gregarious species of open bushed and wooded areas, with game or domestic stock and short grass, up to 3000 m (Br; Ja; MPG; Se). While most abundant in the rift valley and adjacent highlands, flocks of thousands can also appear in almost any part of the country (Br; HH; La).

A marked Oct-Nov influx to the coastal strip coincided with a similar invasion of Magpie Starlings (895) and the only coastal record of Shelley's Starling (888) (EABR 1978). Thousands of Wattled Starlings flew SE from Kilifi Creek 102D in the early morning of 12 Nov 1978 (EABR 1978). Birds have appeared during the night at Ngulia Lodge 101A in Nov and Dec (Br; EABR 1978), but there is no direct evidence that they were attracted to the lights while migrating (Backhurst & Pearson 1977b; D J Pearson pers. comm.). Small, transient numbers at the top of the altitudinal range around Mau Narok 61D are mostly Jul-Dec (Se).

Br suggests that this starling may never be resident in one locality and, while present throughout the year in the semi-arid country around Baringo 50A, St notes fluctuations in abundance. These nomadic movements are probably stimulated by occurrence of abundant insect prey (Br; Brown & Britton 1980; Ja; MPG). Annual non-breeding flocks in the L Victoria basin 48CD 60 (June-Oct) and at Nairobi 75B (Jul-Oct) exhibit a similar seasonality to Se's reports, and may illustrate some more regular migrational pattern (Br).

Breeding squares are with one exception in or E of the rift valley: this breeding range is 76% at 500-2000 m, 76% within the intermediate, 250-1000 mm rainfall regime and 72% arid-semiarid (*cf.* Br). Particularly in the drier areas, breeding and the appearance of wattled males only occur in the temporarily lush periods following heavy rains (Br; Brown & Britton 1980; Lewis 1981e). In arid Tsavo E 101B, La found presence in general to be mostly during the rains.

900 YELLOW-BILLED OXPECKER *Buphagus africanus*
B1077 / MPG1217 / D2C2

Afrotropical: widely but in some areas patchily distributed S of 15°N, E to W Ethiopia, Kenya, E Tanzania, Mozambique and NE South Africa. It is apparently

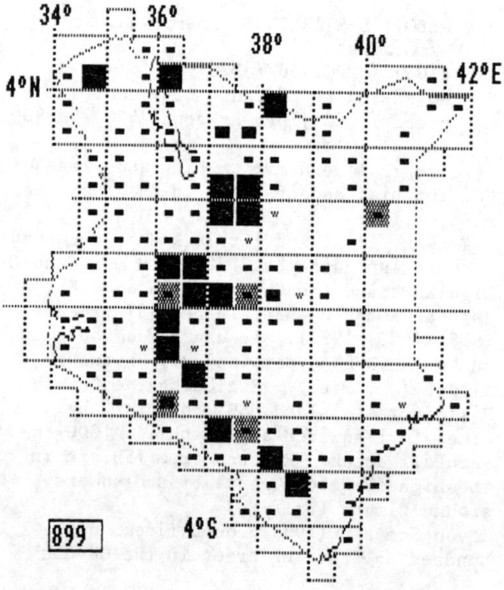

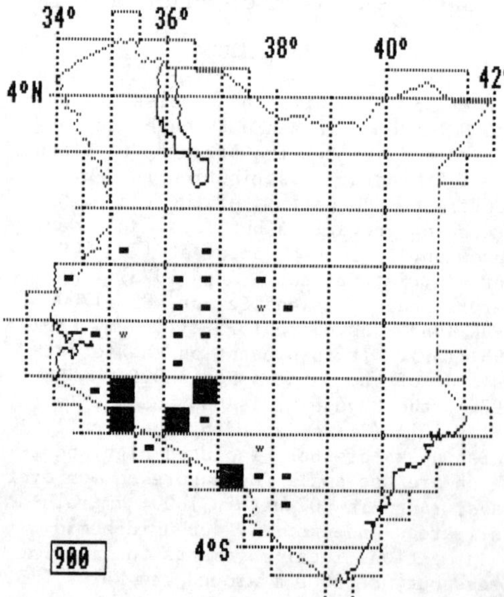

absent from most of the equatorial forest regions, most of the Congo basin and adjacent areas to the N and S, and the dry SW. Distribution in Kenya represents the E limit of the range.

Kenyan: the two oxpeckers are specialised feeders on the ticks Ixodidae and other small ectoparasites of large mammals, both wild and domestic (MPG; HM). They are adapted for this mode of life by having strong, sharp claws to cling to the animals' hides, stiff tail feathers which act as props (as in woodpeckers Picidae) on their hosts' flanks, legs, *etc*., and bulbous bills to facilitate the grasping of prey. They can be harmful in that they enlarge the wounds made by the ectoparasites, and feed on the host animal's blood and tissues (*op. cit*.; Ja; Brown & Britton 1980). They leave areas where there is no game and stock are regularly dipped.

Contrary to Ja's assessment of the Yellow-billed, both species can be common in open habitats, including forest clearings, up to 3000 m (Br). In marked contrast to the Red-billed Oxpecker (901), however, the range of the Yellow-billed is 85% above 1000 m, 78% within the 500+ mm rainfall regime and 81% semiarid-subhumid. The restriction of breeding records to SW areas S of 1°S is of unknown significance. A bird marginally N of the local range edge at semiarid Baringo 50A on 6 Aug (EABR 1983) was presumably a wanderer.

The two oxpeckers form a species-group (HM) and are widely sympatric in Kenya, where the Red-billed is by far the most widely distributed and numerous (Br). They can occur on the same animal, *e.g*. in Nairobi NP 75B (Br; many obs.).

901 RED-BILLED OXPECKER *Buphagus erythrorhynchus*
B1078 / MPG1218 / D2C3

Afrotropical: mainly the E, from E South Africa, N through S Mozambique, Botswana and Zambia, to East Africa, S Sudan, Ethiopia and Somalia. It reaches the W on the borders of Angola and Namibia.

Kenyan: in contrast to the Yellow-billed Oxpecker (900: *q.v*.), this relatively common species could occur in every map square (Br; Se; HH; La; St). Apparent absence from the region around 65B probably reflects lack of observers. Both oxpeckers nest in tree holes, and the Red-billed's breeding squares are 70% within 500+ mm rainfall areas and only 22% arid. 91% of these squares are above 500 m, and there are no breeding records from the apparently suitable coastal lowlands.

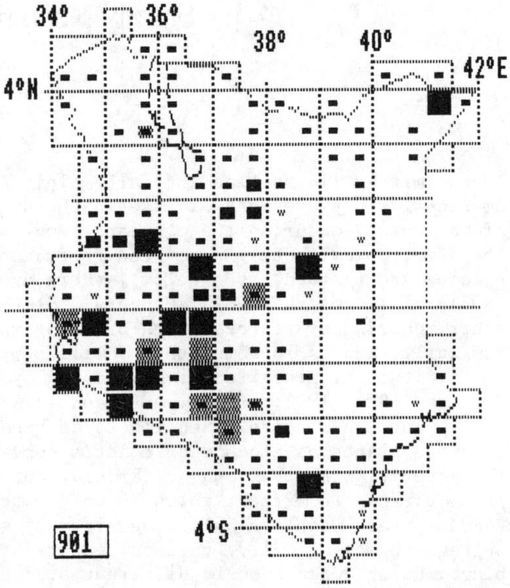

The significance of this apparent breeding distribution is unclear.

The species is generally considered resident (Ja; Be: 74; Se: 61D; La: 101B; DEP pers. obs.: 74BD 75CD 88AD 89AC 100B 101A), and was present around Rhamu 19A (in the vicinity of the Daua riverine woodland) throughout E J Roberts' (*in litt*.) May-Dec study. The breeding distribution could reflect frequency of observers and/or the need for relatively substantial nesting trees. In the latter case, birds in the driest areas would be non-breeding wanderers/migrants. Some movement must occur if the dipping of stock and extermination of large wild animals removes the birds' source of food.

Family Nectariniidae: Sunbirds

Approximately 118 species that are mainly Afrotropical and Oriental, and absent from the Americas. They are most diverse in the Afrotropical Region, where approximately 70 (59% of the total) occur on the mainland, and another four on offshore islands. 35 (50% of the Afrotropical total) have been recorded in Kenya. Despite Br's comment that most species are resident (see below), there are no Kenyan breeding records for the Pygmy (903), Violet-backed (907), Uluguru Violet-backed (908) or Superb (915) Sunbirds, and their records may represent non-breeding wanderers at the extreme limit of their range. Similarly, and as in some other families with members that breed in the highlands, regular altitudinal migration is stimulated by the onset of the wet seasons (Br; see also Britton & Sugg 1973; *cf.* Olive Pigeon (325)). Seasonal movements to flowers occur regularly, and can involve mixed flocks of hundreds of individuals at such particularly favoured plants as Aloes and *Leonotis* spp. (Br). HM group all Afrotropical species into two genera, *Anthreptes* and *Nectarinia*, and note two features of the prolific speciation of *Nectarinia* in Africa, which is only matched by that of the cisticolas (716-742). Whereas these warblers Sylviidae are very similar in appearance to each other but strongly segregated ecologically, various members of *Nectarinia* can often be found on the same plant, being conspicuously differentiated by striking colour, plumage pattern and tail-length characteristics. Secondly, the restriction of the long-tailed species to highland or other non-extensive or arid biomes contrasts with the more widespread and numerous short-tailed species. This suggests the latter to be more successful, while the ranges of the long-tailed species may be relict, *i.e.* the result of unsuccessful competition with the short-tailed members of the family. Other reference: Britton & Britton 1977a.

902 COLLARED SUNBIRD *Anthreptes collaris*
B1080 / MPG1271 / D2C3

Afrotropical: widespread S of 5°N, including the S edge of the Ethiopian highlands, but absent from the dry SW. It reaches 10°N to the N of the Gulf of Guinea. The Pygmy Sunbird (903) replaces it in the N tropics.

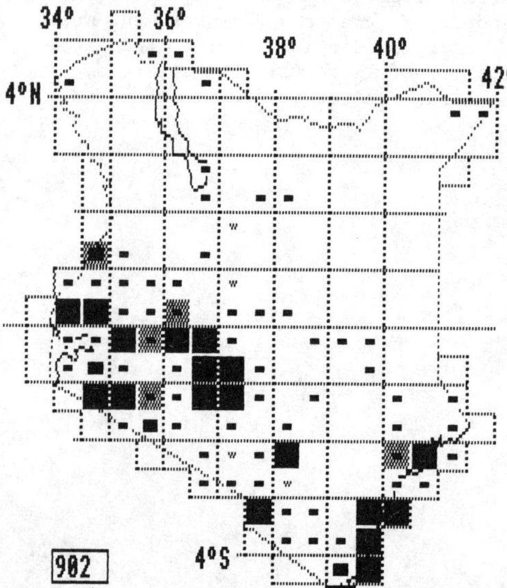

Kenyan: common in a wide variety of bushed, wooded and forested habitats, up to 2800 m (Br; Ja; MPG; Britton & Sugg 1973; Britton & Zimmerman 1979). Range in arid country (23%) is mainly confined to three types of locally moist environment as follows.

The records from the Ethiopian border are either around the Omo R, L Turkana and the S periphery of the Ethiopian highlands 3B 4AD (J S Ash *in litt.*), or the Daua R 19AB (E J Roberts, ADL pers. obs.). The northernmost record from Somalia is in the vicinity of the Juba R immediately NE of 19B (AM). Secondly, this sunbird also occurs on the locally moist peaks of forest islands *e.g.* Mts Kulal 26B (Diamond & Keith 1980), Marsabit 27D 28C (Ja) and Endau 77B (Tennent 1964; A W Diamond *in litt.*), and the Ndoto Mts 39A (Br). Lastly, it is found along rivers, notably the Tana 64B-79C (AGH). Uncommon but resident occurrence in arid Tsavo E 101B is also riverine (La). Excluding these locally moist environments, range in Kenya is 81% within 500+ mm rainfall areas. Restriction in S Somalia to the vicinity of the Juba or Shebelle R or the coast (AM) directly compares to distribution in Kenya E of 39°E.

The Collared Sunbird forms part of a superspecies with the Pygmy Sunbird which, in areas of apparent sympatry, is usually found in drier habitats (HM; Br). See Green Sunbird (904). Other references: Cheke 1971ab.

903 PYGMY SUNBIRD *Anthreptes platurus*
B1086 / MPG1236 / D2C0
NO MAP: post presence 24B 38C 50A; pre presence 13D

Afrotropical: the N tropics, from Senegambia and Mauritania E to N areas of Ethiopia and Somalia, and SE to NE Uganda, SE Sudan and, on its SE limits, NW Kenya.

Kenyan: dry, open, bushed and wooded country of the NW (Br). Ja mentions it at Turkana, and there are specimens taken near Lodwar 13D during Sept 1953 in the National Museum, Nairobi. The only record during recent decades that Br could find from Kenya or E Uganda was in the Ugandan sector of 36D (not mapped). There have also been very occasional Nov records from Kapedo 38C (1979: EABR 1983) and Baringo 50A (St), and a June report from the Uganda/Kenya border E of Moroto 24B (EABR 1983). Of these four squares, three are at 1000-1500 m in 500+ mm rainfall areas. In contrast to the Collared Sunbird (902: *q.v.*), all four squares are arid-semiarid: MPG note that the Pygmy is not attracted to the vicinity of water.

This species is described as a wanderer or migrant by Br and MPG, and is a non-breeding migrant to N Somalia (AM). Although perhaps more common in the remote NW of Kenya than these few records suggest, it it probably still only a non-breeding migrant or wanderer there.

904 GREEN SUNBIRD *Anthreptes rectirostris*
B1087 / MPG1272 / D1C2
NO MAP: post presence 49C 61A; post probable 48D; pre presence 36D 48C 61C

Afrotropical: W and C Africa, from Guinea and Angola, E through the Congo basin to Uganda and, at the E limit of the range, W Kenya.

Kenyan: locally common in forest canopy and edge, up to 2150 m (Br; Ja; MPG; ADL pers. obs.). The six squares are above 1000 m in humid, 1000+ mm rainfall country to the W of the rift valley. There is no confirmed breeding record. This sunbird is closely related to the Collared Sunbird (902) superspecies (HM). Other reference: EABR 1981.

905 PLAIN-BACKED SUNBIRD *Anthreptes reichenowi*
B1088 / MPG1279 / D2C3
NO MAP: post presence 91A 102D 103A 114ABC; post confirmed 102B

Afrotropical: two separate populations in the E: in the coastal lowlands of Kenya and NE Tanzania, and in Mozambique, Zimbabwe and the NE borders of South Africa.

Kenyan: a rather uncommon bird of forest, including remnants and secondary growth, below 500 m in the coastal lowlands (Br; MPG; Britton & Zimmerman 1979). Range extends inland to the lower Tana R 91A (AGH) and the Shimba Hills 114A, and is entirely within semiarid-dry subhumid, 500+ mm rainfall areas.

906 EASTERN VIOLET-BACKED SUNBIRD *Anthreptes orientalis*
B1084 / MPG1274 / D2C3

Afrotropical: a NE species, from E Tanzania, Kenya and NE Uganda to S Sudan, Ethiopia and all but the far NE of Somalia.

Kenyan: dry, bushed and wooded habitats, including open woodland, at 50-1300 m (Br; Ja; La). It is usually in the vicinity of water, and hence only locally common (*op. cit.*; HH; St). Range is 87% arid-semiarid (*cf.* Br), and virtually absent from the highlands and moist W, which define its SW limit. It is also absent from much of the coastal strip, including all of the dry subhumid squares. Allopatric congeners replace it in these moist highland and coastal areas (see below). Apparent

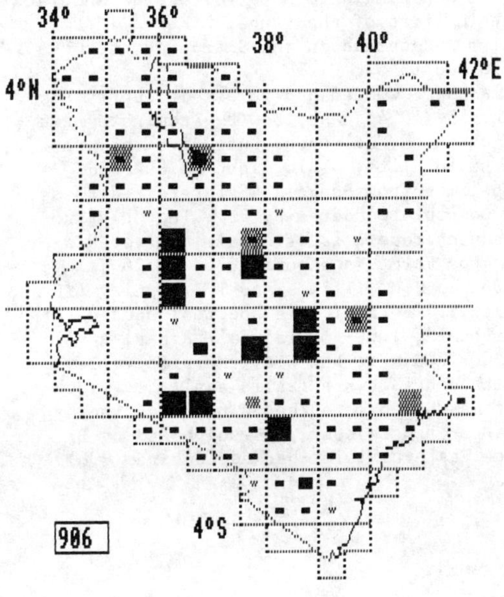

absence from the very arid country around 29 may represent lack of observers. This sunbird is resident in Tsavo E 101B (La) and in 75C 88A 89A (DEP pers. obs.). Breeding occurs throughout the year in the semiarid country around L Baringo 50A, except during exceptionally dry weather (St).

The Eastern Violet-backed Sunbird forms part of a superspecies with the allopatric Violet-backed (907) and Uluguru Violet-backed (908) Sunbirds (HM). Both of these other species favour wooded and forested habitats, whereas the Eastern's range is only 64% well timbered. The Violet-backed Sunbird occurs in the moist W, and both it and the Eastern have been recorded in 37C. This square is on the N edge of the W highlands and, while the Violet-backed record was in high, moist country near Kapenguria (T J Barnley *in litt.*), the Eastern is in lower, drier habitats. Br notes that these two species are unlikely to breed on the same ground, despite reports of sympatry. The Uluguru Violet-backed Sunbird occurs in forests on the S coast 114AB, where the Eastern is absent. Both species have been recorded along the lower Tana R 79C, but the Eastern is in wooded grassland (AGH) whereas the Uluguru frequents the riverine forest.

907 VIOLET-BACKED SUNBIRD *Anthreptes longuemarei*
B1082 / MPG1274 / D1C0
NO MAP: pre presence 37C 48B

Afrotropical: to the N of the equatorial lowland forests, this sunbird extends E from Senegambia to N Zaire, Uganda and, at the E limit of the range, W Kenya. It is also widespread in the S tropics, between 5°S and 15°S.

Kenyan: wooded and forested country (Br). Chapin (1954) described the nominate race from North Kavirondo, but omits the precise locality. This might be the old specimen from Bungoma 48B, referred to as a possible wanderer by Br. The only subsequent report is of a male collected near Kaibos Farm, Kapenguria 37C, on 26 Jul 1966 by J G Williams and T J Barnley (*in litt.*). Both of this species' squares are in humid, 1000+ mm rainfall areas, at 1500-2000 m to the W of the rift valley. Although this species is easily overlooked, it may be no more than a non-breeding wanderer W to Kenya, as suggested by Br. See Eastern Violet-backed Sunbird (906).

908 ULUGURU VIOLET-BACKED SUNBIRD *Anthreptes neglectus*
B1083 / MPG1275 / D1C1
NO MAP: post presence 114AB; pre presence 79C

Afrotropical: the E, from coastal Mozambique, through E Tanzania, to the coastal lowlands of Kenya.

Kenyan: forest and forest edge (Br). The first Kenya record was on 12 Dec 1962 at Makeri 79C on the lower Tana R (Keith 1968); there have been subsequent reports from the Shimba Hills 114A and Jadini Forest 114B (Br; Squire 1976a). These three squares are all below 500 m, in 500+ mm rainfall areas of the coastal lowlands. The species is quite common in the East Usambara Mts of adjacent NE Tanzania (Br), but its status in Kenya is uncertain. It may well be overlooked. See Eastern Violet-backed Sunbird (906).

909 AMANI SUNBIRD *Anthreptes pallidigaster*
B1085 / MPG1278 / D2C3
NO MAP: post presence 103A; post confirmed 102B

Afrotropical: endemic to the Sokoke Forest 102B 103A of coastal Kenya, and to forests on the East Usambara Mts of extreme NE Tanzania. Total range is approximately 150 km² (Britton & Britton 1978b).

Kenyan: a reasonably common but extremely local species, confined to the canopy of *Brachystegia* woodland in Sokoke Forest 102B 103A (Turner 1977; Britton & Britton 1978b; Britton & Zimmerman 1979; Br; Britton *et al.* 1984; EABR 1977, 1983). Birds in adjacent forest habitats (Britton & Zimmerman 1979) are considered local wanderers (Br). The population in the 70 km² of suitable habitat at Sokoke is estimated at 2900-4700 pairs (Britton & Britton 1978b). The two squares are semiarid-dry subhumid, and within the 500+ mm rainfall regime. The species could soon be endangered by forest and woodland destruction (Turner 1977; EABR 1983).

910 VARIABLE SUNBIRD *Nectarinia venusta*
B1128 / MPG1251 / D2C3

Afrotropical: widespread between 10°N and 17°S and extending to 15°N on the W coast and in Ethiopia, but absent from much of the Congo basin and adjacent areas to the N.

Kenyan: common in a wide range of bushed, wooded and forest edge habitats, up to 3000 m (Br; Britton 1979; MPG; Britton & Sugg 1973). Distribution is throughout

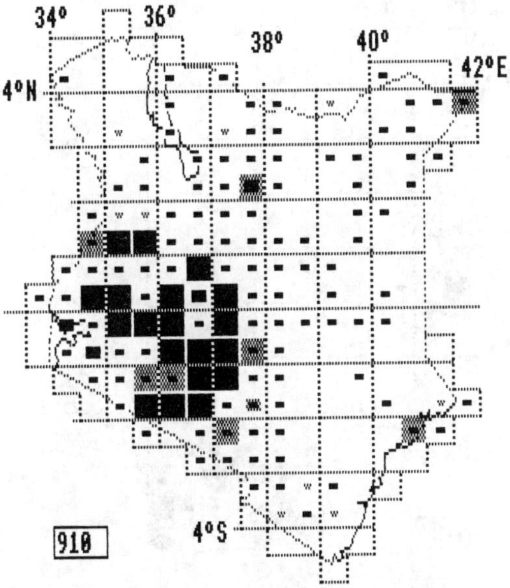

our altitude, rainfall and moisture categories, and there is no obvious reason why it should not occur in every square (but see below). Apparent absence from the E borders and the far NW probably represents paucity of observers, although HH do not list localities W of L Turkana. Absence from the SE interior around 90 may reflect lack of observers, but there are no post-1969 records from the well-known Tsavo E 101B area (La).

Cheke (1976) describes breeding on the coast S of Mombasa 114B and various other observers have submitted records from the coast S of 3°S. Britton & Britton (1977a and *in litt.*), who were resident on the coast for some years, reject all these reports and regard the species as absent from the coastal strip S of Manda Island 91B. We agree with this view. This coastal absence extends S to Dar es Salaam at 6°48'S in Tanzania, where there are occasional records (Br). The species is thus absent from all of the dry subhumid, 1000+ mm rainfall parts of Kenya's coast.

The distinctive white-bellied race *N.v. albiventris* occurs in the NE and E, from Moyale 17A and Mandera 19B S through areas E of 52A and the Tana R 65A-79C, to the coast at Manda Island 91B (Ja; Br; Britton & Britton 1977a; P L Britton, J E Miskell *in litt.*). There are possible old records from Tsavo 89C 101 and Taru 102C (*op. cit.*). Yellow-bellied races occur further W, with intergrades around L Turkana and at Mt Marsabit 27D 28C (HM; Br).

Be described the Variable Sunbird as resident in Narok district 74, DEP (pers. obs.) concurs for 74D 75BD 88A 89A, and Br considers it probably generally so. The fact that its breeding squares are 85% above 1000 m, 67% subhumid-humid and 86% within 500+ mm rainfall areas probably therefore reflects observer cover. Breeding records away from the highlands are from the forest island of Mt Marsabit 27D, Karo-Lola 19B (May: Erlanger 1905, 1907) and Manda Island 91B (May: Ja). P L Britton (*in litt.*) found a nest, apparently of this species, on Manda, and considered the bird resident there. This status may well also apply to Kenya's arid N and E, with breeding during the rains (as at Karo-Lola and Manda), when these areas are at their most inaccessible (*cf.* Crested Lark (543)). Another possibility is that birds move into arid areas during the rains, so that they are resident only in moist or locally moist areas. Some support for this proposition comes from the data of St, who records small, annual numbers after rains in the semarid country around L Baringo 50A. Other references: Williams 1978b; Cunningham-van Someren 1984a.

911 ORANGE-TUFTED SUNBIRD *Nectarinia bouvieri*
B1093 / MPG1247 / D1C2
NO MAP: post presence 48D; pre probable 48D

Afrotropical: a restricted distribution E from Cameroon along the N edge of the Congo basin to Uganda and, at the E limit of the range, W Kenya. To the S of this basin, it also occurs in Congo, Zaire and Angola.

Kenyan: a generally uncommon bird of forest edge and remnants, known only from the Kakamega Forest 48D (Br). It was first recorded there in Dec 1965, and probable breeding is based upon enlarged ovaries in some of the specimens (Ripley & Bond 1971). There have been frequent subsequent records from this locality (Mann 1976), which is at 1700 m in a humid, 1000+ mm rainfall square.

This species forms part of a superspecies with the allopatric Shining Sunbird (912) (HM), which is in the much drier bush country of N Kenya.

912 SHINING SUNBIRD *Nectarinia habessinica*
B1100 / MPG1240 / D2C2

Afrotropical: the NE, from N areas of Ethiopia and Somalia S to Kenya and 3°30'N in Somalia (AM). It probably reaches NE Uganda, but there is no certain record

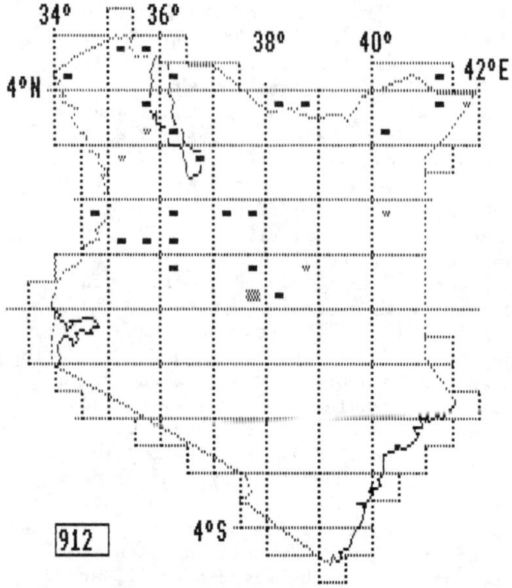

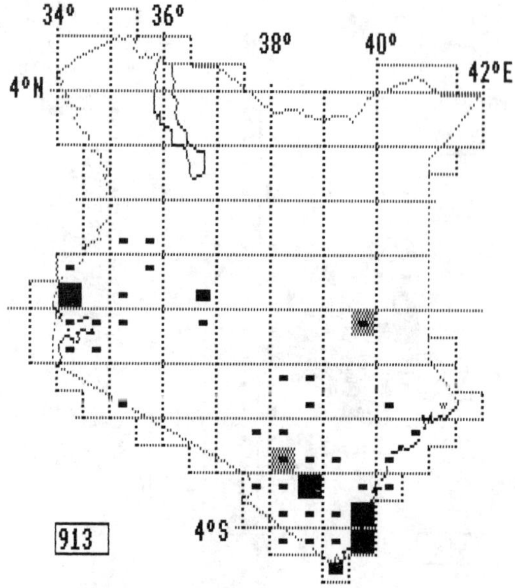

(see Mann 1976).

Kenyan: local and uncommon in bush on rough rocky or sandy ground, at 400-1000 m (Br; HH; EABR). It occurs in the N, SW to the N edges of the highlands at the base of the Kongelai escarpment 37C (EABR 1979), the Kito Pass 37D (ADL pers. obs.), Baringo 50A (St), Isiolo 51D (Tomlinson 1946) and Meru NP 52C (Br). This range is 84% arid-semiarid (*cf.* Br) and 63% within 0-500 mm rainfall areas.

Tomlinson's (1946) note of nest building at Isiolo 51D in Dec 1940 may be the source of MPG's mention of Dec breeding in N Kenya. There are no other breeding records. Dates for localities near the S range edge are throughout the year (St; EABR), and the species is resident in the Kapedo area 38C (EABR 1981, 1983, 1984). See Orange-tufted Sunbird (911).

913 LITTLE PURPLE-BANDED SUNBIRD *Nectarinia bifasciata*
B1092 / MPG1241, 1242 / D2C2

Afrotropical: a S species, widespread between 5°S and 15°S, and extending down the E coast to 30°S. To the E of the Congo basin, it ranges N to Uganda, Kenya and S Somalia.

Kenyan: locally common in a wide variety of well vegetated habitats including bush, thicket, woodland, mangrove and forest edge, mainly S of the equator and below 1500 m (Br; La). It is most frequent in the coastal lowlands, the SE plateau and the L Victoria basin, but has a peculiarly fragmented distribution elsewhere (*op. cit.*; Ja). Range is 82% below 1500 m, 75% within 500+ mm rainfall areas and only 21% arid.

La consider it resident, but Br notes occasional local wanderers to the apparently unsuitable habitat of Kisite Island 114C. It makes considerable movements to temporary food sources (MPG; *e.g.* Ja).

This sunbird forms part of a superspecies with the Violet-breasted Sunbird (914)(HM). Their ranges overlap, but the Violet-breasted is in generally more arid habitats and sympatry is probably more apparent than real (Br).

914 VIOLET-BREASTED SUNBIRD *Nectarinia pembae*
B1114 / MPG1243, 1244 / D2C2

Afrotropical: a restricted distribution in the E and NE, from Pemba Island in NE Tanzania to SE Kenya, S Somalia, and the Kenya/Ethiopia border.

Kenyan: bushed and wooded country, up to approximately 1000 m (Br; Britton & Zimmerman 1979). Apart from a curiously isolated record at Moyale 17A (Ja), range is confined to the coastal and SE lowlands. It is locally common in the coastal strip between Malindi 103A and Kiunga 80C, and in the vicinity of the lower Tana R 78B 79A, but scarce elsewhere (Br; EABR 1978, 1982, 1984). Range is 90% below 500 m and 85% arid-semiarid (*cf.* Little Purple-banded Sunbird (913: *q.v.*)). Br describes it as mainly in low rainfall areas, but there

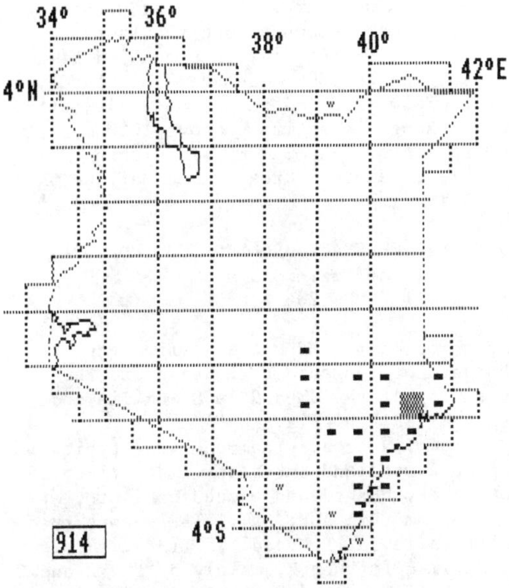

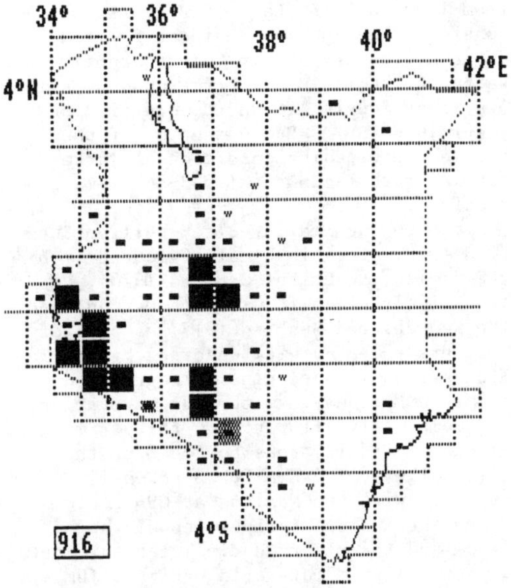

are records from 1000+ mm areas of the coastal strip and none from 0-250 mm squares. The single breeding record is from Ijara 79D (Percy *et al.* 1953). Other reference: Williams 1968.

915 SUPERB SUNBIRD *Nectarinia superba*
B1125 / MPG1237 / D2C0
NO MAP: post presence 48D; pre presence 48C

Afrotropical: W and C Africa, from the lowland forests around the Gulf of Guinea, E through the Congo basin to Uganda and, at the E limit of the range, W Kenya.

Kenyan: in forest canopy and edge, and other adjacent habitats (Br). There is a specimen from Mumias 48C (van Someren 1916) and recent sight records from Kakamega Forest 48D (Br). It is mentioned at Kakamega by Williams (1967), but not by Williams (1968). These two localities are at 1300-1700 m, W of the rift valley and N of the Winam Gulf 60AB. Both squares are in humid, 1000+ mm rainfall country. There is no breeding record, and the few reports (none during the past seven years: EABR) may concern non-breeding wanderers to the E edge of their range.

916 MARIQUA SUNBIRD *Nectarinia mariquensis*
B1107 / MPG1245 / D2C2

Afrotropical: a NE population ranges S from Ethiopia and N Somalia, through East Africa, to S Tanzania. Another population is widespread between 15°S and 25°S. Distribution in Kenya represents the E limit of the range.

Kenyan: locally common in diverse open bushed and wooded habitats (Br; Ja; MPG; AGH). Br gives the altitudinal range as 800-2000 m, with most records at 1000-1500 m (*cf.* Be), but is uncertain of status in the far SE, where there are old records from 400-500 m in Tsavo E 101B (La), and recent reports from below 100 m along the lower Tana R 79C 91A (AGH). Range is throughout our altitude categories, but 84% at 1000-2000 m, with a distinct scarcity of records from the main highland areas 61-63. Br describes the species as most typically in intermediate rainfall areas but, while there are reports from only 5% of the driest, 0-250 mm squares, range is 26% within the wettest, 1000+ mm regime.

Distribution is only 16% arid, virtually entirely in four types of situation. The records from the Lapurr Mts 3D (MacInnes 1933-34), Moyale 17A (ADL pers. obs.) and Takaba 18C (E J Roberts *in litt.*) represent the S limit of the population centred in Ethiopia. The Moyale record was in Jul after rains, whereas the species was not seen there in drier conditions during the previous Mar, and the Takaba record was late June (*op. cit.*). Secondly, the Mariqua is also found in the vicinity of the isolated highlands of Mts Kulal 26B (Diamond & Keith 1980), Nyiru 26D (Ja; Br) and Marsabit 27D (van Someren 1929), and the Ndoto Mts 39A (Br). Thirdly, it has been noted near water at Merti 40D (H Isack *in litt.*), and the Tana R 79C (AGH). Finally, there are the pre-1970 records from Tsavo E 101B (La). DEP (pers. obs.) considers the species resident in 74BD 75D 89A. Occurrence in the apparently suitable, semiarid country around L Baringo 50A is curiously limited to one bird on 1 Apr 1979 (St). MPG's mention of the species in breeding condition at Turkana may refer to MacInnes' (1933-34) female specimen.

Although the Smaller Black-bellied Sunbird (918: *q.v.*) is usually grouped with the similar Beautiful Sunbird (920), HM tentatively place it with the Mariqua and the Red-chested Sunbird (917) as part of a superspecies, a view endorsed by Br. The Red-chested is ecologically segregated from the Mariqua in waterside vegetation, and completely allopatric with the Smaller Black-bellied in areas to the W of the rift valley. The Smaller Black-bellied overlaps with the Mariqua at QSD scale along the E and S peripheries of the highlands and their surrounding plateaux, but is generally in more arid habitats (HM).

917 RED-CHESTED SUNBIRD *Nectarinia erythroceria*
B1098 / MPG1232 / D2C2

Afrotropical: the E, in S Sudan, the E Zaire borders, NW Tanzania, Uganda and, at the E limit of the range, W Kenya.

Kenyan: locally common in waterside habitats to the W of the rift valley (Br; Britton 1978; Ja; MPG). Typically associated with low vegetation cover, it has also adapted to more open gardens and areas with scattered trees near water (*op. cit.*). Most common around the shores and islands of L Victoria, this sunbird also extends up to 1800 m in the W highlands, NE to Saiwa 37C and SE to Loita 74D (Br; Ja). Range is entirely above 1000 m, 91% moist subhumid-humid and 91% within 1000+ mm rainfall areas. See Mariqua Sunbird (916).

918 SMALLER BLACK-BELLIED SUNBIRD *Nectarinia nectarinioides*
B1111 / MPG1234 / D2C2

Afrotropical: a NE species, from the NE borders of Tanzania to E Kenya and the vicinity of the Juba R in S Somalia and extreme SE Ethiopia.

Kenyan: a locally common bird, typically of open woodland but also in a variety of other dry bushed and wooded habitats, at 50-1300 m (Br; AGH; La). It reaches the rift valley at L Magadi 75C (Br), but is otherwise further E, mainly on the E and S peripheries of the C highlands or near permanent water in the E lowlands. It has been recorded along the Tana R 64B-91A (Ja; AGH; Br), and recent reports from the woodland along the Daua R 7C 19AB represent an extension of known range in Kenya (EABR 1984). E J Roberts (*in litt.*) noted it along the Daua in all months but May and Jul of her May-Dec survey. These records correlate with observations along the upper Juba R of S Somalia (of which the Daua is a tributary), from the QSD NE of 19B S to 1°30'N (AM).

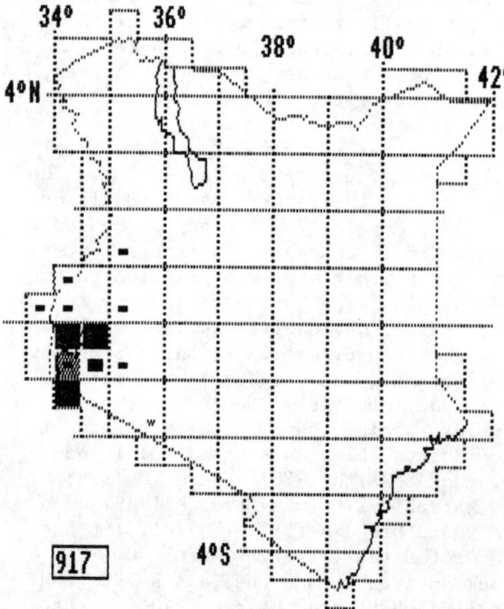

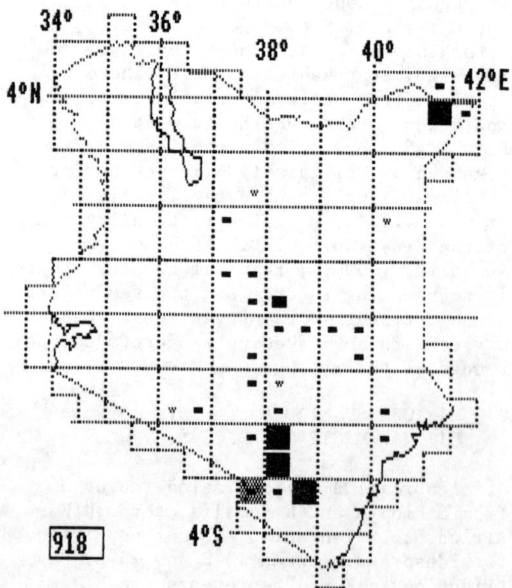

In contrast to the Mariqua Sunbird (916: *q.v.*), range is 95% arid-semiarid. The Smaller Black-bellied's range in arid country is in the vicinity of locally moist environments *e.g.* the major rivers mentioned above, Mt Marsabit 27D, the Ndoto Mts 39A, and waterholes at Wajir 42A. However, even if these arid squares are excluded from the analysis, the range of this species is distinctly drier than that of the Mariqua, being 55% semiarid (Br) as against 38%. The Smaller Black-bellied was formerly common in the woodland of Tsavo E 101B, but after that habitat's virtual elimination by fires and elephant *Loxodonta africana*, La could list only four Nov-Dec records, all of single birds. See Beautiful Sunbird (920).

919 OLIVE-BELLIED SUNBIRD *Nectarinia chloropygia*
B1094 / MPG1257 / D2C2

Afrotropical: the moist lowlands around the Gulf of Guinea, and E throughout the Congo basin to NW Tanzania, Uganda and, on the E limit of the range, W Kenya.

Kenyan: a locally common sunbird of forest edge and occasionally interior, and of moist bush, thicket and cultivation, up to 1750 m (Br; Ja; MPG; Sugg & Sugg 1972). It is frequently near water (*op. cit.*). Ja recorded this species at Sotik 61C, but there were no further records until 1963, when a specimen was taken at Kakamega Forest 48D (Zimmerman & Mumford 1965). There have been many subsequent reports from this latter locality (Mann 1976) and other areas (Sugg & Sugg 1972; Br), and it was probably generally overlooked after Ja's note. Range is entirely in 1000+ m areas to the W of the rift valley. The squares are all moist subhumid-humid and within 1000+ mm rainfall areas.

The Olive-bellied Sunbird forms part of a species-group with the Beautiful Sunbird (920: *q.v.*), which inhabits far drier areas (HM).

920 BEAUTIFUL SUNBIRD *Nectarinia pulchella*
B1116 / MPG1233 / D2C2

Afrotropical: the N tropics, mainly between 8°N and 17°N, from the W coast to Ethiopia. Further S, it reaches most areas of East Africa, where distribution in Kenya represents its E limits. The black-bellied race *N.p. melanogastra* is endemic to Kenya near or S of the equator, and to the dry interior of Tanzania (Br).

Kenyan: a locally common bird of open bushed and wooded habitats at 400-1900 m, most typically in semiarid country at 700-1300 m (Br; Ja; MPG; St). Range is only 19% arid and 7% within the driest, 0-250 mm rainfall areas. Apart from the Turkana area, where it is occasional (HH), this sunbird is virtually absent from the dry N and E. The moister coastal lowlands are below its lower altitudinal limit. DEP (pers. obs.) considers this sunbird resident in 75BCD 88A, and St reports presence throughout the year in the semiarid Baringo area 50A. Absence of recent records from the E limit of the range in arid Tsavo E 101B, where it was formerly plentiful, may

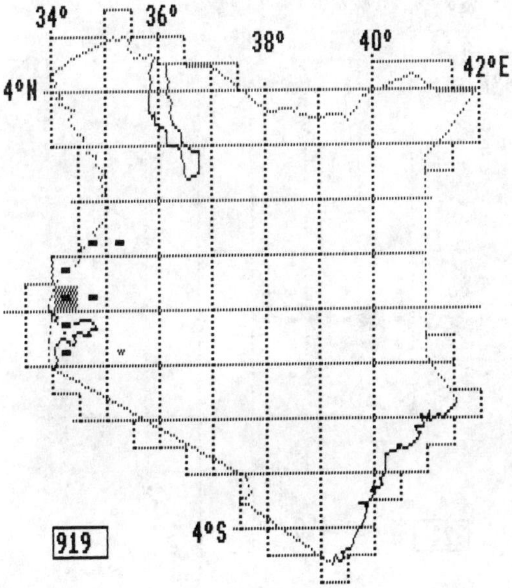

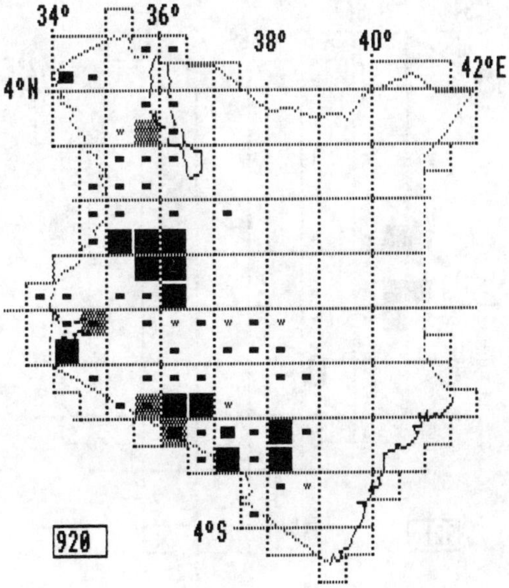

reflect destruction of woodland (La).

The partially black-bellied nominate race occurs SE to Kongelai 37C, the Kerio Valley (approximately 49B), Baringo 50A and Bogoria 50C, and E to Turkana, while the totally black-bellied *N.p. melanogastra* is S and E of the highlands, including drier areas of the L Victoria basin (Br). Investigation of the ranges and possible interbreeding of these two races in the Naivasha-Kedong area 62C is desirable (Br).

In direct contrast to the Olive-bellied Sunbird (919: *q.v.*), the Beautiful extends far into semiarid country (*cf.* Br). The Beautiful's range overlaps that of the Smaller Black-bellied Sunbird (918) at QSD scale along the E periphery of the C highlands and their adjacent plateaux. However, as noted by HM, the Beautiful is generally in less arid areas (57% arid-semiarid as against the Smaller's 95%). See Mariqua Sunbird (916).

921 NORTHERN DOUBLE-COLLARED SUNBIRD *Nectarinia preussi*
B1115 / MPG1256 / D2C2

Afrotropical: a highland species, in Cameroon, the E Zaire-Uganda borders, NE Uganda, S Sudan and Kenya.

Kenyan: locally common in wooded and forested habitats, including bamboo, in the W and C highlands (Br; Ja; Britton & Sugg 1973). Range is 77% moist subhumid-humid and 70% within 1000+ mm rainfall areas. Presence in 51A is on the small isolated highland of Mt Uraguess (Ja). Some degree of wandering or migration is reported by Britton & Sugg (1973), who found the bird abundant at 2800 m on Mt Elgon 36D during Dec 1969, but rare there when few plants were in flower during Jul 1971.

This species displays some degree of sympatry with the similar Eastern Double-collared Sunbird (922)(Br; HM). However, whereas the Northern is mainly at 1700-2800 m and only exceptionally reaches 3400 m, the Eastern is mainly at 1800-3700 m (Br). The ranges of these two species are very similar at QSD scale, except for the Eastern's presence on more forest islands: 26BD 39A 51A 88D 101A.

922 EASTERN DOUBLE-COLLARED SUNBIRD *Nectarinia mediocris*
B1108 / MPG1254 / D2C2

Afrotropical: highlands of the E, from Kenya through E Tanzania to Malawi and Mozambique.

Kenyan: generally common in wooded and forested (usually edge) habitats, including bamboo and tree-heath, mainly at 1800-3700 m but occasionally down to 1400 m (Br; Se; Diamond & Keith 1980). It is present on the forest islands of Mts Kulal 26B, Nyiru 26D (Diamond & Keith 1980; Ja) and Uraguess 51A and the Ndoto Mts 39A, the Chyulu Hills 88D and the Taita Hills 101A (Br). Excluding the localised environments of these isolated highlands, which are too small to be shown on our overlays, range is 73% moist subhumid-humid and entirely within 500+ mm

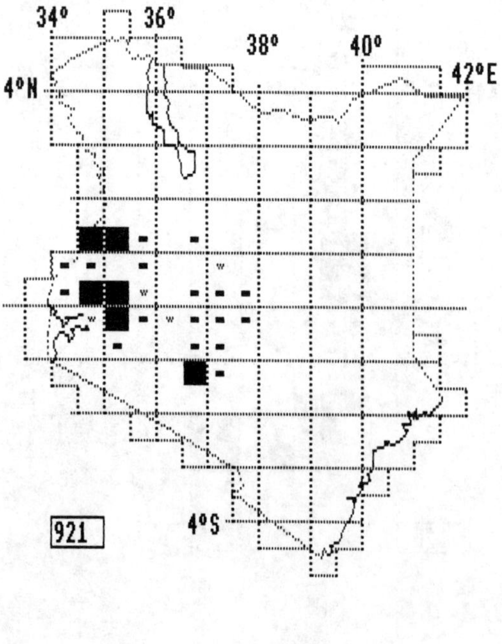

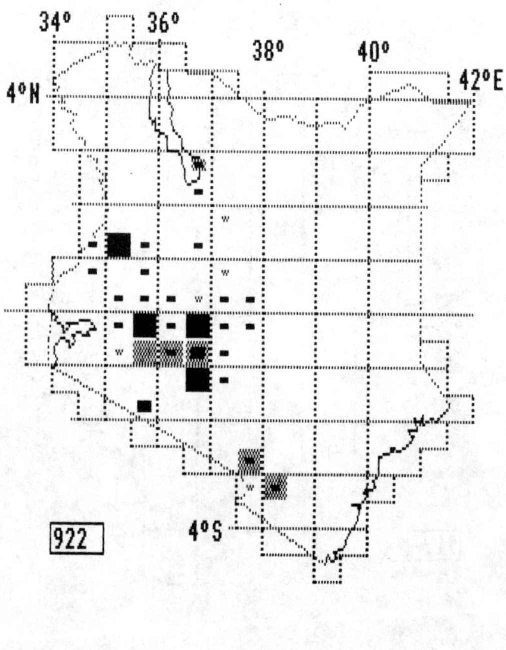

rainfall areas.

Be and Se consider the species resident, but all May-June 'long' rains breeding records are from below 2000 m (Brown & Britton 1980), and seasonal movements are known at 8°S in Tanzania (Br). Presumed wanderers (Br) have descended to 900 m at Elgeyu (approximately 37D 49B) and Useri 100B (Ja). See Northern Double-collared Sunbird (921). Other references: Cheke 1971ab; Williams 1978b.

923 COPPER SUNBIRD *Nectarinia cuprea*
B1096 / MPG1238 / D2C3

Afrotropical: widespread between 10°N and 16°S, E to W Ethiopia, Kenya, W and S Tanzania and Mozambique, but sparse in most of the equatorial forest regions. Distribution in Kenya represents the E limit of the range.

Kenyan: locally common in bushed, wooded and forest edge habitats, up to 1700 m (Br; MPG). It is frequently near water, and characteristic of the papyrus swamps of W Kenya (*op. cit.*; Britton 1978). Range is in 1000+ m, moist subhumid-humid, 1000+ mm rainfall areas to the W of the rift valley.

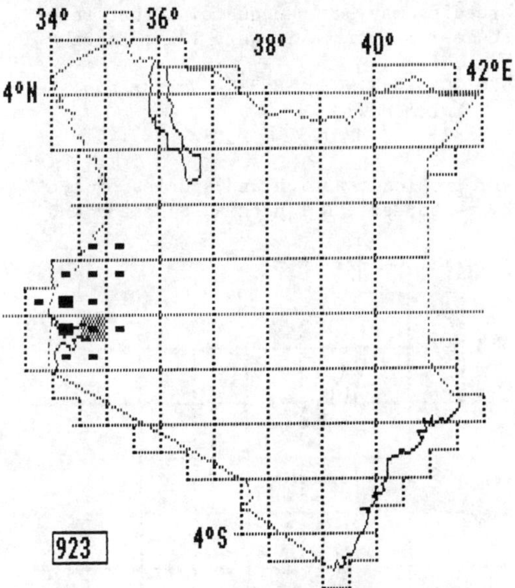

924 TACAZZE SUNBIRD *Nectarinia tacazze*
B1126 / MPG1229 / D2C3

Afrotropical: the highlands of the NE, from NE Tanzania, Kenya and E Uganda to S Sudan and Ethiopia.

Kenyan: this sunbird frequents forest edge, bamboo and shrubby heath at 1800-4200 m in the W and C highlands, where it

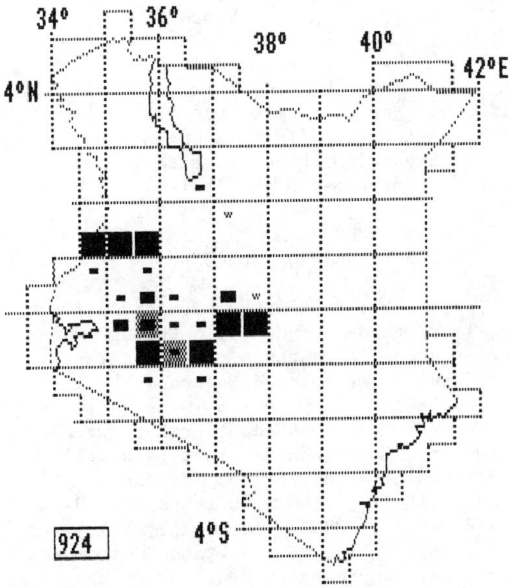

is sometimes common (Br; Ja; Se). It is present on the forest islands of Mt Nyiru 26D (Ja; Diamond & Keith 1980) and the Ndoto Mts 39A (Br). Excluding these isolated highlands, range is 81% moist subhumid-humid and 67% within the 1000+ mm rainfall regime.

All breeding records are above 2100 m (Brown & Britton 1980), and Br considers birds at lower altitudes to be non-breeding, altitudinal migrants that descend during the wet, cool months (*cf*. Olive Pigeon (325) and Golden-winged Sunbird (926)). These months are mainly June-Aug in W Kenya, and this seasonality compares with Britton & Sugg's (1973) note of apparent altitudinal movement on Mt Elgon 36D (*cf*. Northern Double-collared Sunbird (921)).

But Br also points out exceptions to this generalised migration model. To the W of the rift valley, the Tacazze Sunbird was common at 4100 m in the crater of Mt Elgon 36D during June (Ja), resident around 3000 m at Mau Narok 61D (Se), and resident above 2100 m in Narok district 74 (Be). E of the rift, Br also records resident birds around 2000 m at Naro Moru 63A. Three records down to approximately 1650 m at Nairobi 75B presumably represent wanderers (Ja; Belcher 1939; Jan 28: EABR 1981).

The Tacazze Sunbird is very closely related to the Bronze Sunbird (925: *q.v.*), and the two may form part of a superspecies (HM; Br). The Bronze breeds at lower altitudes than the Tacazze (Br), and HM's note that altitudinal segregation is at least partly achieved by the Bronze's total range being mainly below 1800 m broadly compares with Br's estimates, *i.e.* that the Tacazze is normally above 1800 m, while the Bronze is most frequent below 2100 m. The absence of the Bronze from Britton & Sugg's (1973) 2400+ m study area on Mt Elgon 36D agrees with this pattern. Other references: Cheke 1971ab.

925 BRONZE SUNBIRD *Nectarinia kilimensis*
B1103 / MPG1230 / D3C3

Afrotropical: mainly the highlands of the E, from Uganda and Kenya, S through suitable areas of Tanzania and the E Zaire borders, to Malawi and Mozambique. There is an isolated population in Angola.

Kenyan: in forest edge, bushed, wooded and sometimes heath habitats, and well known in highland gardens (Br; Ja). It is common at 1500-2100 m, but scarce and erratic up to 2800 m (*op. cit.*; Be). This is mainly a bird of the W and C highlands and their surrounding plateaux. Absence from more N forest islands, which are occupied by the Tacazze (924: *q.v.*) and Golden-winged (926: *q.v.*) Sunbirds, is notable. To the SE, it is present on the Chyulu Hills 88D (J F M Horne, L L Short *in litt.*), while occurrence in 88C (the Tanzanian sector: J S S Beesley *in litt.*) and 100B represents the periphery of birds on the Kilimanjaro massif.

Be considered this sunbird resident at 1500-2100 m in Narok district 74 and it is similarly so in gardens around 1650 m in Nairobi 75B (ADL pers. obs.), but some degree of movement to lower altitudes, possibly from the higher parts of its range, may occur during the wet seasons (*cf.* Tacazze and Golden-winged Sunbirds). Annual Apr-Jul influxes to the Ruaraka-Kahawa-Ruiru area of 75B (DEP pers. obs.) coincide with the 'long' rains and the succeeding cool season, as does a single Apr-May wanderer to 1000 m in the semiarid Baringo area 50A (St; EABR 1980). A mid-Dec wanderer around 550 m at Voi Safari Lodge 101B was during the 'short' rains, and may have come E from the Kilimanjaro population. La treat this record as unconfirmed, but the seasonality and the observer's experience are such that we map the record here. Interestingly, both this and the Baringo bird appeared in the artificially lush gardens of hotels.

Excluding these two wanderers, the Bronze's range is 83% subhumid-humid and 90% within 500+ mm rainfall areas. The Bronze's breeding squares are 80% above 1500 m and entirely within 500+ mm rainfall areas. This restriction of breeding records to the higher and wetter parts of the total range suggests some degree of post-reproductive dispersal, and this is supported by the influxes and wanderers noted above, although our coverage of breeding may be inadequate. Other references: Stewart & Stewart 1964; Stagg 1973.

926 GOLDEN-WINGED SUNBIRD *Nectarinia reichenowi*
B1119 / MPG1235 / D3C3

Afrotropical: a highland species of the E Zaire borders, N Tanzania, S Uganda and

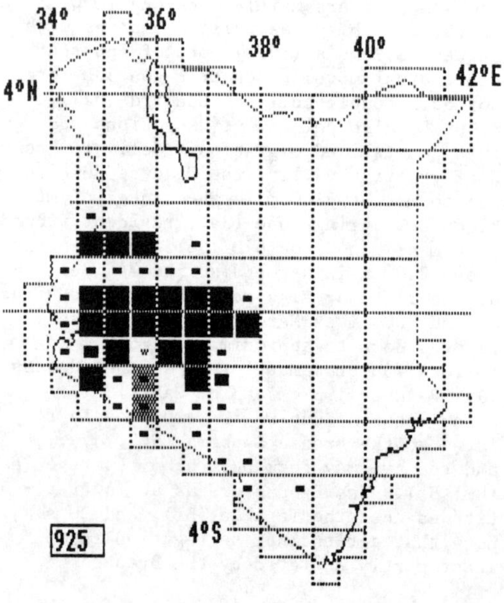

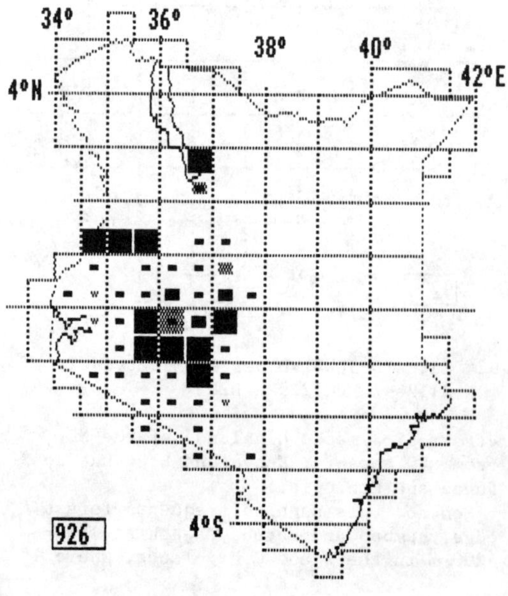

Kenya.

Kenyan: this locally common sunbird frequents a wide variety of well vegetated habitats, including bamboo, forest edge and glades, and gardens, but will also move into open grassland to feed on flowers (Br; Ja; Se; Britton & Sugg 1973; Diamond & Keith 1980). The race *N.r. lathburyi* is endemic to the forest islands of Mt Nyiru 26D, the Mathews Range 39C and Mt Uraguess 51A (Williams 1956; Br; Ja). Breeding birds in the even more isolated forest of Mt Kulal 26B (Diamond & Keith 1980) are presumably also this subspecies. Br notes this race at 2200-2400 m on Nyiru and the Mathews, while the birds of Kulal were between 1800 and 2300 m (*op. cit.*).

The Golden-winged Sunbird shows the clearest evidence of regular altitudinal migration in Kenya's highland sunbirds, but Br presumes *N.r. lathburyi*, being confined to isolated peaks, to be mainly resident. That some altitudinal movement occurs on even these very localised highlands is shown by Pelchen (1977d), who found the species around 1700 m at Tum, at the foot of Mt Nyiru, during very wet weather in Dec 1976-Jan 1977.

The rest of this species' range is occupied by the nominate race, which breeds at 1800-3400 m (exceptionally down to 1600 m), and then makes extensive and regular migrations to lower altitudes during the cool rainy seasons (Br; Brown & Britton 1980). Be considered this species resident above 2100 m on the Mau (approximately 61D 62C), breeding in Nov-Jan, but Se described it as an Aug/Sept-June migrant to around 3000 m at Mau Narok 61D breeding in Apr and Oct. Small numbers that descend annually to 1200 m during Mar-Sept in the L Victoria basin, W to 48C, show moult features and (in the single specimen taken) a lack of gonadal activity consistent with non-breeding migrants (Britton & Sugg 1973; Br). Similar Apr-June influxes occur in the Ruaraka-Kahawa-Ruiru area of 75B, and contrast with the complete lack of Aug-Jan records (DEP pers. obs.). Single wanderers to semiarid country at around 1000 m have also been in wet, cool months, in mid-Apr at Baringo 50A (EABR 1981), and during June (EABR 1979) and Nov (P Le Pelley *in litt.*) at Olorgesaille 75C.

The most conclusive evidence of extensive movements derives from two ringing recoveries. A male ringed during May at 1600 m in Nairobi 75B moved 65 km to 2530 m at South Kinangop 62D before the following Sept (Backhurst 1973b; Britton & Sugg 1974), while a bird ringed at South Kinangop 62D during May 1973 was recovered 101 km away at approximately 1500 m at Mitaboni Location 76A in Apr 1975 (Backhurst 1977).

In addition to the forest islands already mentioned, this sunbird also occurs on the Chyulu Hills 88D (Br). Excluding all of these isolated highlands, and the 50A and 75C wanderers, range is 76% above 1500 m, 84% subhumid-humid and 95% within 500+ mm rainfall areas. The breeding squares are 92% above 1500 m, 78% moist subhumid-humid and entirely within 500+ mm rainfall areas. The differences between these two sets of figures reflect the altitudinal migration to some extent. The Golden-winged Sunbird is broadly sympatric with the Tacazze Sunbird (924)(HM).

927 MALACHITE SUNBIRD *Nectarinia famosa*
B1099 / MPG1227 / D3C2

Afrotropical: mainly the E, from the Ethiopian highlands, S through the highlands of East Africa and the E Zaire borders, to Zimbabwe and the Cape; and N along the W coast to the Orange R. Distribution in Kenya represents the E limit of the range.

Kenyan: locally common in a wide diversity of habitats, from scrub to bamboo and forest edge, and also attracted far into open grasslands by suitable flowers (Br; Ja; MPG; Britton & Sugg 1973). Range is mainly at 1700-3000 m and, as in the Golden-winged Sunbird (926) and other highland congeners, there is regular altitudinal descent during the cool, wet seasons (Br). All breeding records are from above 2000 m (Brown & Britton 1980).

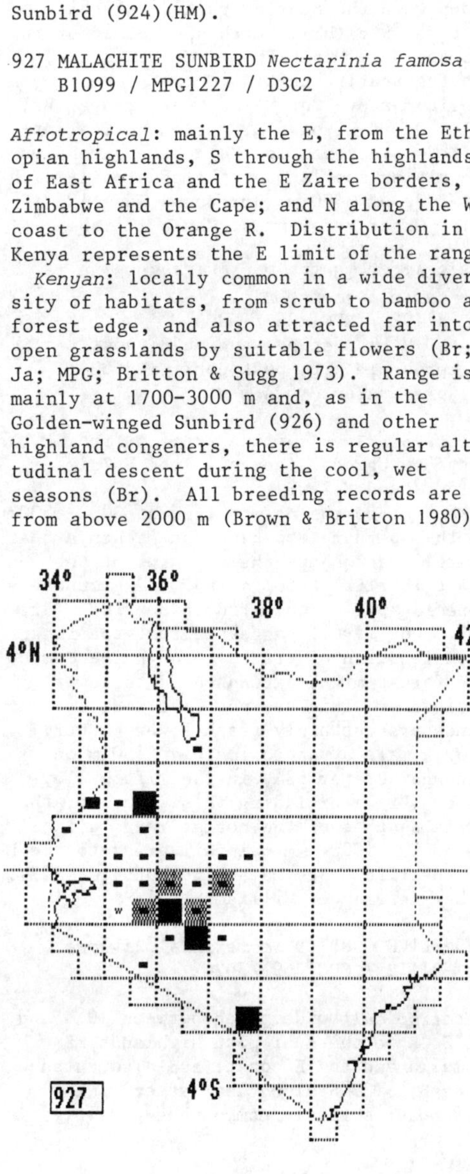

Be considers the Malachite resident at 1800-2700 m on the Mau 61D. At 3000 m around Mau Narok 61D, however, Se found it an annual Sept-Apr visitor, and saw birds 600 m lower near Njoro 61B during other months of the year. This altitudinal migration, together with extensive wandering and wide habitat tolerance, produces an extensive range in the highlands (Br). The Malachite is also found on the forest islands of Mt Nyiru 26D (Diamond & Keith 1980) and the Chyulu Hills 88D. Excluding these two localised highlands, range is 78% above 1500 m and 74% moist subhumid-humid, and all squares receive 500+ mm rainfall.

The Malachite Sunbird forms a species-group with the Scarlet-tufted Malachite Sunbird (928)(HM). Both species occur in the Aberdare Mts 62BD and Mt Kenya 63A, but the Scarlet-tufted is restricted to moorland above the Malachite's range (Br; HM). On Mt Elgon 36D, where the Scarlet-tufted is absent, the Malachite breeds on the alpine moorland at the abnormally high altitude of 3400 m (Britton & Sugg 1973; Br). Other references: Cheke 1971ab.

928 SCARLET-TUFTED MALACHITE SUNBIRD *Nectarinia johnstoni*
B1102 / MPG1228 / D3C3
NO MAP: post presence 62B; post probable 62D; post confirmed 63A; pre presence 62A

Afrotropical: a montane sunbird, from C Kenya to Uganda, Tanzania, the E Zaire borders and Malawi.

Kenyan: common on moorland at 3000-4500 m on the Aberdare Mts 62BD and Mt Kenya 63A, where it frequents the vicinity of the giant lobelias *Lobelia* spp. and groundsels *Senecio* spp. of the Afroalpine flora (Br; Ja). Coe (1961) suggested that the recent proliferation of tracks through the bamboo and forest might extend range to lower altitudes. This has not occurred, but wanderers certainly reach lower country *e.g.* a male at approximately 2150 m on Menengai Crater 62A, in the rift valley floor (Br). Excluding this wanderer, the three squares of the normal range are all in humid, 1000+ mm rainfall areas to the E of the rift. See Malachite Sunbird (927). Other reference: Williams 1951.

929 OLIVE SUNBIRD *Nectarinia olivacea*
B1112 / MPG1269 / D2C2

Afrotropical: widespread between 10°N and 12°S, E to the Ethiopian highlands, S Somalia and the E coast; and also extending S to 33°S along the latter.

Kenyan: a fairly common but possibly

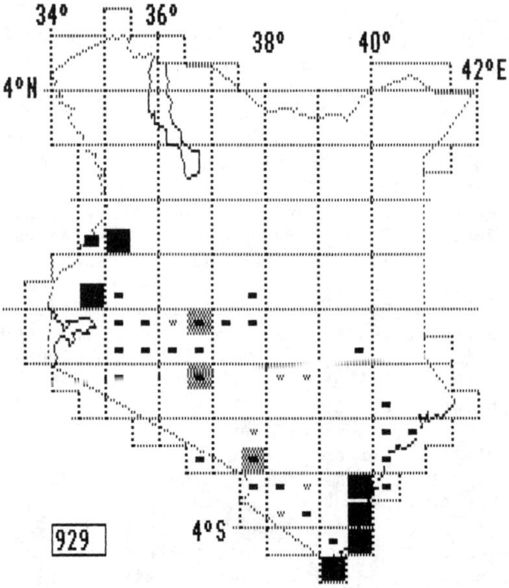

overlooked species, found in bamboo and forest, and modifications of these habitats that retain tall trees, up to 2300 m (Br; Ja; AGH; Britton & Zimmerman 1979). Sessions recorded a single bird, possibly a wanderer, around 3000 m at Mau Narok 61D in Mar (Britton 1979). This sunbird's range is 80% within 500+ mm rainfall areas. Occurrence in arid areas (15%) is in locally moist environments (see below).

The three races present in Kenya occupy differing regions. *N.o. ragazzii* is in the highlands and adjacent plateaux to the W of the rift valley, reaching E to the floor of the latter at Subukia 62A (Br) and possibly Naivasha 62C (race unknown: J Hayes *in litt.*). This race's range is entirely above 1000 m, 80% within 1000+ mm rainfall areas and entirely moist subhumid-humid. It is isolated from and does not intergrade with the other two races, and differs from them in that its females lack pectoral tufts (Br).

N.o. neglecta inhabits the peripheries of the C highlands, E and S to the forest islands of Mt Endau 77B (Tennent 1964), the Chyulu Hills 88D (Br) and Ol Doinyo Orok 87D (race unknown: J Fanshawe *in litt.*). Excluding these isolated highlands, range of this race is 88% above 1000 m, 75% moist subhumid-humid, and entirely within 500+ mm rainfall areas.

N.o. changamwensis occupies the coastal lowlands, inland to the Taita Hills 101A (Br) and the lower Tana R 79C 91A (AGH). A specimen from Koro Kora 65D on the Tana (National Museum Collection, Nairobi) may belong here. This race's range is approximately 86% below 500 m, 79% within 500+ mm rainfall areas and 71% semiarid-dry subhumid. The few occurrences in arid squares are in locally moist environments: along the Tana R 65D 79C (*op. cit.*), and in the Voi R Forest 101B, where it is now apparently extinct (La). Occurrence along the Tana parallels restriction to the woodlands of the lower Juba R of adjacent S Somalia (AM).

930 SCARLET-CHESTED SUNBIRD *Nectarinia senegalensis*
B1122 / MPG1263 / D2C2

Afrotropical: widespread S from 15°N to Namibia, Botswana and E South Africa, E to Ethiopia, S Sudan, E Uganda, Kenya and the E coast, but absent from the equatorial lowland forest regions. Distribution in Kenya represents the E limits of this bird's range: it is replaced in the drier areas of Kenya and in Somalia by Hunter's Sunbird (931: *q.v.*).

Kenyan: a wide range of bushed and wooded habitats in medium-high rainfall areas, up to 2150 m (Br). Distribution is 82% within 500+ mm rainfall areas and only 5% arid (see below).

Occurrence is in three distinct regions. The reports from 6C 16B 17B represent the SE limit of birds in Ethiopia, and presumably refer to the race *N.s. cruentata*. 6C and 17B were in the Ethiopian sectors of the squares (J S Ash *in litt.*) but 16B was at Sololo, Kenya (G C Backhurst *in litt.*). These three squares are shown as

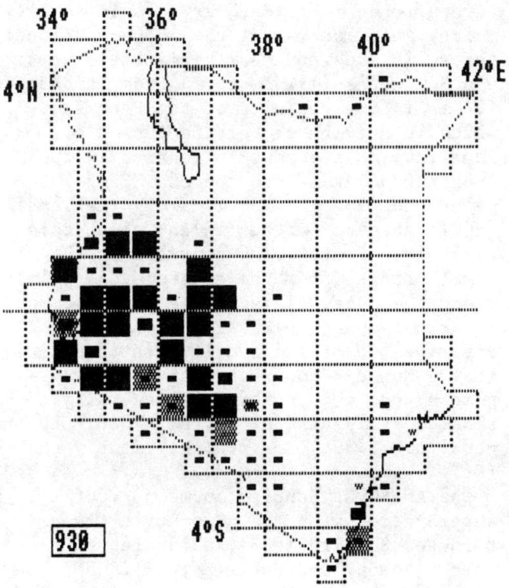

arid on our overlays, but the climatic data may be inaccurate. Sololo 16B is at 500-1000 m, in an apparently 250-500 mm rainfall area. The highlands and their peripheral plateaux are at least mainly occupied by *N.s. lamperti*, which is common (Br) and 92% above 1000 m. The coastal lowlands' race *N.s. gutturalis* is decidedly more local (Ja; Br), and entirely below 500 m. The species' only other arid square is 79C, where *gutturalis* is in the locally lush habitats around the Tana R (AGH).

Be considered this species resident in Narok district 74. DEP (pers. obs.) concurs for 74D 75CD 88D 89AC 100B 101A, but also notes annual influxes to the Ruaraka-Kahawa-Ruiru area of 75B during the Feb-Apr and Sept-Nov dry seasons, that augment the resident population. Breeding squares are 79% subhumid-humid, but whether this may be biased by distribution of observers is unknown.

The Scarlet-chested forms part of a superspecies with the allopatric Hunter's Sunbird (HM). The latter inhabits the drier N and E of Kenya and, apart from most of Turkana (HH) and the far NW, the ranges of these two species, taken together, cover most of the country. The two are very similar in appearance, and often confused (*e.g.* see Hunter's Sunbird). Br doubts that the two occur alongside each other even as wanderers, being separated by habitat preferences in areas of apparent sympatry. The sole known exception to this relationship apparently exists along the lower Tana R 79C 91A, where both have been recorded in wooded grassland that is intermediate in character between the lush riverine habitats and the surrounding arid bush (AGH; Br).

931 HUNTER'S SUNBIRD *Nectarinia hunteri*
B1101 / MPG1264 / D2C2

Afrotropical: the NE, from extreme NE Tanzania, through Kenya and NE Uganda, to S Sudan, S and SE Ethiopia, and Somalia.

Kenyan: usually uncommon in dry, open bushed and wooded habitats at 50-1200 m in the N and E (Br; Ja; La). The N and E edges of the highlands 37C-49D-63B-76 define its SW Afrotropical limit. In squares on the highlands' edge, *e.g.* 37C, this species is in lower, drier areas than the closely related Scarlet-chested Sunbird (930: *q.v.*), which occupies higher, wetter environments. In contrast to the latter, Hunter's range is 91% arid-semiarid, but it is virtually absent from the apparently suitable L Turkana basin (*cf.* HH). However, whereas Br considers that it

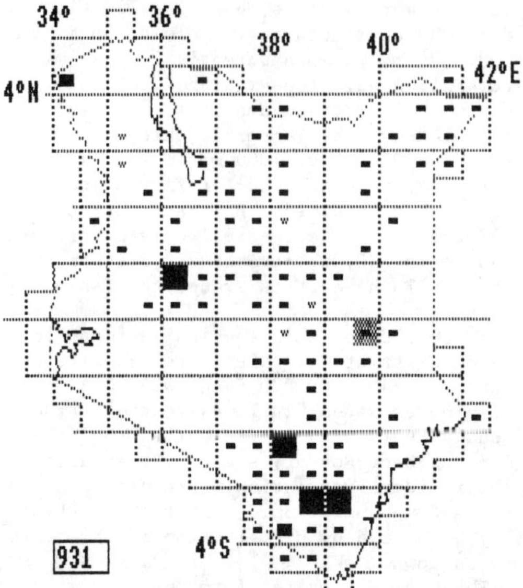

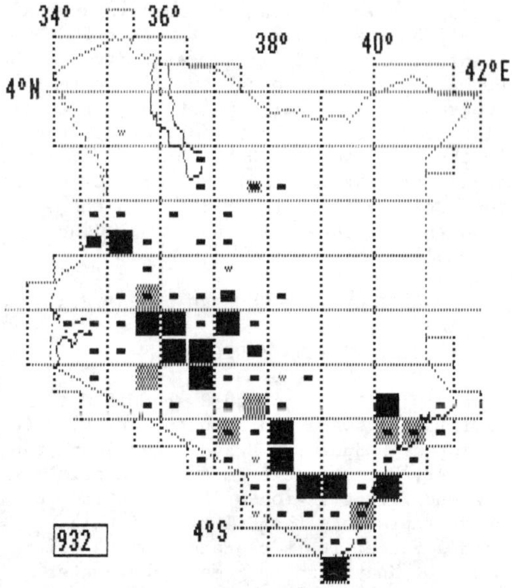

mainly avoids the most arid areas, its range is in fact in 41% of the most arid squares (moisture index below 1.2) and in 42% of those with the least (0-250 mm) rainfall. This sunbird is not found in the coastal strip, where it is replaced by the Scarlet-chested. In Somalia where the latter is absent, Hunter's reaches the coastal strip and the record in 80D, which derives from Somalia, marks the S tip of this coastal area (AM).

La consider Hunter's resident in arid Tsavo E 101B, and DEP (pers. obs.) concurs for 88D 89C 100B 101A. St notes similar status at semiarid Baringo 50A, adding that breeding possibly occurs in all months during abnormally wet years. A report of a single male of this species at Akala 60A (EABR 1981) would imply wandering into unsuitably humid country, over 100 km SW of the normal range. ADL can no longer be certain of the identification, and withdraws the record. Other reference: Lack 1976b.

932 AMETHYST SUNBIRD *Nectarinia amethystina*
B1091 / MPG1261 / D2C2

Afrotropical: widely distributed between 5°S and 12°S, and further S to Mozambique, Botswana, E South Africa and the Cape. To the E of the Congo basin, the N limits of the range are in S Sudan, E Uganda, Kenya and S Somalia.

Kenyan: locally common, typically in woodland but also frequenting a wide range of other well vegetated habitats, including mangroves and forest edge, up to 2200 m (Br; Ja; MPG; Be; Britton & Zimmerman 1979; Diamond & Keith 1980). Range includes the forest islands of Mts Loima 13C (National Museum Collection, Nairobi), Kulal 26B (Diamond & Keith 1980), Nyiru 26D, Marsabit 27D 28C, Uraguess 51A (Ja) and Endau 77B (Tennent 1964; A W Diamond *in litt.*), and the Ndoto Mts 39A (Br). Apart from these isolated highlands, which are too small to be represented on our overlays, range is 83% within 500+ mm rainfall areas and only 9% arid.

Occurrence in arid squares is in locally moist situations, *e.g.* the above-mentioned forest islands and near permanent rivers. van Someren's (1929) record from Mandera 19B is in the vicinity of the Daua R, which is a tributary of the Juba R of Somalia. AM record this sunbird in the Juba's woodlands N to the QSD NE of 19B. Occurrence in arid 79C is along the Tana R (AGH): absence further inland along this river is in surprising contrast to the Somali range (from the equator to 4°30'N) along the Juba. This species is uncommon in arid Tsavo E 101B, again confined to riverine and other locally moist habitats (La). Numbers congregate at flowering plants, but this sunbird is generally considered resident (Be: 74; La: 101B; DEP pers. obs.: 89A), or probably mainly so (Br).

The Amethyst Sunbird forms part of a superspecies with the allopatric Green-throated Sunbird (933), which inhabits forest canopy and edge (HM; Br). HM note

Plate 17. Female Bronze Sunbird at the nest (J.S. Karmali F.R.P.S.). *See page 492*.

Plate 18. Jackson's Widowbird, endemic to central Kenya and northeast Tanzania (J.S. Karmali F.R.P.S.). *See page 523*.

marginal range overlap of these two birds, but consider them mainly ecologically segregated. Rarity of the Amethyst in the apparently suitable L Victoria basin (Br), including the total lack of records from 48, probably reflects this allopatry.

933 GREEN-THROATED SUNBIRD *Nectarinia rubescens*
B1120 / MPG1262 / D2C2
NO MAP: post presence 36D 48AC; post probable 48D 49C

Afrotropical: a C African species, from Cameroon, E throughout the Congo basin and adjacent areas, to Uganda, NW Tanzania and, at the E limit of the range, W Kenya.

Kenyan: locally common in forest canopy and edge, up to 1800 m (Br; Ja; MPG). Range is above 1000 m, to the W of the rift valley. The five squares are all humid and within the 1000+ mm rainfall regime. There is no confirmed breeding record, but Br considers the species probably mainly resident, mentioning a single example of wandering in Uganda. See Amethyst Sunbird (932).

934 GREEN-HEADED SUNBIRD *Nectarinia verticalis*
B1130 / MPG1266 / D2C2

Afrotropical: from Senegambia and Angola, E through the Congo basin and adjacent regions to mainly highland areas of S Sudan, Uganda, Kenya and W, S and NE Tanzania.

Kenyan: locally common in forest edge and clearings and well wooded country at 1200-2400 m, though typically above 1700 m to the E of the rift valley (Br; Ja; MPG; Britton & Sugg 1973; Diamond & Fayad 1979). It is a bird of the highlands and their peripheral plateaux, with range 89% moist subhumid-humid and 78% within 1000+ mm rainfall areas. It does not extend to forest islands. The extent of this sunbird's distribution on either side of the rift valley is clearly related to rainfall. To the W, it occupies 86% of the 1000+ mm squares so that, as well as occurring in the W highlands, it also extends throughout the moist plateau and the L Victoria basin (*cf.* Br). E of the rift, the 1000+ mm regime is largely confined to the true highlands, and the bird is sparse on the surrounding 500-1000 mm plateaux, *e.g.* at Nairobi 75B (EABR 1978; ADL pers. obs.).

Br considers this species apparently mainly resident, and the absence of confirmed breeding records from areas in and E of the rift is probably not significant. Br suggests that occasional birds at Nakuru 62A and Nairobi 75B are wanderers but, at the latter locality, the species appears to be resident in very small numbers near its lower altitudinal limit (P M Allen, ADL pers. obs.). See Blue-throated Brown Sunbird (935).

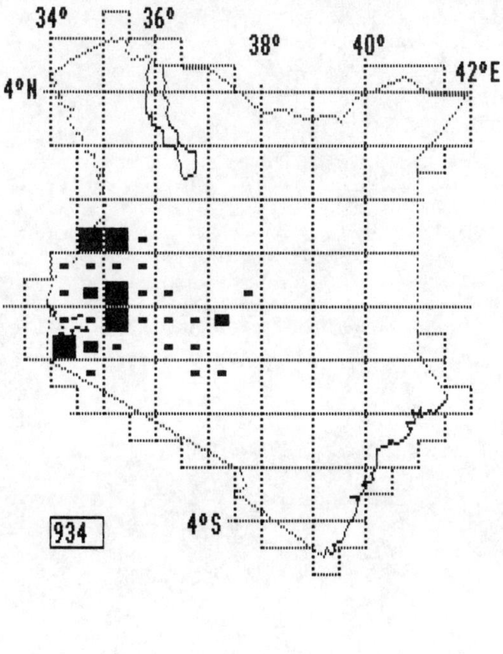

935 BLUE-THROATED BROWN SUNBIRD *Nectarinia cyanolaema*
B1097 / MPG1268 / D1C1
NO MAP: post presence 48D

Afrotropical: W and C Africa, from the lowland forests around the Gulf of Guinea, E through the Congo basin to W Tanzania, Uganda and, at the E limit of the range, W Kenya.

Kenyan: rare in canopy and edge habitats in Kakamega Forest 48D (Br; MPG). This forest is at 1700 m to the W of the rift valley in a humid, 1000+ mm rainfall square. There is no breeding record.

The Blue-throated Brown is related to and often sympatric with the Green-headed Sunbird (934), but more restricted to forest habitats (HM). Other reference: Mann 1976.

936 MOUSE-COLOURED SUNBIRD *Nectarinia veroxii*
B1129 / MPG1265 / D1C3

Afrotropical: the E coastal lowlands, from S Somalia to Cape Province, South Africa.

Kenyan: this sunbird is awarded D1 because of its drab, easily overlooked appearance, and the fact that it is often scarce and local, even in apparently

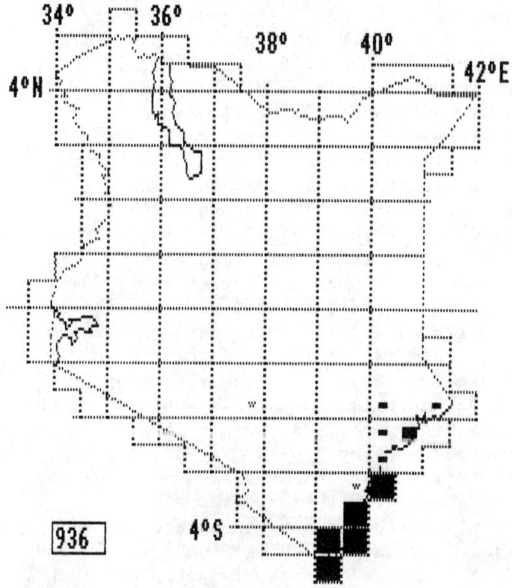

suitable areas (Ja; MPG; Br; *cf.* AM). It inhabits bush, woodland thicket, mangroves and forest edge in the coastal lowlands (Br), and along the lower Tana R 79C 91A (AGH)(hence C3). Occurrence along the Tana parallels distribution in adjacent Somalia, where it extends at least 240 km inland on the lower Juba R (AM).

The sole Kenyan record away from the coastal/SE lowlands is a female, presumably a wanderer, collected in acacias at the foot of Nzaui Hill 76D by Ja. Excluding this bird, range is entirely below 500 m in 500+ mm rainfall areas, and 91% semiarid-dry subhumid. Although at approximately 1200 m, the Nzaui bird was also in semiarid, 500-1000 mm, lowland wooded country. All recorded nests have been on buildings, which provide protection from the weather, probably ensuring successful breeding during the wettest months of the year: there may also be other, less obvious advantages to this choice of site (Britton & Britton 1977a; Brown & Britton 1980).

Family Zosteropidae: White-eyes

The taxonomy of this family is particularly controversial and locally complex, and we follow Br in adhering to the Afrotropical taxa given by HM. The world's 79 species of white-eyes are mainly tropical, being most diverse in the Australasian and Oriental Regions, and absent from the Americas. The family has been notably successful in colonising islands and, of the nine Afrotropical species (11% of the world's total), five are endemic to oceanic islands. The white-eyes are most diverse in the NE Afrotropical, with three species in highland areas of W Ethiopia, Kenya and extreme NE Tanzania.

937 YELLOW WHITE-EYE *Zosterops senegalensis*
B1133 / MPG1219-1221 / D2C2

Afrotropical: widespread between 15°N and 15°S, E to W Ethiopia, Kenya, Tanzania and parts of the coast further S. It is largely absent from the Congo basin. Distribution in Kenya represents the E limit of the range, beyond which it is replaced by the Montane (938) and Abyssinian (939) White-eyes (HM, and see below).

Kenyan: common in wooded and forest habitats, and giant heath, up to 3400 m (Br; Ja; Se; Britton & Sugg 1973; Diamond & Fayad 1979). This white-eye occurs on the forest islands of Mts Loima 13C (National Museum Collection, Nairobi), Nyiru 26D (Ja) and Marsabit 27D, the Ndoto Mts 39A, and the Mathews Range 39C (Br).

Excluding these isolated highlands, which are too small to be represented on our overlays, range is 71% above 1500 m,

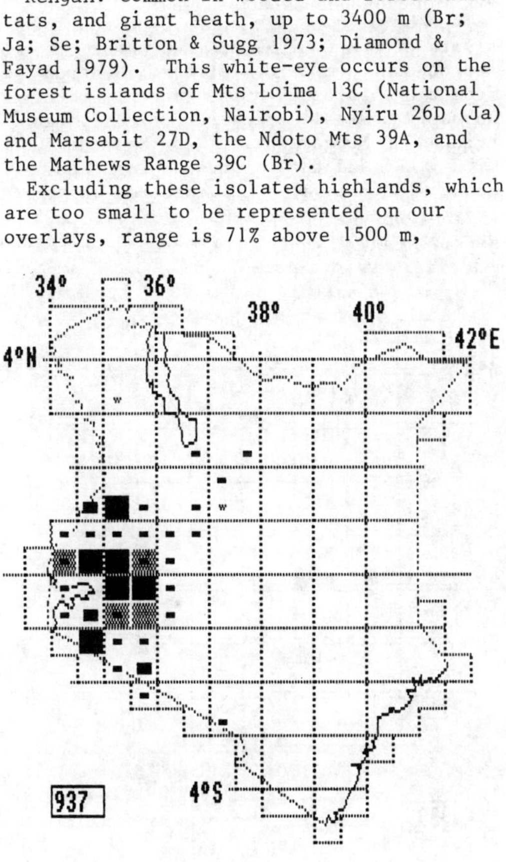

entirely within 500+ mm rainfall areas and 87% subhumid-humid. Range to the W of the rift valley is 72% within the 1000+ mm rainfall regime. Further E, it extends into the 500-1000 mm areas of the W plateau 37D 74, the rift floor 50AC 62AC, and the E plateau 38D 39C 50B, and thence further NE to the forest islands. It is uncommon in the semiarid rift floor around L Baringo 50A (St). There is no suggestion of seasonality at Baringo, and Be (74) and Se (61D) both consider the species resident.

The Yellow White-eye forms part of a species-group with the Montane and Abyssinian White-eyes (HM). The Montane is more strictly a forest bird than the Yellow, and is confined to areas E of the rift. Both the Yellow (Br) and the Montane (Friedmann & Stager 1969, reiterated by Diamond & Keith 1980) have been reported from Mt Nyiru 26D, but whether both actually occur, or this reflects confusion due to recent nomenclatural changes is uncertain. HM's and Br's omission of the Montane on Nyiru suggests the latter and, since its presence there would imply a notable coexistence, we do not map the Montane in 26D.

The Abyssinian White-eye is also mostly E of the rift valley. It reaches the edges of the highlands but, in contrast to the other two species, also extends down to sea level and into far drier country. The Abyssinian may be altitudinally segregated from the Yellow in areas of apparent sympatry because, in the Mau 61D-Narok 74B area, Be lists the former at 1500-1800 m and the latter at 1800-2700 m. Furthermore, in 26D, the Yellow occupies the Mt Nyiru forest island, while the Abyssinian is lower down in the adjacent Horr Valley (Br).

Mixed flocks of Montane and Abyssinian White-eyes are common around 1650 m at Nairobi 75B, and Br considers the two species sympatric there. These two birds also flock together on the Taita Hills 101A (EABR 1982). HM consider the Montane to be generally ecologically segregated from both of the other white-eyes, however, and note that there is no sign of hybridisation in their restricted zones of range overlap.

938 MONTANE WHITE-EYE *Zosterops poliogastra*
B1132 / MPG1222-1225 / D2C3

Afrotropical: highlands of the NE, from the far NE of Tanzania to Kenya and Ethiopia. The assumption that this species includes all the montane forms scattered throughout this range implies the latter to have extended originally to lowland areas between Eritrea and NE Tanzania (HM). Subsequent retraction from the lowlands has left relict populations isolated in highland areas.

Kenyan: common in forest, forest edge and gardens, up to 3000 m (Br; Ja; MPG). Four races occur (Br). The grey-bellied *Z.p. kulalensis* is endemic to the Mt Kulal forest island 26B, where Diamond & Keith (1980) found it common. Probable breeding on Kulal is based upon enlarged testes in some of the type specimens (Williams 1948b). The similarly grey-bellied *Z.p. silvana* has long been considered endemic to the Taita Hills forest island 101A, but was recently also found on the nearby forest island of Mt Kasigau 101D, at 1300 m (EABR 1981). Confirmed breeding of *silvana* in 101A represents MPG's bare mention from Taita. The yellow-bellied race *Z.p. mbuluensis* occurs on the forest islands of Ol Donyo Erok 87D and the Chyulu Hills 88D, while the similarly coloured *Z.p. kikuyuensis* occupies the C highlands.

Excluding the forest islands, which are too localised to appear on our overlays, range in the C highlands is 69% above 1500 m, and entirely within subhumid-humid, 500+ mm rainfall country. It is present in all of the 1000+ mm rainfall squares to the E of the rift. See Yellow White-eye (937).

939 ABYSSINIAN WHITE-EYE *Zosterops abyssinica*
B1131 / MPG1219 / D2C2

Afrotropical: the NE, from NE Tanzania to Kenya, Ethiopia and all but E areas of Somalia. It may occur in NE Uganda (Br).

Kenyan: common in bushed and wooded habitats, and forest edge, up to 1800 m (Br; MPG). 96% of its squares are in or E of the rift valley, but it also extends to the W plateau at Narok 74B (Be; Br). In contrast to the other two white-eyes, the Abyssinian's range is 81% within intermediate 250-1000 mm rainfall areas, and 68% semiarid-dry subhumid.

This bird is mostly known from the lower slopes of the C highlands and their surrounding semiarid plateaux. Records from the apparently suitable semiarid-dry subhumid coastal strip are curiously restricted to Mongeya Hill 102B, Ngomeni 103A, Witu 91AB, and the islands of Lamu and Manda 91B (Br; Ja). Unidentified white-eyes along the lower Tana R 79C 91A (AGH) were presumably this species. The records

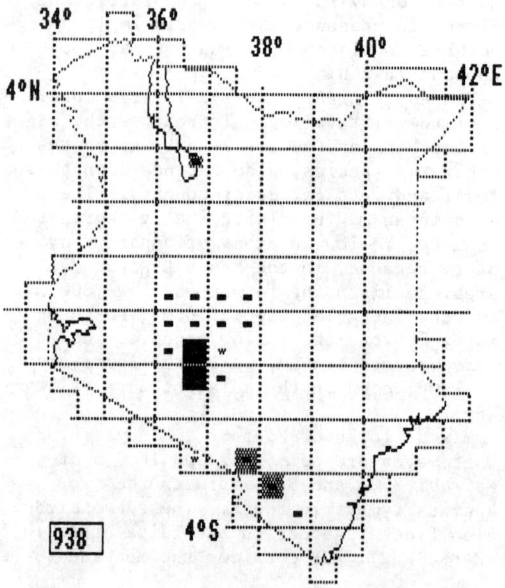

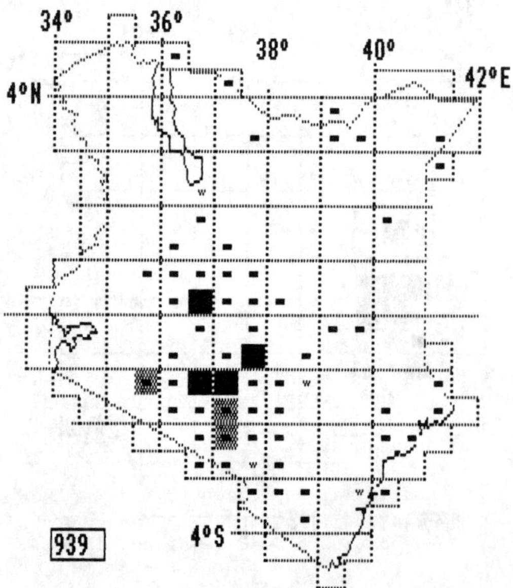

in 80AC derive from Somalia (AM). This species' distribution is only 18% arid. In the arid E, localities include locally moist habitat around Wajir 42A (Br; J E Miskell *in litt.*), and squares containing the Tana R 65AB 79C.

Be considered this white-eye resident in the dry subhumid Narok area 74B and DEP (pers. obs.) concurs for 75CD 89AC 100B 101A. In arid Tsavo E 101B, however, this bird is a sporadic visitor to areas with large trees (La). The records in 19C and 31A are from the Somali sectors of the squares (AM). Reports further W represent the S edge of the Ethiopian population in 4A 5C 17CD (J S Ash *in litt.*), in a stand of mature trees at *c.* 900 m in 15D (F Alexander *in litt.*), and in bush at Moyale 17A on 18 Jul 1984 (F J McCartney, ADL pers. obs.). See Yellow White-eye (937). Other reference: Cunningham-van Someren 1970.

Family Ploceidae: Weavers, Bishops, Widowbirds, Malimbe, Queleas, Sparrows, Petronia, Indigobirds, Whydahs

Approximately 147 species, of which 126 (85%) occur in sub-Saharan Africa, and the remainder (excluding introductions) in the Oriental and Palearctic Regions. The genera *Malimbus*, *Amblyospiza*, *Euplectes*, *Quelea*, *Vidua*, *Hypochera*, *Anomalospiza*, *Sporopipes*, *Plocepasser*, *Pseudonigrita*, *Histurgops*, *Philetairus*, *Bubalornis* and *Dinemellia* are all Afrotropical endemics, as are 48 of the 52 members (92%) of *Ploceus* (HM). Of the 126 species recorded from the Afrotropical Region, 72 (57%) are known in Kenya, where at least 67 (93%) either breed or are presumed to do so. Clarke's Weaver (959) is endemic to the Sokoke Forest area *c.* 102B, but has yet to be found nesting. The Red-headed Quelea (987) is apparently only a non-breeding visitor. The Orange (943) and Yellow-mantled (966) Weavers and the Broad-tailed Paradise Whydah (1011) may be non-breeding wanderers at or beyond the limits of their normal ranges. Following HM, we discuss the parasitic whydahs *Vidua* and indigobirds *Hypochera* along with their host species in the Family Estrildidae (Waxbills *etc.*).

940 BLACK-HEADED WEAVER *Ploceus cucullatus*
B1165 / MPG1312, 1313 / D2C3

Afrotropical: widespread S of 15°N, apart from the drier areas of NE Somalia and the SW.

Kenyan: a common and gregarious weaver of open bushed or wooded country, up to 2500 m, but mainly below 2000 m (Br; Ja; MPG; Britton & Zimmerman 1979). Range is 76% within 500+ mm rainfall areas and only 19% arid, with distribution in the dry N and E mainly confined to locally moist areas (*cf.* HM; AM). The records in 3B 17A and 19A represent the S limit of the Ethiopian highlands population. The report from 3B, which contains the Omo R, is from Ethiopia (J S Ash *in litt.*). 17A represents occurrence in locally moist habitats at Moyale (Br; Dec: J E Miskell *in litt.*), and 19A reflects the Daua riverine woodland (E J Roberts *in litt.*). Presence in 27D is on the moist, isolated Mt Marsabit (Br). Further S, occurrence in arid squares similarly either borders moister country (*e.g.* La; see below), or is along the Tana R 64B-79C (AGH). Extension along the latter parallels virtual restriction to the valleys of the Juba and Shebelle R of adjacent S Somalia (AM).

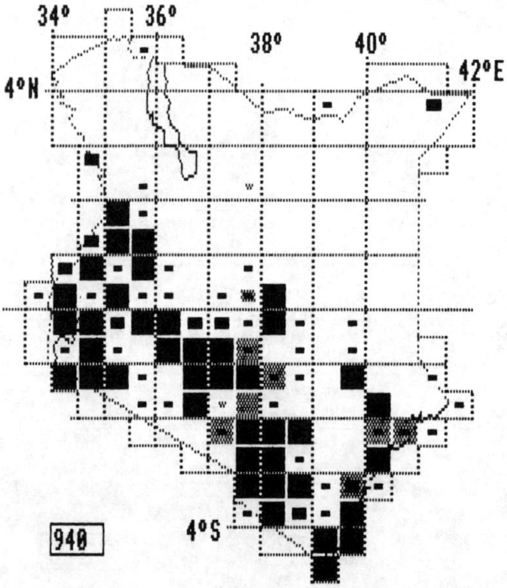

The high proportion of breeding squares (62%) and their uniform distribution through the range S of 2°N suggest a mainly resident species, although frequent nesting near human habitation (Ja; Br) obviously increases the chances of detecting breeding records. Br describes the bird as resident, but also notes seasonal movement into dry country during the rains. This happens in the arid Voi-Tsavo E 101B area where, although a few birds may be resident, numbers increase significantly during the rains, particularly in Nov-Jan (La). Such movements are likely to occur in other arid squares adjacent to moister country *e.g.* 64AC 77B 89BD (see above). In addition to this apparently regular movement, wandering also occurs (Br; AM). See Vieillot's Black Weaver (941). Other reference: Kasoma 1982.

941 VIEILLOT'S BLACK WEAVER *Ploceus nigerrimus*
B1175 / MPG1344 / D3C3
NO MAP: post presence 48C 60B; post probable 48A 49C; post confirmed 48D; pre confirmed 60D

Afrotropical: W and C Africa, from Guinea

S to Angola, E through the Congo basin to W Tanzania, Uganda and, at the E limit of the range, W Kenya.

Kenyan: a conspicuous and gregarious species, locally common in forest edge and clearings, and in wooded country, up to 2000 m (Br; Ja; MPG). Range is in 1000+ m areas to the W of the rift valley. All five squares are moist subhumid-humid, and within the 1000+ mm rainfall regime. The record in 48A is from Uganda (J S Ash *in litt.*), and the species is also known from the Ugandan side of Mt Elgon 36CD (Ja: not mapped). Br considers this weaver probably mainly resident, though occasional individuals at Ng'iya 48C and Kisumu 60B are apparently wandering.

Vieillot's Black Weaver possibly forms part of a superspecies with the allopatric Speke's Weaver (942), and both constitute part of a species-group with the Black-headed Weaver (940)(HM; Br). Speke's is a bird of more open bushed and wooded habitats than Vieillot's, and extends into drier country. Their ranges are separate at QSD scale except in 60B, where both have been reported from Kisumu, although the Vieillot's there was probably a wanderer (Br). Speke's can be sympatric with the Black-headed Weaver but, in Kenya, it more often frequents thorny vegetation (HM). The Black-headed is generally in rather more open habitats than Vieillot's (*op. cit.*; ADL pers. obs.).

942 SPEKE'S WEAVER *Ploceus spekei*
B1181 / MPG1314 / D2C2

Afrotropical: a localised bird of the NE, from highland areas of NE Tanzania, Kenya and Ethiopia to N and E Somalia.

Kenyan: this gregarious and social species inhabits open bushed and wooded country at 1200-2200 m, and is one of the commonest weavers in the Nakuru 62A-Narok 74B-Nairobi 75B area of the highlands (Br; Ja). It is also a common breeding bird in modified habitats around habitation *e.g.* in urban and periurban Nairobi 75B (*op. cit.*; Gichuki 1983). Range is 74% in or E of the rift valley, 80% semiarid-subhumid (*cf.* Vieillot's Black Weaver (941)) and 87% within 500+ mm rainfall areas. A male in Moyale 17A on 27 Mar 1984 (ADL pers. obs.) represents the S limit of birds in Ethiopia. The high proportion of breeding squares (71%) suggests a mainly resident species, which agrees with the opinions of MPG, Be (74), Br and DEP (pers. obs.: 75B). However, Br also mentions wandering.

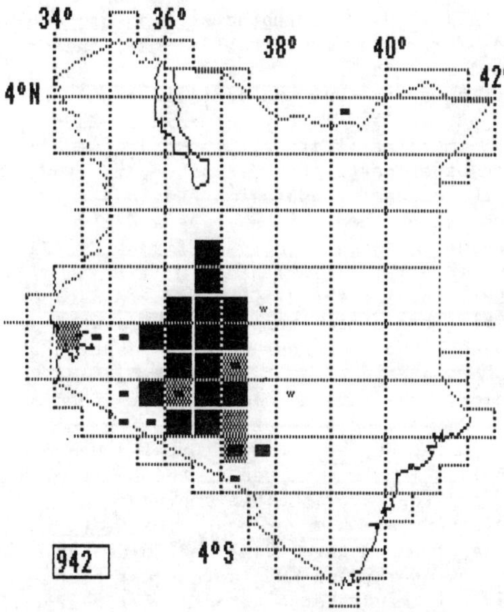

943 ORANGE WEAVER *Ploceus aurantius*
B1158 / MPG1334 / D2C0
NO MAP: post presence 60B

Afrotropical: W and C Africa, from Liberia to Congo, and E into the Congo basin. Further E, the race *P.a. rex* is endemic to the N shores of L Victoria, in NW Tanzania, Uganda and, at the E extremity of the range, W Kenya.

Kenyan: three reports from the L Victoria basin, at Kisumu 60B. The first record, in Mann (1976), omits details. The other two, both of single males, were on the lakeshore on 2 Nov 1981 (EABR 1981) and in lakeside papyrus on 25 Nov 1984 (EABR 1984). This square is moist subhumid, and within the 1000+ mm rainfall regime. The lakeshore is at about 1140 m altitude. There is no breeding record, and Br considers the early record at Kisumu to have been a wanderer. See Taveta Golden Weaver (944).

944 TAVETA GOLDEN WEAVER *Ploceus castaneiceps*
B1163 / MPG1333 / D1C2
NO MAP: post presence 100D; post confirmed 88C 100B; pre presence 101AC

Afrotropical: endemic to border areas of S Kenya and NE Tanzania.

Kenya: locally common, endemic to swamp and riverine habitats below 1500 m on the S borders (Br; Turner 1977; Walters 1978). All nests overhang water, but birds frequently wander to nearby open bushed areas (*op. cit.*; Ja). Abrupt desertion of colonies for weeks or months (Ja; MPG) indicates more extensive movement. Of the five squares, four are semiarid, and all are at 500-1500 m in intermediate, 250-1000 mm rainfall areas. There are records from other parts of E Kenya but, in view of the extensive confusion that has occurred between this species and close congeners, these require confirmation (Ja; MPG; Turner 1977; Br)(hence D1).

HM emphasise the difficulties in establishing taxonomic groupings amongst the weavers, and are unsure of the position of this species. In terms of range and ecology, the Taveta Golden Weaver appears to form part of a superspecies with the allopatric Northern Brown-throated Weaver (948), which is confined to the L Victoria basin. The Taveta Golden's plumage characters are however most similar to those of the Orange (943) and Golden Palm (945) Weavers, and HM also suggest that this trio might constitute a species-group. They are allopatric at QSD scale, with the Orange in the L Victoria basin, and the other two in closely adjacent ranges to the E of the rift valley. HM mention that, in cases of apparent sympatry, the Taveta Golden achieves ecological isolation from the Golden Palm by nesting exclusively in swamp vegetation. See Golden Weaver (946).

945 GOLDEN PALM WEAVER *Ploceus bojeri*
B1162 / MPG1331 / D2C2

Afrotropical: endemic to NE Tanzania, E Kenya, S Somalia and extreme SE Ethiopia.

Kenyan: a common, gregarious species of open bushed and wooded habitats, particularly those with palms, in the coastal strip (Br; Ja). Inland it is a more local bird of riverine habitats below about 1200 m, in the E lowlands and on the E periphery of the C highlands and adjacent plateaux (Br). Extension along the Tana R 64B-79C (AGH) parallels virtual restriction to the valleys of the Juba and Shebelle R of S Somalia (AM). Presence at Sarigo 19A in May 1901 (Erlanger 1905, 1907) is in the same square as the R Daua, and relates to occurrence to 3°N on the Juba, of which the Daua is a tributary. 76% of this weaver's squares contain coast and/or permanent rivers. Squares lacking permanent rivers are mainly in semiarid-dry subhumid, 500+ mm rainfall country.

Br considers this species resident, but also mentions wandering (*cf.* AM). See Taveta Golden Weaver (944) and Golden Weaver (946). Other reference: Turner 1977.

946 GOLDEN WEAVER *Ploceus subaureus*
B1183 / MPG1341 / D2C2

Afrotropical: the E periphery of Africa, N from the Cape through Mozambique and Malawi to E Tanzania and, on the N limits of the range, Kenya.

Kenyan: near water in reed beds and in adjacent bushed and wooded country, below 1300 m (Br; MPG; La). Like the closely related Golden Palm Weaver (945), the Golden is mainly a common bird of the

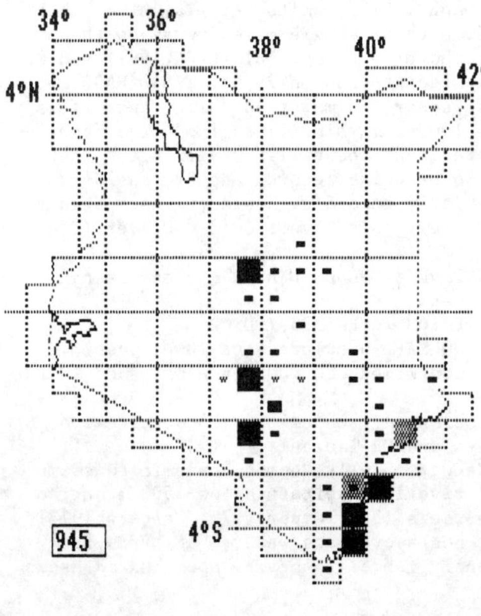

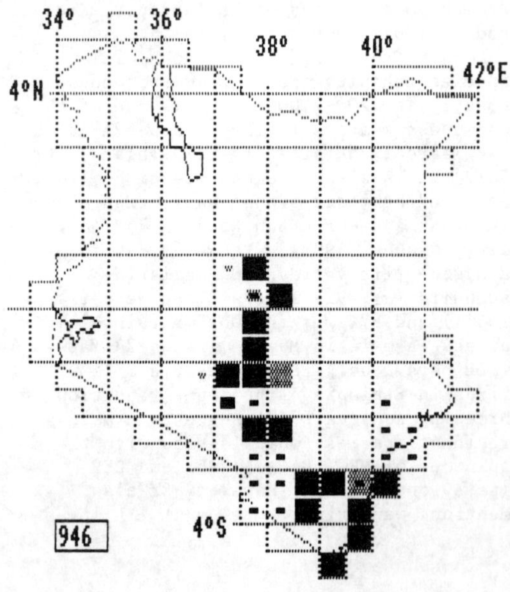

coastal lowlands that also extends more locally inland along rivers to the E edge of the C highlands and plateaux (Br). Range is entirely E of the rift valley, and 84% of the squares are coastal and/or contain permanent rivers. In contrast to the Golden Palm, the Golden is at the N and E limits of its range in Kenya, and penetrates less into dry country, even along rivers. Thus the Golden's range, even including riverine squares in arid country, is still 86% semiarid-subhumid and 76% within 500+ mm rainfall areas. In contrast to the Golden Palm, it is present along only the lower Tana R 79C 91A (AGH), does not extend so far NE along the Uaso Nyiro R 51AB, and is absent from the extreme NE 19.

Br describes the Golden as a resident and a wanderer, and MPG mention large congregations in the non-breeding season. La's description of this species as an uncommon visitor to arid Tsavo E 101B during the rains suggests some more regular movement.

HM place the gregarious Golden Weaver in a superspecies with the solitary Holub's Golden Weaver (947). They consider marginal overlap of these two species' Afrotropical ranges to possibly reflect nonbreeding wanderers, and Br describes them as allopatric. Holub's is segregated from the Golden at greater altitudes, but there is overlap at QSD scale along the E edge of the C highlands and plateaux. The two birds' combined ranges occupy most of Kenya's wetter, 500+ mm rainfall areas.

The Golden Weaver is closely related to and sympatric with the Taveta Golden Weaver (944) and the Golden Palm Weaver (HM; Br).

947 HOLUB'S GOLDEN WEAVER *Ploceus xanthops*
B1189 / MPG1342 / D2C2

Afrotropical: widespread between 5°S and 20°S, E to C Tanzania, Malawi and C Mozambique. Further S, this weaver reaches S Mozambique and E South Africa. Further N, it extends to the E Zaire borders, Uganda and Kenya. Distribution in Kenya represents the N and E limits of the range.

Kenyan: rather local and retiring in moist bushed habitats, usually near water, at 900-2300 m (Br; Ja; MPG; HM). Range is in the highlands and adjacent plateaux, 78% subhumid-humid. All squares receive 500+ mm rainfall except 75C, where occurrence is in the lush foothills of the Nguruman escarpment. The species is resident (Br; MPG; DEP pers. obs.: 75B). See Golden Weaver (946).

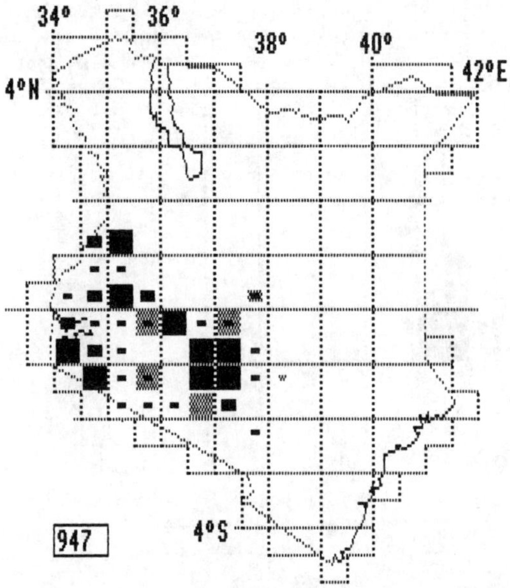

948 NORTHERN BROWN-THROATED WEAVER *Ploceus castanops*
B1164 / MPG1330 / D2C3
NO MAP: post presence 60C; post probable 60A; post confirmed 48AC 60B; pre confirmed 60A

Afrotropical: endemic to Nile catchment areas of N Tanzania, Uganda and W Kenya.

Kenyan: first reported from Kenya by Britton & Harper (1969), this locally common weaver is restricted to papyrus and adjacent habitats in the Winam Gulf 60ABC and at L Kanyaboli 48C and Ukwala 48C (Britton 1978; Br). The record in 48A is from Uganda (J S Ash *in litt.*). The five squares are all at 1000-1500 m in moist subhumid-humid, 1000+ mm rainfall country. It nests solitarily or colonially over water and, in contrast to its congeners, often forages in vegetation overhanging water or on floating waterlily pads (MPG; HM; Britton 1978). See Taveta Golden Weaver (944).

949 YELLOW-BACKED WEAVER *Ploceus melanocephalus*
B1173 / MPG1325 / D2C3

Afrotropical: mainly W and C Africa, in the N tropics from Senegambia to Chad, and SE through the Congo forests to Zambia, Tanzania, Uganda and, at the E limit of the range, W Kenya.

Kenyan: a gregarious and mainly very common bird of reeds, bush and tall grass

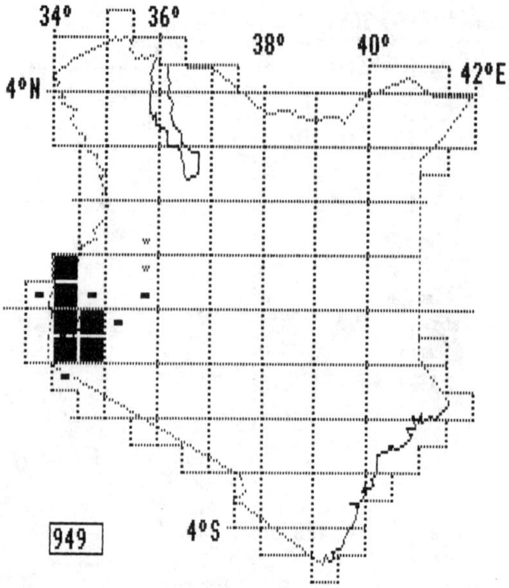

near water, up to 1700 m (Br; Britton 1978; Ja; MPG). Occurrence is mainly W of the rift valley, but it extends E to the latter in the Kerio Valley, between Tot 37D and Chebloch 49D (Br). This range is 69% at 1000-1500 m, 92% moist subhumid-humid, and 92% within 1000+ mm rainfall areas. Br considers this species to be mainly resident, but also notes presumed wanderers in unsuitable country away from water.

HM divide the "masked weavers" into those closely associated with water and those not. The former are a species-group containing the Yellow-backed, Golden-backed (950: *q.v.*), Northern Masked (951), Jubaland (952) and Masked (953) Weavers, while the latter are discussed under Vitelline Masked Weaver (954).

HM mention the possibility of local ecological segregation in the Yellow-backed species-group, but this does not appear to be the case amongst the more closely related members. The particularly closely related Yellow-backed and Golden-backed typically nest in mixed colonies in the W, with no obvious segregation (Ja; Britton 1978; Br). The Northern Masked is geographically allopatric with the Yellow-backed, but shares colonies with the Golden-backed at L Baringo 50A (Br; St). The Jubaland is geographically isolated from the preceeding three species, in the extreme NE.

In contrast, the Masked Weaver occurs throughout Kenya, and is at least broadly sympatric with the other four. HM regard the Masked as the least certain member of this species-group, and point out that, in plumage characters, it is closely similar to the distinctly smaller Little (957) and Slender-billed (958) Weavers. In contrast to the rest of its group, the Masked often moves well away from water (Br; *cf.* AM). At L Baringo 50A, the Masked nests over water alongside Golden-backed and Northern Masked Weavers (St). However, the Masked and the Golden-backed are reproductively allopatric in Kenya to the W of the rift valley, and mainly or even totally so in Tanzania (Br). The Masked is normally resident in Amboseli GR 88C, but most if not all left the area when abnormally heavy rain coincided with an unusual influx of Golden-backed Weavers (Walters 1978). This evacuation of semiarid country at the onset of rain is contrary to the movements noted by Br and La.

950 GOLDEN-BACKED WEAVER *Ploceus jacksoni*
B1171 / MPG1326 / D2C2

Afrotropical: confined to East Africa and adjacent areas of Burundi and S Sudan.

Kenyan: gregarious and locally common, typically associated with swamp and riverine habitats in open bushed or wooded country up to 1800 m (Br; Ja; MPG; St). Normal range is on the plateau to the W of the rift valley, together with the rift valley floor on the N periphery of the highlands (Br; St). This distribution is 88% above 1000 m, 82% subhumid-humid, and entirely within 500+ mm rainfall areas. Within this range, St notes resident status at L Baringo 50A.

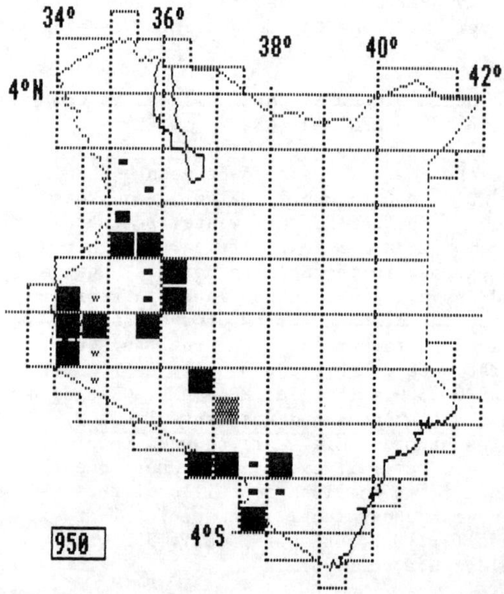

This species can also appear and breed on the E plateau, in the Konza 76C-Namanga 87D-Amboseli 88C-L Jipe 100D area (Br). Of these eight E squares, six are above 1000 m, seven semiarid, and all within the intermediate, 250-1000 mm rainfall regime. Breeding on the E plateau at Nairobi 75B is by formerly captive birds (EABR 1977), and is omitted from the above analysis. Br describes occurrences on the E plateau as erratic and considers movements to semiarid areas as wandering, but this irregular movement to a relatively restricted E area is more irruptive in character. Walters (1978) reports sudden appearance and attempted breeding in Amboseli GR 88C during unusually heavy rains, but whether rainfall generally stimulates these irruptions is unknown. Due to this irruptive nature, this species' range is more extensive and fragmented than that of the very closely related but mainly resident Yellow-backed Weaver (949: *q.v.*). Other reference: Backhurst 1977.

951 NORTHERN MASKED WEAVER *Ploceus taeniopterus*
B1185 / MPG1316 / D2C2
NO MAP: post presence 3B 4A; post confirmed 50AC

Afrotropical: the NE, from NE Zaire to Sudan and extreme SW Ethiopia. Range has apparently extended only recently to the rift valley on the N slopes of the Kenya highlands.

Kenyan: this gregarious weaver is a common resident of reeds and bush around the edge of freshwater L Baringo 50A, and throughout the swamp N of L Bogoria 50C (Br; St; T Stevenson *in litt.*; EABR 1978). These rift lakes are at approximately 960-1000 m, in a semiarid-dry subhumid area with 500-1000 mm rainfall. In the absence of previous records, presence in these areas apparently represents a substantial recent extension of range (Br). Confusion with close congeners is possible, and records from elsewhere in Kenya and Uganda lack confirmation (Br).

Presence in 3B and 4A represents the S limit of Ethiopian birds (J S Ash *in litt.*). Breeding at Sagantal, near L Turkana (MPG) is presumably also in extreme SE Sudan or SW Ethiopia. See Yellow-backed Weaver (949).

952 JUBALAND WEAVER *Ploceus dichrocephalus*
B1166 / MPG1325 / D2C3
NO MAP: post presence 7C 19AB; pre presence 6D

Afrotropical: the NE, largely confined to the valleys of the Daua, Juba and Shebelle R of S Ethiopia, extreme NE Kenya and S Somalia.

Kenyan: gregarious and locally common, restricted to the vicinity of the Daua R 6D 7C 19AB, on Kenya's NE borders (Br; MPG). Br describes it frequenting villages but the only note that we can trace of this habit in Kenya is from D A Turner (*in litt.*), who saw a colony very close to a small village on the banks of the Daua.

This closeness to water appears to be this species' real trait. E J Roberts (*in litt.*), who was resident at Rhamu 19A, found it almost exclusively in lush riverine bush, and her observations agree with this species-group's affinity for water, with Williams & Arlott's (1980) and AM's descriptions of the bird's habitat, and with the observations of T Stevenson and ADL at Rhamu Dimtu 7C and Rhamu 19A in June 1984. All four squares are below 600 m, in arid, 0-250 mm rainfall country. There is no confirmed breeding record, but Roberts (*op. cit.*) found it apparently resident at Rhamu 19A. See Yellow-backed Weaver (949).

953 MASKED WEAVER *Ploceus intermedius*
B1170 / MPG1319 / D2C2

Afrotropical: widespread S of 10°S, though absent from the dry SW. It also extends through East Africa to S Sudan, Ethiopia and all except E areas of Somalia.

Kenyan: this gregarious species occurs in a wide variety of open bushed and wooded habitats, up to 2000 m, but mainly

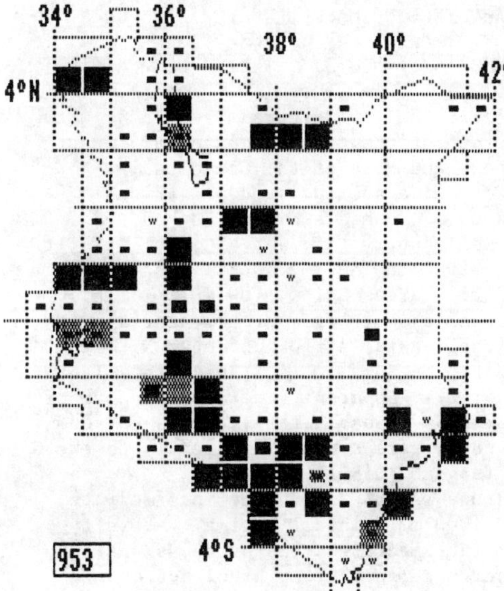

below 1500 m (Br; MPG; Be; HH; La; St). It is peculiarly local but often common (Br). Range is throughout our altitude categories. Br states that it often avoids the wettest country, but there are records from 59% of the 1000+ mm rainfall squares. Reports from only 21% of the most arid squares (moisture index below 1.2) and 30% of those with 0-250 mm rainfall suggest apparently uncommon status in the NE to be real (*cf.* Br). This pattern is perhaps exaggerated by a low frequency of observers and/or utilisation of these very dry areas during and shortly after the rains, when they are temporarily verdant but least accessible (see Crested Lark (543)).

Be (74), Walters (1978: 88C), St (50A), Br and DEP (pers. obs.: 75CD 88AD 89AC 100B 101A) all consider this weaver resident, but Br also mentions wandering, and movement into arid-semiarid country during the rains. The latter, more regular type of movement occurs in arid Tsavo E 101B, where visitors are fairly common during the rains, breeding in woodland near water (La). Br also mentions residence near water in arid-semiarid areas, as at L Turkana (HH) and Baringo 50A (St). The Masked's squares in S Somalia mainly contain coast and/or permanent rivers (AM). This type of localisation can also be seen in Kenya E of 39°30'E, where distribution is mainly in the coastal strip, or along the Daua 19AB and Tana R 65B-91A. See Yellow-backed Weaver (949). Other references: Gichuki 1982; Pomeroy *et al.* 1982.

954 VITELLINE MASKED WEAVER *Ploceus velatus*
B1187 / MPG1317, 1321 / D2C2

Afrotropical: widespread S of 15°N, with the exceptions of the equatorial forest areas, and most of the Congo basin.

Kenyan: a solitary and gregarious weaver of bushed and open bushed and wooded country (Br; Ja; MPG; Be; St). It occurs up to 1800 m, but is most commonly in drier areas below 1400 m (*op. cit.*). Virtual absence from even semiarid areas of the coastal strip continues N to the equator in Somalia (AM) and S into Tanzania (Br). Apparent general absence S of 3°30'S is probably real, and is corroborated by rare status immediately to the N in Tsavo E 101B (La).

Range is 78% within the intermediate, 250-1000 mm rainfall regime. Affinity for intermediate areas produces sparse occurrence in the wet highlands and W plateau, and also in the dry N and NE,

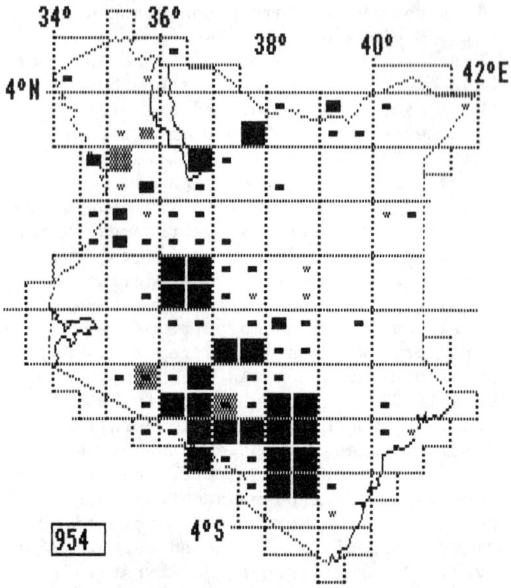

including L Turkana. Distribution is 77% arid-semiarid but, as with rainfall, only 2% in the driest areas (moisture index below 1.2). In the arid E, presence in squares containing the Tana R compares with restriction to the valleys of the Juba and Shebelle R of S Somalia (AM).

Br, Be (74), St (50A) and DEP (pers. obs.: 75CD 88A 89A) all consider this species resident, but Br also notes wandering (*cf.* AM). La's mention of rare visitors to arid Tsavo E 101B during the rains suggests some more regular movement.

HM consider this species and Heuglin's Masked Weaver (955) to form part of a superspecies that constitutes a group of masked weavers not associated with water (see Yellow-backed Weaver (949)). Heuglin's inhabits woodland and is virtually allopatric with the Vitelline (Br), which extends into distinctly drier country (*cf.* HM).

955 HEUGLIN'S MASKED WEAVER *Ploceus heuglini*
B1168 / MPG1318 / D2C2
NO MAP: post presence 36B 37AC; post probable 48A 49A; pre presence 48B; pre confirmed 37C

Afrotropical: replaces the Vitelline Masked Weaver (954: *q.v.*) at lower, moister latitudes in the N tropics. Heuglin's has an apparently rather discontinuous range E from Senegambia to Ghana, Nigeria, N Zaire, S Sudan, Uganda, and at its E limit, W Kenya.

Kenyan: a usually solitary, woodland bird, with very few Kenyan records (MPG; Br). There are old reports from Trans Nzoia 37C (breeding: Stoneham 1944e) and Bungoma 48B (Br), and T J Barnley (*in litt.*) knows it as a Jan-Mar visitor to the Kodich 36B-Kunyao 37A area.

An EANHS Nest Record Card describing breeding at L Baringo 50A is quoted by Br but omitted here, due to the possibility of confusion with another species, the uncharacteristic environment (see below), and the lack of subsequent records despite an observer resident in the area. Recent breeding indications involve a male at a nest 18 km west of Bungoma 48A on 1 June, and one with two or three females, nest building near Soy 49A on 7 Aug (EABR 1980).

In contrast to the range of the Vitelline Masked Weaver, Heuglin's six squares are all at 1000-2000 m to the W of the rift valley, in well-wooded, subhumid-humid, 500+ mm rainfall areas. The three breeding squares are all in humid, 1000+ mm rainfall country. T J Barnley's data (see above) agree with MPG's note of local migration.

956 CHESTNUT WEAVER *Ploceus rubiginosus*
B1180 / MPG1327 / D3C2

Afrotropical: separate populations in the NE and SW. The former is in areas of East Africa to the S and E of L Victoria, N to extreme SE Sudan, Ethiopia and W Somalia. In the SW, it occurs in Namibia and Angola.

Kenyan: gregarious and locally common in open bushed and wooded habitats of drier areas, mainly below 1500 m (Br; Ja; MPG; AGH; HH). Occurrence up to 2000 m is regular, but records at 3000 m are exceptional, *e.g.* a June bird at Mau Narok 61D (Br; Britton 1979). Range is 75% above 500 m and 68% within the intermediate, 250-1000 mm rainfall regime (*cf.* Pomeroy *et al.* 1982).

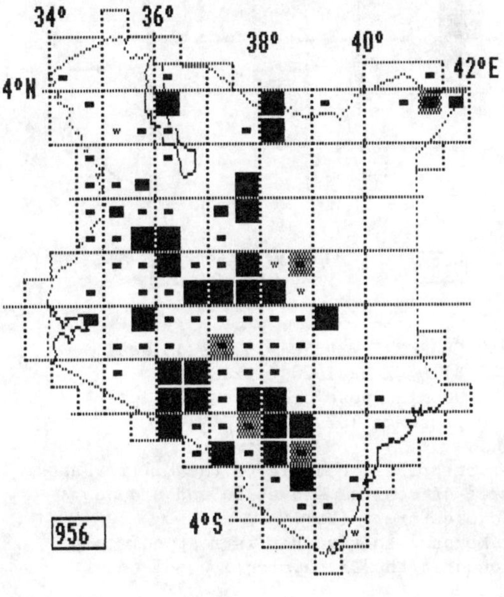

Virtual absence from the E lowlands is corroborated by lack of records from observers resident in 42A and 65B, and by restriction of range in adjacent S Somalia to the valleys of the Juba and Shebelle R (AM). It is similarly restricted in Kenya E of 40°E to the vicinities of the Daua R 7C 18B 19AB and the Tana R 79C 91A (AGH). The single coastal record is from Mombasa 114B (Ja). Virtual absence from even semiarid areas of the coastal strip contrasts strongly with its inland range, which is 40% semiarid. Scarce occurrence in the L Victoria basin and the adjacent moist W plateau 48 60 73 74 is evident on the map, but similar status in the moist highlands is probably obscured at QSD scale by wanderers to abnormal altitudes *e.g.* the Mau Narok 61D bird mentioned above. Presence in arid areas (30% of the range) is mainly confined to squares bordering moister country *e.g.* 52B S to 101D and, in the N, to the vicinity of L Turkana (HH) and the Daua R 7C 18B 19AB.

This weaver's breeding squares are 64% less arid-semiarid and 72% within intermediate, 250-1000 mm rainfall areas (*cf.* Red-billed Quelea (986)). Despite the paucity of observers in more remote regions, particularly during the wet seasons when it breeds (Brown & Britton 1980, see Masked Weaver (953)), the lack of breeding records from the driest environments is probably real (Pomeroy *et al.* 1982; *cf.* HM).

The Chestnut Weaver's extensive movements are stimulated by rainfall, and occurrence is seasonal in most areas (Br). Numbers in the semiarid country around L Baringo 50A fluctuate both monthly and from year to year (St). Thousands usually breed there during the May-Jul cool wet season but, thereafter, dispersal produces uncommon status from Jul to Nov or, in very dry years, complete Aug-Apr absence. This weaver is a fairly common visitor to arid Tsavo E 101B during both wet seasons, but particularly in Dec (La). Occurrence in other arid areas, particularly those lacking extensive permanent water, is probably similarly seasonal. The unusual appearance in 1984 of large flocks in Nairobi gardens 75B (1650 m, moist subhumid) coincided with very dry weather (many obs.). A bird ringed at

Ngulia Lodge 101A was recovered almost five months later 213 km away at Migwani 77A (Backhurst 1977).

957 LITTLE WEAVER *Ploceus luteolus*
B1172 / MPG1324 / D2C2

Afrotropical: the N tropics between 7°N and 17°N, E to Ethiopia, and SE to NW Tanzania, Uganda and, at the SE extremity of the range, Kenya.

Kenyan: locally common and gregarious in dry woodland and open bushed and wooded habitats at 400-1500 m (Br; Ja; MPG; HH; St). With the possible exception of 38D, this species' squares are entirely within or W of the rift valley. *P.l. kavirondensis* is restricted to the L Victoria basin, E to Muhoroni 61A, while the remainder of the Kenyan range from Bogoria 50C N to Turkana is occupied by the nominate race (Br). Due to these contrasting distributions, the environmental characteristics of these two races differ markedly. *P.l. kavirondensis* is mainly at 1000-1500 m, in moist subhumid, 1000+ mm rainfall areas, while nominate birds range from 400-1500 m, and throughout the rainfall and moisture regimes. This weaver is resident at semiarid Baringo 50A (St) and Br considers it generally so, though also mentioning wandering.

The Little Weaver forms part of a superspecies with the Slender-billed Weaver (958)(HM). These two birds exhibit extensive range overlap but remain at least mainly allopatric by means of ecological segregation, in which the Slender-billed is virtually confined to papyrus (Britton 1978; Br; HM; Ja). This relationship is illustrated at L Kanyaboli 48C, where the Little breeds in seasonally inundated woodland, while the Slender-billed breeds in the adjacent swamp (Br). See Yellow-backed Weaver (949).

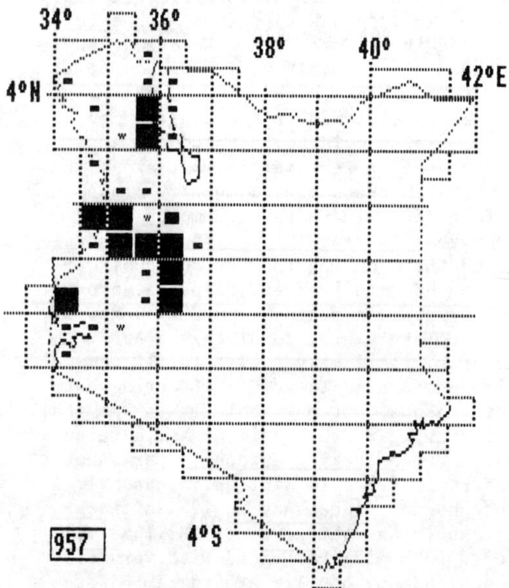

958 SLENDER-BILLED WEAVER *Ploceus pelzelni*
B1179 / MPG1343 / D2C2

Afrotropical: mainly W and C Africa, from the Gulf of Guinea and the Congo basin, E to Tanzania, Uganda and, at the E limit of the range, W Kenya.

Kenyan: a gregarious and locally common bird, virtually restricted to papyrus in the L Victoria basin, up to 1700 m (Br; Britton 1978; Ja; MPG). The record in 48A is from Uganda (J S Ash *in litt.*). Range is above 1000 m, to the W of the rift valley. The squares are all moist subhumid-humid, and within 1000+ mm rainfall areas. Br describes this weaver as mainly resident, but reports an apparent wanderer at Ng'iya 48C. See Little Weaver (957).

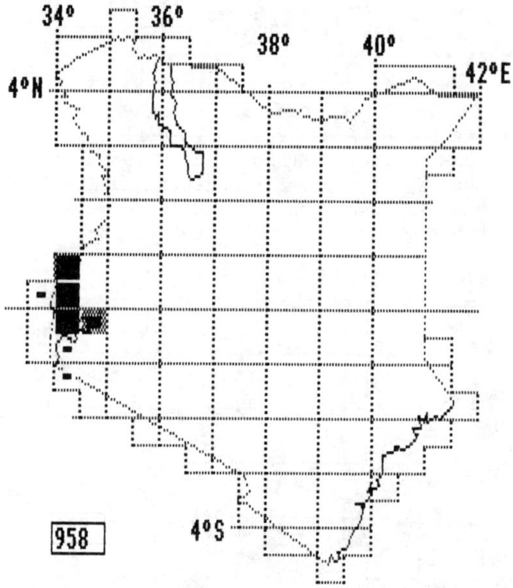

959 CLARKE'S WEAVER *Ploceus golandi*
B1167 / MPG1320 / D2C3
NO MAP: post presence 102B; pre presence 102D

Afrotropical: endemic to the Kilifi-Sabaki area of coastal Kenya, in and around Sokoke Forest (Br).

Kenyan: this weaver is most commonly found in the *Brachystegia* woodland of

Sokoke Forest 102BD, but also occurs in true forest there (Britton & Zimmerman 1979; Br). Often a member of mixed species foraging parties, it also congregates in monospecific flocks of over 100 individuals (*op. cit.*: but see Short & Horne 1985a). Total population is probably less than 2000 pairs (Britton *et al.* 1984). Williams (1957) describes a locality in 102D, but this area may no longer be forested (see Britton & Zimmerman 1979), and all recent records are apparently from 102B. This species is endangered by forest destruction at Sokoke (Turner 1977; Britton & Zimmerman 1979). Range is below 150 m (Br), and the two squares are semiarid-dry subhumid, and within the 500+ mm rainfall regime.

This is an enigmatic species because, despite its apparently accessible and extremely restricted range, it remains little known. The lack of records between the collection of the type specimen in 1913 and the species' rediscovery in 1955 gave rise to the proposal of the type as an aberrant Masked Weaver (953)(Williams 1957; Ja; MPG; Ripley & Bond 1971). Although Clarke's is now reported annually from Sokoke (Mann 1976; EABR), its elusive character persists in that sightings are markedly seasonal, with intervening apparent absences of months' duration, and breeding has never been observed (Br). Most records are Aug-Nov, but there are also reports during Apr, Jul and Dec (Br; Taylor 1984b; Short & Horne 1985a; EABR). This lack of breeding records must be due either to oversight or, in view of the Jan-Mar absence suggested above, to breeding away from Sokoke.

Breeding indications are few. Large flocks that appeared in Sokoke on 16 Aug were mostly immature birds (EABR 1978). A pair on 18 Apr were thought to be nesting nearby (EABR 1979). A flock on 4 Apr included juveniles which, although free-flying, were closely accompanied by adult female birds (Taylor 1984b). Some of these juveniles begged for food from their attendant females, and one was actually fed. Taylor (*op. cit.*) suggested this activity to represent Feb nesting but noted that, since breeding in this month is not common in any other *Ploceus* species (including a forest dweller) in the coastal lowlands, the possibility of a breeding area outside Sokoke cannot be precluded.

Unless Clarke's utilises some extremely localised and obscure breeding site, the sole sufficiently poorly known forest areas between E Tanzania and S Somalia would appear to lie between the Tana R and the Somali border. However, in view of the males' distinctive plumage, the total absence of any record away from Sokoke more strongly suggests true endemism to this forest. Short & Horne (1985a) concur with this latter view, and suggest that breeding may occur in both the Jul-Sept and Jan-Mar periods.

Non-detection of birds breeding in Sokoke likewise appears improbable but Britton *et al.* (1984) suggest that, although gregarious during the off-season, Clarke's may not be so while breeding (*cf.* the EABR 1979 pair). They propose that breeding could be overlooked if the nests are thinly distributed in the tops of tall forest trees, and the birds more or less silent (*cf.* HM).

The final intriguing point concerns Weyns' Weaver *P. weynsi*, which has a far more extensive range in Zaire, Uganda and possibly NW Tanzania, and which forms a superspecies with Clarke's (HM). Although a bird of forest edge and moist bush (Br), Weyns' mirrors Clarke's characteristics and enigmas almost exactly. It is conspicuous and gregarious, but wandering flocks appear erratically and can be apparently absent for periods of months, and breeding is unknown (Br). Other reference: Clancey & Williams 1959.

960 BAGLAFECHT WEAVER *Ploceus baglafecht*
B1159 / MPG1323, 1347-1349 / D3C3

Afrotropical: mainly highlands of the E, from Ethiopia and S Sudan, S through East Africa and the E Zaire borders, to Zambia and Malawi. There is also a population in

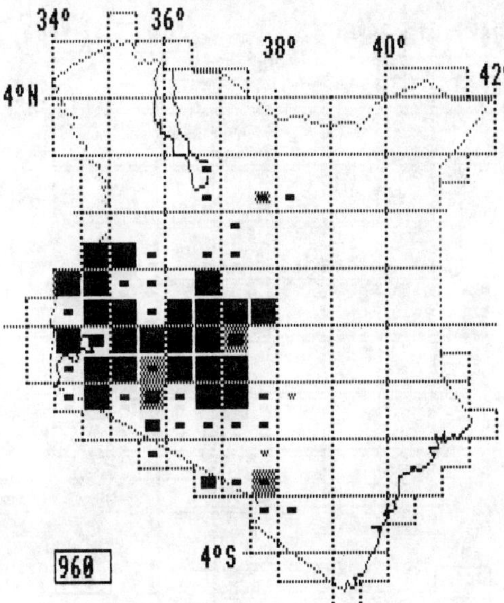

Cameroon. Distribution in Kenya represents the E limit of the range.

Kenyan: this conspicuous species, usually seen in pairs, is common in a wide range of wooded and forest edge habitats, up to 3000 m in the highlands and peripheral plateaux (Br; Ja; MPG; Be; Se; Britton & Sugg 1973; Diamond & Keith 1980). It is well known in gardens, and is present on the forest islands of Mts Kulal 26B (Diamond & Keith 1980), Nyiru 26D (around forest at 2500 m: Pelchen 1977d) and Marsabit 27D 28C (Ja; Br), and the Ndoto Mts 39A, the Mathews Range 39C, the Chyulu Hills 88D and the Taita Hills 101A (Br). Excluding these isolated highlands, which are too small to appear on our overlays, range is 98% above 1000 m, 94% within 500+ mm rainfall areas and 76% subhumid-humid.

Distinctive races occur. Reichenow's Weaver *P.b. reichenowi* is found in most parts of the Kenyan range, but intergrades with the nominate race in the Mt Elgon 36D-Kerio area (approximately 37D 49B) and with Stuhlmann's Weaver *P.b. stuhlmanni* in Bungoma 48B and Siaya 48C 60A (Br). The Baglafecht Weaver is resident (Br; Be (74); Se (61D); ADL and DEP (pers. obs.: 75B)). Other reference: Collias 1978.

961 COMPACT WEAVER *Ploceus superciliosus*
B1184 / MPG1350 / D2C2

Afrotropical: W and C Africa, from the country around the Gulf of Guinea, E through the Congo basin to SE Ethiopia, NW Tanzania, Uganda and, at the E limit of the range, W Kenya.

Kenyan: local and uncommon, foraging singly or in social groups in a wide diversity of habitats, including open grassland, bush, woodland and forest, but mainly nesting in marshes (Br; Ja; MPG). Br notes it up to 1700 m to the N of the Winam Gulf 60AB, but there are subsequent records further S, to the SE of Homa Bay 60D (EABR 1981) and from the Migori R near Lolgorien 73B (A Lohding *in litt.*). Range is above 1000 m, to the W of the rift valley. All squares are moist subhumid-humid, and within the 1000+ mm rainfall regime.

962 SPECTACLED WEAVER *Ploceus ocularis*
B1177 / MPG1337 / D2C2

Afrotropical: widespread S of 5°N, E to Ethiopia, S Sudan, Kenya and the E coast, and absent from the dry SW and most of the Congo basin. Distribution in Kenya represents the E limit of the range.

Kenyan: fairly sparse, usually in pairs, and easily overlooked in the dense undergrowth and creepers of bushed and forested habitats up to 2200 m (Br; Ja; MPG; Be). Range is throughout our altitude categories, 86% within the 500+ mm rainfall regime and only 3% arid. Occurrence in arid country is confined to the vicinity of the lower Tana R 79C (AGH) and, formerly, the Voi R 101B (La). HH mention it as occasional at Ferguson's Gulf 13B, L Turkana. We omit this locality because, in contrast to 79C and 101B, 13B is arid and not adjacent to semiarid country, and because even the

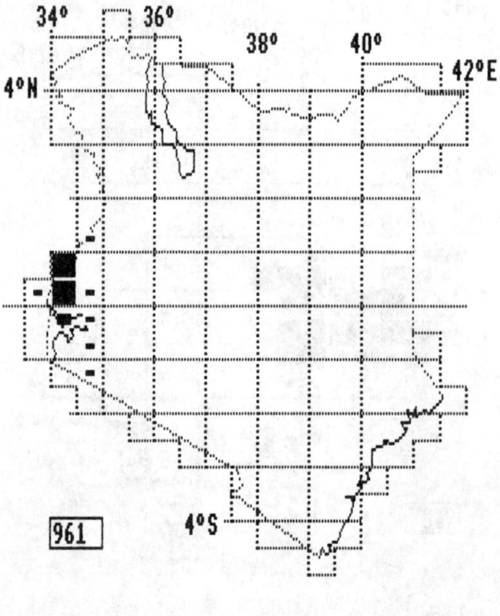

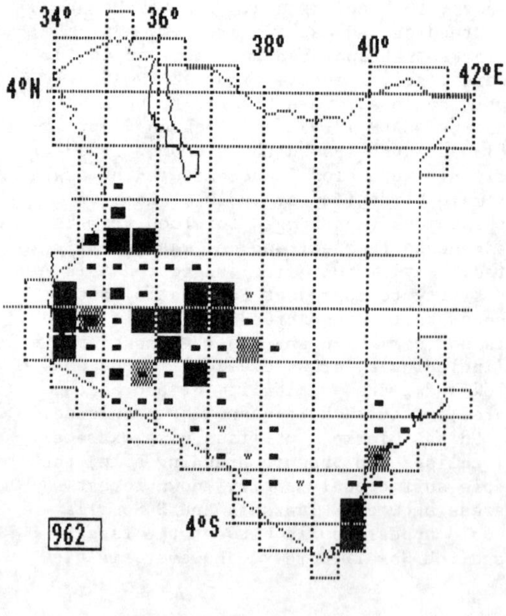

Black-necked Weaver (963: *q.v.*), a bird that extends into drier country than the Spectacled, is not recorded from Turkana. Br also omits this locality.

The Spectacled is considered resident, with different races on either side of the rift valley (Br; Be). A bird in the garden of L Baringo Club 50A on 7 Mar (St) was presumably a wanderer, just beyond the limit of its normal range.

The Spectacled Weaver forms a superspecies with the Black-necked and Black-billed (964) Weavers (HM). All three forage in dense vegetation. Their ranges overlap widely at QSD scale, but sympatry is mainly precluded by habitat preferences (HM; Br). The Spectacled Weaver is typical of moister, lusher vegetation than the Black-necked (*e.g.* La), while the Black-billed is restricted to highland forest. However, Br notes true sympatry between the Black-necked and the Spectacled at the Kenya coast.

963 BLACK-NECKED WEAVER *Ploceus nigricollis*
B1176 / MPG1336 / D2C2

Afrotropical: from the W coast between 15°N and 10°S, E through the Congo basin and adjacent country to East Africa and S areas of Sudan, Ethiopia and Somalia. The Black-necked inhabits the lowland equatorial forests of W and C Africa where the Spectacled Weaver (962: *q.v.*) is virtually absent (HM). In East Africa, where the Spectacled is more widespread, the Black-necked is more typical of woodland and bush.

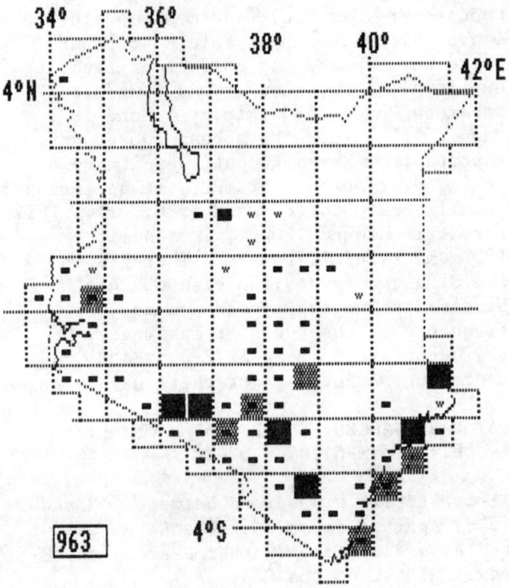

Kenyan: two races, with differing environmental characters. The W and C African nominate race reaches W Kenya. Br lists localities to the N of the Winam Gulf 60AB but, by describing the other race on the NE, E and S edges of the C highlands, presumably implies nominate birds to be general W of the rift valley. This range is entirely above 1000 m, 92% moist subhumid-humid and 83% within 1000+ mm rainfall areas. The nominate race is a locally common resident which, although it may occur in forest, is more usually in forest edge, or in more open thickets and secondary vegetation nearby (Br; Ja; MPG).

P.n. melanoxanthus replaces the nominate race further E, in the rift valley at Magadi 75C 87A and on the E plateau (Br). In contrast to the nominate's range, *melanoxanthus* is 94% at 0-1500 m, 86% arid-semiarid and 88% within the intermediate, 250-1000 mm rainfall regime. *P.n. melanoxanthus* is an uncommon to fairly common resident of drier bushed and wooded areas up to 1300 m (Br; La). It is found in the coastal lowlands S to Mombasa 114B (Br), and inland to the NE, E and S peripheries of the C highlands and their adjacent plateaux. Br mentions this race in river valleys, but it is notably absent from most of the Tana R. Distribution in Kenya E of 40°E compares with S Somalia, where this race occupies the coastal strip, but not the Juba riverine habitats (AM). The small cluster of old records around 39D probably reflects the low frequency of observers, and this species' shy nature.

The record near Lokichokio 2C (J Harjula *in litt.*) is of unknown race. This square is at 500-1000 m, in a semiarid, 500-1000 mm rainfall environment. The total range of the Black-necked Weaver differs from that of the Spectacled Weaver by being peripheral to the highlands, 84% below 1500 m, 23% arid and 36% in 0-500 mm rainfall areas. Other reference: Cunningham-van Someren 1971a.

964 BLACK-BILLED WEAVER *Ploceus melanogaster*
B1174 / MPG1346 / D2C2

Afrotropical: apart from a population in Cameroon, this is a highland forest species of the E, from the E Zaire borders to NW Tanzania, S Sudan, Uganda and, at the E limit of the range, W Kenya.

Kenyan: uncommon in the undergrowth and mid-levels of forest, mainly at 1500-3000 m but also down to 1200 m (Br; Ja; MPG; Se). It is usually seen in pairs. Occurrence is mainly W of the rift valley, but it reaches the E plateau on the Karissia

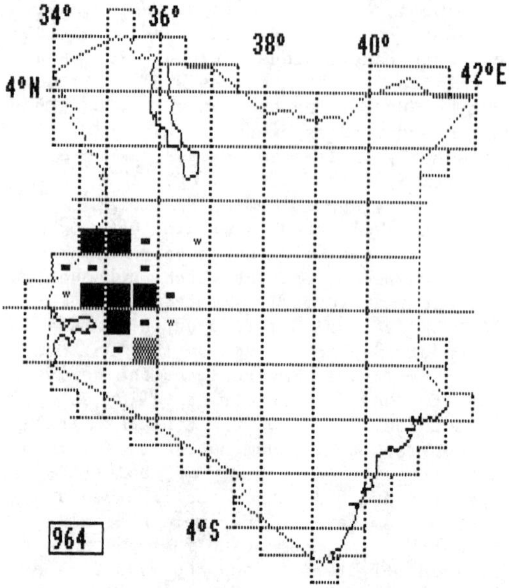

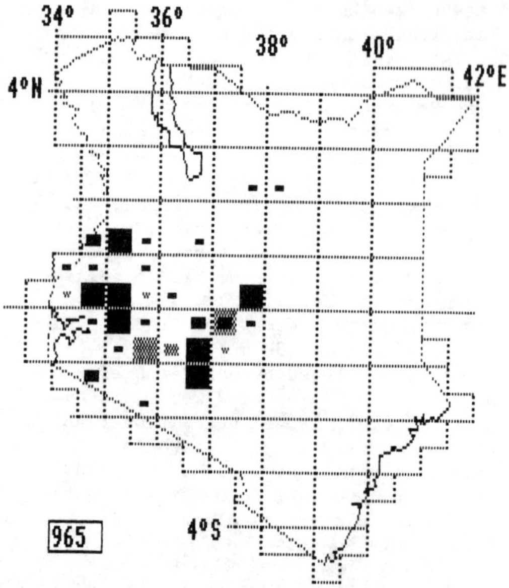

Hills 38D and at Subukia 62A (Br). In contrast to the closely related Spectacled (962: *q.v.*) and Black-necked (963) Weavers, range is 70% above 1500 m, 81% moist subhumid-humid and 76% within 1000+ mm rainfall areas. This weaver is resident (Br; Be).

965 BROWN-CAPPED WEAVER *Ploceus insignis*
B1169 / MPG1351 / D2C2

Afrotropical: known from Cameroon and Angola in the W, and in the E highlands from the E Zaire-Tanzania borders to S Sudan, Uganda and, at the E limit of the range, Kenya.

Kenyan: a fairly common, solitary bird of highland forests and bamboo, including edge and partially modified habitats, at 1200-3000 m (Br; Ja; MPG; Se). It has a distinctive foraging method, crawling on the trunks and branches of trees like a Nuthatch *Sitta europaea* (*op. cit.*; HM). Excluding the forest island of Mt Marsabit 27D 28C (Br), which is too localised to appear on our overlays, range is 79% moist subhumid-humid and 75% within 1000+ mm rainfall areas. Br, Be and Se all consider this weaver resident, although the last also mentions local movements.

966 YELLOW-MANTLED WEAVER *Ploceus tricolor*
B1186 / MPG1353 / D2C1
NO MAP: post presence 48D

Afrotropical: W and C Africa, from the forests around the Gulf of Guinea, E through the Congo basin to Uganda and, at the E extremity of the range, W Kenya.

Kenyan: a forest canopy species that also occurs in more open, edge habitats with tall trees (Br; Ja). It forages by climbing spirally up tree trunks, like a woodpecker (Ja; MPG). Ja mentions this weaver along the upper Yala R 48D, but there were apparently no further records until the 1960s, when it was found in Kakamega Forest 48D (Ripley & Bond 1971; Mann 1976). There have been subsequent reports from Kakamega but, despite being easily overlooked (Ja; MPG), it appears actually rare there (Mann 1976; Br). This forest is around 1700 m, in a humid, 1000+ mm rainfall square. There is no breeding record and, in view of this bird's distinctive appearance and the frequency of observers at Kakamega, the few occurrences (none 1977-84: EABR) may represent wanderers E of their usual range.

967 DARK-BACKED WEAVER *Ploceus bicolor*
B1161 / MPG1335 / D2C2

Afrotropical: widespread between 5°N and 15°S, apart from most of the Congo basin; and also extending further S along the E coast almost to the Cape.

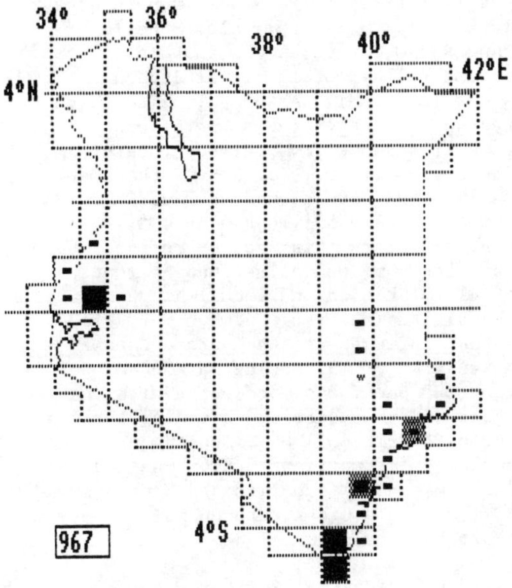

Kenyan: a locally common, solitary weaver that frequents forest and moist woodland with thickets (Br; MPG). Two distinctive races occur. *P.b. mentalis*, which has a greyish mantle, is found up to 1900 m in areas to the W of the rift valley and N of the Winam Gulf 60AB (Br; Ja). Its range is entirely above 1000 m, in humid, 1000+ mm rainfall areas.

P.b. kersteni, with a black mantle, occurs in the coastal lowlands and inland along the Tana R 65B-91A (Br; Britton & Zimmerman 1979). This distribution is entirely below 500 m. Excluding the arid squares 65BD 78B 79C, where occurrence is totally confined to the locally lush habitats around the Tana R (*e.g.* AGH), this lowland range is semiarid-dry subhumid, and within the 500+ mm rainfall regime. Presence along the Tana compares with virtual restriction to the Juba R of adjacent S Somalia (AM). Br considers this species resident. Other reference: Fletcher 1983.

968 RED-HEADED MALIMBE *Malimbus rubricollis*
B1155 / MPG1356 / D2C3
NO MAP: post probable 48D; pre presence 61A

Afrotropical: W and C Africa, from the lowland forests around the Gulf of Guinea, E through the Congo basin to S Sudan, NW Tanzania, Uganda and, at the E limit of the range, W Kenya.

Kenyan: this solitary, forest species is mentioned at Nyarondo 61A by Ja, but all recent records come from Kakamega Forest 48D, where it is not uncommon (ADL pers. obs.). Both of these squares are above 1000 m to the W of the rift valley, and Br mentions 1700 m as the species' upper altitudinal limit. Both squares are humid and within the 1000+ mm rainfall regime. There is no breeding record, but Br considers this malimbe resident. See Red-headed Weaver (969).

969 RED-HEADED WEAVER *Anaplectes rubriceps*
B1135 / MPG1359 / D3C2

Afrotropical: widespread S of 13°N, with the exceptions of the Congo basin and the lowlands bordering the Gulf of Guinea, and the dry areas of E Somalia and the SW.

Kenyan: local and never numerous in bushed and wooded country below 1800 m (Br; MPG; Be; La; St). Range is only 13% below 500 m, and this weaver is absent from the coastal lowlands. This coastal absence extends S to 10°S in Tanzania (Br) and N to 80AB in Somalia (AM).

Br notes that this bird usually avoids the wettest and driest country, though occurring in the L Victoria basin and at L Turkana. Our map does not show aversion to high rainfall (presence in 67% of the non-coastal 1000+ mm squares), but range is only 11% within the driest, 0-250 mm rainfall regime and 26% arid. Occurrence is thus sparse in the N and E. The records from 4A E to 17BD (J S Ash *in litt.*) represent the S limit of birds in Ethiopia, and are in country that may not actually be arid. This weaver is occasional around

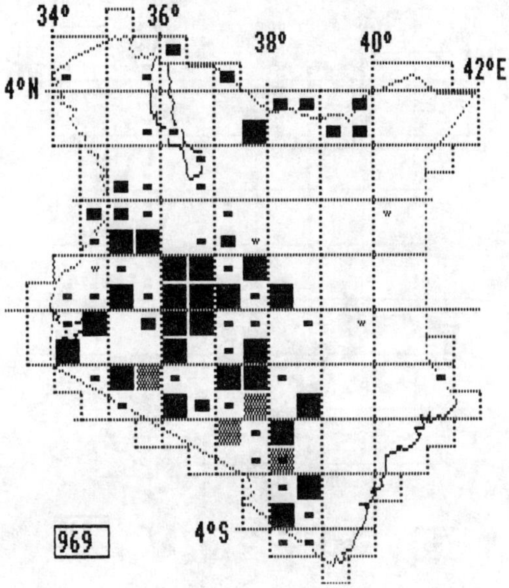

Turkana (HH), and habitat in arid Tsavo E 101B is partly riverine (La). Presence in only two of the squares containing the Tana R (64B 65B) agrees with K L Campbell's (*in litt.*) note of uncommon or local occurrence between 64 and the sea, and with AM's listing of only two inland squares on the Juba R of S Somalia.

This species is considered resident (Br; Be (74); La (101B); St (50A)).

The Red-headed Weaver is often classified in the genus *Malimbus*, of which it is the only non-forest member (HM).

970 GROSBEAK WEAVER *Amblyospiza albifrons*
B1134 / MPG1358 / D2C2

Afrotropical: widespread in many areas S of 10°N, apart from much of the Congo basin and the dry regions of E Somalia and the SW.

Kenyan: this social and locally common species occurs up to 3000 m (Br; Ja). It breeds in wetlands but sometimes forages far from water, in a variety of bushed, wooded and forest edge habitats (*op. cit.*; MPG; Be; Britton 1978).

Range is 88% within 500+ mm rainfall areas, mostly in the highlands and adjacent plateaux, and in the coastal strip. This distribution is mainly subhumid-humid, but also extends to arid-semiarid squares on the E and SE peripheries of the C highlands and at the coast. 71% of these drier squares contain permanent rivers, as does 3B (J S Ash *in litt.*), which represents the S limit of birds in Ethiopia (*cf.* HM's map). Range has recently retracted from Voi swamp in arid 101B (La). Occurrence along the lower Tana R 79C 91A parallels that in S Somalia in the valleys of the lower Juba and Shebelle R (AM). Absence from the higher reaches of these rivers corresponds with general absence from permanent water in the dry interior, *e.g.* at L Turkana and Baringo 50A, and along the Daua R 7C 19AB and the N Uaso Ny'iro R 51AB 52A.

Br considers this species resident but mentions some wandering, and Se reports a single, late Jan individual at the top of the altitudinal range at Mau Narok 61D. Annual influxes to the Ruaraka-Kahawa-Ruiru area of 75B during Mar-June and Nov-Dec (DEP pers. obs.) coincide with the rains and this species' breeding seasons (Brown & Britton 1980), and suggest some more regular movement. Brown-headed males (*A.a. melanota*) occur in and W of 36D-49C-60 (Br; Ja). Other references: Gichuki 1982; Pomeroy *et al.* 1982.

971 SOUTHERN RED BISHOP *Euplectes orix*
B1150 / MPG1363 / D2C2

Afrotropical: a S species, widespread S of 5°S but extending N to Tanzania, Uganda and, at the NE limit of the range, Kenya. It is replaced by the Northern Red Bishop (972: *q.v.*) in the N tropics, and by the Zanzibar Red Bishop (973) in the coastal and adjacent lowlands of the E. Kenya is apparently the only country in which all three birds occur.

Kenyan: a locally common bird of tall grassland, typically in open country near

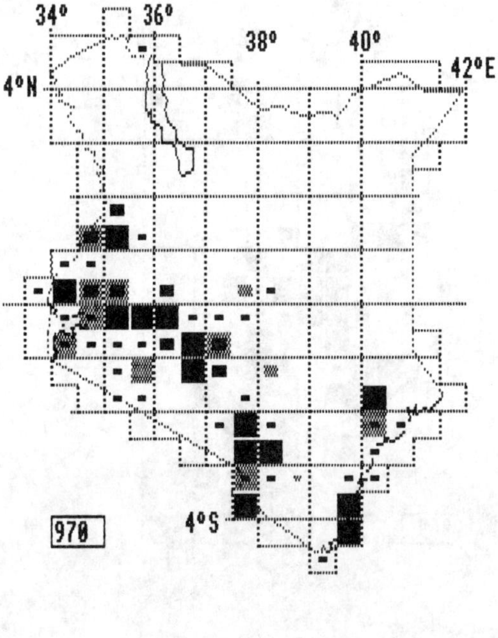

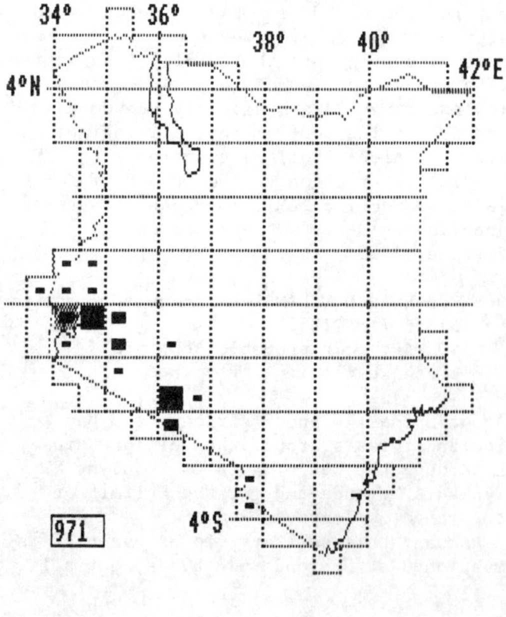

water (Br; Ja). Br gives altitudinal range as mainly at 1000-1500 m, but also down to 600 m. Distribution is 87% within the 500+ mm rainfall regime, and absent from arid areas. Br considers this bishop resident, but also mentions wandering. The latter is presumably illustrated by a lone male in breeding plumage on the NE edge of its range at L Naivasha 62C on 26 Apr (EABR 1981).

The Southern Red Bishop forms a superspecies with the similarly gregarious Northern and Zanzibar Red Bishops (HM). These three species' breeding ranges are allopatric and, where two are found together, they remain separated by minor ecological differences (HM; Br). The Northern is geographically isolated to the N and NE of the other two. The Zanzibar replaces the Southern in the SE lowlands, with overlap at QSD scale in 100BD. The Southern and the Northern are replaced in wetter areas by the Black-winged Red (974) and Black (975) Bishops, which form a species-group with this superspecies (HM).

972 NORTHERN RED BISHOP *Euplectes franciscanus*
B1143 / MPG1363 / D2C1

Afrotropical: the N tropics, E at 5°N-15°N to Ethiopia and NW Somalia, and SE to N Uganda, Kenya and 1°S in Somalia (AM).

Kenyan: very local but occasionally common in tall open or bushed grassland at 600-1800 m (Br; Ja). Distribution is in three distinct areas. An apparently disjunct population of the nominate race occurs on the N edge of the highlands in the L Baringo 50A-L Bogoria 50C area (Br; St). Ja also records this population at Elgeyu (approximately 37D 49B: not mapped), Eldama Ravine 49D, and the N Uaso Ny'iro R (approximately 50BD 51AB: not mapped), and the lack of recent records from these areas suggests a decline. Range in 49D 50AC is above 1000 m in semiarid-subhumid, 500+ mm rainfall country.

The record in 12A is from the Ugandan sector of the square (J S Ash *in litt.*) and represents the limit of the population in that country. This is presumably the nominate race. Presence in 28A is the sole Kenyan record of the distinctively coloured *E.f. pusilla* of Ethiopia and Somalia: two males and three females were found 40 km N of Marsabit on 25 Mar 1978 (Moore 1983; EABR 1982). This race is known in 4B 5B and the square NE of 19B (all entirely within Ethiopia/Somalia: J S Ash *in litt.*, AM), and other occurrences along Kenya's N borders may well be undetected. In contrast to the disjunct southern population, 12A and 28A are at 500-1000 m, in arid-semiarid, 0-500 mm rainfall country.

This species is mainly resident (Br; St), but some degree of wandering is shown by the late Mar flock in 28A which, like the late Apr Southern Red Bishop (971: *q.v.*) wanderer at Naivasha, was during the 'long' rains period.

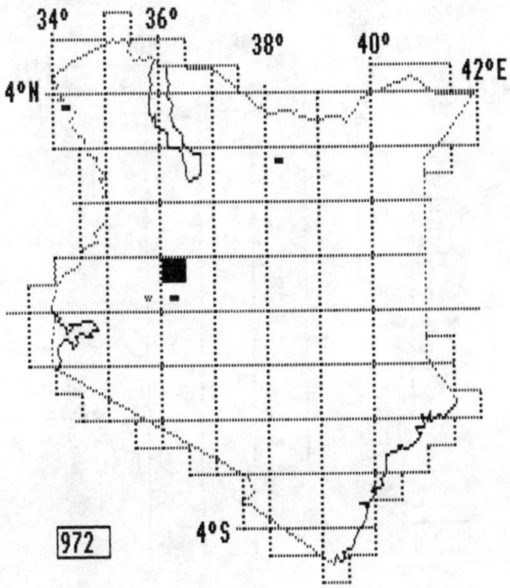

973 ZANZIBAR RED BISHOP *Euplectes nigroventris*
B1149 / MPG1364 / D2C3

Afrotropical: confined to E areas of Mozambique and Tanzania, and SE Kenya.

Kenyan: common in open and bushed grasslands below 1000 m in the coastal and adjacent lowlands, N to Lamu and Manda 91B (Br; Ja; MPG). Range is 83% semiarid-dry subhumid and 83% within 500+ mm rainfall areas. Males with some red on the throat are most frequent inland (Br; Ja).

Br describes this species as both a resident and wanderer, and a record at Kitui 77A (Ja) may be in the latter category. Rare visitors to rivers in arid Tsavo E 101B are mainly in Dec-Jan (La), and the bird has appeared in the Ngulia area of Tsavo W 101A in Jan (D J Pearson *in litt.*). These dates are during and just after the 'short' rains, and compare with rains wandering in the Southern (971: *q.v.*) and Northern (972) Red Bishops.

(map overleaf)

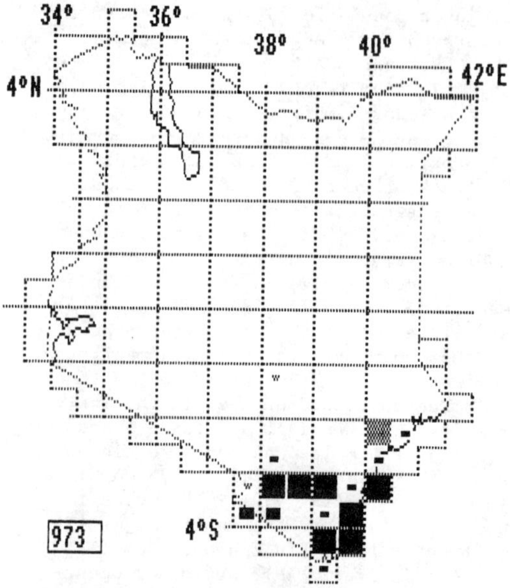

974 BLACK-WINGED RED BISHOP *Euplectes hordeaceus*
B1146 / MPG1365 / D2C2

Afrotropical: between Senegambia and Angola on the W coast, E to SW Ethiopia, Kenya, the E coast and S Mozambique.

Kenyan: locally common in tall bushed grassland, up to 1600 m (Br; Ja; MPG). Two widely separated races occur. The nominate race inhabits the coastal lowlands, inland to the Tiva R flood plain 90A (Br), Taru 102C (Ja) and (presumably, on geographical grounds) L Jipe 100D (P C Lack *in litt.*). This range is 92% below 500 m, 92% semiarid-dry subhumid and 92% within 500+ mm rainfall areas.

E.h. craspedopterus is mainly found W of the rift valley, but also extends NE to Marsabit 27D (Ja; Tomlinson 1946). The record in 24B derives from Uganda (J S Ash *in litt.*). In contrast to the coastal race, this inland race is 89% at 1000-2000 m, 78% moist subhumid-humid and 72% within the 1000+ mm rainfall regime. Br considers this species resident, but the Marsabit record presumably represents wandering, possibly from SW Ethiopia (*cf.* Northern Red Bishop (972)).

The Black-winged Red Bishop forms a species-group with the Black Bishop (975) and the preceding superspecies (HM). The Black-winged is allopatric with the small and similar race *E.g. friederichseni* of the Black in N Tanzania, but often alongside the larger *E.g. ansorgei* in W Kenya (Br; HM). The Black-winged is sympatric with the Zanzibar Red Bishop (973) but, although overlapping extensively at QSD scale, is usually segregated from the Southern Red Bishop (971: *q.v.*) in wetter grassland (HM).

975 BLACK BISHOP *Euplectes gierowii*
B1144 / MPG1366 / D2C2

Afrotropical: three distinctive and separate populations. In the N, *E.g. ansorgei* extends discontinuously from W Ethiopia to

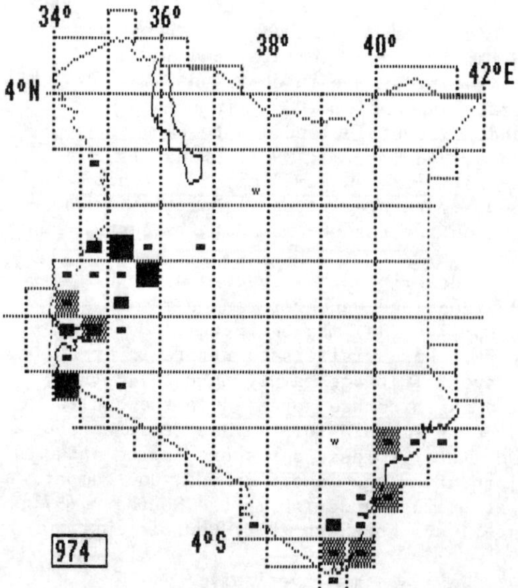

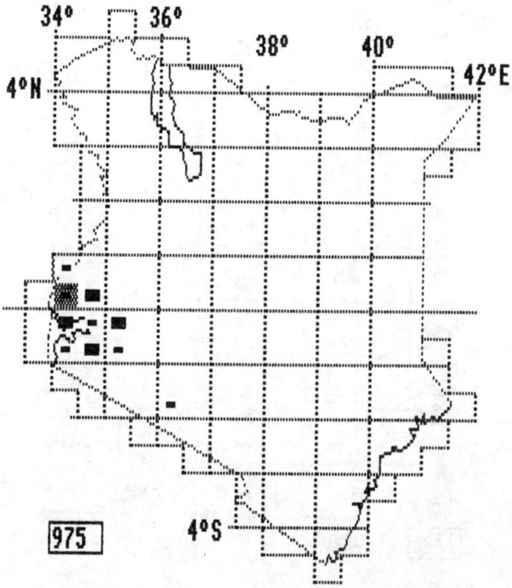

S Sudan, NE Zaire, Uganda and W Kenya. *E. g. friederichseni* is centred in interior N Tanzania, just reaching NE to the Kenyan border at L Natron. The third population is in Angola.

Kenyan: the first confirmed record of this bishop within Kenya was by Britton & Harper (1969). It is now known to be a local and generally uncommon bird of tall and bushed grassland at 700-1600 m, usually near water or swamps (Br; MPG). In all but one case, records refer to *E.g. ansorgei*, which is found to the W of the rift valley. Its range is entirely at 1000-2000 m in moist subhumid-humid, 1000+ mm rainfall areas.

This species has also been observed on the rift valley floor in moist grasslands near the Ewaso Ngiro R 75C during Aug 1979 (specimen) and June 1982 (DEP pers. obs.). This site is at 670 m in 600 mm rainfall country. On geographical grounds, these were most probably the race *E.g. friederichseni*, which Br describes as endemic to interior N Tanzania. This race's type locality is on the Nguruman Escarpment, apparently in the Tanzanian sector of 86B (Ja; Br), though MPG place this locality in Kenya. The area W of L Magadi is not well known, and this race may be regular there. See Southern Red Bishop (971) and Black-winged Red Bishop (974).

976 YELLOW-CROWNED BISHOP *Euplectes afer*
B1137 / MPG1369 / D3C2

Afrotropical: widespread but local S of 15°N, but absent from the dry SW and Somalia. Distribution in Kenya represents the E limit of the range.

Kenyan: very local in swampy and moist grassland at 400-2000 m (Br; Ja; HH; St). Range is mainly in the highlands and adjacent plateaux, being 93% above 1000 m, 71% subhumid-humid and 82% within 500+ mm rainfall areas. The six breeding squares all fall within these altitudinal and rainfall categories. Occurrence around 400 m in marshland at the mouth of the Kerio R 14C, L Turkana (Oct: HH), is in very different country.

Br describes this bishop as a resident and wanderer, and movements are at least partly stimulated by the production of suitable habitat by rainfall (*cf.* Zanzibar Red Bishop (973) and congeners). This is described at L Baringo 50A by St (and EABR), and also occurs in the Nairobi NP area 75B (ADL pers. obs.).

The Yellow-crowned Bishop forms a superspecies with the allopatric Fire-fronted Bishop (977)(HM; Br). These two birds overlap at QSD scale, but the Fire-fronted is mainly in lower and drier country in the E/SE. Other reference: Cunningham-van Someren 1949.

977 FIRE-FRONTED BISHOP *Euplectes diadematus*
B1142 / MPG1368 / D2C2

Afrotropical: endemic to Kenya, NE Tanzania and S Somalia.

Kenyan: local in open bushed habitats, mainly below 1000 m (Br; MPG). It occurs in the E and SE lowlands and on the pla-

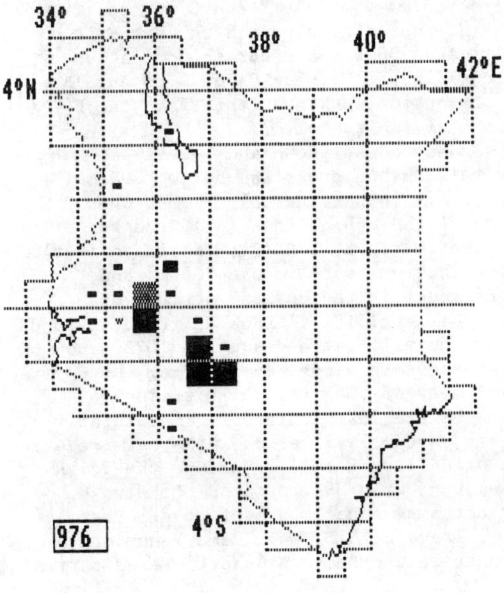

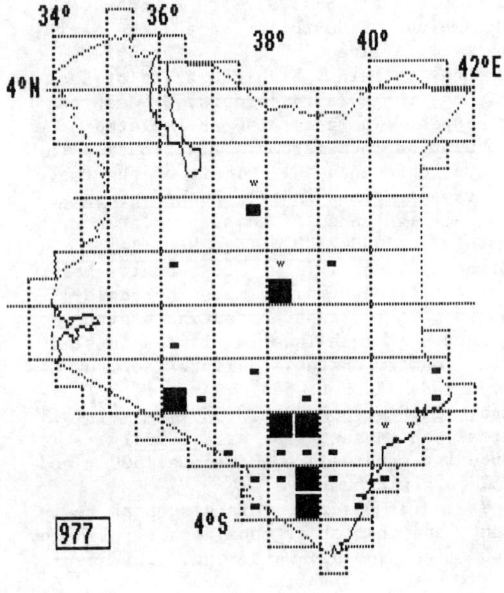

teaux around the C highlands, W to lower altitude, semiarid areas of the rift valley at Baringo 50A (Ja) and Magadi/Olorgesaille 75C. Recent appearance in Tsavo E 101B (La) may reflect opening up of habitat. In contrast to the closely related Yellow-crowned Bishop (976: *q.v.*), range is 72% below 1000 m, 88% arid-semiarid and 85% within the intermediate, 250-1000 mm rainfall regime.

Like the Yellow-crowned, the Fire-fronted makes extensive movements in response to rainfall, and can appear and breed in relatively large numbers after substantial rains in arid-semiarid areas (Br). This type of breeding is particularly known from Meru NP 52C (Br), the Magadi/Olorgesaille area 75C (Cunningham-van Someren 1971h; Mann 1976), and the Tsavo region 89 101 (commoner after the 'short' rains in Tsavo E: La; EABR 1977, 1978, 1983). Of the seven breeding squares, six are below 1000 m, five within 250-500 mm rainfall areas, and all arid-semiarid (*cf.* Yellow-crowned Bishop). Most records are of wanderers (*cf.* AM), but small numbers are present during most months in bushed dunes at dry subhumid Sabaki 103A (Br). A single breeding plumage male at L Naivasha 62C on 26 Apr (EABR 1981) was in moist subhumid country near 2000 m, which is presumably unsuitable, on the same day as a wandering Southern Red Bishop (971). Other reference: Moore 1983.

978 YELLOW BISHOP *Euplectes capensis*
B1141 / MPG1367 / D2C3

Afrotropical: widespread but local in the S tropics at 5°S-15°S, S to Zimbabwe, S Mozambique, E South Africa and the periphery of Africa S of 30°S. Further N, it extends to mainly highland areas of East Africa, the E Zaire borders, S Sudan and Ethiopia. There is also an isolated population in Cameroon. Distribution in Kenya represents the E limit of the range.

Kenyan: common in tall bushed grassland in the highlands and adjacent plateaux, mainly at 1400-2300 m (Br; MPG; Be; Britton & Sugg 1973). Its scarcity below these altitudes in the moist subhumid-humid L Victoria basin contrasts with presence at much lower altitudes in semi-arid country on the E periphery of the C highlands 52C S to 89A. Range is 82% above 1000 m, 86% within 500+ mm rainfall areas, and absent from arid country. Breeding squares are 70% above 1500 m and 75% subhumid-humid.

Be and Br describe this bishop as resident, and the latter considers birds below 1400 m as probably wandering. This cer-

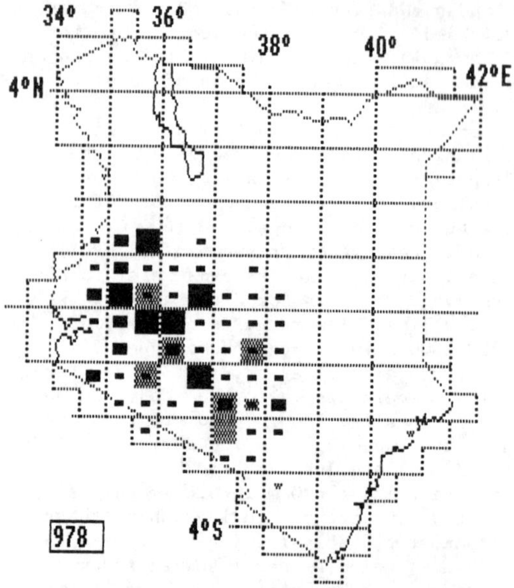

978

tainly appears true of Ja's mention of Lamu 91B, which is the sole record for the coastal lowlands and far E of the normal range. Coastal occurrence is regular S of 7°S in Tanzania however (Br).

HM consider the Yellow Bishop to be intermediate between the typical bishops and the typical widowbirds.

979 WHITE-WINGED WIDOWBIRD *Euplectes albonotatus*
B1138 / MPG1373 / D3C2

Afrotropical: mainly E and C Africa, from the Ethiopian highlands S through East Africa to 30°S. A separate population occurs in the W, from Gabon S to Angola. Distribution in Kenya represents the E limit of the range.

Kenyan: conspicuous and locally common in tall bushed grassland below 2000 m (Br; MPG). It is particularly common on the E edge of the C highlands (Br), and has reached Marsabit 27D (Ja; Tomlinson 1946). Distribution is 86% above 500 m, and occurrence in the coastal strip S of 3°S is occasional (Br). Lack of recent records from the well-watched squares 102BD and 114B suggests a contraction from the E extremity of the range. This widow's squares are 79% within 500+ mm rainfall areas, and only 7% arid. Seasonal presence in arid Tsavo E 101B is around the Voi R (La), and arid 89B contains the Tiva R. Br notes seasonal movement to drier areas *e.g.* Tsavo E 101B, where small numbers appear in the Dec-Jan 'short' rains period

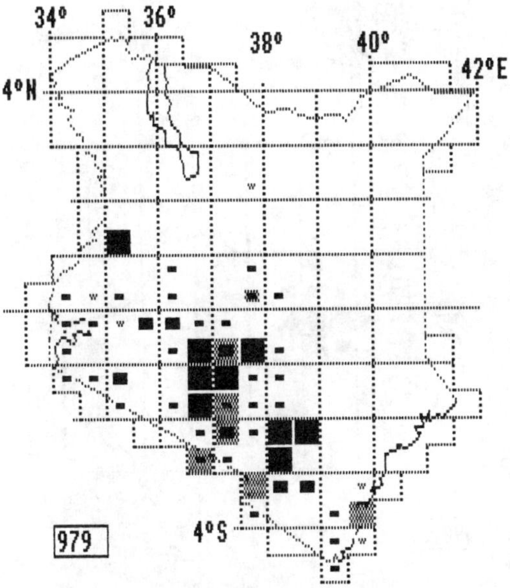

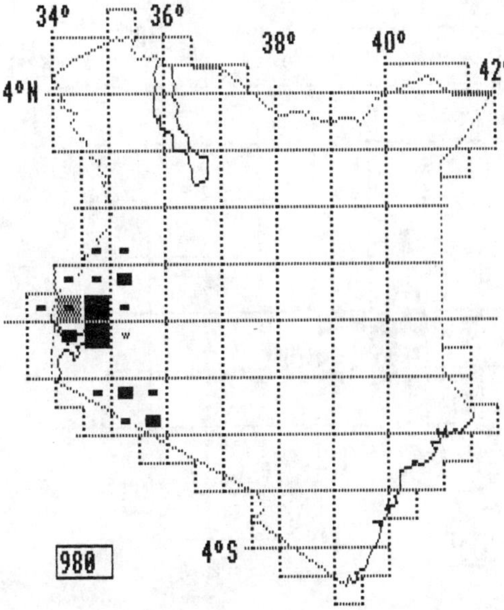

(La). An annual May-Aug influx to the Ruaraka-Kahawa-Ruiru area of 75B (DEP pers. obs.) coincides with this species' main breeding season (Brown & Britton 1980).

HM are tentative in their subdivision of the widowbirds. They place the White-winged in a species-group with the Yellow-mantled Widowbird (980), and quote areas where the latter is ecologically segregated in wetter grasslands. Br specifically describes the Yellow-mantled as a bird of moist grassland, and its range is moister than the White-winged's. The White-winged has only 14% of its breeding squares W of the rift valley, within the range of the Yellow-mantled. This partly reflects observer coverage, but the possibility exists of some degree of allopatry.

980 YELLOW-MANTLED WIDOWBIRD *Euplectes macrourus*
B1148 / MPG1371, 1372 / D3C2

Afrotropical: widespread, from Senegambia E to the Ethiopian highlands, S to Angola in the W and Zimbabwe in the E. Distribution in Kenya represents the E limit of the range.

Kenyan: locally common in open moist grassland, at 1000-1800 m to the W of the rift valley (Br; MPG). In contrast to the White-winged Widowbird (979: *q.v.*), range is 76% within 1000+ mm rainfall country and 87% subhumid-humid. Br considers this bird resident. Birds with black mantles (the Yellow-shouldered Widowbird) occur spasmodically, sometimes alongside yellow-mantled ones (Br; Ja; HM). HM suggest some degree of allopatry with the Fan-tailed Widowbird (984), but note range overlap in Uganda. Similar overlap occurs at QSD scale in Kenya, but Br describes the two species as preferring slightly differing habitats.

981 RED-NAPED WIDOWBIRD *Euplectes ardens*
B1139 / MPG1375, 1376 / D3C2

Afrotropical: throughout the E, from the Ethiopian highlands S almost to the Cape, and to the W coast between Gabon and Angola. Further NW, there are isolated populations in the vicinity of Cameroon and Sierra Leone. Distribution in Kenya represents the E limit of the range.

Kenyan: locally common in tall open or bushed grassland, with two geographically distinct races (Br; Ja). *E.a. suahelica* occupies the highlands at 1500-3000 m (Br; Ja; Be; Se). Its range is 78% subhumid-humid and 86% within 500+ mm rainfall areas. Br considers the species resident and Be thinks the same of this race, but movements are known. Se found it an Oct-Dec visitor in small numbers to its upper altitudinal limit around Mau Narok 61D. An annual Apr-June influx to the Ruaraka-Kahawa-Ruiru area of 75B (DEP pers. obs.) coincides with this species' main breeding season (Brown & Britton 1980).

E.a. tropicus occurs in the Taita area 101ABC (Br; Ja), below 1500 m in an arid-semiarid, 250-1000 mm rainfall area. Jul-

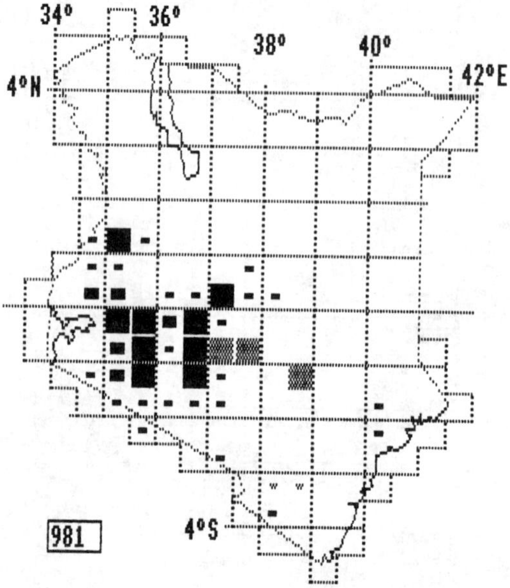

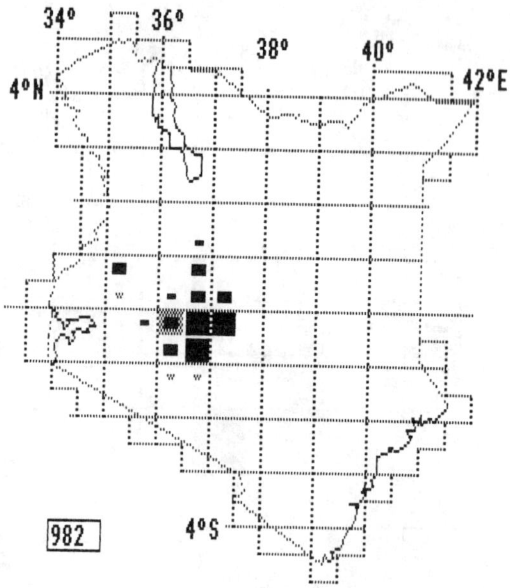

Aug birds around the lower Tana R 79C 91A were not racially identified (AGH), but were presumably also *tropicus*. Br omits this locality, but there appears little chance of confusion if the birds were in breeding plumage.

HM do not consider this species particularly closely related to any of the other widows, and point out its at least partial sympatry with all of them. However, influxes of this species and of Jackson's Widowbird (985) to the Ruaraka-Kahawa-Ruiru area of 75B show little overlap in seasonality (DEP pers. obs.).

982 LONG-TAILED WIDOWBIRD *Euplectes progne*
B1151 / MPG1377 / D3C3

Afrotropical: three separate populations, with *E.p. delamerei* being endemic to the highlands of Kenya. Other races occur in S Mozambique and E South Africa, and from Angola E to S Zaire and Zambia.

Kenyan: a conspicuous and locally common widow that breeds in marshy highland grassland at 1800-2800 m (Br; Ja; MPG). It is found in or E of the rift valley, apart from an apparently isolated breeding population in the Soy 49A-Eldoret 49C area (Br; A G Johnston *in litt.*). Range is entirely in 500+ mm rainfall areas, and 86% subhumid-humid. There may have been a slight contraction of range in the 75AB area. Br suggests that this species is mainly resident, although some wandering may occur. Andersson (1982) has shown that the number of mates attracted by a breeding male depends upon the length of the male's tail.

HM place the Long-tailed in a species-group with the Marsh Widowbird (983) and add that, since the two appear to be ecologically segregated, they may in fact constitute a superspecies. They are virtually allopatric at QSD scale, the Marsh's range is wetter, and Br's altitude estimates imply altitudinal segregation. See Jackson's Widowbird (985).

983 MARSH WIDOWBIRD *Euplectes hartlaubi*
B1145 / MPG1374 / D2C2
NO MAP: post presence 48BD; post confirmed 37C; pre presence 49C

Afrotropical: a broken distribution W from S Tanzania to Angola, and thence N through the Congo basin to Cameroon. An isolated population occurs on the N shores of L Victoria, E to W Kenya.

Kenyan: local and uncommon in moist grassland and swamp edge at 1100-1800 m (Br; Ja; MPG; Tennent 1965). Range is W of the rift valley and N of the Winam Gulf 60AB, and the four squares are all above 1000 m in humid, 1000+ mm rainfall areas. The only breeding records are from Kapsara and Saiwa Swamp 37C, although breeding probably occurs on many coarsely grassed and periodically inundated alluvial flats along the Nzoia R (T J Barnley *in litt.*). See Long-tailed Widowbird (982) and Fan-tailed Widowbird (984).

984 FAN-TAILED WIDOWBIRD *Euplectes axillaris*
B1140 / MPG1370 / D3C2

Afrotropical: mainly the E, from S Sudan, Ethiopia and S Somalia, S through East Africa to 33°S in E South Africa. It extends to the W coast in Angola, and there are isolated populations in Cameroon and Mali.

Kenyan: two separate populations. In the highlands, it is found in tall moist grassland and swamp edge, usually below 1500 m but reaching 2300 m in Uasin Gishu 49AC (Br; Britton 1978). This range is entirely above 1000 m, 85% moist subhumid-humid and 70% within 1000+ mm rainfall areas. Br describes the race *E.a. phoeniceus* as very common in the L Victoria basin, E only to Nandi 49C. Either Br has omitted them, or birds further E in the highlands (many obs.) have not been racially identified. The restriction of highland breeding records to 1000+ mm rainfall areas to the W of the rift valley is of unknown significance.

The race *E.a. zanzibaricus* is reasonably common in dry bushed grassland and bushland in the coastal lowlands N to Lamu 91B, extending inland along the Tana R (Br; AGH), and possibly to L Jipe 100D (P C Lack *in litt.*). This range is 91% below 500 m, 82% semiarid-dry subhumid and 91% within 500+ mm rainfall areas. Presence along the Tana R corresponds to range in the wetlands of the Juba and Shebelle R of adjacent S Somalia (AM). Ja notes range to an altitude of 800 ft (240 m) on the Tana R, which is somewhere in the 64B 65A area (not mapped). In view of this, Br's note of rare occurrences of this coastal race at Murang'a 63C seems plausible, although Ja is unsure of the racial identification. Br draws attention to the virtual allopatry of this species and the Marsh Widowbird (983). See Yellow-mantled Widowbird (980).

985 JACKSON'S WIDOWBIRD *Euplectes jacksoni*
B1147 / MPG1378 / D3C3

Afrotropical: endemic to the highlands of W and C Kenya and the far NE of Tanzania.

Kenyan: conspicuous and locally common in open grassland at 1500-3000 m (Br; Ja; MPG; Be; Se; Turner 1977). Range is entirely within subhumid-humid, 500+ mm rainfall areas. Br considers it resident, but movements occur. Be describes appearance as erratic, and Se notes local migration during the Jan-Mar dry season at the upper limit of the altitudinal range around Mau Narok 61D. Se's mention that most birds are in non-breeding plumage at this time corresponds with DEP's observations in the Ruaraka-Kahawa-Ruiru area of 75B, where there is an annual Nov-Feb influx.

This widow has a unique courtship display (Ja; van Someren 1946, 1958). HM note that the broadly sympatric Long-tailed Widowbird (982) generally breeds in moister grass than this species (*cf.* Br). See Red-naped Widowbird (981).

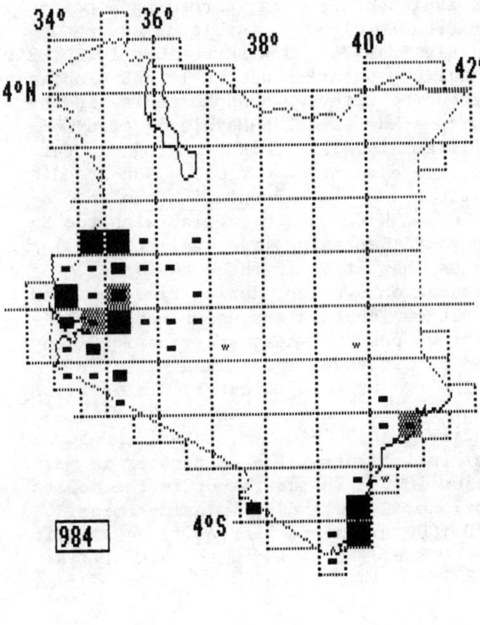

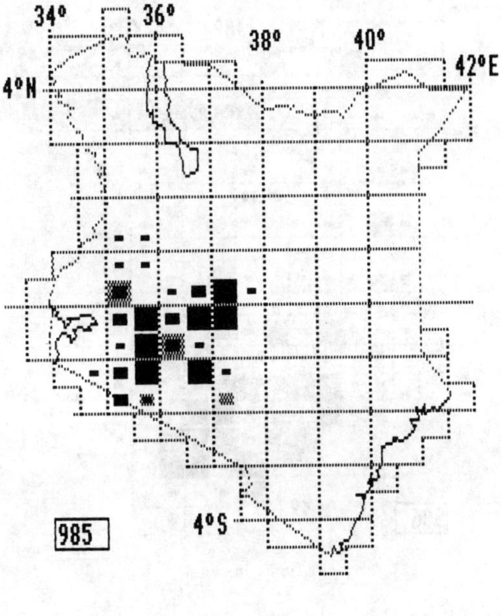

986 RED-BILLED QUELEA *Quelea quelea*
B1193 / MPG1360 / D2C3

Afrotropical: widespread S of 17°N, with the exceptions of the wetter equatorial W coast and the Congo lowlands, and the drier areas of NE and C Somalia.

Kenyan: this quelea has been the subject of more research than any other bird in the Afrotropical Region, because it is the most serious avian agricultural pest. It is nomadic and intensely gregarious, and can attain concentrations of millions of individuals, and devastate cereal crops (MPG; Magor & Ward 1972; Allan 1983). Habitat is principally open, bushed grassland at 500-1500 m, although the total altitudinal range extends from sea level up to 3000 m (Br; AGH). It regularly roosts at altitudes up to 3000 m near Timau 51C (R G Allan *in litt.*), but Se recorded only rare stragglers in non-breeding plumage at 3000 m around Mau Narok 61D. Birds have been recorded in the wettest and driest areas of Kenya, but are most abundant in the semiarid and less arid regimes.

Vast movements occur, but La report some birds in all months in arid Tsavo E 101B. St found similar status in 50A; and DEP (pers. obs.) in 75C 89A. Breeding is stimulated by the appearance of new grass and of insects in areas which have received rain some weeks earlier (Disney & Marshall 1956; Marshall & Disney 1957; Ward 1971). Abundant insect prey appears necessary for the production of young but, when not breeding, the birds can exist on seeds taken from the ground in dry country (*op. cit.*; La). Hence breeding occurs opportunistically wherever and whenever conditions are suitable (*op. cit.*; Allan 1983; Leuthold & Leuthold 1972a), in extremely localised colonies that may contain millions of birds (Brown & Britton 1980; EABR 1977, 1980). These dense breeding concentrations can attract numerous raptors (Leuthold & Leuthold 1972ab).

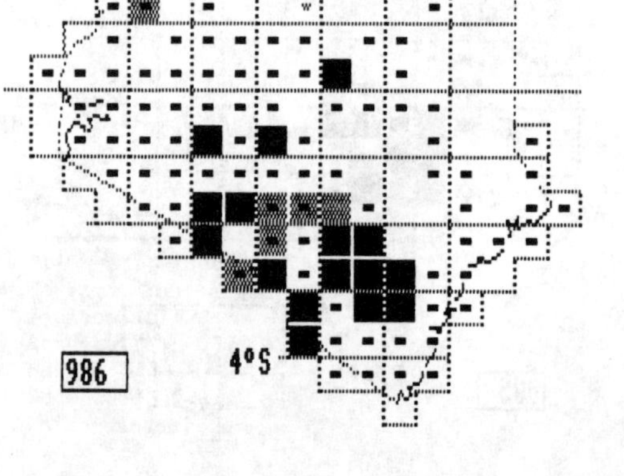

As the Intertropical Convergence Zone (ITCZ) moves S into Kenya Oct-Nov, quelea leave areas at the onset of the rain, when food becomes scarce or absent due to the germination of annual grass seeds (R G Allan *in litt.*). The birds appear to migrate NE to cross the ITCZ, and breed in the temporarily verdant conditions immediately in its wake (Ward 1971; Br). The birds then follow the ITCZ as it moves S, utilising post-rains conditions in more SW areas and apparently breeding more than once at different latitudes (*op. cit.*). Reappearance from the NE and breeding in Kenya occur from Nov onwards, and later in N and then C Tanzania (Br; Ja; La; Leuthold & Leuthold 1972ab; EABR 1977, 1980, 1981).

Allan (1983) maps the breeding and total ranges of three proposed quelea populations in Kenya, which differ slightly in the extent of facial black in the males. A population in S Somalia (AM) and E Kenya apparently causes little agricultural damage at present. Another population in S and SW Kenya and NE Tanzania overlaps with the E birds in a common breeding area between the lower Tana R and the Kenya/Tanzania border. This breeding concentration is clearly visible on our map, mainly from 75CD 87A SE to 100D 101B 102A. A third population occurs in S Ethiopia, extending into Kenya and breeding in the Turkana-Marsabit 27D-Moyale 17 area (*cf.* Williams 1954a). This breeding cluster is also evident on our map. Non-breeding birds from this N population can move S to C and W Kenya, to overlap with the S Kenya-NE Tanzania birds. Allan (1983) notes that it is in these two areas of population overlap, during times of food shortage, that this species can be a serious pest to wheat and rice in C Kenya, and possibly to rice in the W.

The two breeding clusters are in different environments. That in the N is entirely at 500-1000 m, in arid, 0-500 mm rainfall country. The S cluster is mainly below 1000 m (although not in the coastal strip or E lowlands), in arid-semiarid, 250-1000 mm rainfall areas. Overall distribution of breeding with rainfall is similar to that seen in the Chestnut Weaver (956)(Pomeroy *et al.* 1982), with 83% less arid-semiarid, and 80% within the intermediate, 250-1000 mm rainfall regime.

The first serious crop damage in Kenya occurred in Rift Valley Province during 1952 (Williams 1954a). Over 30 years of killing these birds at their roosts and breeding colonies has nowhere lessened the overall problem, and Allan (1983) describes other methods that can produce more economical and effective crop protection. Ja found these birds good to eat, but their considerable biomass has yet to be exploited as a food source.

The Red-billed Quelea forms part of a species-group with the Red-headed (987) and Cardinal (988) Queleas, and the three can flock together during the off-season (HM). These authors state the Red-billed to be replaced by the Red-headed in moister areas, and suggest that they may constitute a superspecies. This contrast can be seen in Kenya, where the Red-billed is most widespread while the Red-headed, on the edge of its range, barely penetrates arid country. HM's description of the Cardinal Quelea as ecologically intermediate between the other two, being sympatric with the Red-billed in the wetter areas of that species' range but only marginally overlapping the Red-headed, precisely fits these birds' habitats (Br) and ranges in Kenya.

987 RED-HEADED QUELEA *Quelea erythrops*
B1192 / MPG1361 / D2C1

Afrotropical: widespread S of 10°N, E to SW Ethiopia, SW Kenya, E Tanzania, Malawi, S Mozambique and E South Africa. It is absent from much of the Congo basin and adjacent areas, and from the dry SW. Apart from a single Aug record in extreme S Somalia 80B (AM), distribution in Kenya represents the E limit of the range.

Kenyan: a distinctly local quelea of moist grassland, mainly below 1400 m in the L Victoria basin, the Magadi area 75C and the coast (Br; Ja). In contrast to the Red-billed Quelea (986: *q.v.*), range is entirely within 250+ mm rainfall areas and only 6% arid. E of the normal range, Ja recorded it from an unspecified area of the Tana R, and there is the single record from Somalia (see above).

The Red-headed is apparently a non-breeding visitor, although breeding plumage males are often seen (Br). Apr-May breeding (the first in East Africa) has recently occurred in Tanzania, at 4-5°S in the N and 7°S in the NE (EABR 1981-1983). Flocks totalling over 1000 birds in the

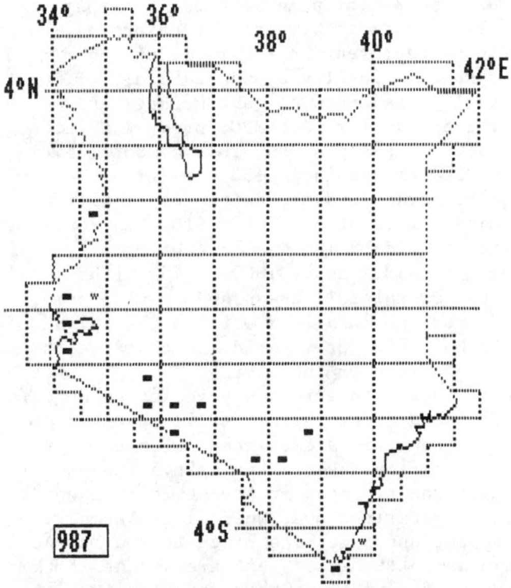

Ramisi 114C area (4°30'S) during Oct (EABR 1981) could have been non-breeding migrants (MPG) from these S areas. Br suggests that other seasonality is probably obscured by migrants from both the N and S tropics. Other reference: Magor & Ward 1972.

988 CARDINAL QUELEA *Quelea cardinalis*
B1191 / MPG1362 / D2C2

Afrotropical: an E species, from S Sudan

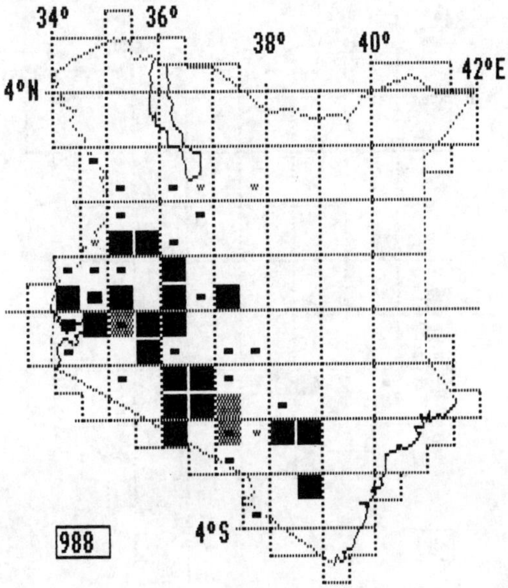

and the S Ethiopian borders, S through East Africa to Zambia. Distribution in Kenya represents the E limit of the range.

Kenyan: locally abundant in dry or moist tall grassland with rank herbage and scattered bush (Br; Ja; MPG; Brown & Britton 1980). Mainly at 800-2000 m, total altitudinal range is 400-3000 m (Br). Se saw rare stragglers around the upper altitudinal limit at Mau Narok 61D. In contrast to the Red-billed Quelea (986: *q.v.*), range is 81% above 1000 m, 75% within 500+ mm rainfall areas and only 8% arid. Occurrence in arid 27D is on the upland of Mt Marsabit (Br). Arid 89B contains the flood plain of the Tiva R.

Br notes that this species often appears to breed after rains, and is absent from arid-semiarid areas except when they are temporarily green after rain. These patterns are seen in the semiarid country around Baringo 50A, where large numbers appear and sometimes breed in the May-Aug rains/moist period, but are absent if the rains fail (St). Other reference: Magor & Ward 1972.

989 PARASITIC WEAVER *Anomalospiza imberbis*
B1136 / MPG1405 / D1C1

Afrotropical: a sporadic distribution in the N tropics from Sierra Leone to Ethiopia, and S through East Africa and the E Zaire borders to S Zaire, Zambia, Zimbabwe and NE South Africa. Distribution in Kenya represents the E limit of the range.

Kenyan: uncommon in moist, open or bushed grassland up to 2000 m (Br; MPG; Williams & Keith 1962; La). Range is 70% at 1000-2000 m, 75% subhumid-humid and 90% within the 500+ mm rainfall regime (*cf.* Br). It is present in most months in the L Victoria basin but is sporadic elsewhere, typically appearing when substantial rains produce a good growth of grass (Williams & Keith 1962; Br), *e.g.* in the Nairobi 75B-Thika 76A area (EABR). This weaver mainly parasitises grassland warblers *e.g.* cisticolas and prinias, which breed during the rains (Brown & Britton 1980; Williams & Keith 1962; HM; Britton & Britton 1971c, 1972a; EABR).

990 GREY-HEADED SPARROW *Passer griseus*
B1206 / MPG1300-1303 / D3C2

Afrotropical: wide-ranging, in most regions S of 18°N apart from the driest areas of E Somalia and the SW.

Kenyan: this common sparrow frequents bushed and wooded country, and human habitation (including Nairobi city centre 75B), up to 3000 m (Br; MPG). Se lists two June stragglers around the upper altitudinal limit at Mau Narok 61D, but found the species more common at 2400 m. Range is throughout our altitude, rainfall and moisture categories, and there appears no reason why this sparrow should not occur in every square (Lewis & Pomeroy 1983a). Br mentions absence from some arid areas, but presence in 56% of the most arid squares (moisture index below 1.2) suggests otherwise. Apparent virtual restriction of breeding to 500+ m, 250+ mm rainfall squares could well reflect low observer

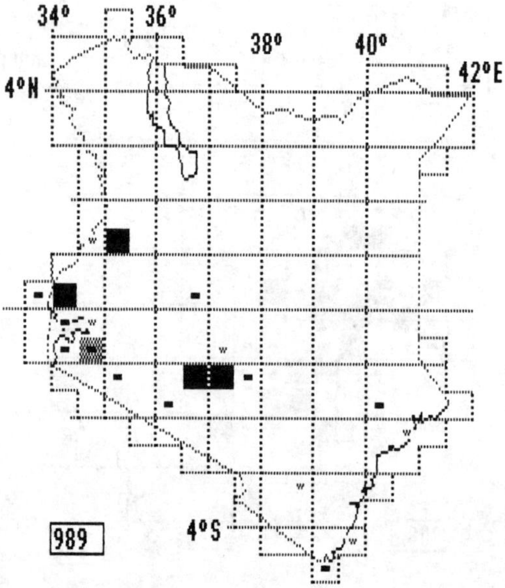

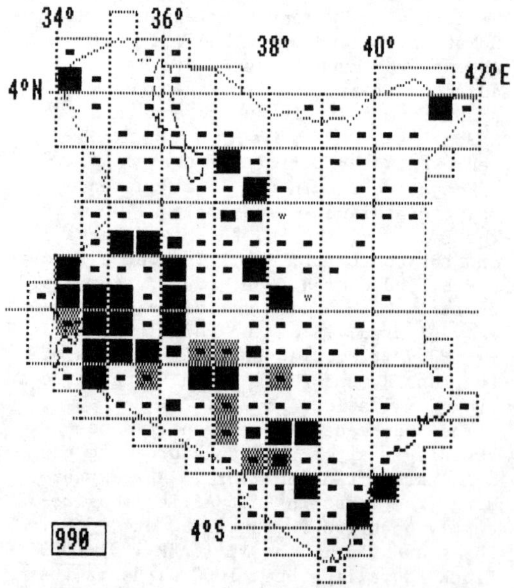

frequencies (Lewis & Pomeroy 1983a). Be considers this species resident in Narok district 74, and DEP (pers. obs.) notes similar status in 74B 75CD 88AD 89AC 100B 101A.

The taxonomy of the Grey-headed Sparrow is complex, with four distinct and mainly allopatric forms in Kenya (Br). Partly for convenience, HM treat these forms as species forming part of a superspecies, although mentioning both interbreeding and ecological segregation in various overlap areas. Br treats them as races, but suggests that the distinctly heavier billed *P.g. gongonensis* (the Parrot-billed Sparrow) probably warrants specific status, and that the others may also do so. Each race exhibits wide habitat diversity (Br). Using Br's racial ranges, we compare their distributions, and summarise specific literature.

Swainson's Sparrow *P.g. swainsonii* is known only from Moyale 17AC (and presumably also 16B), where it is common in the town and the surrounding S foothills of the Ethiopian highlands (Turner & Forbes-Watson 1976b; F J McCartney & ADL pers. obs.). 17AC are at 500-1000 m, in arid, 250-500 mm rainfall country. Whereas Swainson's is in the hills around Moyale, the Parrot-billed Sparrow is lower down on the plains to the S (Turner & Forbes-Watson 1976b). Birds at Lokitaung 3D were identified as the nominate race (MacInnes 1933-34), but are more probably nominate-Swainson's intergrades (Br).

The nominate race (the Grey-headed Sparrow *per se*) is confined to the highlands and moist plateau W of the rift valley, approximately in and W of 37C 49BD 61BD. Its range is above 1000 m, in moist subhumid-humid, 1000+ mm rainfall country (*cf*. Br).

The Swahili Sparrow *P.g. suahelicus* is in SW and S Kenya, in and SW of 60C 74B 88B 100B. Range here is mostly at 1000-2000 m in semiarid-subhumid, 250-1000 mm rainfall country.

The Parrot-billed Sparrow is apparently allopatric with Swainson's (see above) and the nominate race (Br), but otherwise occurs throughout Kenya. Its range is throughout our environmental categories. This range overlaps extensively with that of the Swahili Sparrow and van Someren (1922) records them together at Simba 88B. In C Kenya, the nominate race and the Parrot-billed do not intergrade, and are at least partly ecologically separate. Specific notes on the Parrot-billed come from Turkana (HH), Baringo 50A (St), the lower Tana R 79C 91A (AGH) and Tsavo E 101B (La). All except AGH describe it as common, while La consider it resident. Other reference: Clancey 1959b.

991 RUFOUS SPARROW *Passer motitensis*
B1207 / MPG1294, 1295 / D3C3

Afrotropical: four distinct mainland populations. *P.m. rufocinctus* is endemic to the highlands of Kenya and NE Tanzania, and *P.m. shelleyi* extends from NE Uganda and NW Kenya to S Sudan, SE Ethiopia and NW Somalia. The other populations are in C Sudan and S Africa. Distribution in Kenya represents the E limit of the range.

Kenyan: most of the range is occupied by *P.m. rufocinctus*, which is common in open bushed and wooded country, and around human habitation, in the highlands and adjacent plateaux (Br; Ja; MPG). This race is mainly at 1400-2200 m, although total altitudinal range is 1000-3000 m (Br; Britton 1979). Range is 78% within the 500+ mm rainfall regime, and it extends from humid to arid country.

In contrast, the breeding squares are clustered in a compact area around the true highlands, with 90% above 1500 m, 94% within 500+ mm rainfall areas and 74% subhumid-humid. Lewis & Pomeroy (1983a) mapped this breeding distribution, suggested that it might be real, and requested breeding records from other regions. The sole response to this request reported probable breeding in the Ugandan sector of 36B (J S Ash *in litt*.). Hence, on present evidence, breeding of *P.m. rufocinctus* appears to be confined to the central part of its restricted Kenyan (and

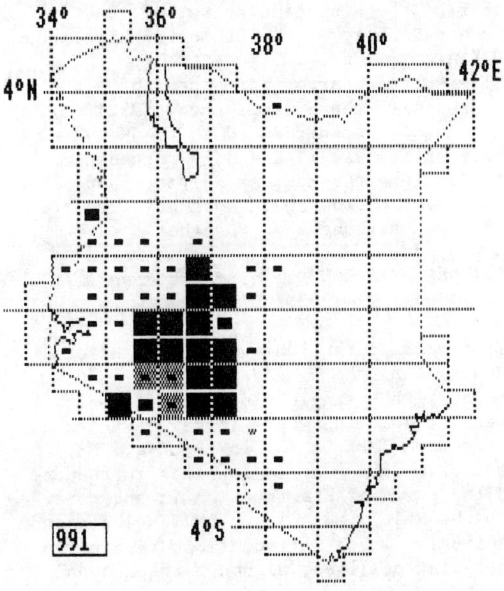

presumably NE Tanzanian) range. Non-breeding birds evidently disperse peripherally to lower altitudes on the moist W plateau 36D 37CD S to 73B, and to much lower and drier country with permanent rivers in the NE 51B 52A (*cf.* Ja) and in the SE 89C. These last three squares are at 500-1000 m in semiarid, 250-1000 mm rainfall areas. All three contain permanent rivers.

Several points support this proposal. This species' breeding is conspicuous, and can even occur on buildings (Br). W areas such as 37C 48D 49ACD 60B 61A 74A either have resident birdwatchers or are visited very frequently, so that observer cover can hardly be a factor. On the W edge of the breeding area in 61D, Sessions first recorded the bird at Mau Narok in 1972, and it bred in succeeding years (Britton 1979). This might show a slight W expansion of the breeding area. Be considered this sparrow resident in Narok district 74, and DEP (pers. obs.) found similar status in 74BD and 75. Both of these one degree squares are within the observed breeding area but, even there, the species shows evidence of movement. Annual Jul-Sept influxes to the Ruaraka-Kahawa-Ruiru area of 75B (DEP pers. obs.) are after the main breeding season (Brown & Britton 1980), and may demonstrate the dispersal of non-breeding birds that we propose.

P.m. shelleyi is known in Kenya from a single record near the Kerio R in the N of the Kerio Valley (approximately 37D 49B: not mapped)(Ja; Br). These two squares are at 1000-1500 m in a subhumid, 500+ mm rainfall area. This race has also been reported from the Ethiopian sector of 16A, on the S periphery of the Ethiopian highlands (J S Ash *in litt.*).

The Rufous Sparrow forms part of a superspecies with the House Sparrow (992)(HM), which has been introduced to the Kenya coast at Mombasa 114B and is spreading inland along the Mombasa-Nairobi road. These two sparrows have both been reported from 89C, although not together.

992 HOUSE SPARROW *Passer domesticus*
B1204 / MPG1293 / D2C2

Afrotropical: the indigenous *P.d. niloticus* occurs along the Nile Valley S to 13°N in Sudan, with a single old record from the N coast of Somalia, where it may have been introduced (MPG; AM). In the E and NE, the Asian race *P.d. indicus* was introduced to the ports of Khartoum (an inland port, via the Nile), Mogadishu (AM), Mombasa and Zanzibar. These introductions remain relatively localised, although the Kenyan

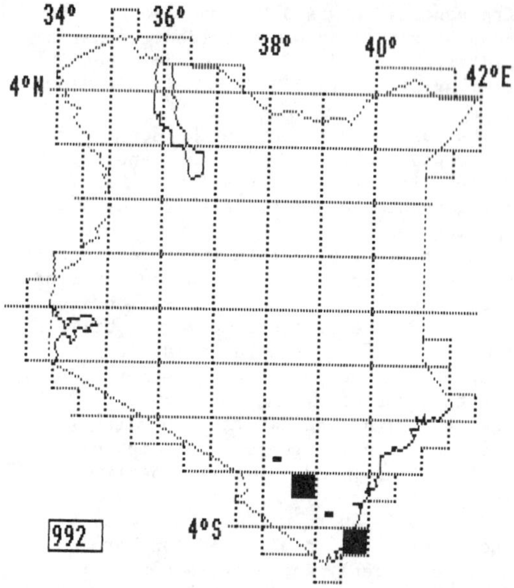

population is showing signs of expansion (see below). Similar introduction of the species to South Africa at Durban and East London has resulted in widespread colonisation N to at least N Namibia, Zambia and C Mozambique. There are recent reports from Senegambia.

Kenyan: Lewis (1983c) summarised the status of this species in Kenya up to the end of 1982; all unreferenced material in this account derives from this source. The House Sparrow was introduced to Mombasa 114B early in this century, probably as an accidental passenger on cargo ships. It remained uncommon there, mainly in the old town and docks, until approximately the mid-1970s. From 1979 to 1982, it spread inland from Mombasa Island 114B to Moi International Airport 114B, Mariakani 102C and Voi 101B, and became generally more common in Mombasa. Repeated breeding occurred at Voi. The Mombasa population continued common in 1983, and the first certain records from Mtito Andei 89C came in late Dec 1983 (EABR 1983) and in 1984 (Gomez de Silva 1984).

Range extension from Mombasa 114B has been along the main Mombasa-Nairobi 75B road and rail links. We map this species' four squares to emphasise this linearity. Given this bird's familiarity with man and human constructions, range extension by accidental transport in trains or road haulage vehicles seems possible (*cf.* ship-borne introduction). This expansion has moved out from the dry subhumid, 1000+ mm rainfall coastal strip to arid-semiarid,

250-500 mm rainfall areas, and to an altitude of 730 m at Mtito Andei 89C. Whether this trend will continue to Nairobi 75B and the highlands is unknown. These areas are occupied by the closely related Rufous Sparrow (991: *q.v.*), and there is yet to be a record of these two species together. Other references: Ash & Colston 1981; Lewis & Pomeroy 1983a; Lewis 1984a.

993 SOMALI SPARROW *Passer castanopterus*
B1203 / MPG1298 / D2C2

Afrotropical: the NE, with two separate populations. The nominate race occurs in N and C Somalia, whilst *P.c. fulgens* is found in the L Turkana basin of N Kenya and adjacent SW Ethiopia.

Kenyan: a local bird of open country at 400-900 m on the N plateau (Br), S to Kapedo 38C. This range is 86% arid (*cf.* Br) and 90% within 0-500 mm rainfall areas (Lewis & Pomeroy 1983a). Br lists Ferguson's Gulf 13B as the only locality W of L Turkana but there are now further records in this region (EABR 1979, 1983; P J Squelch *in litt.*), W to Lodakach 2D (T Romig *in litt.*). The records in 3B 4A are from Ethiopia (J S Ash *in litt.*).

The nominate race is common around human habitation in Somalia (MPG; HM; AM), but P L Britton (*in litt.*) thinks that Br's mention of a similar association for *P.c. fulgens* in Kenya may simply be based on the erroneous assumption that the two races behave in the same manner. We can trace only two certain instances of occurrences in settlements in Kenya. The bird is resident in Kapedo village 38C, where it is attracted to spilt flour (J Harjula, T Stevenson *in litt.*; EABR 1983, 1984). The only other record is of a flock in the centre of Laisamis 39B on 11 June 1984 (EABR 1984). In contrast, HH specifically note presence inland from L Turkana, and other records appear to describe similarly sparsely inhabited country (Tomlinson 1946; Lewis 1981d; EABR 1978, 1979, 1983). Williams & Arlott's (1980) mention of tree-lined wadis is certainly more appropriate in some cases (ADL pers. obs.).

P.c. fulgens presumably breeds regularly in N Kenya, but there are very few records. Although apparently resident at Kapedo 38C, actual nests have yet to be found. Breeding condition males with black bills are seen there (T Stevenson pers. comm.), a dead juvenile has been found, and hatching apparently occurs in June (J Harjula *in litt.*). van Someren (1929) saw juveniles in a flock at Marsabit 27D 28C in Jul. A pair were feeding young in a solitary tree nest, near a wadi near Korr 27D on 23 May 1981 (Lewis 1981d). These three May-Jul dates are during or soon after the 'long' rains, when these dry areas are temporarily lush. The dearth of breeding records suggests that nesting is generally away from habitation, during wet/moist conditions when N Kenya is at its most inaccessible (*cf.* Crested Lark (543))(Lewis & Pomeroy 1983a).

Br suggests wandering of this species, contrasting earlier accounts of common occurrence at Ferguson's Gulf 13B and Central Island 14AC with normally rare/absent status there. Tomlinson (1946) reports abundance E of North Horr 15C during dry weather, but disappearance after rain. J Harjula (*in litt.*) relates fluctuations from 0-200 birds at Kapedo 38C. The population in this latter village has increased from small numbers during 1977-82, to hundreds in 1983 (J Harjula *in litt.*; EABR 1981, 1983, 1984). MPG describe similar increase around an artificial food source in Somalia. Other reference: Lewis 1984a.

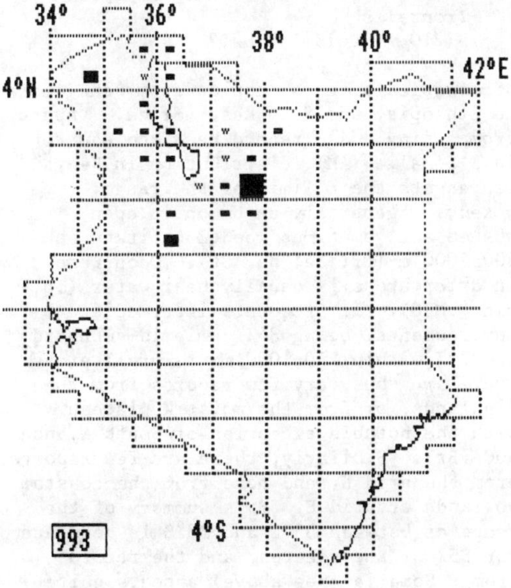

994 CHESTNUT SPARROW *Passer eminibey*
B1205 / MPG1306 / D2C2

Afrotropical: the NE, from Sudan and Ethiopia S to the E of L Victoria to NE Uganda, Kenya and interior Tanzania. There are recent records between the equator and 1°N in S Somalia (AM).

Kenyan: locally common in open wooded habitats up to 2000 m, especially near settlements or swampy ground (Br; Ja; MPG). It is most common in and W of the rift

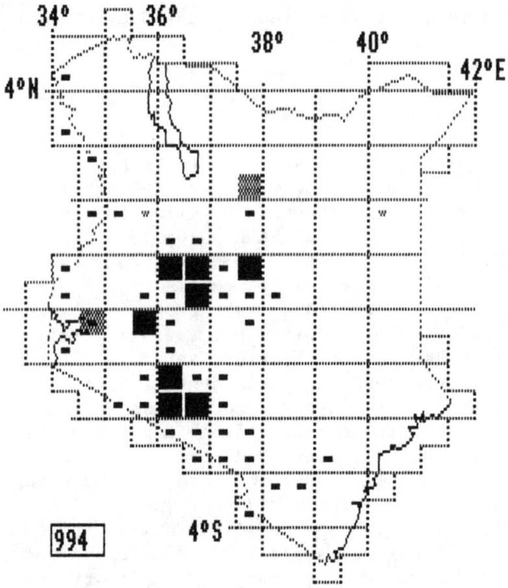

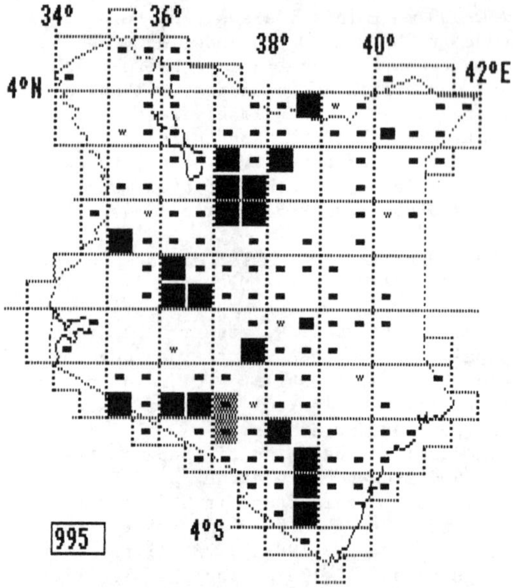

valley, but more local further E (Br). Range is 80% above 1000 m, but mainly peripheral to the true highlands (Lewis & Pomeroy 1983a). There are few records from the E lowlands which, apart from the recent Somali records mentioned above, define this bird's E limit. Absence from the coastal lowlands continues S into Tanzania, apart from a Jan flock at 6°48'S (Br). Distribution is 77% within intermediate, 250-1000 mm rainfall areas and 82% semiarid-subhumid, and thus generally very sparse in the dry N and E.

The Chestnut Sparrow is considered resident in Narok district 74 (Be) and in 75C (DEP pers. obs.), but Br suggests that most birds wander extensively. This latter view is supported by observations in the semiarid Baringo area 50A, where the species is always present, but definitely more numerous in wet periods (St; T Stevenson *in litt.*). Other reference: Payne 1969.

995 YELLOW-SPOTTED PETRONIA *Petronia pyrgita*
B1208 / MPG1308 / D2C3

Afrotropical: sporadically distributed along the S edge of the Sahara around 15°N, and S through S Sudan, Ethiopia and Somalia to NE Uganda, Kenya and NE Tanzania.

Kenyan: widespread in small numbers in dry bushed and wooded habitats, below 1800 m but mainly at 500-1500 m (Br; Ja; MPG; AGH; La; St). Primarily associated with trees, it also frequents rocks (*op. cit.*; HM). Range is 82% arid-semiarid (*cf.* Br). It is virtually absent from the highlands, and from the moist W plateau with the exception of the L Victoria basin at Kisumu 60B and South Nyanza 60C (Br; P L Britton *in litt.*). There are no reports from four of the five dry subhumid, 1000+ mm rainfall coastal squares, and it is completely absent from the coastal strip S of Malindi 103A (Br). The species is generally considered resident (Br; La; St; DEP pers. obs.).

996 SPECKLE-FRONTED WEAVER *Sporopipes frontalis*
B1210 / MPG1311 / D2C2

Afrotropical: the N tropics from Senegambia to Ethiopia, and S to East Africa. Apart from a single Jan record near the equator in S Somalia (AM), distribution in Kenya represents the E limit of the range.

Kenyan: generally uncommon in open bushed and sometimes wooded habitats at 400-2000 m (Br; Ja; Be; St). Occurrence in drier areas is usually near water (*op. cit.*; MPG). It frequents intermediate environments, being 84% semiarid-subhumid and 79% within 250-1000 mm rainfall areas. There are thus very few records from the highlands, or from the moist W plateau with the notable exception of South Nyanza 60C (Br). Similarly, there are few reports from the arid N, and none from the coastal lowlands or arid E. Br's summary of the range as between 35°E and 38°E is accurate for 85% of the squares, and the record from S Somalia (see above) appears quite

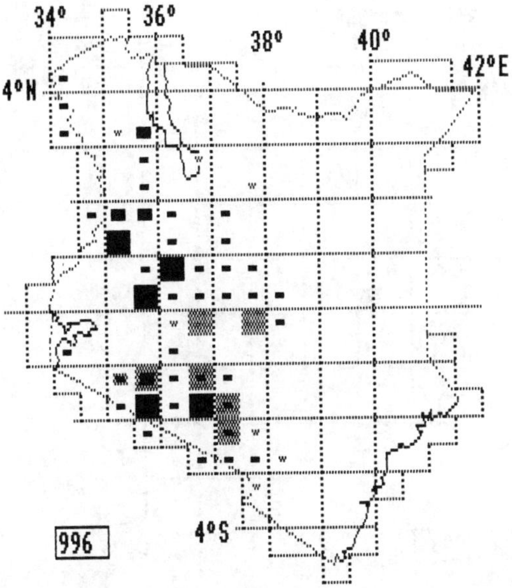

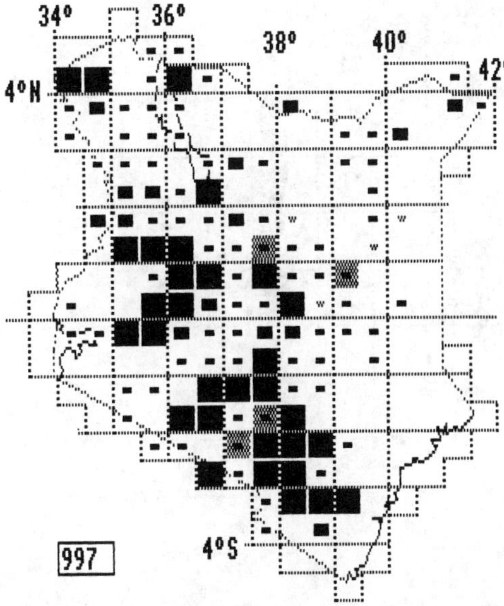

isolated.

Be considers this weaver resident in Narok district 74, and DEP (pers. obs.) concurs for 74B and 75D. Br describes it as both resident and wandering however, and small, permanent numbers in the semiarid Baringo area 50A undergo increases of unknown origin (St).

997 WHITE-BROWED SPARROW WEAVER *Plocepasser mahali*
B1199 / MPG1287 / D3C3

Afrotropical: two separate populations. In the NE, this sparrow weaver extends from N Uganda and extreme NE Tanzania to S Sudan, Kenya, Ethiopia and Somalia. In the S, it is found in S Tanzania, and is widespread S of 15°S with the exception of the dry SW.

Kenyan: noisy, conspicuous and locally numerous in dry wooded and open wooded country, mainly at 400-1400 m, but also reaching 1900 m (Br; Ja; MPG; HH; La; St). It is often around habitation (*op. cit.*). Range is throughout our rainfall and moisture categories although, on the moist W plateau, it is mainly confined to the Winam Gulf 60AB-Nyanza Rift 61A area. Affinity for the locally moister habitats of arid country is shown by restriction to woodland in the L Turkana basin (HH), and (partly) to riverine growth in Tsavo E 101B (La). Absence from the coastal lowlands and from the E borders S of 3°N compares with absence W of the Juba R in adjacent S Somalia (AM).

This sparrow weaver is considered resident at semiarid Baringo 50A (St) and in arid Tsavo E 101B (La), and E J Roberts (*in litt.*) recorded it throughout her May-Dec study of the Daua riverine at Rhamu 19A. Br thinks it mostly resident, but also mentions wandering. Although finding it always present in 75B 88D 89AC 100B 101A, DEP (pers. obs.) notes an annual Oct-Dec influx to the Ruaraka-Kahawa-Ruiru area of 75B, which is during the non-breeding season (Brown & Britton 1980).

The White-browed Sparrow Weaver forms part of a species-group with the similarly colonial Donaldson-Smith's (998) and Chestnut-crowned (999) Sparrow Weavers (HM). The range of Donaldson-Smith's is enclosed by that of the White-browed (HM), and overlap probably reflects wanderers and/or flocks moving to drink (Br). Donaldson-Smith's inhabits generally drier and more open country than the White-browed (Br). The Chestnut-crowned is on the edge of its range and little known in Kenya. Its few squares are mainly moister than those of the White-browed, and HM suggest that it may prefer greater altitudes. Other references: Collias & Collias 1978ab.

998 DONALDSON-SMITH'S SPARROW WEAVER *Plocepasser donaldsoni*
B1198 / MPG1289 / D2C2

Afrotropical: restricted to N Kenya, the S borders of Ethiopia and a very localised population in S Somalia. The Somali population is 1°E of 54B (AM), and the two

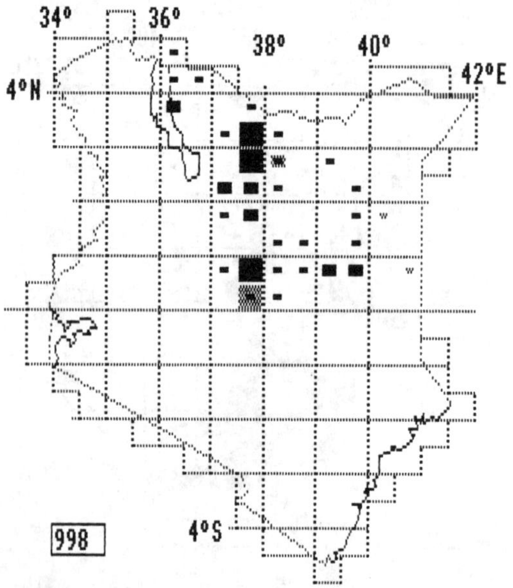

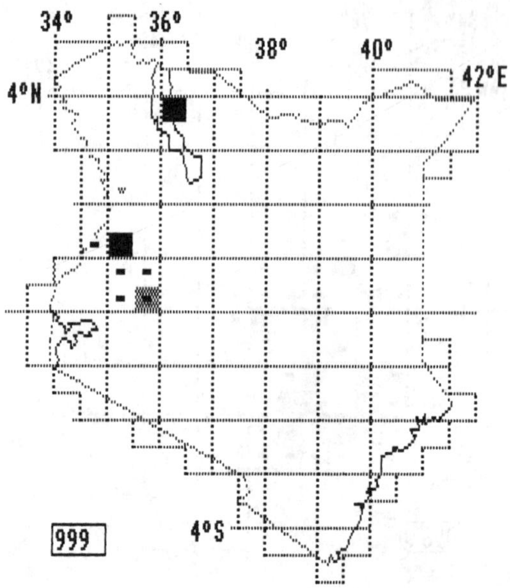

ranges were presumably originally continuous. In contrast to the White-browed Sparrow Weaver (997: *q.v.*), this Somali population is W of the Juba R (AM).

Kenyan: locally common in open and often rocky semidesert with scattered bushes and trees (Br; Ja; MPG). Range is E of the rift valley apart from the N Turkana area, where the lake forms its W limit (HH). Br describes this species in arid-semiarid country at 400-1500 m, and distribution is 83% arid, 90% below 1000 m and 90% within 0-500 mm rainfall areas (*cf.* White-browed Sparrow Weaver). It is resident, and particularly abundant in parts of Wajir district (approximately 41 42)(Br).

999 CHESTNUT-CROWNED SPARROW WEAVER *Plocepasser superciliosus*
B1200 / MPG1288 / D2C1

Afrotropical: the N tropics between 8°N and 15°N, E to Ethiopia, Uganda and, at the SE limit of the range, NW Kenya.

Kenyan: local, uncommon and unobtrusive in dry bushed and open wooded country at 400-1800 m (Br; Ja; MPG; EABR 1984). It is known mainly from the W highlands and their N periphery, N to the Karasuk Hills 25C (Br). Of the seven squares in this area, all are above 1000 m in 500+ mm rainfall areas to the W of the rift valley, and six are moist subhumid-humid. Br considers it resident in this area, and breeding is known at 1800 m on the Kamassia Hills 49D in Aug (Ja), and at Cheperaria 37C in June (T J Barnley *in litt.*). Apart from 25C (see above), we can trace no records for N Kenya to the W of the rift valley beyond MPG's note of breeding in NW Kenya, and T J Barnley's (*in litt.*) mention of increasing abundance N of 37C.

Several pairs nesting on 9 Oct 1984 at Kampi ya Kavuli 14A (EABR 1984) were in the rift valley and to the E of L Turkana. This is apparently a marked extension of range and, in contrast to the areas mentioned above, is in arid, 0-250 mm rainfall country at about 400 m altitude. See White-browed Sparrow Weaver (997).

1000 GREY-HEADED SOCIAL WEAVER *Pseudonigrita arnaudi*
B1201 / MPG1291 / D3C2

Afrotropical: the E, from S areas of Sudan, Ethiopia and (very locally) Somalia, to N Uganda, Kenya and C and E Tanzania.

Kenyan: locally common in bushed and open wooded habitats, mainly at 500-1400 m but occasionally reaching 1900 m (Br; MPG). It is often in areas of sandy soil (*op. cit.*). Range is 75% in or E of the rift valley, 86% semiarid-subhumid and 84% within intermediate, 250-1000 mm rainfall areas (*cf.* Br). It is absent from the W and C highlands, but widespread on the peripheral, lower rainfall plateaux to the E of the rift. More sporadic distribution to the W of the rift apparently reflects the greater extent of 1000+ mm rainfall country there. The record in 16A is from Ethiopia (J S Ash *in litt.*), in the S foothills of the highlands. The overlay data for this

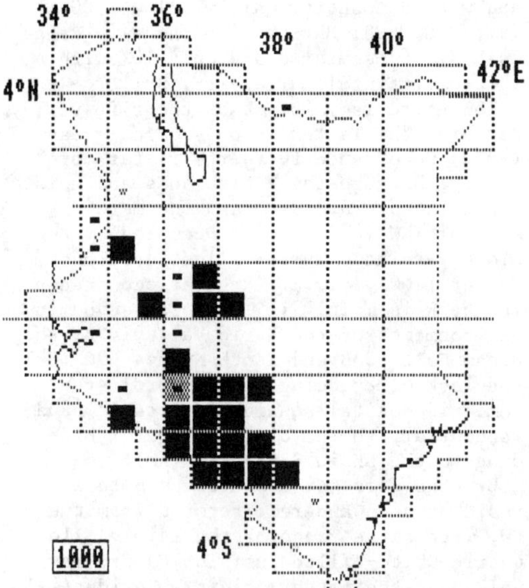

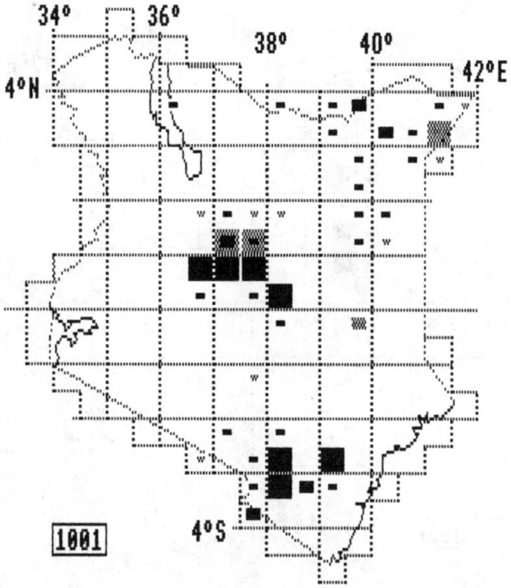

Ethiopian square may be incorrect, and the only other QSD in arid country is Tsavo E 101B, from which range has recently withdrawn (La).

The high proportion (69%) of breeding squares suggests a resident species, and this view is endorsed by DEP (pers. obs.) for 75CD, and by Collias & Collias (1978c) and Br. Occasional non-breeding visitors to the range edge in the semiarid Baringo area 50A have been in Jan, May and June-Sept (St).

The Grey-headed Social Weaver forms a species-group with the similarly colonial Black-capped Social Weaver (1001)(HM). These two birds can be sympatric, but the Black-capped is mainly in lower, drier country (Br). Other references: Collias & Collias 1980; Collias 1984.

1001 BLACK-CAPPED SOCIAL WEAVER *Pseudonigrita cabanisi*
B1202 / MPG1292 / D2C2

Afrotropical: S Ethiopia, the SW borders of Somalia, Kenya and NE Tanzania.

Kenyan: locally common in dry, open, bushed and wooded habitats at 200-1300 m (Br; MPG). It is found to the E of the rift valley, with the exception of specimens taken at Koobi Fora 14A, L Turkana in June (National Museum Collection, Nairobi). This species is mainly (89%) allopatric at QSD scale with the Grey-headed Social Weaver (1000: *q.v.*), in the lowlands to the NE and E of the latter. In contrast to the Grey-headed's range, the Black-capped is 94% arid-semiarid (*cf.* Br) and 74% within 0-500 mm rainfall areas. The only records from Somalia are in 19C and 31A (AM).

MPG and Br consider this species resident, although the latter mentions rare, seasonal, non-breeding flocks in the Tsavo E area 101B (La).

1002 WHITE-BILLED BUFFALO WEAVER *Bubalornis albirostris*
B1194 / MPG1284 / D2C2

Afrotropical: mainly the N tropics, across Africa between 10°N and 17°N. There is an apparently isolated population in SE Sudan, NE Uganda and NW Kenya.

Kenyan: locally common in dry, bushed and open wooded country at 400-1200 m in the NW (Br; Ja; HH; St). It is mainly N of the W highlands, but penetrates further S in the Kerio Valley 49BD and the main rift valley 50A (Br; St). Range is 76% arid-semiarid and 69% within intermediate, 250-1000 mm rainfall areas. Breeding records are also mainly within these categories. Br considers it both resident and wanderer. In the semiarid Baringo area 50A, birds frequent some of the breeding colonies throughout the year (St).

The White-billed Buffalo Weaver forms a superspecies with the similarly colonial Red-billed Buffalo Weaver (1003)(HM). These two are mainly allopatric, but some overlap does occur (Br; HM). This may in part be due to non-breeding wanderers, but true sympatry (apparently without interbreeding)

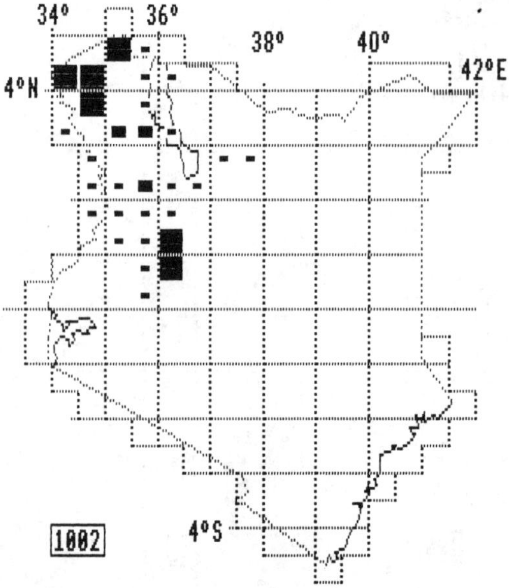

occurs in the Kerio Valley 49BD (*op. cit.*).

1003 RED-BILLED BUFFALO WEAVER *Bubalornis niger*
B1195 / MPG1285 / D3C2

Afrotropical: a population in the E and NE, from S Sudan, Ethiopia and Somalia to Kenya and interior E Tanzania. Another race occurs in Angola, Zambia, Namibia, Botswana, Zimbabwe and South Africa.

Kenyan: generally common in open bushed and wooded country, mainly below 1500 m (Br; MPG; AGH; HH; La; *contra* Ja). Range is 87% arid-semiarid and only 10% within 1000+ mm rainfall areas (*cf.* Br).

This bird is mainly E of the rift valley, reaching the latter in only a few areas (Br). Occurrence further W is far more patchy, because the W highlands and moist W plateau are avoided, and the more suitable NW lowlands are occupied by the closely-related White-billed Buffalo Weaver (1002: *q.v.*). Regular occurrence in the W is mainly in the moist subhumid environments of the Kerio Valley 49B, the Winam Gulf 60AB and South Nyanza 60C (Br). The lack of breeding records from even semiarid coastal squares is notable, and is contrary to the distribution of breeding to the NE in S Somalia (AM).

Br considers this species as both a resident and wanderer: records from the NW, within the range of the mainly allopatric White-billed Buffalo Weaver, may well have the latter origin. Residential status has been reported from Tsavo E 101B (La), although the bird is in fact commonest there during the rains (Lack 1985), and from 75C 88AD 89AC 100B 101A (DEP pers. obs.). Other reference: Leuthold 1970a.

1004 WHITE-HEADED BUFFALO WEAVER *Dinemellia dinemelli*
B1196 / MPG1286 / D3C3

Afrotropical: the E and NE, from S Sudan, Ethiopia and Somalia to East Africa E of L Victoria.

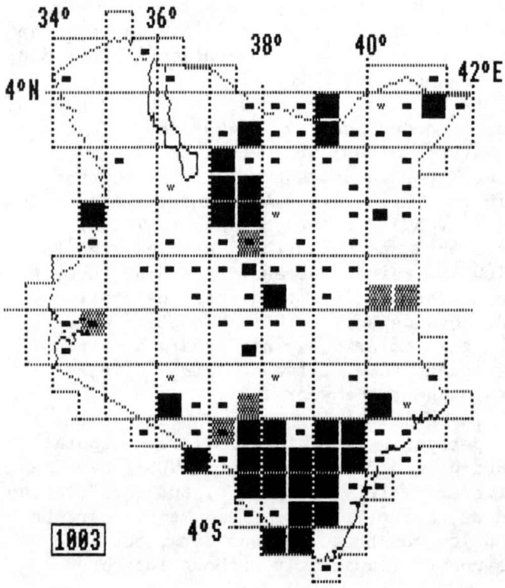

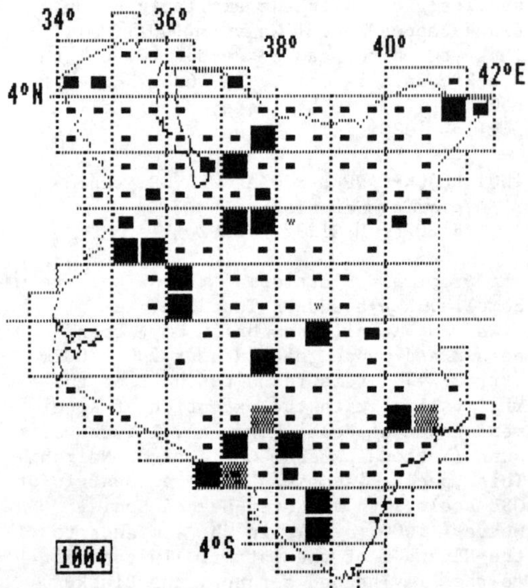

Kenyan: a common, colonial and conspicuous species of open bushed and wooded habitats below 1400 m (Br; Ja; MPG; AGH; HH; La; St). Range is 86% arid-semiarid and only 6% within the 1000+ mm rainfall regime. It is thus scarce in the highlands and the moist W plateau (Br), which form part of the W edge of the range. It is absent from the coastal lowlands S of Ngomeni 103A (Br). The absence of breeding records from even semiarid coastal squares contrasts with S Somalia (AM; *cf.* Red-billed Buffalo Weaver (1003)).

Br describes the White-headed as probably mainly resident, and this status is reported from Tsavo E 101B (La), and from 75D 88AD 89AC 100B 101A (DEP pers. obs.). Lack (1985) reports a June-Aug peak in Tsavo E 101B, and Br mentions some wandering. See Pygmy Falcon (142).

Family Estrildidae: Waxbills, Twinspots, Firefinches, Negrofinches, Cordon-bleus, Mannikins etc.

Around 124 species that are Afrotropical, Oriental and Australasian. 67 (54%) are found in the Afrotropical Region, where the genus *Amadina* and 93% of the brightly coloured "true waxbills" are endemic (HM). The other estrildid group, the more sombre mannikins and silver-bills, is also extensively Oriental and Australasian. 35 (52% of the Afrotropical total) occur in Kenya. All breed or are thought to do so, although the Cordon-bleu (1026) and Magpie Mannikin (1043) are at the limit of their range and known from very few records. Following HM, we discuss the parasitic indigobirds and whydahs of the Ploceidae alongside their estrildid hosts.

1005 GREY-HEADED NEGROFINCH *Nigrita canicapilla*
B1246 / MPG1386 / D2C2

Afrotropical: from the lowland forests around the Gulf of Guinea, E through the Congo basin to S Sudan, Uganda and, at the E limits of the range, the highlands of Kenya and NE Tanzania.

Kenyan: a locally common bird of the edges and more open interior of forest, at 1700-3350 m in the W and C highlands (Br; Ja; MPG; Se; HM; Diamond & Fayad 1979). Range is 72% within the 1000+ mm rainfall regime and 86% moist subhumid-humid. Different races are resident either side of the rift valley (Ja; Br) and, particularly in terms of numbers of breeding squares, that to the E appears less well covered and hence possibly less common.

Although diverse in plumage features, the negrofinches are similar ecologically, being associated with shrubs and trees in and around forest (HM). The three Kenyan species can occur together in feeding flocks at Kakamega Forest 48D (Br; EABR 1978). Some diffuse ecological segregation is known in E Zaire, where the Chestnut-breasted Negrofinch (1006) is often at low levels in primary forest while the Grey-headed reaches higher levels in more open habitats (HM; *cf.* Ja).

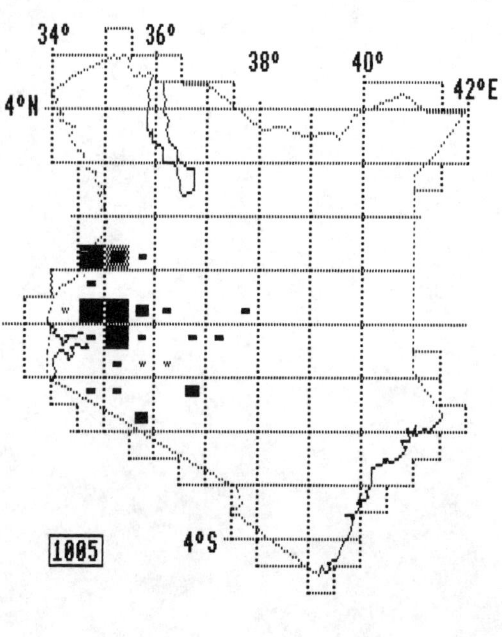

1006 CHESTNUT-BREASTED NEGROFINCH *Nigrita bicolor*
B1245 / MPG1387 / D2C2
NO MAP: post presence 48D

Afrotropical: W and C Africa, from the lowland forests around the Gulf of Guinea, E through the Congo basin to S Sudan, Uganda and, at the E limit of the range, W Kenya.

Kenyan: a forest edge species, known from rare sightings in Kakamega Forest 48D (Mann 1976; Br; EABR 1978, 1981). This forest is at 1700 m in a humid, 1000+ mm rainfall square. There is no breeding record, but Br considers it resident and forest birds of this type rarely wander far. See Grey-headed Negrofinch (1005).

1007 WHITE-BREASTED NEGROFINCH *Nigrita fusconota*
B1247 / MPG1388 / D2C2
NO MAP: post probable 48D; pre presence 36D

Afrotropical: W and C Africa, from the lowland forests around the Gulf of Guinea to the Congo basin, Uganda and, at the E extremity of the range, W Kenya.

Kenyan: there is an old record from Mt Elgon 36D (Ja), but all recent reports of this forest edge species have been in the Kakamega Forest 48D, where it is reasonably common (Br; MPG; ADL pers. obs.). Both of these localities are at 1500-2000 m in humid, 1000+ mm rainfall squares. Probable breeding in 48D reflects regular sightings at Kakamega (many obs.), and Br's note of resident status. See Grey-headed Negrofinch (1005).

1008 GREEN-WINGED PYTILIA *Pytilia melba*
B1256 / MPG1410 / D2C3

Afrotropical: widespread S of 15°N, with the exceptions of the equatorial lowland forest areas of the W, and the dry areas of E Somalia and the SW.

Kenyan: a distinctive and generally common bird in the lower levels of dry scrub, bush and wooded habitats, up to 1800 m but mainly below 1400 m (Br; Ja; MPG; AGH; HH; La; St). Range is 69% less arid-semiarid and 66% within the intermediate, 250-1000 mm rainfall regime. This pytilia is absent from the coastal strip from Malindi 103A S to 6°48'S in Tanzania (Br), and scarce in humid areas of the W highlands and plateau E of 34°30'E. Br considers it resident, and similar reports come from Tsavo E 101B (La) and 74D 75C 89A (DEP pers. obs.).

The Green-winged Pytilia forms part of a species-group with the Orange-winged Pytilia (1009)(HM). In contrast to the Green-winged, the Orange's few squares are mainly restricted to the higher rainfall areas, and it is more a bird of trees than thicket and scrub (HM; Br). See Paradise Whydah (1010) and Pin-tailed Whydah (1032).

1009 ORANGE-WINGED PYTILIA *Pytilia afra*
B1255 / MPG1409 / D1C1

Afrotropical: widespread S of 5°S, apart from the dry SW. Further N, it extends through the E Zaire borders and E Tanzania to Kenya, N Uganda, S Sudan and Ethiopia. Distribution in Kenya represents the E limit of the range.

Kenyan: very scarce and local in bushed and open wooded country up to 1800 m (Br). In contrast to the Green-winged Pytilia (1008: *q.v.*), seven of the Orange's nine squares are within the 500+ mm rainfall regime. Range is in two distinct areas, in the SE lowlands and on the SE edge of the C highlands. The latter consists solely of old records from Murang'a 63C and Kikuyu 75B (Ja), which are in 1000+ m, moist subhumid squares. Range in the SE lowlands is inland to Voi 101B (La) and N to Lamu 91B (Ja). Of these seven squares, six are below 500 m and five arid-semiarid.

The only recent Kenyan records of this species are of single birds in dry subhumid 102D and 114B (P L Britton *in litt.*). There is no breeding record, but Br considers the bird resident. This pytilia is very little known and local, and easily overlooked. Whether the preponderance of old records reflects an actual decline or the modern disinclination for specimens is unknown. See Broad-tailed Paradise Whydah (1011).

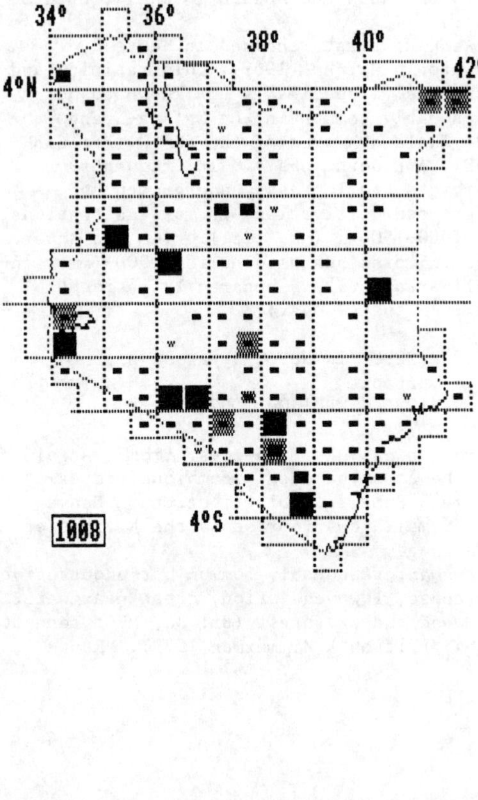

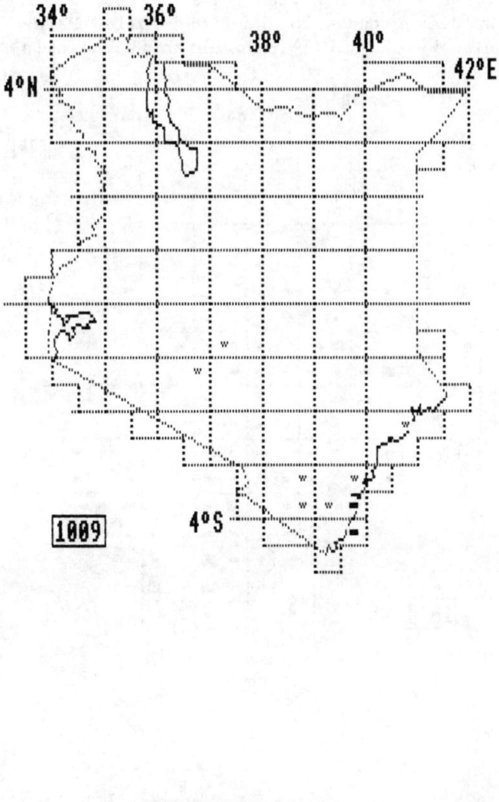

1010 PARADISE WHYDAH *Vidua paradisaea*
B1218 / MPG1444 / D3C2

Afrotropical: widespread S of 10°S, apart from the dry SW. Further N it extends through E Tanzania and Kenya to NE Uganda, S Sudan, Ethiopia and W Somalia.

Kenyan: a locally common bird of open bushed and wooded habitats, typically in dry country below 1400 m but also up to 2000 m (Br; HH; St). A record nearer 3000 m around Mau Narok 61D (Britton 1979) may have been a wanderer.

This whydah parasitises the Green-winged Pytilia (1008), and their distributions show many similarities (HM; Br). They are both distributed broadly throughout Kenya, and Ja's and Br's notes of the whydah in the E of the country only are incorrect. The pytilia is apparently resident in even arid areas, but the whydah may only visit these during wet periods. This is seen in arid Tsavo E 101B, where the whydah only appears, in variable numbers, after rains (La). The same is true at semiarid Baringo 50A (T Stevenson *in litt.*), but DEP (pers. obs.) recorded resident status in semiarid 89A. This whydah also compares with its pytilia host by being 69% within intermediate, 250-1000 mm rainfall areas and 67% less arid-semiarid. Both the parasite and the host are thinly distributed in the humid W highlands and plateau 48 49 60 61, and apparently absent from the S coastal 114.

The Paradise Whydah forms part of a species-group with the Broad-tailed Paradise Whydah (1011)(HM), which is known in Kenya from a single record.

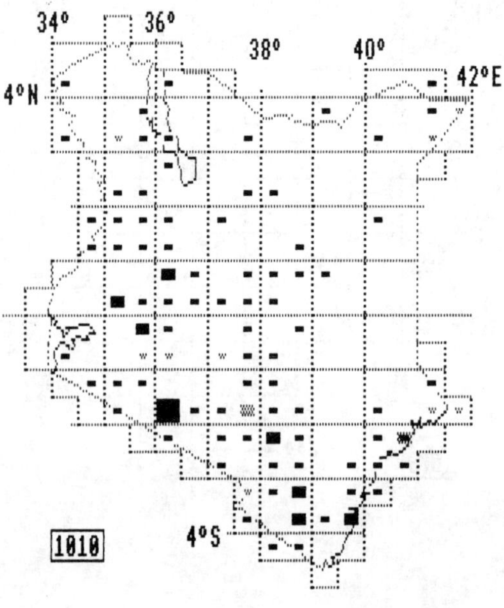

1011 BROAD-TAILED PARADISE WHYDAH *Vidua obtusa*
B1217 / MPG1445 / D3C0
NO MAP: pre presence 63B

Afrotropical: a S tropics species, widespread S of 5°S apart from the dry SW. Further N, it reaches S and W Tanzania, the E Zaire borders and SW Uganda, and there is a single specimen from C Kenya.

Kenyan: the single record of this distinctive, woodland species is a male taken at Chuka 63B on 21 Feb 1947 (National Museum Collection, Nairobi; Br). The locality is around 1200 m, in a moist subhumid, 1000+ mm rainfall square. The individual was presumably wandering, although the locality is immediately NE of the highland range of this parasite's host, the Orange-winged Pytilia (1009). See Paradise Whydah (1010).

1012 BROWN TWINSPOT *Clytospiza monteiri*
B1221 / MPG1389 / D2C2
NO MAP: post presence 48C 60A; pre confirmed 48C

Afrotropical: W and C Africa, with populations N and S of the Congo basin forests. To the N, it extends E from Cameroon to the Central African Republic, N Zaire, S Sudan, Uganda and, at the E limit of the range, W Kenya. The S population is found between Gabon and Angola in the W, E to E Zaire.

Kenyan: first recorded in Kenya by Britton & Harper (1969), this retiring and unobtrusive species is now known to be reasonably common in the Ng'iya-Nyangweso 48C-Akala 60A area of the far W (Br; EABR 1981; ADL pers. obs.). It frequents thickets, tall grass, undergrowth and overgrown cultivation, in moist situations at 1000-1500 m (*op. cit.*). The two squares are in moist subhumid-humid, 1000+ mm rainfall areas. It is apparently resident (Br; ADL pers. obs.).

1013 PETERS' TWINSPOT *Hypargos niveoguttatus*
B1235 / MPG1406 / D2C2

Afrotropical: an E species, from E Angola to the coast between Mozambique and extreme S Somalia. Distribution in Kenya and Somalia 80A represents the N limit of the range.

Kenyan: reasonably common but unobtrusive in dense, low vegetation, often near water in woodland or forest (Br; Ja; MPG; Tennent 1964; Britton & Zimmerman 1979). Range

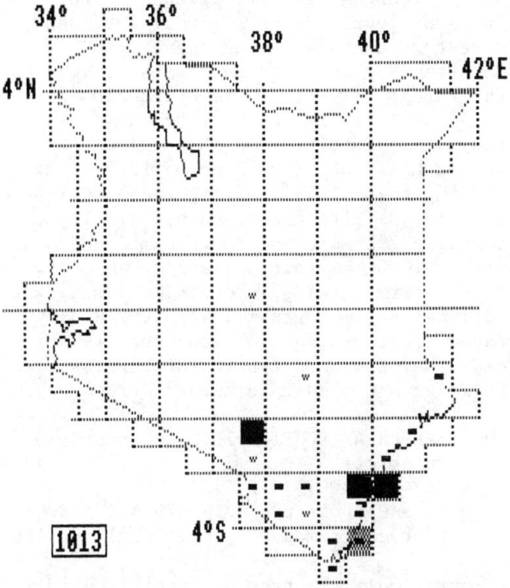

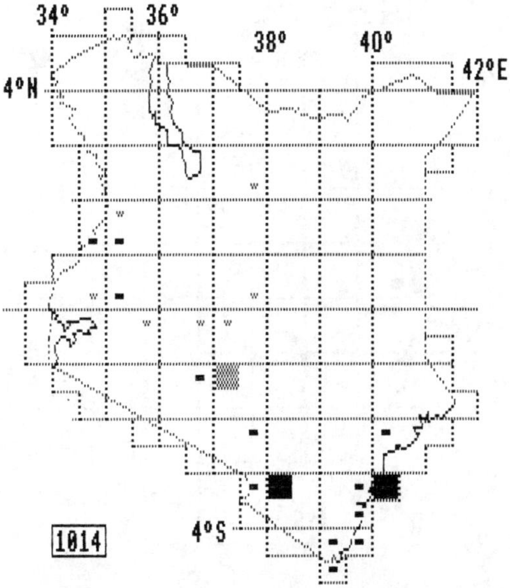

reaches 2000 m but is mainly below 1000 m in the coastal lowlands (Br), inland to Voi 101B (La) and N to 80A, which is the sole locality in Somalia (AM; Ash 1983). It is more local further inland, with recent reports from the Kibwezi area 88B, and older records from the forest island of Mt Endau 77B (Tennent 1964) and highland forest at Meru 51D (Ja).

A W Diamond (*in litt.*) failed to relocate the species during four days on Endau in 1979, and continued presence there and at Meru, both on the N limit of the Afrotropical range, is uncertain. Overall range is entirely E of the rift valley, 72% below 1000 m and 93% less arid-dry subhumid. Br considers this bird resident.

1014 GREEN-BACKED TWINSPOT *Mandingoa nitidula*
B1242 / MPG1407 / D2C2

Afrotropical: widespread between 8°N and 10°S, E to S Ethiopia, Kenya and coastal Tanzania. Further S, it extends through Malawi, Mozambique and Zimbabwe to E South Africa. Distribution in Kenya represents the E limit of the range.

Kenyan: an uncommon bird of moist thickets and undergrowth in forest and forest edge, up to 2150 m (Br; Ja; MPG; Britton & Zimmerman 1979; EABR 1977). As with the two other twinspots (1012, 1013), D2 is awarded because, although retiring and unobtrusive, these birds are readily identifiable when seen and of sufficient interest to be reported.

The Green-backed inhabits moist country in the coastal lowlands N to the lower Tana R 91A, and inland to the highlands and the forest island of Mt Marsabit 27D (Ja). Excluding the latter mountain, which is too small to feature on the overlays, range is entirely within the 500+ mm rainfall regime and absent from arid country. The relatively high proportion of pre-1970 squares in the interior may reflect less common status than in the SE (*cf.* Peters' Twinspot (1013)) or a decline.

1015 RED-HEADED BLUEBILL *Spermophaga ruficapilla*
B1259 / MPG1391 / D2C2

Afrotropical: W and C Africa, from the S and E peripheries of the Congo basin, E to NW Tanzania, S Sudan, Uganda and, at the E limit of the range, Kenya. There is also an isolated population in the East Usambara Mts of NE Tanzania.

Kenyan: locally common but retiring in highland forest undergrowth and edge, up to 2150 m (Br; MPG; Sugg & Sugg 1972). It is mainly W of the rift valley, but there is also an old record from Meru 51D and a recent sight record at Embu 63C (Br). Range is entirely above 1000 m in moist subhumid-humid, 1000+ mm rainfall areas. The species is resident (Br).

(map overleaf)

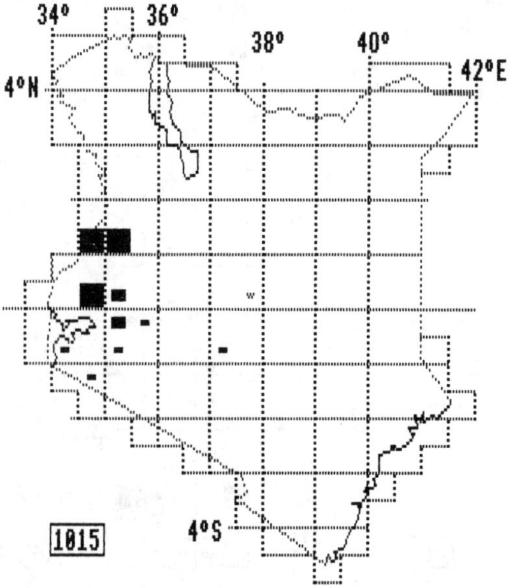

1016 ABYSSINIAN CRIMSON-WING *Cryptospiza salvadorii*
B1224 / MPG1399 / D2C2

Afrotropical: an E montane species, in the highlands of W Ethiopia, S Sudan, Uganda, Kenya and NE Tanzania.

Kenyan: common but retiring in the undergrowth and edges of highland forest and bamboo, at 1500-3000 m (Br; Ja; MPG; Se; Britton & Sugg 1973). There were occasional records from the grounds of the (now) National Museum, Nairobi 75B, when forest remnants survived there (van Someren 1939). This species is present on the forest islands of Mts Nyiru 26D (Diamond & Keith 1980) and Uraguess 51A (Ja), the Ndoto Mts 39A (Br) and the Chyulu Hills 88D (van Someren 1939). Excluding these localised highlands, which are too small to appear on our overlays, range is 76% moist subhumid-humid and 67% within 1000+ mm rainfall areas.

This crimson-wing is considered resident (Br; Be; Se). Confusion with the Red-faced Crimson-wing *C. reichenovii* is very easy (HM; Se), and reports of this species in Kenya are almost certainly erroneous.

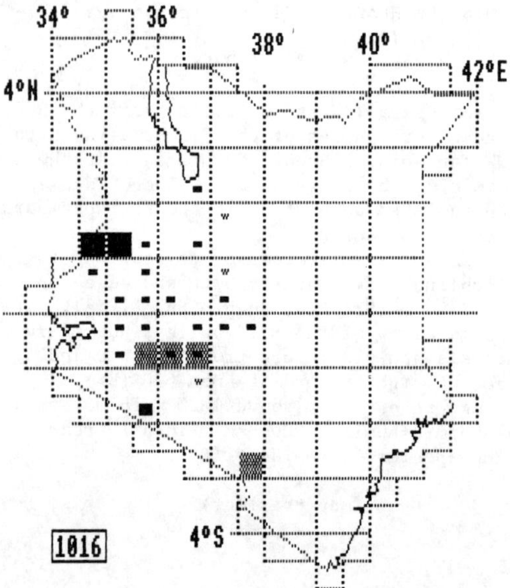

1017 BAR-BREASTED FIREFINCH *Lagonosticta rufopicta*
B1240 / MPG1414 / D1C2
NO MAP: post presence 47D 48C 60B; post confirmed 60A; pre confirmed 48C

Afrotropical: a N tropics species, mainly at 5°N to 10°N, from Sierra Leone E to S Sudan and SE to NW Uganda. Birds in E Uganda and W Kenya represent the SE extremity of the range. This population has been considered disjunct (Br), but may not be so (EABR 1983). A further population to the S of the Congo basin may not belong to this species.

Kenyan: first recorded in Kenya by Britton & Harper (1969), and now known to extend from Siaya district 47D 48C to the N shores of the Winam Gulf 60AB (Br; ADL pers. obs.). It is locally common in bushed and wooded habitats, typically in low cover near water, up to 1400 m (Br). In contrast to the closely related Red-billed Firefinch (1018), the four squares are all above 1000 m in moist subhumid-humid, 1000+ mm rainfall country. Br considers it resident.

The firefinches are a link between the twinspots and related genera (1012-1016) and the typical waxbills (1024-1038), and the Bar-breasted forms part of a species-group with the Red-billed Firefinch (HM). These authors describe the Bar-breasted from moister, more vegetated habitats than the Red-billed, which agrees both with our overlay data and Br's habitat summaries. Sympatry has apparently been induced by these birds' (especially the Red-billed's) affinity for human habitation (HM), and they commonly occur on the same ground in the Siaya area 48C 60A (Britton & Harper 1969; Br).

1018 RED-BILLED FIREFINCH *Lagonosticta senegala*
B1241 / MPG1413 / D2C3

Afrotropical: widespread S of 20°N, with the exceptions of many of the equatorial lowland forest areas of the W, and the drier areas of E Somalia and the SW.

Kenyan: most of the Kenyan range is occupied by *L.s. ruberrima*, which is a common and approachable bird of the moist highlands, plateaux and coastal lowlands (Br; Ja; MPG; St). It is found in a wide diversity of open bushed, wooded and grassy habitats up to 2200+ m, and is often around cultivation and human habitation (*op. cit.*). Range is throughout our altitude categories, 75% within 500+ mm rainfall areas and only 13% arid. Distribution E of 39°E compares with that in Somalia S of 5°N, where most of the bird's squares are coastal or contain permanent rivers (AM). Br considers this bird generally resident, and the lack of breeding records from the coastal lowlands presumably represents poor observer coverage. DEP (pers. obs.) describes it as resident in 75BC 88D 89AC 100B 101A, but notes annual June-Sept influxes to the Ruaraka-Kahawa-Ruiru area of 75B.

Two other populations occur on Kenya's N borders, in the vicinity of permanent rivers on the S periphery of the Ethiopian highlands. Records from around the Omo R 3B 4A (HH; J S Ash *in litt.*) presumably refer to *L.s. brunneiceps* (MPG). Range in this area is below 1000 m in arid, 0-500 mm rainfall areas. Records around the Daua R 7C 19AB (E J Roberts, ADL pers. obs.) apparently refer to *L.s. somaliensis* (Br). This race's range is below 1000 m in arid, 0-250 mm rainfall areas. E J Roberts (*in litt.*) recorded it during all months of her May-Dec survey at Rhamu 19A.

The Red-billed Firefinch can be sympatric with Jameson's (1021), the African (1022) and the Black-bellied (1023) Firefinches (Br). See Bar-breasted Firefinch (1017) and Red-billed Firefinch Indigobird (1019).

1019 RED-BILLED FIREFINCH INDIGOBIRD *Hypochera chalybeata*
B1211 / MPG1435, 1436 / D2C2

Afrotropical: widespread S of 20°N, apart from the equatorial lowland forests, most of the Congo basin and adjacent areas, and the dry SW.

Kenyan: rather uncommon in open bushed and wooded habitats up to 1800 m (Br). It parasitises the Red-billed Firefinch (1018) (HM; Br) and their distributions show many similarities, including affinity for habitation (Ja; ADL pers. obs.). They both occupy the moister areas of the country, or are localised near permanent rivers elsewhere. The latter mode of occurrence is illustrated by the report from the Ethiopian sector of 3B (J S Ash *in litt.*) which contains the Omo R. The firefinch also has a population in this area. The indigobird may also be expected around the Daua R firefinch population 7C 19AB, although AM found this parasite unaccountably rare in adjacent S Somalia. 65B (D J Pearson *in litt.*) contains the Tana R, but

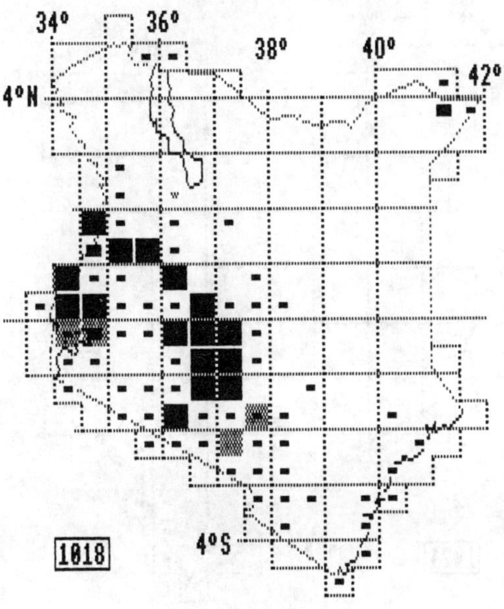

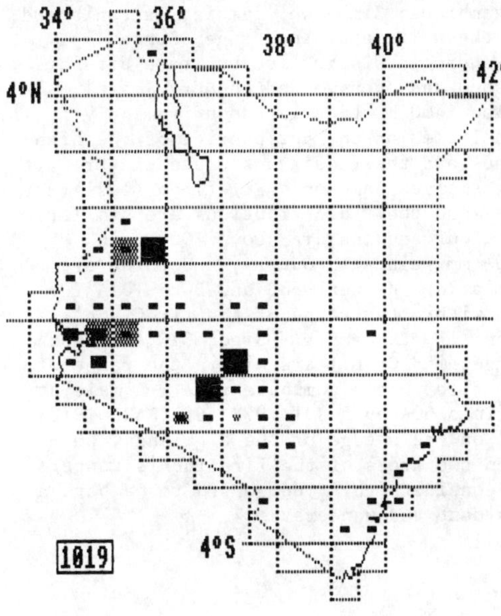

we have no record for the firefinch in this area. In view of the latter's penetration inland along the Juba R of S Somalia (AM), our coverage of its range along the Tana may be inadequate. Br mentions the firefinch on the Tana, but omits localities.

The indigobird's red-billed race *H.c. amauropteryx* inhabits the coastal lowlands (Br) and, like the firefinch, is apparently isolated from the population in the interior by the dry country between Tsavo 101 and the coastal strip. This coastal range is below 500 m in semiarid-dry subhumid, 500+ mm rainfall areas. The white-billed *H.c. centralis* occupies the highlands and surrounding plateaux (Br), and is 92% above 1000 m, 92% within 500+ mm rainfall areas and entirely semiarid-humid.

Br describes this indigobird as a resident that also wanders, but more regular movements seem to occur. Near the edge of the range in the semiarid country around Baringo 50A, small, annual Apr-Sept numbers (St) coincide with cooler, moister conditions and the Apr-Jul breeding season (Brown & Britton 1980). In contrast, DEP (pers. obs.) found it resident in semiarid 75C, but with an annual June-Sept absence in the Ruaraka-Kahawa-Ruiru area of moist subhumid 75B.

1020 JAMESON'S FIREFINCH INDIGOBIRD *Hypochera purpurascens*
B1213 / MPG1438 / D2C0
NO MAP: post presence 50A; pre presence 88B 101A; pre probable 37C

Afrotropical: the E, from South Africa to Mozambique, Zimbabwe, Zambia, Tanzania and, at the N limit of the range, Kenya.

Kenyan: a distinctive but rare bird of bushed and open wooded country, mainly below 1400 m, that is known from only five localities on the peripheries of the highlands and their adjacent plateaux (Br). It parasitises Jameson's Firefinch (1021)(HM; Br), and their distributions are similar. All four squares are above 1000 m in 500+ mm rainfall country, and three are semiarid. Kibwezi 88B and Bura 101A (Br) are within the range of the firefinch's race *L.r. taruensis*. Sigor 37C (HM) and Kongelai 37C (Br) are within the range of the firefinch's nominate race. A male at Baringo 50A on 5 Jul 1978 (St; EABR 1978) was on the N edge of the highlands, between two areas of the firefinch's range. Br considers this indigobird to be both a resident and wanderer.

1021 JAMESON'S FIREFINCH *Lagonosticta rhodopareia*
B1238 / MPG1411, 1412 / D1C2

Afrotropical: widespread S of 5°S, apart from the dry SW. Further N, it extends through E Tanzania to Kenya, NE Uganda, S Sudan and Ethiopia. There is an isolated population in Mali. Distribution in Kenya represents the E limit of the range.

Kenyan: a reasonably common but unobtrusive bird that favours low, dense cover in bushed and open wooded country up to 1400+ m (Br; MPG). Two races occur. *L.r. taruensis* is S of 1°S, in the coastal strip between Lamu 91B and Bamburi 102D 114B, along the Tana R N to Bura 78B, and on the S and SE peripheries of the C plateau (Br; AGH). This range is 68% below 1000 m, 68% semiarid (*cf.* Br, and the nominate race's range), and 59% within 500+ mm rainfall areas. As in the Red-billed Firefinch (1018: *q.v.*), the coastal and interior populations are apparently separate.

The nominate race occurs on the N edge of the highlands from the Wamba 51A-Isiolo 51D area (Br) NW to Sigor 37C (HM) and Moroto 24B (not mapped) in NE Uganda (Br). Of the four squares in the Isiolo 51D area, all are at 500-1500 m within the 250+ mm rainfall regime, whilst three are semiarid. 37C is omitted from this analysis because the locality, Sigor, is in atypically low (approximately 1100 m) and dry country near the square's NE corner. Probable breeding in 37C is based upon parasitism by Jameson's Firefinch Indigobird (1020: *q.v.*) at Sigor (HM). There is no confirmed

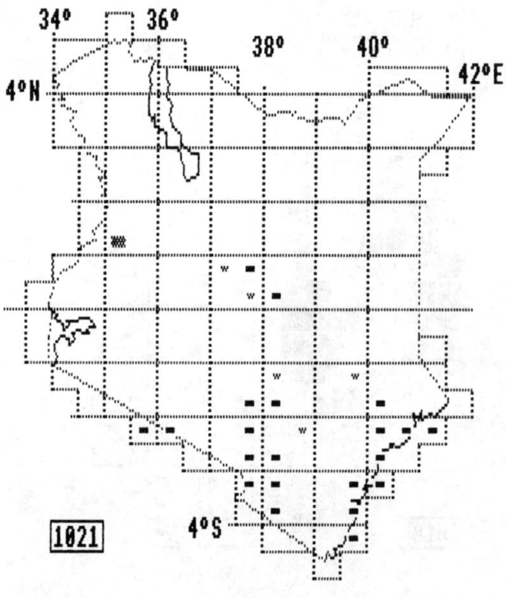

breeding record for this species, but Br considers it a resident and wanderer.

Jameson's Firefinch forms a superspecies with the Black-bellied Firefinch (1023) (HM). The two are completely allopatric even at QSD scale, the Black-bellied being in much moister country in the far W (Br). This superspecies forms part of a species-group with the African Firefinch (1022), which is easily confused with Jameson's in the field (HM; see La). Jameson's and the African are mainly allopatric, the former being on lower altitude and/or drier ground where their ranges coincide (Br; HM). The ranges of the African and the Black-bellied overlap completely at QSD scale, and, from Br's descriptions of their habitats, they could occur together, but this has yet to be established.

1022 AFRICAN FIREFINCH *Lagonosticta rubricata*
B1239 / MPG1411 / D2C2

Afrotropical: in the E, from Ethiopia and S Sudan, through the highlands of East Africa and the E Zaire borders, to S Zaire, Malawi, Mozambique and E South Africa. It also occurs in the W, around the Gulf of Guinea. Distribution in Kenya represents the E limit of the range.

Kenyan: an uncommon bird of low cover in bushed, usually moist, highland areas up to 2000 m (Br; MPG). In contrast to Jameson's Firefinch (1021: *q.v.*), range is 87% above 1000 m, 81% within the 500+ mm rainfall regime (*cf.* Br), and extensively in humid areas. The sole arid square is 77B, where the bird is on the locally moist Mt Endau forest island (A W Diamond *in litt.*). This species is both resident and a wanderer (Br). See Red-billed Firefinch (1018).

1023 BLACK-BELLIED FIREFINCH *Lagonosticta rara*
B1237 / MPG1426 / D2C3
NO MAP: post presence 48ACD 60A; pre probable 48B

Afrotropical: a fragmentary distribution between 5°N and 10°N, from Sierra Leone E to the N edge of the Congo basin. Further SE, it extends from S Sudan and NE Zaire to Uganda and, on the SE extremity of the range, W Kenya.

Kenyan: an uncommon and peculiarly local bird of low cover in bushed and open wooded country up to 1700 m (Br; MPG). Range is above 1000 m, to the W of the rift valley and N of the Winam Gulf 60AB. In contrast to Jameson's Firefinch (1021: *q.v.*), all five squares are moist subhumid-humid and within the 1000+ mm rainfall regime. There is no confirmed breeding record, but Br regards it as resident. See Red-billed Firefinch (1018).

1024 RED-CHEEKED CORDON-BLEU *Uraeginthus bengalus*
B1261 / MPG1431 / D3C3

Afrotropical: the N tropics, S of 15°N and N of the equatorial lowland forests, E to Ethiopia. Further S, it extends through East Africa and S Somalia to S Zaire and

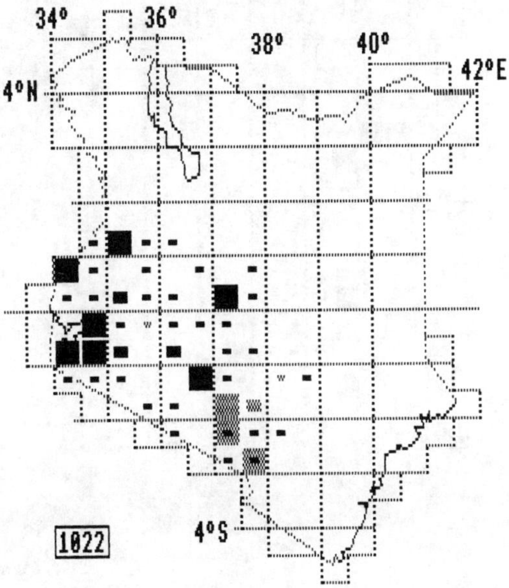

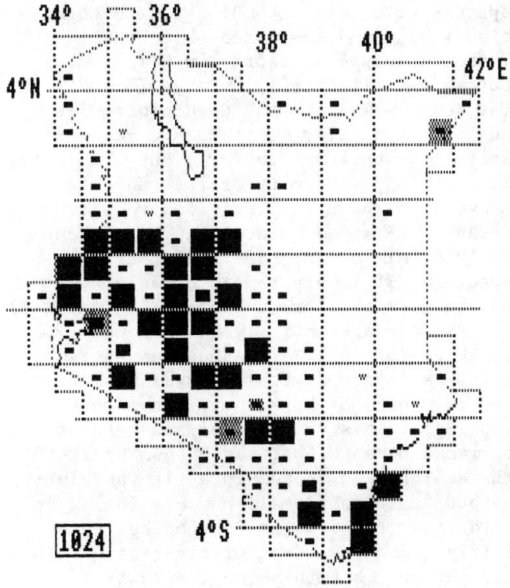

Zambia. It is replaced in other areas of the S tropics by the Cordon-bleu (1026).

Kenyan: usually common in a wide variety of bushed and wooded habitats, up to 2300+ m (Br; Ja; La; St). It is approachable, and often around human habitation (*op. cit.*; MPG). Range is 70% within the 500+ mm rainfall regime, mainly in the wetter highlands, plateaux and coastal lowlands (*cf.* Br). Occurrence in arid country (20% of its squares) is mainly in various locally moist situations, but notably absent from the vicinity of L Turkana (Br; HH). 16A (J S Ash *in litt.*) and Moyale 17AC represent the S limit of the population centred in the moist Ethiopian highlands. The records in 19 are the only ones N of the equator in Somalia (AM), and reflect the SE limit of the Ethiopian population. Presence in 27D 28C is on the locally moist highland of Marsabit, and 42A represents the moister environment around Wajir (Br).

S of the equator, this cordon-bleu enters arid country in the locally lusher habitats along the Tana R 78B 79C 91A (AGH). There is also a well-defined range edge in the strip of arid squares 64D S to 101D bordering the semiarid plateau. In one of these arid squares, Tsavo E 101B, this species is confined to locally moist riverine vegetation and gardens (La). Br thinks this species generally resident, and this view is shared by La for 101B, and by DEP (pers. obs.) for 74D 75CD 88AD 89AC 100B 101A. The only mention of movement comes from MPG, who note temporary local disappearances.

The Red-cheeked Cordon-bleu forms a species-group with the Blue-capped Cordon-bleu (1025) and the Cordon-bleu (HM). These three are sympatric in C Tanzania but mainly allopatric in Kenya (Br; HM). The Red-cheeked and the Cordon-bleu frequent very similar habitats and may constitute a superspecies (Br), but the latter is on the NE extremity of its range in Kenya, and known from only one record.

The Blue-capped Cordon-bleu is in generally drier country than the other two species. It is apparently broadly sympatric with the Cordon-bleu in NE Tanzania, but the two are unlikely to be found together (Br; HM). The Blue-capped is mainly allopatric with the Red-cheeked, and is uncommon where their ranges overlap (Br). This relationship is seen at semiarid Baringo 50A, where both breed but the Red-cheeked is common while the Blue-capped is in only small numbers (St). In arid Tsavo E 101B, the Red-cheeked is a fairly common resident of riverine growth and gardens, but the Blue-capped is a sporadic visitor in small numbers (La). Both species are resident in the Olorgesaille area of semiarid 75C (DEP pers. obs.), and sympatry is also known at Wajir 42A (Br). Some instances of sympatry amongst the cordon-bleus probably result from human influence (HM), and feral Blue-cappeds have bred at 1650 m in moist subhumid Nairobi 75B (not mapped)(Br).

1025 BLUE-CAPPED CORDON-BLEU *Uraeginthus cyanocephalus*
B1262 / MPG1432 / D3C2

Afrotropical: a more localised bird than the other two cordon-bleus, confined to extreme SE Sudan, S Ethiopia, S Somalia, Kenya and Tanzania.

Kenyan: locally common in drier bushed and wooded country, up to 1300 m (Br; Ja; MPG). In contrast to the Red-cheeked Cordon-bleu (1024: *q.v.*), range is 80% arid-semiarid, 78% within intermediate, 250-1000 mm rainfall areas, and only 11% below 500 m. It is mainly found on the semiarid plateaux to the N and S of the highlands (*cf.* Br), with relatively few localities in the arid N and E. Sparse occurrence S of 2°N in the E lowlands correlates with absence S of that latitude in S Somalia (AM). Br's summaries of this bird's Afrotropical and Kenyan ranges are incorrect (see Nikolaus 1979, AM).

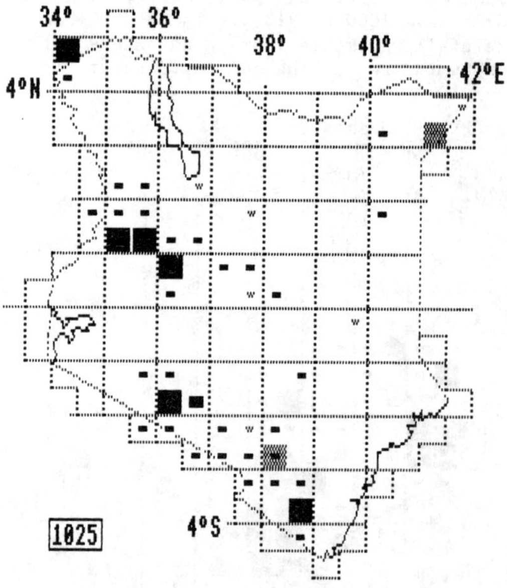

1026 CORDON-BLEU *Uraeginthus angolensis*
B1260 / MPG1430 / D2C1
NO MAP: pre presence 100D

Afrotropical: the S tropics, widespread S of 5°S with the exception of the dry SW. Further N, it extends through E Tanzania to the N extremity of the range on the S Kenya border.

Kenyan: the single record is from L Jipe 100D (Br), at approximately 900 m in a semiarid, 500-1000 mm rainfall square. The lack of subsequent records from this reasonably regularly visited area suggests an extralimital wanderer. See Red-cheeked Cordon-bleu (1024).

1027 PURPLE GRENADIER *Uraeginthus ianthinogaster*
B1263 / MPG1433 / D3C2

Afrotropical: the E and NE, from E Tanzania to Kenya, NE Uganda, S Sudan, Ethiopia and Somalia.

Kenyan: locally common in open parts of bushed and wooded habitats, up to 2300 m (Br; Ja; MPG; Be; La; St). Range is 76% at 500-2000 m and only 23% arid. The distribution has three components. This is mainly a bird of the highlands and adjacent plateaux. It is distinctive, and coverage in these areas is virtually complete with the notable exceptions of 48 and 49. There is no obvious reason for the shortage of records from these one degree squares, but we believe it to be real because some of the constituent QSDs receive regular coverage. In the same area, Br notes the contrast between its being widespread in Siaya 48C 60A but absent from adjacent E Uganda.

The squares between 15 and 19 (several obs.) are on the S periphery of the population centred in Ethiopia. These areas account for most of this species' arid and 0-250 mm rainfall percentages but, as noted previously, our overlay data may be incorrect here. The bird's rarity in the lowlands S of 2°N and E of 39°E corresponds with very scattered distribution in S Somalia (AM). 79A is the species' type locality (Ja), and 67C 80A are from Somalia (AM). The Purple Grenadier is not uncommon on the coast in S Somalia (AM), but the sole record from the Kenyan coast is a party of two males and three females in a garden at Shela 91B (J Hayes *in litt.*).

This species is considered resident by Br, Be (74), La (101B), St (50A) and DEP (pers. obs.: 74BD 75BCD 89A). See Straw-tailed Whydah (1028).

1028 STRAW-TAILED WHYDAH *Vidua fischeri*
B1214 / MPG1443 / D3C2

Afrotropical: the E and NE, from C Tanzania N to Kenya, NE Uganda, extreme SE Sudan, Ethiopia and Somalia.

Kenyan: uncommon in dry bushed habitats below 1600 m (Br; Ja; MPG). It parasitises the Purple Grenadier (1027)(MPG; Cunningham-van Someren 1973c; Br), and their distributions show many similarities (see HM, and AM). The whydah's range is 70% arid-semiarid and 72% within the intermediate, 250-1000 mm rainfall regime (*cf.* Br).

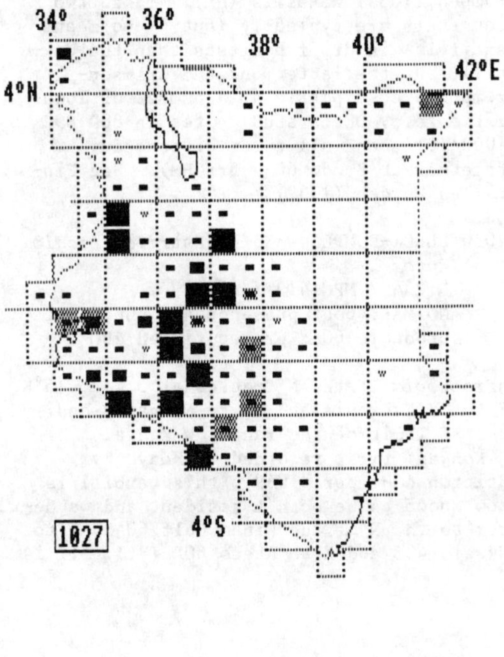

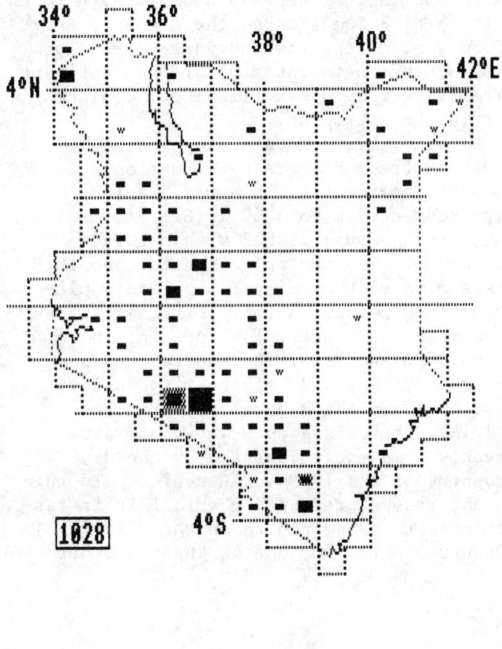

Like the grenadier's, range can be divided into three broad regions. The whydah is mainly a bird of the highlands and surrounding plateaux, although it avoids the true highlands *e.g.* 61BD 62BD 63AC and is less widespread to the W of the rift valley (Ja; Br). There is a scattering of records on the N borders, from 2AC (Nikolaus 1979) E to 19, but distribution in this area is rather more widespread than the grenadier's *e.g.* at L Turkana and in the vicinity of 30. Finally, the whydah and the grenadier are both rare in the SE lowlands. There is only a single inland record of the former from Garissa 65B (Br), and it is completely absent from within 150 km of the Kenyan border S of 2°N in Somalia (AM). The single record from the Kenyan coastal strip was near Lamu 91C (DEP pers. obs.). Whereas the whydah is known from many squares on the Somali coast (AM), the only other reports from coastal East Africa are at 6°48'S in Tanzania (Br; EABR 1981, 1982).

This whydah is present throughout the year at semiarid Olorgesaille 75C (DEP pers. obs.) and Baringo 50A (St), although the latter observer also notes apparent influxes. It is an irregular and sporadic visitor to the E edge of its main range in arid Tsavo E 101B (La). Br assumes that occasional birds in the C highlands are wanderers, and the same possibly also applies to birds beyond the grenadier's range in the N (see above).

1029 WAXBILL *Estrilda astrild*
B1226 / MPG1418 / D2C3

Afrotropical: widespread S of 10°N, E to Ethiopia, S Somalia and the E coast, apart from some of the lowland forest areas of the W, and the S interior of the continent. The Black-rumped Waxbill (1030) replaces it in the N tropics.

Kenyan: this common and approachable species frequents grassy situations, usually near water, in a wide variety of environments up to 3000 m (Br; Ja; MPG; Se; AGH; La). Range is 82% within 500+ mm rainfall areas and only 4% arid, and this is a bird of the moister highlands, plateaux and coastal strip. Different races occur in the coastal lowlands inland to L Jipe 100D and Voi 101B, and in the interior to either side of the rift valley (Br). As in the Red-billed Firefinch (1018), the coastal population is apparently completely isolated by the dry country of the 101D-102AC region. In this area, recent presence in arid 101B is restricted to a Dec record on the Voi R (La). Occurrence in 79C, one of the two other

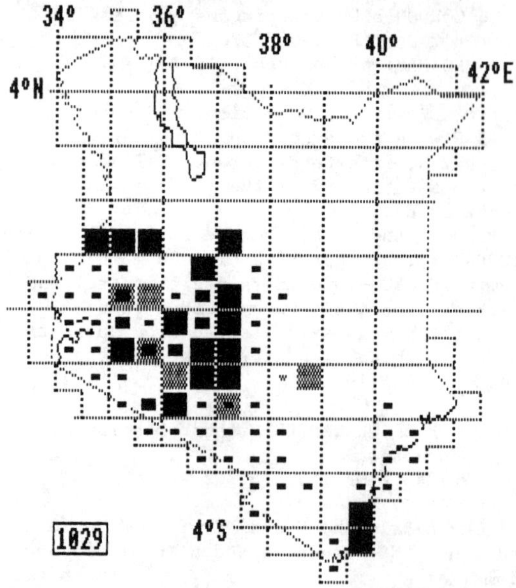

arid squares, is similarly along the Tana R (AGH). This extension along the Tana parallels distribution in S Somalia, which is confined to the Juba and Shebelle R (AM).

Br describes this waxbill as a resident and wanderer, and the former status is also noted for Narok district 74 (Be), the top of the altitudinal range at Mau Narok 61D (Se) and 75B 88D 89C 100B 101A (DEP pers. obs.).

The Waxbill forms part of a species-group with the Black-rumped and Crimson-rumped (1031) Waxbills (HM). These two congeners are typically (but by no means entirely) in drier habitats than the Waxbill, but the latter and the Crimson-rumped are sympatric at a number of localities *e.g.* DEP's study sites in 88D 89C 100B 101A, and all three birds occur together at Bondo 60A (Br; HM). See Pin-tailed Whydah (1032).

1030 BLACK-RUMPED WAXBILL *Estrilda troglodytes*
B1234 / MPG1419 / D2C3
NO MAP: post presence 48C 60B; post probable 60A; pre confirmed 48C

Afrotropical: the N tropics at 5°N to 15°N, E to NE Ethiopia and SE to N Uganda and, at the SE limit of the range, W Kenya.

Kenyan: first recorded in Kenya by Britton & Harper (1969), this waxbill is now known to be both a resident and wanderer to the N of the Winam Gulf 60AB, N to Ukwala 48C and E to Ahero 60B (Br). It is locally common (Br), and a Jan flock of 100+ occurred at Ahero 60B (Mann 1976). This is typically a bird of dry scrub and cultivation (see HM), but it also inhabits moist thicket, swamp edge and rice fields in Kenya (Br; ADL pers. obs.). Br mentions range up to 1400 m, and the three squares are all above 1000 m in moist subhumid-humid, 1000+ mm rainfall country. See Waxbill (1029). Other reference: Harper & Harper 1972.

1031 CRIMSON-RUMPED WAXBILL *Estrilda rhodopyga*
B1233 / MPG1420 / D2C2

Afrotropical: the E and NE, from Sudan, Ethiopia and S Somalia to East Africa, the E Zaire borders, Malawi and the far NW of Mozambique.

Kenyan: a reasonably common and approachable waxbill of grassy situations in bushed and wooded country (Br; MPG; La; St). Range is 89% above 500 m, and only 19% arid and 2% within 0-250 mm rainfall areas. Br mentions it up to 1800 m, but Be estimates 2280 m and there are records from 60% of our 2000+ m squares.

Overall distribution is in two broad regions. Most of the range is in the moister plateaux, and in three dry subhumid squares of the S coastal strip (Br). Apart from the latter, it is virtually absent below 500 m in the E lowlands. Secondly, occurrence in arid country is in three distinct areas. The records along the N borders from 3AB E to 19A (several obs.) represent the S edge of populations centred in moister areas in Sudan and Ethiopia. These squares may not actually be arid. Further S, 27D 28C represents presence on the locally moist Mt Marsabit (Br). S of the equator, occurrence in arid environments is restricted to a well defined strip of squares 89B S to 101D, on the edge of semiarid country.

In one of these S arid squares, Tsavo E 101B, this waxbill is an annual Dec-Mar visitor to locally moist habitats including the vicinity of rivers (La). This movement into arid country during and after the 'short' rains would seem more regular than Br's note of wandering. Br also describes it as resident, and this view is endorsed by Be for subhumid 74 and St for semiarid 50A. See Waxbill (1029), Black-cheeked Waxbill (1035) and Pin-tailed Whydah (1032).

1032 PIN-TAILED WHYDAH *Vidua macroura*
B1216 / MPG1441 / D3C3

Afrotropical: very widespread S of 15°N, apart from N Somalia and the driest areas of the SW.

Kenyan: a common and distinctive bird of open bushed and wooded country, mainly below 2300 m but also reaching 3000 m (Br; Ja; MPG; AGH). Range is only 20% arid and 8% within 0-250 mm rainfall areas, and occurrence in the arid N and E is restricted to three types of habitat. It is found around permanent water on both sides of L Turkana (HH), along the Daua R 7C 19AB (E J Roberts, ADL pers. obs.), and along the Tana R 65B 79AC 91A (AGH). Occurrence along these two E rivers corresponds with

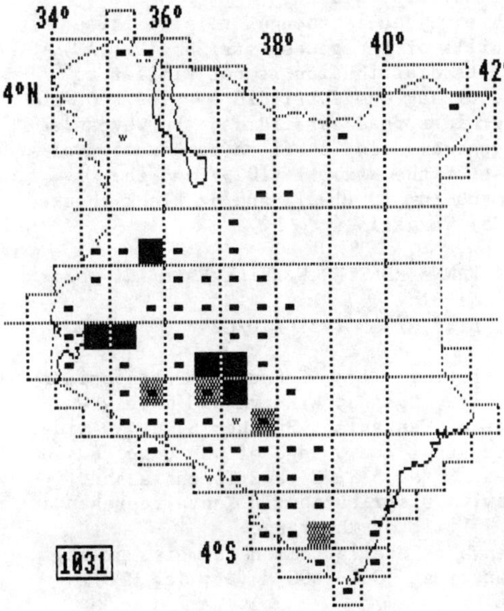

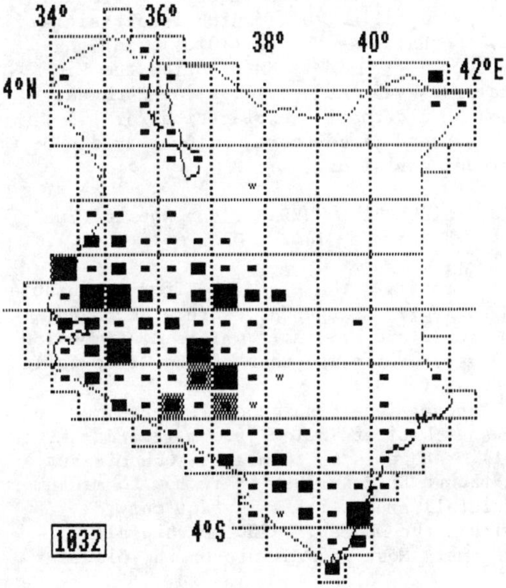

virtual restriction to the Juba and Shebelle R valleys of adjacent S Somalia (AM). The second habitat in arid country is on the locally moist Mt Marsabit 27D, where Br describes it as regular but we have only pre-1970 records. Finally, it is found in arid country in S Kenya which is adjacent to semiarid areas.

In Tsavo E 101B, one of these arid S squares, this whydah is a visitor during and soon after the rains, particularly to lush riverine habitats (La). MPG also mention seasonal movements, and St observed constantly fluctuating numbers throughout the year in the semiarid country around L Baringo 50A. DEP recorded presence throughout the year at his study sites in semiarid 75C and dry subhumid 74B. However, he noted movements in the more intensively studied Ruaraka-Kahawa-Ruiru area of moist subhumid 75B, where an annual Jan-May influx corresponds with the dry non-breeding season (Brown & Britton 1980), while a June-Aug diminution in numbers may also occur. Birds around 3000 m at Mau Narok 61D were in Oct (Britton 1979).

This whydah parasitises the Waxbill (1029) and perhaps other congeners (Ja; HM; Br). Range S of 2°N broadly corresponds to those of the Waxbill, the Crimson-rumped Waxbill (1031) and other more localised, moist country congeners. Presence of the whydah in some squares where waxbills are apparently absent might reflect the former's greater detectability. N of 2°N, the Crimson-rumped Waxbill also occurs at Marsabit 27D 28C and around the Daua R 7C 19AB, but the whydah's range around L Turkana is more extensive than those of any waxbills except the Green-winged Pytilia (1008)(which is parasitised by the Paradise Whydah (1010)), and the Silver-bill (1044). HH describe the Pin-tailed Whydah as occasional at Turkana, and an alternative explanation for its widespread range there would be as a non-breeding visitor.

1033 STEEL-BLUE WHYDAH *Vidua hypocherina*
B1215 / MPG1442 / D3C2

Afrotropical: the E and NE, from C Tanzania N to Kenya, NE Uganda, extreme SE Sudan, Ethiopia and W Somalia. This distribution is very similar to that of the Straw-tailed Whydah (1028).

Kenyan: local and generally uncommon in bushed habitats below 1400 m (Br; Ja; MPG; St). Range is 63% semiarid-subhumid but, although Br describes it in low to medium rainfall areas, it is in fact only 8% within the lowest, 0-250 mm rainfall regime. Most records are on the plateaux

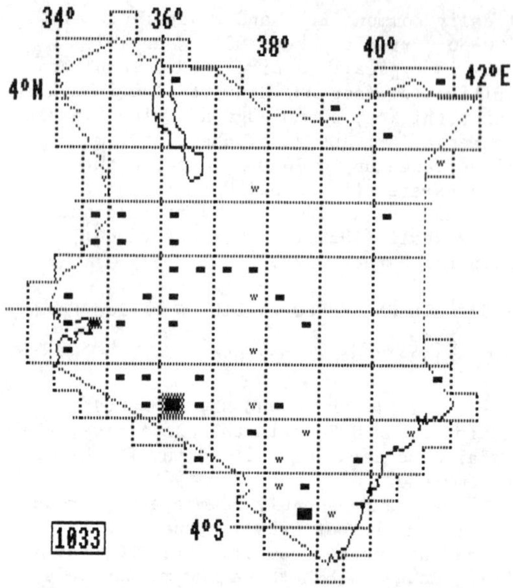

around the highlands, with far smaller numbers in the lowlands of the NE and SE. This is possibly a true pattern because this species is conspicuous and, further E, there have been few records from S Somalia (AM). Records from the Tana R (Ja; Br) omit precise localities.

The only possible references to movement are Ja's mention of frequent occurrence in the Tsavo-Voi 101 area during the Apr 'long' rains period, and St's note that most records in the semiarid Baringo area 50A are of breeding plumage birds in the moist, cool May-Sept season.

This whydah is thought to parasitise waxbills of the genus *Estrilda* (HM; AM), but its distribution is not similar to that of any one estrildid species. Of the more widespread estrildids, the whydah's range seems best to fit a combination of those of the Waxbill (1029) and the Crimson-rumped (1031) and/or Black-cheeked (1035) Waxbills (*cf.* HM).

1034 FAWN-BREASTED WAXBILL *Estrilda paludicola*
B1231 / MPG1422 / D2C2

Afrotropical: from S Sudan and Ethiopia to NE Zaire, Uganda, W Kenya and mainly W areas of Tanzania. Further S, it is widespread from the S edge of the Congo basin forest S to 15°S, E to S Tanzania and Malawi. Distribution in Kenya represents the E limit of the range.

Kenyan: locally common in moist grassy habitats at 1000-2000 m (Br; Ja; MPG).

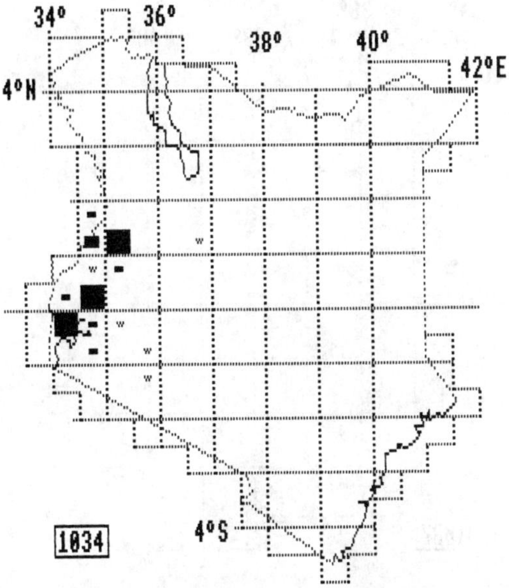

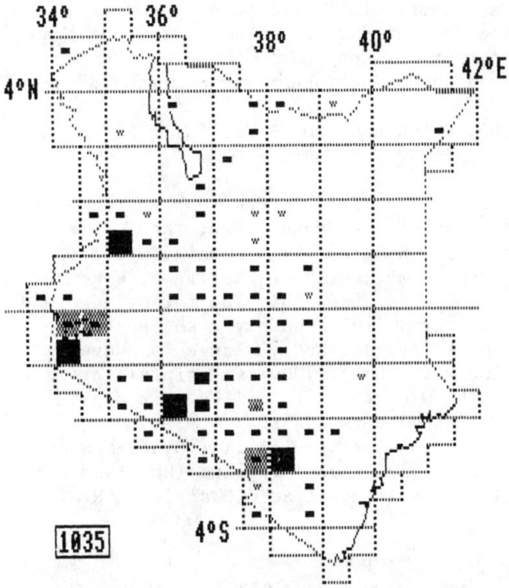

With the exception of the Karissia Hills 38D (Br), distribution is confined to areas W of the rift valley. The squares are 78% moist subhumid-humid and 85% within the 1000+ mm rainfall regime. Ja mentions presence E to the Mau (mapped in 61BD) and Br quotes the Narok R (mapped in 61D 74B). These records and 38D (see above) are at least one QSD E of all others that we know of, and may imply a retraction from the E limit of the range. Br considers this bird resident. Other reference: Britton & Britton 1971c.

1035 BLACK-CHEEKED WAXBILL *Estrilda erythronotus*
B1228 / MPG1427 / D2C2

Afrotropical: occupies two separate areas. The population in the E and NE consists of two distinctively plumaged subspecies, the Black-chinned Waxbill *E.e. delamerei* and the Black-cheeked Waxbill *E.e. charmosyna*. Their combined ranges extend through East Africa and NE to extreme SE Sudan, Ethiopia and W Somalia. These two races occur either side of a line from NE Uganda SE to the S Kenya borders in the Taveta-Tsavo area 100-101. They meet and possibly overlap E and SE of Nairobi 75B, and further research in this area may show that they are in fact separate species. Another population exists S of 15°S, from S Angola to Zimbabwe and South Africa.

Kenyan: reasonably common but never numerous in scrub, thicket and open wooded habitats below 1600 m (Br; Ja; MPG). Range is 76% within the intermediate, 250-1000 mm rainfall regime (*cf*. Br) and only 5% humid. This waxbill is mainly found on the plateaux around the highlands, and in the lower country of the N. It is absent from the coastal lowlands, and virtually so in those of the E (see below). Further E, in S Somalia, there are only five, scattered QSDs (including 19C (AM)).

Both races occur in the Thika area 76A (Plumb 1978b), and Br described the Black-chinned NE to 48C 60AB 74BD 76A 88D 100B. Using this range summary, the Black-chinned is entirely above 1000 m, mainly within the 500+ mm rainfall regime, and absent from arid environments. The Black-cheeked race is in distinctly lower and drier country (Ja; MPG; HM; Br), being mainly arid-semiarid and within the intermediate 250-1000 mm rainfall regime, and descending to 100 m at Bura 78B (Br). Although the Black-cheeked race extends well out into the arid country of the N *e.g.* around 27A, occurrence in the arid E is virtually confined to squares adjacent to semiarid areas, from 52B S to 101D (see above). In Tsavo E 101B, one of these E arid squares, this race was formerly common in riverine habitats, while three recent records have been in the rains or post-rains months of Jan and May (La). DEP (pers. obs.) finds this race resident in the Olorgesaille area of semiarid 75C.

This species' range is quite similar to that of the Crimson-rumped Waxbill (1031), and the two can be sympatric (HM). In general, however, the Black-cheeked Waxbill (and particularly its moister country, Black-chinned race) is more arboreal than most of its congeners (HM; Br). Other reference: Cunningham-van Someren 1978d.

1036 BLACK-CROWNED WAXBILL *Estrilda nonnula*
B1230 / MPG1425 / D2C3

Afrotropical: E from Cameroon and Gabon to N areas of the Congo basin, the E Zaire borders, S Sudan, NW Tanzania, Uganda and, at the E extremity of the range, W Kenya.

Kenyan: a common bird of forest edge and moist grass in secondary bush, up to 2200 m (Br; Ja; MPG). Range is above 1000 m in humid, 1000+ mm rainfall areas to the W of the rift valley. Br considers this species resident.

This and the very similar Black-headed Waxbill (1037) are siblings (HM), and completely allopatric at QSD scale in Kenya (Br).

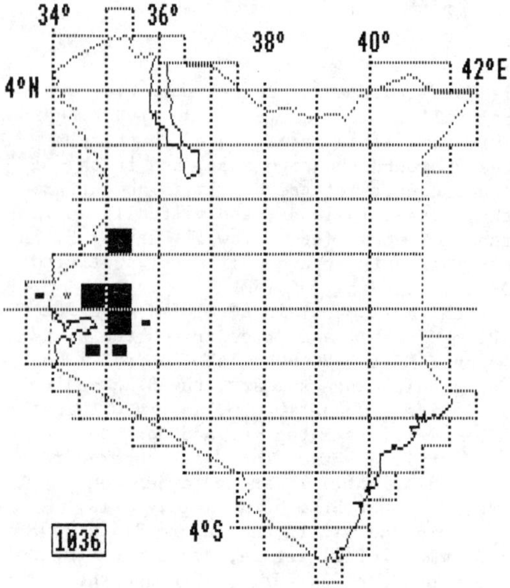

1037 BLACK-HEADED WAXBILL *Estrilda atricapilla*
B1227 / MPG1424 / D2C3

Afrotropical: mainly W and C Africa, on the W, S and E edges of the Congo basin. Disjunct populations in SW Uganda and on Mt Elgon and in the C highlands in Kenya represent the E extremities of the range.

Kenyan: locally common on the edges of highland forest and bamboo at 2100-3300 m (Br; Ja; MPG; Britton & Sugg 1973; EABR 1977). The two populations are on Mt Elgon 36D 48AB, and to the E of the rift valley on the Aberdare Mts 62BD 75B and Mt Kenya

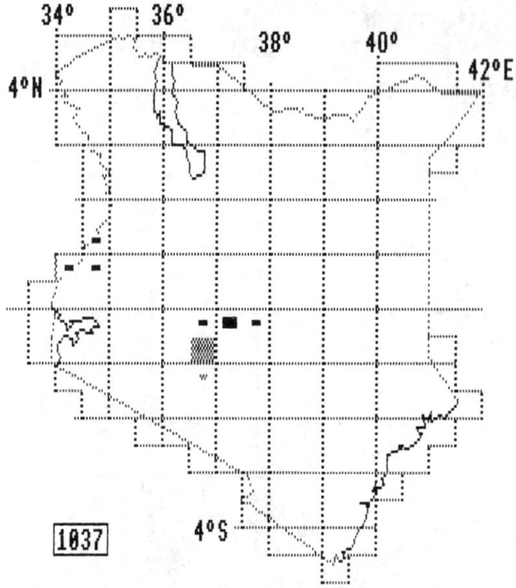

63AB. This range is 75% humid and 87% within the 1000+ mm rainfall regime. Br describes this waxbill as resident. See Black-crowned Waxbill (1036).

1038 YELLOW-BELLIED WAXBILL *Estrilda melanotis*
B1229 / MPG1417 / D2C2

Afrotropical: mainly the E, from Ethiopia and S Sudan, S through mostly highland areas of East Africa and the E Zaire bor-

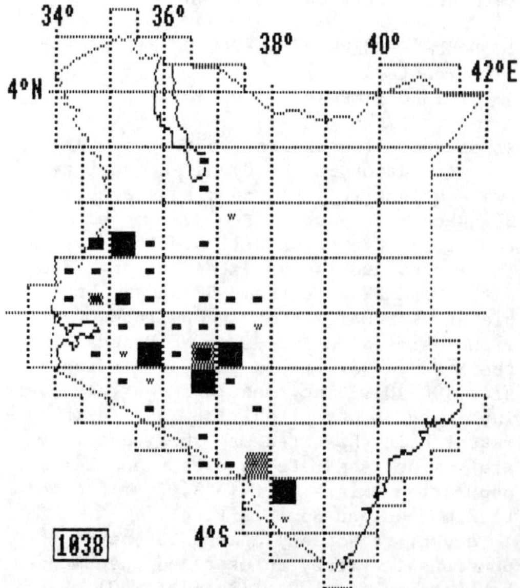

ders, to Zambia, Malawi, Mozambique and E South Africa S to the Cape. There is also an isolated race in Angola. Distribution in Kenya represents the E limit of the range.

Kenyan: fairly common but never numerous in moist thickets and forest edges and glades, at 900-3000 m (Br; MPG; Se; Britton & Sugg 1973; Diamond & Keith 1980). It occurs on the forest islands of Mts Kulal 26B (Diamond & Keith 1980) and Nyiru 26D, the Ndoto Mts 39A (Br), the Chyulu Hills 88D and the Taita Hills 101AC. Excluding these locally moist highlands, which are too small to be represented on our overlays, range is 89% subhumid-humid and 91% within 500+ mm rainfall areas.

Br considers this bird resident. Se found similar status at the top of the altitudinal range around Mau Narok 61D, although the species appeared most common when apparently breeding, in Aug.

1039 ZEBRA WAXBILL *Amandava subflava*
B1219 / MPG1421 / D2C2

Afrotropical: locally distributed S of 15°N, apart from many of the equatorial lowland forest areas, and the dry regions of Somalia and the SW. Distribution in Kenya represents the E limit of the range.

Kenyan: an uncommon bird of tall grassed habitats up to 2200 m, usually near permanent water in high rainfall areas (Br; Ja; MPG; EABR 1978). It is most common to the W of the rift valley (Br; Mann 1976), where the nominate race is above 1000 m in moist subhumid-humid, 1000+ mm rainfall areas.

Further E, *A.s. clarkei* is scarce and locally distributed around the C highlands, and is also known from Witu 91A in the coastal lowlands (Ja; Br). This E range is 78% semiarid-subhumid, 89% within the 500+ mm rainfall regime and, apart from Witu 91A, also above 1000 m. Br describes this species as a resident and wanderer.

1040 QUAILFINCH *Ortygospiza atricollis*
B1249 / MPG1403 / D2C2

Afrotropical: widespread but local S of 15°N, with the exceptions of most of the equatorial lowland forest areas, and of the dry areas of Somalia and the SW. Distribution in Kenya represents the E limit of the range.

Kenyan: locally common in open grassy habitats on dry, moist or flooded ground up to 3000 m (Br; Ja; MPG; Se; St). It is typically near water on short grass flats (Br). Range is in the highlands and peripheral, non-arid plateaux, 90% above 1000 m and 84% within the 500+ mm rainfall regime.

Br describes this bird as a resident and wanderer. Appearance in three widely separated areas seems to have been during or after wet seasons. These reports involve flocks in June and Aug 1980 at semiarid Baringo 50A (St), presence in Dec-Mar and Nov-Dec around 3000 m at humid Mau Narok 61D (Se), and a single Jan bird in the species' only arid square, Tsavo E 101B (La).

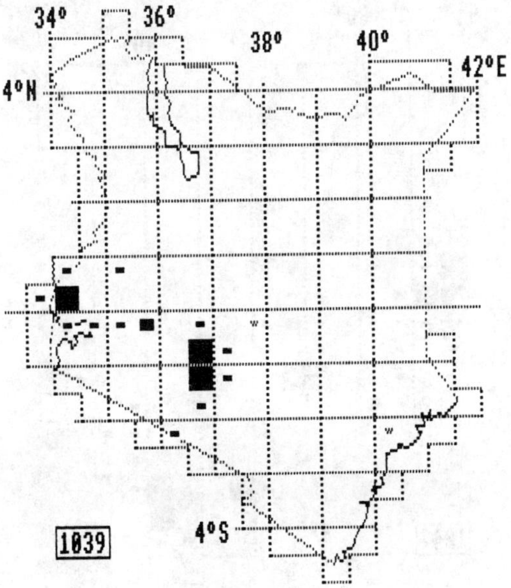

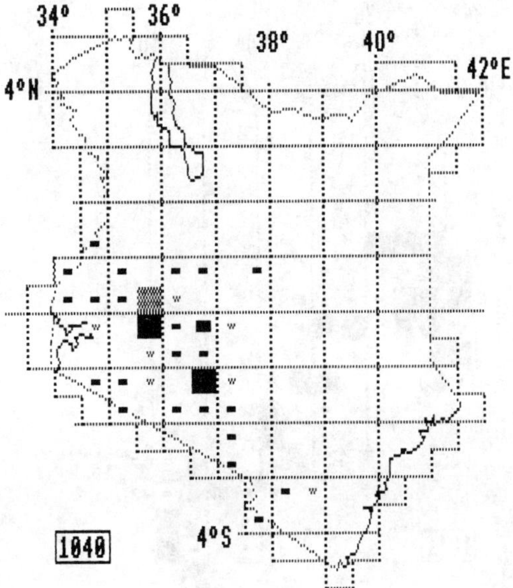

1041 BRONZE MANNIKIN *Lonchura cucullata*
B1266 / MPG1379 / D2C2

Afrotropical: widespread S of 10°N, with the exceptions of the dry areas of Somalia and the SW. Distribution in Kenya represents the E limit of the range (see AM).

Kenyan: a common species of open grassy situations up to 2000 m, typically in open bushed and wooded country (Br; Ja; MPG; AGH). It is approachable, and often around human habitation (*op. cit.*; La). Different races occur either side of the rift valley (Br). Range is 88% within 500+ mm rainfall areas, and virtually restricted to the moister highlands, plateaux and coastal strip. The few (7%) arid squares are all adjacent to semiarid terrains, and specific reports mention locally moist environments, *i.e.* the Mt Endau forest island in 77B (A W Diamond *in litt.*), around the lower Tana R in 79C (AGH), and in riverine vegetation and around habitation in Tsavo E 101B (La).

La consider this mannikin resident and Br describes it as generally so, but wandering and more regular movements occur. Wanderers include an early Mar individual at semiarid Baringo 50A (St), and a late Dec bird around 3000 m at Mau Narok 61D (Britton 1979). DEP (pers. obs.) found the species throughout the year in the Ruaraka-Kahawa-Ruiru area of moist subhumid 75B, but also noted an annual Apr-Jul influx which corresponds to the peak breeding season (Brown & Britton 1980).

The Bronze Mannikin forms a species-group with the Black and White (1042) and Magpie (1043) Mannikins (HM). The Magpie is on the edge of its range in Kenya, and known from only two localities. The Black and White is often in rather moister and less open habitats than the Bronze (HM), but the two flock together in some areas and are truly sympatric at Mombasa 114B (Ja; Br; AGH).

1042 BLACK AND WHITE MANNIKIN *Lonchura bicolor*
B1265 / MPG1380, 1381 / D2C2

Afrotropical: widespread S of 10°N, apart from the dry areas of E Somalia and the SW. Birds S of a line through C Angola, S Zaire, C Tanzania, C Kenya and the Ethiopia-Somalia borders have chestnut backs, while those further N have black.

Kenyan: locally common in open forest and forest edge, moist secondary bush, and woodland with tall grass and thicket, up to 2000 m (Br; Ja; MPG). It is mainly found in the moister highlands, plateaux and coastal strip.

The black-backed Black and White Mannikin *L.b. poensis* occurs W of the rift valley, and is entirely above 1000 m, 94% moist subhumid-humid and 81% within 1000+ mm rainfall areas.

The Rufous-backed Mannikin *L.b. nigriceps* is E of the rift, throughout our altitude categories, 80% semiarid-subhumid and 87% within 500+ mm rainfall areas. Occurrence in arid country includes the S periphery of the Ethiopian highlands at Moyale 17A (Ja), the locally moist forest islands of Mts Marsabit 27D (Ja) and Endau 77B (A W Diamond *in litt.*), and the vicinity of the

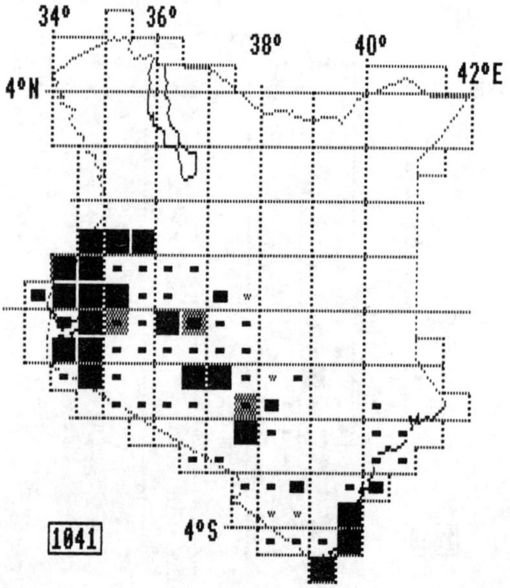

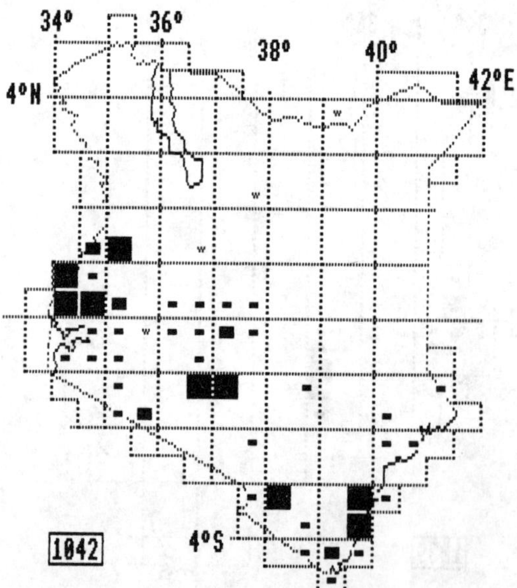

lower Tana R 79C (AGH). Presence along the lower Tana corresponds with restriction to the lower Juba R and 80A in S Somalia (AM). That the Rufous-backed Mannikin is apparently less of a forest bird than the black-backed race (Ja; MPG) presumably reflects the former's drier range. See Bronze Mannikin (1041).

1043 MAGPIE MANNIKIN *Lonchura fringilloides*
B1267 / MPG1382 / D2C0
NO MAP: pre presence 100B 114C

Afrotropical: a patchy distribution around the Gulf of Guinea from Senegambia to N Angola, E to S Sudan, W Uganda and E Tanzania. Further SE, it extends to Zambia, Malawi, Zimbabwe, Mozambique and E South Africa. Two old records on Kenya's S borders presumably represent wanderers just E or N of their normal range in NE Tanzania.

Kenyan: usually scarce in forest edge and moist secondary bush (Br). The old records are from Taveta 100B and Vanga 114C (Ja). These two squares are below 1500 m in semiarid-dry subhumid, 500+ mm rainfall areas. See Bronze Mannikin (1041).

1044 SILVER-BILL *Lonchura malabarica*
B1269 / MPG1383 / D2C2

Afrotropical: mainly the N tropics at 10°N to 20°N, E to Ethiopia and N Somalia; and also SE to S Sudan, NE Uganda, NE Tanzania, Kenya and all except the far S of Somalia.

Kenyan: locally common in dry, open bushed and wooded country, mainly at 400-1300 m but also reaching 1600 m in SW Kenya (Br; MPG; HH). Range is 89% arid-semiarid, mainly on the plateaux to the N and S of the highlands, and in the far NE. This species is virtually absent from the moist W highlands and plateau, so that the rift valley forms the W limit of the Afrotropical range between 2°N and 1°S. On this W range edge, St has occasional records from the semiarid rift floor at Baringo 50A in Aug 1980.

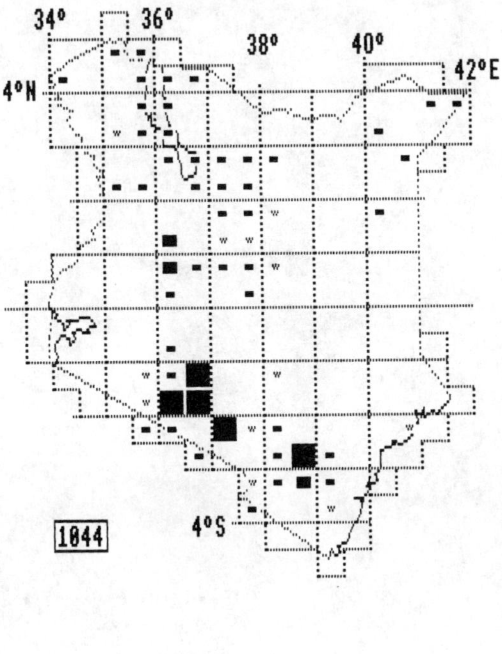

Distribution in the far E reflects that in S Somalia, where range is S to 1°N (AM). Hence there are reports S to Wajir 42A (Br; J E Miskell *in litt.*), but only a single coastal record, at semiarid Lamu 91B (Ja), further S. In the arid SE, rare individuals in Tsavo E 101B are almost always restricted to the vicinity of water, with the exception of a flock formerly resident around Voi Safari Lodge (La). The latter apparently constitutes the only published note of association with habitation in Kenya, although ADL (pers. obs.) has seen flocks in the grounds of Magadi Club 75C, and this affinity is a pronounced trait in Sudan (MPG) and Somalia (AM; *cf.* Somali Sparrow (993)). La describe this bird as resident in Tsavo E 101B, and DEP (pers. obs.) finds similar status in the Olorgesaille area of 75C. Br considers it a resident and wanderer.

The Silver-bill and the Grey-headed Silver-bill (1045) are mainly allopatric, with the latter in generally moister country, but mixed flocks are regular in areas where their ranges overlap (Br). These two are the dry country representatives of *Lonchura*, the mannikins (1041-1043) all being confined to high rainfall areas. See Pin-tailed Whydah (1032).

1045 GREY-HEADED SILVER-BILL *Lonchura griseicapilla*
B1268 / MPG1384 / D2C2

Afrotropical: a restricted range in the NE, from S Tanzania to Kenya, S Sudan and S Ethiopia. Although it reaches the E borders, there is no certain record for Uganda (Br).

Kenyan: generally uncommon in bushed and open wooded habitats at 100-1650 m (Br; La; St; EABR 1983). Range is mainly on the plateaux N and S of the highlands. In contrast to the Silver-bill (1044: *q.v.*), this distribution is 73% semiarid-subhumid, 84% within intermediate, 250-1000 mm rainfall areas, and virtually absent from the driest, 0-250 mm regime. The few occurrences in arid country include a flock in Banya Laga 4C (HH's only record from the Turkana area), presence on the locally

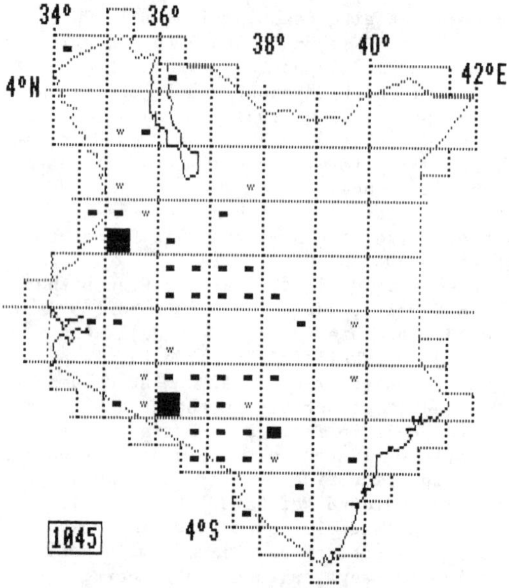

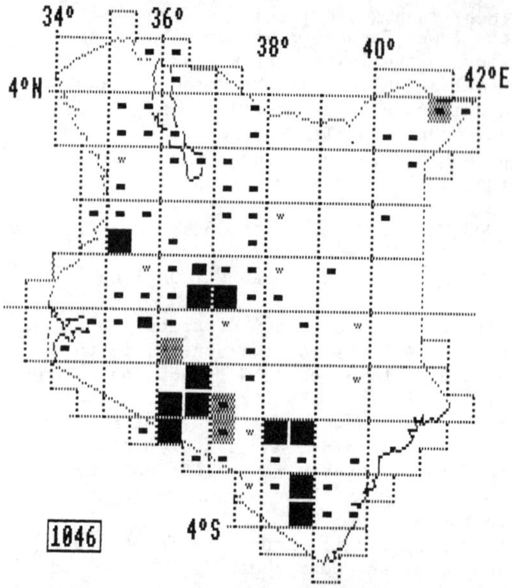

moist Mt Marsabit 27D (Ja; Tomlinson 1946), and records near the Tana R at Garissa 65B (National Museum Collection, Nairobi) and Bura 78B (Br). The only mentions of breeding come from West Pokot (mapped in 37C: T J Barnley *in litt.*) and L Magadi 75C (M E J Gore *in litt.*).

DEP (pers. obs.) finds this silver-bill resident in 75CD, and Br notes similar status with wandering. Records near the E limit of the range in arid Tsavo E 101B are mostly Dec-May, however (La). This seasonality compares with occurrences further E, at Garissa 65B in Jan (see above) and on Galana Ranch 90D in Mar (D J Pearson, ADL pers. obs.). Wanderers above the normal altitudinal limit, around 1900 m near L Naivasha 62C, were in Feb (National Museum Collection, Nairobi).

1046 CUT-THROAT *Amadina fasciata*
B1264 / MPG1402 / D2C2

Afrotropical: the N tropics at 10°N to 17°N, E to Ethiopia and N Somalia; and S through S Sudan, NE Uganda, Kenya, S Somalia and E Tanzania to Mozambique, Zimbabwe, Botswana and NE South Africa.

Kenyan: locally common in dry and reasonably open bushed and wooded habitats below 1300 m (Br; Ja; MPG; HH; La; St). Range is 75% arid-semiarid (*cf.* Br) and 62% within the intermediate, 250-1000 mm rainfall regime. It is noticeably scarce in the driest areas of the NE (around 29), and in the wet highlands and W plateau. Scarce occurrence in the far SE may correspond to apparently similar status in adjacent S Somalia (AM). 102D is one of only four coastal squares from which this bird has been reported in East Africa and Somalia (Br; AM).

DEP (pers. obs.) finds this species resident in the Olorgesaille area of 75C, and Br thinks it generally so. Wandering is also known (Br; HM), but La's note that it is a visitor to arid Tsavo E 101B, in variable numbers and mainly Dec-June, suggests some more regular movement.

Family Fringillidae: Buntings, Seed-eaters, Canaries

678 species, world-wide but most diverse in the Americas. Only 40 (6%) are found in the Afrotropical Region. Of the 19 (47% of the Afrotropical total) recorded in Kenya, 16 (84%) breed or are thought to do so, the Ortolan Bunting (1049) is a Palearctic vagrant, and the Brown-rumped Bunting (1052) and Streaky-headed Seed-eater (1061) appear to be wanderers at the extremity of their ranges. The occurrence in Kenya of Cretzschmar's Bunting is unsubstantiated.

1047 CINNAMON-BREASTED ROCK BUNTING *Emberiza tahapisi*
B1278 / MPG1476 / D2C2

Afrotropical: widespread S of 15°N with the exceptions of the lowland equatorial forest regions, most of Somalia, and the dry SW. Distribution in Kenya represents the E limit of the range.

Kenyan: local and generally uncommon on rocky ground in a wide variety of habitats at 400-3000 m (Br; Ja; MPG; Se; HH; St). Br describes it as typical of rocky hills and boulder-clad slopes in open wooded country. With the exception of the L Turkana basin, range is mainly above 500 m and absent from the most arid (moisture index below 1.2), 0-250 mm rainfall country. This bunting is however widespread at Turkana (HH), and occurrence in arid country can be related to locally moister environments. In the N, these are the vicinity of Turkana, the S foothills of the Ethiopian highlands 16A (J S Ash *in litt.*), and Mt Marsabit 27D (Ja). S of the equator, presence in arid environments is restricted to squares adjacent to extensive semiarid areas, from 89B S to 101D. In one of these squares, Tsavo E 101B, occurrence is mainly near water and seasonal (La, see below).

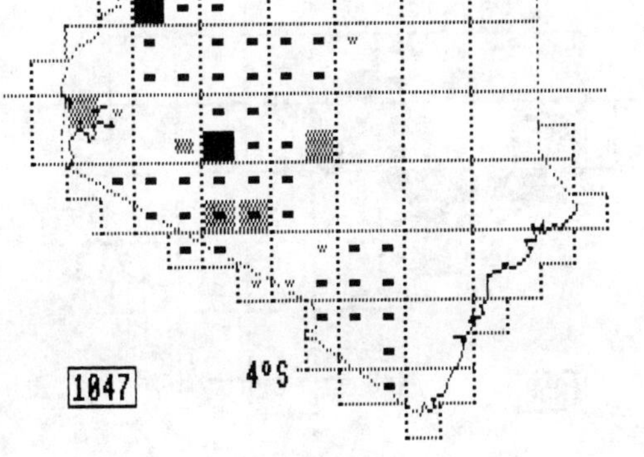

This bunting is considered resident in semiarid country at Baringo 50A (St), 88D 89C 100B and 101A, and in moist subhumid 75B (DEP pers. obs.). Br describes it as both a resident and wanderer, and two Nov individuals around the top of the altitude range at humid Mau Narok 61D (Se) presumably belong in the latter category. Occurrence in Tsavo E 101B, one of the S arid squares mentioned above, suggests more regular movement into dry country. The bird is a sporadic visitor to the area, but mainly in Dec-Feb (La).

The Cinnamon-breasted Rock Bunting forms part of a species-group with the House Bunting (1048)(HM). These two species both favour rocky ground but may be segregated by minor habitat preferences (HM). In Kenya, the House Bunting is confined to the far N, almost entirely in arid country, but overlaps with the Cinnamon-breasted at QSD scale in the L Turkana basin.

1048 HOUSE BUNTING *Emberiza striolata*
B1277 / MPG1477 / D2C1

Afrotropical: the N tropics, at 15°N to 21°N along the S edges of the Sahara across the whole continent. It extends further S to N areas of Somalia and Ethiopia and, on the S extremity of the range, also reaches S Ethiopia, N Kenya and probably extreme SE Sudan (see below).

Kenyan: local but sometimes common in dry, stony habitats at 400-800 m in N Kenya (Br; MPG). Br mentions affinity for human habitation in Kenya, but this statement may have been erroneously based on the species' habits in other countries (P L Britton *in litt.*). We have only seen it in wild country and can find no specific note of association with dwellings in

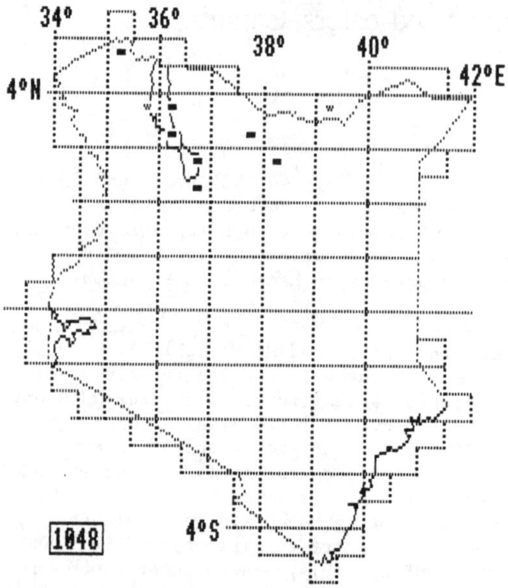

Kenya. Br quotes Moyale 17A as a locality, but ADL did not observe it in the town there.

Range is 88% arid (*cf.* Br), and completely within the 0-500 mm rainfall regime. A record at Ferguson's Gulf 13B has been the only one to the W of L Turkana (Br), but the bird was reasonably common near the Sudan border at Kamathia 3A in Feb 1984 (EABR 1984). There is no breeding record, but Br considers this bunting resident, and also mentions wandering. See Cinnamon-breasted Rock Bunting (1047).

1049 ORTOLAN BUNTING *Emberiza hortulana*
B1275 / MPG1472 / D1CO
NO MAP: pre presence 50A

Afrotropical: this Palearctic species winters in the N tropics, S to around 12°N in the W and 8-9°N in the E.

Kenyan: despite MPG's implication that there have been more, the single known record is of one collected at L Baringo 50A on 15 Oct 1910 (Ja). This area is near 1000 m, in a semiarid, 500-1000 mm rainfall square. The lack of other records indicates a vagrant S of its normal range, and the date suggests "overshooting" of this range while on S passage. Other reference: Ba.

\- CRETZSCHMAR'S BUNTING *Emberiza caesia*
B- / MPG1473 / D2CO
NO CONFIRMED RECORD

Afrotropical: this Palearctic species winters S to around 14°N in Sudan and N Ethiopia.

Kenyan: Moreau (1972) noted the only record S of 14°N as being in dry country at the N end of the rift valley in Kenya. Mann (1976) was unable to trace the origin of this report, and Br considers it unsubstantiated. We agree with this view.

1050 GOLDEN-BREASTED BUNTING *Emberiza flaviventris*
B1273 / MPG1469 / D3C2

Afrotropical: widespread S of 5°S, apart from the dry SW. Further N it extends to East Africa and S Sudan, with an isolated population at 10°N to 15°N along the S margin of the Sahara. Distribution in Kenya represents the NE and E limits of the range, beyond which it is replaced by the Somali Golden-breasted Bunting (1051).

Kenyan: moderately common in open wooded habitats at 1400-2200 m (Br; Ja; MPG; Be). It inhabits the C highlands and their peripheral plateaux but is less widespread to the W of the rift valley, particularly on the moist W plateau 48 60. Range is 87% within the 500+ mm rainfall regime, and absent from arid country. This bunting is resident in Narok district 74 (Be), and in 74D 75D (DEP pers. obs.). Br considers it generally so, but also mentions wandering. A late Feb individual around

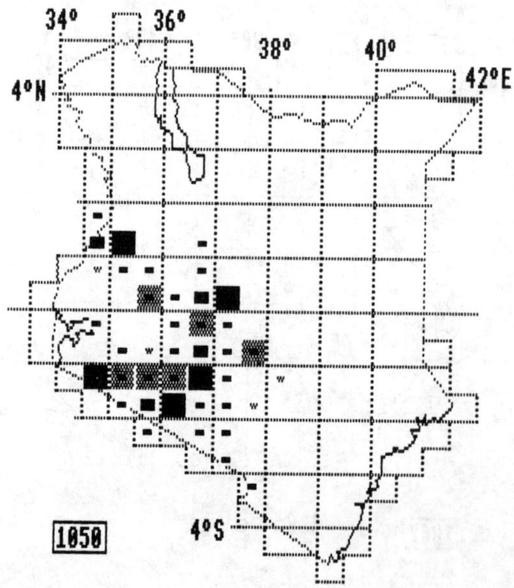

3000 m at Mau Narok 61D (Britton 1979) is presumably in the latter category.

The Golden-breasted Bunting forms a superspecies with the very similar Somali Golden-breasted Bunting, which replaces it allopatrically in lower and drier country (HM; Br). The few squares in which both of these species occur are mostly on the edge of the highlands, with large variations in altitude. This superspecies forms part of a species-group with the Brown-rumped Bunting (1052)(HM), which is known in Kenya from very few records and which may be a wanderer beyond its usual range.

1051 SOMALI GOLDEN-BREASTED BUNTING *Emberiza poliopleura*
B1276 / MPG1470 / D2C2

Afrotropical: the NE, from the NE borders of Tanzania to Kenya, NE Uganda, extreme SE Sudan (see below), Ethiopia and most of Somalia.

Kenyan: reasonably common in open bushed and wooded country up to 1200 m (Br; Ja; MPG; HH; La). In contrast to the Golden-breasted Bunting (1050: *q.v.*), range is 70% within the 0-500 mm rainfall regime and 91% arid-semiarid (*cf.* Br). It is virtually absent from the moist subhumid-humid highlands and their peripheral moist plateaux, which form the SW limit of its Afrotropical range. Occurrences immediately S of the highlands are in the semiarid floor of the rift valley at Kedong 75A and Magadi-Olorgesaille 75C (Br).

Presence in extreme SE Sudan is shown by

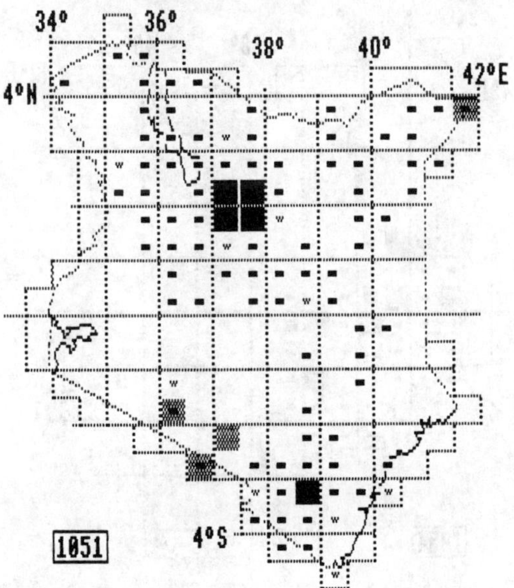

the record in 2C, which was on the Kenya-Sudan border (G Nikolaus *in litt.*), and by that in 3A, which was just S of this border at Kamathia (D J Pearson, ADL pers. obs.). Absence from the E borders S of 2°N corresponds with distribution in S Somalia which, between 2°30'N and 0°30'S is confined to the coastal strip (AM). This bunting reaches the Kenya coast at and N of Malindi 103A (Ja), but is confined to the coastal hinterland further S (including 114C)(Br). Absence from the coast N of 91C may be real because the Garsen 91A-Lamu 91B area is reasonably well known, and this bunting is also absent N to 0°30'S in Somalia (AM).

Br considers this bird resident, and reports of this status also come from semiarid areas at Baringo 50A (St) and Olorgesaille 75C (DEP pers. obs.), and from arid Tsavo E 101B (La). Br also mentions wandering, and three early May birds at 1830 m on W Laikipia 50A presumably fall in this category (Short & Horne 1985b). Birds have been seen at dawn around Ngulia Lodge 101A and one was caught at the lights there on a night in Nov, but this species is not a known migrant (Backhurst & Pearson 1977b; EABR 1980). The concentration of breeding squares 27CD 39AB represents data from a single observer in the Kaisut Desert, during a brief period just after the 'long' rains (Lewis 1981e). Since it is at least mainly resident, this bunting presumably breeds in all such areas of the dry N and E when they are temporarily green and at their most inaccessible after wet weather (*cf.* Brown & Britton 1980). Our coverage of breeding is hence very poor (*cf.* Crested Lark (543), Superb Starling (890)).

1052 BROWN-RUMPED BUNTING *Emberiza forbesi*
B1274 / MPG1471 / D1C0
NO MAP: RECORDS IMPRECISELY LOCATED

Afrotropical: a relatively scarce and local species of the N tropics, from Senegambia E to Ethiopia, and SE to N Uganda and, at the SE extremity of the range, NW Kenya.

Kenyan: this bunting favours open bushed and wooded country (Br). van Someren (1922) lists "north Kerio River and Simu River, Elgon", and these localities are reiterated by Ja. MPG do not mention the bird in Kenya, but Mann (1976) describes van Someren's (1922) two localities as Kenyan and Br follows suit by describing them in the NW. However, the Simu R is apparently wholly within Uganda 36CD, as Br's gazetteer confirms. Thus the sole Kenyan record is from the N Kerio R, presumably somewhere in 37D 49BD. This probably represents a wanderer SE of its normal range, although the species could easily be overlooked.

1053 YELLOW-CROWNED CANARY *Serinus canicollis*
B1282 / MPG1454 / D2C3

Afrotropical: mainly the E, from the highlands of S Sudan, Ethiopia, East Africa, the E Zaire borders and Malawi, S to Zimbabwe, E South Africa and the Cape. There is also a population in Angola. Distribution in Kenya represents the E limit of the range.

Kenyan: this common species favours moorland, open bushed country and the edges of forest and bamboo at 1600-4300 m (Br; MPG; Se; Britton & Sugg 1973). It is found in the W and C highlands, and on the forest island of Mt Loima 13C (National Museum Collection, Nairobi). Excluding this localised highland, which is too small to appear on our overlays, range is 89% moist subhumid-humid and 72% within the 1000+ mm rainfall regime.

Both Be and Se regard this canary as resident, and Br suggests that it may be mainly so. But large flocks occur (Ja; MPG; Se), and these are seasonal around Mau Narok 61D (Br). Altitudinal migration is known in NE Tanzania (Br).

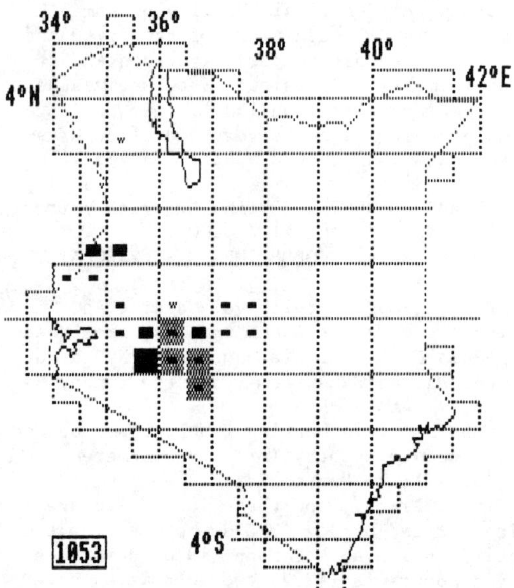

1054 AFRICAN CITRIL *Serinus citrinelloides*
B1283 / MPG1464 / D2C3

Afrotropical: the E, from the Ethiopian highlands, through moderate to high altitude areas of S Sudan, East Africa and the E Zaire borders, S to Malawi. Distribution in Kenya represents the E limit of the range.

Kenyan: common in a variety of fairly open habitats, including bamboo and forest edge, moist secondary bush and gardens, at 1100-3300 m (Br; MPG; Se; Britton & Sugg 1973). Range is in the highlands and peripheral plateaux, 77% subhumid-humid and 87% within the 500+ mm rainfall regime (*cf*. Br). Se and Br consider this bird resident, although the latter also mentions wandering. DEP (pers. obs.) recorded it during all months in the Ruaraka-Kahawa-Ruiru area of 75B, but also noted an annual Aug-Dec influx there that may correspond to the main non-breeding season (Brown & Britton 1980).

This species has a complex taxonomy within Kenya. The black-faced race, *S.c. kikuyuensis*, is endemic to the highlands from 51D 75B W to 49C 61AC 60C (Br). The grey-faced *S.c. brittoni*, first described by Traylor (1970a), is endemic to W Kenya, from 60A 48D 49C N to 36D 37C, at 1300-2800 m (Britton 1971a; Br). Using these ranges, the two races can be seen to differ environmentally: *kikuyuensis* in subhumid-humid, 500+ mm rainfall areas, and *brittoni* in humid, 1000+ mm. The lack of intergradation where these two races meet in W Kenya, and the occurrence of singing males

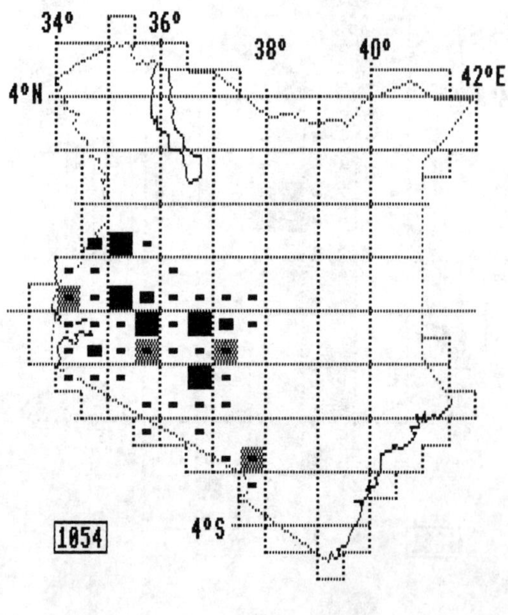

of both in the same garden at Nandi Hills 49C, suggests that they are reproductively isolated, *i.e.* that these two 'races' may in fact be separate species within a superspecies (Br). Another grey-faced race occurs to the S of these two (Br).

HM regard the African Citril and the Papyrus Canary (1055) as part of a species-group. Their ranges overlap completely at QSD scale and Britton (1971a) notes actual overlap at 1300 m near Siaya 48C, but they are segregated locally by differing habitat preferences (Br).

1055 PAPYRUS CANARY *Serinus koliensis*
B1287 / MPG1451 / D2C3
NO MAP: post presence 60A; post probable 60B; post confirmed 48C; pre confirmed 60B

Afrotropical: endemic to papyrus swamps in NE Zaire, S and E Uganda and W Kenya (Br).

Kenyan: this locally common species often forages in cleared areas adjacent to papyrus swamps, but returns to the papyrus to roost (Britton 1971a, 1978; Br; MPG). It is found only in the vicinity of the N shores of the Winam Gulf 60AB, E to Kisumu 60B, and in isolated areas of papyrus along streams in Siaya 48C 60A, N to Ukwala (Britton 1971a; Br). The three squares are all at 1000-1500 m in moist subhumid-humid, 1000+ mm rainfall country. Br considers this canary resident. See African Citril (1054).

1056 YELLOW-FRONTED CANARY *Serinus mozambicus*
B1290 / MPG1448 / D2C2

Afrotropical: widespread S of 15°N, with the exception of most of the equatorial forest areas and the dry SW, E to W Ethiopia, SE Sudan, W Kenya and the E coast. It also extends N from E Tanzania to the SE lowlands of Kenya. Distribution in Kenya represents the E limit of the range.

Kenyan: common in open bushed and wooded habitats up to 2300 m (Br; Ja; MPG). We received records of this canary from most of the moister parts of Kenya but, from personal observations and consultations with the literature, it became evident that widespread confusion occurs between sight records of this species and the very similar White-bellied Canary (1057), and possibly also the less similar Brimstone Canary (1059)(Lewis 1982i). This particularly affects the Yellow-fronted in the rift valley and on the high plateau to the E. Br mentions the Yellow-fronted in the Arusha-Moshi area of NE Tanzania, but records from the nearby 87D and 100B, al-

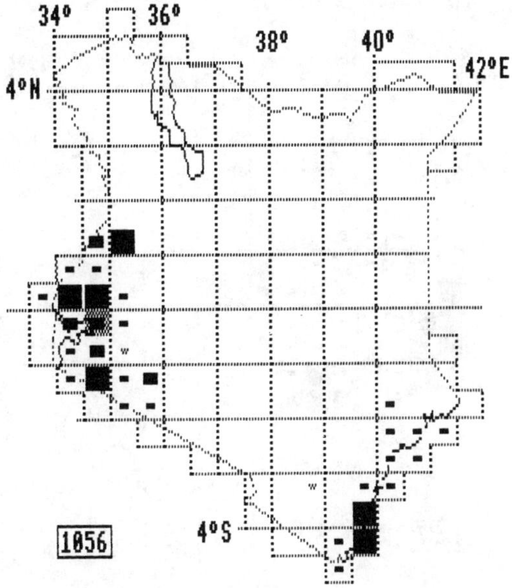

though possibly correct, were still discarded pending further positive identifications.

The Yellow-fronted hence occurs in two separate areas. The nominate race occupies the coastal lowlands, inland at least to arid Tsavo E 101B, where it formerly occurred around houses (La), and to the lower Tana R 79C 91A (AGH). This distribution is 69% semiarid-dry subhumid, and entirely below 500 m. All squares except Tsavo E 101B (see above) are within the 500+ mm rainfall regime. *S.m. barbatus* is found to the W of the rift valley, where occurrence is 90% at 1000-2000 m, 90% moist subhumid-humid and 80% within 1000+ mm rainfall areas. Br considers this canary a resident and wanderer.

The Yellow-fronted Canary forms a superspecies with the White-bellied Canary (HM). These two are mainly allopatric, with the White-bellied in drier environments (Br; HM). This superspecies forms part of a species-group with the Yellow-rumped Seed-eater (1058)(HM) and their ranges overlap extensively.

1057 WHITE-BELLIED CANARY *Serinus dorsostriatus*
B1285 / MPG1449 / D2C2

Afrotropical: the E and NE, from N and E Tanzania to Kenya, NE Uganda, SE Sudan, Ethiopia and W areas of Somalia.

Kenyan: locally common in drier, open bushed and wooded country, typically at 400-1400 m but also reaching 1900 m in

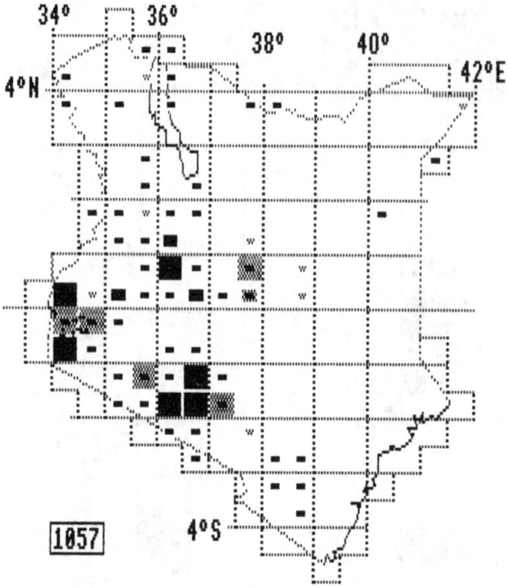

Narok 74 and 2650 m on Mt Nyiru 26D (Br; Ja; MPG; Be; HH; La; St). In contrast to the Yellow-fronted Canary (1056: *q.v.*), the White-bellied is 69% within intermediate, 250-1000 mm rainfall areas (*cf.* Br) and absent from the coastal 500+ mm rainfall zone. The only records from the lowlands E of 39°E are from Mandera 19B (van Someren 1929), 31A (Somalia: AM) and Wajir 42A (J E Miskell *in litt.*). This virtual absence from the E interior is paralleled in adjacent S Somalia where, S of 31A, the bird is confined to the coastal strip (AM). Absence from the coastal lowlands of Kenya, which are occupied by the closely related Yellow-fronted Canary, contrasts with occurrence near the Somali coast where the Yellow-fronted is absent.

Range is 27% arid, mainly in locally moister situations. It occurs near permanent water around the Omo R 3B 4A and L Turkana (HH), near the Daua R at Mandera 19B, and in the vicinity of Wajir 42A. The records in 15B (F Alexander *in litt.*) and 16A (J S Ash *in litt.*) are from the S foothills of the Ethiopian highlands. Other arid squares are on the peripheries of semiarid areas *e.g.* 25B 39D 52BD 89D 101BD. In one of the latter, Tsavo E 101B, this species occurs only rarely and after rains (La). However, this canary is described as resident by Be (74), Br, DEP (pers. obs.: 74D 75CD) and St (50A). Br also mentions wandering.

1058 YELLOW-RUMPED SEED-EATER *Serinus atrogularis*
B1280 / MPG1459 / D2C2

Afrotropical: widespread S of 5°S, apart from most of the dry SW. Further N it extends through East Africa to S Sudan, Ethiopia and S Somalia.

Kenyan: locally common but irregularly distributed in open bushed and wooded country up to 2000 m (Br; Ja; MPG; Be). Range is only 14% arid and 2% within the lowest, 0-250 mm rainfall regime (*cf.* Br). Occurrence in arid areas is mainly in locally moister situations, as on the S periphery of the Ethiopian highlands at Moyale 17AC (Br), around the forest island of Mt Marsabit 27D 28AC (Ja), and in the vicinity of Wajir 42A (Br). All arid squares S of the equator are adjacent to semiarid country. Presence in locally moister situations in these S areas includes the forest island of Mt Endau 77B (A W Diamond *in litt.*) and around the lower Tana R 79C (AGH). It is rare, but not seasonal, in arid Tsavo E 101B, usually near water or in locally moist gardens (La). Range E of 40°E parallels that in adjacent S Somalia, *i.e.* near the coast or permanent water (AM).

There are no breeding records from the coastal lowlands, but Br thinks this bird resident, as do Be for 74, and DEP (pers. obs.) for 74D 75BCD. MPG note local movements and Br mentions wandering. Occasional small flocks near the local range edge at semiarid Baringo 50A were during exceptionally heavy rains in 1978 (St). See

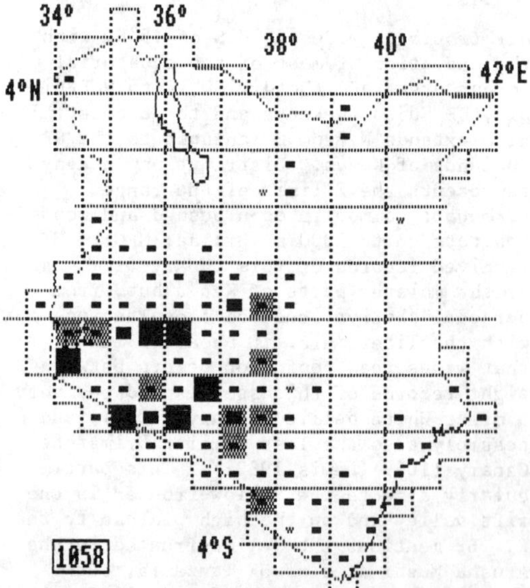

Yellow-fronted Canary (1056).

1059 BRIMSTONE CANARY *Serinus sulphuratus*
B1293 / MPG1450 / D2C2

Afrotropical: the S and E, widespread S of 10°S but absent from the dry SW. Further N, it extends to S Tanzania, the E Zaire borders, NW Tanzania, Uganda and, at the NE limits of the range, Kenya and Mt Kilimanjaro.

Kenyan: often common in moist, open, bushed habitats with scattered trees, up to 2400 m in the highlands and adjacent plateaux (Br; Ja; MPG; HM). Confusion between this species and other congeners is possible (see Yellow-fronted Canary (1056)), and records from 89C SE to 114B have been discarded on the basis of information from P L Britton, R A M McVicker and the literature. The type locality of Mt Kilimanjaro (approximately 100A) is on the SE periphery of the Kenyan range (*cf.* Br). Distribution is 95% above 1000 m, 78% subhumid-humid and 93% within 500+ mm rainfall areas. This canary is considered resident by Br and Be.

It is closely related to the Grosbeak Canary (1060), which replaces it in the drier N and E, with overlap at QSD scale on the NE and SE peripheries of the highlands. HM place these two birds in a species-group, while Br considers them allopatric members of a superspecies.

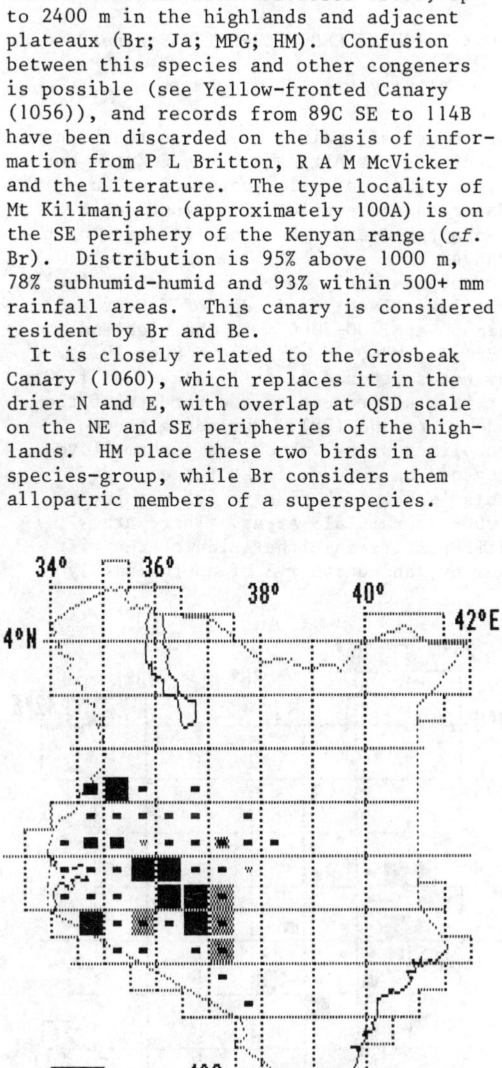

1060 GROSBEAK CANARY *Serinus donaldsoni*
B1284 / MPG1452, 1453 / D2C1

Afrotropical: the NE, from the drier areas of Tanzania to Kenya, S and SE Ethiopia, and Somalia. *S.d. buchanani*, which may be an incipient species, occurs in Tanzania and NE to the Tana R in Kenya (Br). Areas further N and NE are occupied by the nominate race.

Kenyan: scarce in open bushed and wooded country, mainly at 400-1300 m but also reaching 200-1600 m (Br; MPG; HH). Apparently more common status near the rift valley probably represents the frequency of observers recording a scarce but noteworthy bird. Range is 83% arid-semiarid (*cf.* Br), and absent from the coastal lowlands. The rift valley forms the W limit of the Afrotropical range with the sole exception of Maji Moto 74B (Br). Records from L Turkana are on the E shores and Central Island 14C (HH). *S.d. buchanani* extends SW from Meru 51D and Garissa 65B, and the nominate race is further N and NE (Br). DEP (pers. obs.) found this canary resident in the Olorgesaille area of 75C, while Br describes it as a resident and wanderer. See Brimstone Canary (1059).

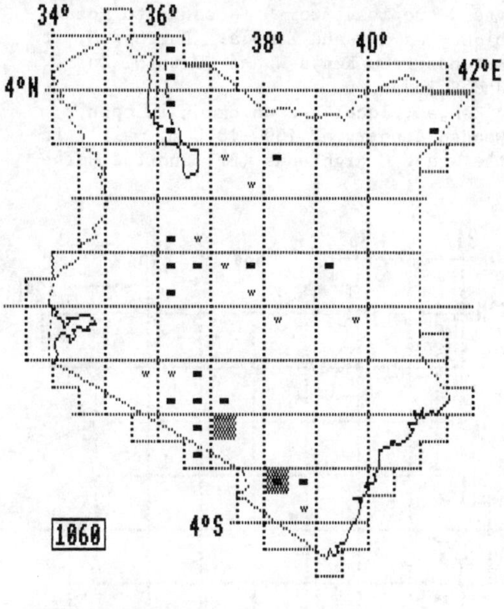

1061 STREAKY-HEADED SEED-EATER *Serinus gularis*
B1286 / MPG1456 / D1C0
NO MAP: post presence 37A; pre presence 48B

Afrotropical: three isolated populations. In the N tropics, this bird is found to the N of the lowland forests, from Ghana E to Nigeria, Cameroon, S Sudan, NE Uganda and, at its SE limit, NW Kenya. The other populations are in Angola, and in the SE from Zimbabwe to the Cape. The distribution of the closely related Stripe-breasted Seed-eater (1062) links these three populations together.

Kenyan: a bird of dry woodland at medium altitudes, known in Kenya from only two records (Br). Ja quotes Mangiki 48B, and there is a recent sighting from the Suam R at Nakuijit 37A (Br). These two squares are at 1500-2000 m, in subhumid-humid, 500+ mm rainfall country. There is no breeding record. Although this seed-eater may well be overlooked, the very few records probably also reflect very small numbers of individuals at or beyond the limit of their normal range.

The Streaky-headed Seed-eater forms part of a species-group with the Stripe-breasted Seed-eater, and the two appear to be little differentiated ecologically (HM; Br).

1062 STRIPE-BREASTED SEED-EATER *Serinus reichardi*
B1291 / MPG1456 / D1C1

Afrotropical: range is most extensive in the S tropics, from S Tanzania to Mozambique, Malawi and Zambia. Further N, it extends from Kenya W and NW to NE Zaire and S Sudan.

Kenyan: local and uncommon in open wooded country at 1600-1800 m, mainly in the W and C highlands (Br; Short & Horne 1985b; EABR). Excluding the isolated peaks of Mts Nyiru 26D and Ololokwe 51B (Br), which are too small to feature on our overlays, range is 78% subhumid-humid and entirely within the 500+ mm rainfall regime. These parameters are similar to those of the Streaky-headed Seed-eater (1061: *q.v.*). Ja mentions Jul breeding at Elgeyu (approximately 37D 49BD: not mapped). The high proportion of pre-1970 squares (55%) probably represents the species' scarce status and D1 characteristics.

1063 THICK-BILLED SEED-EATER *Serinus burtoni*
B1281 / MPG1462 / D2C3

Afrotropical: mainly the E, in the highlands of East Africa and the E Zaire borders. There are also populations in Cameroon and Angola. Distribution in Kenya represents the N and E limits of the range.

Kenyan: fairly common but unobtrusive in the lower levels and edge of forest and bamboo at 1700-3000 m in the highlands (Br; Ja; MPG; Se; Britton & Sugg 1973). Br notes it exceptionally as low as 1200 m. It is also present in the forests of the Ndoto Mts 39A (Br). Excluding these localised peaks, which are too small to feature on our overlays, range is 74% moist subhumid-humid and 67% within 1000+ mm rainfall areas. There are different races either side of the rift valley, and another S of approximately

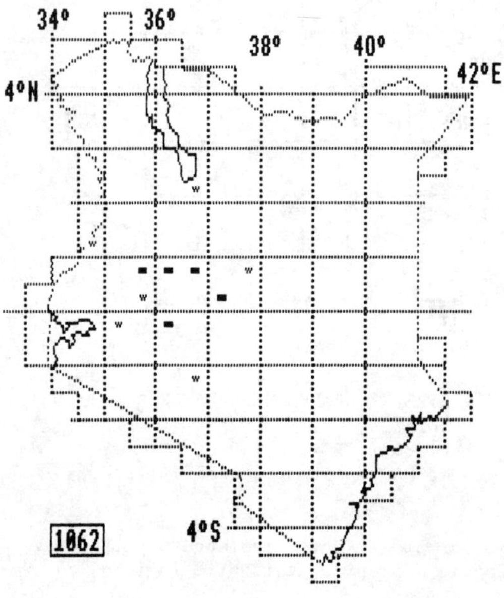

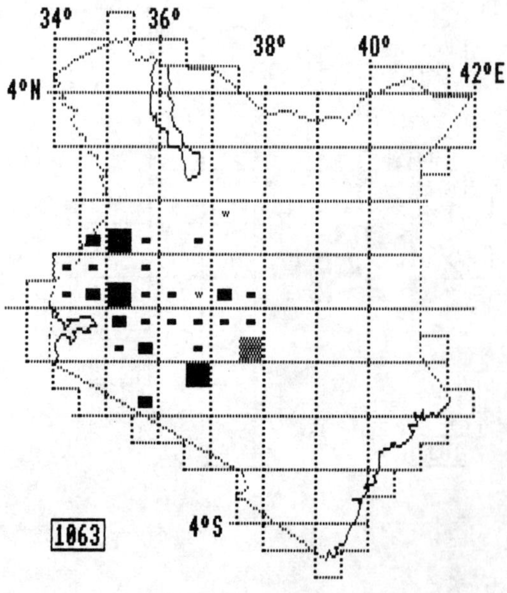

61CD in the W (Ja; Br). This species is considered resident (Br; Se).

The Thick-billed Seed-eater forms part of a species-group with the Streaky Seed-eater (1064)(HM). They are widely sympatric but, whereas the Thick-billed is more retiring and associated with forests, the Streaky prefers more open vegetation and has a greater altitudinal range (HM; Br).

1064 STREAKY SEED-EATER *Serinus striolatus*
B1292 / MPG1461 / D3C3

Afrotropical: highlands of the E and NE, from Ethiopia and S Sudan to East Africa, the E Zaire borders and Malawi. Distribution in Kenya represents the E limit of the range.

Kenyan: common and well known in a great variety of habitats, including forest edge, gardens, scrub, secondary bush and alpine moorland, at 1300-4300 m in the highlands (Br; Ja; MPG; Se; Britton & Sugg 1973). It is present on the forest islands of Mt Nyiru 26D (Br), Mt Endau 77B and the Taita Hills 101A. Excluding these localised peaks, which are too small to feature on our overlays, range is 90% within the 500+ mm rainfall regime, and absent from arid country. Widespread extension into semiarid areas, particularly on the S periphery of the C highlands, clearly distinguishes this bird's range from that of the Thick-billed Seed-eater (1063: *q.v.*). Be (74), Se (61D) and Br all consider this species resident, and Br also mentions wandering. DEP (pers. obs.), whilst observing it throughout the year in 74D and 75B, also notes annual Apr-May and Aug-Oct influxes to the Ruaraka-Kahawa-Ruiru area of 75B, a moist subhumid square.

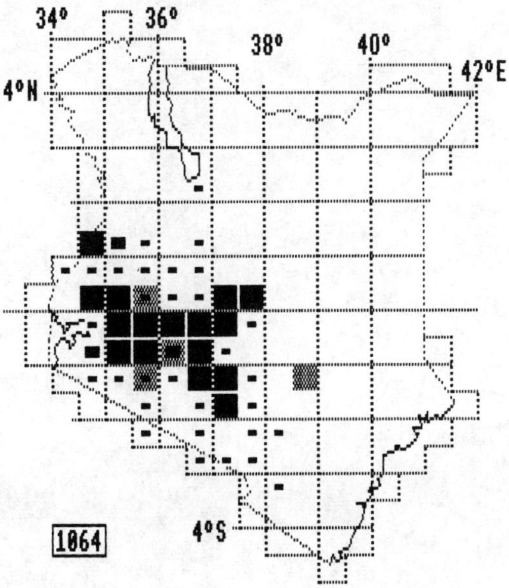

1064

1065 ORIOLE FINCH *Linurgus olivaceus*
B1279 / MPG1463 / D2C3

Afrotropical: a highland species, mainly in the E, from S Sudan to East Africa, the E Zaire borders and Malawi. There is another population in Cameroon. Distribution in Kenya represents the NE limit of the range.

Kenyan: uncommon and somewhat retiring in the edges and interiors of highland forest, mainly at 1700-3000 m (Br; Ja; MPG). It is found E of the rift valley on Mt Kenya 51D 63A, but is more widespread to the W, from Molo 61B NW to Mt Elgon 36D (Br). Range is entirely in moist subhumid-humid, 1000+ mm rainfall areas. It does not extend to forest islands. Br considers this species resident.

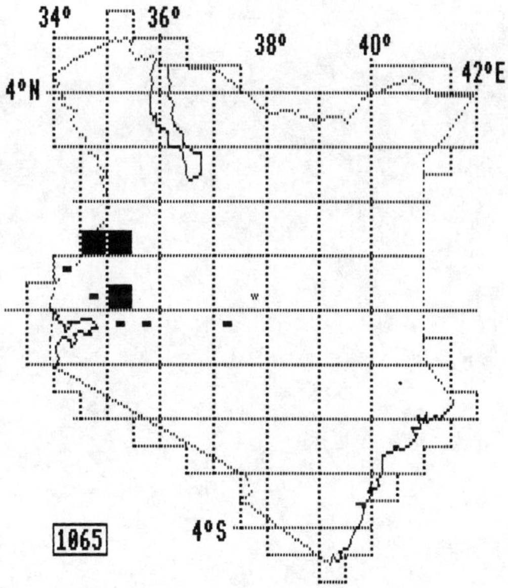

1065

Glossary

(Further information on many of these terms is to be found in Campbell & Lack (1985)).

Afrotropical: one of the six zoogeographical regions of the world, comprising Africa to the south of the Sahara, but excluding Madagascar (see Fig. 1). Formerly called Ethiopian Region.

Allopatric: term used to describe two or more closely-related species whose distributions, and especially breeding distributions, do not overlap. See Sympatric, Superspecies.

Arid: areas with sparse vegetation, mainly along drainage lines. See Moisture.

Brood parasite: see Nest Parasite.

Bushed, bushland: having an appreciable number of woody plants, which are mainly shrubs with some trees. This vegetation covers at least 20% of the ground surface and is often fairly dense.

Carbonate lake: applies to water bodies with very high concentration of this ion, the so-called 'soda lakes', *e.g.* Nakuru 62A, Magadi 75C.

Coast: in this Atlas refers to the actual coastline and the adjacent land, up to some 20-40 km inland.

Common: terms of relative abundance used in this Atlas include rare, uncommon, common and abundant. They have no precise meanings.

Congener: belonging to the same genus.

Conspecific: belonging to the same species (*q.v.*).

Crepuscular: active in the half-light of dawn and dusk.

Dimorphic: having two morphs (*q.v.*).

Disjunct distribution: a distribution comprising two or more parts, geographically isolated from each other.

Ectoparasite: parasite living on the outside of an animal, *e.g.* ticks.

Endemic: confined to a specified area or habitat, *e.g.* endemic to Kenya, or to Mt. Kulal; or to papyrus swamps in central Africa.

Eutrophic: a lake or other water body is eutrophic if it is relatively rich in nutrients; if nutrients are few, and productivity therefore low, it is termed oligotrophic.

Exotic: introduced by man from another place, as with Feral Pigeon (323) and House Sparrow (992) in Kenya.

Extralimital: in this Atlas, refers to birds outside of their normal range. See Vagrant, Wanderer.

Feral: describes an animal originally kept captive but subsequently established in the wild, *e.g.* Yellow-collared Lovebird (348) at Naivasha 62C and Nairobi 75B.

Forest: vegetation dominated by tall trees, which form a more or less closed canopy.

Forest island: an area of forest isolated by more open habitats from other forests. Usually refers to forests separated by a considerable distance, such as those on Mts Marsabit 27D 28C and Endau 77B. However, deforestation in the highlands and at the coast has produced islands of forest where formerly it was extensive.

Grassland: vegetation with few or no trees and bushes (if these are present, terms such as wooded grassland, bushed grassland are used).

Highlands: an ill-defined area of central southern Kenya, especially the area over and above 1800 m from Nairobi 75B to Eldoret 49C and Narok 74B to Nyahururu 50C.

Humid: moist areas capable of supporting forest, which sometimes extends to the subhumid zone. See Moisture.

Hybrid; hybridisation: a bird whose parents are of two different species; the process involved. An example is the two Paradise Flycatchers *Terpsiphone rufiventer* and *T. viridis* (813, 812), which are known to hybridise in western Kenya.

Intra-African: migrants breeding in one place and then moving to another, within the Afrotropical Region.

Introduced: see Exotic.

Kleptoparasitism: in this Atlas the forcible and often violent theft of food items from other birds. Also called Piracy.

Local (of distribution): usually refers to a species occurring sparsely in widely-scattered populations, *e.g.* White-eyed Kestrel (158), Square-tailed Drongo (567).

'Locally common (*etc.*)' is applied to cases where a bird is only common in relatively few places.

Migration: typically, the movement of a whole population from one place to another. Where the migration is altitudinal, as with the Golden-winged Sunbird (926), the distances moved may be quite short. At the other extreme, Eurasian Swallows (553) migrate from the USSR and Europe through tropical Africa to the south of the continent. See also Afrotropical, Intra-African, Palearctic, Wanderer.

Moisture: in this Atlas, a climatic term embracing arid, semiarid, subhumid and humid zones (see Fig. 7 and Table 4).

Morph: Form: thus the Little Egret (35) has white and dark morphs. See Dimorphic, Polymorphic.

Nest parasite: a bird whose eggs are laid in the nest of another species which then rears the young produced. All African cuckoos and honeyguides are nest parasites, as are the indigobirds and most if not all whydahs. Also known as brood parasite.

Nominate race(more correctly, nominotypical): the type race of a species, having the same subspecific and specific names, *e.g.* the race *Nectarinia pulchella pulchella* of the Beautiful Sunbird (920).

Oligotrophic: see Eutrophic.

Oversummering: term to describe individuals of Palearctic migrant species, usually subadults, which remain in the tropics whilst most of the species is breeding in Eurasia.

Palearctic: the zoogeographical region north of the Afrotropical (*q.v.*), consisting of north Africa, Europe and Asia north of the Himalayas.

Passage: a bird 'on passage' is migrating.

Passage migrant: a bird which passes through Kenya, for example when migrating from the Palearctic Region to southern Africa *e.g.* Lesser Grey Shrike (863).

Phase: another term for morph (*q.v.*).

Polymorphic: having more than one morph (*q.v.*).

Population: the population of a species is usually taken to mean all of the individuals at a particular time and place.

Race: see Subspecies.

Range: the overall area within which a species is usually found. Few species occur at all places within their overall range, due to their habitat preferences.

Raptor: predatory bird, bird of prey, usually taken to include owls as well as the vultures, hawks, eagles, falcons and allies.

Rare: uncommon. See Common.

Resident: a species found in a specified area throughout the year, and usually breeding there, but not necessarily. Thus the Marabou (50) is present throughout the year in Nairobi 75B, and the Lesser Flamingo (60) at Lake Nakuru 62A, but they normally do not breed there. In these cases the individuals making up the population may change with time.

Riparian: associated with river banks.

Savanna: a vague term covering most non-forested, non-arid parts of Kenya, from grassland to woodland (*q.v.*).

Semiarid: a category in the moisture range (*q.v.*), supporting grassland, bushland, shrubland and woodland vegetation.

Sibling: in this Atlas, species which are very closely-related but usually allopatric (*q.v.*).

Soda lake: see Carbonate Lake.

Sparse: term of relative abundance: see Common.

Species: groups of interbreeding natural populations that are reproductively isolated from other such groups (Mayr 1969).

Species-group: a group of species which, whilst closely-related, are less close than those belonging to a superspecies (*q.v.*). The members of a species-group are often sympatric, at least for part of their distributions.

Subhumid: see Moisture.

Subspecies: morphologically distinct populations within a species, capable of interbreeding amongst themselves. This happens commonly where there distributions adjoin. The term Race is synonymous.

Superspecies: two or more species which are believed to have evolved recently from a common ancestor. Although reproductively isolated they are still allopatric (*q.v.*), or largely so. See Species-group.

Sympatric: having overlapping distributions, and occurring together. See Allopatric.

Vagrant: a bird outside its normal range (*q.v.*). See Extralimital, Wanderer.

Wanderer: (a) in this Atlas, generally refers to birds that are outside their normal ranges, but not sufficiently so to be called vagrants. (b) the meaning given to this term by Britton (1980a), *i.e.* birds making relatively limited movements, often varying from year to year as a result of rainfall patterns.

Wintering: usually applies to Palearctic migrants spending most of the period November-March in Kenya *e.g.* Nightingale (645)(*cf.* Passage).

Wooded: having an appreciable number of woody plants, which are mainly trees with some shrubs. This vegetation typically covers between 2 and 20% of the land surface. In this Atlas, can also refer to woodland *sensu stricto* (*q.v.*).

Woodland: vegetation type, usually in semiarid or subhumid areas, with trees covering at least 20% of the ground surface. Trees may be fairly short (usually 5-10 m) and the ground layer is usually dominated by grasses.

Zoogeographical region: one of the six regions of the world, each with a distinct fauna. They are: Neotropical, Nearctic, Palearctic, Afrotropical, Oriental and Malagasy (Madagascar).

Appendix – Computer production of the maps

by A.R.D.Taylor

The distribution maps in this book were produced on a microcomputer and printed with a dot matrix printer. The following notes are intended to offer guidance to those contemplating distribution mapping by computer as well as those wishing to know the mapping method used for this book.

Microcomputers are equipped with graphics systems allowing the representation of any image on the monitor screen, provided that the image can be built up of dots, rather like a tapestry image. The detail present in this image depends upon the microcomputer used, but should not be less than 256 points (horizontally) x 256 (vertically). A BBC Model B microcomputer was used in a screen resolution of 640 points x 256 points. Given this limitation, planning how the map should be laid out and the number of divisions the map should be composed of follows as the next step.

Kenya has a convenient shape, being rather rectangular in outline. The available bird record data was suitable for recording every half degree square of Kenya's area, giving a total of 228 valid squares. Counting across the map from west to east, there are a maximum of 17 squares, while from north to south there are 21. Each square could have 1 of 10 possible descriptions for a bird species, i.e. absent, pre-1970 presence, post-1970 presence, etc. For final printing therefore, the map symbols for each square must be sufficiently distinct. Each square (symbol) had to occupy 16 monitor screen dots horizontally x 8 dots vertically, while leaving a space of 1 dot on all sides of the symbol for visual clarity of separation. A large sheet of graph paper was used to lay out the Kenya outline in terms of squares. The national boundaries were drawn in then the map was coded into a Basic language program for screen reproduction.

The transfer of this map into Basic was the most time-consuming step, a series of DRAW and MOVE statements had to be used to recreate the national boundary and the grid lines. Once this had been done and the map was correctly drawn on screen together with symbol positions, the program could be used whenever the map had to be drawn. In this way, memory did not need to be reserved for the disc storage of each of the maps, only for the bird distribution data.

Data for disc storage was read off the original hand drawn record maps using a screen editor (called 'Birdmap') by pressing user-defined function keys labelled with each symbol. The data for each bird was stored on disc as a single 'string' record. Each 'string' can be 256 bytes long, for this atlas it was convenient that there were a maximum of 228 squares per map. Therefore, each square could be stored as one byte, a very efficient form of storage. The screen editor could later be used to update or correct bird records at any time, simply by recalling data from disc.

Printing of the bird atlas used a program similar to 'Birdmap' called 'Birdprin'. This program read the data created by 'Birdmap' and converted the symbols into their final form for printing.

Finally, for analysis of the ecoclimatic distribution characteristics of each species, the program 'Birdstat' was used to compare the data held on disc for each bird and relate this to a lookup table of ecoclimatic characteristics for each of the mapping squares.

'BIRDMAP'

```
>L.
   10 REM**********BIRDMAP***************
   20 REM  FOR BIRD ATLAS OF KENYA
   30 REM  BY DOUG TAYLOR, JULY 1985
   40 ON ERROR CLOSE £0:*ACCESS Z.* L
   50 record$="":*OPT1,1
   60 *TV0,1
   70 MODE4:VDU4:COLOUR128:CLS:display=0
   80 PRINT TAB(15,10);"OPTIONS"
   90 PRINT'''Enter a new birdmap                    - 1"
  100 PRINT'''Search for and display a map           - 2"
  110 A$=GET$
  120 IF A$="1" THEN PROCENTER
  130 IF A$="2" THEN PROCDISPLAY
  140 GOTO 60
  150 END
  160  DEFPROCDRAWMAP
  170 COLOUR 129: GCOL 0,0
  180 CLS
  190 MOVE600,130:DRAW636,130:DRAW636,166:DRAW672,166:DRAW672,238:DRAW708,238:DRAW708,274:
DRAW744,274:DRAW744,310:DRAW780,310:DRAW780,346:DRAW816,346:DRAW816,382:DRAW780,382:DRAW7
80,454:DRAW744,454:DRAW744,670:DRAW780,670:DRAW780,706
  200 DRAW816,706:DRAW816,778:DRAW780,778:DRAW780,814:DRAW672,814:DRAW672,778:DRAW492,778:
DRAW492,814:DRAW420,814:DRAW420,850:DRAW348,850:DRAW348,886:DRAW312,886:DRAW312,850:DRAW2
40,850:DRAW240,706:DRAW276,706:DRAW276,562:DRAW240,562
  210 DRAW240,526:DRAW204,526:DRAW204,418:DRAW240,418:DRAW240,382:DRAW276,382:DRAW276,346:
DRAW348,346:DRAW348,310:DRAW420,310:DRAW420,274:DRAW492,274:DRAW492,202:DRAW528,202:DRAW5
28,166:DRAW600,166:DRAW600,130:MOVE484,202:DRAW672,202
  220 MOVE232,346:DRAW816,346:MOVE232,634:DRAW816,634:MOVE232,490:DRAW816,490:MOVE232,778:
DRAW816,778:MOVE240,858:DRAW240,850:MOVE384,858:DRAW384,310:MOVE528,822:DRAW528,202:MOVE6
72,822:DRAW672,202:MOVE816,786:DRAW816,778:MOVE244,414
  230 PLOT21,500,270:PLOT21,508,250:PLOT21,492,238:PLOT21,496,234:PLOT21,504,234:PLOT21,51
2,226:PLOT21,616,154:DRAW620,166:DRAW624,158:DRAW632,162:DRAW628,170:DRAW636,175:DRAW640,
173:DRAW648,194:DRAW660,214:DRAW664,230:DRAW652,234
  240 DRAW664,234:DRAW672,246:DRAW668,250:DRAW672,258:DRAW672,250:DRAW684,258:DRAW680,266:
DRAW688,274:DRAW688,298:DRAW708,306:DRAW704,310:DRAW716,306:DRAW732,318:DRAW736,326:DRAW7
32,326:DRAW728,334:DRAW740,334:DRAW736,350:DRAW744,342
  250 DRAW748,354:DRAW748,342:DRAW760,350:DRAW764,346:DRAW784,366:DRAW784,370:PLOT21,744,4
26:PLOT21,744,690:PLOT21,772,718:PLOT21,812,774:PLOT21,760,774:PLOT21,728,798:PLOT21,660,
770:PLOT21,652,754:PLOT21,644,746:PLOT21,640,738
  260 PLOT21,632,746:PLOT21,620,746:PLOT21,596,742:PLOT21,576,746:PLOT21,572,754:PLOT21,56
8,750:PLOT21,560,754:PLOT21,532,750:PLOT21,468,802:PLOT21,452,810:PLOT21,388,812:PLOT21,3
76,826:PLOT21,364,822:PLOT21,360,826:PLOT21,352,822
  270 PLOT21,344,834:PLOT21,348,846:PLOT21,336,850:PLOT21,328,846:PLOT21,324,850:PLOT21,26
8,822:PLOT21,240,798:PLOT21,248,790:PLOT21,244,778:PLOT21,252,774:PLOT21,248,770:PLOT21,2
56,770:PLOT21,252,762:PLOT21,272,754:PLOT21,264,742
  280 PLOT21,276,722:PLOT21,284,718:PLOT21,284,706:PLOT21,288,698:PLOT21,300,694:PLOT21,30
0,682:PLOT21,312,670:PLOT21,304,666:PLOT21,312,646:PLOT21,308,642:PLOT21,316,626:PLOT21,3
12,610:PLOT21,296,594:PLOT21,300,582:PLOT21,280,574
  290 PLOT21,276,554:PLOT21,268,546:PLOT21,256,542:PLOT21,244,526:PLOT21,240,510:PLOT21,23
6,498:PLOT21,240,482:PLOT21,236,454:PLOT21,236,418:PLOT21,244,414:DRAW248,426:DRAW244,434
:DRAW244,442:DRAW252,454:DRAW264,450:DRAW264,442
  300 DRAW272,450:DRAW268,458:DRAW276,462:DRAW284,458:DRAW292,458:DRAW288,462:DRAW292,466:
DRAW300,466:DRAW292,478:DRAW280,478:DRAW272,474:DRAW260,474:DRAW264,470:DRAW256,462:DRAW2
56,470:DRAW244,478:DRAW248,482:DRAW240,486:DRAW236,494
  310 MOVE380,822:DRAW372,802:DRAW376,790:DRAW372,778:DRAW372,750:DRAW376,746:DRAW376,742:
```

```
DRAW380,750:DRAW380,734:DRAW388,730:DRAW396,726:DRAW396,706:DRAW408,707:DRAW416,698:DRAW
416,690:DRAW420,682:DRAW420,670:DRAW424,662:DRAW440,670
 320 DRAW440,698:DRAW432,698:DRAW416,710:DRAW416,722:DRAW404,734:DRAW404,738:DRAW400,750
:DRAW404,762:DRAW396,774:DRAW400,790:DRAW396,798:DRAW396,810
 330 MOVE312,346:DRAW312,850:MOVE456,274:DRAW456,814:MOVE600,130:DRAW600,778:MOVE492,274
:DRAW708,274:MOVE240,418:DRAW780,418:MOVE276,562:DRAW744,562:MOVE276,706:DRAW780,706
 340  ENDPROC
 350  DEFPROCLOCATECHR
 360 COLOUR128:PRINT"Bird nr: ";number;"  Name: ";birdname$:COLOUR129
 370 P=0:*FX225,48
 380 VDU23,241,1,2,4,8,16,32,64,128:VDU23,242,129,66,36,24,24,36,66,129:VDU23,243,137,74
,44,24,255,44,74,137:VDU23,244,3,7,14,28,56,112,224,192:VDU23,245,195,231,126,60,60,126,
231,195
 390 VDU23,246,219,255,126,255,255,126,255,219:VDU23,247,131,71,46,28,56,116,226,193:VDU
23,248,131,79,46,28,255,124,234,201:VDU23,249,203,239,126,60,255,126,239,203
 400 VDU5
 410 MOVE316,882:PROCCHR:FORN=244TO388STEP36:MOVEN,846:PROCCHR:NEXT:FORN=244TO460STEP36:
MOVEN,810:PROCCHR:NEXT:FORN=676TO748STEP36:MOVEN,810:PROCCHR:NEXT:FORN=244TO784STEP36:MO
VEN,774:PROCCHR:NEXT
 420 FORN=244TO784STEP36:MOVEN,738:PROCCHR:NEXT:FORN=280TO748STEP36:MOVEN,702:PROCCHR:NE
XT:FORN=280TO712STEP36:MOVEN,666:PROCCHR:NEXT:FORN=280TO712STEP36:MOVEN,630:PROCCHR:NEXT
:FORN=280TO712STEP36:MOVEN,594:PROCCHR:NEXT
 430 FORN=244TO712STEP36:MOVEN,558:PROCCHR:NEXT:FORN=208TO712STEP36:MOVEN,522:PROCCHR:NE
XT:FORN=208TO712STEP36:MOVEN,486:PROCCHR:NEXT:FORN=208TO748STEP36:MOVEN,450:PROCCHR:NEXT
:FORN=244TO748STEP36:MOVEN,414:PROCCHR:NEXT
 440 FORN=280TO784STEP36:MOVEN,378:PROCCHR:NEXT:FORN=352TO748STEP36:MOVEN,342:PROCCHR:NE
XT:FORN=424TO712STEP36:MOVEN,306:PROCCHR:NEXT:FORN=496TO676STEP36:MOVEN,270:PROCCHR:NEXT
:FORN=496TO640STEP36:MOVEN,234:PROCCHR:NEXT
 450 FORN=532TO640STEP36:MOVEN,198:PROCCHR:NEXT:MOVE604,162:PROCCHR:ENDPROC
 460  DEFPROCCHR
 470 P=P+1
 480 IF display<>1 THENB$=GET$
 490 IF display=1 THEN B$=MID$(record$,P,1)
 500 FORM=240 TO 249
 510    IFVAL(B$)=M-240 THEN VDUM
 520    NEXT
 530 MOVE640,896:VDU240,240,240:MOVE640,896:GCOL0,3:PRINT;P:GCOL0,0
 540 IF display<>1 THENrecord$=record$+B$
 550  ENDPROC
 560  DEFPROCEDIT:VDU23,240,255,255,255,255,255,255,255,255
 570 VDU4:COLOUR128:PRINTTAB(0,0);"INPUT SQUARE NUMBER TO BE CHANGED":INPUT square:VDU11
:PRINT"CONTENTS=";:VDU(240+VAL(MID$(record$,square,1))):PRINT
 575 PRINT"PRESS CORRECT SYMBOL FOR THE SQUARE":B$=GET$:GCOL0,3:VDU240+VAL(B$):record$=L
EFT$(record$,square-1)+B$+MID$(record$,square+1)
 580 PRINT"ANY MORE? (Y/N)":B$=GET$:PRINTTAB(0,1);"";TAB(0,2);STRING$(56,""):IFB$="Y"THE
NGOTO560
 590  ENDPROC
 600  DEFPROCENTER:VDU23,240,255,255,255,255,255,255,255,255
 610 *FX12,14
 620 record$=""
 630 CLS:PRINT'''"WHAT IS THE BIRD NAME?":INPUT birdname$
 640 PRINT"THE BIRD NUMBER?":INPUT number:IF number>1400 THEN PRINT"REPEAT ";:GOTO640
 650  PROCDRAWMAP
 660  PROCLOCATECHR
 670 *FX12,0
 680 MOVE0,125:PRINT"PRESS 'S' TO SAVE RECORD OR 'E' TO EDIT OR SPACE BAR TO RESTART":T$
=GET$:IFT$="S"THEN PROCSAVE
 690 IFT$="E"THEN PROCEDIT
 700 IFT$<>" "THENGOTO680
 710  ENDPROC
 720  DEFPROCDISPLAY:VDU23,240,0,0,0,0,0,0,0,0
 730 CLS:T$="": IF record$<>"" THEN PRINT'''"Do you want the current map printed?      (Ye
s/No)":T$=GET$:IF T$="Y"THENPROCPRINT
```

```
 740 PRINT'''"ENTER THE BIRD NUMBER":INPUT number:B%=number
 745 PRINT'''"Print (P) or display (D) ?":IF GET$="P" PROCPRINT
 750 IF number< =700 THEN BIRDFILE$="Z.RECORD1":*DRIVE0
 760 IF number> 700 THEN number=number-700:BIRDFILE$="Z.RECORD2":*DRIVE2
 770 X=OPENIN BIRDFILE$:PTR£X=number*252:INPUT£X,record$:CLOSE£0
 780 number=VAL(MID$(record$,229,4)):birdname$=MID$(record$,233,18):record$=LEFT$(record
$,228)
 790  PROCDRAWMAP
 800 display=1
 810  PROCLOCATECHR
 820 MOVE0,125:PRINT"PRESS 'S' TO SAVE RECORD OR 'E' TO EDIT OR SPACE BAR TO RESTART":T$
=GET$:IFT$="S"THEN PROCSAVE
 830 IFT$="E"THEN PROCEDIT
 840 IFT$<>" "THENGOTO820
 850  ENDPROC
 860  DEFPROCSAVE:N=4-LEN(STR$(number)):N$=STR$(number)+STRING$(N," "):N=18-LEN(birdname
$):IFN>0 THENbirdname$=birdname$+STRING$(N," ")
 870 record$=record$+N$+LEFT$(birdname$,18)
 875 point=number
 880 IF point<=700 THEN BIRDFILE$="Z.RECORD1":*DRIVE0
 890 IF point>700 THEN point=point-700:BIRDFILE$="Z.RECORD2":*DRIVE2
 900 *ACCESS Z.*
 910 X=OPENUP BIRDFILE$:PTR£X=point*252:PRINT£X,record$:CLOSE£0
 920 *ACCESS Z.* L
 930 ENDPROC
 940  DEFPROCPRINT
 950 *EXEC PRINT
 960 ENDPROC
```

'BIRDPRIN'

```
>L.
  10 record$="":*OPT1,1
  20 *TV0,1
  30 *FX5,1
  40  MODE0:VDU4:CLS:IF record$=""THEN PROCDISPLAY
  50 PRINT TAB(25,10);"REPEAT (Y/N) ?":A$=GET$:IF A$="Y" PROCDISPLAY
  60 CHAIN"BIRDMAP":END
  70  DEFPROCDRAWMAP:GCOL0,129:GCOL 0,0:CLG
  80 MOVE600,130:PLOT21,636,130:PLOT21,636,166:PLOT21,672,166:PLOT21,672,238:PLOT21,708,
238:PLOT21,708,274:PLOT21,744,274:PLOT21,744,310:PLOT21,780,310:PLOT21,780,346:PLOT21,81
6,346:PLOT21,816,382:PLOT21,780,382:PLOT21,780,454
  90 PLOT21,744,454:MOVE744,670:PLOT21,780,670:PLOT21,780,706:MOVE638,130:PLOT21,638,166
:MOVE710,238:PLOT21,710,274:MOVE746,274:PLOT21,746,310:MOVE782,310:PLOT21,782,346:MOVE81
8,346:PLOT21,818,382:MOVE782,382:PLOT21,782,454:MOVE782,670
 100 PLOT21,782,706:MOVE530,786:PLOT21,530,202:MOVE386,858:PLOT21,386,310:MOVE780,706:PL
OT21,816,706:PLOT21,816,778:PLOT21,780,778:PLOT21,780,814:PLOT21,672,814:MOVE672,778:PLO
T21,492,778:PLOT21,492,814:PLOT21,420,814:PLOT21,420,850
 110 PLOT21,348,850:PLOT21,348,886:PLOT21,312,886:PLOT21,312,850:PLOT21.240,850:PLOT21,2
40,706:PLOT21,276,706:PLOT21,276,562:PLOT21,240,562:MOVE818,706:PLOT21,818,786:MOVE782,7
78:PLOT21,782,814:MOVE674,822:PLOT21,674,166:MOVE494,778
 120 PLOT21,494,814:MOVE422,814:PLOT21,422,850:MOVE350,850:PLOT21,350,886:MOVE314,886:PL
OT21,314,850:MOVE242,858:PLOT21,242,706:MOVE278,706:PLOT21,278,562:MOVE240,562:PLOT21,24
0,526:PLOT21,204,526:PLOT21,204,418:PLOT21,240,418
 130 PLOT21,240,382:PLOT21,276,382:PLOT21,276,346:PLOT21,348,346:PLOT21,348,310:PLOT21,4
20,310:PLOT21,420,274:PLOT21,492,274:PLOT21,492,202:PLOT21,528,202:PLOT21,528,166:PLOT21
,600,166:PLOT21,600,130:MOVE484,202:PLOT21,672,202:MOVE206,526
 140 PLOT21,206,418:MOVE242,418:PLOT21,242,382:MOVE278,382:PLOT21,278,346:MOVE350,346:PL
OT21,350,310:MOVE422,310:PLOT21,422,274:MOVE494,274:PLOT21,494,202:MOVE530,202:PLOT21,53
0,166:MOVE602,166:PLOT21,602,130:MOVE242,562:PLOT21,242,528
 150 MOVE268,346:PLOT21,816,346:MOVE268,634:PLOT21,752,634:MOVE176,490:PLOT21,816,490:MO
VE232,778:PLOT21,816,778:MOVE240,858:PLOT21,240,850:MOVE384,858:PLOT21,384,310:MOVE528,7
86:PLOT21,528,202:MOVE672,822:PLOT21,672,166:MOVE816,786
```

```
 160 PLOT21,816,778:MOVE312,346:PLOT21,312,850:MOVE456,274:PLOT21,456,814:MOVE600,130:PL
OT21,600,778:MOVE492,274:PLOT21,708,274:MOVE240,418:PLOT21,780,418:MOVE276,562:PLOT21,74
4,562:MOVE276,706:PLOT21,780,706:MOVE314,346:PLOT21,314,886
 170 MOVE458,274:PLOT21,458,814:MOVE602,130:PLOT21,602,778:MOVE244,414
 180 MOVE244,414
 190 PLOT21,500,270:PLOT21,508,250:PLOT21,492,238:PLOT21,496,234:PLOT21,504,234:PLOT21,5
12,226:PLOT21,616,154:DRAW620,166:DRAW624,158:DRAW632,162:DRAW628,170:DRAW636,175:DRAW64
0,173:DRAW648,194:DRAW660,214:DRAW664,230:DRAW652,234
 200 DRAW664,234:DRAW672,246:DRAW668,250:DRAW672,258:DRAW672,250:DRAW684,258:DRAW680,266
:DRAW688,274:DRAW688,298:DRAW708,306:DRAW704,310:DRAW716,306:DRAW732,318:DRAW736,326:DRA
W732,326:DRAW728,334:DRAW740,334:DRAW736,350:DRAW744,342
 210 DRAW748,354:DRAW748,342:DRAW760,350:DRAW764,346:DRAW784,366:DRAW784,370:PLOT21,744,
426:PLOT21,744,690:PLOT21,772,718:PLOT21,812,774:PLOT21,760,774:PLOT21,728,798:PLOT21,66
0,770:PLOT21,652,754:PLOT21,644,746:PLOT21,640,738
 220 PLOT21,632,746:PLOT21,620,746:PLOT21,596,742:PLOT21,576,746:PLOT21,572,754:PLOT21,5
68,750:PLOT21,560,754:PLOT21,532,750:PLOT21,468,802:PLOT21,452,810:PLOT21,388,812:PLOT21
,376,826:PLOT21,364,822:PLOT21,360,826:PLOT21,352,822
 230 PLOT21,344,834:PLOT21,348,846:PLOT21,336,850:PLOT21,328,846:PLOT21,324,850:PLOT21,2
68,822:PLOT21,240,798:PLOT21,248,790:PLOT21,244,778:PLOT21,252,774:PLOT21,248,770:PLOT21
,256,770:PLOT21,252,762:PLOT21,272,754:PLOT21,264,742
 240 PLOT21,276,722:PLOT21,284,718:PLOT21,284,706:PLOT21,288,698:PLOT21,300,694:PLOT21,3
00,682:PLOT21,312,670:PLOT21,304,666:PLOT21,312,646:PLOT21,308,642:PLOT21,316,626:PLOT21
,312,610:PLOT21,296,594:PLOT21,300,582:PLOT21,280,574
 250 PLOT21,276,554:PLOT21,268,546:PLOT21,256,542:PLOT21,244,526:PLOT21,240,510:PLOT21,2
36,498:PLOT21,240,482:PLOT21,236,454:PLOT21,236,418:PLOT21,244,414:DRAW248,426:DRAW244,4
34:DRAW244,442:DRAW252,454:DRAW264,450:DRAW264,442
 260 DRAW272,450:DRAW268,458:DRAW276,462:DRAW284,458:DRAW292,458:DRAW288,462:DRAW292,466
:DRAW300,466:DRAW292,478:DRAW280,478:DRAW272,474:DRAW260,474:DRAW264,470:DRAW256,462:DRA
W256,470:DRAW244,478:DRAW248,482:DRAW240,486:DRAW236,494
 270 MOVE380,822:DRAW372,802:DRAW376,790:DRAW372,778:DRAW372,750:DRAW376,746:DRAW376,742:
DRAW380,750:DRAW380,734:DRAW388,730:DRAW396,726:DRAW396,706:DRAW408,707:DRAW416,698:DRAW
416,690:DRAW420:682:DRAW420,670:DRAW424,662:DRAW440,670
 280 DRAW440,698:DRAW432,698:DRAW416,710:DRAW416,722:DRAW404,734:DRAW404,738:DRAW400,750
:DRAW404,762:DRAW396,774:DRAW400,790:DRAW396,798:DRAW396,810
 290 MOVE240,166:DRAW312,166:DRAW312,206:DRAW240,206:DRAW240,166:VDU5:MOVE244,198:PRINTS
TR$(new%):VDU4
 300 ENDPROC
 310  DEFPROCLOCATECHR:P=0:*FX225,48
 320 PRINTTAB(0,0);"Bird nr: ";number;" Name: ";birdname$
 330 VDU23,230,0,0,0,2,1,0,0,0:VDU23,231,0,0,0,160,64,0,0,0:VDU23,232,0,0,21,10,21,10,0,
0,:VDU23,233,0,0,80,168,80,168,0,0:VDU23,234,85,170,85,170,85,170,85,170:VDU23,235,85,17
0,85,170,85,170,85,170:VDU23,236,0,0,0,7,7,0,0,0
 340 VDU23,237,0,0,0,224,224,0,0,0:VDU23,238,0,0,31,31,31,31,0,0:VDU23,239,0,0,248,248,2
48,248,0,0:VDU23,240,255,255,255,255,255,255,255,255:VDU23,241,255,255,255,255,255,255,2
55,255:VDU23,242,0,0,21,15,23,10,0,0
 350 VDU23,243,0,0,80,232,240,168,0,0:VDU23,244,85,170,85,175,87,170,85,170:VDU23,245,85
,170,85,234,245,170,85,170:VDU23,246,85,170,95,191,95,191,85,170:VDU23,247,85,170,253,25
0,253,250,85,170:VDU23,248,56,108,108,56,0,0,0,0:VDU5
 360 MOVE176,794:PRINT"4";CHR$(248);"N":MOVE428,208:PRINT"4";CHR$(248);"S":MOVE220,890:P
RINT"34";CHR$(248):MOVE368,890:PRINT"36";CHR$(248):MOVE512,854:PRINT"38";CHR$(248):MOVE6
56,854:PRINT"40";CHR$(248):MOVE796,818:PRINT"42";CHR$(248);"E"
 370 MOVE316,882:PROCCHR:FORN=244TO388STEP36:MOVEN,846:PROCCHR:NEXT:FORN=244TO460STEP36:
MOVEN,810:PROCCHR:NEXT:FORN=676TO748STEP36:MOVEN,810:PROCCHR:NEXT:FORN=244TO784STEP36:MO
VEN,774:PROCCHR:NEXT
 380 FORN=244TO784STEP36:MOVEN,738:PROCCHR:NEXT:FORN=280TO748STEP36:MOVEN,702:PROCCHR:NE
XT:FORN=280TO712STEP36:MOVEN,666:PROCCHR:NEXT:FORN=280TO712STEP36:MOVEN,630:PROCCHR:NEXT
:FORN=280TO712STEP36:MOVEN,594:PROCCHR:NEXT
 390 FORN=244TO712STEP36:MOVEN,558:PROCCHR:NEXT:FORN=208TO712STEP36:MOVEN,522:PROCCHR:NE
XT:FORN=208TO712STEP36:MOVEN,486:PROCCHR:NEXT:FORN=208TO748STEP36:MOVEN,450:PROCCHR:NEXT
:FORN=244TO748STEP36:MOVEN,414:PROCCHR:NEXT
 400 FORN=280TO784STEP36:MOVEN,378:PROCCHR:NEXT:FORN=352TO748STEP36:MOVEN,342:PROCCHR:NE
XT:FORN=424TO712STEP36:MOVEN,306:PROCCHR:NEXT:FORN=496TO676STEP36:MOVEN,270:PROCCHR:NEXT
:FORN=496TO640STEP36:MOVEN,234:PROCCHR:NEXT
```

```
 410 FORN=532TO640STEP36:MOVEN,198:PROCCHR:NEXT:MOVE604,162:PROCCHR:ENDPROC
 420 DEFPROCCHR:P=P+1:B$=MID$(record$,P,1)
 430 FORM=228 TO 247STEP2:IFVAL(B$)*2=M-228 ANDVAL(B$)=OTHENVDUM;M+1
 440   IFVAL(B$)*2=M-228 ANDVAL(B$)<> OTHENVDU238,239,8,8:GCOL4,0:VDU238,239,8,8:GCOLO,0:
VDUM;M+1
 450   NEXT:ENDPROC
 460  DEFPROCDISPLAY:CLS:T$=""
 470 PRINT"BRITTON NUMBER IS: ";B%;TAB(50)"* ENTER NEW NUMBER *":INPUT new%:CLS
 480 number=B%
 490 IFnumber<=700 THEN BIRDFILE$="Z.RECORD1":*DRIVEO
 500 IFnumber>700 THEN number= number-700:BIRDFILE$="Z.RECORD2":*DRIVE2
 510 X=OPENIN BIRDFILE$:PTR£X=number*252:INPUT£X,record$:CLOSE£0
 520 number=VAL(MID$(record$,229,4)):birdname$=MID$(record$,233,18):record$=LEFT$(record
$,228):PROCDRAWMAP:PROCLOCATECHR:IF PAGE=&1B00 THEN CALL&1900 ELSE CALL&1500
 530  ENDPROC
```

'BIRDSTAT'

```
>L.
   10 REM"*****BIRDSTAT PROGRAM*****
   15 *TVO,1
   16 *FX6
   20 MODEO
   25 PRINT"START NUMBER":INPUTstart:PRINT"FINISH NUMBER":INPUTfinish
   30 DIMrecord$(10)
   40 DIMMAP$(228),BREEDING$(24),ALLREC$(24)
   50 FORI%=start TO finish STEP10
   60   PROCLOAD
   61   FORK%=1TO10:birdnum$=MID$(record$(K%),229,4):birdname$=MID$(record$(K%),233,18):r
ecord$=LEFT$(record$(K%),228)
   62    IFbirdnum$="    "AND K%<>10 THENNEXTK%
   63   IFbirdnum$="    "AND K%=10 THENNEXTI%
   70 PROCREAD
   80 PROCCALC
   90 PROCDISPLAY
   95 IFK%<10 THEN NEXTK%
  100 NEXTI%
  110 END
  120 DEFPROCREAD:A%=0:B%=0:FORN%=1TO24:BREEDING$(N%)="":ALLREC$(N%)="":NEXT
  130 FORJ%=1TO229
  140   IF J%=229 THEN ENDPROC
  150   IFMID$(record$,J%,1)="0"THENNEXT
  160 A%=A%+1
  170 IFMID$(record$,J%,1)<>"1"ANDMID$(record$,J%,1)<>"4"THEN B%=B%+1
  180 FORN=1TOLEN(MAP$(J%))
  190   T$=MID$(MAP$(J%),N,1)
  200   IFMID$(record$,J%,1)<>"1"ANDMID$(record$,J%,1)<>"4"THEN BREEDING$(ASC(T$)-64)=BRE
EDING$(ASC(T$)-64)+T$
  210   ALLREC$(ASC(T$)-64)=ALLREC$(ASC(T$)-64)+T$
  220   NEXT:NEXT
  230 ENDPROC
  240 DEFPROCCALC
  250 FORN%=1TO24
  260   IFB%>0 THENBREEDING$(N%)=STR$(100*(LEN(BREEDING$(N%)))/B%)
  270   IFA%>0 THENALLREC$(N%)=STR$(100*(LEN(ALLREC$(N%)))/A%)
  280   NEXT
  290 ENDPROC
  300 DEFPROCDISPLAY:VDU2:VDU1,27,1,15
  305 IFK%=1 THENPRINTSPC(18);"       ALTITUDE       RAINFALL      MOISTURE INDEX     HIGH
LOW   FRESH       SALT"
  310 IFK%=1 THENPRINT"NR    BIRDNAME";TAB(23);" A  B  C  D  E..F  G  H  I..J  K  L  M
N  O..P  Q..R..S  T  U..V  W  X":PRINT"------------------------------------------------
-----------------------------------------"
```

```
 320 PRINTbirdnum$;' ';birdname$;:FORM%=1TO24:PRINTTAB(M%*3+21);STR$(INT(VAL(BREEDING$(M
%))));:NEXT:PRINTTAB(96);"Br:";B%:PRINTbirdnum$;" ";SPC(18);:FORM%=1TO24:PRINTTAB(M%*3+2
1);STR$(INT(VAL(ALLREC$(M%))));:NEXT:PRINTTAB(96);"Tt:";A%
 325 PRINT"-----------------------------------------------------------------------------
---------------------"
 330 VDU3:ENDPROC
 340 DEFPROCLOAD:*DRIVE0
 345 IFI%<>start THEN360
 350 Y=OPENIN"LOOKUP":FORN%=1TO228:INPUT£Y,MAP$(N%):NEXT:CLOSE£Y
 360 IFI%<=690 THEN *DRIVE0
 365 IFI%>690 THEN *DRIVE2
 370 IF I%<=690 THEN X=OPENIN"Z.RECORD1"
 375 IF I%>690 THEN X=OPENIN"Z.RECORD2"
 380 FORnumber=I%TOI%+9
 390   point=number
 400   IF number>700 THEN point=number-700
 410   PTR£X=point*252
 420   INPUT£X,record$((number-I%)+1)
 425   NEXT
 460 CLOSE£0
 470 ENDPROC
```

Bibliography

a: abbreviations for frequently cited references

AGH Andrews, P., C.P. Groves & J.F.M. Horne 1975. Ecology of the lower Tana river flood plain (Kenya). J. E.Afr.nat.Hist.Soc. & natn.Mus. 151: 1-31.

AM Ash, J.S. & J.E. Miskell 1983. Birds of Somalia: their habitat, status and distribution. Scopus spec. Suppl.1.

Ba Backhurst, G.C., P.L. Britton & C.F. Mann 1973. The less common Palearctic migrant birds of Kenya and Tanzania. J.E.Afr.nat.Hist.Soc. & natn.Mus. 140: 1-38.

Be Betts, F.N. 1966. Notes on some resident breeding birds of southwest Kenya. Ibis 108: 513-530.

Br Britton, P.L. (ed.) 1980a. Birds of East Africa: their habitat, status and distribution. Nairobi: East African Natural History Society.

BUN Brown, L.H., E.K. Urban & K. Newman 1982. The birds of Africa. London: Academic Press.

EABR East African Bird Report. Scopus. Years are given if three or less issues are quoted.

HM Hall, B.P. & R.E. Moreau 1970. An atlas of speciation in African passerine birds. London: British Museum (Natural History).

HH Hopson, A.J. & J. Hopson 1975. Preliminary notes on the birds of the Lake Turkana area. Kitale: cyclostyled.

Ja Jackson, F.J. 1938. The birds of Kenya Colony and the Uganda Protectorate. London: Gurney & Jackson.

La Lack, P.C., W. Leuthold & C. Smeenk 1980. Checklist of the birds of Tsavo East National Park, Kenya. J.E.Afr.nat.Hist.Soc. & natn.Mus. 170: 1-25.

MPG Mackworth-Praed, C.W. & C.H.B. Grant 1957, 1960. African handbook of birds. Series I, vols. 1 & 2. Birds of eastern and north eastern Africa. London: Longmans, Green & Co.

Se Sessions, P.H.B. 1966. Notes on the birds of Lengetia Farm, Mau Narok. J.E.Afr.nat.Hist.Soc. & natn.Mus. 26: 18-48.

Sn Snow, D.W. (ed.) 1978. An atlas of speciation in African non-passerine birds. London: British Museum (Natural History).

St Stevenson, T. 1980. The birds of Lake Baringo. Nairobi: Sealpoint Publicity.

b: non-abbreviated references

Akeley, C.E. 1912. Flamingoes of Lake Hannington, Africa. Am.Mus.Jl. 12: 304-308.

Alamargot, J. 1978. Observations ornithologiques dans le sud-ouest de L'Ethiopie (Province de Guemou-Gofa) en 1973-1974. France: Institut d'Elevage et de Medecine Veterinaire des Pays Tropicaux.

Allan, R.G. 1983. The strategy for protecting crops from the depredations of quelea birds in Kenya. Proc. Ninth Bird Control Semin. BGSU, Bowling Green, Ohio.

Allen, P.M. 1969. Some birds seen at Castle Forest, September, 1969. Unpubl. MS. kept by the East African Natural History Society, Nairobi.

Allen, P.M. 1971a. A visit to Ragati Forest. EANHS Bull. 1971: 46.

Allen, P.M. 1971b. Trips to Baringo Lodge. *ibidem* 1971: 125-126.

Allen, P.M. 1972. Field trip to Mtara Estate. *ibidem* 1972: 128.

Allen, P.M. 1973a. Camp at the Arboretum, Chagaik Estate, Kericho. *ibidem* 1973: 22.

Allen, P.M. 1973b. Camp at Bushwhackers. *ibidem* 1973: 44-45.

Allen, P.M. 1973c. Field trip to Magadi, 29th-30th July. *ibidem* 1973: 126.

Allen, P.M. 1973d. Camp on Mr. C.H. Mayers' farm, Kedong, 14-16 September. *ibidem* 1973: 141.

Allen, P.M. 1974a. Expedition to Chemomi Estate, Nandi Hills, Kenya 8th-10th February 1974. *ibidem* 1974: 39.

Allen, P.M. 1974b. Camp in the grounds of Falloden Cottage, Ol Kalou. *ibidem* 1974: 109.

Allen, P.M. 1975. Ololokwe Camp. *ibidem* 1975: 4.

Allen, P.M. 1976. A visit to Mutomo. *ibidem* 1976: 77-78.

Allen, P.M. 1983. Society camp at Bushwhackers 15-17 July, 1983. *ibidem* 1983: 76-77.

Allen, P.M. & M.P. Clifton 1972. Aggressive behaviour of Kori Bustard *Otis kori*. *ibidem* 1972: 188-189.

Andersson, M. 1982. Female choice selects for extreme tail length in a widowbird. Nature 299: 818-820.

Andrewartha, H.G. & L.C. Birch 1954. The distribution and abundance of animals. Chicago: University of Chicago Press.

Andrews, P., C.P. Groves & J.F.M. Horne 1975. Ecology of the lower Tana river flood plain (Kenya). J.E.Afr.nat.Hist. Soc. & natn.Mus. 151: 1-31. (see Bibliography a).

Angwin, D. 1968a. Rapids Estate, near Mitubiri, 16/6/68. Unpubl. MS. kept by the East African Natural History Society, Nairobi.

Angwin, D. 1968b. Lake Naivasha 14/7/68 checklist of birds. *ibidem*.

Angwin, D. 1968c. Checklist of birds seen on the Nairobi falls trip 31/8/68. *ibidem*.

Angwin, D. 1970. Magadi Road to Campi ya Bibi, 1st May 1970. *ibidem*.

Angwin, D. 1971a. Camp on the Wei Wei River. EANHS Bull. 1971: 179-180.

Angwin, D. 1971b. Kindaruma Dam, Tana River, Central Province. EANHS camp, July 31st-August 2nd 1971. Unpubl. MS. kept by the East African Natural History Society, Nairobi.

Angwin, D. 1971c. Red-wing Starlings and Klipspringers. EANHS Bull. 1971: 182-183.

Angwin, D. 1977. Great Spotted Cuckoo. *ibidem* 1977: 108.

Angwin, D. 1980. Natural history notes from Kakamega Forest, western Kenya. *ibidem* 1980: 27-34.

Angwin, D. 1981. Report on the third annual Nairobi bird census. *ibidem* 1981: 2-5.

Angwin, D. 1982. Report on the fourth annual Nairobi bird census. *ibidem* 1982: 37-40.

Angwin, D. & J. Angwin 1968. Camp to Elemborasha Hill 14-15th September 1968; checklist of birds. Unpubl. MS. kept by the East African Natural History Society, Nairobi.

Angwin, E.B, 1977. Food storage in Indian House Crows. EANHS Bull. 1977: 131.

Angwin, E.B. 1980. Feathering the nest. *ibidem* 1980: 66.

Anonymous 1912. From Saxony to Njoro. J. E.Afr. & Uganda nat.Hist.Soc. 4: 131.

Anonymous 1913. Nesting of some E.African birds. *ibidem* 6: 56-57.

Anonymous 1943. *Psittacula krameri scopoli*. Parrakeet. J.E.Afr.nat.Hist.Soc. 3 & 4 (77 & 78): 286-287.

Anonymous 1970. Sands Farm, Athi River, 23rd May 1970. Unpubl. MS. kept by the East African Natural History Society, Nairobi.

Anonymous 1980. Report on the second annual Nairobi bird census. EANHS Bull. 1980: 2-5.

Anonymous 1985a. Crow population set to be reduced. The Daily Nation 7634:3.

Anonymous 1985b. Ringing news. EANHS Bull. 1985: 93.

Anonymous (n.d.)a. List of birds seen at Hunter's Lodge and in Kiboko Game Reserve. Unpubl. MS. kept by the East African Natural History Society, Nairobi.

Anonymous (n.d.)b. Bird list - Rowallan Scout Camp, Nairobi. *ibidem*.

Anonymous (n.d.)c. Preliminary checklist of birds collected by the Cambridge Pathological Expedition, 26 July-Aug.24, 1977. *ibidem*.

Archer, A.L. 1978. Marabous breeding at Kiboko - Kenya. EANHS Bull. 1978: 106-107.

Archer, A.L. 1979. Somali Bee-eater *Merops revoilii* in Tsavo West. Scopus 3: 58-59.

Archer, A.L. 1985. Protective behaviour in a nestling Red-chested Cuckoo *Cuculus solitarius*. *ibidem* 9: 52-53.

Archer, G.F. & E.M. Godman 1937, 1961. The birds of British Somaliland and the Gulf of Aden. Edinburgh & London: Gurney & Jackson, and Oliver & Boyd.

Ash, J.S. 1974. The Boran Cisticola in Ethiopia. Bull.Br.Orn.Club 94: 24-26.

Ash, J.S. 1976. Bird-ringing in Ethiopia report No. 5 1969-1975. Tech.Rep.1, Med. Ecol.Div., U.S.Nav.Med.Res.Unit 5: 1-17.

Ash, J.S. 1977. First known breeding of the Ruddy Shelduck *Tadorna ferruginea* south of the Sahara. Bull.Br.Orn.Club 97: 56-59.

Ash, J.S. 1980. The Lesser Golden Plover *Pluvialis dominica* in northeast Africa and the southern Red Sea. Scopus 4: 64-66.

Ash, J.S. 1981a. Desert Wheatears *Oenanthe deserti* in Ethiopia and Somalia. *ibidem* 5: 35-36.

Ash, J.S. 1981b. Spring passage of Whimbrel *Numenius phaeopus* and other waders off the coast of Somalia. *ibidem* 5: 71-76.

Ash, J.S. 1981c. A new race of Scaly Babbler *Turdoides squamulatus* from Somalia. Bull.Br.Orn.Club 101: 399-403.

Ash, J.S. 1983. Over fifty additions of birds to the Somalia list including two hybrids, together with notes from Ethiopia and Kenya. Scopus 7: 54-79.

Ash, J.S. 1985a. Midwinter observations from Djibouti. *ibidem* 9: 43-49.

Ash, J.S. 1985b. Two additions to the Somalia list: Greater Frigatebird *Fregata minor* and Indian House Crow *Corvus splen-*

dens. *ibidem* 9: 108-110.

Ash, J.S. 1985c. Birds, including a hybrid, new to Uganda. *ibidem* 9: 133-137.

Ash, J.S. & O. Ashford 1977. Great Black-headed Gulls *Larus ichthyaetus* and Red-necked Phalaropes *Phalaropus lobatus* inland in Ethiopia. J.E.Afr.nat.Hist.Soc. & natn.Mus. 31: 1-3.

Ash, J.S. & P.R. Colston 1981. A House x Somali Sparrow *Passer domesticus* x *P. castanopterus* hybrid. Bull.Br.Orn.Club 101: 291-294.

Ash, J.S. & J.E. Miskell 1983. Birds of Somalia: their habitat, status and distribution. Scopus spec.Suppl.1 (see Bibliography a).

Ash, J.S. & D.E. Pomeroy 1981. Mapping schemes in the Afrotropical Region. Ibis 123: 552-553.

Ash, J.S. & G.E. Watson 1974. *Locustella naevia* in Ethiopia. Bull.Br.Orn.Club 94: 39-40.

Backhurst, D.E.G. 1985. Falcons and eagles in Tsavo West National Park. EANHS Bull. 1985: 91-92.

Backhurst, D.E.G., D.J. Pearson & D.K. Richards 1984. A Kenya record of the Black-eared Wheatear *Oenanthe hispanica*. Scopus 8: 50-51.

Backhurst, G.C. 1968. Report on bird ringing in East Africa for 1966-1967. J.E.Afr.nat.Hist.Soc. & natn.Mus.1(116): 61-65.

Backhurst, G.C. 1969a. A record of *Gallinago stenura* from Kenya. Bull.Br.Orn. Club 89: 95-96.

Backhurst, G.C. 1969b. Bird ringing report 1967-68. J.E.Afr.nat.Hist.Soc. & natn.Mus. 3(118): 217-225.

Backhurst, G.C. 1970. Bird ringing report 1968-1969. *ibidem* 1(119): 16-26.

Backhurst, G.C. 1971a. A sight record of a new bird for Kenya. EANHS Bull. 1971: 40.

Backhurst, G.C. 1971b. East African bird ringing report 1969-1970. J.E.Afr.nat. Hist.Soc.& natn.Mus. 123:1-14.

Backhurst, G.C. 1972. East African bird ringing report 1970-1971. *ibidem* 136: 1-16.

Backhurst, G.C. 1973a. Scarce Palearctic migrants in Kenya. EANHS Bull. 1973: 113-114.

Backhurst, G.C. 1973b. East African bird ringing report 1971-1972. J.E.Afr.nat. Hist.Soc. & natn.Mus. 144: 1-15.

Backhurst, G.C. 1974. East African bird ringing report 1972-73, 1973-74. *ibidem* 146: 1-9.

Backhurst, G.C. 1977. East African bird ringing report 1974-77. *ibidem* 163: 1-10.

Backhurst, G.C. 1981a. Bird migration spectacular. Swara 4: 28-31.

Backhurst, G.C. 1981b. East African bird ringing report 1977-1981. J.E.Afr.nat. Hist.Soc. & natn.Mus. 174: 1-19.

Backhurst, G.C. 1985a. Ringing news. EANHS Bull. 1985: 65.

Backhurst, G.C. 1985b. Desert Wheatear *Oenanthe deserti* in Kenya. Scopus 9: 140-141.

Backhurst, G.C. & D.E.G. Backhurst 1970. A preliminary checklist of East African birds. Nairobi: cyclostyled.

Backhurst, G.C. & P.L. Britton 1969. A record of *Calidris subminuta* from Kenya. Bull.Br.Orn.Club 89: 121.

Backhurst, G.C., P.L. Britton & C.F. Mann 1973. The less common Palearctic migrant birds of Kenya and Tanzania. J. E.Afr.nat.Hist.Soc. & natn.Mus. 140: 1-38 (see Bibliography a).

Backhurst, G.C., M.G. Kelsey & D.J. Pearson 1986. Ringing and migration at Ngulia, Tsavo, autumn 1984. Scopus 8: 128-130.

Backhurst, G.C. & D.J. Pearson 1976. Savi's Warbler in Tsavo, an addition to the Kenya list. EANHS Bull. 1976: 21-22.

Backhurst, G.C. & D.J. Pearson 1977a. Southward migration at Ngulia, Tsavo, Kenya 1976/77. Scopus 1: 12-17.

Backhurst, G.C. & D.J. Pearson 1977b. Ethiopian Region birds attracted to the lights of Ngulia Safari Lodge, Kenya. *ibidem* 1:98-103.

Backhurst, G.C. & D.J. Pearson 1979. Southward migration at Ngulia, Tsavo, Kenya 1978/79. *ibidem* 3: 19-25.

Backhurst, G.C. & D.J. Pearson 1980. Southward migration at Ngulia, Tsavo, Kenya 1979/80. *ibidem* 4: 14-20.

Backhurst, G.C. & D.J. Pearson 1981. Ringing and migration at Ngulia, Tsavo, November 1980-January 1981. *ibidem* 5: 28-30.

Backhurst, G.C. & D.J. Pearson 1983a. Ringing and migration at Ngulia, Tsavo, October to December 1982. *ibidem* 6: 144-145.

Backhurst, G.C. & D.J. Pearson 1983b. Ringing and migration at Ngulia, Tsavo, autumn 1983.*ibidem* 7: 136-138.

Backhurst, G.C. & D.J. Pearson 1984. The timing of the southward night migration of Palearctic birds over Ngulia, south-east Kenya. Proc.Fifth Pan-Afr.orn. Congr. 361-369.

Bailey, H.P. 1979. Semi-arid climates: their definition and distribution. In A.E. Hall, G.H. Cannell & H.W. Lawton (eds.), Agriculture in semi-arid environments, p. 73-97. Berlin: Springer-Verlag.

Bailey, R.S., R. Pocklington & P.R. Willis 1968. Storm-petrels *Oceanodroma* spp. in the Indian Ocean. Ibis 110: 27-34.

Baker, N.E. 1984. Recent coastal records of the White-starred Forest Robin *Pogonocichla stellata* in Tanzania. Scopus 8: 51-52.
Baker, W.J. & P.M. Allen 1978. Silvery-cheeked Hornbills. EANHS Bull. 1978: 56-58.
Bally, P.R.O. 1943. An accident to a stork. J.E.Afr.nat.Hist.Soc. 3 & 4 (77 & 78): 283-284.
Bannerman, D.A. 1910. On a collection of birds made by Mr. A.B. Percival in British East Africa. Ibis Ninth Ser. 4 (16): 676-709.
Bartholomew, G.A. & C.J. Pennycuick 1973. The flamingo and pelican populations of the Rift Valley lakes in 1968-69. E.Afr. Wild.Jl. 11: 189-198.
Bednall, D.K. 1956. Black-browed Albatross at Mombasa. Ibis 98: 138.
Bednall, D.K. 1958. 'Drumming' by Swifts. J.E.Afr.nat.Hist.Soc. 2 (99): 8.
Bednall, D.K. 1959a. A Spotted Eagle in Nairobi? *ibidem* 3 (100): 144.
Bednall, D.K. 1959b. Migrant records - August/September 1958. *ibidem* 3 (100): 145.
Bednall, D.K. 1959c. A migration of Lesser Kestrels. *ibidem* 4 (101): 176.
Bednall, D.K. 1959d. Migrant records - January to May 1959. *ibidem* 4 (101): 177-178.
Bednall, D.K. 1959e. Late migration of Hobbies. *ibidem* 4 (101): 180.
Bednall, D.K. 1959f. Drumming of swifts. *ibidem* 4 (101): 180.
Belcher, C.F. 1939. Some Kinangop sunbirds. J.E.Afr. & Uganda nat.Hist.Soc. 3 (64): 167-171.
Belcher, C.F. 1943. Field notes from Kenya. Ibis Fourteenth Ser. 6: 91-93.
Bengtsson, E. 1975. The Shining-blue Kingfisher *Alcedo quadribrachys* in Kenya. EANHS Bull. 1975: 31.
Bennun, L.A. 1985. The Spotted Ground Thrush *Turdus fischeri fischeri* at Gede in coastal Kenya. Scopus 9: 97-107.
Benson, C.W. 1945. Observations from Kenya Colony, Uganda and Tanganyika Territory. Ibis 87: 90-95.
Benson, C.W. 1968. Singing by the Chiff-Chaff in Kenya. J.E.Afr.nat.Hist.Soc. & natn.Mus. 2 (117): 163.
Benson, C.W., R.K. Brooke, R.J. Dowsett & M.P.S. Irwin 1970. Notes on the birds of Zambia. Arnoldia (Rhod.) 4(40): 1-59.
Benson, C.W. & M.P.S. Irwin 1964. The migrations of the pitta of eastern Africa. N.Rhod.Jl. 5: 465-475.
Benson, C.W. & M.P.S. Irwin 1974. The significance of records of the Common Sandpiper breeding in East Africa. Bull.Br. orn.Club 94: 20-21.
Berg-Schlosser, G. 1979. Dünnschnabelmöwe *Larus genei* am Nakuru-see/Kenia. Vogelwarte 30: 145-146.
Best, J.R. 1977. A large hirundine roost in Uasin Gishu. EANHS Bull. 1977: 39-40.
Betts, F.N. 1966. Notes on some resident breeding birds of southwest Kenya. Ibis 108: 513-530 (see Bibliography a).
Beverley, P.E. 1976. Observations at Makalia Falls, Nakuru National Park. EANHS Bull. 1976: 129-130.
Binford, L.C. & D.A. Zimmerman 1974. Rufous-bellied Heron in Kenya. Bull.Br. Orn.Club 94: 101-102.
Binks, A. 1984. Swallow-tailed Kites in Meru National Park. EANHS Bull. 1984: 59.
Black, H.L. 1970. A record of the African Pitta *Pitta angolensis* from the Kenya coast. EANHS Bull. 1970: 64.
Blankenship, L., C.R. Field & I.S.C. Parker 1971. Swallow-tailed Kites *Elanus riocourii* in the Kedong Valley. *ibidem* 1971: 187.
Blayney Percival, A. 1914. East African Snipe. J.E.Afr. & Uganda nat.Hist.Soc. 8: 141-142.
Blayney Percival, A. 1916. Bird and snake. *ibidem* 10: 127.
Blayney Percival, A. 1917. Migration notes. *ibidem* 11: 186.
Bleazard, S. 1982. Lammergeyer in Amboseli. EANHS Bull. 1982: 61.
Blencowe, E.J. 1960a. Hinde's Pied Babbler in the Embu district. J.E.Afr. nat.Hist.Soc. 6 (103): 248.
Blencowe, E.J. 1960b. Report on bird ringing -1960. *ibidem* 7 (104): 300-301.
Blencowe, E.J. 1962a. Abdim's Storks breeding on cliffs. J.E.Afr.nat.Hist. Soc. & Coryndon Museum 1 (105): 64.
Blencowe, E.J. 1962b. Grey Plover on Lake Nakuru. *ibidem* 1 (105): 64.
Blencowe, E.J. 1962c. Pink footed Puffback breeding in Kakamega Forest. *ibidem* 1 (105): 65.
Blencowe, E.J. 1962d. Bird ringing report for 1960/61. *ibidem* 1 (105): 74.
Blencowe, E.J. 1963. The extension of range of Hildebrandt's Starling *Spreo hildebrandti* (Cabanis). *ibidem* 2 (106): 75.
Blom-Bjorner, J. 1945a. Breeding season of *Sagittarius serpentarius*. J.E.Afr. nat.Hist.Soc. 3 & 4 (83 & 84): 158.
Blom-Bjorner, J. 1945b. Harrier-Eagle eating a toad. *ibidem* 3 & 4 (83 & 84): 158-159.
Blom-Bjorner, J. 1945c. *Circaetus cinerascens*. *ibidem* 3 & 4 (83 & 84): 159.
Blom-Bjorner, J. 1945d. *Pernis apivorus apivorus*. *ibidem* 3 & 4 (83 & 84): 159.
Boothroyd, B. 1986. A second record of Jouanin's Petrel *Bulweria fallax* from

Kenya. Scopus 10: 28-29.

Bowles, R.V. 1967. Migration of Lesser Spotted Eagle. J.E.Afr.nat.Hist.Soc. & natn.Mus. 2 (114): 87.

Bowles, V.V. 1974. Wednesday morning bird walks. EANHS Bull. 1974: 119-121.

Brass, D. 1981. Backyard birds: Bendera Lane, Nairobi. *ibidem* 1981: 98-99.

Briffett, C. 1982a. Camp at Saiwa Swamp, 11th-13th December 1981. *ibidem* 1982: 15-17.

Briffett, C. 1982b. Society weekend at Fisherman's Camp, Naivasha. *ibidem* 1982: 110-112.

Britton, H.A. 1980. Eggs from East Africa in the collection of the Western Foundation of Vertebrate Zoology, California. Scopus 4: 79-83.

Britton, H.A. & P.L. Britton 1976. Records section (birds). EANHS Bull. 1976: 52-61.

Britton, P.L. 1969. Some records of albinism in East Africa. Bull.Br.Orn.Club 89: 169-171.

Britton, P.L. 1970a. The immature plumage of two African warblers. *ibidem* 90: 26-28.

Britton, P.L. 1970b. Two new shrikes for Kenya. *ibidem* 90: 133-134.

Britton, P.L. 1970c. Some non-passerine bird weights from East Africa. *ibidem* 90: 142-144, 152-154.

Britton, P.L. 1971a. Two sympatric canaries, *Serinus koliensis* and *S. citrinelloides*, in western Kenya. Auk 88: 911-914.

Britton, P.L. 1971b. On the apparent movements of *Cossypha natalensis*. Bull. Br.Orn.Club 91: 137-144.

Britton, P.L. 1972. Weights of African bulbuls (Pycnonotidae). Ostrich 43: 23-42.

Britton, P.L. 1973. Seasonal movements of the black cuckoo shrikes *Campephaga phoenicea* & *C. flava*, especially in East Africa. Bull.Br.Orn.Club 93: 41-48.

Britton, P.L. 1974a. Sandwich Tern on the Kenya coast. EANHS Bull. 1974: 30-31.

Britton, P.L. 1974b. A new gull for Kenya - *Larus ichthyaetus*? *ibidem* 1974: 59.

Britton, P.L. 1974c. Further Frigate Birds on the Kenya coast. *ibidem* 1974: 93.

Britton, P.L. 1974d. Broad-billed Sandpipers and Herring Gulls wintering on the north Kenya coast. *ibidem* 1974: 112-113.

Britton, P.L. 1974e. The Garsen heronry on the Tana River. *ibidem* 1974: 129-130.

Britton, P.L. 1974f. Relative biomass of Ethiopian and Palearctic passerines in west Kenya habitats. Bull.Br.Orn.Club 94: 108-113.

Britton, P.L. 1975. Dark-coloured egrets on the Kenya coast. EANHS Bull. 1975: 64-65.

Britton, P.L. 1976. Letter to the Editor: The identity of small Palearctic gulls in East Africa. *ibidem* 1976: 37-38.

Britton, P.L. 1977a. A Kenya record of *Cercococcyx montanus* at sea level. Scopus 1: 23-24.

Britton, P.L. 1977b. Status and identification of East African terns. *ibidem* 1: 29-34.

Britton, P.L. 1977c. Further records of the Sandwich Tern *Sterna sandvicensis* on the north Kenya coast. *ibidem* 1: 48.

Britton, P.L. 1977d. First African records of two Malagasy sea-birds. Bull.Br.Orn. Club. 97: 54-56.

Britton, P.L. 1977e. The Madagascar Pratincole *Glareola ocularis* in Africa. Scopus 1: 94-97.

Britton, P.L. 1978. Seasonality, density & diversity of birds of a papyrus swamp in western Kenya. Ibis 120: 450-466.

Britton, P.L. 1979. Notes on the birds of Lengetia Farm, Mau Narok, from the diaries of the late P.H.B. Sessions. Scopus 3: 54-57.

Britton, P.L. (ed) 1980a. Birds of East Africa: their habitat, status and distribution. Nairobi: East African Natural History Society (see Bibliography a).

Britton, P.L. 1980b. *Ficedula* Flycatchers in East Africa. Scopus 4: 21-22.

Britton, P.L. 1980c. A clarification of the status of the Red-chested Cuckoo *Cuculus solitarius* in lowland areas of East Africa. *ibidem* 4: 71.

Britton, P.L. 1981. Wilson's Storm Petrel *Oceanites oceanicus* in Kenya. *ibidem* 5: 31.

Britton, P.L. 1982a. Giant Petrel *Macronectes* sp. in Tanzania. *ibidem* 6: 42.

Britton, P.L. 1982b. Eurasian Cuckoo *Cuculus canorus* oversummering in East Africa. *ibidem* 6: 45-46.

Britton, P.L. & H.A. Britton 1971a. Somali Bee-eater in Tsavo East National Park. EANHS Bull. 1971: 71.

Britton, P.L. & H.A. Britton 1971b. Black-billed Barbet and Chestnut-fronted Shrike breeding in Kenya. *ibidem* 1971: 126-127.

Britton, P.L. & H.A. Britton 1971c. Some interesting Kenya breeding records. *ibidem* 1971: 165-166.

Britton, P.L. & H.A. Britton 1972a. An earlier breeding record of the Parasitic Weaver. *ibidem* 1972: 18.

Britton, P.L. & H.A. Britton 1972b. A Kenya record of the Southern Carmine Bee-eater *Merops nubicus nubicoides*. *ibidem* 1972: 120-121.

Britton, P.L. & H.A. Britton 1973a. Collared Flycatcher and Red-footed Falcon in western Kenya. *ibidem* 1973: 94-95.

Britton, P.L. & H.A. Britton 1973b. Broad-billed Sandpipers on the Kenya coast. *ibidem* 1973: 146-147.

Britton, P.L. & H.A. Britton 1973c. The Fiery-necked Nightjar in Kenya. *ibidem* 1973: 158-159.

Britton, P.L. & H.A. Britton 1974a. Pomarine Skua on the Kenya coast. *ibidem* 1974: 4-5.

Britton, P.L. & H.A. Britton 1974b. White-faced Tree Ducks nesting on islets off the Kenya coast. *ibidem* 1974: 5-6.

Britton, P.L. & H.A. Britton 1974c. Migratory Pygmy Kingfishers in coastal Kenya. *ibidem* 1974: 128.

Britton, P.L. & H.A. Britton 1976a. The nest and eggs of *Tauraco fischeri*. Bull. Br.Orn.Club 96: 134-136.

Britton, P.L. & H.A. Britton 1976b. Records section (Birds). EANHS Bull. 1976: 52-61.

Britton, P.L. & H.A. Britton 1977a. Sunbirds nesting inside buildings at the Kenya coast. Scopus 1: 68-70.

Britton, P.L. & H.A. Britton 1977b. An April fall of Palearctic migrants at Ngulia. *ibidem* 1: 109-110.

Britton, P.L. & H.A. Britton 1978a. The Malindi Pipit *Anthus melindae* in coastal Kenya. Ibis 120: 215-219.

Britton, P.L. & H.A. Britton 1978b. Notes on the Amani Sunbird *Anthreptes pallidigaster* including a description of nest and eggs. Scopus 2: 102-103.

Britton, P.L. & H.A. Britton 1979. Phalaropes in coastal Kenya. *ibidem* 3: 58.

Britton, P.L. & H.A. Britton 1985. Shoebill *Balaeniceps rex*: a deletion from the Kenya avifauna. *ibidem* 9: 50.

Britton, P.L., H.A. Britton & M.A.C. Coverdale 1980. The avifauna of Mrima Hill, south Kenya coast. *ibidem* 4: 73-78.

Britton, P.L. & L.H. Brown 1971. Breeding seabirds at the Kiunga islands, Kenya. Ibis 113: 364-366.

Britton, P.L. & L.H. Brown 1974. The status and breeding behaviour of East African Lari. Ostrich 45: 63-82.

Britton, P.L. & W.P.H. Duffus 1974. Great Black-headed Gull at Malindi. EANHS Bull. 1974: 51-52.

Britton, P.L. & J.F. Harper 1969. Some new distributional records for Kenya. Bull.Br.Orn.Club 89: 162-165.

Britton, P.L. & T.O. Osborne 1976. The race of *Sterna bergii* in Kenya. *ibidem* 96: 132-134.

Britton, P.L. & G.B. Rathbun 1978. Two migratory thrushes and the African Pitta in coastal Kenya. Scopus 2: 11-17.

Britton, P.L., S.N. Stuart & D.A. Turner 1984. East African endangered species. Proc.Fifth Pan-Afr.orn.Congr.: 679-688.

Britton, P.L. & M.St.J. Sugg 1973. Birds recorded on Kimilili track, Mount Elgon. J.E.Afr.nat.Hist.Soc. & natn.Mus. 143: 1-7.

Britton, P.L. & M.St.J. Sugg 1974. Birds recorded on the Kimilili track, Mount Elgon, Kenya - a postscript. EANHS Bull. 1974: 34.

Britton, P.L. & D.A. Zimmerman 1979. The avifauna of Sokoke Forest, Kenya. J.E. Afr.nat.Hist.Soc. & natn.Mus. 169: 1-15.

Brooke, R.K. 1969. *Apus berliozi* Ripley, its races and siblings. Bull.Br.Orn. Club 89: 11-16.

Brooke, R.K. 1971. The eastern and southern populations of the Mottled Spinetail. *ibidem* 91: 134-135.

Brooke, R.K. 1972. On the breeding and migratory status of *Apus berliozi bensoni*. *ibidem* 92: 114.

Brooke, R.K. & S.A. Parker 1984. The African bird collections of S.A. White of South Australia. Scopus 8: 33-36.

Brown, E.S. & C.F. Dewhurst 1970a. List of birds seen in Mandera region. Unpubl. MS. kept by the East African Natural History Society, Nairobi.

Brown, E.S. & C.F. Dewhurst 1970b. Birds seen in Wajir area. *ibidem*.

Brown, E.S. & C.F. Dewhurst 1971. N.F.D. EANHS Bull. 1971: 17-19.

Brown, G.H.H. 1963a. A first record of the Imperial Eagle *Aquila heliaca*, Savigny, in Kenya. J.E.Afr.nat.Hist. Soc. & Coryndon Mus. 2 (106): 71.

Brown, G.H.H. 1963b. The Marabou Stork, *Leptoptilos crumeniferus* (Lesson), killing live prey. *ibidem* 2 (106): 71-72.

Brown, G.H.H. 1963c. Note on the Pearl-spotted Owlet *Glaucidium perlatum* (Vieillot). *ibidem* 2 (106): 76.

Brown, G.H.H. 1963d. A note on the Violet Wood-Hoopoe *Phoeniculus damarensis*, (Grant). *ibidem* 2 (106): 76-77.

Brown, L.H. 1952-3. On the biology of the large birds of prey of the Embu District, Kenya Colony. Ibis 94: 577-620 & 95: 74-114.

Brown, L.H. 1955a. Supplementary notes on the biology of the large birds of prey of Embu District, Kenya Colony. *ibidem* 97: 38-64, 183-221.

Brown, L.H. 1955b. The breeding of Lesser and Greater Flamingoes in East Africa. J.E.Afr.nat.Hist.Soc. 5 (97): 159-162.

Brown, L.H. 1958. The breeding of the Greater Flamingo *Phoenicopterus ruber* at Lake Elmentaita, Kenya Colony. Ibis 100: 388-420.

Brown, L.H. 1959. Flamingoes in Africa. Ostrich Suppl. 3: 79-83.

Brown, L.H. 1960. The African Fish Eagle *Haliaeetus vocifer* especially in the Kavirondo Gulf. Ibis 102: 285-297.

Brown, L.H. 1963. Observations on East African birds of prey. E.Afr.Wildl.Jl. 1: 5-17.

Brown, L.H. 1965a. Redwinged Starlings of Kenya. J.E.Afr.nat.Hist.Soc. & natn. Mus. 1 (110): 41-56.

Brown, L.H. 1965b. Observations on Verreaux's Eagle Owl *Bubo lacteus* (Temminck) in Kenya. *ibidem* 2 (111): 101-107.

Brown, L.H. 1966. Observations on some Kenya eagles. Ibis 108: 531-572.

Brown, L.H. 1969. A first breeding record for the Southern Banded Snake Eagle *Circaetus fasciolatus* in Kenya. *ibidem* 111: 390-391.

Brown, L.H. 1970a. African birds of prey. London: Collins.

Brown, L.H. 1970b. Recent new breeding records for Kenya: Ovampo Sparrowhawk *Accipiter ovampensis*, Narina's Trogon *Apaloderma narina*, & Red-tailed Ant-thrush *Neocossyphus rufus*. Bull.Br.Orn. Club 90: 2-6.

Brown, L.H. 1970c. A Nairobi record of the African Pitta *Pitta angolensis*. EANHS Newsl. 63.

Brown, L.H. 1971a. A record of the Short-eared Owl *Asio flammeus* (Pontoppidan) from Kenya. EANHS Bull. 1971: 77-78.

Brown, L.H. 1971b. *Cuculus poliocephalus* at Watamu. *ibidem* 1971: 117.

Brown, L.H. 1971c. The relation of the Crowned Eagle *Stephanoaetus coronatus* and some of its prey animals. Ibis 113: 240-243.

Brown, L.H. 1971d. The flamingoes of Lake Nakuru. New Scientist and Sci.Jl. 51: 97-101.

Brown, L.H. 1972a. The breeding behaviour of the African Harrier Hawk *Polyboroides typus* in Kenya. Ostrich 43: 169-175.

Brown, L.H. 1972b. Natural longevity in wild Crowned Eagles *Stephanoaetus coronatus*. Ibis 114: 263-265.

Brown, L.H. 1973a. The mystery of the flamingoes. Nairobi: East African Publishing House.

Brown, L.H. 1973b. Unusual seabird records off Watamu, Kenya. EANHS Bull. 1973: 39-40.

Brown, L.H. 1973c. Notes from Watamu. *ibidem* 1973: 83-84.

Brown, L.H. 1973d. A Honey Buzzard at Karen. *ibidem* 1973: 86.

Brown, L.H. 1974a. Probable Wedge-tailed Shearwater *Puffinus pacificus* off Watamu. *ibidem* 1974: 23.

Brown, L.H. 1974b. Is poor breeding success a reason for the rarity of Ayres' Hawk-eagle? Ostrich 45: 145-146.

Brown, L.H. 1974c. A record of two young reared by Verreaux's Eagle. *ibidem* 45: 146-147.

Brown, L.H. 1974d. The races of the European Snake Eagle *Circaetus gallicus*. Bull.Br.Orn.Club 94: 126-128.

Brown, L.H. 1975a. A Frigate Bird in Mida Creek. EANHS Bull. 1975: 14.

Brown, L.H. 1975b. Breeding of Stuhlmann's Starling and Narina's Trogon. *ibidem* 1975: 44-45.

Brown, L.H. 1977a. The White-winged Dove *Streptopelia reichenowi* in SE Ethiopia, comparisons with other species, and a field key for identification. Scopus 1: 107-109.

Brown, L.H. 1977b. The status, population structure, and breeding dates of the African Lammergeyer. Raptor Res. 11 (3): 49-80.

Brown, L.H. 1978. Wahlberg's Eagle in northeastern Africa and the first record for Somalia. Scopus 2: 19-20.

Brown, L.H. 1980. The African Fish Eagle. Cape Town: Purnell.

Brown, L.H. 1982. The prey of the Crowned Eagle *Stephanoaetus coronatus* in central Kenya. Scopus 6: 91-94.

Brown, L.H. & P.L. Britton 1980. The breeding seasons of East African birds. Nairobi: East African Natural History Society.

Brown, L.H. & B.E. Brown 1974. The relative numbers of migrant and resident rollers in eastern Kenya. Bull.Br.Orn. Club 94: 126-128.

Brown, L.H. & B.E. Brown 1979. The behaviour of the Black Sparrowhawk *Accipiter melanoleucus*. Ardea 67: 77-95.

Brown, L.H. & G.H.H. Brown 1963. Recent records of the Lammergeyer, *Gypaetus barbatus* (Linn), in western Kenya. J.E. Afr.nat.Hist.Soc. & Coryndon Mus. 2 (106): 72.

Brown, L.H. & G. Bursell 1968. A first breeding record of the Cuckoo Falcon in Kenya. J.E.Afr.nat.Hist.Soc. & natn. Mus. 1 (116): 49-51.

Brown, L.H. & T.J. Cade 1972. Age classes and population dynamics of the Bateleur and African Fish Eagle. Ostrich 43: 1-16.

Brown, L.H. & P.R.A. Davey 1978. Natural longevity, as determined by plumage variation, in Ayre's Hawk-eagle (*Hieraaetus dubius*). Bokmakierie 30: 27-31.

Brown, L.H., V. Gargett & P. Steyn 1977. Breeding success in some African eagles related to theories about sibling aggression and its effects. Ostrich 48:

65-71.
Brown, L.H. & J.B.D. Hopcraft 1973. Population structure and dynamics in the African Fish Eagle *Haliaeetus vocifer* (Daudin) at Lake Naivasha, Kenya. E.Afr. Wildl.Jl. 11: 255-269.
Brown, L.H. & K.B. Newman 1974. "Anting" in African passerine birds. Ostrich 45: 194-195.
Brown, L.H., D. Powell-Cotton & J.B.D. Hopcraft 1973. The breeding of the Greater Flamingo and Great White Pelican in East Africa. Ibis 115: 352-374.
Brown, L.H. & A. Root 1971. The breeding behaviour of the Lesser Flamingo *Phoeniconaias minor*. *ibidem* 113: 147-172.
Brown, L.H., E.K. Urban & K. Newman 1982. The birds of Africa. London: Academic Press (see Bibliography a).
Bruce, M. 1972. Bird bath at Njoro. EANHS Bull. 1972: 204.
Bruce, M. & D. Curtis 1976. Visit to Mill Farm, Njoro, 21-22 February 1976. *ibidem* 1976: 63-64.
Bruce, N. 1972. Black-headed Gulls off Kenya coast. *ibidem* 1972: 42.
Bullock, D.A. & K. Bullock 1976. Madagascar Bee-eater *Merops superciliosus* nesting on Tana River. *ibidem* 1976: 4.
Burke, V.E.M. 1965. A count of Crowned Cranes (*Balearica regulorum* (Bennett)) in the Kisii district, Kenya. J.E.Afr. nat.Hist.Soc. & natn.Mus. 2(111):162-163.
Burke, V.E.M. & L.H. Brown 1970. Observations on the breeding of the Pink-backed Pelican *Pelecanus rufescens*. Ibis 112: 499-512.
Burrell, J.H. & R. Abel 1976. A not so 'extinct' thrush on the Kenya coast. EANHS Bull. 1976: 32-33.
Campbell, A.L. 1971a. Camp at the coast 16th-20th October 1971. *ibidem* 1971: 197-198.
Campbell, A.L. 1971b. A petrel at Lake Rudolph, Kenya. *ibidem* 1971: 200-201.
Campbell, A.L. 1972a. Caspian Plover *Charadrius asiaticus* at the coast. *ibidem* 1972: 65.
Campbell, A.L. 1972b. Visit to Kisumu heronry. *ibidem* 1972: 143-144.
Campbell, A.L. 1973. Marina weekend. *ibidem* 1973: 161-162.
Campbell, B. & E. Lack 1985. A dictionary of birds. London: T. & A.D. Poyser.
Campbell, I. & A.L. Campbell 1975. Lesser Spotted Eagle in Kenya. EANHS Bull. 1975: 65.
Campbell, K.L.I. 1977. Observations on the Fishing Owl *Scotopelia peli* on the Tana River, Kenya. *ibidem* 1977: 36-37.
Cambridge Pathological Expedition 1977. Preliminary checklist of birds collected by the Cambridge Pathological Expedition 26 July-August 24, 1977. Unpubl. MS. kept by the East African Natural History Society, Nairobi.
Candy, M. 1984. Habits and breeding biology of the Great Blue Turaco *Corythaeola cristata*. J.E.Afr.nat.Hist.Soc. & natn.Mus. 180: 1-19.
Carlson, A. 1986a. A comparison of birds inhabiting pine plantation and indigenous forest in a tropical mountain area. Biol.Conserv. in press.
Carlson, A. 1986b. Group territoriality in the Rattling Cisticola, *Cisticola chiniana*. Oikos in press.
Carswell, M. & D.E. Pomeroy 1984. A bird atlas for Uganda. EANHS Bull. 1984: 96-99.
Carver, J. & J. Carver 1979. Peter's Finfoot nest. *ibidem* 1979: 113-114.
Chambers, R.J.H. 1960. Flamingoes on the Lokippi salt lake. J.E.Afr.nat.Hist. Soc. 5 (102): 209.
Chapin, J.P. 1939, 1954. The Birds of the Belgian Congo. Bull.Am.Mus.nat.Hist. 75.
Cheke, R.A. 1971a. Temperature rhythms in African montane sunbirds. Ibis 113: 500-506.
Cheke, R.A. 1971b. Feeding ecology and significance of interspecific territoriality of African montane sunbirds (Nectariniidae). Revue Zool. & Bot.Afr. 84: 50-64.
Cheke, R.A. 1972. Birds of the Cherangani montane forests and their parasites. E.Afr.Wildl.Jl. 10: 245-249.
Cheke, R.A. 1976. Notes on a nesting female Variable Sunbird. Bull.Br.Orn. Club 96: 5-8.
Cheke, R.A. 1978. Records of birds and their parasites from the Cherangani Mountains, Kenya. E.Afr.Wildl.Jl. 16: 61-64.
Cheke, R.A. & P.J. Taylor 1969. A checklist of the birds of the Cherangani Hills, recorded by the Oxford University Expedition in 1969. Unpubl. cyclostyled MS. kept by the East African Natural History Society, Nairobi.
Clancey, P.A. 1959a. A new race of the Weaver *Ploceus cucullatus* (Muller) from East Africa. Bull.Br.Orn.Club 79: 41-42.
Clancey, P.A. 1959b. Notes on some Grey-headed Sparrows from Kenya Colony. *ibidem* 79: 131-132.
Clancey, P.A. 1974. On the validity of range of *Lamprotornis corruscus mandanus* van Someren, 1921. *ibidem* 94: 113-116.
Clancey, P.A. & J.G. Williams 1959. On the unknown female dress and specific relationships of *Ploceus golandi* (Clarke). Ibis 101: 247-248.
Clifton, M.P. 1976. Records section. EANHS Bull. 1976: 80-82.

Clifton, M.P. 1977. A report on a visit to Kibwezi Forest. *ibidem* 1977: 76-78.

Cockett, R.E. 1976. Some observations on the Spotted Eagle Owl *Bubo africanus*. *ibidem* 1976: 78-80.

Coe, M.J. 1961. Notes on *Nectarinia johnstoni johnstoni* on Mt. Kenya. Ostrich 32: 101-103.

Collias, N.E. 1978. Building of one nest by two male Reichenow's Weavers. Scopus 2: 101.

Collias, N.E. 1984. Social behaviour of the Grey-capped Social Weaver *Pseudonigrita arnaudi* in East Africa. Proc. Fifth Pan-Afr.orn.Congr. 779-786.

Collias, N.E. & E.C. Collias 1977. Weaverbird nest aggregation and evolution of the compound nest. Auk 94: 50-64.

Collias, N.E. & E.C. Collias 1978a. Cooperative breeding behaviour in the White-browed Sparrow Weaver. *ibidem* 95: 472-484.

Collias, N.E. & E.C. Collias 1978b. Survival and intercolony movement of White-browed Sparrow Weavers *Plocepasser mahali* over a two-year period. Scopus 2: 75-76.

Collias, N.E. & E.C. Collias 1978c. Behaviour of the Grey-headed Social Weaver. EANHS Bull. 1978: 102-105.

Collias, N.E. & E.C. Collias 1980. Behaviour of the Grey-capped Social Weaver (*Pseudonigrita arnaudi*) in Kenya. Auk 97: 213-226.

Collins, C.T. 1980. Notes on the food of the Horus Swift *Apus horus* in Kenya. Scopus 4: 10-13.

Collins, D. 1975. Camp at Fallodon Farm, Ol Kalou. 15th-16th February 1975. EANHS Bull. 1975: 36-37.

Collins, S.J.K. 1958. List of birds seen on EANHS ramble, 17th August 1958. Wakeford's Farm and Lukenia. Unpubl. MS. kept by the East African Natural History Society, Nairobi.

Colston, P.R. 1971. Additional non-passerine bird weights from East Africa. Bull.Br.Orn.Club 91: 110-111.

Colston, P.R. 1972. African passerine bird weights. *ibidem* 92: 115-116.

Congreve, W.M. 1913. The nesting habits of some East African birds. J.E.Afr. & Uganda nat.Hist.Soc. 6: 44-47.

Cooper, J.E., C.S. Dewhurst & A.D. Irvin 1971. The Ring-necked Parrakeet *Psittacula krameri* in Nairobi National Park. EANHS Bull. 1971: 173-174.

Copley, H. 1942. Note on a Gadwall collected at Kisumu. J.E.Afr.nat.Hist. Soc. 4 & 5 (73 & 74): 223.

Copley, H. 1944. Occasional notes. *ibidem* 5 & 6 (79 & 80): 395.

Copley, H. 1945. African Harrier-Hawks hunting weaver fledglings. *ibidem* 3 & 4 (83 & 84): 161.

Coverdale, M.A.C., J. Hancock & D.J. Pearson 1983. Unusual December-January breeding at the Garsen heronry, Kenya. Scopus 7: 49-50.

Coverdale, M.A.C. & D.J. Pearson 1984. A Broad-billed Prion *Pachyptila vittata* on the Kenya coast. *ibidem* 8: 43-44.

Craddock, G. 1970. List of birds seen within a 5-mile radius of the Tea Hotel, Kericho. Unpubl. MS. kept by the East African Natural History Society, Nairobi.

Cramp, S. & K.E.L. Simmons (eds.) 1977-. The birds of the western Palearctic. Oxford: OUP.

Crowe, T.M. & A.A. Crowe 1982. Patterns of distribution, diversity and endemism in Afrotropical birds. J.Zool.Lond. 198: 417-442.

Cunningham-van Someren, G.R. 1949. *Euplectes taha* in Kenya. Ibis 91: 345-346.

Cunningham-van Someren, G.R. 1969. Escapes of *Psittacula krameri* and *Agapornis* spp. breeding in Kenya. Bull.Br.Orn.Club 89: 137-139.

Cunningham-van Someren, G.R. 1970. On *Prodotiscus insignis* (Cassin) parasitising *Zosterops abyssinica* Guerin. *ibidem* 90: 129-131.

Cunningham-van Someren, G.R. 1971a. Is *Ploceus nigricollis melanoxanthus* (Cabanis) monogamous or polygamous? EANHS Bull. 1971: 96-97.

Cunningham-van Someren, G.R. 1971b. Cuckoos at Watamu. *ibidem* 1971: 116.

Cunningham-van Someren, G.R. 1971c. On the nesting of the 'Chestnut-banded' or 'Magadi Plover' *Charadrius pallidus venustus* Fischer & Reichenow, at Lake Nakuru. *ibidem* 1971: 130-132.

Cunningham-van Someren, G.R. 1971d. Cuckoos along the Magadi road. *ibidem* 1971: 132.

Cunningham-van Someren, G.R. 1971e. Palm-nut Vulture *Gypohierax angolensis* (Gmelin). *ibidem* 1971: 135-136.

Cunningham-van Someren, G.R. 1971f. Breeding of White-throated Bee-eaters *Merops albicollis*. *ibidem* 1971: 162-163.

Cunningham-van Someren, G.R. 1971g. Notes from Karen. *ibidem* 1971: 199-200.

Cunningham-van Someren, G.R. 1971h. *Euplectes diadematus* Fischer & Reichenow; courtship, nest and eggs. Bull.Br.Orn. Club 91: 119-121.

Cunningham-van Someren, G.R. 1972a. Lesser Frigate Bird off Watamu-Mida Creek. EANHS Bull. 1972: 30.

Cunningham-van Someren, G.R. 1972b. A second frigate bird off Watamu. *ibidem* 1972: 48.

Cunningham-van Someren, G.R. 1972c. Casts or pellets of the Barn Owl *Tyto alba*. *ibidem* 1972: 76-77.

Cunningham-van Someren, G.R. 1972d. A Honey Buzzard *Pernis apivorus* at Kiambu. *ibidem* 1972: 155-156.

Cunningham-van Someren, G.R. 1972e. Some White-throated Bee-eaters are resident in Kenya. *ibidem* 1972: 157.

Cunningham-van Someren, G.R. 1973a. *Tringa hypoleucos* breeding in East Africa. Bull.Br.Orn.Club 93: 39-40.

Cunningham-van Someren, G.R. 1973b. Unusual nest sites. EANHS Bull. 1973: 85.

Cunningham-van Someren, G.R. 1973c. *Vidua fischeri* (Reichenow) parasitic on *Granatina ianthinogaster* (Reichenow). J. E.Afr.nat.Hist.Soc. & natn.Mus 139: 1-4.

Cunningham-van Someren, G.R. 1973d. Calling and flight display assembly of male trogons. EANHS Bull. 1973: 81-83.

Cunningham-van Someren, G.R. 1974. At sea off Watamu. *ibidem* 1974: 19-20.

Cunningham-van Someren, G.R. 1975a. Breeding of Stuhlmann's Starling *Poeoptera stuhlmanni*. *ibidem* 1975: 12-13.

Cunningham-van Someren, G.R. 1975b. A note on the Lake Magadi Avocets and other birds. *ibidem* 1975: 90-92.

Cunningham-van Someren, G.R. 1975c. *Agapornis fischeri* Reichenow in Kenya? Auk 92: 370-371.

Cunningham-van Someren, G.R. 1976. Bird notes from Karen. EANHS Bull. 1976: 134-136.

Cunningham-van Someren, G.R. 1977a. On the nest of the Brown-chested Alethe. *ibidem* 1977: 6.

Cunningham-van Someren, G.R. 1977b. Late news from the rift valley swamp. *ibidem* 1977: 116-117.

Cunningham-van Someren, G.R. 1977c. Birds of the Nguruman Range, Kenya. Rep.1, Sect.Orn., Div.nat.Sci., natn.Mus Kenya: cyclostyled.

Cunningham-van Someren, G.R. 1977d. The European Wryneck *Jynx torquilla* in East Africa. Scopus 1: 54.

Cunningham-van Someren, G.R. 1978a. North Nandi Forest. Skins collected by Lorio Lukiro, V.C. Fayad and Dr. A.W. Diamond 3-9/1/78. Unpubl. MS. kept by the East African Natural History Society, Nairobi.

Cunningham-van Someren, G.R. 1978b. Nguruman Range. Skins collected by Lorio Luko and V.C. Fayad 17-24/1/78. *ibidem.*

Cunningham-van Someren, G.R. 1978c. A further note on the rift valley swamps. EANHS Bull. 1978: 60-62.

Cunningham-van Someren, G.R. 1978d. Additional notes on the Black-faced Waxbill. *ibidem* 1978: 87.

Cunningham-van Someren, G.R. 1979. Joint National Museum-Wien Museum ecological survey of the North Nandi Forest. *ibidem* 1979: 35-40.

Cunningham-van Someren, G.R. 1980a. Checklist of the birds of the North Nandi Forest. Report No.2, January 1980. Sect. Orn., Div.nat.Sci., natn.Mus.Kenya: cyclostyled.

Cunningham-van Someren, G.R. 1980b. North Nandi expedition. EANHS Bull. 1980: 6-9.

Cunningham-van Someren, G.R. 1980c. Weaver nesting colonies on the Magadi road. *ibidem* 1980: 64-66.

Cunningham-van Someren, G.R. 1981. Dry season troubles. *ibidem* 1981: 26-27.

Cunningham-van Someren, G.R. 1982. Prey of the Black-headed Heron at Njoro. *ibidem* 1982: 75-77.

Cunningham-van Someren, G.R. 1984a. Strange behaviour of a Variable Sunbird at Miotoni. *ibidem* 1984: 75-78.

Cunningham-van Someren, G.R. 1984b. Birds and the drought at Miotoni, Karen. *ibidem* 1984: 100-103.

Cunningham-van Someren, G.R. & D.K. Richards 1977a. Note on the Ewaso Ngiro swamp. *ibidem* 1977: 32-34.

Cunningham-van Someren, G.R. & D.K. Richards 1977b. A note on the flora and fauna of two seasonal swamps in the rift valley. *ibidem* 1977: 96-104.

Cunningham-van Someren, G.R. & H. Schifter 1981. New races of montane birds from Kenya and southern Sudan. Bull.Br.Orn. Club 101: 347-352 & 355-363.

Curry-Lindahl, K. 1981. Bird migration in Africa. Movements between six continents. London & New York: Academic Press.

Cyrus, D. & N. Robson 1980. Bird atlas of Natal. Pietermaritzburg: University of Natal Press.

Davey, F.A. 1983. Verreaux's Eagles. EANHS Bull. 1983: 11.

Davey, P.R.A. 1978a. Wahlberg's Eagle. *ibidem* 1978: 56.

Davey, P.R.A. 1978b. Pel's Fishing Owl. *ibidem* 1978: 56.

Davey, P.R.A. 1980. Random bird notes. *ibidem* 1980: 52.

Davey, P.R.A. 1982a. Notes from Masinga Dam, Tana River. *ibidem* 1982: 12.

Davey, P.R.A. 1982b. General report on Lake Turkana - 24th April-6th May 1982. *ibidem* 1982: 78-79.

Davey, P.R.A. 1982c. Albinism in Blue-eared Glossy Starling. *ibidem* 1982: 77.

Davey, P.R.A. 1983a. Red-billed Firefinch Indigobird. *ibidem* 1983: 75.

Davey, P.R.A. 1983b. Letter to the Editor. *ibidem* 1983: 75.

Davey, P.R.A. 1983c. Letter to the Editor. *ibidem* 1983: 91.

Davey, P.R.A. 1984. Young African Fish

Eagle - killing and eating a terrapin. *ibidem* 1984: 57.
Davey, P.R.A. & G.M. Davey 1980. Swallow-tailed Kites breeding at Porr, Lake Turkana. *ibidem* 1980: 47-49.
Davey, P.R.A. & J. Harrison 1976. Notes on migrant Eurasian duck in the shooting bag at Ol Bolossat Lake, Kenya. WAGBI, Rossett: cyclostyled.
Deane, A.J. 1971. Letter to the Editor. EANHS Bull. 1971: 118-119.
Dewhurst, C.F. 1971. A Lammergeyer *Gypaetus barbatus* seen in the Nairobi area. *ibidem* 1971: 133.
Dewhurst, C.F. & L.D.C. Fishpool 1984. The Swallow-tailed Kite *Chelictinia riocourii* breeding in the Kedong Valley, Kenya. Scopus 8: 25-26.
Diamond, A.W. 1977. Birds caught in Malaba Forest, Kakamega, December 1977. Unpubl. MS. kept by the East African Natural History Society, Nairobi.
Diamond, A.W. 1983. The Black and White Flycatcher *Bias musicus* in Kenya. Scopus 7: 22.
Diamond, A.W. & V.C. Fayad 1979. Preliminary comparisons between the avifaunas of the North Nandi and Kakamega Forests. *ibidem* 3: 93-100.
Diamond, A.W. & A.C. Hamilton 1980. The distribution of forest passerine birds and Quaternary climatic change in tropical Africa. J.Zool.Lond. 191: 379-402.
Diamond, A.W. & G.S. Keith 1980. Avifaunas of Kenya forest islands. 1 - Mt. Kulal. Scopus 4: 49-55.
Dillingham, I.H. & R.E. Moreau 1961. Musophagidae in northwestern Tanganyika Territory. Ibis 103a: 294-295.
Disney, H.J.De S. & A.J. Marshall 1956. A contribution to the breeding biology of the weaver-finch *Quelea quelea* (Linnaeus) in East Africa. Proc.Zool.Soc.Lond. 127: 379-387.
Dittami, J.P. 1981. Observations on wintering wheatears in burned grass areas at Lake Nakuru, Kenya. Vogelwarte 31: 177-178.
Dittami, J.P. 1983. Notes on Blue-eared Glossy Starlings *Lamprotornis chalybaeus* at Nakuru, Kenya. Scopus 7: 37-39.
Dittami, J.P. & V. Haas 1982. A family of White Storks *Ciconia ciconia* at Lake Nakuru: the first breeding record for Kenya? *ibidem* 6: 70.
Dobbs, C.M. 1914. On Crested Cranes at Kericho. J.E.Afr. & Uganda nat.Hist. Soc. 8: 138-140.
Donnelly, A.V. 1976. They are at it again! EANHS Bull. 1976: 62-63.
Donnelly, A.V. 1978. Notes from Watamu, coastal Kenya. *ibidem* 1978: 122-125.
Donnelly, A.V. & P. Donnelly 1977. Carmine Bee-eaters roosting up the Mida Creek. *ibidem* 1977: 61-62.
Dowsett, R.J. 1972a. Geographical variation in *Pseudhirundo griseopyga*. Bull. Br.Orn.Club 92: 97-100.
Dowsett, R.J. 1972b. Is the bulbul *Phyllastrephus placidus* a good species? *ibidem* 92: 132-138.
Dowsett, R.J. 1975. Lesser Spotted Eagle in Tsavo National Park. EANHS Bull. 1975: 14-15.
Dowsett, R.J. 1977. The distribution of some falcons and plovers in East Africa. Scopus 1: 73-78.
Dowsett, R.J. 1980. The migration of coastal waders from the Palearctic across Africa. Gerfaut 70: 3-35.
Dowsett, R.J. & F. Dowsett-Lemaire 1978. The voice of the Scarce Swift *Schoutedenapus myoptilus*. Scopus 2: 51-52.
Dowsett-Lemaire, F. 1979. The imitative range of the song of the Marsh Warbler *Acrocephalus palustris*, with special reference to imitations of African birds. Ibis 121: 453-468.
Dowsett-Lemaire, F. & R.J. Dowsett 1978a. Vocal mimicry of the lark *Mirafra hypermetra* as a possible species-isolating mechanism. Bull.Br.Orn.Club 98: 140-144.
Dowsett-Lemaire, F. & R.J. Dowsett 1978b. The Boran Cisticola *Cisticola bodessa* in Kenya, and its possible affinities. Scopus 2: 29-33.
Duffus, W.P.H. & A.E. Butterworth 1977. A record of the European Pochard *Aythya ferina* from Lake Ol Bolossat, Kenya. *ibidem* 1: 21-22.
Dyer, A. 1974. Letter to the Editor. EANHS Bull. 1974: 59-60.
Dyer, R. 1971. Partial albino starling. *ibidem* 1971: 161.
Dyer, R. & T. Dyer 1981. Ngare Ndare. Checklist of birds seen and heard during the EANHS weekend camp Easter 1981 17-20th April. Unpubl. MS. kept by the East African Natural History Society, Nairobi.
Dyson, W.G. 1971. A Kenya nest record for the Ethiopian Swallow. EANHS Bull. 1971: 38-39.
East African Bird Report. Scopus. (see Bibliography a).
East African Natural History Society/ Kenya Museum Society Wednesday Birdwalk. Card index (unpubl.) of birds seen in the Nairobi area.
Eddy, C.T. 1978a. Lambwe Valley checklist. Unpubl. MS. kept by the East African Natural History Society, Nairobi.
Eddy, C.T. 1978b. A checklist of birds seen at Mosocho between September 1975 and January 1978. *ibidem*.
Edwards, J.K. 1973. Silver bird *Empidor-*

nis semipartitus at Karen. EANHS Bull. 1973: 87.

Ellis, M. 1971. Chestnut-banded Plover at Baringo. *ibidem* 1971: 182.

Emlen, S.T. 1979. Fiscal attacks upon White-fronted Bee-eaters. Scopus 3: 101-102.

Emlen, S.T. 1981. Altruism, kinship and reciprocity in the White-fronted Bee-eater. In R.D. Alexander & D. Tinkle (eds.). Natural selection and social behaviour: recent research and new theory, p. 217-230. New York: Chiron Press.

Erard, C. 1974. The problem of the Boran Cisticola. Bull.Br.Orn.Club 94: 26-38.

Erlanger, C.F. von 1905. Beiträge zur Vogelfauna Nordostafrikas. J.Orn. 1905: 42-158, 433-499, 670-756.

Erlanger, C.F. von 1907. Beiträge zur Vogelfauna Nordostafrikas. *ibidem* 1-58.

Evans, S.M., M.A. Cantrell & A. Cram (eds.) 1973. Report of the University of Newcastle upon Tyne Exploration Society's Expedition to Kenya 1972. Cyclostyled.

Fanshawe, J.H. 1985. Bustards: an account of Paul Goriup's lecture to the Society. EANHS Bull. 1985: 9-11.

Fayad, V.C. & C.C. Fayad 1977a. A Grasshopper Warbler *Locustella naevia* from Kenya. Scopus 1: 84-86

Fayad, V.C. & C.C. Fayad 1977b. A Grasshopper Warbler from Nguruman Escarpment, Kenya. EANHS Bull. 1977: 104-105.

Fayad, V.C. & C.C. Fayad 1980. An ecological survey of the Nguruman Forest, Kenya. Nairobi: private publication.

Firmin, A. 1944. Verreaux's Eagle breeding in Kenya. J.E.Afr.nat.Hist.Soc. 1 & 2 (81 & 82): 95-96.

Fletcher, P.C. 1983. Breast markings in the Dark-backed Weaver at Kakamega Forest. EANHS Bull. 1983: 63-64.

Fogden, M.P.L. 1963. Early autumn migrants in coastal Kenya. Ibis 105: 112-113.

Fogden, M.P.L. 1964. The reproductive behaviour and taxonomy of Hemprich's Gull *Larus hemprichii*. *ibidem* 106: 299-320.

Forbes-Watson, A.D. 1966. A further note on Reef Herons in East Africa. J.E.Afr. nat.Hist.Soc. & natn.Mus. 3 (112): 233.

Forbes-Watson, A.D. 1971. Skeleton check-list of East African birds. Nairobi: cyclostyled.

Forbes-Watson, A.D. 1972a. East African Frigate Birds. EANHS Bull. 1972: 64.

Forbes-Watson, A.D. 1972b. Birds naturalised in East Africa. *ibidem* 1972: 144-146.

Forbes-Watson, A.D. 1977a. Notes on the field identification of East African honeyguides (Indicatoridae). Scopus 1: 17-20.

Forbes-Watson, A.D. 1977b. African Harrier-hawk *Polyboroides typus* feeding on carrion. *ibidem* 1: 22.

Forbes-Watson, A.D. 1977c. Drongo *Dicrurus adsimilis* using animate perch. *ibidem* 1: 25.

Forbes-Watson, A.D. 1977d. Maximum carrying potential of a Black-shouldered Kite *Elanus caeruleus*. *ibidem* 1: 44.

Forbes-Watson, A.D. (n.d.). Kakamega. Unpubl. MS. kept by the East African Natural History Society, Nairobi.

Forbes-Watson, A.D. & D.A. Turner 1973. A visit to Taveta, Lake Jipe and the Teita Hills. EANHS Bull. 1973: 156-157.

Forbes-Watson, N.M. 1963. Notes on the African Hobby, *Falco cuvieri* Smith. J. E.Afr.nat.Hist.Soc. & Coryndon Mus. 2 (106): 74-75.

Frank, L.G., R.M. Jackson, J.E. Cooper & M.C. French 1977. A survey of chlorinated hydrocarbon residues in Kenyan birds of prey. E.Afr.Wildl.Jl. 15: 295-304.

Frere, P.J. 1982. Verreaux's Eagle *Aquila verreauxi* eating carrion. Scopus 6: 101.

Frere, P.J. 1984. White-browed Coucal *Centropus superciliosus* robbing nest of Grosbeak Weaver *Amblyospiza albifrons* of eggs. *ibidem* 8: 48.

Friedmann, H. 1930. A lark new to science from north-central Kenya Colony. Auk 47: 418-419.

Friedmann, H. 1930, 1937. Birds collected by the Childs Frick expedition to Ethiopia and Kenya Colony. Bull.U.S.nat. Mus. 153: 1930, 1-516; 1937, 1-506.

Friedmann, H. & G.S. Keith 1968. First specimen of *Otus scops turanicus* (Loudon) from Africa. Bull.Br.Orn.Club 88: 112.

Friedmann, H. & A. Loveridge 1937. Notes on the ornithology of tropical East Africa. Bull.Mus.comp.Zool., Harv. 81: 1-413.

Friedmann, H. & K.E. Stager 1967. Results of the 1966 Cheney Expedition to the Samburu District, Kenya. Ornithology. Contr.Sci., Los Angeles 130: 1-34.

Friedmann, H. & K.E. Stager 1969. Results of the 1968 Avil Expedition to Mt. Nyiru, Samburu District, Kenya. Ornithology. *ibidem* 174: 1-30.

Fry, C.H. 1981. The residential status of the Madagascar Bee-eater *Merops superciliosus* in Africa. Scopus 5: 41-45.

Fry, C.H. 1984. The Bee-eaters. London: T. & A.D. Poyser.

Fry, C.H., P.L. Britton & J.F.M. Horne 1974. Lake Rudolph and the Palearctic exodus from East Africa. Ibis 116: 44-51.

Fry, C.H. & J.F.M. Horne 1972a. Black-headed Gulls at Lake Rudolph. EANHS Bull. 1972: 138-139.

Fry, C.H. & J.F.M. Horne 1972b. Kentish Plover at Lake Rudolph: a new species for Kenya. *ibidem* 1972: 139.

Gardiner, N. 1975. Bird notes from Dar es Salaam. *ibidem* 1975: 86-87.

Gerhart, J.D. 1971. A Lammergeyer record. *ibidem* 1971: 173.

Gerhart, J.D. 1980. Finfoot eating snake. Scopus 4: 69.

Gerhart, J.D., F. Ng'weno and D. Angwin 1979. Report on the first annual Nairobi bird census. EANHS Bull. 1979: 10-13.

Gerhart, J.D. & R.O. Paxton 1980. The Green Crombec *Sylvietta virens* in Kenya. Scopus 4: 47.

Gichuki, C.M. 1983. Some aspects of the ecology of Speke's Weaver *Ploceus spekei* (Heuglin) in Nairobi, Kenya. Unpubl. M.Sc. thesis, University of Nairobi.

Gichuki, N.N. 1982. The ecology and pest status of some granivorous birds in eastern Machakos District, Kenya. *ibidem.*

Gichuki, N.N. 1984. The pest status of some granivorous birds in eastern Machakos District, Kenya. Proc.Fifth Pan-Afr.orn.Congr. 371-386.

Gichuki, N.N. & D.E. Pomeroy 1983. The populations, diet, and pest status of two doves of the genus *Streptopelia* in eastern Machakos District, Kenya. Kenya J. Sci.Technol.Ser. B 4 (2): 97-110.

Gilpin, S. 1980. List of birds, EANHS day trip, Harries' Farm, Thika, 10th February 1980. Unpubl. MS. kept by the East African Natural History Society, Nairobi.

Gilpin, S. 1981. Field trip to Ruiru Dam. EANHS Bull. 1981: 11-12.

Glover, P.E., W.P. Langridge, E.M. Dougall & L.E.D. Wateridge 1970. A list of birds from the Mau Mara area of Masailand. Nairobi: Kenya National Parks.

Goddard, M. 1971. A week's birding at Malindi. EANHS Bull. 1971: 20-21.

Goddard, M. 1972. Egyptian Plover in Kenya. *ibidem* 1972: 14.

Gomez de Silva, H. 1984. The House Sparrow in Kenya. *ibidem* 1984: 113.

Gough, F.B. 1952. Notes on the nesting of the Hadada *Hagedashia hagedash erlangeri* Neum.J.E.Af.nat.Hist.Soc. 1 (91): 52.

Grant, C.H.B. 1915. On a collection of birds from British East Africa and Uganda, presented to the British Museum by Capt. G.S. Cozens. Parts I, II & III. Ibis Tenth Ser. (3) 1: 1-76, 235-315, 400-472.

Grant, L.V.J. 1985. A record of the Lesser Frigatebird *Fregata ariel* from Kenya. Scopus 9: 110-111.

Grant, L.V.J. & A.D. Lewis 1984. Breeding of the Ethiopian Swallow *Hirundo aethiopica* in interior Kenya. *ibidem* 8: 67-72.

Granvik, H. 1923. Contributions to the knowledge of East African ornithology. Birds collected by the Swedish Mount Elgon Expedition 1920. J.Orn. 71: 1-280.

Granvik, H. 1934. The ornithology of north western Kenya Colony with special regard to the Suk and Turkana districts. Revue Zool. & Bot.Afr. 25: 1-190.

Greenham, R. & L.M. Greenham 1975. Blackhead Plover nesting. EANHS Bull. 1975: 66-67.

Gregory, A.R. 1971. Crowned Hawk-Eagle *Stephanoaetus coronatus*. *ibidem* 1971: 62.

Groh, G. 1979. Little Gull *Larus minutus* at Lake Turkana: a new bird for Kenya. Scopus 3: 26.

Grumbley, T. 1978. EANHS Ol Lolokwe camp 24th/27th March 1978. Unpubl. MS. kept by the East African Natural History Society, Nairobi.

Grzimek, M. & B. Grzimek 1960. Flamingoes censused in East Africa by aerial photography. J.Wildl.Management 24: 215-217.

Guillet, A. & R.W. Furness 1985. Energy requirements of a Great White Pelican (*Pelecanus onocrotalus*) population and its impact on fish stocks. J.Zool.Lond. (A) 205: 573-583.

Gurney, G.H. 1909. Notes on a collection of birds made in British East Africa. Ibis Ninth Ser. (3) 11: 484-532.

Haas, V. 1982. Social organisation and role of helpers in Anteater Chats (*Myrmecocichla aethiops*) and Mourning Wheatears (*Oenanthe lugens schalowi*). Abstr.XVIII Congr.int.orn. 204, Moskau, Nauka.

Haas, V. 1985. Feeding rate reduction of mother Anteater Chats in the presence of helpers. BES-Workshop on Cooperative Breeding: Cambridge 29/3-1/4 1985.

Haas, V., U. Häussler & R. Stick 1981. Turtle Dove *Streptopelia turtur* in Kenya. Scopus 5: 128.

Haas, V., I.G. Marshall & A.D. Lewis 1982. Records of the Grasshopper Buzzard *Butastur rufipennis* at Nakuru, rift valley, Kenya, and a note on its prey. *ibidem* 6: 101-102.

Haas, V. & E. Nickel 1982. Breeding of Magpie Starlings *Speculipastor bicolor* in Kenya. *ibidem* 6: 41.

Haas, V. & K.H. Siebenrock 1983. Successful breeding of Greater Flamingo *Phoenicopterus ruber* at Lake Nakuru, Kenya. *ibidem* 7: 91-92.

Hale, W.H. 1958. Wild life seen above forest line on Mt. Kenya, 26th-28th February, 1958. J.E.Afr.nat.Hist.Soc. 2 (99): 15-16.

Hall, B.P. & R.E. Moreau 1970. An atlas of speciation in African passerine birds. London: British Museum (Natural History). (see Bibliography a).

Hamel, Rev. P. 1971. Letter to the Editor. EANHS Bull. 1971: 204-205.

Hamilton, A.C. 1981. The quaternary history of African forests: its relevance to conservation. Afr.J.Ecol. 19: 1-6.

Harper, J.F. 1974. Kisumu heronry. EANHS Bull. 1974: 79.

Harper, J.F. & L. Harper 1972. Black-rumped Waxbills nesting in Kenya. *ibidem* 1972: 77.

Harper, J.F. & L. Harper 1974a. Where do all the House Martins go? *ibidem* 1974: 113-115.

Harper, J.F. & L. Harper 1974b. Cherangani Hills - Chiffchaff and Black Kite. *ibidem* 1974: 129.

Harper, J.F. & L. Harper 1974c. White-eyed Pochard at Marsabit. *ibidem* 1974: 163.

Harrison, C. 1982. An atlas of the birds of the western Palaearctic. London: Collins.

Harrison, P. 1983. Seabirds: an identification guide. Beckenham: Croom Helm.

Hartley, N.D.R. 1974. Birds seen at Lake Hannington 16-18 Aug. 1974 EANHS camp. Unpubl. MS. kept by the East African Natural History Society, Nairobi.

Hartley, N.D.R. 1980. Natural history weekend at Timau. EANHS Bull. 1980: 105-107.

Harvey, W.G. 1972. A note on Black-headed Gulls in Dar es Salaam. *ibidem* 1972: 41.

Hayes, J. 1970. The Selengai Camp. EANHS Newsl. 4-7.

Hayes, J. 1978. Camp at Kilamata, Engari Ngiro Swamp. 15/16th July 1978. EANHS Bull. 1978: 115-117.

Hayes, J. & J.A. Kerr 1980. Birds seen at Buffalo Camp and environs, Mara Block 60, 4 to 7 April, 1980. Unpubl. MS. kept by the East African Natural History Society, Nairobi.

Hegner, R.E. 1980. Group dispersal in Hemprich's Hornbill *Tockus hemprichii*. Scopus 4: 67-68.

Hegner, R.E., S.T. Emlen, N.J. Demong & C.E. Miller 1979. Helpers at the nest in the White-fronted Bee-eater. *ibidem* 3: 9-13.

Henley, - 1943. Field notes on Secretary Bird. J.E.Afr.nat.Hist.Soc. 1 & 2 (75 & 76): 123.

Heinzel, H., R. Fitter & J. Parslow 1974. The birds of Britain and Europe with North Africa and the Middle East. London: Collins.

Hinde, S.L. 1898. On birds observed near Machakos Station, in British East Africa. Ibis Seventh Ser. (4) 16: 576-586.

Hinde, S.L. 1900. On further collections of British-East-African birds. *ibidem* Seventh Ser (6) 23: 492-500.

Hockey, P.A.R. & A.A. Ferrar (eds.) 1985. Guidelines for the Bird Atlas of southern Africa. Pretoria: CSIR.

Hofer, A. 1982. Some facts about the breeding biology of the Ground Hornbill. EANHS Bull. 1982: 10-12.

Hooker, T. & B.I. Hooker 1969. Duetting. In R.A. Hinde (ed.), Bird Vocalizations, Cambridge: Cambridge University Press.

Hopson, A.J. 1974a. A further record of the Kentish Plover at Lake Rudolph. EANHS Bull. 1974: 18.

Hopson, A.J. 1974b. Egyptian Plovers in the Lake Rudolph area. *ibidem* 1974: 131.

Hopson, A.J., R. de Vos, P. Magnusson & L. Johansson 1975. An occurrence of *Larus melanocephalus* the Mediterranean Gull at Lake Rudolph, Kenya. *ibidem* 1975: 65-66.

Hopson, A.J. & J. Hopson 1973. More Broad-billed Sandpipers at Lake Rudolph. *ibidem* 1973: 52.

Hopson, A.J. & J. Hopson 1974a. A Buff-breasted Sandpiper *Tryngites subruficollis* at Kerio Bay, Lake Rudolph. *ibidem* 1974: 17-18.

Hopson, A.J. & J. Hopson 1974b. Sooty Gulls at Lake Rudolph. *ibidem* 1974: 131-132.

Hopson, A.J. & J. Hopson 1975a. Preliminary notes on the birds of the Lake Turkana area. Kitale: cyclostyled (see Bibliography a).

Hopson, A.J. & J. Hopson 1975b. On the occurrence of two rare gulls *Larus leucophthalmus* White-eyed Gull and *Larus ichthyaetus* Great Black-headed Gull, at Lake Turkana. EANHS Bull. 1975: 83-84.

Hopson, J. & A.J. Hopson 1972. Broad-billed Sandpiper at Lake Rudolph. *ibidem* 1972: 170-171.

Hopson, J. & A.J. Hopson 1974. Further records of the Broad-billed Sandpiper *Limicola falcinellus* from Lake Rudolph. *ibidem* 1974: 33.

Hopson, J. & A.J. Hopson 1977. Breeding Carmine and White-throated Bee-eaters in the vicinity of Ferguson's Gulf, Turkana. *ibidem* 1977: 2-3.

Horne, J.F.M. 1971. A further pitta record. *ibidem* 1971: 28.

Horne, J.F.M. & L.L. Short 1977. First record of the Turtle Dove *Streptopelia turtur* in Kenya. Scopus 1: 50.

Horne, J.F.M. & L.L. Short 1981. Birds seen while casually barbetting 6-13 Dec. 1981, Koru (Homa Lime Co.). Unpubl. MS. kept by the East African Natural History Society, Nairobi.

Horwitz, R.J. & J. Anderson 1976. A

record of *Falco dickinsoni* from Kenya. EANHS Bull. 1976: 48-49.

Houston, D.C. 1975. Ecological isolation of African scavenging birds. Ardea 63: 55-64.

Howell, K.M., C.A. Msuya & S.N. Stuart 1980. The Red-chested Cuckoo *Cuculus solitarius* in coastal Tanzania. Scopus 4: 44.

Hudson, J.R. 1945. Pale specimen of Yellow-vented Bulbul (*Pycnonotus tricolor fayi*). J.E.Afr.nat.Hist.Soc. 3 & 4 (83 & 84): 161-162.

Hudson, J.R. 1973. Corncrake at Kabete in 1946. EANHS Bull. 1973: 112.

Huels, T.R. 1981. Cooperative breeding in the Golden-breasted Starling *Cosmopsarus regius*. Ibis 123: 539-542.

Huels, T.R. 1982. Cooperative feeding of conspecific and *Clamator jacobinus* young by *Turdoides rubiginosus*. Scopus 6: 33-35.

Hull, R.L. 1941, 1942. The names of some birds in several Nyanza languages. J.E. Afr. & Uganda nat.Hist.Soc. 1 (70): 55 and 2 & 3 (71 & 72): 125-126.

Hull, R.L. 1944. Nesting of the Red-throated Rock-martin, *Ptyonoprogne R. rufigula*. *ibidem* 1 & 2 (81 & 82): 94.

Hurxthal, L.M. 1979. Breeding biology of the Ostrich *Struthio camelus massaicus* Neumann in Nairobi Park. Unpubl. Ph.D. thesis, University of Nairobi.

Irvine, G.C. & D. Irvine 1970. Diani Forest - October 1970. EANHS Newsl.: 42-43.

Irvine, G.C. & D. Irvine 1974. Down on the Tana. EANHS Bull. 1974: 158-159.

Irvine, G.C. & D. Irvine 1977a. Down at Diani - October 1976. *ibidem* 1977: 85-86.

Irvine, G.C. & D. Irvine 1977b. Notes on birds seen in Diani Forest, Kenya. *ibidem* 1977: 106-108.

Irvine, G.C. & D. Irvine 1978. A visit to Lake Solai. *ibidem* 1978: 100-102.

Irvine, G.C. & D. Irvine 1980a. Diani bush. *ibidem* 1980: 10-11.

Irvine, G.C. & D. Irvine 1980b. The Palm Swift. *ibidem* 1980: 11-12.

Irvine, G.C. & D. Irvine 1981. Things are pretty dull birdwise these days. *ibidem* 1981: 100-103.

Isack, H.A. (n.d.). Birds of Kangaru School (Embu), including a few birds from the neighbouring areas. Unpubl. MS. kept by the East African Natural History Society, Nairobi.

Isack, H.A. (n.d.). Birds of Merti. Unpubl. MS. kept by the East African Natural History Society, Nairobi.

Isenmann, P. 1978. Bemerkungen zur Ausdehnung des Winterareals der Lachmöwe (*Larus ridibundus*) in Ost-Afrika und ihre vermutliche Ursache. Vogelwarte 29: 279-280.

Jackson, F.J. 1898. On a collection of birds from Witu, British East Africa. Ibis Seventh Ser. (4) 13: 133-141.

Jackson, F.J. 1899, 1902. List of birds obtained in British East Africa. Parts 1-3. Ibis Seventh Ser.(5) 20: 587-639; Eighth Ser.(2) 1: 33-96 & 8: 611-643.

Jackson, F.J. 1905. Notes on the East African species of *Macronyx* and *Tmetothylacus*. *ibidem* Eighth Ser.(5) 17: 101-104.

Jackson, F.J. 1910. East Africa and Uganda francolins. J.E.Afr. & Uganda nat.Hist.Soc. 1: 7-23.

Jackson, F.J. 1911. Game birds of the East Africa and Uganda Protectorates. *ibidem* 2: 60-74.

Jackson, F.J. 1926. Notes on the Game Birds of Kenya and Uganda. London: Williams & Norgate.

Jackson, F.J. 1938. The birds of Kenya Colony and the Uganda Protectorate. London: Gurney & Jackson (see Bibliography a).

Jenkin, P.M. 1929. Biology of lakes in Kenya. Nature 124: 574.

Jenkin, P.M. 1957. The filter-feeding and food of flamingoes (*Phoenicopteri*). phil. Trans.R.Soc. B240: 401-493.

Jensen, J.V. & J. Kirkeby 1980. The birds of the Gambia. Århus: Aros Nature Guides.

Johnston, H. 1902. On the occurrence of *Balaeniceps rex* on Lake Victoria. Ibis Eighth Ser.(2): 334-336.

Jones, M. 1983. Papyrus: a new fuel for the third world. New Scientist 99: 418-421.

Jouanin, C. 1957. Les Procellariides melaniques signales en Mer d'Oman. L'Oiseau & le Revue fr.d'Orn. 27: 12-27.

Kaestner, P. & J. Kaestner 1978. Violet Wood-Hoopoe raiding nest of Speckle-fronted Weaver. Scopus 2: 24.

Kahl, M.P. 1968. Recent breeding records of storks in eastern Africa. J.E.Afr. nat.Hist.Soc. & natn.Mus. 1 (116): 67-72.

Kahl, M.P. 1983. Breeding displays of the African Spoonbill *Platalea alba*. Ibis 125: 324-338.

Karmali, J.S. 1979. A weekend at Lake Baringo. EANHS Bull. 1979: 48-50.

Kasoma, P.M.B. 1982. Some aspects of the ecology of the black-headed weaver (*Ploceus cucullatus bohndorfii*, Reich.). Unpubl. M.Sc. thesis, Makerere University.

Kean-Hammerson, D. 1972. Advance weather indicators. EANHS Bull. 1972: 32-33.

Kear, J. & N. Duplaix-Hall (eds.) 1975. Flamingoes. Berkhamsted: T. & A.D. Poyser.

Keating, M.I. & C. Foxall 1980. Adventure Trails camp. EANHS Bull. 1980: 36-37.

Keith, G.S. 1964. A new subspecies of *Spreo albicapillus* (Blyth) from Kenya. Bull.Br.Orn.Club 84: 162-163.

Keith, G.S. 1968. Notes on birds of East Africa, including additions to the avifauna. Am.Mus.Novit. 2321.

Keith, G.S. & A. Twomey 1968. New distributional records of some East African birds. Ibis 110: 537-548.

Kent, D.F. 1981. EANHS field trip near Ngong Hills. EANHS Bull. 1981: 54.

Kenworthy, J.M. 1966. Temperature conditions in the tropical highland climates of East Africa. E.Afr.geogr.Rev. 4: 4-6.

Kreuger, R. 1970. First finding of the egg of *Circaetus fasciolatus*. Ibis 112: 117-118.

Kruuk, H. 1967. The competition for food between vultures (Aegypiinae) in East Africa. Ardea 55: 171-193.

Kutilek, M.J. 1974. Notes on Kittlitz's sandplover and the blacksmith plover at Lake Nakuru. E.Afr.Wildl.Jl. 12: 87-91.

Lack, D. 1968. The sequence in European bird lists. Ibis 110: 208-209.

Lack, P.C. 1975a. Records of 'less common Palaearctic migrants' in Tsavo National Park East 1974-75. EANHS Bull. 1975: 87-88.

Lack, P.C. 1975b. Range expansion of the Quail Plover. *ibidem* 1975: 110.

Lack, P.C. 1976a. The first recorded nest of the Pink-breasted Lark *Mirafra poecilosterna* (Reichenow). Bull.Br.Orn.Club 96: 111-112.

Lack, P.C. 1976b. Nesting of Hunter's Sunbird. EANHS Bull. 1976: 49.

Lack, P.C. 1976c. The status of the Taita Falcon. *ibidem* 1976: 103-104.

Lack, P.C. 1977a. The status of Friedmann's Bush-Lark *Mirafra pulpa*. Scopus 1: 34-39.

Lack, P.C. 1977b. A small heronry at Lake Jipe, Tsavo. *ibidem* 1: 82.

Lack, P.C. 1980. The habitats and feeding stations of the land-birds of Tsavo East National Park, Kenya. Unpubl. D.Phil. thesis, University of Oxford.

Lack, P.C. 1983. The movements of Palearctic landbird migrants in Tsavo East National Park, Kenya. J.anim.Ecology 52: 513-524.

Lack, P.C. 1985. The ecology of the landbirds of Tsavo East National Park, Kenya. Scopus 9: 2-23, 57-96.

Lack, P.C., W. Leuthold & C. Smeenk 1980. Checklist of the birds of Tsavo East National Park, Kenya. J.E.Afr.nat.Hist. Soc. & natn.Mus. 170: 1-25 (see Bibliography a).

Lack, P.C. & D.L.J. Quicke 1978. Dietary notes on some Kenyan birds. Scopus 2: 86-91.

Lambert, F.R. 1984. Birds at ant swarms in Kenya and southern Sudan. *ibidem* 8: 31-32.

Leakey, L.S.B. 1943a. Notes on the Falconidae in the Coryndon Memorial Museum. J.E.Afr.nat.Hist.Soc. 1 & 2 (75 & 76): 103-122.

Leakey, L.S.B. 1943b. Note on Gadwall. *ibidem* 1 & 2 (75 & 76): 127.

Leakey, L.S.B. 1943c. Mackinder's Owl. *ibidem* 3 & 4 (77 & 78): 284.

Lee, H.J. 1959. Grey Phalarope at Lake Nakuru. *ibidem* 4 (101): 175.

Lee, H.J. (n.d.). Migrant records 1959/60 & 1960/61. Unpubl. MS. kept by the East African Natural History Society, Nairobi.

Lee, H.J. 1962a. Unusual birds at the coast. J.E.Afr.nat.Hist.Soc. & Coryndon Mus. 1 (105): 66.

Lee, H.J. 1962b. Early arrival of Caspian Plover. *ibidem* 1 (105): 66.

Lee, H.J. 1963. Use of a live perch by the Red-winged Starling, *Onychognathus morio* (Linn.). *ibidem* 2 (106): 75.

Lee, H.J. 1971. Ethiopian Swallow *Hirundo aethiopica*. EANHS Bull. 1971: 167.

Leisler, B., G. Heine & K.H. Siebenrock 1983. Einnischung und interspezifische Territorialität überwinternder Steinschmatzer (*Oenanthe isabellina*, *O. oenanthe*, *O. pleschanka*) in Kenia. J. Orn. 124: 393-413.

Leith, H. 1972. Modelling the primary productivity of the world. Nature & Resour. 8: 5-10.

Leuthold, W. 1970a. Albino buffalo weaver. E.Afr.Wildl.Jl. 8: 205.

Leuthold, W. 1970b. Some breeding data on Somali ostrich. *ibidem* 8: 206.

Leuthold, W. 1971a. Crowned Hawk-Eagle *Stephanoaetus coronatus*. EANHS Bull. 1971: 62.

Leuthold, W. 1971b. Somali Bee-eater in Tsavo National Park. *ibidem* 1971: 97.

Leuthold, W. 1973a. Is the Somali Bee-eater extending its range? *ibidem* 1973: 101.

Leuthold, W. 1973b. Less common Palearctic migrants. *ibidem* 1973: 111-112.

Leuthold, W. 1974. Inland observations of the Terek Sandpiper. *ibidem* 1974: 153.

Leuthold, W. 1977. Notes on the breeding biology of the Ostrich *Struthio camelus* in Tsavo East National Park, Kenya. Ibis 119: 541-544.

Leuthold, W. & B. Leuthold 1972a. Blutschnabelweber (*Quelea quelea*) als Beute von Grief - und Stelzvögeln. Vogelwarte 26: 352-354.

Leuthold, W. & B. Leuthold 1972b. Queleas and attendant predators. EANHS Bull.

1972: 77-78.

Leuthold, W. & B. Leuthold 1974. Recent observations of migratory shore and water birds in Tsavo East National Park. *ibidem* 1974: 72-74.

Lewis, A.D. 1978. Sighting of Mackinder's Eagle Owl. *ibidem* 1978: 107.

Lewis, A.D. 1981a. Red-footed Falcon *Falco vespertinus* at Nairobi. Scopus 5: 57.

Lewis, A.D. 1981b. A record of the Collared Flycatcher *Ficedula albicollis* from Nairobi. *ibidem* 5: 60.

Lewis, A.D. 1981c. The past and present status and distribution of the Grey-crested Helmet Shrike *Prionops poliolopha*. *ibidem* 5: 66-70.

Lewis, A.D. 1981d. The Somali Sparrow *Passer castanopterus*: a breeding record for East Africa. *ibidem* 5: 83.

Lewis, A.D. 1981e. Birds of the Kaisut Desert, northern Kenya. EANHS Bull. 1981: 60-64.

Lewis, A.D. 1981-1982. Field identification of the genus *Cisticola* (Aves) in Kenya: parts 1-3. *ibidem* 1981: 108-114; 1982: 2-9, 28-37.

Lewis, A.D. 1982a. Pelican migration in the Kenya rift valley. Scopus 6: 40.

Lewis, A.D. 1982b. Further records of the Grey-crested Helmet Shrike. *ibidem* 6: 47-48.

Lewis, A.D. 1982c. Threat display of the Yellow-breasted Apalis *Apalis flavida*. *ibidem* 6: 73-74.

Lewis, A.D. 1982d. Form and function of the duetting of the Yellow-breasted Apalis *Apalis flavida*. *ibidem* 6: 95-100.

Lewis, A.D. 1982e. The Trumpeter Hornbill *Bycanistes bucinator* in Nairobi. *ibidem* 6: 102-103.

Lewis, A.D. 1982f. The breeding of the Rufous-chested Swallow *Hirundo semirufa* in East Africa. *ibidem* 6: 103-105.

Lewis, A.D. 1982g. Field trip to Hell's Gate, Naivasha. EANHS Bull. 1982: 14-15.

Lewis, A.D. 1982h. The bird atlas for Kenya: a request for information on place names. *ibidem* 1982: 84.

Lewis, A.D. 1982i. Possible widespread misidentification of canaries *Serinus* in Kenya. *ibidem* 1982: 101-103.

Lewis, A.D. 1982j. Further notes on Cisticolas. *ibidem* 1982: 104-105.

Lewis, A.D. 1983a. Why doesn't the Yellow-breasted Apalis duet at Kenyatta University College? *ibidem* 1983: 2-4.

Lewis, A.D. 1983b. Informal outing to the top of the Mua Hills. *ibidem* 1983: 18-19.

Lewis, A.D. 1983c. Range extension and population increase of the House Sparrow in Kenya. Scopus 7: 23-26.

Lewis, A.D. 1983d. A record of the Grey-crested Helmet Shrike near Naivasha, Kenya. *ibidem* 7: 26-27.

Lewis, A.D. 1983e. Old records of some scarce or little-known species from Kenya. *ibidem* 7: 89-90.

Lewis, A.D. 1983f. Unusual prey of Lizard Buzzard *Kaupifalco monogrammicus*. *ibidem* 7: 92.

Lewis, A.D. 1983g. A leucistic Common Bulbul *Pycnonotus barbatus*. *ibidem* 7: 94-95.

Lewis, A.D. 1983h. Further records of the white-throated form of the Pied Wheatear *Oenanthe pleschanka* from Kenya. *ibidem* 7: 96.

Lewis, A.D. 1983-4. Ten great places to see birds in Kenya. EANHS Bull. 1983: 65-71, 83-87; 1984: 71-72.

Lewis, A.D. 1984a. Further notes on sparrows *Passer* in Kenya and Somalia. *ibidem* 1984: 22-23.

Lewis, A.D. 1984b. Society excursion to Potha, Machakos. *ibidem* 1984: 34.

Lewis, A.D. 1984c. Grey Hornbills accompanying baboons. *ibidem* 1984: 58.

Lewis, A.D. 1984d. Notes on the ranges of three species in northern and eastern Kenya. Scopus 8: 27-28.

Lewis, A.D. 1984e. Hinde's Pied Babbler *Turdoides hindei* south of Machakos, Kenya. *ibidem* 8: 48-49.

Lewis, A.D. 1985. A visit to Mt Suswa. EANHS Bull. 1985: 66-67.

Lewis, A.D., I.J.P. Loefler & D.J. Pearson 1984. Four-banded Sandgrouse *Pterocles quadricinctus* in northwest Kenya. Scopus 8: 46-48.

Lewis, A.D. & D.J. Pearson 1981a. Stork and raptor migration in South Nyanza, Kenya. *ibidem* 5: 51.

Lewis, A.D. & D.J. Pearson 1981b. Raptor and stork migration at Namanga, southern Kenya. *ibidem* 5: 83-84.

Lewis, A.D. & D.E. Pomeroy 1982a. A bird atlas for Kenya. EANHS Bull. 1982: 64-70.

Lewis, A.D. & D.E. Pomeroy 1982b. A bird atlas for Kenya: further progress. *ibidem* 1982: 86-91.

Lewis, A.D. & D.E. Pomeroy 1983a. A bird atlas for Kenya: further progress II. *ibidem* 1983: 25-31.

Lewis, A.D. & D.E. Pomeroy 1983b. A bird atlas for Kenya: a progress report. Scopus 7: 27-30.

Lewis, A.D. & D.E. Pomeroy 1984. A bird atlas for Kenya: where do we now stand? EANHS Bull. 1984: 2-9.

Ligon, J.D. & S.H. Ligon 1978. The communal social system of the Green Wood Hoopoe in Kenya. Living Bird 17: 159-197.

Lind, E.M. & M.E.S. Morrison 1974. East

African vegetation. London: Longman.
Loefler, I.J.P. 1984. Breeding of Verreaux's Eagle Owl prevented by crows. EANHS Bull. 1984: 105-106.
Lohding, A. 1985. A mixed bird party and a snake. *ibidem* 1985: 23.
Long, R.C. 1963. Little Swifts at Mombasa, Kenya Colony. Ostrich 34: 49.
Loveridge, A. 1916. Natural history notes from British East Africa. J.E.Afr. & Uganda nat.Hist.Soc. 10: 108-124.
Loveridge, A. 1917a. A natural history expedition through the Kedong Valley, B.E.A. *ibidem* 11: 157-167.
Loveridge, A. 1917b. Natural history notes from British East Africa. *ibidem* 11: 170-185.
Loveridge, A. 1922. Notes on East African birds (chiefly nesting habits and stomach -contents) collected 1915-1919. Proc.zool.Soc.Lond. 1922: 837-862.
Loveridge, A. (n.d.). Records of European and other migrants in East Africa. Unpubl. MS. kept by the East African Natural History Society, Nairobi.
Lowis, R.A. 1974. Letter to the Editor. EANHS Bull. 1974: 123.
Lynes, H. 1930. Review of the genus *Cisticola*. Ibis (12) 6, Suppl.
Lynn-Allen, B.G. 1942. A few notes on birds in the northern Lake Rudolph area. J.E.Afr.nat.Hist.Soc. 4 & 5 (73 & 74): 220-222.
Maas, M. 1979. Abdim's Stork breeding in Kisii? EANHS Bull. 1979: 66-67.
Maas, M. 1983. White-faced Scops Owl breeding in Kisii. *ibidem* 1983: 32.
Macdonald, J.D. 1956. A new species of lark from Kenya. Bull.Br.Orn.Club 76: 70-72.
MacInnes, D.G. 1932-33. Nesting habits of some East African birds. J.E.Afr. & Uganda nat.Hist.Soc. 47 & 48: 128-135.
MacInnes, D.G. 1933-34. Notes on the birds of Turkana Province. *ibidem* 1 & 2 (51 & 52): 24-50.
MacInnes, D.G. 1946. European Wigeon, *Anas penelope* Linn. J.E.Afr.nat.Hist. Soc. 1 & 2 (85 & 86): 67.
MacInnes, D.G. 1948. Migratory Kestrels. Nature E.Afr. 5: 3-5.
Mackworth-Praed, C.W. 1917. A collection of birds from two Districts of British East Africa. Ibis Tenth Ser. (5) 3: 362-421.
Mackworth-Praed, C.W. & C.H.B. Grant 1957, 1960. African handbook of birds. Series I, vols. 1 & 2. Birds of eastern and north eastern Africa. London: Longmans, Green & Co. (see Bibliography a).
Maclean, G.L. 1985. Roberts' birds of southern Africa. Cape Town: The John Voelcker Book Fund.
Madge, S.G. 1974. A probable Sandwich Tern on the Kenya coast. EANHS Bull. 1974: 30.
Madge, S.G. & G.R. Cunningham-van Someren 1975. Black-throated Honeyguide and Abyssinian Scimitarbill. *ibidem* 1975: 130-131.
Madge, S.G. & D.K. Richards 1976. Some Kenya bird notes. *ibidem* 1976: 33-34.
Magor, J.I. & P. Ward 1972. Illustrated descriptions, distribution maps, and bibliography of the species of *Quelea* (weaverbirds: Ploceidae). Trop.Pest Bull. 1.
Mann, C.F. 1971. The first record of the Black-winged Pratincole for Kenya. EANHS Bull. 1971: 76-77.
Mann, C.F. 1972. The occurrence of the Grey Apalis *Apalis cinerea* on Mt. Marsabit. *ibidem* 1972: 168-169.
Mann, C.F. 1973a. Two new bird breeding records for Kenya. *ibidem* 1973: 39.
Mann, C.F. 1973b. Some recent raptor records from East Africa. *ibidem* 1973: 96-98.
Mann, C.F. 1974a. A second Pomarine Skua on the Kenya coast. *ibidem* 1974: 31.
Mann, C.F. 1974b. A record of the Collared Flycatcher *Ficedula albicollis semitorquata* in western Kenya. *ibidem* 1974: 78.
Mann, C.F. 1974c. Some bird notes from the Mara. *ibidem* 1974: 78-79.
Mann, C.F. 1974d. Songs of two Kenya turdids. Auk 91: 166-167.
Mann, C.F. 1976. Some recent changes in our knowledge of bird distribution in East Africa. J.E.Afr.nat.Hist.Soc. & natn.Mus. 157: 1-24.
Mann, C.F. 1980. Notes on the avifaunas of the Kakamega and the Nandi Forests. Scopus 4: 97-99.
Mann, C.F. & P.L. Britton 1971a. The identification of Frigate Birds. EANHS Bull. 1971: 185-186.
Mann, C.F. & P.L. Britton 1971b. Identification of Frigate Birds - a postscript. *ibidem* 1971: 207-208.
Mann, C.F. & P.L. Britton 1972. Naturalised birds on the Kenya coast. *ibidem* 1972: 181-182.
Mann, C.F., P.J. Burton & I. Lennerstedt 1978. A re-appraisal of the systematic position of *Trichastoma poliothorax* (Timaliinae, Muscicapidae). Bull.Br. Orn.Club 98: 131-140.
Marsh, C., S. Nichools & D. Parker 1975. Birds of the Tana River Primate Game Reserve; species abundance and habitat preferences. Unpubl. MS. kept by the East African Natural History Society, Nairobi.
Marshall, A.J. & H.J.de S. Disney 1957. Experimental induction of the breeding

season in a xerophilous bird. Nature 180: 647-649.

Mathews, T. 1973. Boat trip through southern Kenya and Tanzania waters to Mafia Island. EANHS Bull. 1973: 18-20.

Mayr, E. 1969. Principles of Systematic Zoology. New York: McGraw Hill.

McCartney, R. & N. Rottcher 1983. Chat about Pembroke House. EANHS Bull. 1983: 72-73.

McKay, K.S.N. 1960. Cape Teal *Anas capensis* on Lake Rudolph. Ostrich 31: 106.

McLachlan, G.R. & R. Liversidge 1978. Roberts' Birds of South Africa. Cape Town: The John Voelcker Bird Book Fund.

McVicker, R.A.M. 1982. Exotic birds in Mombasa. EANHS Bull. 1982: 74.

Meadows, B.S. 1974. A June record of the Madagascar Pratincole on the Kenya coast. *ibidem* 1974: 102-103.

Meadows, B.S. 1976. Breeding of the Half-collared Kingfisher *Alcedo semitorquatus* in Kenya. *ibidem* 1976: 28-29.

Meadows, B.S. 1977a. The food of a Malachite Kingfisher *Alcedo cristata* holding a territory on a fishless river. Scopus 1: 24-25.

Meadows, B.S. 1977b. Exceptional numbers of Black-tailed Godwits *Limosa limosa* wintering at Lake Naivasha during 1976/77. *ibidem* 1: 49-50.

Meadows, B.S. 1978. On the wintering of the Wigeon *Anas penelope* in the Kenya highlands. *ibidem* 2: 97.

Meadows, B.S. 1979. Preliminary report on the January range and abundance of Palearctic ducks in East Africa. *ibidem* 3: 29-31.

Meadows, B.S. 1980. Mapping the distribution of Palearctic wildfowl in East Africa - 1980 progress report. *ibidem* 4: 89-90.

Meadows, B.S. 1984. Numbers and seasonality of filter-feeding ducks in Kenya. Proc.Fifth Pan-Afr.orn.Congr. 441-459.

Meadows, B.S., C.F. Mann & D. Cape 1975. Observations on the Shining-blue Kingfisher *Alcedo quadribrachys* on the Lugosida River, Kakamega Forest. EANHS Bull. 1975: 32-33.

Meadows, B.S. & D.J. Pearson 1981. White-eyed Pochards *Aythya nyroca* in central Kenya during early 1981. Scopus 5: 77-79.

Meadows, B.S. & J.K. Weston 1974. Icterine Warbler and Chiffchaff on Mount Kenya. EANHS Bull. 1974: 95.

Meadows, B.S., J.K. Weston & D. Weston 1974. Spotted Redshanks in Kenya during March 1974. *ibidem* 1974: 103.

Mearns, E.A. 1914. Descriptions of three new African weaver-birds of the genera *Estrilda* and *Granatina*. J.E.Afr. & Uganda nat.Hist.Soc. 8: 134-138.

Meeth, P. 1977. Herring Gulls *Larus argentatus* in Kenya. Ardea 65: 88.

Meinertzhagen, R. 1922. Notes on some birds from the Near East and from Tropical East Africa. Ibis Eleventh Ser.(4) 1: 1-74.

Meinertzhagen, R. 1930. Nicoll's birds of Egypt. London: Hugh Rees.

Meinertzhagen, R. 1937. Some notes on the birds of Kenya Colony, with especial reference to Mount Kenya. Ibis Fourteenth Ser. 1: 731-760.

Meinertzhagen, R. 1944. The breeding of the Bearded Vulture in Kenya Colony and the movements of the European Swallow and Swift. *ibidem* 86: 549.

Meinertzhagen, R. 1950. A visit to Magadi, in Kenya, in March 1949. *ibidem* 92: 148-149.

Meinertzhagen, R. 1958. Greater Flamingoes in Kenya. *ibidem* 100: 624.

Meneghetti, F. 1943. Notes on the nest of *Tchitrea viridis viridis*. J.E.Afr.nat. Hist.Soc. 3 & 4 (77 & 78): 285-286.

Meneghetti, F. 1944. Further notes on the nesting of Paradise Flycatchers and on parasitism by Klaas' Cuckoc. *ibidem* 1 & 2 (81 & 82): 96.

Merz, A. 1980. Marabou Stork kills Lesser Flamingo. EANHS Bull. 1980: 97-98.

Miskell, J.E. 1977. Cooperative feeding of young at the nest by Fischer's Starling *Spreo fischeri*. Scopus 1: 87-88.

Miskell, J.E. (n.d.). Kenya records from von Erlanger's bird collection. Unpubl. MS.

Miskell, J.E. & J.S. Ash 1985. Gillett's Lark *Mirafra gilletti* new to Kenya. Scopus 9: 53-54.

Miskell, J.E. & J.D. Gerhart 1977. N.F.D., Garissa-Madogashe-Habaswein-Isiolo, 24-27 Sept. 1977. Unpubl. MS. kept by the East African Natural History Society, Nairobi.

Modha, M.L. & M.J. Coe 1969. Notes on the breeding of the African Skimmer *Rhynchops flavirostris* on Central Island, Lake Rudolph. Ibis 111: 593-598.

Moore, R.D. 1979. Spike-heeled Lark *Chersomanes albofasciata* at Amboseli: a new bird for Kenya. Scopus 3: 103.

Moore, R.D. 1981. Long-tailed Skuas *Stercorarius longicaudus* at the Kenya coast. *ibidem* 5: 79.

Moore, R.D. 1982. Where to watch birds in Kenya. Nairobi: Transafrica Press.

Moore, R.D. 1983. The first record of the northeastern race of the Northern Red Bishop *Euplectes franciscanus* for Kenya. Scopus 7: 23.

Moreau, R.E. 1935. The breeding biology

of certain East African hornbills (Bucerotidae). J.E.Afr. & Uganda nat.Hist. Soc. 1 & 2 (57 & 58): 1-28.

Moreau, R.E. 1938. Bird-migration over the northwestern part of the Indian Ocean, the Red Sea, and the Mediterranean. Proc. zool.Soc.Lond. A108 (1): 1-26.

Moreau, R.E. 1944. Mount Kenya: a contribution to the biology and bibliography. J.E.Afr.nat.Hist.Soc. 1 & 2 (81 & 82): 61-92.

Moreau, R.E. 1958. Some aspects of the Musophagidae. Ibis 100: 67-112, 238-270.

Moreau, R.E. 1966. The bird faunas of Africa and its islands. London: Academic Press.

Moreau, R.E. 1969. The Sooty Falcon *Falco concolor* Temminck. Bull.Br.Orn.Club 89: 62-67.

Moreau, R.E. 1972. The Palearctic-African bird migration systems. London: Academic Press.

Morrison, A.F. 1945. Notes on the occurrence of migrant wheatears at Nairobi. J.E.Afr.nat.Hist.Soc. 3 & 4 (83 & 84): 116-119.

Morrison, A.F. 1946. Nest of the Grey-capped Swamp Warbler. *ibidem* 1 & 2 (85 & 86): 71-72.

Mulder, J. & J.de Ridder 1985. Kelp Gull in Kenya in January 1984. Dutch Birding 7: 138-139.

Muringo, C. 1980. Observation of a pair of Pygmy Falcon at their nest. EANHS Bull. 1980: 51.

Myall, E. & R. Myall 1979. Weekend excursion to Salt Lick Lodge. *ibidem* 1979: 93-95.

Ng'weno, F. 1971. Frigate Birds. *ibidem* 1971: 168.

Ng'weno, F. 1972. Letter to the Editor. *ibidem* 1972: 206.

Ng'weno, F. 1973. Honeyguide host. *ibidem* 1973: 53.

Ng'weno, F. 1975. Bird notes. *ibidem* 1975: 17.

Ng'weno, F. 1980. Some observations on birds in the Jadini Forest. *ibidem* 1980: 81-82.

Nickalls, E.M. 1983. Notes on Fiscal Shrike and Klaas' Cuckoo. *ibidem* 1983: 74.

Nikolaus, G. 1979. Notes on some birds new to the south Sudan. Scopus 3: 68-73.

Nikolaus, G. 1980. An experiment to attract migrating birds with car headlights in the Chyulu Hills, Kenya. *ibidem* 4: 45-46.

Nikolaus, G. 1982. Further notes on some birds new to south Sudan. *ibidem* 6: 1-4.

Nikolaus, G. 1984. Further notes on birds new or little known in the Sudan. *ibidem* 8: 38-42.

Nikolaus, G. 1987. Distribution atlas of Sudan's birds with notes on habitat and status. Bonner zool.monogrn. 25: 1-322.

Nikolaus, G. & D.M. Hamed 1984. Distinct status changes of certain Palearctic migrants in the Sudan. Scopus 8: 36-38.

Norris, C.E. 1970. Selengai Camp, 21/22 June 1970. EANHS Newsl.: 22.

Norris, C.E. 1971a. Bird notes from Meru National Park. EANHS Bull. 1971: 38.

Norris, C.E. 1971b. Movement of migrants. *ibidem* 1971: 123-124.

Norris, C.E. 1971c. Breeding at Lake Rudolph. *ibidem* 1971: 130-131.

Norris, C.E. 1971d. Ring-necked Parrakeet *Psittacula krameri*. *ibidem* 1971: 200.

Norris, C.E. 1972. Martial Eagle killing a young Grant's Gazelle. *ibidem* 1972: 129-130.

Norris, C.E. 1974. A Bittern in Nairobi National Park. *ibidem* 1974: 163-164.

North, M.E.W. 1936. Breeding habits of the Crested Wattled Plover (*Sarciphorus tectus latifrons*). J.E.Afr. & Uganda nat.Hist.Soc. 3 & 4 (59 & 60): 132-145.

North, M.E.W. 1939. Field notes on certain raptorials and water-birds in Kenya Colony. Ibis Fourteenth Ser. 3: 487-507, 617-643.

North, M.E.W. 1940. Field notes on Abdim's Stork in two Kenya Provinces. J.E.Afr. & Uganda nat.Hist.Soc. 1 & 2 (66 & 67): 1-5.

North, M.E.W. 1942a. A field guide to the scavenging birds of Kenya. *ibidem* 2 & 3 (71 & 72): 61-88.

North, M.E.W. 1942b. The nesting of some Kenya Colony hornbills. Ibis Fourteenth Ser. 6: 499-508.

North, M.E.W. 1943. The breeding of the Marabou Stork in East Africa. *ibidem* 85: 190-198.

North, M.E.W. 1944. Some East African birds of prey. *ibidem* 86: 117-138.

North, M.E.W. 1947. Breeding of the Egyptian Vulture in Kenya Colony. *ibidem* 89: 662-663.

North, M.E.W. 1948. The Lammergeyer in Kenya Colony. *ibidem* 90: 138-141.

North, M.E.W. 1951. Opportunism in the use of burnt ground for breeding. *ibidem* 93: 466-467.

North, M.E.W. 1959a. Ring-necked Doves drinking at Wajir. J.E.Afr.nat.Hist. Soc. 3 (100): 142-143.

North, M.E.W. 1959b. The great heronry of Garsen on the river Tana. *ibidem* 4 (101): 159-160.

North, M.E.W. 1959c. A lucky escape for a swift. *ibidem* 4 (101): 176.

North, M.E.W. 1963. Breeding of the Black-headed Herons at Nairobi, Kenya, 1958-62. J.E.Afr.nat.Hist.Soc. & Coryndon Mus. 2 (106): 33-63.

North, M.E.W. 1966. A Reef Heron at Lake Nakuru, Kenya. J.E.Afr.nat.Hist.Soc. & natn.Mus. 3 (112): 231-232.

North, M.E.W. (n.d.). Waterfowl records, Ziwa Estates, Hoey's Bridge, Kenya. Unpubl. MS. kept by the East African Natural History Society, Nairobi.

North, M.E.W. & D.S. McChesney 1964. More voices of African birds. (Long-playing record). Cornell University, Laboratory of Ornithology.

Oberholser, H.C. 1945. Birds of the White-Fuller Expedition to Kenya, East Africa. Scient.Publs. Cleveland Mus.nat.Hist. 4 (3): 43-122.

Ogilvie-Grant, W.R. 1901. On the birds collected by Mr. J.J. Harrison between Zeila and Lakes Rudolph and Baringo, Eastern Africa. Ibis Eighth Ser. (1) 2: 278-299.

Oliver, P.J., D.L. Davenport, A. Gibbs & C.E. Wheeler 1971. Chiffchaffs in Kenya. EANHS Bull. 1971: 78.

Olson, D.K., J. Chism & D.P. Harding (n.d.). A checklist of the Birds of A.D.C. Mutara Ranch, Laikipia District, Kenya. Unpubl. MS.

OS-C 1977. Some recent records of Palearctic birds in Kenya and Tanzania. Scopus 1: 39-43, 78-81.

OS-C 1979. Distribution mapping scheme for East Africa. *ibidem* 3:32.

OS-C 1981. Check-list of the Birds of Kenya. Nairobi: Ornithological Sub-Committee of the East African Natural History Society.

Osborne, R.M. & A.M. Alio 1984. A bird-watching safari in North-Eastern Province. Parts 1 & 2. EANHS Bull. 1984: 65-71, 88-96.

Oulton, J.T. 1938. A note on the Crowned Eagle (*Spizaetus coronatus*). J.E.Afr. & Uganda nat.Hist.Soc. 5 (61): 189.

Owre, O.T. 1966. The Crowned Crane at Lake Rudolph. Bull.Br.Orn.Club 86: 54-56.

Owre, O.T. 1967. The Reef Heron, *Egretta schistacea* Ehrenb. in interior East Africa. J.E.Afr.nat.Hist.Soc. & natn. Mus. 2 (114): 61-63.

Owre, O.T. & D.R. Paulson 1968. Records of Falconiformes from the Lake Rudolph area, Kenya. Bull.Br.Orn.Club 88: 151-152.

Parker, I.S.C. 1982. Two observations of nesting in the eastern race of the Green Ibis *Bostrychia olivacea akeleyorum*. Scopus 6: 20.

Parker, I.S.C. 1984. Shoebill *Balaeniceps rex*: a deletion from the Kenya avifauna. *ibidem* 8: 79.

Parsons, B.T. 1959. Some Makueni birds. J. E.Afr.nat.Hist.Soc. 4 (101): 164-166.

Parsons, B.T. 1969. A record of *Oceanodroma leucorhoa* from Kenya. Bull.Br. Orn.Club 89: 120-121.

Parsons, J. 1976. Counts of Ciconiiformes nesting in three colonies near Kisumu - 27 April 1976. EANHS Bull. 1976: 61-62.

Parsons, J. 1977. The effect of predation by Fish Eagles on the breeding success of various Ciconiiformes nesting near Kisumu, Kenya. J.nat.Hist. 11: 337-353.

Payne, R.B. 1967. *Lybius guifsobalito* occurs in Kenya. Ostrich 38: 283.

Payne, R.B. 1969. Nest parasitism and display of Chestnut Sparrows in a colony of Grey-capped Social Weavers. Ibis 111: 300-307.

Pearson, D.J. 1972. Black-headed Gulls. EANHS Bull. 1972: 67-68.

Pearson, D.J. 1973. On the wintering of the Sprosser *Luscinia luscinia* and the Marsh Warbler *Acrocephalus palustris* in Kenya. *ibidem* 1973: 140.

Pearson, D.J. 1974. The timing of wing moult in some Palearctic waders wintering in East Africa. Wader Stud.Gr. Bull. 12: 6-12.

Pearson, D.J. 1977a. Slender-billed Gulls *Larus genei* and other Palearctic gulls at Lake Turkana, December 1976. Scopus 1: 45-47.

Pearson, D.J. 1977b. The first year moult of the Common Sandpiper *Tringa hypoleucos* in Kenya. *ibidem* 1: 89-94.

Pearson, D.J. 1978. The genus *Sylvia* in Kenya and Uganda. *ibidem* 2: 63-71.

Pearson, D.J. 1979a. The races of Red-tailed Shrike *Lanius isabellinus* occurring in East Africa. *ibidem* 3: 74-78.

Pearson, D.J. 1979b. A bird distributional mapping scheme for East Africa. EANHS Bull. 1979: 71-72.

Pearson, D.J. 1980. Northward spring passage of Palearctic passerines across Tsavo. Scopus 4: 25-28.

Pearson, D.J. 1981a. The wintering and moult of Ruffs *Philomachus pugnax* in the Kenyan rift valley. Ibis 123: 158-182.

Pearson, D.J. 1981b. Red-footed Falcon *Falco vespertinus* in Tsavo. Scopus 5:57.

Pearson, D.J. 1981c. The identity of two *Ficedula* flycatchers recently collected in Kenya. *ibidem* 5: 59-60.

Pearson, D.J. 1981d. Spring falls of Palearctic passerines at Mtito Andei, Kenya. *ibidem* 5: 80-81.

Pearson, D.J. 1982. The migration and wintering of Palearctic *Acrocephalus* warblers in Kenya and Uganda. *ibidem* 6: 49-59.

Pearson, D.J. 1983a. A Blacksmith/Spur-winged Plover *Vanellus armatus* x *V. spinosus* hybrid? *ibidem* 7: 93-94.

Pearson, D.J. 1983b. Requests for records of Palearctic migrants. *ibidem* 7: 102-103.

Pearson, D.J. 1984a. The moult of the Little Stint *Calidris minuta* in the Kenyan rift valley. Ibis 126: 1-15.

Pearson, D.J. 1984b. The Nightingale *Luscinia megarhynchos*, the Sprosser *Luscinia luscinia* and the Irania *Irania gutturalis* in Kenya. Scopus 8: 18-23.

Pearson, D.J. 1984c. A Pintail Snipe *Gallinago stenura* at Naivasha. *ibidem* 8: 45-46.

Pearson, D.J. 1984d. Some counts of wintering waders on the south Kenya coast. *ibidem* 8: 93-95.

Pearson, D.J. & G.C. Backhurst 1973. The head plumage of Eastern Yellow-headed Yellow Wagtails wintering in Nairobi, Kenya. Ibis 115: 589-591.

Pearson, D.J. & G.C. Backhurst 1976a. Palearctic passerine migration at Kariobangi, Nairobi. EANHS Bull. 1976: 23-28.

Pearson, D.J. & G.C. Backhurst 1976b. The southward migration of Palearctic birds over Ngulia, Kenya. Ibis 118: 78-105.

Pearson, D.J. & G.C. Backhurst 1978. Southward migration at Ngulia, Tsavo, Kenya 1977/1978. Scopus 2: 42-47.

Pearson, D.J. & G.C. Backhurst 1979. The white-throated form of the Pied Wheatear *Oenanthe pleschanka* in Kenya. *ibidem* 3: 79-80.

Pearson, D.J. & G.C. Backhurst 1983. Moult in the River Warbler *Locustella fluviatilis*. Ringing and Migration 4: 227-230.

Pearson, D.J., G.C. Backhurst & D.E.G. Backhurst 1979. Spring weights and passage of Sedge Warblers *Acrocephalus schoenobaenus* in central Kenya. Ibis 121: 8-19.

Pearson, D.J., H.A. Britton & P.L. Britton 1978. Substantial wintering populations of the Basra Reed Warbler *Acrocephalus griseldis* in eastern Kenya. Scopus 2: 33-35.

Pearson, D.J. & P.L. Britton 1980. Arrival and departure of Palearctic waders on the Kenya coast. *ibidem* 4: 84-88.

Pearson, D.J. & M.A.C. Coverdale 1984. Wedge-tailed Shearwaters *Puffinus pacificus* off Ras Ngomeni, Kenya. *ibidem* 8: 24.

Pearson, D.J. & G.J. Jobson 1979. River Warblers in song in Kenya. *ibidem* 3: 59-60.

Pearson, D.J. & A.D. Lewis 1980. Occurrence of a Black Tern *Childonias niger* at Dandora, Kenya. *ibidem* 4: 95-96.

Pearson, D.J., A.D. Lewis & D.A. Turner 1980. The Gambaga Flycatcher *Muscicapa gambagae* in northern Kenya. *ibidem* 4:96.

Pearson, D.J. & B.S. Meadows 1979. Lesser Spotted Eagles *Aquila pomarina* in Kenya during 1978/79, with comments on the identification of the species. *ibidem* 3: 48-53.

Pearson, D.J., J.H. Phillips & G.C. Backhurst 1970. Weights of some Palearctic waders wintering in Kenya. Ibis 112: 199-208.

Pearson, D.J. & T. Stevenson 1980. A survey of wintering Palearctic waders in the southern part of the Kenya rift valley. Scopus 4: 59-63.

Peirce, M.A. & G.C. Backhurst 1970. Observations on the haematozoa found in birds from the Northern Frontier District of Kenya. E.Afr.Wildl.Jl. 8: 208-212.

Peirce, M.A., G.C. Backhurst & D.E.G. Backhurst 1977. Haematozoa of East African birds. III. Three years' observations on the blood parasites of birds from Ngulia. *ibidem* 15: 71-79.

Peirce, M.A. & J.E. Cooper 1977. Haematozoa of East African birds. V. Blood parasites of birds of prey. *ibidem* 15: 213-216.

Pelchen, H. 1977a. A definite sighting of the Taita Falcon. EANHS Bull. 1977: 105-106.

Pelchen, H. 1977b. A Taita Falcon *Falco fasciinucha* in Tsavo. Scopus 1: 83-84.

Pelchen, H. 1977c. Some birds observed during a visit to Mt. Sabachi. Unpubl. MS. kept by the East African Natural History Society, Nairobi.

Pelchen, H. 1977d. Some birds observed during a visit to Mt. Nyiru. *ibidem*.

Pelchen, H. 1977e. Birds observed on and at Lake Bogoria. *ibidem*.

Pennycuick, C.J. 1972. Soaring behaviour and performance of some East African birds, observed from a motor-glider. Ibis 114: 178-218.

Pennycuick, C.J. & G.A. Bartholomew 1973. Energy budget of the lesser flamingo (*Phoeniconaias minor* Geoffroy). E.Afr. Wildl.Jl. 11: 199-207.

Percival, A.B. 1911. European migrants in British East Africa. J.E.Afr. & Uganda nat.Hist.Soc. 2: 101-106.

Percival, A.B. 1914a. Anderson's Peon. *ibidem* 8: 151.

Percival, A.B. 1914b. Distribution of game birds. *ibidem* 8: 151-152.

Percy, Lord R.C., H.E. Percy & M.W. Ridley 1953. The water-holes at Ijara, Northern Province, Kenya. J.E.Afr.nat.Hist.Soc. 1 (93): 2-14.

Pitman, C.R.S. 1956. Remarks on the nidification of the Kenya Hill Chat *Pinarochroa sordida ernesti* Sharpe. Bull.Br. Orn.Club 76: 107.

Pitman, C.R.S. 1965a. The nest and eggs of the Striped Crake *Porzana marginalis* Hartlaub. *ibidem* 85: 32-40.

Pitman, C.R.S. 1965b. The nesting, eggs and young of the Saddle-bill Stork *Ephi-*

ppiorhynchus senegalensis (Shaw). *ibidem* 85: 70-80.

Plumb, W.J. 1978a. The Lesser Golden Plover in Kenya. Scopus 2: 72-73.

Plumb, W.J. 1978b. Thika as a locality in the overlap zone between the two races of the Black-faced Waxbill *Estrilda erythronotus*. EANHS Bull. 1978: 85-86.

Plumb, W.J. 1978c. Co-operative feeding of young by adult and sub-adult Red-knobbed Coots. *ibidem* 1978: 88.

Plumb, W.J. 1978d. Co-operative feeding of young by White-crowned Shrikes. *ibidem* 1978: 89.

Plumb, W.J. 1979. Observations on Hinde's Babbler *Turdoides hindei*. Scopus 3: 61-67.

Pollard, J.R.T. 1944. Photographs of the African Lammergier, *Gypaetus barbatus meridionalis*. J.E.Afr.nat.Hist.Soc. 1 & 2 (81 & 82): 94.

Pollard, J.R.T. 1946a. Nest and eggs of the Kenya Hill-Chat *Pinarochroa sordida ernesti* Sharpe. Ibis 88: 520-521.

Pollard, J.R.T. 1946b. Nesting of the African Hoopoe, *Upupa africana* Bechstein, in Kenya Colony. *ibidem* 88: 521.

Pollard, J.R.T. 1946c. Note on a nest of Verreaux's Eagle in Kenya Colony. *ibidem* 88: 521-522.

Pomeroy, D.E. 1975. Letter to the Editor. EANHS Bull. 1975: 135.

Pomeroy, D.E. 1977. Marabous associated with vulture parties in East Africa. Scopus 1: 103-106.

Pomeroy, D.E. 1978a. Seasonality of Marabou Storks *Leptoptilos crumeniferus* in eastern Africa. Ibis 120: 313-321.

Pomeroy, D.E. 1978b. Counts of Marabou Storks *Leptoptilos crumeniferus* in relation to their movements in eastern Africa. Scopus 2: 92-96.

Pomeroy, D.E. 1979. Nesting of Marabou Storks in the Makindu-Kiboko-Kibwezi area. EANHS Bull. 1979: 114-116.

Pomeroy, D.E. 1981. Composition of Kenya's avifauna. *ibidem* 1981: 30-31.

Pomeroy, D.E. 1982. Lack of seasonality in song by some birds of the Nairobi area. Scopus 6: 105-107.

Pomeroy, D.E. 1986. The Marabou in Kenya. *ibidem* 10: 1-9.

Pomeroy, D.E. in press, a. A preliminary report on the avifauna of south-west Marsabit District, northern Kenya. UNESCO/IPAL Tech.Rep.

Pomeroy, D.E. in press, b. Land bird populations in East Africa. Afr.J.Ecol.

Pomeroy, D.E. in press, c. Using East African atlas data for ecological studies. Ann.Zool.Fennica.

Pomeroy, D.E. & A.D. Lewis 1983. Estimating the number of species likely to occur in Kenya bird atlas squares. Scopus 7: 40-43.

Pomeroy, D.E. & A.D. Lewis 1987. Bird species richness in tropical Africa: some comparisons. Biol.Conserv. 70: 11-28.

Pomeroy, D.E., A.D. Lewis & A. Mutere 1982. Use of the 'Bird Atlas for Kenya' data for ecological studies. EANHS Bull. 1982: 105-109.

Pomeroy, D.E. & C. Muringo 1984. Studies of birds in a semi-arid area of Kenya. I. The use of transects to assess the effects of bush-clearance on bird populations. Proc.Fifth Pan-Afr.orn.Congr. 179-199.

Pomeroy, D.E. & B. Tengecho 1980. Probable breeding of Trumpeter Hornbills in Tsavo West National Park. EANHS Bull. 1980: 91.

Pomeroy, D.E. & B. Tengecho 1982a. The importance of dams. A study of birds of the Thika Road area, central Kenya. *ibidem* 1982: 77-85.

Pomeroy, D.E. & B. Tengecho 1982b. Studies of birds in a semi-arid area of Kenya. II. Bird parties in two woodland areas. Scopus 6: 25-32.

Pomeroy, D.E. & B. Tengecho 1986. Studies of birds in a semi-arid area of Kenya. III. The use of 'Timed Species Counts' for studying regional avifauna. J.trop. Ecol. 2: 231-247.

Porter, R.F., I. Willis, S. Christensen & B.P. Nielsen 1976. Flight identification of European raptors. Berkhamsted: T. & A.D. Poyser.

Powiss, G. & T. Wacher 1982. Report on a walking and camel safari in Tsavo East National Park. EANHS Bull. 1982: 54-60.

Pratt, D.J. & M.D. Gwynne (eds.) 1977. Rangeland management and ecology in East Africa. London: Hodder & Stoughton.

Preston, I. 1976. Whale-headed Stork. EANHS Bull. 1976: 131-132.

Prigogine, A. 1978a. Note sur les petits indicateurs de la forêt de Kakamega. Gerfaut 68: 87-89.

Prigogine, A. 1978b. Le statut du Loriot de Percival, *Oriolus percivali*, et son hybridation avec *Oriolus larvatus* dans l'est africain. *ibidem* 68: 253-320.

Rader, F. 1974. Field trip to Lake Hannington 16-18th August 1974. EANHS Bull. 1974: 138-141.

Rand, A.L. 1958. Notes on African Bulbuls. Fieldiana, Zool. 35 (6): 145-220.

Rathbun, G.B. 1978. The African Pitta at Gedi Ruins, Kenya. Scopus 2: 7-10.

Reichenow, A. 1900-1903. Die Vogel Afrikas. Neudamm: J. Neumann.

Reid, C. & C. Ryall 1985. Notes from south-east Kenya. EANHS Bull. 1985: 57-58.

Reid, G.D.F. 1984. The Kenya Bird Atlas: the results. *ibidem* 1984: 117-118.
Reid, G.D.F. 1985. Excursion to Bushwhackers' Camp, Kibwezi. *ibidem* 1985: 39.
Reyer, H.-U. 1980. Flexible helper structure as an ecological adaptation in the Pied Kingfisher (*Ceryle rudis rudis* L.). Behav.Ecol.Sociobiol. 6: 219-227.
Reynolds, F.L. 1960a. Bird ringing in East Africa. J.E.Afr.nat.Hist.Soc. 5 (102): 198-199.
Reynolds, F.L. 1960b. Notes on attempts to trap waders and duck for ringing, East Africa 1959/60. *ibidem* 7 (104): 301-303.
Reynolds, F.L. 1960c. Cinnamon-chested Bee-eater (*Melittophagus oreobates* Sharpe). *ibidem* 7 (104): 306.
Reynolds, J.F. 1972a. Somali Ostrich in Nairobi National Park. EANHS Bull. 1972: 182-183.
Reynolds, J.F. 1972b. Feeding association between Marsh Sandpiper and Hottentot Teal. *ibidem* 1972: 189.
Reynolds, J.F. 1972c. On the nesting habits of White-throated Bee-eaters *Merops albicollis*. *ibidem* 1972: 116-120.
Reynolds, J.F. 1973a. Some observations on Bustards in Nairobi National Park. *ibidem* 1973: 98-99.
Reynolds, J.F. 1973b. Some bird records from Meru and Nairobi National Parks. *ibidem* 1973: 112-113.
Reynolds, J.F. 1974. Palearctic birds in East Africa. Br.Birds 67: 70-76.
Reynolds, J.F. 1975a. Migrants near Garissa. EANHS Bull. 1975: 88-89.
Reynolds, J.F. 1975b. Nest building of Striped Swallows *Hirundo abyssinica*. *ibidem* 1975: 90.
Reynolds, J.F. 1975c. Redshanks at Lake Magadi. *ibidem* 1975: 111.
Reynolds, J.F. 1977. Pratincole *Glareola pratincola* catching flies at ground level. Scopus 1: 23.
Reynolds, J.F. 1980. Sympatry of Blacksmith and Spur-winged Plovers *Vanellus armatus* and *spinosus* at Amboseli. *ibidem* 4: 43.
Ridley, M.W. 1953. Grey Phalarope in Kenya. J.E.Afr.nat.Hist.Soc. 1 (93): 17.
Ridley, M.W., B.L. Moss & Lord R.C. Percy 1955. The food of flamingoes in Kenya Colony. *ibidem* 5 (97): 147-158.
Ridley, M.W. & Lord R.C. Percy 1953. Notes on the birds of Lake Elmentaita, Kenya Colony. Proc.Univ.Durham phil. Soc. 12 (8): 103-118.
Ripley, S.D. 1966. A notable owlet from Kenya. Ibis 108: 136-137.
Ripley, S.D. & G.M. Bond 1971. Systematic notes on a collection of birds from Kenya. Smithsonian Contr.Zool. 111: 1-21.
Roberts, E.J. 1985. A remarkable journey: bicycle safari in North-Eastern Province. EANHS Bull. 1985: 16-19.
Robertson, P., J. Hopson & A.J. Hopson 1974. Skuas at Lake Rudolph. *ibidem* 1974: 31-32.
Rolfe, J.G. & D.J. Pearson 1973. Some recent records of Palearctic migrants from eastern Uganda. *ibidem* 1973: 62-66.
Rudnai, J. 1982. Easter weekend camp at Maralal. *ibidem* 1982: 84-86.
Ryves, V.W. 1959. An Emerald Cuckoo's foster parents. J.E.Afr.nat.Hist.Soc. 4 (101): 175.
Schluter, D. & S. Harcourt 1984. Sighting of Pel's Fishing Owl at Bushwhackers'. EANHS Bull. 1984: 26
Schmidl, D. (n.d.). Contributional notes to the distribution of birds in East Africa. Unpubl. MS. kept by the East African Natural History Society, Nairobi.
Schüz, E. 1973. Some remarks on the Lammergeyer *Gypaetus barbatus* in East Africa. Bull.Br.Orn.Club 93: 161-163.
Schwab, L.T. 1979. Some nesting data on the Nyanza Swift *Apus niansae*. Scopus 3: 26-27.
Serle, W. 1943. Notes on East African birds. Ibis 85: 55-82.
Sessions, P.H.B. 1966. Notes on the birds of Lengetia Farm, Mau Narok. J.E.Afr. nat.Hist.Soc. & natn.Mus. 26: 18-48. (see Bibliography a).
Sessions, P.H.B. 1972. Observations on Mackinder's Eagle Owl *Bubo capensis mackinderi* Sharpe. *ibidem* 138: 1-20.
Sessions, P.H.B. 1974a. Letter to the Editor. EANHS Bull. 1974: 133-134.
Sessions, P.H.B. 1974b. Letter to the Editor. *ibidem* 1974: 165-166.
Sessions, P.H.B. 1975. Uncommon migrants on Mau Narok. *ibidem* 1975: 40-41.
Sessions, P.H.B. (n.d.). Notes of birds observed on Lengetia Farm, Mau Narok, mostly between August 1960 and August 1961; with special reference to status, numbers and voice. Unpubl. MS. kept by the East African Natural History Society, Nairobi.
Sharpe, R.B. 1930. The birds of Marsabit Mountain, Kenya Colony. Bateleur 2: 44-47, 106.
Sharrock, J.T.R. 1976. The atlas of breeding birds of Britain and Ireland. Calton, England: T. & A.D. Poyser.
Sheppard, D.M. 1954. Bird notes from Molo - 1. The dam. J.E.Afr.nat.Hist. Soc. 3 (95): 80-81.
Sheppard, D.M. 1958. Bird notes from Molo - 2. The garden. *ibidem* 2 (99): 6-8.
Short, L.L. & J.F.M. Horne 1980. Vocal and other behaviour of the Green Barbet

in Kenya. Ostrich 51: 219-229.

Short, L.L. & J.F.M. Horne 1982a. Vocal and other behaviour of Kenyan Black-collared Barbets *Lybius torquatus*. Ibis 124: 27-43.

Short, L.L. & J.F.M. Horne 1982b. 'White-tailed' White-headed Barbet in central western Kenya. Scopus 6: 40-41.

Short, L.L. & J.F.M. Horne 1983. The relationship of male Lesser Honeyguides *Indicator minor* with duetting barbet pairs. Bull.Br.Orn.Club 103: 25-32.

Short, L.L. & J.F.M. Horne 1984. Aspects of duetting in some ground barbets. Proc.Fifth Pan-Afr.orn.Congr. 729-744.

Short, L.L. & J.F.M. Horne 1985a. Notes on some birds of the Arabuko-Sokoke Forest. Scopus 9: 117-126.

Short, L.L. & J.F.M. Horne 1985b. Notes on some birds of Ol Ari Nyiro, Laikipia Plateau. *ibidem* 9: 137-140.

Short, L.L., J.F.M. Horne & J-P vande Weghe 1983. Aberrantly plumaged barbets. *ibidem* 7: 10-14.

Siemens, L.A. 1981. Crowned Hornbills nesting near Eldoret. EANHS Bull. 1981: 65-66.

Siemens, L.A. 1983. Field notes from AUEA Biology Department. *ibidem* 1983: 71.

Siemens, L.A. 1984. Field notes from AUEA Biology Department. *ibidem* 1984: 27-28.

Siemens, L.A. 1985. Notes from north-west Kenya, Nakuru and Masai Mara. *ibidem* 1985: 58-59.

Simpson, V.R. 1971a. Sighting of a Piapiac in Kenya. *ibidem* 1971: 99.

Simpson, V.R. 1971b. An interesting morning spent at Ahero. *ibidem* 1971: 127-129.

Simpson, V.R. 1971c. Letter to the Editor. *ibidem* 1971: 203-204.

Sinclair, J.C. 1981. Pintado Petrel *Daption capense* in Kenya. Scopus 5: 77.

Skinner, N.J. 1971. A nesting record of the Crested Lark in Kenya. EANHS Bull. 1971: 98.

Skinner, N.J. 1972. Birds on the lava fields of Teleki's Volcano. *ibidem* 1972: 164-166.

Skinner, N.J. 1974. Checklist of the birds of the Kibwezi Forest Reserve. Unpubl. MS. kept by the East African Natural History Society, Nairobi.

Skinner, N.J. 1975. Bird checklist; Professor Eric Lucas' farm, Ol Kalou. *ibidem*.

Skinner, N.J. 1977. Checklists of birds in Nairobi National Park. EANHS Bull. 1977: 38-39.

Skinner, N.J. 1978. Notes on breeding of the Red-chested Cuckoo in Nairobi. Scopus 2: 59-62.

Smalley, M.E. 1983. A major Lesser Kestrel *Falco naumanni* roost in Kenya. *ibidem* 7: 44-47.

Smart, J.B. 1949. Records of migratory birds. Nature E.Afr. 2: 5.

Smart, J. 1952. A brief note on some birds seen around Malindi 9-18/12/50. J.E.Afr.nat.Hist.Soc. 1 (91): 52-54.

Smart, J.B. 1965. A breeding record for the Sooty Tern in Kenya. J.E.Afr.nat. Hist.Soc. & natn.Mus. 1 (110): 69-70.

Smart, J.B. 1967. Report on birdringing for 1961-1966. *ibidem* 2 (114): 47-51.

Smart, J.B. & G.C. Backhurst 1967. A list of the birds recorded on the EANHS Easter Camp at Maralal and Lake Kisima 24-27 March 1967. Unpubl. MS. kept by the East African Natural History Society, Nairobi.

Smart, J.B. & A.D. Forbes-Watson 1971. Occurrence of the Asiatic Dowitcher in Kenya. EANHS Bull. 1971: 74-75.

Smeenk, C. 1974. Comparative-ecological studies of some East African birds of prey. Ardea 62: 1-97.

Smeenk, C. & N. Smeenk-Enserink 1975. Observations on the Pale Chanting Goshawk *Melierax poliopterus*, with comparative notes on the Gabar Goshawk *Micronisus gabar*. *ibidem* 63: 93-115.

Snow, D.W. (ed.) 1978. An atlas of speciation in African non-passerine birds. London: British Museum (Natural History). (see Bibliography a).

Spindlow, D.D. 1974. Two exciting Kenya bird spots. EANHS Bull. 1974: 9.

Spoerry, A. 1976. Notes on Carmine Bee-eaters seen nesting in Ileret on the 8th May 1974. *ibidem* 1976: 126-129.

Spratt, J. 1963. Countryman's Letter. Africana 1 (8): 26.

Squire, J.E. 1972. Bird list from Loiengalani, L. Rudolph. Unpubl. MS. kept by the East African Natural History Society, Nairobi.

Squire, J.E. 1976a. Some observations on the forests south of Mombasa. EANHS Bull. 1976: 76-77.

Squire, J.E. 1976b. Some observations during spring passage in Tsavo West. *ibidem* 1976: 108.

Squire, J.E. 1976c. List from Tana River below Garissa 10-17/8/76. Unpubl. MS. kept by the East African Natural History Society, Nairobi.

Squire, J.E. 1976d. Tinderet Forest 25/6/76. *ibidem*.

Squire, J.E. 1976e. Tinderet Forest 11-12/12/76. *ibidem*.

Squire, J.E. 1976f. List from forest along the Kaptegat River, Kericho. *ibidem*.

Squire, J.E. 1977. White-crowned Starlings near Maikona. EANHS Bull. 1977: 3-4.
Squire, J.E. (n.d.). List from Saiwa Swamp. Unpubl. MS. kept by the East African Natural History Society, Nairobi.
Squire, J.E. (n.d.). Miscellaneous records. *ibidem*.
Squire, J.E. (n.d.). Bird list from Mt. Elgon Nat. Park and its immediate vicinity. *ibidem*.
Squire, J.E. (n.d.). Witu Forest 17/8/76. *ibidem*.
Squire, J.E. (n.d.). Additions to the lists in Williams' National Parks book. *ibidem*.
Stagg, S.J. 1972. Ground Hornbills in Nairobi. EANHS Bull. 1972: 62.
Stagg, S.J. 1973. Parasitic birds of Kenton College, Nairobi. *ibidem* 1973: 53-54.
Standish-King, J. 1970. Birds seen on Mt. Suswa, September 26th and 27th 1970. Unpubl. MS. kept by the East African Natural History Society, Nairobi.
Standish-King, J. 1972a. Black-headed Gulls? EANHS Bull. 1972: 87.
Standish-King, J. 1972b. Heronry at Njoro. *ibidem* 1972: 129.
Start, A.N. 1971. A study of mixed species bird flocks in a highland forest area in Kenya, East Africa. Unpubl. report, National Museum Nairobi, and Department of Zoology, University of Aberdeen, Scotland.
Start, J. & H. Start 1978. Nesting record of the Silvery-cheeked Hornbill. EANHS Bull. 1978: 125-127.
Steinbacher, J. 1977. Nachweise von paläarktischen Möwen und Enten als Wintergaste in Kenya. Vogelwarte 29: 65.
Stevenson, T. 1980. The birds of Lake Baringo. Nairobi: Sealpoint Publicity. (see Bibliography a).
Stevenson, T. 1982a. Further records of the Slender-billed Gull *Larus genei* at Lake Turkana. Scopus 6: 22-23.
Stevenson, T. 1982b. Large numbers of Palearctic ducks in the Omo Delta, Lake Turkana. *ibidem* 6: 72
Stevenson, T. 1983a. Kestrel *Falco tinnunculus* feeding on a bat. *ibidem* 7: 93.
Stevenson, T. 1983b. A second record of the Nubian Shrike *Lanius nubicus* in East Africa. *ibidem* 7: 97-98.
Stewart, D.R.M. & J. Stewart 1964. Anting behaviour in the Bronze Sunbird. J.E. Afr.nat.Hist.Soc. & Coryndon Mus. 5 (109): 92.
Stoneham, H.F. 1944a. Frigate Bird near Malindi. J.E.Afr.nat.Hist.Soc. 5 & 6 (79 & 80): 395.
Stoneham, H.F. 1944b. Late date for European Storks. *ibidem* 5 & 6 (79 & 80): 395.
Stoneham, H.F. 1944c. Steppe Eagle in Trans-Nzoia. *ibidem* 5 & 6 (79 & 80): 395.
Stoneham, H.F. 1944d. First Kenya record of Temminck's Stint. *ibidem* 5 & 6 (79 & 80): 396.
Stoneham, H.F. 1944e. Heuglin's Weaver. *ibidem* 5 & 6 (79 & 80): 396.
Stott, K. & J. Selsor (n.d.). Birds of the Bushwhackers-Kibwezi area. Unpubl. MS. kept by the East African Natural History Society, Nairobi.
Stuart, S.N. 1981. A comparison of the avifaunas of seven east African forest islands. Afr.J.Ecol. 19: 133-151.
Sugg, M.St.J. 1974. Mensural and moult data from a breeding colony of Pied Kingfishers. Ostrich 45: 227-234.
Sugg, M.St.J. & G. Sugg 1970. A weekend on the Loita Plains. EANHS Newsl: 16-18.
Sugg, M.St.J. & G. Sugg 1971. A nest record of the Snowy-headed Robin Chat. EANHS Bull. 1971: 163.
Sugg, M.St.J. & G. Sugg 1972. Notes on some birds recorded in South Nyanza, Kenya. *ibidem* 1972: 38-40.
Survey of Kenya 1970. National Atlas of Kenya. Nairobi: Survey of Kenya.
Sutton, J., A. Sutton, C.F. Dewhurst & L. M. Dewhurst 1984. A confirmed breeding record for the Swallow-tailed Kite in the rift valley of Kenya. EANHS Bull. 1984: 12.
Swank, W.G. 1977. Food of three upland game birds in Selengei Area, Kajiado District, Kenya. E.Afr.Wildl.Jl. 15: 99-105.
Swegen, H. 1984. Breeding at Kisumu heronry 1982. EANHS Bull. 1984: 112.
Taylor, P.B. 1978. Two records of Knot *Calidris canutus* from the Kenyan coast. Scopus 2: 97-98.
Taylor, P.B. 1981a. Red-footed Falcons *Falco vespertinus* in Tsavo West National Park. *ibidem* 5: 125-126.
Taylor, P.B. 1981b. First East African record of the Red-necked Stint *Calidris ruficollis*. *ibidem* 5: 126-128.
Taylor, P.B. 1982a. Storm-Petrels Hydrobatidae off the Kenya coast. *ibidem* 6: 13-16.
Taylor, P.B. 1982b. Pectoral Sandpiper *Calidris melanotos* at Mombasa. *ibidem* 6: 21-22.
Taylor, P.B. 1982c. Slender-billed Gull *Larus genei* at Mombasa. *ibidem* 6: 23-24.
Taylor, P.B. 1983. A melanistic Common Snipe *Gallinago gallinago* near Nairobi: an example of 'Sabine's Snipe'? *ibidem* 7: 50-51.
Taylor, P.B. 1984a. A roost of Lesser Kestrels *Falco naumanni* on the Athi Plains, Kenya. *ibidem* 8: 26.
Taylor, P.B. 1984b. Recent Kenya records

of Clarke's Weaver *Ploceus golandi* and an indication of its breeding season. *ibidem* 8: 28-29.

Tennent, J.R.M. 1961. Kitui District, June 1960 to April 1961. Unpubl. MS. kept by the East African Natural History Society, Nairobi.

Tennent, J.R.M. 1962. Notes on the migration and dispersal of birds at Kitui. J. E.Afr.nat.Hist.Soc. & Coryndon Mus. 1 (105): 41-42.

Tennent, J.R.M. 1964. The birds of Endau Mountain in the Kitui District of Kenya. Ibis 106: 1-6.

Tennent, J.R.M. 1965. Notes on the birds of Kakamega Forest. J.E.Afr.nat.Hist. Soc. & natn.Mus. 2 (111): 95-100.

Tennent, J.R.M. (n.d.). Dates of first and last viewings of migrant birds. Unpubl. MS. kept by the East African Natural History Society, Nairobi.

Thomsen, P. & P. Jacobsen 1979. The birds of Tunisia. Copenhagen: Thomsen.

Thomsett, S. 1981. Some observations on the Bat Hawk *Macheiramphus alcinus*. Scopus 5: 56.

Tomlinson, W. 1946. Bird notes chiefly from the Northern Frontier District of Kenya. J.E.Afr.nat.Hist.Soc. 1 & 2 (85 & 86): 1-17, and 5 (89): 225-250.

Toschi, A. 1946a. Coryndon Museum Expedition to the Mau Forest. Part III: Birds collected on the expedition. *ibidem* 3 & 4 (87 & 88): 81-100.

Toschi, A. 1946b. Notes on some little known nesting birds from Olorgasailie. *ibidem* 3 & 4 (87 & 88): 134-137.

Traylor, M.A. 1967a. A new species of *Cisticola*. Bull.Br.Orn.Club 87: 45-48.

Traylor, M.A. 1967b. *Cisticola aberdare* is a good species. *ibidem* 87: 137-141.

Traylor, M.A. 1970a. A new race of *Serinus citrinelloides*. *ibidem* 90: 83-86.

Traylor, M.A. 1970b. East African *Bradornis*. Ibis 112: 513-531.

Trevor, S. & P.C. Lack 1976. Great Spotted Cuckoo parasitising Superb Starling. EANHS Bull. 1976: 50.

Tuck, G.S. & H. Heinzel 1978. A field guide to the seabirds of Britain and the World. London: Collins.

Tuite, C.H. 1978. The lesser flamingo (*Phoeniconaias minor*, Geoffroy): aspects of its ecology and behaviour in the eastern Rift Valley of Kenya and northern Tanzania. Unpubl. Ph.D. thesis, University of Bristol.

Tuite, C.H. 1979. Population size, distribution and biomass density of the Lesser Flamingo in the Eastern Rift Valley, 1974-1976. J.Appl.Ecol. 16: 765-775.

Turner, D.A. 1961. Observations from Mombasa area. Unpubl. MS. kept by the East African Natural History Society, Nairobi.

Turner, D.A. 1973a. Birds observed in the Kitovu Forest, Taveta November 2-3, 1973. *ibidem*.

Turner, D.A. 1973b. Birds observed at Lake Jipe, Tsavo West. Nov. 3-4, 1973. *ibidem*.

Turner, D.A. 1973c. Birds observed in the Teita Hills forest, November 4-5, 1973. *ibidem*.

Turner, D.A. 1976. A record of the Abyssinian Long-eared Owl *Asio otus* from Mount Kenya. EANHS Bull. 1976: 8.

Turner, D.A. 1977. Status and distribution of East African endemic species. Scopus 1: 2-11.

Turner, D.A. 1978. The heronry at Lake Jipe. *ibidem* 2: 47-48.

Turner, D.A. 1979. Greenbuls of the Taita Hills, S.E. Kenya. *ibidem* 3: 27-28.

Turner, D.A. 1980. The Madagascar Squacco Heron *Ardeola idae* in East Africa, with notes on its field identification. *ibidem* 4: 42-43.

Turner, D.A. 1981. The East African distributional mapping scheme - a progress report. *ibidem* 5: 38-40.

Turner, D.A. 1982. The status and distribution of the Arabian Bustard *Otis arabs* in north-eastern Africa, and its possible occurrence in northern Kenya. *ibidem* 6: 20-21.

Turner, D.A. 1985. On the claimed occurrence of the Spike-heeled Lark *Chersomanes albofasciata* in Kenya. *ibidem* 9: 142.

Turner, D.A. & A.D. Forbes-Watson 1974. The Long-legged Buzzard *Buteo rufinus* in East Africa. EANHS Bull. 1974: 2-4.

Turner, D.A. & A.D. Forbes-Watson 1976a. Status of the White-crowned Starling *Spreo albicapillus* (Blyth) in Kenya. Bull.Br.Orn.Club 96: 58.

Turner, D.A. & A.D. Forbes-Watson 1976b. Additions to the Kenya avifauna. EANHS Bull. 1976: 4.

Turner, D.A. & J.D. Gerhart 1971. 'Foot-wetting' by incubating African Skimmers *Rynchops flavirostris*. Ibis 113: 244.

Turner, D.A., T. Stevenson & D.K. Richards 1984. Kenya breaks big-day birding marathon record. EANHS Bull. 1984: 59-60.

Turner, D.A. & D.A. Zimmerman 1979. Field identification of Kenya greenbuls. Scopus 3: 33-47.

Tyler, S.J. & S.J. Ormerod 1986. Interactions between resident and migratory wagtails *Motacilla* spp. in Ethiopia - an ecological conundrum. *ibidem* 10: 10-19.

Udvardy, M.D.F. 1981. An overview of grid-based atlas works in ornithology. In C.J. Ralph & J.M. Scott (eds.),

Studies in Avian Biology, 6., p.103-109. U.S.A.: Cooper Ornithological Society.
Ulfstrand, S. 1973. Proportions of Palearctic birds in some East African habitats. Vogelwarte 27: 137-141.
United States Board on Geographic Names 1964. Kenya Official Standard Names Gazetteer No.78. Washington: United States Department of the Interior.
Urban, E.K. & L.H. Brown 1971. A Checklist of the Birds of Ethiopia. Addis Ababa: Haile Selassie I University Press.
Ussher, A. 1944. Nesting habits of Red-throated Rock-Martin. (*Ptyonoprogne R. rufigula*). J.E.Afr.nat.Hist.Soc. 5 & 6 (79 & 80): 399.
van Lawick-Goodall, J. & H. van Lawick-Goodall 1966. Use of tools by the Egyptian Vulture *Neophron percnopterus*. Nature 212: 1468-1469.
van Someren, R.A.L. & V.G.L. van Someren 1911. Studies of birdlife in Uganda. London: John Bale, Sons & Danielson.
van Someren, V.D. 1946. The dancing display and courtship of Jackson's Wydah (*Coliuspasser jacksoni* Sharpe). J.E.Afr. nat.Hist.Soc. 3 & 4 (83 & 84): 131-141.
van Someren, V.D. 1947. Katamayu river and rain forest - 23/2/47. Nature E.Afr. 1: 14-15.
van Someren, V.D. 1958. A bird watcher in Kenya. London and Edinburgh: Oliver & Boyd.
van Someren, V.G.L. 1914. Breeding of game birds. J.E.Afr. & Uganda nat.Hist. Soc. 8: 152.
van Someren, V.G.L. 1916. A list of birds collected in Uganda and British East Africa, with notes on their nesting and other habits. Parts I & II. Ibis Tenth Ser.(4) 2: 193-252, & 3: 373-472.
van Someren, V.G.L. 1918a. Migration of birds. J.E.Afr. & Uganda nat.Hist.Soc. 12: 218-240.
van Someren, V.G.L. 1918b. Notes on a collection of birds from Lamu and district, made by Mr H.J. Allen Turner in April 1916. *ibidem* 12: 249-261.
van Someren, V.G.L. 1918c. *Pitta angolensis longipennis* (Reichenow). *ibidem* 13: 279-280.
van Someren, V.G.L. 1919a. Notes of interest on some East African birds. *ibidem* 14: 414.
van Someren, V.G.L. 1919b. European Crane. *ibidem* 14: 416-417.
van Someren, V.G.L. 1921. On a collection of birds from Turkanaland. *ibidem* 16: 3-38.
van Someren, V.G.L. 1922. Notes on the birds of East Africa. Novit.Zool. 29: 1-246.
van Someren, V.G.L. 1925-35. The birds of Kenya and Uganda. Vols.I & II. J.E. Afr. & Uganda nat.Hist.Soc. 21-58.
van Someren, V.G.L. 1929. Notes on the birds of Jubaland and the northern frontier, Kenya. *ibidem* 35: 100-146.
van Someren, V.G.L. 1930. New races of birds from eastern Africa. *ibidem* 37: 193-197.
van Someren, V.G.L. 1931. Catalogue of the European and Asiatic migrants to Kenya and Uganda, with a brief outline on the subject of migration of birds. *ibidem* spec.Suppl. 4.
van Someren, V.G.L. 1932. Birds of Kenya and Uganda, being addenda and corrigenda to my previous paper in Novitates Zoologicae 29, 1922. Novit.Zool. 37: 252-380.
van Someren, V.G.L. 1939. Reports on the Coryndon Museum Expedition to the Chyulu Hills. Part 2. The birds of the Chyulu Hills. J.E.Afr. & Uganda nat.Hist.Soc. 1 & 2 (62 & 63): 15-129.
van Someren, V.G.L. (n.d.). Birds of Emali-Garabani. Unpubl. suppl. to van Someren, V.G.L. 1939.
van Someren, V.G.L. 1943. Migratory ducks in Kenya Colony. Ibis 85: 345.
van Someren, V.G.L. 1944. Nesting of *Apus caffer streubelii* (Hartlaub). *ibidem* 86: 98.
van Someren, V.G.L. 1945. Further records of Temminck's Stint in Kenya. J.E.Afr. nat.Hist.Soc. 3 & 4 (83 & 84): 163.
van Someren, V.G.L. 1948. *Cisticola distincta*. Ibis 90: 478.
van Someren, V.G.L. 1956. Days with birds. Fieldiana: Zool. 38.
van Someren, V.G.L. & R.A.L. van Someren 1917. Provisional check list of birds of East Africa and Uganda. J.E.Afr. & Uganda nat.Hist.Soc. spec.Suppl. 1.
Vareschi, E. 1978. The ecology of Lake Nakuru (Kenya). I. Abundance and feeding of the Lesser Flamingo. Oecologia 32: 11-35.
Vareschi, E. & J. Jacobs 1985. The ecology of Lake Nakuru. VI. Synopsis of production and energy flow. *ibidem* 65: 412-424.
Vincent, J. 1947. Remarks on *Cisticola distincta* Lynes. Ibis 89: 650-652.
Voous, K.H. 1948. Notes on some Kenya birds. *ibidem* 90: 472-473.
Voous, K.H. 1965. Summer records of *Larus fuscus* in Mombasa, Kenya. Ardea 53: 83.
Voous, K.H. 1977. List of recent Holarctic bird species. London: Ibis.
Wallace, D.I.M. 1955. The mixing of the races of the Yellow Wagtail in Kenya. Br.Birds 48: 337-340.
Wallace, D.I.M. 1975. Rare and unusual Palearctic migrant birds in Central Kenya in 1953. EANHS Bull. 1975: 24-31.
Walters, J.R. 1977. Bird-watching in

Amboseli. *ibidem* 1977: 58-61.
Walters, J.R. 1978. The appearance of the Golden-backed Weaver in Amboseli. Scopus 2: 26-27.
Ward, P. 1971. The migration patterns of *Quelea quelea* in Africa. Ibis 113: 275-297.
Watson, R.M., T. Singh & I.S.C. Parker 1970. The diet of duck and coot on Lake Naivasha. E.Afr.Wildl.Jl. 8: 131-144.
Whitaker, B.K. 1958a. Some notes on birds seen in the Teldet Farms area, Elgon, Kenya, between Nov.22nd and Feb.6th 1958. Unpubl. MS. kept by the East African Natural History Society, Nairobi.
Whitaker, B.K. 1958b. Birds seen Katcheleba district of Suam Jan.17-20, 1958. *ibidem*.
White, C.M.N. 1960. Some inter relationships in the Larks of the *Mirafra africana* group. Bull.Br.Orn.Club 80: 7-10.
White, C.M.N. 1962. A revised check-list of African shrikes ... etc. Lusaka: Government Printer.
White, C.M.N. 1965. A revised checklist of African Non-Passerine Birds. *ibidem*.
Wickler, W. 1973. Artunterschiede im Duettgesang zwischen *Trachyphonus d'arnaudii usambiro* und den anderen Unterarten von *T. d'arnaudii*. J.Orn. 114: 123-128.
Wiley, A.J. 1945. Nest of the Grey-capped Warbler (*Eminia lepida*). J.E.Afr.nat. Hist.Soc. 3 & 4 (83 & 84): 162.
Williams, A.A.E. 1978a. Notes on *Tockus* Hornbills breeding at Lake Baringo, Kenya. Scopus 2: 21-23.
Williams, A.A.E. 1978b. Accidental parasitization of an Eastern Double-collared Sunbird by a Variable Sunbird. *ibidem* 2: 25-26.
Williams, J.G. 1947. Kikuyu Escarpment, 20/4/47. Nature E.Afr. 2: 21-22.
Williams, J.G. 1948a. Bird notes. Saker Falcon (*Falco cherrug cherrug*) in Kenya Colony. Nature E.Afr. 8: 1-2.
Williams, J.G. 1948b. A new race of White-eye from Kenya Colony. Bull.Br.Orn.Club 68: 101-102.
Williams, J.G. 1951. *Nectarinia johnstoni*: a revision of the species, together with data on plumage, moults and habits. Ibis 93: 579-595.
Williams, J.G. 1952. American Pectoral Sandpiper in Kenya Colony. *ibidem* 94: 538.
Williams, J.G. 1953. Spotted Redshank in Kenya. J.E.Afr.nat.Hist.Soc. 1(93): 17.
Williams, J.G. 1954a. The quelea threat to Africa's grain crops. E.Afr.agric.Jl.: 133-136.
Williams, J.G. 1954b. The identification of Kenya birds of prey in flight. Part 1, vultures. J.E.Afr.nat.Hist.Soc. 3 (95): 78-79.
Williams, J.G. 1955. The identification of Kenya birds of prey in flight. Part 2, falcons, hobbies, kestrels and Pygmy Falcon. *ibidem* 5 (97): 165-167.
Williams, J.G. 1956. A new Golden-winged Sunbird from Kenya. Bull.Br.Orn.Club 76: 136-139.
Williams, J.G. 1957. The re-discovery and status of *Ploceus golandi*. Ibis 99: 123-124.
Williams, J.G. 1960a. The recognition of Temminck's Stint in the field. J.E.Afr. nat.Hist.Soc. 5 (102): 220.
Williams, J.G. 1960b. Sabine's Spinetail (*Chaetura sabini* Gray). *ibidem* 7 (104): 305-306.
Williams, J.G. 1963. A field guide to the birds of East and Central Africa. London: Collins.
Williams, J.G. 1966. A new species of swallow from Kenya. Bull.Br.Orn.Club 86: 40.
Williams, J.G. 1967. A field guide to the national parks of East Africa. London: Collins.
Williams, J.G. 1968. Safari Journal. London: Collins.
Williams, J.G. & N. Arlott 1980. A field guide to the birds of East Africa. London: Collins.
Williams, J.G. & H. Friedmann 1965. The Pygmy Honey-guide *Indicator pumilio* Chapin in East Africa. Bull.Br.Orn.Club 85: 21-22.
Williams, J.G. & G.S. Keith 1962. A contribution to our knowledge of the Parasitic Weaver *Anomalospiza imberbis*. *ibidem* 82: 141-142.
Winterbottom, J.M. 1974. The Cape Teal. Ostrich 45: 110-132.
Wolf, D.E. 1984. Ayres' Hawk Eagle *Hieraaetus dubius* feeding on a fruit bat. Scopus 8: 44.
Wolf, D.E. (n.d.).Thompson's Falls area. Unpubl. MS. kept by the East African Natural History Society, Nairobi.
Wood, D. 1974. A Bittern at Lake Naivasha. EANHS Bull. 1974: 104.
Woosnam, R.B. 1912. New localities for Hubbard's Francolin. J.E.Afr. & Uganda nat.Hist.Soc. 4: 135.
Zimmerman, D.A. 1967. *Agapornis fischeri, Lybius guifsobalito,* and *Stiphrornis erythrothorax* in Kenya. Auk 84: 594-595.
Zimmerman, D.A. 1970. An earlier nesting record of *Circaetus fasciolatus* in Kenya. Ibis 112: 264.
Zimmerman, D.A. 1972. The avifauna of the Kakamega Forest, western Kenya, including a bird population study. Bull.Am.Mus. nat.Hist. 149: 255-340.

Zimmerman, D.A. 1978a. A second record of Dickinson's Kestrel in southern Kenya. Scopus 2: 73-74.

Zimmerman, D.A. 1978b. Mascarene Martins in Kenya. *ibidem* 2: 74-75.

Zimmerman, D.A. & R.E. Mumford 1965. First specimens of three species of birds from Kenya. Auk 82: 282-283.

Index of scientific names

Index of English names